"Dr. Appasani has put together a collection of papers from investigators around the world. These works describe the application of single-particle Cryo-EM and Cryo-electron tomography to a wide range of molecular and cellular systems. They provide an overview of the technologies driving the field - from sample preparation to image processing and provide insights into the challenges specific to different types of systems - be they membrane-associated structures, nucleic acid processing molecular machines or fibrous molecular scaffolds. Finally, a number of chapters explore the rapidly expanding potential of this new technology to drug discovery and its application to the most pressing problems in biomedical research."

— **Lee Makowski, PhD.,** Professor of Bioengineering, Chemistry and Chemical Biology,
Northeastern University & Visiting Professor of Neurology
at Harvard Medical School, Boston, **USA**

"As a relative new-comer to the field of cryo-EM, I thoroughly enjoyed reading this comprehensive book. It not only ensures a solid foundation by reviewing the history and underlying principles of the techniques but also describes practical guidelines and technological advances for facility design, sample preparation, data acquisition, and data analysis. A collection of fascinating structural discoveries are shared into the biology of various enzymes, membrane proteins, filaments, and cellular machinery. Importantly, therapeutic applications of these findings are highlighted and herald a bright future for the use of cryo-EM and cryo-ET in drug discovery."

— **Lizelle Lubbe, PhD.,** Scientist, University of Cape Town, **SOUTH AFRICA**

"The cryo-electron microscopy field is advancing at an accelerating pace, both in technology and applications, and too many papers are being published for a researcher. Therefore, this book is a very useful guidepost for both beginners and advanced cryo-EM researchers, covering basic principles, such as Fourier Shell Correlation, to many biological applications."

— **Masahide Kikkawa, MD., PhD.,** Professor of Cell Biology & Anatomy,
The University of Tokyo, **JAPAN**

"The revolutionary Cryo-EM has transformed the field of structural biology by prompting the revisiting of numerous protein complex structures. This book by Appasani, showcases a refined collection that explores the most recent and vital structures solved by Cryo-EM, providing a comprehensive resource in one convenient place."

— **Prince Tiwari, PhD.** Assistant Professor of Biosciences and Bioengineering,
Indian Institute of Technology-Roorkee, **INDIA**

Cryo-Electron Microscopy in Structural Biology

Cryo-electron microscopy, in combination with tomography, has emerged as a new technology for visualizing molecular structures at a resolution beyond even 1 Å. Using this technology has revealed the native molecular details of viruses, membranes, enzymes, ribosomes, and cells. This comprehensive volume brings together authoritative overviews of these methods from structural and biological perspectives. It is a must-have for researchers and graduate students, as well as those working in industry, primarily in the areas of biophysics, structural biology, crystallography, and genomics.

Key Features

- Focuses on the applications of cryo-EM to structural biology
- Documents the importance of cryo-EM/ET approaches in studying the structural determinants of cellular organelle and membrane protein biochemistry
- Reviews the applications of high-resolution structures of viruses
- Emphasizes structural insights of nuclear and gene machineries
- Includes a section focused entirely on the applications of cryo-EM/ET in drug discovery and therapeutic development

KRISHNARAO APPASANI is an award-winning Indian-born American scientist, educator, and science communicator and the chief executive officer of GeneExpression Systems, a global conference–producing organization in the biomedical and physical sciences. He is an elected life fellow of the Royal Society of Biologists, London, UK. In 2022, he was bestowed with an *Outstanding Science Communicator Award* by the Ministry of Science & Technology, Government of India. Appasani is also the editor of *RNAi* (2005), *MicroRNAs* (2007), *Epigenomics* (2012), *GWAS* (2016); *Optogenetics* (2017), *Genome Editing* (2019), and *Single Molecule Science* (2022), all published by Cambridge University Press. Some of these have been translated into Japanese, Chinese, and Arabic.

Cryo-Electron Microscopy in Structural Biology

From Structural Insights to Tomography and Drug Discovery

Edited by
Krishnarao Appasani, PhD.
Gene Expression Systems, Inc.

Foreword by
Gregory A. Petsko, DPhil.
Harvard Medical School and Brigham & Women's Hospital

CRC Press
Taylor & Francis Group
Boca Raton London New York

CRC Press is an imprint of the
Taylor & Francis Group, an **informa** business

Designed cover image: Courtesy of Dr. Agniezka Obarska, Max Planck Institute of Biophysics, Germany, and Dr. Shyamal Mosalaganti of the University of Michigan, USA

First edition published 2025
by CRC Press
2385 NW Executive Center Drive, Suite 320, Boca Raton FL 33431

and by CRC Press
4 Park Square, Milton Park, Abingdon, Oxon, OX14 4RN

CRC Press is an imprint of Taylor & Francis Group, LLC

Library of Congress Cataloging-in-Publication Data
Names: Appasani, Krishnarao, 1959- editor.
Title: Cryo-electron microscopy in structural biology : from structural insights to tomography and drug discovery / edited by Krishnarao Appasani.
Description: First edition. | Boca Raton, FL : CRC Press, 2025. | Includes bibliographical references and index.
Identifiers: LCCN 2024013223 (print) | LCCN 2024013224 (ebook) | ISBN 9781032352534 (hbk) | ISBN 9781032352572 (pbk) | ISBN 9781003326106 (ebk)
Subjects: LCSH: Cytology--Technique. | Electron microscopy. | Electron microscopy--Technique. | Molecular structure.
Classification: LCC QH583 .C796 2025 (print) | LCC QH583 (ebook) | DDC 571.6--dc23/eng/20240805
LC record available at https://lccn.loc.gov/2024013223
LC ebook record available at https://lccn.loc.gov/2024013224

ISBN: 9781032352534 (hbk)
ISBN: 9781032352572 (pbk)
ISBN: 9781003326106 (ebk)

DOI: 10.1201/9781003326106

Typeset in Times New Roman
by KnowledgeWorks Global Ltd.

Dedicated to

Donald L. D. Caspar, PhD *(1927–2021), an American crystallographer, who coined the term* structural biology, *and who was a legendary professor who had trained many structural biologists around the world while he was at Harvard, Brandeis, and Florida State University. The editor was inspired by his discussions when they met many times at the Rosenstiel Award Ceremonies at Brandeis University, Waltham, Massachusetts, USA.*

My late friend ***Oliver C. Steinbach, PhD*** *(1968–2022), a German-American biochemist, who led pharmaceutical, diagnostic, and imaging products while he was at Altana Pharma AG (Germany and USA), Philips Research (The Netherlands, and USA), TransCode Therapeutics, Inc. (USA), Imagion Biosystems, Inc. (USA), and Roche (USA). A true friend who strived to be ethical always and who lived up to extremely high standards, who had compassion and hiked with me during the Covid epidemic isolation period when the writing of this book began.*

Krishna hiking with Oliver, who was standing on a rock at California's Anza Borrego Desert State Park, Hellhole Canyon, on March 21, 2021.

Contents

Editor ... xiii
List of contributors ... xv
Foreword ... xix
Preface ... xxi

SECTION I Cryo-EM in Cell Biology: Structure Determination of Protein Assemblies and Cellular Organelle

Chapter 1 Introduction on Cryogenic Electron Microscopy in Structural Biology ... 3
Krishnarao Appasani

Chapter 2 Structural Dynamics of Proteasome Regulation by Cryo-EM ... 25
Youdong Mao

Chapter 3 Cryo-EM Structures of a Ubiquitin-Specific Protease in Complex with Substrate ... 45
Martin Rennie

Chapter 4 Structural Biology of Protein Kinases and Their Regulation ... 55
Samrat Mitra, Pranita Ray, Prateeka Borar, and Smarajit Polley

Chapter 5 The Life and Times of p53: Structural Advancements to Decipher the Guardian of the Genome ... 69
Maria J. Solares, Samantha Berry, and Deborah F. Kelly

SECTION II Technology Platform and Evolving Technologies

Chapter 6 Single-Particle Cryo-EM: What Happens Inside the Black Box? ... 87
Sayan Bhakta and Marin van Heel

Chapter 7 Streptavidin Affinity Grids in the Study of Macromolecular Complexes Using Cryo-Electron Microscopy ... 110
Nils Marechal and Patrick Schultz

Chapter 8 Running a Cost-Effective Modern Cryo-EM Facility for SPA Analysis ... 121
Lijia Jia

Chapter 9 Computational Methods for Bimolecular Structure Modeling for Cryo-EM ... 131
Javad Baghirov, Xiao Wang, Genki Terashi, Shu Li, and Daisuke Kihara

Chapter 10 Artificial Intelligence-Driven Data Acquisition for Single-Particle Cryo-EM and Three-Dimensional Electron Diffraction ... 142
Koji Yonekura

Chapter 11 From Icy Specimens to Digital Insights: Integrating Sample Preparation, Data Analysis, and IT Infrastructure 147

Eric Shell, Karine Minari, and Vitor Hugo Balasco Serrão

SECTION III Cryo-EM in Virology and Enzymology: High-Resolution Structures of Chaperonins and Enzymes

Chapter 12 Cryo-EM Structures of Viral and Bacterial Chaperonins 169

Olga S. Sokolova and Tatiana B. Stanishneva-Konovalova

Chapter 13 Cryo-EM Structure of Monkeypox Virus DNA Polymerase Holoenzyme 176

Yi Shi and George F Gao

Chapter 14 Determining the High-Resolution Structure of the [NiFe]-hydrogenase Huc by Native Host Purification and Cryo-EM 187

Hari Venugopal, Ashleigh Kropp, Chris Greening, and Rhys Grinter

Chapter 15 Membrane Rupture and Lipid–Protein Interaction of γ-Hemolysin from *Staphylococcus aureus* Using Cryo-EM 200

Suman Mishra, Anupam Roy, Arnab Chatterjee, and Somnath Dutta

SECTION IV Cryo-EM in the Study of Membrane Biology: Probing the Structures of Membrane Proteins

Chapter 16 Historical Evolution and Perspective on Cryo-EM Analysis of Membrane Proteins 213

Youzhong Guo

Chapter 17 Single-Particle Cryogenic Electron Microscopy Unfolds Distinct Membrane Transporter Folds: Focus on Autoinducer-2 Exporter Family 227

Radhika Khera

Chapter 18 Rhodopsin Dimers in the Retinal Rod Outer Segment 243

Sahil Gulati and Krzysztof Palczewski

Chapter 19 Cryo-EM Structures of the Fungal GPCR Dimer Ste2 254

Vaithish Velazhahan and Christopher G. Tate

Chapter 20 Human V-ATPases: Assembly, Function, and Regulation 268

Elizabeth Fosuah, Qingpeng Lin, and Tian-Min Fu

Chapter 21 Structure and Function of the Intracellular Calcium Channel Ryanodine Receptor 275

Marco C. Miotto

Chapter 22 Cryo-EM of Mitochondrial Membrane Proteins 287
Swati Yadav, Shaileshanand Jha, and Kutti R. Vinothkumar

SECTION V Cryo-EM in the Study of Molecular Biology: Structural Insights of Gene Machineries

Chapter 23 Imaging Structurally Dynamic Ribosomes with Cryogenic Electron Microscopy 305
Samantha M. Webster, Mira B. May, Barrett M. Powell, and Joseph H. Davis

Chapter 24 Cryo-EM in the Study of Ribosome Biogenesis 319
Ankita Arora and Prem S. Kaushal

Chapter 25 Cryo-EM Structures of Ribosomes from Human Pathogens: Implications for Drug Targeting 335
Ankit Dhur, Aneek Banerjee, Krishnamoorthi Srinivasan, and Jayati Sengupta

SECTION VI Cryo-EM in the Study of Neurobiology and Skin—Neuroreceptors to Ion Channels and Synaptic Transmission

Chapter 26 Cryo-EM Structural Features of Glutamate Receptors Involved in Synaptic Neurotransmission or Brain Pathology 349
Nebojša Bogdanović

Chapter 27 Cryo-EM Structures of Systemic Amyloid Fibrils 367
Binh An Nguyen, Maria del Carmen Fernandez-Ramirez, Shumaila Afrin, and Lorena Saelices

Chapter 28 Structural Insights into the 5-HT_{3A} Receptor Revealed by Cryogenic Electron Microscopy (Cryo-EM) 377
Zhuowen Li, Nikhil Bharambe, and Sandip Basak

Chapter 29 Molecular Cryo-EM in the Study of Skin Tissue 385
Lars Norlén

SECTION VII Applications of Cryo-ET and Cryo-EM in the Biopharmaceutical Industry

Chapter 30 Emerging Technologies in Cryo-Electron Tomography 395
Kanika Khanna

Chapter 31 Cryo-Electron Tomographic Imaging of Membrane Proteins from Mitochondria and Chloroplasts 407
Tofayel Ahmed and Antara Saha

Chapter 32 Functional Studies of the Actin Cytoskeleton by Cryogenic Electron Tomography 418

Dorit Hanein and Niels Volkmann

Chapter 33 Cryo-EM in Drug Discovery.. 429

Youxin Kong, Maxime Killer, Madhumati Sevvana, and Chiara Rapisarda

Chapter 34 Applications of Single-Particle Analysis and Cryo-ET in Drug Discovery 446

Qing Yao

Index... 459

Editor

Krishnarao Appasani trained as a molecular cell biologist, earning Masters and Doctoral degrees in Biochemistry and Molecular biology. He accepted a Lectureship at Boston University and taught cell and molecular biology courses to undergraduates and graduate students. He was a post-doc in the laboratory of Nobel Laureate, the late Prof. H. Gobind Khorana at the Massachusetts Institute of Technology. Subsequently, he joined the faculty of Harvard Medical School and directed the Thoracic Oncology Laboratory at the Brigham &Women's Hospital and the Dana-Farber Cancer Institute in Boston. During this time, he completed an MBA at Bryant University, Smithfield, RI, and then moved to Perkin Elmer and directed the Microarrays group. Dr Appasani has published about 60 research papers and abstracts in international journals, especially on Gene Expression. He was awarded the Outstanding Investigator Award by the American Federation for Medical Research, Washington D.C., competed globally for the Ranbaxy Research Medal, and was honored with the 1997-Ranbaxy Research Award in the field of Applied Medical Sciences. At Perkin Elmer, Dr Appasani organized a global Biomics Seminar series to educate scientists and to initiate and develop collaborations with researchers in academia and other industries. For about 15 years he ran a multi-million-dollar knowledge management enterprise and served as an international authority in bridging Academia and Industry by fostering inter-disciplinary and translational science. In 2016 Dr Appasani was inducted as a Life Fellow of the Royal Society of Biology, England, for his global efforts in promoting science education.

List of contributors

Shumaila Afrin
Peter O'Donnell Jr. Brain Institute
University of Texas–Southwestern Medical Center
Dallas, Texas

Tofayel Ahmed
Saha Institute of Nuclear Physics
Homi Bhabha National Institute of Mumbai
Kolkata, West Bengal, India

Krishnarao Appasani
GeneExpression Systems, Inc.
Carlsbad, California

Ankita Arora
UNESCO-DBT–Regional Centre for Biotechnology
NCR Biotech Science Cluster
Faridabad, Haryana, India

Javad Baghirov
Purdue University
West Lafayette, Indiana

Aneek Banerjee
CSIR-Indian Institute of Chemical Biology
&
Academy of Scientific and Innovative Research (AcSIR)
Ghaziabad, Uttar Pradesh, India

Sandip Basak
Nanyang Technological University
Singapore

Samantha Berry
Pennsylvania State University
University Park, Pennsylvania

Sayan Bhakta
Laboratório Nacional de Nanotecnologia (LNNano)
Centro Nacional de PesquisaemEnergia e Materiais (CNPEM)
Campinas, Brazil

Nikhil Bharambe
Nanyang Technological University
Singapore

Nebojsa Bogdanovic
Florida State University
Tallahassee, Florida

Prateeka Borar
Bose Institute, Unified Academic Campus
Kolkata, West Bengal, India

Arnab Chatterjee
Indian Institute of Science
Bangalore, India

Joseph H. Davis
Massachusetts Institute of Technology
Cambridge, Massachusetts

Ankit Dhur
CSIR–Indian Institute of Chemical Biology
Kolkata, West Bengal, India

Somnath Dutta
Indian Institute of Science
Bangalore, India

Maria del Carmen Fernandez-Ramirez
Peter O'Donnell Jr. Brain Institute
University of Texas–Southwestern Medical Center
Dallas, Texas

Elizabeth Fosuah
College of Medicine, The Ohio State University
Columbus, Ohio

Tianmin Fu
College of Medicine, The Ohio State University
Columbus, Ohio

George F. Gao
Chinese Academy of Sciences–Institute of Microbiology
Chaoyang District, Beijing, China

Chris Greening
Monash University
Clayton, VIC, Australia

Rhys Grinter
Monash University
Clayton, VIC, Australia

Sahil Gulati
Gatan, Inc.
Pleasanton, California

Youzhong Guo
Virginia Commonwealth University
Richmond, Virginia

Dorit Hanein
University of California Santa Barbara
Santa Barbara, California

Shaileshanand Jha
National Centre for Biological Sciences
Tata Institute of Fundamental Research
Bangalore, India

Lijia Jia
University of Texas Health Sciences Center
San Antonio, Texas

Prem S. Kaushal
UNESCO-DBT–Regional Centre
for Biotechnology
Faridabad, Haryana, India

Deb Kelly
Structural Oncology LLC.
State College, Pennsylvania

Kanika Khanna
Gladstone Institutes at UCSF
San Francisco, California

Radhika Khera
Max Planck Institute of Biophysics
Frankfurt am Main, Germany
&
University of California Berkeley
Berkeley, California

Daisuke Kihara
Purdue University
West Lafayette, Indiana

Maxime Killer
Sanofi
Vitry-sur-Seine, France

Youxin Kong
Sanofi
Vitry-sur-Seine, France

Ashleigh Kropp
Monash University
Clayton, VIC, Australia

Lidia Kurochkina
Lomonosov Moscow State University
Moscow, Russia

Shu Li
Purdue University
West Lafayette, Indiana

Zhuowen Li
Nanyang Technological University
Singapore

Qingpeng Lin
University of California San Diego
San Diego, California

Youdong Mao
Peking University
Beijing, China

Nils Marechal
IGBMC–Institute of Genetics and Molecular
and Cellular Biology
Illkirch Cedex, France

Mira May
Massachusetts Institute of Technology
Cambridge, Massachusetts

Karine Minari
University California–Santa Cruz
Santa Cruz, California

Marco C. Miotto
Columbia University Vagelos College of
Physicians and Surgeons
New York City, New York

Suman Mishra
Indian Institute of Science
Bangalore, India

Samrat Mitra
Bose Institute, Unified Academic Campus
Kolkata, West Bengal, India

BinhAn Nguyen
Peter O'Donnell Jr. Brain Institute
University of Texas–Southwestern Medical
Center
Dallas, Texas

Lars Norlen
Karolinska Institute
Stockholm, Sweden

Krzysztof (Kris) Palczewski
University of California Irvine
Irvine, California

Smarajit Polley
Bose Institute, Unified Academic
Campus
Kolkata, West Bengal, India

Barrett M. Powell
Massachusetts Institute of Technology
Cambridge, Massachusetts

Chiara Rapisarda
Sanofi
Vitry-sur-Seine, France

Pranita Ray
Bose Institute, Unified Academic Campus
Kolkata, West Bengal, India

Martin Rennie
University of Glasgow
Glasgow, United Kingdom

Anupam Roy
Indian Institute of Science
Bangalore, India

Lorena Saelices
Peter O'Donnell Jr. Brain Institute
University of Texas–Southwestern Medical Center
Dallas, Texas

Antara Saha
Saha Institute of Nuclear Physics
Homi Bhabha National Institute of Mumbai
Kolkata, West Bengal, India

Patrick Schultz
IGBMC–Institute of Genetics and Molecular and Cellular Biology
IllkirchCedex, France

Jayati Sengupta
CSIR-Indian Institute of Chemical Biology
Kolkata, West Bengal, India
&
Academy of Scientific and Innovative Research (AcSIR)
Ghaziabad, Uttar Pradesh, India

Vitor Hugo Balasco Serrão
University California–Santa Cruz
Santa Cruz, California

Madhumati Sevvana
Sanofi, Inc.
Cambridge

Eric Shell
University California–Santa Cruz
Santa Cruz, California

Yi Shi
Chinese Academy of Sciences–Institute of Microbiology
Chaoyang District, Beijing, China

Maria J. Solares
Perelman School of Medicine-University of Pennsylvania
Philadelphia, Pennsylvania

Olga S. Sokolova
Lomonosov Moscow State University
Moscow, Russia

Krishnamoorthi Srinivasan
CSIR-Indian Institute of Chemical Biology
Kolkata, West Bengal, India
&
Academy of Scientific and Innovative Research (AcSIR)
Ghaziabad, Uttar Pradesh, India

Tatyana B. Stanishneva-Konovalova
LomonosovMoscow State University
Moscow, Russia

Christopher G. Tate
MRC Laboratory of Molecular Biology
University of Cambridge
Cambridge, United Kingdom

Genki Terashi
Purdue University
West Lafayette, Indiana

Marin van Heel
Laboratório Nacional de Nanotecnologia (LNNano)
Centro Nacional de PesquisaemEnergia e Materiais (CNPEM)
Campinas, Brazil

Vaithish Velazhahan
MRC Laboratory of Molecular Biology
Gonville and Caius College, University of Cambridge
Cambridge, United Kingdom

Hari Venugopal
Monash University
Clayton, VIC, Australia

Kutti R. Vinothkumar
National Centre for Biological Sciences
Tata Institute of Fundamental Research
Bangalore, India

Niels Volkmann
University of California Santa Barbara
Santa Barbara, California

Xiao Wang
Purdue University
West Lafayette, Indiana

Samantha M. Webster
Massachusetts Institute of Technology
Cambridge, Massachusetts

Swati Yadav
National Centre for Biological Sciences
Tata Institute of Fundamental Research
Bangalore, India

Qing Yao
Gilead Sciences, Inc.
Foster City, California

Koji Yonekura
RIKEN SPring-8 CenterSayo, Hyogo, Japan
&
Tohoku University, Katahira Campus
Katahira, Aoba-Ku, Sendai, Japan

Foreword

Cryo-Electron Microscopy in Structural Biology: From Structural Insights to Tomography and Drug Discovery provides a detailed overview of the techniques of cryoEM. I am thankful to Krishnarao Appasani for making the extraordinary effort to prepare this timely volume on the rapidly expanding cryoEM technique used to probe structural biology.

We all knew it would happen eventually; we just didn't think it would happen so soon. Those of us who came into structural biology in the 1960s, when the field was just getting started, had seen small molecule structure determination go from a field in its own right to a routine analytical tool, automated enough for almost anyone to do it, expected protein structure determination to become commoditized in a similar way at some point. We just didn't expect we'd see it in our professional lifetimes.

But that, of course, is exactly what happened in the 1990s and early 2000s. What drove this radical change was a perfect storm of user-friendly software development, area detector technology, synchroton radiation sources, low temperature to minimize radiation damage from the intense synchrotron beam; and the ability to clone and express virtually any protein of interest. Indeed, so powerful had these technologies become that, by the turn of the millennium, in an almost unparalleled example of bandwagon-hopping (onto the "omics" revolution), scientists in several countries had proposed, and embarked up on, the Structural Genomics Initiative, an effort to determine a three-dimensional structure of every protein in an organism – or, in some manifestations, at least one member of every protein fold family - via experimental or computational approaches or a combination of both. (Full disclosure: I was against this idea at the time, still am, and am convinced I was right.)

It was protein crystallography (and, later, protein structure prediction) that was at the center of these efforts, because by 2000, just as had happened in small molecule crystallography, the structure determination of proteins had become easy enough that people with little or no formal training in the discipline could import it into their laboratories. One result, as probably should have been expected, is that a number of completely or partially incorrect structures were published by relatively inexperienced people who encountered technical problems they could not even identify, much less deal with.

Still, the requirement for proteins to crystallize somewhat limited the application of these cookie-cutter methods to single proteins and relatively small protein complexes (though the ribosome crystallized, and it was solved, all 79 different proteins and 4 ribosomal RNAs of it), and preparations that were conformationally homogeneous.

And then, along came cryo-electron microscopy (cryoEM), the subject of this book. Electron microscopy itself is about as old as protein crystallography: Ernst Ruska and Max Knoll, a physicist and an electrical engineer, respectively, from the University of Berlin, created the first electron microscope (it had 400x magnification) in 1931. (James Sumner accomplished the first crystallization of an enzyme, urease, in 1926, and Dorothy Hodgkin and J.D. Bernal published the first X-ray diffraction photographs of a protein crystal, that of the enzyme pepsin, in 1934.) But it wasn't until 1975 that Richard Henderson and Nigel Unwin published the first EM structure of a protein, the 7A resolution structure of the purple membrane protein from *Halobacterium halobium*. That breakthrough was possible by the use of tilted, unstained images and the fact that the protein formed an ordered two-dimensional lattice in the membrane, allowing Fourier reconstruction techniques similar to those used in crystallography.

Though herculean effort was eventually able to push the purple membrane EM structure to near atomic resolution, for a long time high-resolution electron microscopy of proteins was limited to those that could form a lattice, or that had many copies in different but symmetrical orientations such as viruses. Right around the time that structural genomics was consuming a lot of the energy and resources of the structural biology community, a handful of scientists were developing the exact analog of the technological advances that had fueled the protein crystallography explosion: user-friendly software for processing, sorting, categorizing and analyzing images; powerful new detectors to make acquisition of thousands of high signal-to-noise images rapid; and new EM instruments that were able to collect data automatically from specimens frozen in vitreous ice at liquid-helium temperatures. These breakthroughs meant that single particle structures could be determined if the particle was large enough (which enabled the structure determination of big multi-protein complexes), even though it might have several conformations in solution. And the absence of the need for crystals, whether two- or three-dimensional, vastly expanded the range of what could be studied especially since large amounts of material were not required.

By 2015, single particle cryoEM had begun to rival protein crystallography as the method of choice for structural studies of large proteins (by large I mean around 250,000 MW or bigger) and multiprotein complexes, and resolutions of ~3A for such structures were becoming common. Now, almost 10 years later, it has clearly become the method of choice in such cases. In 2023 there were about 14,000 new depositions in the Protein Data Bank. More than 4500 of these were determined by cryoEM. A little more than twice that were determined by crystallography, but that number has been fairly constant for the past eight years, whereas the number of cryoEM structures is growing by more than 500 every year. At present rate of growth, it will surpass the yearly crystal structure output by around 2030.

This wonderful book will give you an introduction to the techniques of cryoEM, but more importantly will provide you with a buffet of applications, in many areas of biology, to sample from. It can be read sequentially, or dipped into almost at random as you desire (I must confess a personal preference for the latter).

It seems fair to ask, as one reads it, where structural biology is going now. The book offers a glimpse of that, but let me give my own perspective: I think the next frontier, which is almost upon us, is high resolution structure determination *in situ* – that is, in living cells - using the cryoEM variant called cryo-electron tomography (cryoET). The brainchild of Wolfgang Baumeister, cryoET is basically medical tomography applied to thin slices from frozen cells. The slices are imaged by fairly standard cryoEM methods, and the tilted section images are processed in reciprocal space to reconstruct a three-dimensional image. Currently, the method is limited to about 1-4 nM (10-40A) resolution, but there is optimism that this can be bettered considerably, and probably rather soon. The cryoET version of cryoEM promises the Holy Grail of structural biology: the ability to observe protein structures as they function in their native habitat. The problem, of course, is knowing what one is looking at in the dense environment of the cell, where proteins are as crowded as New Year's revelers at Times Square in New York City. But can there be any doubt that this limitation, too, will be overcome, somehow?

All experimental science can be thought of as an attempt to overcome the limitations of the human eye. If our eyes were like that of Superman, who possesses microscopic vision, we could just peer at proteins in solution or inside the cell and watch them in glorious atomic detail. CryoEM has given us our first pair of Superman's eyes. If you want to see what they reveal, read on.

Manchester-by-the-Sea, Massachusetts
August 2024

Gregory A. Petsko, DPhil.
Professor of Neurology
Ann Romney Center for Neurological Diseases
Harvard Medical School and
Brigham & Women's Hospital
Boston, Massachusetts, USA

Preface

"The important thing in science is not so much to obtain new facts as to discover new ways of thinking about them."

—Sir William Lawrence Bragg, FRS, 1915 Nobel Prize -winning Australia-born British physicist (1890–1971), who discovered atomic arrangements of crystal structure by means of X-rays

As the famous Austrian-British philosopher of science Karl R. Popper (1902–1994) believed, *"before we can find the answers, we need the power to ask new questions; in other words, we need new technology."* A key example of such a technological advance is the recent development of *cryogenic electron microscopy (cryo-EM).* This breakthrough technology has integrated disciplines such as cell biology, biochemistry, biophysics, and physiology. It also incorporates chemistry, physics, mathematics, computational biology, and engineering, as well as approaches from biotechnology, nanotechnology, and optics. One can now understand how cellular components and machineries operate molecule by molecule, which has opened up the possibility of visualizing whole new worlds. Cryo-electron microscopy and other related live-cell imaging techniques, among the most exciting tools available in *structural biology* today, offer powerful new ways to elucidate bimolecular structure and function, in terms of both revealing mechanisms of action on an atomic level and tracking the behavior of molecules in living cells. Among the most valuable achievements strengthening biological understanding are combinations between super-resolution techniques such as cryogenic electron tomography (cryo-ET) and cryo-EM. With the amazing developments both approaches are taking, combined 3D reconstruction will have a large impact in the future.

The major challenge with microscopy, even until recent years, was re-creating the fuzzy 2D images as sharp 3D structures. Therefore, three scientists—Richard Henderson (MRC Laboratory of Molecular Biology, UK), Jacques Dubochet (University of Lausanne, Switzerland) and Joachim Frank (Columbia University)—developed the cryo-EM that transformed the field of structural biology. In the last decade, the scientific community has witnessed several breakthroughs in the field of structural biology by adopting cryo-EM and cryo-ET to solve 3D structures of cytoskeletal proteins and membrane proteins, ion channels, receptors, ribosomes, and many more. In recent years, innovative artificial intelligence and machine learning–based software tools also revolutionized the the prediction of thousands of protein structures.

Cryo-Electron Microscopy in Structural Biology is intended for those working in the fields of structural biology, cell biology, neurobiology, molecular biology, genomics, biotechnology, drug discovery, and molecular medicine. There are a number of books already available covering structural biology, cryo-electron microscopy, and cryo-ET. They include: J. Frank (1992), *Electron Tomography: Three-Dimensional Imaging with Transmission Electron Microscope*, Plenum Press, UK; J. Frank (2006), *Electron Tomography: Methods for Three-Dimensional Visualization of Structures in the Cell*, 2nd ed., Springer Press, USA; J. Frank (2006), *Three-Dimensional Electron Microscopy of Macromolecular Assemblies: Visualization of Biological Molecules in Their Native State*, Oxford University Press, UK; A. Cavalier, D. Spehner, and B. M. Humbel (2009), *Handbook of Cryo-Preparation Methods for EM*, CRC Press, USA; G. J. Jensen (2010), *Cryo-EM Part B: 3-Dimensional Reconstruction*, Part 2, in *Methods in Enzymology*, vol. 482, Elsevier Science–Academic Press, USA; J. Frank (2013), *Electron Tomography, Methods for 3-D Visualization of Structures in the Cell*, Springer Press, USA; R. A. Crowther (2016), *The Resolution Revolution: Recent Advances in Cryo-EM*, in *Methods in Enzymology*, vol. 579, Academic Press, USA; J. Frank (2018), *Single-Particle Cryo-Electron Microscopy: The Path Toward Atomic Resolutions, Selected Papers*, World Scientific Publishing company, Pvt. Ltd, Singapore; T. Muller-Reichert and G. Pigino (2019). *Three-Dimensional Electron Microscopy*, in *Methods in Cell Biology*, vol. 152, Academic Press, USA; T. Gronen and B. L. Nannenga (2021), *Cryo-EM Methods and Protocols*, in *Methods in Molecular Biology*, vol. 2215, Humana Press, USA; R. M. Glaeser, E. Nogales and Wah Chiu (2021), *Single Particle Cryo-EM of Biological Macromolecules*, IOP Publishing Ltd., UK; S. M. Lea and P. A. Williams (2022), *Current Opinion in Structural Biology: Cryo-Electron Microscopy-2022*. Elsevier Ltd, The Netherlands, and J. Frank (2024), *Novel Developments in Cryo-EM of Biological Molecules*, Jenny Stanford Publishing Pte. Ltd, Singapore, Routledge Press.

Most of the above books primarily focus on methods, lab protocols, and selected papers, except the recent book by Glaeser and Nogales, which discusses details of the field from bimolecular structure determination and its technological perspectives. The recent book by Frank discusses the principles of SPA, data collection, and highlights of emerging computational, microfluidic-based, and the latest time-resolved cryo-EM methods. This book differs in that it is the first text completely devoted to describing the applications of cryo-EM and cryo-ET in the fields of cell biology, virology, enzymology, membrane biology, molecular biology, neurobiology, and skin biology and emphasizing their importance to the biopharmaceutical industry.

The goal of this book is to serve as a reference for graduate students, postdoctoral researchers, and primary investigators, as well as to provide an explanatory analysis for executives and scientists from biotechnology and pharmaceutical companies. My hope is that this volume will serve as a prologue to the field for newcomers and as a reference for those already active in the field. Most importantly, this book serves as a bridge between the basic science of cryo-electron microscopy and structural biology and its diverse applications in areas such as vaccine development and drug discovery. The chapters listed in the present volume discuss insights that have been revealed about molecular mechanisms, structures, and functions by both cryo-EM and cryo-ET techniques. Many topics are covered in this text, including enzymes, motor proteins, membrane channels, mitochondria, ribosomes, and other key organelles and molecules of current interest. An introduction by the editor provides a brief review of the key principles along with a historical overview. The final part of the introduction provides a discussion of future perspectives, including the latest applications and the relevance of computational biology tools to drug design and discovery.

Many people have contributed to making my involvement in this project possible. I thank my teachers for their guidance and mentorship, as well as their excellent teaching, which have helped me to bring about this educational enterprise. Each chapter is a self-contained treatise by authorities in the field, and I am extremely grateful to all of the contributors, without whose commitment this project would not have been possible. Many people have had a hand in the preparation of this book. Each chapter has been passed back and forth between the authors for criticism and revision. Each contributor is responsible for their own chapter content; the editor and publisher are not responsible for discrepancies of any kind. I thank the reviewers, who have made the hours spent putting together this volume worthwhile. I am indebted to the staff of the Taylor & Francis publishing group of the CRC Press, and in particular Chuck Crumly for his generosity and efficiency throughout the editing of this book. He truly understands the urgency and need for this volume. I also extend my appreciation to Kara Roberts for her excellent cooperation during the development of this volume. I want to thank Dr Gregory Petsko, Professor at Harvard University and 2023-US National Medal of Science Awardee for his kindness in writing the Foreword to this book. Last, but not least, I thank Shyamala (Sham) Appasani for her understanding and support during the development of this volume.

This book, *Cryo-Electron Microscopy in Structural Biology*, is the eleventh in the series *Gene Expression and Regulation* that I have produced in the last two decades. A portion of the royalties will be contributed to the Dr. Appasani Foundation, a nonprofit organization devoted to bringing social change through the education of youth in developing nations; and The MINDS Foundation (**M**ental **I**llness and **N**eurological **D**isorder**S**), which is committed to taking a grassroots approach to providing high-quality mental health care and educational awareness in rural India.

Krishnarao Appasani, PhD.
San Diego, California, USA

Section I

Cryo-EM in Cell Biology

Structure Determination of Protein Assemblies and Cellular Organelle

1 Introduction on Cryogenic Electron Microscopy in Structural Biology

Krishnarao Appasani

The light microscope opened the 1st gate to microcosm.

The electron microscope opened the 2nd gate to microcosm.

What will we find opening the 3rd gate?

—**Ernst August Friedrich Ruska**
1986 Nobel Prize–winning German physicist and electrical engineer (1906–1988), who contributed fundamental work in electron optics and designed the first electron microscope

It seems now the cryogenic electron microscope opened the 3rd gate to microcosm and brought a resolution revolution.

INTRODUCTION

Structural studies of biomolecules allow access to an incredible wealth of molecular information. Most importantly, previously *"unanswerable"* questions in the physical, chemical, and biological sciences can now be answered. The field of *structural biology* can be divided into two general areas: (1) improvements in microscopy methodologies (technology development) and (2) use of these methodologies to address important scientific questions in fundamental biological research (applied research). Over the past decades, cryogenic electron microscopy (cryo-EM) and cryogenic electron tomography (cryo-ET) research has fostered excellent collaboration and interdisciplinary research with input from biology, chemistry, physics, and engineering sciences.

Scientists are progressively aiming to discover and describe the secrets in the index of the book of life. In order to understand complex molecular interactions, a novel technological platform is needed. Techniques such as X-ray diffraction, electron microscopy (EM), nuclear magnetic resonance (NMR), and several others allow researchers to uncover part of this information; however, they lack the needed sensitivity, the resolution to unravel the details of individual molecular events. However, with cryo-EM and cryo-ET one can resolve the structures of biomolecules at greater resolution.

This introductory chapter begins from the initial identification of crystal structures by X-ray crystallography to the discovery and applications of cryo-EM in many areas of biology, including structural biology.

HISTORICAL PERSPECTIVES FROM CRYSTAL STRUCTURE TO X-RAY CRYSTALLOGRAPHY

Abbé Haüy, professor of the humanities at the University of Paris, France, was the founder of the *"structure of the crystal science"* during the last decades of the 18th century. For his systematic observation and pioneering research on crystals, he will forever be remembered as a distinguished scientist and the *"Father of Crystallography."* Haüy was hailed by Henry James Brooke in his book *Familiar Introduction to Crystallography*, published in 1823. Since then, many successors have worked in this field and unlocked the secrets of crystal science (Adams, 1918).

In order to study the arrangement of all of the atoms in crystals, we needed a new kind of wave—one with a shorter wavelength that could penetrate surfaces to show not just the atoms on the outside but also the atoms within the crystal or any biological protein. That was made possible by the discovery of X-rays in 1895 by Wilhelm Conrod Röntgen, a German engineer and physicist from the University of Würzburg. For this breakthrough, he received the inaugural Nobel Prize for physics in 1901. It remained an open question whether these rays consisted of corpuscles or of waves. Max Theodr Felix von Laue, another German physicist, showed for the first time in 1912 that X-rays were a kind of wave that could be diffracted by a crystal. For this discovery he received the Nobel Prize for physics in 1914.

To extend von Laue's discovery, William Henry Bragg, a British-born physicist, and his son William Lawrence Bragg, an Australian-born British physicist from the University of Leeds, first analyzed the atomic arrangement—that is, a sort of atomic fingerprint—of crystals of common salt by means of X-rays (Bragg, 1912; Bragg and Bragg, 1913). This seminal work heralded a revolution in the scientific understanding of crystals and their atomic arrangements, for which they shared the physics Nobel Prize in 1915. Following this they both pursued research that led to a new branch of science called *X-ray crystallography*. Max von Laue's and Bragg and Bragg's discoveries thus laid the foundations for understanding the structure of several molecules of interest in biology, chemistry, and materials science.

William Henry Bragg had produced some stalwart X-ray crystallographers, including his son William Lawrence Bragg, John Desmond Bernal, William T. Astbury, Dorothy Crowfoot Hodgkin, and many more. In fact, John Bernal and his colleagues at the University of Cambridge

DOI: 10.1201/9781003326106-2

pioneered the use of X-ray crystallography in studying large biomolecules such as proteins (Bernal, 1939). Later, Max Ferdinand Perutz, an Austrian-born British scientist, joined John Bernal and Bragg at Cambridge. In later years, John Kendrew and Francis Crick worked under Perutz and learned the use of X-ray crystallography. William Lawrence Bragg, popularly known as *one of the fathers of X-ray crystallography*, worked as the director of the Cavendish Laboratory, fostered the use of X-ray crystallography, and mentored a whole group of scientists.

THE BIRTH OF STRUCTURAL BIOLOGY

In 1926, James Batcheller Sumner, an American biochemist, succeeded in crystallizing the enzyme urease (which breaks down urea into ammonia and carbon dioxide) and showed that an enzyme is a protein. In 1929, John Howard Northrop, an American biochemist at the Rockefeller Institute for Medical Research, crystallized the pepsin, trypsin, and chymotrypsin enzymes, which are active in the digestive process (Northrop, 1929). For their discoveries Sumner and Northrop shared the Nobel Prize in chemistry in 1946 along with Wendell Stanley. In 1934, John Bernal and Dorothy Crowfoot Hodgkin from the Crystallographic Laboratory in Cambridge, England, also discovered the pepsin crystals and showed the X-ray diffraction patterns (Bernal and Crowfoot, 1934). These studies opened the research area of *protein crystallography*. During mid-1930s, Dorothy Crowfoot Hodgkin independently developed the X-ray crystal structure of insulin while she was at the University of Oxford (Crowfoot, 1935). For this seminal work she was awarded the Nobel Prize in chemistry in 1964.

In 1937, John Bernal gave an ambitious assignment to Max Perutz to determine the structure of haemoglobin, the protein in red blood cells that is responsible for transporting oxygen. After many years, Perutz and his colleagues finally resolved the structure of haemoglobin (Bernal et al., 1938; Perutz, 1942; Bragg and Perutz, 1954; Perutz, 1954; Perutz et al., 1960). At the same time, John Cowdery Kendrew had focused on the crystallography of myoglobin, a muscle protein, which serves as an intracellular storage site for oxygen (Kendrew et al., 1958). For the remarkable achievement of identifying the crystal structure of haemoglobin and myoglobin, Perutz and Kendrew were awarded the Nobel Prize for chemistry in 1962. It is worth mentioning that Linus Pauling, an American chemist at the California Institute of Technology in Pasadena, was also involved in developing the structure of biomolecules. For his seminal work on chemical bonds, he was honoured with the Nobel Prize for chemistry in 1954.

In the early 1950s, Maurice Wilkins, Rosalind Franklin's groups at the King's College, London, and Francis Crick and James Watson at the University of Cambridge first resolved the double-helical structure of deoxyribonucleic acid using X-ray diffraction studies (Franklin and Gosling, 1953; Watson and Crick, 1953; Wilkins et al., 1953). For this ground-breaking contribution, Crick, Watson, and Wilkins were awarded the Nobel Prize in medicine and physiology in 1962. The determination of the 3D structure of DNA, myoglobin, and hemoglobin using X-ray crystallography in the 1950s marked the beginning of structural biology, and a new field of molecular biology evolved and flourished.

The single-minded devotion of William Henry Bragg, Max Perutz, John Kendrew, and Dorothy Crowfoot Hodgkin, all Nobel laureates, played a pivotal role in the creation of this new and important branch of biology, *structural biology*. Recently, Sir John Meurig Thomas, a professor and Master of Peterhouse in Cambridge, has referred to them as the *"architects of structural biology"* (Thomas, 2020).

Donald L. D. Caspar, an American crystallographer, who discovered the structure of tobacco mosaic virus along with Rosalind Franklin of the King's College, London, coined the terms *"structural biology"* and *"structural molecular biology"* for the first time in the 1960s.

FROM LIGHT MICROSCOPY TO ELECTRON MICROSCOPY

The idea of seeing objects through multiple lenses to attain maximum magnification was initiated in the late 1500s by two Dutch opticians, Zacharias Janssen and Hans Janssen, a father and son team. However, Antony Van Leeuwenhoek, a Dutch scientist and one of the pioneers of microscopy, became the first to create and use a microscope for scientific purposes. He used his new microscope to analyse insects and eventually to discover bacteria, earning him the title *"Father of the Microscope."* By the middle of the 17th century, simple microscopes had become popular and were widely used to discover many things in biology. In 1644, Giovanni Battista Odierna had produced the first insights into tissue anatomy (Stafleu and Cowan, 1976; Solares and Kelly, 2022). By the mid-1660s, Robert Hooke published the work *Micrographia* and coined the term *"cell"* (Hooke, 1665).

During the late 1920s, light microscopes were approaching the theoretical limits of their resolution due to the wavelength of light. The conception that it was possible to construct a microscope that used electrons instead of light was first realized when Ernst Ruska was able to build a short-focus magnetic lens using a magnetic coil. Thus, the first transmission electron microscope (TEM) was invented in 1931 by Ernst August Friedrich Ruska, a German physicist and electrical engineer at the Technical University of Berlin under the supervision of Max Knoll. This opened the door to many further instrumental developments that eventually led to the revolutionary elucidation of molecular architectures at an atomic scale (Ruska and Knoll, 1931). By 1938, Ernst Ruska had produced a prototype with a magnification of 30,000. In the late 1960s, this technique was used to deduce the structures of helical fibres (DeRosier and Klug, 1968). For this breakthrough, Ruska shared the physics Nobel Prize in 1986 with Gerd Binnig and Heinrich Rohrer, who developed a scanning tunneling microscope while they were at the IBM Research Laboratories in Zürich, Switzerland. Both of these microscopes would eventually enable humans to see atoms for the first time and would dramatically increase our knowledge of the cells that make up all plant and animal life.

The digital microscope was developed in 1986 by Japanese scientists, which many claim revolutionized the field of microscopy. They created a way to transfer the image from under the microscope to a computer for instant analysis. Nowadays, microscopes have built-in, high-definition monitors, eliminating the need for an external computer to view images.

RISE OF MICROSCOPY TO CRYOGENIC ELECTRON MICROSCOPY

The major challenge with microscopy, even up to recent years, was re-creating the fuzzy 2D images as sharp 3D structures. Over the course of nearly two decades, three researchers from three countries independently created a technique for generating a 3D structure of the protein at an atomic level using an electron microscope. Their technique used vitrification to cool a sample to cryogenic temperatures, thereby allowing the biomolecules to retain their shape in a vacuum. This approach was called *cryogenic electron microscopy* (cryo-EM). The three scientists—Richard Henderson, Jacques Dubochet, and Joachim Frank—led the transformation of the cryo-EM field from *"blobology"* to respectable atomic structures.

In 1975, Richard Henderson, a British molecular biologist, alongside his professional colleague Peter Nigel Tripp Unwin, a neuroscientist from the Medical Research Council Laboratory of Molecular Biology from the University of Cambridge, United Kingdom, had published a 7 Å structure of the membrane protein bacteriorhodopsin (Henderson and Unwin, 1975). This seminal paper demonstrated that it was possible to resolve many of the transmembrane helices in their target protein. With his success in the field, Henderson kept pushing the resolution of bacteriorhodopsin to new heights. In the 1990s, using this new cryogenic method, Henderson and colleagues resolved a high-resolution density map of bacteriorhodopsin to 3.5 Å (Henderson et al., 1990). After that, the atomic barrier was no longer just a dream for cryo-EM.

Jacques Dubochet, a Swiss biophysicist from the University of Lausanne, was the second pioneer; his work was fundamental in adding the *"cryo"* to cryo-EM. He pursued methods based on cryogenic cooling to create a clean form of vitreous ice. The sample freezing process can often damage biological samples because ice crystals may shear fragile protein bonds. By developing a method that cools water faster than ice crystals can form, it was possible to suspend particles in a protective layer of amorphous ice, a process termed *"vitrification"*. Dubochet used this technique to publish the first cryo-EM images of viruses (Dubochet and McDowall, 1981; Adrian et al., 1984; Al-Amoudi et al., 2005). In parallel to Dubochet's work that was initiated at the European Molecular Biology Laboratory in Germany, Robert Glaeser at the Lawrence Berkeley National Laboratory in the USA, Friedrich Zemlin at the Fritz-Haber Institute in Germany, and Fujiyoshi Yoshinori from Japan were experimenting with frozen-hydrated specimens (Dubochet, 2011; Kühlbrandt, 2014, 2022).

Joachim Frank, a German-American biophysicist who did significant work while at the Wadsworth Center in Albany, is currently at Columbia University in New York. From 1975 through 1986 he developed an image-processing method in which the electron microscope's fuzzy two-dimensional images are analyzed and merged to reveal a sharp three-dimensional structure (Frank, 1975; Frank and Al-Ali, 1975; van Heel and Frank, 1981). Most importantly, he developed the single-particle reconstruction approach and applied it to the study of ribosome structure and function (Radermacher et al., 1987; Frank et al., 1991; Srivastava et al., 1995; Frank et al., 2000).

Since 2013, as a consequence of the superior performance of the new cameras, a large number of cryo-EM maps have been obtained at near-atomic resolution (2 to 4 Å). Cryo-EM has become a mainstream technique of structural biology (Nogales, 2016). Single-particle cryo electron microscopy is now being used to solve macromolecular structures at high resolution. This technique was featured on the cover of *Nature Methods* magazine as the *"Method of the Year 2015"* (Method of the Year 2015, 2016; Stuart et al., 2016; Subramaniam, 2019). As the cryo-EM technique *"moved biochemistry into a new era,"* the three pioneers Richard Henderson, Jacques Dubochet, and Joachim Frank were awarded the Nobel Prize for chemistry in 2017 *"for developing cryo-electron microscopy for the high-resolution structure determination of biomolecules in solution"* (Cressey and Callaway, 2017).

Cryo-electron microscopy has become more and more attractive for scientists coming from the structural biology and drug discovery fields. The resolution of proteins can be determined to better than 1.5 Å resolution with the more powerful microscope (Frechin et al., 2023). In recent years, cryo-EM became a powerful tool for public health in determining the structures of Zika virus and SARS-CoV-2. With the advancement of these microscopes, we can now see items at the atomic level and up to 10,000,000 times more closely than the human eye can see.

CRYO-EM IN THE STUDY OF CELLULAR MEMBRANE BIOLOGY

EM was utilized to initiate the structural analysis of cell membranes in the late 1950s (Sjostrand et al., 1958; Robertson, 1960). It was J. David Robertson who identified the dark electron-dense bands as the headgroups and proteins associated with two opposite lipid monolayers (Robertson, 1959, 1960). As mentioned earlier, in the mid-1970s Richard Henderson and Unwin first identified the 3D structure of a membrane protein, bacterial rhodopsin, at 7 Å (Henderson and Unwin, 1975). The first atomic-level, high-resolution structure of a membrane protein, the photosynthesis reaction center from the photosynthetic bacterium *Rhodopseudomonas viridis* was solved by X-ray crystallography by Deisenhofer and colleagues in the early 1980s by growing 3D crystals (Deisenhofer et al., 1985). By refining methods, Henderson and his colleagues had determined the atomic resolution of bacteriorhodopsin to be 3.5 Å (Henderson et al., 1990).

In the early 2010s, using 3D crystallography, many high-resolution membrane protein structures were determined through X-ray crystallography. David Julius, an American physiologist from the University of California–San Francisco, had determined the first atomic resolution single-particle cryo-EM structure of a transient receptor potential cation channel subfamily V member 1 (TRPV1), also known as the capsaicin receptor and the vanilloid receptor 1 (Liao et al., 2013). For this breakthrough on taste receptors, he was awarded the 2021 Nobel Prize in medicine or physiology, shared with Ardem Patapoutian, a Lebanese-born American scientist from Scripps Research in La Jolla, California. Recently, Tian Hua's group from the Shanghai Tech University in China has revealed the cryo-EM structures of human bitter taste receptor TAS2R46 complexed with chimeric mini-G protein gustducin (Xu et al., 2022).

Single-particle cryo-EM has opened up new avenues of investigation in G-protein coupled receptor structure biology, including the study of transient intermediates, continuous motions, and structure determination of flexible complexes. The high-resolution three-dimensional atomic structural information about membrane proteins is extraordinarily beneficial in understanding their molecular mechanism of biochemical and physiological or pathological functions.

HISTORICAL ASPECTS OF MITOCHONDRIAL BIOLOGY AND CRYO-EM

Historically, the mitochondria were first discovered in 1857 by Swiss physiologist and the first light microscope pioneer Rudolf Albert von Kolliker, who named them *"sarcosomes"*. In 1894, German pathologist Richard Altman referred to the observed granules as *"bioblasts"* (Altman, 1890). Later, in 1898, German microbiologist Carl Benda observed these organelles as *"thread granules"* all over the eukaryotic cell cytoplasm and called them *"mitochondria"* (Benda, 1898). These mitochondria were first visualized as cylindrical bodies in the 1950s using transmission electron microscopy (TEM) by the Belgian-American cell biologist and medical doctor Albert Claude and his colleagues (Claude and Fullam, 1945). The first high-resolution electron micrographs were published in 1952–1953 independently by George Palade and Sjostrand, and Palade is the one who found that the mitochondrion is surrounded by a membrane, which is folded to form ridges inside the mitochondrion; he named these ridges cristae mitochondriales (Palade, 1952; Sjostrand, 1953). Because of the production of energy molecules by the mitochondrion, they were popularly known as the *"powerhouse of the cell"* (Siekevitz, 1957).

Mitochondria are membrane-bound organelles, present in the cytoplasm of all eukaryotic cells, that produce adenosine triphosphate (ATP), the main energy molecule used by the cell. Anatomically, mitochondria are surrounded by a double-membrane system, consisting of inner and outer mitochondrial membranes separated by an intermembrane space. The inner membrane forms numerous folds called cristae, which extend into the interior (or matrix) of the organelle (Ernster and Schatz, 1981). They play a major role in breaking down nutrients and generating energy-rich molecules for the cell. Many of the biochemical reactions involved in cellular respiration take place within the mitochondria. The mitochondrial proteins are spread across the matrix, inner membrane, inner membrane space, and outer membrane to carry out many vital processes, including fatty acid metabolism and oxidative phosphorylation.

Over the last century, several biochemists from many countries, including Hans Krebs (German-British), Otto Meyerhof (German), Herman Kalckar (Danish-American), Efraim Racker (Austrian-American), Paul Boyer (American), Fritz Lippmann (German-American), Britton Chance (American), Mildred Cohn (American), David Green (American), Lars Ernster (Swedish), Albert Lehninger (American), Peter Mitchell (British), John Walker (British), and several others contributed to the growth of the new field of *bioenergetics* (Racker, 1980; Madeira, 2018).

Cryo-EM has been successfully used in the study of organelle biology, including mitochondria and chloroplasts, which is one of the subjects of this book.

HISTORICAL ASPECTS OF RIBOSOMAL BIOLOGY AND CRYO-EM

Ribosomes were first observed in bacterial viruses (bacteriophages) by Salvador Luria and his colleagues in 1943. Later they were shown as *"small particulate components"* attached to the endoplasmic reticulum in the rat liver cellular cytoplasm by Romanian-born American cell biologist George Palade in 1955 (Luria et al., 1943; Palade, 1955). In 1958, Howard Dintzis coined the term *"ribosome"* for purified microsomes devoid of membrane fragments (Roberts, 1958). A ribosome is an intercellular structure made of both RNA and protein, and it is the site of protein synthesis in the cell. Throughout the history of electron microscopy, ribosomes have served as an ideal subject for imaging and technological development, which in turn has driven our understanding of *"ribosomal biology"*. The ribosome, which is responsible for protein translation in all domains of life, is a massive ribo-nucleoprotein complex ranging in size from ~2.5 MDa in bacteria to ~4 MDa in eukaryotes (Ben-Shem et al., 2011; Davis and Williamson, 2017). The prokaryotic molecular machines or ribosomes sediment at 70S and consist of 30S and 50S subunits, whereas eukaryotic ones are more complex, sediment at 80S, and consist of 40S and 60S subunits. These subunits are formed by interwoven RNA helices and ribosomal proteins that assemble through a dynamic, multi-step biogenesis pathway. Ribosome biogenesis involves a series of steps, including transcription of pre-ribosomal RNA, processing, post-transcriptional modifications, and folding of rRNA.

Over the past five decades, work from the laboratories of Masayasu Nomura, a Japanese-American from the University of Wisconsin–Madison, later at the University of California–Irvine; Harry Noller, an American biochemist from the University of California–Santa Cruz; James Lake, an American molecular biologist from the University of California–Los Angeles; Debi P. Burma (my co-mentor), an Indian biochemist from the Banaras Hindu University; Peter Moore, an American chemist from Yale University; Alexander Spirin, a Russian biochemist from the Institute of Protein Research, Russia; Knud Nierhaus, a German biochemist from the Max Planck Institute of Molecular Genetics; and many other scientists around the world had contributed and provided insights into ribosome biogenesis and error-free protein synthesis (Nomura et al., 1969; Moore et al., 1974; Lake, 1976; Raziuddin et al., 1979; Rohl and Nierhaus, 1982; Moazed and Noller, 1989; Spirin, 1999).

However, biogenesis remained poorly understood mainly because of the lack of high-resolution crystal structure images. Those were achieved by the groups of H. G. Wittmann and Ada Yonath from the Max Planck Institute and Weizmann Institute of Science, Israel, Thomas Steitz from Yale University, and Venkatraman (Venki) Ramakrishnan from the University of Utah and later the MRC Laboratory of Molecular Biology of the University of Cambridge (Wittmann, 1983; Yonath and Wittmann, 1989; Steitz, 2008; Schmeing and Ramakrishnan, 2009). Over the past 15 years, technical advances in cryo-EM have produced ever-higher resolution structures that led to new insights into ribosome structure and function (Mitra and Frank, 2006; Bai et al., 2015).

The structure of the ribosomal large 50S subunit at atomic resolution of 2.4 Å was first identified in 2000 by Tom Steitz's group using X-ray crystallography (Ban et al., 2000). The crystal structure of the 50S subunit from *Deinococcus radiodurans* was determined at a resolution of 3.1 Å by Ada Yonath's group (Harms et al., 2001). Similarly, the structure of the ribosomal small subunit 30S from *Thermus thermophilus* was determined at an atomic resolution of 3.3 Å and 3 Å independently by Ramakrishnan's and Yonath's groups respectively (Wimberly et al., 2000; Pioletti et al., 2001). The complete crystal structure of the 70S ribosome from *Thermus thermophilus* was elucidated in 2001 by Harry Noller's group at a resolution of 5.5 Å (Yusupov et al., 2001). These crystal studies helped us to understand the binding site of several antibiotics. The crystal structure of the eukaryotic 80S was first identified much later in 2011 in *Saccharomyces cerevisiae* at a resolution of 3 Å (Ben-Shem et al., 2011). The first cryo-EM structure image of the bacterial ribosome was achieved at a resolution of ~40 Å by Joachim Frank and his colleagues (Frank et al., 1991), later achieved at a resolution of 23 Å by Marin van Heel's group (Stark et al., 1995).

With the further development of direct detectors, cryo-EM saw a significant improvement in resolution. This *"Resolution Revolution"* heralded a golden era for cryo-EM, as the technique saw a marked increase in usage globally. Now that it is possible to obtain cryo-EM maps with resolution of less than 2 Å, we can analyze minute structural details of biomolecules. High-resolution cryo-EM structures for human small subunit processome, nucleosome, and spliceosome are also available (Bai et al., 2021; Singh et al., 2021; Ehara et al., 2022; Tholen et al., 2022).

Cryo-EM has been successfully used in the study of molecular machines such as ribosomes, which is one of the subjects of this book.

CRYO-EM IN STRUCTURAL VIROLOGY

Study of Human Viruses

Cryo-EM has been extensively used in the past few years to determine the structure of many coronaviruses and their variants in order to develop antiviral drugs. Andrew Ward's group from Scripps Research analysed the structural features of SARS-CoV-2 spike protein in order to develop a vaccine candidate (Bangaru et al., 2020). Priyamvada Acharya's group from Duke University revealed the spike structure, conformation, and antigenicity of SARS-CoV-2 (Gobeil et al., 2021), whereas Bing Chen's group from the Harvard Medical School studied the infectivity and immune evasion of SARS-CoV-2 variants (Cai et al., 2020, 2021; Zhang et al., 2021). David Veesler's group from the University of Washington used cryo-EM to study antibody responses against SARS-CoV-2 Omicron sublineages (Park et al., 2022), whereas Sriram Subramaniam's group from the University of British Columbia and Jonathan Abraham's group from Harvard Medical School studied antibody evasion and the cryo-EM structure of the spike protein–ACE2 complex from the SARS-CoV-2 Omicron variant (Mannar et al., 2022; Nabel et al., 2022). Recently, the cryo-EM structure of hepatitis C virus E1E2 glycoprotein complex and Crimean-Congo hemorrhagic fever virus was also determined (de la Pena et al., 2022; Mishra et al., 2022).

Cryo-EM in the Study of Bacterial Viruses

Bacteriophage T4 is one of the seven *Escherichia coli* phages, and it is the most well-studied member of Myoviridae, the most complex family of tailed phages. T4 assembly is divided into three independent pathways: the head, the tail, and the long tail fibers (Goldberg et al., 1994). The structure of bacteriophage T4 has been extensively studied using various techniques, including X-ray crystallography and cryo-EM, providing a structural model of phage T4 at near atomic resolution. T4 isometric head cryo-EM structure was revealed at 3.3 Å, its tail tube structure was obtained at 3.4 Å resolution, and its contracted tail (head–tail joining region) structure was determined at 3.4 Å (Aksyuk et al., 2009; Chen et al., 2017; Zheng et al., 2017). Bacteriophage P22 is a member of the family Podoviridae that infects *Salmonella enterica* and *Salmonella typhimurium*; its virion cryo-EM structure was also determined (McNulty et al., 2018).

CRYOGENIC ELECTRON TOMOGRAPHY (CRYO-ET)

Another approach to obtaining 3D structures of macromolecular assemblies using electron microscopy is called tomography, popularly known as *electron tomography* (cryo-ET), in which a series of images is collected, with each image taken at a different tilt relative to the direction of the incident electron beam. Electron tomography has become a standard technique with applications in cell biology, structural biology, and materials science. During the early 1990s, work on cryo-ET was initiated by Wolfgang Baumeister from the Max Planck Institute for Biochemistry in Martinsried, Germany (Dierksen et al., 1992, 1993). Since then, it has evolved steadily into a powerful tool for *structural cell biology* or *visual proteomics*. It has unique potential to generate molecular maps of cellular landscapes at near-atomic resolution. Ongoing developments in technology and methodology hold promise to push the boundaries further toward achieving complete maps and resolution of *in situ* structures (Bäuerlein and Baumeister, 2021; Baumeister, 2022). Visualizing the real-world interplay of proteins in their native environment (*visual proteomics*) is a difficult task, but now scientists are hoping to understand it through cryo-ET (Eisenstein, 2023).

Using cryo-ET, one can produce 3D pictures (tomograms) of complex objects such as asymmetric viruses, cellular organelles, or whole cells from a series of tilted cryo-EM images (Martynowycz and Gonen, 2018; Baumeister, 2022). Perhaps the melding of both cryo-EM and ET technologies will enable the determination of cryo-EM structures at a resolution beyond even 1 Å, which was thought to be a near impossible quest to embark upon. Very recently, researchers have applied cryo-ET to gain insights into processes such as photosynthesis and protein production. Using cryo-ET, the nuclear pore puzzle has been solved independently by four different groups, and researchers have mapped the structure of the nuclear pore complex, an assembly of 30 different proteins that controls access to the cell nucleus. This is all part of the emerging field of visual proteomics (Bley et al., 2022; Fontana et al., 2022; Mosalaganti et al., 2022; Petrovic et al., 2022; Zhu et al., 2022). Cryo-ET has been successfully used in the study of the human immunodeficiency virus-1 maturation and assembly process (Qu et al., 2021). In addition, Benjamin Engel's group from the University of Basel, Switzerland, revealed the protein architecture of cilia from *Chlamydomonas* using *in situ* cryo-ET (van den Hoek et al., 2022). Overall, cryo-ET provides holistic views of the inner spaces of cells and therefore has huge discovery potential.

Cryo-EM tomography combined with cryo-FIB (focused ion beam) sample preparation is a new trend for *in situ* membrane protein structural biology. In the last two years, two notable reports of a hybrid method using SPA and cryo-ET have been developed (Fukuda et al., 2023). One method is known as TYGRESS (tomography-guided 3D reconstruction of subcellular structures), and the other is hStA (hybrid subtomogram averaging) (Sanchez et al., 2020; Song et al., 2020). Using subtomogram averaging, the structure of the SARS-CoV-2 spike protein has been obtained (Ke et al., 2020; Turonová et al., 2020). To be efficient in tilt-series image acquisition, the beam image-shift electron cryo-tomography (BISECT) method has recently been developed (Bouvette et al., 2021). The BISECT method applies a beam tilt–induced image-shift approach combined with SPA. In addition, two other beam image-shift cryo-ET methods, parallel cryo-electron tomography (PACE-tomo) and multishot tomography (Eisenstein et al., 2022; Khavnekar et al., 2022) have been developed. Both PACE-tomo and multi-shot tomography have been used to acquire tilt-series images of not only purified specimens but also focused ion beam (FIB) milled cellular specimens. Ion beam milling and cryo-ET were also used to obtain structures of cardiac and skeletal muscle thin filaments (Wang et al., 2022).

Recently, the ELI-TriScope platform, a new generation of integrated cryo-CLEM system, was developed by Fei Sun from the Chinese Academy of Sciences, Beijing. It provides researchers with co-localized electron (E), light (L), and ion (I) beams. Using this, his group studied the centrioles that sort and separate chromosomes during cell division (Li et al., 2023). Cryo-EM holography is another emerging technology for high-resolution structure determination of a single protein molecule (Longchamp et al., 2017; Appasani, 2022). In a different context, to study X chromosome inactivation, time-resolved structured illumination microscopy (a kind of sequential pulse localization imaging over time) was developed by Neil Brockdorff's group from Oxford University. This study revealed key principles of noncoding RNA Xist (X-inactive specific transcript) RNA spreading (Rodermund et al., 2021). Cryo-EM and related technologies are listed in Figure 1.1.

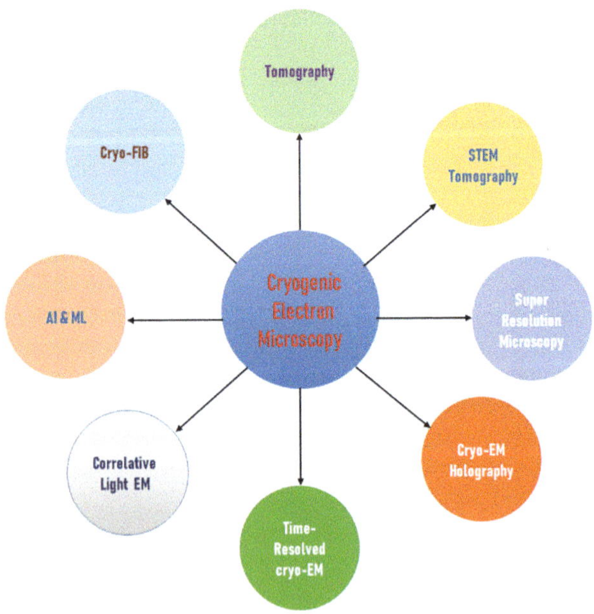

FIGURE 1.1 Cryo-EM and related technologies.

ARTIFICIAL INTELLIGENCE (AI)

AI is the buzz of this decade, and it seems to penetrate every aspect of our daily lives, including the structural biology research field. The most well-known AI program is *AlphaFold2* (simply referred as AlphaFold), a neural-network-based approach and a machine-learning algorithm for protein structure prediction. This was developed by Demis Hassabis and his colleagues at the Alphabet-Google's Deep Mind company in London (Senior et al., 2020; Jumper et al., 2021). In addition, David Baker's group from the University of Washington has provided an alternative machine-learning algorithm and explored network architectures based on the DeepMind framework. The method is called *RoseTTA* fold, which can solve challenging X-ray crystallography and cryo-EM modeling problems and generate accurate models of protein-protein complexes (Baek et al., 2021). In addition, the *RoseTTA* deep-learning method also helped to systematically identify and build structures for the protein complexes that mediate key processes in eukaryotes (Humphreys et al., 2021). The potential applications of this software in the study of biology are listed in Figure 1.2.

AI-driven software has been recognized as the *"Breakthrough of the Year 2021"* by *Science* magazine based on two fronts: the protein folding scientific achievement and enabling of future research. *"AI-driven advances offer a view of the dance of life as never seen before, a panorama that will forever change biology and medicine (Service, 2021)."*

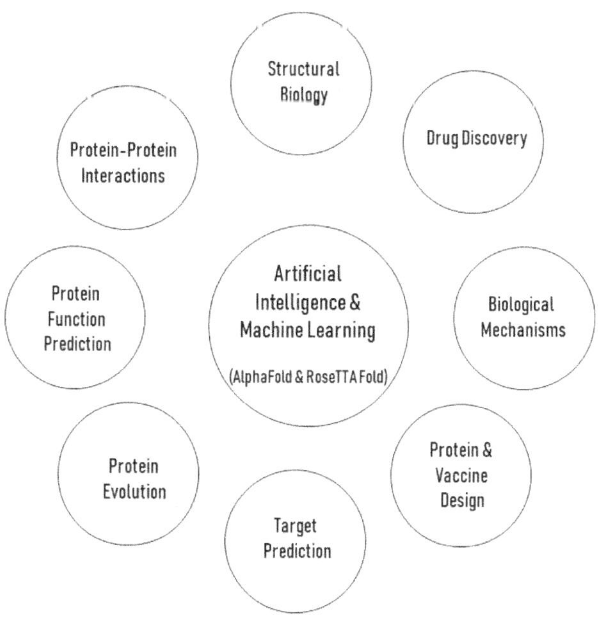

FIGURE 1.2 Applications of artificial intelligence and machine-learning tools in the study of various fields of biology. (Adopted and modified from Yang et al. (2023)).

The pros and cons of AI tools are listed below:

Pros:

A. AlphaFold can accurately predict the 3D structure of soluble proteins, given that the templates from the protein database bank are adequate.
B. Prediction accuracy depends on the accuracy of the 3D structure template and the similarity of the common structural modules and the sequence.
C. Using AlphaFold, it is now possible to predict almost two-thirds of the target structures to an accuracy that is comparable to the experimental results.

Cons:

A. New AI tools have to be developed to analyze and predict conformational changes in proteins specially to solve structures of polymorphic assemblies and protein–nucleic acid complexes.
B. The present tools are not suitable yet for better prediction of protein folding and function and to facilitate protein engineering (Moore et al., 2022).
C. AI might also play an important role in predicting the 3D structure of membrane proteins; however, it cannot be successful until an adequate database of accurate experimental membrane protein structures, including protein–lipid interactions, is available.

Recently, Baker's team has developed *"top-down"* design of protein architectures with reinforcement learning, and their computed protein design data reached a close agreement with the experimentally determined cryo-EM structure of icosahedral viral capsids (Lutz et al., 2023).

COMMERCIAL AND LOGISTICS PERSPECTIVES

Several thousand research papers have already been published in top journals, including *Nature, Nature Biotechnology, Nature Methods, Lancet, Science, Cell, Cell Metabolism, EMBO Journal, Journal of Molecular Biology, PNAS, Journal of Structural Biology*, and others, indicating that the field of cryo-EM and cryo-ET research is growing at an exponential rate within academia, bio-tech, and pharmaceutical industries.

Infrastructure Development and Core Facilities

Both cryo-EM and cryo-ET are still young technologies with a limited workforce and very low access to the necessary equipment and expertise. Therefore, to train a new workforce, and to continually improve the technology while also providing new research discoveries, in 2018 the National Institutes of Health developed an initiative with $129.5 million dollars to support a few core facilities in the United States. These are at SLAC at Stanford University, New York Structural Biology Center at Columbia University,

University of Colorado at Boulder, Oregon Health & Science University in Portland, and University of Wisconsin at Madison, in addition to its own facilities at NIDDK and the National Cancer Institute. Additionally, four multi-year grants for the creation of instructional and hands-on training in cryo-EM methodology have also been awarded to the California Institute of Technology, Pasadena, California; Yale University, New Haven, Connecticut; the University of Utah, Salt Lake City, Utah; and Purdue University, West Lafayette, Indiana. Many other universities, such as MIT, Harvard, the University of Pennsylvania, and others have formed their own core facilities with funding from private endowments.

Cost of the Instrumentation

Generally, a top cryo-EM machine costs about $7 million to install and set up in an appropriate room. The next parameter is operational costs that include staff and utility expenses that can rise to $10,000 per day. Developing a full workflow for cryo-ET can cost upwards of $10.7 million.

Competition

The most popular cryo-EM models are Glacios 2, 130 Krios machines available from Thermo Fisher Scientific, headquartered in Waltham, Massachusetts. The other commercial sellers for cryo-EM include Hitachi (SEM-SU9000) and JEOL (CRYOARM300), but the greatest market share was held by Thermo Fisher Scientific (Titan Krios 300 kV). Many other players supply related instruments and software; these include Olympus Corporation, OPTIKA, Leica Microsystems, ZEISS International, Nikon Instruments Inc., Thorlabs, Inc., Oxford Instruments, Hamamatsu Photonics K.K., NanoFocus AG, HORIBA, Ltd., and Gatan, among others.

Growth of Publications

Early X-ray crystallography experiments were supported through government-owned billion-dollar synchrotron facilities that provided a number of protein structures. However, the structures determined through cryo-EM and deposited in the Protein Data Bank are doubling every year.

Market Trends

The Data Bridge Market Research Firm based in London analysed the growing trends in the global cryo-EM market and forecasted trends for 2023 to 2030. According to their analysis, the market will reach $1,533.08 million by 2030, with a compound annual growth rate of 8.7%. Most of the manufacturers are continuously updating their microscopes with new AI and automation tools in order to get imaging faster and with less sample preparation. It is expected that these new capabilities will drive the global cryo-EM market growth. In terms of market share, the U.S. is expected to dominate due to the rising technological advancements in microscopy, growing R&D investments, and the launch of new core facilities in the bio-pharmaceutical industry. The European Union is the next big market for cryo-EM since many universities and research institutes are installing new core facilities for biomedical science research. It is also true in China, where a number of research institutes are setting up cryo-EM for research as well to develop therapeutics.

SCOPE OF THIS BOOK

This text consists of 34 chapters, grouped into 7 sections, and many of the aforementioned applications are described in various sections of the book, which are summarized as follows. The applications of cryo-EM in various subdisciplines of biology covered in this book are listed in Figure 1.3.

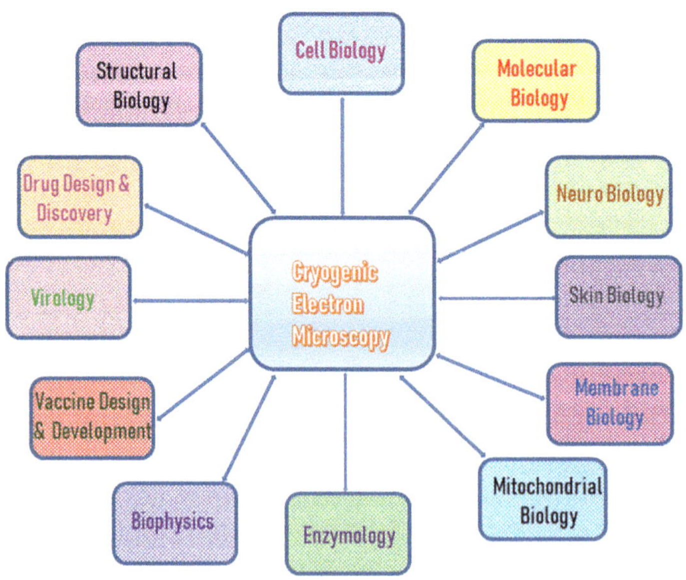

FIGURE 1.3 Applications of cryogenic electron microscopy in various subdisciplines of biology.

SECTION I. CRYO-EM IN CELL BIOLOGY—STRUCTURE DETERMINATION OF PROTEIN ASSEMBLIES AND CELLULAR ORGANELLES

This part consists of five chapters, including this introduction.

The fate of the cellular proteome is elegantly regulated by the ubiquitin–proteasome system. The broad importance of the proteasome in life sciences is exemplified by its roles in the cell cycle, gene expression, antigen presentation, immunity regulation and tumorigenesis, and neurodegenerative diseases (Coux et al., 1996; Ciechanover and Kwon 2015). In recent years the atomic dissection of the innermost workings of the proteasome using cryo-EM and advanced machine-learning approaches have aided in the study of the functional dynamics of molecular machines and have provided implications for drug discovery. In Chapter 2, Yudong Mao has summarized the structural biology of the proteasome and its many conformers using time-resolved cryo-EM. This chapter also narrates proteasome regulation by various ubiquitylation machineries and other proteasome-interacting proteins in general. These studies will undoubtedly drive biotechnological expansion toward the development of novel therapeutics for treating many traditionally incurable diseases.

Ubiquitin-specific protease 1 (USP1) is an enzyme participating in DNA repair that is upregulated in several cancers and is a potential anticancer target (Groelly et al., 2023). Structural characterization of USP1 is crucial to understanding how it is involved in DNA repair pathways and to aid structure-guided drug design of USP1 inhibitors. In Chapter 3, Martin Rennie summarizes the structural basis through which USP1-UAF1 recognizes and deubiquitinates the FANCI-FANCD2 heterodimer. To understand how USP1 deubiquitinates the FANCI-FANCD2 heterodimer at the molecular level, Rennie reconstituted the enzyme–substrate complex and determined its structure by cryo-EM to 2.5–3.7 Å. This work highlights the potential of cryo-EM in the analysis of large enzymes with large substrates.

The protein kinase superfamily is one of the largest gene families, coding for more than 500 kinases (Hanks et al., 1988). Almost all, if not all, cellular signaling events involve phosphorylation that makes kinases important nodes of intervention in much pathology such as cancer, immune disorders, infectious diseases, and cardiovascular and neurological disorders (Brognard and Hunter, 2011). Kinases have remained the center of attraction to biologists, chemists, and chemical biologists alike for many decades. In Chapter 4, Mitra et al. have nicely discussed multiple aspects of kinase structure and activation mechanisms. In addition, they have summarized the applications of cryo-EM analysis and cryo-ET methods to study macromolecular structure and functional relationships, especially in deciphering kinase regulation. The atomic structures unravelled the inner workings of kinase-signaling complexes and ultimately help to develop designer drugs.

Every day, cells are exposed to different stressors. Although the sources range from internal metabolic errors to external radiation exposure, all of these factors can lead to DNA insults culminating in apoptosis (Chatterjee and Walker, 2017). Through evolution, cells have developed robust DNA damage and repair pathways, with the tumor suppressor protein TP53 (p53) serving as the *"guardian of the genome"* (Levine and Oren, 2009). Decades of research have focused on the role of the p53 gene in healthy cells and disease progression. P53 is the most deregulated protein in cancer. In recent years, efforts to study small proteins on the order of p53 have come to life using cryo-EM. In Chapter 5, Solares et al. discuss the historical findings for p53 structure–function studies and demonstrate the unlimited potential to extend cryo-EM studies for small proteins. In addition, Solares et al. explore mutational features by modeling flexible states of the protein based on higher-resolution work. They have utilized biological, technological, and computational advancement tools to better understand the structural properties of p53.

SECTION II. TECHNOLOGY PLATFORM AND EVOLVING TECHNOLOGIES

This part consists of six chapters.

In Chapter 6, Bhakta et al. have reviewed the historical perspective, from the discovery of bacterial cells by Leuwenhoek to the first discovery of EM by Ruska, and then the present cryo-EM. This chapter also highlights the visualization of individual biological molecules in their hydrated state, freeze-plunging specimen-preparation techniques, camera normalization approaches, associated algorithms, data acquisition, and software issues that gave rise to today's routine cryo-EM use (Sass et al., 1989; van Heel and Frank, 1981, Sass et al., 1989; Henderson et al., 1990). Bhakta et al. in this chapter argue that there have also been incoherent methodological deviations that created a swamp of confusion. The main purpose of this chapter is to look back to correct persistent errors and missed opportunities that entered the cryo-EM procedures. In this review chapter, van Heel and his colleagues urge a new generation of cryo-EM researchers to critically look behind their graphic user interfaces (GUIs) to understand that cryo-EM is not a *"black box"*. Finally, van Heel and his colleagues critically review some basic principles of single-particle cryo-EM and warn about procedures and metrics that have gone astray and continue to generate problems.

In Chapter 7, Nils Marechal and Patrick Schultz have summarized different strategies based on affinity layers, which not only allow sequestering the sample far from disruptive interfaces, but also permit work with a reduced sample amount and tight control of sample behavior. In this context, they have developed streptavidin affinity grids in the study of macromolecular complexes using cryo-EM. Moreover, streptavidin grids promise to be of major interest for the study of DNA binding complexes and for developing methods for on-grid purification. Many cryo-EM specimen preparation, image acquisition, data storage, processing, map model building, and software systems have been developed to speed up the determination of about 100,000 protein structures. In recent years, more and more cryo-EM labs or core facilities have been built in academic and biopharma settings. In Chapter 8, Lijia Jia has discussed very elegantly how to set up and run a cost-effective modern cryo-EM facility. Cryo-EM has emerged as a new technology for structural biology, offering distinct advantages over previous approaches.

However, despite cryo-EM's remarkable success, it often suffers from the relatively low resolution of density maps, which poses challenges for accurate structure modeling. To address this need, Daisuke Kihara's group members have been developing a series of computational tools, especially the state-of-the-art protein structure modeling method DeepMainmast (Terashi et al., 2022). In Chapter 9, Bhagirov et al. summarize the application of various computer tools to determine the structure of many biomolecules. In Chapter 10, Koji Yonekura documents artificial intelligence–driven data acquisition and three-dimensional electron diffraction methods to reduce the substantial human involvement required by the conventional control of electron microscopes and their mechanical precision.

Establishing a state-of-the-art cryo-EM facility represents a pivotal enterprise at the intersection of structural biology's advanced imaging techniques and high-volume data management. It also needs meticulous planning and integration of diverse components. From sample preparation to data analysis, storage, and IT infrastructure, every facet plays a crucial role in enabling ground-breaking discoveries in the field of structural biology. Implementing high-performance computing resources, network connectivity, and advanced visualization tools is essential for managing the data-intensive processes involved. In Chapter 11, Shell et al. have provided an overview of the multifaceted process of building a cryo-EM facility that includes sample preparation, data analysis, and the information technology infrastructure. In particular, this chapter help researchers and managers gain the expertise necessary to establish cryo-EM facilities, thereby increasing the availability of this vital technique across the globe.

SECTION III. CRYO-EM IN VIROLOGY AND ENZYMOLOGY—HIGH-RESOLUTION STRUCTURES OF CHAPERONINS AND ENZYMES

This part consists of four chapters.

The study of the mechanisms underlying the process of protein folding in the cell remains one of the major challenges in biochemistry and cell biology. Protein folding assisting complexes, known as heat shock proteins (HSPs), have been discovered in bacteriophages and many other organisms (Georgopoulos et al., 1972; Tissiéres et al., 1974). Proteins such as GroEL and GroES show similar architecture to HSPs, therefore they were allocated to the family of chaperones termed *chaperonins*. Chaperonins are multimeric protein complexes that assist protein folding in an ATP-dependent way. Recently, several chaperonins were also found in the genomes of bacteriophages (Kurochkina et al., 2012). In Chapter 12, Sokolova and Stanishneva-Konovalova have described the cryo-EM structures of phage and bacterial chaperones and discussed the similarities and differences among members of the growing family of chaperonins.

Monkeypox virus (MPXV), an orthopoxvirus genus, belongs to the Poxviridae family and causes infection to humans. This is a global burden according to the World Health Organization, and currently a total of 90,656 cases have been reported. Monkeypox virus is a large, enveloped, double-strand DNA virus that infects a wide range of host species and replicates exclusively in the cytoplasm (Falendysz et al., 2017). The mature poxvirus virions enter cells either by macropinocytosis and release from the endosome compartment or by direct fusion with the cell plasma membrane. During DNA replication, a few viral proteins are known to be involved, including the DNA polymerase holoenzyme. The MPXV polymerase holoenzyme consists of three protein subunits, F8, A22, and E4 (Peng et al., 2023). The structural studies of this enzyme help us to develop therapeutic targets to control the infection in humans. Shi and Gao utilized the power of cryo-EM to resolve the structure of the holoenzyme/DNA complex at a resolution of ~2.8 Å, and this is summarized in Chapter 13. Additionally, the EM map described in this chapter shows the key structural features of all proteins and DNA elements.

Over the past decade, cryo-electron microscopy has emerged as a central technique for the structural analysis of macromolecules, particularly large complexes and membrane proteins (Fukuda et al., 2023). Because of the difficulty of producing many of these macromolecular complexes in recombinant systems, many researchers have purified them from their native hosts, often gaining new biochemical insights as a result. In Chapter 14, Venugopal et al. have described a novel affinity-tag-based native purification method for isolating the Huc complex from *Mycobacterium smegmatis*. This bacterium uses Huc, an enzyme (forms an 833 kDa complex) that oxidises the trace quantities of H_2 present in the atmosphere as an energy source, allowing it to live on air. These authors have used cryo-EM and determined the structure of Huc, at an exceptional resolution of 1.52 Å (Grinter et al., 2023). Reconstruction of the Huc structure from cryo-EM images allowed them to reveal that the enzyme forms a C4 symmetrical complex, with a flexible membrane anchoring tail. Hopefully, this novel affinity-tag-based strategy will serve as a practical example of how this methodology can be used to determine the structure and function of other challenging macromolecular complexes.

Staphylococcus aureus is the leading cause of nosocomial and community-acquired infections, with an annual killing rate of over 100,000 people globally (Howden et al., 2023). Pore-forming toxins (PFTs) constitute the largest class of bacterial virulence factors (Los et al., 2013). PFTs are amphipathic proteins that are secreted as water-soluble monomers, which oligomerize on host cellular membranes and subsequently puncture holes by forming a transmembrane pore complex. *S. aureus* employs several such PFTs, including α-hemolysin, γ-hemolysin, Panton-Valentine leukocidin, and leukocidins AB and ED. However, how this PFT interacts with lipids in its native environment remains largely unexplained. In Chapter 15, Mishra et al. have employed cryo-EM for structural characterization of γ-hemolysin in a lipid membrane environment, elucidated the native oligomeric state of γ-hemolysin in liposomes, and predicted the amino acid residues engaged in binding to the membrane. They resolved an octahedral superassembly of γ-hemolysin oligomers to 3.5 Å and observed the molecular interactions stabilizing these superassemblies. These studies further help to aid therapeutic interventions for toxin–membrane interaction and pore formation.

SECTION IV. CRYO-EM IN THE STUDY OF MEMBRANE BIOLOGY—PROBING THE STRUCTURES OF MEMBRANE PROTEINS

This part consists of seven chapters.

The biomembrane is an essential component of living cells. Until the 1970s we did not fully understand the presence of integral or membrane proteins. In the early 1970s, Singer and Nicolson proposed the *fluid-mosaic model* to explain the structure of the lipid bilayer and demonstrated the location of integral membrane proteins (Singer and Nicolson, 1972). Since then we have learned that the cell membrane consists primarily of lipid bilayers, membrane proteins, and other lipid-based substances. Membrane proteins are distinct macromolecules that exist as either integral, peripheral, or lipid-anchored. The integral membrane proteins include transporters, channels, and receptors. They play essential roles in facilitating the transfer and transport of molecules and substances across the membranes. Most importantly, they are involved in biological processes like cell signaling, cell adhesion, sense perception, metabolism, signal transduction, and photosynthesis. To study the structure and function of membrane proteins one has to consider the lipid environment, especially how the lipid bilayer constituents play a role in folding, assembly, and stability. Membrane proteins are difficult to crystallize in two-dimensional or three-dimensional structures, and they are difficult to stabilize following extraction from the cell membrane using detergents.

The field of *membrane protein structural biology* is entering a golden age due to recent technological advancements such as single-particle cryo-EM technologies and the creation of detergent-free methods for sample preparation. Apart from sample preparation using detergent or detergent-free methods, another concern is the host cells of membrane proteins those overexpressed heterogeneously in many organisms, including *E. coli*, yeast, insect, CHO cells, and HEK 293 cells.

In Chapter 16, Youzhong Guo summarizes the details of membrane-embedded proteins in the lipid bilayers and elaborates on the use of various lipid detergent and purification methods. This chapter also covers representative cases or model studies of structure determination of six membrane proteins: (1) bacteriorhodopsin, (2) mammalian aquaporin, (3) human patched 1 protein, (4) *E. coli* mechanosensitive channel protein, (5) *E. coli* multi-efflux transporter, AcrB protein, and (6) the human Slo1 channel. This chapter also covers the recently used native cell membrane nanoparticles (NCMN) system, which deduced the high-resolution cryo-EM structure of the full-length human platelet integrin αIIbβ3, a challenging membrane protein and a drug target for anticoagulation therapy (Adair et al., 2023). Additionally, Guo summarizes the cryo-focused ion beam (cryo-FIB), an emerging technology platform to study membrane proteins using cryo-EM tomography (Berger et al., 2023).

The prokaryotic signaling mechanism was first discovered in *Streptococcus pneumonia* and bioluminescence *Vibrio fischeri* (Tomasz, 1965; Nealson et al., 1970). This signaling/communication is mediated by pheromones or signaling molecules known as autoinducers. The autoinducer-2 exporter (AI-2 exporter) family is a large prokaryotic secondary active transporter family principally described in bacteria and archaea (Rettner and Saier, Jr., 2010). Autoinducer-2, a universal molecule for both intra- and interspecies communication, is involved in the regulation of biofilm formation, virulence, motility, chemotaxis, and antibiotic resistance. AI-2 acts as a universal signaling molecule in prokaryotes and is involved in quorum sensing (Bassler and Losick, 2006).

Although Single Particle Analysis (SPA) has greatly advanced the *membrane protein structure biology* arena, optimizations are time-consuming, and certain steps in the workflow are still not completely streamlined. In Chapter 17, Khare has systematically described the steps in the current single-particle workflow. She adopted this methodology to *E. coli* autoinducer-2 exporters (TqsA and YdiK), novel members of the membrane transporter family, and obtained their first high-resolution structures at 3.35 Å and 2.80 Å respectively (Khera et al., 2022). Additionally, she concludes by highlighting the importance of complementary biophysical, genetic, and computational studies that work together with SPA to provide an overall representation of the functioning of the membrane protein superfamily.

G protein–coupled receptors (GPCRs) are a large family of cell membrane receptors that play crucial roles in many cellular processes and respond to various external signals, including light, ions, lipids, small molecules, peptides, and proteins (Kobilka, 2007; Mustafi and Palczewski, 2009). GPCRs can bind to various G protein subtypes, such as Gs,

Gi/o, Gq, and G12/13, to set in motion specific physiological functions, including taste, vision, and synaptic transmission (Flock et al., 2017). GPCRs are encoded as single polypeptide chains that have an extracellular N-terminus, an intracellular C-terminus, and seven helical transmembrane (TM1-TM7) regions linked by three extracellular loops and three intracellular loops (Molday, 1998). The first cryo-EM structure of a GPCR-G protein complex was determined in 2017; since then, cryo-EM has revolutionized GPCR structural biology (Liang et al., 2017; García-Nafría et al., 2018; Chen et al., 2021).

In Chapter 18, Gulati and Palczewski have examined dimerization in rhodopsin protein, which presents in the rod photoreceptor cells of the retina. Interestingly, rhodopsin is both a retinal-binding protein and a GPCR involved in visual transduction (Khorana, 1993). In this chapter, they summarize the structural insights on GPCR dimerization (Gulati et al., 2017). The study of *GPCR biology* and its dimerization using cryo-EM is a promising area of research that has the potential to lead to new therapeutic strategies for various diseases.

GPCRs present in many organisms, including the yeast *Saccharomyces cerevisiae,* and they help the fungi in sensing their environment by detecting a wide range of signals. Ste2 is a fungal GPCR that is essential for mating in the yeast. Since no structures of fungal GPCRs are available, it was difficult to understand their function. Despite major attempts from several laboratories to crystallize Ste2 over the past two decades, no structures of Ste2 could be determined (Naider and Becker, 2020). Thus, Velazhahan and Tate developed new methodologies and strategies to determine the first high-resolution structure of Ste2 as described in Chapter 19. They developed a novel engineered Gpa1 protein to stabilize the active-state Ste2 complex for structural studies. They also reasoned that cryo-EM would be the best method for determining the structure of Ste2 since cryo-EM eliminates the need for crystal formation. By employing the above strategies, they were finally successful in obtaining cryo-EM images of Ste2 at a resolution of 2 Å (Velazhahan et al., 2021). This chapter also covers the cryo-EM studies on Ste2 dimerization.

Vacuolar-type adenosine triphosphatases (V-ATPases) are large protein complexes in many cellular organelles, including lysosomes, endosomes, and the Golgi apparatus in eukaryotic cells (Vasanthakumar and Rubinstein, 2020). These play a crucial role in maintaining the acidic pH of these cellular organelles and are essential for various cellular processes. Dysregulation of V-ATPases can impact cellular and organismal functions and have been associated with many diseases, including osteoporosis, renal tubular acidosis, cancer, and neurodegeneration (Hinton et al., 2009). In order to understand the molecular mechanism of the human V-ATPases and to reveal their assembly, function, and regulation, cryo-EM structures have been deduced to atomic scale by Fosuah et al., and this is reported in Chapter 20. In addition, this chapter also describes how V-ATPase activity was regulated by small molecule inhibitors and binding proteins through structural analysis of V-ATPase complexes.

Ryanodine receptors (RyRs) are large ion channels located in the sarcoplasmic reticulum and endoplasmic reticulum of cells. These ion channels are essential in intracellular calcium signaling pathways, and they are involved in many cellular processes, such as muscle contraction, synaptic transmission, proliferation, differentiation, and neuronal plasticity (Zalk et al., 2007). In fact, mutations in these ion channels lead to dysregulated Ca^{2+} homeostasis and cause several muscle disorders. Three distinct isoforms of RyRs have been identified: RyR1, RyR2, and RyR3. Despite the great progress made in the past decade, many aspects of *RyR biology* remain to be explored (Hadiatullah et al., 2022). Therefore, Miotto nicely summarizes in Chapter 21 how cryo-EM can be used extensively to investigate the structure and function of RyRs. Additionally, he has elaborated the whole process involving RyR purification, cryo-EM sample preparation, data acquisition, image processing, and atomic model building. Cryo-EM has helped his group to obtain RyR structures at near-atomic resolution at 2.58 Å (Miotto et al., 2022). This chapter also highlights the structures of pathological RyR variants that helped us to understand the disease mechanisms. Hopefully, the *cryo-EM/ET resolution revolution* will allow them in a few years to understand many of the physiological and pathological mysteries around these channel proteins from ligand gating to disease origins.

Mitochondrial membrane protein complexes have been extensively studied using biochemical, biophysical, and structural studies due to their high abundance, stability, and role in energy generation. Understanding the bioenergetic mechanisms of these proteins, especially the enzymes involved in the respiratory electron transport chain, are crucial. The recent cryo-EM structure studies helped us to understand not only individual enzyme complexes but also supercomplexes of multiple different proteins. In Chapter 22, Yadav et al. nicely reviews the structural and functional aspects of mammalian F_1F_O ATP synthase in detail. This chapter also summarizes the architecture of the higher-order oligomers of the ATP synthases using cryo-EM. Using the cryo-ET technique, the role of ATP synthases in cristae formation and maintenance has been also demonstrated. The field of *mitochondrial biology* is quite fascinating to researchers since the functional defects in several mitochondrial proteins have been linked to health problems such as cancer, aging, obesity, cardiovascular diseases, and neurodegenerative disorders (Ebanks and Chakrabarti, 2022).

SECTION V. CRYO-EM IN THE STUDY OF MOLECULAR BIOLOGY—STRUCTURAL INSIGHTS INTO GENE MACHINERIES

This part consists of three chapters describing the cryo-EM structures of ribosomes.

Ribosomes are macromolecular assemblies that are sites for protein synthesis or translation in all cells. These are protein factories that read the messenger RNA (mRNA) sequence and translate that genetic code into a specified string of amino acids, which grow into long chains that fold to form proteins. Ribosomes are composed of two subunits, the large and the small subunit, both of which consist of ribosomal RNA (rRNA) molecules and a variable number of ribosomal proteins. Overall, these are large and dynamic thermal ratchet molecular motors at the heart of information processing in cells.

In Chapter 23, Webster et al. have provided a detailed historical perspective of the development of electron microscopy (EM) with respect to *ribosome biology* in understanding the life cycle of dynamic ribosomes or gene machines. This chapter focuses on the study of ribosomes, especially in eukaryotes. Study of these gene machines could lead to the development of better antibiotics against life-threatening diseases (Ramakrishnan, 2018). As described early in this chapter, for the past several decades EM has contributed to our understanding of ribosome structure and function relationships in both eukaryotes and prokaryotes. This chapter also emphasizes the use of cryo-EM in the field of *ribosomology* and addresses key issues, including ribosome biogenesis, translation, and sequestration, along with their associated cofactors, and their distributions in diseased states. Recently, cryo-EM structures of mitoribosomes has revealed those involved in membrane-tethered mitochondrial protein synthesis and translational termination (Desai et al., 2020; Itoh et al., 2021; Saurer et al., 2023).

Bacterial *Escherichia coli* small subunit (30S) consists of 16S ribosomal RNA (r-RNA) 1542 nucleotides long and 21 ribosomal proteins (r-proteins). The large subunit (50S) contains two rRNAs, 2904-nucleotide-long 23S rRNA, 120-nucleotide-long 5S rRNA, and 33 r-proteins. In Chapter 24, Arora et al. have summarized the biogenesis of prokaryotic ribosomes and the early work performed on the *in vitro* assembly intermediates. Additionally, this chapter covers the recent cryo-EM complexes of the biogenesis factors and their role in incrementally driving the conformations into fully mature subunits. These atomic resolution structural studies of the assembly intermediates may be explored as a new drug target in the future.

The ribosomes of pathogenic organisms are the major target for antibiotics since most antibiotics target the protein translational machinery (Poehlsgaard and Douthwaite, 2005). Understanding the structural aspects of these molecular machines and their interaction with various drugs can provide novel insights into the working of these drugs. In addition, they help us understand the evolution of drug resistance, and they provide avenues to develop new targeted drugs. Since it is ideal to decode the structure of ribosomes from pathogenic organisms, Dhur et al. have summarized the structural details of ribosomes from several species of pathogens using cryo-EM in Chapter 25. This chapter also covers the cryo-EM structures of ribosomes from both prokaryotic and eukaryotic pathogens.

Studying ribosomes *in situ* via electron microscopy is typically accomplished using cryo-electron tomography (cryo-ET), often preceded by the focused ion beam (FIB) milling method (Asano et al., 2016; Xing et al., 2023; Young and Villa, 2023). Additionally, translation machinery could be visualized through time-resolved cryo-EM (Kaledhonkar et al., 2019).

SECTION VI. CRYO-EM IN THE STUDY OF NEUROBIOLOGY AND SKIN—NEURORECEPTORS TO ION CHANNELS AND SYNAPTIC NEUROTRANSMISSION

The next four chapters in this section discuss the importance of cryo-EM in the fields of skin biology and neurobiology.

Ion channels are integral membrane proteins that form aqueous pores that enable ions to rapidly diffuse along their electrochemical gradient. This ion flux precisely controls diverse physiological processes in electrically excitable cells. In the mammalian central nervous system, ion channels orchestrate the signalling between neurons and other cells that shapes synaptic transmission, plasticity, and information processing. Among the most widely studied and functionally critical channels mediating brain function are glutamate receptors, which transduce the majority of excitatory neurotransmission. Milestone discoveries made in the 1950s and 1960s on the electrophysiology of the glutamate receptors, and molecular cloning experiments done in the early 1990s, helped us to understand *receptor physiology* (Armstrong et al., 1998; Collingridge et al., 2009). Recent advances in cryo-EM have provided unprecedented atomic-resolution snapshots of glutamate receptor architecture in multiple conformational states (Creekmore et al., 2021).

In Chapter 26, Bogdanovic describes the cross-disciplinary research integrating genetics, biochemistry, electrophysiology, and structural biology to unveil the elegant intricacy of glutamate receptor structure–function relationships that underlie synaptic transmission in the brain. In addition, this chapter also highlights the importance of the cryo-EM structures of ion channels and their involvement in numerous neurological diseases, including Alzheimer, autism, and many neuropsychiatric disorders. Finally, Bogdanovic suggests that upcoming methods such as new super-resolution imaging, spatial proteomics, multi-omics linked to physiology, and structure-based drug design coupled with patient-derived neurons may help to address synaptic computation, plasticity, learning and memory, and overall treatment of brain disorders.

Systemic amyloidoses are degenerative disorders caused by the misfolding and aggregation of proteins into amyloid fibrils that deposit in diverse tissues, causing organ failure and death. Organs commonly affected include the heart, gastrointestinal tract, kidney, liver, nerves, and eye (Gertz and Dispenzieri, 2020). The term *amyloid* was coined by German pathologist Rudolf Virchow in 1854 to refer to starch-like deposits in liver, and later its composition was discovered by Friedreich and Kekulé in 1859

(Virchow, 1854; Friedreich and Kekulé, 1859). Forty-two proteins have been identified and implicated in the formation of both localized and systemic amyloidosis, including Aβ, tau, and α-synuclein.

During the mid-2010s, the structures of dozens of amyloid fibrils were identified with the use of cryo-EM (Fitzpatrick et al., 2017; Sawaya et al., 2021). It is very important to mention that cryo-EM was used to deduce amyloid fibril structures from the systemic amyloidosis patient samples and Aβ42 filaments from Alzheimer patients' brains (Schmidt et al., 2019; Nguyen et al., 2022; Steinebrei et al., 2022; Yang et al., 2022). In Chapter 27, Binh An Nguyen et al. present historical perspectives and show the mechanism of formation of amyloid fibrils in systemic amyloidoses. In addition, this chapter describes how the cryo-EM technique was adopted to study the structures of amyloid fibrils from systemic amyloidosis. Finally, this chapter highlights the description of cryo-EM structures of amyloid fibrils extracted from the organs of patients suffering from systemic amyloidosis.

The 5-hydroxytryptamine$_3$ (5-HT$_3$) receptor is a member of the Cys-loop family of ligand-gated ion channels, which includes nicotinic acetylcholine (nACh), GABA$_A$, and glycine receptors (Reeves and Lummis, 2002). In the late 1950s, the 5-HT$_3$R was initially identified as a serotonin receptor found in the submucosal neurons of the guinea pig intestine. Serotonin receptors are a large family of neurotransmitter receptors located in the nervous system. They respond to chemical signals such as serotonin (5-HT) and its derivatives to transfer nerve impulses across neurons. Electrophysiological experiments furthur suggested that 5-HT$_3$ are indeed ion channels (Derkach et al., 1989).

Basak and his colleagues first reported the cryo-EM structure of apo 5-HT$_{3A}$R in the absence of nanobodies, which showed an overall architecture similar to that of the nanobody-bound crystal structure (Basak et al., 2018). In Chapter 28, Li et al. summarize the *biology of 5-HT$_{3A}$R serotonin receptors*, which includes their cryo-EM studies in greater detail. These studies helped us to understand the structures of both the apo and serotonin-bound states of 5-5-HT$_3$, demonstrating conformational transitions upon agonist binding. This chapter also covered the analysis of 5-HT$_{3A}$R structures within lipid environments, offering novel insights into the asymmetric opening of the receptor.

Human skin serves to uphold homeostasis by preventing water loss from the body as well as by preventing permeation of exogenous substances into the body. The major physical barrier is situated in the skin's topmost epidermal layer, the stratum corneum (Norlén and Al-Amoudi, 2004). Norlen has utilized cryo-EM during epidermal differentiation and has observed that the skin's barrier structure undergoes five apparent maturation stages, as described in Chapter 29.

Cryo-EM of vitreous sections allows for investigation directly *in situ* of the molecular architecture of skin (Norlén et al., 2003). In addition, Norlen gives an overview of the cryo-EM of vitreous sections procedure and its analysis using atomic-detail molecular dynamics modeling and electron microscopy simulations. Since these studies were obtained from healthy skin, it may now be possible to explore differences in lipid structure that could be the basis of barrier malfunction in skin diseases such as atopic dermatitis, psoriasis, and the ichthyoses.

SECTION VII. EMERGING TECHNOLOGIES AND APPLICATIONS OF CRYO-ET AND CRYO-EM IN THE BIOPHARMACEUTICAL INDUSTRY

This part consists of five chapters that highlight the applications of cryo-ET in solving high-resolution structures.

The development of new techniques in light and electron microscopy have propelled the discovery of novel cell biological events. At the forefront of these advancements, cryo-ET is a powerful technique that provides three-dimensional insights into cellular components and organelles. While cryo-EM excels in determining the average structure of purified macromolecules, cryo-ET takes it further to reveal the intricacies of whole cells, their components, and even tissues (Khanna and Villa, 2022; Turk and Baumeister, 2020; Hong et al., 2023).

In Chapter 30, Khanna has detailed the recent developments and breakthroughs in the cryo-ET field, highlighting their potential to enhance resolution, efficiency, and accessibility. Among the groundbreaking technologies, cryo-focused ion beam (cryo-FIB) milling stands out as an innovative approach enabling precision thinning of cellular specimens. Complementing this is cryo-correlative light and electron microscopy (cryo-CLEM), which facilitates the correlation of molecular-level intricacies with the broader cellular context. This chapter also discusses the hardware and software developments that play an essential role in cryo-ET data acquisition. Additionally, Khanna has illustrated a few examples of how cryo-ET alone as well as combined with other techniques has resolved the high-resolution structures of many biological macromolecules. Finally, this chapter focuses on recent work that has revealed a dynamic model of the human nuclear pore complex by combining AI-based structure prediction with single-particle cryo-EM maps, cryo-FIB-ET coupled with subtomogram analysis and integrative modeling (Mosalaganti et al., 2022).

Mitochondria generate energy molecules in eukaryotic cells, whereas chloroplasts in plant cells, algae, and cyanobacteria are responsible for generating energy and producing sugar molecules by converting CO_2 to carbohydrates through a complex process known as photosynthesis (Cooper and Hausman, 2013). Chloroplasts contain an extensive network of thylakoid membranes, which are arranged in stacks known as *grana* and surrounded by fluid called *stroma* (Albertsson, 2001).

Yadav et al. in Chapter 22 of this book's section IV described the structures of mitochondrial proteins using cryo-EM. However, in Chapter 31 of this section, Ahmed and Saha have detailed the cryo-ET structures of mitochondrial electron transport chain complexes, mitochondrial ATP synthases, photosystems and associated light-harvesting complexes, and chloroplast ATP synthases (Perkins et al., 1998; Daum and Kühlbrandt, 2011). High-resolution cryo-EM studies have provided detailed insights into the functional mechanism of these protein complexes. Additionally, this chapter provides historical details about mitochondria and chloroplasts and discusses the topography of various membrane protein complexes in these two organelles using mainly the results obtained from conventional TEM and cryo-ET studies. This chapter also highlights the recent application of biological cryo-FIB milling and cryo-ET, that has proven to be successful in mining thylakoid membranes and the association of organelle membrane proteins at molecular level (Marko et al., 2007).

The actin cytoskeleton is pivotal in orchestrating cell migration and cellular morphology across diverse eukaryotic organisms. The diverse architectural arrangements of the actin cytoskeleton contribute to various cellular functions, including protrusion, adhesion, contraction, and retraction (Fäßler et al., 2023). High-resolution structures of actin filaments and assemblies are crucial to deciphering the underlying molecular mechanisms. In recent years, cryo-ET has significantly advanced our knowledge of the structural biology of actin organization within cells at a nanoscale level (van der Heide et al., 2007). In Chapter 32, Hanein and Volkmann describe actin-filament organization by generating hundreds of tomograms. They also discuss various strategies for controlling the cellular environment and the role of factors that play a critical part in understanding the dynamic behavior of the actin cytoskeleton. Despite the impressive progress attained through these approaches, their suitability for cell biology studies is often constrained by several factors that are discussed in detail in this chapter. This chapter also describes cryo-ET complementary techniques such as cryo-CLEM and cryo-FIB milling methods to get thin sections that help us to understand the nanoarchitecture of the actin cytoskeleton (Mahamid et al., 2015; Klein et al., 2021).

The development of drugs to treat human diseases usually requires the modulation of proteins or RNA by either biologics or new chemical entities (de la Torre and Albericio, 2022). Since the old-fashioned drug discovery and optimization cycle is very time-consuming, today the pharmaceutical industry employs knowledge-driven structural biology techniques such as cryo-EM (Sader et al., 2020).

In Chapter 33, Kong et al. discuss the impact that cryo-EM has had on the drug discovery process, from small and large molecule structure to its specific application to membrane proteins and vaccine design. Cryo-EM has shown immense potential in the development of new drugs based on enzyme complexes, neurodegenerative filaments, GPCRs, and transporters. This chapter provides several examples of the cryo-EM structures of many molecules that are ultimately developed as potential drugs to treat human diseases. Interestingly, this chapter also highlights the development of novel vaccines for combating the recent Covid-19 epidemic. Less than one year after the discovery of the Covid-19 virus and its variants, efficacious vaccines were developed at an unprecedented speed because of the structures obtained by cryo-EM (Wrapp et al., 2020). In industrial settings, cryo-EM and cryo-ET help to determine the structures of large protein complexes like G protein–coupled receptors (GPCRs), membrane proteins, antibody complexes, and proteolysis targeting chimeras (Paiva and Crews, 2019).

In addition to the use of cryo-EM in drug discovery, in Chapter 34, Yao details the adoption of both cryo-EM and cryo-ET techniques in all stages of early drug development, including (1) identification of novel drug candidates, (2) elucidation of drug–target complex structures, (3) analysis of drug effects, and (4) exploration of drug delivery systems. In fact, visualizing the structures obtained from both cryo-EM and cryo-ET helped biopharmaceutical researchers to discover structure-based drug design and vaccine development (Earl and Subramaniam, 2016; Cai et al., 2020). This chapter highlights few examples of the use of cryo-EM structures in the development of Covid-19 mRNA vaccine BNT162b2 and Moderna's mRNA1273 (Corbett et al., 2020), as well as the development of drugs for obesity and blood glucose control based on the cryo-EM structure of glucagon-like-peptide-1(GLP-1) receptor (Wang et al., 2023).

FUTURE PERSPECTIVES OF STRUCTURAL BIOLOGY

Because of the difficulties of specimen preparation and the high costs of using cryo-EM for biomolecule structure determination, we need a people's cryo-EM. Scientists hope to bring the revolutionary microscope to the masses (Hand, 2020). Efforts are underway at Hitachi High-Technologies Corporation, Japan, to develop a new and cheaper cryo-EM that operates at 30 kV with cryo-transfer holder, primarily to observe frozen samples of living cells (Usukura et al., 2018).

Recently Richard Henderson, Christopher Russo, and colleagues from the Laboratory of Molecular Biology in Cambridge, UK, have developed a new cryo-EM that operates at 100 keV (McMullan et al., 2023). This new instrument features advances in electron optics, detection, and processing that make structure determination fast and simple at a fraction of current costs. They have determined structures of 11 different specimens, ranging in size from 140 kDa to 2 MDa. Hopefully, this new instrument will expand experimental structural biology to a wide range of previously intractable problems and will be more affordable to many labs once it is commercially available (McMullan et al., 2023).

The recent advances in computational biology (AI/ML)–based protein structure prediction mark the beginning of a new era in structural biology. Surely these algorithms will accelerate biomedical science research to speed up drug discovery and foster biotechnology applications. However, on the experimental side, a combination of cell biology tools such as correlative light and electron microscopy (cryo-CLEM), single particle analysis (SPA), cryo-FIB milling, and cryo-ET promises to bridge cell biology and structural biology to discover the structural and functional intricacies of proteins (Young and Villa, 2023).

A long-term goal of structural biology remains the visualization of molecular structures in their natural context, often referred to as in-cell or *in situ* structure determination. As mentioned earlier in this chapter and by other contributors in this book, methods such as mass spectrometry in conjunction with cryo-EM and cryo-ET will complement the use of AI tools and algorithms (Cramer, 2021). The combination of AI-based algorithms with cryo-EM and cryo-ET helps us to map large protein complexes. However, we have to wait and see how this approach can be further applied to understand large protein assemblies.

Based on the cryo-EM/ET high-resolution structures, scientists can now accelerate the path to new treatments and therapies for diseases such as Alzheimer's, Parkinson's, and Huntington's, as well as cancers, AIDS, malaria, Zika, Covid-19, and more. Cryo-EM has the potential to affect every corner of medicine, from protein and cellular engineering to neurological diseases and vaccine development. The high-resolution structures of coronaviruses deduced from cryo-EM/ET and the mRNA technology platform helped many biopharmaceutical companies to develop vaccines for the Covid-19 pandemic.

Finally, we must wait to see how the combination of structural biology with *"artificial intelligence will change the transformative research in the 21st century that impacts science and society (Cramer, 2021)."*

ACKNOWLEDGEMENTS

Due to space constraints, I have limited the citation of references but have not intentionally omitted the work of others.

REFERENCES

Adair, B. D., Xiong, J. P., Yeager, M. and Arnaout, M. A. (2023). Cryo-EM structures of full-length integrin αIIbβ3 in native lipids. *Nature Communications*, **14**, 4168.

Adams, F. D. (1918). Haüy, the "father of crystallography". *American Mineralogist*, **3**, 131–132.

Adrian, M. Dubochet, J., Lepault, J. and McDowall, A. W. (1984). Cryo-electron microscopy of viruses. *Nature*, **308**, 32–36.

Aksyuk, A., Leiman, P. G., Kurochkina, L. and Rossmann, M. G. (2009). The tail sheath structure of bacteriophage T4: A molecular machine for infecting bacteria, *European Molecular Biology Organization Journal*, **28**, 821–829.

Al-Amoudi, A., Studer, D. and Dubochet, J. (2005). Cutting artefacts and cutting process in vitreous sections for cryo-electron microscopy. *Journal of Structural Biology*, **150**, 109–121.

Albertsson, P. Å. (2001). A quantitative model of the domain structure of the photosynthetic membrane. *Trends in Plant Sciences*, **6**, 349–354.

Altman, R. (1890). Die Elementarorganismen und ihre Beziehungen zu den Zellen [The cellular organelles and their relations to cells], p. 125. Veit & Co., Leipzig, Germany.

Appasani, K. (2022). Introduction on Single-Molecule Science. In Single Molecule Science, Appasani, K. (ed), pp. 3–19, Cambridge University press, Cambridge, UK.

Armstrong, N., Sun, Y., Chen, G.-Q. and Gouaux, E. (1998). Structure of a glutamate-receptor ligand-binding core in complex with kainate. *Nature*, **395**, 913–917.

Asano, S., Engel, B. D. and Baumeister, W. (2016). In situ cryo-electron tomography: A Post-reductionist approach to structural biology. *Journal of Molecular Biology*, **428**, 332–343.

Baek, M., Dimaio, F., Anishchenko, I. and Baker, D. (2021). Accurate prediction of protein structures and interactions using a three-track neural network. *Science*, **373**, 871–876.

Bai, R., Wan, R., Wang, L., Xu, K., Zhang, Q., Lei, J. and Shi, Y. (2021). Structure of the activated human minor spliceosome. *Science*, **371**, 1220.

Bai, X. C., McMullan, G. and Scheres, S. H. (2015). How cryo-EM is revolutionizing structural biology. *Trends in Biochemical Sciences*, **40**, 49–57.

Ban, N., Nissen, P., Hansen, J., Moore, P. B. and Steitz, T. A. (2000). The complete atomic structure of the large ribosomal subunit at 2.4 A resolution. *Science*, **289**, 905–920.

Bangaru, S., Ozorowski, G., Turner, H. L. and Ward, A. W. (2020). Structural analysis of full-length SARS-CoV-2 spike protein from an advanced vaccine candidate. *Science*, **370**, 1089–1094.

Basak, S., Gicheru, Y., Rao, S., Sansom, M. S. P. and Chakrapani, S. (2018). Cryo-EM reveals two distinct serotonin-bound conformations of full-length 5-HT(3A) receptor. *Nature*, **563**, 270–274.

Bassler, B. L. and Losick, R. (2006). Bacterially speaking. *Cell*, **125**, 237–246.

Bäuerlein, F. J. B. and Baumeister, W. (2021). Towards visual Proteomics at high resolution. *Journal of Molecular Biology*, **433**, 167187.

Baumeister, W. (2022). Cryo-electron tomography: A long journey to the inner space of cells. *Cell*, **185**, 2649–2652.

Ben-Shem, A., Garreau de Loubresse, N., Melnikov, S., Jenner, L., Yusupova, G. and Yusupov, M. (2011). The structure of the eukaryotic ribosome at 3.0 Å resolution. *Science*, **334**, 1524–1529.

Benda, C. (1898) "Ueber die Spermatogenese der Vertebraten und höherer Evertebraten, II. Theil: Die Histiogenese der Spermien" (On spermatogenesis of vertebrates and higher invertebrates, Part II: The histogenesis of sperm), *Archiv für Anatomie und Physiologie*, **73**, 393–398.

Berger, C., Premaraj, N., Ravelli, R. B. G., Knoops, K., Lopez-Iglesias, C. and Peters, P. J. (2023). Cryo-electron tomography on focused ion beam lamellae transforms structural cell biology. *Nature Methods*, **20**, 499–511.

Bernal, J. D. (1939). Structure of Proteins, *Nature*, **143**, 663–667.

Bernal, J. D. and Crowfoot, D. (1934). X-ray photographs of crystalline pepsin. *Nature*, **133**, 794–795.

Bernal, J. D., Fankuchen, I. and Perutz, M. (1938). An X-ray study of Chymotrypsin and Haemoglobin. *Nature*, **141**, 523–524.

Bley, C. J., Nie, S., Mobbs, G. W., et al. (2022). Architecture of the cytoplasmic face of the nuclear pore. *Science*, **376**, 1174.

Bouvette, J., Liu, H.-F., Du, X. and Bartesaghi, A. (2021) Beam image-shift accelerated data acquisition for near-atomic resolution single-particle cryo-electron tomography. *Nature Communications*, **12**, 1957.

Bragg, W. L. (1912). The diffraction of short electromagnetic waves by a crystal. *Proceedings of the Cambridge Philosophical Society*, **XVIL**, Part 1, 43–57.

Bragg, W. H. and Bragg, W. L. (1913). The reflection of X-rays by crystals. *Proceedings of the Royal Society A, Mathematical, Physical and Engineering Sciences*, 428–438. The Royal Society Publishing.

Bragg, W. L. and Perutz, M. (1954). The structure of haemoglobin. VI. Fourier projections on the 010 plane. *Proceedings of the Royal Society*, **A225**, 315–329.

Brognard, J. and Hunter, T. (2011). Protein kinase signalling networks in cancer. *Current Opinions in Genetics & Development*, **21**, 4–11.

Cai, Y., Zhang, J., Xiao, T. and Chen, B. (2021). Structural basis for enhanced infectivity and immune evasion of SARS-CoV-2 variants. *Science*, **373**, 342348.

Cai, Y., Zhang, J., Xiao, T., Peng, H., Sterling, S. M., Walsh Jr., R. M., Rawson, S., Rits-Volloch, S. and Chen, B. (2020). Distinct conformational states of SARS-CoV-2 spike protein. *Science*, **369**, 1586–1592.

Chatterjee, N. and Walker, G. C. (2017). Mechanisms of DNA damage, repair, and mutagenesis: DNA Damage and Repair. *Environmental and Molecular Mutagenesis*, **58**, 235–263.

Chen, Q., Plasencia, M., Li, Z., Mukherjee, S., Patra, D., Chen, C.-L., Klose, T., Yao, X.-Q., Kossiakoff, A. A. and Chang, L. (2021). Structures of rhodopsin in complex with G-protein-coupled receptor kinase 1. *Nature*, **595**, 600–605.

Chen, Z., Sun, L., Zhang, Z. and Rao, V. B. (2017). Cryo-EM structure of the bacteriophage T4 isometric head at 3.3-Å resolution and its relevance to the assembly of icosahedral viruses. *Proceedings of the National Academy of Sciences United States of America*, **114**, E8184–E8193.

Ciechanover, A. and Kwon, Y. T. (2015). Degradation of misfolded proteins in neurodegenerative diseases. Therapeutic targets and strategies. *Experimental Molecular Medicine*, **47**, e147.

Claude, A. and Fullam, E. F. (1945). An electron microscope study of isolated mitochondria. Methods and preliminary results. *Journal of Experimental Medicine*, **81**, 51.

Collingridge, G. L., Olsen, R. W., Peters, J. and Spedding, M. (2009). A nomenclature for ligand-gated ion channels. *Neuropharmacology*, **56**, 2–5.

Cooper, G. M. and Hausman, R. E. (2013). The cell: A molecular approach, Sixth edition. Sinauer Associates Publishers Inc., Sunderland, Massachusetts, USA.

Corbett, K. S., Edwards, D. K., Leist, S. R., et al. (2020). SARS-CoV-2 mRNA vaccine design enabled by prototype pathogen preparedness. *Nature*, **586**, 567–571.

Coux, O., Tanaka, K. and Goldberg, A. L. (1996). Structure and functions of the 20S and 26S proteasomes. *Annual Reviews of Biochemistry*, **65**, 801–847.

Cramer, P. (2021). AlphaFold2 and the future of structural biology. *Nature Structural & Molecular Biology*, **28**, 704–705.

Creekmore, B. C., Chang, Y. W. and Lee, E. B. (2021). The cryo-EM effect: Structural biology of neurodegenerative disease aggregates. *Journal of Neuropathology and Experimental Neurology*, **80**, 514–529.

Cressey, D. and Callaway, E. (2017). Cryo-electron microscopy wins' chemistry Nobel. *Nature*, **550**, 167.

Crowfoot, D. (1935). X-ray single crystal photographs of insulin. *Nature*, **135**, 591–592.

Daum, B. and Kühlbrandt, W. (2011). Electron tomography of plant thylakoid membranes. *Journal of Experimental Botany*, **62**, 2393–2402.

Davis, J. H. and Williamson, J. R. (2017). Structure and dynamics of bacterial ribosome biogenesis. *Philosophical Transactions of the Royal Society of London. Series B, Biological Sciences*, **372**, 20160181. http://dx.doi.org/10.1098/rstb.2016.0181.

Deisenhofer, J., Epp, O., Miki, K., Huber, R. and Michel, H. (1985). Structure of the protein subunits in the photosynthetic reaction centre of *Rhodopseudomonas viridis* at 3Å resolution. *Nature*, **318**, 618–624.

de la Pena, A. T., Sliepen, K., Eshun-Wilson, L. and Ward, A. B. (2022). Structure of hepatitis C virus E1E2 glycoprotein complex. *Science*, **378**, 263–269.

de la Torre, B. G. and Albericio, F. (2022). The Pharmaceutical Industry in 2021. An analysis of FDA drug approvals from the perspective of molecules. *Molecules*, **27**, 1075.

Derkach, V., Surprenant, A. and North, R. A. (1989). 5-HT_3 receptors are membrane ion channels. *Nature*, **339**, 706–709.

DeRosier, D. J. and Klug, A. (1968). Reconstruction of three-dimensional structures from electron micrographs. *Nature*, **217**, 130–134.

Desai, N. Yang H., Chandrasekaran, V., Kazi, R., Minczuk, M. and Ramakrishnan, V. (2020). Elongational stalling activates mitoribosome-associated quality control. *Science*, **370**, 1105–1110.

Dierksen, K., Typke, D., Hegerl, R. and Baumeister, W. (1993). Towards automatic electron tomography. II. Implementation of autofocus and low-dose procedures. *Ultramicroscopy* **49**, 109–120.

Dierksen, K., Typke, D., Hegerl, R., Koster, A. J. and Baumeister, W. (1992) Towards automatic electron tomography. *Ultramicroscopy*, **40**, 71–87.

Dubochet, J. (2011). Cryo-EM – The first thirty years. *Journal of Microscopy*, **245**, 221–224.

Dubochet, J., Adrian, M., Chang, J.-J., Homo, J.-C., Lepault, J., McDowall, A. W. and Schultz, P. (1988), Cryo-electron microscopy of vitrified specimens. *Quarterly Reviews of Biophysics*, **21**, 129–228.

Dubochet, J. and McDowall, A. W. (1981). Vitrification of pure water for electron microscopy. *Journal of Microscopy*, **124**, 3–4.

Earl, L. A. and Subramaniam, S. (2016). Cryo-EM of viruses and vaccine design. *Proceedings of the National Academy of Sciences United States of America*, **113**, 8903–8905.

Ebanks, B. and Chakrabarti, L. (2022). Mitochondrial ATP Synthase is a target of oxidative stress in neurodegenerative diseases. *Frontiers in Molecular Biosciences*, **9**, 1–7.

Ehara, H., Kujirai, T., Shirouzu, M., Kurumizaka, H. and Sekine, S. (2022). Structural basis of nucleosome disassembly and reassembly by RNAPII elongation complex with FACT. *Science*, **377**, 1169.

Eisenstein, F., Yanagisawa, H., Kashihara, H. and Danev, R. (2022). Parallel cryo-electron tomography on in situlamellae. *bioRxiv*. **10**.1101/2022.04.07.487557.

Eisenstein, M. (2023). Catching proteins at play: The method revealing the cell's inner mysteries. *Nature*, **621**, 646–648.

Ernster, L. and Schatz, G. (1981). Mitochondria: A Historical Review. *Journal of Cell Biology*, **91**, 227s–255s.

Fäßler, F., Javoor, M. G. and Schur, F. K. (2023). Deciphering the molecular mechanisms of actin cytoskeleton regulation in cell migration using cryo-EM. *Biochemical Society Transactions*, **51**, 87–99.

Falendysz, E. A., Lopera, J. G., Doty, J. B., et al. (2017). Characterization of Monkeypox virus infection in African rope squirrels (*Funisciurus sp.*). *Public Library of Sciences (PLOS) in Neglected Tropical Diseases* **11**, e0005809.

Fitzpatrick, A. W. P., Falcon, B., He, S., Murzin, A. G., Murshudov, G., Garringer, H. J., Crowther, R. A., Ghetti, B., Goedert, M. and Scheres, S. H. W. (2017). Cryo-EM structures of tau filaments from Alzheimer's disease. *Nature*, **547**, 185–190.

Flock, T., Hauser, A. S., Lund, N., Gloriam, D. E., Balaji, S. and Babu, M. M. (2017). Selectivity determinants of GPCR-G-protein binding. *Nature*. **545**, 317–322.

Fontana, P., Dong, Y., Pi, X., et al. (2022). Structure of cytoplasmic ring of nuclear pore complex by integrative cryo-EM and AlphaFold. *Science*, **376**, 1178.

Frank. J. (1975). A practical resolution criterion in optics and electron microscopy. *Optik*, **43**, 25–34.

Frank, J. and Al-Ali, L. (1975). Signal-to-noise ratio of electron micrographs obtained by cross correlation. *Nature*, **256**, 376–379.

Frank, J., Penczek, P., Grassucci, R. and Srivastava, S. (1991). Three-dimensional reconstruction of the 70S *Escherichia coli* ribosome in ice: The distribution of ribosomal RNA. *Journal of Cell Biology*, **115**, 597–605.

Frank, J., Penczek, P., Grassucci, R. A., Heagle, A., Spahn, C. M. T. and Agrawal, R. K. (2000). Cryo-electron microscopy of the translational apparatus: Experimental evidence for the paths of mRNA, tRNA, and the polypeptide chain. In The Ribosome: Structure, Function, Antibiotics, and Cellular Interactions, Garrett, R. A., Douthwaite, S. R., Liljas, A., Matheson, A. T., Moore, P. B., Noller, H. F. (eds). American Society for Microbiology Press, Washington, DC.

Franklin, R. E. and Gosling, R. G. (1953). Molecular configuration in sodium thymonucleate. *Nature*, **171**, 740–741.

Frechin, L., Holvec, S., von Loeffelholz, O., Hazemann, I. and Klaholz, B. P. (2023). High-resolution cryo-EM performance comparison of two latest-generation cryo electron microscopes on the human ribosome. *Journal of Structural Biology*, **215**, 107905.

Friedreich, N. and Kekulé, A. (1859). Zur Amyloidfrage. *Archives für Pathology. Anatomy und Physiology und für Klinical Medicine*, **16**, 50–65.

Fukuda, Y., Stapleton, K. and Kato, T. (2023). Progress in spatial resolution of structural analysis by cryo-EM. *Microscopy*, **72**, 135–143.

García-Nafría, J., Lee, Y., Bai, X., Carpenter, B. and Tate, C. G. (2018). Cryo-EM structure of the adenosine A2A receptor coupled to an engineered heterotrimeric G protein. *elife*, **7**, e35946.

Georgopoulos, C. P., Hendrix, R. W., Kaiser, A. D. and Wood, W. B. (1972). Role of the host cell in bacteriophage morphogenesis: Effects of a bacterial mutation on T4 head assembly. *Nature New Biology*, **239**, 38–41.

Gertz, M. A. and Dispenzieri, A. (2020). Systemic amyloidosis recognition, prognosis, and therapy: A systematic review. *Journal of the American Medical Association*, **324**, 79–89.

Gobeil, S. M. C., Janowska, K., McDowell, S. and Acharya, P. (2021). Effect of natural mutations of SARS-CoV-2 on spike structure, confirmation, and antigenicity. *Science*, **373**, 641.

Goldberg, E., Grinius, L. and Letellier, L. (1994). Recognition, Attachment, and Injection. In Molecular Biology of Bacteriophage T4, Karam, J. D. (ed), pp. 347–356. American Society for Microbiology, Washington, DC.

Grinter, R., Kropp, A., Venugopal, H., et al. (2023). Structural basis for bacterial energy extraction from atmospheric hydrogen. *Nature*, **615**, 541–547.

Groelly, F. J., Fawkes, M., Dagg, R. A., Blackford, A. N. and Tarsounas, M. (2023). Targeting DNA damage response pathways in cancer. *Nature Reviews in Cancer*, **23**, 78–94.

Gulati, S., Jastrzebska, B., Banerjee, S., Placeres, A. L., Miszta, P., Gao, S., Gunderson, K., Tochtrop, G. P., Filipek, S., Katayama, K., Kiser, P. D., Mogi, M., Stewart, P. L. and Palczewski, K. (2017). Photocyclic behavior of rhodopsin induced by an atypical isomerization mechanism. *Proceedings of the National Academy of Sciences United States of America*, **114**, E2608–E2615.

Hadiatullah, H., He, Z. and Yuchi, Z. (2022). Structural insight into ryanodine receptor channelopathies. *Frontiers in Pharmacology*, **13**, 897494.

Hand, E. (2020). We need a people's cryo-EM. Scientists hope to bring revolutionary microscope to the masses - Cheap cryo–electron microscope could accelerate discovery of protein structures. *Science*, Jan 23. https://www.science.org/content/article/we-need-people-s-cryo-em-scientists-hope-bring-revolutionary-microscope-masses; doi: 10.1126/science.aba9954.

Hanks, S. K., Quinn, A. M. and Hunter, T. (1988). The protein kinase family: Conserved features and deduced phylogeny of the catalytic domains. *Science*, **241**, 42–52.

Harms, J., Schluenzen, F., Zarivach, R., Bashan, A., Gat, S., Agmon, I., Bartels, H., Franceschi, F. and Yonath, A. (2001). High resolution structure of the large ribosomal subunit from a mesophilic eubacterium. *Cell*, **107**, 679–688.

Henderson, R., Baldwin, J. M., Ceska, T. A., Zemlin, F., Beckmann, E., & Downing, K. H. (1990). Model for the structure of bacteriorhodopsin based on high-resolution electron cryo-microscopy. *Journal of Molecular Biology*, **213**, 899–929.

Henderson, R. and Unwin, P. N. T. (1975). Three-dimensional model of purple membrane obtained by electron microscopy. *Nature*, **257**, 28–32.

Hinton, A., Bond, S. and Forgac, M. (2009). V-ATPase functions in normal and disease processes. *Pflugers Archives*, **457**, 589–598.

Hooke, R. (1665). Micrographia, or, Some Physiological Descriptions of Minute Bodies Made by Magnifying Glasses, with Observations and Inquiries Thereupon. Royal Society, London. http://ttp.royalsociety.org/ttp/ttp.html?id=a9c4863d-db77-42d1-b294-fe66c85958b3&type=book.CrossRefGoogleScholar

Hong, Y., Song, Y., Zhang, Z. and Li, S. (2023). Cryo-Electron Tomography: The resolution revolution and a surge *in situ* virological discoveries. *Annual Review of Biophysics*, **52**, 339–360.

Howden, B. P., Giulieri, S. G., Wong Fok Lung, T., et al. (2023). *Staphylococcus aureus* host interactions and adaptation. *Nature Reviews in Microbiology*, **21**, 380–395.

Humphreys, I. R., Pei, J., Baek, M., et al. (2021). Computed structures of core eukaryotic protein complexes. *Science*, **374**, 1340.

Itoh, Y., Andrell, J., Choi, A., Richter, U. and Amunts, A. (2021). Mechanism of membrane-tethered mitochondrial protein synthesis. *Science*, **371**, 846–849.

Jumper, J., Evans, R., Pritzel, A., et al. (2021). Highly accurate protein structure prediction with AlphaFold. *Nature*, **596**, 583–589.

Kaledhonkar, S., Fu, Z., Caban, K., Li, W., Chen, B., Sun, M., Gonzalez Jr., R. L., and Frank, J. (2019). Late steps in bacterial translation initiation visualized using time-resolved cryo-EM. *Nature*, **570**, 400–404.

Ke, Z., Oton, J., Qu, K. and Briggs, J. A. G. (2020) Structures and distribution of SARS-CoV-2 spike proteins on intact virions. *Nature*, **588**, 498–502.

Kendrew, J. C., Bodo, G., Dintzis, H. M., Parrish, R. G., Wyckoff, H. and Phillips, D. C. (1958). A three-dimensional model of the myoglobin molecule obtained by x-ray analysis. *Nature*, **181**, 662–666.

Khanna, K. and Villa, E. (2022). Revealing bacterial cell biology using cryo-electron tomography. *Current Opinion in Structural Biology*, **75**, 102419.

Khavnekar, S., Wan, W., Mujunder, P., Wietrzynski, W., Erdmann, P. S. and Plitzko, J. M. (2022). Multishot tomography for high-resolution *in situ* subtomogram averaging. *bioRxiv*. https://doi.org/10.1101/2022.04.10.487763.

Khera, R., Mehdipour, A. R., Bolla, J. R., Kahnt, J., Welsch, S., Ermler, U., Muenke, C., Robinson, C. V., Hummer, G., Xie, H. and Michel, H. (2022). Cryo-EM structures of pentameric autoinducer-2 exporter from *Escherichia coli* reveal its transport mechanism. *European Molecular Biology Organization (EMBO) Journal*, **41**, e109990.

Khorana, H. G. (1993). Two light-transducing membrane proteins: Bacteriorhodopsin and the mammalian rhodopsin. *Proceedings of the National Academy of Sciences United States of America*, **90**, 1166–1171.

Klein, S., Wimmer, B. H., Winter, S. L., Kolovou, A., Laketa, V. and Chlanda, P. (2021). Post-correlation on-lamella cryo-CLEM reveals the membrane architecture of lamellar bodies. *Communications Biology*, **4**, 137.

Kobilka, B. K. (2007). G protein coupled receptor structure and activation. *Biochimica et Biophysica Acta (BBA)-Biomembranes*, **1768**, 794–807.

Kühlbrandt, W. (2014). The resolution revolution. *Science*, **343**, 1443–1444.

Kühlbrandt, W. (2022). Forty years in cryoEM of membrane proteins. *Microscopy*, **71**(S1), i30–i50.

Kurochkina, L. P., Semenyuk, P. I., Orlov, V. N., Robben, J., Sykilinda, N. N. and Mesyanzhinov, V. V. (2012). Expression and functional characterization of the first bacteriophage-encoded chaperonin. *Journal of Virology*, **86**, 10103–10111.

Lake, J. A. (1976). Ribosome structure determined by electron microscopy of *Escherichia coli* small subunits, large subunits and monomeric ribosomes. *Journal of Molecular Biology*, **105**, 131–139.

Levine, A. and Oren, M. (2009). The first 30 years of P53: Growing ever more complex. *Nature Reviews Cancer*, **9**, 749–758.

Li, S., Wang, Z., Jia, A., et al. (2023). ELI trifocal microscope: A precise system to prepare target cryo-lamellae for *in situ* cryo-ET study. *Nature Methods*, **20**, 276–283.

Liang, Y. L., Khoshouei, M., Radjainia, M., et al. (2017). Phase-plate cryo-EM structure of a class B GPCR-G-protein complex. *Nature* **546**, 118–123.

Liao, M., Cao, E., Julius, D. and Cheng, Y. (2013). Structure of the TRPV1 ion channel determined by electron cryo-microscopy. *Nature*, **504**, 107–112.

Longchamp, J. N., Rauschenbachb, S., Abbb, S., Eschera, C., Latychevskaiaa, T., Kernb, K. and Finka, H. W. (2017). Imaging proteins at the single-molecule level. *Proceedings of the National Academy of Sciences United States of America*, **114**, 1474–1479.

Los, F. C. O., Randis, T. M., Aroian, R. V. and Ratner, A. J. (2013). Role of pore-forming toxins in bacterial infectious diseases. *Microbiology and Molecuar Biology Reviews*, **77**, 173–207.

Luria, S. E., Delbruck, M. and Anderson, T. F. (1943). Electron microscope studies of bacterial viruses. *Journal of Bacteriology*, **46**, 57–77.

Lutz, I. D., Wang, S., Norn, C., et al. (2023). Top-down design of protein architectures with reinforcement learning. *Science*, **380**, 266–273.

Madeira, V. M. (2018). Overview of mitochondrial bioenergetics, *Methods in Molecular Biology*, **1782**, 1–6.

Mahamid, J., Schampers, R., Persoon, H., Hyman, A. A., Baumeister, W. and Plitzko, J. M. (2015). A focused ion beam milling and lift-out approach for site-specific preparation of frozen-hydrated lamellas from multicellular organisms. *Journal of Structural Biology*, **192**, 262–269.

Mannar, D., Saville, J., Zhu, X., et al. (2022). SARS-Co-V-2 Omicron variant: Antobody evasion and cryo-EM structure of spike protein-ACE2 complex. *Science*, **375**, 760–764.

Marko, M., Hsieh, C., Schalek, R., Frank, J. and Mannella, C. (2007). Focused-ion-beam thinning of frozen-hydrated biological specimens for cryo-electron microscopy. *Nature Methods*, **4**, 215–217.

Martynowycz, M. W. and Gonen, T. (2018). From electron crystallography of 2D crystals to MicroED of 3D crystals. *Current Opinion in Colloidal Interface Science*, **34**, 9–16.

McMullan, G., Naydenova, K., Mihaylov, D., et al. (2023). Structure determination by cryoEM at 100 keV. *Proceedings of the National Academy of Sciences United States of America*, **120**, e2312905120, 1–8.

McNulty, R., Cardone, G., Gilcrease, E., Baker, T. S., Casjens, S. R. and Johnson, J. E. (2018). Cryo-EM elucidation of the structure of bacteriophage P22 virions after genome release. *Biophysical Journal*, **114**, 1295–1301.

Method of the Year 2015. *Nature Methods*. 2016; **13**, 1.

Miotto, M. C., Weninger, G. Dridi, H., Yuan, Q., Liu, Y., Wronska, A., Melville, Z., Sittenfeld, L., Reiken, S. and Marks, A. R. (2022). Structural analyses of human ryanodine receptor type 2 channels reveal the mechanisms for sudden cardiac death and treatment. *Science Advances*, **8**, eabo1272.

Mishra, A. K., Hellert, J. H. Freitas, N., et al. (2022). Structural basis of synergistic neutralization of Crimean-Congo hemorrhagic fever virus by human antibodies. *Science*, **375**, 104–19.

Mitra, K. and Frank, J. (2006). Ribosome Dynamics: Insights from atomic structure modeling into cryo-electron microscopy maps. *The Annual Review of Biophysics and Biomolecular Structure*, **35**, 299–317.

Moazed, D. and Noller, H. F. (1989). Intermediate states in the movement of transfer RNA in the ribosome. *Nature*, **342**, 142–148.

Molday, R. S. (1998). Photoreceptor membrane proteins, phototransduction, and retinal degenerative diseases. The Friedenwald Lecture. *Investigative Ophthalmology and Visual Science*, **39**, 2491–2513.

Moore, P. B., Engelman, D. M. and Schoenborn, B. P. (1974). Asymmetry in the 50S ribosomal subunit of *Escherichia coli*. *Proceedings of the National Academy of Sciences United States of America*, **71**, 172–176.

Moore, P. B., Hendrickson, W. A., Henderson, R. and Brunger, A. T. (2022). The protein-folding problem: Not yet solved. *Science*, **375**, 507.

Mosalaganti, S., Obarska-Kosinska, A., Siggel, M., et al. (2022). AI-based structure prediction empowers integrative structural analysis of human nuclear pores. *Science*, **376**, 1–15.

Mustafi, D. and Palczewski, K. (2009). Topology of class A G protein-coupled receptors: Insights gained from crystal structures of rhodopsins, adrenergic and adenosine receptors. *Molecular Pharmacology*, **75**, 1–12.

Nabel, K. G., Clark, S. A., Shankar, S., et al. (2022). Structural basis for continued antibody evasion by the SARS-Co-V-2 receptor binding domain. *Science*, **375**, 282.

Naider, F. and Becker, J. M. (2004). The α-factor mating pheromone of *Saccharomyces cerevisiae*: A model for studying the interaction of peptide hormones and G protein-coupled receptors. *Peptides*, **25**, 1441–1463.

Nealson, K. H., Platt, T. and Hastings, J. W. (1970). Cellular control of the synthesis and activity of the bacterial luminescent system. *Journal of Bacteriology*, **104**, 313–322.

Nguyen, B. A., Afrin, S., Singh, V., Ahmed, Y., Pedretti, R., Fernandez-Ramirez, M. D. C., Benson, M. D., Sawaya, R. M., Cao, Q., Boyer, D., Pope, A., Wydorski, P. M., Chhapra, F., Eisenberg, D. S. and Saelices, L. (2022). Structural polymorphism of amyloid fibrils in cardiac ATTR amyloidosis revealed by cryo-electron microscopy. *bioRxiv* 2022.06.21.496949. https://doi.org/10.1101/2022.06.21.496949.

Nogales, E. (2016). The development of cryo-EM into a mainstream structural biology technique. *Nature Methods*, **13**, 24–27.

Nomura, M., Mizushima, S., Ozaki, M., Traub, P. and Lowry, C. V. (1969). Structure and function of ribosomes and their molecular components. *Cold Spring Harbor Symposia on Quantitative Biology*, **34**, 49–61.

Norlén, L. and Al-Amoudi, A. (2004). Stratum corneum keratin structure, function, and formation: The cubic rod-packing and membrane templating model. *Journal of Investigative Dermatology*, **123**, 715–732.

Norlén, L., Al-Amoudi, A. and Dubochet, J. (2003). A cryo-transmission electron microscopy study of skin barrier formation. *Journal of Investigative Dermatology*, **120**, 555–560.

Northrop, J. H. (1929). Crystalline pepsin. *Science*, **69**, 580.

Paiva, S.-L. and Crews, C. M. (2019). Targeted protein degradation: Elements of PROTAC design. *Current Opinion in Chemical Biology*, **50**, 111–119.

Palade, G. E. (1952). An electron microscope study of the mitochondrial structure. *Anatomical Records*, **114**, 427–451.

Palade, G. E. (1955). A small particulate component of the cytoplasm. *Journal of Biophysical and Biochemical Cytology*, **1**, 59–68.

Park, Y. J., Pinto, D., Walls, A. C., et al. (2022). Imprinted antibody responses against SARS-Co-V-2 Omicron sublineages. *Science*, **378**, 619–627.

Peng, Q., Xie, Y., Kuai, L., et al. (2023). Structure of monkeypox virus DNA polymerase holoenzyme. *Science*, **379**, 100–105.

Perkins, G. A., Song, J. Y., Tarsa, L., Deerinck, T. J., Ellisman, M. H. and Frey, T. G. (1998). Electron tomography of mitochondria from brown adipocytes reveals crista junctions. *Journal of Bioenergy and Biomembranes*, **30**, 431–442.

Perutz, M. (1942). X-ray analysis of haemoglobin. *Nature*, **149**, 491–494.

Perutz, M. F. (1954). The structure of haemoglobin. III. direct determination of the molecular transform. *Proceedings of the Royal Society*, **A225**, 264–286.

Perutz, M. F., Rossmann, M. G., Cullis, A. F., Muirhead, H., Wil, G. and North, A. C. T. (1960). Structure of Haemoglobin: A three-dimensional Fourier synthesis at 5.5 A resolution. *Nature*, **185**, 416–422.

Petrovic, S., Samanta, D., Perriches, T. and Hoelz, A. (2022). Architecture of the linker-scaffold in the nuclear pore. *Science*, **376**, 1175.

Pioletti, M., Schlunzen, F., Harms, J., Zarivach, R., Gluhmann, M., Avila, H., Bashan, A., Bartels, H., Auerbach, T., Jacobi, C. Hartsch, T., Yonath, A. and Franceschi, F. (2001). Crystal structures of complexes of the small ribosomal subunit with tetracycline, edeine and IF3. *The European Molecular Biology Organization Journal*, **20**, 1829–1839.

Poehlsgaard, J. and Douthwaite, S. (2005). The bacterial ribosome as a target for antibiotics. *Nature Reviews Microbiology*, **3**, 870–881.

Qu, K., Ke Z., Zila, V. and Briggs, J. A. G. (2021). Maturation of the matrix and viral membrane of human immunodeficiency virus-1. *Science*, **373**, 700–704.

Racker, E. (1980). From Pasteur to Mitchell: A hundred years of bioenergetics. *Federation Proceedings*, **39**, 210–215.

Radermacher, M., Wagenknecht, T., Verschoor, A. and Frank, J. (1987). Three-dimensional structure of the large ribosomal subunit from Escherichia coli. *European Molecular Biology Organization Journal*, **6**, 1107–1114.

Ramakrishnan, V. (2018). Gene Machine: The Race to Decipher the Secrets of the Ribosome. Basic Books, New York, NY, USA.

Raziuddin, Chatterji, D., Ghosh, S. and Burma, D. P. (1979). Site of action of RNase I on the 50 S ribosome of *Escherichia coli* and the association of the enzyme with the partially degraded subunit. *Journal of Biological Chemistry*, **254**, 10575–10578.

Reeves, D. C. and Lummis, S. C. R. (2002). The molecular basis of the structure and function of the 5-HT_3 receptor: A model ligand-gated ion channel. *Molecular Membrane Biology*, **19**, 11–26.

Rettner, R. E. and Saier Jr., M. H. (2010). The autoinducer-2 exporter superfamily. *Microbial Physiology*, **18**, 195–205.

Roberts, R. B. (1958). Introduction, In Microsomal Particles and Protein Synthesis. Roberts, R. B. (ed), pp. vii–viii. Pergamon Press, New York.

Robertson, J. D. (1959). The ultrastructure of cell membranes and their derivatives. *Biochemical Society Symposium Proceedings*, **16**, 3–43.

Robertson, J. D. (1960). Membranen und Membranmodelle: A Molecular Theory of Cell Membrane Structure. Bargmann, W. et al. (eds), pp. 159–160. Springer-Verlag, Berlin-Heidelberg.

Rodermund, L., Coker, H., Oldenkamp, R., et al. (2021). Time-resolved structured illumination microscopy reveals key principles of Xist RNA spreading. *Science*, **372**, 1167.

Rohl, R. and Nierhaus, K. H. (1982). Assembly map of the large subunit (50S) of Escherichia coli ribosomes. *Proceedings of the National Academy of Sciences of the United States of America*, **79**, 729–733.

Ruska, E. and Knoll, M. (1931). Die magnetische Sammelspule für schnelle Elektronenstrahlen. *Z. Techn. Physik* **12**, 389–400 and 448.

Sader, K., Matadeen, R., Castro Hartmann, P., Halsan, T. and Schlichten, C. (2020). Industrial cryo-EM facility setup and management. *Acta Crystallography D Structural Biology*, **76**, 313–325.

Sanchez, R. M., Zhang, Y., Chen, W., Dietrich, L. and Kudryashev, M. (2020) Subnanometer-resolution structure determination in situ by hybrid subtomogram averaging – single particle cryo-EM. *Nature Communications*, **11**, 3709.

Sass, H. J., Beckmann, E., Büldt, G., Dorset, D., Rosenbusch, J. P., van Heel, M., Zeitler, E., Zemlin, F. and Massalski, A. (1989). Densely packed ß-structure at the protein-lipid interface of porin is revealed by high-resolution cryo-electron microscopy, *Journal of Molecular Biology*, **209**, 171–175.

Saurer, M., Leibundgut, M., Nadimpalli, H. P., et al. (2023). Molecular basis of translation termination at noncanonical stop codons in human mitochondria. *Science*, **380**, 531–536.

Sawaya, M. R., Hughes, M. P., Rodriguez, J. A., Riek, R. and Eisenberg, D. S. (2021). The expanding amyloid family: Structure, stability, function, and pathogenesis. *Cell*, **184**, 4857–4873.

Schmeing, T. M. and Ramakrishnan, V. (2009). What recent ribosome structures have revealed about the mechanism of translation. *Nature*, **461**, 1234–1242.

Schmidt, M., Wiese, S., Adak, V., Engler, J., Agarwal, S., Fritz, G., Westermark, P., Zacharias, M. and Fändrich, M. (2019). Cryo-EM structure of a transthyretin-derived amyloid fibril from a patient with hereditary ATTR amyloidosis. *Nature Communications*, **10**, 5008.

Senior, A. W., Evans, R., Jumper, J., et al. (2020). Improved protein structure prediction using potentials from deep learning. *Nature*, **577**, 706–710.

Service, R. F. (2021). Protein structures for all.: AI-powered predictions reveal the shapes of proteins by the thousands. *Science*, **374**, 1427.

Siekevitz, P. (1957). Powerhouse of the cell. *Scientific American*, **197**, 131–144.

Singer, S. J. and Nicolson, G. L. (1972). The fluid mosaic model of the structure of cell membranes. *Science*, **175**, 720–731.

Singh, S., Broeck, A. V., Miller, L., Chaker-Margot, M. and Kline, S. (2021). Nucleolar maturation of the human small subunit processome. *Science*, **373**, 1216.

Sjostrand, F. S. (1953). Electron microscopy of mitochondria and cytoplasmic double membranes. *Nature*, **171**, 30–32.

Sjostrand, F. S., Andersson-Cedergren. E. and Dewey, M. M. (1958). The ultrastructure of the intercalated discs of frog, mouse and guinea pig cardiac muscle. *Journal of Ultrastructure Research*, **1**, 271–297.

Solares, M. J. and Kelly, D. F. (2022). Harnessing the Power of Structural Oncology. *Microscopy Today*, **30**, 10–17.

Spirin, A. S. (1999). Ribosomes. Springer Publication, New York, NY, USA.

Song, K., Shang, Z., Fu, X., et al. (2020). In situ structure determination at nanometer resolution using TYGRESS. *Nature Methods*, **17**, 201–208.

Srivastava, S., Verschoor, A., Radermacher, M., Grassucci, R. and Frank, J. (1995). Three-dimensional reconstruction of mammalian 40S ribosomal subunit embedded in ice. *Journal of Molecular Biology*, **245**, 461–466.

Stafleu, F. A. and Cowan, R. S. (1976). Taxonomic Literature, vol. 1. A-G Bohn, Scheltema & Holkema, Utrecht. https://www.biodiversitylibrary.org/item/103414.GoogleScholar

Stark, H., Mueller, F., Orlova, E. V., Schatz, M., Dube, P., Erdemir, T., Zemlin, D., Brimacombe, R. and van Heel, M. (1995). The 70S *Escherichia coli* ribosome at 23 A resolution: Fitting the ribosomal RNA. *Structure*, **3**, 815–821.

Steitz, T. A. (2008). A structural understanding of the dynamic ribosome machine. *Nature Reviews in Molecular Cell Biology*, **9**, 242–253.

Steinebrei, M., Gottwald, J., Baur, J., Röcken, C., Hegenbart, U., Schönland, S. and Schmidt, M. (2022). Cryo-EM structure of an ATTRwt amyloid fibril from systemic non-hereditary transthyretin amyloidosis. *Nature Communications*, **13**, 6398.

Stuart, D. I., Subramaniam, S. and Abrescia, N. G. A. (2016). The cryo-EM revolution: Fueling the next phase. *Nature Methods*, **13**, 607–608.

Subramaniam, S. (2019). The cryo-EM revolution: Fueling the next phase. *International Union of Crystallography Journal*, **6**, 1–2.

Terashi, G., Wang, X., Maddhuri Venkata Subramaniya, S. R., Tesmer, J. J. and Kihara, D. (2022). Residue-wise local quality estimation for protein models from cryo-EM maps. *Nature Methods*, **19**, 1116–1125.

Tissiéres, A., Mitchell, H. K. and Tracy, U. M. (1974). Protein synthesis in salivary glands of *Drosophila melanogaster*: Relation to chromosome puffs. *Journal of Molecular Biology*, **84**, 389–398.

Tholen, J., Razew, M., Weis, F. and Galej, W. (2022). Structural basis of branch site recognition by the human spliceosome. *Science*, **375**, 50–56.

Thomas, J. M. (2020). Architects of Structural Biology. Oxford University Press, UK.

Tomasz, A. (1965). Control of the competent state in *Pneumococcus* by a hormone-like cell product: An example for a new type of regulatory mechanism in bacteria. *Nature*, **208**, 155–159.

Turk, M. and Baumeister, W. (2020). The promise and the challenges of cryo-electron tomography. *Federation of European Biological Sciences Letters*, **594**, 3243–3261.

Turonová, B., Sikora, M., Schürmann, C., and Beck, M. (2020). *In situ* structural analysis of SARS-CoV-2 spike reveals flexibility mediated by three hinges. *Science*, **370**, 203–208.

Usukura, J., Narita, A., Matsumoto, T. and Ose, Y. (2018). Cryo-electron microscope developed for simultaneous STEM, SEM imaging and its application to biological samples. Hitachi High-Technologies Corporation Scientific instrument news, Technical magazine of Electron Microscope and Analytical Instruments. 10, 1–9.

van den Hoek, H., Klena, N., Jordan, M. and Engel, B. D. (2022). *In situ* architecture of the ciliary base reveals the stepwise assembly of intraflagellar transport trains. *Science*, **377**, 543–548.

van der Heide, P., Xu, X. P., Marsh, B. J., Hanein, D. and Volkmann, N. (2007). Efficient automatic noise reduction of electron tomographic reconstructions based on iterative median filtering. *Journal of Structural Biology*, **158**, 196–204.

van Heel, M. and Frank, J. (1981). Use of multivariate statistics in analysing the images of biological macromolecules. *Ultramicroscopy*, **6**, 187–194.

Vasanthakumar, T. and Rubinstein, J. L. (2020). Structure and roles of V-type ATPases. *Trends in Biochemical Sciences*, **45**, 295–307.

Velazhahan, V., Ma, N., Pándy-Szekeres, G., Kooistra, A. J., Lee, Y., Gloriam, D. E., Vaidehi, N. and Tate, C. G. (2021). Structure of the class D GPCR Ste2 dimer coupled to two G proteins. *Nature*, **589**, 148–153.

Virchow, R. (1854). Zur Cellulose-Frage. Archives für pathology. *Anatomy und Physiology und für Klinical Medicine*, **6**, 416–426.

Wang, J. Y., Wang, Q. W., Yang, X. Y., Yang, W., Li, D. R., Jin, J. Y., Zhang, H. C. and Zhang, X. F. (2023). GLP-1 receptor agonists for the treatment of obesity: Role as a promising approach. *Frontiers in Endocrinology*, **14**, 1085799.

Wang, Z., Grange, M., Pospich, S., Wagner, T., Kho, A. L., Gautel, M. and Rausner, S. (2022). Structures from intact myofibrils reveal mechanism of thin filament regulation through nebulin. *Science*, **375**, 738.

Watson, J. D. and Crick, F. H. C. (1953). Molecular structure of nucleic acids; a structure for deoxyribose nucleic acid. *Nature*, **171**, 737–738.

Wilkins, M. H. F., Stokes, A. R. and Wilson, H. R. (1953). Molecular structure of deoxypentose nucleic acids. *Nature*, **171**, 738–740.

Wimberly, B. T., Brodersen, D. E., Clemons Jr., W. M., Morgan-Warren, R. J., Carter, A. P., Vonrhein, C., Hartsch, T. and Ramakrishnan, V. (2000). Structure of the 30S ribosomal subunit. *Nature*, **407**, 327–339.

Wittman, H. G. (1983). Architecture of prokaryotic ribosome. *Annual Reviews of Biochemistry*, **52**, 35–65.

Wrapp, D., Wang, N., Corbett, K. S., et al. (2020). Cryo-EM structure of the 2019-nCoV spike in the prefusion conformation. *Science*, **367**, 1260–1263.

Xing, H., Taniguchi, R., Khusainov, I., Kreysing, J. P., Welsch, S., Turonova, B. and Beck, M. (2023). Translation dynamics in human cells visualized at high resolution reveal cancer drug action. *Science*, **381**, 70.

Xu, W., Wu, L., Liu, S., et al. (2022). Structural basis for strychnine activation of human bitter taste receptor TAS2R46. *Science*, **377**, 1298–1304.

Yang, Y., Arseni, D., Zhang, W., et al. (2022). Cryo-EM structures of amyloid-β 42 filaments from human brains. *Science*, **375**, 167–172.

Yang, Z., Zeng, X., Zhao, Y. and Chen, R. (2023). AlphaFold2 and its applications in the fields of biology and medicine. *Signal Transduction and Targeted Therapy*, **8**, 115.

Yonath, A. and Wittmann, H. G. (1989). Challenging the three dimensional structure of ribosomes. *Trends in Biochemical Sciences*,**14**, 329–335.

Young, L. N. and Villa, E. (2023). Bringings structure to cell biology with cryo-electron tomography. *Annual Reviews of Biophysics*, **52**, 573–595.

Yusupov, M. M., Yusupova, G. Z., Baucom, A., Lieberman, K., Earnest, T. N., Cate, J. H. and Noller, H. F. (2001). Crystal structure of the ribosome at 5.5 A resolution. *Science*, **292**, 883–896.

Zalk, R., Lehnart, S. E. and Marks, A. R. (2007). Modulation of the ryanodine receptor and intracellular calcium. *Annual Rviews of Biochemistry*, **76**, 367–385.

Zhang, J., Xiao, T., Cai, Y. and Chen, B. (2021). Membrane fusion and immune evasion by the spike protein of SARS-Co-V-2 delta variant. *Science*, **374**, 1353–1360.

Zheng, W., Wang, F., Taylor, N. M. I. and Egelman, E. H. (2017). Refined cryo-EM structure of the T4 tail tube: Exploring the lowest dose limit. *Structure*, **25**, 1436–1441.

Zhu, X., Huang, G., Zeng, C., et al. (2022). Structure of the cytoplasmic ring of the *Xenopus laevis* nuclear pore complex. *Science*, **376**, 1177.

2 Structural Dynamics of Proteasome Regulation by Cryo-EM

Youdong Mao

PROLOGUE

This chapter is dedicated to the memory of Prof. Dr. Alfred Lewis Goldberg (September 3, 1942 – April 18, 2023) for his seminal contributions to the discovery of the proteasome system.

INTRODUCTION: UBIQUITIN-PROTEASOME SYSTEM IN A NUTSHELL

Intracellular proteins are dynamically synthesized and degraded to maintain homeostasis. Among more than 600 protease genes in humans, only the proteasome universally governs the controlled death of proteins, without sequence specificity, to recycle amino acids for the synthesis of new proteins by the ribosome. The proteasome removes virtually any misfolded proteins that are toxic or pathogenic for protein quality control, cleaves inactive proteins for functional activation, and degrades functional proteins that become detrimental when cellular function no longer needs them under certain circumstances. The broad importance of the proteasome in life sciences is exemplified by its roles in the cell cycle, gene expression, antigen presentation, immunity regulation, and tumorigenesis (Coux, Tanaka et al., 1996; Voges, Zwickl et al., 1999; Mayer, 2000; Tanaka, 2009; Ciechanover and Kwon, 2015; Finley, Chen et al., 2016; Collins and Goldberg, 2017; Bard, Goodall et al., 2018; Hnia, Clausen et al., 2019; Meyer-Schwesinger, 2019; Finley and Prado, 2020). The 26S proteasome mediate ubiquitin-dependent selective degradation of proteins in cells, in which one or two 19S regulatory particles (RP, also known as Proteasome Activator 700 or PA700) are capped on the 20S core particle (CP). The doubly RP-capped proteasome was also termed 30S in certain literature (Sahu, Mali et al., 2021). The proteasome can exist in several compositionally distinct forms of holocomplex assembly known as 20S (CP), 26S (RP-CP), 30S (RP-CP-RP), PA28-CP, PA200-CP, and even PA200-CP-RP hybrids (Savulescu and Glickman, 2011; Toste Rego and da Fonseca, 2019; Xie, Metcalfe et al., 2019; Guan, Wang et al., 2020; Mao, 2021). The 26S proteasome is the most popular form in cells.

The CP houses peptidase activities and is composed of 14 α-type and 14 β-type protein subunits that are assembled into four heptameric rings stacking into a cylindrical barrel in a $\alpha_{1\text{-}7}\beta_{1\text{-}7}\beta_{1\text{-}7}\alpha_{1\text{-}7}$ arrangement, with an approximately two-fold symmetry (Arrigo, Tanaka et al., 1988; Lowe, Stock et al., 1995; Groll, Ditzel et al., 1997; Dong, Zhang et al., 2019). In contrast to seven distinct α-type or β-type subunits in eukaryotes, there is only one gene for either α-type or β-type subunits in prokaryotic cells that assembles into a homo-heptameric ring. In eukaryotes, each β-ring houses three threonine residues at the N-termini of β1, β2, and β5, constituting proteolytically active sites in the catalytic chamber. Positioned on both sides of the chamber are the hetero-heptameric α-rings controlling substrate entry into the CP interior. Opening of an axial channel in the α-ring, also known as the CP gate, is allosterically controlled by the RP docking onto the CP.

The CP may be assembled into three alternative tissue-specific subtypes known as immunoproteasome, thymoproteasome, and spermatoproteasome that fine-tune their catalytic activities to meet tissue-specific functional needs. In lymphoid tissues or other interferon-γ (IFN-γ) stimulated cells, three constitutively expressed β-subunits are substituted with the subtype subunits β1i, β2i, and β5i to assemble the immunoproteasome (Ferrington and Gregerson, 2012; Murata, Takahama et al., 2018). Thymus-specific thymoproteasome (Murata, Sasaki et al., 2007) and testis-specific spermatoproteasome (Uechi, Hamazaki et al., 2014; Zhang, Jiang et al., 2021) contain subunits β5t and α4s, which replace β5 and α4, respectively. Curiously, different subtypes of β-subunits can be mixed into a single proteasome assembly that permits differential regulation of enzymatic properties (Dahlmann, Ruppert et al., 2000).

For most proteins targeted to the proteasome, covalent modification of ubiquitin to surface lysine residues is catalyzed by a cascade of three enzymes: the ubiquitin-activating E1 enzyme, the ubiquitin-conjugating E2 enzyme, and the substrate-selective E3 ubiquitin ligase (Figure 2.1) (Ciehanover, Hod et al., 1978; Hershko, Ciechanover et al., 1979; Ciechanover, Elias et al., 1980; Ciechanover, Heller et al., 1980; Hershko, Ciechanover et al., 1980). Ubiquitin consists of 76 amino acids and is highly conserved in eukaryotes (Goldstein, Scheid et al., 1975; Wilkinson, Urban et al., 1980). Human cells contain a few E1 enzymes, about 40 members of E2 enzymes, and at least 600 E3 ligases (Hua and Vierstra, 2011; Buetow and Huang, 2016; Zheng and Shabek, 2017). The hierarchical cascade of ubiquitylation enzymes allows parallel targeting of myriad substrates to a reservoir of proteasomes. Polyubiquitin chains, which form via all possible lysine residues (Lys6, Lys11, Lys27, Lys29, Lys33, Lys48 and Lys63) as well as the N-terminal Met1 of

DOI: 10.1201/9781003326106-3

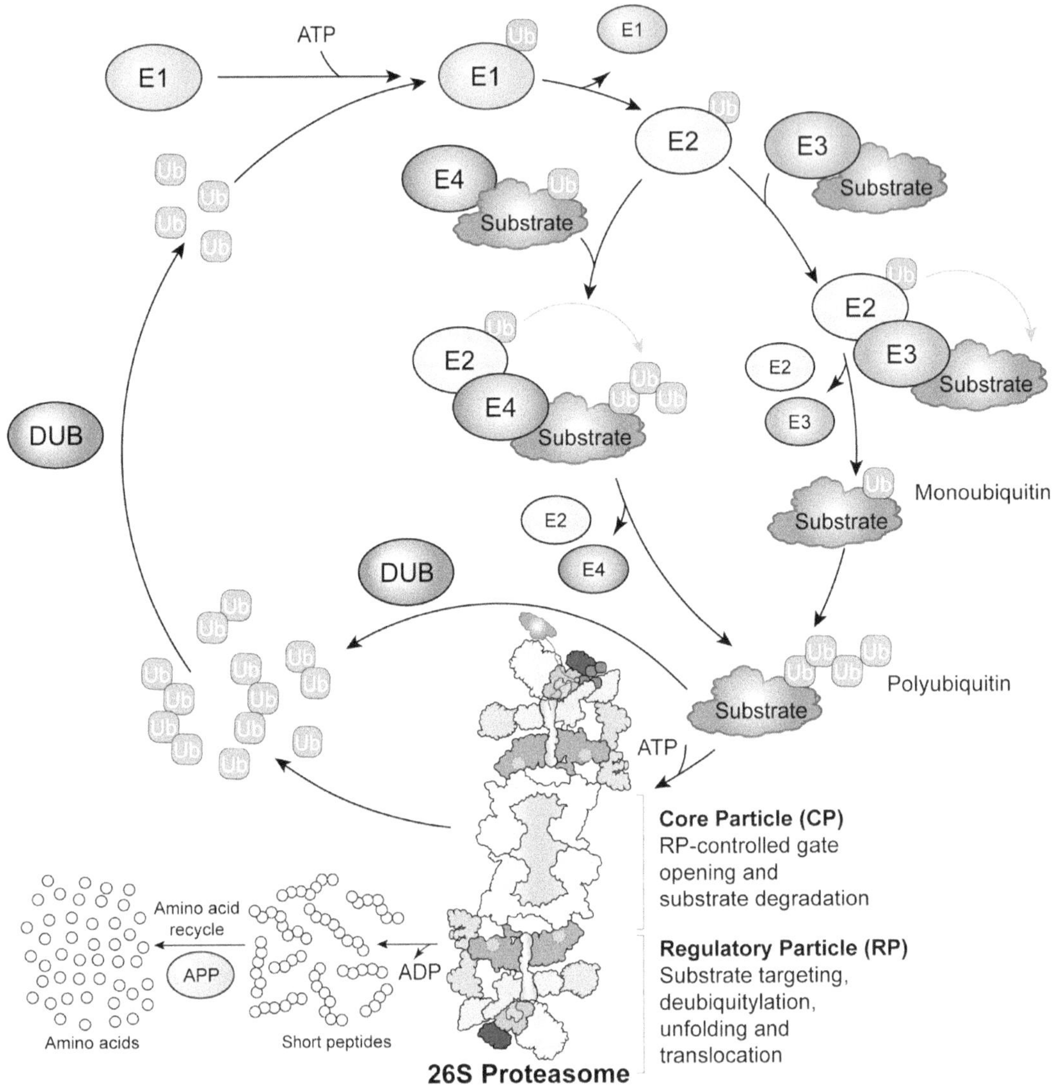

FIGURE 2.1 Overview of the ubiquitin-proteasome system. Ubiquitylation refers to the covalent conjugation of ubiquitin to lysine residues exposed on the surface of targeted substrates. Ubiquitin is activated by the El enzyme in an ATP-dependent fashion. Activated ubiquitin is transferred from El to E2. A E3 ubiquitin ligase transfers the ubiquitin from E2 to the substrate and catalyzes the isopeptide bond formation between ubiquitin and substrate. A specific type of ubiquitin ligase (E4) elongates ubiquitin chains after monoubiquitin or polyubiquitin with little or no substrate recognition. The formation of Lys48-linked polyubiquitin chains, the main signal for proteasomal degradation, is mediated by successive cycles of ubiquitin conjugation. Monoubiquitylation could be sufficient for proteasome targeting in certain scenarios. The monoubiquitin and polyubiquitin are removed from the substrate by the deubiquitylating enzymes (DUBs) associated with the proteasome. After being translocated into the proteasome, substrates are cleaved into short peptides, which are further broken down to amino acids by aminopeptidases (APPs). Released free polyubiquitin molecules are further recycled by cytosolic DUBs for used in ubiquitylation reactions.

ubiquitin, are the predominant degradation signals *in vivo* (Komander and Rape, 2012). The compactness and flexibility of the polyubiquitin chains are dependent of the lysine-linkage type and serve as a "ubiquitin code" to control the selectivity and specificity of substrate targeting to the proteasome. Lys48 is the most abundant linkage type used by proteasomal degradation in cells, followed by Lys11 and Lys63 in yeast (Xu, Duong et al., 2009) and Lys63, Lys29, and Lys11 in humans (Dammer, Na et al., 2011). Many E3 ubiquitin ligases, such as UBE3C/HUL5, UBE3A/E6AP, UFD4, HUWE1, PARKIN, anaphase-promoting complex/cyclosome (APC/C), SCF^{CDC4}, and even some E2 enzymes have been suggested to transiently associate with the proteasome, presumably to fine-tune its function by unknown mechanisms (Xie and Varshavsky, 2000; Leggett, Hanna et al., 2002; Xie and Varshavsky, 2002; Sakata, Yamaguchi et al., 2003; Crosas, Hanna et al., 2006; Demartino and Gillette, 2007; Martinez-Noel, Galligan et al., 2012; Kuhnle, Martinez-Noel et al., 2018).

The RP consists of at least 18 subunits and regulates substrate recognition, recruitment, unfolding, and translocation in an ATP-dependent manner (Figure 2.2). The RP is made of two subcomplexes named the lid and base (Glickman, Rubin et al., 1998). The lid subcomplex is assembled into a horseshoe-like architecture, forming an elaborate bundle by the C-terminal α-helical domain of each lid subunit except RPN15 (SEM1) (Estrin, Lopez-Blanco et al., 2013). The base subcomplex is centered on the hetero-hexameric motor of AAA-ATPase (ATPases Associated with diverse cellular Activities) (consisting of subunits RPT1–RPT6) that harnesses the chemical energy of ATP hydrolysis to generate mechanical work, unfolding and translocating substrate into the proteolytic chamber for degradation. Recognition of a ubiquitylated substrate is mediated by three ubiquitin receptors RPN1 (PSMC2/S7) (Shi, Chen et al., 2016), RPN10 (PSMD4/S5a) (Deveraux, Ustrell et al., 1994; van Nocker, Sadis et al., 1996), and RPN13 (ADRM1) (Husnjak, Elsasser et al., 2008; Schreiner, Chen et al., 2008). Ubiquitylated substrates can also be recruited to the AAA-ATPase unfoldase by substrate-shuttling factors consisting of ubiquitin-like (UBL) and ubiquitin-associated (UBA) domains, such as RAD23, DSK2, and DDI1, which are reversibly associated with the three stoichiometric ubiquitin receptors (Elsasser, Chandler-Militello et al., 2004; Elsasser and Finley, 2005; Zhang, Chen et al., 2009).

There are more than 100 deubiquitylating enzymes (DUBs) functioning upstream of the proteasome to stabilize the proteome, reverse improper ubiquitylation, and avoid unnecessary degradation (Leznicki and Kulathu, 2017). Most of these DUBs do not directly interact with the proteasome except USP14/Ubp6 and UCH37/UCHL5, which are reversibly recruited to the proteasome by binding RPN1 and RPN13, respectively (Verma, Chen et al., 2000; Borodovsky, Kessler et al., 2001; Hamazaki, Iemura et al., 2006; Qiu, Ouyang et al., 2006; Yao, Song et al., 2006; Lee, Lu et al., 2016). However, the 26S proteasome contains a stoichiometric DUB subunit RPN11 (PSMD14/POH1) that instead removes ubiquitin chains *en bloc* for promoting substrate degradation rather than stabilizing substrates. As the sole universal terminal machinery for degrading virtually any protein substrates, the proteasome appears to be designed with exceptional versatility and adaptability. Its functional expandability appears to be achieved by dynamically organizing an ever-expanding network of proteasome-associated proteins, such as substrate-shuttling factors, DUBs, ubiquitin ligases, and other pathway-dependent or tissue-specific regulatory factors, which elegantly program

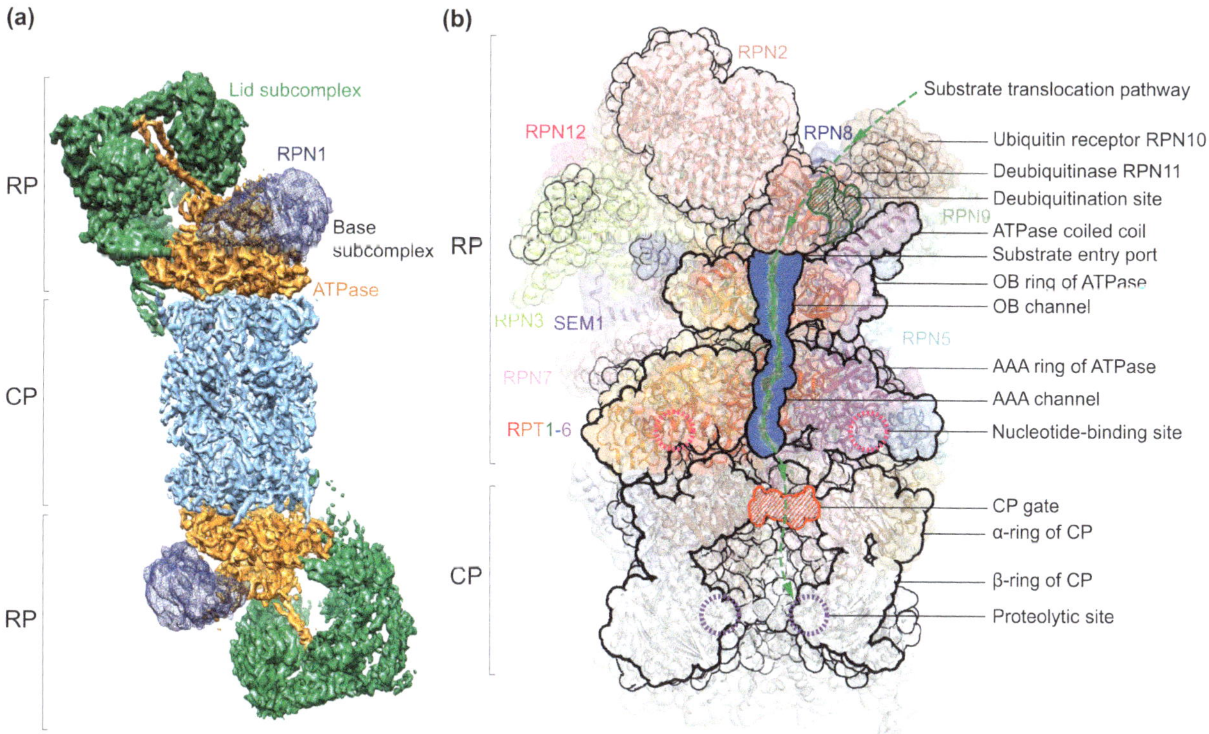

FIGURE 2.2 Overall architecture of the 26S proteasome. (a) The cryo-EM reconstruction of the doubly capped 26S proteasome. The CP is shown as cyan. (b) Illustrative anatomy of the RP-CP subcomplex structure, with black silhouettes representing the central cross-section along the ATPase channel colored opaque blue. Dashed green curve illustrates the substrate-translocation pathway across the channel to the proteolytic sites in the CP chamber. Only the upper half of the CP is shown.

the proteasome activity for cellular needs (Finley and Prado, 2020; Mao, 2021).

In the absence of ubiquitylation, a protein normally remains intact without interactions with the 26S proteasome. However, the 19S RP can be replaced with several ATP-independent activators, typically PA28 (11S Reg) and PA200 (Blm10 in yeast), to activate the proteasome and induce ubiquitin-independent degradation under specific circumstances (Whitby, Masters et al., 2000; Stadtmueller and Hill, 2011; Huber and Groll, 2017; de la Pena, Goodall et al., 2018; Zhu, Wang et al., 2018; Dong, Zhang et al., 2019; Toste Rego and da Fonseca, 2019; Xie, Metcalfe et al., 2019; Guan, Wang et al., 2020; Chen, Wang et al., 2021; Opoku-Nsiah, de la Pena et al., 2022; Zhao, Makhija et al., 2022). For instance, cytosolic PA28αβ activates immunoproteasome and plays a role in antigen presentation by the major histocompatibility complex (MHC) class I (Dick, Ruppert et al., 1996). By contrast, PA28γ (REGγ) and PA200 are nuclear proteins. PA200-20S appears to catalyze ubiquitin-independent degradation of acetylated histones during DNA repair and spermatogenesis (Qian, Pang et al., 2013; Mandemaker, Geijer et al., 2018). Even without replacing the 19S RP, ubiquitylation can be bypassed via certain proteasome-interacting factors to effect ubiquitin-independent degradation. For example, specific C-terminal degrons, sometimes with the assistance of substrate-shuttling proteins from the Ubiquilin family, can mediate ubiquitin-independent degradation by the 26S proteasome (Makaros, Raiff et al., 2023). In the nucleus, a proteosome-associated protein named midnolin uses its Catch domain to target nuclear proteins for degradation in the 26S proteasome in a ubiquitin-independent fashion, bypassing the ubiquitylation system (Gu, Nardone et al., 2023). In the absence of any proteasome activators, degradation of unfolded proteins may still be accomplished by 20S alone, which can even degrade ubiquitin tags under hypoxia and in a failing heart (Sahu, Mali et al., 2021).

CHRONICLE OF STRUCTURAL BIOLOGY OF THE PROTEASOME

The proteasome is unequivocally one of the most dynamic holoenzyme complexes with respect to conformational motion and compositional heterogeneity. For this reason, the atomic structure of 26S was not determined until four decades after its first discovery in the 1970s (Etlinger and Goldberg, 1977). Likely because of the abundance of the proteasome in cells, the human CP was one of the first protein complexes imaged by negative-staining electron microscopy half a century ago, long before its function was characterized (Figure 2.3) (Harris, 1968). The elucidation of the proteasome structure and function was propelled by methodology innovations in biophysics and biochemistry. In contrast to 26S, the CP is conformationally stable and became the first component solved by X-ray crystallography (Lowe, Stock et al., 1995; Groll, Ditzel et al., 1997). The crystal structure of the human CP was determined much later, due to its difficulties in expression and purification from human cells (Harshbarger, Miller et al., 2015; Schrader, Henneberg et al., 2016). Crystal structures of the isolated CP or immunoproteasome CP were also determined for several species (Groll, Ditzel et al., 1997; Unno, Mizushima et al., 2002; Huber, Basler et al., 2012, Harshbarger, Miller et al., 2015). All these CP structures present a closed gate in the center of their α-rings. The first yeast CP with an open gate in the α-ring was determined by X-ray crystallography in the presence of activation by the PA26 (11S) regulator from *Trypanosoma brucei* (Whitby, Masters et al., 2000), whereas the crystal structure of mammalian 11S illustrated an asymmetric PA28$\alpha_4\beta_3$ assembled from four PA28α and three PA28β subunits (Huber and Groll, 2017).

Although the structure of the RP is highly dynamic and completely evaded X-ray crystallography, most RP subunits or their homologs were at least in part solved by X-ray crystallography or nuclear magnetic resonance (NMR) in isolation (Schreiner, Chen et al., 2008; Zhang, Hu et al., 2009; Zhang, Wu et al., 2009; Riedinger, Boehringer et al., 2010; Boehringer, Riedinger et al., 2012; He, Kulkarni et al., 2012; Pathare, Nagy et al., 2012, 2014; Worden, Padovani et al., 2014; Shi, Chen et al., 2016). Electron microscopy reconstruction can overcome structural dynamics of the RP and became the only option to obtain the molecular architecture of the 26S proteasome in the 1990s (Peters, Cejka et al., 1993; Frank, 2006). Following the first cryo-EM reconstruction of the 26S proteasome from *Drosophila* at 1–2 nm resolution in 2009 (Nickell, Beck et al., 2009),

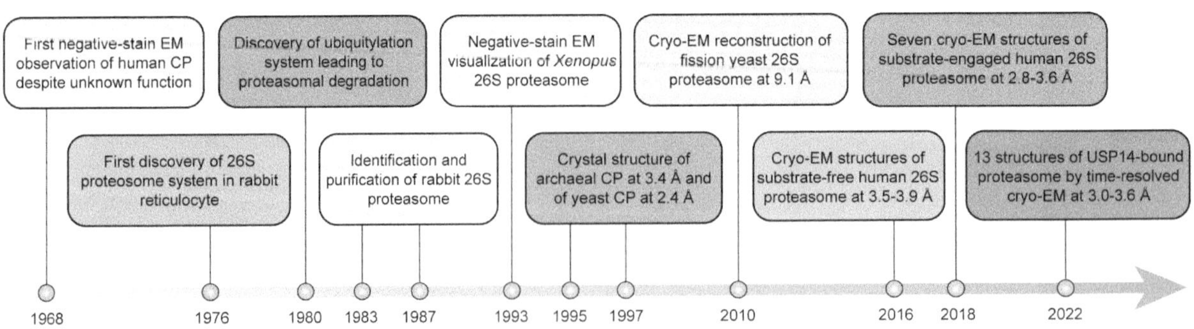

FIGURE 2.3 Chronicle of the elucidation of structure–function relationship of the 26S proteasome.

the cryo-EM structure of *Schizosaccharomyces pombe*26S proteasome was presented at 9.1 Å resolution (Bohn, Beck et al., 2010). Because the resolution was insufficient for direct assignment of subunit organization, further integration with cross-linking mass spectrometry was used to infer the subunit architecture of the RP (Beck, Unverdorben et al., 2012; da Fonseca, He et al., 2012; Lander, Estrin et al., 2012; Lasker, Forster et al., 2012). By docking the crystal or homology-modeled structures of individual RP subunits into an 8 Å density map, the first pseudo-atomic model for the yeast 26S proteasome was achieved (Beck, Unverdorben et al., 2012). Further cryo-EM analysis of the substrate-free yeast proteasome revealed three distinct conformational states (designated s1, s2 and s3), which were hypothetically proposed as the substrate-accepting, commitment, and translocating states, respectively (Matyskiela, Lander et al., 2013; Sledz, Unverdorben et al., 2013; Unverdorben, Beck et al., 2014). Meanwhile, a 9 Å map of the yeast proteasome in the presence of substrate was reconstructed, but the density of substrate was not observed (Matyskiela, Lander et al., 2013). The low resolution of these studies precludes understanding of structural mechanisms for ATP-dependent degradation by the 26S proteasome (Forster, Unverdorben et al., 2013).

The first high-resolution cryo-EM structures of the substrate-free 26S proteasome were reported in 2016 and were obtained from functional holoenzymes purified from human cells by three independent studies (Chen, Wu et al., 2016; Huang, Luan et al., 2016; Schweitzer, Aufderheide et al., 2016). All three studies reported virtually the same conformation of the proteasome in a resting, inactive state at 3.5–3.9 Å resolution exhibiting closed CP gates (referred to as state S_A by Chen, Wu et al., 2016). Only one of the studies also reported three alternative states (S_B, S_C, and S_D) at moderate resolution, among which only state S_D exhibits open CP gates indicative of an activated conformation (Chen, Wu et al., 2016). The discovery of state S_D revealed the insertion of five RPT C-terminal tails into the CP, albeit at a moderate resolution (6–8 Å) (Chen, Wu et al., 2016). Meanwhile, the free RP complex was analyzed by cryo-EM in states presumably prior to its assembly with the CP (Lu, Wu et al., 2017). Surprisingly, in the free RP, the AAA-ATPase ring exhibited continuous motion of a broad conformational landscape, captured in seven conformational states at ~9 Å resolution in the AAA-ATPase but at 4.6 Å for the rest of the free RP (Lu, Wu et al., 2017). The initial breakthrough in solving human 26S proteasome structures was followed by cryo-EM structures of the yeast counterparts (Ding, Fu et al., 2017; Wehmer, Rudack et al., 2017). Further studies expanded the structural ensemble of substrate-free 26S into six coexisting conformations for both human and yeast, providing increasingly higher resolution for the key features in the activated S_D-compatible conformations (Eisele, Reed et al., 2018; Zhu, Wang et al., 2018; Ding, Xu et al., 2019).

The first atomic structures of the substrate-engaged human 26S proteasome in seven functional states by cryo-EM were reported at an average resolution of 2.8–3.6 Å (Figure 2.4) (Dong, Zhang et al., 2019), whereas the yeast counterparts in four substrate-engaged states were analyzed at 4.2–4.7 Å (de la Pena, Goodall et al., 2018). The former employed a "nucleotide-substitution" strategy that dilutes ATP with ATPγS in a time-dependent manner, which was expected to partially stall as many intermediate substrate-bound conformations in action as possible (Dong, Zhang et al., 2019). In contrast, the latter inactivated the yeast Rpn11 with the inhibitor ortho-phenanthroline in an attempt to stall the proteasome in a translocation-competent state prior to the Rpn11-catalyzed deubiquitylation reaction (de la Pena, Goodall et al., 2018). Despite the notable differences in the strategy of capturing substrate-engaged 26S proteasome, the substrate-bound AAA-ATPase structures in states 5D, 5T, and 4D of the yeast proteasome resemble their human counterparts in states E_{D1} and E_{D2} within the limits of their reconstruction quality and resolution, suggesting a conserved hand-over-hand mechanism of substrate translocation in the proteasome (de la Pena, Goodall et al., 2018; Dong, Zhang et al., 2019). Further analysis of the human proteasome by a deep-learning-based 3D classification method with improved accuracy revealed more than a dozen conformational states of the substrate-engaged RP and discovered several initially missed states (Wu, Chen et al., 2022). Following this success, time-resolved cryo-EM data were further combined with the deep-learning-based 3D classification (Wu, Chen et al., 2022) to examine the mutual regulation of USP14 and the human 26S proteasome in the act of polyubiquitylated substrate degradation (Zhang, Zou et al., 2022). This work advanced the revelation of proteasome complexity to another level. By determining 13 coexisting conformations and tracking their time-dependent changes, the findings illuminate parallel kinetic pathways of USP14-regulated structural transition during substrate degradation, with one pathway rejecting substrate by USP14-catalyzed deubiquitylation and the other promoting substrate degradation by energetically favoring conformers of processive translocation (Zhang, Zou et al., 2022).

In addition to decoding the canonical working mechanism of 26S proteasome, applications of cryo-EM have recently allowed the determination of the CP bound to an endogenous proteasome inhibitor named PI31/Fub1 (Jespersen, Ehrenbolger et al., 2022; Rawson, Walsh et al., 2022) and of the CP bound to a nuclear transport factor AKIRIN2 (de Almeida, Hinterndorfer et al., 2021), and they have also captured intermediate assemblies of the CP during its biogenesis (Schnell, Walsh et al., 2021; Walsh, Rawson et al., 2023). Notably, PI31 was found to directly bind all six active sites in the CP (Rawson, Walsh et al., 2022), whereas its ortholog in microsporidia was observed to bind the dormant spore 20S proteasome in a similar mode, implicating a tissue-dependent role of PI31. These studies highlight the feasibility of cryo-EM to analyze interactions of the proteasome with various cellular proteins that may be difficult for traditional X-ray crystallography.

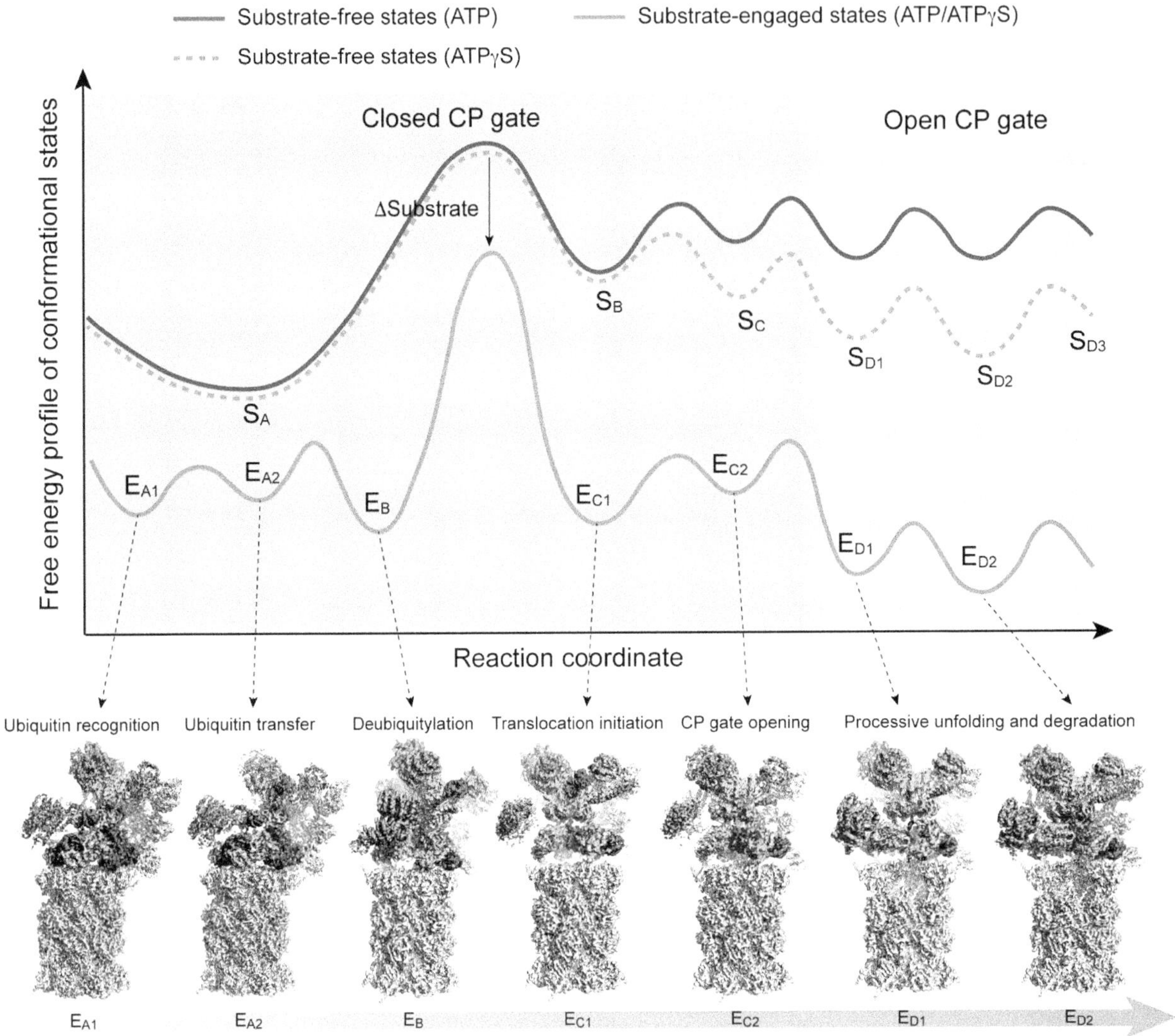

FIGURE 2.4 Schematic illustration of the energy landscape of 26S proteasome regulation by substrates and nucleotides. The replacement of ATP with slowly hydrolyzed ATPγS enriches the conformational states corresponding to an open CP gate. The engagement of a substrate with the proteasome further stabilizes these states but also induces more substates corresponding to a closed CP gate. Seven cryo-EM maps of the substrate-engaged *Homo sapiens* 26S proteasome at 2.8–3.6 Å resolution capture the key intermediate steps of substrate processing and provide important insights into the chemical cycle of proteasome-mediated degradation (Dong, Zhang et al., 2019). Dashed arrows map the seven cryo-EM structures to the local energy minima at the energy landscape.

CANONICAL UBIQUITIN-MEDIATED PATHWAYS OF PROTEASOMAL DEGRADATION

Although ubiquitin-independent degradation can be carried out by either 20S or 26S proteasome in a pathway-dependent or tissue-specific manner, the functional roles of ubiquitin-mediated degradation by the 26S proteasome appear to be universal and broadly observed. Thus, it is plausible to consider ubiquitin-mediated degradation as the canonical pathway of the proteasome. Growing evidence suggests that three modes of coordinated ATP hydrolysis in the AAA-ATPase motor regulate distinct steps of substrate processing in the proteasome, including ubiquitin recognition (states E_{A1} and E_{A1}), substrate deubiquitylation (state E_B), initiation of substrate translocation (states E_{C1} and E_{C2}), and processive substrate translocation (states E_{D1} and E_{D2}) (de la Pena, Goodall et al., 2018; Dong, Zhang et al., 2019; Wu, Chen et al., 2022; Zhang, Zou et al., 2022). Together with recent time-resolved cryo-EM studies, more than 30 conformers resolved for the human proteasome in the act of substrate degradation characterize a nearly complete cycle of substrate processing in the 26S proteasome under various regulatory conditions.

Ubiquitin-mediated substrate commitment. Most proteins targeted for proteasomal degradation are recognized through substrate-conjugated polyubiquitin chains that mediate protein–protein interactions with the ubiquitin receptors (Figure 2.5a and b) (Kulathu and Komander, 2012; Yau and Rape 2016). The three ubiquitin receptors are organized around the substrate entrance of the AAA-ATPase motor, surrounding the stoichiometric DUB RPN11. Such a geometric arrangement would allow them to recognize a diverse variation of polyubiquitin chains of distinct linkage types, including those branched tree-like chains or hybrid linkage types (Martinez-Fonts, Davis et al., 2020; Wu, Chen et al., 2022). In addition to the ubiquitin receptor sites of high affinity, the proteasome also harbors several low-affinity ubiquitin-binding sites, in the vicinity of the primary sites, to assist the ubiquitin recruitment, inter-subunit transferring, and chain remodeling (Wu, Chen et al., 2022). The initial ubiquitin binding to the ubiquitin receptor RPN1 and to the N-terminal coiled coils of RPT4/PRT5 near RPN10 were visualized in the cryo-EM structures of the proteasome in state E_{A1} and E_{A2} (Dong, Zhang et al., 2019; Wu, Chen et al., 2022; Zhang, Zou et al., 2022).

An essential requirement for substrate commitment to the proteasome is an unstructured initiation region in the substrate, which was visualized to bind the pore loops of the AAA-ATPase in the cryo-EM structure of state E_B (Figure 2.5c), and which was interpreted as the starting point of translocation initiation and was synchronized with the RPN11-catalyzed deubiquitylation reaction (Dong, Zhang et al., 2019). Such an initiation region must be of sufficient length, allowing simultaneous binding of ubiquitin and a stretch of substrate residues to a ubiquitin receptor and the AAA-ATPase in the base subcomplex, respectively (Prakash, Tian et al., 2004; Takeuchi, Chen et al., 2007; Zhao, Zhang et al., 2010; Dong, Zhang et al., 2019). Despite a lack of visualization of the complete ubiquitin chains bound to the proteasome, these structural snapshots suggest an intriguing model for consecutive steps in initial ubiquitin recognition, leading to substrate commitment with the AAA-ATPase motor (Dong, Zhang et al., 2019). Interestingly, localization of a protein with an unstructured initiation region of sufficient length around the proteasome can be sufficient to target it for degradation even without ubiquitylation (Murakami, Matsufuji et al., 1992; Janse, Crosas et al., 2004; Erales and Coffino, 2014). Consistently, deleting such an initiation region can prevent a protein from being degraded (Elsasser and Finley, 2005; Fishbain, Prakash et al., 2011; Heinen, Acs et al., 2011). Variations in the length and composition of this initiation region can regulate the half-life of substrates due to changes of its binding affinity with the proteasome (van der Lee, Lang et al., 2014; Fishbain, Inobe et al., 2015).

Consequential Bifurcation of Deubiquitylation

Once ubiquitin receptors deliver a polyubiquitylated substrate to the proteasome, the proteasome must decide on two opposite consequences of deubiquitylation of substrates. If the ubiquitin tag is removed from the substrate upon or after substrate commitment, the DUBs in the proteasome promote delivery of substrate to the proteolytic chamber by ridding substrate translocation of the steric hindrance of ubiquitin chains. If instead the ubiquitin tag is removed early prior to substrate commitment, the substrate then dissociates from the proteasome and avoids being degraded. Unlike most DUBs incapable of binding the proteasome, the isopeptidase activities of all three proteasome-associated DUBs (RPN11, USP14/Ubp6, UCH37/UCHL5) are rather low in isolation; by incorporating into the proteasome, they can be dramatically enhanced. In the cryo-EM structure of the substrate-engaged proteasome in state E_B, the Insert-1 loop of RPN11 adopts a β-hairpin conformation and forms a four-stranded β-sheet from one side with the C-terminal strand of ubiquitin and from the other side with a segment of the RPT5 N-loop (residues 99-119) emanating from the oligonucleotide- or oligosaccharide-binding (OB) ring (Figure 2.6a and b). This β-sheet places the isopeptide bond right next to the zinc ion in the active site of RPN11 (Dong, Zhang et al., 2019), offering a geometric explanation for how RPN11-mediated ubiquitylation is optimally activated and coupled to substrate commitment with the AAA-ATPase motor (Worden, Dong et al., 2017).

In contrast to the JAMM metalloprotease RPN11, USP14, or its yeast ortholog Ubp6 is a cysteine-dependent DUB that primarily functions in the proteasome via reversible association with RPN1 and the AAA-ATPase motor (Borodovsky, Kessler et al., 2001; Hung, Klumpe et al., 2022; Zhang, Zou et al., 2022). Binding the proteasome activates USP14/Ubp6 ~300- to 800-fold (Leggett, Hanna et al., 2002; Lee, Lee et al., 2010). The N-terminal UBL domain of USP14 binds the T2 site of RPN1, whereas its catalytic USP domain binds the OB and AAA rings near RPN11 (Figure 2.6c) (Leggett, Hanna et al., 2002; Aufderheide, Beck et al., 2015; Bashore, Dambacher et al., 2015; Huang, Luan et al., 2016; Shi, Chen et al., 2016). Activation of USP14 or Ubp6 results from the refolding of the BL1 loop into a β-hairpin by direct interaction with the OB loops of mostly RPT1, which removes steric occlusion for ubiquitin binding (Hung, Klumpe et al., 2022; Zhang, Zou et al., 2022). The active sites of USP14 and RPN11 are only ~38 Å apart. Binding of peptide-proximal ubiquitin moiety allosterically antagonizes ubiquitin binding to RPN11 and decouples the ATPase activity from RPN11-catalyzed deubiquitylation (Hanna, Hathaway et al., 2006; Bashore, Dambacher et al., 2015; Zhang, Zou et al., 2022). Early studies have observed seemingly self-contradicting effects. Notably, USP14 stimulates the proteasomal ATPase activity and CP gate opening but inhibits substrate degradation, and curiously it enhances the ubiquitin ligase UBE3C/Hul5 association with the proteasome (Leggett, Hanna et al., 2002; Peth, Besche et al., 2009; Aufderheide, Beck et al., 2015; Bashore, Dambacher et al., 2015; Kim and Goldberg, 2017; Kuo and Goldberg, 2017). Recently, time-resolved cryo-EM analysis visualized two parallel pathways of conformational transitions of the proteasome, thus reconciling the early contradictory observations (Zhang, Zou et al., 2022). It turned out one pathway stimulates the ATPase activity for substrate degradation when substrate commitment precedes USP14 activation, and the other inhibits substrate engagement when USP14 activation precedes substrate engagement. Essentially,

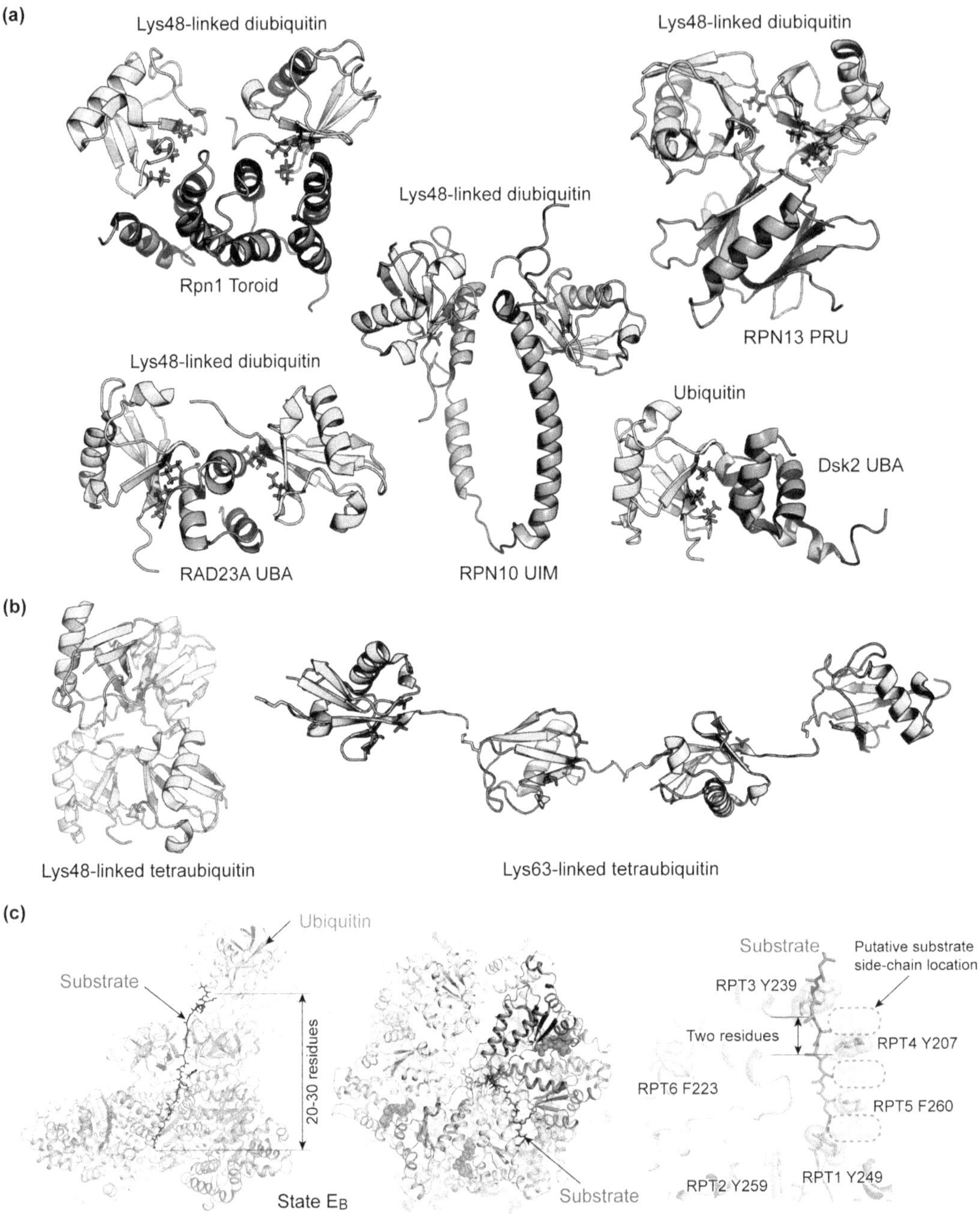

FIGURE 2.5 Proteasomal recognition of ubiquitin codes and initiation region of substrates. (a) Distinct modes of interaction with ubiquitin are shown for each receptor. Amino acids of each ubiquitin-binding domain that contact ubiquitin are highlighted. The structures correspond to PDB IDs as follows: Lys48-linked diubiquitin-bound yeast Rpn1 toroid segment (2N3V) (Shi, Chen et al., 2016), Lys48-linked diubiquitin-bound human RPN10 UIM (2KDE) (Zhang, Wang et al., 2009), Lys48-linked diubiquitin-bound human RPN13 PRU (5YMY) (Liu, Dong et al., 2019), Lys48-diubiquitin-bound human RAD23A/hHR23A UBA (1ZO6) (Varadan, Assfalg et al., 2005), and ubiquitin-bound yeast Dsk2 UBA (1WR1) (Ohno, Jee et al., 2005). PRU, Pleckstrin-like receptor for ubiquitin domain; UBA, ubiquitin-associated domain; UBL, ubiquitin-like domain; UIM, ubiquitin-interacting motif; VWA, von Willebrand factor A domain; UBA, ubiquitin-associated domain. (b) Structures of tetraubiquitin chains with two different linkages. The PDB IDs for the Lys48- and Lys63-linked tetraubiquitin chains are 2O6V and 3HM3, respectively (Eddins, Varadan et al., 2007; Datta, Hura et al., 2009). (c) The initiation region of substrate recognized by the AAA-ATPase channels. The ATPase structure in the substrate-bound proteasome in state E_B (PDB ID: 6MSE) is shown from the perspective of side view (left) and top view (middle) (Dong, Zhang et al., 2019). The right insert shows the close-up view of the substrate interaction with the pore-1 loops, with the aromatic residues in the pore loop highlighted by the transparent sphere representations. The putative location of the substrate side chains intercalating with these pore-loop aromatic side chains are marked by dashed red boxes.

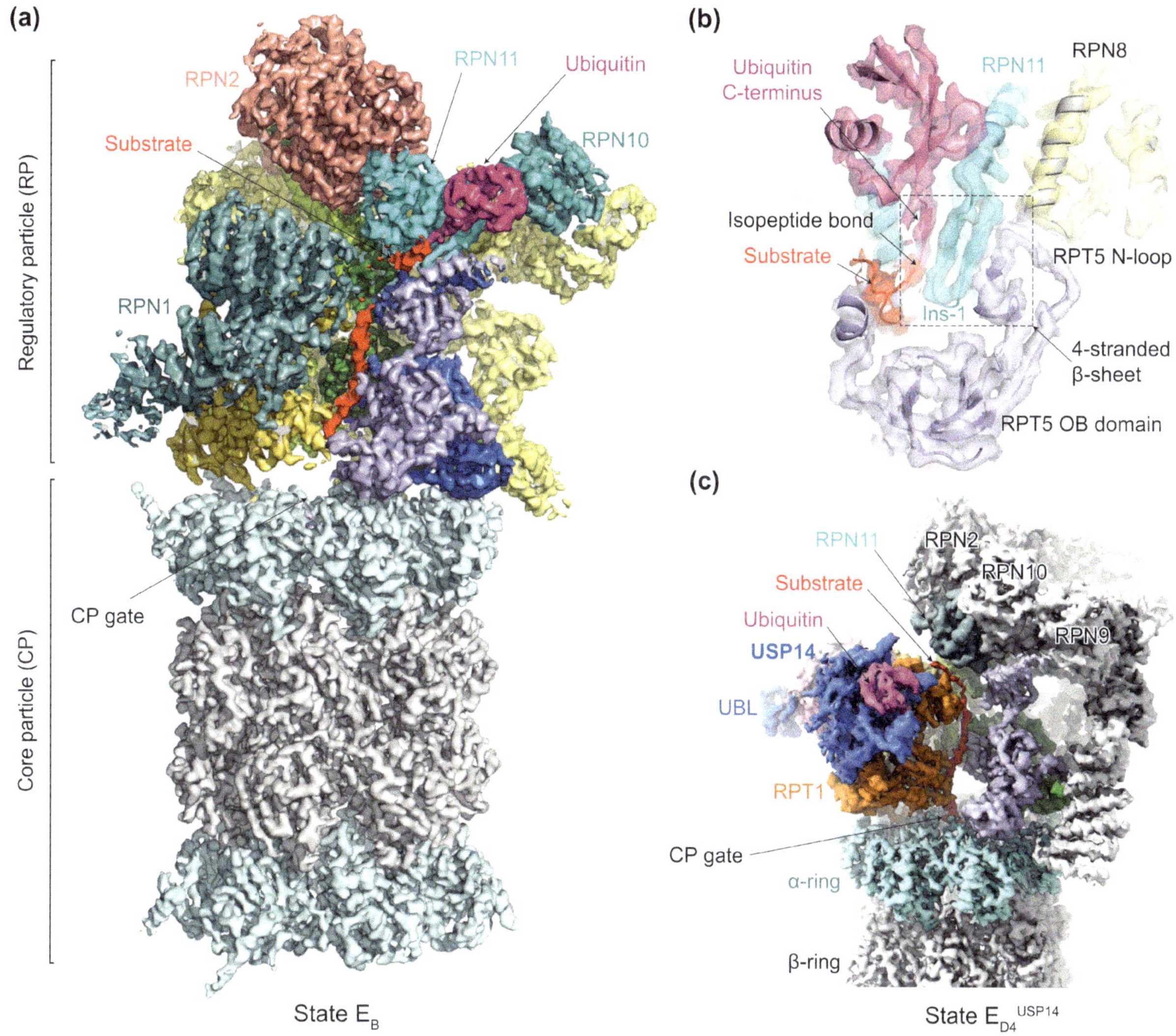

FIGURE 2.6 Mechanism of substrate deubiquitylation by the 26S proteasome. (a) Cryo-EM density maps of substrate-bound human proteasome in state E_B at 3.3 Å (Dong, Zhang et al., 2019). The density of RPT1 is omitted to show the substrate bound inside the AAA channel. (b) A close-up view of the quaternary interface around the scissile isopeptide bond between ubiquitin and the substrate lysine in E_B. The cryo-EM density is rendered as a transparent surface, superimposed with the cartoon representation of the atomic model. (c) Side view of the cryo-EM density map of the substrate-engaged USP14-proteasome complex in state E_{D4}^{USP14} (Zhang, Zou et al., 2022). Compared to state E_{D2}^{USP14}, USP14 is rotated about 30° to dock onto the AAA domain of RPT1. To visualize the substrate density inside the AAA-ATPase motor, the density of RPT5 is omitted.

the bidirectional allostery of USP14 and the proteasome kinetically introduces three regulatory checkpoints for proteasome function, at the steps of ubiquitin recognition, initiation of substrate translocation, and ubiquitin chain recycling (Zhang, Zou et al., 2022).

UCH37/UCHL5 is another cysteine-dependent DUB reversibly associated with the proteasome in higher eukaryotes, such as *Homo sapiens* and *Drosophila melanogaster*. Its ortholog Uch2 was found in *Schizosaccharomyces pombe* but not in *Saccharomyces cerevisiae* (Lam, Xu et al., 1997; Holzl, Kapelari et al., 2000). UCH37 consists of two domains, an N-terminal catalytic domain representative of the UCH family and a fibrous C-terminal tail (Burgie, Bingman et al., 2012). Binding of its C-terminal region to the N-terminal DEUBAD domain of RPN13 activates UCH37 in the proteasome (Yao, Song et al., 2006; Sahtoe, van Dijk et al., 2015; Vander Linden, Hemmis et al., 2015). UCH37 selectively cleaves Lys48-containing branched chains for promoting proteasomal degradation (Deol, Crowe et al., 2020; Du, Babik et al., 2022). The debranching activity and selectivity toward branched chains is notably enhanced by binding RPN13 in the proteasome (Deol, Crowe et al., 2020). Upon association with RPN13, a loop sequestering the catalytic cysteine (Cys88) in the free UCH37 is remodeled to expose the catalytic triad (Cys88, His164, and Asp179) (Hamazaki, Iemura et al., 2006; Qiu, Ouyang et al., 2006; Yao, Song et al., 2006; Burgie, Bingman et al., 2012; Sahtoe, van Dijk et al., 2015; Vander Linden, Hemmis et al., 2015).

In the free RPN13, the DEUBAD domain interacts with the PRU domain and sterically occludes its ubiquitin-binding site. Association of RPN13 with RPN2 in the proteasome opens up the PRU domain for ubiquitin binding, allowing activation of UCH37 (Chen, Lee et al., 2010; Lu, Nowicka et al., 2017).

Substrate unfolding and translocation. The proteasomal AAA-ATPases come from the classic clade of the AAA+ superfamily. Most cryo-EM structures of substrate-engaged AAA+ ATPase complex resemble the architecture of substrate–pore loop interactions observed in the proteasome (de la Pena, Goodall et al., 2018; Dong, Zhang et al., 2019; Zhang, Zou et al., 2022). These studies consistently proposed the hypothetical hand-over-hand model of substrate translocation, despite lack of detailed evidence for structural snapshots of ATPases within a cycle of single-nucleotide exchange (Puchades, Sandate et al., 2020). The initial proposal of the hand-over-hand model assumes the sequential symmetric hydrolysis of ATP around the AAA-ATPase ring, similar to that in a typical AAA-ATPase homohexamer like Yme1 and ClpXP (Puchades, Sandate et al., 2020). In this model, the pore loop of one ADP-bound ATPase subunit is dissociated from the translocating front of the substrate, while its adjacent nucleotide-exchanging ATPase subunit makes a large outward rotation in attempt to re-associate with the substrate at the entry end of the substrate after ATP binding (Figure 2.7). During this interval, the other four substrate-bound ATPase subunits are allosterically driven and rotate together inward to move the substrate forward.

However, the proteasomal AAA-ATPase motor exhibits its functional asymmetry, implicating an asymmetric

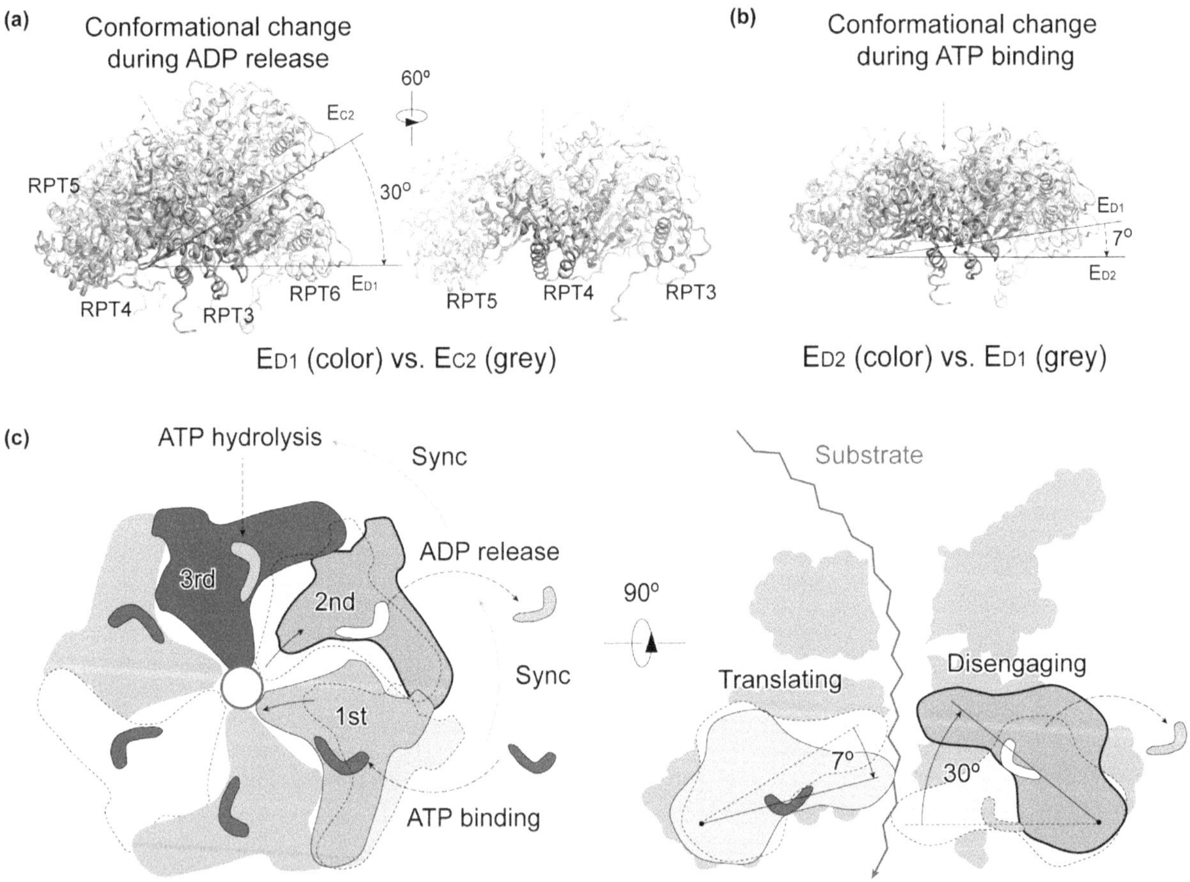

FIGURE 2.7 Mechanism of processive substrate translocation in the proteasome (PDB IDs: 6MSH, 6MSJ, 6MSK) (Dong, Zhang et al., 2019). (a) Structural comparison between E_{D1} (color) and E_{C2} (grey), by using the large AAA subdomain of RPT5 to align the AAA ring structures in distinct states. A 30° out-of-plane rotation is observed during RPT5 disengagement from substrate. The right panel, viewed from a perspective rotated vertically against the left panel, shows that out-of-plane rotation in RPT5 is more prominently amplified in its counterclockwise neighboring ATPases than in its clockwise neighbors. Red arrows mark the center of the AAA ring. (b) Structural comparison between E_{D2} (color) and E_{D1} (grey) in which the large AAA subdomain of RPT5 is used to align the two AAA-ring structures together. A 7° out-of-plane rotation of Rpt5 is observed during RPT5 reassociation with the substrate. (c) Schematic illustrates the mechanism for processive substrate translocation that involves synchronization of nucleotide processing in three adjacent ATPases, i.e., ATP binding, ADP release, and ATP hydrolysis (left), creating vertical rotations in the ATPases that cooperatively propel the substrate (right).

mechanism of substrate-engaged ATPase heterohexamer (Rubin, Glickman et al., 1998; Erales, Hoyt et al., 2012; Beckwith, Estrin et al., 2013; Fang, Hon et al., 2022; Zhang, Zou et al., 2022). Growing evidence points to the asymmetric interactions of the ATPase motor with various proteasome-associated factors, including USP14/Ubp6 and ZFAND5 that regulate differential activation of the proteasome (Hung, Klumpe et al., 2022; Zhang, Zou et al., 2022; Lee, Zhu et al., 2023). Recently, six conformers of the substrate-engaged ATPase ring within a cycle of single-nucleotide exchange during initiation of substrate translocation were resolved by application of the deep-learning-enhanced 3D classification method (Wu, Chen et al., 2022). These structures indicate two consecutive substeps involved. A substep of faster dissociation of the ADP-releasing ATPase subunit from the forefront of substrate movement near the CP gate is followed by another substep of slower reassociation of the ATP-exchanged ATPase subunit with substrate near the entrance of the OB ring. It was speculated that kinetic variation of the second substep could accommodate certain probabilistic events such as substrate sliding or inchworm motion. During this transient interval, two or more substrate-dissociated ATPase subunits result in weaker gripping of substrate that could hypothetically permit intermittent substrate translocation driven by its own conformational entropy, in addition to ATPase motor-driven movement. Consistently, recent computational modeling found such kinetic substeps to be crucial in recapitulating the dynamics of substrate translocation *in silico* (Fang, Hon et al., 2022). This mechanism can potentially reconcile the cryo-EM observation of two substrate residues gripped between adjacent pore loops and other experimental observation of substrate translocation of more than two residues per ATP hydrolyzed (Fei, Bell et al., 2020).

Processive degradation. The proteasome degrades substrates into short peptides in a processive fashion and normally does not release substrate until it is fully degraded (Dick, Moomaw et al., 1991; Akopian, Kisselev et al., 1997; Nussbaum, Dick et al., 1998). However, an exception was observed in the activation of many cellular proteins with partial degradation that is hypothetically terminated at degradation-stop signals, though the consensus signature of such degradation-stop signals remains elusive (Saeki, Kudo et al., 2009; Zhang, Lv et al., 2013; Chui, Okondo et al., 2019; Sandstrom, Mitchell et al., 2019). The Thr1 residues in β1, β2, and β5 subunits possess caspase-like (acidic), trypsin-like (basic), and chymotrypsin-like (hydrophobic) activities, respectively. In these β-subunits, an open cleft referred to as the S1 ("specificity") pocket harbors the active site at the N-terminal Thr1 embraced by a catalytic triad. The side chain Oγ of Thr1 provides the catalytic nucleophile attack of the carbonyl carbon in a peptide bond at the C-terminal of the P1 residue of the substrate (Lowe, Stock et al., 1995; Seemuller, Lupas et al., 1995). Opening of the CP gate was found to allosterically lower the activation energy for the nucleophilic attack for peptide bond cleavage at the active site (Wu, Chen et al., 2022).

Peptide products released from substrate degradation exhibit a length distribution ranging from 4 to 25 residues and an average length of 7 to 9 residues (Wenzel, Eckerskorn et al., 1994; Ehring, Meyer et al., 1996; Kisselev, Akopian et al., 1998; Nussbaum, Dick et al., 1998). How does the CP regulate the peptide length of degraded substrates? As observed in the cryo-EM reconstructions of the substrate-engaged human proteasome, a dozen noncatalytic substrate-binding sites inside the CP chamber provide a possible answer (Dong, Zhang et al., 2019). Of note, all these substrate-binding sites are heavily decorated with aromatic residues, consistent with the hallmark of substrate-binding sites inside the OB and AAA channels. These sites can stabilize an unfolded polypeptide segment with a length ranging from 3 to 10 residues and are located in a range of 4 to 25 residues distant from the three catalytic sites in β1, β2, and β5 (Dong, Zhang et al., 2019), consistent with the experimental observation of the product length variation. The auxiliary noncatalytic substrate-binding sites can anchor the unstructured substrates during their proteolysis, inhibit refolding of substrates in the CP, and thus maintain processivity of degradation. In support of this mechanism, NMR studies have shown that the CP interior stabilizes unfolded conformations of substrates (Ruschak, Religa et al., 2010), and long peptides were degraded faster than short peptides (Dolenc, Seemuller et al., 1998).

AN INTEGRATIVE MECHANISM OF PROTEASOME REGULATION

Systematic analysis of cryo-EM structures of substrate-engaged proteasome in more than 30 conformational states under various conditions has now provided extensive insights into a unified mechanism for proteasomal regulation through a multilevel hierarchy of protein–protein interaction networks, in which all subunit conformations are allosterically coupled (Figure 2.8) (de la Pena, Goodall et al., 2018; Dong, Zhang et al., 2019; Wu, Chen et al., 2022; Zhang, Zou et al., 2022). The first and outermost level of functional regulation is via dynamic interactions between ubiquitin signals and an array of ubiquitin receptors, DUBs, and other proteasome-interacting proteins. This is manifested as ubiquitin-binding sites recognizing both ubiquitin signals and UBL proteins and transmitting such protein–protein interactions into allosteric regulation of AAA-ATPase conformations that control the CP gate. Binding of the UBL domain of USP14 or tetra-ubiquitin chain alone can stimulate proteasome activation (Kim and Goldberg, 2018; Ding, Xu et al., 2019; Jonsson, Htet et al., 2022). Ubiquitin binding and remodeling destabilize the resting state of the proteasome and promote its transition to downstream translocation-competent (E_D-like) states. A polyubiquitin chain binding multiple sites in the proteasome could have profound allosteric and conformational selection effects (Jonsson, Htet et al., 2022).

The second level of proteasome regulation lies between the AAA-ATPase unfoldase and the rest of the proteasome.

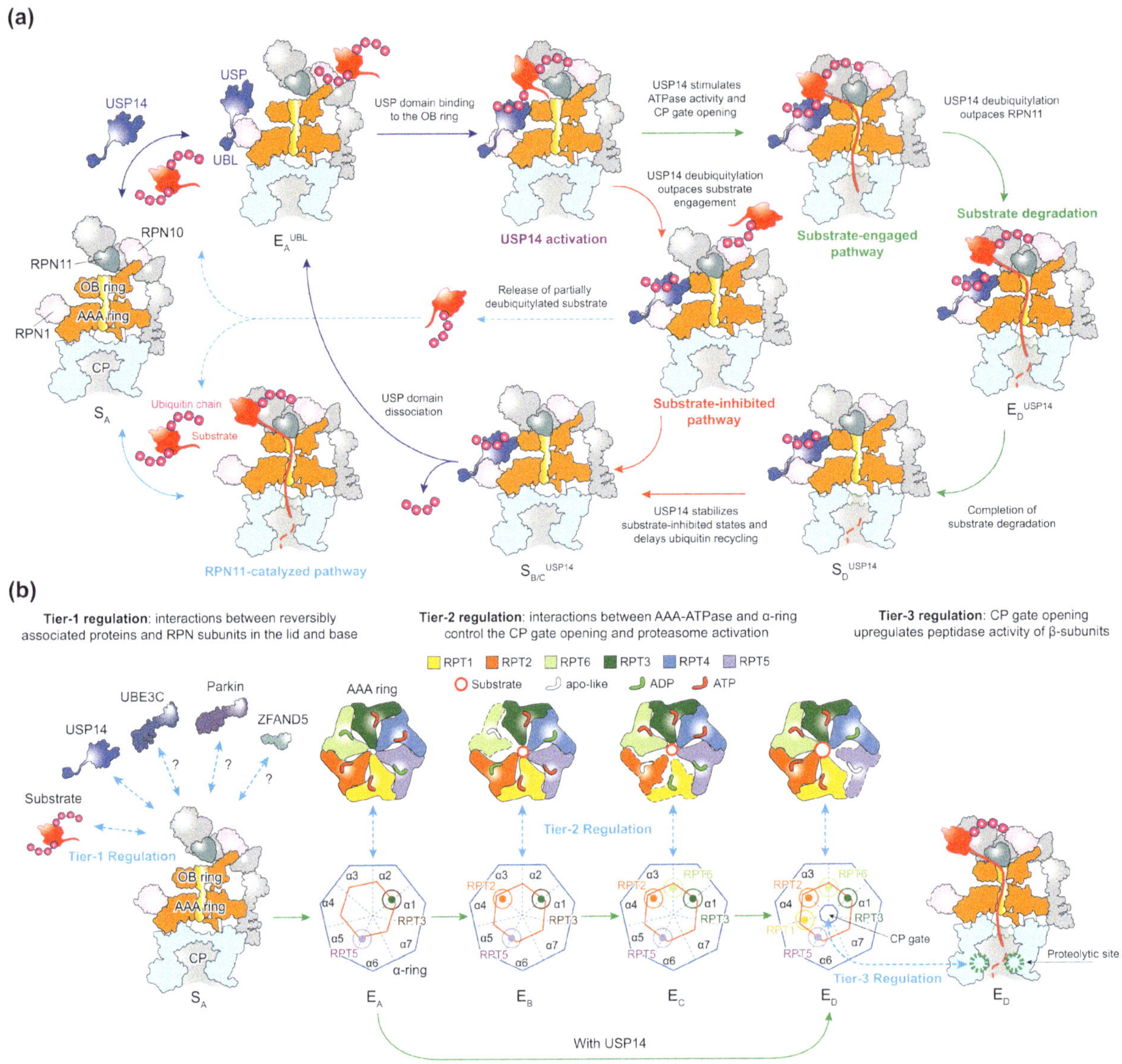

FIGURE 2.8 Complexity of proteasome regulation. (a) Mechanistic modelof USP14-mediated regulation of proteasome function (Zhang, Zou et al., 2022). USP14 binding to the RPN1 and RPT1 subunits of the proteasome activates USP14, whereas ubiquitin-substrate conjugates recruited to the proteasome's ubiquitin receptors facilitate ubiquitin recognition by USP14. USP14 binding creates two parallel state-transition pathways of the proteasome. Along the substrate-inhibited pathway (red arrows), which has RPN11 blocking the substrate entrance of the OB ring before any substrate insertion takes place, USP14 trims ubiquitin chains and releases the substrate from the proteasome, thus preventing substrate degradation. Along the substrate-engaged pathway (blue arrows), a substrate has already been inserted into the ATPase ring, and RPN11 narrows down on the OB ring but does not block substrate translocation through the OB ring. (b) Illustration explaining hierarchical allosteric regulation of proteasome function composed of three tiers of regulatory pairs (Wu, Chen et al., 2022). In the presence of USP14, the proteasome may skip states E_B and E_C and switch directly to state E_D for processive degradation.

The AAA-ATPase motor bridges the architectural gap between the CP and RPN subunits, including the lid subcomplex and the ubiquitin receptors in the base. Distinct dynamic modes of coordinated ATP hydrolysis appear to regulate the intermediate steps of substrate processing, in which initiation of substrate unfolding is allosterically coupled with CP gating (Dong, Zhang et al., 2019; Zhang, Zou et al., 2022). The conformation of the lid subcomplex and ubiquitin receptors allosterically regulate the insertion of the C-terminal tails of AAA-ATPases into the inter-subunit surface pockets of the α-ring (Dong, Zhang et al., 2019; Greene, Goodall et al., 2019). The importance of activating the proteasome via the regulation of AAA-ATPase is highlighted by the direct interactions of USP14 and ZFAND5 with the AAA-ATPase motor, which allosterically stabilize the translocation-compatible states or promote structural transition into these states (Zhang, Zou et al., 2022; Lee, Zhu et al., 2023).

The last and innermost level of proteasome regulation is between the CP gate in the α-ring and the catalytic sites of the β-subunits. Opening of the CP gate allosterically upregulates the proteolytic activity of catalytic β-subunits by lowering the energetic barrier of nucleophilic attack (Wu, Chen et al., 2022). The hierarchical allosteric regulation may occur mutually with both RPs and allow the conformational state of one RP to influence that of the other RP in the doubly capped proteasome (Chen, Wu et al., 2016; Zhu, Wang et al., 2018; Wu, Chen et al., 2022). The proteasome structure and function are expected to have been optimally evolved in response to extremely complicated microenvironmental challenges. This advanced design principle is manifested as various phenotypic effects, such as UBL-induced proteasome activation (Kim and Goldberg, 2018; Collins and Goldberg, 2020) and long-range allosteric communications in the proteasome (Kleijnen, Roelofs et al., 2007; Haselbach, Schrader et al., 2017; Rennella, Huang et al., 2020; Yu, Yu et al., 2020).

PERSPECTIVE

The past decade has seen several breakthrough studies in the elucidation of structural mechanisms of the proteasome at the atomic level. Continuous improvements of biochemical and cryo-EM methods have yielded atomic information critical for understanding the functional dynamics of the proteasome. Not only have the conformational states of the proteasome been extensively mapped, but also their correspondence to the intermediate steps of substrate processing and a nearly complete cycle of protein degradation by the proteasome have been visualized at the atomic level. A unified mechanistic model of proteasome function is now emerging. These advancements lay an essential foundation for further understanding how the proteasome decodes the ubiquitin signals and regulates many cellular functions of hitherto unknown mechanisms. The lack of other cellular regulators of the proteasome in previous structural studies, such as UBE3A, UBE3C, and PARKIN, limited our understanding of proteasome regulation under specific physiological conditions. The previously reported structures used substrates mainly modified by Lys63-linked ubiquitin chains (de la Pena, Goodall et al., 2018; Dong, Zhang et al., 2019; Zhang, Zou et al., 2022). These model substrates have limited the structural insights into diverse ubiquitin chain interactions with the proteasome. Replacing the model substrates with Lys48-linked chains or other linkage types, including branched chains, is anticipated in future studies.

The proteasome is a highly valuable drug target, with three FDA-approved drugs (bortezomib, carfilzomib, and ixazomib) currently in clinical use for treating multiple myeloma and mantle cell lymphoma (Moreau, Pylypenko et al., 2011; Huber and Groll, 2012; Rentsch, Landsberg et al., 2013; Cromm and Crews, 2017). Several other FDA-approved drugs, such as thalidomide, lenalidomide, and pomalidomide, work as bispecific molecular glues to hijack the Cullin-RING ligase 4-cereblon ($CRL4^{CRBN}$) E3 ligase and induce degradation of two specific B cell transcription factors, Ikaros family zinc finger proteins 1 and 3 (IKZF1 and IKZF3) by the proteasome in multiple myeloma cells (Fischer, Bohm et al., 2014; Gandhi, Kang et al., 2014; Kronke, Udeshi et al., 2014; Lu, Middleton et al., 2014; Petzold, Fischer et al., 2016). Several clinical trials are currently underway to test proteolysis-targeting chimera (PROTAC) compounds as potential drugs that act in a way similar to molecular glue to induce the degradation of cancer-pivoting targets, such as androgen and estrogen receptors for treating prostate and breast cancers, respectively (Bondeson and Crews, 2017; Bekes, Langley et al., 2022). Many viruses attack the host by injecting a viral protein that also works in a molecular glue-like mechanism to hijack the host E3 ligases to degrade host proteins by the proteasome that get in the way of viral replication (Bialek, Collawn et al., 2023). A notorious example is the human immunodeficiency virus type 1 (HIV-1) that uses its accessory factors Vif, Vpu, and Vpr to counteract host defenses by hijacking Cullin-RING E3 ligases (CUL5, CUL1, and CUL4A) to degrade APOBEC3 family proteins and other restriction factors (Jager, Kim et al., 2011). CertainE3 ligases hijacked by virus, such as UBE3A/E6AP, were found to associate with the proteasome with elusive functions and mechanisms. More recently, bispecific compounds directly targeting certain proteasomal subunits and bypassing any E3 ligases have been found to induce the degradation of intracellular proteins, illuminating the possibility of targeted degradation harnessing the proteasome itself (Bashore, Prakash et al., 2023). Alternatively, PROTACs inducing degradation of proteasomal subunit RPN13 also exhibited anticancer activity for multiple myeloma (Lu, Sabbasani et al., 2021). Both multispecific drugs and targeted degradation present new paradigms of therapeutic discovery, the advantages of which are yet to be fully exploited (Deshaies, 2020).

Despite the recent breakthrough studies in elucidating the structure–function relationship of the proteasome from the lens of functional dynamics, major missing gaps of mechanistic understanding lie in the interplay between numerous ubiquitylation pathways and the working machineries of the proteasome. In addition, the exact biological functions of the ATP-independent activators and ubiquitin-independent proteasomal pathways remain incompletely understood (Bialek, Collawn et al., 2023), though their roles are expected to be not as universal as the 26S proteasome (Kocberber, Willemsen et al., 2023). Deeper understanding of proteasome regulation by various ubiquitylation machineries and other proteasome-interacting proteins in general and in tissue-specific conditions will undoubtedly drive biotechnological expansion of targeted protein degradation and further development of novel therapeutics for treating many traditionally incurable diseases. Addressing these problems poses greater technical challenges due to the transient nature of the interactions of these cellular proteins with the proteasome. Another major gap is the understanding of how the proteasome selectively splices certain polypeptides for optimal antigen presentation (Hanada, Yewdell et al., 2004; Vigneron, Stroobant et al., 2004; Liepe, Marino et al., 2016; Paes, Leonov et al., 2019). Understanding the mechanism of proteasome-mediated protein splicing might help develop it into a system of targeted protein editing, a potential technology that is envisioned to rival gene editing for widespread biomedical applications.

ACKNOWLEDGMENTS

The authors thank S. Zou, S. Zhang, D. Yin, and other members of the Mao laboratory for discussions and comments. This work was funded in part by grants from the National Natural Science Foundation of China (12125401) to Y. M.

REFERENCES

Akopian, T. N., Kisselev, A. F. and Goldberg, A. L. (1997). Processive degradation of proteins and other catalytic properties of the proteasome from Thermoplasma acidophilum. *J Biol Chem* **272**(3): 1791–1798.

Arrigo, A. P., Tanaka, K., Goldberg, A. L. and Welch, W. J. (1988). Identity of the 19S 'prosome' particle with the large multifunctional protease complex of mammalian cells (the proteasome). *Nature* **331**(6152): 192–194.

Aufderheide, A., Beck, F., Stengel, F., Hartwig, M., Schweitzer, A., Pfeifer, G., Goldberg, A. L., Sakata, E., Baumeister, W. and Forster, F. (2015). Structural characterization of the interaction of Ubp6 with the 26S proteasome. *Proc Natl Acad Sci USA* **112**(28): 8626–8631.

Bard, J. A. M., Goodall, E. A., Greene, E. R., Jonsson, E., Dong, K. C. and Martin, A. (2018). Structure and function of the 26S proteasome. *Annu Rev Biochem* **87**: 697–724.

Bashore, C., Dambacher, C. M., Goodall, E. A., Matyskiela, M. E., Lander, G. C. and Martin, A. (2015). Ubp6 deubiquitinase controls conformational dynamics and substrate degradation of the 26S proteasome. *Nat Struct Mol Biol* **22**(9): 712–719.

Bashore, C., Prakash, S., Johnson, M. C., Conrad, R. J., Kekessie, I. A., Scales, S. J., Ishisoko, N., Kleinheinz, T., Liu, P. S., Popovych, N., Wecksler, A. T., Zhou, L., Tam, C., Zilberleyb, I., Srinivasan, R., Blake, R. A., Song, A., Staben, S. T., Zhang, Y., Arnott, D., Fairbrother, W. J., Foster, S. A., Wertz, I. E., Ciferri, C. and Dueber, E. C. (2023). Targeted degradation via direct 26S proteasome recruitment. *Nat Chem Biol* **19**(1): 55–63.

Beck, F., Unverdorben, P., Bohn, S., Schweitzer, A., Pfeifer, G., Sakata, E., Nickell, S., Plitzko, J. M., Villa, E., Baumeister, W. and Forster, F. (2012). Near-atomic resolution structural model of the yeast 26S proteasome. *Proc Natl Acad Sci USA* **109**(37): 14870–14875.

Beckwith, R., Estrin, E., Worden, E. J. and Martin, A. (2013). Reconstitution of the 26S proteasome reveals functional asymmetries in its AAA+ unfoldase. *Nat Struct Mol Biol* **20**(10): 1164–1172.

Bekes, M., Langley, D. R. and Crews, C. M. (2022). PROTAC targeted protein degraders: The past is prologue. *Nat Rev Drug Discov* **21**(3): 181–200.

Bialek, W., Collawn, J. F. and Bartoszewski, R. (2023). Ubiquitin-dependent and independent proteasomal degradation in host-pathogen interactions. *Molecules* **28**(18): 6740.

Boehringer, J., Riedinger, C., Paraskevopoulos, K., Johnson, E. O., Lowe, E. D., Khoudian, C., Smith, D., Noble, M. E., Gordon, C. and Endicott, J. A. (2012). Structural and functional characterization of Rpn12 identifies residues required for Rpn10 proteasome incorporation. *Biochem J* **448**(1): 55–65.

Bohn, S., Beck, F., Sakata, E., Walzthoeni, T., Beck, M., Aebersold, R., Forster, F., Baumeister, W. and Nickell, S. (2010). Structure of the 26S proteasome from schizosaccharomyces pombe at subnanometer resolution. *Proc Natl Acad Sci USA* **107**(49): 20992–20997.

Bondeson, D. P. and Crews, C. M. (2017). Targeted protein degradation by small molecules. *Annu Rev Pharmacol Toxicol* **57**: 107–123.

Borodovsky, A., Kessler, B. M., Casagrande, R., Overkleeft, H. S., Wilkinson, K. D. and Ploegh, H. L. (2001). A novel active site-directed probe specific for deubiquitylating enzymes reveals proteasome association of USP14. *EMBO J* **20**(18): 5187–5196.

Buetow, L. and Huang, D. T. (2016). Structural insights into the catalysis and regulation of E3 ubiquitin ligases. *Nat Rev Mol Cell Biol* **17**(10): 626–642.

Burgie, S. E., Bingman, C. A., Soni, A. B. and Phillips, G. N., Jr. (2012). Structural characterization of human Uch37. *Proteins* **80**(2): 649–654.

Chen, J., Wang, Y., Xu, C., Chen, K., Zhao, Q., Wang, S., Yin, Y., Peng, C., Ding, Z. and Cong, Y. (2021). Cryo-EM of mammalian PA28αβ-iCP immunoproteasome reveals a distinct mechanism of proteasome activation by PA28αβ. *Nat Commun* **12**(1): 739.

Chen, S., Wu, J., Lu, Y., Ma, Y. B., Lee, B. H., Yu, Z., Ouyang, Q., Finley, D. J., Kirschner, M. W. and Mao, Y. (2016). Structural basis for dynamic regulation of the human 26S proteasome. *Proc Natl Acad Sci USA* **113**(46): 12991–12996.

Chen, X., Lee, B. H., Finley, D. and Walters, K. J. (2010). Structure of proteasome ubiquitin receptor hRpn13 and its activation by the scaffolding protein hRpn2. *Mol Cell* **38**(3): 404–415.

Chui, A. J., Okondo, M. C., Rao, S. D., Gai, K., Griswold, A. R., Johnson, D. C., Ball, D. P., Taabazuing, C. Y., Orth, E. L., Vittimberga, B. A. and Bachovchin, D. A. (2019). N-terminal degradation activates the NLRP1B inflammasome. *Science* **364**(6435): 82–85.

Ciechanover, A., Elias, S., Heller, H., Ferber, S. and Hershko, A. (1980). Characterization of the heat-stable polypeptide of the ATP-dependent proteolytic system from reticulocytes. *J Biol Chem* **255**(16): 7525–7528.

Ciechanover, A., Heller, H., Elias, S., Haas, A. L. and Hershko, A. (1980). ATP-dependent conjugation of reticulocyte proteins with the polypeptide required for protein degradation. *Proc Natl Acad Sci USA* **77**(3): 1365–1368.

Ciehanover, A., Hod, Y. and Hershko, A. (1978). A heat-stable polypeptide component of an ATP-dependent proteolytic system from reticulocytes. *Biochem Biophys Res Commun* **81**(4): 1100–1105.

Ciechanover, A. and Kwon, Y. T. (2015). Degradation of misfolded proteins in neurodegenerative diseases: Therapeutic targets and strategies. *Exp Mol Med* **47**: e147.

Collins, G. A. and Goldberg, A. L. (2017). The logic of the 26S proteasome. *Cell* **169**(5): 792–806.

Collins, G. A. and Goldberg, A. L. (2020). Proteins containing ubiquitin-like (Ubl) domains not only bind to 26S proteasomes but also induce their activation. *Proc Natl Acad Sci USA* **117**(9): 4664–4674.

Coux, O., Tanaka, K. and Goldberg, A. L. (1996). Structure and functions of the 20S and 26S proteasomes. *Annu Rev Biochem* **65**: 801–847.

Cromm, P. M. and Crews, C. M. (2017). The proteasome in modern drug discovery: Second life of a highly valuable drug target. *ACS Cent Sci* **3**(8): 830–838.

Crosas, B., Hanna, J., Kirkpatrick, D. S., Zhang, D. P., Tone, Y., Hathaway, N. A., Buecker, C., Leggett, D. S., Schmidt, M., King, R. W., Gygi, S. P. and Finley, D. (2006). Ubiquitin chains are remodeled at the proteasome by opposing ubiquitin ligase and deubiquitinating activities. *Cell* **127**(7): 1401–1413.

da Fonseca, P. C., He, J. and Morris, E. P. (2012). Molecular model of the human 26S proteasome. *Mol Cell* **46**(1): 54–66.

Dahlmann, B., Ruppert, T., Kuehn, L., Merforth, S. and Kloetzel, P. M. (2000). Different proteasome subtypes in a single tissue exhibit different enzymatic properties. *J Mol Biol* **303**(5): 643–653.

Dammer, E. B., Na, C. H., Xu, P., Seyfried, N. T., Duong, D. M., Cheng, D., Gearing, M., Rees, H., Lah, J. J., Levey, A. I., Rush, J. and Peng, J. (2011). Polyubiquitin linkage profiles in three models of proteolytic stress suggest the etiology of Alzheimer disease. *J Biol Chem* **286**(12): 10457–10465.

Datta, A. B., Hura, G. L. and Wolberger, C. (2009). The structure and conformation of Lys63-linked tetraubiquitin. *J Mol Biol* **392**(5): 1117–1124.

de Almeida, M., Hinterndorfer, M., Brunner, H., Grishkovskaya, I., Singh, K., Schleiffer, A., Jude, J., Deswal, S., Kalis, R., Vunjak, M., Lendl, T., Imre, R., Roitinger, E., Neumann, T., Kandolf, S., Schutzbier, M., Mechtler, K., Versteeg, G. A., Haselbach, D. and Zuber, J. (2021). AKIRIN2 controls the nuclear import of proteasomes in vertebrates. *Nature* **599**(7885): 491–496.

de la Pena, A. H., Goodall, E. A., Gates, S. N., Lander, G. C. and Martin, A. (2018). Substrate-engaged 26S proteasome structures reveal mechanisms for ATP-hydrolysis-driven translocation. *Science* **362**(6418): aav0725.

Demartino, G. N. and Gillette, T. G. (2007). Proteasomes: Machines for all reasons. *Cell* **129**(4): 659–662.

Deol, K. K., Crowe, S. O., Du, J., Bisbee, H. A., Guenette, R. G. and Strieter, E. R. (2020). Proteasome-bound UCH37/UCHL5 debranches ubiquitin chains to promote degradation. *Mol Cell* **80**(5): 796–809 e799.

Deshaies, R. J. (2020). Multispecific drugs herald a new era of biopharmaceutical innovation. *Nature* **580**(7803): 329–338.

Deveraux, Q., Ustrell, V., Pickart, C. and Rechsteiner, M. (1994). A 26 S protease subunit that binds ubiquitin conjugates. *J Biol Chem* **269**(10): 7059–7061.

Dick, L. R., Moomaw, C. R., DeMartino, G. N. and Slaughter, C. A. (1991). Degradation of oxidized insulin B chain by the multiproteinase complex macropain (proteasome). *Biochemistry* **30**(10): 2725–2734.

Dick, T. P., Ruppert, T., Groettrup, M., Kloetzel, P. M., Kuehn, L., Koszinowski, U. H., Stevanovic, S., Schild, H. and Rammensee, H. G. (1996). Coordinated dual cleavages induced by the proteasome regulator PA28 lead to dominant MHC ligands. *Cell* **86**(2): 253–262.

Ding, Z., Fu, Z., Xu, C., Wang, Y., Wang, Y., Li, J., Kong, L., Chen, J., Li, N., Zhang, R. and Cong, Y. (2017). High-resolution cryo-EM structure of the proteasome in complex with ADP-AlFx. *Cell Res* **27**(3): 373–385.

Ding, Z., Xu, C., Sahu, I., Wang, Y., Fu, Z., Huang, M., Wong, C. C. L., Glickman, M. H. and Cong, Y. (2019). Structural snapshots of 26S proteasome reveal tetraubiquitin-induced conformations. *Mol Cell* **73**(6): 1150–1161 e1156.

Dolenc, I., Seemuller, E. and Baumeister, W. (1998). Decelerated degradation of short peptides by the 20S proteasome. *FEBS Lett* **434**(3): 357–361.

Dong, Y., Zhang, S., Wu, Z., Li, X., Wang, W. L., Zhu, Y., Stoilova-McPhie, S., Lu, Y., Finley, D. and Mao, Y. (2019). Cryo-EM structures and dynamics of substrate-engaged human 26S proteasome. *Nature* **565**(7737): 49–55.

Du, J., Babik, S., Li, Y., Deol, K. K., Eyles, S. J., Fejzo, J., Tonelli, M. and Strieter, E. (2022). A cryptic K48 ubiquitin chain binding site on UCH37 is required for its role in proteasomal degradation. *Elife* **11**: e76100.

Eddins, M. J., Varadan, R., Fushman, D., Pickart, C. M. and Wolberger, C. (2007). Crystal structure and solution NMR studies of Lys48-linked tetraubiquitin at neutral pH. *J Mol Biol* **367**(1): 204–211.

Ehring, B., Meyer, T. H., Eckerskorn, C., Lottspeich, F. and Tampe, R. (1996). Effects of major-histocompatibility-complex-encoded subunits on the peptidase and proteolytic activities of human 20S proteasomes. Cleavage of proteins and antigenic peptides. *Eur J Biochem* **235**(1–2): 404–415.

Eisele, M. R., Reed, R. G., Rudack, T., Schweitzer, A., Beck, F., Nagy, I., Pfeifer, G., Plitzko, J. M., Baumeister, W., Tomko, R. J., Jr. and Sakata, E. (2018). Expanded coverage of the 26S proteasome conformational landscape reveals mechanisms of peptidase gating. *Cell Rep* **24**(5): 1301–1315 e1305.

Elsasser, S., Chandler-Militello, D., Muller, B., Hanna, J. and Finley, D. (2004). Rad23 and Rpn10 serve as alternative ubiquitin receptors for the proteasome. *J Biol Chem* **279**(26): 26817–26822.

Elsasser, S. and Finley, D. (2005). Delivery of ubiquitinated substrates to protein-unfolding machines. *Nat Cell Biol* **7**(8): 742–749.

Erales, J. and Coffino, P. (2014). Ubiquitin-independent proteasomal degradation. *Biochim Biophys Acta* **1843**(1): 216–221.

Erales, J., Hoyt, M. A., Troll, F. and Coffino, P. (2012). Functional asymmetries of proteasome translocase pore. *J Biol Chem* **287**(22): 18535–18543.

Estrin, E., Lopez-Blanco, J. R., Chacon, P. and Martin, A. (2013). Formation of an intricate helical bundle dictates the assembly of the 26S proteasome lid. *Structure* **21**(9): 1624–1635.

Etlinger, J. D. and Goldberg, A. L. (1977). A soluble ATP-dependent proteolytic system responsible for the degradation of abnormal proteins in reticulocytes. *Proc Natl Acad Sci USA* **74**(1): 54–58.

Fang, R., Hon, J., Zhou, M. and Lu, Y. (2022). An empirical energy landscape reveals mechanism of proteasome in polypeptide translocation. *Elife* **11**: e71911.

Fei, X., Bell, T. A., Jenni, S., Stinson, B. M., Baker, T. A., Harrison, S. C. and Sauer, R. T. (2020). Structures of the ATP-fueled ClpXP proteolytic machine bound to protein substrate. *Elife* **9**: e52774.

Ferrington, D. A. and Gregerson, D. S. (2012). Immunoproteasomes: Structure, function, and antigen presentation. *Prog Mol Biol Transl Sci* **109**: 75–112.

Finley, D., Chen, X. and Walters, K. J. (2016). Gates, channels, and switches: Elements of the proteasome machine. *Trends Biochem Sci* **41**(1): 77–93.

Finley, D. and Prado, M. A. (2020). The Proteasome and Its network: Engineering for adaptability. *Cold Spring Harb Perspect Biol* **12**(1): a033985.

Fischer, E. S., Bohm, K., Lydeard, J. R., Yang, H., Stadler, M. B., Cavadini, S., Nagel, J., Serluca, F., Acker, V., Lingaraju, G. M., Tichkule, R. B., Schebesta, M., Forrester, W. C., Schirle, M., Hassiepen, U., Ottl, J., Hild, M., Beckwith, R. E., Harper, J. W., Jenkins, J. L. and Thoma, N. H. (2014). Structure of the DDB1-CRBN E3 ubiquitin ligase in complex with thalidomide. *Nature* **512**(7512): 49–53.

Fishbain, S., Inobe, T., Israeli, E., Chavali, S., Yu, H., Kago, G., Babu, M. M. and Matouschek, A. (2015). Sequence composition of disordered regions fine-tunes protein half-life. *Nat Struct Mol Biol* **22**(3): 214–221.

Fishbain, S., Prakash, S., Herrig, A., Elsasser, S. and Matouschek, A. (2011). Rad23 escapes degradation because it lacks a proteasome initiation region. *Nat Commun* **2**: 192.

Forster, F., Unverdorben, P., Sledz, P. and Baumeister, W. (2013). Unveiling the long-held secrets of the 26S proteasome. *Structure* **21**(9): 1551–1562.

Frank, J. (2006). Three-dimensional electron microscopy of macromolecular assemblies: Visualization of biological molecules in their native state. Oxford, New York, Oxford University Press.

Gandhi, A. K., Kang, J., Havens, C. G., Conklin, T., Ning, Y., Wu, L., Ito, T., Ando, H., Waldman, M. F., Thakurta, A., Klippel, A., Handa, H., Daniel, T. O., Schafer, P. H. and Chopra, R. (2014). Immunomodulatory agents lenalidomide and pomalidomide co-stimulate T cells by inducing degradation of T cell repressors Ikaros and Aiolos via modulation of the E3 ubiquitin ligase complex CRL4(CRBN.). *Br J Haematol* **164**(6): 811–821.

Glickman, M. H., Rubin, D. M., Coux, O., Wefes, I., Pfeifer, G., Cjeka, Z., Baumeister, W., Fried, V. A. and Finley, D. (1998). A subcomplex of the proteasome regulatory particle required for ubiquitin-conjugate degradation and related to the COP9-signalosome and eIF3. *Cell* **94**(5): 615–623.

Goldstein, G., Scheid, M., Hammerling, U., Schlesinger, D. H., Niall, H. D. and Boyse, E. A. (1975). Isolation of a polypeptide that has lymphocyte-differentiating properties and is probably represented universally in living cells. *Proc Natl Acad Sci USA* **72**(1): 11–15.

Greene, E. R., Goodall, E. A., de la Pena, A. H., Matyskiela, M. E., Lander, G. C. and Martin, A. (2019). Specific lid-base contacts in the 26s proteasome control the conformational switching required for substrate degradation. *Elife* **8**: e49806.

Groll, M., Ditzel, L., Lowe, J., Stock, D., Bochtler, M., Bartunik, H. D. and Huber, R. (1997). Structure of 20S proteasome from yeast at 2.4 A resolution. *Nature* **386**(6624): 463–471.

Gu, X., Nardone, C., Kamitaki, N., Mao, A., Elledge, S. J. and Greenberg, M. E. (2023). The midnolin-proteasome pathway catches proteins for ubiquitination-independent degradation. *Science* **381**(6660): eadh5021.

Guan, H., Wang, Y., Yu, T., Huang, Y., Li, M., Saeed, A., Perculija, V., Li, D., Xiao, J., Wang, D., Zhu, P. and Ouyang, S. (2020). Cryo-EM structures of the human PA200 and PA200-20S complex reveal regulation of proteasome gate opening and two PA200 apertures. *PLoS Biol* **18**(3): e3000654.

Hamazaki, J., Iemura, S., Natsume, T., Yashiroda, H., Tanaka, K. and Murata, S. (2006). A novel proteasome interacting protein recruits the deubiquitinating enzyme UCH37 to 26S proteasomes. *EMBO J* **25**(19): 4524–4536.

Hanada, K., Yewdell, J. W. and Yang, J. C. (2004). Immune recognition of a human renal cancer antigen through post-translational protein splicing. *Nature* **427**(6971): 252–256.

Hanna, J., Hathaway, N. A., Tone, Y., Crosas, B., Elsasser, S., Kirkpatrick, D. S., Leggett, D. S., Gygi, S. P., King, R. W. and Finley, D. (2006). Deubiquitinating enzyme Ubp6 functions noncatalytically to delay proteasomal degradation. *Cell* **127**(1): 99–111.

Harris, J. R. (1968). Release of a macromolecular protein component from human erythrocyte ghosts. *Biochim Biophys Acta* **150**(3): 534–537.

Harshbarger, W., Miller, C., Diedrich, C. and Sacchettini, J. (2015). Crystal structure of the human 20S proteasome in complex with carfilzomib. *Structure* **23**(2): 418–424.

Haselbach, D., Schrader, J., Lambrecht, F., Henneberg, F., Chari, A. and Stark, H. (2017). Long-range allosteric regulation of the human 26S proteasome by 20S proteasome-targeting cancer drugs. *Nat Commun* **8**: 15578.

He, J., Kulkarni, K., da Fonseca, P. C., Krutauz, D., Glickman, M. H., Barford, D. and Morris, E. P. (2012). The structure of the 26S proteasome subunit Rpn2 reveals its PC repeat domain as a closed toroid of two concentric alpha-helical rings. *Structure* **20**(3): 513–521.

Heinen, C., Acs, K., Hoogstraten, D. and Dantuma, N. P. (2011). C-terminal UBA domains protect ubiquitin receptors by preventing initiation of protein degradation. *Nat Commun* **2**: 191.

Hershko, A., Ciechanover, A., Heller, H., Haas, A. L. and Rose, I. A. (1980). Proposed role of ATP in protein breakdown: Conjugation of protein with multiple chains of the polypeptide of ATP-dependent proteolysis. *Proc Natl Acad Sci USA* **77**(4): 1783–1786.

Hershko, A., Ciechanover, A. and Rose, I. A. (1979). Resolution of the ATP-dependent proteolytic system from reticulocytes: A component that interacts with ATP. *Proc Natl Acad Sci USA* **76**(7): 3107–3110.

Hnia, K., Clausen, T. and Moog-Lutz, C. (2019). Shaping striated muscles with ubiquitin proteasome system in health and disease. *Trends Mol Med* **25**(9): 760–774.

Holzl, H., Kapelari, B., Kellermann, J., Seemuller, E., Sumegi, M., Udvardy, A., Medalia, O., Sperling, J., Muller, S. A., Engel, A. and Baumeister, W. (2000). The regulatory complex of Drosophila melanogaster 26S proteasomes. Subunit composition and localization of a deubiquitylating enzyme. *J Cell Biol* **150**(1): 119–130.

Hua, Z. and Vierstra, R. D. (2011). The cullin-RING ubiquitin-protein ligases. *Annu Rev Plant Biol* **62**: 299–334.

Huang, X., Luan, B., Wu, J. and Shi, Y. (2016). An atomic structure of the human 26S proteasome. *Nat Struct Mol Biol* **23**(9): 778–785.

Huber, E. M., Basler, M., Schwab, R., Heinemeyer, W., Kirk, C. J., Groettrup, M. and Groll, M. (2012). Immuno- and constitutive proteasome crystal structures reveal differences in substrate and inhibitor specificity. *Cell* **148**(4): 727–738.

Huber, E. M. and Groll, M. (2012). Inhibitors for the immuno- and constitutive proteasome: Current and future trends in drug development. *Angew Chem Int Ed Engl* **51**(35): 8708–8720.

Huber, E. M. and Groll, M. (2017). The Mammalian Proteasome Activator PA28 Forms an Asymmetric alpha(4)beta(3) Complex. *Structure* **25**(10): 1473–1480 e1473.

Hung, K. Y. S., Klumpe, S., Eisele, M. R., Elsasser, S., Tian, G., Sun, S., Moroco, J. A., Cheng, T. C., Joshi, T., Seibel, T., Van Dalen, D., Feng, X. H., Lu, Y., Ovaa, H., Engen, J. R., Lee, B. H., Rudack, T., Sakata, E. and Finley, D. (2022). Allosteric control of Ubp6 and the proteasome via a bidirectional switch. *Nat Commun* **13**(1): 838.

Husnjak, K., Elsasser, S., Zhang, N., Chen, X., Randles, L., Shi, Y., Hofmann, K., Walters, K. J., Finley, D. and Dikic, I. (2008). Proteasome subunit Rpn13 is a novel ubiquitin receptor. *Nature* **453**(7194): 481–488.

Jager, S., Kim, D. Y., Hultquist, J. F., Shindo, K., LaRue, R. S., Kwon, E., Li, M., Anderson, B. D., Yen, L., Stanley, D., Mahon, C., Kane, J., Franks-Skiba, K., Cimermancic, P., Burlingame, A., Sali, A., Craik, C. S., Harris, R. S., Gross, J. D. and Krogan, N. J. (2011). Vif hijacks CBF-beta to degrade APOBEC3G and promote HIV-1 infection. *Nature* **481**(7381): 371–375.

Janse, D. M., Crosas, B., Finley, D. and Church, G. M. (2004). Localization to the proteasome is sufficient for degradation. *J Biol Chem* **279**(20): 21415–21420.

Jespersen, N., Ehrenbolger, K., Winiger, R. R., Svedberg, D., Vossbrinck, C. R. and Barandun, J. (2022). Structure of the reduced microsporidian proteasome bound by PI31-like peptides in dormant spores. *Nat Commun* **13**(1): 6962.

Jonsson, E., Htet, Z. M., Bard, J. A. M., Dong, K. C. and Martin, A. (2022). Ubiquitin modulates 26S proteasome conformational dynamics and promotes substrate degradation. *Sci Adv* **8**(51): eadd9520.

Kim, H. T. and Goldberg, A. L. (2017). The deubiquitinating enzyme Usp14 allosterically inhibits multiple proteasomal activities and ubiquitin-independent proteolysis. *J Biol Chem* **292**(23): 9830–9839.

Kim, H. T. and Goldberg, A. L. (2018). UBL domain of Usp14 and other proteins stimulates proteasome activities and protein degradation in cells. *Proc Natl Acad Sci USA* **115**(50): E11642–E11650.

Kisselev, A. F., Akopian, T. N. and Goldberg, A. L. (1998). Range of sizes of peptide products generated during degradation of different proteins by archaeal proteasomes. *J Biol Chem* **273**(4): 1982–1989.

Kleijnen, M. F., Roelofs, J., Park, S., Hathaway, N. A., Glickman, M., King, R. W. and Finley, D. (2007). Stability of the proteasome can be regulated allosterically through engagement of its proteolytic active sites. *Nat Struct Mol Biol* **14**(12): 1180–1188.

Kocberber, Z., Willemsen, N. and Bartelt, A. (2023). The role of proteasome activators PA28alphabeta and PA200 in brown adipocyte differentiation and function. *Front Endocrinol (Lausanne)* **14**: 1176733.

Komander, D. and Rape, M. (2012). The ubiquitin code. *Annu Rev Biochem* **81**: 203–229.

Kronke, J., Udeshi, N. D., Narla, A., Grauman, P., Hurst, S. N., McConkey, M., Svinkina, T., Heckl, D., Comer, E., Li, X., Ciarlo, C., Hartman, E., Munshi, N., Schenone, M., Schreiber, S. L., Carr, S. A. and Ebert, B. L. (2014). Lenalidomide causes selective degradation of IKZF1 and IKZF3 in multiple myeloma cells. *Science* **343**(6168): 301–305.

Kuhnle, S., Martinez-Noel, G., Leclere, F., Hayes, S. D., Harper, J. W. and Howley, P. M. (2018). Angelman syndrome-associated point mutations in the Zn(2+)-binding N-terminal (AZUL) domain of UBE3A ubiquitin ligase inhibit binding to the proteasome. *J Biol Chem* **293**(47): 18387–18399.

Kulathu, Y. and Komander, D. (2012). Atypical ubiquitylation - the unexplored world of polyubiquitin beyond Lys48 and Lys63 linkages. *Nat Rev Mol Cell Biol* **13**(8): 508–523.

Kuo, C. L. and Goldberg, A. L. (2017). Ubiquitinated proteins promote the association of proteasomes with the deubiquitinating enzyme Usp14 and the ubiquitin ligase Ube3c. *Proc Natl Acad Sci USA* **114**(17): E3404–E3413.

Lam, Y. A., Xu, W., DeMartino, G. N. and Cohen, R. E. (1997). Editing of ubiquitin conjugates by an isopeptidase in the 26S proteasome. *Nature* **385**(6618): 737–740.

Lander, G. C., Estrin, E., Matyskiela, M. E., Bashore, C., Nogales, E. and Martin, A. (2012). Complete subunit architecture of the proteasome regulatory particle. *Nature* **482**(7384): 186–191.

Lasker, K., Forster, F., Bohn, S., Walzthoeni, T., Villa, E., Unverdorben, P., Beck, F., Aebersold, R., Sali, A. and Baumeister, W. (2012). Molecular architecture of the 26S proteasome holocomplex determined by an integrative approach. *Proc Natl Acad Sci USA* **109**(5): 1380–1387.

Lee, B. H., Lee, M. J., Park, S., Oh, D. C., Elsasser, S., Chen, P. C., Gartner, C., Dimova, N., Hanna, J., Gygi, S. P., Wilson, S. M., King, R. W. and Finley, D. (2010). Enhancement of proteasome activity by a small-molecule inhibitor of USP14. *Nature* **467**(7312): 179–184.

Lee, B. H., Lu, Y., Prado, M. A., Shi, Y., Tian, G., Sun, S., Elsasser, S., Gygi, S. P., King, R. W. and Finley, D. (2016). USP14 deubiquitinates proteasome-bound substrates that are ubiquitinated at multiple sites. *Nature* **532**(7599): 398–401.

Lee, D., Zhu, Y., Colson, L., Wang, X., Chen, S., Tkacik, E., Huang, L., Ouyang, Q., Goldberg, A. L. and Lu, Y. (2023). Molecular mechanism for activation of the 26S proteasome by ZFAND5. *Mol Cell* **83**(16): 2959–2975 e2957.

Leggett, D. S., Hanna, J., Borodovsky, A., Crosas, B., Schmidt, M., Baker, R. T., Walz, T., Ploegh, H. and Finley, D. (2002). Multiple associated proteins regulate proteasome structure and function. *Mol Cell* **10**(3): 495–507.

Leznicki, P. and Kulathu, Y. (2017). Mechanisms of regulation and diversification of deubiquitylating enzyme function. *J Cell Sci* **130**(12): 1997–2006.

Liepe, J., Marino, F., Sidney, J., Jeko, A., Bunting, D. E., Sette, A., Kloetzel, P. M., Stumpf, M. P., Heck, A. J. and Mishto, M. (2016). A large fraction of HLA class I ligands are proteasome-generated spliced peptides. *Science* **354**(6310): 354–358.

Liu, Z., Dong, X., Yi, H. W., Yang, J., Gong, Z., Wang, Y., Liu, K., Zhang, W. P. and Tang, C. (2019). Structural basis for the recognition of K48-linked Ub chain by proteasomal receptor Rpn13. *Cell Discov* **5**: 19.

Lowe, J., Stock, D., Jap, B., Zwickl, P., Baumeister, W. and Huber, R. (1995). Crystal structure of the 20S proteasome from the archaeon T. acidophilum at 3.4 A resolution. *Science* **268**(5210): 533–539.

Lu, G., Middleton, R. E., Sun, H., Naniong, M., Ott, C. J., Mitsiades, C. S., Wong, K. K., Bradner, J. E. and Kaelin, W. G., Jr. (2014). The myeloma drug lenalidomide promotes the cereblon-dependent destruction of Ikaros proteins. *Science* **343**(6168): 305–309.

Lu, X., Nowicka, U., Sridharan, V., Liu, F., Randles, L., Hymel, D., Dyba, M., Tarasov, S. G., Tarasova, N. I., Zhao, X. Z., Hamazaki, J., Murata, S., Burke, T. R., Jr. and Walters, K. J. (2017). Structure of the Rpn13-Rpn2 complex provides insights for Rpn13 and Uch37 as anticancer targets. *Nat Commun* **8**: 15540.

Lu, X., Sabbasani, V. R., Osei-Amponsa, V., Evans, C. N., King, J. C., Tarasov, S. G., Dyba, M., Das, S., Chan, K. C., Schwieters, C. D., Choudhari, S., Fromont, C., Zhao, Y., Tran, B., Chen, X., Matsuo, H., Andresson, T., Chari, R., Swenson, R. E., Tarasova, N. I. and Walters, K. J. (2021). Structure-guided bifunctional molecules hit a DEUBAD-lacking hRpn13 species upregulated in multiple myeloma. *Nat Commun* **12**(1): 7318.

Lu, Y., Wu, J., Dong, Y., Chen, S., Sun, S., Ma, Y. B., Ouyang, Q., Finley, D., Kirschner, M. W. and Mao, Y. (2017). Conformational landscape of the p28-bound human proteasome regulatory particle. *Mol Cell* **67**(2): 322–333 e326.

Makaros, Y., Raiff, A., Timms, R. T., Wagh, A. R., Gueta, M. I., Bekturova, A., Guez-Haddad, J., Brodsky, S., Opatowsky, Y., Glickman, M. H., Elledge, S. J. and Koren, I. (2023). Ubiquitin-independent proteasomal degradation driven by C-degron pathways. *Mol Cell* **83**(11): 1921–1935 e1927.

Mandemaker, I. K., Geijer, M. E., Kik, I., Bezstarosti, K., Rijkers, E., Raams, A., Janssens, R. C., Lans, H., Hoeijmakers, J. H., Demmers, J. A., Vermeulen, W. and Marteijn, J. A. (2018). DNA damage-induced replication stress results in PA200-proteasome-mediated degradation of acetylated histones. *EMBO Rep* **19**(10): e45566.

Mao, Y. (2021). Structure, dynamics and function of the 26S proteasome. *Subcell Biochem* **96**: 1–151.

Martinez-Fonts, K., Davis, C., Tomita, T., Elsasser, S., Nager, A. R., Shi, Y., Finley, D. and Matouschek, A. (2020). The proteasome 19S cap and its ubiquitin receptors provide a versatile recognition platform for substrates. *Nat Commun* **11**(1): 477.

Martinez-Noel, G., Galligan, J. T., Sowa, M. E., Arndt, V., Overton, T. M., Harper, J. W. and Howley, P. M. (2012). Identification and proteomic analysis of distinct UBE3A/E6AP protein complexes. *Mol Cell Biol* **32**(15): 3095–3106.

Matyskiela, M. E., Lander, G. C. and Martin, A. (2013). Conformational switching of the 26S proteasome enables substrate degradation. *Nat Struct Mol Biol* **20**(7): 781–788.

Mayer, R. J. (2000). The meteoric rise of regulated intracellular proteolysis. *Nat Rev Mol Cell Biol* **1**(2): 145–148.

Meyer-Schwesinger, C. (2019). The ubiquitin-proteasome system in kidney physiology and disease. *Nat Rev Nephrol* **15**(7): 393–411.

Moreau, P., Pylypenko, H., Grosicki, S., Karamanesht, I., Leleu, X., Grishunina, M., Rekhtman, G., Masliak, Z., Robak, T., Shubina, A., Arnulf, B., Kropff, M., Cavet, J., Esseltine, D. L., Feng, H., Girgis, S., van de Velde, H., Deraedt, W. and Harousseau, J. L. (2011). Subcutaneous versus intravenous administration of bortezomib in patients with relapsed multiple myeloma: A randomised, phase 3, non-inferiority study. *Lancet Oncol* **12**(5): 431–440.

Murakami, Y., Matsufuji, S., Kameji, T., Hayashi, S., Igarashi, K., Tamura, T., Tanaka, K. and Ichihara, A. (1992). Ornithine decarboxylase is degraded by the 26S proteasome without ubiquitination. *Nature* **360**(6404): 597–599.

Murata, S., Sasaki, K., Kishimoto, T., Niwa, S., Hayashi, H., Takahama, Y. and Tanaka, K. (2007). Regulation of CD8+ T cell development by thymus-specific proteasomes. *Science* **316**(5829): 1349–1353.

Murata, S., Takahama, Y., Kasahara, M. and Tanaka, K. (2018). The immunoproteasome and thymoproteasome: Functions, evolution and human disease. *Nat Immunol* **19**(9): 923–931.

Nickell, S., Beck, F., Scheres, S. H., Korinek, A., Forster, F., Lasker, K., Mihalache, O., Sun, N., Nagy, I., Sali, A., Plitzko, J. M., Carazo, J. M., Mann, M. and Baumeister, W. (2009). Insights into the molecular architecture of the 26S proteasome. *Proc Natl Acad Sci USA* **106**(29): 11943–11947.

Nussbaum, A. K., Dick, T. P., Keilholz, W., Schirle, M., Stevanovic, S., Dietz, K., Heinemeyer, W., Groll, M., Wolf, D. H., Huber, R., Rammensee, H. G. and Schild, H. (1998). Cleavage motifs of the yeast 20S proteasome beta subunits deduced from digests of enolase 1. *Proc Natl Acad Sci USA* **95**(21): 12504–12509.

Ohno, A., Jee, J., Fujiwara, K., Tenno, T., Goda, N., Tochio, H., Kobayashi, H., Hiroaki, H. and Shirakawa, M. (2005). Structure of the UBA domain of Dsk2p in complex with ubiquitin molecular determinants for ubiquitin recognition. *Structure* **13**(4): 521–532.

Opoku-Nsiah, K. A., de la Pena, A. H., Williams, S. K., Chopra, N., Sali, A., Lander, G. C. and Gestwicki, J. E. (2022). The YPhi motif defines the structure-activity relationships of human 20S proteasome activators. *Nat Commun* **13**(1): 1226.

Paes, W., Leonov, G., Partridge, T., Chikata, T., Murakoshi, H., Frangou, A., Brackenridge, S., Nicastri, A., Smith, A. G., Learn, G. H., Li, Y., Parker, R., Oka, S., Pellegrino, P., Williams, I., Haynes, B. F., McMichael, A. J., Shaw, G. M., Hahn, B. H., Takiguchi, M., Ternette, N. and Borrow, P. (2019). Contribution of proteasome-catalyzed peptide cis-splicing to viral targeting by CD8(+) T cells in HIV-1 infection. *Proc Natl Acad Sci USA* **116**(49): 24748–24759.

Pathare, G. R., Nagy, I., Bohn, S., Unverdorben, P., Hubert, A., Korner, R., Nickell, S., Lasker, K., Sali, A., Tamura, T., Nishioka, T., Forster, F., Baumeister, W. and Bracher, A. (2012). The proteasomal subunit Rpn6 is a molecular clamp holding the core and regulatory subcomplexes together. *Proc Natl Acad Sci USA* **109**(1): 149–154.

Pathare, G. R., Nagy, I., Sledz, P., Anderson, D. J., Zhou, H. J., Pardon, E., Steyaert, J., Forster, F., Bracher, A. and Baumeister, W. (2014). Crystal structure of the proteasomal deubiquitylation module Rpn8-Rpn11. *Proc Natl Acad Sci USA* **111**(8): 2984–2989.

Peters, J. M., Cejka, Z., Harris, J. R., Kleinschmidt, J. A. and Baumeister, W. (1993). Structural features of the 26 S proteasome complex. *J Mol Biol* **234**(4): 932–937.

Peth, A., Besche, H. C. and Goldberg, A. L. (2009). Ubiquitinated proteins activate the proteasome by binding to Usp14/Ubp6, which causes 20S gate opening. *Mol Cell* **36**(5): 794–804.

Petzold, G., Fischer, E. S. and Thoma, N. H. (2016). Structural basis of lenalidomide-induced CK1alpha degradation by the CRL4(CRBN) ubiquitin ligase. *Nature* **532**(7597): 127–130.

Prakash, S., Tian, L., Ratliff, K. S., Lehotzky, R. E. and Matouschek, A. (2004). An unstructured initiation site is required for efficient proteasome-mediated degradation. *Nat Struct Mol Biol* **11**(9): 830–837.

Puchades, C., Sandate, C. R. and Lander, G. C. (2020). The molecular principles governing the activity and functional diversity of AAA+ proteins. *Nat Rev Mol Cell Biol* **21**(1): 43–58.

Qian, M. X., Pang, Y., Liu, C. H., Haratake, K., Du, B. Y., Ji, D. Y., Wang, G. F., Zhu, Q. Q., Song, W., Yu, Y., Zhang, X. X., Huang, H. T., Miao, S., Chen, L. B., Zhang, Z. H., Liang, Y. N., Liu, S., Cha, H., Yang, D., Zhai, Y., Komatsu, T., Tsuruta, F., Li, H., Cao, C., Li, W., Li, G. H., Cheng, Y., Chiba, T., Wang, L., Goldberg, A. L., Shen, Y. and Qiu, X. B. (2013). Acetylation-mediated proteasomal degradation of core histones during DNA repair and spermatogenesis. *Cell* **153**(5): 1012–1024.

Qiu, X. B., Ouyang, S. Y., Li, C. J., Miao, S., Wang, L. and Goldberg, A. L. (2006). hRpn13/ADRM1/GP110 is a novel proteasome subunit that binds the deubiquitinating enzyme, UCH37. *EMBO J* **25**(24): 5742–5753.

Rawson, S., Walsh, R. M., Jr., Velez, B., Schnell, H. M., Jiao, F., Blickling, M., Ang, J., Bhanu, M. K., Huang, L. and Hanna, J. (2022). Yeast PI31 inhibits the proteasome by a direct multisite mechanism. *Nat Struct Mol Biol* **29**(8): 791–800.

Rennella, E., Huang, R., Yu, Z. and Kay, L. E. (2020). Exploring long-range cooperativity in the 20S proteasome core particle from Thermoplasma acidophilum using methyl-TROSY-based NMR. *Proc Natl Acad Sci USA* **117**(10): 5298–5309.

Rentsch, A., Landsberg, D., Brodmann, T., Bulow, L., Girbig, A. K. and Kalesse, M. (2013). Synthesis and pharmacology of proteasome inhibitors. *Angew Chem Int Ed Engl* **52**(21): 5450–5488.

Riedinger, C., Boehringer, J., Trempe, J. F., Lowe, E. D., Brown, N. R., Gehring, K., Noble, M. E., Gordon, C. and Endicott, J. A. (2010). Structure of Rpn10 and its interactions with polyubiquitin chains and the proteasome subunit Rpn12. *J Biol Chem* **285**(44): 33992–34003.

Rubin, D. M., Glickman, M. H., Larsen, C. N., Dhruvakumar, S. and Finley, D. (1998). Active site mutants in the six regulatory particle ATPases reveal multiple roles for ATP in the proteasome. *EMBO J* **17**(17): 4909–4919.

Ruschak, A. M., Religa, T. L., Breuer, S., Witt, S. and Kay, L. E. (2010). The proteasome antechamber maintains substrates in an unfolded state. *Nature* **467**(7317): 868–871.

Saeki, Y., Kudo, T., Sone, T., Kikuchi, Y., Yokosawa, H., Toh-e, A. and Tanaka, K. (2009). Lysine 63-linked polyubiquitin chain may serve as a targeting signal for the 26S proteasome. *EMBO J* **28**(4): 359–371.

Sahtoe, D. D., van Dijk, W. J., El Oualid, F., Ekkebus, R., Ovaa, H. and Sixma, T. K. (2015). Mechanism of UCH-L5 activation and inhibition by DEUBAD domains in RPN13 and INO80G. *Mol Cell* **57**(5): 887–900.

Sahu, I., Mali, S. M., Sulkshane, P., Xu, C., Rozenberg, A., Morag, R., Sahoo, M. P., Singh, S. K., Ding, Z., Wang, Y., Day, S., Cong, Y., Kleifeld, O., Brik, A. and Glickman, M. H. (2021). The 20S as a stand-alone proteasome in cells can degrade the ubiquitin tag. *Nat Commun* **12**(1): 6173.

Sakata, E., Yamaguchi, Y., Kurimoto, E., Kikuchi, J., Yokoyama, S., Yamada, S., Kawahara, H., Yokosawa, H., Hattori, N., Mizuno, Y., Tanaka, K. and Kato, K. (2003). Parkin binds the Rpn10 subunit of 26S proteasomes through its ubiquitin-like domain. *EMBO Rep* **4**(3): 301–306.

Sandstrom, A., Mitchell, P. S., Goers, L., Mu, E. W., Lesser, C. F. and Vance, R. E. (2019). Functional degradation: A mechanism of NLRP1 inflammasome activation by diverse pathogen enzymes. *Science* **364**(6435): eaau1330.

Savulescu, A. F. and Glickman, M. H. (2011). Proteasome activator 200: The heat is on. *Mol Cell Proteomics* **10**(5): R110.006890.

Schnell, H. M., Walsh, R. M., Jr., Rawson, S., Kaur, M., Bhanu, M. K., Tian, G., Prado, M. A., Guerra-Moreno, A., Paulo, J. A., Gygi, S. P., Roelofs, J., Finley, D. and Hanna, J. (2021). Structures of chaperone-associated assembly intermediates reveal coordinated mechanisms of proteasome biogenesis. *Nat Struct Mol Biol* **28**(5): 418–425.

Schrader, J., Henneberg, F., Mata, R. A., Tittmann, K., Schneider, T. R., Stark, H., Bourenkov, G. and Chari, A. (2016). The inhibition mechanism of human 20S proteasomes enables next-generation inhibitor design. *Science* **353**(6299): 594–598.

Schreiner, P., Chen, X., Husnjak, K., Randles, L., Zhang, N., Elsasser, S., Finley, D., Dikic, I., Walters, K. J. and Groll, M. (2008). Ubiquitin docking at the proteasome through a novel pleckstrin-homology domain interaction. *Nature* **453**(7194): 548–552.

Schweitzer, A., Aufderheide, A., Rudack, T., Beck, F., Pfeifer, G., Plitzko, J. M., Sakata, E., Schulten, K., Forster, F. and Baumeister, W. (2016). Structure of the human 26S proteasome at a resolution of 3.9 A. *Proc Natl Acad Sci USA* **113**(28): 7816–7821.

Seemuller, E., Lupas, A., Stock, D., Lowe, J., Huber, R. and Baumeister, W. (1995). Proteasome from Thermoplasma acidophilum: A threonine protease. *Science* **268**(5210): 579–582.

Shi, Y., Chen, X., Elsasser, S., Stocks, B. B., Tian, G., Lee, B. H., Shi, Y., Zhang, N., de Poot, S. A., Tuebing, F., Sun, S., Vannoy, J., Tarasov, S. G., Engen, J. R., Finley, D. and Walters, K. J. (2016). Rpn1 provides adjacent receptor sites for substrate binding and deubiquitination by the proteasome. *Science* **351**(6275): aad9421.

Sledz, P., Unverdorben, P., Beck, F., Pfeifer, G., Schweitzer, A., Forster, F. and Baumeister, W. (2013). Structure of the 26S proteasome with ATP-gammaS bound provides insights into the mechanism of nucleotide-dependent substrate translocation. *Proc Natl Acad Sci USA* **110**(18): 7264–7269.

Stadtmueller, B. M. and Hill, C. P. (2011). Proteasome activators. *Mol Cell* **41**(1): 8–19.

Takeuchi, J., Chen, H. and Coffino, P. (2007). Proteasome substrate degradation requires association plus extended peptide. *EMBO J* **26**(1): 123–131.

Tanaka, K. (2009). The proteasome: Overview of structure and functions. *Proc Jpn Acad Ser B Phys Biol Sci* **85**(1): 12–36.

Toste Rego, A. and da Fonseca, P. C. A. (2019). Characterization of fully recombinant human 20S and 20S-PA200 proteasome complexes. *Mol Cell* **76**(1): 138–147 e135.

Uechi, H., Hamazaki, J. and Murata, S. (2014). Characterization of the testis-specific proteasome subunit alpha4s in mammals. *J Biol Chem* **289**(18): 12365–12374.

Unno, M., Mizushima, T., Morimoto, Y., Tomisugi, Y., Tanaka, K., Yasuoka, N. and Tsukihara, T. (2002). The structure of the mammalian 20S proteasome at 2.75 A resolution. *Structure* **10**(5): 609–618.

Unverdorben, P., Beck, F., Sledz, P., Schweitzer, A., Pfeifer, G., Plitzko, J. M., Baumeister, W. and Forster, F. (2014). Deep classification of a large cryo-EM dataset defines the conformational landscape of the 26S proteasome. *Proc Natl Acad Sci USA* **111**(15): 5544–5549.

van der Lee, R., Lang, B., Kruse, K., Gsponer, J., Sanchez de Groot, N., Huynen, M. A., Matouschek, A., Fuxreiter, M. and Babu, M. M. (2014). Intrinsically disordered segments affect protein half-life in the cell and during evolution. *Cell Rep* **8**(6): 1832–1844.

van Nocker, S., Sadis, S., Rubin, D. M., Glickman, M., Fu, H., Coux, O., Wefes, I., Finley, D. and Vierstra, R. D. (1996). The multiubiquitin-chain-binding protein Mcb1 is a component of the 26S proteasome in Saccharomyces cerevisiae and plays a nonessential, substrate-specific role in protein turnover. *Mol Cell Biol* **16**(11): 6020–6028.

Vander Linden, R. T., Hemmis, C. W., Schmitt, B., Ndoja, A., Whitby, F. G., Robinson, H., Cohen, R. E., Yao, T. and Hill, C. P. (2015). Structural basis for the activation and inhibition of the UCH37 deubiquitylase. *Mol Cell* **57**(5): 901–911.

Varadan, R., Assfalg, M., Raasi, S., Pickart, C. and Fushman, D. (2005). Structural determinants for selective recognition of a Lys48-linked polyubiquitin chain by a UBA domain. *Mol Cell* **18**(6): 687–698.

Verma, R., Chen, S., Feldman, R., Schieltz, D., Yates, J., Dohmen, J. and Deshaies, R. J. (2000). Proteasomal proteomics: Identification of nucleotide-sensitive proteasome-interacting proteins by mass spectrometric analysis of affinity-purified proteasomes. *Mol Biol Cell* **11**(10): 3425–3439.

Vigneron, N., Stroobant, V., Chapiro, J., Ooms, A., Degiovanni, G., Morel, S., van der Bruggen, P., Boon, T. and Van den Eynde, B. J. (2004). An antigenic peptide produced by peptide splicing in the proteasome. *Science* **304**(5670): 587–590.

Voges, D., Zwickl, P. and Baumeister, W. (1999). The 26S proteasome: A molecular machine designed for controlled proteolysis. *Annu Rev Biochem* **68**: 1015–1068.

Walsh, R. M., Jr., Rawson, S., Schnell, H. M., Velez, B., Rajakumar, T. and Hanna, J. (2023). Structure of the preholoproteasome reveals late steps in proteasome core particle biogenesis. *Nat Struct Mol Biol* **30**: 1516–1524.

Wehmer, M., Rudack, T., Beck, F., Aufderheide, A., Pfeifer, G., Plitzko, J. M., Forster, F., Schulten, K., Baumeister, W. and Sakata, E. (2017). Structural insights into the functional cycle of the ATPase module of the 26S proteasome. *Proc Natl Acad Sci USA* **114**(6): 1305–1310.

Wenzel, T., Eckerskorn, C., Lottspeich, F. and Baumeister, W. (1994). Existence of a molecular ruler in proteasomes suggested by analysis of degradation products. *FEBS Lett* **349**(2): 205–209.

Whitby, F. G., Masters, E. I., Kramer, L., Knowlton, J. R., Yao, Y., Wang, C. C. and Hill, C. P. (2000). Structural basis for the activation of 20S proteasomes by 11S regulators. *Nature* **408**(6808): 115–120.

Wilkinson, K. D., Urban, M. K. and Haas, A. L. (1980). Ubiquitin is the ATP-dependent proteolysis factor I of rabbit reticulocytes. *J Biol Chem* **255**(16): 7529–7532.

Worden, E. J., Dong, K. C. and Martin, A. (2017). An AAA motor-driven mechanical switch in Rpn11 controls deubiquitination at the 26S proteasome. *Mol Cell* **67**(5): 799–811 e798.

Worden, E. J., Padovani, C. and Martin, A. (2014). Structure of the Rpn11-Rpn8 dimer reveals mechanisms of substrate deubiquitination during proteasomal degradation. *Nat Struct Mol Biol* **21**(3): 220–227.

Wu, Z., Chen, E., Zhang, S., Ma, Y. and Mao, Y. (2022). Visualizing conformational space of functional biomolecular complexes by deep manifold learning. *Int J Mol Sci* **23**(16): 8872.

Wu, Z., Chen, E., Zou, S., Hou, Y., Zhang, S., Wang, W. L., Dong, Y. and Mao, Y. (2022). Hidden dynamics of proteasome autoregulation discovered by cryo-EM data-driven deep learning. *bioRxiv*: 2020.2012.2022.423932.

Xie, S. C., Metcalfe, R. D., Hanssen, E., Yang, T., Gillett, D. L., Leis, A. P., Morton, C. J., Kuiper, M. J., Parker, M. W., Spillman, N. J., Wong, W., Tsu, C., Dick, L. R., Griffin, M. D. W. and Tilley, L. (2019). The structure of the PA28-20S proteasome complex from Plasmodium falciparum and implications for proteostasis. *Nat Microbiol* **4**(11): 1990–2000.

Xie, Y. and Varshavsky, A. (2000). Physical association of ubiquitin ligases and the 26S proteasome. *Proc Natl Acad Sci USA* **97**(6): 2497–2502.

Xie, Y. and Varshavsky, A. (2002). UFD4 lacking the proteasome-binding region catalyses ubiquitination but is impaired in proteolysis. *Nat Cell Biol* **4**(12): 1003–1007.

Xu, P., Duong, D. M., Seyfried, N. T., Cheng, D., Xie, Y., Robert, J., Rush, J., Hochstrasser, M., Finley, D. and Peng, J. (2009). Quantitative proteomics reveals the function of unconventional ubiquitin chains in proteasomal degradation. *Cell* **137**(1): 133–145.

Yao, T., Song, L., Xu, W., DeMartino, G. N., Florens, L., Swanson, S. K., Washburn, M. P., Conaway, R. C., Conaway, J. W. and Cohen, R. E. (2006). Proteasome recruitment and activation of the Uch37 deubiquitinating enzyme by Adrm1. *Nat Cell Biol* **8**(9): 994–1002.

Yau, R. and Rape, M. (2016). The increasing complexity of the ubiquitin code. *Nat Cell Biol* **18**(6): 579–586.

Yu, Z., Yu, Y., Wang, F., Myasnikov, A. G., Coffino, P. and Cheng, Y. (2020). Allosteric coupling between alpha-rings of the 20S proteasome. *Nat Commun* **11**(1): 4580.

Zhang, D., Chen, T., Ziv, I., Rosenzweig, R., Matiuhin, Y., Bronner, V., Glickman, M. H. and Fushman, D. (2009). Together, Rpn10 and Dsk2 can serve as a polyubiquitin chain-length sensor. *Mol Cell* **36**(6): 1018–1033.

Zhang, F., Hu, M., Tian, G., Zhang, P., Finley, D., Jeffrey, P. D. and Shi, Y. (2009). Structural insights into the regulatory particle of the proteasome from *Methanocaldococcus jannaschii*. *Mol Cell* **34**(4): 473–484.

Zhang, F., Wu, Z., Zhang, P., Tian, G., Finley, D. and Shi, Y. (2009). Mechanism of substrate unfolding and translocation by the regulatory particle of the proteasome from *Methanocaldococcus jannaschii*. *Mol Cell* **34**(4): 485–496.

Zhang, N., Wang, Q., Ehlinger, A., Randles, L., Lary, J. W., Kang, Y., Haririnia, A., Storaska, A. J., Cole, J. L., Fushman, D. and Walters, K. J. (2009). Structure of the s5a:k48-linked diubiquitin complex and its interactions with rpn13. *Mol Cell* **35**(3): 280–290.

Zhang, S., Zou, S., Yin, D., Zhao, L., Finley, D., Wu, Z. and Mao, Y. (2022). USP14-regulated allostery of the human proteasome by time-resolved cryo-EM. *Nature* **605**(7910): 567–574.

Zhang, Z., Lv, X., Yin, W. C., Zhang, X., Feng, J., Wu, W., Hui, C. C., Zhang, L. and Zhao, Y. (2013). Ter94 ATPase complex targets K11-linked ubiquitinated Ci to proteasomes for partial degradation. *Dev Cell* **25**(6): 636–644.

Zhang, Z. H., Jiang, T. X., Chen, L. B., Zhou, W., Liu, Y., Gao, F. and Qiu, X. B. (2021). Proteasome subunit α4s is essential for formation of spermatoproteasomes and histone degradation during meiotic DNA repair in spermatocytes. *J Biol Chem* **296**: 100130.

Zhao, J., Makhija, S., Zhou, C., Zhang, H., Wang, Y., Muralidharan, M., Huang, B. and Cheng, Y. (2022). Structural insights into the human PA28-20S proteasome enabled by efficient tagging and purification of endogenous proteins. *Proc Natl Acad Sci USA* **119**(33): e2207200119.

Zhao, M., Zhang, N. Y., Zurawel, A., Hansen, K. C. and Liu, C. W. (2010). Degradation of some polyubiquitinated proteins requires an intrinsic proteasomal binding element in the substrates. *J Biol Chem* **285**(7): 4771–4780.

Zheng, N. and Shabek, N. (2017). Ubiquitin Ligases: Structure, Function, and Regulation. *Annu Rev Biochem* **86**: 129–157.

Zhu, Y., Wang, W. L., Yu, D., Ouyang, Q., Lu, Y. and Mao, Y. (2018). Structural mechanism for nucleotide-driven remodeling of the AAA-ATPase unfoldase in the activated human 26S proteasome. *Nat Commun* **9**(1): 1360.

3 Cryo-EM Structures of a Ubiquitin-Specific Protease in Complex with Substrate

Martin Rennie

INTRODUCTION

USP1 IS INVOLVED IN DNA REPAIR AND IS A POTENTIAL ANTICANCER TARGET

Uncontrolled chemical modification of DNA impairs DNA replication and transcription, which is toxic to cells. Recognition of this DNA damage by protein complexes within the nucleus triggers signaling cascades to repair the DNA (Groelly et al., 2023). Ubiquitin-specific protease 1 (USP1) is involved in regulation of the DNA damage response, being a well-established component of the Fanconi anemia (FA) pathway (Nijman et al., 2005) and of translesion synthesis (TLS) (Huang et al., 2006). The FA pathway facilitates repair of cross-links between the two DNA strands, while TLS bypasses sites of DNA damage during DNA replication. USP1 expression is upregulated in several cancers, frequently correlating with poor prognosis (Ma et al., 2019; Sonego et al., 2019; Xu et al., 2019). Small molecule inhibitors against USP1 have been developed and highlight its potential as a target for cancer treatment (Chen et al., 2011; Dexheimer et al., 2014; Liang et al., 2014). Structural characterization of USP1 is crucial to understanding how it is involved in DNA repair pathways and to aid structure-guided drug design of USP1 inhibitors.

MONO-UBIQUITINATED FANCI-FANCD2 COMPLEX IS A SUBSTRATE OF USP1

USP1 deubiquitinates DNA clamps to regulate their protein–protein and protein–DNA interactions (Alcón et al., 2020; Kaszubowski and Trakselis, 2022; Rennie et al., 2020; Tan et al., 2020a; Wang et al., 2020), as well as oncogenic proteins to reduce their degradation (Sonego et al., 2019; Williams et al., 2011). A well-studied USP1 substrate is the FANCI-FANCD2 heterodimer, which is involved in the FA pathway (Nijman et al., 2005; Sims et al., 2007). This complex binds DNA in a sequence-independent manner, with mono-ubiquitination driving clamping of the complex onto DNA (Figure 3.1a) (Alcón et al., 2020; Rennie et al., 2020; Tan et al., 2020a; Wang et al., 2020). The FANCI and FANCD2 subunits are paralogs, with the mono-ubiquitination occurring on specific lysines, K523 and K561, respectively. Formation of the isopeptide bond between the C-terminus of ubiquitin and these lysines is catalyzed by the FA core complex (Rajendra et al., 2014; van Twest et al., 2017; Wang et al., 2021). This mono-ubiquitination reaction is stimulated by binding of DNA to FANCI-FANCD2 (Sato et al., 2012) and phosphorylation of FANCI (Tan et al., 2020b). Clamping of the FANCI-FANCD2 heterodimer onto DNA can be achieved by mono-ubiquitination of the FANCI (Lemonidis et al., 2022) or FANCD2 subunit (Alcón et al., 2020; Rennie et al., 2020; Tan et al., 2020a; Wang et al., 2020), or phosphorylation of FANCI (Sijacki et al., 2022). However, mono-ubiquitination of the FANCD2 subunit is essential for tightest DNA binding (Rennie et al., 2020). Preventing FANCD2 mono-ubiquitination by a K561R point mutation increases cell susceptibility to DNA interstrand cross-linking reagents (Garcia-Higuera et al., 2001), highlighting the importance of this event in DNA repair.

Interestingly, the mono-ubiquitinated lysines of FANCI-FANCD2 are buried within the interface between the subunits in the non-ubiquitinated state (Joo et al., 2011). This interface is mediated by interactions across the N-terminal domains (NTDs) and helical domains (HDs) of each subunit (Figure 3.1a) (Alcón et al., 2020; Joo et al., 2011; Wang et al., 2020). The FA core complex restructures the interface, exposing the ubiquitination sites and creating a new interface between the C terminal domains (CTDs) of each subunit (Wang et al., 2021). Mono-ubiquitination acts to stabilize this closed conformation, with the ubiquitin on one subunit interacting with the NTD of the opposite subunit, suggesting that the partner subunit serves as the "reader" of the ubiquitin signal (Figure 3.1a) (Alcón et al., 2020; Wang et al., 2020). Deubiquitination of FANCI-FANCD2 by USP1 presumably unlocks the clamp from DNA, allowing other proteins to access the DNA. USP1-knockouts, like FANCD2 K561R mutants, have increased susceptibility to interstrand cross-linkers (Oestergaard et al., 2007); thus both putting on the mono-ubiquitin signal and removing it appear to be important for DNA repair.

USP1 targets FANCI-FANCD2 through several distinct features. The USP1 protein is encoded by one of around 100 human deubiquitinase genes that can be subdivided into seven families, the largest being the ubiquitin-specific protease (USP) family (Clague et al., 2019). USP1 has a USP domain which contains the catalytic residues and binds ubiquitin. Structurally the USP domain is reminiscent of a right hand, with fingers, palm, and thumb subdomains (Figure 3.1b) (Rennie et al., 2021). This domain

DOI: 10.1201/9781003326106-4

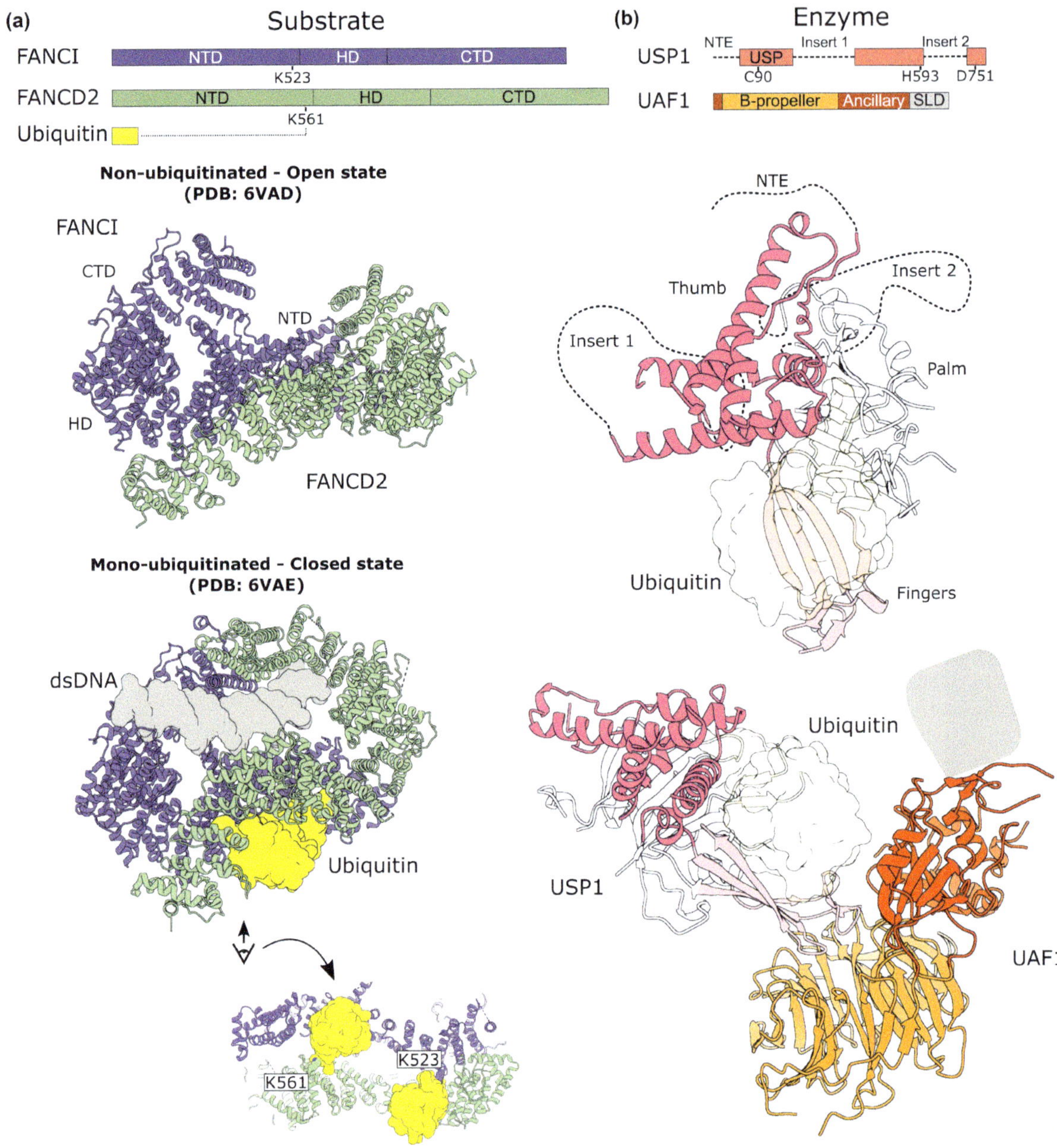

FIGURE 3.1 Structures of the FANCI-FANCD2 substrate and USP1-UAF1 enzyme. (a) The substrate consists of two components with ubiquitin conjugated to one or both lysines highlighted (*top*). Non-ubiquitinated FANCI-FANCD2 heterodimer is open (*middle*; PDB: 6VAD (Wang et al., 2020)), while mono-ubiquitinated heterodimer is closed with the ubiquitin of each subunit interacting with the other subunit (*bottom*; PDB: 6VAE (Wang et al., 2020)). Domains of FANCI are indicated. (b) The enzyme consists of two components (*top*). USP1 is shown as a cartoon, ubiquitin as a transparent volume, and the disordered regions, deleted from the crystallized construct, are indicated as dashed lines (*middle*; PDB: 7AY2 (Rennie et al., 2021)). The USP1-UAF1 complex is mediated by the fingers of USP1 (*bottom*; PDB: 7AY2 (Rennie et al., 2021)). The SLD domain, removed in the crystallized construct, is indicated by a grey box.

is interspaced by three large, disordered regions (residues 1-75, 227-422, 602-744). The N-terminal extension (NTE; 1-75) has been implicated in substrate targeting (Arkinson et al., 2018), insertion 1 (227-422) contains nuclear localization signals (Garcia-Santisteban et al., 2012), and insertion 2 (602-744) contains an autocleavage site (Huang et al., 2006). Insertion 1 and a shorter loop (465-483) play a role in auto-inhibition of USP1 (Dharadhar et al., 2021). To relieve this auto-inhibition a cofactor, USP1-associated protein 1 (UAF1), binds through the fingers subdomain of USP1 (Figure 3.1b) (Cohn et al., 2007; Rennie et al., 2021). UAF1 also plays a role in substrate binding (Rennie et al., 2021; Yang et al., 2011). The primary goal of the work described in this chapter was to determine the structural basis through which USP1-UAF1 recognizes and deubiquitinates the FANCI-FANCD2 heterodimer.

EXPERIMENTAL APPROACH

We had used crystallography for determining structures of truncated USP1-UAF1 (Figure 3.1b; PDB: 7AY0, 7AY2) (Rennie et al., 2021). However, experiments with the full-length enzyme or enzyme with substrate did not yield crystals. At a molecular weight of ~300 kDa, the FANCI-FANCD2 size is favorable for cryo-EM single particle analysis, so we turned our focus to this approach. The non-ubiquitinated FANCI and FANCD2 subunits could be expressed and purified separately and reconstituted, or co-expressed and co-purified, similarly to the enzyme subunits, USP1 and UAF1. However, insect cells rather than bacteria were required for high yields of the full-length human orthologs of each of these four proteins. We opted for a reconstitution approach, purifying each of the protein components separately and recombining them (Figure 3.2a) to give us greater control over the system. With the proteins in hand, two key challenges were preparing the natively mono-ubiquitinated FANCI-FANCD2 substrate and trapping the enzyme–substrate complex by preventing the catalytic reaction.

Preparation of the Substrate

Producing K561 mono-ubiquitinated FANCD2 or K523 mono-ubiquitinated FANCI in high yields and purity is not trivial. Efficient and specific mono-ubiquitination of FANCI and FANCD2 requires binding of DNA to the heterodimer (Sato et al., 2012), phosphorylation of FANCI (Tan et al., 2020b), and the 1.1 MDa FA core complex (Rajendra et al., 2014; van Twest et al., 2017; Wang et al., 2021). Despite these cumbersome requirements, the underlying components of this enzymatic reaction are an E2 ubiquitin-conjugating enzyme, UBE2T, and an E3 ligase, FANCL, one of the FA core complex subunits. FANCL binding to UBE2T rewires UBE2T to optimize transfer of ubiquitin onto the specifically targeted lysines (Chaugule et al., 2020). An allosteric conduit on UBE2T mediating this rewiring was engineered to drastically improve the efficiency of *in vitro* mono-ubiquitination of FANCI and FANCD2, whilst also retaining specificity (Chaugule et al., 2020, 2019). Together with a truncation of FANCL, this mutated UBE2T allowed us to efficiently mono-ubiquitinate purified FANCI or FANCD2, without the need for DNA or FANCL incorporation into the FA core complex. With this procedure established, we could reconstitute the FANCI-FANCD2 heterodimer in various mono-ubiquitination states (on FANCD2, on FANCI, or on both) (Rennie et al., 2020).

The FANCI-FANCD2 heterodimer with ubiquitin on FANCD2 (FANCI-FANCD2^{Ub}) is readily deubiquitinated *in vitro* by USP1 but not by other, more generic, USPs (Rennie et al., 2020). This suggested features of USP1-UAF1 may allow efficient deubiquitination of FANCI-FANCD2^{Ub}. Therefore we focused on this state as our substrate for USP1 cryo-EM analysis.

Preventing Substrate Turnover

Chemical modification of the substrate or the enzyme can be employed to prevent substrate turnover by the enzyme. For the former, nonhydrolyzable nucleotide triphosphate analogs have been extensively used for ATPases and GTPases. Nonhydrolyzable analogs of the isopeptide bond between ubiquitin and target lysines have also been

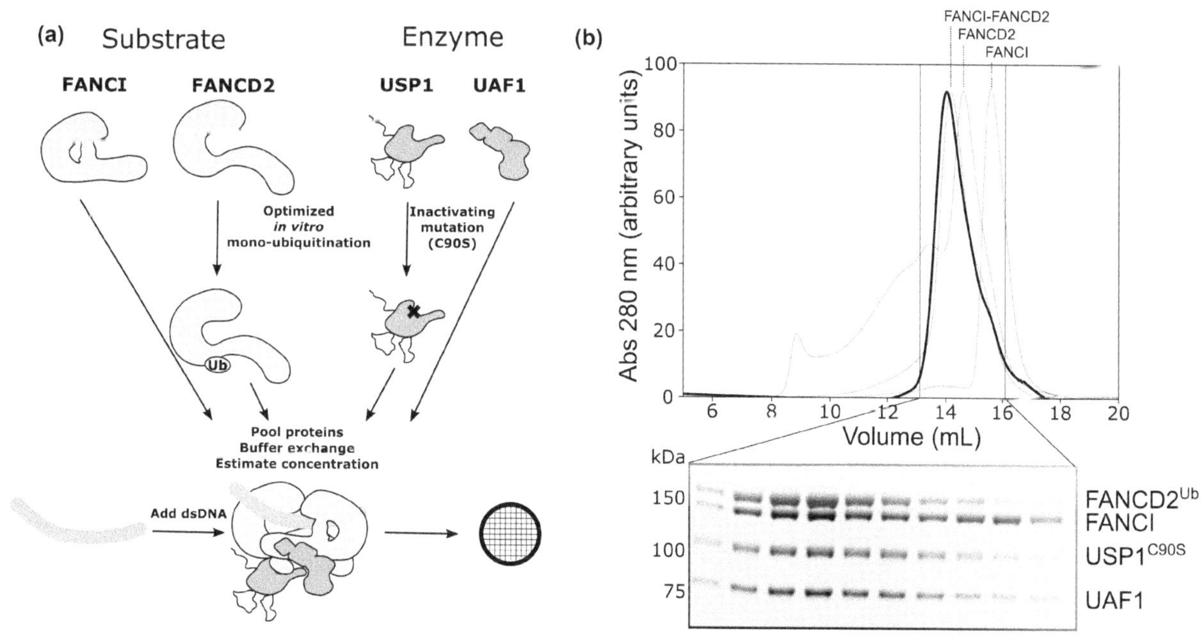

FIGURE 3.2 Preparation of the USP1^{C90S}-UAF1-FANCI-FANCD2^{Ub} complex. (a) Schematic representation for assembly of the enzyme–substrate complex. (b) Size-exclusion chromatography of the resulting complex confirming co-elution of all four protein components. (Adapted from Rennie et al., 2021.)

developed (Shanmugham et al., 2010; Yang et al., 2016; Yin et al., 2000) and successfully used to capture the SAGA deubiquitinase with mono-ubiquitinated histone substrate (Morgan et al., 2016). However, mutation of the deubiquitinase active site is also feasible through site-directed mutagenesis. Mutating the active site cysteine to serine or alanine greatly reduces isopeptide hydrolysis but does not prevent ubiquitin binding (Morrow et al., 2018).

For full-length USP1, we expressed both the C90A and C90S active site mutants in insect cells. However, the C90A mutant copurified with a contaminant consistent with a ubiquitin or ubiquitin-like protein. We attributed this to the very tight affinity of ubiquitin for cysteine to alanine active site mutations of deubiquitinases (Morrow et al., 2018). Therefore, we focused on the C90S mutant, which we were able to purify similar to the wild-type. We also included a mutation of USP1 in insert 2 that prevents autocleavage (G670A,G671A). We confirmed that when combined with UAF1, this mutant was inactive on the FANCI-FANCD2^{Ub} substrate even at equimolar concentrations of enzyme and substrate.

Assembling the Complex and Optimizing the Grids

To confirm assembly of the enzyme–substrate complex, we used NativePAGE and size-exclusion chromatography (Figure 3.2b). Initially, cryo-EM grids were prepared directly from size-exclusion chromatography purified complex. However, these grids had low particle density (Figure 3.3a),

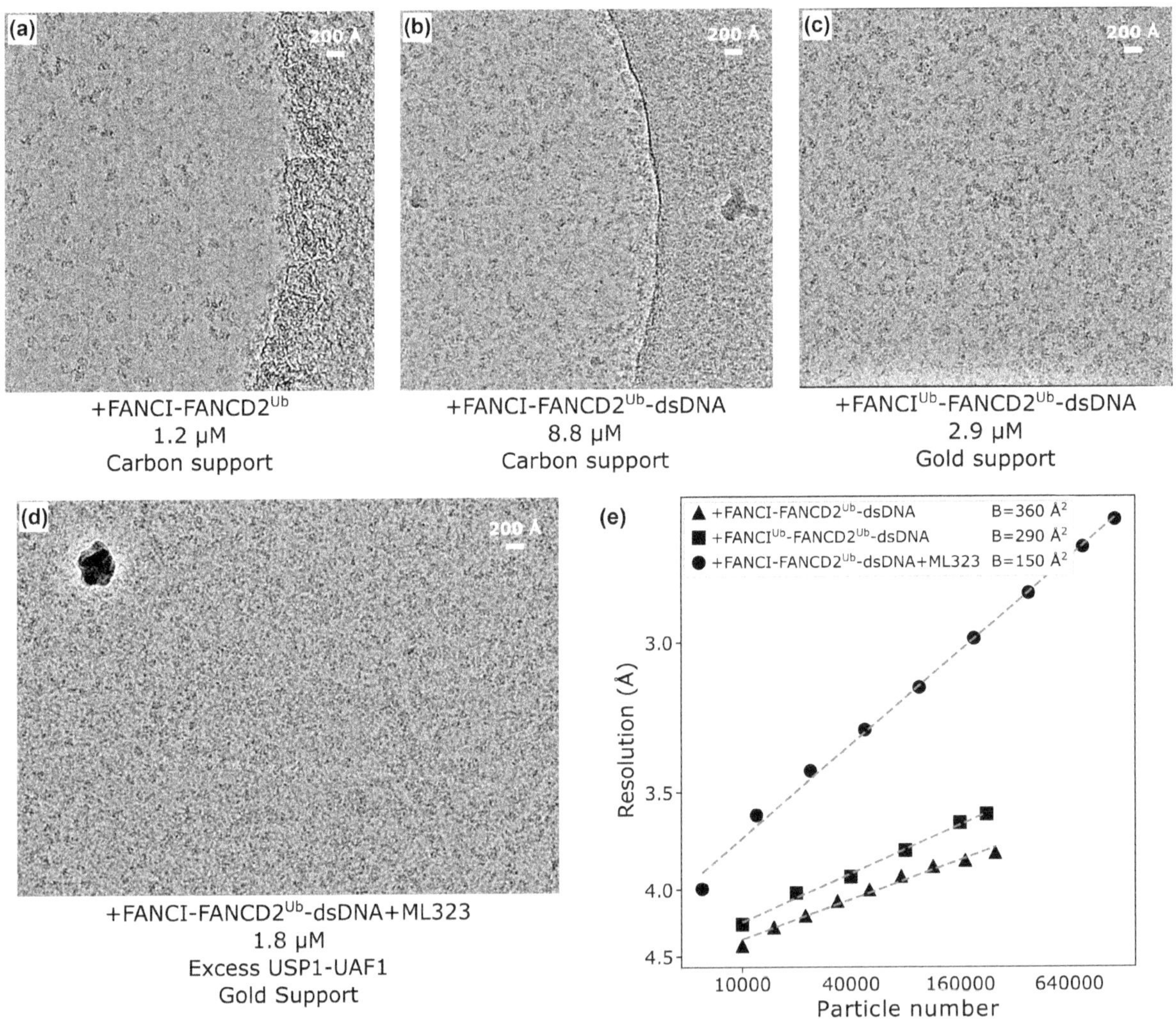

FIGURE 3.3 Example micrographs of USP1^{C90S}-UAF1 bound to different substrates. (a) With FANCI-FANCD2^{Ub} substrate applied to C-Flat 1.2/1.3 grids and collected on a CRYOARM/DE-64 system at the Scottish Centre for Macromolecular Imaging (SCMI). (b) With FANCI-FANCD2^{Ub}-dsDNA substrate applied to Quantifoil 1.2/1.3 grids and collected on a CRYOARM/DE-64 system at SCMI. (c) With FANCI^{Ub}-FANCD2^{Ub}-dsDNA substrate applied to UltrAuFoil 1.2/1.3 grids and collected on a CRYOARM/DE-64/Omega energy filter system at SCMI. (d) With FANCI-FANCD2^{Ub}-dsDNA substrate and excess ML323 inhibitor and excess USP1-UAF1 applied to UltrAuFoil 1.2/1.3 grids and collected on a Krios/K3/GIF energy filter system at eBIC, Diamond Light Source. All micrographs are at 2.5 µm defocus and ~1 Å/px. (e) B-factor plots computed from datasets corresponding to panels (b)–(d) using ResLog in cryoSPARC (with homogenous refinement).

and regions with very thin ice resulted in complex dissociation. It was subsequently found to be adequate, and more convenient, to reconstitute the complex 1:1:1:1 from snap-frozen aliquots of the protein components, and then buffer exchange into EM-suitable buffer using a desalting column. This facilitated preparation of the complex at higher concentrations without the need for a reconcentration step. Initial reconstructions suggested the FANCI-FANCD2 complex remained in a clamped conformation; therefore double-stranded DNA (dsDNA) was added following buffer exchange to further increase the size of the complex. Higher complex concentration and addition of dsDNA yielded grids with high particle density (Figure 3.3b), which enabled determination of the structure to a global resolution of 3.7 Å (Rennie et al., 2021).

Further Datasets

With the experience of producing grids of USP1^{C90S}-UAF1-FANCI-FANCD2^{Ub}-dsDNA and knowledge of the particle shapes, we sought an inhibitor-bound enzyme–substrate structure (Rennie et al., 2022) and USP1-UAF1 bound to the di-mono-ubiquitinated substrate (FANCI^{Ub}-FANCD2^{Ub}; currently unpublished). We used NativePAGE and microscale thermophoresis to confirm our reconstitution of both variations of the enzyme–substrate complex. Earlier cryo-EM samples (Figure 3.3a and b) were prepared using carbon supports, but for these two samples (Figure 3.3c and d) we used gold supports (Russo and Passmore, 2016). Both of the gold support datasets surpassed the carbon support resolutions, even when accounting for the same number of particles in B-factor plots (Rosenthal and Henderson, 2003) (Figure 3.3e).

Data Processing and Model Building

Movies were processed using cryoSPARC (Punjani et al., 2017), with all datasets using a similar pipeline (Figure 3.4). Motion correction and CTF estimation was performed on all movies, followed by removal of micrographs with poor CTF fit or those missing the holes. Blob-picking of the resulting micrographs yielded 2D classes suitable for *ab initio* reconstruction and for template picking. *Ab initio* reconstruction yielded one structure with well-resolved secondary structure. Template-picked particles were taken straight to

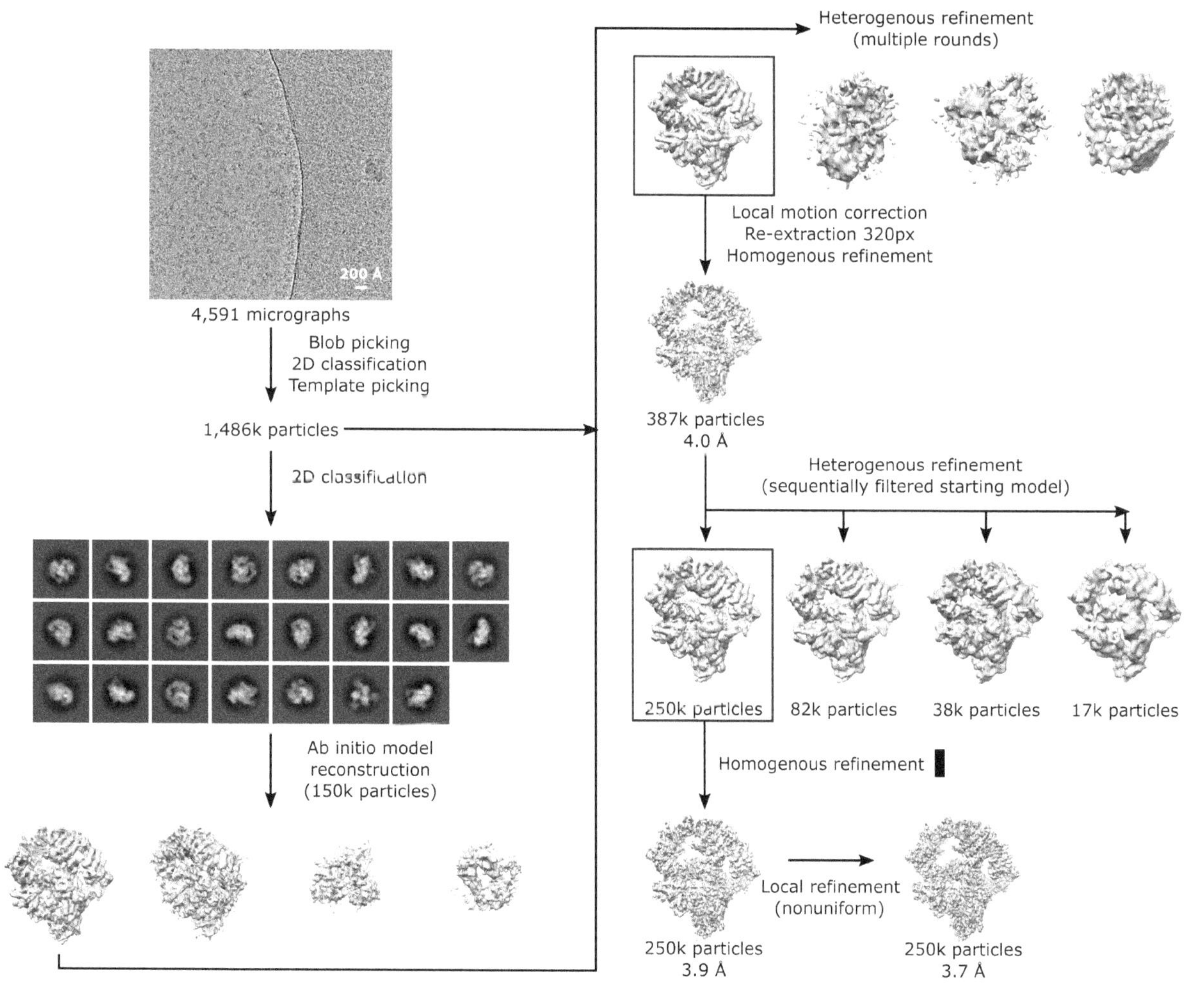

FIGURE 3.4 Typical processing workflow of the cryo-EM data. (Reproduced from Rennie et al., 2021.)

heterogenous refinement, i.e., without 2D classification, and multiple rounds of heterogenous refinement were performed. Initial models used for these heterogenous refinement steps included the target structure, generated from *ab initio* reconstruction, and junk classes, generated from non-target particles, or sequentially low-pass filtered classes of the target structure (30, 20, 15, 12 Å). Particles falling into the highest resolution class of the target structure were used for homogenous refinement and nonuniform refinement (Punjani et al., 2020). Nonuniform refinement yielded improved maps compared to homogenous refinement, but significant variability remained as identified by 3D variability analysis (Punjani and Fleet, 2021), particularly with respect to the bound DNA (Rennie et al., 2021). Focused reconstruction on different subsections improved maps, and atomic models were built into consensus maps and focused maps.

For atomic model building into the consensus map from the first high-resolution dataset we obtained (EMD-11934), we started with existing experimental models of the enzyme and substrate separately. Crystal structures of USP1 (PDB: 7AY2 (Rennie et al., 2021)) and UAF1 (5K1A (Li et al., 2016)) as well as cryo-EM models of FANCI-FANCD2^{Ub} (PDB: 6VAF (Wang et al., 2020)) were docked into the map. Iterative manual model building in COOT (Emsley et al., 2010) and automated real-space refinement in phenix (Afonine et al., 2018) were then performed. In particular, an apparent sequence register error was corrected for residues 1019–1037 in the FANCD2 subunit of the FANCI-FANCD2^{Ub} initial model (PDB: 6VAF). The AlphaFold model of FANCD2 exhibited the same sequence register for residues 1019–1037 as our model, providing further support that we had correctly remodeled the region. Modeling into subsequent consensus and focused maps was performed similarly to the initial dataset except using as a starting model the enzyme substrate complex derived from the initial dataset (PDB: 7AY1 (Rennie et al., 2021)). AlphaFold models (Jumper et al., 2021) were also used to guide modeling when they became available through the AlphaFold Protein Structure Database.

RESULTS AND DISCUSSION

Insight into USP1-mediated Deubiquitination

The structures, together with deubiquitination assays and AlphaFold-multimer (Evans et al., 2022) predictions, provide insight into the multifaceted way USP1-UAF1 deubiquitinates FANCI-FANCD2 (Figure 3.5). In the enzyme–substrate complex USP1 clasps the ubiquitin, whereas in the substrate-only complex (i.e. FANCI-FANCD2^{Ub} without USP1-UAF1) ubiquitin interacts extensively with FANCI (Wang et al., 2020) (Figure 3.5a). As such, when USP1-UAF1 engages FANCI-FANCD2^{Ub} the ubiquitin attached to K561 undergoes large-scale conformational changes. Further, the isopeptide bond between FANCD2 and ubiquitin becomes threaded through the active site of USP1 (Figure 3.5b). The remainder of FANCI-FANCD2 is in the closed conformation, still clamped around dsDNA. However, the N-terminus of FANCD2 exhibits more flexibility than the substrate alone, seemingly in an intermediate between the FANCI-FANCD2 ubiquitinated and non-ubiquitinated states (Figure 3.5c). This suggests the FANCI-FANCD2 complex may be beginning to return to the open conformation in the state that we trapped.

Interestingly, an extensive interface forms between the accessory subunits of the enzyme and substrate—UAF1, which does not contain the active site, and FANCI, which does not contain the isopeptide bond. The interface buries ~1600 Å^2 and consists of two regions: The SLD of UAF1 interacts with a SIM-like motif on FANCI (Figure 3.5d), and the β-propeller domain of UAF1 interacts with a loop on FANCI (Figure 3.5e). The latter loop can be phosphorylated to promote ubiquitination and inhibit deubiquitination (Cheung et al., 2017; Tan et al., 2020b). Together with biophysical data (Rennie et al., 2021), this suggests the interaction with UAF1 is regulated by FANCI phosphorylation.

Inserts 1 and 2 of USP1 are not visible in any of the cryo-EM maps we have obtained, and so they are likely unstructured and/or not tightly associated with the USP domain of USP1. The NTE is also highly disordered. However, before we explored cryo-EM to characterize the interaction, alanine scanning mutagenesis and biochemical assays had pinpointed a short region of the NTE of USP1 as critical for FANCD2 deubiquitination (Arkinson et al., 2018). AlphaFold-multimer predictions of USP1^{NTE} and FANCD2 identified the same region of the NTE as interacting with the FANCD2 HD (Rennie et al., 2022). Within the cryo-EM maps, weak density at this site was visible and consistent with the AlphaFold models (Figure 3.5f). As such, the NTE helps mediate targeting of the ubiquitin on FANCD2.

USP1 and Inhibitors

As USP1 is a potential anticancer target with multiple established small molecule inhibitors, one of the goals was to identify the inhibitor binding site. However, the first structures of USP1 we obtained gave no obvious insight into where the inhibitors, thought to be allosteric, might bind. To experimentally determine the binding site, we sought to use the USP1^{C90S}-UAF1-FANCI-FANCD2^{Ub}-dsDNA complex as a scaffold, adding in the inhibitor ML323 (Liang et al., 2014) that was commercially available. We found that even in the presence of ML323, the enzyme–substrate complex was able to form. We took this six-component complex—four protein subunits, one dsDNA, one small molecule—through for cryo-EM analysis and were able to determine a consensus structure to 2.7 Å. Within these maps there was no obvious density for the inhibitor; however, near the interface between the palm and thumb subdomains of USP1, there were distinct breaks in the mainchain density where in previous structures the density was contiguous. Focusing on the USP1 subunit and classifying around the palm and thumb, we identified two states, both resolved to 2.5 Å, one with clear density for inhibitor and one without inhibitor.

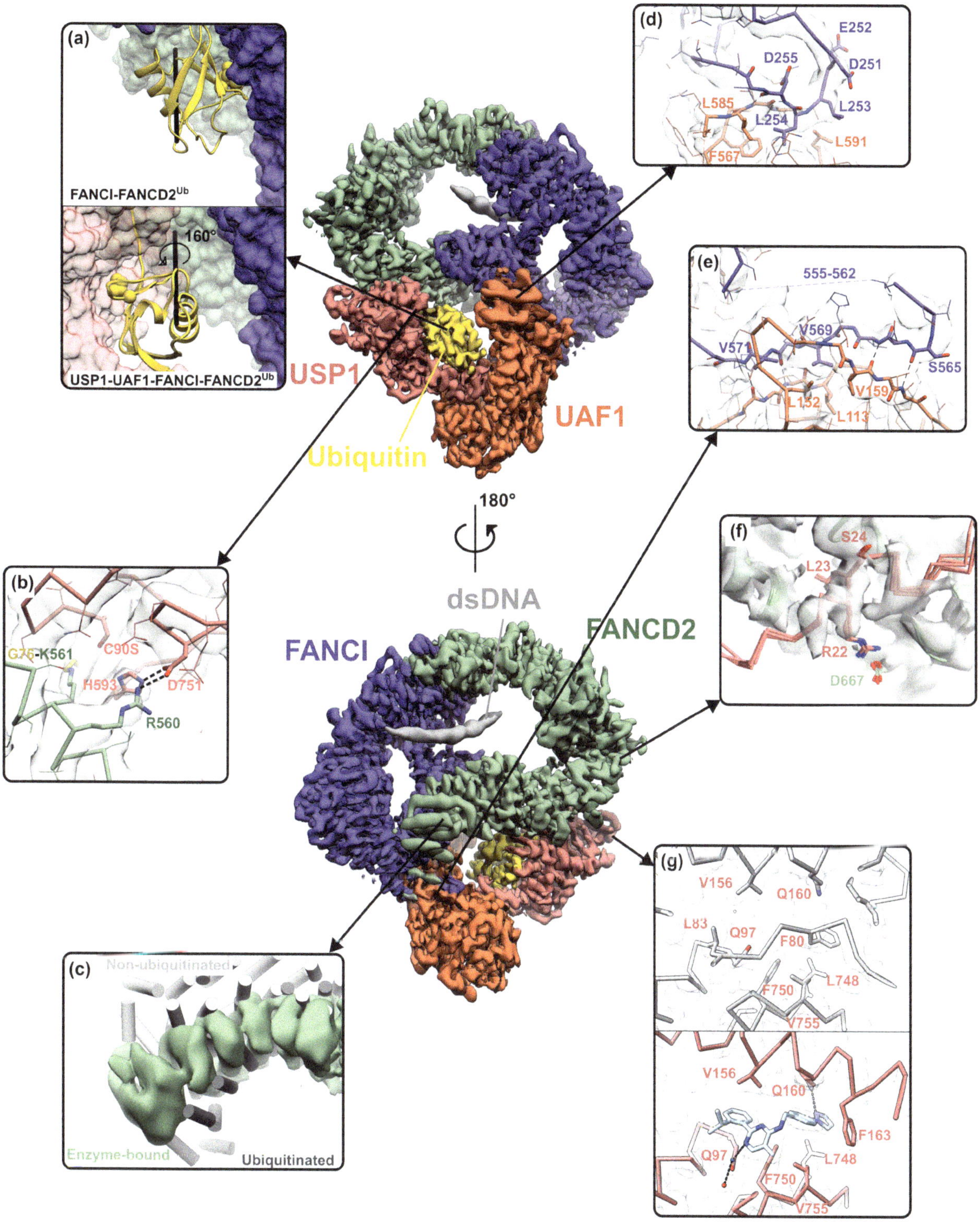

FIGURE 3.5 Cryo-EM reconstructions of USP1^{C90S}-UAF1-FANCI-FANCD2^{Ub} reveal multifaceted control of deubiquitination. (a) Comparison of substrate-alone (*top*; PDB: 6VAF (Wang et al., 2020)) with the enzyme-bound state (*bottom*; PDB: 7AY1 (Rennie et al., 2021)). Structures were aligned by the FANCI subunit. (b) The active site of USP1 with the ubiquitin^{G76}-FANCD2^{K561} isopeptide bond. (c) Density of FANCD2^{NTD} from the reconstruction (green), with models of the equivalent region from non-ubiquitinated (PDB: 6VAD) and ubiquitinated (PDB: 6VAF) heterodimers. Structures aligned by the FANCD2 subunit. (d) UAF1^{SLD} interacts with FANCI. (e) UAF1$^{\beta\text{-propeller}}$ interacts with FANCI. (f) USP1^{NTE} interacts with FANCD2^{HD}. (g) Comparison of the ML323 binding site in the absence (*top*; grey) and presence of the inhibitor (*bottom*; colored). (Adapted from Rennie et al., 2021 and Rennie et al., 2022.)

The ML323 inhibitor binds to a strikingly cryptic site, replacing part of the hydrophobic core of USP1 (Figure 3.5g). A pair of β-strands near the active site are displaced, resulting in increased disorder in this region. Inhibitor binding also leads to much more subtle rearrangements in the active site that likely flip the catalytic histidine, preventing it from efficiently deprotonating the catalytic cysteine and thereby inhibiting USP1.

Final Remarks

Our cryo-EM samples were prepared by reconstituting the components, having previously identified suitable buffers to snap-freeze aliquots of FANCI, FANCD2, USP1, or UAF1. This approach gave us considerable control in tweaking the grid preparations. For example, the inhibitor of USP1 was expected to disrupt the complex somewhat based on biophysical data, so we used an excess of USP1-UAF1 to drive the large FANCI-FANCD2 substrate toward the USP1-UAF1-bound state. We have been able to prepare and determine structures of several related complexes without modification of the protein purifications.

For removing junk particles from cryo-EM datasets, we used heterogenous refinement exclusively. We first removed obvious junk, using classes distinct from the target. Then using sequentially low-pass filtered classes of the target, we rationalized that low-resolution particles or junk with shapes similar to the target would be removed. We found this approach retained more particles, and in more orientations, than including a 2D classification step and yielded clearer maps.

Of the three datasets that resolved to better than 4 Å, we found that gold support grids yielded higher-resolution structures than carbon supports (Figure 3.3). However, we must note that each dataset was a slightly different complex, and gold supports were used for the later datasets when we were more experienced with cryo-EM. Nevertheless, other evidence also suggests gold supports yield higher-resolution structures (Danev et al., 2021).

Assembling USP1 into the large enzyme–substrate complex improved tractability for cryo-EM analysis and allowed us to unambiguously determine the binding site of a small molecule inhibitor. This approach might be a useful alternative to other methods for increasing particle size for cryo-EM, such as binding megabodies (Uchański et al., 2021).

We focused on preventing turnover of the enzyme–substrate complex to yield a homogenous sample for cryo-EM, but this might not even be necessary for some enzyme–substrate pairs. Impressively, the Pavletich group captured the FA core complex enzyme with FANCI-FANCD2 substrate during the mono-ubiquitination reaction (Wang et al., 2021). They set up grids at early time points of the reaction (~20% ubiquitination) and classified the resulting mixture of enzyme and enzyme–substrate particles *in silico*. They were able to classify multiple states of enzyme substrate, suggesting a complex energy landscape for the reaction. Similarly, cryo-EM samples from actively turning over proteasome have also been studied (De la Peña et al., 2018). These examples, where the signal is sufficient to allow classification of the states, highlight the potential of cryo-EM in the analysis of large enzymes with large substrates.

REFERENCES

Afonine PV., Poon BK, Read RJ, Sobolev OV., Terwilliger TC, Urzhumtsev A, Adams PD. 2018. Real-space refinement in PHENIX for cryo-EM and crystallography. *Acta Crystallogr Sect D Struct Biol* **74**:531–544. doi:10.1107/S2059798318006551

Alcón P, Shakeel S, Chen ZA, Rappsilber J, Patel KJ, Passmore LA. 2020. FANCD2–FANCI is a clamp stabilized on DNA by monoubiquitination of FANCD2 during DNA repair. *Nat Struct Mol Biol* **27**:240–248. doi:10.1038/s41594-020-0380-1

Arkinson C, Chaugule VK, Toth R, Walden H. 2018. Specificity for deubiquitination of monoubiquitinated FANCD2 is driven by the N-terminus of USP1. *Life Sci Alliance* **1**:e201800162. doi:10.1101/430124

Chaugule VK, Arkinson C, Rennie ML, Kamarainen O, Toth R, Walden H. 2020. Allosteric mechanism for site-specific ubiquitination of FANCD2. *Nat Chem Biol* **16**:291–301. doi:10.5956/jriet.31.753

Chaugule VK, Arkinson C, Toth R, Walden H. 2019. Enzymatic preparation of monoubiquitinated FANCD2 and FANCI proteins. *Methods Enzymol* **618**:73–104. doi:10.1016/bs.mie.2018.12.021

Chen J, Dexheimer TS, Ai Y, Liang Q, Villamil MA, Inglese J, Maloney DJ, Jadhav A, Simeonov A, Zhuang Z. 2011. Selective and cell-active inhibitors of the USP1/UAF1 deubiquitinase complex reverse cisplatin resistance in non-small cell lung cancer cells. *Chem Biol* **18**:1390–1400. doi:10.1016/j.chembiol.2011.08.014

Cheung RS, Castella M, Abeyta A, Gafken PR, Tucker N, Taniguchi T. 2017. Ubiquitination-linked phosphorylation of the FANCI S/TQ cluster contributes to activation of the Fanconi anemia I/D2 complex. *Cell Rep* **19**:2432–2440. doi:10.1016/j.celrep.2017.05.081

Clague MJ, Urbé S, Komander D. 2019. Breaking the chains: Deubiquitylating enzyme specificity begets function. *Nat Rev Mol Cell Biol* **20**:338–352. doi:10.1038/s41580-019-0099-1

Cohn MA, Kowal P, Yang K, Haas W, Huang TT, Gygi SP, D'Andrea AD. 2007. A UAF1-containing multisubunit protein complex regulates the fanconi anemia pathway. *Mol Cell* **28**:786–797. doi:10.1016/j.molcel.2007.09.031

Danev R, Belousoff M, Liang YL, Zhang X, Eisenstein F, Wootten D, Sexton PM. 2021. Routine sub-2.5 Å cryo-EM structure determination of GPCRs. *Nat Commun* **12**:4333. doi:10.1038/s41467-021-24650-3

De la Peña AH, Goodall EA, Gates SN, Lander GC, Martin A. 2018. Substrate-engaged 26S proteasome structures reveal mechanisms for ATP-hydrolysis–driven translocation. *Science* **362**:eaav0725. doi:10.1126/science.aav0725

Dexheimer TS, Rosenthal AS, Luci DK, Liang Q, Villamil MA, Chen J, Sun H, Kerns EH, Simeonov A, Jadhav A, Zhuang Z, Maloney DJ. 2014. Synthesis and structure-activity relationship studies of N -benzyl-2-phenylpyrimidin-4-amine derivatives as potent USP1/UAF1 deubiquitinase inhibitors with anticancer activity against nonsmall cell lung cancer. *J Med Chem* **57**:8099–8110. doi:10.1021/jm5010495

Dharadhar S, Dijk WJ Van, Scheffers S, Fish A, Sixma TK. 2021. Insert L1 is a central hub for allosteric regulation of USP1 activity. *EMBO Rep* **22**:e51749. doi:10.15252/embr.202051749

Emsley P, Lohkamp B, Scott WG, Cowtan K. 2010. Features and development of Coot. *Acta Crystallogr Sect D Biol Crystallogr* **66**:486–501. doi:10.1107/S0907444910007493

Evans R, O'Neill M, Pritzel A, Antropova N, Senior AW, Green T, Žídek A, Bates R, Blackwell S, Yim J, Ronneberger O, Bodenstein S, Zielinski M, Bridgland A, Potapenko A, Cowie A, Tunyasuvunakool K, Jain R, Clancy E, Kohli P, Jumper J, Hassabis D. 2022. Protein complex prediction with AlphaFold-Multimer. *bioRxiv*. doi: doi:10.1101/2021.10.04.463034

Garcia-Higuera I, Taniguchi T, Ganesan S, Meyn MS, Timmers C, Hejna J, Grompe M, D'Andrea AD. 2001. Interaction of the Fanconi anemia proteins and BRCA1 in a common pathway. *Mol Cell* **7**:249–262. doi:10.1016/S1097-2765(01)00173-3

Garcia-Santisteban I, Zorroza K, Rodriguez JA. 2012. Two nuclear localization signals in USP1 mediate nuclear import of the USP1/UAF1 complex. *PLoS One* **7**:e38570. doi:10.1371/journal.pone.0038570

Groelly FJ, Fawkes M, Dagg RA, Blackford AN, Tarsounas M. 2023. Targeting DNA damage response pathways in cancer. *Nat Rev Cancer* **23**:78–94. doi:10.1038/s41568-022-00535-5

Huang TT, Nijman SMB, Mirchandani KD, Galardy PJ, Cohn MA, Haas W, Gygi SP, Ploegh HL, Bernards R, D'Andrea AD. 2006. Regulation of monoubiquitinated PCNA by DUB autocleavage. *Nat Cell Biol* **8**:339–347. doi:10.1038/ncb1378

Joo W, Xu G, Persky NS, Smogorzewska A, Rudge DG, Buzovetsky O, Elledge SJ, Pavletich NP. 2011. Structure of the FANCI-FANCD2 complex: Insights into the Fanconi anemia DNA repair pathway. *Science* **333**:312–316. doi:10.1126/science.1205805

Jumper J, Evans R, Pritzel A, Green T, Figurnov M, Ronneberger O, Tunyasuvunakool K, Bates R, Žídek A, Potapenko A, Bridgland A, Meyer C, Kohl SAA, Ballard AJ, Cowie A, Romera-Paredes B, Nikolov S, Jain R, Adler J, Back T, Petersen S, Reiman D, Clancy E, Zielinski M, Steinegger M, Pacholska M, Berghammer T, Bodenstein S, Silver D, Vinyals O, Senior AW, Kavukcuoglu K, Kohli P, Hassabis D. 2021. Highly accurate protein structure prediction with AlphaFold. *Nature* **596**:583–589. doi:10.1038/s41586-021-03819-2

Kaszubowski JD, Trakselis MA. 2022. Beyond the lesion: Back to high fidelity DNA synthesis. *Front Mol Biosci* **8**:811540. doi:10.3389/fmolb.2021.811540

Lemonidis K, Rennie ML, Arkinson C, Chaugule VK, Clarke M, Streetley J, Walden H. 2022. Structural and biochemical basis of interdependent FANCI-FANCD2 ubiquitination. *EMBO J* **42**:e111898. doi:10.15252/embj.2022111898

Li H, Lim KS, Kim H, Hinds TR, Jo U, Mao H, Weller CE, Sun J, Chatterjee C, D'Andrea AD, Zheng N. 2016. Allosteric activation of ubiquitin-specific proteases by β-propeller proteins UAF1 and WDR20. *Mol Cell* **63**:249–260. doi:10.1016/j.molcel.2016.05.031

Liang Q, Dexheimer TS, Zhang P, Rosenthal AS, Villamil MA, You C, Zhang Q, Chen J, Ott CA, Sun H, Luci DK, Yuan B, Simeonov A, Jadhav A, Xiao H, Wang Y, Maloney DJ, Zhuang Z. 2014. A selective USP1-UAF1 inhibitor links deubiquitination to DNA damage responses. *Nat Chem Biol* **10**:298–304. doi:10.1038/nchembio.1455

Ma A, Tang M, Zhang L, Wang B, Yang Z, Liu Yun, Xu G, Wu L, Jing T, Xu X, Yang S, Liu Y. 2019. USP1 inhibition destabilizes KPNA2 and suppresses breast cancer metastasis. *Oncogene* **38**:2405–2419. doi:10.1038/s41388-018-0590-8

Morgan MT, Haj-Yahya M, Ringel AE, Bandi P, Brik A, Wolberger C. 2016. Structural basis for histone H2B deubiquitination by the SAGA DUB module. *Science* **351**:725–728. doi:10.1126/science.aac5681

Morrow ME, Morgan MT, Clerici M, Growkova K, Yan M, Komander D, Sixma TK, Simicek M, Wolberger C. 2018. Active site alanine mutations convert deubiquitinases into high-affinity ubiquitin-binding proteins. *EMBO Rep* **19**:e45680. doi:10.15252/embr.201745680

Nijman SMB, Huang TT, Dirac AMG, Brummelkamp TR, Kerkhoven RM, D'Andrea AD, Bernards R. 2005. The deubiquitinating enzyme USP1 regulates the fanconi anemia pathway. *Mol Cell* **17**:331–339. doi:10.1016/j.molcel.2005.01.008

Oestergaard VH, Langevin F, Kuiken HJ, Pace P, Niedzwiedz W, Simpson LJ, Ohzeki M, Takata M, Sale JE, Patel KJ. 2007. Deubiquitination of FANCD2 is required for DNA crosslink repair. *Mol Cell* **28**:798–809. doi:10.1016/j.molcel.2007.09.020

Punjani A, Fleet DJ. 2021. 3D variability analysis: Resolving continuous flexibility and discrete heterogeneity from single particle cryo-EM. *J Struct Biol* **213**:107702. doi:10.1016/j.jsb.2021.107702

Punjani A, Rubinstein JL, Fleet DJ, Brubaker MA. 2017. CryoSPARC: Algorithms for rapid unsupervised cryo-EM structure determination. *Nat Methods* **14**:290–296. doi:10.1038/nmeth.4169

Punjani A, Zhang H, Fleet DJ. 2020. Non-uniform refinement: Adaptive regularization improves single-particle cryo-EM reconstruction. *Nat Methods* **17**:1214–1221. doi:10.1038/s41592-020-00990-8

Rajendra E, Oestergaard VH, Langevin F, Wang M, Dornan GL, Patel KJ, Passmore LA. 2014. The genetic and biochemical basis of FANCD2 monoubiquitination. *Mol Cell* **54**:858–869. doi:10.1016/j.molcel.2014.05.001

Rennie ML, Arkinson C, Chaugule VK, Toth R, Walden H. 2021. Structural basis of FANCD2 deubiquitination by USP1-UAF1. *Nat Struct Mol Biol* **28**:356–364. doi:10.1038/s41594-021-00576-8

Rennie ML, Arkinson C, Chaugule VK, Walden H. 2022. Cryo-EM reveals a mechanism of USP1 inhibition through a cryptic binding site. *Sci Adv* **8**:eabq6353. doi:10.1126/sciadv.abq6353

Rennie ML, Lemonidis K, Arkinson C, Chaugule VK, Clarke M, Streetley J, Spagnolo L, Walden H. 2020. Differential functions of FANCI and FANCD2 ubiquitination stabilize ID2 complex on DNA. *EMBO Rep* **21**:e50133. doi:10.15252/embr.202050133

Rosenthal PB, Henderson R. 2003. Optimal determination of particle orientation, absolute hand, and contrast loss in single-particle electron cryomicroscopy. *J Mol Biol* **333**:721–745. doi:10.1016/j.jmb.2003.07.013

Russo CJ, Passmore LA. 2016. Ultrastable gold substrates: Properties of a support for high-resolution electron cryomicroscopy of biological specimens. *J Struct Biol* **193**:33–44. doi:10.1016/j.jsb.2015.11.006

Sato K, Toda K, Ishiai M, Takata M, Kurumizaka H. 2012. DNA robustly stimulates FANCD2 monoubiquitylation in the complex with FANCI. *Nucleic Acids Res* **40**:4553–4561. doi:10.1093/nar/gks053

Shanmugham A, Fish A, Luna-Vargas MPA, Faesen AC, Oualid F El, Sixma TK, Ovaa H. 2010. Nonhydrolyzable ubiquitin-isopeptide isosteres as deubiquitinating enzyme probes. *J Am Chem Soc* **132**:8834–8835. doi:10.1021/ja101803s

Sijacki T, Alcón P, Chen ZA, McLaughlin SH, Shakeel S, Rappsilber J, Passmore LA. 2022. The DNA-damage kinase ATR activates the FANCD2-FANCI clamp by priming it for ubiquitination. *Nat Struct Mol Biol* **29**:881–890. doi:10.1038/s41594-022-00820-9

Sims AE, Spiteri E, Sims RJ, Arita AG, Lach FP, Landers T, Wurm M, Freund M, Neveling K, Hanenberg H, Auerbach AD, Huang TT. 2007. FANCI is a second monoubiquitinated member of the Fanconi anemia pathway. *Nat Struct Mol Biol* **14**:564–567. doi:10.1038/nsmb1252

Sonego M, Pellarin I, Costa A, Vinciguerra GLR, Coan M, Kraut A, D'Andrea S, Dall'Acqua A, Castillo-Tong DC, Califano D, Losito S, Spizzo R, Couté Y, Vecchione A, Belletti B, Schiappacassi M, Baldassarre G. 2019. USP1 links platinum resistance to cancer cell dissemination by regulating Snail stability. *Sci Adv* **5**:eaav3235. doi:10.1126/sciadv.aav3235

Tan W, van Twest S, Leis A, Bythell-Douglas R, Murphy VJ, Sharp M, Parker MW, Crismani WM, Deans AJ. 2020a. Monoubiquitination by the human Fanconi Anemia core complex clamps FANCI:FANCD2 on DNA in filamentous arrays. *Elife* **9**:e54128. doi:10.7554/elife.54128

Tan W, van Twest S, Murphy VJ, Deans AJ. 2020b. ATR-mediated FANCI phosphorylation regulates both ubiquitination and deubiquitination of FANCD2. *Front Cell Dev Biol* **8**:2. doi:10.3389/fcell.2020.00002

Uchański T, Masiulis S, Fischer B, Kalichuk V, López-Sánchez U, Zarkadas E, Weckener M, Sente A, Ward P, Wohlkönig A, Zögg T, Remaut H, Naismith JH, Nury H, Vranken W, Aricescu AR, Pardon E, Steyaert J. 2021. Megabodies expand the nanobody toolkit for protein structure determination by single-particle cryo-EM. *Nat Methods* **18**:60–68. doi:10.1038/s41592-020-01001-6

van Twest S, Murphy VJ, Hodson C, Tan W, Swuec P, O'Rourke JJ, Heierhorst J, Crismani W, Deans AJ. 2017. Mechanism of ubiquitination and deubiquitination in the Fanconi anemia pathway. *Mol Cell* **65**:247–259. doi:10.1016/j.molcel.2016.11.005

Wang R, Wang S, Dhar A, Peralta C, Pavletich NP. 2020. DNA clamp function of the monoubiquitinated Fanconi anaemia ID complex. *Nature* **580**:278–282. doi:10.1038/s41586-020-2110-6

Wang S, Wang R, Peralta C, Yaseen A, Pavletich NP. 2021. Structure of the FA core ubiquitin ligase closing the ID clamp on DNA. *Nat Struct Mol Biol* **28**:300–309. doi:10.1038/s41594-021-00568-8

Williams SA, Maecker HL, French DM, Liu J, Gregg A, Silverstein LB, Cao TC, Carano RAD, Dixit VM. 2011. USP1 deubiquitinates ID proteins to preserve a mesenchymal stem cell program in osteosarcoma. *Cell* **146**:918–930. doi:10.1016/j.cell.2011.07.040

Xu X, Li S, Cui X, Han K, Wang J, Hou X, Cui L, He S, Xiao J, Yang Y. 2019. Inhibition of ubiquitin specific protease 1 sensitizes colorectal cancer cells to DNA-damaging chemotherapeutics. *Front Oncol* **9**:1406. doi:10.3389/fonc.2019.01406

Yang K, Li G, Gong P, Gui W, Yuan L, Zhuang Z. 2016. Chemical protein ubiquitylation with preservation of the native cysteine residues. *ChemBioChem* **17**:995–998. doi:10.1002/cbic.201600042

Yang K, Moldovan GL, Vinciguerra P, Murai J, Takeda S, D'Andrea AD. 2011. Regulation of the Fanconi anemia pathway by a SUMO-like delivery network. *Genes Dev* **25**:1847–1858. doi:10.1101/gad.17020911

Yin L, Krantz B, Russell NS, Deshpande S, Wilkinson KD. 2000. Nonhydrolyzable diubiquitin analogues are inhibitors of ubiquitin conjugation and deconjugation. *Biochemistry* **39**:10001–10010. doi:10.1021/bi0007019

4 Structural Biology of Protein Kinases and Their Regulation

Samrat Mitra, Pranita Ray, Prateeka Borar, and Smarajit Polley

INTRODUCTION

Phosphorylation is one of the most abundant posttranslational modifications found on proteins. Protein kinases transfer the γ-phosphate moiety from ATP onto Ser, Thr, Tyr, His, Asp, Arg, Lys, and Cys residues in the cognate protein substrates. Most of the eukaryotic protein kinases (EPKs) are Ser/Thr/Tyr kinases (STYKs) that phosphorylate the respective namesake residues. A small number of kinases, however, defy this residue-level specificity; they are capable of phosphorylating both Ser/Thr and Tyr residues, making them dual-specificity kinases. Phosphomodification imparts novel physico-chemical properties to the protein substrates, leading to downstream signaling consequences otherwise unprecedented (Figure 4.1). This generalized transformative property of kinases led to their recognition as molecular switches.

Protein kinases populate about 2% of the human genome and code for more than 500 kinases, making the protein kinase superfamily one of the largest gene families in humans (Hanks et al., 1988). It is estimated that they phosphorylate about 30% of the total proteome (Cohen, 2000). Almost all, if not all cellular signaling events involve phosphorylation, which makes kinases important nodes of intervention in many pathologies such as cancer, immune disorders, infectious diseases, and cardiovascular and neurological disorders (Figure 4.1) (Brognard and Hunter, 2011; Kung and Jura, 2019; Lahiry et al., 2010; Torkamani et al., 2009; Zarrin et al., 2021).

All EPKs contain a catalytic domain, typically 250 to 300 amino acids long (Hanks et al., 1988; Manning et al., 2002a; Manning et al., 2002b). The catalytic domain represents a fairly conserved fold broadly separated into a β-strand–rich N-lobe and a helix-rich C-lobe (Knighton et al., 1991). They often contain other noncatalytic regulatory/scaffolding domains beyond the kinase catalytic domain. Kinases are found in every subcellular location, which further suggests their ubiquitous involvement in cellular functions (Figure 4.1). They are known to undergo posttranslational modifications that could play pivotal regulatory roles in their biology. Many EPKs are pleiotropic in nature, but many of them are highly specific in their choice of substrates in a context-dependent manner (Schrofelbauer and Hoffmann, 2011; Ubersax and Ferrell, 2007). It is not rare to find kinases that work in conjunction with scaffolding proteins, thereby owing their biological roles to the respective complexes (Figure 4.1) (Schrofelbauer et al., 2012). A plethora of scientific disciplines and techniques, including structural, biochemical, biophysical, chemical, genetic, chemical genetics, and computational approaches have been used extensively to help us understand the nuances of EPKs (Hunter, 2015, 2022). We will try to provide a glimpse of where we stand today and what to look for in the coming days to gain a detailed understanding of kinase biology.

CLASSIFICATION OF KINASES AND THE CATALYTIC DOMAIN

The catalytic domains of EPKs have to perform three rudimentary jobs: bind and orient phosphate donor ATP/GTP along with divalent cations like Mg^{2+}/Mn^{2+}; bind and orient phosphate acceptor protein/peptide; and transfer the γ-phosphate from ATP/GTP to the acceptor residue of the protein/peptide substrates. As these basic jobs are the same for all such kinases, it is expected that they would show high sequence similarity. This feature was used to classify the kinome into 10 broad groups, 90 families, and 145 subfamilies (Hanks and Hunter, 1995; Hanks et al., 1988; Manning et al., 2002b). These groups are: AGC (protein kinase A, G, and C families), CAMK (calcium- and calmodulin-regulated kinases), CK1 (CK1, TTBK and VRK families) CMGC (CDK, MAPK, GSK3, and CLK families), Other (unique kinases that do not match with other groups), STE (yeast STE7, STE11, and STE20), TK (tyrosine kinase), TKL (tyrosine kinase–like), RGC (receptor guanylate cyclases that generate second messenger cGMP but harbor catalytically inactive kinase domain), and APKs (atypical, harbor protein kinase activity but with little sequence similarity to known EPKs). Sequence comparison also led to the identification of 12 subdomains in the kinase domain (Hanks et al., 1988).

The first high-resolution information on the topology of the core kinase structure came from the structure of PKA in 1991 (Figure 4.2a) (Knighton et al., 1991). This structure was a milestone in our understanding of the kinase structure–function relationship. It provided a 3D description of the kinase fold and elucidated the features that make a kinase active (Figure 4.2b and c). Subsequent structural studies on various different kinases bolstered this view of an active kinase conformation and deciphered the inactive conformations. It is broadly accepted that active conformations of protein kinases encompass a tiny portion in the conformational space, if not a single well-constructed conformation; however, inactive kinases represent a variety of conformations (De Bondt et al., 1993). Nonetheless, the

DOI: 10.1201/9781003326106-5

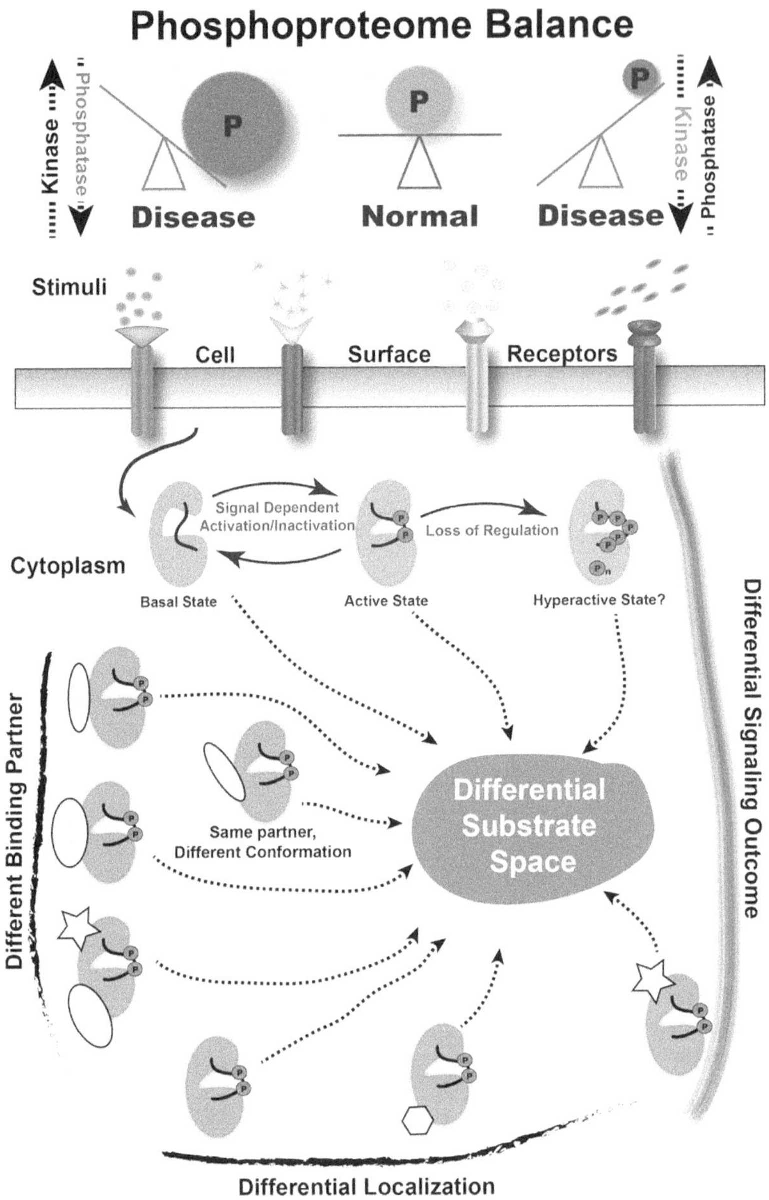

FIGURE 4.1 Schematic depiction of kinase signaling in cells: A balanced phosphoproteome status is maintained in a healthy cell or organism. This balance is maintained by fine-tuned activities of kinases and phosphatases. Dysregulation of this fine-tuning toward either hyper- or hypo-phosphorylation of the proteome may cause havoc. Cell surface receptors transmit extracellular stimuli-induced signals that often end up activating kinases. Kinases may remain in basal states that only get activated as and when required. Kinases may undergo phosphorylation by themselves or by an upstream kinase, leading to their activation or hyper-activation (also to inhibition). The same kinase may phosphorylate different pools of substrates depending upon the phosphorylation status of the kinase, a change in binding partner, or a change in subcellular localization, which may lead to differential signaling outcomes.

well-folded inactive conformations also do not vary a great deal; they often maintain most of the structural features of an active kinase with small exceptions that ultimately lead to their inactivity (Figure 4.2d–f) (Huse and Kuriyan, 2002; Ngo et al., 2007; Nolen et al., 2004; Wang et al., 2014).

As mentioned earlier, kinases are highly homologous owing to the same steps of catalysis they all follow in phosphorylating their respective substrates. Massive sequence analyses were undertaken in the late 1980s, exploiting the computational power and sequence information available at that time before the publication of the crystal structure of PKA in 1991, to delineate subdomains in the primary sequences of protein kinases. Release of the PKA structure, the availability of more primary sequences of kinases, and improvements in computational tools allowed for refining these subdomains and put them in the context of kinase activity in terms of their role in performing the catalytic functions. A brief description of

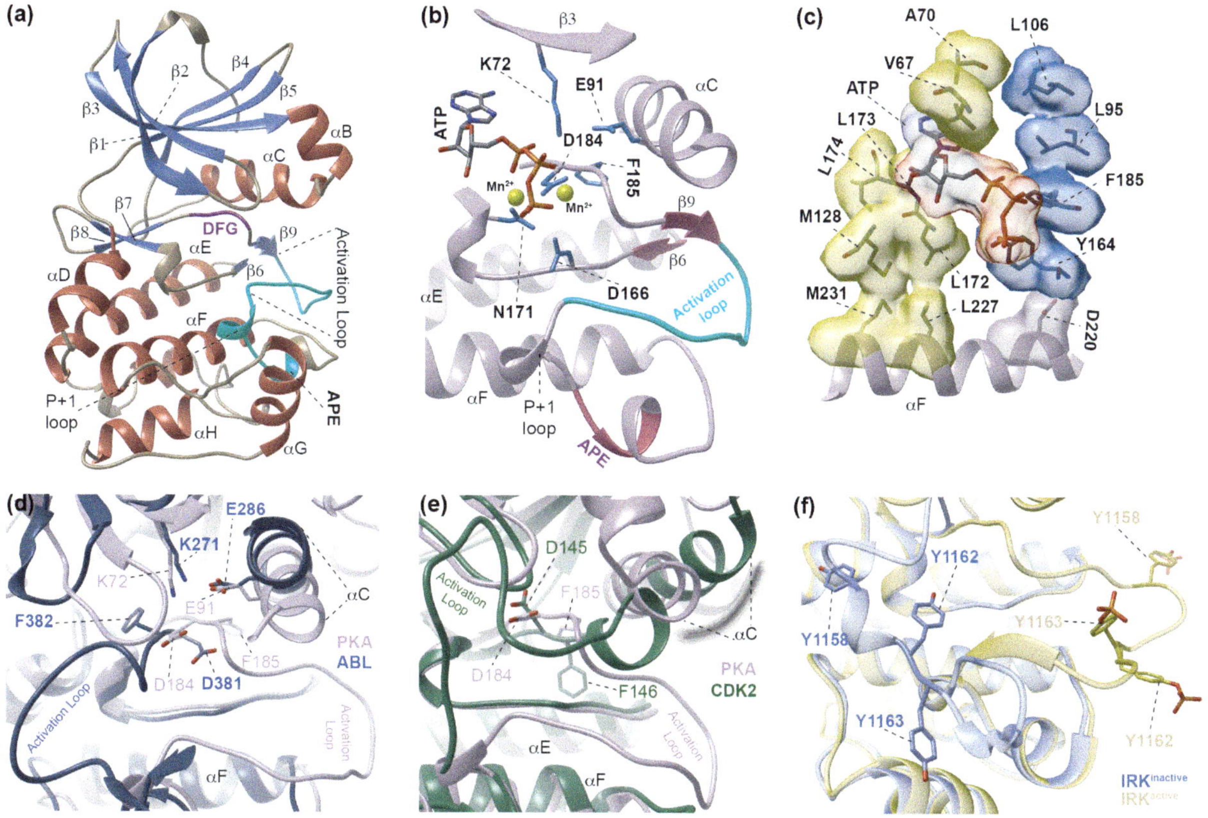

FIGURE 4.2 Depiction of the structural elements in a kinase domain. Molecular graphics were created using UCSF Chimera (Pettersen et al., 2004). (a) The active kinase fold composed of β-strand–rich N-lobe and α-helix–rich C-lobe. β-strands are shown in blue, whereas α-helices are shown in hot pink. Position of the DFG-loop is shown in purple, activation segment is shown in cyan. Position of the APE-loop is also labeled (PDB ID 1ATP). (b) Conserved residues and their positions along with ATP and Mn^{2+} are shown. K72, D184, N171 stabilize Mn^{2+}-ATP. D166 is the catalytic base. D184 and F185 are in "DFG in" conformation (PDB ID 1ATP). (c) Hydrophobic spine formation is shown that is completed by the presence of ATP. The R-spine is shown in blue while the C-spine is shown in yellow, D220 on the αF-helix is shown in mauve (PDB ID 1ATP). (d) Comparison of "DFG in" (active PKA, mauve) and "DFG out" (inactive ABL, blue, PDB ID 1IEP) (Nagar et al., 2002) is shown. Notice the change in position of the activation loops in the two kinases. (e) Comparison of active conformation of PKA (mauve) and inactive conformation of CDK2 (green, PDB ID 1HCK) due to movement of the αC-helices. Both of them harbor "DFG in" conformation, but the activation loop positions in the two kinases are very different. (f) Activation loop phosphorylation in IRK makes it active (yellow, PDB ID 1IR3) (Hubbard, 1997) by relieving the autoinhibitory conformation of the activation loop in the unphosphorylated state (blue, PDB ID 1IRK) (Hubbard et al., 1994). Three tyrosine residues that undergo phosphorylation are shown in both situations depicting massive rearrangement of the activation loop.

these subdomains and their relevant features appears in Table 4.1. The amino acid residue numbers used in the table and in the text here corresponds to the primary sequence of PKA (Figure 4.2b).

The crystal structure of PKA revealed the two-lobed arrangement of the kinase domains: the N-lobe rich in β-strands and the C-lobe consisting primarily of α-helices (Figure 4.2a) (Nolen et al., 2004; Taylor et al., 1993, 1992, 2021). Both these lobes, however, have their structural elements, i.e., β-strands and α-helices, connected by loops of different lengths. While these structural elements are well conserved, they often contain insertions of varying lengths and compositions in-between that are specific for different individual kinases (Sicheri et al., 1997; Zhang et al., 1994). The 12 subdomains of a kinase fold are described here in brief with special emphasis on their features related to kinase activity/active fold.

The first four subdomains starting from the N-terminal of the kinase fold, I through IV, are located in the small/N-lobe. The rest of the subdomains (VIA–XI) constitute the large/C-lobe, while subdomain V connects the two lobes. Subdomain I is the first one from the N-terminal of the kinase domain. It contains the glycine-rich loop, harboring the consensus **Gly-x-Gly-x-x-Gly-x-Val** motif encompassing the strands β1 and β2 that folds into a β-turn-β structure. This structure forms a lid or flap covering the ATP to anchor the α- and β-phosphates of the nucleotide. Subdomain II consists of the β3 strand followed by a small helix in PKA. This particular helix is not conserved in many other kinases; it is found in some like PKA but not

TABLE 4.1
Kinase subdomains and their features *(residue numbers correspond to PKA sequence)*

Subdomains	Features	Location and Function
I	Contains **G-x-G-x-x-G-x-V** Folds into β-turn-β structure Encompass β1 and β2 Covers and anchors the α- & β-phosphates of ATP	• Located in the N-lobe • Predominantly anti-parallel β-sheet • Binding and orientation of ATP/GTP
II	Spans β3, followed by a small helix (not conserved, not found in many cases) Contains the invariant **K72** within β3 Mutation of this K renders kinases largely inactive This K anchors ATP by interacting with α- & β-phosphates of ATP	
III	Spans the αC-helix **E91** (well conserved) is located in this helix Stabilize interaction between K72 and ATP	
IV	Spans the β4-strand, no invariant residue	
V	Spans β5-strand and αD-helix β5-strand is highly hydrophobic Gatekeeper residue is located in the β5-strand	Connects the N- and C-lobe
VIA	Spans the large hydrophobic αE-helix No direct contact with ATP or peptide structure Mainly provides structural support	• Located in the C-lobe • Predominantly α-helical • Binding of the substrate peptide and aiding phosphotransfer
VIB	Spans strands β6 & β7, and the loop connecting the two Consensus motif **HRDLKP-x-N** is present D166 and N171 are invariant D166 is the catalytic base N171 stabilizes the catalytic loop by hydrogen bonding with backbone carbonyl of D166 N171 chelates the secondary Mg^{2+} that holds the α- & γ-phosphates of ATP K168 neutralizes the negative charge on the γ-phosphate	
VII	Spans strands β8 & β9, and the loop connecting the two Contains the highly conserved DFG motif D184 chelates the primary Mg^{2+} that holds the β- & γ-phosphates of ATP "DFG in" and "DFG out" conformations are noticed in active and inactive kinase structures, respectively	
VIII	Spans a long loop post β9-strand to αF-helix Plays major role in substrate peptide recognition Contains highly conserved APE motif Well-conserved E208 stabilizes the large lobe by forming an ion pair with R280 in subdomain XI Contains the activation loop and the P+1 loop	
IX	Contains the αF-helix Well-conserved D220 stabilizes the catalytic loop by forming hydrogen bonds with the backbone amides of H & R of the HRD motif in subdomain VIB	
X	Contains the αG-helix and variable insertion up to αH-helix of subdomain XI Poor sequence conservation	
XI	Spans to the C-terminal end of the kinase domain from subdomain X Contains αH-helix and αI-helix	

necessarily in the vast majority of them. Subdomain II harbors one of the most significant residues, a Lys residue (K72 in PKA) that anchors ATP by interacting with the α- and β-phosphates (Figure 4.2b). The effect of this Lys on kinase activity and function has been widely studied. In general, kinases become inactive when the Lys is mutated, although in some cases they have been shown to retain some activity (~5% or less as compared to their wild-type counterpart).

All these studies concentrated on the effect of this residue on the activity of the kinase without much attention given to its effect on the specificity of the kinase, if any. In other words, the effort to connect the phosphotransfer efficiency and desired specificity of the kinase has been inadequate. Apart from interaction with ATP, K72 makes another important contact within the kinase polypeptide—it forms an important salt bridge (Figure 4.2b) with another well-conserved residue, E91, that is located in the αC-helix in subdomain III, the relative positioning of which with respect to the catalytic cleft modulates the active or inactive status of a kinase (Figure 4.2e). This salt bridge between K72 and E91 stabilizes the interaction between K72 and ATP and provides the correct architectural framework for efficient kinase activity. Movement of the αC-helix has also been correlated with the inactive and active states of kinases. The αC-helix must be positioned inward so that E91 is poised to communicate with K72 in a manner conducive to proper orientation of ATP, leading to efficient catalysis. Meddling with this critical interaction between these two residues in various ways is detrimental to kinase activity; movement of the αC-helix away from the appropriate spatial position would cause the loss of this pivotal interaction and thus inactivation of the kinase (Huse and Kuriyan, 2002; Kovacs et al., 2015). The following region, subdomain IV, spans mostly the strand β4 that does not contain any conserved or invariant residues.

The next subdomain, V, physically connects the N-lobe with the C-lobe and is composed of the β5-strand and αD-helix, wherein the β5-strand is composed of highly hydrophobic residues. One of the most important residues in this subdomain that is directly connected to the regulation of kinase function is a large hydrophobic residue called the gatekeeper that restricts the entry of other small molecule ligands, except cognate phosphate-donor ATP, into the active site cleft. Plant receptor-like kinases (RLKs) contain tyrosine at this gatekeeper position, phosphorylation of which was shown to cross-talk with the activation loop and regulate its phosphorylation status, which in turn dictates the specificity of the kinase (Bhattacharya et al., 2019).

Subdomain VIA contains a long hydrophobic helix, denoted the αE-helix, that marks the beginning of the C-lobe. This helix mostly provides structural support to the overall kinase fold and does not make any direct contact with ATP or the substrate peptide. The next two subdomains, VIB and VII, are very important because they are home to a number of well-conserved/invariant residues related to catalysis. Subdomain VIB consists of the strands β6 and β7 and the loop connecting those two strands. It harbors the consensus motif **His-Arg-Asp-Leu-Lys-Pro-x-Asn (HRDLKP-x-N)**, where D166 and N171 are invariant among kinases. D166 is considered the catalytic base that accepts the proton from the substrate hydroxyl group and orchestrates the nucleophilic attack. The invariant N171 stabilizes the activation loop by making hydrogen bonds with the backbone carbonyl of D166 in the same motif. Furthermore, N171 chelates the secondary Mg^{2+} that coordinates with the α- and γ-phosphates of ATP. In addition, K168 neutralizes the negative charges on the γ-phosphate of ATP.

Subdomain VII, which spans the strands β8 and β9 and the loop connecting them, houses another highly conserved motif, **Asp-Phe-Gly** or **DFG**. The conformation and relative orientation of this motif have been found to be vital to the activity of a kinase (Figure 4.2a and b). Phe in the DFG motif has been found to have been replaced with other residues in some kinases, e.g., Leu in IKK1 and IKK2, which makes the motif DLG without altering its significance in the kinase activity. Aspartate in DFG (D184 in PKA) chelates the primary Mg^{2+} that secures the β- and γ-phosphates of ATP poised for donating the terminal phosphate. The conformation and relative orientation of the DFG motif have been widely implicated in the active state of a kinase. Careful and rigorous analysis of this motif in the available kinase crystal structures has revealed that this DFG could exist in two broadly classified conformational states: "DFG in" and "DFG out" (Figure 4.2d). It is prudent to call them a collection of similar conformational states rather than distinct single conformations.

Kinase structural biology has come a long way since the appearance of the first X-ray crystal structure of PKA in 1991. Several hundred high-resolution structures of kinases have been deposited since then. This collection of structures captures kinases in many different forms, states, stages of catalysis, bound or free of substrates or inhibitors or interaction partners. They offer detailed structural frameworks of kinase activation and catalysis, and possible means of shutting them off. The relative orientation of DFG is a major determinant of the active status of a kinase. "DFG in" conformation assumes an ensemble of conformations where Mg^{2+}-binding Asp (D184 in PKA) in this motif is directed into the active site cleft to coordinate the primary-Mg^{2+} poised for catalysis (Figure 4.2b). This conformation is considered the canonical active conformation of kinases, even though backbone torsion angles of this motif fall in a more restricted region of the Ramachandran plot corresponding to the left-handed helices (Huse and Kuriyan, 2002). The inactive kinase conformations, on the other hand, have been shown to harbor this motif in "DFG out" state, where the same Asp is steered away from the active site. However, this conformational flip of DFG entails a change of the backbone torsion angles from a more restricted to a more allowed region of the Ramachandran plot corresponding to the β-strands (Huse and Kuriyan, 2002; Taylor and Kornev, 2011). It is thus reasonable to conclude that flipping the DFG motif controls switching between active and inactive conformations of a kinase. It's also worth mentioning that the identity of the DFG+1 residue dictates the specificity of the residue to be phosphorylated by a kinase (Chen et al., 2014).

The next subdomain, VIII, spans a long loop post the β9-strand and up to the αF-helix and is a major contributor to substrate peptide recognition. It contains the activation loop and P+1 loop, along with the highly conserved **Ala-Pro-Glu, APE** motif (Figure 4.2b). The well-conserved E208 of the APE motif forms an ion pair with R280 in subdomain XI that provides stability to maintain the structural integrity of the large/C-lobe. The αF-helix is part of the next subdomain, IX, which harbors another well-conserved Asp (D220 in PKA). This Asp helps stabilize the catalytic loop by forming H-bonds with the backbone amides of His (Y164) and Arg (R165) of the HRD motif in subdomain VIB. Subdomain X contains the αG-helix followed by an insertion of variable length in individual kinases that span to the αH-helix of subdomain XI. Subdomain X shows poor sequence conservation. The C-terminal boundary of the kinase domain is ascribed to subdomain XI, consisting of helices αH and αI.

HYDROPHOBIC SPINES CONNECT THE N- AND C-LOBES

Sequence analysis and correlation with high-resolution structures have revealed how well-conserved, mostly hydrophilic residues contribute to kinase activity. It was still unclear what role hydrophobic residues played in ascribing the active state to a kinase and how N- and C-lobes communicated with each other to support productive catalysis. Rigorous comparative analysis of high-resolution structures of kinases in both active and inactive conformations revealed that the core of the kinase domain is composed of three major elements: a long hydrophobic helix, αF, that traverses the C-lobe, and two spines, each made up of hydrophobic residues from both small and large lobes (Figure 4.2c) (Hu et al., 2015; Kornev et al., 2006; Taylor and Kornev, 2011). These spine-forming residues are discontinuous in the primary sequence of these kinases, and sequence alignment alone would not be able to place them in spatial motifs that are formed only in the 3D space. Comparison of active and inactive conformations further revealed that this three-component hydrophobic core is stable yet dynamic. Every active EPK structure represented a spatially contiguous hydrophobic spine made up of two residues each from the N- (L95, L106) and C-lobes (Y164, F185) (Figure 4.2c). In the inactive kinase structures, however, this spine was broken, which resulted in its being called the regulatory or R-spine. These four residues of the R-spine are discontinuously spaced: L95 is located in the catalytically crucial αC-helix, L106 in the strand β4, Y164 in the H(Y)RD motif in the β6, and F185 in the DFG motif in the loop connecting β8 and β9.

The regulatory spine is primarily involved in substrate binding, assembly of which depends upon the conformation of the activation loop. In other words, the continuity of the R-spine depends upon the active conformation of the activation loop, thus making it a signature feature of the kinase active state. Further analysis revealed the presence of another such spine that runs parallel to the R-spine, consisting of residues from the both the N- and C-lobes. Interestingly, this second spine is spatially incomplete (discontinuous) irrespective of the active status of the kinase; it gets completed only upon incorporation of the adenine ring of ATP at a certain position. This spine is referred to as the catalytic or C-spine and is composed of eight hydrophobic residues from both lobes: V57(β2), part of the ATP-binding loop, A70 (β3, same strand that harbors K72), and M128 (αD) from the N-lobe, and L172 (β7), L173 (β7), I174 (β7), L227 (αF), and M231 (αF) from the C-lobe. Clearly, the catalytic spine is well grounded on the helix αF (Figure 4.2c). The regulatory spine, on the other hand, is anchored to the αF-helix by an interaction between Y164 (of the catalytic loop H(Y)RD motif) and a conserved Asp (D220) in that helix. In summary, the hydrophobic αF-helix functions as the anchoring stem that connects the two lobes, and the juxtaposition of these spine residues in the presence of ATP is a major determinant of productive catalysis (Figure 4.2c) (Taylor and Kornev, 2011).

ACTIVATION MECHANISMS

Kinases often switch between inactive and active states (Figure 4.2d-f). In the previous sections we have described the sequence and conformational requirements of a kinase to be considered active. In biochemical terms, in most cases it denotes phosphorylation of one or more residues in the activation loop region (Zhang et al., 1995). In such cases, the activation loop collapses into the active site when unphosphorylated, which deters binding of both the phosphate donor (ATP) and acceptor (substrate peptide) amenable to fruitful catalysis (Hubbard et al., 1994). Phosphorylation of the activation loop at desired site(s) in the same kinase induces conformational changes in the loop that releases it from the active site and makes it move away from the catalytic center to facilitate binding of ATP and the substrate peptide, leading to productive phosphotransfer (Hubbard, 1997). The insulin receptor tyrosine kinase domain represents such an intriguing scenario (Figure 4.2f). Its kinase domain remains autoinhibited by a sequence originating from its activation segment. The activation segment contains three Tyr residues, one of which (Y1162) hinders ATP binding, which percolates into a conformation of the activation segment that disallows substrate peptide binding. *Trans*-autophosphorylation of those three tyrosine residues disrupts autoinhibitory interactions in the ATP-binding cleft that in turn stabilize the activation segment in a conformation conducive to proper placement of the substrate peptide (Huse and Kuriyan, 2002).

It is intriguing that enzymes that catalyze phosphorylation are themselves phosphoproteins, and specific phosphorylation can be considered critical for their activation. The position and number of phosphorylations in the activation loop do not follow any consensus and are kinase specific. In case of multisite phosphorylation in the activation loop, one of the residues assumes the central position, and its phosphate forms an ion pair with a conserved Arg (R165 in PKA) in the HRD motif in the catalytic loop. In the case of PKA, despite having two Thr residues (T197 and T201) in the activation loop, only one, T197, gets phosphorylated and forms an ion pair with R165. It has been proposed that this ion pair may allosterically influence proper positioning of the DFG motif required for catalysis. However, phosphorylation of the central position does not necessarily restrict its interaction exclusively with the catalytic loop; it may extend well beyond into other regions in the N-lobe. For example, in case of ERK2 (MAPK), there are two residues in its activation loop where it undergoes phosphorylation. Of these two, the Thr183 in its phosphorylated state (pT183) assumes the central position and forms contacts with the αC-helix. Such interactions lead to closing in of the N- and C-lobes, promoting activation. In addition, phosphorylation of the Thr183 induces homodimerization of ERK2 and thus nuclear localization by propagating conformational changes in the MAPK-specific C-terminal extension segment of the kinase.

Activation loop phosphorylation may occur via a variety of mechanisms. It may be autophosphorylation or phosphorylation by an upstream kinase, or it may achieve autophosphorylation by binding to a cognate binding partner (Lochhead, 2009). It is easier to perceive phosphorylation in *trans* by an upstream kinase when the substrate-kinase is in inactive conformation and the activation loop extends into the active site of the activating kinase that is already in the active form. It is also understandable that a similar, but not identical, mechanism would be at play when the kinase undergoes *trans*-autophosphorylation wherein the two kinase domains are brought into proximity by homo-oligomerization to exchange their activation segments with each other. This mechanism avoids a paradoxical situation if *trans*-autophosphorylation had followed the enzyme:substrate relationship, as in the case of involvement of a *trans*-acting upstream kinase. If the inactive molecule was presented as the cognate substrate to its active counterpart, the amino acid sequence in the activation segment of the inactive kinase must conform to the requirement of the substrate consensus sequence of the kinase. It may lead to a situation wherein the active kinase may face challenges in maintaining its phosphorylation specificity (Oliver et al., 2007).

In the activation segment exchange model, however, this does not pose any such challenge to the kinase as long as the exchange happened successfully, as is found in the crystal structures of many kinases like Chk2 (PDB ID 2CN5) (Pike et al., 2008), STK10 (PDB ID 2J7T), DAPK3 (PDB ID 2J90), SLK (PDB ID 2J51), and IRE1α (PDB ID 3P23) (Ali et al., 2011). In this model, interactions of the exchanged segment with the body of its counterpart stabilizes the active conformation poised for catalysis even in the absence of phosphorylation. Local dynamicity of the target segment in one kinase creates the appropriate spatial arrangement for phosphate transfer from ATP of the opposing kinase domain. There are examples of mixed modes of activation as well. The mechanism that helps achieve high local concentration supports rapid amplification of kinase activity. IKK2 requires phosphorylation of two Ser residues at positions 177 and 181 to become active; the unphosphorylated state is deemed inactive or poorly active. It was shown that recombinantly purified IKK2 can achieve phosphorylation at these two sites through *trans*-autophosphorylation via concentration-dependent oligomerization. Later, it was shown that an upstream kinase TAK1 primes IKK2 by phosphorylating S177, and this singly phosphorylated IKK2 then undergoes autophosphorylation at S181 to generate fully active doubly phosphorylated IKK2 (Polley et al., 2013; Zhang et al., 2014).

A distinct mechanism of activation is witnessed in the case of Cdk2 that exists by itself in an inactive conformation (PDB ID 1HCK) (Schulze-Gahmen et al., 1996) and is activated in a stepwise manner. First, it binds to a partner cyclin that induces conformational changes in Cdk2 to achieve a partially activated state by virtue of attaining some of the structural hallmarks of an active kinase with concomitant exposure of a Thr residue (T160) to the solvent (PDB ID 1FIN) (Jeffrey et al., 1995). The solvent-exposed T160 is then phosphorylated by a group of kinases called Cdk-activating kinases that render Cdk2 fully active (PDB ID 1JST) (Russo et al., 1996). A different mechanism is observed in the case of Src kinase. Docking of SH2 and SH3 domains at the back of the kinase domain renders the kinase inactive. Removal of SH2/SH3 docking either by binding of these domains to another recognition partner or by removal of the SH2-docking phosphorylation in the kinase itself releases the kinase from this constrained, inhibited state, which can achieve the active conformation (PDB ID 1FMK and 1Y57) (Cowan-Jacob et al., 2005; Xu et al., 1997).

An unusual activation mechanism is adopted by three of the four EGFR kinases (EGFR or ErbB1, HER2 or ErbB2, HER4 or ErbB4; HER3 or ErbB3 is a pseudokinase) of the receptor tyrosine kinase family. They do require phosphorylation of the activation segment for full activity, and their juxtamembrane segment plays an activating role for the kinase domain. The crystal structures of EGFR (PDB ID 3GOP & 1M14) (Red Brewer et al., 2009; Stamos et al., 2002), HER2 (3PP0) (Aertgeerts et al., 2011) and HER4 (3BCE) (Qiu et al., 2008) reveal that these kinases form an asymmetric dimer in which one kinase represents the active state and acts as the activator kinase. The inactive counterpart (receiver kinase) undergoes significant conformational changes upon its association with the activator kinase that convert the inactive receiver kinase into an active kinase.

There are curious instances of autophosphorylation in kinases as well. For example, DYRKs and GSK3β undergo autophosphorylation at Tyr that transform them into Ser/Thr kinases. Nascent DYRKs and GSK3β behave as Tyr kinases, and cotranslational autophosphorylation at Tyr occur intramolecularly (in *cis*) when the kinase is yet to mature. Once these kinases are autophosphorylated at Tyr in the activation loop and are matured, they specifically phosphorylate Ser/Thr residues intermolecularly, i.e., in *trans* (Kinstrie et al., 2010; Lochhead et al., 2006; Lochhead et al., 2005). Furthermore, the two states of the kinase before and after Tyr autophosphorylation display differential drug sensitivity.

INHIBITORS: STRATEGIES AND EXAMPLES

There are many different kinds of inhibitors available for kinases (Knox et al., 2021; Martin et al., 2017; Noble et al., 2004), and many of them have been approved (Anderson et al., 2023; Wang et al., 2014) to be used in the clinic (Table 4.2). One of the most challenging problems in deriving ATP-binding cleft-directed kinase inhibitors is the fact that the inhibitor has to overcome competition from millimolar-level concentrations of cellular ATP.

Type I inhibitors target the well-ordered "DFG in" state of the kinase. For example, Dastinib, used in CML and ALL patients, inhibits ABL, EphB4, and MYT1. Type II inhibitors, on the other hand, bind to and stabilize inactive

TABLE 4.2
FDA-approved kinase inhibitor drugs

Commercial Name	Target Kinase	Inhibitor	Mode of Inhibition	Disease
GLEEVEC	c-Src, Bcr-abl, c-KIT	IMATINIB		CML and ALL that are Ph⁺, GIST, HES, CEL
NEXAVAR	BRAF Kinase domain b3aC loop deletion mutant, VEGFR, PDGFR	SORAFENIB		Liver carcinoma, advanced renal carcinoma and differentiated thyroid carcinoma
VANFLYTA	FLT3	QUIZARTINIB (AC220)		AML
ICLUSIG	FGFR1, ROR1	PONATINIB		CML and Ph⁺ ALL
SUTENT	ITK	SUNITINIB		RCC and imatinib-resistant gastrointestinal stromal tumor
TASIGNA	Bcr-Abl	NILOTINIB		CML
SCEMBLIX	Bcr-Abl allosteric	ASCIMINIB		Ph⁺ myeloid leukemia
BOSULIF	**EGFR Bcr-Abl**	**BOSUTINIB**	**DFG OUT**	**CML**
BOSULIF	**MYT1, HER3 pseudokinase**	**BOSUTINIB**	**DFG IN**	**CML**
	Src family kinase HCK, STK17B	QUERCETIN		Breast cancer
TARCEVA	EGFR	ERLOTINIB		NSCLC and pancreatic cancer
	MYT1	SARACATINIB (AZD0530)		Idiopathic pulmonary fibrosis
ROZLYTREK	ALK	ENTRECTINIB		ROS1-positive NSCLC and NTRK fusion-positive solid tumors
TYKERB	EGFR, Her2/neu	LAPATINIB		Breast cancer
IRESSA	EGFR	GEFITINIB		NSCLC
CAPRELSA	VEGFR, EGFR RET-kinase	VANDETANIB		Symptomatic or progressive medullary thyroid cancer
SPRYCEL	Abl, EphB4	DASATINIB		CML and ALL
XALKORI	ALK, c-MET	CRIZOTINIB		ALK or ROS1-positive NSCLC, as well as ALK-positive ALCL and inflammatory myofibroblastic tumor
Xeljanz, Jaquinus, Tofacinix	JAK family	Tofacitinib		Rheumatoid arthritis, psoriatic arthritis, ankylosing spondylitis, polyarticular course juvenile idiopathic arthritis, and ulcerative colitis
Imbruvica	BTK	Ibrutinib	Irreversible covalent modification	MCL, CLL
Calquence	BTK	Acalabrutinib		MCL, non-Hodgkin lymphoma, CLL, SLL
Velexbru	BTK	Tirabrutinib	Irreversible covalent modification	CLL, Rheumatoid arthritis, autoimmune disorders
Jaypirca	BTK	Pirtobrutinib		MCL
Brukinsa	BTK	Zanubrutinib		CLL, MCL, MZL
Cotellic	MEK	Cobimetinib	Allosteric	Melanoma

CML: chronic myelogenous leukemia; ALL: acute lymphocytic/lymphoblastic leukemia; Ph⁺: Philadelphia chromosome (+)ve; GIST: gastrointestinal stromal tumors; HES: hypereosinophilic syndrome; CEL: chronic eosinophilic leukemia; AML: acute myeloid leukemia; NSCLC: non-small cell lung cancer; RCC: renal cell carcinoma; ALCL: anaplastic large cell lymphoma; MCL: mantle cell lymphoma; MZL: marginal zone lymphoma.

conformations of kinases. There are many examples of this class of inhibitors, like Gleevec (imatinib, the first kinase inhibitor to be used in the clinic), which was found to inhibit Bcr-Abl and was used in Philadelphia chromosome positive CML, ALL, and some other types of cancers. The development of imatinib is also quite intriguing. The lead compound was originally identified as a PKC (Ser/Thr kinase) inhibitor in high-throughput screening. The lead compound was then systematically developed into a tyrosine kinase–specific inhibitor with higher conformational specificity and water solubility. Even though type II inhibitors are more specific than their type I counterparts, they also show nonspecific kinase inhibition, and they have failed to inhibit mutant kinases with larger gatekeeper residues. Type III inhibitors are supposed to overcome such problems. One such example is a highly selective inhibitor of MEK kinase, PD184532, that binds to a hydrophobic cleft close to the ATP-binding pocket without affecting ATP binding yet achieving remarkable inhibition of the kinase. A major obstacle in designing an inhibitor other than the ATP-binding pocket in kinases is the lack of proper clefts or grooves on a relatively flat kinase surface. Traditional small molecule drug design approaches primarily rely upon the availability of such pockets. It is known that kinases adopt diverse allosteric mechanisms of regulation. A type IV inhibitor is expected to bind to regions outside the ATP-binding cleft and the phosphoacceptor site to be considered a truly allosteric inhibitor. Several such molecules are known that exploit a variety of ways adopted by the respective kinases to elicit their biological activities. Notable examples are reported for ABL kinase, PDK1, CDK2, EGFR, etc. These inhibitors are, however, quite inefficient, likely due to their destination being a flat protein surface. Nonetheless, they provide avenues to inhibit a kinase in a truly allosteric manner. Another means to achieve high selectivity as well as potency is to develop bivalent or bisubstrate analogues. Both of these inhibitors, called type V inhibitors, thrive on the ATP-binding pockets and either the substrate-binding pocket (bisubstrate) or any other site on the kinase besides the substrate-binding region (bivalent).

Though yet to be fully realized, we believe it will be more advantageous to search for novel, activation state–specific and context-dependent interaction surfaces on kinases, either with themselves or with their cognate binding partners away from the active sites, to engineer designer drugs against kinases. These interaction surfaces can be identified using techniques like cryo-EM or cryo-ET because they present the structures of biologically relevant kinases without the need for highly diffracting single crystals.

CRYO-EM AND CRYO-ET IN DECIPHERING KINASE REGULATION

Single-particle cryo-electron microscopy (cryo-EM) analysis and cryo-electron tomography (cryo-ET) are the methods of choice to understand macromolecular structure–function relationships in a more real-life manner than in other structural methods. An understanding of the structural aspects of the canonical kinase fold, both active and inactive, primarily came from X-ray crystallography, which needs well-diffracting single crystals. Multidomain kinases and kinases that require scaffolding or regulatory partners are scarce and often refractory to crystallization. Cryo-EM techniques can solve this problem to a great extent (Cheng et al., 2016; Chua et al., 2022; Nogales and Scheres, 2015). Furthermore, it can enable us to capture these kinase complexes *in situ* (Watanabe et al., 2020).

For example, more than 50% of human kinases are known to be client proteins of Hsp90, and many of them require Cdc37 for maturation. A structural understanding of this dependence on Hsp90 and Cdc37 remained elusive until recently, owing to the dynamic nature of Hsp90–client complexes and due to challenges in reconstituting such complexes. Extraordinary advancements in detector technologies, the advent of highly efficient computers, and improved data processing algorithms allow the community to address many structural problems that would be intractable otherwise. In 2016, the cryo-EM structure of the asymmetric Hsp90:Cdc37:Cdk4 complex was published to provide the first impression of how the Hsp90:Cdc37 duo enables maturation of client kinases. Many flexible loops in Hsp90 were shown to be stabilized due to kinase binding. This structure also revealed that Cdc37 destabilizes the kinase domain and Hsp90 traps the kinase in an open conformation, thereby protecting the kinase in a partially unfolded state (Verba and Agard, 2017; Verba et al., 2016). Subsequently, the cryo-EM structure of Hsp90:Cdc37:CRaf:PP5 was determined, which provided the structural basis for understanding Hsp90's role in regulating protein homeostasis by activating PP5 followed by scaffolding it in a manner poised to dephosphorylate the Craf kinase (Jaime-Garza et al., 2023).

The human epidermal growth factor receptor (HER or ErbB) family of RTKs, consisting of EGFR or HER1, HER2, HER3, and HER4, are heavily implicated in many cancers, including breast, lung, and gastroesophageal. EGFR and HER2 are more potent oncoproteins than HER3 and HER4. These receptors undergo ligand-induced specific homo- and hetero-dimerization to initiate a downstream phospho-signaling cascade that promotes cell proliferation, growth, and differentiation. For example, HER2 does not have any known ligand, nor can it undergo ligand-induced homodimerization. Nevertheless, it forms clinically relevant heterodimers with EGFR and HER3 (pseudokinase). Cryo-EM was employed to determine the structure of pro-oncogenic HER2:HER3 heterodimer in complex with growth factor neuregulin-1β (HER2:HER3:NRG1β) (Diwanji et al., 2021). In this study, another tripartite complex involving an oncogenic extracellular domain mutant of HER2, S310F (HER2S310F:HER3:NRG1β) was resolved by cryo-EM. Structure of the mentioned tripartite complex along with the HER2-directed therapeutic antibody trastuzumab Fab fragment Herceptin (HER2S310F:HER3:NRG1β:Herceptin). The structures revealed that HER2:HER3 forms a heart-shaped asymmetric where density for the HER3 dimerization arm in domain II is unresolved, indicating that the HER3 dimerization arm might be dispensable for this dimerization. This study also revealed how the HER2S310F mutation stabilized this heterodimer, and the structural basis of trastuzumab binding to the HER2S310F:HER3:NRG1β (Figure 4.3a). Furthermore, this study deciphered the mechanism by which HER2 oncogenic mutant (S310F) exploits the intrinsic dynamicity of the HER2:HER3 interaction interface to give rise to uncontrolled HER2 kinase activity.

Cryo-EM analyses also helped in deciphering the structural basis of pathogenesis caused by the leucine-rich

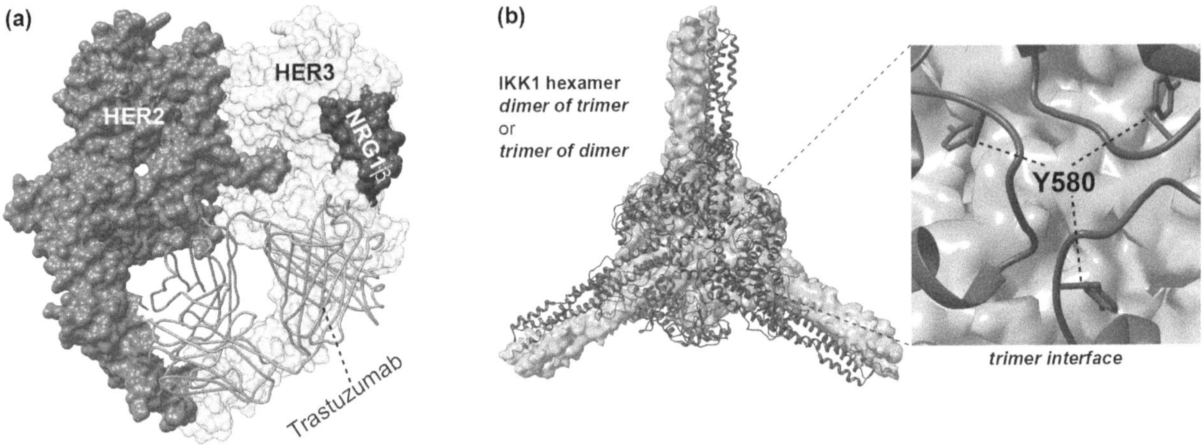

FIGURE 4.3 Cryo-EM analyses of kinases. (a) Cryo-EM–derived structure of the HER2-HER3-NRG1β bound to therapeutic Fab, Trastuzumab is shown. NRG1β-binds asymmetrically to HER3 only (EMDB ID 23918, PDB ID 7MN8). (b) Cryo-EM–aided X-ray structure of human IKK1/α depicting novel hexameric (trimer of dimers) arrangement unprecedented in similar kinases (EMDB ID 8439, PDB ID 5EBZ). Trimeric-interface residue is shown, mutation of which caused systemic immunodeficiency and immature bone development in patients.

repeat kinase 2 (LRRK2), a 286 kDa polypeptide consisting of seven distinct domains. It is a commonly mutated gene associated with both familial and sporadic Parkinson disease (PD), and increased LRRK2 kinase activity has been reported in idiopathic PD brain tissue. Mutations in LRRK2 have also been reported in Crohn disease and Hansen disease. Despite such relevance in disease biology, structure of the full length LRRK2 remained an enigma until 2017, when the first structure of the full-length protein was reported, albeit at a very low resolution (~24 Å). Since then, a series of single-particle cryo-EM and *in situ* cryo-ET structural studies were published that helped us to understand the structural basis of LRRK2 function (Leschziner and Reck-Peterson, 2021). These studies revealed that the catalytic half of LRRK2 is sufficient for filament formation, and conformation of the LRRK2 kinase domain regulates its interaction with the microtubule (Deniston et al., 2020). It was also shown that LRRK2 places its GTPase domain closer to the microtubule surface, making the kinase domain exposed to the cytoplasm (Watanabe et al., 2020). Interestingly, a pathogenic mutant of LRRK2 formed a right-handed double helical oligomer around a left-handed microtubule composed of 11, 12, or 13 protofilaments (Watanabe et al., 2020). Furthermore, while the close conformation favors its oligomerization on the microtubule, its open conformation limits LRRK2 filament formation in cells. These studies indicated that LRRK2 oligomers pose a physical barrier for the microtubule-associated motor proteins, and that kinase inhibitors that block the active closed conformation and stabilize the inactive open conformation might have therapeutically beneficial effects (Deniston et al., 2020; Snead et al., 2022).

One of the applications of cryo-EM that is of particular interest in drug design is its ability to quickly deliver structures bound to small molecules that are potential drug/lead compounds (Passos et al., 2020). Traditional soaking/co-crystallization approaches are far more tedious than this approach. It is possible to capture the drug-bound complexes, binding modes, and salient differences in binding attributes of different molecules that would aid in improving specificity and efficacy. In other words, a path to developing better drugs may be adopted based on real structural data rather than assumptions (Scapin et al., 2018).

Another aspect of cryo-EM that can transform our understanding of kinase signaling complexes (and other signaling complexes) is its ability to identify and determine structures of heterogenous samples (Bai et al., 2015; Fernandez-Leiro and Scheres, 2016; Lyumkis, 2019; Nogales and Scheres, 2015; Scheres, 2016). This heterogeneity may be compositional, stoichiometric, or conformational. The cryo-EM data is collected on the ensemble of particles present in the bulk of the sample. Computational "purification" allows us to classify them differently, thereby separating them from each other. These different classes may then be treated as different complexes, and each of them may provide useful information orthogonal to each other. The most populous species might give rise to the highest-resolution structure of a particular state of a complex, but it need not be the most important state relevant to its functionality. In other words, the most relevant state to decipher the biology of a complex need not be the most abundant species in a heterogenous sample. For example, Inhibitor of -κB kinase (IKK)1/α and IKK2/β are highly homologous kinases that play a pivotal role in the activation of noncanonical/alternative and canonical NF-κB activation crucial for immune signaling in metazoans. Human IKK2 X-ray crystal structures revealed multiple conformational states of its dimer in the crystal and helped reveal its activation mechanism through autophosphorylation via concentration-dependent oligomerization (Polley et al., 2013). Though IKK1 shares more than 50% of its identity with the IKK2 primary sequence, and preliminary bioinformatics analyses indicated similar domain architectures, X-ray structural models of IKK2 (PDB ID 4E3C) (Polley et al., 2013) failed to generate a molecular replacement solution to the IKK1 X-ray diffraction data. Cryo-EM analysis provided a low-resolution map of the IKK1-dimer that then provided the solution to the X-ray diffraction data when used as the search model in molecular replacement (Figure 4.3b) (Polley et al., 2016). It revealed that both the tertiary and quaternary arrangements of IKK1 differed significantly from those of IKK2. IKK1 was shown to arrange itself in a hexameric fashion that could be described as a dimer of trimers or a trimer of dimers. Interestingly, we found about 2% of the particles represented the same hexameric structure in the cryo-EM data even though the dimeric structures (the trimer of which formed the hexamer) were overwhelmingly the most abundant species. The hexameric arrangement unearthed a trimer interface that was unprecedented in such kinases (Figure 4.3b) (Polley et al., 2016). Most importantly, one of the residues (Y580) that were important for forming this interface turned out to be extremely crucial in human immune biology. A patient with combined immunodeficiency and a history of recurrent pneumonias who failed to gain weight and height (ranked below third percentile) at childhood, suffered from recurrent infections by *Staphylococcus aureus* as well as fungal infections. She lacked secondary lymphoid tissues like palpable lymph nodes and the Waldeyer's ring and showed skeletal abnormalities, including incomplete fusions of bones. However, a targeted NGS analysis for a panel of genes could not identify any candidate variant among the 264 genes linked to immunodeficiencies. Surprisingly, whole-exome sequencing identified a point mutation in the CHUK gene that encodes for the IKK1 kinase. This mutation was found to be an autosomal recessive missense mutation altering Y580 to C (Y580C) (Bainter et al., 2021)—the same residue that was identified by our cryo-EM and X-ray crystallography analyses to be crucial for the hexameric arrangement of IKK1 (Figure 4.3b).

Last but not least, time-resolved freezing (vitrification), popularly known as time-resolved cryo-EM, holds promise for pinpointing the kinetic aspects of the multicomponent macromolecular assembly process that is intractable in a classical structural biology approach that deals primarily with steady-state macromolecular assemblages (Chen et al., 2015; Dandey et al., 2020; Kaledhonkar et al., 2018, 2019).

Kinases have remained the center of attraction for biologists, chemists, and chemical biologists for many decades. They are lucrative drug targets in many diseases. Single-particle cryo-EM and *in situ* cryo-ET techniques will continue to unravel the inner workings of kinase-signaling complexes to enable us to more effectively develop designer drugs against the enzymes or assemblies that contain these kinases (Attwood et al., 2021; Ferguson and Gray, 2018; Kontzias et al., 2012; Nussinov and Tsai, 2013; Wang et al., 2014; Zarrin et al., 2021).

ACKNOWLEDGEMENTS

We believe despite the great deal that we know about kinases, it is still the tip of the iceberg. The field of protein kinases is a vast one, and many labs around the globe have enriched the field. We acknowledge every original contributor, and we apologize to those who were not cited here due to lack of space.

Research in SP lab is primarily supported by the DBT Wellcome Trust India Alliance (IA/I/15/1/501852), Bose Institute, and DST-SERB, Government of India (IPA/2020/000414). SM and PB acknowledge Bose Institute, and PR acknowledges the University Grants Commission, Government of India, for fellowship support. Authors thank other members of the SP lab for helpful discussions. Molecular graphics and analyses were performed using UCSF Chimera, developed by the Resource for Biocomputing, Visualization and Informatics at the University of California, San Francisco, supported by the NIH P41-GM103311.

REFERENCES

Aertgeerts, K., Skene, R., Yano, J., Sang, B.C., Zou, H., Snell, G., Jennings, A., Iwamoto, K., Habuka, N., Hirokawa, A., Ishikawa, T., Tanaka, T., Miki, H., Ohta, Y., Sogabe, S., 2011. Structural analysis of the mechanism of inhibition and allosteric activation of the kinase domain of HER2 protein. J Biol Chem 286, 18756–18765.

Ali, M.M., Bagratuni, T., Davenport, E.L., Nowak, P.R., Silva-Santisteban, M.C., Hardcastle, A., McAndrews, C., Rowlands, M.G., Morgan, G.J., Aherne, W., Collins, I., Davies, F.E., Pearl, L.H., 2011. Structure of the Ire1 autophosphorylation complex and implications for the unfolded protein response. EMBO J 30, 894–905.

Anderson, B., Rosston, P., Ong, H.W., Hossain, M.A., Davis-Gilbert, Z.W., Drewry, D.H., 2023. How many kinases are druggable? A review of our current understanding. Biochem J 480, 1331–1363.

Attwood, M.M., Fabbro, D., Sokolov, A.V., Knapp, S., Schioth, H.B., 2021. Trends in kinase drug discovery: targets, indications and inhibitor design. Nat Rev Drug Discov 20, 839–861.

Bai, X.C., McMullan, G., Scheres, S.H., 2015. How cryo-EM is revolutionizing structural biology. Trends Biochem Sci 40, 49–57.

Bainter, W., Lougaris, V., Wallace, J.G., Badran, Y., Hoyos-Bachiloglu, R., Peters, Z., Wilkie, H., Das, M., Janssen, E., Beano, A., Farhat, K.B., Kam, C., Bercich, L., Incardona, P., Villanacci, V., Bondioni, M.P., Meini, A., Baronio, M., Abarzua, P., Parolini, S., Tabellini, G., Maio, S., Schmidt, B., Goldsmith, J.D., Murphy, G., Hollander, G., Plebani, A., Chou, J., Geha, R.S., 2021. Combined immunodeficiency with autoimmunity caused by a homozygous missense mutation in inhibitor of nuclear factor Sci Immunol 6, eabf6723.

Bhattacharya, A., Paul, A., Chakrabarti, D., DasGupta, M., 2019. Gatekeeper-activation loop cross-talk determines distinct autoactivation states of symbiosis receptor kinase. Biochemistry 58, 2419–2431.

Brognard, J., Hunter, T., 2011. Protein kinase signaling networks in cancer. Curr Opin Genet Dev 21, 4–11.

Chen, B., Kaledhonkar, S., Sun, M., Shen, B., Lu, Z., Barnard, D., Lu, T.M., Gonzalez, R.L., Jr., Frank, J., 2015. Structural dynamics of ribosome subunit association studied by mixing-spraying time-resolved cryogenic electron microscopy. Structure 23, 1097–1105.

Chen, C., Ha, B.H., Thevenin, A.F., Lou, H.J., Zhang, R., Yip, K.Y., Peterson, J.R., Gerstein, M., Kim, P.M., Filippakopoulos, P., Knapp, S., Boggon, T.J., Turk, B.E., 2014. Identification of a major determinant for serine-threonine kinase phosphoacceptor specificity. Mol Cell 53, 140–147.

Cheng, A., Tan, Y.Z., Dandey, V.P., Potter, C.S., Carragher, B., 2016. Strategies for automated CryoEM data collection using direct detectors. Methods Enzymol 579, 87–102.

Chua, E.Y.D., Mendez, J.H., Rapp, M., Ilca, S.L., Tan, Y.Z., Maruthi, K., Kuang, H., Zimanyi, C.M., Cheng, A., Eng, E.T., Noble, A.J., Potter, C.S., Carragher, B., 2022. Better, Faster, Cheaper: Recent Advances in Cryo-Electron Microscopy. Annu Rev Biochem 91, 1–32.

Cohen, P., 2000. The regulation of protein function by multisite phosphorylation–a 25 year update. Trends Biochem Sci 25, 596–601.

Cowan-Jacob, S.W., Fendrich, G., Manley, P.W., Jahnke, W., Fabbro, D., Liebetanz, J., Meyer, T., 2005. The crystal structure of a c-Src complex in an active conformation suggests possible steps in c-Src activation. Structure 13, 861–871.

Dandey, V.P., Budell, W.C., Wei, H., Bobe, D., Maruthi, K., Kopylov, M., Eng, E.T., Kahn, P.A., Hinshaw, J.E., Kundu, N., Nimigean, C.M., Fan, C., Sukomon, N., Darst, S.A., Saecker, R.M., Chen, J., Malone, B., Potter, C.S., Carragher, B., 2020. Time-resolved cryo-EM using Spotiton. Nat Methods 17, 897–900.

De Bondt, H.L., Rosenblatt, J., Jancarik, J., Jones, H.D., Morgan, D.O., Kim, S.H., 1993. Crystal structure of cyclin-dependent kinase 2. Nature 363, 595–602.

Deniston, C.K., Salogiannis, J., Mathea, S., Snead, D.M., Lahiri, I., Matyszewski, M., Donosa, O., Watanabe, R., Bohning, J., Shiau, A.K., Knapp, S., Villa, E., Reck-Peterson, S.L., Leschziner, A.E., 2020. Structure of LRRK2 in Parkinson's disease and model for microtubule interaction. Nature 588, 344–349.

Diwanji, D., Trenker, R., Thaker, T.M., Wang, F., Agard, D.A., Verba, K.A., Jura, N., 2021. Structures of the HER2-HER3-NRG1beta complex reveal a dynamic dimer interface. Nature 600, 339–343.

Ferguson, F.M., Gray, N.S., 2018. Kinase inhibitors: the road ahead. Nat Rev Drug Discov 17, 353–377.

Fernandez-Leiro, R., Scheres, S.H., 2016. Unravelling biological macromolecules with cryo-electron microscopy. Nature 537, 339–346.

Hanks, S.K., Hunter, T., 1995. Protein kinases 6. The eukaryotic protein kinase superfamily: kinase (catalytic) domain structure and classification. FASEB J 9, 576–596.

Hanks, S.K., Quinn, A.M., Hunter, T., 1988. The protein kinase family: conserved features and deduced phylogeny of the catalytic domains. Science 241, 42–52.

Hu, J., Ahuja, L.G., Meharena, H.S., Kannan, N., Kornev, A.P., Taylor, S.S., Shaw, A.S., 2015. Kinase regulation by hydrophobic spine assembly in cancer. Mol Cell Biol 35, 264–276.

Hubbard, S.R., 1997. Crystal structure of the activated insulin receptor tyrosine kinase in complex with peptide substrate and ATP analog. EMBO J 16, 5572–5581.

Hubbard, S.R., Wei, L., Ellis, L., Hendrickson, W.A., 1994. Crystal structure of the tyrosine kinase domain of the human insulin receptor. Nature 372, 746–754.

Hunter, T., 2015. Discovering the first tyrosine kinase. Proc Natl Acad Sci USA 112, 7877–7882.

Hunter, T., 2022. A journey from phosphotyrosine to phosphohistidine and beyond. Mol Cell 82, 2190–2200.

Huse, M., Kuriyan, J., 2002. The conformational plasticity of protein kinases. Cell 109, 275–282.

Jaime-Garza, M., Nowotny, C.A., Coutandin, D., Wang, F., Tabios, M., Agard, D.A., 2023. Hsp90 provides a platform for kinase dephosphorylation by PP5. Nat Commun 14, 2197.

Jeffrey, P.D., Russo, A.A., Polyak, K., Gibbs, E., Hurwitz, J., Massague, J., Pavletich, N.P., 1995. Mechanism of CDK activation revealed by the structure of a cyclinA-CDK2 complex. Nature 376, 313–320.

Kaledhonkar, S., Fu, Z., Caban, K., Li, W., Chen, B., Sun, M., Gonzalez, R.L., Jr., Frank, J., 2019. Late steps in bacterial translation initiation visualized using time-resolved cryo-EM. Nature 570, 400–404.

Kaledhonkar, S., Fu, Z., White, H., Frank, J., 2018. Time-resolved cryo-electron microscopy using a microfluidic chip. Methods Mol Biol 1764, 59–71.

Kinstrie, R., Luebbering, N., Miranda-Saavedra, D., Sibbet, G., Han, J., Lochhead, P.A., Cleghon, V., 2010. Characterization of a domain that transiently converts class 2 DYRKs into intramolecular tyrosine kinases. Sci Signal 3, ra16.

Knighton, D.R., Zheng, J.H., Ten Eyck, L.F., Ashford, V.A., Xuong, N.H., Taylor, S.S., Sowadski, J.M., 1991. Crystal structure of the catalytic subunit of cyclic adenosine monophosphate-dependent protein kinase. Science 253, 407–414.

Knox, J., Joly, N., Linossi, E.M., Carmona-Negron, J.A., Jura, N., Pintard, L., Zuercher, W., Roy, P.J., 2021. A survey of the kinome pharmacopeia reveals multiple scaffolds and targets for the development of novel anthelmintics. Sci Rep 11, 9161.

Kontzias, A., Kotlyar, A., Laurence, A., Changelian, P., O'Shea, J.J., 2012. Jakinibs: a new class of kinase inhibitors in cancer and autoimmune disease. Curr Opin Pharmacol 12, 464–470.

Kornev, A.P., Haste, N.M., Taylor, S.S., Eyck, L.F., 2006. Surface comparison of active and inactive protein kinases identifies a conserved activation mechanism. Proc Natl Acad Sci USA 103, 17783–17788.

Kovacs, E., Zorn, J.A., Huang, Y., Barros, T., Kuriyan, J., 2015. A structural perspective on the regulation of the epidermal growth factor receptor. Annu Rev Biochem 84, 739–764.

Kung, J.E., Jura, N., 2019. Prospects for pharmacological targeting of pseudokinases. Nat Rev Drug Discov 18, 501–526.

Lahiry, P., Torkamani, A., Schork, N.J., Hegele, R.A., 2010. Kinase mutations in human disease: interpreting genotype-phenotype relationships. Nat Rev Genet 11, 60–74.

Leschziner, A.E., Reck-Peterson, S.L., 2021. Structural biology of LRRK2 and its interaction with microtubules. Mov Disord 36, 2494–2504.

Lochhead, P.A., 2009. Protein kinase activation loop autophosphorylation in cis: overcoming a Catch-22 situation. Sci Signal 2, pe4.

Lochhead, P.A., Kinstrie, R., Sibbet, G., Rawjee, T., Morrice, N., Cleghon, V., 2006. A chaperone-dependent GSK3beta transitional intermediate mediates activation-loop autophosphorylation. Mol Cell 24, 627–633.

Lochhead, P.A., Sibbet, G., Morrice, N., Cleghon, V., 2005. Activation-loop autophosphorylation is mediated by a novel transitional intermediate form of DYRKs. Cell 121, 925–936.

Lyumkis, D., 2019. Challenges and opportunities in cryo-EM single-particle analysis. J Biol Chem 294, 5181–5197.

Manning, G., Plowman, G.D., Hunter, T., Sudarsanam, S., 2002a. Evolution of protein kinase signaling from yeast to man. Trends Biochem Sci 27, 514–520.

Manning, G., Whyte, D.B., Martinez, R., Hunter, T., Sudarsanam, S., 2002b. The protein kinase complement of the human genome. Science 298, 1912–1934.

Martin, M.P., Endicott, J.A., Noble, M.E.M., 2017. Structure-based discovery of cyclin-dependent protein kinase inhibitors. Essays Biochem 61, 439–452.

Nagar, B., Bornmann, W.G., Pellicena, P., Schindler, T., Veach, D.R., Miller, W.T., Clarkson, B., Kuriyan, J., 2002. Crystal structures of the kinase domain of c-Abl in complex with the small molecule inhibitors PD173955 and imatinib (STI-571). Cancer Res 62, 4236–4243.

Ngo, J.C., Gullingsrud, J., Giang, K., Yeh, M.J., Fu, X.D., Adams, J.A., McCammon, J.A., Ghosh, G., 2007. SR protein kinase 1 is resilient to inactivation. Structure 15, 123–133.

Noble, M.E., Endicott, J.A., Johnson, L.N., 2004. Protein kinase inhibitors: insights into drug design from structure. Science 303, 1800–1805.

Nogales, E., Scheres, S.H., 2015. Cryo-EM: A unique tool for the visualization of macromolecular complexity. Mol Cell 58, 677–689.

Nolen, B., Taylor, S., Ghosh, G., 2004. Regulation of protein kinases; controlling activity through activation segment conformation. Mol Cell 15, 661–675.

Nussinov, R., Tsai, C.J., 2013. Allostery in disease and in drug discovery. Cell 153, 293–305.

Oliver, A.W., Knapp, S., Pearl, L.H., 2007. Activation segment exchange: a common mechanism of kinase autophosphorylation? Trends Biochem Sci 32, 351–356.

Passos, D.O., Li, M., Jozwik, I.K., Zhao, X.Z., Santos-Martins, D., Yang, R., Smith, S.J., Jeon, Y., Forli, S., Hughes, S.H., Burke, T.R., Jr., Craigie, R., Lyumkis, D., 2020. Structural basis for strand-transfer inhibitor binding to HIV intasomes. Science 367, 810–814.

Pettersen, E.F., Goddard, T.D., Huang, C.C., Couch, G.S., Greenblatt, D.M., Meng, E.C., Ferrin, T.E., 2004. UCSF Chimera–a visualization system for exploratory research and analysis. J Comput Chem 25, 1605–1612.

Pike, A.C., Rellos, P., Niesen, F.H., Turnbull, A., Oliver, A.W., Parker, S.A., Turk, B.E., Pearl, L.H., Knapp, S., 2008. Activation segment dimerization: a mechanism for kinase autophosphorylation of non-consensus sites. EMBO J 27, 704–714.

Polley, S., Huang, D.B., Hauenstein, A.V., Fusco, A.J., Zhong, X., Vu, D., Schrofelbauer, B., Kim, Y., Hoffmann, A., Verma, I.M., Ghosh, G., Huxford, T., 2013. A structural basis for IkappaB kinase 2 activation via oligomerization-dependent trans auto-phosphorylation. PLoS Biol 11, e1001581.

Polley, S., Passos, D.O., Huang, D.B., Mulero, M.C., Mazumder, A., Biswas, T., Verma, I.M., Lyumkis, D., Ghosh, G., 2016. Structural basis for the activation of IKK1/alpha. Cell Rep 17, 1907–1914.

Qiu, C., Tarrant, M.K., Choi, S.H., Sathyamurthy, A., Bose, R., Banjade, S., Pal, A., Bornmann, W.G., Lemmon, M.A., Cole, P.A., Leahy, D.J., 2008. Mechanism of activation and inhibition of the HER4/ErbB4 kinase. Structure 16, 460–467.

Red Brewer, M., Choi, S.H., Alvarado, D., Moravcevic, K., Pozzi, A., Lemmon, M.A., Carpenter, G., 2009. The juxtamembrane region of the EGF receptor functions as an activation domain. Mol Cell 34, 641–651.

Russo, A.A., Jeffrey, P.D., Pavletich, N.P., 1996. Structural basis of cyclin-dependent kinase activation by phosphorylation. Nat Struct Biol 3, 696–700.

Scapin, G., Potter, C.S., Carragher, B., 2018. Cryo-EM for small molecules discovery, design, understanding, and application. Cell Chem Biol 25, 1318–1325.

Scheres, S.H., 2016. Processing of structurally heterogeneous Cryo-EM data in RELION. Methods Enzymol 579, 125–157.

Schrofelbauer, B., Hoffmann, A., 2011. How do pleiotropic kinase hubs mediate specific signaling by TNFR superfamily members? Immunol Rev 244, 29–43.

Schrofelbauer, B., Polley, S., Behar, M., Ghosh, G., Hoffmann, A., 2012. NEMO ensures signaling specificity of the pleiotropic IKKbeta by directing its kinase activity toward IkappaBalpha. Mol Cell 47, 111–121.

Schulze-Gahmen, U., De Bondt, H.L., Kim, S.H., 1996. High-resolution crystal structures of human cyclin-dependent kinase 2 with and without ATP: bound waters and natural ligand as guides for inhibitor design. J Med Chem 39, 4540–4546.

Sicheri, F., Moarefi, I., Kuriyan, J., 1997. Crystal structure of the Src family tyrosine kinase Hck. Nature 385, 602–609.

Snead, D.M., Matyszewski, M., Dickey, A.M., Lin, Y.X., Leschziner, A.E., Reck-Peterson, S.L., 2022. Structural basis for Parkinson's disease-linked LRRK2's binding to microtubules. Nat Struct Mol Biol 29, 1196–1207.

Stamos, J., Sliwkowski, M.X., Eigenbrot, C., 2002. Structure of the epidermal growth factor receptor kinase domain alone and in complex with a 4-anilinoquinazoline inhibitor. J Biol Chem 277, 46265–46272.

Taylor, S.S., Knighton, D.R., Zheng, J., Sowadski, J.M., Gibbs, C.S., Zoller, M.J., 1993. A template for the protein kinase family. Trends Biochem Sci 18, 84–89.

Taylor, S.S., Knighton, D.R., Zheng, J., Ten Eyck, L.F., Sowadski, J.M., 1992. Structural framework for the protein kinase family. Annu Rev Cell Biol 8, 429–462.

Taylor, S.S., Kornev, A.P., 2011. Protein kinases: evolution of dynamic regulatory proteins. Trends Biochem Sci 36, 65–77.

Taylor, S.S., Wu, J., Bruystens, J.G.H., Del Rio, J.C., Lu, T.W., Kornev, A.P., Ten Eyck, L.F., 2021. From structure to the dynamic regulation of a molecular switch: a journey over 3 decades. J Biol Chem 296, 100746.

Torkamani, A., Verkhivker, G., Schork, N.J., 2009. Cancer driver mutations in protein kinase genes. Cancer Lett 281, 117–127.

Ubersax, J.A., Ferrell, J.E., Jr., 2007. Mechanisms of specificity in protein phosphorylation. Nat Rev Mol Cell Biol 8, 530–541.

Verba, K.A., Agard, D.A., 2017. How Hsp90 and Cdc37 lubricate kinase molecular switches. Trends Biochem Sci 42, 799–811.

Verba, K.A., Wang, R.Y., Arakawa, A., Liu, Y., Shirouzu, M., Yokoyama, S., Agard, D.A., 2016. Atomic structure of Hsp90-Cdc37-Cdk4 reveals that Hsp90 traps and stabilizes an unfolded kinase. Science 352, 1542–1547.

Wang, Q., Zorn, J.A., Kuriyan, J., 2014. A structural atlas of kinases inhibited by clinically approved drugs. Methods Enzymol 548, 23–67.

Watanabe, R., Buschauer, R., Bohning, J., Audagnotto, M., Lasker, K., Lu, T.W., Boassa, D., Taylor, S., Villa, E., 2020. The in situ structure of parkinson's disease-linked LRRK2. Cell 182, 1508–1518 e1516.

Xu, W., Harrison, S.C., Eck, M.J., 1997. Three-dimensional structure of the tyrosine kinase c-Src. Nature 385, 595–602.

Zarrin, A.A., Bao, K., Lupardus, P., Vucic, D., 2021. Kinase inhibition in autoimmunity and inflammation. Nat Rev Drug Discov 20, 39–63.

Zhang, F., Strand, A., Robbins, D., Cobb, M.H., Goldsmith, E.J., 1994. Atomic structure of the MAP kinase ERK2 at 2.3 A resolution. Nature 367, 704–711.

Zhang, J., Clark, K., Lawrence, T., Peggie, M.W., Cohen, P., 2014. An unexpected twist to the activation of IKKbeta: TAK1 primes IKKbeta for activation by autophosphorylation. Biochem J 461, 531–537.

Zhang, J., Zhang, F., Ebert, D., Cobb, M.H., Goldsmith, E.J., 1995. Activity of the MAP kinase ERK2 is controlled by a flexible surface loop. Structure 3, 299–307.

5 The Life and Times of p53

Structural Advancements to Decipher the Guardian of the Genome

Maria J. Solares, Samantha Berry, and Deborah F. Kelly

INTRODUCTION

The Guardian of the Genome: A Call to Action

Every day, cells are exposed to different stressors. Although the sources range from internal metabolic errors to external irradiation exposure, all of these factors can lead to DNA insults culminating in apoptosis (Chatterjee & Walker, 2017; Hakem, 2008). The accumulation of genotoxic events may lead to severe outcomes at the cellular level, especially in cases where genetic predispositions elevate the risk of disease (Bouwman & Jonkers, 2012; Wolters & Schumacher, 2013). Through evolution, cells have developed robust DNA damage and repair pathways with the tumor suppressor protein TP53 (p53) serving as the "guardian of the genome."

A well-characterized function of p53 involves cell cycle arrest (Chen, 2016; Vousden & Prives, 2009). In healthy cells, the concentration of p53 is maintained at relatively low levels until needed for the repair process (Ozaki & Nakagawara, 2011). These regulatory mechanisms are mediated by a plethora of proteins, including mouse double minute 2 (MDM2). As an E3 ligase, MDM2 ubiquitinates p53 for destruction to maintain homeostasis through an autoregulatory feedback loop (Haupt et al., 1997; Honda et al., 1997; Kubbutat et al., 1997). Upon detecting double-strand breaks in the DNA helix, ataxia telangiectasia mutated (ATM) kinase is activated and phosphorylates p53 (Chen, 2012). The phosphorylation of p53 can impede MDM2 binding activity and prevent degradation-linked ubiquitination (Böttger et al., 1999). During DNA damage, a protein called alternate reading frame (ARF) is transcribed and directly inhibits MDM2 (Kung & Weber, 2022). This series of checks and balances allows p53 to accumulate in the cell as needed and transcriptionally activate downstream factors like p21 (El-Deiry, 1998; El-Deiry et al., 1993). As p21 is produced, G_1 arrest in the cell cycle occurs as p21 binds to cyclin E/Cdk2 and cyclin D/Cdk4 complexes (Deng et al., 1995; Wade Harper, 1993; Xiong et al., 1993); this initiates a signaling cascade that silences genes critical for DNA replication and cell-cycle progression (Noda et al., 1994; Xiong et al., 1993). It has also been reported that p21 interacts with proliferating cell nuclear antigen (PCNA), which in turn inhibits DNA replication *in vitro* (Luo et al., 1995). Further, p53 may also arrest cell cycle progression at the G_2/M phase through the inhibition of other checkpoints by p53-activated partners, or even bind directly to the promoters of these same checkpoint proteins to repress transcription (Clair et al., 2004; Hermeking et al., 1997; Ohki et al., 2000; Tanaka et al., 2000). Additional research highlights the redundancy found in arresting the cell cycle at these two checkpoints. The absence of p21 does not prevent G_1 arrest, signifying that there are several p53-induced factors in the cell that participate in cell cycle arrest (Brugarolas et al., 1995). This is also observed with other well-characterized p53-related factors found to inhibit G_2/M phase transition (Chen, 2016). Cell-cycle arrest can also be reversed if DNA repair is accomplished. Cycle reentry usually occurs by achieving homeostatic concentrations of p53, where ATM and ATR are suppressed, MDM2 is transcribed/activated, and excess p53 is degraded (Chen, 2012; Kung & Weber, 2022).

If DNA repair cannot be accomplished, p53 initiates cell death. Apoptosis is achieved through the induction of several p53-dependent genes: BH3 domain-only proapoptotic proteins (ex.: Puma, Bax, Bak, etc.); death receptors (ex.: Fas, Dr4, Killer/Dr5, etc.); and apoptosis execution factors (ex.: Apaf1, caspase 6, etc.) (Riley et al., 2008; Tait & Green, 2010). Each of these apoptotic proteins participates in either the extrinsic or intrinsic apoptosis pathways (Tait & Green, 2010). The latter is targeted both directly and indirectly by p53. One of the vital steps to the intrinsic apoptosis pathway is the permeabilization of the mitochondrial outer membrane (MOMP). BH2 domain–only proapoptotic proteins like Bax and Bak form complexes with other mitochondrial proteins to oligomerize and form pores on the outer organelle's membrane (Eskes et al., 2000; Hsu et al., 1997; Wei et al., 2000). In turn, cytochrome c is released into the cell, which recruits and creates the apoptosome complex that induces downstream cell death (Goldstein et al., 2000; Wei et al., 2000). Other BH3 domain–only proteins like Puma also play an important part by targeting other stages of MOMP (Jeffers et al., 2003). In fact, research suggests that multiple target genes work together to execute cell death (Tait & Green, 2010).

As a coup de grâce, p53 translocalizes to the mitochondria to initiate transcriptional-independent apoptotic mechanisms. Reports suggest that p53 can interact directly with Bax and Bak inhibitors to promote permeabilization inside the mitochondria (Marchenko & Moll, 2007; Mihara et al., 2003). However, other research indicates that transcriptional activity is unnecessary to promote apoptosis and that cell death can be initiated by direct targeting of p53 to signal peptides (Vaseva et al., 2012). As the judge, jury,

DOI: 10.1201/9781003326106-6

and executioner, p53 holds the proverbial power of life and death at the cellular level.

Cut the Brakes: Effects of p53 Mutations in Disease Development and Progression

When so much responsibility is placed on a single protein, what happens when errors ensue? If cells lose their ability to combat genotoxic events and prevent the transmission of faulty genes to their daughter cells, the organism's health is in jeopardy. Mutations in *TP53* have been mapped to an extensive list of human disorders. Certain germline mutations cause Li-Fraumeni syndrome, wherein patients are at risk of developing spontaneous cancer growths throughout life (Royds & Iacopetta, 2006). Most other germline mutations cause major developmental abnormalities, if not death *in utero* (Armstrong et al., 1995; Sah et al., 1995). Somatic mutations have also been linked to neurodegenerative diseases, and they are being studied as disease markers in both Alzheimer and Parkinson disease (Lanni et al., 2007; Szybińska & Leśniakx, 2017; Wolfrum et al., 2022). However, through the years, errors in p53 often take center stage in oncology research.

p53 is often the primary line of defense against cancer induction. It can suppress abnormal cell growth and replication effects observed in tumorigenesis (Sullivan et al., 2018; Vousden & Lu, 2002). As such, it is not surprising that it is among the most deregulated proteins in cancer. Around 50% of all cancer cases will exhibit a random, missense mutation in the *TP53* gene (Hollstein et al., 1991; Kandoth et al., 2013; Olivier et al., 2010). In general, cutting the brakes to the cell cycle can spell trouble, yet in cancer, the repercussions are devastating. Perturbing p53 leads not only to increased tumor growth, but also to increased chemoresistance, expedited metastasis, and poor patient prognosis.

Although every codon in *TP53* has been reported to contain mutations, there are seven prevalent mutations directly linked to cancer (Kandoth et al., 2013; Walker et al., 1999). These "hotspot" mutations are generally associated with heightened disease progression and worse patient prognosis. Since these seven hotspots occur in the important DNA-binding domain region of p53, they are thought to destabilize the protein's ability to interact with DNA (Hollstein et al., 1991). Further, they have been reported to have a dominant negative effect on wild-type (WT) p53 complexes (Willis et al., 2004).

Years of research on *TP53* suggests that not all mutations are equivalent (Muller & Vousden, 2014). Each hotspot has a specific disease profile (Bamford et al., 2004; Barretina et al., 2012; Bartek et al., 1990; Gaidano et al., 1991; Liu & Bodmer, 2006; Markl & Jones, 1998; Moneo et al., 2007; Moore et al., 2001; Tseng et al., 1999; Yin et al., 1992) (Table 5.1). Some of these changes, for example, have been

TABLE 5.1
Reported hotspot mutations that affect p53 function

Hotspot Mutation	Impact on p53	Topography	References	Biological Impact	References
R175H	Conformational	Biliary tract, Bones, Brain, Breast, Colon, Hematop. system, Lung, Ovary, Pancreas, Stomach, Urinary tract	50–53	Proliferation, Genomic Instability, Migration, Invasion, Colony Formation, Drug Resistance, Architecture Disruption, Stem Cell Differentiation	60-82
G2455		Brain, Colon, Colorectum, Liver, Conformational Pancreas, Hematop. System, Pancreas, Stomach	51–53	Architecture Disruption	82
R248Q	Conformational	Bladder, Bones, Brain, Breast, Corpus Uteri, Esophagus, Hematop. System, Kidney, Lung, Lymph nodes	51, 53, 54	Migration, Invasion, Drug Resistance	65, 83
R248W	DNA Contact	Brain, Breast, Colon, Esophagus, Hematop. System, Kidney, Lung, Lymph 51, 53, 55, 56 Nodes, Skin		Proliferation, Genomic Instability, Migration, Colony Formation, Drug Resistance, Architecture Disruption	61, 63, 70, 72, 74, 82, 84–87
R2495	Conformational	Breast, Colorectum, Heart/Medulla, Liver, Lung	51–53, 57	Proliferation, Migration	70, 73, 78, 88
R273H	DNA Contact	Biliary Tract, Brain, Colon/Colorectum, Corpus Uteri, Hematop. System, Lung, Lymph Nodes, Ovary, Pancreas, Skin, Thyroid	50–53, 58	Proliferation, Genomic Instability, Migration, Invasion, Colony Formation, Drug Resistance	60–62, 70, 71, 73, 81, 82, 87, 89, 90
R282H	Conformational	Lung	59	Proliferation, Colony Formation	90

Note: Seven hotspot mutations are listed along with their impact on the p53 protein, tissue topology, and biological function.

considered worse for patients, when they are linked to "gain of function" phenotypes (Adorno et al., 2009; Agapova et al., 1996; Bossi et al., 2006; Buganim et al., 2006; Di Agostino et al., 2006; Dong et al., 2009, 2013; Freed-Pastor et al., 2012; Hsiao et al., 1994; Kalo et al., 2012; Lang et al., 2004; Liu et al., 2011; Loging & Reisman, 1999; Matas, 2001; Muller et al., 2009, 2013; Noll et al., 2012; Pugacheva et al., 2002; Sarig et al., 2010; Scian et al., 2004; Shi et al., 2002; Song et al., 2007, p. 200; Tsang et al., 2005; Vaughan, Frum, et al., 2012; Vaughan, Singh, et al., 2012; Weisz et al., 2004; Yan et al., 2008; Yeudall et al., 2012; Yi et al., 2012; Yoshikawa et al., 2010; Zhang et al., 2011, p. 53). Some mutations have been found to increase immune evasion and to enhance motility in xenograft models and *in vitro* systems (Adorno et al., 2009; Grugan et al., 2013; Muller et al., 2009, 2013; Muller & Vousden, 2014; Sauer et al., 2010; Wang et al., 2013). It is not yet well understood why certain point mutations create such a wide diversity of outcomes. Collectively, the consensus opinion is that p53, mutant or WT, is a good candidate for targeted chemotherapeutics.

Deciphering Domains of the p53 Protein

Major barriers to fully utilizing p53 as a targeted therapy are its cryptic structure and its myriad oligomerization states (Figure 5.1). The primary structure of the p53 protein is composed of three broad regions: an N-terminal domain (NTD), a DNA-binding domain (DBD) or core domain, and a C-terminal domain (CTD) (Figure 5.1a). As the functional center of the protein, the DBD coordinates binding with p53-response elements in double-stranded DNA (Joerger & Fersht, 2010). Since all hotspot mutations occur in the DBD, it is an important region to understand in terms of disease relevance and functionality.

The first look into p53's core was achieved in 1994. Yunje Cho and his coworkers were able to crystallize a single DBD, in the form of a monomer, bound to a DNA duplex (Cho et al., 1994) (Figure 5.1b). Researchers found that the DBD can be described in a few key regions: two large loops (L2 and L3) followed by a short loop-sheet-helix motif (L1). L2 and L3 are stabilized by a zinc ion, which

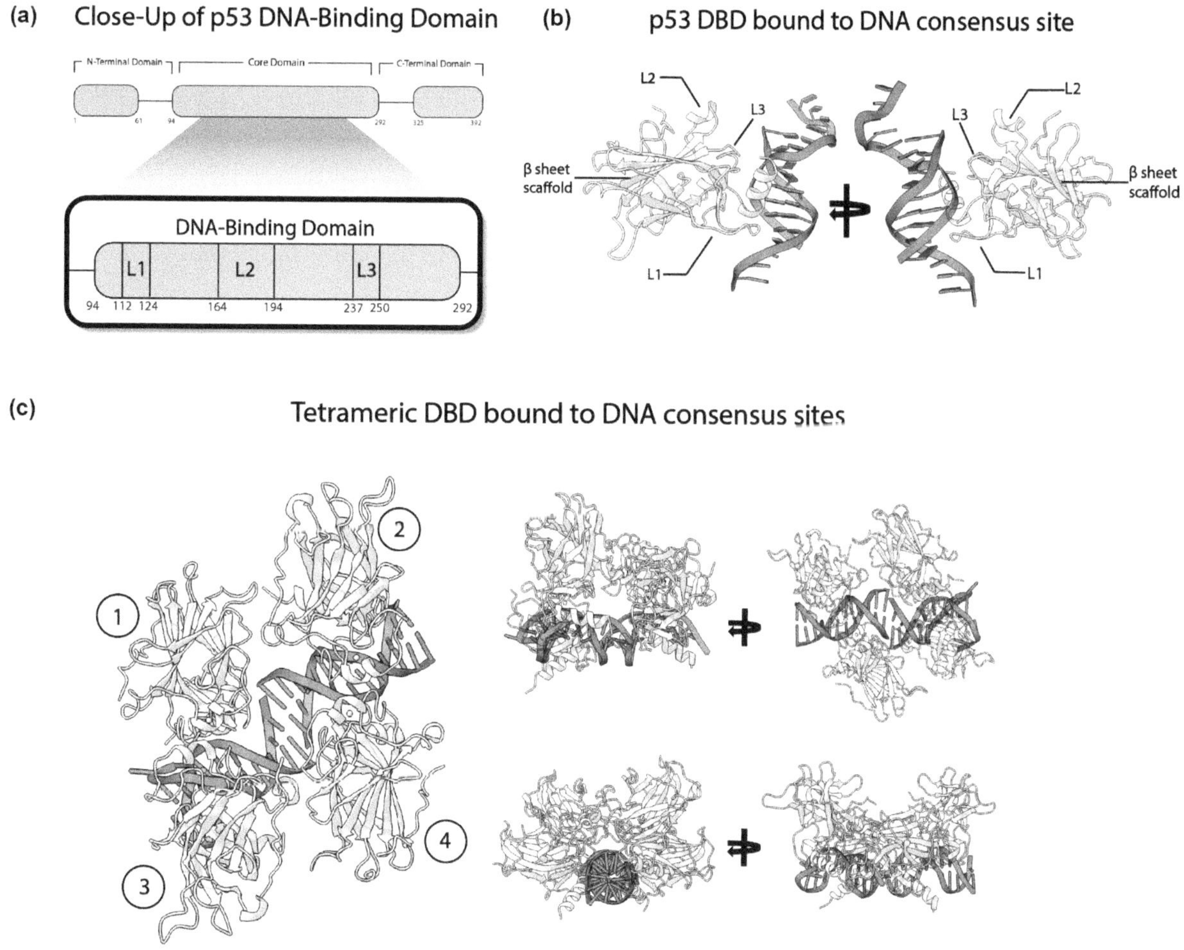

FIGURE 5.1 Structural elements of the p53 DBD. (a) The primary sequence of p53 is comprised of the N-terminal domain (NTD), core domain that consists of the DNA-binding domain (DBD), and the C-terminal domain (CTD). The DBD (residues 94–292) has three featured sites (L1–L3) that foster DNA engagement. (b) Crystal structure of the DBD interacting with DNA [Cho et al]. (c) Crystal structure of the DBD tetramer (1–4) bound to DNA containing a double-stranded break [C. Kitayner].

in turn coordinates DNA binding. They were also able to determine the residues and regions involved in forming the protein–DNA interface. Their findings would then serve as the basis for other structural studies.

Although this model was able to advance knowledge of p53 structure and function, there were still many questions regarding the DBD. First, it was still an incomplete look. The functional unit of p53 is understood to be a tetramer that assembles as a "dimer of dimers." Structural information on how that assembly came together was still missing. Second, the structural basis and mechanism the functional unit used to recognize DNA targets remained unknown. It took over a decade for these questions to be answered.

Malka Kitayner and her collaborators published a more complete model. Unlike Cho's structure, Kitayner's models depict how four core domains of human p53 assemble around two DNA half-sites to form the functional tetramer (Figure 5.1c) (Kitayner et al., 2006). Researchers also crystallized p53-DNA complexes with different DNA dodecamers. A total of four crystal structures were resolved from the study. The team found that within the dimer, each monomer interacted with each other and formed sequence-specific contacts with symmetry-related regions of each of the half-sites. To form tetramers, dimers associated through specific "stabilization networks" caused by protein–protein and base-stacking interactions. Kitayner was also able to deduce how p53 recognized different DNA patterns. A cohort of different residues was identified for each of the half-sites that were resolved, and how each interacted with the DNA backbone. Thanks to these advances, there's a wealth of structural information available on this region. As an example, high-resolution crystal structures are available for the DBD bound to DNA (Chen et al., 2013; Cho et al., 1994; Kitayner et al., 2006), regulatory proteins (Derbyshire, 2002; Gorina & Pavletich, 1996; Joo et al., 2002), viral oncoproteins (Lavin & Gueven, 2006), and select DBD mutants (Joerger et al., 2004, 2005, 2006).

Form Follows Function: Flexible Parts of p53

In contrast to the structural information available for the DBD of p53, the NTD and CTD are less well characterized. The NTD contains important regulatory regions for p53, including two transactivation domains (TADs). A wide array of regulatory proteins, including MDM2, interact with a TAD region to either activate or degrade p53 (Marine et al., 2006; Momand et al., 2000). The CTD serves a similar function, although consensus upon its mechanism of action has not yet been established. Generally, it is understood that this region interacts with other regulatory proteins to impede DNA binding, perhaps in a tissue-specific manner (Hamard et al., 2013). Equally important, the CTD contains the oligomerization domain (OD). This region allows for the coordination of separate oligomers to come together in forming a quaternary structure. Although some protein–protein and protein–DNA interactions are strong enough to hold the tetramer complex together, as seen in Kitayner's models, the precise manner in which oligomerization occurs remains a mystery.

In terms of structural features, the NTD and CTD are highly flexible (Bell et al., 2002; Joerger & Fersht, 2007). This intrinsic disorder may aid the complex function of p53 by conveying to it the ability to interact with many diverse partners (Dunker et al., 2005). However, the inherent disorder in these regions limits structural analysis at the atomic scale. Limited success has been achieved into reconstructing some of these regions. For instance, the p53 tetramerization domain has been resolved through X-ray crystallography and nuclear magnetic resonance (NMR) (Jeffrey et al., 1995; Mittl et al., 1998), and certain regions of the NTD like TAD1/TAD2 domains have also been structurally evaluated (Botuyan et al., 1997; Lee et al., 2000). Unfortunately, the greater picture of the full-length arrangement of the p53 protein has been elusive.

An added layer of p53's complexity is due to its oligomerization states. Although *TP53* encodes for a single monomeric unit, the p53 protein typically exists as dimers sequestered in the cell (Nicholls et al., 2002). These dimers, once activated, oligomerize around DNA and form the tetrameric unit (Nicholls et al., 2002). As such, the tetramer is considered the functional unit of p53 and is involved in the transcription of downstream activators of DNA-repair and apoptotic signaling molecules (McLure & Lee, 1998). Moreover, the variability of oligomeric states can limit structural analysis caused by heterogeneity in samples.

It is imperative to seek the full structure of p53. Resolving the structure can help us discern the true mechanism of action for the NTD and CTD. Further, disease-related regulation could be hidden in these unknown regulatory regions, waiting to be exploited as the next treatment. Yet, to resolve flexible proteins like p53, new techniques must be employed. X-ray crystallography yields high-resolution structures of stable domains, but it is limited in its ability to resolve more flexible subunits and protein structures. NMR has a different set of advantages and limitations. Although it is better suited to discern flexible structures, the size of many protein oligomers is out of range for the current state of the technique. A third technique that can better handle flexible macromolecules is ideal.

From Blob to Beauty: The Path to Atomic Resolution

Cryo-electron microscopy (EM) was developed in parallel as an alternative to X-ray crystallography. By allowing scientists to bypass the crystallization step, it is now possible to observe flexible proteins and domains in a near-native frozen environment. Protein samples are preserved in thin layers of vitreous ice that somewhat protect their fragile peptide bonds during imaging steps. In its infancy, cryo-EM was unable to deliver atomic-level models like its other structural counterparts. Not until the recent "resolution revolution" came into view.

The research of three scientists served as pillars to transform the cryo-EM field from "blobology" to respectable atomic structures. Early on, EM imaging was performed on larger macromolecules embedded in heavy metal salts or resins. In 1975, Richard Henderson, alongside his partner Nigel Unwin, published a 7 Å structure of the membrane protein bacteriorhodopsin (Unwin & Henderson, 1975). This seminal paper demonstrated that it was possible to resolve many of the transmembrane helices in their target protein. With his success in the field, Henderson kept pushing the resolution of bacteriorhodopsin to new heights. He and his team developed novel ways to collect images and diffraction patterns to retain the true conformation of visualized proteins (Shen, 2018).

Jacques Dubochet is the second pillar as his work was fundamental in adding the "cryo" to cryo-EM. He pursued methods based on cryogenic cooling to create a clean form of vitreous ice. The sample freezing process can often damage biological samples because ice crystals may shear fragile protein bonds. By developing a method that cools water faster than ice crystals can form, it was possible to suspend particles in a protective layer of amorphous ice, a process termed vitrification. Dubochet used this technique to publish the first cryo-EM images of viruses (Adrian et al., 1984; Dubochet & McDowall, 1981). In the 1990s, using these new cryogenic methods, Henderson and colleagues resolved a high-resolution density map of bacteriorhodopsin to 3.5 Å (Henderson et al., 1990). The atomic barrier was no longer just a dream for cryo-EM.

There was still more to tackle in cryo-EM than just sample preparation. Joachim Frank and others were hard at work on data analysis techniques that would give rise to single-particle reconstruction strategies. To reveal atomic features of individual protein particles, it was necessary to mathematically reduce the noise in EM micrographs. Frank accomplished this by computationally aligning particle images via cross-correlation strategies, molecular symmetry notwithstanding (Frank, 1975; van Heel & Frank, 1981). These particle features could then be classified based on orientation values and used to produce image reconstructions of the original 3D objects. This method was tested and proven using asymmetrical bacterial ribosomes (Radermacher et al., 1986).

Richard Henderson, Jacques Dubochet, and Joachim Frank, three key figures behind the revolution, were awarded the Nobel Prize in chemistry in 2017. We would be remiss not to mention also the years of community building and collaborative efforts between life scientists, engineers, and industry teams to develop new tools such as direct electron detectors and advanced software systems. Thanks to these efforts, it is possible to image and process data at unprecedented structural detail. The field once known as "blobology" has rapidly grown, with the Electron Microscopy Data Bank containing more than 26,500 entries to date (EMDB; http://emdb-empiar.org).

Shining a Spotlight on p53 with Cryo-EM

Research waits for nothing, not even a Nobel Prize. Structural biologists had already begun adapting the "resolution revolution" technology to image smaller and more flexible proteins even as the aforementioned papers were being published. The same applied to p53. The promise of deciphering p53 structures using cryo-EM sparked the interest of the structural biology community (Figure 5.2).

The same year Kitayner was working on her crystallographic structures, one of the earliest first looks into the p53 tetrameric structure using cryo-EM was achieved by Andrei Okorokov and colleagues (Okorokov et al., 2006). Following the initial findings that p53 has a low level of ATPase activity, it was determined that the addition of ATP, ADP, and dADP can stabilize p53–DNA interactions (Okorokov & Milner, 1999). In fact, the ATP/ADP molecular switch can serve as a regulatory mechanism for specific and nonspecific DNA binding. More importantly, this endogenous feature of p53 could enable researchers to stabilize a tetramer in a single conformation to be resolved using cryo-EM.

After expressing, purifying, and untagging ATP-stabilized murine p53 proteins, samples were plunge-frozen on holey carbon films. Images were collected on a Tecnai F20 at 200kV, low electron dose conditions (~20 e/Å^2). Under these conditions, researchers determined their high-resolution cutoff was about 7 Å, which was cutting edge for the era. Through further analysis and reconstruction, they determined that a dihedral symmetry (D2), where four subunits are related by three twofold axes, was appropriate when comparing reprojections to 2D averages. Their final reconstruction landed at 13.7 Å (Figure 5.2a).

Since they used D2 symmetry, Okorokov's final reconstruction was in an interesting formation. ATP-stabilized murine p53 tetramer turned out to look like a hollow, skewed cube with eight vertices/nodes connected by linkers, with four of these nodes being slightly larger. The 3D volume of the larger nodes was experimentally determined to be ~25 kDa and could easily accommodate a murine or human DBD. The sum of the density of all regions also came up to total ~185 kDa of mass, similar to the molecular weight of a p53 tetramer (~174 kDa).

Although the first reconstruction of a complete murine p53 is a low-resolution model, it was possible to discern key features. In this ATP-stabilized form, neither the NTD nor the CTD were found to be interacting directly with the DBD. Instead, through automated fitting, previously resolved OD, TAD1, TAD2 crystal structures were assigned together to the smaller nodes. This 3D organization implies that the NTD region of a monomer interacts with the CTD of another, almost, Okorokov notes, "as twins facing each other and holding hands" to form dimer structures (Okorokov et al., 2006). This was further proved *in vitro* through co-immunoprecipitation assay.

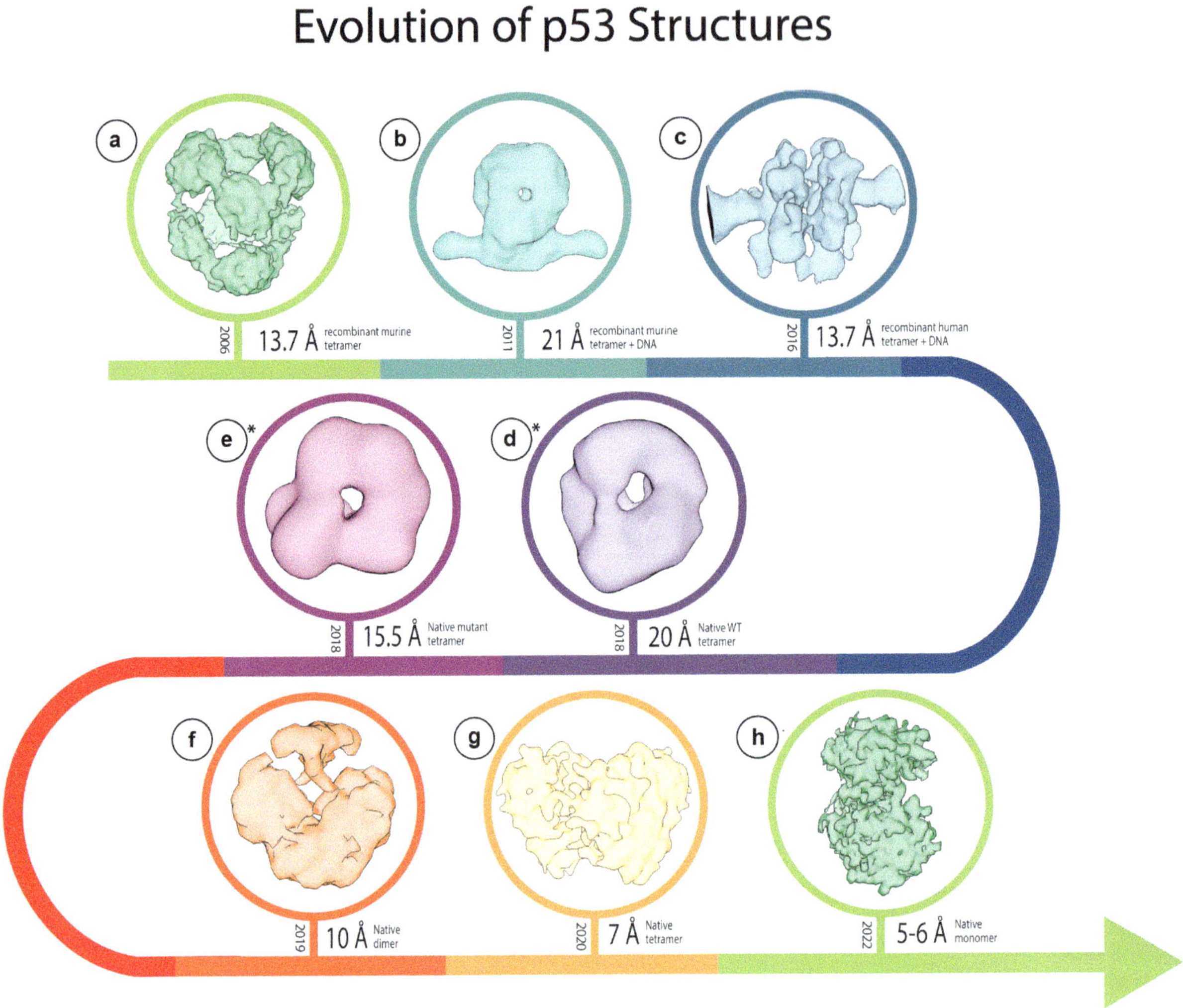

FIGURE 5.2 EM structural analysis of p53 through the ages. (a) First murine recombinant p53 cryo-EM structure lacking DNA. (b) Murine recombinant cryo-EM structure of p53 bound to DNA. (c) Truncated human p53 tetramer bound to DNA. (d) Wild-type human p53 tetramer isolated from cancer cells.* (e) Mutated human p53 tetramer isolated from cancer cells.* (f) Wild-type human p53 dimer isolated from brain cancer cells. (g) Wild-type human p53 tetramer bound to DNA and isolated from brain cancer cells. (h) Wild-type full-length human p53 monomer isolated from brain cancer cells. *Indicates negative stain structures.

Understanding Molecular Differences in p53 Structures

Differences arose in the biological interpretations of crystallographic structures versus cryo-EM structures of p53. It was difficult to reconcile the structural symmetry differences from Kitayner's tetramer on DNA (C2) (Kitayner et al., 2006) compared to Okorokov's tetramer lacking DNA (D2) (Okorokov et al., 2006). Researchers needed to avoid speculation and had to present the experimental data that precisely demonstrated which interactions were biologically relevant. As a guide, reasearchers knew to study the accepted functional form of p53, which is its tetramer state bound to DNA. There are specific and well-documented DNA sequences called response elements (REs) to which p53 binds (Harris et al., 2009; Kern et al., 1991; Tokino et al., 1994). As previously noted, these regions are recognized and bound to by the protein's DBD, and it is generally accepted that each RE can be bound to by several DBDs (Cho et al., 1994; Kitayner et al., 2006). Biologically speaking, two dimer cores can come together at a single RE in a manner consistent with twofold symmetry (C2). Outside of the DBD, DNA binding properties have also been attributed to the CTD (Kim & Deppert, 2006; Laptenko et al., 2015). Several reports have noted that the CTD usually binds nonspecific elements and can often stabilize whole protein–DNA interactions through ill-defined spatial arrangements (Laptenko et al., 2015).

In 2011, Ricardo Aramayo investigated the first view of full-length murine p53 bound to DNA using EM techniques (Aramayo et al., 2011). Researchers purified recombinant murine p53 and combined it with an excess of DNA fragments containing 13 consecutive REs. Although this large number of RE repetitions does not naturally occur, researchers believed that it could ensure reliable complex formation for easier visualization; the largest amount of consecutive REs found in nature is four repetitions in the *mdm2* gene (Aramayo et al., 2011). In conjunction with biochemical experiments and negative stain imaging, samples were vitrified on holey carbon films and examined at 200 kV under low-dose conditions (~10–20 e/Å^2).

The micrographs revealed "beads on a string" in which each "bead" was identified as a tetramer interwoven by DNA "strings." (Aramayo et al., 2011) Single-particle reconstructions yielded a ~21 Å structure imposing a C2 symmetry (Figure 5.2b). The resulting density map was best described by researchers as a "globular mass attached to a stem." (Aramayo et al., 2011) Just like his last model, four larger-density nodes were arranged in an apparent pseudo D2 symmetry with two openings found on each "face." More interestingly, it was determined that the DNA was only bound to the side of the tetramer and did not "weave" through the complex. Aramayo used Kitayner's models to guide his interpretations of the map. The region of the map closest to the DNA density was in good agreement with the predetermined structure. The other core domains distant from the DNA density also fit two core domains but did not participate in any DNA bonding. Aramayo concludes that the most likely model of engagement is that one single core from each dimer engages the RE (Aramayo et al., 2011).

Aramayo further suggested that since only two DBDs bind to DNA, this can explain the dominant negative effect hotspot mutations impose in disease (Aramayo et al., 2011). Biochemical and molecular research has indicated that a single missense mutation in one allele of *p53* can dominate and even negate a WT allele (Willis et al., 2004). In this case, one perturbed core in a dimer is sufficient in destabilizing the complex and causing the detrimental downstream effects seen in human disease. In fact, Aramayo notes, one mutated core can inactivate 75% of p53 tetramers (Aramayo et al., 2011).

It is hard to overlook the variables at play in cryo-EM that can affect biological structures. Some of these variables can be as simple as molecules falling into preferred orientations or disrupted protein organizations when exposed to an extended air–water interface during vitrification. This is not counting the sheering forces exercised on molecules when blotting and freezing samples. As the cryo-EM field grew and advanced, it became increasingly clear that new adaptations had to be made to image fragile biological complexes and to control for these variables.

But what if researchers could have a support structure that could control these variables? In 2016, Thomas Martin and colleagues set out to create a DNA-origami support structure that could do three things: (1) protect particles from adopting uncontrolled orientations to get more accurate viewing angles, (2) protect particles from the harsh conditions experienced during sample preparation, and (3) protect particles from strong degradation caused by electron bombardment (Martin et al., 2016). Their final design was akin to a birdcage with double arrays of dsDNA serving as bars. Following a honeycomb motif, their cage's dimensions were 31.7 nm × 26.4 nm × 26.2 nm, with a hollow inside measuring 13.6 nm × 13.2 nm. They also gave their cage a "flag" of 10 parallel helices as an added feature, so that top and bottom views could be easily recognized. Finally, they left single-stranded DNA (ssDNA) overhangs with exposed aromatic bases, as previous research had noted that this kind of design had preferred orientation in ice layer creations (Martin et al., 2016). As such, the height of their cage (~260 Å) could ensure that the ice layer was thicker than the expected size of their target complex.

Martin recognized that a good candidate to test this cage on was p53. Due to the DBD's inherent nature of binding specific REs, the cage's interior was designed to hold a "perch" made of a p53 binding element. Researchers also noted that shifting the DNA perch sequence one base pair upward would cause the complex to rotate 34° along the axis; in short, by shifting the sequence upward one base pair at a time, they could generate several controlled views of p53. This could be maximized by using a tilted stage. Finally, the size of the DNA–p53 complex was also considered. Although it could be considered a limitation, researchers chose to use a recombinant human p53 transcript that lacked a CTD. This ensured that the tetramer bound to DNA (<120 Å) could be encapsulated safely within the ice layer created by the "cage."

Their birdcage was successful. After some perfecting of the blotting conditions, they were finally able to collect images on an FEI Titan Krios microscope at 300 kV with a K2 Summit detector. Initial structures proposed that the final model would not have a D2 symmetry as proposed by earlier work. Instead, final maps that were resolved to ~15 Å showed an apparent C2 symmetry, much like that found in previous X-ray crystallography models, albeit with additional densities in the front and back of the complexes (Figure 5.2c) (Chen et al., 2013; Kitayner et al., 2006). The models showed local translational symmetry along the DNA axis. The tetramerization domain, they argue, is most likely found in the extra densities, and most likely binds to the opposite side of the DNA helix. As a bonus, the authors add in their paper, their structure "settles an ongoing debate about the symmetry of the p53 tetramer bound to DNA" (Martin et al., 2016).

P53 Modifications Come into Focus

These breakthroughs collectively changed the field. Okorokov, Aramayo, Martin, and their respective collaborators helped demonstrate the strength of the technology. Their findings on the folding of tetrameric p53 can be used as a launchpad by other scientists to answer other important questions on p53. There is a level of flexibility and transcendence to having research not be modeled solely behind "*what does full-length, high resolution p53 look like?*" Although still important, instead current research focuses more on the biological impacts of structure on the function of p53. Yet, these questions cannot be solved through recombinant models. The answers lie in studying proteins straight from the source: patient cells.

Although cryo-EM is a technology that can certainly "view" anything, compositional and conformational heterogeneity limit the technique. Samples are frozen to limit movements of flexible regions, but flexible diverse particles may still exist regardless of how many regions are removed to stabilize the target proteins. Another much less recognized limitation is the quantity and quality of particles. For a high-resolution cryo-EM structure, thousands of particles must be processed. Therein, a high concentration of sample must be used. Traditionally, recombinant proteins technology can overcome this limitation. Yet, yeast and bacterial expression systems are just that: model systems. By relying on model systems to guide product development and drug discovery, research omits important processes that only occur in human cells or diseased states.

As an example, posttranslational modifications (PTMs) serve an important role within the biological workings of a cell. Each of these small moieties serves to fine-tune reactions. Since p53 is involved in a wide range of cell signaling cascades, it is tightly regulated by PTMs. Regulation patterns can be so sensitive as to be one molecule away from switching p53's role from an apoptotic factor to proteasomal degradation, as seen in mono- and polyubiquitination, respectively (Brooks & Gu, 2011). These are natural and healthy occurrences within cells to prevent premature cell cycle arrest and apoptosis. Other factors aside from ubiquitination that are important to p53 function are phosphorylation and acetylation (Figure 5.3b and c) (Brooks & Gu, 2003, p. 53; Dai & Gu, 2010; Hornbeck et al., 2012). Yet, their roles within p53 structural research are understudied. These exquisite molecular tags can often increase heterogeneity in samples, making it more difficult to obtain the pristine image samples structural biologists yearn to understand.

Notably, years of clinical data suggest that hotspot mutations lead to increased cell metabolism and metastatic signaling (Koulgi et al., 2013; Lepre et al., 2017; Walker et al., 1999, p. 199). However, the impact of these mutations on p53 structure is still a work in progress. Previous literature classifies the seven "hotspots" as either "DNA contact" or "conformational" mutations (Figure 5.3d) (Cho et al., 1994). The impact of these mutations has been modeled in the DNA binding domain, but whole-length conformational changes have not been documented (Joerger et al., 2006). Further, if and how these mutations affect the PTM regulation of p53 through conformational changes is even less studied. Abnormal modification patterns could be at the root of many of the different phenotypes observed in oncology. The field of molecular subtyping is underexplored in series of deadly cancers, and even more barren when it comes to p53 (Collisson et al., 2019; Zhao et al., 2019).

Structural Features of Natively-Sourced p53 Protein

Although cells in homeostasis exhibit low concentrations of p53, it is often overexpressed in cancer cells. In 2018, Yanping Liang and colleagues exploited this advantage to initiate the exploration of natively sourced p53 through negative stained, single-particle EM analysis (Liang et al., 2019). By using two overexpressing cell models, one with WT p53 and the other with a truncated, mutated version, they reconstructed low-resolution models for each. The team reported a WT p53 map resolved to 20 Å and a mutated WT map to 15 Å (Figure 5.2d and e) (Liang et al., 2019). The structures were in good agreement with previous DBD models as well, demonstrating C2 symmetry. This first proof of concept was an avalanche.

The study of new materials to visualize fragile proteins did not stop with Martin's work. Nick Alden and colleagues demonstrated the use of custom-designed substrates to enhance protein particle capture (Alden et al., 2019). His system relied on forgoing materials that only weakly interacted with biological components and focused on using silicon nitride (SiN). This chemical composition also improved the ice thickness inconsistencies and improved support material (Murarka, 2003). Unsurprisingly, the p53 cryo-EM structure was resolved to ~10 Å (Figure 5.2f) (Alden et al., 2019). What was surprising was their determination of extra density that could not be attributed to the protein itself. What did fit this density, and in a biologically relevant matter, was ubiquitin. Alden's results demonstrated the power of using new materials to protect delicate native p53 assemblies and resolve structures with PTMs present.

These advances moved full steam ahead in 2020. A highly reproducible approach to extract, purify, and enrich molecules from cells had been evading researchers. Without the traditional his-tags, the creation of p53 density maps relied on excessive amounts of data, which is typically unfeasible. The same messy nature of p53 that prevent its crystallization, however, could be exploited to harvest it without the use of recombinant tags. Maria Solares discovered that by adapting IMAC (immobilized metal affinity chromatography) strategies, she could capture phosphorylation-rich p53 from the nuclear material of cancer cells (Solares et al., 2020). One of the advantages of SiN is its ability to be functionalized. By layering thin films of Ni-NTA coatings, the same IMAC reagent, partially purified p53 complexes could interact with the microchips to preserve it during cryo-EM sample preparation. This last step can also aid in further enriching and selecting for native p53 proteins.

Enrichment and purification efforts paid off. Monomers (~50 kDa), even today, are just at the edge of being too small to be reconstructed by cryo-EM. Nevertheless, the team purified natively sourced p53 to determine the full-length monomer in sub-nanometer range (Solares et al., 2020). The map of ~70 Å in size easily accommodated the well-known X-ray structure of a single DBD (40 Å) but also could accommodate a theoretical NTD domain calculated through molecular modeling servers (Solares et al., 2020). Some smaller unoccupied densities could accommodate linking residues between the NTD and DBD. These reconstructions served to further identify tetramer assemblies in samples (Figure 5.2g). The lower half of the map was interpreted with the same DBD structure, but in tetrameric form, and the upper half contained the theoretical NTD model. Two smaller, unaccounted densities were

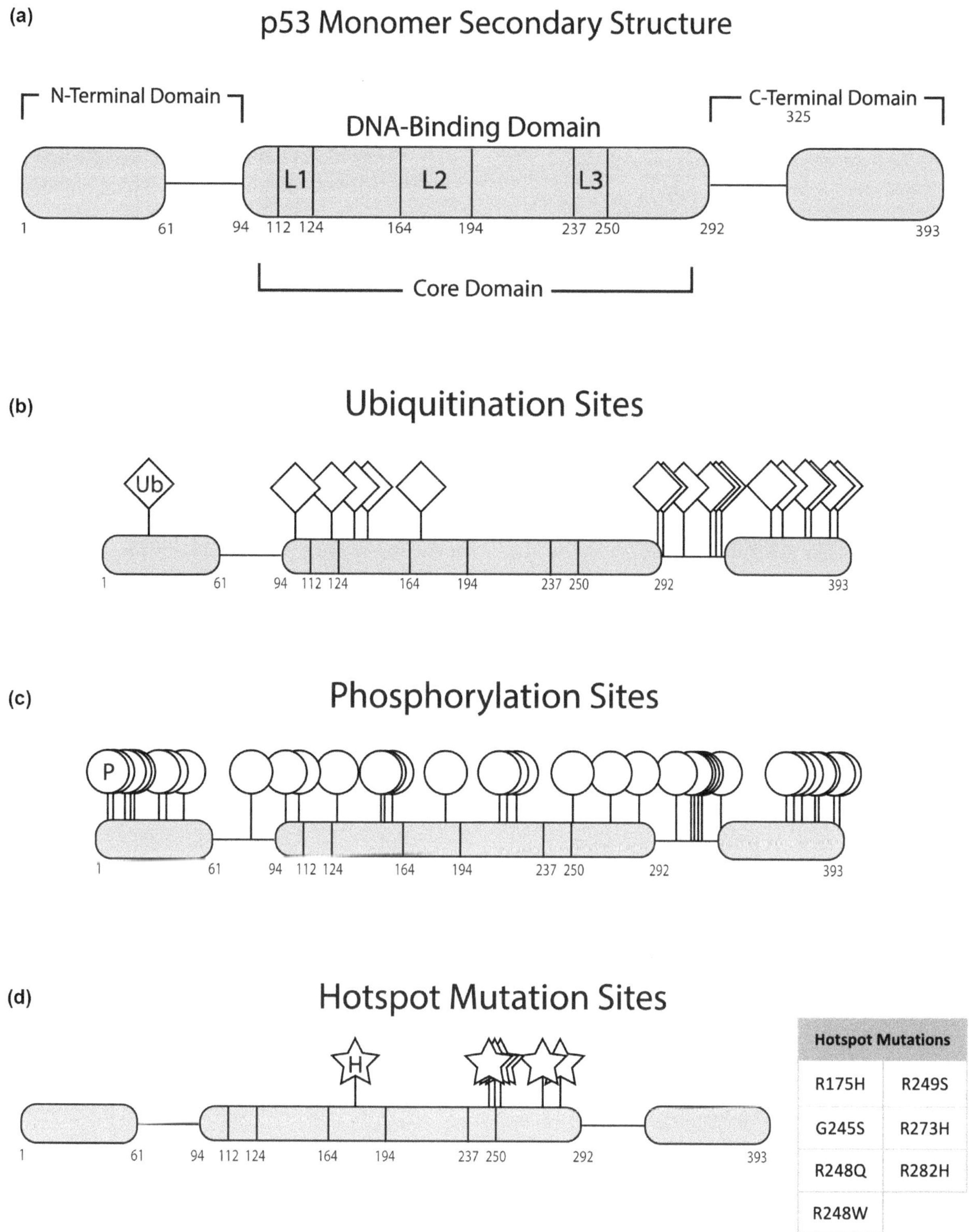

FIGURE 5.3 Posttranslational modifications (PTMs) are prevalent throughout the p53 primary sequence. (a) Schematic of the p53 primary sequence showing features of the N-terminal domain (NTD), core DBD, and C-terminal domain (CTD). The NTD transactivation domains 1/2 (TAD domains) are located in the N-terminus while the adjacent L1-L3 sequences are located in the DBD, to help p53 engage DNA. The CTD contains oligomerization domains (OD). (b, c) Prevalent PTMs, including ubiquitination and phosphorylation sites, are distributed through the p53 structure. (d) Disease-related hotspot mutations are generally clustered in the DBD of the p53 structure and include R175H, G245S, R248Q, R248W, R249S, R273H, and R282H.

attributed to ubiquitin modifications. Just as Alden discovered, cryo-EM can help piece together the biological story of some of these proteins. In the case of this tetramer, the glioblastoma cells that were used produce native WT p53. The ubiquitin moiety was resolved in this structure adjacent to residue K24, a known site that signals DNA repair. Further, a DNA fragment within the tetramer exhibited features associated with dsDNA breaks. As discussed earlier, increasing the number of ubiquitin moieties can decrease the likelihood that p53 can signal appropriate DNA repair. It is likely that within these cancer cells, p53 is appropriately signaling for DNA repair but is modified abnormally to maintain or even progress disease states (Dai & Gu, 2010).

Further work into the monomer proved effective. A new model refined to ~5–6 Å depicted density regions that were attributed to not just an NTD and DBD but also a CTD (Figure 5.2h) (Solares & Kelly, 2022). Protein prediction servers were able to construct an initial structure that could then be rebuilt into the density map using rigid-body refinement and flexible fitting techniques. The final model provides a new starting point into novel molecular interpretations of the full-length p53 structure. As many of the "hotspot" mutations reside in the DBD, the lack of complete structural models had not been a perceived simulation for researchers. Questions remained regarding changes in global conformations upon the introduction of single-point mutations. Using the new full-length p53 structure, Solares employed structure prediction software servers to calculate models for all the hotspot mutations (Solares & Kelly, 2022). Results showed that even subtle missense mutations could disrupt the electrostatic identity of the molecules (Figure 5.4).

Conformational mutations (R175H, G245S, R248Q, R249S, and R282H), for example, change the surface charge potentials of the protein. Introducing amino acids that shift the overall positive charge to a more negative value can destabilize the DNA binding pocket and potentially cause the loss of the key zinc ion. Additionally, DNA contact mutations (R273H and R248W) occur in residues that anchor key regions of p53 to the grooves of DNA. WT p53 has arginine residues that protrude and interact with DNA; however, the introduction of tryptophan can reduce the "reach" of p53 and its ability to engage DNA. Both amino acid changes, however, also affect the electrostatic profiles of the monomer and could potentially impede binding with the DNA backbone. The rippling effects of these mutations were evident in theoretical models, and initial modeling could shape experimental models in the future.

DISCUSSION/CONCLUSIONS

In terms of understanding the structure and function of p53, several researchers have become the new pillars in the field. Although Cho laid the initial groundwork by resolving the first structure of p53's core, it was Kitayner who was able to illuminate the dimerization and tetramerization process using X-ray crystallography. These structural efforts were magnified in turn through the revolution of cryo-EM by Nobel laureates Henderson, Dubochet, and Frank. With improved cryo-EM technology, Okorokov and Aramayo further probed p53's structure, revealing key features of the tetramer and how it binds to DNA. Additionally, Martin ended the p53 symmetry debate by applying his unique "birdcage" method for protein capture. The field of

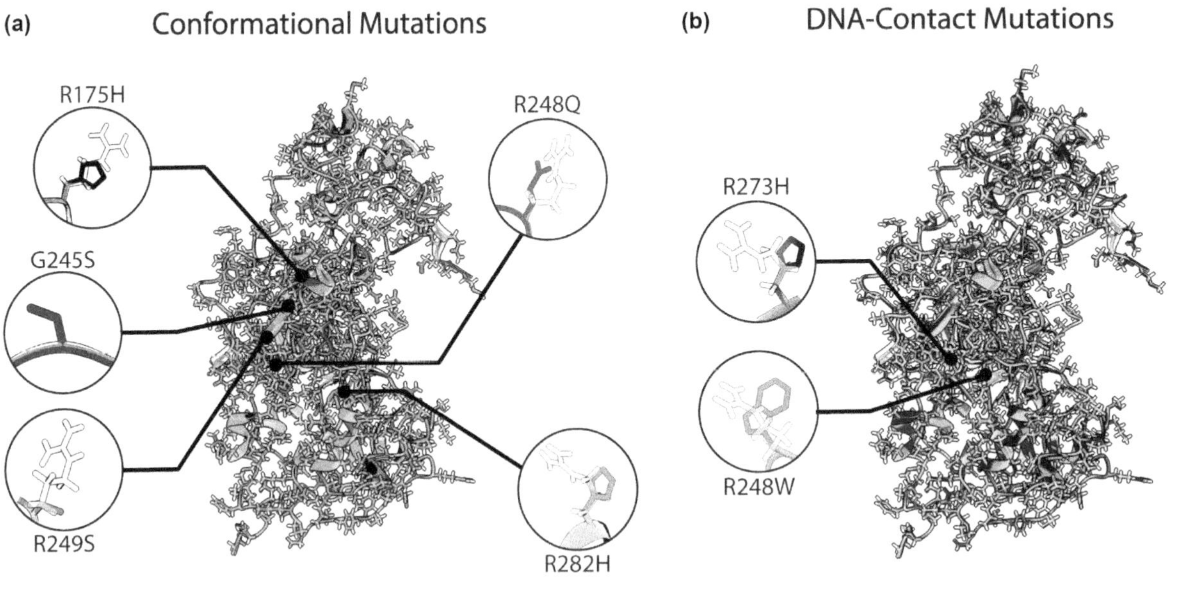

FIGURE 5.4 Molecular models of p53 hotspot mutations reveal changes in the DBD properties. (a) Overlay of molecular models containing conformational mutations with the individual hotspot mutation magnified. WT p53 model is shown in white for comparison. (b) Overlay of molecular models containing DNA-contact mutations with the individual hotspot mutation magnified. WT p53 model is shown in white for comparison.

biological sample extraction and preservation was furthered by Liang and Alden. Subsequently, Solares was finally able to explore mutational features by modeling flexible states of the protein based on higher-resolution work. It took a combination of biological, technological, and computational advancements to improve our current understanding of the structural properties of p53.

The result of the work done by the aforementioned individuals, alongside many others, has opened frontiers for researchers. Although the "moon race" to the highest-resolution, full-length structure continues, researchers can now go further into understanding the structure and function of one of the most insidious cancer proteins. As we pinpoint and assess these alterations, we may vanquish the myth of the undruggable target. Complete models and a fuller understanding of p53 can aid scientists in designing better targeted therapeutics and repair systems.

ACKNOWLEDGMENTS

This work was supported by the National Institutes of Health and the National Cancer Institute [R01CA193578 to DFK] and the Center for Structural Oncology at the Huck Institutes of the Life Sciences, Pennsylvania State University.

REFERENCES

Adorno, M., Cordenonsi, M., Montagner, M., Dupont, S., Wong, C., Hann, B., Solari, A., Bobisse, S., Rondina, M. B., Guzzardo, V., Parenti, A. R., Rosato, A., Bicciato, S., Balmain, A., & Piccolo, S. (2009). A Mutant-p53/smad complex opposes p63 to empower TGFβ-induced metastasis. *Cell*, *137*(1), 87–98. https://doi.org/10.1016/j.cell.2009.01.039

Adrian, M., Dubochet, J., Lepault, J., & McDowall, A. W. (1984). Cryo-electron microscopy of viruses. *Nature*, *308*(5954), 32–36. https://doi.org/10.1038/308032a0

Agapova, L. S., Ilyinskaya, G. V., Turovets, N. A., Ivanov, A. V., Chumakov, P. M., & Kopnin, B. P. (1996). Chromosome changes caused by alterations of p53 expression. *Mutation Research/Fundamental and Molecular Mechanisms of Mutagenesis*, *354*(1), 129–138. https://doi.org/10.1016/0027-5107(96)00062-0

Alden, N. A., Varano, A. C., Dearnaley, W. J., Solares, M. J., Luqiu, W. Y., Liang, Y., Sheng, Z., McDonald, S. M., Damiano, J., McConnell, J., Dukes, M. J., & Kelly, D. F. (2019). Cryo-EM-on-a-chip: Custom-designed substrates for the 3D analysis of macromolecules. *Small*, *15*(21), 1900918. https://doi.org/10.1002/smll.201900918

Aramayo, R., Sherman, M. B., Brownless, K., Lurz, R., Okorokov, A. L., & Orlova, E. V. (2011). Quaternary structure of the specific p53–DNA complex reveals the mechanism of p53 mutant dominance. *Nucleic Acids Research*, *39*(20), 8960–8971. https://doi.org/10.1093/nar/gkr386

Armstrong, J. F., Kaufman, M. H., Harrison, D. J., & Clarke, A. R. (1995). High-frequency developmental abnormalities in p53-deficient mice. *Current Biology*, *5*(8), 931–936. https://doi.org/10.1016/S0960-9822(95)00183-7

Bamford, S., Dawson, E., Forbes, S., Clements, J., Pettett, R., Dogan, A., Flanagan, A., Teague, J., Futreal, P. A., Stratton, M. R., & Wooster, R. (2004). The COSMIC (Catalogue of Somatic Mutations in Cancer) database and website. *British Journal of Cancer*, *91*(2), 355–358. https://doi.org/10.1038/sj.bjc.6601894

Barretina, J., Caponigro, G., Stransky, N., Venkatesan, K., Margolin, A. A., Kim, S., Wilson, C. J., Lehár, J., Kryukov, G. V., Sonkin, D., Reddy, A., Liu, M., Murray, L., Berger, M. F., Monahan, J. E., Morais, P., Meltzer, J., Korejwa, A., Jané-Valbuena, J., … Garraway, L. A. (2012). The Cancer Cell Line Encyclopedia enables predictive modelling of anticancer drug sensitivity. *Nature*, *483*(7391), 603–607. https://doi.org/10.1038/nature11003

Bartek, J., Iggo, R., Gannon, J., & Lane, D. P. (1990). Genetic and immunochemical analysis of mutant p53 in human breast cancer cell lines. *Oncogene*, *5*(6), 893–899.

Bell, S., Klein, C., Müller, L., Hansen, S., & Buchner, J. (2002). P53 contains large unstructured regions in its native state. *Journal of Molecular Biology*, *322*(5), 917–927. https://doi.org/10.1016/S0022-2836(02)00848-3

Bossi, G., Lapi, E., Strano, S., Rinaldo, C., Blandino, G., & Sacchi, A. (2006). Mutant p53 gain of function: Reduction of tumor malignancy of human cancer cell lines through abrogation of mutant p53 expression. *Oncogene*, *25*(2), 304–309. https://doi.org/10.1038/sj.onc.1209026

Böttger, V., Böttger, A., Garcia-Echeverria, C., Ramos, Y. F., van der Eb, A. J., Jochemsen, A. G., & Lane, D. P. (1999). Comparative study of the p53-mdm2 and p53-MDMX interfaces. *Oncogene*, *18*(1), 189–199. https://doi.org/10.1038/sj.onc.1202281

Botuyan, M. V. E., Momand, J., & Chen, Y. (1997). Solution conformation of an essential region of the p53 transactivation domain. *Folding and Design*, *2*(6), 331–342. https://doi.org/10.1016/S1359-0278(97)00047-3

Bouwman, P., & Jonkers, J. (2012). The effects of deregulated DNA damage signalling on cancer chemotherapy response and resistance. *Nature Reviews Cancer*, *12*(9), 587–598. https://doi.org/10.1038/nrc3342

Brooks, C. L., & Gu, W. (2003). Ubiquitination, phosphorylation and acetylation: The molecular basis for p53 regulation. *Current Opinion in Cell Biology*, *15*(2), 164–171. https://doi.org/10.1016/S0955-0674(03)00003-6

Brooks, C. L., & Gu, W. (2011). P53 regulation by ubiquitin. *FEBS Letters*, *585*(18), 2803–2809. https://doi.org/10.1016/j.febslet.2011.05.022

Brugarolas, J., Chandrasekaran, C., Gordon, J. I., Beach, D., Jacks, T., & Hannon, G. J. (1995). Radiation-induced cell cycle arrest compromised by p21 deficiency. *Nature*, *377*(6549), 552–557. https://doi.org/10.1038/377552a0

Buganim, Y., Kalo, E., Brosh, R., Besserglick, H., Nachmany, I., Rais, Y., Stambolsky, P., Tang, X., Milyavsky, M., Shats, I., Kalis, M., Goldfinger, N., & Rotter, V. (2006). Mutant p53 protects cells from 12-*O*-tetradecanoylphorbol-13-acetate–induced death by attenuating activating transcription factor 3 induction. *Cancer Research*, *66*(22), 10750–10759. https://doi.org/10.1158/0008-5472.CAN-06-0916

Chatterjee, N., & Walker, G. C. (2017). Mechanisms of DNA damage, repair, and mutagenesis: DNA damage and repair. *Environmental and Molecular Mutagenesis*, *58*(5), 235–263. https://doi.org/10.1002/em.22087

Chen, J. (2012). The roles of MDM2 and MDMX phosphorylation in stress signaling to p53. *Genes & Cancer*, *3*(3–4), 274–282. https://doi.org/10.1177/1947601912454733

Chen, J. (2016). The cell-cycle arrest and apoptotic functions of p53 in tumor initiation and progression. *Cold Spring Harbor Perspectives in Medicine*, *6*(3), a026104. https://doi.org/10.1101/cshperspect.a026104

Chen, Y., Zhang, X., Dantas Machado, A. C., Ding, Y., Chen, Z., Qin, P. Z., Rohs, R., & Chen, L. (2013). Structure of p53 binding to the BAX response element reveals DNA unwinding and compression to accommodate base-pair insertion. *Nucleic Acids Research*, *41*(17), 8368–8376. https://doi.org/10.1093/nar/gkt584

Cho, Y., Gorina, S., Jeffrey, P. D., & Pavletich, N. P. (1994). Crystal structure of a p53 tumor suppressor-DNA complex: Understanding tumorigenic mutations. *Science*, *265*(5170), 346–355. https://doi.org/10.1126/science.8023157

Clair, S. St., Giono, L., Varmeh-Ziaie, S., Resnick-Silverman, L., Liu, W., Padi, A., Dastidar, J., DaCosta, A., Mattia, M., & Manfredi, J. J. (2004). DNA damage-induced downregulation of Cdc25C is mediated by p53 via two independent mechanisms. *Molecular Cell*, *16*(5), 725–736. https://doi.org/10.1016/j.molcel.2004.11.002

Collisson, E. A., Bailey, P., Chang, D. K., & Biankin, A. V. (2019). Molecular subtypes of pancreatic cancer. *Nature Reviews Gastroenterology & Hepatology*, *16*(4), 207–220. https://doi.org/10.1038/s41575-019-0109-y

Dai, C., & Gu, W. (2010). p53 post-translational modification: Deregulated in tumorigenesis. *Trends in Molecular Medicine*, *16*(11), 528–536. https://doi.org/10.1016/j.molmed.2010.09.002

Deng, C., Zhang, P., Wade Harper, J., Elledge, S. J., & Leder, P. (1995). Mice lacking p21CIP1/WAF1 undergo normal development, but are defective in G1 checkpoint control. *Cell*, *82*(4), 675–684. https://doi.org/10.1016/0092-8674(95)90039-X

Derbyshire, D. J. (2002). Crystal structure of human 53BP1 BRCT domains bound to p53 tumour suppressor. *The EMBO Journal*, *21*(14), 3863–3872. https://doi.org/10.1093/emboj/cdf383

Di Agostino, S., Strano, S., Emiliozzi, V., Zerbini, V., Mottolese, M., Sacchi, A., Blandino, G., & Piaggio, G. (2006). Gain of function of mutant p53: The mutant p53/NF-Y protein complex reveals an aberrant transcriptional mechanism of cell cycle regulation. *Cancer Cell*, *10*(3), 191–202. https://doi.org/10.1016/j.ccr.2006.08.013

Dong, P., Karaayvaz, M., Jia, N., Kaneuchi, M., Hamada, J., Watari, H., Sudo, S., Ju, J., & Sakuragi, N. (2013). Mutant p53 gain-of-function induces epithelial–mesenchymal transition through modulation of the miR-130b–ZEB1 axis. *Oncogene*, *32*(27), 3286–3295. https://doi.org/10.1038/onc.2012.334

Dong, P., Xu, Z., Jia, N., Li, D., & Feng, Y. (2009). Elevated expression of p53 gain-of-function mutation R175H in endometrial cancer cells can increase the invasive phenotypes by activation of the EGFR/PI3K/AKT pathway. *Molecular Cancer*, *8*(1), 103. https://doi.org/10.1186/1476-4598-8-103

Dubochet, J., & McDowall, A. (1981). Vitrification of pure water for electron microscopy. *Journal of Microscopy*, *124*(3), 3–4.

Dunker, A. K., Cortese, M. S., Romero, P., Iakoucheva, L. M., & Uversky, V. N. (2005). Flexible nets. The roles of intrinsic disorder in protein interaction networks. *FEBS Journal*, *272*(20), 5129–5148. https://doi.org/10.1111/j.1742-4658.2005.04948.x

El-Deiry, W. S. (1998). Regulation of p53 downstream genes. *Seminars in Cancer Biology*, *8*(5), 345–357. https://doi.org/10.1006/scbi.1998.0097

El-Deiry, W. S., Tokino, T., Velculescu, V. E., Levy, D. B., Parsons, R., Trent, J. M., Lin, D., Mercer, W. E., Kinzler, K. W., & Vogelstein, B. (1993). WAF1, a potential mediator of p53 tumor suppression. *Cell*, *75*(4), 817–825. https://doi.org/10.1016/0092-8674(93)90500-P

Eskes, R., Desagher, S., Antonsson, B., & Martinou, J.-C. (2000). Bid induces the oligomerization and insertion of bax into the outer mitochondrial membrane. *Molecular and Cellular Biology*, *20*(3), 929–935. https://doi.org/10.1128/MCB.20.3.929-935.2000

Frank, J. (1975). Averaging of low exposure electron micrographs of non-periodic objects. *Ultramicroscopy*, *1*(2), 159–162. https://doi.org/10.1016/S0304-3991(75)80020-9

Freed-Pastor, W. A., Mizuno, H., Zhao, X., Langerød, A., Moon, S.-H., Rodriguez-Barrueco, R., Barsotti, A., Chicas, A., Li, W., Polotskaia, A., Bissell, M. J., Osborne, T. F., Tian, B., Lowe, S. W., Silva, J. M., Børresen-Dale, A.-L., Levine, A. J., Bargonetti, J., & Prives, C. (2012). Mutant p53 disrupts mammary tissue architecture via the mevalonate pathway. *Cell*, *148*(1–2), 244–258. https://doi.org/10.1016/j.cell.2011.12.017

Gaidano, G., Ballerini, P., Gong, J. Z., Inghirami, G., Neri, A., Newcomb, E. W., Magrath, I. T., Knowles, D. M., & Dalla-Favera, R. (1991). p53 mutations in human lymphoid malignancies: Association with Burkitt lymphoma and chronic lymphocytic leukemia. *Proceedings of the National Academy of Sciences of the United States of America*, *88*(12), 5413–5417. https://doi.org/10.1073/pnas.88.12.5413

Goldstein, J. C., Waterhouse, N. J., Juin, P., Evan, G. I., & Green, D. R. (2000). The coordinate release of cytochrome c during apoptosis is rapid, complete and kinetically invariant. *Nature Cell Biology*, *2*(3), 156–162. https://doi.org/10.1038/35004029

Gorina, S., & Pavletich, N. P. (1996). Structure of the p53 tumor suppressor bound to the ankyrin and SH3 domains of 53BP2. *Science*, *274*(5289), 1001–1005. https://doi.org/10.1126/science.274.5289.1001

Grugan, K. D., Vega, M. E., Wong, G. S., Diehl, J. A., Bass, A. J., Wong, K. K., Nakagawa, H., & Rustgi, A. K. (2013). A common p53 mutation (R175H) activates c-Met receptor tyrosine kinase to enhance tumor cell invasion. *Cancer Biology & Therapy*, *14*(9), 853–859. https://doi.org/10.4161/cbt.25406

Hakem, R. (2008). DNA-damage repair; the good, the bad, and the ugly. *The EMBO Journal*, *27*(4), 589–605. https://doi.org/10.1038/emboj.2008.15

Hamard, P.-J., Barthelery, N., Hogstad, B., Mungamuri, S. K., Tonnessen, C. A., Carvajal, L. A., Senturk, E., Gillespie, V., Aaronson, S. A., Merad, M., & Manfredi, J. J. (2013). The C terminus of p53 regulates gene expression by multiple mechanisms in a target- and tissue-specific manner in vivo. *Genes & Development*, *27*(17), 1868–1885. https://doi.org/10.1101/gad.224386.113

Harris, C. R., DeWan, A., Zupnick, A., Normart, R., Gabriel, A., Prives, C., Levine, A. J., & Hoh, J. (2009). P53 responsive elements in human retrotransposons. *Oncogene*, *28*(44), 3857–3865. https://doi.org/10.1038/onc.2009.246

Haupt, Y., Maya, R., Kazaz, A., & Oren, M. (1997). Mdm2 promotes the rapid degradation of p53. *Nature*, *387*(6630), 296–299. https://doi.org/10.1038/387296a0

Henderson, R., Baldwin, J. M., Ceska, T. A., Zemlin, F., Beckmann, E., & Downing, K. H. (1990). Model for the structure of bacteriorhodopsin based on high-resolution electron cryo-microscopy. *Journal of Molecular Biology*, *213*(4), 899–929. https://doi.org/10.1016/S0022-2836(05)80271-2

Hermeking, H., Lengauer, C., Polyak, K., He, T.-C., Zhang, L., Thiagalingam, S., Kinzler, K. W., & Vogelstein, B. (1997). 14-3-3sigma is a p53-Regulated Inhibitor of G2/M Progression. *Molecular Cell*, *1*(1), 3–11. https://doi.org/10.1016/S1097-2765(00)80002-7

Hollstein, M., Sidransky, D., Vogelstein, B., & Harris, C. (1991). P53 mutations in human cancers. *Science*, *253*(5015), 49. https://doi.org/10.1126/science.1905840

Honda, R., Tanaka, H., & Yasuda, H. (1997). Oncoprotein MDM2 is a ubiquitin ligase E3 for tumor suppressor p53. *FEBS Letters*, *420*(1), 25–27. https://doi.org/10.1016/S0014-5793(97)01480-4

Hornbeck, P. V., Kornhauser, J. M., Tkachev, S., Zhang, B., Skrzypek, E., Murray, B., Latham, V., & Sullivan, M. (2012). PhosphoSitePlus: A comprehensive resource for investigating the structure and function of experimentally determined post-translational modifications in man and mouse. *Nucleic Acids Research*, *40*(Database issue), D261–D270. https://doi.org/10.1093/nar/gkr1122

Hsiao, M., Low, J., Dorn, E., Ku, D., Pattengale, P., Yeargin, J., & Haas, M. (1994). Gain-of-function mutations of the p53 gene induce lymphohematopoietic metastatic potential and tissue invasiveness. *The American Journal of Pathology*, *145*(3), 702–714.

Hsu, Y.-T., Wolter, K. G., & Youle, R. J. (1997). Cytosol-to-membrane redistribution of Bax and Bcl-X_L during apoptosis. *Proceedings of the National Academy of Sciences*, *94*(8), 3668–3672. https://doi.org/10.1073/pnas.94.8.3668

Jeffers, J. R., Parganas, E., Lee, Y., Yang, C., Wang, J., Brennan, J., MacLean, K. H., Han, J., Chittenden, T., Ihle, J. N., McKinnon, P. J., Cleveland, J. L., & Zambetti, G. P. (2003). Puma is an essential mediator of p53-dependent and -independent apoptotic pathways. *Cancer Cell*, *4*(4), 321–328. https://doi.org/10.1016/S1535-6108(03)00244-7

Jeffrey, P. D., Gorina, S., & Pavletich, N. P. (1995). Crystal structure of the tetramerization domain of the p53 tumor suppressor at 1.7 angstroms. *Science*, *267*(5203), 1498–1502. https://doi.org/10.1126/science.7878469

Joerger, A. C., Allen, M. D., & Fersht, A. R. (2004). Crystal structure of a superstable mutant of human p53 core domain. *Journal of Biological Chemistry*, *279*(2), 1291–1296. https://doi.org/10.1074/jbc.M309732200

Joerger, A. C., Ang, H. C., & Fersht, A. R. (2006). Structural basis for understanding oncogenic p53 mutations and designing rescue drugs. *Proceedings of the National Academy of Sciences*, *103*(41), 15056–15061. https://doi.org/10.1073/pnas.0607286103

Joerger, A. C., Ang, H. C., Veprintsev, D. B., Blair, C. M., & Fersht, A. R. (2005). Structures of p53 cancer mutants and mechanism of rescue by second site suppressor mutations. *Journal of Biological Chemistry*, *280*(16), 16030–16037. https://doi.org/10.1074/jbc.M500179200

Joerger, A. C., & Fersht, A. R. (2007). Structure–function–rescue: The diverse nature of common p53 cancer mutants. *Oncogene*, *26*(15), 2226–2242. https://doi.org/10.1038/sj.onc.1210291

Joerger, A. C., & Fersht, A. R. (2010). The tumor suppressor p53: From structures to drug discovery. *Cold Spring Harbor Perspectives in Biology*, *2*(6), a000919. https://doi.org/10.1101/cshperspect.a000919

Joo, W. S., Jeffrey, P. D., Cantor, S. B., Finnin, M. S., Livingston, D. M., & Pavletich, N. P. (2002). Structure of the 53BP1 BRCT region bound to p53 and its comparison to the Brca1 BRCT structure. *Genes & Development*, *16*(5), 583–593. https://doi.org/10.1101/gad.959202

Kalo, E., Kogan-Sakin, I., Solomon, H., Bar-Nathan, E., Shay, M., Shetzer, Y., Dekel, E., Goldfinger, N., Buganim, Y., Stambolsky, P., Goldstein, I., Madar, S., & Rotter, V. (2012). Mutant p53R273H attenuates the expression of phase 2 detoxifying enzymes and promotes the survival of cells with high ROS levels. *Journal of Cell Science*, *125*, 5578–5586. https://doi.org/10.1242/jcs.106815

Kandoth, C., McLellan, M. D., Vandin, F., Ye, K., Niu, B., Lu, C., Xie, M., Zhang, Q., McMichael, J. F., Wyczalkowski, M. A., Leiserson, M. D. M., Miller, C. A., Welch, J. S., Walter, M. J., Wendl, M. C., Ley, T. J., Wilson, R. K., Raphael, B. J., & Ding, L. (2013). Mutational landscape and significance across 12 major cancer types. *Nature*, *502*(7471), 333–339. https://doi.org/10.1038/nature12634

Kern, S. E., Kinzler, K. W., Bruskin, A., Jarosz, D., Friedman, P., Prives, C., & Vogelstein, B. (1991). Identification of p53 as a sequence-specific DNA-binding protein. *Science*, *252*(5013), 1708–1711. https://doi.org/10.1126/science.2047879

Kim, E., & Deppert, W. (2006). The versatile interactions of p53 with DNA: When flexibility serves specificity. *Cell Death & Differentiation*, *13*(6), 885–889. https://doi.org/10.1038/sj.cdd.4401909

Kitayner, M., Rozenberg, H., Kessler, N., Rabinovich, D., Shaulov, L., Haran, T. E., & Shakked, Z. (2006). Structural basis of DNA recognition by p53 tetramers. *Molecular Cell*, *22*(6), 741–753. https://doi.org/10.1016/j.molcel.2006.05.015

Koulgi, S., Achalere, A., Sharma, N., Sonavane, U., & Joshi, R. (2013). QM-MM simulations on p53-DNA complex: A study of hot spot and rescue mutants. *Journal of Molecular Modeling*, *19*(12), 5545–5559. https://doi.org/10.1007/s00894-013-2042-2

Kubbutat, M. H. G., Jones, S. N., & Vousden, K. H. (1997). Regulation of p53 stability by Mdm2. *Nature*, *387*(6630), 299–303. https://doi.org/10.1038/387299a0

Kung, C.-P., & Weber, J. D. (2022). It's getting complicated—A fresh look at p53-MDM2-ARF triangle in tumorigenesis and cancer therapy. *Frontiers in Cell and Developmental Biology*, *10*, 818744. https://doi.org/10.3389/fcell.2022.818744

Lang, G. A., Iwakuma, T., Suh, Y.-A., Liu, G., Rao, V. A., Parant, J. M., Valentin-Vega, Y. A., Terzian, T., Caldwell, L. C., Strong, L. C., El-Naggar, A. K., & Lozano, G. (2004). Gain of function of a p53 hot spot mutation in a mouse model of Li-Fraumeni syndrome. *Cell*, *119*(6), 861–872. https://doi.org/10.1016/j.cell.2004.11.006

Lanni, C., Uberti, D., Racchi, M., Govoni, S., & Memo, M. (2007). Unfolded p53: A potential biomarker for Alzheimer's disease. *Journal of Alzheimer's Disease*, *12*(1), 93–99. https://doi.org/10.3233/JAD-2007-12109

Laptenko, O., Shiff, I., Freed-Pastor, W., Zupnick, A., Mattia, M., Freulich, E., Shamir, I., Kadouri, N., Kahan, T., Manfredi, J., Simon, I., & Prives, C. (2015). The p53 C terminus controls site-specific DNA binding and promotes structural changes within the central DNA binding domain. *Molecular Cell*, *57*(6), 1034–1046. https://doi.org/10.1016/j.molcel.2015.02.015

Lavin, M. F., & Gueven, N. (2006). The complexity of p53 stabilization and activation. *Cell Death & Differentiation*, *13*(6), 941–950. https://doi.org/10.1038/sj.cdd.4401925

Lee, H., Mok, K. H., Muhandiram, R., Park, K.-H., Suk, J.-E., Kim, D.-H., Chang, J., Sung, Y. C., Choi, K. Y., & Han, K.-H. (2000). Local structural elements in the mostly unstructured transcriptional activation domain of human p53. *Journal of Biological Chemistry*, *275*(38), 29426–29432. https://doi.org/10.1074/jbc.M003107200

Lepre, M. G., Omar, S. I., Grasso, G., Morbiducci, U., Deriu, M. A., & Tuszynski, J. A. (2017). Insights into the effect of the G245S single point mutation on the structure of p53 and the binding of the protein to DNA. *Molecules*, *22*(8). https://doi.org/10.3390/molecules22081358

Liang, Y., Dearnaley, W. J., Alden, N. A., Solares, M. J., Gilmore, B. L., Pridham, K. J., Varano, A. C., Sheng, Z., Alli, E., & Kelly, D. F. (2019). Correcting errors in the BRCA1 warning system. *DNA Repair, 73*, 120–128. https://doi.org/10.1016/j.dnarep.2018.11.009

Liu, K., Ling, S., & Lin, W.-C. (2011). TopBP1 mediates mutant p53 gain of function through NF-Y and p63/p73. *Molecular and Cellular Biology, 31*(22), 4464–4481. https://doi.org/10.1128/MCB.05574-11

Liu, Y., & Bodmer, W. F. (2006). Analysis of P53 mutations and their expression in 56 colorectal cancer cell lines. *Proceedings of the National Academy of Sciences of the United States of America, 103*(4), 976–981. https://doi.org/10.1073/pnas.0510146103

Loging, W. T., & Reisman, D. (1999). Elevated expression of ribosomal protein genes L37, RPP-1, and S2 in the presence of mutant p531. *Cancer Epidemiology, Biomarkers & Prevention, 8*(11), 1011–1016.

Luo, Y., Hurwitz, J., & Massagué, J. (1995). Cell-cycle inhibition by independent CDK and PCNA binding domains in p21Cip1. *Nature, 375*(6527), 159–161. https://doi.org/10.1038/375159a0

Marchenko, N. D., & Moll, U. M. (2007). The role of ubiquitination in the direct mitochondrial death program of p53. *Cell Cycle, 6*(14), 1718–1723. https://doi.org/10.4161/cc.6.14.4503

Marine, J.-C., Francoz, S., Maetens, M., Wahl, G., Toledo, F., & Lozano, G. (2006). Keeping p53 in check: Essential and synergistic functions of Mdm2 and Mdm4. *Cell Death & Differentiation, 13*(6), 927–934. https://doi.org/10.1038/sj.cdd.4401912

Markl, I. D., & Jones, P. A. (1998). Presence and location of TP53 mutation determines pattern of CDKN2A/ARF pathway inactivation in bladder cancer. *Cancer Research, 58*(23), 5348–5353.

Martin, T. G., Bharat, T. A. M., Joerger, A. C., Bai, X., Praetorius, F., Fersht, A. R., Dietz, H., & Scheres, S. H. W. (2016). Design of a molecular support for cryo-EM structure determination. *Proceedings of the National Academy of Sciences, 113*(47). https://doi.org/10.1073/pnas.1612720113

Matas, D. (2001). Integrity of the N-terminal transcription domain of p53 is required for mutant p53 interference with drug-induced apoptosis. *The EMBO Journal, 20*(15), 4163–4172. https://doi.org/10.1093/emboj/20.15.4163

McLure, K. G., & Lee, P. W. K. (1998). How p53 binds DNA as a tetramer. *The EMBO Journal, 17*(12), 3342–3350. https://doi.org/10.1093/emboj/17.12.3342

Mihara, M., Erster, S., Zaika, A., Petrenko, O., Chittenden, T., Pancoska, P., & Moll, U. M. (2003). P53 has a direct apoptogenic role at the mitochondria. *Molecular Cell, 11*(3), 577–590. https://doi.org/10.1016/S1097-2765(03)00050-9

Mittl, P. R. E., Chène, P., & Grütter, M. G. (1998). Crystallization and structure solution of p53 (residues 326–356) by molecular replacement using an NMR model as template. *Acta Crystallographica Section D Biological Crystallography, 54*(1), 86–89. https://doi.org/10.1107/S0907444997006550

Momand, J., Wu, H.-H., & Dasgupta, G. (2000). MDM2—Master regulator of the p53 tumor suppressor protein. *Gene, 242*(1–2), 15–29. https://doi.org/10.1016/S0378-1119(99)00487-4

Moneo, V., Serelde, B. G., Fominaya, J., Leal, J. F. M., Blanco-Aparicio, C., Romero, L., Sánchez-Beato, M., Cigudosa, J. C., Tercero, J. C., Piris, M. A., Jimeno, J., & Carnero, A. (2007). Extreme sensitivity to Yondelis (Trabectedin, ET-743) in low passaged sarcoma cell lines correlates with mutated p53. *Journal of Cellular Biochemistry, 100*(2), 339–348. https://doi.org/10.1002/jcb.21073

Moore, P. S., Sipos, B., Orlandini, S., Sorio, C., Real, F. X., Lemoine, N. R., Gress, T., Bassi, C., Klöppel, G., Kalthoff, H., Ungefroren, H., Löhr, M., & Scarpa, A. (2001). Genetic profile of 22 pancreatic carcinoma cell lines. Analysis of K-ras, p53, p16 and DPC4/Smad4. *Virchows Archiv: An International Journal of Pathology, 439*(6), 798–802. https://doi.org/10.1007/s004280100474

Muller, P. A. J., Caswell, P. T., Doyle, B., Iwanicki, M. P., Tan, E. H., Karim, S., Lukashchuk, N., Gillespie, D. A., Ludwig, R. L., Gosselin, P., Cromer, A., Brugge, J. S., Sansom, O. J., Norman, J. C., & Vousden, K. H. (2009). Mutant p53 drives invasion by promoting integrin recycling. *Cell, 139*(7), 1327–1341. https://doi.org/10.1016/j.cell.2009.11.026

Muller, P. A. J., Trinidad, A. G., Timpson, P., Morton, J. P., Zanivan, S., van den Berghe, P. V. E., Nixon, C., Karim, S. A., Caswell, P. T., Noll, J. E., Coffill, C. R., Lane, D. P., Sansom, O. J., Neilsen, P. M., Norman, J. C., & Vousden, K. H. (2013). Mutant p53 enhances MET trafficking and signalling to drive cell scattering and invasion. *Oncogene, 32*(10), 1252–1265. https://doi.org/10.1038/onc.2012.148

Muller, P. A. J., & Vousden, K. H. (2014). Mutant p53 in cancer: New functions and therapeutic opportunities. *Cancer Cell, 25*(3), 304–317. https://doi.org/10.1016/j.ccr.2014.01.021

Murarka, S. P. (2003). Silicon dioxide, nitride, and oxynitride. In K. H. Jürgen Buschow, R. W. Cahn, M. C. Flemings, B. Ilschner, E. J. Kramer, S. Mahajan, P. Veyssière (Eds.). *Encyclopedia of Materials: Science and Technology* (pp. 1–14). Elsevier. https://doi.org/10.1016/B0-08-043152-6/01895-7

Nicholls, C. D., McLure, K. G., Shields, M. A., & Lee, P. W. K. (2002). Biogenesis of p53 involves cotranslational dimerization of monomers and posttranslational dimerization of dimers: Implications on the dominant negative effect. *Journal of Biological Chemistry, 277*(15), 12937–12945. https://doi.org/10.1074/jbc.M108815200

Noda, A., Ning, Y., Venable, S. F., Pereira-Smith, O. M., & Smith, J. R. (1994). Cloning of senescent cell-derived inhibitors of DNA synthesis using an expression screen. *Experimental Cell Research, 211*(1), 90–98. https://doi.org/10.1006/excr.1994.1063

Noll, J. E., Jeffery, J., Al-Ejeh, F., Kumar, R., Khanna, K. K., Callen, D. F., & Neilsen, P. M. (2012). Mutant p53 drives multinucleation and invasion through a process that is suppressed by ANKRD11. *Oncogene, 31*(23), 2836–2848. https://doi.org/10.1038/onc.2011.456

Ohki, R., Nemoto, J., Murasawa, H., Oda, E., Inazawa, J., Tanaka, N., & Taniguchi, T. (2000). Reprimo, a new candidate mediator of the p53-mediated cell cycle arrest at the G2 phase. *Journal of Biological Chemistry, 275*(30), 22627–22630. https://doi.org/10.1074/jbc.C000235200

Okorokov, A. L., & Milner, J. (1999). An ATP/ADP-Dependent Molecular Switch Regulates the Stability of p53-DNA Complexes. *Molecular and Cellular Biology, 19*(11), 7501–7510. https://doi.org/10.1128/MCB.19.11.7501

Okorokov, A. L., Sherman, M. B., Plisson, C., Grinkevich, V., Sigmundsson, K., Selivanova, G., Milner, J., & Orlova, E. V. (2006). The structure of p53 tumour suppressor protein reveals the basis for its functional plasticity. *The EMBO Journal, 25*(21), 5191–5200. https://doi.org/10.1038/sj.emboj.7601382

Olivier, M., Hollstein, M., & Hainaut, P. (2010). TP53 mutations in human cancers: Origins, consequences, and clinical use. *Cold Spring Harbor Perspectives in Biology, 2*(1), a001008. https://doi.org/10.1101/cshperspect.a001008

Ozaki, T., & Nakagawara, A. (2011). Role of p53 in cell death and human cancers. *Cancers*, *3*(1), 994–1013. https://doi.org/10.3390/cancers3010994

Pugacheva, E. N., Ivanov, A. V., Kravchenko, J. E., Kopnin, B. P., Levine, A. J., & Chumakov, P. M. (2002). Novel gain of function activity of p53 mutants: Activation of the dUTPase gene expression leading to resistance to 5-fluorouracil. *Oncogene*, *21*(30), 4595–4600. https://doi.org/10.1038/sj.onc.1205704

Radermacher, M., Wagenknecht, T., Verschoor, A., & Frank, J. (1986). A new 3-D reconstruction scheme applied to the 50S ribosomal subunit of *E. coli*. *Journal of Microscopy*, *141*(1), RP1–RP2. https://doi.org/10.1111/j.1365-2818.1986.tb02693.x

Riley, T., Sontag, E., Chen, P., & Levine, A. (2008). Transcriptional control of human p53-regulated genes. *Nature Reviews Molecular Cell Biology*, *9*(5), 402–412. https://doi.org/10.1038/nrm2395

Royds, J. A., & Iacopetta, B. (2006). p53 and disease: When the guardian angel fails. *Cell Death & Differentiation*, *13*(6), 1017–1026. https://doi.org/10.1038/sj.cdd.4401913

Sah, V. P., Attardi, L. D., Mulligan, G. J., Williams, B. O., Bronson, R. T., & Jacks, T. (1995). A subset of p53-deficient embryos exhibit exencephaly. *Nature Genetics*, *10*(2), 175–180. https://doi.org/10.1038/ng0695-175

Sarig, R., Rivlin, N., Brosh, R., Bornstein, C., Kamer, I., Ezra, O., Molchadsky, A., Goldfinger, N., Brenner, O., & Rotter, V. (2010). Mutant p53 facilitates somatic cell reprogramming and augments the malignant potential of reprogrammed cells. *Journal of Experimental Medicine*, *207*(10), 2127–2140. https://doi.org/10.1084/jem.20100797

Sauer, L., Gitenay, D., Vo, C., & Baron, V. T. (2010). Mutant p53 initiates a feedback loop that involves Egr-1/EGF receptor/ERK in prostate cancer cells. *Oncogene*, *29*(18), 2628–2637. https://doi.org/10.1038/onc.2010.24

Scian, M. J., Stagliano, K. E. R., Ellis, M. A., Hassan, S., Bowman, M., Miles, M. F., Deb, S. P., & Deb, S. (2004). Modulation of gene expression by tumor-derived p53 mutants. *Cancer Research*, *64*(20), 7447–7454. https://doi.org/10.1158/0008-5472.CAN-04-1568

Shen, P. S. (2018). The 2017 nobel prize in chemistry: Cryo-EM comes of age. *Analytical and Bioanalytical Chemistry*, *410*(8), 2053–2057. https://doi.org/10.1007/s00216-018-0899-8

Shi, X.-B., Nesslinger, N. J., Deitch, A. D., Gumerlock, P. H., & deVere White, R. W. (2002). Complex functions of mutantp53 alleles from human prostate cancer. *The Prostate*, *51*(1), 59–72. https://doi.org/10.1002/pros.10072

Solares, M. J., Jonaid, G. M., Luqiu, W. Y., Liang, Y., Evans, M. C., Dearnaley, W. J., Sheng, Z., & Kelly, D. F. (2020). Microchip-based structure determination of disease-relevant p53. *Analytical Chemistry*, *92*(23), 15558–15564. https://doi.org/10.1021/acs.analchem.0c03599

Solares, M. J., & Kelly, D. F. (2022). Complete models of p53 better inform the impact of hotspot mutations. *International Journal of Molecular Sciences*, *23*(23), 15267. https://doi.org/10.3390/ijms232315267

Song, H., Hollstein, M., & Xu, Y. (2007). P53 gain-of-function cancer mutants induce genetic instability by inactivating ATM. *Nature Cell Biology*, *9*(5), 573–580. https://doi.org/10.1038/ncb1571

Sullivan, K. D., Galbraith, M. D., Andrysik, Z., & Espinosa, J. M. (2018). Mechanisms of transcriptional regulation by p53. *Cell Death & Differentiation*, *25*(1), 133–143. https://doi.org/10.1038/cdd.2017.174

Szybińska, A., & Leśniakx, L. (2017). P53 dysfunction in neurodegenerative diseases—The cause or effect of pathological changes? *Aging and Disease*, *8*(4), 506. https://doi.org/10.14336/AD.2016.1120

Tait, S. W. G., & Green, D. R. (2010). Mitochondria and cell death: Outer membrane permeabilization and beyond. *Nature Reviews Molecular Cell Biology*, *11*(9), 621–632. https://doi.org/10.1038/nrm2952

Tanaka, H., Arakawa, H., Yamaguchi, T., Shiraishi, K., Fukuda, S., Matsui, K., Takei, Y., & Nakamura, Y. (2000). A ribonucleotide reductase gene involved in a p53-dependent cell-cycle checkpoint for DNA damage. *Nature*, *404*(6773), 42–49. https://doi.org/10.1038/35003506

Tokino, T., Thiagalingam, S., El-Deiry, W. S., Waldman, T., Kinzler, K. W., & Vogelstein, B. (1994). P53 tagged sites from human genomic DNA. *Human Molecular Genetics*, *3*(9), 1537–1542. https://doi.org/10.1093/hmg/3.9.1537

Tsang, W., Ho, F. Y. F., Fung, K., Kong, S., & Kwok, T. (2005). P53-R175H mutant gains new function in regulation of doxorubicin-induced apoptosis. *International Journal of Cancer*, *114*(3), 331–336. https://doi.org/10.1002/ijc.20818

Tseng, J. E., Rodriguez, M., Ro, J., Liu, D., Hong, W. K., & Mao, L. (1999). Gender differences in p53 mutational status in small cell lung cancer. *Cancer Research*, *59*(22), 5666–5670.

Unwin, P. N. T., & Henderson, R. (1975). Molecular structure determination by electron microscopy of unstained crystalline specimens. *Journal of Molecular Biology*, *94*(3), 425–440. https://doi.org/10.1016/0022-2836(75)90212-0

van Heel, M., & Frank, J. (1981). Use of multivariates statistics in analysing the images of biological macromolecules. *Ultramicroscopy*, *6*(1), 187–194. https://doi.org/10.1016/S0304-3991(81)80197-0

Vaseva, A. V., Marchenko, N. D., Ji, K., Tsirka, S. E., Holzmann, S., & Moll, U. M. (2012). P53 Opens the mitochondrial permeability transition pore to trigger necrosis. *Cell*, *149*(7), 1536–1548. https://doi.org/10.1016/j.cell.2012.05.014

Vaughan, C. A., Frum, R., Pearsall, I., Singh, S., Windle, B., Yeudall, A., Deb, S. P., & Deb, S. (2012). Allele specific gain-of-function activity of p53 mutants in lung cancer cells. *Biochemical and Biophysical Research Communications*, *428*(1), 6–10. https://doi.org/10.1016/j.bbrc.2012.09.029

Vaughan, C. A., Singh, S., Windle, B., Sankala, H. M., Graves, P. R., Andrew Yeudall, W., Deb, S. P., & Deb, S. (2012). P53 mutants induce transcription of NF-κB2 in H1299 cells through CBP and STAT binding on the NF-κB2 promoter and gain of function activity. *Archives of Biochemistry and Biophysics*, *518*(1), 79–88. https://doi.org/10.1016/j.abb.2011.12.006

Vousden, K. H., & Lu, X. (2002). Live or let die: The cell's response to p53. *Nature Reviews Cancer*, *2*(8), 594–604. https://doi.org/10.1038/nrc864

Vousden, K. H., & Prives, C. (2009). Blinded by the light: The growing complexity of p53. *Cell*, *137*(3), 413–431. https://doi.org/10.1016/j.cell.2009.04.037

Wade Harper, J. (1993). The p21 Cdk-interacting protein Cip1 is a potent inhibitor of G1 cyclin-dependent kinases. *Cell*, *75*(4), 805–816. https://doi.org/10.1016/0092-8674(93)90499-G

Walker, D. R., Bond, J. P., Tarone, R. E., Harris, C. C., Makalowski, W., Boguski, M. S., & Greenblatt, M. S. (1999). Evolutionary conservation and somatic mutation hotspot maps of p53: Correlation with p53 protein structural and functional features. *Oncogene*, *18*(1), 211–218. https://doi.org/10.1038/sj.onc.1202298

Wang, W., Cheng, B., Miao, L., Mei, Y., & Wu, M. (2013). Mutant p53-R273H gains new function in sustained activation of EGFR signaling via suppressing miR-27a expression. *Cell Death & Disease*, *4*(4), e574. https://doi.org/10.1038/cddis.2013.97

Wei, M. C., Lindsten, T., Mootha, V. K., Weiler, S., Gross, A., Ashiya, M., Thompson, C. B., & Korsmeyer, S. J. (2000). tBID, a membrane-targeted death ligand, oligomerizes BAK to release cytochrome c. *Genes & Development*, *14*(16), 2060–2071. https://doi.org/10.1101/gad.14.16.2060

Weisz, L., Zalcenstein, A., Stambolsky, P., Cohen, Y., Goldfinger, N., Oren, M., & Rotter, V. (2004). Transactivation of the *EGR1* gene contributes to mutant p53 gain of function. *Cancer Research*, *64*(22), 8318–8327. https://doi.org/10.1158/0008-5472.CAN-04-1145

Willis, A., Jung, E. J., Wakefield, T., & Chen, X. (2004). Mutant p53 exerts a dominant negative effect by preventing wild-type p53 from binding to the promoter of its target genes. *Oncogene*, *23*(13), 2330–2338. https://doi.org/10.1038/sj.onc.1207396

Wolfrum, P., Fietz, A., Schnichels, S., & Hurst, J. (2022). The function of p53 and its role in Alzheimer's and Parkinson's disease compared to age-related macular degeneration. *Frontiers in Neuroscience*, *16*, 1029473. https://doi.org/10.3389/fnins.2022.1029473

Wolters, S., & Schumacher, B. (2013). Genome maintenance and transcription integrity in aging and disease. *Frontiers in Genetics*, *4*. https://doi.org/10.3389/fgene.2013.00019

Xiong, Y., Hannon, G. J., Zhang, H., Casso, D., Kobayashi, R., & Beach, D. (1993). P21 is a universal inhibitor of cyclin kinases. *Nature*, *366*(6456), 701–704. https://doi.org/10.1038/366701a0

Yan, W., Liu, G., Scoumanne, A., & Chen, X. (2008). Suppression of inhibitor of differentiation 2, a target of mutant p53, is required for gain-of-function mutations. *Cancer Research*, *68*(16), 6789–6796. https://doi.org/10.1158/0008-5472.CAN-08-0810

Yeudall, W. A., Vaughan, C. A., Miyazaki, H., Ramamoorthy, M., Choi, M.-Y., Chapman, C. G., Wang, H., Black, E., Bulysheva, A. A., Deb, S. P., Windle, B., & Deb, S. (2012). Gain-of-function mutant p53 upregulates CXC chemokines and enhances cell migration. *Carcinogenesis*, *33*(2), 442–451. https://doi.org/10.1093/carcin/bgr270

Yi, L., Lu, C., Hu, W., Sun, Y., & Levine, A. J. (2012). Multiple roles of p53-related pathways in somatic cell reprogramming and stem cell differentiation. *Cancer Research*, *72*(21), 5635–5645. https://doi.org/10.1158/0008-5472.CAN-12-1451

Yin, Y., Tainsky, M. A., Bischoff, F. Z., Strong, L. C., & Wahl, G. M. (1992). Wild-type p53 restores cell cycle control and inhibits gene amplification in cells with mutant p53 alleles. *Cell*, *70*(6), 937–948. https://doi.org/10.1016/0092-8674(92)90244-7

Yoshikawa, K., Hamada, J., Tada, M., Kameyama, T., Nakagawa, K., Suzuki, Y., Ikawa, M., Monsur Hassan, N. M., Kitagawa, Y., & Moriuchi, T. (2010). Mutant p53 R248Q but not R248W enhances in vitro invasiveness of human lung cancer NCI-H1299 cells. *Biomedical Research*, *31*(6), 401–411. https://doi.org/10.2220/biomedres.31.401

Zhang, Y., Yan, W., & Chen, X. (2011). Mutant p53 disrupts MCF-10A cell polarity in three-dimensional culture via epithelial-to-mesenchymal transitions. *Journal of Biological Chemistry*, *286*(18), 16218–16228. https://doi.org/10.1074/jbc.M110.214585

Zhao, L., Lee, V. H. F., Ng, M. K., Yan, H., & Bijlsma, M. F. (2019). Molecular subtyping of cancer: Current status and moving toward clinical applications. *Briefings in Bioinformatics*, *20*(2), 572–584. https://doi.org/10.1093/bib/bby026

Section II

Technology Platform and Evolving Technologies

6 Single-Particle Cryo-EM
What Happens Inside the Black Box?

Sayan Bhakta and Marin van Heel

HISTORICAL INTRODUCTION

Microscopy provides direct visual access to the miniscule biological world. The first observation of monocellular organisms by van Leeuwenhoek with his single-lens optical microscopes (van Leeuwenhoek, 1677) was a sensational landmark in our understanding of the biological world. With gradual improvements made to light microscopes, we eventually reached the resolution limits imposed by the wavelength of light. Whereas optical microscopy is a good tool for revealing the spatial distribution of large cellular organelles, surpassing this resolution limitation requires the use of radiation with a (much) shorter wavelength, such as that of accelerated electrons, postulated by De Broglie (De Broglie, 1924). The invention of the electron microscope (**EM**) (Ruska & Knoll, 1931) opened the door to many important further instrumental developments that eventually led to the revolutionary elucidation of molecular architectures at an atomic scale (Kühlbrandt, 2014). Those developments in EM took some 70 years to mature, because of the many challenges to overcome. First, in the 1960s and 1970s, the challenge was to improve the inherently low contrast of biological samples, often not at all visible in the EM, using protocols like negative staining with heavy metals (Brenner & Horne, 1959; Van Bruggen et al., 1960). Next came the better understanding of the information obtained in EM images, and how to exploit that information in digital form, using the "software machines" that were developed for the ever-more powerful computing resources that became available. The first 2D and 3D reconstruction algorithms were developed for regular aggregates of molecules like helical fibers (DeRosier & Klug, 1968; DeRosier & Moore, 1970); icosahedral viruses (Crowther, 1971); two-dimensional crystals (Unwin & Henderson, 1975); and three-dimensional reconstruction from images recorded by tilting the sample (Hoppe et al., 1974; Oettl et al., 1983; Harauz et al., 1987; Radermacher et al., 1987). These techniques were developed for use with the available high-contrast, negative staining specimen preparation approaches. In the high vacuum of the EM, such specimen preparation led to structural degradation due to the loss of the bound water of the biological molecules, and they typically also showed a strong orientation relative to the support film.

In the 1980s and 1990s, an important focus in biological EM was on maintaining the structural integrity of biological specimens by keeping them hydrated in the high-vacuum environment of the EM. Life arose in aqueous environments, and removing water from biological samples normally destroys the sample's structural integrity. Early landmark successes in cryogenic EM were achieved by low-dose imaging of 2D crystals like purple membrane and porins that are sometimes found naturally in cell membranes, samples of which could be embedded in sugars for structural preservation (Unwin & Henderson, 1975; Sass et al., 1989; Henderson et al., 1990). It was especially the remarkable discovery of the ***vitreous-ice*** state of water that enabled the direct visualization of individual biological complexes in their hydrated state within the EM (Fernández-Morán, 1960; Taylor & Glaser, 1976; Chanzy et al., 1976; Brüggeller & Mayer, 1980; Dubochet & McDowall, 1981; Jeng & Chiu 1984; Chiu et al.,1986), and of the freeze-plunging specimen-preparation technique (Adrian et al., 1984; Dubochet et al., 1988) that gave rise to today's cryo-EM routine.

With the advent of a practical cryogenic EM specimen preparation came a new challenge: to avoid tilting altogether and exploit random orientations of single particles within the vitreous sample to achieve 3D reconstruction based on thousands or millions of extremely noisy molecular images. A universal zero-tilt method was what was called for, to assign Euler angles to class averages of aligned molecular images, found by unsupervised classification (Van Heel, 1984a; 1984b; Borland & van Heel, 1990) in a compressed eigenvector space (Van Heel & Frank, 1981). The real-space angular reconstitution approach was proposed in 1987 (Van Heel, 1987a) and first came to routine operation in 1994 (Orlova & van Heel, 1994; Van Heel et al., 1992; Van Heel et al., 1997). This zero-tilt approach to generate three-dimensional (3D) structures of biological complexes from their 2D cryo-EM projection images (or *class averages*) superseded tilt-based approaches because of its experimental simplicity. The zero-tilt approach, in combination with projection matching or angular reconstitution, is now the norm [Van Heel, 1984b; 1987a], although that may not always be transparent to the users of black-box programs.

In the decades after the advent of the cryo-EM technique, we have seen a continuous stream of technical and methodological improvements in sample preparation, electron optics, direct electron recording of images, the recording and alignment of "movies" (Kunath et al., 1984), and especially in data analysis. Very important also were the interpretation and validation tools that emerged over the decades. For example, prior to the first direct visualization of the surfaces of 3D reconstructions some 40 years ago, the best tool available was to create physical 3D models by cutting out balsawood sections and gluing them together

DOI: 10.1201/9781003326106-8

(Van Heel, 1983). We now have advanced visualization tools like Chimera (Pettersen et al., 2004) to simulate surface views or even to simulate the "docking" of different 3D structures. All these developments, combined, brought a revolution in resolution of cryo-EM (Kühlbrandt 2014). The purpose of this chapter is not so much to review all that is currently happening in the cryo-EM field, but rather to look back to correct persistent errors and missed opportunities that entered the cryo-EM procedures.

THE ABC OF SINGLE-PARTICLE CRYO-EM

Cryo-EM single-particle analysis (SPA) consists of an array of different procedures, many of which include iterative refinement cycles. Metrics to assess the success of such refinement cycles are paramount to the success of the overall approach. In general, one can state that angular reconstitution approaches (Van Heel, 1987a; Afanasyev et al., 2017) are less prone to reference bias than the projection matching approaches (Van Heel, 1984b; Harauz & Van Heel, 1986b). We here follow the alignment-by-classification (ABC) pipeline (Dube et al., 1993; Afanasyev et al., 2017), a procedure that is largely reference-free and is thus relatively insensitive to reference bias, an effect also known as "Einstein from noise" (Schatz & Van Heel, 1990). The basic overall ABC procedures covered in this review include (Figure 6.1): camera normalization of the input images, CTF correction of the images (or their movie frames), movie-frame alignments, automatic particle selection, and selected particle alignment and classification. This is followed by 3D reconstruction(s) of the structure(s) based on the obtained 2D class averages whereby the Euler angle assignments are made by angular reconstitution (Van Heel, 1987a) or by projection matching (Van Heel, 1984b), or later variants thereof. These cryo-EM procedures are important and deserve extra attention. Many of these methodologies were pioneered within the IMAGIC system, yet in various black boxes the original procedures have been given new names that obscure their purpose. We highlight critical points that exist in general in single-particle cryo-EM procedures and explain the underlying principles.

In the first phase of the ABC pipeline, the micrographs (typically "movies") are preprocessed to normalize the digital camera's output (**CAMERA CORRECTION**), to correct anisotropic magnification present in the images, and to determine the local contrast transfer function (CTF) in different areas of the micrographs (**CTF AND ANISOTROPIC MAGNIFICATION**). Modern-day cryo-EM cameras produce movies, that is, a sequence of "movie frames," from the same area of the sample. These movies are the best source for finding the local CTFs, and the CTF-corrected movie frames are then aligned with respect to each other and summed. (**MOVIE ALIGNMENT**) The resulting movie averages are then used to find the "particles" from the micrographs (**PARTICLE PICKING**). Unsupervised classification of the particles yields similar projection views in similar orientations, which are then averaged to produce still relatively noisy class-averages, a procedure called alignment by classification. The aligned averaged particles obtained by this zero-tilt approach are then merged into a single 3D volume, exploiting the different orientations of the molecular complexes revealed by angular reconstitution. A great advantage of the zero-tilt approach is that it is amenable to 4D approaches where different structural-functional states of the complex, present in the same sample, can be studied separately (**4D PROCESSING**).

CAMERA CORRECTION

In the age of silver-halide film, each film was used only once, developed individually, scanned to provide a digital image, and then discarded. Modern direct electron detectors (DEDs) (Faruqi & Henderson, 2007), in contrast, are fundamentally different since the same sensor is used to record thousands of images in every data-collection experiment. That means that during the collection of a large cryo-EM dataset with, say, 10,000 images, we collect for each individual pixel of the 4K-by-4K image, a vector of 10,000 measurements. That implies that each pixel in the 2D sensor can be seen as an individual measurement device used to collect thousands of measurements. Each of these pixel vectors, here of length 10,000, characterize an individual pixel in terms of its sensitivity and average output. In the simplest case, in which the camera properties remain constant over time, it suffices to calculate a corrected image by subtracting the average image from each measured image and divide that by the standard deviation image (Afanasyev et al., 2015). That is so simple and straightforward that a referee commented, "This is a case of: Why didn't someone think of this before."

When the digital camera is not stable over time, or more specifically, the software by the camera manufacturer regularly resets the device to compensate for instabilities, one needs a more sophisticated (multivariate) approach to diagnose the problem. The effect of the fixed pattern of the digital camera significantly complicates the downstream movie-alignment process (as discussed in (**MOVIE ALIGNMENT**) and thus sometimes ends up ruining the whole cryo-EM experiment. The running costs and depreciation costs of a modern cryo-EM instrument is of the order of US$5000 per day. We have diagnosed camera instability normalization problems with two high-end cryo-EMs at the Biozentrum in Basel, Switzerland, that rendered those EMs useless for two years, a depreciation loss of over US$5 million, not including the costs to the Institute's science of not having access to those crucial instruments! The importance of the camera normalization approach is underestimated, given the relatively low number of citations that this (free-download) paper collected.

Camera correction is not restricted to cryo-EM data but is a generally applicable principle. When we analyzed a large dataset of Mars images collected with the NASA *Curiosity* rover (Afanasyev et al., 2015), we managed to *a posteriori* improve the interpretable 2D resolution of those *Curiosity* images by a factor of ~10 at no cost, for a project with an estimated overall cost of US$2.5 billion. The improvement in data collection by this camera

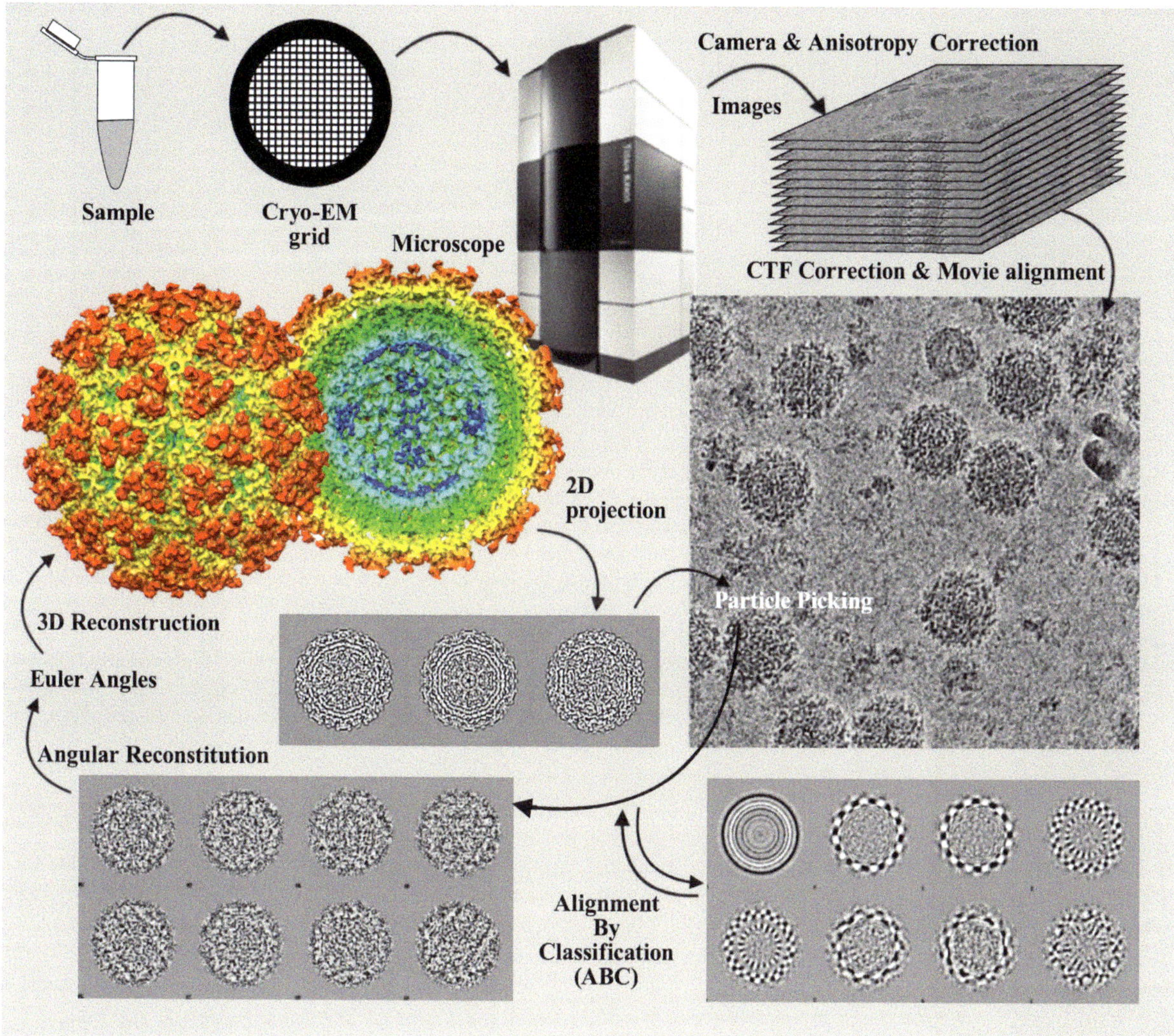

FIGURE 6.1 Overview of cryo-EM and ABC single-particle analysis. The purified sample is vitrified on a cryo-EM grid, rapidly cooled to liquid nitrogen temperature, and imaged in a cryogenic electron microscope. The recorded images are processed in multiple steps: camera correction, anisotropic magnification, and CTF correction, etc. High-resolution micrographs are recorded as a sequence of movie frames that must be aligned with respect to each other. The molecular images (particles) are then detected and extracted from the preprocessed micrographs. An unsupervised alignment by classification of the particles yields "class averages." Euler angles are assigned by angular reconstitution to the members of each class, and a first 3D reconstruction is calculated. The procedures are refined iteratively, repeating the particle picking but now using projection images from the first 3D as references in order to refine the whole process and harvest atomic information.

normalization can be assessed "at the source" by reliable metrics such as the Fourier Ring Correlation (FRC) and Fourier Ring Information (FRI). It is important to have good metrics for assessing each procedural phase so as to reach an optimal cryo-EM result.

CTF AND ANISOTROPIC MAGNIFICATION

The electron microscope is primarily a phase-contrast instrument (Scherzer, 1949). Largely the same rules apply for phase contrast in EM as for the phase-contrast microscope in light optics (Zernike, 1942), the main difference being that in EM the phase contrast is not a specific hardware device, but is rather created "virtually" by the combination of spherical aberration and defocus of the magnetic lenses. The most important aspect of an imaging device is its linearity. Linearity implies that we can predict the output of a linear imaging system (Figure 6.2) if we know the transmission function of the object (in real space) and the (phase-contrast) transfer function of the imaging device (in Fourier space). As is shown in Figure 6.2, the linear imaging procedure is mathematically formalized as the result of multiplying the Fourier transform of the object by the CTF and Fourier transforming that product back to real space.

Unfortunately, fundamental problems were introduced in early (biological) CTF publications, flaws that have persisted until the present day. In the first place, in the much-cited paper

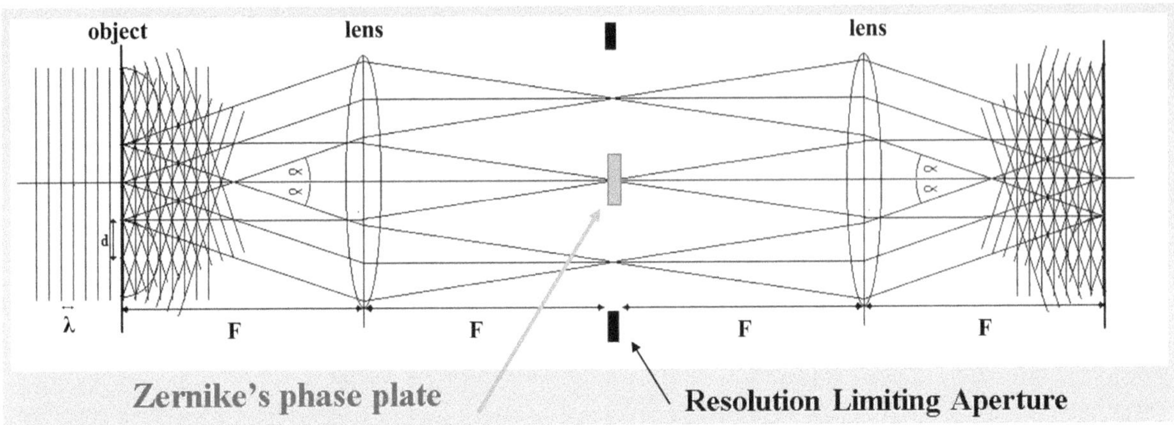

FIGURE 6.2 Symbolic representation of a linear imaging system. The 4F symbolic imaging system (four focal distances "F") contains all elements necessary for linear imaging compressed to a necessary minimum: It consists of two Fourier transform lenses that create a one-to-one image of the object in the imaging plane (modified from: Van Heel, 1980–2022). The object plane (left) and the image plane (right) are in "real space," whereas the back-focal plane of the first lens is in Fourier space, where phase plates, the CTF, and/or the limiting aperture are located, and to be multiplied to the Fourier transform of the object wave. The second lens then Fourier-transforms that product back to real space. A Zernike phase plate rotates the relative phase of the illumination ("zero-order beam") with respect to the beams diffracted by the object. This phase plate is what makes the phase information in the object visible as density variations in the image. In the electron microscope, the phase-contrast transfer function (PhCTF) normally fulfills the role of the phase plate in light microscopy. Note that if the illumination is not strictly coherent, the illumination that is focused in the back-focal plane becomes a disk or a Gaussian, complicating the calculation of the CTF by a convolution over the image of the source.

by Erickson and Klug (Erickson & Klug, 1971; Erickson, 1973), the EM amplitude contrast of the object is assumed to be proportional to the phase contrast of the same object over all spatial frequencies—at least that is how that original paper is interpreted today (Frank, 2006). If that proportionality were indeed true, one single transfer function would suffice to describe how the linear imaging device generates an output image. In reality, however, amplitude contrast and phase contrast are two separate properties of the complex transmission function of the object, and these are associated with different physical properties of the sample (Van Heel, 1978; Van Heel et al. 2000) and are covered by two different transfer functions: the amplitude or intensity transfer function (ITF), and the phase-contrast transfer function (PhCTF) (Van Heel, 1978). In their most correct form, these can be combined into a single complex transfer function, operating on an object described by a single complex function (which is not implemented in current programs). That assumed amplitude-to-phase proportionality is fictitious and leads to many different proportionality constants (between, say, 5% and 35% amplitude contrast versus phase contrast), summarized in Mindell & Grigorieff (2003) and Frank (2006). However, the error that emerges from that incorrect assumption is that the defocus values that the CTF determination programs produce becomes incorrect. Assuming that direct proportionality introduces a cosine term that shifts the zero-crossings of the resulting "hybrid CTF" compared to the correct, sine PhCTF. Experimental results assessed with such a "standard" yet incorrect mathematical model will yield invalid results in comparative studies like Marabini et al., 2015. Popular CTF-finding programs like CTFFIND4 (Rohou & Grigorieff, 2015) that are often used in GUIs like Relion (Scheres, 2012) also suffer from this proportionality problem.

A further CTF problem in EM was the coverage of partial coherence in Erickson (1973) and in Frank (1973), which are both problematic for different reasons. The latter paper especially approximates a convolution-like operation in Fourier space (an integration over the image of the source in the back-focal plane) by a multiplication with an envelope function (moreover, a note-added-in-press points at an error in the paper that the author never corrects). Such an envelope function cannot shift the zeroes of the CTF and can also not dampen the CTF to a zero value. In the Van Heel (1978) paper, partial coherence is treated correctly, but the paper did not explicitly refute the flaws in Frank (1973). The necessary partially coherent Fourier-space CTF convolutions are implemented as multiplications in the conjugate space in IMAGIC-4D (Van Heel et al., 2012). A better appreciation of the effects of partial coherence may expand the applicability of cheaper, non-FEG electron sources in EM in high-resolution cryo-EM.

Full dataset CTF correction (FDS-CTF) procedures as proposed in Afanasyev et al. (2017) probably are the fastest and most efficient CTF correction procedures available for large datasets. The approach, summarized in Figure 6.3, starts off with the anisotropic magnification correction (Afanasyev et al., 2017) without which the CTF parameters found become only approximate, and the "blurriness" of

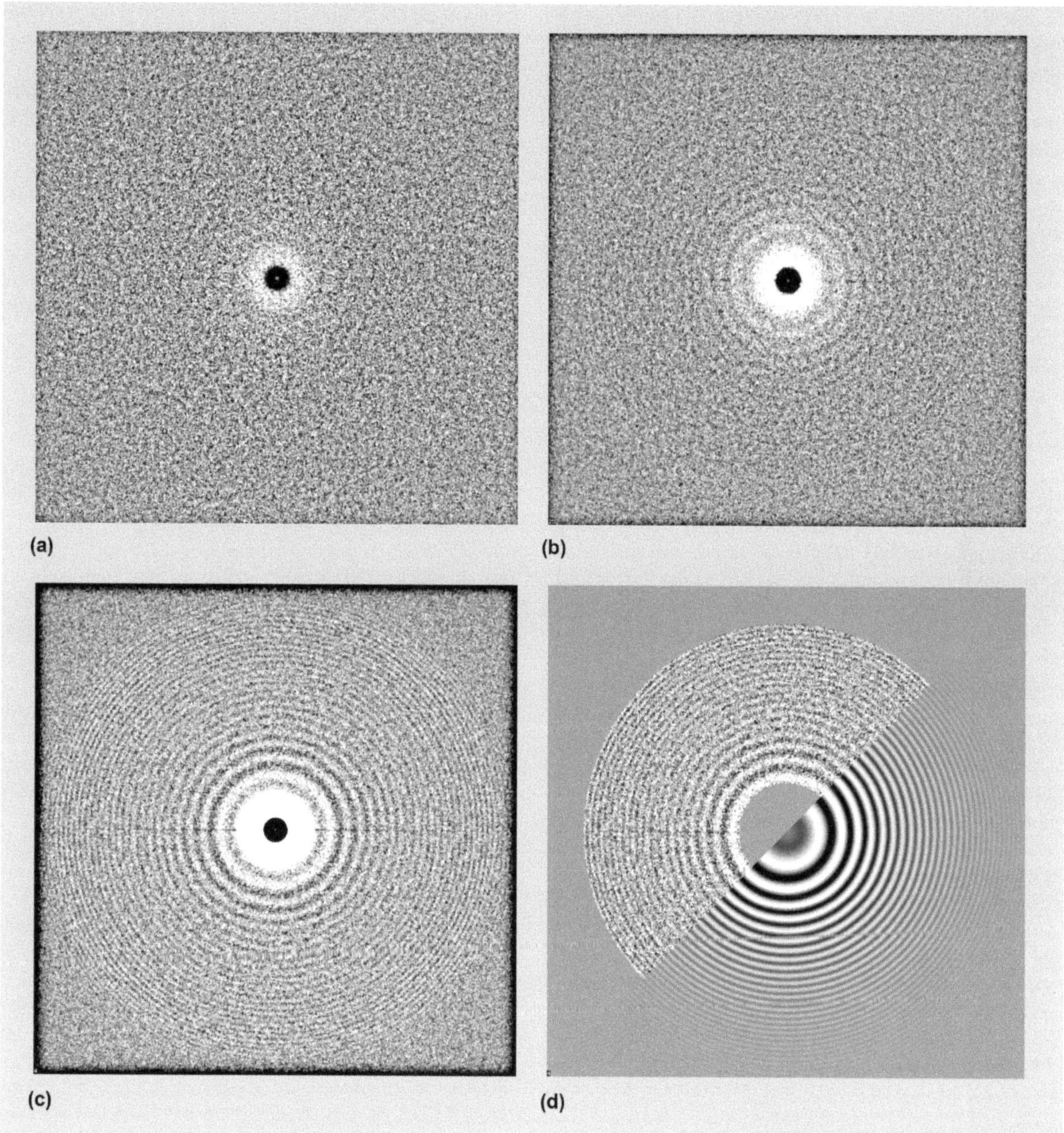

FIGURE 6.3 Full-dataset CTF correction: of a SARS-CoV-2 dataset (EMPIAR 10723). (a) Spectrum of a 1/12th movie cut from a full movie ("one micrograph"). A total of 24,000 such spectra entered into this CTF analysis. (b) Spectrum of the full movie ("one micrograph"). Since this corresponds to 12 times the area of the spectrum shown in (a), CTF rings (Thon rings) become visible. There are 2,000 such micrograph spectra in this preliminary analysis. (c) Class average of 1/12th spectra. The 24,000 1/12th spectra were MSA-classified into 600 class averages each, thus containing the sum of ~100 such 1/12th spectra. One of these 1/12th spectra is shown here. (d) CTF fitting to 1/12th spectra class averages. In short: The precision with which the CTF parameters of the 1/12th movie spectra as shown in (a) is given by the quality of the CTF fit as shown in (d). The CTF parameters are then conveyed to the original 1/12th movie for CTF correction. Further details in main text.

larger single-particle complexes increases as a function of the distance to the center of the structure. The FDS-CTF approach is superior to any CTF procedure that is based on correcting a single micrograph (or movie-micrograph) because the CTF information of the full dataset containing thousands of micrographs or (movie-micrographs) is integrated into the analysis on a subframe basis. That implies that even if the subframe (movie-) micrographs are very small, the local CTF precision is based on the integrated local information from everywhere in the full dataset.

As is illustrated in Figure 6.3, the full-frame movie-micrographs are cut into overlapping submovies (12, in this case) from the full movie. The advantage is that the 1/12th submovies can then each be treated independently in terms of their defocus parameters and in terms of their local movie-frame alignments. Although the spectrum of a 1/12th submovie is fully invisible (Figure 6.3a), when 12 such subspectra from one micrograph are summed, Thon rings start to appear (Figure 6.3b). Much better than averaging 12 such subspectra is to classify all 1/12th submovie spectrum patches (from the *full* dataset) and to then sum the 1/12th movie-spectrum patches that are members of the same class (Figure 6.3c). The amazing precision in CTF determination (Figure 6.3d) then applies to each member of the corresponding class (Figure 6.3a), where one, visually, does not recognize any Thon rings at all!

MOVIE ALIGNMENT

During an electron exposure aimed at collecting enough information on an EM sample, the sample will suffer from continuous (local) movement due to many factors like the electron exposure itself charging up or heating up the sample locally, or due to mechanical or electronic instabilities of the microscope. The idea of collecting sequences of individual micrographs to be treated as "movie frames" arose already in the 1980s, a methodology pioneered by Wolfgang Kunath and coworkers (Kunath et al., 1984). Aligning individual frames with respect to each other, and subsequently averaging them to a single micrograph with a better overall reproducible resolution was a methodology pioneered in the Fritz Haber Institute of the Max Planck Society. The technology was first implemented using the first "electron counting" cameras with a total pixel count of only 256 × 256 pixels. The approach was reinvented for use in cryo-EM (Campbell et al., 2012) without citing the original work. With new cameras with a field of view of ~4096 × 4096 pixels, movie alignment algorithms are now performed more locally by, for example, subdividing the full field of view into, say, 16 or 25 overlapping submovies covering the full field of view (see, for example, Afanasyev et al., 2017).

The precision of a movie alignment is often hampered by the presence of fixed-pattern high-frequency component in the images. The routine camera-correction method discussed above minimizes the spurious correlation due to such fixed background pattern in the transducer chip and electronics. Spurious correlations from a fixed background pattern are characterized by a "zero" shift in the translational alignment and are due to the highest-frequency components in the data, associated by the line and column patterns in the imaging chip. These spurious correlations can thus be significantly suppressed by the movie-alignment program by low-pass filtering and by suppressing zero shifts. The quality improvement by the movie alignment can be also assessed immediately at the source by reliable metrics such as amplitude spectrum measurements and Fourier ring information (FRI)—see below (**Fourier Ring/Shell Correlation and Fourier Ring/Shell Information**). Just as with the camera correction, it is important to use metrics for the quality of each procedure performed rather than to wait for final-result metrics. In terms of speed of processing, much can be done in a single night using "gaming" computers with, say, eight cores or more and large memories (32–64 gigabytes) using an appropriate program like the IMAGIC-4D, Align-Movie-Frames program (Afanasyev et al., 2017). Very important for such operations is the use of fast, large M.2 SSD disk drives because the speed of I/O operations is a main limiting factor for movie-alignment operations.

PARTICLE PICKING

The selection of individual molecular images from noisy electron micrographs is a challenging task since these projection images are extremely noisy and are often not even visible in the raw micrographs. The noise level of the micrographs is so high that conventional image segmentation approaches (Wikipedia, Image segmentation) such as, for example, edge detection, stand no chance of success because the noise level already makes the whole molecular image invisible, let alone its edges. Yet these approaches continue to emerge in the literature, ignoring such obvious principles of particle picking. Detection of particles from the micrographs becomes easier when some *a priori* knowledge of the molecule is available that one can exploit to look for specific views of a particle. However, such reference-based approaches are prone to reference bias (see discussion).

One of the most sensitive and reference-free particle-picking approaches is based on local variance or the local modulation of the images. The presence of an object at a certain point of the image contributes to the local variance or modulation that adds to the background noise present everywhere. This thus generates a peak in the local variance in the micrograph due to particle's presence. The local variance or modulation can be calculated exploiting the speed of the FFT algorithm (Van Heel, 1982). Band-pass filtering the input data to emphasize the relevant spatial frequencies that best discriminate the particles from the background is of paramount importance. Strangely enough, almost none of the scores of papers on particle picking mention this aspect of particle picking. The presence of strong low-frequency components like density ramps, edges of the grid-hole, and ice crystals may corrupt the results of local variance/modulation calculations and are thus best removed. The average densities of the images are set to zero since modulations caused by phase contrast in the EM have a zero average. An example of modulation-based particle picking is shown in Figure 6.4. Our preferred overall procedure is to first pick the particles by modulation picking, where no reference bias is possible, and to later use the alignment by classification (ABC) approach to focus on the specific properties of the particles being studied and to get new particle-related references for a new round of particle

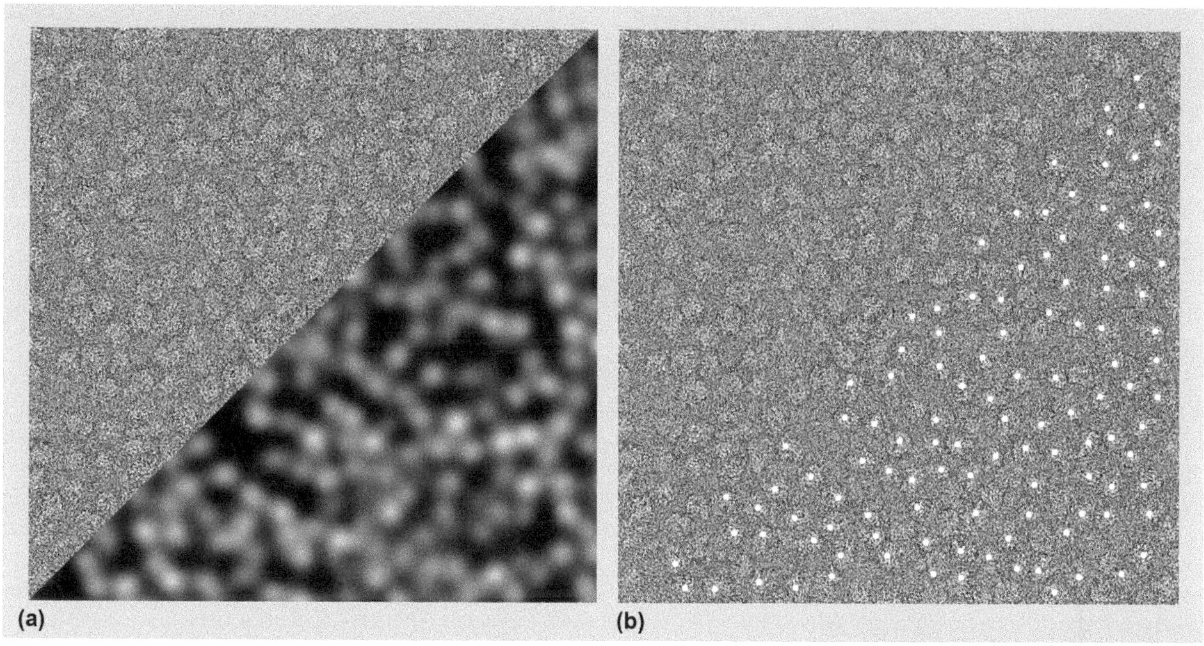

FIGURE 6.4 Reference-free particle picking based on local modulations. (a) Top left corner is the original micrograph (taken from dataset: EMPIAR 10443) and the bottom right corner is the variance/modulation image. (b) Top left corner is again the original micrograph and the bottom right corner shows the positions of the picked particles selected from that area.

picking and analysis (Van Heel, 2013; Afanasyev et al., 2017). The variance-based particle picking approach has been criticized (Nicholson & Glaeser, 2001) by authors who did not quite understand its principles and who presented the output results in an upside-down orientation compared to the input image. Simple statistical selection criteria and advanced MSA classification of the picked particles (discussed below) is a routine selection procedure for focusing exclusively on statistically significant particles rather than on impurities like ice crystals. This variance/modulation particle picking approach, developed more than 40 years ago, remains one of the most powerful and objective procedures for locating particles in cryo-EM images without needing prior information on the molecule. Finally, a word of warning against AI-based particle picking: AI picking is being "trained" by what humans see as the object. However, in routine operations, like in Afanasyev et al., (2017), particles are picked that are not even visible to the human eye and thus cannot serve as training examples.

MSA EIGENVECTOR DATA COMPRESSION

The introduction of eigenvector data compression (Van Heel & Frank, 1981) and unsupervised classification (Van Heel, 1984a, 1989) to electron microscopy was the beginning of a paradigm change in the analysis of macromolecular structures. This multivariate statistical analysis (MSA) revolution resulted from a six-weeks' visit of one of the current authors (MvH) to the laboratory of J. Frank in Albany in 1979. MvH had finished his degree in theoretical optics and became a full-time EM researcher in the Biochemistry Department of the University of Groningen. During his short Albany visit, the historical emergence of the MSA approach was triggered by his social contacts with a French colleague, Jean-Pierre Bretaudière (JP) as described in his obituary written after his untimely passing (Van Heel, 2009). The MSA method that JP was using was "correspondence analysis" (Wikipedia, Correspondence analysis), which stemmed from the school of French mathematicians around Jean-Paul Benzécri (Wikipedia, Jean-Paul Benzécri) and was specifically designed for histogram data. It is a powerful approach, based on the χ^2-metric, especially designed for use in the social sciences but not designed for signals that have a zero average value, which applies to many signals in science, including phase-contrast microscopy. A new, more flexible version of MSA was defined for signals using the modulation distance metric, combining the advantages of principal components analysis (PCA) and *analyse factorielle des correspondances* (AFC). Quoting from our MSA review (Van Heel et al., 2016), "For high-resolution EM work … the χ^2-metric must be considered obsolete since the publication of (Borland & van Heel, 1990)." Nevertheless, where Frank in his book (Frank, 2006) copies large chunks of the MSA mathematics from Borland & van Heel (1990), he continues to praise correspondence analysis and its χ^2-metric that had already been refuted for signal processing applications and had been superseded by the modulation metric (Borland & van Heel, 1990), which Frank does not even mention in his version of the story.

Let us not linger on the historical and theoretical aspects of MSA, but let us first explain what is actually done (for more details, see the Van Heel et al. (2016) review, a free download). The first phase in such MSA procedures is to compress the huge amounts of input information that result from cryo-EM data collection into a manageable number of main eigenvectors of the dataset. A raw dataset can contain some two terabytes (2 TB) of raw information, from which we want to extract, say, a 3D reconstruction volume of 256^3 bytes (~16 MB). To calculate that 3D structure we will need, again roughly speaking, 256 projection images of the structure with each 256 × 256 byte ≈ 64 kB, that is, a total of ~16 MB. To reduce the raw dataset from the 2 TB level to ~16 MB represents a data reduction of around a factor of 100,000. This necessary data reduction is best performed in an unsupervised way, in the compressed space representing the main sources of variance existing in the full dataset.

The basic principle of the MSA data compression by eigenvectors method is as follows: If we have a large number of measurements, say 10,000 images of each 256 × 256 = 65,536 pixel values, each of those images can be represented as a single point in a 65,536-dimensional space (64K space). The density of each pixel is thereby used as the coordinate along each of the 64K unit vectors spanning that 64K-dimensional "hyperspace." The 10,000 measurements together form a "cloud" of 10,000 points in that 64K-dimensional space. That data cloud is specific for the set of measurements we have performed, and from which cloud we want to extract all the relevant systematic information. The shape of the cloud is not uniform, but it has a specific shape; certain directions describe more variations in the dataset than other directions.

The origin of the orthogonal 64K space corresponds to an image where all 64K pixels contain a zero value. We can simply rotate the coordinate system of that 64K space the way we can rotate a three-dimensional coordinate system. However, a pure rotation of the full 64K orthogonal coordinate system does not give us any compressed form of information. The family of PCA approaches aims at finding the "principal components" of the dataset, which will become the new unit vectors of the new coordinate system. The trick is to order these unit vectors such that the first unit vector will describe the direction in the 64K space that describes the main variances in the dataset. The second most important direction (unit vector) is placed such that it describes most of the remaining variances in the data cloud, which means that that second unit vector must be orthogonal to the first. The third unit vector is then placed such that it is orthogonal to both unit vectors 1 and 2 and best describes the remaining variance of the dataset. The new unit vectors of the 64K space are sorted by the amount of variance they describe. The new unit vectors are called the eigenvectors of the dataset. Very often just tens of eigenvectors (often called eigenimages, eigenvolumes, or eigenspectra, depending on the application) suffice to describe the properties of the dataset.

We have already seen an example of the massive data compression exploited in finding the CTF parameters of a large dataset discussed in **(CTF AND ANISOTROPIC MAGNIFICATION)** and visualized in Figure 6.3. Only about 1% of the variation seen in the local-area amplitude spectrum of Figure 6.3a is due to the practically invisible CTF Thon rings in that local amplitude spectrum. Because the CTF information over the full size of the collected dataset is integrated into, say, some 30 eigenspectra, the precision of the CTF parameter assignment is indeed unsurpassable (Figure 6.3a).

MSA CLASSIFICATION

Having compressed the information content of the 10,000-member data cloud in the 64K-dimensional dataspace into a 10,000-member data cloud in a ***subspace*** spanned by only a few tens of eigenvectors, it becomes much more feasible to compare all 10,000 members of the cloud to each other in a reasonable amount of time. In that subspace, only the important variance directions of the cloud are taken into account. We can thus now compare all 10,000 members with each other for the purpose of subdividing them into smaller groups (subclouds) based on the mutual similarity of the individual class members. For methodological consistency, the classification procedure used to merge individual members into subclouds must be based on the same variance-based philosophy as used for the eigenvector data compression (Van Heel, 1984a, 1989; Borland & Van Heel, 1990). One of the best-known ways to achieve that is through the Ward criterion (Wikipedia, Ward's method; Van Heel, 1984a) where the merging of two classes (subclouds) is selected by the Ward criterion: minimizing added intraclass variance. The moving-elements postprocessor (Van Heel, 1989; Borland & Van Heel, 1990) is an important extension of Ward's method of hierarchical ascendant classification (HAC) by allowing individual members of classes to move to another class, there they in retrospect feel more at home using the same intraclass variance criterion.

For very large datasets, it can be advantageous to start a Ward-criterion classification using a suboptimal K-means preclassification strategy, where a predefined number (K) of randomly selected elements are used as first classification seeds. Each of the elements in the dataset is thus first assigned to one of the K seeds based on its closest direct distance. After this fast but simplistic classification, the procedures enter a true HAC where the variance-based criteria soon stabilize the initial disparities introduced by the K-means procedures (Van Heel, 1989). A moving-elements postprocessor then provides the final classification.

2D PARTICLE ALIGNMENT

The raw cryo-images are extremely noisy to minimize the electron-bombardment damage to the biological molecules in the cooled sample. Averaging procedures are needed to extract the repetitive information present in, say, 2D crystals (Unwin & Henderson, 1975), which is a relatively simple operation because of the repetitive nature of a 2D-crystal sample. In single-particle analysis, however, the averaging procedure is far more complicated since the individual molecular complexes are not held in a fixed orientation (by their crystal neighbors). The isolated "particles" are projection images through the molecules in the sample, in three different rotational orientations (Euler angles) and two degrees of translational freedom in the image plane (x- and y-directions). The third direction (the z-direction) is in the direction of the electron beam and represents the defocus of that particular particle. In single-particle analysis (SPA), the number of parameters is thus much higher than with the earlier crystallography-inspired approaches, and a single average no longer suffices. The averaging procedure thus must be multidimensional, and that translates to the need to characterize a single molecular projection image over multiple dimensions, that is, over multiple eigenvectors/eigenimages.

When a large dataset of selected single particles is first processed by MSA eigenvector data compression the first, say, 10 eigenvectors will cover a certain percentage of the total variance present in the unaligned dataset. The goal of our iterative ABC procedure (Afanasyev et al., 2017) is to find the best overall alignment, namely the alignment that maximizes the variance (sum of eigenvalues) in the lower eigenimages (Dube et al., 1993; Van Heel et al., 2016). Multireference alignment procedures (MRA) (Van Heel & Stöffler-Meilicke, 1985) are used traditionally to align noisy originals to projection images of earlier 3D reconstructions. The iterative ABC procedures (Afanasyev et al., 2017) represent a more advanced approach in which the intermediate alignment assessments are performed on class averages. This latter approach has a better perspective to achieve optimal results in terms of maximizing the variance covered by the first eigenvectors.

Note that the concept of what the variance is, depends on the metric applied (the modulation metric is the preferred one). Simpler (multi) reference alignment procedures are more prone to bias toward the chosen references(s) (Boekema et al., 1986), causing the problem also known as "Einstein from noise" (Schatz 1992). The ABC approach is probably the best current approach for focusing the variance information in a dataset into the main eigenvectors of the system. However, the very concept of "information" is not well defined, and classical metrics like the "information capacity" of a channel (Shannon, 1949) were found to be inconsistent and not suitable for harvesting new information. Information remains a primary focus of our current research (Van Heel & Schatz, 2020; Van Heel, 2024b).

TILTING THE SAMPLE HOLDER?

In 2D electron crystallography (Unwin & Henderson, 1975), the 3D information results from tilting the crystal to achieve different projection views of the 2D crystallographic array of strictly aligned unit cells of the crystal. This tilting is a macroscopic, mechanical tilting of the sample holder that is typically restricted to a range of −60° to +60°. Tilting the sample of individual molecules was also tried in early 3D imaging of individual molecular complexes (Hoppe et al., 1974; Oettl et al., 1983). That mechanical tilting approach only had very limited success for a number of reasons: the dried negatively stained samples were flattened against the supporting carbon film; the mechanical tilting of the sample holder introduced serious positioning errors; the noise level in the data did not allow for any constructive refinement; etc. To improve on the limitations of the single-tilt axis approach, a random conical tilt (RCT) approach emerged from the group of Hoppe in Martinsried. It was presented in a lecture by Radermacher in 1978 at the Toronto International EM meeting and was later implemented in practice (Radermacher et al., 1987). The approach consists of using an untilted micrograph with many particles in the same (preferred) orientation and averaging the same particles from another micrograph taken under a, say, 45° tilt. The untilted micrograph serves to find the rotational orientation on the individual complexes in the plane of the sample. No-tilt alternatives already existed that exploited the "natural tilt" of some forms of regular arrays that incorporate multiple copies of a molecule in different 3D orientations. Such aggregates occur, for example, in fibers with helical symmetry (DeRosier & Moore, 1970) or in icosahedral viruses with 60 copies of the unit cell in different orientations (Crowther, 1971). Although such assemblies also suffered from a flattening onto the support film, the resulting 3D reconstructions were much more reliable than single-tilt (tomographic) reconstructions of asymmetric molecular complexes such as ribosomes. With the advent of the superior vitreous-ice specimen-preparation approach (Adrian et al., 1984), the SPA situation changed entirely. Sample tilting remains important for samples with little redundancy, like thin sections of entire cells in electron tomography but the resolutions achieved from such tilt experiments are normally significantly worse than those achieved by cryo-EM SPA.

ANGULAR RECONSTITUTION: THE ZERO-TILT APPROACH

Angular reconstitution was proposed in 1987 as a general method to *a posteriori* find the relative orientations of the 2D projection directions of a 3D structure. With the advent of vitreous-ice specimen preparation (Adrian et al., 1984) and of MSA eigenvector data compression and unsupervised classification (Borland & van Heel, 1990; Van Heel, 1984a; Van Heel et al., 2016), not knowing the molecular projection orientations of the now noise-free 2D class averages became the next problem. To reconstruct the 3D structures associated with these projection images, their relative orientations must first be determined. This problem did not previously exist in medical computer tomography or in other mechanical tilt-based 3D-reconstruction approaches. To solve the problem of not knowing those relative orientations *a priori*, two different methods were developed to find the necessary orientation information from the projection images themselves, that is, *a posteriori* to the data collection.

The first solution is projection matching, in which one simply assigns random Euler angles to the noise-free class-average images and then counts on gradually refining the Euler angle assignments iteratively by multireference

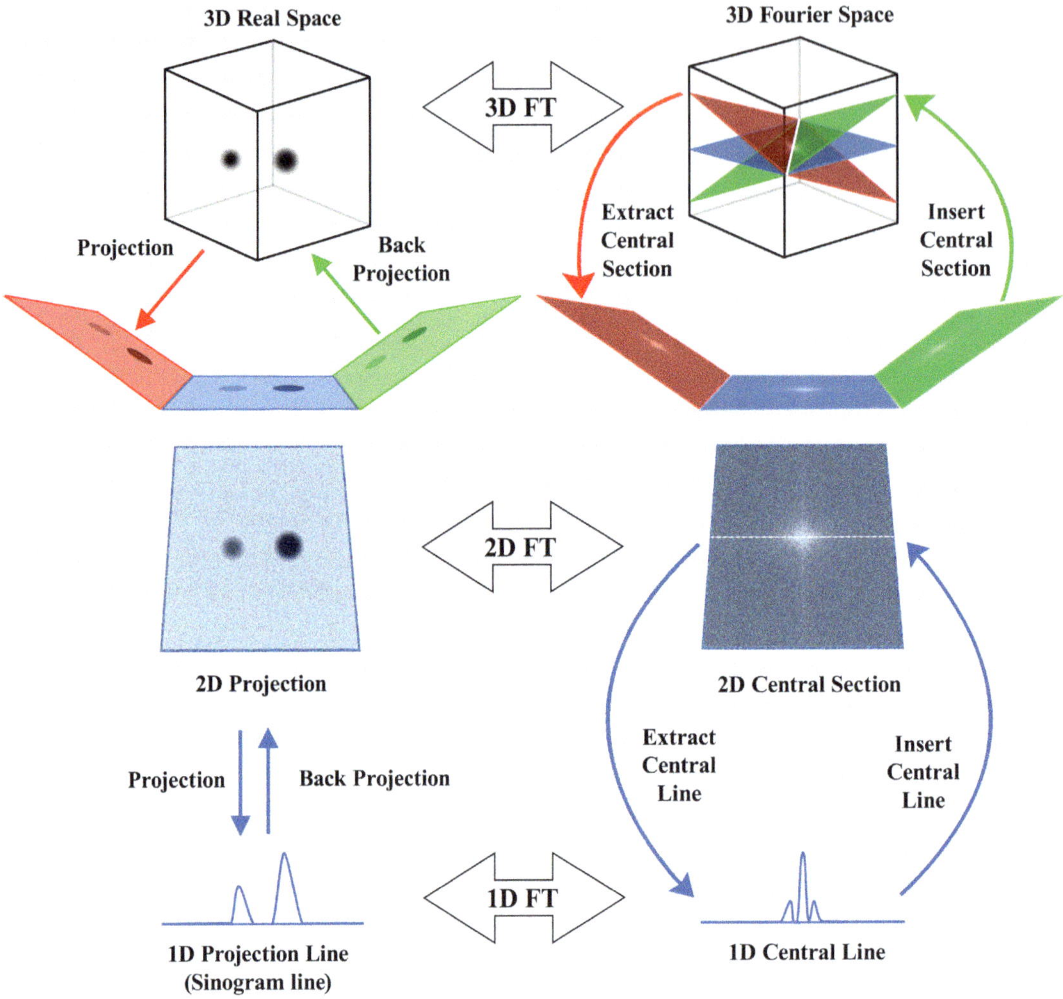

FIGURE 6.5 Central sections, common lines, and angular reconstitution. This is an elaborated, visual version of the more symbolic, abstract original illustration presented previously, Figure 2 of Van Heel (1987a). The full information of a 3D dataset can be represented either in 1D, 2D, or 3D form. Moreover, the data can be represented either in real space or in Fourier space. Given the availability of the necessary full dataset (and the necessary orientational parameters) in any of these six forms, all other forms can be uniquely derived from the chosen one. Finding the necessary orientation parameter is performed by angular reconstitution, which requires finding the common line projections (CLP) of all "noise-free" class averages, which are the projection images of the 3D structure to be determined (Van Heel, 1987a). Those CLPs are extracted from sinograms (collections of all line projections) calculated from the class averages. The nature of (iterative) angular reconstitution Euler-angle determination is explained in more detail in Figure 6.6.

alignments with respect to reprojections of the intermediate 3D reconstructions achieved (Van Heel 1984b). Projection matching is a relatively crude method for starting up a fresh 3D reconstruction; it is better suited for a final refinement of the projection directions (for example, see Van Heel et al. 2000) than for the very first model generation. When a similar structure is already available from previous analyses, projection matching can be used in a refinement mode straight away. Most current popular programs only use the projection matching approach and rely on the availability of a starting model to initialize the processing.

The second and methodologically more powerful approach is angular reconstitution (AR), which is the quintessential zero-tilt method (Van Heel, 1987a). To explain the AR method, one needs to first explain the different concepts of what a 3D dataset is in its various manifestations (Figure 6.5). This illustration, a more visual elaboration of the original diagram shown in Van Heel (1987a), shows the equivalency of six different ways of representing the same data: three ways in real space and three ways in Fourier space. Note that all operations in this diagram represent continuous mathematical operations (like a continuous Fourier transform, the CFT) but are used as discrete operators (like the discrete Fourier transform, the DFT) (Van Heel, 2024b).

The top (3D) tier of Figure 6.5 shows the equivalency of a 3D volume in real space and in 3D Fourier space. We can Fourier transform the data from 3D real space to 3D Fourier space, back and forth, many times using a 3D DFT that will not lead to any loss of information; the data representation in two spaces contains exactly the same information. Dropping down one tier, we change from the 3D level to the 2D level. In real space (left) we project one 3D volume into many Euler directions to obtain many 2D projections of that volume. Thinking simplistically, as discussed in **(MSA EIGENVECTOR DATA COMPRESSION)**, to calculate a 3D structure of size 256 × 256 × 256 voxels, we need roughly 256 projection images of the structure with each containing 256 × 256 pixels. The inverse is also true: The 3D volume can produce ~256 independent 2D projections of the structure. On the right-hand side of Figure 6.5, the 2D Fourier transforms of the 2D projections represent central sections through the 3D Fourier transform (DeRosier & Klug, 1968). Extracting and inserting central sections in Fourier space is equivalent to creating 2D projections from the 3D volume and back-projecting 2D projection through the 3D volume. When the 2D projections are related to each other by a single tilt axis, as is the case in this diagram, in Fourier space that tilt axis represents a common line between the corresponding central sections (marked in white).

Dropping one further tier down, each of the 256 projection images (or 2D class averages) can be converted to a sinogram (Wikipedia, Tomographic reconstruction; Van Heel, 1987a). The sinogram is sometimes called a Radon transform (Radon, 1917), but in single-axis medical CT, its discrete version is mostly called a sinogram for a clear visual reason (Figure 6.6). The basic angular reconstitution procedure is explained in the legend of Figure 6.6, consisting

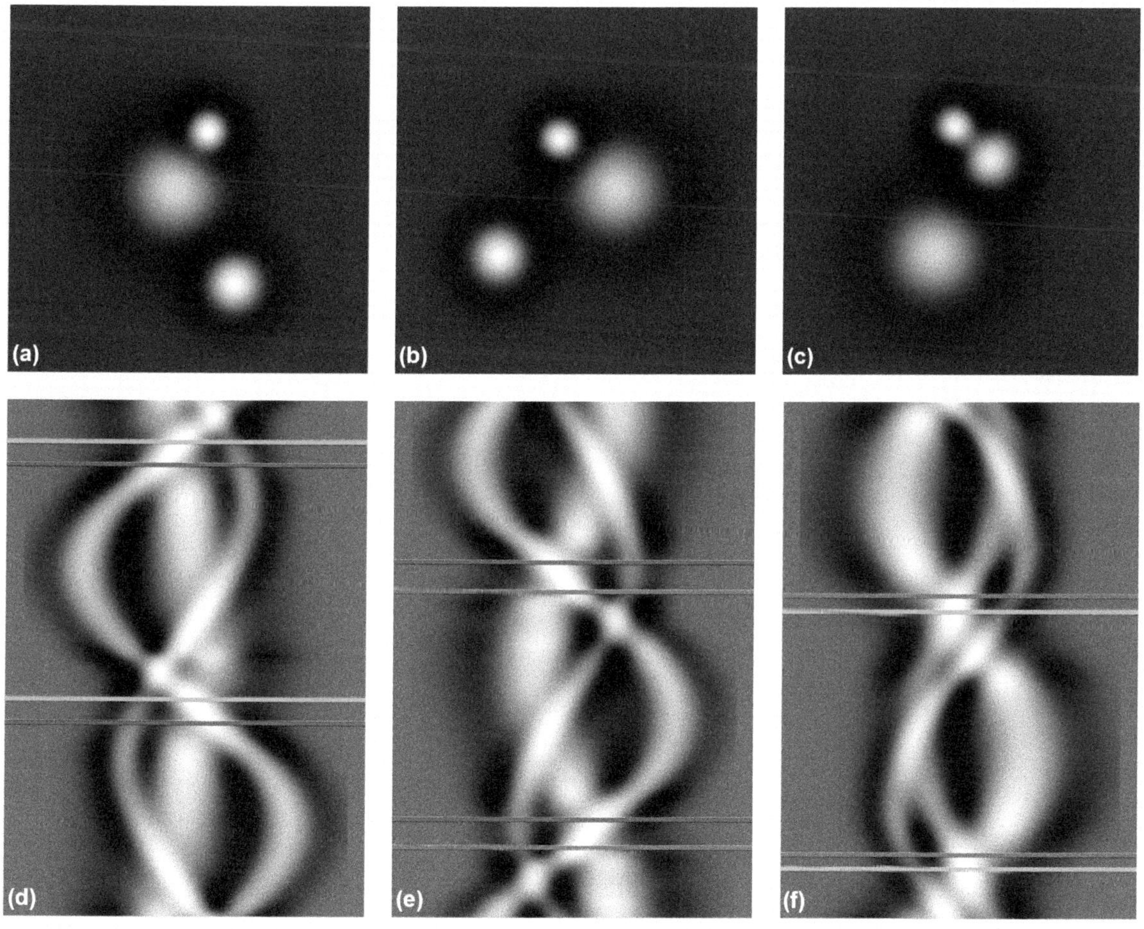

FIGURE 6.6 Angular reconstitution: matching all the common-line projections. A simple three-dimensional test object was created consisting of three spheres which was band-pass filtered to remove its very low-frequency and very high-frequency components. That 3D model was projected in three random orientation directions (A-B-C). Each of these 2D projections was further projected into all possible 1D projection directions, each forming a classical sinogram (D-E-F). The trick is to search for all possible pairs of line projections that two 2D projections share. Thus, through an exhaustive search of the common line projections (CLPs) between projections A and B, we find two "red" CLPs in the sinogram D (of A) and two "red" CLPs in the sinogram E (of B). These CLPs come in pairs because the sinograms here cover the full 360° range of possible line projections; the pairs are always 180° apart. The green pairs of CLPs are those between projections A and C (D and F sinograms), and the blue CLP pairs are those between projections B and C (E and F sinograms).

of the Euler angle assignment for three projections of an asymmetric sample (Van Heel, 1987a). In practice, one starts up a 3D reconstruction with a larger number of different class averages in an iterative procedure where an anchor set of projections from the previous iteration is used to rapidly converge to quasi-atomic resolution (Afanasyev et al., 2017). During these iterative refinements, one gradually refines various parameters, such as the centering of the particle images that must remain in synch with the corresponding center of the 3D structure as a whole. Note that a long controversy existed between the RCT and the angular reconstitution approaches, which was decided in favor of the latter (Van Heel et al., 1997).

4D PROCESSING

In 2D electron crystallography, like in 3D X-ray crystallography (XRC), the *a priori* assumption is that all unit cells are the same and that we can thus average all the unit cells in the crystal to obtain the noise-free average. Any free movement of a side chain, for example, will blur that part of the unit cell of the macromolecule we are studying. Indeed, in XRC one will often proteolytically cleave off such loose ends to obtain better crystals that diffract to a higher resolution. Such cleaving may yield a more rigid crystal, but that may also leave the biological molecules in a state that is farther from their physiological state. Single biological complexes embedded in vitreous ice reside in a quasi–*in vivo* aqueous environment with fewer unnatural constraints and can thus yield more reliable, more physiologically relevant 3D structures.

Zero-tilt approaches like angular reconstitution or projection-matching refinements are not limited to only one 3D reconstruction from the dataset of 2D images. The approach is equally applicable to datasets containing multiple 3D structures within the same sample—in other words, a 4D dataset. The first example of such 4D processing revealed an unexpected separation of two different states of the *E. coli* ribosome, in complex with RF 3, during termination of protein synthesis (Klaholz et al., 2004). One of the most intriguing and biologically most relevant aspects of single-particle cryo-EM is indeed this 4D analysis. The MSA eigenvector data compression and unsupervised MSA classification are very powerful tools in 4D analysis. One can compare all 3D structures, stored in a large 4D stack with each other, and thus achieve a better understanding of the dynamical behavior of the 3D structures.

VALIDATION, RESOLUTION, AND INFORMATION

For assessing measurements in general, it is fundamental to have sound, objective quality metrics. However, that objectivity must often give way to strong personal beliefs that "mine-is-better-than-yours." Handwaving arguments like "my metric is more conservative than yours" cannot be part of a scientific argument. Beware, however; defining reliable metrics is not a problem specific to cryo-EM. Some metrics that are in routine use, like the signal-to-noise ratio (SNR), are not the information metrics that people believe them to be (Van Heel & Schatz, 2020) due to assumptions about the properties of signal and noise that were made in the early days of information theory (Shannon, 1949). As a consequence, all secondary metrics based on the assumption that the ***SNR*** is a valid ***information metric***, like the ***Spectral Signal to Noise Ratio*** ("***SSNR***") or the "***detective quantum efficiency***" (***DQE***), used to assess the quality of direct-electron detectors ***are incorrect*** [Van Heel & Schatz 2020]. That also affects cryo-EM since the quality of the expensive direct electron cameras in use in the field are assessed (and priced) based on their DQE (Van Heel & Schatz, 2023).

The original concepts of resolution in optics are based on the principle of the diffraction limit: the resolution of optical instruments like telescopes and microscopes, typically presented as the size of the resolution-limiting (Figure 6.2) numerical aperture (NA), known as the Abbe criterion (Wikipedia, Ernst Abbe). This classic linear instrumental resolution always remains the same, whether or not the instrument is in use or is stored in a cupboard, and whether or not its illumination is switched on. This instrumental resolution is often confused with the resolution achieved as the result of a data collection experiment on a specific sample. The latter is the results resolution one wants to optimize to obtain good cryogenic electron microscopical results. Cryo-EM methods emerged against the scientific background of X-ray crystallography (XRC) in which the primary quality metrics are R-factors and phase residuals. In XRC one iteratively refines an atomic model such that its power spectrum approaches the experimentally determined diffraction pattern. This iterative procedure is repeated until those metrics converge to a minimum.

In XRC, one can only measure the diffraction pattern, that is, the amplitudes of the sample; the phases of the sample, which are the most important part of the information, must be retrieved by indirect, iterative methods. This X-ray procedure is inherently nonlinear, meaning that we cannot simply add two experimental 3D, results independently obtained from the same biological complex, to produce a better final 3D result. These XRC linearity problems were not so clear in the early 1980s when the first metrics were inherited from XRC. They have now come closer to the forefront, but the structural biology community still appears largely oblivious to this fact (Henderson et al., 2012). In spite of the more direct and complete nature of cryo-EM data when compared to classical X-ray diffraction data, XRC with its early atomic-resolution results was intuitively considered to have the better resolution metrics. In reality, the metrics in XRC are difficult to define because of the nonlinear methodologies used. XRC actually also has its share of unprovoked errors, such as the DPR in electron microscopy. For example, the R-factor, which is defined as:

$$R = \frac{\sum|F_{obs} - F_{calc}|}{\sum|F_{obs}|} \quad (6.1)$$

is normalized by the observed amplitudes $\Sigma|\boldsymbol{F}_{obs}|$ only, and not by the calculated $\Sigma|\boldsymbol{F}_{calc}|$. Multiplying all $\boldsymbol{F}_{calc}$ by any nonzero constant does not change the calculated final structure, but it does dramatically change the R-factor. The R-factor must thus be saved by some hidden normalization that is not part of its definition.

In (electron) microscopy, we measure both the amplitudes and the phases of the sample, and that allows us to incorporate the full information, that is, both the real and the imaginary part of the data into the quality metrics. Unfortunately, the nonlinearity of XRC was already inherited by such phase-oriented metrics as the EM phase residual of Crowther (1971) and the differential phase residual (DPR) of Frank et al. (1981). Directly after the DPR was proposed as a quality metric for SPA, it was superseded by the Fourier ring correlation (FRC) in 2D (Van Heel et al. 1982) and the Fourier shell correlation (FSC) for 3D data (Harauz & Van Heel, 1986a). For assessing the resolution of 2D-crystal analyses, Saxton & Baumeister (1982) had proposed a "spatial frequency correlation function"; unfortunately, no formulas were provided. The authors did correctly suggest a 2σ-significance threshold to account for the noise statistics of the correlation that changes with the number of sampling points in each spatial-frequency ring. The next section (on DPRs) is more about human behavior than about science and may be skipped without interrupting the logical flow of this review. More details on these historical issues—which continue to be relevant—are discussed in a separate review (Van Heel, 2024a).

THE SAGA OF THE DIFFERENTIAL PHASE RESIDUAL

Apart from the problems that the DPR (as a phase residual) inherited from XRC, the specific phase-residual definition introduced in Frank et al. (1981) represented an additional, unprovoked error. The applied normalization (by the sum of the amplitudes of the two objects being compared) means that multiplying either input image by a constant—which does not change their information content—changes the DPR value completely. With the publication of the all-inclusive FRC Van Heel et al. 1982, one would expect the flawed DPR to soon disappear from the literature. However, the opposite was the case: The publication of the FRC triggered a ferocious response on issues like why the DPR is more "conservative" than the FRC. That reluctance to appreciate and accept the clear advantages of the FRC made a more formal discussion on DPR problems necessary. In the paper that followed (Van Heel, 1987b), the good, the bad, and the ugly aspects of the DPR were discussed, and corrected phase residuals were proposed that avoided those errors. Nevertheless, as was clearly stated in that paper, the corrected phase residuals of Van Heel (1987b), like the DPRX, still suffered from their XRC inheritance of being based on incomplete data coverage, and they could not match the quality of the FRC (Van Heel et al. 1982) and FSC (Harauz & Van Heel, 1986a) metrics. The only purpose of those corrected phase residuals was to correct the flawed DPR. In spite of explaining in Van Heel (1987b) the flaws of the chosen DPR definition (Frank et al., 1981), and showing that the DPR random-noise limit was 103.8° and not the suggested 90°, and in spite of suggesting simple corrections to the DPR, its proponents never took ownership of the problem. This attitude of avoiding a real discussion is illustrated by some truly weird examples:

Example 1:

Frank and coworkers in Liao & Frank (2010), a paper in which they continue discussing the DPR some 25+ years after it was refuted, cite the van Heel & Schatz, 2005 paper only once: "in contrast, in (van Heel and Schatz, 2005) it is argued strongly in favor of the criterion based on the σ." That van Heel & Schatz, 2005 paper, however, is not about σ but about the introduction of the ½ bit criterion that superseded the σ criterion as the FRC-FSC threshold; the paper also listed errors that had accumulated in literature, including the DPR errors. To correctly quote van Heel & Schatz, 2005 from its conclusion: "...We suggest using the 1/2-bit threshold curve as a general-purpose indicator of interpretable resolution in FSC curves." Such "citations" are misleading since the original research states exactly the opposite of what it is quoted for.

Example 2:

Again, on the issue of resolution metrics in a different review paper (Frank, 2009), it is stated, some 25 years after the DPR was refuted, that "Initial controversies (Orlova et al., 1997; Malhotra, 1998) have given way to the acceptance of FSC = 0.5 as criterion, which was initially introduced by Boettcher et al. (1997) in a study of the hepatitis B virus capsid. (This criterion is, incidentally, quite consistent with the criterion DPR = 45° earlier used.)" Another false citation; Orlova et al. (1997) actually states: "fixed phase residual thresholds should be replaced by a significance threshold curve that also depends on the number of pixels/voxels in any particular Fourier-space ring (2D) or shell (3D)." The Frank (2009) review also avoids citing the van Heel & Schatz, 2005 paper where it is stated that: "There is no scientific justification for the use of fixed FSC threshold criteria (such as the "0.5" criterion) as is emphasized by our theoretical considerations and our model experiments. All fixed-valued threshold criteria are implicitly or explicitly the result of flawed assumptions on the statistical behaviour of the noise and the signal components of the data."

The next section is about updated FRC/FSC metrics and about new and more advanced quality metrics to assess the harvesting of new information. Yet, there too, we still need to first remove some historically accumulated debris from the playing field.

FOURIER RING/SHELL CORRELATION AND FOURIER RING/SHELL INFORMATION

Here we need to take a step back to the basics and explain the original ideas behind the FRC/FSC metrics. These ideas have almost gotten lost in the course of the many decades since they were first proposed (Van Heel et al. 1982). The field has suffered from a poor level of understanding of the basic statistics and basic signal processing issues, in combination with an aggressive dissemination of misinformation as shown in the previous section. To avoid this historic muddling, let first us establish a firm basis upon which we can update the FRC/FSC metrics and introduce the new FRI/FSI metrics, where the C for Correlation has been superseded by the I for Information.

Preliminaries: Formulas in Rings and Shells in Fourier Space

The first mention of the Fourier shell correlation (FSC) in the literature was in Harauz & Van Heel (1986a). This is the 3D version of the first 2D Fourier ring correlation (FRC) metric (Van Heel et al. 1982) and the 3D version had already been announced together with the FRC. In formula form:

$$\boldsymbol{FRC}_{12}(r_i)=\frac{\sum_{r\in r_i}\boldsymbol{F}_1(\boldsymbol{r})\cdot\boldsymbol{F}_2^*(\boldsymbol{r})}{\sqrt{\sum_{r\in r_i}\left|\boldsymbol{F}_1(\boldsymbol{r})\right|^2\cdot\sum_{r\in r_i}\left|\boldsymbol{F}_2(\boldsymbol{r})\right|^2}} \tag{6.2}$$

In this formula, all summations are over rings in 2D Fourier space (or over shells in 3D Fourier space). In the first paper (Van Heel, 1982b), the FRC was used to compare 2D EM results with projected, established XRC results, over all spatial frequencies. The FSC (Harauz & Van Heel (1986a)) was first used to compare the quality of 3D reconstruction algorithms by comparing 3D reconstruction results with the known 3D test object, again, over all spatial frequencies. The normalization in the FRC/FSC formulas makes the different types of data comparable by dividing with the standard deviation of each of the two sets of data, in each of the resolution rings or shells $(\sum_{r\in r_i}\left|\boldsymbol{F}(\boldsymbol{r})\right|^2)^{1/2}$. Note that the FRC/FSC metrics are general 2D/3D data processing metrics that are not at all tied to cryo-EM or even to structural biology. This is a very important linearity property that is not shared by, for example, the R-factor or phase residual in XRC or in EM! This fundamental insight is lacking in just about every paper of resolution in cryo-EM or in XRC. This also means that no pure cryo-EM or structural biology arguments may be used to assess the quality of the FRC or FSC as a metric. Coupling the FSC to a 2D alignment procedure and calling that a "gold-standard FSC" reveals ignorance of the relevant science. The FRC/FSC are metrics defined for comparing two independently collected images or volumes. What those two images or volumes actually are, and how they were acquired, has nothing to do with the quality of the FRC/FSC as a resolution metric. A ruler can measure the length of a pencil or the width of a melting ice cube, neither of which has anything to do with the quality of the ruler as a metric. If your living room is too hot, don't blame that on your gold-standard thermometer; it is more effective to call your central-heating/air-conditioning repair person!

What is also often not understood—although clearly explained in the original papers—is that summation over a ring or shell of the complex function $\Sigma\ (\boldsymbol{F}_1{\cdot}\boldsymbol{F}_2{}^*)$ always yields a real value because the function $(\boldsymbol{F}_1{\cdot}\boldsymbol{F}_2{}^*)$ exhibits *Hermitian symmetry* (contains only Friedel pairs). The added "$Re\left\{\Sigma\boldsymbol{F}_1\cdot\boldsymbol{F}_2^*\right\}$" notation in Frank (2006), while explicitly citing the original FRC formulation of Van Heel et al. (1982), is mathematically redundant. Such lack of understanding is thus communicated to and perpetuated by others; see for example Cardone et al. (2013), implicitly suggesting there was something wrong with the original mathematical formulation of the FRC.

Why Are ALL Fixed-Value FRC-FSC Thresholds Incorrect?

The first reason why all fixed-value FRC-FSC thresholds are incorrect for 2D and 3D data is due to the number of sampling points being considered in the correlation calculation. The correlation statistics of two data vectors in a ring or a shell in Fourier space are considered: When one calculates the FRC, the number of sampling points in that ring is proportional to the radius of that ring $\boldsymbol{R}^1$, and in the case of the FSC the number of sampling points in a shell is proportional to $\boldsymbol{R}^2$. The statistics of the FRC and FSC are thus different, and therewith the threshold curves to be applied are different. In fact, in the 1D signal processing of electronic engineering (Shannon, 1949), a comparable threshold applies, but there the Fourier correlation (FC) of $\boldsymbol{R}^0 = 1$ is independent of the distance to the Fourier space origin (Van Heel & Schatz, 2020). Those differences by themselves already justify different names for what look like identical formulas. The same type of difference is relevant when the 3D structures to be compared have a high degree of imposed symmetry; these effects have all been proven extensively both theoretically and experimentally (Van Heel & Schatz, 2005, 2017).

Let us, for didactical reasons, assume that there is no signal at all in the data and the individual measurements thus consist of pure random noise. In that case the FRC/FSC will oscillate randomly around the zero-correlation line whereby the amplitudes of those random fluctuations decrease with the distance to the origin of Fourier space. The reason its amplitude decreases is that that amplitude is proportional to *the square root* of the number of $\boldsymbol{K}$ sampling points in each ring/shell. This is not an opinion, it is a mathematically proven fact known as the central limit theorem, which refutes all fixed-value resolution thresholds (0.5; 0.143; 0.3, 45°, etc.) for 2D or 3D data. Model calculations have also exhaustively confirmed those established facts (Van Heel & Schatz, 2005, 2017). The paper by Saxton and Baumeister (1982) on the spatial frequency correlation function (SFCF) did not contain formulas, but the authors did correctly specify the expected random-noise limits. All existing (and future) fixed-value resolution thresholds for 2D or 3D data are thus incorrect.

The second and equally fundamental mathematical reason why fixed-value thresholds for resolution metrics are wrong is the assumption that the signal $\boldsymbol{S}$ and the noise $\boldsymbol{N}$ are independent and that thus the cross-products between signal and noise can be left out of the equation, or in mathematical terms:

$$\sum_{n} S(x) \cdot N(x) = 0 \tag{6.3}$$

However, not only is it true that for the signal vector $\boldsymbol{S}(\boldsymbol{r})$, the noise vector $\boldsymbol{N}(\boldsymbol{r})$ is just an uncorrelated random vector, the inverse is also true: for the noise vector $\boldsymbol{N}(\boldsymbol{r})$ the signal vector $\boldsymbol{S}(\boldsymbol{r})$ is just an uncorrelated random vector. Since, however, the inner product between two random vectors of length $\boldsymbol{n}$ is not zero, but is rather proportional to $\sqrt{\boldsymbol{n}}$, equation (6.3) is wrong in this case. The equation is valid if and only if the vectors $\boldsymbol{S}(\boldsymbol{r})$ and $\boldsymbol{N}(\boldsymbol{r})$ are orthogonal. That orthogonality does not happen just by chance (it would need to be enforced by a numerical Gram–Schmidt orthogonalization). *A priori* assuming orthogonality profoundly messes up all that follows (Van Heel & Schatz, 2005, 2017). We have added a copy of the Appendix of the 2017 paper to this review for the reader's convenience. This assumption, as is clearly shown in the Appendix, refutes the 0.143 FSC threshold proposed in Rosenthal & Henderson (2003). That paper fails between formulas (A2) and (A3), stating: "Assuming signal and noise are uncorrelated, and for data on the same scale, the above expression (A2) may be written as follows…(A3)," whereupon the authors assume equation (6.3) to apply. The consequences of this failure are detailed below, and in Van Heel (2024a).

The third (and oldest) reason why fixed-valued resolution thresholds are wrong is that they are based on the flawed formula relating the signal-to-noise ratio (SNR) (Frank & Al-Ali, 1975) of a measurement to its cross-correlation coefficient (CCC), with a formula taken directly from Bershad & Rockmore (1974):

$$SNR = \frac{CCC}{(1 - CCC)} \tag{6.4}$$

The Bershad & Rockmore formula had a number of strict *a priori* restrictions attached, thus making it inapplicable in practice (Van Heel & Schatz, 2005, 2017), but Frank & Al-Ali (1975) broadened its application to real-life electron microscopy. Since the SNR is defined as the square of the signal over the square of the noise, the SNR is a definite positive entity, whereas the CCC is defined between $-1 \leq CCC \leq +1$. When the signal is very small, the CCC starts to oscillate around zero. When a negative value is inserted in equation (6.4), the SNR becomes negative, beyond its definition range; in other words, equation (6.4) is wrong and should be retracted from the literature. This flawed formula is used routinely in arguments for a fixed-value resolution threshold, arguments that were neither coherent nor mathematical. The argument used in Frank & Al-Ali (1975) to generalize the Bershad & Rockmore (1974) formula included the claim that 2D images are large and that thus the "large N" requirement of the 1D formulation in the original work could be dropped. This argument, which is invalid in Fourier space, therefore already had the denial of the R dependence ("fixed threshold") built in as a postulate and could/can no longer be used to prove the R independence. Any paper referring to the flawed Frank & Al-Ali (1975) paper to support fixed resolution thresholds directly enters into a circular argumentation (Van Heel, 2024b). Such theoretical arguments, unfortunately, fall on deaf ears for those not trained in the logics of mathematics or theoretical physics.

Is the Half-Bit FSC Threshold Still the Half-Bit Threshold Criterion?

What took us a bit longer to realize (Van Heel & Schatz, 2020) is that we had tailored the 1/2-bit FSC criterion (Van Heel & Schatz, 2005) based on Shannon's channel information capacity (Shannon, 1949), which itself is based on an SNR defined as "S^2/N^2," with the additional assumption that the S^2 and N^2 are independent of each other. That implies that Shannon's channel capacity suffers from the same *a priori* orthogonality assumption as expressed in equation (6.3). To cut a long story short, it was not the $FSC_{½\text{-bit}}$ nor the $FSC_{1\text{-bit}}$ criterion that changed, but rather the definition of the bit! We had originally assessed the FSC while considering the SNR to be an established information metric. However, we only then realized that in fact it is the other way around: The SNR is not a useful metric, whereas the CCC, in contrast, is always a real measurable physical entity. The SNR is—at best—useful in the extreme case of a very high SNR where we also already deterministically know the analytical input signal $\boldsymbol{S}$. Even then, the SNR metric can only serve to measure the loss of information in a 1D channel and can never serve as a metric to assess the collection of new information. With the CCC, in contrast, we can always define the amount of information harvested in an experiment (Van Heel & Schatz, 2020) and in all dimensions, including 1D, 2D, and 3D. In other words, the new information bit is based on measurable CCCs, and that new bit supersedes Shannon's earlier information bit, which was based on a nonmeasurable SNR metric. (Note that information bits are continuous entities, not to be confused with storage bits that can only assume the values 0 or 1). Therewith, the classic SNR, as a metric upon which to base the concept of information, became superfluous.

And What About Those New Information Metrics?

When the FSC/FRC curves get very close to their −1 or +1 limits, they no longer reflect the similarity level between the two measurements correctly. A 1% difference in value between an $FSC(r_i)$ of 0.99 and one of 0.999 does not correctly represent their huge significance difference (Van Heel & Schatz, 2020). To compensate for this effect, we apply a Fisher transform (Fisher, 1915; Wikipedia, Fisher), to the Fourier correlation (FC), the FRC, or the FSC. The resulting function is called the Fourier shell information (FSI) or Fourier ring information (FRI):

$$FSI(r_i) = K \cdot log_2 \left\{ \frac{1 + FSC(r_i)}{1 - FSC(r_i)} \right\} \tag{6.5}$$

In fact, we again have three versions of this information metric: the 1D version is simply called the Fourier Information (FI). The factor K of the FI, the FRI, and the FSI is a function of R^0 (=1) in 1D cases, of R^1 in 2D cases, and of R^2 in 3D cases (Van Heel & Schatz, 2020).

What Is the Story of Local Resolution and Local Information Metrics?

Holding on to refuted metrics and fixed-value resolution thresholds in cryo-EM has created significant collateral damage. What is the matter? The resolution thresholds in 2D and 3D, FRC and FSC, as we have seen above, are necessarily functions of R^1 or R^2, respectively (Van Heel & Schatz, 2005, 2017). That is true for any size box, and it remains incompatible with any fixed-value resolution threshold. When looking at the local 3D resolution, that is, the resolution within a small volume contained within a much larger ("global") volume, assuming a fixed threshold value necessarily leads to erroneous results that must be corrected with flawed and ad hoc threshold-value decisions (Figure 6.7). This section on local resolution and information is about

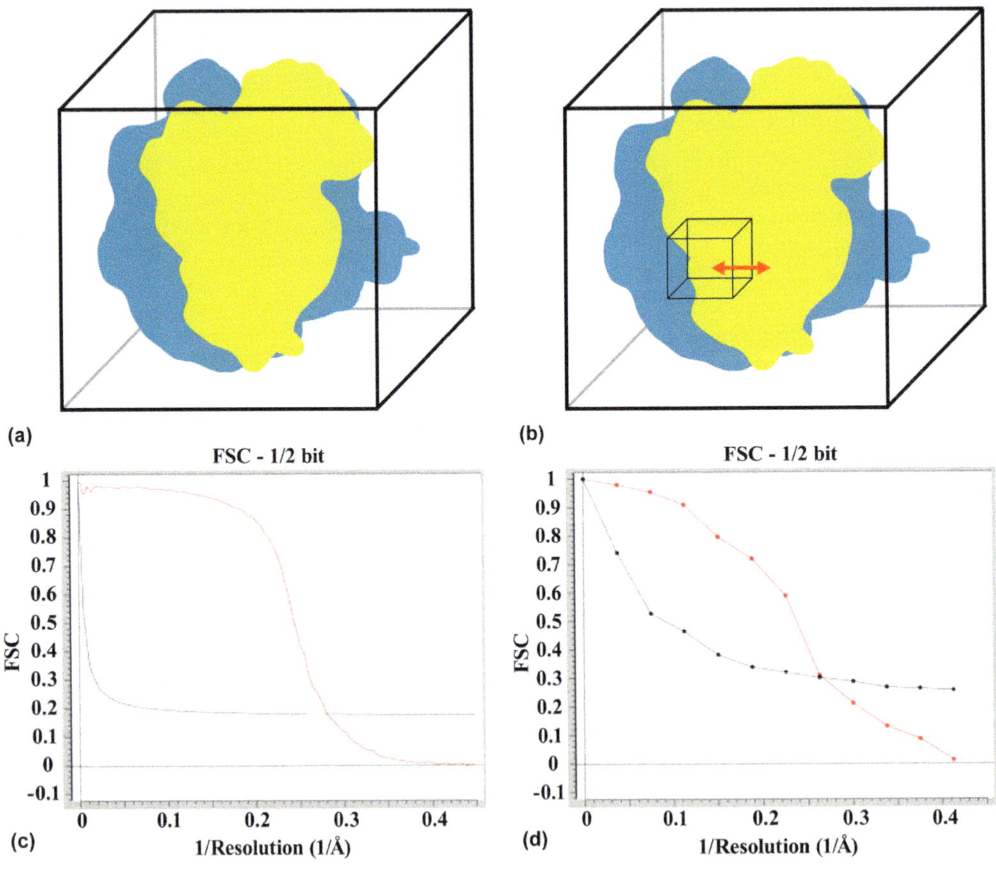

FIGURE 6.7 Global resolution versus local resolution. (a) Rather than looking at global resolution, it makes more sense to assess local resolutions in running subvolumes (b). (c) A practical global resolution assessment FSC (360^3 voxels, 0.6 Gaussian mask) compared with (d) a local-volume FSC (24^3 voxels, 0.6 Gaussian mask). The appropriate ½-bit threshold curves make the global and local resolution values comparable, without the need to introduce different fixed thresholds. This allows for such metrics to be incorporated in automatic procedures such as those needed for determining local resolution and information. The curves show a high-resolution ½-bit cutoff at ~1/0.27 = 3.7 Å, although the corresponding threshold levels differ from the global FSC to the local FSCs. (c) and (d) were modified from (Van Heel & Schatz, 2017).

a more fundamental issue than just whether a local metric is any good. The issue requires more serious attention than we are giving it here. The polemics that have dominated resolution discussions over decades have overshadowed the underlying science, namely, that real resolution/information metrics must be local (see Discussion, below).

DISCUSSION

Linearity

Cryo-EM can now achieve resolution levels that are competitive with those classically achieved in X-ray crystallography, and that for complexes that are not even amenable to XRC. The separation of different conformational states of 3D structures from the same sample (4D cryo-EM), pioneered some 20 years ago (Klaholz et al., 2004), has become an important tool in understanding structure–function relations of biological complexes. No comparable XRC approach appears to be available with a similar level of flexibility. For decades, XRC had been the dominant technique for determining atomic coordinates of proteins and their complexes, and cryo-EM had much to learn from XRC. However, especially now that cryo-EM achieves quasi-atomic resolution levels, it is time to reevaluate the historical limitations of XRC with respect to cryo-EM in terms of methodological purity. XRC is a nonlinear structural biology approach where only the amplitudes of the crystallized samples are measured and the phases are then approximated iteratively. The nonlinearity also applies to the XRC quality metrics, which therefore are per-definition inferior to the linear metrics in cryo-EM like the FRC and FSC. Unfortunately, the majority of the structural biology field has not digested the importance of this statement, and most continue to adhere to long-since-refuted "gold-standard" metrics. For example, the fixed threshold FSC_{143} is justified by its historical connection to XRC quality in Rosenthal & Henderson (2003). However, the underlying logic must here be reversed: By complying with nonlinear XRC metrics, that FSC_{143} metric no longer complies with the linearity requirements of the original FSC and should no longer be called an FSC (see Appendix).

The Damage that Bad Metrics Cause

What is clear to anyone who understands the importance of the issues discussed in this review is that cryo-EM's inheritance of nonlinear XRC-based information metrics like R-factors or phase residuals will slowly disappear because those metrics throw away half of the available information, and they are not applicable outside a narrow, specific field anyway. The same is true for all fixed-value resolution thresholds in cryo-EM. For example, in electron ptychography (Bosch et al., 2023), in super-resolution light microscopy (Banterle et al., 2013), in X-ray imaging, in ultrasound imaging, or actually in any field where the need is to distinguish between instrumental and results resolution, the linear FRC- or FSC-type metrics will prevail. However, the question remains what actual damage is done by using incorrect metrics like the fixed-threshold FSC_{143}. The proponents of such metrics always tell us it does not really matter, but they refuse to withdraw their flawed work or even to correctly cite the research criticizing it; e.g., Van Heel & Schatz (2005). One significant type of damage occurs when incorrect metrics are used as a target function to iteratively optimize the quality of the results. The result can then only move toward optimizing that specific metric, whether it is correct or not. The typical reaction of cryo-EM scientists is to blame the FSC metric for problems encountered, ignoring the fact that the inconsistent "gold-standard" FSC143 procedures they are following violate the FSC definition! The consequence of these nonlinear lines of thinking was a whole cottage industry of xxx-FSC_{yyy} metrics that must all be refuted for the same reason (Van Heel, 2024b).

Is Local Resolution the Fundamental One?

In the previous section, global resolution metrics were compared to local resolution metrics, and we argued that in combination with a correct 1/2-bit FSC threshold criterion, one threshold serves all purposes (Van Heel & Schatz 2005, 2017) and there is no need for different thresholds for different purposes as suggested by Cardone et al. (2013): "One number does not fit all." The need for two different thresholds is purely the result of insisting on erroneous fixed-value resolution thresholds. As was also shown in Figure 6.7, the same threshold criterion must be used (but not the same numerical threshold value) to define the resolution correctly, and one then indeed finds comparable resolution level locally as well as globally. In fact, the preferred metric now is the local information density (LID) metric, which also includes the local 1/2-bit threshold (Van Heel & Schatz, 2017). To measure global resolutions, one must mask the two 2D images or two 3D volumes by a smooth function, like a Gaussian function with a half width of, say, 0.6, to comply with the requirements of the symmetric-sampling theorem (Van Heel & Schatz, 2020; Van Heel, 2024b). However, what is relevant here will be the overall size of the object to be masked by a smooth mask that must be larger than the object. What comes beyond that smooth mask is no longer relevant to the object's global resolution. That "global" adjective is therefore often more confusing than helpful since the global resolution implicitly assumes that the object is embedded in an appropriate-size sampling frame and that the densities in that volume are slowly apodized to zero prior to reaching the edges of the sampling area or volume. The local resolution suffers much less from such ambiguities since there one is always within the 1D, 2D, or 3D boundaries of the object, and one complies with the need to mask off the environment by multiplying with a Gaussian mask following the symmetric-sampling theorem requirements (Van Heel & Schatz, 2020; Van Heel,

2024b). Moreover, interesting objects are typically not uniform in information content, and the variations within an object are mostly the focus of our interest. Finally, the comparative quality assessment between two volumes of different sizes will often be facilitated by comparing local subvolumes of identical size of, say, 32^3 voxels with each other, irrespective of the actual size of the two volumes.

The Einstein from Noise Issue

The reference bias problem in cryo-EM is not a resolution metric issue but rather an issue of alignment with respect to reference images or volumes. A classical experiment was presented in the PhD thesis of Michael Schatz (Schatz, 1992) where a dataset of hundreds of 2D random-noise images were aligned using cross-correlation procedures with respect to an image of Albert Einstein. The total sum of the aligned random-noise images yielded a perfect copy of the Einstein reference image because of the reference bias caused by the alignment algorithm. To avoid such reference bias, a number of reference-free alignment algorithms were proposed, such as Dube et al. (1993), where an MSA-based alignment by classification (ABC) was proposed to prove the13-fold (***C***13) symmetry of a bacteriophage SPP1 portal protein, without using a 13-fold symmetric reference image, which would have automatically pulled the result toward 13-fold symmetry; see also Afanasyev et al. (2017). Reference bias has been diagnosed in the classic case of Mao et al. (2013) with Sodroski as senior author. The authors managed to get papers published in top journals with unrealistic claims that simply could not be true with the technique of those days. The chaotic controversies that existed in cryo-EM around resolution criteria and reference bias were exploited by those authors in an atmosphere of "anything goes" by using reference images that were created from a predetermined 3D model. In their defense against a barrage of criticism (Mao et al., 2013; Henderson, 2013; Subramaniam, 2013; Van Heel, 2013), Mao et al. cited an unreferenced "critical FSC" and claimed that (especially MvH) did not know how to interpret an FSC. A "reverse-engineering" manuscript that Afanasyev and Van Heel wrote to pinpoint how those authors had manipulated their data (that the authors took two years to deposit and that did not contain realistic data) was not accepted by the journal that had originally published their work since it was deemed not interesting for the general public, and that manuscript was never published. Unfortunately, because that issue was not properly closed off with a retraction by the journals or the authors, others were recently found doing the same as described in Huntington et al. (2023).

Perspective and Future

Cryo-EM has become a competitive industry operated around hi-tech equipment that costs ~5000 Euros per day in depreciation costs alone and requires a team of about 10 people with different expertises to operate. Setting up a cryo-EM facility takes a personal administrational investment of many years that may be seen as unproductive years in a scientific career. To set up a new facility may take a decade, and the finances for such an endeavor must be guaranteed on a long-term basis. Now that there are some 1000 high-end cryogenic electron microscopes installed worldwide, hopefully the access to such high-level instrumentation has become less problematic. Nevertheless, in a country like Brazil, achieving the first such facility in the Latin American continent was a real milestone. It is very important to realize that the availability of such advanced instrumentation in conjunction with a highly skilled personnel infrastructure may be the key to combating pandemics such as the recent Covid-19 event. The question is not whether an emerging economy like Brazil can afford to have such equipment but whether it can afford *not* to have such an infrastructure and a think tank of specialists around it.

A Final Positive Twist

Through the years, we and our group members have been frustrated by the continuous unscientific attacks on our work. People who only cite their own gold standard alignment FSC yet never understood the universal concept of the FSC have undoubtedly cost us grant income. We have simply stopped reacting to such abuse. The unpleasant details are discussed in a separate paper (Van Heel, 2024a). One good thing that came out of all this: We were forced to repeatedly think about the theoretical basis for the science of data collection. We thus found that the very foundation of sampling science established by Claude Shannon and others was flawed from the beginning (Shannon, 1949) and must be superseded by the symmetric sampling concept (Van Heel & Schatz, 2020, 2023; Van Heel, 2024b). This new insight changes the perspective on what measurements are in the physical sciences in general; it is not restricted to specific cryo-EM or X-ray crystallography issues.

CONCLUSIONS

Single-particle cryo-EM has become an essential tool for visualizing biological macromolecules. Over the decades of its development, cryo-EM has matured to a level that it now plays a crucial role in the development of, for example, vaccines to combat the Covid-19 pandemic. Many good things can and have been said about cryo-EM, and especially about its superb specimen-preparation methodology. However, there have also been methodological deviations that created severe confusion. Confusion around validation procedures and inappropriate nomenclatures, like calling processes "reference-free" when they clearly depend on reference choices, have even made it possible to bend results toward a desired, predetermined goal.

To counter such confusions, we have here emphasized original research methodological ideas that continue to be more science-based and intrinsically faster than most popular ones. Proudly promoted "free" software distributions are de facto illusions backed up by strong commercial interests. Ignoring the costs of developing and maintaining cutting-edge scientific software is counterproductive to the advancement of the cryo-EM community! The most shocking results from our review of unethical behavior involving cryo-EM are presented in a separate paper (Van Heel, 2024a). Above all, we urge newcomers to cryo-EM to critically study and understand the scientific tools they are using. Cryo-EM is not a black box; understanding the underlying science saves time and money and helps humanity.

ACKNOWLEDGEMENTS

We thank Michael Schatz and Ralf Schmidt for comments and technical and software assistance, Rodrigo Portugal for comments. MvH thanks Harold Erickson for discussing the problems of partial coherence and of the proportionality of the phase and amplitude contrast in cryo-EM.

REFERENCES

Adrian M, Dubochet J, Lepault J, McDowall AW (1984). Cryo-electron microscopy of viruses. *Nature* **308**, 32–36.

Afanasyev P, Ravelli RBG, Matadeen R, De Carlo S, Van Duinen G, Alewijnse B, Peters P, Abrahams J-P, Portugal RV, Schatz M, Van Heel M (2015). A posteriori correction of camera characteristics from large image data sets. *Sci. Rep* **5**, 10317.

Afanasyev P, Seer-Linnemayr C, Ravelli RBG, Matadeen R, De Carlo S, Alewijnse B, Portugal RV, Pannu NS, Schatz M, Van Heel M (2017). Single-particle cryo-EM using alignment by classification (ABC): The structure of *Lumbricus terrestris* haemoglobin. *IUCrJ* **4**, 678–694.

Banterle N, Bui KH, Lemke EA, Beck M (2013). Fourier ring correlation as a resolution criterion for super-resolution microscopy. *J Struct Biol* **183**, 363–367.

Bershad NJ, Rockmore AJ (1974). On estimating signal-to-noise ratio using the sample correlation coefficient. *IEEE Trans. Inf. Theory* **20**, 112–113.

Boekema EJ, Berden JA, van Heel M (1986). Structure of mitochondrial F1-ATPase studied by electron microscopy and image processing. *Bioch. Biophys. Acta* **851**, 353–360.

Boettcher B, Wynne SA, Crowther RA (1997). Determination of the fold of the core protein of hepatitis B virus by electron cryomicroscopy. Nature **386**, 88–91.

Borland L, van Heel M (1990). Classification of image data in conjugate representation spaces. *J Opt Soc Am* **A7**, 601–610.

Bosch C, Diaz A, Holler M, Guizar-Sicairos M, Aidukas T, Pacureanu A, Mueller E, Peddie CJ, Collinson L, Zhang Y, Menzel A (2023). 3D-Imaging of synapses in neuronal tissues with synchrotron X-ray ptychography. *BioRxiv* **2023**, 2023–11.

Brenner S, Horne RW (1959). A negative staining method for high resolution electron microscopy of viruses. *Biochim Biophys Acta* **34**, 103.

Brüggeller P, Mayer E (1980). Complete vitrification in pure liquid water and dilute aqueous solutions. *Nature* **288**, 569–571.

Campbell MG, Cheng A, Brilot AF, Moeller A, Lyumkis D, Veesler D, Pan J, Harrison SC, Potter CS, Carragher B, Grigorieff N (2012). Movies of ice-embedded particles enhance resolution in electron cryo-microscopy. *Structure* **20**, 1823–1828.

Cardone G, Heymann JB, Steven AC (2013). One number does not fit all: Mapping local variations in resolution in cryo-EM reconstructions. *J Struct Biol* **184**, 226–236.

Chanzy H, Franc JM, Herbage D (1976). High-angle electron diffraction of frozen-hydrated collagen. *BiochEM J* **153**, 139–140.

Chiu W, Downing KH, Dubochet J, Glaeser RM, Heide HG, Knapek E, Kopf DA, Lamvik MK, Lepault J, Robertson JD, Zeitler E, Zemlin F, International Experimental Study Group (1986). Cryoprotection in electron microscopy. *J Microsc (Oxford)* **141**, 385–391.

Crowther RA (1971). Procedures for three-dimensional reconstruction of spherical viruses by Fourier synthesis from electron micrographs. *Phil Trans Roy Soc Lond B* **261**, 221–230.

De Broglie L (1924). Recherches sur la théorie des quanta. *Annales de Physique (1925)* **10**, 22–128.

DeRosier DJ, Klug A (1968). Reconstruction of three-dimensional structures from electron micrographs. *Nature* **217**, 130–134.

DeRosier DJ, Moore PB (1970). Reconstruction of three-dimensional images from electron micrographs of structures with helical symmetry. *J Mol Biol* **52**, 355–369.

Dube P, Tavares P, Lurz R, van Heel M (1993). Bacteriophage SPP1 portal protein: A DNA pump with 13-fold symmetry. *EMBO J* **15**, 1303–1309.

Dubochet J, Adrian M, Chang J-J, Homo J-C, Lepault J, McDowall AW, Schultz P (1988). Cryo-electron microscopy of vitrified specimens. *Q Rev Biophys* **21**, 129–228.

Dubochet J, McDowall AW (1981). Vitrification of pure water for electron microscopy. *J of Microscopy (Oxford)* **124**, RP3–4.

Erickson HP, Klug A (1971). Measurement and compensation of defocusing and aberrations by Fourier processing of electron micrographs. *Phil Trans R Soc Lond B* **261**, 105–118.

Erickson HP (1973). The Fourier transform of an electron micrograph-first order and second order theory of image formation. *Adv Optic Electr Microsc* **5**, 163–199.

Faruqi AR, Henderson R (2007). Electronic detectors for electron microscopy. *Curr Opin Struct Biol* **17**, 549–555.

Fernández-Morán H (1960). Low temperature preparation techniques for electron microscopy of biological specimens based on rapid freezing with liquid helium II. *Ann N Y Acad Sci* **85**, 689–713.

Fisher RA (1915). Frequency distribution of the values of the correlation coefficient in samples of an indefinitely large population. *Biometrika* 10 (1915) 507–521.

Frank J, Al-Ali L (1975). Signal-to-noise ratio of electron micrographs obtained by cross-correlation. *Nature* **256**, 376–379.

Frank J (1973). The envelope of electron microscopic transfer functions for partially coherent illumination. *Optik* **38**, 519–536.

Frank J, Verschoor A, Boublik M (1981). Computer averaging of electron micrographs of 40 S ribosomal subunits. *Science* **214**, 1353–1355.

Frank J (2006). Three-dimensional electron microscopy of macromolecular assemblies: visualization of biological molecules in their native state. Oxford university press.

Frank J (2009). Single-particle reconstruction of biological macromolecules in electron microscopy–30 years. *Quarterly Reviews of Biophysics* **42**, 139–158.

Harauz G, Van Heel M (1986a). Exact filters for general-geometry three-dimensional reconstruction. *Optik* **73**, 146–156.

Harauz G, Van Heel M (1986b). Direct 3D reconstruction from orientations with initially unknown angles. *Patter Recognition in Practice* **II**, 279–288.

Harauz G, Borland L, Bahr GF, Zeitler E, Van Heel M (1987). Three-dimensional reconstruction of a human metaphase chromosome from electron micrographs. *Chromosoma* **95**, 366–374.

Henderson R, Baldwin JM, Ceska TA, Zemlin F, Beckman E, Downing KH (1990). Model for the structure of bacteriorhodopsin based on high-resolution electron cryomicroscopy. *J Mol Biol* **213**, 899–929.

Henderson R, Sali A, Baker ML, Carragher B, Devkota B, Downing KH, Egelman EH, Feng Z, Frank J, Grigorieff N, Jiang W, Ludtke SJ, Medalia O, Penczek PA, Rosenthal PB, Rossmann MG, Schmid MF, Schröder GF, Steven AC, Stokes DL, Westbrook JD, Wriggers W, Yang H, Young J, Berman HM, Chiu W, Kleywegt GJ, Lawson CL (2012). Outcome of the first electron microscopy validation task force meeting. *Structure* 20, 205–214.

Henderson R (2013). Avoiding the pitfalls of single particle cryo-electron microscopy: Einstein from noise. *Proc Natl Acad Sci* **110**, 18037–18041.

Hoppe W, Gassmann J, Hunsmann N, Schramm HJ, Sturm M (1974). Three-dimensional reconstruction of individual negatively stained yeast fatty acid synthetase molecules from tilt-series in the electron microscope. *Hoppe-Seyler's Z Physiol ChEM* **355**, 1483–1487.

Huntington JA, Faille A, Ustok FI. (2023). Serious Issues with Cryo-EM Structures of Human Prothrombinase- A Case of Einstein from Noise? https://doi.org/10.31219/osf.io/rd263

Klaholz BP, Myasnikov AG, Van Heel M (2004). Visualization of release factor 3 on the ribosome during termination of protein synthesis. *Nature* 427, 862–865.

Jeng T-W, Chiu W (1984). Quantitative assessment of radiation damage in a thin protein crystal. *J Microsc (Oxford)* **136**, 35–44.

Kühlbrandt W (2014). The resolution revolution. *Science* **343**, 1443–1444.

Kunath W, Weiss K, Sack-Kongehl H, Kessel M, Zeitler E (1984). Time-resolved low-dose microscopy of glutamine synthetase molecules. *Ultramicroscopy* **13**, 241–252.

Liao HY, Frank J (2010). Definition and estimation of resolution in single-particle reconstructions. *Structure* **18**, 768–775.

Malhotra A, Penczek P, Agrawal RK, Gabashvili IS, Grassucci RA, Jünemann R, Burkhardt N, Nierhaus KH, Frank J (1998). Escherichia coli 70 S ribosome at 15 A resolution by cryo-electron microscopy: localization of fMet-tRNAfMet and fitting of L1 protein. J Mol Biol. 280(1), 103–16.

Mao Y, Wang L, Gu C, Herschhorn A, Désormeaux A, Finzi A, Xiang S-H, Sodroski GJ (2013). Molecular architecture of the uncleaved HIV-1 envelope glycoprotein trimer. *Proc Natl Acad Sci* **110**, 12438–12443.

Marabini R, Carragher B, Chen S, Chen J, Cheng A, Downing KH, Frank J, Grassucci RA, Heymann JB, Jiang W, Jonic S, Liao HY, Ludtke SJ, Patwari S, Piotrowski AL, Quintana A, Sorzano COS, Stahlberg H, Vargas J, Voss NR, Chiu W, Carazo JM (2015). CTF challenge: Result summary. *J Struct Biol* **19**, 348–359.

Mindell JA, Grigorieff N (2003). Accurate determination of local defocus and specimen tilt in electron microscopy. *J Struct Biol* **142**, 334–347.

Nicholson WV, Glaeser RM (2001). Review: Automatic particle detection in electron microscopy. *J Struct Biol* **133**, 90–101.

Oettl H, Hegerl R, Hoppe W (1983). Three-dimensional reconstruction and averaging of 50S ribosomal subunits of Escherichia coli from electron micrographs. *J Molec Biol* **163**, 431–450.

Orlova EV, van Heel M. (1994). Angular reconstitution of macromolecules with arbitrary point group symmetry. *Proceedings ICEM 13-Paris 1994*, **1**, 507–508.

Orlova EV, Dube P, Harris JR, Beckman E, Zemlin F, Markl J, Van Heel M (1997). Structure of Keyhole Limpet Hemocyanin Type 1 (KLH1) at 15Å resolution by electron cryomicroscopy and angular reconstitution. *J Mol Biol* **271**, 417–437.

Pettersen EF, Goddard TD, Huang CC, Couch GS, Greenblatt DM, Meng EC, Ferrin TE (2004). UCSF chimera-a visualization system for exploratory research and analysis. *J Comput ChEM* **25**, 1605–1612.

Radermacher M, Wagenknecht T, Verschoor A, Frank J (1987). Three-dimensional structure of the large ribosomal subunit from Escherichia coli. *EMBO J* **6**, 1107–1114.

Radon J. (1917). Über die Bestimmung von Funktionen durch ihre Integralwerte langs gewissez Mannigfaltigheiten, Ber. Verh. Sachs. Akad. Wiss. Leipzig, Math Phys Klass, 69.

Rohou A, Grigorieff N (2015). CTFFIND4: Fast and accurate defocus estimation from electron micrographs. *J Struct Biol* **192**, 216–221.

Rosenthal PB, Henderson R (2003). Optimal determination of particle orientation, absolute hand, and contrast loss in single-particle electron cryomicroscopy. *J Mol Biol* **333**, 721–745.

Ruska E, Knoll M (1931). Die magnetische Sammelspule für schnelle Elektronenstrahlen. *Z Techn Physik* **12**, 389–400 and 448.

Sass HJ, Beckmann E, Büldt G, Dorset D, Rosenbusch JP, van Heel M, Zeitler E, Zemlin F, Massalski A (1989). Densely packed ß-structure at the protein-lipid interface of porin is revealed by high-resolution cryo-electron microscopy. *J Mol Biol* **209**, 171–175.

Saxton WO, Baumeister W (1982). The correlation averaging of a regularly arranged bacterial cell envelope protein. *J Microsc* **127**, 127–138.

Schatz M (1992). PhD thesis: Invariante Klassifizierung elektronenmikroskopischer Aufnahmen von eiseingebetteten biologischen Makromolekulen. Thesis, Freie Universität Berlin. Deutsche Hochschulschriften 452, Verlag Hänsel-Hohenhausen, Egelsbach/Köln/New York.

Schatz M, Orlova EV, Dube P, Stark H, Zemlin F, Van Heel M (1997). Angular reconstitution in 3D electron microscopy: Practical and technical aspects. *Scan Microsc* **11**, 179–194.

Schatz M, Van Heel M (1990). Invariant classification of molecular views in electron micrographs. *Ultramicroscopy* **32**, 255–264.

Scheres SHW (2012). RELION: Implementation of a Bayesian approach to cryo-EM structure determination. *J Struct Biol* **180(3)**, 519–530.

Scherzer O (1949). The theoretical Resolution limit of the electron microscope. *J Appl Phys* **20**, 20–29.

Shannon CE (1949). Communication in the presence of noise. *Proc. of the IRE* **37**, 10–21.

Subramaniam S (2013). Structure of trimeric HIV-1 envelope glycoproteins. *Proc. Natl. Acad. Sci* **110**, E4172–E4174.

Taylor KA, Glaser RM (1976). Electron microscopy of frozen hydrated biological specimens. *J Ultrastruct Res* **55**, 448–546.

Unwin PNT, Henderson R (1975). Molecular structure determination by electron microscopy of unstained crystalline specimens. *J Mol Biol* **94**, 425–440.

Van Bruggen EFJ, Wiebenga EH, Gruber M (1960). Negative-staining electron microscopy of proteins at pH values below their isoelectric points. Its application to hemocyanin. *Bioch Biophys Act* **42**, 171.

Van Heel M (1978). On the imaging of relatively strong objects in partially coherent illumination in optics and electron optics. *Optik* **49**, 389–408.
Van Heel M. (1980–2022). *Principles of Phase Contrast (Electron) Microscopy*. Lecture notes.
Van Heel M, Frank J (1981). Use of multivariate statistics in analyzing the images of biological macromolecules. *Ultramicroscopy* **6**, 187–194.
Van Heel M (1982). Detection of objects in quantum-noise-limited images. *Ultramicros* **7**, 331–342.
Van Heel M, Keegstra W, Schutter W, van Bruggen EFJ (1982). Arthropod hemocyanin structures studied by image analysis. In: Life Chemistry Reports, Suppl. 1, "The Structure and Function of Invertebrate Respiratory Proteins", EMBO workshop, Leeds, Ed. E.J. Wood, 69–73..
Van Heel M (1983). Stereographic representation of three-dimensional density distributions. *Ultramicroscopy* **11**, 307–314.
Van Heel M (1984a). Multivariate statistical classification of noisy images (randomly oriented biological macromolecules). *Ultramicroscopy* **13**, 165–183.
Van Heel M (1984b). Three-dimensional reconstruction from projections with unknown angular relationships. *Proc. Eighth Europ. Conr. on Electr. Microsc., Budapest*. 1347–1348.
Van Heel M (1987a). Angular reconstitution: A posteriori assignment of projection directions for 3D reconstruction. *Ultramicroscopy* **21**, 111–124.
Van Heel M (1987b). Similarity measures between images. *Ultramicroscopy* **21**, 95–100.
Van Heel M (1989). Classification of very large electron microscopical image data sets. *Optik* **82**, 114–126.
Van Heel M, Winkler HP, Orlova EV, Schatz M (1992). Structure analysis of ice-embedded single particles. *Scanning Microscopy Suppl* **6**, 23–42.
Van Heel M, Stöffler-Meilicke M (1985). Characteristic views of *E. coli* and *B. stearothermophilus* 30S ribosomal subunits in the electron microscope. *EMBO J* **4**, 2389–2395.
Van Heel M, Orlova EV, Harauz G, Stark H, Dube P, Zemlin F, Schatz M (1997). Angular reconstitution in three-dimensional electron microscopy: Historical and theoretical aspects. *Scan Microsc* **11**, 195–210.
Van Heel M, Gowen B, Matadeen R, Orlova EV, Finn R, Pape T, Cohen D, Stark H, Schmidt R, Schatz M, Patwardhan A. (2000). Single-particle electron cryo-microscopy: Towards atomic Resolution. *Q Rev Biophys*, **33(4)**, 307–369.
Van Heel M, Schatz M (2005). Fourier shell correlation threshold criteria. *J Struct Biol* **151**, 250–262.
Van Heel M (2009). Jean-Pierre bretaudière *obituary*: The early days of multivariate statistics in electron microscopy. Appendix to: *Open J Stat* **6** (2016) 735–739.
Van Heel M, Portugal R, Rohou A, Linnemayr C, Bebeacua C, Schmidt R, Grant T, Schatz M (2012). Four-dimensional cryo-electron microscopy at quasi-atomic resolution: IMAGIC 4D. International Tables for Crystallography. Vol. F, Chapter 19.9, 624–628.
Van Heel M (2013). Finding trimeric HIV-1 envelope glycoproteins in random noise. *Proc. Natl. Acad. Sci* **110**, E4175–E4177.
Van Heel M, Portugal R, Schatz M (2016). Multivariate statistical analysis of large datasets: Single particle electron microscopy. *Open J Stat* **6**, 701–739.
Van Heel M, Schatz M. (2017). Reassessing the Revolution's Resolutions. *BioRxiv*
Van Heel M, Schatz M. (2020). Information: to Harvest, to Have and to Hold. *arXiv*.
Van Heel M, Schatz M. (2023). Systems, devices, components, and methods for optimizing information and data acquisition, transmission, processing, and analysis. US Patent App. 18/024,225
Van Heel M (2024a). Unethical distortions and deceptions in cryo-EM, in preparation.
Van Heel M (2024b). Symmetric Sampling and the Discrete Laws of Physics (1) Refuting the Classical Shannon-Nyquist Sampling Theorem. https://www.researchsquare.com/article/rs-3783624/v1
van Leeuwenhoek A (1677). Observation, communicated to the publisher ck, in a Dutch letter of the 9 October 1676 here in English: Concerning little animals by him observed in rain-well-sea and snow water; as also in water wherein pepper had lain infused. *Phil Trans* **12**, 821–831.
Zernike F (1942). Phase contrast, a new method for the microscopic observation of transparent objects part II. *Physica* **9**, 686–974.
Wikipedia, ***Image segmentation***, (September, 2023).
Wikipedia, ***Jean-Paul Benzécri***, (September, 2023).
Wikipedia, ***Correspondence analysis***, (September, 2023).
Wikipedia, ***Fisher transformation***, (16 Apr. 2020)
Wikipedia, ***Ward's method***, (September, 2023).
Wikipedia, ***Tomographic reconstruction***, (September, 2023).
Wikipedia, ***Ernst Abbe***, (September, 2023)

Appendix

The Difference between Uncorrelated and Orthogonal Vectors

Marin van Heel and Michael Schatz

The gold standard used for assessing the quality of 3D reconstructions in cryo-EM is the Fourier shell correlation metric (FSC) applied between two independently determined volumes (Harauz & van Heel, 1986a). The *a priori* assumption made for such comparisons in theoretical papers is that the two volumes (Figure S1a–b) contain identical 3D information (the signal $\boldsymbol{S}$, Figure S1d–e), but that each volume is deteriorated by a different realization of an additive zero-mean random noise ($\boldsymbol{N}_i$, Figure S1f–g). In our simple test, $\boldsymbol{S}$ is the structure of a giant hemoglobin (Afanasyev et al., 2017), but that is immaterial to this model experiment. The two volumes, $\boldsymbol{A} = \boldsymbol{S} + \boldsymbol{N}_1$ (Figure S1a) and $\boldsymbol{B} = \boldsymbol{S} + \boldsymbol{N}_2$ (Figure S1b), are generated using the same starting volume $\boldsymbol{S}$ and adding two random-noise volumes $\boldsymbol{N}_1$ and $\boldsymbol{N}_2$, respectively. Important to the experiment is that we know all these components *a priori*, such that we can assess all cross-correlation components separately, which is a luxury one never has in real experiments. From the 3D Fourier transforms of these 3D densities ($\boldsymbol{A}$ and $\boldsymbol{B}$; transforms denoted in italics) all correlations are calculated: $\boldsymbol{A}\cdot\boldsymbol{B}^* = (\boldsymbol{S} + \boldsymbol{N}_1)\cdot(\boldsymbol{S}^* + \boldsymbol{N}_2^*) = \boldsymbol{S}^2 + \boldsymbol{S}\cdot\boldsymbol{N}_2^* + \boldsymbol{N}_1\cdot\boldsymbol{S}^* + \boldsymbol{N}_1\cdot\boldsymbol{N}_2^*$. The following FSC curves are generated (function of radius $\boldsymbol{r}$):

1. A $\boldsymbol{FSC}_{AB}$ (Figure S1c) between the two noisy volumes $\boldsymbol{A}$ and $\boldsymbol{B}$,
2. The ***signal-versus-noise*** correlation $\boldsymbol{FSC}_{SN1}$ between $\boldsymbol{S}$ and $\boldsymbol{N}_1$ (Figure S1i),
3. The ***signal-versus-noise*** correlation $\boldsymbol{FSC}_{SN2}$ between $\boldsymbol{S}$ and $\boldsymbol{N}_2$ (Figure S1j), and,
4. The ***noise-versus-noise*** correlation: $\boldsymbol{FSC}_{N1N2}$, between $\boldsymbol{N}_1$ and $\boldsymbol{N}_2$ (Figure S1h).

The $\boldsymbol{FSC}_{SN1}$, $\boldsymbol{FSC}_{SN2}$, and $\boldsymbol{FSC}_{N1N2}$ curves are normalized only by the number of voxels in the shells and are directly comparable. The $\boldsymbol{FSC}_{AB}$ (Figure S1c) has a typical FSC appearance, indicating that realistic levels of noise have been added to the signal $\boldsymbol{S}$ to create the noisy volumes $\boldsymbol{A}$ and $\boldsymbol{B}$. The noise-versus-noise $\boldsymbol{FSC}_{N1N2}$ (Figure S1h) oscillates around the zero mark over all spatial frequencies with its modulation increasing close to the origin, where Fourier shells contain fewer voxels. The correlation fluctuations between the signal $\boldsymbol{S}$ and each of the noise terms $\boldsymbol{N}_1$ and $\boldsymbol{N}_2$ are significantly larger than are the correlations between the noise terms $\boldsymbol{N}_1$ and $\boldsymbol{N}_2$ in the relevant frequency ranges. Whenever the signal $\boldsymbol{S}$ and the uncorrelated noise $\boldsymbol{N}_i$ are of similar size, the predominant noise component consists of the cross-terms ($\boldsymbol{S}\cdot\boldsymbol{N}_2^* + \boldsymbol{N}_1\cdot\boldsymbol{S}^*$) rather than of the smaller ($\boldsymbol{N}_1\cdot\boldsymbol{N}_2^*$) correlations. Resolution criteria must compare the overall signal component to the influence of all three noise-related cross-terms. It is at this level that a fundamental error is often made in the literature, namely: It is assumed that because the signal and the noise are independent or uncorrelated, their inner products $\boldsymbol{S}\cdot\boldsymbol{N}_2^*$ and $\boldsymbol{N}_1\cdot\boldsymbol{S}^*$ are zero and may be dropped from the equations!

However, uncorrelated in this context means only that the inner product between these vectors is zero as an expectation value—that is, as the average over an infinite number of repeats of the experiment (symbolically: $<\boldsymbol{S}\cdot\boldsymbol{N}> = 0$). Assuming that each of the individual $\boldsymbol{S}\cdot\boldsymbol{N}$ inner products is zero implies, in contrast, that each of these vector pairs are orthogonal. The latter assumption—prolific in the EM literature—is untenable, as is clearly illustrated in our model experiment. Thus, although the statement that the signal $\boldsymbol{S}$ and the noise $\boldsymbol{N}$ are independent or uncorrelated is by itself correct, the standard interpretation in cryo-EM is to therefore take those cross-products out of the equations. This is equivalent to stating that all these vectors are orthogonal, which is fundamentally wrong.

Note that there are various definitions of *orthogonal*, *uncorrelated*, or *independent vectors*, and this confusion extends to most fields of science. The same confusion exists in the use of these concepts in our daily language. In mathematics, for example, when the inner product of two vectors is zero, those vectors are called *orthogonal*. In statistics, on the other hand, the same *orthogonal* nomenclature is often used for when the expectation value of the inner product is zero. One must thus clearly define exactly what one means with any such statistical statement, and one must stay within a consistent set of definitions.

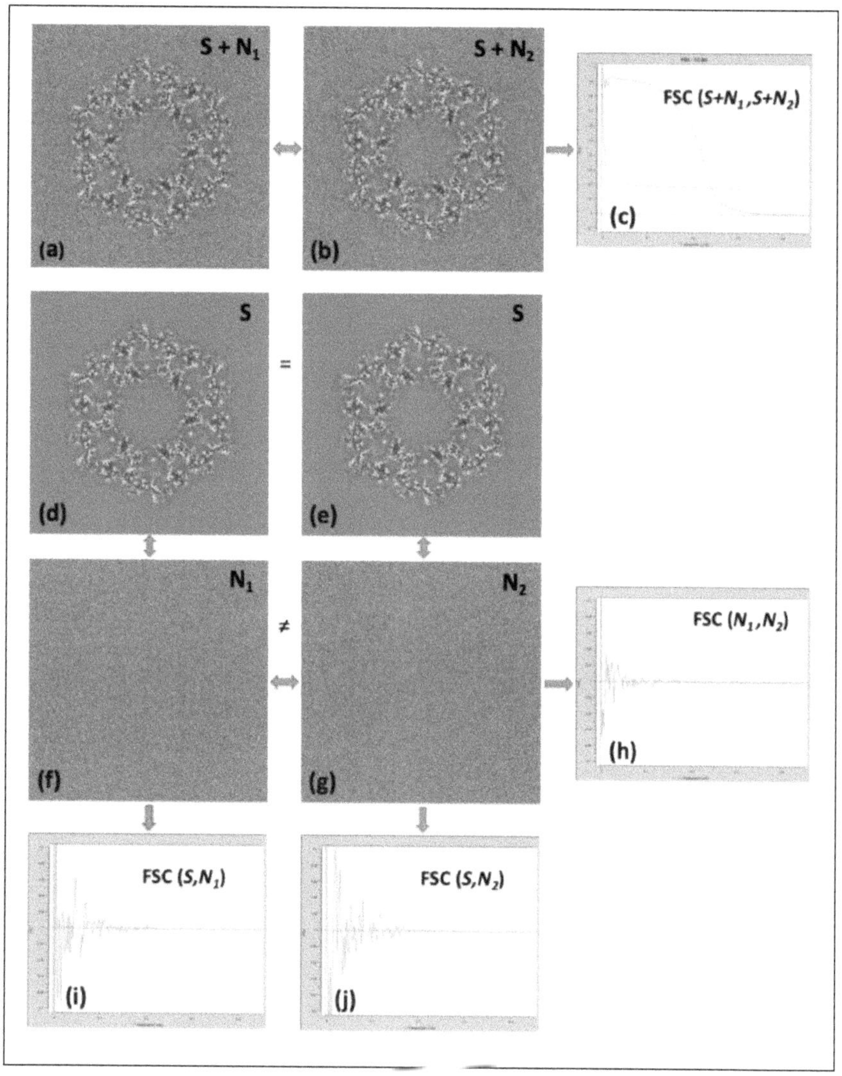

FIGURE A1.1. Fourier shell correlations between signal and noise volumes.

a) Section through test volume $\boldsymbol{A}$, the sum of noise-free signal volume $\boldsymbol{S}$ (d) and random noise volume $\boldsymbol{N}_1$ (f). Shown is section 151 (of 360).
b) Corresponding section through test volume $\boldsymbol{B}$, the sum of signal volume $\boldsymbol{S}$ (d) and a second random noise volume $\boldsymbol{N}_2$ (g).
c) FSC between the two noisy volumes $\boldsymbol{A}$ and $\boldsymbol{B}$.
d) Noise-free signal volume $\boldsymbol{S}$ (an arbitrary cryo-EM 3D reconstruction). (Shown is section 151 (of 360).
e) Copy of the same noise-free signal volume $\boldsymbol{S}$ (d).
f) Corresponding section through noise volume $\boldsymbol{N}_1$.
g) Corresponding section through noise volume $\boldsymbol{N}_2$.
h) FSC (not normalized) between noise volumes $\boldsymbol{N}_1$ and $\boldsymbol{N}_2$.
i) FSC (not normalized) between signal volume $\boldsymbol{S}$ and noise volume $\boldsymbol{N}_1$.
j) FSC (not normalized) between signal volume $\boldsymbol{S}$ and noise volume $\boldsymbol{N}_2$.

7 Streptavidin Affinity Grids in the Study of Macromolecular Complexes Using Cryo-Electron Microscopy

Nils Marechal and Patrick Schultz

INTRODUCTION

Electron microscopists seeking structural information from single-particle analysis have benefited from the resolution revolution, during which improved microscopes, direct electron detectors, and image analysis software were used to reach atomic resolution from large datasets of molecular images. A remaining major challenge is to preserve the structural integrity of the specimen in order to study its organization and conformational dynamics in close to solution conditions. The most important step is to conserve specimen hydration within the microscope's vacuum. Key techniques were identified in the 1980s to vitrify the specimen, to transfer and image the sample in a frozen hydrated state, and to avoid specimen contamination from external sources (Dubochet et al., 1988).

A crucial stage during the specimen preparation process is the formation of a 20 to 50 nm thick aqueous film that contains the molecules of interest. Such a thin film is a prerequisite to retrieving high-resolution structural information from transmission electron microscopy images (Wu et al., 2016), but its preparation is difficult to control. The most popular approach consists of depositing a drop of solution containing the biomolecule of interest onto the electron microscopy support and to blot the excess solution with filter paper. This process involves several high-energy steps, including exposure of the specimen to potentially denaturing surface effects and strong shearing forces (Alex et al., 2018; Glaeser, 2016; Glaeser and Han, 2017). Many fragile molecular complexes did not survive this step, and they needed chemical cross-linking to be stabilized. The concentration of the sample often did not match the number of particles depicted in the projection of the thin layer. Protein concentrations as low as 0.1 mG/mL yielded closely packed protein distributions while they should have occupied only 0.01% of the volume of the solution. Electron tomography experiments clearly showed that all proteins were concentrated on one surface of the thin layer, most probably by interacting with the air–water interface and thus reducing the surface tension by compromising their structural integrity (D'Imprima et al., 2019). High-speed interference contrast microscopy was used to analyze the fluid behavior during the blotting process, and this revealed that fiber irregularities contributed to irregular ice thickness and that macromolecular complexes could be exposed to severe shear forces during blotting (Armstrong et al., 2020).

Several directions were explored to reduce the effects of specimen blotting and surface tension. Elaborating on the very early method of spraying micrometric droplets onto the grid surface (Dubochet et al., 1988), the Carrager group proposed a similar spraying approach coupled to a self-whipping device consisting of epitaxially grown nanowires (Tilak et al., 2012). This very elegant method has the advantage of considerably reducing the time the specimen experiences the proximity of the air–water interface (a typical experiment can be performed in less than 75 ms). In addition, filter paper blotting is not required anymore, thus reducing the shear forces and ice thickness irregularities. The second family of approaches consists in using solid supports, such as continuous carbon films or graphene oxide flakes, in order to adsorb the sample on a less denaturing surface than the air–water interface. Both methods proved useful for several molecular systems, but neither proved to be a universal solution.

Alternatively, the air–water interface can be shielded by the use of surfactants such as lipids or detergents (Frederik et al., 1989). Several nonionic detergents, such as beta octyl glucoside, DDM, NodinetP40, or CHAPS/CHAPSO, proved useful to protect fragile specimens from partial denaturation but with the drawback of requiring very high specimen concentration (Chen et al., 2019). This makes sense since preventing biomolecules from interacting with the air–water interface also reduces the number of imaged molecules, which can only be restored by increasing specimen concentration. Despite the success of this method, a compromise has thus to be found between the risk of denaturing the sample and the local increase in concentration at the interface needed to collect a sufficiently large image dataset suitable for high-resolution structure determination.

Early biophysical studies showed that protein denaturation can be prevented by spreading a protective lipid monolayer that self-organizes at the air–water interface (Fromherz, 1971; Peters and Fromherz, 1975). Further inspired by the successful two-dimensional (2D) crystallization of membrane proteins and the potential to solve protein structures by using crystallographic data analysis methods (Unwin and Henderson, 1975), the Kornberg group developed 2D crystallization methods adapted to soluble proteins (Uzgiris and Kornberg, 1983). They chemically synthesized a functionalized lipid by coupling dinitrophenyl (DNP) to the hydrophilic moiety of an amphipathic phospholipid molecule. Once spread at the air–water interface, the lipid-bound ligand

DOI: 10.1201/9781003326106-9

is solvent accessible and can interact with IgG molecules directed against DNP, placed in the solution below the lipid monolayer. The IgG molecules were oriented and concentrated at the lipid-protected interface and were found to form single-molecule-thick 2D crystals in favorable conditions.

This innovative method has been adapted to other ligands and target combinations, such as novobiocine for the gyrase B subunit (Lebeau et al., 1990) or ribonucleotide reductase bound to its effector (Ribi et al., 1987), and it was extended to the use of charged lipid monolayers (Darst et al., 1991; Schultz et al., 1990). To deliver a more universally available method, lipids coupled to nickel-chelating lipids were developed and used to crystallize histidine-tagged protein complexes (Bischler et al., 1998; Kubalek et al., 1994). But one of the most versatile systems turned out to be the 2D crystallization of streptavidin on biotinylated lipid monolayers (Darst et al., 1991).

STREPTAVIDIN AFFINITY GRIDS FOR CRYO-ELECTRON MICROSCOPY: A DECADE OF DEVELOPMENT

The streptavidin affinity grids are based on mounting a 2D streptavidin crystal on an electron microscopy support. Placing the streptavidin molecules under a biotinylated lipid monolayer concentrates the streptavidin tetramer on the lipid surface and favors their subsequent 2D crystallization (Darst et al., 1991). According to the C222 space group of the crystals, each streptavidin tetramer has two biotin binding pockets occupied at the lipid monolayer interface, which leaves two biotin binding sites available on the solvent-facing surface of the crystal, thus opening the possibility to bind macromolecular targets through a variety of approaches (Figure 7.1a and b).

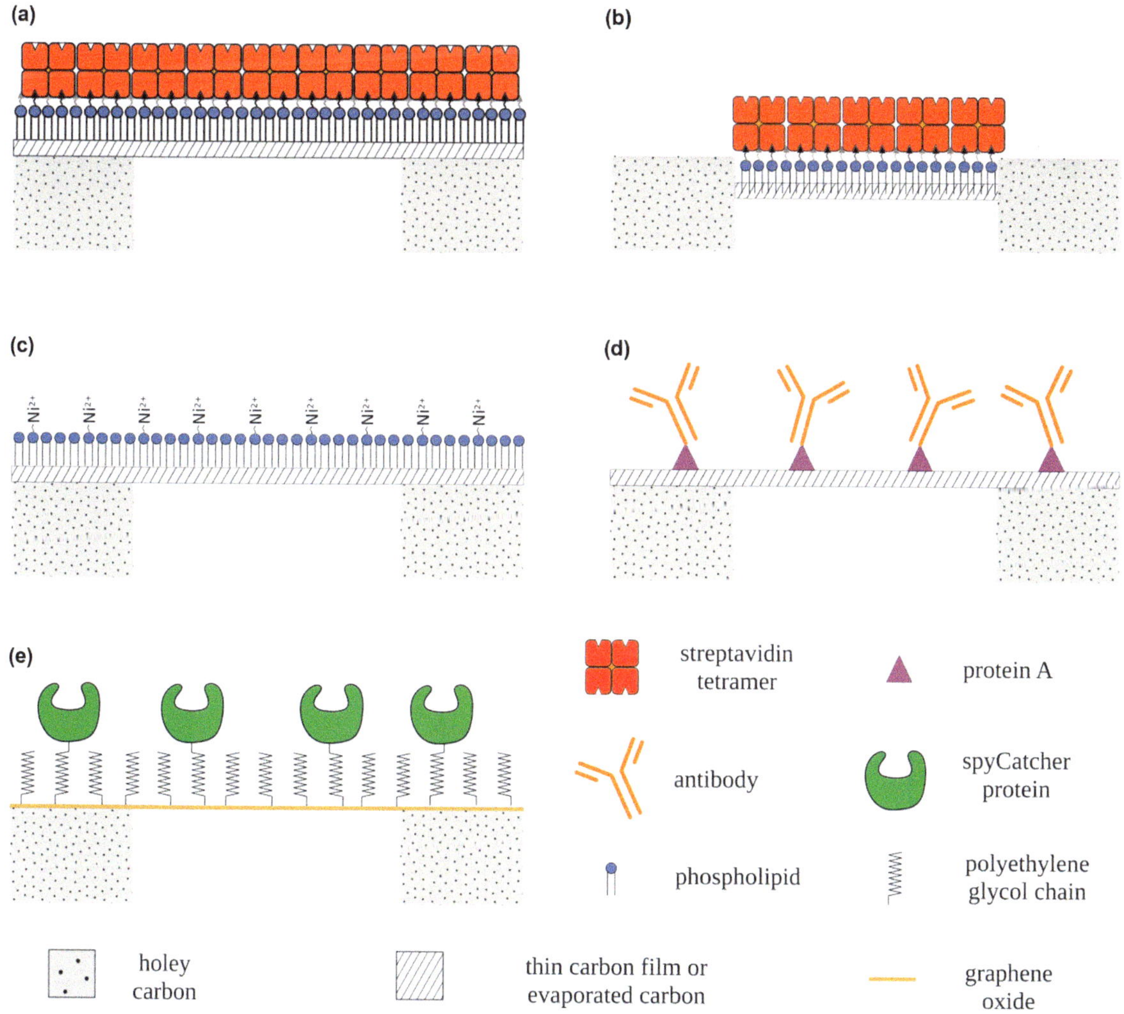

FIGURE 7.1 Architecture of several class of affinity grids. Schematic representation of the four currently existing types of affinity grids: (a) streptavidin affinity grid design as first established by Darst and coworkers (1991) and Crucifix and coworkers in 2004, and (b) as developed by Han and coworkers in 2016. (c) Architecture of a functionalized lipid monolayer grid exposing a Ni-NTA moiety (Kelly et al., 2008). (d) Design of antibody affinity grids using protein A to anchor the antibody (Yu et al., 2016). (e) Representation of a SpyCatcher affinity grid (Wang *et al*., 2020). All these grids are based on holey- or lacey-carbon grids, on which the affinity film is assembled.

The binding of labeled macromolecular targets onto the surface of 2D streptavidin crystals takes advantage of the streptavidin–biotin interaction, which is the strongest known noncovalent protein–ligand association, with a dissociation constant (K_d) of ~10^{-15} M (Green, 1962, 1963; Melamed and Green, 1963). In its pioneering study, the Darst group captured biotinylated ferritin on the streptavidin crystals, thus showing that the solvent-facing biotin binding sites are indeed accessible. In 2004, Schultz and coworkers used the 2D streptavidin crystals to bind a biotinylated double-stranded DNA and showed that this system can be used to capture RNA polymerase I. They further demonstrated that the periodic signal of the crystal lattice can be removed from the micrograph by masking the diffraction peaks in Fourier space. They showed that it is not only possible to bind RNA polymerase I to the immobilized DNA template, but also to reconstruct a 3D model of the enzyme (Crucifix et al., 2004). Similarly, the recruitment of biotinylated liposomes onto 2D streptavidin crystals was proposed as a strategy to study membrane proteins in their lipid environment by cryo-electron microscopy (Wang et al., 2008; Wang and Sigworth, 2010). In these studies, the 2D streptavidin crystals were formed at the surface of a drop containing the sample to be captured, then decorated crystals were transferred onto an electron microscopy grid and processed for cryo-EM (Figure 7.1a). Despite the considerable opportunities offered by the functionalized streptavidin crystal support, the complexity, the poor reproducibility, and the limited surface of 2D crystals, in addition to the required Fourier filtering to remove the lattice signal, restrained the use of streptavidin grids to a few methodological studies (Figure 7.2). In 2016, the Glaeser group revisited the 2D streptavidin crystal preparation method. Instead of being transferred onto the electron microscopy grid from a drop, the crystal was directly grown onto a holey carbon grid (Han et al., 2016). Briefly, the biotinylated lipid monolayer is transferred on a hydrophobic holey carbon grid from a simplified Langmuir-Blodgett trough. A streptavidin solution is then loaded on the grid, and crystallization occurs within a few tens of minutes. Remarkably, a large number of 2D crystals form, and they can even be dried in the presence of a protective trehalose layer. A few nanometers of amorphous carbon is finally evaporated onto the exposed lipid tails to rigidify the assembly and prevent lipid oxidation (Figure 7.1b). The streptavidin crystal, sandwiched between a layer of dehydrated trehalose and the carbon-stabilized lipid monolayer, remains active for months, making it possible to prepare many grids in advance, and to use them whenever required. This development removed a methodological hurdle, and since then a growing body of research using streptavidin affinity grids has been published, as illustrated in Figure 7.2.

However, manufacturing the streptavidin affinity grids based on the above mentioned procedure remained challenging: It involved repetitive grid handling, causing mechanical stress, and suffered from a lack of control of the crystallization conditions—i.e., streptavidin concentration, temperature, and humidity. Furthermore, the crystallization process is highly sensitive to the properties of the holey carbon film. Taken together, these factors led to large variations in batch quality. An optimized streptavidin affinity grid preparation procedure (see Annexe) was proposed in a research study

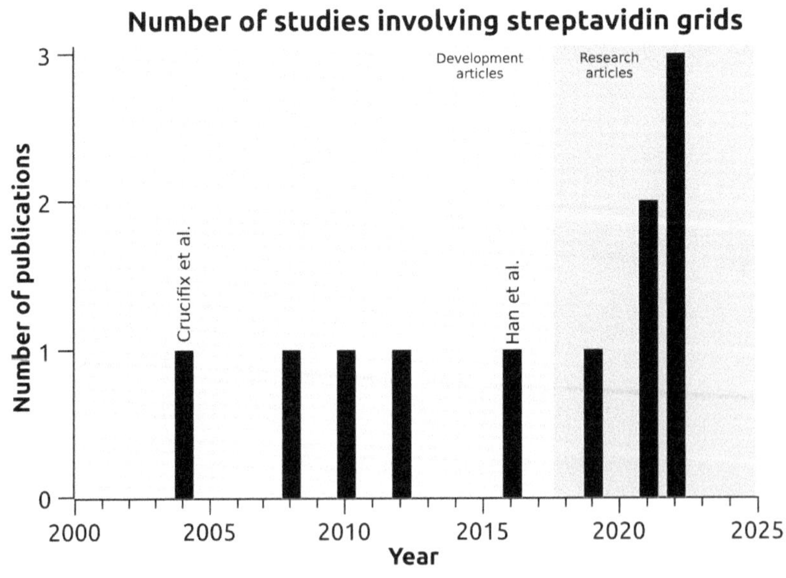

FIGURE 7.2 Occurrence of streptavidin affinity grids in literature. Bar chart illustrating the appearance and evolution of streptavidin affinity grids since the first article published in 2004 (Crucifix et al., 2004) until 2023. The light grey area represents the method development phase. The dark grey area corresponds to the period when streptavidin grids appeared in research studies.

focusing on the structure of bovine thyroglobulin (Marechal et al., 2022). Modifications to the Glaeser protocol include: (1) the standardization of the grid surface properties, making use of a liquid hexane treatment; (2) the fine-tuning of the streptavidin concentration during crystallization, and (3) the use of a sitting drop crystallization plate for better control of the relative humidity and crystallization kinetics. Using a crystallization plate also reduces mechanical damage generated by repetitive handling of the grid with tweezers, and it limits surface-tension-induced stress at the washing step. Together, these optimized steps improve in-batch and batch-to-batch reproducibility.

A VERSATILE PLATFORM FOR A VARIETY OF SAMPLE CAPTURE STRATEGIES

A variety of affinity grids have been developed and can be sorted into four categories: (1) affinity surfaces based on monolayers of Ni-NTA functionalized lipids (Figure 7.1c) used to bind his-tagged macromolecules (Kelly et al., 2008); (2) antibody-functionalized, carbon-coated grids (Figure 7.1d) to capture a complex of interest from cell extract (Yu et al., 2016); (3) SpyCatcher-functionalized graphene oxide grids (Figure 7.1e) for covalent binding and high-resolution determination of Spy-tagged proteins as small as 150 kDa (Wang et al., 2020); and (4) streptavidin affinity grids. If each of these approaches has its own advantages, they are either highly specialized (1, 3), require modification of the protein of interest (1, 3), or require to have sample-specific antibodies available (2). Streptavidin grids are remarkably versatile because they can accommodate many of the assets of the three other categories, as soon as the protein of interest, a binding partner, a small molecule ligand, or a selective antibody has the ability to bind to streptavidin. The capture of a protein of interest on the streptavidin crytal can be achieved in two ways: direct attachment of the target macromolecule to the affinity surface, or through interaction with an intervening binding partner.

Direct binding. As demonstrated using apoferritin in the early developments (Darst et al., 1991), random chemical biotinylation of the sample can be performed to trigger direct attachment to the 2D streptavidin crystal. This method enables enrichment of the imaging field with particles at concentrations as low as 50 nM (Figure 7.3e and g, left panel).

Since the sample binds to the crystal with femtomolar affinity, chemical biotinylation provides similar particle distribution in negative staining and cryo conditions, thus making it possible to identify optimal sample preparation parameters (biotinylation rate, sample concentration, or incubation time) in stained conditions and using a standard room-temperature electron microscope before switching to more demanding cryo-preparation for more elaborate cryo-EMs (Figure 7.4).

Furthermore, random biotinylation of protein primary amines—i.e., N-terminus and lysine residues—is likely to induce a more uniform distribution of particle orientations and thus can be used as a strategy to overcome the problem of preferred orientation (Han et al., 2012). This is well illustrated by structural studies carried out on yeast RNA polymerase I (yRNAP1). In this work, Lahiri and coworkers performed random biotinylation of yRNAP1 and its subsequent capture on streptavidin grids to overcome the strong preferred orientation impairing high-resolution reconstruction of yRNAP1 stalled on a DNA lesion (Lahiri et al., 2019; Xu et al., 2017). *In vitro* chemical modification of the sample requires the biotinylated target to be separated from the excess of reagent; otherwise the biotin-containing compound competes with the sample for binding to the affinity surface. This separation step can be performed by dialysis, size exclusion, or ultrafiltration, and is therefore mainly applicable to samples available in large amounts. Otherwise, a challenging but promising and elegant approach could be to cotranslationally incorporate biocytins, a chimeric, nonnatural biotin–lysine amino acid at different positions on the target protein during *in vivo* protein synthesis (Dougherty and Van Arnam, 2014). If sample characteristics prevent the use of random biotinylation, a fusion peptide with affinity for streptavidin can be grafted to the protein of interest by genetic engineering, thus allowing the purified sample to be captured on the grid surface without further modification (Figure 7.3f). A variety of these tags are available, with affinity ranging from µM to fM scales; the avitag, specifically biotinylated by BirA (Beckett et al., 1999), provides high affinity (K_d = ~10^{-15} M) and site-specific functionalization of the target protein. The streptavidin binding peptide (SBP-tag) and Strep-tag II bind streptavidin with K_d as low as ~1 µM and 2.5 nM, respectively (Keefe et al., 2001; Schmidt and Skerra, 2007). The angular distribution of the particles may be restricted when the peptide tag is placed in a non-flexible part of the target. This effect can be overcome by tagging multiple subunits of the macromolecular complex, by introducing a flexible linker between the tag and the target, or by making use of electron tomography approaches combined with sub-tomogram averaging (Myasnikov et al., 2013; Noble et al., 2018).

Partner-based binding. An intermediate molecule that binds both to the streptavidin crystal and to the macromolecular target can be used. This intermediate can be synthetic, such as biotinylated DNA (Figure 7.3a) to capture DNA binding complexes (Crucifix et al., 2004), or liposomes/nanodiscs enriched with biotinylated lipids (Figure 7.3b) to anchor membrane proteins (Wang et al., 2008; Wang and Sigworth, 2010). Any biotinylated small molecule ligand could also be used, such as biotinylated nicotinamide-adenine-dinucleotide to capture NAD-binding complexes (Figures 7.3c and 7.6). This binding intermediate can also be a biotinylated/tagged antibody or antibody derivative with selectivity for the macromolecular target (Figure 7.3d) or a genetically engineered peptide tag (i.e., HA, FLAG). This latter solution makes it possible to capture untagged macromolecular complexes from endogenous sources.

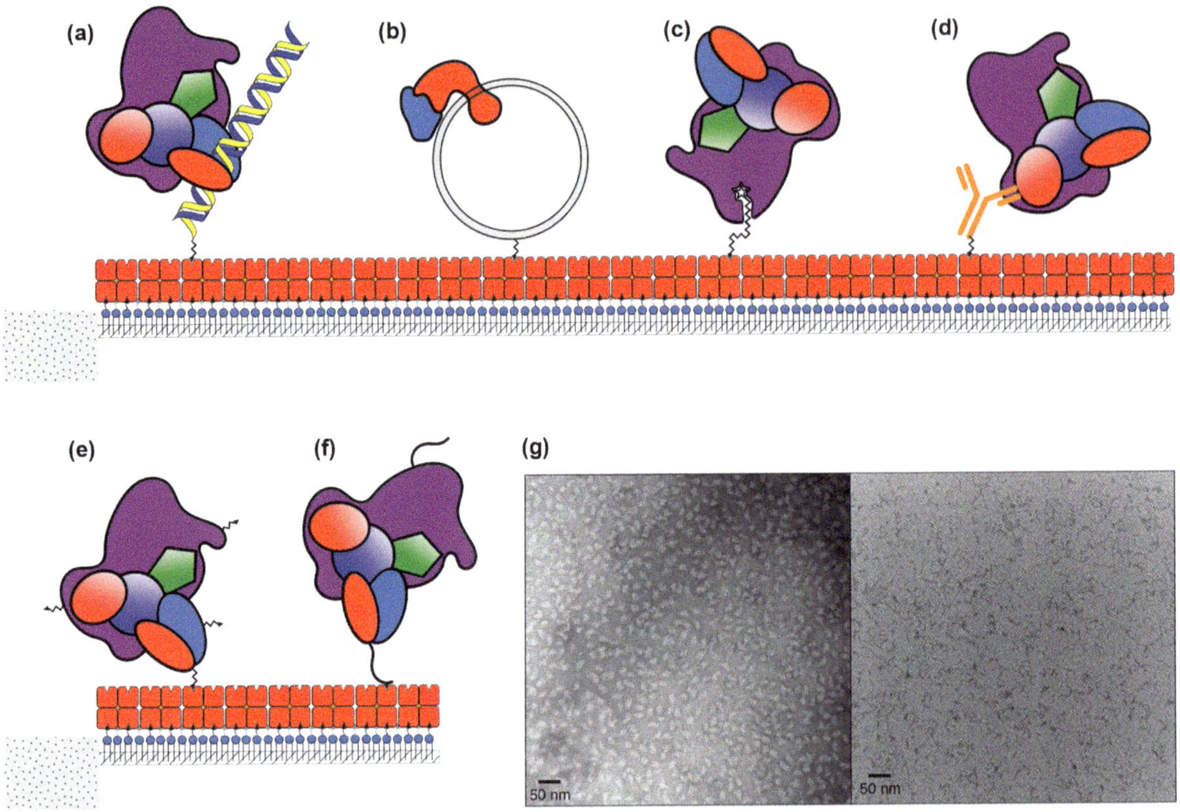

FIGURE 7.3 Strategies for direct or indirect sample capture on streptavidin affinity grids. Schematic representation of the capture of a multiprotein complex onto the streptavidin crystal (a–f). Macromolecular complexes can interact with the affinity surface indirectly by the intermediate of a partner, such as (a) specific biotinylated DNA, (b) a liposome or nanodisc doped with biotinylated lipid, (c) a small molecule ligand coupled to a biotin moiety, or (d) a tagged, biotinylated antibody. Direct capture of the complex of interest onto the streptavidin film can be achieved by (e) random, chemical biotinylation or (f) by genetically engineering a component of the complex to add a peptide tag providing affinity for streptavidin. (g) Left panel shows an electron micrograph of a negatively stained streptavidin crystal decorated with biotinylated thyroglobulin. To obtain this particle distribution, the streptavidin grid was incubated for 20 minutes, 4°C, on a 10 μL drop of biotinylated thyroglobulin solution at 50 nM, in a sealed sitting drop crystallization plate. Right panel represents a positively stained streptavidin crystal decorated with biotinylated double-stranded DNA.

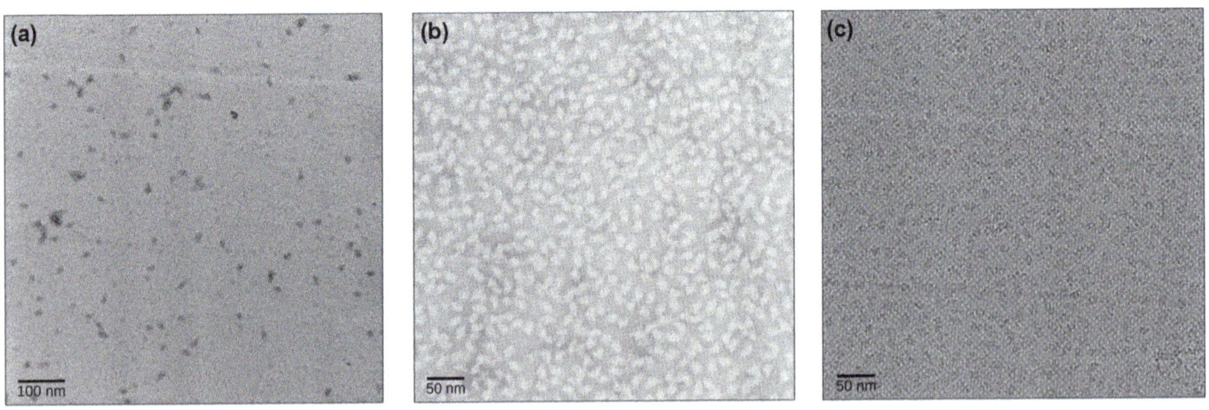

FIGURE 7.4 Streptavidin grids provide scalable particle distribution in negative stain and frozen hydrated conditions. Electron micrographs of two-dimensional streptavidin crystals after 20 minutes incubation at 4°C, with (a) 10 μL, 75 nM solution of unmodified thyroglobulin or (b, c) biotinylated thyroglobulin. Panels (a) and (b) show stained streptavidin grids. Panel (c) presents a plunge-frozen, hydrated specimen. Only sparse protein binding is visible when the affinity grid is incubated with unmodified specimen (a), suggesting a low level of nonspecific binding at this thyroglobulin concentration. Particle distribution is similar in negative staining (b) and cryo (c) conditions, making it possible to carry out many optimization steps at room temperature in dehydrated conditions.

STREPTAVIDIN GRIDS, A PLATFORM TO DEVELOP ON-GRID PURIFICATION

For high-resolution structural biology studies, the optimal sample needs to be highly purified, conformationally and compositionally homogenous, and stable in a buffer compatible with the biophysical method used for analysis. While electron microscopy can deal with reasonable composition and conformation heterogeneity, a pure, stable sample in a low-background buffer is a prerequisite.

Streptavidin affinity grids are intrinsically coupled with a selectivity step that enriches the biomolecule of interest onto the grid, thus introducing an additional on-grid purification step. Any biotinylated binding partner can thus be used to specifically recruit the target macromolecules from an intricate mixture. In cases when the target complex is heterogenous with one sub-stoichiometric subunit, tagging this subunit will enrich onto the grid the subpopulation of complexes containing the full set of components (Figure 7.3d and f).

In a standard purification workflow, the target particle is usually isolated from cellular or nuclear extracts using an affinity resin, typically composed of a target binding partner immobilized on a polymer matrix. A binding buffer, usually containing detergents and high ionic strength, is used to prevent sample precipitation/aggregation and to limit nonspecific binding of contaminants onto the matrix. Furthermore, a mobile washing buffer is used to remove the unbound material and adjust interaction specificity. To be used as a suitable affinity purification step, the streptavidin grids have to withstand such harsh conditions. We observed that, once formed, the 2D streptavidin crystals are resistant to detergents commonly used in purification schemes such as Triton X-100, CHAPS, CHAPSO, dodecyl maltoside, NP-40, or ß-octylglucoside, and they tolerate repetitive washing steps. The crystals can also withstand at least 2 mol/L of NaCl and 50% glycerol. These properties make it possible to carry out on-grid purification involving sequential steps of target binding, washing, and buffer exchange. If on-grid purification has not been performed on streptavidin grids yet, it was recently carried out with SpyCatcher grids using a flow cell (Ramlaul et al., 2023). Unlike standard chromatographic purification methods, since most actions can be carried out on the electron microscopy support, on-grid purification reduces sample loss at each purification step. It has the advantage of dramatically limiting exposition of the sample to liquid–gas interfaces, of reducing dilution-driven complex dissociation (i.e., during size exclusion) and of avoiding sample concentration steps that can lead to complex aggregation or precipitation. Considering the small size of the grid and its capacity to concentrate the sample on its surface, a very small amount of starting material can be used, which makes such an approach particularly suitable for low-abundant macromolecular complexes. On-grid purification also strongly shortens the steps between cell lysis and vitrification, which is of critical importance for short transient and purification-sensitive complexes that can be captured on-grid directly from cell lysate or partially purified extracts. Finally, since the sample is immobilized on the grid, the final buffer composition can be tuned to better fit cryo-EM imaging needs without diluting the target, as would result from any desalting or buffer exchange step in a standard purification pipeline.

PROCESSING STREPTAVIDIN AFFINITY GRID DATA

The only difference in processing images collected on streptavidin affinity grids compared to standard single-particle cryo-EM data analysis (SPA) is the impact of the periodic background introduced by the streptavidin crystal lattice on downstream 2D particle alignments and 3D image refinement. When using standard 2D classification procedures, the crystal periodicity dramatically overcomes the signal from the macromolecule of interest, thus leading to models representing the streptavidin lattice (Figure 7.5a).

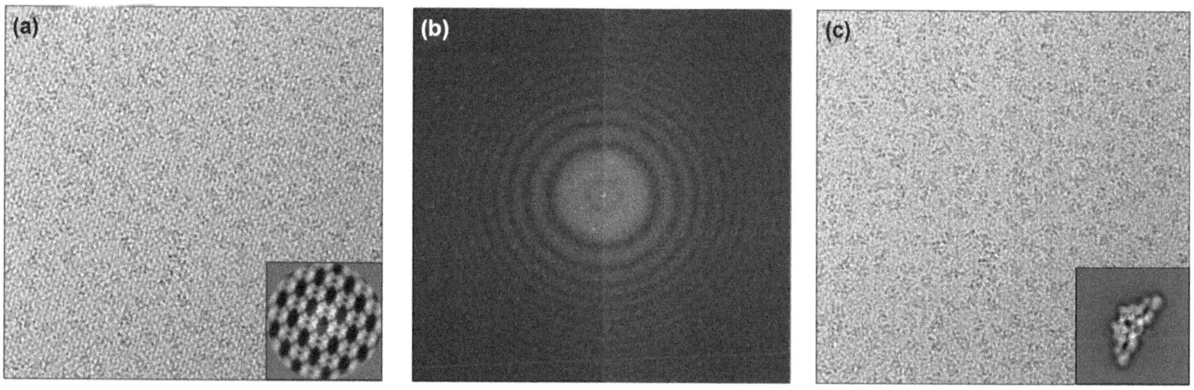

FIGURE 7.5 Lattice subtraction is a prerequisite to data processing. (a) Cryo-electron micrograph showing the distribution of biotinylated particle on the streptavidin crystal. The insert shows a two-dimensional average of the streptavidin lattice. (b) The left half represents the power spectrum of the Fourier transform computed from the micrograph shown in panel (a) in which the diffraction spots are generated by the periodic signal of the streptavidin crystal. The right half shows the same power spectrum after spot subtraction using an in-house Python script. (c) The corresponding micrograph after lattice subtraction in Fourier space, followed by reverse Fourier transform. The insert shows a representative class average obtained after two-dimensional particle alignment and classification.

The crystal lattice can, however, be subtracted from the dose-weighted, drift-corrected micrograph by Fourier filtering of the diffraction spots followed by reverse Fourier transform (Crucifix et al., 2004), as illustrated in Figure 7.5b and c. The subtracted images are funneled into a standard SPA pipeline and can make use of deep learning tools for particle selection (Figure 7.5c). Once a refined 3D model is computed, the lattice signal can be reintroduced—for instance, at the particle polishing step—without interfering with particle alignment, since Euler angles attributed to each individual particle are already close to their optimal values (Marechal et al., 2022).

STREPTAVIDIN AFFINITY GRIDS TO STUDY GENE EXPRESSION REGULATION

Streptavidin affinity grids promise to be a method of choice to perform cryo-EM structural analysis of macromolecular mechanisms involved in regulated gene expression, taking advantage of biotinylated nucleic acids onto which functional complexes can be assembled. Several companies propose to incorporate biotinylated nucleotides into synthetic DNA primers for biotin labeling of long DNA or plasmids by PCR amplification or by using the restriction-ligation process. The biotin-labeled DNA can then be used to capture DNA-binding macromolecular complexes onto the 2D streptavidin crystals and perform a controlled biochemical reaction on grid. This method was first used as a proof of concept to decorate streptavidin grids with RNA polymerase I (Crucifix et al., 2004). In this pilot work, Schultz and coworkers first functionalized the streptavidin surface with a 450 base-pair, double-stranded DNA. In a second step, the RNA polymerase I was captured on-grid through its interaction with DNA. Alternatively, the biotinylated DNA can also be incorporated into the macromolecular complex prior to its binding to the affinity grid. Such a strategy was chosen to solve the structure of the polycomb repressive complex 2 (PRC2) bound to a nucleosome core particle (NCP) (Kasinath et al., 2021). In this study, the authors mixed the NCP, assembled onto a biotinylated DNA, with a large molar excess of PRC2 in order to promote the formation of the PRC2–NCP complex. After binding on the streptavidin grid, the excess of PRC2 was washed out, resulting in the on-grid enrichment of the assembled PRC2–NCP complex. Another advantage of this method is that the capture of the target is driven by streptavidin–biotin affinity rather than the target–DNA association.

Recently, the proteinaceous streptavidin surface was used to better understand the mechanisms underlying chromatin condensation and to prevent charge repulsion between the electron microscopy support and the biological sample (Zhang et al., 2022). Chromatin-like structures made of tetra- and dodecanucleosome condensates are affected by charge repulsion with glow-discharged carbon. To overcome this issue, the authors immobilized biotin-labeled polynucleosomes onto streptavidin crystals and were able to capture chromatin-like condensates.

SETTING UP SAMPLE CAPTURE AND METHOD LIMITATIONS

If the streptavidin affinity grid approach sounds like a magic bullet, one needs to pay attention to the design of the capture experiment and to the limitations of the method. The binding strategy, the sample concentration, the accessibility and distance of the streptavidin binding effector must be carefully tuned. For the most challenging cases, capture of the complex of interest on the streptavidin crystal may require multiple rounds of optimization. Here are a few suggestions to start setting up a streptavidin grid system:

1. One should always try to select the binding system with the strongest affinity for streptavidin. If streptavidin affinity is used in your purification pipeline, prefer SBP tag to strep-tags since it will permit a better capture on the crystal. Make sure you get rid of any trace of biotin used for elution. If compatible with your purification, the avi-tag, once biotinylated by the BirA enzyme *in vitro* or in the expression system, is the peptide tag with the highest affinity.
2. If you chose random, chemical biotinylation of your sample, it is advised to conjugate about one biotin per particle. This will induce a better angular distribution of the sample and prevent artifacts potentially provoked by over-biotinylation. Methods to monitor biotin incorporation can be found in these two articles: Lahiri et al. (2019), Marechal et al. (2022). Briefly, the degree of biotinylation can be measured using a comparative pull-down experiment: The sample is labeled with different molar excesses of reagent, or left unlabeled for the negative control. Then, biotinylated samples are mixed with an excess of streptavidin-conjugated resin or with resin blocked by biotin. Later, the supernatants are analyzed by SDS-PAGE. The conditions where most proteins were pulled down by unblocked resin correspond to a biotinylation ratio close to 1 biotin per protein. A complete pull-down is the sign of too high a degree of biotinylation. Blocked resin is necessary to measure sample nonspecific binding on the resin matrix. A variety of biotinylation reagents with linker length ranging from a few Å to several nanometers are commercially available (Figure 7.6).
3. When trying to capture a multicomponent complex, the streptavidin-binding moiety should always be the limiting factor, and its concentration should be in the 50 to 500 nM range, in order to ensure sufficient particle distribution while preventing dilution-driven particle dissociation. In cases where some components need micromolar concentration, it is advised to add some detergents to the buffer in order to prevent nonspecific binding of the sample to the streptavidin crystal. As

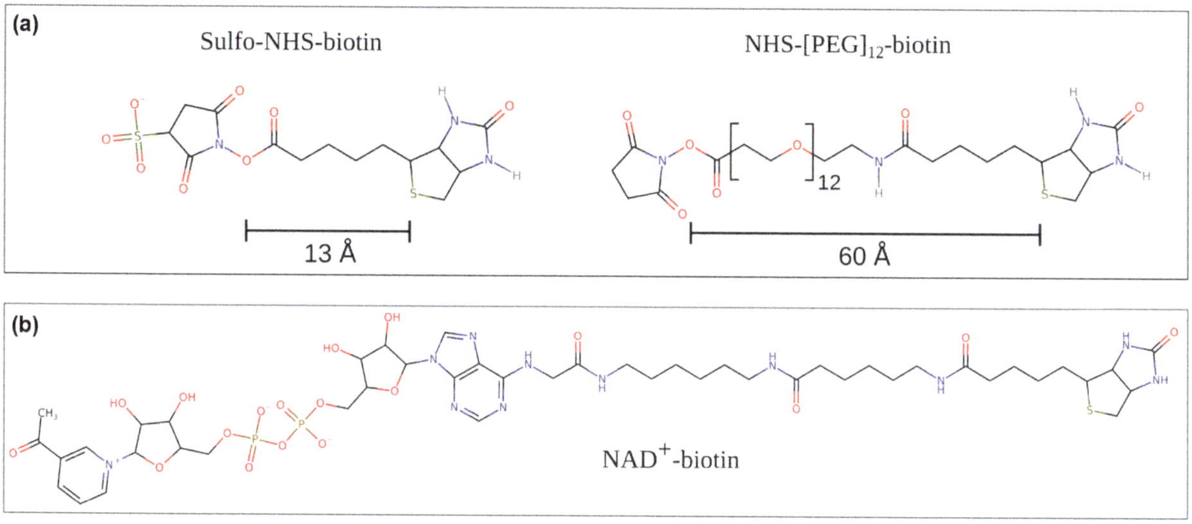

FIGURE 7.6 A few examples of biotinylation compounds and biotinylated ligands. (a) Sulfo-NHS-biotin and NHS-$[PEG]_{12}$-biotin are two commercially available compounds used to chemically add a biotin moiety onto primary amines, such as amino-terminal function and lysines in proteins. The distances between the primary amine and the biotin group after labeling are of ~13 Å and ~60 Å for Sulfo-NHS-biotin and NHS-$[PEG]_{12}$-biotin, respectively. (b) An example of a small molecule ligand (nicotinamide adenine dinucleotide) coupled to a biotin group, which can be used to capture a receptor of this ligand onto the streptavidin grid.

a general rule, a 75 to 100 nM concentration of a biotinylated target is sufficient to entirely saturate the crystal surface after a few minutes (Figure 7.4).

4. The streptavidin binding moiety must be accessible at the surface of the macromolecule of interest. It also needs to be either flexible or conjugated to a flexible part of the protein of interest in order to permit a sufficiently large angular distribution. Most importantly, the binding moiety should be distant enough from the particle core to prevent binding-induced macromolecular complex dissociation.
5. Based on literature and on experiments performed in our laboratory, the streptavidin affinity grids approach is not suitable, to date, to study macromolecules smaller that 200 kDa, whose dimensions are too small to be visualized on the 5 nm periodic streptavidin array.

CONCLUSION

Streptavidin affinity grids permit fine-tuning of sample distribution on their surface, allow us to work with small amounts of macromolecular complexes, and can accommodate a large variety of sample capture strategies. Their properties are particularly useful for addressing the issues of preferred orientation and macromolecule denaturation at the air–water interface. Moreover, streptavidin grids promise to be of major interest for the studies of DNA binding complexes and to develop methods for on-grid purification. We emphasize that their use is about to quickly spread over the structural cryo-EM community, as supported by the growing body of research articles using this technique and published after Glaeser's group protocol in 2016. However, the use of this sample preparation method is currently limited by the need to produce affinity grids in-house, which limits their utilization to a few specialized groups. The development of commercially available streptavidin grids could overcome this limitation in the future. Further development will also be needed to facilitate processing of data collected on streptavidin affinity grid specimens: software to subtract the lattice signal on-the-fly, to compute diffraction-based statistics, to assess the image quality, and to clean up the dataset to get rid of poor imaging areas.

ANNEXE

Preparation of Streptavidin affinity Grids

If this protocol was used in your research, please cite: Formation of thyroid hormone revealed by a cryo-EM structure of native bovine thyroglobulin. Nils Marechal, Banyuhay P. Serrano, Xinyan Zhang & Charles J. Weitz. Nat. Com. 2022. doi.org/10.1038/s41467-022-30082-4.

This procedure is optimized to obtain proper crystallization results at a temperature ranging from 18°C to 22°C. Lower temperature may result in partial crystallization, while higher temperature will result in mosaic crystals. If working at a different temperature, it is recommended to adjust streptavidin solution concentration.

Material

- HPLC grade chloroform
- HPLC grade liquid hexane
- 35 mm glass petri dishes
- 35 mm disposable plastic petri dishes

- 35 mm and 8 cm filter papers
- Glass Pasteur pipette
- Quantifoil R2/1, R2/2, R3.5/1, or R4/2, Cu or Cu/Rh, 200 or 300 mesh EM grids
- Anti-capillary tweezers (Dumont N5AC)
- Streptavidin (1 mg/mL, NEB N7021S)
- SAG subphase buffer (50 mM HEPES pH 7.5, 150 mM NaCl, 10% trehalose)
- SAG wash buffer (10 mM HEPES pH 7.5, 50 mM NaCl, 10% trehalose)
- SAG Rehydration buffer (50 mM HEPES pH 7.5, 150 mM NaCl)
- Lipid buffer (chloroform:methanol:H2O 65:35:8)
- 16:0 biotinyl PE lipid stock solution (10 mg/mL, Avanti Polar Lipids 870285P)
- Castor oil (Sigma 259853)
- 0,5 µL glass syringe
- 24-well sitting drop crystallization plate (Cryschem M Plate, Hampton Research)
- Siliconized glass coverslip
- Vaseline in a syringe (VWR 28908.290)
- Microscope slide
- Parafilm
- 1 timer
- Carbon evaporator
- Must work by flash or have a shutter
- Carbon coil or rod
- Must reach at least 5.10^{-5} Torr

Manufacturing

Day 0 – Grid cleaning and preparation (~ 5 hours)

- Wear gloves and work under a fume hood when manipulating chloroform!
- Load two to three 35 mm Wattman filter papers into a 35 mm glass petri dish.
- Dispose 24 holey carbon grids, carbon facing up, onto filter paper.
- Use a glass Pasteur pipette to delicately pour chloroform onto the filter paper, until grids start to move. Make sure papers are fully soaked.
- Close the petri dish, and incubate 3 to 4 hours, then open cover and let chloroform completely evaporate.
- Evaporate a "dark grey" (~20–50 nm thick) layer of carbon onto the holey foil (this step can also be done on day 1).

Day 1 – grid aging (~ 2 hours)

- Wear gloves and work under a fume hood when manipulating liquid hexane!
- Transfer manually each grid, carbon facing up, onto a clean filter paper placed in a new 35 mm glass petri dish.
- With a glass Pasteur pipette, softly cover the grids with liquid hexane. Make sure grids are covered with enough chemicals (3–4 mL should be enough), and do not float.
- Close the petri dish and incubate for 10 minutes.
- Pipette-out liquid hexane (tilt the petri dish to remove the last excess), then leave grids and filter papers to dry for about 60 minutes.

Day 1 – Setting up work area (~ 30 minutes)

- With anti-capillary tweezers, grab each grid so that the carbon side faces the straight tip of the tweezers. Thus, the bent tip should face the metal mesh side.
- Align the tweezers (still holding grids, carbon side facing down)
- Grease the crystallization plate wells with vacuum grease for later sealing (can be done during streptavidin stock thawing).
- Add 0.5 mL ddH2O in all the reservoirs. Make sure not to wet the pedestals.

Day 1 – Streptavidin solution preparation (~ 30 minutes)

- On ice, thaw 1 mg/mL NEB streptavidin aliquots.
- Dilute aliquots to 0.15 mg/mL with ice-cold SAG subphase buffer.
- Equilibrate SAG subphase buffer stock at 18°C–22°C.

Day 1 – Preparing lipid monolayer (~15 minutes)

- Dilute the 10 mg/mL lipid solution to 1 mg/mL using lipid dilution solution. Vortex to homogenize. Leave on ice until use.
- Fill up a 35 mm disposable petri dish cover with SAG subphase buffer (until meniscus reaches the top).
- Pipette-in 11 µL castor oil after having cut the extremity of a 20 µL pipette tip to improve aspiration of the viscous solution.
- Load the castor oil onto the surface of the SAG subphase buffer, in the middle. Let the castor oil spread for 5 to 10 minutes.
- Prepare a large parafilm surface at the right of the petri dish. Place the crystallization plate at the right of the parafilm.
- After this, be ready to start the crystallization step within 2 minutes. Using the Hamilton syringe, softly deposit the 1 mg/mL lipid solution onto the surface of the castor oil, in the middle, until you can see a circular lipid monolayer of ~1.5 cm diameter.
- For each grid, prepare a row of 3 drops (50 µL) of SAG subphase buffer. Prepare a maximum of 12 rows at once.

Day 1 – Crystallization (~ 50–60 minutes). Process by rows of 4 wells. Preparing a row takes about 3 to 5 minutes

- (1) Add 10 µL 0.15 mg/mL ice-cold streptavidin solution to each of 4 sitting drop supports.

- (2) With the carbon facing down, briefly touch the middle of the lipid monolayer with the grids. Make sure not to touch the castor oil on the side.
- (3) To get rid of excess, unbound lipids, subsequently touch 3 drops of subphase buffer.
- (4) Using a bent filter paper, briefly blot-off excess buffer from the side of the grid.
- (5) Immediately transfer the grid onto the streptavidin solution, carbon facing down.
- (6) Seal the wells with siliconized coverslips.
- (7) Start a timer. Each rows should incubate at 18°C–22°C, without shaking, for 25–30 minutes
- (8) Repeat steps 1–7 for the 5 remaining rows.

Day 1 – Washing and drying (~ 60 minutes). Process by row of 4 wells

- (1) Once a row is done, get rid of the coverslips with dedicated tweezers.
- (2) Inject 100 µL SAG wash buffer under each grid.
- (3) Remove the wash buffer until the grid floats at the same level as the sitting drop support edges.
- Repeat steps 2 and 3 two additional times. In the final repetition, do not remove the wash buffer.
- (4) With anticapillary tweezers, fish the grid floating onto the drop. Make sure the bent tip is facing the crystal side and the straight tip is on the metal mesh side. Make sure you grab the grid from an edge.
- (5) Lift up the grid from the drop horizontally, as slowly as possible; the grid surface should ideally be parallel to the buffer surface.
- (6) With a bent filter paper, blot off excess buffer from the edge of the grid.
- (7) Put the grid onto a flat, 8 cm filter paper, carbon (crystal) facing up.
- (8) Repeat steps 2–7 for the 5 remaining rows.
- Leave the grids dry for at least 40 minutes. Protect grids from light and dust.

Day – 1 Back coating and storage (~ 60 minutes)

- Flatten parafilm onto a glass microscope slide.
- Transfer dry grids onto the slide, flipping down the carbon side. Tighten grids as much as you can.
- With the grids placed as far as possible from the carbon source and at a pressure of ~6.10–5 mbar, evaporate 0.5–2 nm carbon onto the back of the grids. Whatever carbon evaporator you use, ideally limit the number of flashes to 9 or fewer.
- Store the grids in a desiccator, using silica gel, until use.

REFERENCES

Alex J.N., Hui, W., Venkata P.D., Zhening, Z., Yong Zi, T., Clinton S.P. and Bridget, C. (2018), "Reducing effects of particle adsorption to the air-water interface in cryo-EM", *Nature Methods*, Vol. 15 No. 10, pp. 793–795, doi: 10.1038/s41592-018-0139-3.

Armstrong, M., Han, B.G., Gomez, S., Turner, J., Fletcher, D.A. and Glaeser, R.M. (2020), "Microscale fluid behavior during cryo-EM sample blotting", *Biophysical Journal*, Vol. 118 No. 3, pp. 708–719, doi: 10.1016/j.bpj.2019.12.017.

Beckett, D., Kovaleva, E., Petter, S. and Schatz, P.J. (1999), "A minimal peptide substrate in biotin holoenzyme synthetase-catalyzed biotinylation", *Protien Science*, Vol. 8, pp. 921–929.

Bischler, N., Balavoine, F., Milkereit, P., Tschochner, H., Mioskowski, C. and Schultz, P. (1998), "Specific interaction and two-dimensional crystallization of histidine tagged yeast RNA polymerase I on nickel-chelating lipids", *Biophysical Journal*, Vol. 74 No. 3, pp. 1522–1532, doi: 10.1016/S0006-3495(98)77864-6.

Chen, J., Noble, A.J., Kang, J.Y. and Darst, S.A. (2019), "Eliminating effects of particle adsorption to the air/water interface in single-particle cryo-electron microscopy: Bacterial RNA polymerase and CHAPSO", *Journal of Structural Biology: X*, Vol. 1 No. January, p. 100005, doi: 10.1016/j.yjsbx.2019.100005.

Crucifix, C., Uhring, M. and Schultz, P. (2004), "Immobilization of biotinylated DNA on 2-D streptavidin crystals", *Journal of Structural Biology*, Vol. 146 No. 3, doi: 10.1016/j.jsb.2004.02.001.

D'Imprima, E., Floris, D., Joppe, M., Sánchez, R., Grininger, M. and Kühlbrandt, W. (2019), "Protein denaturation at the air-water interface and how to prevent it", *ELife*, Vol. 8, pp. 1–18, doi: 10.7554/eLife.42747.

Darst, S.A., Ahlers, M., Meller, P.H., Kubalek, E.W., Blankenburg, R., Ribi, H.O., Ringsdorf, H., et al. (1991), "Two-dimensional crystals of streptavidin on biotinylated lipid layers and their interactions with biotinylated macromolecules", *Biophysical Journal*, Vol. 59 No. 2, pp. 387–396, doi: 10.1016/S0006-3495(91)82232-9.

Dougherty, D.A. and Van Arnam, E.B. (2014), "In vivo incorporation of non-canonical amino acids by using the chemical aminoacylation strategy: A broadly applicable mechanistic tool", *ChemBioChem*, Vol. 15 No. 12, pp. 1710–1720, doi: 10.1002/cbic.201402080.

Frederik, P.M., Stuart, M.C., Bomans, P.H. and Busing, W.M. (1989), "Phospholipid, nature's own slide and cover slip for cryo-electron microscopy", *Journal of Microscopy*, Vol. 153 No. 1, pp. 81–92.

Fromherz, P. (1971), "A new technique for investigating lipid protein films", *BBA - Biomembranes*, Vol. 225 No. 2, pp. 382–387, doi: 10.1016/0005-2736(71)90235-5.

Glaeser, R.M. (2016), "Proteins, interfaces, and cryo-EM grids", *Current Opinion in Colloid & Interface Science*, Vol. 25 No. 3, pp. 289–313, doi: 10.1016/j.cocis.2017.12.009.

Glaeser, R.M. and Han, B.-G. (2017), "Opinion: hazards faced by macromolecules when confined to thin aqueous films", *Biophysics Reports*, Vol. 3 No. 1–3, pp. 1–7, doi: 10.1007/s41048-016-0026-3.

Green, N.M. (1962), "Avidin", *Advances in Protein Chemistry*, Vol. 29 No. C, pp. 85–133, doi: 10.1016/S0065-3233(08)60411-8.

Green, N.M. (1963), "Avidin. 3. the nature of the biotin-binding site", *The Biochemical Journal*, Vol. 89, pp. 599–609, doi: 10.1042/bj0890599.

Han, B.G., Walton, R.W., Song, A., Hwu, P., Stubbs, M.T., Yannone, S.M., Arbeláez, P., et al. (2012), "Electron microscopy of biotinylated protein complexes bound to streptavidin monolayer crystals", *Journal of Structural Biology*, Vol. 180 No. 1, pp. 249–253, doi: 10.1016/j.jsb.2012.04.025.

Han, B.-G., Watson, Z., Kang, H., Pulk, A., Downing, K., Cate, J. and Glaeser, R. (2016), "Long shelf-life streptavidin support-films suitable for electron microscopy of biological macromolecules", *bioRxiv*, doi: 10.1101/054452.

Dubochet, J., Adrian, M., Chang, J.-J., Homo, J.-C., Lepault, J., McDowall, A. W. and Schultz, P. (1988), "Cryo-electron microscopy of vitrified specimens", *Quarterly Reviews of Biophysics*, Vol. 21 No. 2, pp. 129–228, doi: 10.1017/S0033583500004297.

Kasinath, V., Beck, C., Sauer, P., Poepsel, S., Kosmatka, J., Faini, M., Toso, D., et al. (2021), "JARID2 and AEBP2 regulate PRC2 in the presence of H2AK119ub1 and other histone modifications", *Science*, Vol. 371 No. 6527, doi: 10.1126/science.abc3393.

Keefe, A.D., Wilson, D.S., Seelig, B. and Szostak, J.W. (2001), "One-step purification of recombinant proteins using a nanomolar-affinity streptavidin-binding peptide, the SBP-tag", *Protein Expression and Purification*, Vol. 23 No. 3, pp. 440–446, doi: 10.1006/prep.2001.1515.

Kelly, D.F., Dukovski, D. and Walz, T. (2008), "Monolayer purification: A rapid method for isolating protein complexes for single-particle electron microscopy", *Proceedings of the National Academy of Sciences of the United States of America*, Vol. 105 No. 12, pp. 4703–4708, doi: 10.1073/pnas.0800867105.

Kubalek, E.W., Le Grice, S.F.J. and Brown, P.O. (1994), "Two-dimensional crystallization of histidine-tagged, HIV-1 reverse transcriptase promoted by a novel nickel-chelating lipid", *Journal of Structural Biology*, Vol. 113, pp. 117–123, doi: 10.1006/jsbi.1994.1039.

Lahiri, I., Xu, J., Han, B.G., Oh, J., Wang, D., DiMaio, F. and Leschziner, A. (2019), "3.1Å structure of yeast RNA polymerase II elongation complex stalled at a cyclobutane pyrimidine dimer lesion solved using streptavidin affinity grids", BioRxiv, doi: 10.1101/585547.

Lebeau, L., Regnier, E., Schultz, P., Wang, J.C., Mioskowski, C. and Oudet, P. (1990), "Two-dimensional crystallization of DNA gyrase B subunit on specifically designed lipid monolayers", *FEBS Letters*, Vol. 267 No. 1, pp. 38–42, doi: 10.1016/0014-5793(90)80282-N.

Marechal, N., Serrano, B.P., Zhang, X. and Weitz, C.J. (2022), "Formation of thyroid hormone revealed by a cryo-EM structure of native bovine thyroglobulin", *Nature Communications*, Vol. 13 No. 1, doi: 10.1038/s41467-022-30082-4.

Melamed, M.D. and Green, N.M. (1963), "Avidin. 2. Purification and composition", *The Biochemical Journal*, Vol. 89, pp. 591–599, doi: 10.1042/bj0890591.

Myasnikov, A.G., Afonina, Z.A. and Klaholz, B.P. (2013), "Single particle and molecular assembly analysis of polyribosomes by single- and double-tilt cryo electron tomography", *Ultramicroscopy*, Vol. 126, pp. 33–39, doi: 10.1016/j.ultramic.2012.12.009.

Noble, A.J., Dandey, V.P., Wei, H., Brasch, J., Chase, J., Acharya, P., Tan, Y.Z., et al. (2018), "Routine single particle CryoEM sample and grid characterization by tomography", *ELife*, Vol. 7, pp. 1–42, doi: 10.7554/eLife.34257.

Peters, J. and Fromherz, P. (1975), "Interaction of electrically charged lipid monolayers with malate dehydrogenase", *BBA - Biomembranes*, Vol. 394 No. 1, pp. 111–119, doi: 10.1016/0005-2736(75)90209-6.

Ramlaul, K., Feng, Z., Canavan, C. and Garrido, N.D.M. (2023), "A 3D-printed flow-cell for on-grid purification of electron microscopy samples directly from lysate", Journal of Structural Biology 215 107999.

Ribi, H.O., Reichard, P. and Kornberg, D. (1987), "Two-dimensional crystals of enzyme-effector complexes: Ribonucleotide reductase at 18-A resolution", *Biochemistry*, Vol. 26 No. 1987, pp. 7974–7979.

Schmidt, T.G.M. and Skerra, A. (2007), "The Strep-tag system for one-step purification and high-affinity detection or capturing of proteins", *Nature Protocols*, Vol. 2 No. 6, pp. 1528–1535, doi: 10.1038/nprot.2007.209.

Schultz, P., Célia, H., Riva, M., Darst, S.A., Colin, P., Kornberg, R.D., Sentenac, A., et al. (1990), "Structural study of the yeast RNA polymerase A", *Journal of Molecular Biology*, Vol. 216 No. 2, pp. 353–362, doi: 10.1016/s0022-2836(05)80326-2.

Tilak, J., Patrick, S., John, C., Bridget, C. and Simon S.P. (2012), "Spotiton: A prototype for an integrated inkjet dispense and vitrification system for cryo-TEM", *Journal of Structural Biology*, Vol. 179 No. 1, pp. 68–75, doi: 10.1016/j.jsb.2012.04.020.

Unwin, P.N.T. and Henderson, R. (1975), "Molecular structure determination by electron microscopy of unstained crystalline specimens", *Journal of Molecular Biology*, Vol. 94 No. 3, doi: 10.1016/0022-2836(75)90212-0.

Uzgiris, E.E. and Kornberg, R.D. (1983), "Two-dimensional crystallization technique for imaging macromolecules, with application to antigen-antibody-complement complexes", *Nature*, Vol. 301 No. 5896, pp. 125–129, doi: 10.1038/301125a0.

Wang, F., Liu, Y., Yu, Z., Li, S., Feng, S., Cheng, Y. and Agard, D.A. (2020), "General and robust covalently linked graphene oxide affinity grids for high-resolution cryo-EM", *Proceedings of the National Academy of Sciences of the United States of America*, Vol. 117 No. 39, pp. 24269–24273, doi: 10.1073/pnas.2009707117.

Wang, L., Ounjai, P. and Sigworth, F.J. (2008), "Streptavidin crystals as nanostructured supports and image-calibration references for cryo-EM data collection", *Journal of Structural Biology*, Vol. 164 No. 2, pp. 190–198, doi: 10.1016/j.jsb.2008.07.008.

Wang, L. and Sigworth, F.J. (2010), *Liposomes on a Streptavidin Crystal. A System to Study Membrane Proteins by Cryo-EM, Methods in Enzymology*, Vol. 481, Elsevier Masson SAS, doi: 10.1016/S0076-6879(10)81007-9.

Wu, S., Armache, J.P. and Cheng, Y. (2016), "Single-particle cryo-EM data acquisition by using direct electron detection camera", *Microscopy (Oxford, England)*, Vol. 65 No. 1, pp. 35–41, doi: 10.1093/jmicro/dfv355.

Xu, J., Lahiri, I., Wang, W., Wier, A., Cianfrocco, M.A., Hare, A.A., Dervan, P.B., et al. (2017), "Strucutral basis for eukaryotic transcription-coupled DNA repair initiation", *Nature*, Vol. 551 No. 7682, pp. 653–657, doi: 10.1038/nature24658.

Yu, G., Li, K. and Jiang, W. (2016), "Antibody-based affinity cryo-EM grid", *Methods*, Vol. 100, doi: 10.1016/j.ymeth.2016.01.010.

Zhang, M., Díaz-Celis, C., Onoa, B., Cañari-Chumpitaz, C., Requejo, K.I., Liu, J., Vien, M., et al. (2022), "Molecular organization of the early stages of nucleosome phase separation visualized by cryo-electron tomography", *Molecular Cell*, Vol. 82 No. 16, pp. 3000–3014.e9, doi: 10.1016/j.molcel.2022.06.032.

8 Running a Cost-Effective Modern Cryo-EM Facility for SPA Analysis

Lijia Jia

INTRODUCTION

Cryo-electron microscopy (cryo-EM) has become more and more attractive for scientists coming from the structural biology and drug discovery fields. The resolution of proteins can be determined to better than 1.5 Å resolution with the more powerful microscope, which has benefited from improvements to the instrument and to direct electron detectors (Cheng et al., 2016; Wu, Armache and Cheng, 2016; Naydenova et al., 2019; Fréchin et al., 2022). High-resolution cryo-EM maps can be built for atomic models and can provide details about the protein machine working mechanisms that can be used to optimize drug molecules (Renaud et al., 2018; Yip et al., 2020; Zhang et al., 2021; Robertson, Meyerowitz and Skiniotis, 2022). Hence, more and more cryo-EM labs have been built in research universities and drug companies. With the seamless integration of cryo-EM data collection and processing software, getting a high-resolution cryo-EM structure within one or two weeks is becoming more common (Zivanov et al., 2018; Tegunov and Cramer, 2019; Sharov et al., 2021). Soon cryo-EM will be treated as a standard scientific research procedure. Its applications will not be limited to structural biology; cryo-EM will also perform research services like genome sequencing. This chapter focuses mainly on the routine workflow for the operations of a cryo-EM facility, focusing especially on the use of a cost-effective 200 kV microscope, which can produce sub 3 Å structure resolution when equipped with modern electron detector and emery filter. Besides the core microscope parts, the sample preparation equipment, data storage server, and data processing GPU workstation are also important features of a one-stop service cryo-EM facility.

THE CRYO-EM SPECIMEN PREPARATION

Preparing a sample for imaging by EM mainly includes two processes, negative staining (Moyes, Reynolds and Breakwell, 2009; De Carlo and Harris, 2011; Zhang et al., 2013) and cryogenic freezing (Dobro et al., 2010; Hands-Portman and Bakker, 2022). For a new cryo-EM project without previous information from EM analysis, one usually must use negative staining to visualize a sample and then decide if the sample can be used for cryo-EM imaging. EM users should bear in mind that the sample must be biochemically pure and homogenous in protein molecular composition and conformation in order to be considered for cryo-EM analysis.

Solution buffer conditions can significantly alter the sample conformational variations and cause compositional heterogeneity (Drulyte et al., 2018; Basanta et al., 2022). The optimal buffer salt concentration should be lower than 500 mM, but always try to keep it as low as possible. It should not contain reagents such as glycerol or DMSO that will produce thick ice during grid preparation; to keep the sample homogenous in such reagent buffers, a careful optimization of grid freezing conditions may help produce a high-quality EM structure, but it is always difficult for sub–200 kDa protein complexes (Wu et al., 2012; Liu et al., 2018; Liu et al., 2019). Chemical cross-linking is a powerful tool for reducing sample heterogeneity, especially for huge protein complexes with a molecular weight higher than 500 kDa. The GRAFIX method is frequently adopted by EM researchers, which can dramatically stabilize the complex compositional purity (Kastner et al., 2008; Stark, 2010). Chemical cross-linking, however, may introduce artifacts such as the chemical cross-linking reagent glutaraldehyde, trapping the protein complex in a form of non-native status. Any other approaches to functionally stabilizing the complex can also be considered for EM sample preparation, such as some protein partners, substrate, inhibitor, ligand nucleic acid, or antibody (Yu et al., 2016; Yu, Li and Jiang, 2016; Yue et al., 2022).

Once the sample is ready for EM imaging, the first important and limiting factor for cryo-EM analysis is grid preparation. During imaging, the electron beam strikes the protein sample, and transmitted electron signals are recorded on the detector (CCD or now often DDD) (Cheng et al., 2016; McMullan, Faruqi and Henderson, 2016). To minimize damage to the biological specimen, imaging is performed under high vacuum, 1.6e–08 mPa, and liquid nitrogen temperatures; the protein sample must also be embedded in thin vitreous ice, which permits enough of the transmitted electron signal to be recorded for 3D reconstruction of the protein particles. The EM grid usually has a diameter of 3 mm (Figure 8.1b), and it is made from a metal mesh such as copper, gold, or nickel. Copper and gold mesh coated with copper and gold supporting film are most commonly used for biological specimen preparation. Different commercial mesh size grids—200, 300, or 400 meshes or squares—can be chosen. Sample concentration is the most important factor when choosing the grid types; based on our experience, gold grids usually need a finer mesh than copper grids. Grids should usually be kept in a low-humidity environment, such as a temperature- and humidity-controlled cabinet (Figure 8.1a).

DOI: 10.1201/9781003326106-10

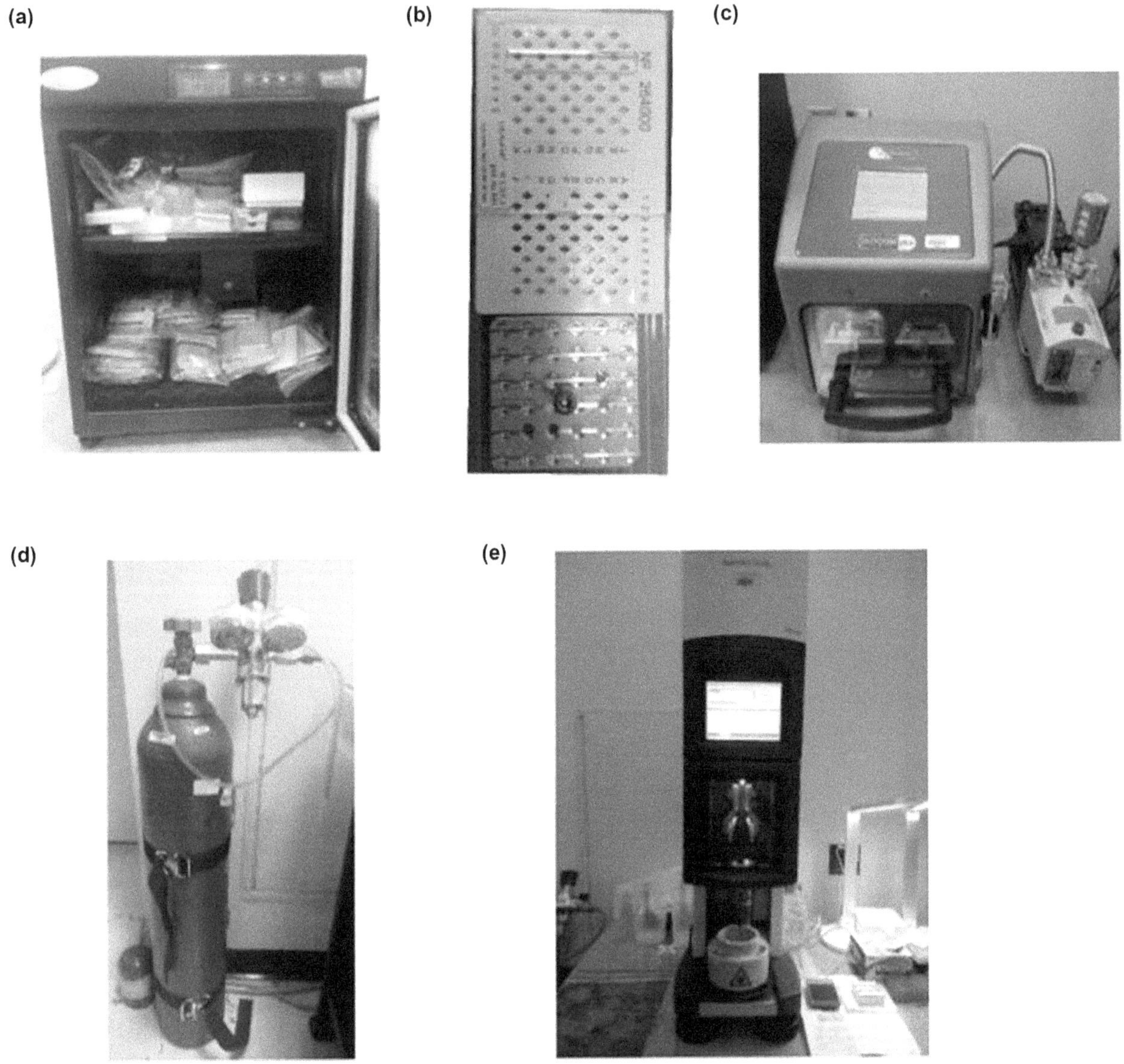

FIGURE 8.1 Cryo-EM sample preparation. (a) Cryo-EM grid storage cabinet with temperature and humidity control; humidity should be kept under 40%. (b) Cryo-EM grid box at the top and grids on glow-discharge holder at the bottom. (c) Cryo-EM glow discharge system to create a hydrophilic surface on the grid before plunge freezing. (d) Ethane storage tank. The purity of the ethane should be higher than 99.9%. (e) Vitrobot instrument.

Before sample vitrification, the grid surfaces need to be made hydrophilic using glow discharge or plasma cleaning machines. The power level is usually 20 mA, and the time ranges from 20 s to several minutes, depending on the specimen. There are some commercial instruments that can be used for this process, such as GloQube glow discharge system by EMS (Figure 8.1c), Solarus Plasma Cleaner by Gatan, and easiGlow glow discharge system by PELCO. Plunge-freezing the sample in liquid ethane cryogen (Figure 8.1d) will cause it to be rapidly frozen in a vitreous ice state and will produce the thin ice needed to optimize the blot force and blot time. At present, blotting and plunging can be done with simple commercial devices, such as Vitrobot (Figure 8.1e) by ThermoFisher (Iancu et al., 2006), EM GP by Leica (Hands-Portman and Bakker, 2022), and Cryoplunge by Gatan. All of these can be set up with a low temperature (<8°C) and 100% humidity when freezing the sample. These freezing apparatuses have different features that may be beneficial for working with certain challenging samples. Usually, Vitrobot uses a double-sided model to blot the sample, which gives rise to gradients of ice thickness; this offers advantages for beginning sample screening and finalizing and optimizing the preferred ice thickness for the sample. Single-sided blotting, such as that provided by EM GP, makes ice thickness uniform across the whole grid. This is very acceptable for difficult samples, with enough to prepare several grids if one uniform ice thickness is enough to produce ample data for 3D map reconstruction. After the grid is frozen in the liquid ethane, it is manually transferred into the grid box pre-immersed in the liquid nitrogen. It is important to always keep the grid under the liquid nitrogen level, or else ice contamination will occur. The room containing the plunging instruments should be maintained at a proper humidity level (<20% or lower).

MICROSCOPE SETUP AND IMAGE ACQUISITION

At present, there are two major manufacturers of cryo-electron microscopes, ThermoFisher Scientific and JEOL. Both offer high-performance microscope products, 200 kV or 300 kV, for single-particle analysis (SPA) in the life sciences, and they can approach near-atomic resolution without any obstacle if equipped with a direct electron detector (ThermoFisher Falcon 3 or Falcon 4 or Gatan K2 or K3) and an energy filter (ThermoFisher Selectris series or Gatan BioQuantum K3 or BioContinuum imaging filter). Considering a trade-off between performance and cost, a 200 kV microscope is a perfect option, such as the ThermoFisher Talos Arctica or Glacios (Figure 8.2a), or the JEOL CRYO ARM. After the microscope installation is done by the engineer, a full microscope alignment and experimental setup needs to be conducted to ensure the microscope performs well. Usually a standard sample, such as apoferritin or aldolase, can approach 2–2.5 Å (Kim et al., 2018).

Before the SPA image acquisition, which involves targeting the holes and then imaging at a high magnification, four microscope magnification beam parameters need to be finalized. These are usually categorized as grid, square, hole, and exposure (Figure 8.3a). These terms are typically derived from EM grid properties. Grid means the field of view contains part of the grid area using the lowest microscope magnification; the square field of view contains one whole grid square; the hole parameter can capture at least one foil hole and the edges of neighboring holes, and the exposure or data acquisition parameter captures only part of one foil hole. For each sample data collection session setup, the square and hole magnification values need to be adjusted for the grid types, which can differ in mesh number and hole size. Based on the grids commonly used at present, the square and hole magnifications are several hundred and several thousand times. The most time-consuming beam setting is the exposure parameter, also called data acquisition, which will produce the final EM data used for the 3D map reconstruction. Generally, SPA experiments use a low-dose model to collect the data to avoid damaging the sample, so a focusing step with the same beam parameters as data acquisition is used to search for a focus point on the foil instead of foil hole before applying defocus to data acquisition. With a direct electron detector, a single electron hitting the sensor will deposit the energy, which is called the counting model (Li et al., 2013). The counting model will record frames, with each exposed frame having an electron number counting response, so the electron dose rate should be lower, otherwise the electrons will be lost during the counting process. At present, the data acquisition beam dose rate is usually between 3 and 9 e^-/pix/s and the total dose between 40 and 80 e^-/Å, and this applies to most SPA projects (Li et al., 2013; Wang et al., 2021).

With the development of advanced hardware and software applications, the accuracy and high throughput of SPA data collection has progressed rapidly, especially the beam image shift data collection strategy, auto-stigmatism, and auto-coma calibration and correction, which speed up the process by a factor of three to four compared to the previous stage shift and manual microscope beam correction method (Cheng et al., 2018; Wu et al., 2019, p. 107396; Cash et al., 2020; Bouvette et al., 2021; Efremov and Stroobants, 2021; Peck, Fay and Strauss, 2022). Most cryo-EM facilities use automated data collection software to acquire stable and recurrent EM data, including SerialEM (Mastronarde, 2003, 2005), Leginon (Carragher et al., 2000; Suloway et al., 2005, 2009; Cheng et al., 2021), EPU (Thompson et al., 2019) (Figure 8.2c), and Latitude S (Gatan). SerialEM and Leginon are free open-source software programs and are the most popular, particularly SerialEM. EPU was developed by ThermoFisher; it is mainly used with the Falcon camera, and it can also be integrated into the Gatan K2 and K3 cameras, but at a relatively high cost. Latitude software was developed by Gatan for use with K2 and K3 camera image acquisition. Except for Leginon, which is based on the Linux operating system, these programs operate in the Windows environment. Leginon also has some unique functions that distinguish it from the others, such as centralized management of multiple microscopes and EM projects. This provides users with live feedback and remote collaboration during data collection. Appion, a software program wrapper compatible with a number of other packages, can be used for preliminary data analysis and filtering as a consequence of its integration with the Leginon database and web server (Lander et al., 2009; Noble and Stagg, 2015; Stagg and Mendez, 2018).

Whichever data collection software is used, setting up the microscope for data collection usually includes the next several procedures. (1) Loading a gold cross-grating grid, then finalizing the grid, square, hole, and exposure magnifications; figuring out the exposure parallel beam intensity with the diffraction model under eucentric focus; measuring the dose rate in the vacuum and calculating the exposure time. (2) After fixing the exposure beam parameters, continue to the beam tilt pivot point X and Y, and center the beam with beam shift and rotation center adjustment. Using the automated correction functions embedded in the data collection software, perform stigmatism and coma correction. (3) Image shift alignment of the four-beam magnification; thus under exposure magnification, first use the stage shift to center an obvious marker, such as an ice crystal or latex spheres in a cross-grating grid, then successively lower the magnification from

This is an Adobe® Illustrator® File that was saved without PDF Content.
To Place or open this file in other applications, it should be re-saved from Adobe Illustrator with the "Create PDF Compatible File" option turned on. This option is in the Illustrator Native Format Options dialog box, which appears when saving an Adobe Illustrator file using the Save As command.

This is an Adobe® Illustrator® File that was saved without PDF Content.
To Place or open this file in other applications, it should be re-saved from Adobe Illustrator with the "Create PDF Compatible File" option turned on. This option is in the Illustrator Native Format Options dialog box, which appears when saving an Adobe Illustrator file using the Save As command.

This is an Adobe® Illustrator® File that was saved without PDF Content.
To Place or open this file in other applications, it should be re-saved from Adobe Illustrator with the "Create PDF Compatible File" option turned on. This option is in the Illustrator Native Format Options dialog box, which appears when saving an Adobe Illustrator file using the Save As command.

This is an Adobe® Illustrator® File that was saved without PDF Content.
To Place or open this file in other applications, it should be re-saved from Adobe Illustrator with the "Create PDF Compatible File" option turned on. This option is in the Illustrator Native Format Options dialog box, which appears when saving an Adobe Illustrator file using the Save As command.

This is an Adobe® Illustrator® File that was saved without PDF Content.
To Place or open this file in other applications, it should be re-saved from Adobe Illustrator with the "Create PDF Compatible File" option turned on. This option is in the Illustrator Native Format Options dialog box, which appears when saving an Adobe Illustrator file using the Save As command.

This is an Adobe® Illustrator® File that was saved without PDF Content.
To Place or open this file in other applications, it should be re-saved from Adobe Illustrator with the "Create PDF Compatible File" option turned on. This option is in the Illustrator Native Format Options dialog box, which appears when saving an Adobe Illustrator file using the Save As command.

This is an Adobe® Illustrator® File that was saved without PDF Content.
To Place or open this file in other applications, it should be re-saved from Adobe Illustrator with the "Create PDF Compatible File" option turned on. This option is in the Illustrator Native Format Options dialog box, which appears when saving an Adobe Illustrator file using the Save As command.

This is an Adobe® Illustrator® File that was saved without PDF Content.
To Place or open this file in other applications, it should be re-saved from Adobe Illustrator with the "Create PDF Compatible File" option turned on. This option is in the Illustrator Native Format Options dialog box, which appears when saving an Adobe Illustrator file using the Save As command.

FIGURE 8.2 Cryo-EM data collection instruments. (a) Microscope room with Glacios microscope. (b) Server room, one TFS Falcon 4 temporary data storage server and one 180 TB data storage server, one GPU data processing workstation with 4x RTX A6000 graphic cards. (c) Microscope computer control room. Three popular data collection programs are configured for use: EPU, Leginon, and SerialEM. For Leginon, one web server computer, one Leginon application computer, and one web viewer computer are shown in the computer room. One Leginon database server is shown in the server room. For SerialEM, two monitors attached to one computer are shown; during data collection, one monitor is for the main application, and the other is used for a SerialEM dummy application, which is used for target selection to avoid conflict with the SerialEM main application.

This is an Adobe® Illustrator® File that was saved without PDF Content.
To Place or open this file in other applications, it should be re-saved from Adobe Illustrator with the "Create PDF Compatible File" option turned on. This option is in the Illustrator Native Format Options dialog box, which appears when saving an Adobe Illustrator file using the Save As command.

This is an Adobe® Illustrator® File that was saved without PDF Content.
To Place or open this file in other applications, it should be re-saved from Adobe Illustrator with the "Create PDF Compatible File" option turned on. This option is in the Illustrator Native Format Options dialog box, which appears when saving an Adobe Illustrator file using the Save As command.

This is an Adobe® Illustrator® File that was saved without PDF Content.
To Place or open this file in other applications, it should be re-saved from Adobe Illustrator with the "Create PDF Compatible File" option turned on. This option is in the Illustrator Native Format Options dialog box, which appears when saving an Adobe Illustrator file using the Save As command.

This is an Adobe® Illustrator® File that was saved without PDF Content.
To Place or open this file in other applications, it should be re-saved from Adobe Illustrator with the "Create PDF Compatible File" option turned on. This option is in the Illustrator Native Format Options dialog box, which appears when saving an Adobe Illustrator file using the Save As command.

This is an Adobe® Illustrator® File that was saved without PDF Content.
To Place or open this file in other applications, it should be re-saved from Adobe Illustrator with the "Create PDF Compatible File" option turned on. This option is in the Illustrator Native Format Options dialog box, which appears when saving an Adobe Illustrator file using the Save As command.

This is an Adobe® Illustrator® File that was saved without PDF Content.
To Place or open this file in other applications, it should be re-saved from Adobe Illustrator with the "Create PDF Compatible File" option turned on. This option is in the Illustrator Native Format Options dialog box, which appears when saving an Adobe Illustrator file using the Save As command.

This is an Adobe® Illustrator® File that was saved without PDF Content.
To Place or open this file in other applications, it should be re-saved from Adobe Illustrator with the "Create PDF Compatible File" option turned on. This option is in the Illustrator Native Format Options dialog box, which appears when saving an Adobe Illustrator file using the Save As command.

This is an Adobe® Illustrator® File that was saved without PDF Content.
To Place or open this file in other applications, it should be re-saved from Adobe Illustrator with the "Create PDF Compatible File" option turned on. This option is in the Illustrator Native Format Options dialog box, which appears when saving an Adobe Illustrator file using the Save As command.

FIGURE 8.3 Cryo-EM data collection, processing, and model building. (a) Cryo-EM data collection workflow. A whole view is taken for the grid, called Atlas; individual squares are then targeted and images taken. A micrograph consisting of several holes from each square is taken; for each individual hole, a final acquisition image is taken. (b) Cryo-EM data processing workflow. First protein particles are picked out from the acquisition image as seen encircled in red. Next, 2D classification is performed to select good particles. In the third step, an initial 3D model is created from good 2D classified particles and following a 3D classification. Finally, the good 3D class goes through homogenous refinement, and resolution is reported by the FCS curve. (c) Atomic model building. Here an initial homologous PDB structure is obtained from either the PDB database or prediction software like AlphaFold. Initial models are docked into the final map using the fit-in-map function of ChimeraX (*left panel*). This is followed by visual inspection at the residue level and manual model building in Coot (*right panel*).

exposure to grid, at each magnification only using image shift to center the marker, keeping the exposure centered. (4) Setting up the foil hole targeting for image acquisition, first take the atlas picture of the whole grid, then choose the squares with optimal ice thickness. Find the eucentic height for each chosen square, and use the hole template to locate the foil hole position. With all the holes position mapped on the square, the software will finalize the image shift range to take the final image. Attribute to the beam tilt astigmatism and coma correction. Even a group of 6 × 6 (total 36) holes can be collected by the image shift method; this means after the autofocus measurement, beam shifting to the target hole position and then coupling image shift to project the image on the camera center without any movement of the stage. Additionally, if the energy filter is used for data acquisition, under the vacuum or empty hole, choose the proper energy filter slit width and zero-loss peak centering.

CRYO-EM DATA STORAGE AND PROCESSING

At present, cryo-EM data is collected in the movies format using the direct detector (Campbell et al., 2012; Ripstein and Rubinstein, 2016). This inevitably yields a huge amount of data; for a Glacios 200 kV equipped with a Falcon 4 camera, a 3 to 6 TB movies dataset is captured over a 24-hour day. Thus, storing and going over this large amount of data is a management challenge, and storage servers with several hundred TB of capacity are now required for the facility (Patwardhan and Lawson, 2016; Baldwin et al., 2018). With the acquisition of a Falcon 4 or K3 camera, the vendor will provide a temporary storage server with tens of TB accompanying the camera installation, and a fiber optic cable connecting the camera and storage server for rapid data storage. The facility also needs to build one or several large-volume storage servers with several hundred TB of capacity for long-term data storage, usually at least one month's worth, and to install fiber optic cable to bridge the temporary and long-term (facility) servers. Each movie file is big enough to reach around 400 MB in the TIF format or 600 MB for EER or even 1.5 GB for MRC. Users transferring the data to their own storage devices can send one 8 TB hard drive to the facility during each data collection session, or they can remotely transfer the data using the Globus system (Ananthakrishnan et al., 2015). The facility needs to set up the Globus server on any computer that can access the facility server (Figure 8.2b).

During data collection, online data processing can begin when using Leginon (Cheng et al., 2021) or Cryosparc (Figure 8.2c) (Punjani et al., 2017), a data processing software program. Data quality can also be monitored live through a conventional first two steps process, movie frame alignment and contrast transfer function (CTF) estimation, for single-particle data. This rapid feedback from motion-corrected micrograph or CTF values can provide the users with information on whether data collection should continue or stop. Next we will outline the general data processing pipeline for single-particle data processing by the facility, particularly for occasional users who need assistance from facility staff to get the EM structure instead of doing it by themselves.

1. Before processing the single-particle data, it is essential to ensure that a parallel GPU workstation with at least a 4 GPU card is purchased; an example configuration is a Centos 7.9 system with NVIDIA drive version 470.74, four RTX A6000 cards, one 8T SDD and 40T SATA HDD (Figure 8.2b). EM data processing software is also needed, such as Cryosparc, Relion, MotionCorr (Zheng et al., 2017), ctffind4 (Rohou and Grigorieff, 2015), or Gctf (Zhang, 2016). Cryosparc and Relion are EM data processing packages that include all the necessary data processing applications; MotionCorr and ctffind4 are the main single-run motion correction and CTF estimation programs. Users can choose whichever they want. Allowing the GPU workstation to access the movies data storage server promotes data processing efficiency and also decreases the burden of storage space volume.
2. The first preprocessing step of single-particle data processing is motion correction; this is completed by a modern algorithm aligning individual movie frames against each other. This ultimately corrects the blur effects of beam-induced motion and ensures the high-resolution signal will be recovered and enabled to achieve near-atomic resolution of the structure. With advances in GPU applications for movie alignment, at least 200 movie frames can be finished in one hour; this facilitates the efficiency of data preprocessing when it is done in parallel with CTF estimation, the second step of data preprocessing. CTF estimation will help to assess the micrograph quality; a higher CTF fitting value means better data quality and potentially higher resolution of the structure.
3. The next stage of data processing is to verify and crop individual particle images out from the motion-corrected micrographs, also called particle extraction (Figure 8.3b). A variety of approaches can be applied to identify and pick out the particles, ranging from fully manual picking to fully automated. The two methods can be combined by first accurately picking the particles manually from a subset of micrographs, the best 2D classes from following 2D classifications, with manually picked particles supplying the template for automatic particle picking from the entire dataset (Figure 8.3b). Cryosparc and Relion both have the two particle picking approaches integrated as a streamlined workflow. After the particles have been identified, they are then extracted from the

micrograph with a manually set box size—usually 2 times bigger than the particle size—to generate a maximal particle stack particle stack; particle numbers can reach several millions.

4. The extracted particles are the initial images of target protein 2D projection with different orientation, using power mathematical processing, aligning each individual particle image and averaging the similar particles together to improve the signal. These averaged new images are 2D classes. To sort out the similar particles, processing the differences in x,y translation, in-plane rotation ultimately groups the same particles into same class. These increased signal-to-noise 2D classes are available to do 3D voxel reconstruction.
5. With 3D reconstruction serving as reference model and classification, 2D classes represent 2D projections of the original 3D object. The relative orientations of the particle images are determined by projection-matching algorithms, which a reference 3D model is compared against from particle images. This process is calculated iteratively, and the convergence 3D model will be the new reference for a new round of orientation assignments. This process cycles many times until the orientation assignment values are not significantly changed. At this point, also described as iteration convergence, the final map has been generated relatively accurately. Biological samples always exhibit divergent conformations when they carry out their functions. These conformation states can be classified through the same processing approaches but several 3D reference models are needed. This type of 3D classification is usually applied when processing single-particle data. More recent research studies reveal that sample compositional and conformational heterogeneity can be distinguished, and they demonstrate the power of these tools in elucidating the mechanism of protein function (Figure 8.3d).
6. Although the 3D classification has helped to assign orientation values to particle images, this still needs to be further refined and validated in case the overfitting problem has occurred. The so-called gold standard approach divides the whole particles into two halves and each half is processed independently to generate a half map; these are then correlated against each other to estimate the final resolution (Figure 8.3b). People call these correlations Fourier shell correlations or FSCs. Cross-correlating the value of the half maps as a function of spatial frequency, a cutoff threshold of 0.143 is the most commonly used criterion to report the nominal resolution of a final reconstruction cryo-EM map (Diebolder et al., 2015; van Heel and Schatz, 2005; Penczek, 2020). This eventual map is used to build the protein model.

CRYO-EM MAP MODEL BUILDING

Cryo-EM map model building is much like conventional crystal map model building by using Coot for *de novo* building if the resolution is <3.5 Å. For some lower-resolution maps, using already known crystal structures, homology model prediction or artificial intelligence prediction, such as the famous AlphaFold (Jumper et al., 2021) and Rosetta (DiMaio, 2017; Du et al., 2021), for fitting into the map by Chimera (Pettersen et al., 2021) is the first labor-intensive step work (Figure 8.3c). Although docking the atom model into the cryo-EM map with Chimera is feasible, it still needs human intervention to adjust the local model map deviation by Coot (Figure 8.2c) (Emsley and Cowtan, 2004). Phenix (Liebschner et al., 2019), a canonical crystallography resolving and refinement package, has developed a robust application, phenix.real_space_refine, to refine a cryo-EM model map in real space, with a working resolution range from 1 to 6 Å. The refinement optimization can be achieved using simulated annealing, morphing, rigid-body refinement, and systematic side-chain improvement. The refined model can be validated using MolProbity tools (Williams et al., 2018) or the wwPDB validation system; the validation information includes covalent geometry, non-bonded interactions, atom outliers, and Ramachandran plot. Any outliers and errors need correction; the model can be submitted to PDB after correction, and the cryo-EM map should also be submitted to EMDB.

NEW TECHNOLOGY AND OUTLOOK

When preparing grids for freezing protein samples, several different types of grids can be optimized to get better particle distribution in the foil holes. Generally, gold mesh–covered gold film can decrease the sample motion induced by the beam compared to carbon mesh–covered carbon film, which is a good option for getting better resolution. Some protein samples have a severe orientation preference; strategies for alleviating this phenomenon include adding detergent to the sample before freezing or collecting data with a 30- or 45-degree stage tilt. Another layer of ultra-thin film made of graphene or its derivatives converging on the grid can decrease the orientation preference; it can also overcome the air–water interface problem during sample blotting, which can cause the protein complex to denature and hinder the ability to obtain higher resolutions (Naydenova, Peet and Russo, 2019; Han et al., 2020; Fan and Sun, 2022). From the standpoint of facility operations, each kind of grid should be tried and optimized when preparing samples for users. Another new sample preparation approach is unblotting, in which only a sub-nanoliter volume of sample is deposited directly on the grid surface, which is then vitrified by cryogen. Recent research has shown this technique can provide controlled and efficient sample preparation (Jain et al., 2012; Razinkov et al., 2016; Rima et al., 2022).

Through a great number of experiments using either crystallography or cryo-EM, more than 100,000 unique protein structures have been determined. This provides a huge dataset for deep learning techniques to explore protein structure prediction. Several software programs have been developed based on neural networks to predict mono protein three-dimensional structures without available homologous structures. AlphaFold is the most famous one; it can accurately predict the backbone of an unknown protein structure from the protein sequence. The model can be used for cryo-EM map model docking. Recently, the neural network has benefited from more robust software that integrates the AlphaFold application into auto model building. EMBuild (He et al., 2022), for example, can build a model from a map with a resolution between 4.0 and 8.0 Å. ModelAngelo (Jamali, Kimanius and Scheres, 2022) can approximate manual building for cryo-EM maps with resolution better than 3.5 Å. Users need only provide the cryo-EM map and the protein sequences; the software models individual protein subunits into the map without human intervention. The accuracy is amazing, and it also saves a lot of time and effort. For a cryo-EM facility, using these AI prediction and auto modeling programs can significantly reduce the manager's labor-intensive and time-consuming model building workload.

Compared to a 300 kV microscope, the drawbacks of a 200 kV microscope are the lack of lens spherical aberration correction, poor thick ice penetration, and reduced ability to maintain a stable column vacuum. But it has recently been shown that a field emission gun electron source 200 kV cryo-EM with a modern direct electron detector can still obtain cryo-EM maps with sub 3.0 Å resolution. And the cost of building a 200 kV microscope is only one-half to two-thirds the cost of building a 300 kV setup. Therefore, building a 200 kV microscope is very feasible for a new facility and enough for performing single-particle cryo-EM projects.

One important factor to consider is data storage. Larger and faster electron counting cameras are now coming onto the market, such as the Falcon 4i and the K3. They can produce vastly more data than earlier models. Consequently, a petabyte (PB) capacity storage system should now be included in facility plans. The data file format produced is movie stacks, but after motion correction, the file is transformed to a single MRC file, which is nearly 10 times smaller. Thereafter, online data preprocessing, motion correction, and CTF estimation will relieve the storage server burden. CTF values can also help in filtering the data according to quality. This is very acceptable, because from the end users' point of view, the data processing usually starts from the motion-corrected micrograph with the highest quality. There are now several cryo-EM data processing programs for online processing, such as WARP (Tegunov and Cramer, 2019), CryoSPARC, and Relion. In addition, data collection software, such as Leginon and EPU, have integrated the online processing application into the package, so preprocessing will begin automatically when the data collection is set up.

Given all these advancements in instruments and automation, cryo-EM will become a standard technical resource in protein biochemistry, much like next-generation sequencing service, which is used in most genomic, transcriptomic, and epigenetic research.

REFERENCES

Ananthakrishnan, R. et al. (2015) 'Globus platform-as-a-service for collaborative science applications', *Concurrency and computation: practice & experience*, 27(2), pp. 290–305. Available at: https://doi.org/10.1002/cpe.3262.

Baldwin, P.R. et al. (2018) 'Big data in cryoEM: automated collection, processing and accessibility of EM data', *Current opinion in microbiology*, 43, pp. 1–8. Available at: https://doi.org/10.1016/j.mib.2017.10.005.

Basanta, B. et al. (2022) 'A case for glycerol as an acceptable additive for single-particle cryoEM samples', *Acta crystallographica. Section D, Structural biology*, 78(Pt 1), pp. 124–135. Available at: https://doi.org/10.1107/S2059798321012110.

Bouvette, J. et al. (2021) 'Beam image-shift accelerated data acquisition for near-atomic resolution single-particle cryo-electron tomography', *Nature communications*, 12(1), p. 1957. Available at: https://doi.org/10.1038/s41467-021-22251-8.

Campbell, M.G. et al. (2012) 'Movies of ice-embedded particles enhance resolution in electron cryo-microscopy', *Structure (London, England: 1993)*, 20(11), pp. 1823–1828. Available at: https://doi.org/10.1016/j.str.2012.08.026.

Carragher, B. et al. (2000) 'Leginon: an automated system for acquisition of images from vitreous ice specimens', *Journal of structural biology*, 132(1), pp. 33–45. Available at: https://doi.org/10.1006/jsbi.2000.4314.

Cash, J.N. et al. (2020) 'High-resolution cryo-EM using beam-image shift at 200 keV', *IUCrJ*, 7(Pt 6), pp. 1179–1187. Available at: https://doi.org/10.1107/S2052252520013482.

Cheng, A. et al. (2016) 'Strategies for automated CryoEM data collection using direct detectors', *Methods in enzymology*, 579, pp. 87–102. Available at: https://doi.org/10.1016/bs.mie.2016.04.008.

Cheng, A. et al. (2018) 'High resolution single particle cryo-electron microscopy using beam-image shift', *Journal of structural biology*, 204(2), pp. 270–275. Available at: https://doi.org/10.1016/j.jsb.2018.07.015.

Cheng, A. et al. (2021) 'Leginon: New features and applications', *Protein science: a publication of the protein society*, 30(1), pp. 136–150. Available at: https://doi.org/10.1002/pro.3967.

De Carlo, S. and Harris, J.R. (2011) 'Negative staining and cryo-negative staining of macromolecules and viruses for TEM', *Micron (Oxford, England: 1993)*, 42(2), pp. 117–131. Available at: https://doi.org/10.1016/j.micron.2010.06.003.

Diebolder, C.A. et al. (2015) 'Conical Fourier shell correlation applied to electron tomograms', *Journal of structural biology*, 190(2), pp. 215–223. Available at: https://doi.org/10.1016/j.jsb.2015.03.010.

DiMaio, F. (2017) 'Rosetta structure prediction as a tool for solving difficult molecular replacement problems', *Methods in molecular biology (Clifton, N.J.)*, 1607, pp. 455–466. Available at: https://doi.org/10.1007/978-1-4939-7000-1_19.

Dobro, M.J. et al. (2010) 'Plunge freezing for electron cryomicroscopy', *Methods in enzymology*, 481, pp. 63–82. Available at: https://doi.org/10.1016/S0076-6879(10)81003-1.

Drulyte, I. et al. (2018) 'Approaches to altering particle distributions in cryo-electron microscopy sample preparation', *Acta crystallographica. Section D, Structural biology*, 74(Pt 6), pp. 560–571. Available at: https://doi.org/10.1107/S2059798318006496.

Du, Z. et al. (2021) 'The trRosetta server for fast and accurate protein structure prediction', *Nature protocols*, 16(12), pp. 5634–5651. Available at: https://doi.org/10.1038/s41596-021-00628-9.

Efremov, R.G. and Stroobants, A. (2021) 'Coma-corrected rapid single-particle cryo-EM data collection on the CRYO ARM 300', *Acta crystallographica. Section D, Structural biology*, 77(Pt 5), pp. 555–564. Available at: https://doi.org/10.1107/S2059798321002151.

Emsley, P. and Cowtan, K. (2004) '*Coot*: model-building tools for molecular graphics', *Acta crystallographica section D biological crystallography*, 60(12), pp. 2126–2132. Available at: https://doi.org/10.1107/S0907444904019158.

Fan, H. and Sun, F. (2022) 'Developing graphene grids for cryoelectron microscopy', *Frontiers in molecular biosciences*, 9, p. 937253. Available at: https://doi.org/10.3389/fmolb.2022.937253.

Fréchin, L. et al. (2022) 'High-resolution cryo-EM performance comparison of two latest-generation cryo electron microscopes on the human ribosome', *Journal of structural biology*, 215(1), p. 107905. Available at: https://doi.org/10.1016/j.jsb.2022.107905.

Han, Y. et al. (2020) 'High-yield monolayer graphene grids for near-atomic resolution cryoelectron microscopy', *Proceedings of the national academy of sciences of the United States of America*, 117(2), pp. 1009–1014. Available at: https://doi.org/10.1073/pnas.1919114117.

Hands-Portman, I. and Bakker, S.E. (2022) 'Customising the plunge-freezing workflow for challenging conditions', *Faraday discussions*, 240(0), pp. 44–54. Available at: https://doi.org/10.1039/d2fd00060a.

He, J. et al. (2022) 'Model building of protein complexes from intermediate-resolution cryo-EM maps with deep learning-guided automatic assembly', *Nature communications*, 13(1), p. 4066. Available at: https://doi.org/10.1038/s41467-022-31748-9.

Iancu, C.V. et al. (2006) 'Electron cryotomography sample preparation using the Vitrobot', *Nature protocols*, 1(6), pp. 2813–2819. Available at: https://doi.org/10.1038/nprot.2006.432.

Jain, T. et al. (2012) 'Spotiton: a prototype for an integrated inkjet dispense and vitrification system for cryo-TEM', *Journal of structural biology*, 179(1), pp. 68–75. Available at: https://doi.org/10.1016/j.jsb.2012.04.020.

Jamali, K., Kimanius, D. and Scheres, S.H.W. (2022) 'A graph neural network approach to automated model building in cryo-EM maps', arXiv. Available at: http://arxiv.org/abs/2210.00006 (Accessed: 15 January 2023).

Jumper, J. et al. (2021) 'Highly accurate protein structure prediction with AlphaFold', *Nature*, 596(7873), pp. 583–589. Available at: https://doi.org/10.1038/s41586-021-03819-2.

Kastner, B. et al. (2008) 'GraFix: sample preparation for single-particle electron cryomicroscopy', *Nature methods*, 5(1), pp. 53–55. Available at: https://doi.org/10.1038/nmeth1139.

Kim, L.Y. et al. (2018) 'Benchmarking cryo-EM single particle analysis workflow', *Frontiers in molecular biosciences*, 5, p. 50. Available at: https://doi.org/10.3389/fmolb.2018.00050.

Koh, A. et al. (2022) 'Routine collection of high-resolution cryo-EM datasets using 200 KV transmission electron microscope', *JoVE*, e63519. Available at: https://doi.org/10.3791/63519.

Lander, G.C. et al. (2009) 'Appion: an integrated, database-driven pipeline to facilitate EM image processing', *Journal of structural biology*, 166(1), pp. 95–102. Available at: https://doi.org/10.1016/j.jsb.2009.01.002.

Li, X. et al. (2013) 'Electron counting and beam-induced motion correction enable near-atomic-resolution single-particle cryo-EM', *Nature methods*, 10(6), pp. 584–590. Available at: https://doi.org/10.1038/nmeth.2472.

Liebschner, D. et al. (2019) 'Macromolecular structure determination using X-rays, neutrons and electrons: recent developments in *Phenix*', *Acta crystallographica section D structural biology*, 75(10), pp. 861–877. Available at: https://doi.org/10.1107/S2059798319011471.

Liu, Y. et al. (2018) 'Near-atomic cryo-EM imaging of a small protein displayed on a designed scaffolding system', *Proceedings of the national academy of sciences of the United States of America*, 115(13), pp. 3362–3367. Available at: https://doi.org/10.1073/pnas.1718825115.

Liu, Y., et al. (2019) 'A 3.8 Å resolution cryo-EM structure of a small protein bound to an imaging scaffold', *Nature Communication*, 10, 1864. Available at: https://doi.org/10.1038/s41467-019-09836-0.

Mastronarde, D.N. (2003) 'SerialEM: a program for automated tilt series acquisition on tecnai microscopes using prediction of specimen position', *Microscopy and microanalysis*, 9(S02), pp. 1182–1183. Available at: https://doi.org/10.1017/S1431927603445911.

Mastronarde, D.N. (2005) 'Automated electron microscope tomography using robust prediction of specimen movements', *Journal of structural biology*, 152(1), pp. 36–51. Available at: https://doi.org/10.1016/j.jsb.2005.07.007.

McMullan, G., Faruqi, A.R. and Henderson, R. (2016) 'Direct electron detectors', *Methods in enzymology*, 579, pp. 1–17. Available at: https://doi.org/10.1016/bs.mie.2016.05.056.

Moyes, R. B., Reynolds, J., & Breakwell, D. P. (2009). Preliminary staining of bacteria: negative stain. Current protocols in microbiology, Appendix 3. Available at: https://doi.org/10.1002/9780471729259.mca03fs15.

Naydenova, K. et al. (2019) 'CryoEM at 100 keV: a demonstration and prospects', *IUCrJ*, 6(Pt 6), pp. 1086–1098. Available at: https://doi.org/10.1107/S2052252519012612.

Naydenova, K., Peet, M.J. and Russo, C.J. (2019) 'Multifunctional graphene supports for electron cryomicroscopy', *Proceedings of the national academy of sciences of the United States of America*, 116(24), pp. 11718–11724. Available at: https://doi.org/10.1073/pnas.1904766116.

Noble, A.J. and Stagg, S.M. (2015) 'Automated batch fiducial-less tilt-series alignment in Appion using Protomo', *Journal of structural biology*, 192(2), pp. 270–278. Available at: https://doi.org/10.1016/j.jsb.2015.10.003.

Patwardhan, A. and Lawson, C.L. (2016) 'Databases and Archiving for CryoEM', *Methods in enzymology*, 579, pp. 393–412. Available at: https://doi.org/10.1016/bs.mie.2016.04.015.

Peck, J.V., Fay, J.F. and Strauss, J.D. (2022) 'High-speed high-resolution data collection on a 200 keV cryo-TEM', *IUCrJ*, 9(Pt 2), pp. 243–252. Available at: https://doi.org/10.1107/S2052252522000069.

Penczek, P.A. (2020) 'Reliable cryo-EM resolution estimation with modified Fourier shell correlation', *IUCrJ*, 7(Pt 6), pp. 995–1008. Available at: https://doi.org/10.1107/S2052252520011574.

Pettersen, E.F. et al. (2021) 'UCSF ChimeraX : structure visualization for researchers, educators, and developers', *Protein science*, 30(1), pp. 70–82. Available at: https://doi.org/10.1002/pro.3943.

Punjani, A. et al. (2017) 'cryoSPARC: algorithms for rapid unsupervised cryo-EM structure determination', *Nature methods*, 14(3), pp. 290–296. Available at: https://doi.org/10.1038/nmeth.4169.

Razinkov, I. et al. (2016) 'A new method for vitrifying samples for cryoEM', *Journal of structural biology*, 195(2), pp. 190–198. Available at: https://doi.org/10.1016/j.jsb.2016.06.001.

Renaud, J.-P. et al. (2018) 'Cryo-EM in drug discovery: achievements, limitations and prospects', *Nature reviews. Drug discovery*, 17(7), pp. 471–492. Available at: https://doi.org/10.1038/nrd.2018.77.

Rima, L. et al. (2022) 'cryoWriter: a blotting free cryo-EM preparation system with a climate jet and cover-slip injector', *Faraday discussions*, 240(0), pp. 55–66. Available at: https://doi.org/10.1039/d2fd00066k.

Ripstein, Z.A. and Rubinstein, J.L. (2016) 'Processing of cryo-EM movie data', *Methods in enzymology*, 579, pp. 103–124. Available at: https://doi.org/10.1016/bs.mie.2016.04.009.

Robertson, M.J., Meyerowitz, J.G. and Skiniotis, G. (2022) 'Drug discovery in the era of cryo-electron microscopy', *Trends in biochemical sciences*, 47(2), pp. 124–135. Available at: https://doi.org/10.1016/j.tibs.2021.06.008.

Rohou, A. and Grigorieff, N. (2015) 'CTFFIND4: Fast and accurate defocus estimation from electron micrographs', *Journal of structural biology*, 192(2), pp. 216–221. Available at: https://doi.org/10.1016/j.jsb.2015.08.008.

Sharov, G. et al. (2021) 'Using RELION software within the Scipion framework', *Acta crystallographica. Section D, Structural biology*, 77(Pt 4), pp. 403–410. Available at: https://doi.org/10.1107/S2059798321001856.

Stagg, S.M. and Mendez, J.H. (2018) 'Processing apoferritin with the Appion pipeline', *Journal of structural biology*, 204(1), pp. 85–89. Available at: https://doi.org/10.1016/j.jsb.2018.06.009.

Stark, H. (2010) 'GraFix: stabilization of fragile macromolecular complexes for single particle cryo-EM', *Methods in enzymology*, 481, pp. 109–126. Available at: https://doi.org/10.1016/S0076-6879(10)81005-5.

Suloway, C. et al. (2005) 'Automated molecular microscopy: the new Leginon system', *Journal of structural biology*, 151(1), pp. 41–60. Available at: https://doi.org/10.1016/j.jsb.2005.03.010.

Suloway, C. et al. (2009) 'Fully automated, sequential tilt-series acquisition with Leginon', *Journal of structural biology*, 167(1), pp. 11–18. Available at: https://doi.org/10.1016/j.jsb.2009.03.019.

Tegunov, D. and Cramer, P. (2019) 'Real-time cryo-electron microscopy data preprocessing with Warp', *Nature methods*, 16(11), pp. 1146–1152. Available at: https://doi.org/10.1038/s41592-019-0580-y.

Thompson, R.F. et al. (2019) 'Collection, pre-processing and on-the-fly analysis of data for high-resolution, single-particle cryo-electron microscopy', *Nature protocols*, 14(1), pp. 100–118. Available at: https://doi.org/10.1038/s41596-018-0084-8.

van Heel, M. and Schatz, M. (2005) 'Fourier shell correlation threshold criteria', *Journal of structural biology*, 151(3), pp. 250–262. Available at: https://doi.org/10.1016/j.jsb.2005.05.009.

Wang, C.-H. et al. (2021) 'Sub-3 Å cryo-EM structures of necrosis virus particles via the use of multipurpose TEM with electron counting camera', *International journal of molecular sciences*, 22(13). Available at: https://doi.org/10.3390/ijms22136859.

Williams, C.J. et al. (2018) 'MolProbity: more and better reference data for improved all-atom structure validation', *Protein science: a publication of the protein society*, 27(1), pp. 293–315. Available at: https://doi.org/10.1002/pro.3330.

Wu, C. et al. (2019) 'High-quality, high-throughput cryo-electron microscopy data collection via beam tilt and astigmatism-free beam-image shift', *Journal of structural biology*, 208(3), p. 107396. Available at: https://doi.org/10.1016/j.jsb.2019.09.013.

Wu, S. et al. (2012) 'Fabs enable single particle cryoEM studies of small proteins', *Structure (London, England: 1993)*, 20(4), pp. 582–592. Available at: https://doi.org/10.1016/j.str.2012.02.017.

Wu, S., Armache, J.-P. and Cheng, Y. (2016) 'Single-particle cryo-EM data acquisition by using direct electron detection camera', *Microscopy (Oxford, England)*, 65(1), pp. 35–41. Available at: https://doi.org/10.1093/jmicro/dfv355.

Yip, K.M. et al. (2020) 'Atomic-resolution protein structure determination by cryo-EM', *Nature*, 587(7832), pp. 157–161. Available at: https://doi.org/10.1038/s41586-020-2833-4.

Yu, G. et al. (2016) 'Antibody-based affinity cryoelectron microscopy at 2.6-Å resolution', *Structure (London, England: 1993)*, 24(11), pp. 1984–1990. Available at: https://doi.org/10.1016/j.str.2016.09.008.

Yu, G., Li, K. and Jiang, W. (2016) 'Antibody-based affinity cryo-EM grid', *Methods (San Diego, Calif.)*, 100, pp. 16–24. Available at: https://doi.org/10.1016/j.ymeth.2016.01.010.

Yue, Y. et al. (2022) 'Structural insight into apelin receptor-G protein stoichiometry', *Nature structural & molecular biology*, 29(7), pp. 688–697. Available at: https://doi.org/10.1038/s41594-022-00797-5.

Zhang, K. (2016) 'Gctf: real-time CTF determination and correction', *Journal of structural biology*, 193(1), pp. 1–12. Available at: https://doi.org/10.1016/j.jsb.2015.11.003.

Zhang, L. et al. (2013) 'Optimized negative-staining electron microscopy for lipoprotein studies', *Biochimica et biophysica acta*, 1830(1), pp. 2150–2159. Available at: https://doi.org/10.1016/j.bbagen.2012.09.016.

Zhang, X. et al. (2021) 'Evolving cryo-EM structural approaches for GPCR drug discovery', *Structure (London, England: 1993)*, 29(9), pp. 963–974.e6. Available at: https://doi.org/10.1016/j.str.2021.04.008.

Zheng, S.Q. et al. (2017) 'MotionCor2: anisotropic correction of beam-induced motion for improved cryo-electron microscopy', *Nature methods*, 14(4), pp. 331–332. Available at: https://doi.org/10.1038/nmeth.4193.

Zivanov, J. et al. (2018) 'New tools for automated high-resolution cryo-EM structure determination in RELION-3', *eLife*, 7. Available at: https://doi.org/10.7554/eLife.42166.

9 Computational Methods for Bimolecular Structure Modeling for Cryo-EM

Javad Baghirov, Xiao Wang, Genki Terashi, Shu Li, and Daisuke Kihara

INTRODUCTION

The three-dimensional (3D) structures of macromolecules, such as proteins, DNA, RNA, and their complexes, play a crucial role in understanding the mechanisms of their biological functions. The 3D structures also facilitate various applications of biomolecules, including drug design (Warner et al., 2018) and biomolecular engineering (Churkin et al., 2018; Huang et al., 2016).

Cryogenic electron microscopy (cryo-EM) is an emerging technology for structural biology. An obvious advantage of cryo-EM over the conventional method, X-ray crystallography, is that it does not require crystals of samples, which opens the door for determining structures that were difficult for X-ray crystallography, such as large macromolecular complexes and membrane proteins. As a result, in the past few years, the number of structures determined by cryo-EM has been increasing at an exponential pace.

Despite the great success of cryo-EM in structural biology, there are weaknesses, which include the relatively low resolution of density maps it can determine. Building structure models from a density map becomes nontrivial as the resolution decreases. When the resolution of the map is about 3 Å or worse, modeling becomes difficult, particularly for map regions with low local density. To address this issue, several computational methods for protein structure modeling have been developed (Alnabati & Kihara, 2019). These methods include those that trace the main chain through local points with high-density values in the map using graph-based approaches (Hryc & Baker, 2022; Terashi et al., 2020; Terashi & Kihara, 2018), a method that uses a structure fragment library (Song et al., 2013; Wang et al., 2016), or structural prediction (Zhang et al., 2022). Additionally, some methods use deep learning to identify local map patterns that correspond to amino acids (He & Huang, 2021; Pfab et al., 2021).

For DNA and RNA structure modeling, there are not many modeling methods for cryo-EM, reflecting the limited availability of 3D structures of DNA and RNA in the public databases, PDB and EMDB (Berman et al., 2002). Existing related methods for RNA structure modeling for cryo-EM include RCrane (Keating & Pyle, 2012) and auto-DRRAFTER (Kappel et al., 2020). RCrane is an interactive software program that helps users iteratively select phosphate, sugar, and base positions from candidates and build the full atomic structure from the selections. Auto-DRRAFTER utilized the experimental secondary structure information to predict RNA structure via FARFAR2 (Watkins et al., 2020). Then it used the fragments from the predicted structure and the density information to build the full structure. Additionally, Phenix. map_to_model is a fragment-based structure modeling tool mainly developed for protein structure modeling, but it can also support nucleotide structure modeling with limited performance.

To address the need for accurate modeling tools for cryo-EM, we have been developing a series of tools (https://kiharalab.org/emsuites/). The suite includes tools for molecular structure fitting to cryo-EM, called VESPER (Han et al., 2021), a protein structure validation score, DAQ (Nakamura et al., 2023; Terashi et al., 2022), a map sharpening method for assisting structure modeling, EM-GAN (Raghavendra et al., 2023), the state-of-the-art protein structure modeling method, DeepMainmast (Terashi et al., 2023), and a DNA/RNA structure modeling method, Cryo-READ (Wang et al., 2023). Among these tools, in this article, we focus our discussion on DeepMainmast and CryoREAD (Figure 9.1).

DeepMainmast is a novel deep learning–based method for *de novo* protein main-chain tracing. It uses deep learning to identify Cα atom positions and amino acid types in cryo-EM density maps. By incorporating an effective main-chain tracing approach called the Vehicle Routing Problem Solver (Psaraftis, 1988), DeepMainmast can also model multichain protein complexes accurately. DeepMainmast outperforms existing methods, including highly accurate protein structure prediction methods like AlphaFold2 (AF2) (Jumper et al., 2021).

Cryo-READ (Cryo-EM DNA/RNA *de novo* atomic structure modeling) is the first deep learning–based method for fully automated *de novo* DNA/RNA structure modeling. It has a similar framework to that of DeepMainmast: It employs a deep neural network to identify potential positions of phosphates, sugars, and nucleobases in cryo-EM maps. Cryo-READ then connects sugar positions to form the backbone structure and maps the nucleic acid sequence along the backbone by considering predicted base types. This automated approach significantly improves the efficiency of DNA/RNA structure determination from medium-resolution cryo-EM maps.

DOI: 10.1201/9781003326106-11

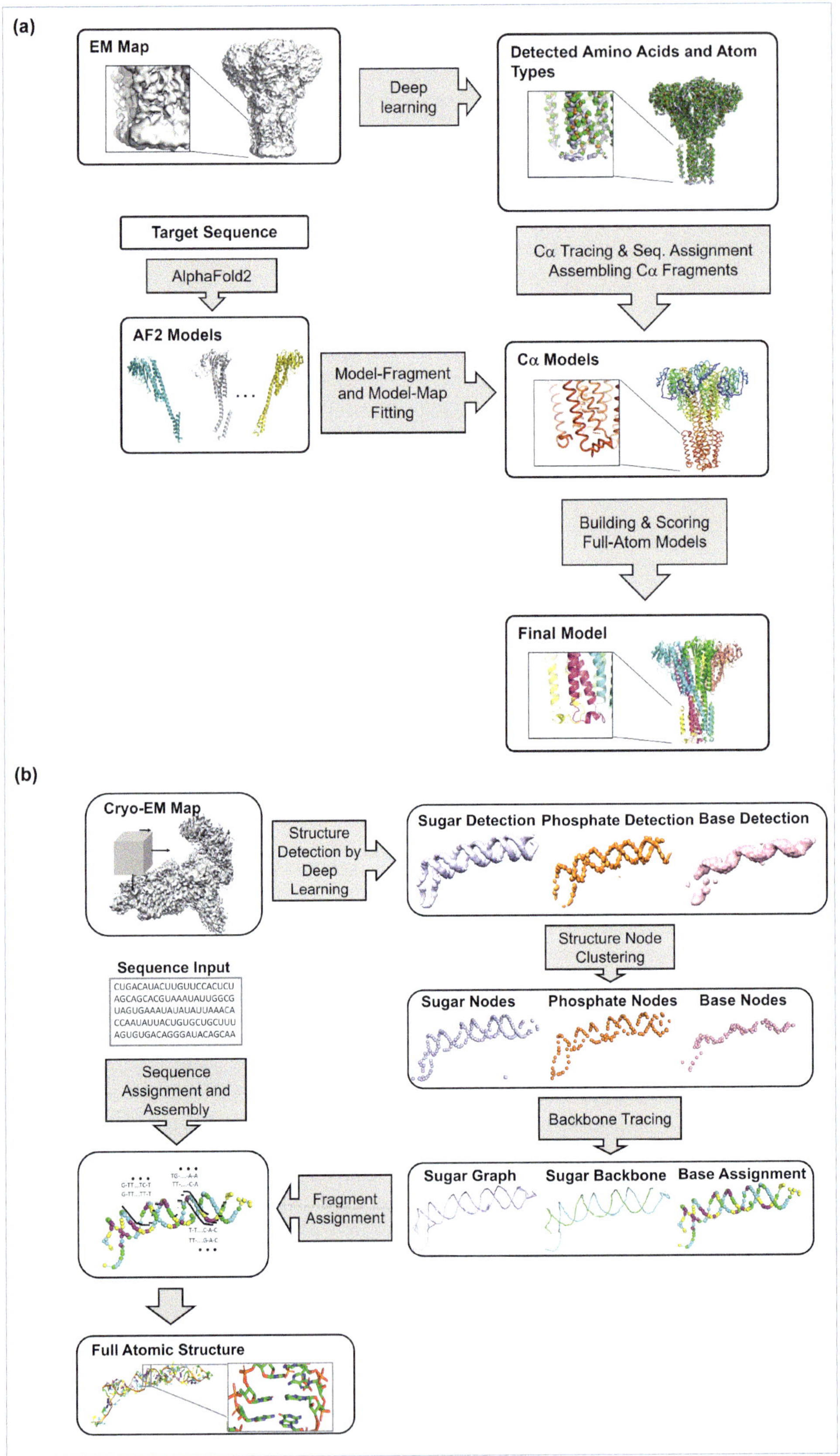

FIGURE 9.1 Overview of the structure modeling software. (a) DeepMainmast. (b) CryoREAD.

DeepMainmast

DeepMainmast Algorithm Overview

DeepMainmast is a protein main-chain modeling protocol that aims to construct accurate protein structure models from cryo-EM density maps of a resolution up to around 5 Å. Beyond 5 Å, automated modeling is difficult because the deep learning used in the modeling pipeline cannot reliably detect structural features, including atom positions and amino acid types, which are needed for modeling. DeepMainmast has two components, a workflow of tracing density, called DeepMainmast(base), and the incorporation of AF2 models.

The DeepMainmast(base) protocol starts by using a deep learning–based method called Emap2sf to detect amino acids and atoms in the given EM map. Emap2sf deploys a U-net architecture (Huang et al., 2020) to compute probabilities for 20 amino acid types and six atom types (N, Cα, C, O, Cb, Others) for each grid point in the density map. Grid points with high probabilities for Cα are clustered using the mean shift algorithm (Carreira-Perpinan, 2006; Terashi & Kihara, 2018), generating representative points named Local Dense Points (LDPs). These LDPs are then connected to form Cα paths using a Vehicle Routing Problem (VRP) Solver (Dantzig & Ramser, 1959), which explores optimal routes to connect LDPs with minimum total costs under specified constraints. The cost between two LDPs is based on the distance and the lowest probability of main-chain atoms along the path. The closer the distance between two connected nodes to the ideal Cα-Cα distance (3.8Å), the more likely the path is accurate. If Emap2sf does not predict a path between two nodes, then the two nodes likely do not have a connection between them.

Next, Cα paths are aligned with the target sequence using the Smith-Waterman Dynamic Programming (DP) algorithm. We used the DAQ score (Terashi et al., 2022) computed from the Emap2sf output to define the matching score between a Cα position in the path to each amino acid type. The DAQ score computes the probability of an amino acid's presence at a specific location within an EM map from local density patterns. Typically, 1000 to 50,000 Cα path-sequence alignments, which we refer to as Cα fragments, are generated for a single-chain protein.

DeepMainmast integrates a predicted protein structure by AF2. Although AF2 shows higher accuracy than existing methods, it is not always accurate. Particularly, it is often the case where AF2 predicts individual domains correctly but places them in a wrong relative orientation. Therefore, a model from AF2 is integrated only when the model has consistent local regions with the Cα fragments traced from the EM map. The advantage of using an AF2 model is that it can fill missing structures in the Cα fragments due to local low map resolution, such as loops and tails in the structure.

Up to this step, we have libraries of Cα fragments, which are traced from the density, and fragments that come from the AF2 model. We combine these two libraries and identify the best combinations of fragments for building the whole protein structure (or structure complex). This process is computed with a Constraint Programming (CP) Solver since it can be considered an optimization problem under a set of constraints. The posed constraints are to assure that combined fragments generate physically possible, protein-like conformations.

Finally, the Cα models are then subjected to full-atom structure building using PULCHRA (Rotkiewicz & Skolnick, 2008), and missing structure regions, if any, are filled and refined with the Rosetta-CM (Song et al., 2013) protocol. Finally, full-atom models are evaluated and ranked based on DAQ (Terashi et al., 2022) and the DOT map-structure fitting score (Han et al., 2021) to select the best models.

DeepMainmast showed higher accuracy in both single-chain protein structures and protein complexes than existing methods, including predicted structures by AF2. See the original paper (Terashi et al., 2023) for more details on the benchmark results.

Availability

DeepMainmast is available on four different platforms: Using the Google Colab notebook website, on a web server, using the CodeOcean capsule, or downloading the code and running it locally on a user's computer. Here is the list of URLs for the four releases:

- Google Colab Notebook
  - Modeling single protein chain: https://bit.ly/dmmsinglechain
  - Modeling multichain complex: https://bit.ly/dmmmultichain
- Web server: https://em.kiharalab.org/algorithm/DeepMainMast
- CodeOcean Capsule: https://codeocean.com/capsule/4294191
- Source code at GitHub: https://github.com/kiharalab/DeepMainMast

Using DeepMainmast on Google Colab

Google Colab is a cloud-based platform provided by Google that allows users to run code and execute computations using GPU (graphics processing unit) or TPU (tensor processing unit) resources. In Google Colab, you can create and run .ipynb, which are interactive documents that contain both code and rich-text elements. These notebooks are stored on your Google Drive, making it easy to share and collaborate with others. Google Colab offers both a free and a paid option. The free option provides access to a limited number of resources, including a GPU, for a restricted amount of time. The paid option, known as Google Colab Pro, offers more powerful hardware, longer session times, and priority access to resources.

We implemented two versions of DeepMainmast, one for modeling a single-chain protein and the other for multichain protein complexes. Both versions have their own notebooks and can be accessed from the links noted above. We use the single-chain notebook and the F420-reducing [NiFe] hydrogenase (EMD-2513, determined at 3.36 Å) for the demonstration of this section. The map of EMD-2513 can be downloaded from the EMDB entry page (https://www.ebi.ac.uk/emdb/EMD-2513). For modeling proteins for EMD-2513, we also need the protein sequence file, which can be obtained from the corresponding PDB entry page (PDB ID: 4ci0, https://www.rcsb.org/structure/4ci0. Click the Download Files button and choose "FASTA Sequence"). Figure 9.2a and b indicates the corresponding places of steps explained below.

1. Visit the Google Colab notebook. Here we use the single-chain notebook shown above.
2. On the notebook, first we select and connect GPUs. To connect to a GPU machine in the Google Colab notebook, click the Connect button on the top right corner. GPU availability in Google Colab

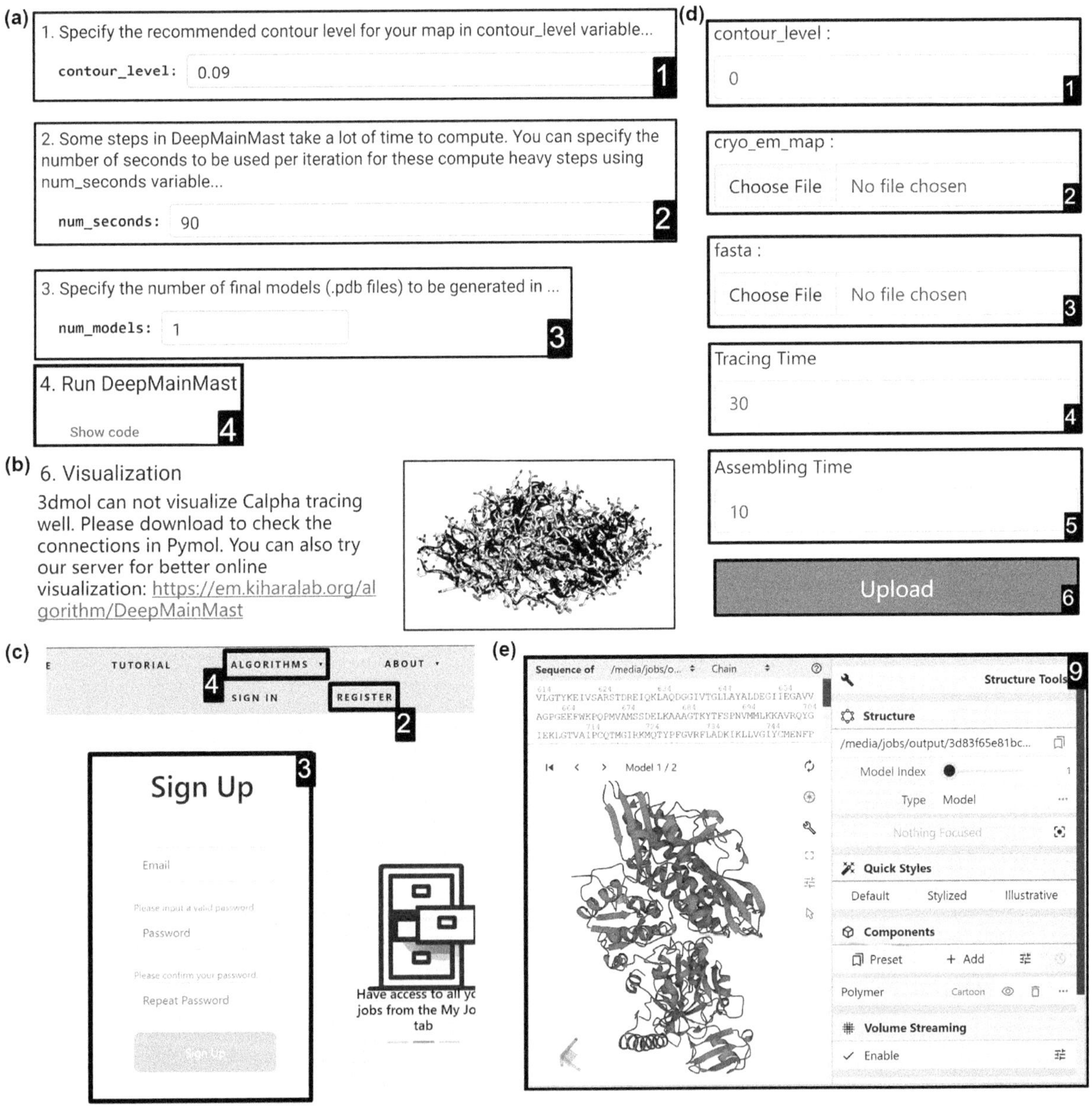

FIGURE 9.2 Running DeepMainmast on Google Colab and the web server. (a) Configuring parameters and Execute DeepMainmast sections of the Google Colab page. (b) Visualization of the model on the Google Colab page. (c) The sign-up page for the web server. (d) The DeepMainmast web server job submission form. (e) The DeepMainmast web server results page for the input map of EMD-2513.

is subject to demand, so users may need to wait for access or upgrade to the pro version. For this example, any GPU is sufficient.

3. Subsequently, we run the Cell button next to each section to execute the step.
   a. Install dependencies. This is to install libraries (components running DeepMainmast code).
   b. Input files. Upload the input cryo-EM map file (.map and or .mrc), the protein sequence file (.fasta), and the predicted structures by AF2 (.pdb). Uploading an AF2 model would increase the accuracy, but its use remains optional if you don't have it. In order to get an AF2 model, you need to run Colabfold separately at the Google Colab website, https://colab.research.google.com/github/sokrypton/ColabFold/blob/main/AlphaFold2.ipynb.
   c. Configure parameters.
      i. Contour Level: The contour level is applied to the map to determine the map region to include in the computation. For our example of EMD-2513, we use a contour level of 0.09, which can be found by locating the map in the EMDB database (https://www.ebi.ac.uk/emdb/EMD-2513).
      ii. Number of seconds limit: Tracing and assembling steps in DeepMainmast is the most expensive part of the computation. You can specify the number of seconds to be used per iteration for these compute-heavy steps. If the value is too low for your map, DeepMainmast may not work correctly. The recommended number of seconds for new maps is 90 seconds, however, please increase this number if DeepMainmast gives an error. For our example, we used 90 seconds.
      iii. Number of final models: Specify the number of final models (.pdb files) to be generated in the /results/ranked folder. Increasing the number of models might make the final model more accurate, but it will also result in a longer computation time. If you are using the free version of Google Colab or are inputting a large map, please specify 1 or 2 for the model count; otherwise, you can specify more to achieve higher accuracy. For this example, we only specified 1 model.
4. Execute DeepMainmast: Execute the DeepMainmast pipeline sequentially with the specified configuration and input.
5. Visualize the model before running structure refinement by Rosetta: Visualize predicted models before the application of RosettaCM (Song et al., 2013), which refines the model structure by filling missing regions in the model (if any) and builds full-atom structures. This process can take a long time, and thus it may be worthwhile to see the structure before going to the next step.
6. Execute Rosetta: Execute RosettaCM to obtain the final full-atom models.
7. Scoring and ranking: Score the predicted protein structure models and rank them according to the DAQ and DOT scores.
8. Visualization: Visualize the final ranked structures. You can choose which structure to visualize by altering the rank field (1 being the most accurate, 2 being the second best, and so on).

Using DeepMainmast Web Server

The web server provides another user-friendly way to use DeepMainmast. It would be easier to use than Google Colab because users only need to upload input files with a couple of parameters to the web server. The web server is also faster than Google Colab because it omits the AF2 pipeline and the structure refinement and full-atom building step with RosettaCM.

Submitting a DeepMainmast Job

To submit a job (i.e., a modeling computation request), users will first need to make an account, activate it, fill in the necessary submission forms, and then submit the job (Figure 9.2c). The user will then get an email letting them know that the job has been completed.

For this demonstration, we use the same input files as we used in the previous section for Google Colab. Submitting a job on the website consists of seven steps. The details are explained below (Figure 9.2).

1. Visit the website at https://em.kiharalab.org
2. Click on the Register button located in the top-right corner of the website. Alternatively, you can use https://em.kiharalab.org/users/registerpage to directly access the sign-up page.
3. Complete the registration process by providing the required information and activating your account. You will need to verify your email address to activate the account. The email will usually arrive instantly, but it might also take up to 1–2 hours.
4. After logging in, on the navigation bar, locate and click on the DeepMainmast option.
5. You will be directed to a page where you need to input specific parameters for your job (Figure 9.2d):
   i. Contour Level: You can specify your desired contour level for the map. For our example, we use a contour level of 0.09. Read the Google Colab section above on how to find the contour level for a map.
   ii. CryoEM Map: Upload the map file (.mrc or .map files only)

iii. Fasta File: Upload the corresponding FASTA file containing the protein sequence information (.fasta or .txt files only).
iv. Tracing Time: Enter the CPU time (minutes) for main-chain tracing. Main-chain tracing involves determining the path or conformation of the protein backbone in a protein structure. This is done by identifying the positions of the Cα atoms and tracing the Cα path. Our example uses 30 minutes for the tracing time, and that is the recommended amount for multichain models.
v. Assembling Time: Enter the CPU time (minutes) for the fragment-assembly. Fragment-assembly involves assembling Cα fragments using the CP Solver and combining Cα models built under different parameter combinations using the CP Solver. Our example uses 10 minutes for the assembling time and that is the recommended amount for multichain models.
vi. Once you have filled in all the required information and uploaded the necessary files, submit the job.
vii. If you are uploading a large file, you can monitor the progress bar at the bottom of the page to track the upload progress.

6. After submission, you will be redirected to the jobs table. This table will display all the jobs you have submitted, with the most recent one appearing at the top. The status of each job will be noted next to it, and you will be notified by email once the job has been completed.
7. Locate the DeepMainmast job you just scheduled and click on the View button to access the 3D outputs generated by the algorithm. If there is no output, then the job is most likely still running or waiting for server resources.

Modeling Results Page

In this section, we will show how the .pdb files are presented (Figure 9.2e). The .pdb file can be seen in the PDBe Molstar plugin. The raw .pdb file can also be downloaded by clicking the Download Outputs button found on the top left of the Results Visualization tab. See Note 1 for information about interacting with the plugin, and see Note 2 for general information about the results page. If you have any questions, check Note 3 for information about the FAQ page. If you wish to share the job with others, you can use the social media sharing options. Alternatively, you can also just copy the link to the job and share it manually.

Code Ocean Capsule for DeepMainmast

Code Ocean is a capsule of codes, which includes the code itself and the environment for running the code, such as libraries that the code depends on. It is used to easily test and run your code online and download the entire capsule and run it locally on a user's computer. To run the code, here is a short instruction:

1. First, you need to create an account on Code Ocean using academic credentials and login into your account.
2. Then, navigate to the Code Ocean link.
   a. Single chain (https://codeocean.com/capsule/4294191)
   b. Multichain (https://codeocean.com/capsule/0749800)
3. To make a reproducible run, i.e., run the code on an example input provided by us, click on the Reproducible Run button in the top right corner. This will start running the code on our example, and the results will be generated in the results folder at the bottom right after the execution is complete.
4. To run the code on an input of your choice, first go to our capsule and click on the Edit Capsule button in the top right corner. This will make a copy of our capsule which you can edit. Follow the instructions in README about how to upload and run your input in this copied capsule by reading the readme file present in the respective Code Ocean capsules.

Case Study of Protein Modeling with DeepMainmast Web Server

Figure 9.3 is the modeling result obtained using the DeepMainmast web server. We used the A, B, and C chains of the F420-reducing [NiFe] hydrogenase map (Allegretti et al., 2014), which corresponds to three different subunits of the enzyme: F420-reducing hydrogenase, subunit alpha, F420-reducing hydrogenase, subunit gamma, and F420-reducing hydrogenase, subunit beta. The modeling was quite successful for this map. 95.6% of Cα atoms were correctly placed and 93.0% of the Cα positions were correctly modeled and had correct amino acid type assignment (Figure 9.3c). The dashed regions in the model indicate where the 2–3 amino acids were missing due to the locally low resolution of the map.

CryoREAD

CryoREAD Algorithm Overview

CryoREAD is software for nucleic acid structural modeling from a cryo-EM map of up to around 5 Å. CryoREAD uses the same framework as DeepMainmast. Given a cryo-EM map, the probability of phosphate, sugar, and base at each grid position of an input density map is computed by a deep neural network. The architecture of the network is a cascaded two-stage U-Net (Huang et al., 2020; Ronneberger et al., 2015), trained with deep supervision. During the inference stage, the network scans the input cryo-EM map with a 64^3 Å^3 box with a stride of 163 Å and produces probabilities for protein, sugar, phosphate, and bases. Proteins are also identified in the map to avoid modeling nucleic acids

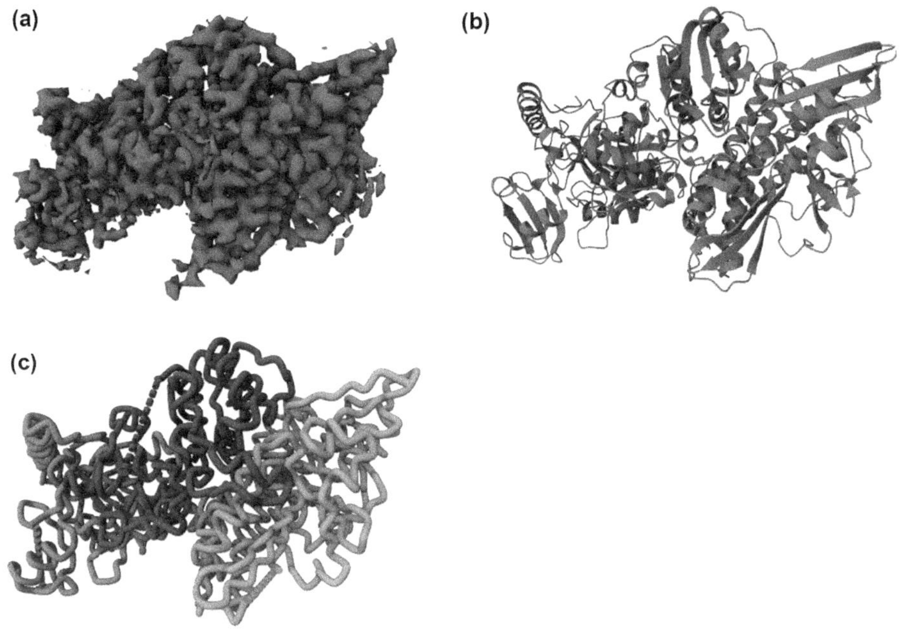

FIGURE 9.3 The output from the DeepMainmast web server for the F420-reducing [NiFe] hydrogenase map (EMD-2513). (a) The map visualized at the author-recommended contour level, 0.09. The resolution of the map is 3.36 Å. (b) The A, B, and C chains of the native structure of the protein deposited by the authors to PDB. PDB ID is 4ci0. (c) The Cα atom model built by the DeepMainmast web server.

in that map region. Identified grids for sugar, phosphate, and base are clustered to produce their representative points (LDPs), respectively. Then, sugar LDPs are connected to construct backbone paths of nucleic acids in the map using the VRP Solver.

The next step is to assign nucleic acid sequences to sugar nodes in the paths. The sequence is matched to each path according to predicted base types computed at each sugar position. Subsequently, the sequence-mapped fragments are assembled into the entire structure by the CP Solver, which maximizes the sum of the probability score of sugar nodes with the assigned bases while satisfying constraints that are required for the consistency of overlapped path segments and nucleic acid sequences.

Once the fragment assembly is complete and a backbone structure is constructed, the atomic structure is built by adding the closest phosphate nodes to sugar nodes and the closest base nodes to each sugar–phosphate pair. The structure is further refined with phenix.real_space_refine (Afonine et al., 2018) and all_atom_ refine (Emsley & Cowtan, 2004) in Coot.

CryoREAD showed higher accuracy than two existing methods, Phenix (Afonine et al., 2018) and auto-DRRAFTER (Kappel et al., 2020). See the original paper (Wang et al., 2023) for more details on the benchmark results.

Availability of CryoREAD

CryoREAD is available for use on three different platforms. Using the Google Colab notebook website, on a web server, or downloading the code and running it locally. Here is the list of URLs:

- Google Colab Notebook: https://colab.research.google.com/github/kiharalab/CryoREAD/blob/main/CryoREAD.ipynb
- Web server: https://em.kiharalab.org/algorithm/CryoREAD
- Source code at GitHub: https://github.com/kiharalab/CryoREAD

Using CryoREAD on Google Colab

The steps to use the Google Colab Notebook are shown in Figure 9.4a–c. For this demonstration, we use the EM map of the primary microRNA, in a complex with the human Drosha and DGCR8 (MP/RNA complex) (EMD-21051) (Partin et al., 2020). This map is determined at 3.7 Å. It can be downloaded from the EMDB entry page (https://www.ebi.ac.uk/emdb/EMD-21051). We also need the RNA sequence file, which can be obtained from the corresponding PDB entry page (PDB ID: 6V5B, https://www.rcsb.org/structure/6V5B).

1. Visit the Google Colab Notebook site to get started.
2. Click the Connect button on the top right corner to connect to a GPU machine. For more information, read the previous DeepMainmast section.
3. Click the Run Cell button next to each of the sections below.

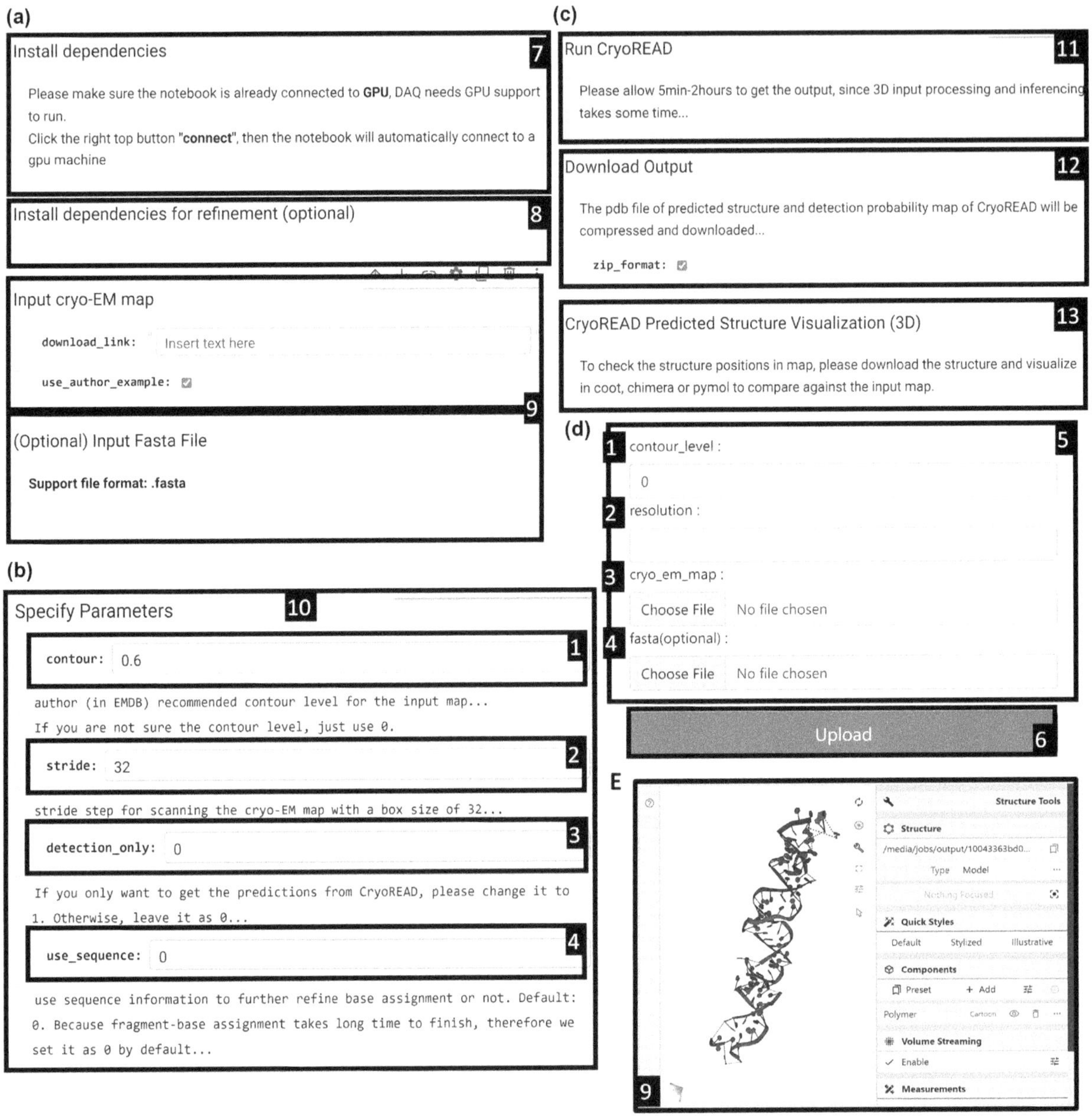

FIGURE 9.4 Running CryoREAD on Google Colab and the web server. (a) Configuring parameter inputs of the CryoREAD sections of the Google Colab page. (b) Configuration parameters of CryoREAD sections of the Google Colab page. (c) Visualization of the model on the Google Colab page. (d) The CryoREAD web server job submission form. (e) The CryoREAD web server results page for the input map of EMD-21051.

4. Install dependencies: Install libraries and download the models/scripts.
5. Install dependencies for Refinement (optional but recommended): This step is needed for running phenix.real_space_refine (Afonine et al., 2018) in Phenix and all_atom_refine (Emsley & Cowtan, 2004) in Coot to refine the structure model. For a large structure with more than 1000 nucleotides, the refinement step may take several hours and could exceed the allocated computational time. This step can be skipped but is recommended to obtain a more accurate structure.
   a. Input files: Upload the cryo-EM map file (.map and or .mrc) and the sequence file (.fasta). The sequence file is optional. If you do not upload the sequence file, CryoREAD models only the backbone structure without the base type assignment.
   b. Specify parameters:
      i. Contour Level. The contour level we use for this example is 0.6, following the

author-recommended contour level of this EM map as shown in the EMDB entry.

ii. Stride. Stride is the step size for scanning the cryo-EM map with a box size of 64^3 Å^3. Suggested values are 16, 32, or 48. Higher values can reduce the computation time but may lead to unreliable results. The stride we use in this example is 32.

iii. Detection_only. Set it to 1 if you only want to see the grid-wise predictions of sugar, phosphate, and base positions, instead of full-atom structure, otherwise, leave it at 0. We leave it at 0 for this example of EMD-21051.

iv. Use_sequence. The default is set at 0 for efficiency purposes. If you have corresponding sequences for the input map, please set it to 1 to utilize sequence information for a more accurate sequence assignment. For this example of EMD-21051, we leave it at 0 for a faster computation.

v. Resolution. The resolution of the input cryo-EM maps is required for the last refinement by Phenix. If you specify 0, the refinement step will be skipped. We use a resolution of 3.7 Å for the current example. For a large map that contains more than 1000 nucleotides, set the resolution to 0 to skip the refinement step because of the wall time limit in Colab. If you use the paid version of Colab, please use the resolution information to improve the quality of the output structure.

c. Run CryoREAD. Execute the CryoREAD pipeline sequentially with the specified configuration and input.

d. Download Output. The .pdb file of the predicted structure and detection probability map of CryoREAD will be compressed and downloaded. You can check the zip_format box if you want to download a tar.gz format file.

e. CryoREAD Predicted Structure Visualization (3D). Visualize the final .pdb structure.

Using CryoREAD Web Server

Similar to DeepMainmast, the web server provides a user-friendly way to use CryoREAD. It is easier to use than Google Colab because there is no need to download any libraries or connect to a GPU. All of the computation is done on the server side. Additionally, unlike Google Colab, there is no wall time limit, so the web server includes the refinement step and outputs the full-atom structure for all job submissions.

Submitting a Job to the CryoREAD Web Server

We use the same example here as we did in the Google Colab section above. All the parameters are the same. Users first need to register. To register, sign in and navigate to the CryoREAD job submission form. Follow the steps described in the DeepMainmast section. Once registered, follow the steps below to submit a job. You will be directed to a page where you input specific parameters for your job (Figure 9.4d):

1. Contour Level: Specify the desired contour level for the input map.
2. Resolution: Resolution of the cryo-EM map. This is required for refinement, which is the last step.
3. CryoEM Map: Upload the map file (.mrc or .map files only).
4. Fasta File (Optional): Upload the corresponding FASTA file containing the protein sequence information (.fasta or .txt files only). The nucleotide sequences should only include 'A', 'U', 'T', 'C', 'G' characters.

The proceeding steps include accessing the jobs table and viewing the results. Both of these steps are explained in the DeepMainmast section and apply to CryoREAD as well. Figure 9.4e shows the model generated by CryoREAD viewed on the web server.

Case Study of CryoREAD

Figure 9.5 shows the modeling result of the primary microRNA in complex with the human Drosha and DGCR8 from the EM map determined at 3.7 Å (EMD-21051) (Figure 9.5a). The Drosha–DGCR8 complex in complex with primary microRNA (MP/RNA complex) in the active state represents a critical step in the generation of mature miRNAs, which are essential for post-transcriptional gene regulation in human cells. The map shown here includes the entire complex, and miRNA is a part of it. That is why the overall shape is different from the shape of the microRNA. The structure of the microRNA (Figure 9.5b) is very well modeled by CryoREAD (Figure 9.5c). The accuracy for the backbone recall is 92.5% and for the sequence recall is 79.5%. Backbone recall measures the fraction of backbone atoms that are correctly modeled. Sequence recall measures the fraction of nucleotides that are correctly modeled with the correct base types.

NOTES

- *Viewing the .pdb structure:* The .pdb structure is visualized using the PDBe Molstar javascript plugin. The structure can be rotated using the left mouse button and can be translated using the right mouse button. Clicking a part of the structure using the right mouse button will center and zoom in to the part that you clicked. The structure can be viewed in full screen by clicking the full-screen

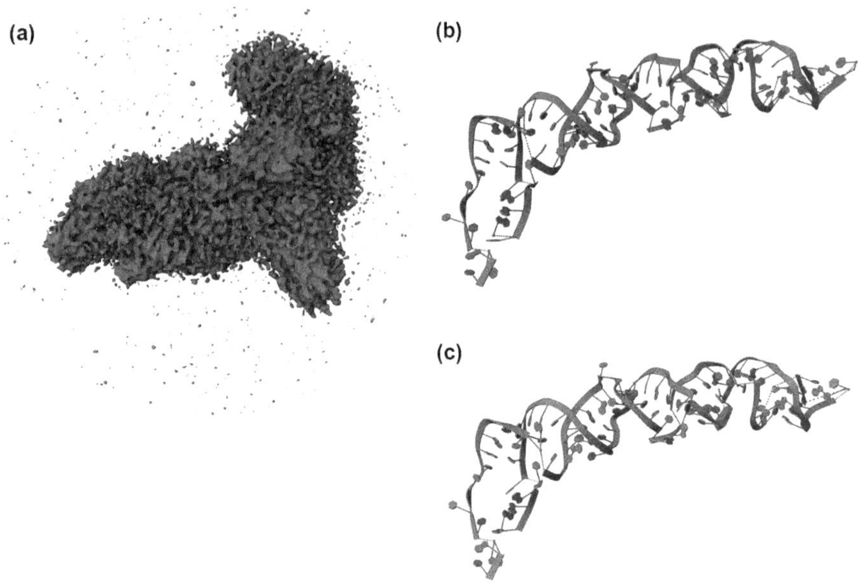

FIGURE 9.5 The output from the CryoREAD web server for the map of the primary microRNA in complex with the human Drosha and DGCR8 (EMD-21051). (a) The map visualized at the author-recommended contour level, 0.6. The resolution of the map is 3.7 Å. (b) The A, B, and C chains of the native structure of the protein deposited by the authors to PDB. PDB ID is 6v5b. (c) The model structure output from the CryoREAD web server.

button on the top right section of the plugin. You can also take a screenshot by clicking the screenshot button located in the same area.

- The results page contains three tabs: Results Visualization, Output Logs, and Job Configuration. The Results Visualization contains the visualized pdb structures, which you can interact with using the Molstar plugin. The Output Logs contains the log file that can be used to monitor the progress of the respective algorithm. Finally, the Job Configuration tab contains the configuration information you supplied when scheduling the job.
- If you have any questions, check out the FAQ page in the About drop-down menu. Alternatively, you can use https://em.kiharalab.org/faqq to directly access the FAQ page. If you don't see your question listed, feel free to submit it.

SUMMARY

In this chapter, we described how to use DeepMainmast for protein modeling and CryoREAD for DNA/RNA structure modeling for cryo-EM maps in a couple of different platforms. We focused on how to use the tools because this is important for structural biology researchers. Readers are referred to the original papers (Terashi et al., 2023; Wang et al., 2023) for details of the algorithms and benchmark results. Biomolecular structure determination by cryo-EM involves the use of many computational tools, and structure modeling is one of the steps where researchers would spend a substantial amount of time. We hope the two tools we introduced here will be a useful assist to structural biologists.

ACKNOWLEDGMENTS

This work was partly supported by the National Institutes of Health (R01GM133840 and 3R01GM133840-02S1) and the National Science Foundation (CMMI1825941, MCB1925643, IIS2211598, DMS2151678, DBI2146026, and DBI2003635).

REFERENCES

Afonine, P. V., Poon, B. K., Read, R. J., Sobolev, O. V., Terwilliger, T. C., Urzhumtsev, A., & Adams, P. D. (2018). Real-space refinement in PHENIX for cryo-EM and crystallography. *Acta Crystallographica Section D: Structural Biology*, *74*(6), 6. https://doi.org/10.1107/S2059798318006551

Allegretti, M., Mills, D. J., McMullan, G., Kühlbrandt, W., & Vonck, J. (2014). Atomic model of the F420-reducing [NiFe] hydrogenase by electron cryo-microscopy using a direct electron detector. *Elife*, *3*, e01963.

Alnabati, E., & Kihara, D. (2019). Advances in structure modeling methods for cryo-electron microscopy maps. *Molecules*, *25*(1), 82.

Berman, H. M., Battistuz, T., Bhat, T. N., Bluhm, W. F., Bourne, P. E., Burkhardt, K., Feng, Z., Gilliland, G. L., Iype, L., & Jain, S. (2002). The protein data bank. *Acta Crystallographica Section D: Biological Crystallography*, *58*(6), 899–907.

Carreira-Perpinan, M. A. (2006). Acceleration strategies for Gaussian mean-shift image segmentation. *Proceedings of the 2006 IEEE Computer Society Conference on Computer Vision and Pattern Recognition*, *1*, 1160–1167. https://doi.org/10.1109/CVPR.2006.44

Churkin, A., Retwitzer, M. D., Reinharz, V., Ponty, Y., Waldispühl, J., & Barash, D. (2018). Design of RNAs: Comparing programs for inverse RNA folding. *Briefings in Bioinformatics*, *19*(2), 350–358.

Dantzig, G. B., & Ramser, J. H. (1959). The truck dispatching problem. *Management Science*, *6*(1), 80–91.

Emsley, P., & Cowtan, K. (2004). Coot: Model-building tools for molecular graphics. *Acta Crystallographica Section D: Biological Crystallography*, *60*(12), 2126–2132.

Han, X., Terashi, G., Christoffer, C., Chen, S., & Kihara, D. (2021). VESPER: Global and local cryo-EM map alignment using local density vectors. *Nature Communications*, *12*(1), 2090.

He, J., & Huang, S.-Y. (2021). Full-length de novo protein structure determination from cryo-EM maps using deep learning. *Bioinformatics*, *37*(20), 3480–3490.

Hryc, C. F., & Baker, M. L. (2022). Beyond the backbone: The next generation of pathwalking utilities for model building in CryoEM density maps. *Biomolecules*, *12*(6), 773.

Huang, H., Lin, L., Tong, R., Hu, H., Zhang, Q., Iwamoto, Y., Han, X., Chen, Y.-W., & Wu, J. (2020). UNet 3+: A Full-Scale Connected UNet for Medical Image Segmentation. *ICASSP 2020 - 2020 IEEE International Conference on Acoustics, Speech and Signal Processing (ICASSP)*, 1055–1059. https://doi.org/10.1109/ICASSP40776.2020.9053405

Huang, P.-S., Boyken, S. E., & Baker, D. (2016). The coming of age of de novo protein design. *Nature*, *537*(7620), 320–327.

Jumper, J., Evans, R., Pritzel, A., Green, T., Figurnov, M., Ronneberger, O., Tunyasuvunakool, K., Bates, R., Žídek, A., & Potapenko, A. (2021). Highly accurate protein structure prediction with AlphaFold. *Nature*, *596*(7873), 583–589.

Kappel, K., Zhang, K., Su, Z., Watkins, A. M., Kladwang, W., Li, S., Pintilie, G., Topkar, V. V., Rangan, R., & Zheludev, I. N. (2020). Accelerated cryo-EM-guided determination of three-dimensional RNA-only structures. *Nature Methods*, *17*(7), 699–707.

Keating, K. S., & Pyle, A. M. (2012). RCrane: Semi-automated RNA model building. *Acta Crystallographica Section D: Biological Crystallography*, *68*(8), 985–995.

Nakamura, T., Wang, X., Terashi, G., & Kihara, D. (2023). DAQ-score database: Assessment of map–model compatibility for protein structure models from cryo-EM maps. *Nature Methods*, *20*(6), 775–776. https://doi.org/10.1038/s41592-023-01876-1

Partin, A. C., Zhang, K., Jeong, B.-C., Herrell, E., Li, S., Chiu, W., & Nam, Y. (2020). Cryo-EM Structures of Human Drosha and DGCR8 in Complex with Primary MicroRNA. *Molecular Cell*, *78*(3), 411–422.e4. https://doi.org/10.1016/j.molcel.2020.02.016

Pfab, J., Phan, N. M., & Si, D. (2021). DeepTracer for fast de novo cryo-EM protein structure modeling and special studies on CoV-related complexes. *Proceedings of the National Academy of Sciences*, *118*(2), e2017525118.

Psaraftis, H. N. (1988). Dynamic Vehicle Routing Problems. In *Vehicle Routing: Methods and Studies* (pp. 223–248). North-Holland.

Ronneberger, O., Fischer, P., & Brox, T. (2015). U-Net: Convolutional Networks for Biomedical Image Segmentation. In N. Navab, J. Hornegger, W. M. Wells, & A. F. Frangi (Eds.), *Medical Image Computing and Computer-Assisted Intervention – MICCAI 2015* (pp. 234–241). Springer International Publishing. https://doi.org/10.1007/978-3-319-24574-4_28

Rotkiewicz, P., & Skolnick, J. (2008). Fast procedure for reconstruction of full-atom protein models from reduced representations. *Journal of Computational Chemistry*, *29*(9), 1460–1465.

Song, Y., DiMaio, F., Wang, R. Y.-R., Kim, D., Miles, C., Brunette, T., Thompson, J., & Baker, D. (2013). High-resolution comparative modeling with RosettaCM. *Structure*, *21*(10), 1735–1742. https://doi.org/10.1016/j.str.2013.08.005

Terashi, G., Kagaya, Y., & Kihara, D. (2020). MAINMASTseg: Automated map segmentation method for Cryo-EM density maps with symmetry. *Journal of Chemical Information and Modeling*, *60*(5), 2634–2643. https://doi.org/10.1021/acs.jcim.9b01110

Terashi, G., & Kihara, D. (2018). De novo main-chain modeling for EM maps using MAINMAST. *Nature Communications*, *9*(1), 1. https://doi.org/10.1038/s41467-018-04053-7

Terashi, G., Wang, X., Maddhuri Venkata Subramaniya, S. R., Tesmer, J. J., & Kihara, D. (2022). Residue-wise local quality estimation for protein models from cryo-EM maps. *Nature Methods*, *19*(9), 1116–1125.

Terashi, G., Wang, X., Prasad, D. et al. DeepMainmast: integrated protocol of protein structure modeling for cryo-EM with deep learning and structure prediction. Nat Methods 21, 122–131 (2024).

Wang, R. Y.-R., Song, Y., Barad, B. A., Cheng, Y., Fraser, J. S., & DiMaio, F. (2016). Automated structure refinement of macromolecular assemblies from cryo-EM maps using Rosetta. *ELife*, *5*, e17219. https://doi.org/10.7554/eLife.17219

Wang, X., Terashi, G., & Kihara, D. (2023). CryoREAD: De novo structure modeling for nucleic acids in cryo-EM maps using deep learning. *Nature Methods*, *20*, 1739–1747.

Watkins, A. M., Rangan, R., & Das, R. (2020). FARFAR2: Improved de novo Rosetta prediction of complex global RNA folds. *Structure*, *28*(8), 963–976. e6.

Zhang, X., Zhang, B., Freddolino, P. L., & Zhang, Y. (2022). CR-I-TASSER: Assemble protein structures from cryo-EM density maps using deep convolutional neural networks. *Nature Methods*, *19*(2), 195–204.

10 Artificial Intelligence-Driven Data Acquisition for Single-Particle Cryo-EM and Three-Dimensional Electron Diffraction

Koji Yonekura

We have introduced AI-driven technologies for data acquisition in single-particle cryo-EM and 3D ED and developed the software named yoneoLocr (You Only Navigate EM Once to Locate in Real-time). It performs well in facilitating data collection for both applications. Its performance is remarkably high in locating the field of view for acquiring molecular images, while further improvements may be possible for diffraction measurements. This chapter serves as an introduction to our AI-driven advancements.

INTRODUCTION

Single-particle cryo-electron microscopy (cryo-EM) eliminates the need for crystallizing biological macromolecules, which was previously essential in X-ray crystallography. Instead, it involves freezing protein solutions in a condition close to a physiological ion environment. It has been widely used for elucidating detailed three-dimensional (3D) structures of relatively larger protein molecules with molecular weights over 100 kDa.

Three-dimensional electron diffraction (3D ED) or electron 3D crystallography exploits scattering power that is four to five orders of magnitude stronger than X-rays, enabling the determination of atomic-resolution structures from tiny crystals where X-ray diffraction is not feasible. This technique can be applied to a diverse range of samples, from small proteins and polypeptides to various low- to medium-sized molecular compounds, including drug candidates and molecular materials.

Both methods require extensive data collection to achieve high-quality structural analysis. Particularly for single-particle cryo-EM, large amounts of image data are needed, making automated data collection essential. However, limitations in the conventional control of electron microscopes and their mechanical precision have led to substantial human involvement. We have developed AI-driven technologies for data acquisition in single-particle cryo-EM and 3D ED (Yonekura et al., 2021), which we will introduce in this chapter.

SINGLE-PARTICLE CRYO-EM

The sample support for the transmission electron microscope (TEM) consists of a circular metal mesh with a diameter of 3 mm. Typically, a mesh with grid sizes ranging from several tens to around 100 µm, containing 200–400 grid squares, is used. A carbon support film with circular holes of approximately 1–2 µm in diameter is attached to the mesh. Protein solution is then applied to the support film, and excess solution is removed using filter paper. Rapid freezing results in the formation of a thin layer of amorphous ice in the holes of the support film (the uppermost and the second rows in Figure 10.1). This frozen sample is transferred into the vacuum of the TEM column while maintaining low temperature, and two-dimensional projection images of molecules are captured under liquid nitrogen cooling. The standard procedure for single-particle analysis in cryo-EM involves reconstructing the three-dimensional structure from the obtained huge numbers of molecular images (Figure 10.1).

The advancements have continued even after the Nobel Prize in chemistry was awarded in 2017 to this method. In 2018 we introduced the first cryo-electron microscope equipped with a cold field emission (CFE) gun (JEOL CRYO ARM 300) that generates a high-coherence electron beam (Hamaguchi et al., 2019; Maki-Yonekura et al., 2021). Our research highlighted the crucial role of coherence in the electron beam for achieving high-resolution single-particle analysis (Hamaguchi et al., 2019). Following this demonstration, in 2020, 1.25 Å and 1.22 Å resolution structures of apoferritin were reported utilizing the coherence beams (Nakane et al., 2020; Yip et al., 2020). More recently, we showed the visualization of most hydrogen atoms in proteins, along with the precision to measure charges and distinguish types of chemical bonds (Maki-Yonekura et al., 2023).

In conventional automated imaging, the selection of a field of view is conducted by calculating a correlation between the acquired image of each carbon hole (the second row in Figure 10.1) and a pre-prepared reference image

DOI: 10.1201/9781003326106-12

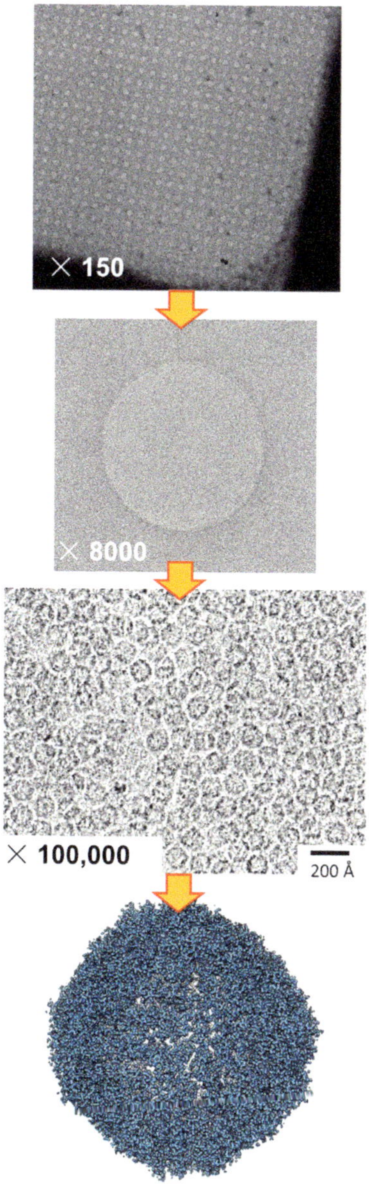

FIGURE 10.1 Flow diagram for single-particle cryo-EM. The sample stage is moved to one of the carbon holes on a low-magnification (×150) image, and the field of view is adjusted based on the hole image taken at a magnification of ×8000. Then, a high-magnification (×100,000) image is captured for image processing. A reconstruction of apoferritin calculated from many collected images is shown at the bottom (Maki-Yonekura et al., 2023).

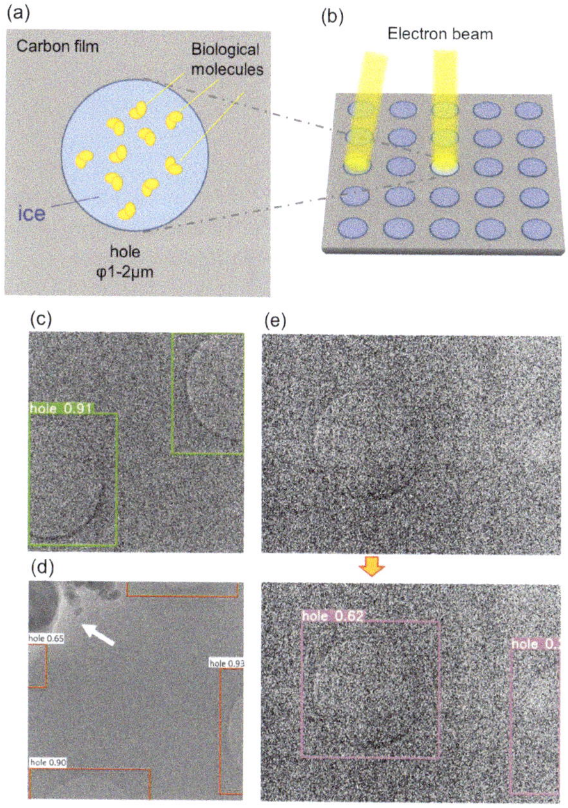

FIGURE 10.2 Illustration of ice-embedded samples for single-particle cryo-EM and AI-based detection of holes in a carbon support film (Yonekura et al., 2021). (a) Biological macromolecules are embedded in thin amorphous ice inside carbon holes. (b) After mechanical adjustment of the sample stage, the beam is deflected to capture images from neighboring holes (in this example, a 5×5 array) around the center hole. (c) Two holes are successfully detected. The larger hole is chosen for stage positioning. (d) A challenging example showing that contaminated ice crystals and debris (white arrows) are present and only small fractions of multiple holes are visible within the field of view. (e) Low-contrast hole image due to thick ice. In (d) and (e), conventional correlation-based alignment fails, whereas AI accurately detects carbon holes even from such images. The numbers in the figure indicate the confidence of identification (1.0 corresponds to 100%).

of an ideal hole. The resulting shifts are then applied to adjust the position of the TEM sample stage. However, the signal-to-noise ratio of the cryo-EM image is low, and factors such as the presence of thick ice, ice crystals, debris, or the inclusion of small portions of multiple holes in the field further exacerbate the situation (Figure 10.2d and e). In such cases, broad peaks or multiple false peaks may appear in the correlation coefficient, making it extremely challenging to obtain the correct solution.

In the widely adopted high-throughput approach, multiple images are captured from a regular array of 3 × 3 or 5 × 5 holes (Figure 10.2b). Here beam shifts are employed to cover the neighboring holes around the center hole once the sample stage is mechanically adjusted to the center hole. Additionally, a small electron beam can be used to capture images from several regions of each hole. These strategies help reduce the mechanical movement of the stage, minimize stage drifts after movement, and facilitate high-speed imaging. However, misalignment in the initial stage positioning can lead to severe consequences. In certain cases, the data from over 100 consecutive images sometimes become unusable.

It is clear that correlation is not well-suited for these tasks. We have developed an alternative approach using AI image recognition (Yonekura et al., 2021). To begin with, we performed annotations by delineating carbon holes with rectangles and labeling them as "hole" using GUI software. This annotation process encompassed all the holes present in 300 images, including challenging cases where correlation-based alignment was unsuccessful, in order to create a labeled dataset for training. This dataset was utilized to train the You Only Look Once (YOLO) image recognition package, a well-established tool in this field. Following this, we created a Python script to calculate the vectors required for bringing a detected hole to the center of the field of view based on the detection.

This procedure was incorporated into scripts of SerialEM (Mastronarde, 2005), an automated data collection program. Alignment failure was eliminated even in difficult cases as long as holes were visually identifiable (Figures 10.2c–e). Importantly, this approach significantly reduces the amount of electron dose required for capturing alignment images, thereby reducing sample damage. We have also confirmed that it functions similarly for images captured at different magnifications or under different conditions, indicating a high degree of versatility (Yonekura et al., 2021).

THREE-DIMENSIONAL ELECTRON DIFFRACTION

In the case of rotational acquisition of electron diffraction patterns from microcrystals (Nannenga et al., 2014; Yonekura et al., 2015; van Genderen et al., 2016), similar challenges arise. As mentioned, this technique allows for atomic-resolution structural determination from tiny crystals that are typically not amenable to other measurements. In recent years, there has been growing interest, particularly in fields such as organic synthesis, drug discovery, and material sciences. See a recent review in Yonekura et al. (2023).

For this approach, various forms of microcrystals, including those in aqueous or organic solvents and powders, are placed on the same carbon film used for single-particle cryo-EM and introduced into the microscope column. The sample stage is moved to the positions of candidate microcrystals identified in low-magnification images, and after confirming crystallinity at low dose, continuous rotation diffraction patterns are captured from only those with good diffraction patterns (Figure 10.3). Although structural determination can be achieved even with data from a small number of crystals, it is typically effective for improving data quality to merge diffraction data from dozens of crystals per sample. A typical rotational data acquisition under standard conditions takes about 1.5–2 minutes. Therefore, a standard automated procedure involves registering the positions of selected candidate crystals and performing

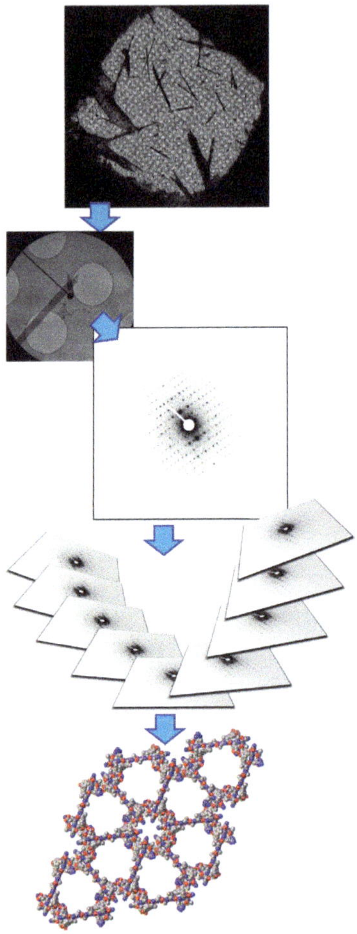

FIGURE 10.3 Flow diagram for 3D ED. A microcrystal is selected from low-magnification images, and a still diffraction pattern is checked. Electron diffraction patterns are acquired from a well-diffracting crystal while rotating the sample stage. The structure of a polypeptide related to the neurodegenerative disease ALS (Sekiyama et al., 2022) is shown at the bottom.

a batch of rotational data acquisition later (Takaba et al., 2020, 2021).

However, as mentioned in the previous section for single-particle cryo-EM, returning precisely to the registered crystal positions cannot be guaranteed. Deviations from the specified positions are common. Continuing automated capture of rotational diffraction patterns under such conditions can accumulate unusable or poor-quality data, leading to a decrease in the quality of the analysis. Adjustments of sample position before data acquisition are essential. Just as in single-particle analysis, the pre-AI step involves correlation calculations with reference images of an ideal carbon hole. With the variations in shapes and appearances of microcrystals, preparing representative reference images is not an effective strategy. Here again, introducing AI is the most suitable choice.

In this data acquisition, there are two additional labor-intensive steps: identifying the positions of microcrystals from low-magnification images and confirming crystallinity before rotation of the sample stage. For challenging samples, even experienced users may need several days to select suitable crystals for measurement and collect a sufficient number of rotational datasets. Automating these operations using machine learning was considered to be feasible for transitioning to AI (the second row of Figure 10.3).

The difficulty of AI-based identification for microcrystals is higher compared to standard carbon holes. Therefore, various types of samples, including proteins, peptides, drugs, organic semiconductors, and organic materials, were labeled in images with microcrystals for alignment purposes and subjected to similar machine learning (Yonekura et al., 2019). Ice crystals, which are often present as contaminants in cryo-EM observations, were labeled separately to ensure their detection. Simultaneously, the quality of diffraction patterns from these samples was also included in the training data. Annotations were made based on the diffraction pattern quality. Figure 10.4 illustrates the identification results and diffraction pattern evaluations obtained from the training for each crystal image (Yonekura et al., 2021).

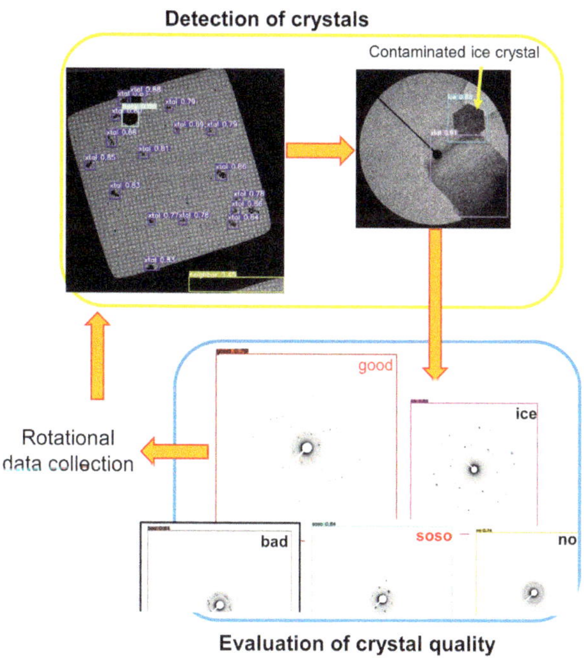

FIGURE 10.4 AI-controlled data collection of electron diffraction patterns (Yonekura et al., 2021). First, crystal positions are identified from low-magnification images. Crystals that are too thick are labeled as "thick," and those blocked by a grid bar while rotating the sample stage are labeled as "edge." Positions of crystals identified as "xtal" are automatically registered to a list. Based on these identified positions, individual crystals are aligned to the center of the field of view. Contaminated ice crystals are labeled as "ice" and can be excluded from the list. Finally, the crystallinity is evaluated from a still diffraction pattern and categorized as "good," "ice," "soso," "bad," or "no." "No" indicates the absence of detected diffraction spots. Rotational data acquisition is performed only for well-diffracting crystals.

Furthermore, the ability to identify crystals from low-magnification images was introduced. Crystals that were considered too thick for measurement or that might be blocked by grid bars during high tilting were labeled separately for exclusion (Yonekura et al., 2021). Additional machine learning was also conducted to exclude crystals located on other squares around the target square of the mesh in the center of the view.

In this way, four distinct machine-learning processes were implemented for diffraction measurements. In the new procedure, including crystal selection from low-magnification images, excluding crystals on neighboring squares, crystal alignment, crystallinity testing, and rotational acquisition of diffraction patterns from good crystals, AI can be applied to more than several hundred registered crystals. Experienced users can initiate fully automated data collection with minimal preparation time (Figure 10.4).

yoneoLocr

As described above, for the automated data collection in single-particle analysis and 3D ED, a Python script that provides instructions to the electron microscope was developed with five implemented modes: "hole" for carbon hole identification in single-particle analysis, "xtal" for crystal alignment, "diff" for diffraction pattern evaluation, "lowmagxtal" for microcrystal position identification, and "square" for mesh grid identification. Each mode demonstrated good values for evaluation metrics such as confusion matrices, precision, and recall (Yonekura et al., 2021).

The software containing these distinct modes was named yoneoLocr (You Only Navigate EM Once to Locate in Real time) and was released as open-source along with control scripts for the electron microscope (Yonekura et al., 2021). The author has also developed GUI software called ParallEM (Hamaguchi et al., 2019; Yonekura et al., 2019; Takaba et al., 2020; Maki-Yonekura et al., 2021; Takaba et al., 2021) to assist in the control of JEOL electron microscopes. For diffraction measurements, a combination of ParallEM and SerialEM was employed, incorporating various modes from yoneoLocr. Both yoneoLocr and ParallEM can be downloaded from the following websites:

yoneoLocr: https://github.com/YonekuraLab/yoneoLocr

ParallEM: https://github.com/YonekuraLab/ParallEM

CONCLUSION

The implementation of AI enables advanced machine control based on captured images without the need for dedicated algorithms. YOLO, originally designed for real-time video analysis, proves efficient in image recognition and operates at high speeds. As demonstrated in this chapter, it can respond to acquired data within 0.1 seconds or less, ensuring real-time responsiveness without causing delays in device control. It exerts minimal computational load and can run on the GPU of the camera control computer for display purposes. Being an independent program, it can be used not only with SerialEM (Mastronarde, 2005) but also with other software. The accuracy of carbon hole identification achieved with the current training data is remarkably high. On the other hand, further enhancements in accuracy are sought for diffraction measurements, and the application of generative AI for advanced improvement is under consideration.

ACKNOWLEDGEMENTS

I thank Saori Maki-Yonekura, Hisashi Naitow, Hasuku Hamaguchi, and Kiyofumi Takaba for their help in developing the AI software. This work was partly supported by JST-Mirai Program Grant Number JPMJMI20G5 (to K. Y.), and Research Support Project for Life Science and Drug Discovery (Basis for Supporting Innovative Drug Discovery and Life Science Research (BINDS)) from AMED under Grant Number JP22ama121006 (to K.Y.).

REFERENCES

Hamaguchi, T., Maki-Yonekura, S., Naitow, H., Matsuura, Y., Ishikawa, T. and Yonekura, K. (2019). A new cryo-EM system for single particle analysis. *J. Struct. Biol.*, **207**, 40–48.

Maki-Yonekura, S., Hamaguchi, T., Naitow, H., Takaba, K. and Yonekura, K. (2021). Advances in cryo-EM and ED with a cold-field emission beam and energy filtration — Refinements of the CRYO ARM 300 system in RIKEN SPring-8 center—. *Microscopy*, **70**, 232–240.

Maki-Yonekura, S., Kawakami, K., Takaba, K., Hamaguchi, T. and Yonekura, K. (2023). Measurement of charges and chemical bonding in a cryo-EM structure. *Commun. Chem.*, **6**, 98.

Mastronarde, D.N. (2005). Automated electron microscope tomography using robust prediction of specimen movements. *J. Struct. Biol.*, **152**, 36–51.

Nakane, T., Kotecha, A., Sente, A., McMullan, G., Masiulis, S., Brown, P. M. G. E., Grigoras, I. T., Malinauskaite, L., Malinauskas, T., Miehling, J., Uchański, T., Yu, L., Karia, D., Pechnikova, E. V., de Jong, E., Keizer, J., Bischoff, M., McCormack, J., Tiemeijer, P., Hardwick, S. W., Chirgadze, D. Y., Murshudov, G., Aricescu, A. R. and Scheres, S. H. W. (2020). Single-particle cryo-EM at atomic resolution. *Nature*, **587**, 152–156.

Nannenga, B. L., Shi, D., Leslie, A. G. W. and Gonen, T. (2014). High-resolution structure determination by continuous-rotation data collection in MicroED. *Nat. Methods*, **11**, 927–930.

Sekiyama, N., Takaba, K., Maki-Yonekura, S., Akagi, K., Ohtani, Y., Imamura, K., Terakawa, T., Yamashita, K., Yonekura, K., Kodama, T. S. and Tochio, H. (2022). ALS mutations in the TIA-1 prion-like domain trigger highly condensed pathogenic structures. *Proc. Natl. Acad. Sci. USA*, **119**, e2122523119.

Takaba, K., Maki-Yonekura, and Yonekura, K. (2020). Collecting large datasets of rotational electron diffraction with ParallEM and SerialEM. *J. Structl. Biol.*, **211**, 107549.

Takaba, K., Maki-Yonekura, S., Inoue, S., Hasegawa, T. and Yonekura, K. (2021). Protein and organic-molecular crystallography with 300kV electrons on a direct electron detector. *Front. Mol. Biosci.*, **7**, 612226.

van Genderen, E., Clabbers, M. T. B., Das, P. P., Stewart, A., Nederlof, I., Barentsen, K. C., Portillo, Q., Pannu, N. S., Nicolopoulos, S., Gruene, T. and Abrahams, J. P. (2016). *Ab initio* structure determination of nanocrystals of organic pharmaceutical compounds by electron diffraction at room temperature using a Timepix quantum area direct electron detector. *Acta Crystallogr. A*, **72**, 236–242.

Yip, K. M., Fischer, N., Paknia, E., Chari, A. and Stark, H. (2020). Atomic-resolution protein structure determination by cryo-EM. *Nature*, **587**, 157–161.

Yonekura, K., Ishikawa, T. and Maki-Yonekura, S. (2019). A new cryo-EM system for electron 3D crystallography by eEFD. *J. Structl. Biol.*, **206**, 243–253.

Yonekura, K., Kato, K., Ogasawara, M., Tomita, M. and Toyoshima, C. (2015). Electron crystallography of ultra-thin 3D protein crystals: Atomic model with charges. *Proc. Natl. Acad. Sci. USA*, **112**, 3368–3373.

Yonekura, K., Maki-Yonekura, S., Naitow, H., Hamaguchi, T. and Takaba, K. (2021). Machine learning-based real-time object locator/evaluator for cryo-EM data collection. *Commun. Biol.*, **4**, 1044.

Yonekura, K., Maki-Yonekura, S. and Takaba, K. (2023). Applications and limitations of electron 3D crystallography. *Structure*, 31, 1328–1334.

11 From Icy Specimens to Digital Insights

Integrating Sample Preparation, Data Analysis, and IT Infrastructure

Eric Shell, Karine Minari, and Vitor Hugo Balasco Serrão

INTRODUCTION

Cryogenic electron microscopy (cryo-EM) has significantly advanced our understanding of the molecular world and continues to contribute to breakthroughs in various scientific disciplines. This method is now established as a major tool for structural biology (Alewijnse et al., 2017; Cheng et al., 2015; Kühlbrandt, 2014; Mills, 2021; Walsh et al., 2022). Cryo-EM-related methodologies underwent key advances in the past decade. The breakthrough development of direct electron detectors (Brilot et al., 2012; Faruqi & McMullan, 2011; Li et al., 2013; Milazzo et al., 2011; Walsh et al., 2022), improvements in the optics and mechanics of the electron microscope (Mills, 2021), and the optimization of graphics processing unit (GPU)–based image-analytical processes (Grant et al., 2018; Punjani et al., 2017; Scheres, 2012; Tang et al., 2007) are the three most critical advances.

Cryo-EM can explore a diverse array of biological samples while maintaining them in conditions closely resembling their native state. Its ability to visualize biological and nanoscale structures in their native state makes it an indispensable tool for researchers worldwide. Cryo-EM can reveal atomic or near-atomic details of purified macromolecules or protein complexes through single-particle analysis (SPA) (Chari & Stark, 2023; Guaita et al., 2022). In addition, recent advances in subtomogram averaging (STA) methods of cryo-electron tomography (cryo-ET) support the visualization of cellular structures (such as organelles, membranes, and large molecular machinery) to near-atomic resolutions and within the native cellular context (Wan & Briggs, 2016; Pyle & Zanetti, 2021; Ni et al., 2022). Additionally, it is necessary to mention the improvements and advances in microelectron diffraction (microED), which also established a new realm in structural determination using a transmission electron microscope (Bu & Nannenga, 2021a, 2021b; Martynowycz et al., 2023; Mu et al., 2021).

With an increasing number of researchers turning to cryo-EM for structural analysis, a bottleneck has developed in gaining access to high-end cryo-EM instruments due to their high operational costs and the scarcity of cryo-EM expertise. In response to this challenge, governments worldwide have made significant investments in the establishment of large national cryo-EM centers (Chua et al., 2022; Walsh et al., 2022; Zimanyi et al., 2022). A number of these centers offer free access to academic users for data collection and user training. Furthermore, smaller-scale cryo-EM centers have also been established through collaborations among institutions or universities with support from various funding sources.

In-house cryo-EM facilities have the flexibility to align with specific research projects, operating on a smaller scale and fostering close ties with nearby research laboratories (Mills, 2021; Walsh et al., 2022). With a smaller user base, the staff can gain familiarity with individual projects and their developments and offer valuable suggestions and guidance to users in pursuing their scientific objectives. While most local cryo-EM centers offer standard protocols for data collection and user training, the formal training process may need to be customized to align with a particular user's strengths and weaknesses. However, there is no guideline or proper training on establishing a local cryo-EM facility. Furthermore, there are very few papers that discuss the management and operation of cryo-EM facilities, and none deeply explains the need for IT infrastructure or data management, which is required to establish a successful facility. Here, we present the requirements to establish a fully operational in-house cryo-EM facility, including performance evaluation of all tasks—particularly of the IT services. This will help researchers and managers gain the expertise needed to establish cryo-EM facilities, thereby increasing the availability of this vital technique across the globe.

WHAT SHOULD I DO BEFORE GETTING IN THE MICROSCOPE?

Cryo-EM involves immobilizing a beam-sensitive biological sample in vitreous ice, allowing for high-resolution imaging in its native state. This is achieved by cryo-protecting the samples at liquid nitrogen temperatures, where radiation damage takes longer to manifest.

The workflow for a cryo-EM-related project is similar regardless of the imaging method. The initial and critical step in this workflow is to obtain a suitable sample. The sample of interest can be a biological macromolecule (such as a protein, nucleic acid, molecular complexes, or virus) or a nanomaterial. The sample can be purified and prepared in aqueous conditions, in order to purify the specimen and obtain a homogenous and stable sample in a buffer solution. Once the sample passes biophysical and biochemical

DOI: 10.1201/9781003326106-13

quality and activity assessments, the samples are ready for grid preparation. Before preparing the cryogenically preserved samples, it is essential to test sample quality using negative-stain electron microscopy (NSEM). NSEM is a rapid, less-costly electron microscopic method that helps an experimenter rule out issues such as aggregation, disintegration, and heterogeneity. It can be conducted using low-kV instruments such as the ThermoFisher Scientific (TFS) Talos L120C or the JEOL JEM-1400 at a significantly lower cost than cryo-EM instruments (Figure 11.1a–c).

Once sample quality is deemed suitable for the continuity of the project, a vitrification session is scheduled and planned. All that is needed is a thin layer of the sample deposited onto an EM grid, typically made of carbon, gold, or graphene. This grid serves as the substrate for imaging. After the sample is placed on top of the grid, the vitrification

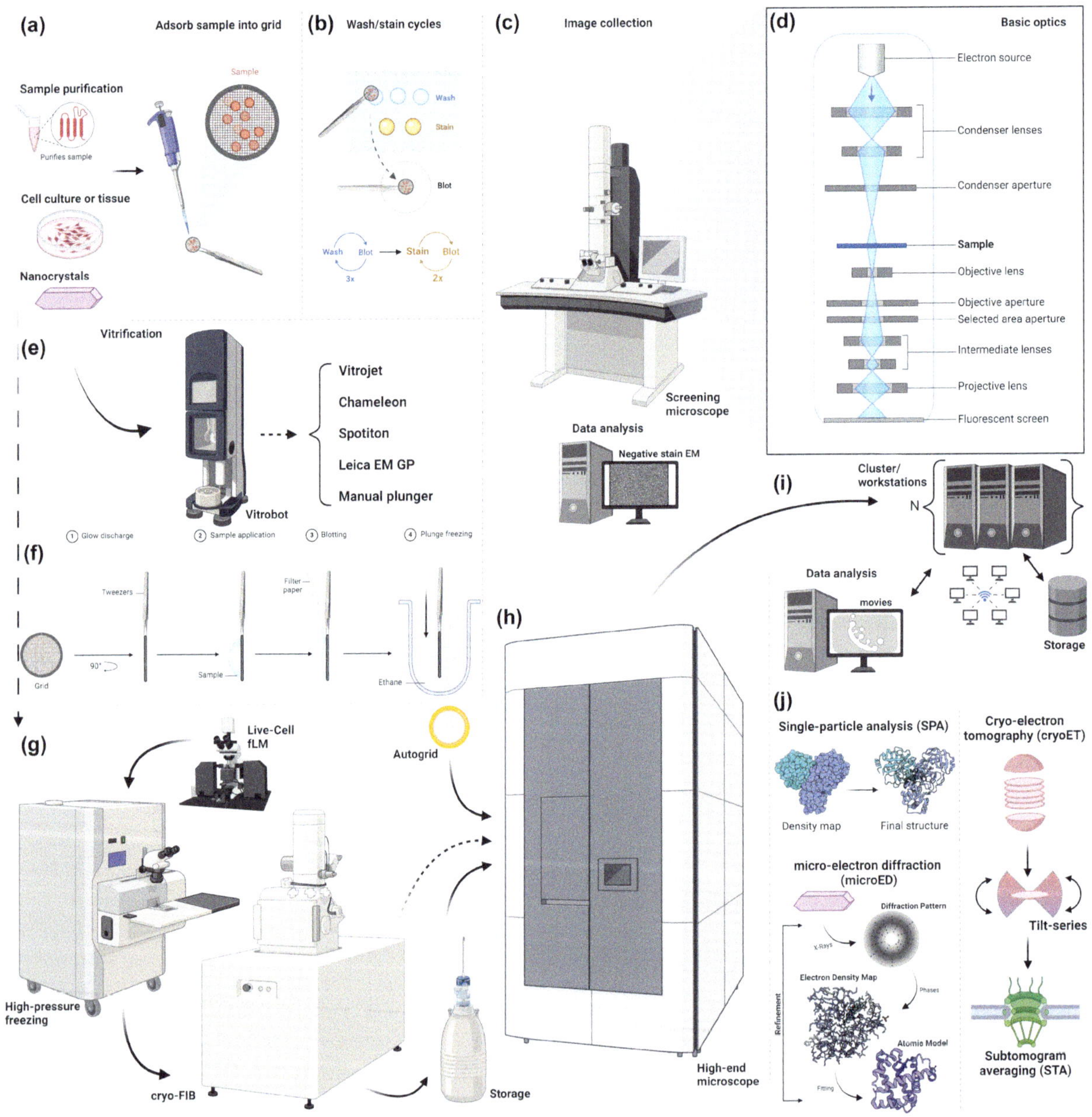

FIGURE 11.1 Cryo-EM general workflow. (a) Different samples can be used for imaging methods using a transmission electron microscope (TEM). Purified macromolecules, cells, tissues, or microcrystals can be prepared in different fashions. (b) Negatively stained sample preparation helps to quickly achieve sample quality using a (c) side-entry TEM. (d) The basic optic is very similar to modern cryo-TEM. (e) Vitrification can happen using different cryo-plungers, but the process is very similar (f). If the samples are cells, tissues, or crystals, it can use a different route (g) in order to preserve the specimen during the vitrification. (h) Once prepared, the samples can be stored long-term or imaged in high-end cryo-TEM. (i) Large datasets are *in-live* transferred to storage and (j) processed to determine the structure of interest by using SPA, STA, or microED. Created with BioRender.

happens by rapidly plunging the grid containing the specimen into liquid ethane, propane, or other mixtures at cryogenic temperatures (approximately −196°C). This process vitrifies the sample, trapping it in a thin layer of amorphous ice, which helps preserve its native structure. After sample vitrification, it can be stored in cryogenic conditions until a screening session is scheduled on the microscope.

This step is still one of the bottlenecks in the field, and numerous research groups are dedicated to enhancing this procedure, as evidenced in recent reviews by several authors (Han et al., 2022; Mills, 2021; Walsh et al., 2022; Weissenberger et al., 2021). Although cryo-EM sample preparation is frequently highlighted as one of the technique's key advantages compared to other structural approaches—a minimal sample volume and concentration requirement for protein and the absence of crystals for structure determination—it is important to not misunderstand the potential drawbacks of low protein concentrations, such as the risk of destabilizing interactions among subunits of protein complexes due to the weak binding affinities and spontaneous dissociation, which may result in aggregation or structural heterogeneity due to stochastic subunit dissociation (Chari & Stark, 2023; Guaita et al., 2022; Kastner et al., 2008; Mills, 2021; Sgro & Costa, 2018; Vénien-Bryan & Fernandes, 2023; Walsh et al., 2022; Xu & Dang, 2022).

Secondly, the process of forming amorphous ice on the grids is inherently complex. Achieving a vitreous ice film of the desired thickness with evenly distributed particles across the holes remains a significant challenge, and it is highly dependent on the specific sample being studied (Dobro et al., 2010; Dubochet et al., 1988; Dubochet & McDowall, 1981; Glaeser et al., 2016). Most current vitrification methods lack high reproducibility, and even minor adjustments to the settings can dramatically impact the results. Screening these grids is an inefficient task, as the entire grid must be meticulously searched to identify a limited number of potentially suitable areas for data collection. Several ways of performing the vitrification are currently on the market from different vendors (TFS Vitrobot Mark IV, CryoSol Vitrojet, SPT LabTech Spotition-based Chameleon, One-side blotting Leica EM GPs or homemade manual plungers) (Figure 11.1e). These pieces of equipment have a common goal but employ different methods for cryo-preserving samples, which may vary depending on the sample.

What if I Have a Crystal?

Micro electron diffraction (microED) is a method employed to gather electron diffraction data using a cryo-EM microscope. The sample preparation methodologies can vary depending on the crystal type. When dealing with protein samples, maintaining crystal hydration is of utmost importance. Consequently, a sample preparation procedure resembling that of SPA cryo-EM is used, where protein microcrystals are vitrified in liquid ethane, ensuring they are preserved in a frozen hydrated state within the microscope. Microcrystals or nanocrystals are applied onto a carbon-coated electron microscopy grid, such as Quantifoil, and excess solution is removed by blotting with filter paper. Subsequently, the crystal-loaded EM grid is promptly vitrified in liquid ethane and either introduced into the cryo-EM or stored under liquid nitrogen for future analysis. While the protocols outlined here emphasize standard sample preparation methods, it is worth noting that other techniques, including crystal fragmentation and cryo-focused ion beam (cryo-FIB) milling, can also be employed. Interested readers are encouraged to consult other sources for comprehensive details on implementing these alternative methods for sample preparation (Bu & Nannenga, 2021a, 2021b; Martynowycz et al., 2023).

What About In-Situ Analysis?

Another groundbreaking method that is rapidly developing involves the ability to observe macromolecules and organelles in the cell environment, or *in situ,* by using cryo-ET and subtomogram averaging (STA) to obtain high-resolution structures. However, the sample preparation workflow is extremely challenging. To illustrate, a typical tomogram typically covers only about 0.01% of the total volume of a cell. Another challenge lies in the physical properties of electrons, as the apparent mean free path of electrons is approximately 280 nm in a cryo-ET-suitable microscope (Guaita et al., 2022; Rice et al., 2018; Wan & Briggs, 2016). Consequently, many regions within a cell are too thick to be directly imaged.

To address these challenges, techniques such as cryo-correlative light and electron microscopy (cryo-CLEM) (Briegel et al., 2010; Guaita et al., 2022; Hylton & Swulius, 2021; Jun et al., 2019; Sartori et al., 2007; Schwartz et al., 2007; Sexton et al., 2021; Tuijtel et al., 2019; van Driel et al., 2009) and cryo-focused ion beam (FIB) milling (Lam & Villa, 2021; Marko et al., 2007; Rigort et al., 2012; Wagner et al., 2020) have been developed and can be employed in combination (Guaita et al., 2022). Cryo-CLEM integrates light microscopy and cryo-ET, allowing for the localization of cellular structures or fluorescently labeled molecules of interest (Hampton et al., 2017; Lam & Villa, 2021; Wagner et al., 2020). Cryo-FIB milling, on the other hand, utilizes a high-current gallium ion beam to create thin sections through cells, referred to as lamellae, which are typically around 80 to 300 nm thick.

The techniques of cryo-FIB have become integral in sample preparation for tomography, allowing samples, whether they are cryogenically preserved or at room temperature, to be thinned to a suitable thickness for observation and data collection in the cryo-EM operating at 200 or 300 kV. To achieve this, the cryo-FIB setup should include a cold stage, enabling the cutting of thin lamellae that are typically 150 nm or less in thickness, thus facilitating their visualization for tomography purposes. This technique plays a crucial role in visualizing intricate intracellular details when it is not feasible to isolate cellular components without causing damage (Marko et al., 2007; Rigort et al., 2012; Schaffer et al., 2017; Zachs et al., 2020). Classical cryo-FIB techniques primarily rely on the conventional plunge-vitrification method. However, the depth of vitrification achievable with this approach, where no crystalline ice forms, is only suitable for vitrification of specimens that are a few microns thick (~10 to 20 μm) (Dubochet et al., 1988). Newer approaches, such as cryo-FIB-lift-out (Lam & Villa, 2021; Mahamid et al., 2015; Parmenter et al., 2016; Rubino et al., 2012; Schaffer et al., 2019; Wagner et al., 2020) and the Waffle method (Kelley et al., 2022), take advantage of high-pressure freezing to achieve uniform vitrification (Figure 11.1g). In the *in-situ* lift-out method, the biological material is initially exposed using a cryo-ultramicrotome. This process allows for the use of cryo-fluorescence light microscopy (cryo-fLM) to identify areas of interest and facilitates the use of a gallium cryo-FIB to create a lamella. Finally, the resulting lamella is mounted on a support grid using a micromanipulator, where it can be further thinned before imaging.

For all the possible sample preparation methods, the prepared cryo grids can be safely stored in cryogenic conditions using large and stable liquid nitrogen (LN2) dewars before the imaging session. Once the time comes for your screening session, there are a few factors that should be considered. Your facility may employ either a conventional side-entry or an automatic grid loading system in its microscopes. However, the fundamental principle shared by all grid loading procedures remains consistent: the secure placement and transfer of vitrified grids into the microscope, all while preserving them at cryogenic temperatures.

To prepare the side-entry cryo-holder (Fischione 2550, Gatan 626, or Elsa cryo-holder, for example) and its associated workstation, the first step is to cool and stabilize them at cryogenic temperatures, using a temperature probe attached to the holder. The procedure requires carefully transferring a grid box to the workstation and then transferring a cryo grid from the grid box onto the tip of the cryo-holder. After securing the clip ring over the grid and covering the holder with a shutter to shield the grid from exposure to the surrounding environment, the specimen is ready to be inserted into the microscope. Grid exchange on a microscope with a side-entry holder is a slow process, and throughput is limited (typically 3–10 grids per day, depending on ambient conditions and user expertise). Consequently, only a restricted number of conditions can be screened in a typical working day. Microscopes equipped with autoloaders simplify the grid exchange process, but grids must be clipped and loaded into cartridges, which can be costly and nonreusable. The cartridge loading step also requires skill and precision to prevent potential jamming of the autoloader, which can lead to extended instrument downtime (Figure 11.1c).

On the other hand, modern cryo-transmission electron microscopes (cryo-TEMs) employ an automated loading system that provides enhanced stability compared to conventional side-entry holders. To utilize autoloading, grids need to be assembled into "auto-grid" cartridges before they are introduced into the microscope. Auto-grids are assembled within dedicated workstations by placing cryo-EM grids onto a clip ring and securely fastening them together using a C-shaped spring, often referred to as a C-clip. Subsequently, the autogrid is meticulously placed into one of the available slots within the autoloader cassette. Finally, the cassette is loaded into the TEM for further use. Therefore, it is usually recommended that grids be clipped and loaded by facility staff or a select group of trusted users. Grids arriving already clipped into cartridges should be closely inspected before loading them into the microscope. It is important to note that clipping grids into cartridges has the potential to damage the grids, which is why some centers may require users to clip their own grids to minimize liability. Autoloaders, however, do have the advantage that if a suitable grid is identified during the screening process, it is straightforward to transfer the same clipped grid to a high-end instrument for high-quality data collection. Unclipped grids can also be "rescued" from a side-entry holder, but this task typically requires a high level of skill and carries the risk of accumulating ice contamination on the grid during the transfer (Figure 11.1h).

OKAY, AND NOW WHAT?

To verify the suitability of the biological sample for cryo-EM data collection, one needs to ensure that the sample is free from issues like aggregation, disintegration, or heterogeneity. This aim is typically achieved through NSEM. A side-entry transmission electron microscope (TEM) with a lanthanum hexaboride (LaB_6) filament electron source operating at 120 kV is well-suited for this purpose. The microscope should have capabilities for low-dose data collection and be equipped with an appropriate electron detector conducive to data analysis, such as a rapid-readout CMOS detector. These microscopes are notably smaller, more cost-effective, and require less maintenance, making them less operationally demanding. Excellent examples include the TFS Talos L120C, Tecnai F20, and T12 models, or JEOL JEM-1400, 1230, or other series, which can be paired with an integrating (noncounting) 4k × 4k 16 MP CMOS detector (e.g., the Gatan Rio Camera, the TFS-Ceta, or the TemCam-XF416 from TVIPS).

For cryogenic imaging sessions, the responsibility for aligning and configuring the microscopes for each session usually lies with the facility staff. Some facility managers strongly emphasize that users should refrain from any contact with the microscope lenses, as this could necessitate realignment, which can be a time-consuming process. The setup of microscopes for optimal data collection quality can be an intricate process that involves several iterative steps. Once the microscope is correctly aligned and ready for operation, there is usually a series of meticulous steps that must be followed to ensure ideal conditions for data collection. Some facilities have adopted the use of checklists (Alewijnse et al., 2017; Nasir, 2010). While conceptually simple, these checklists are invaluable as they ensure no crucial step is overlooked and no data collection session is wasted on acquiring suboptimal data. Even if users are present during the setup, they may not have the necessary knowledge of data quality or collection parameters. These checklists can be employed for individual sessions or incorporated into daily, weekly, and monthly task schedules. Most of the setup work in the cryo-EM facility is handled by the staff, including tasks like grid loading, optimizing microscope and camera performance, configuring the conditions and parameters for image acquisition, setting up automated data collection, and monitoring the progress of data collection. The facility staff also handles the on-the-fly data transfer to users. Users' primary responsibilities involve assessing the quality of a cryo grid and selecting areas for data collection.

For high-end transmission electron microscope (TEM) sessions, typically 8 to 11 grids per eight hours of a regular screening session are loaded into the microscope and subsequently screened to identify the one most suitable for high-resolution data collection. Most facilities report that this initial screening process consumes approximately four to eight hours of microscope time. The goal is to identify a grid and select targets so that data can be automatically collected during the subsequent 16 to 36 hours. Sessions in most facilities last either 24 or 48 hours, so a significant portion of high-end microscope time is dedicated to the screening process. Once data collection begins, a throughput of a range in between 22,000-to-135,000x high-magnification movies in 24 hours is reasonable when automated data collection software is employed.

Remote operation of the microscopes enables 24/7 access to high-end instruments and is desirable for both local and remote access by facility staff. After a cryo-EM grid is loaded into the instrument by staff, two major user tasks remain for a typical high-resolution cryo-EM data collection session: identifying appropriate areas of the sample to form an imaging queue for that sample, and periodically monitoring imaging progress once the queue is submitted. Both tasks are carried out using data acquisition software packages like Leginon (Suloway et al., 2005), SerialEM (Mastronarde, 2005), and TFS/FEI Co. EPU or EPU-2. All three software packages can be run remotely over the internet. Remote monitoring can be conveniently accomplished through the web viewing interface automatically installed with the Leginon server or using other virtual desktop interfaces such as TeamViewer GmbH session using the institution VPN access, depending on the network design (Figure 11.1h–i and Figure 11.2).

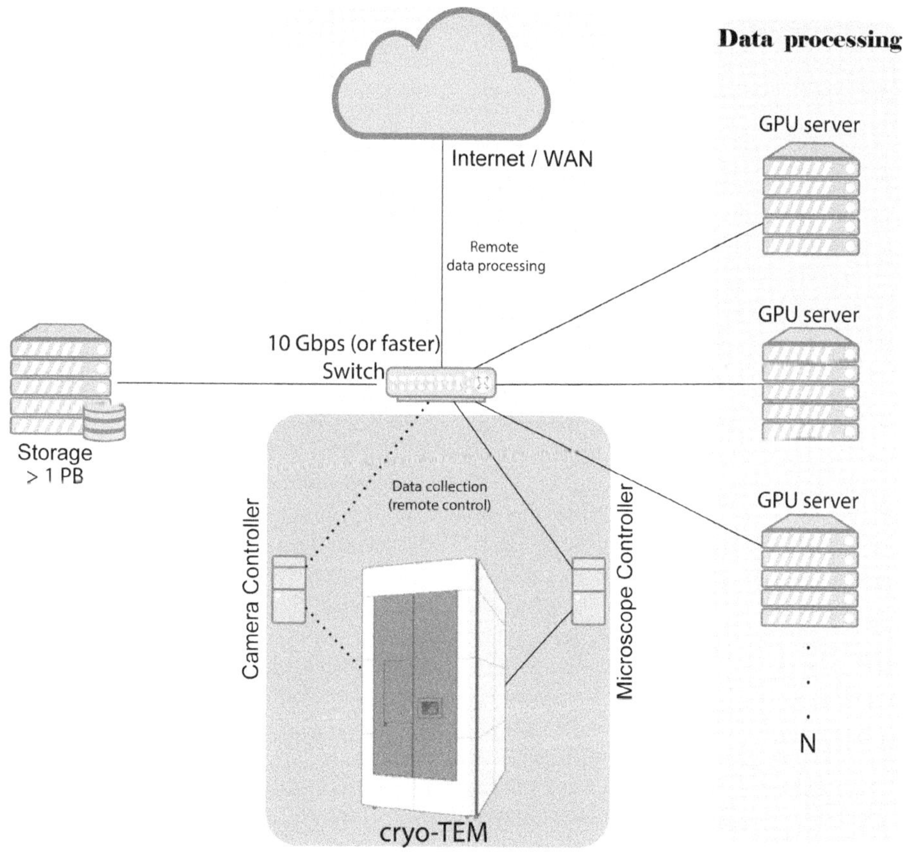

FIGURE 11.2 Minimal network configuration for a small facility. Each device involved in the microscopy data acquisition or transfer is physically connected to the local 10 Gbps (or better) network switch, including storage servers, GPU servers, and the microscope or camera controller (some configurations may have a separate camera controller). Any additional storage and data processing nodes would need to be connected to the 10 Gbps switch. Remote access to the camera and microscope controllers allows 24/7 operation and helps to mitigate issues during the data collection. Additionally, remote access to the GPU server helps several users to be connected and perform the data processing and analysis without physical restrictions, which can be beneficial for institutions or users that are not on-site.

To facilitate user participation, software like TeamViewer GmbH allows remote login to the workstation hosting data collection software. With this software, users can join the facility staff in the image acquisition session, screen for suitable grids among those loaded onto the microscope, and ultimately choose a grid for data collection queuing. During the setup of automated data collection, users can remotely select or add areas of interest, typically holes with ice within a specific thickness range, with assistance from facility staff as needed. While automated data collection is in progress, the facility staff will periodically monitor the process and assess data quality through tasks like motion correction and contrast transfer function (CTF) estimation via '*on-the-fly*' data processing (more details in section 3). Users are encouraged to remotely monitor the data collection process as well and report any potential issues or interruptions to the cryo-EM staff. For users who can provide access to their network drive or cloud storage, cryo-EM staff can upload data on the fly. Users can immediately process the data and notify the facility staff if they believe changes are needed to the imaging parameters or if the grid needs replacement.

Currently, most users opt for remote data collection sessions as they can obtain high-quality data without the need to travel to the facility. After a data collection session, some facilities provide a report summarizing collection parameters and data quality, ideally generated automatically.

Similarly, cryo-ET data collection requires careful consideration of various critical parameters to optimize both data quality and data acquisition efficiency. For cryo-ET tilt-series data collection, one should aim to strike a balance between maximizing signal and resolution while minimizing radiation damage and image blurring. For optimal tilt series data collection, certain specifications are essential. These include the use of a field emission gun (FEG) operating at either 200 or 300 kV, the presence of a suitable detector, a sample holder or stage capable of maintaining cryogenic temperatures, and the inclusion of an energy filter designed for thick samples.

Almost invariably, cryo-ET tilt series often involve thicker samples than those used in single-particle experiments. Employing 200 to 300 kV FEGs enhances the signal by allowing electrons with higher energy to penetrate thicker samples effectively. Moreover, energy filters constitute essential hardware components in the context of cryo-ET data collection. An energy filter selectively removes inelastically scattered electrons, effectively enhancing image contrast. This feature proves especially valuable for cryo-ET specimens because, in addition to the naturally thicker samples commonly investigated by cryo-ET, the tilt configuration also results in greater thickness as the beam traverses the sample layer nonperpendicularly when tilted. As the electron beam penetrates the specimen, a portion of electrons undergoes inelastic scattering due to random interactions with atoms within the sample. With increasing specimen thickness, the electron pathway also extends, resulting in a greater proportion of inelastically scattered electrons. Consequently, this accumulation of inelastically scattered electrons leads to blurry images with diminished contrast.

Additionally, the preference for direct electron detectors over other camera types is due to their significantly higher sensitivity and faster readout capacity. This advantage permits the collection of movies with multiple frames per second, as opposed to capturing a single image. Movie-based data acquisition facilitates the correction of beam-induced particle motion, resulting in sharper images.

The combination of image collection angles and their sequential order is commonly referred to as the tilt-scheme. Recent advances in data collection design allow a dose-symmetric tilt-scheme. This scheme commences with a low tilt angle and then alternates between progressively positive and negative tilts. The primary advantage of this tilt-scheme is its ability to maximize the retention of high-resolution information within the resulting tomogram, which is especially beneficial for subsequent subtomogram averaging. Additionally, this scheme may offer advantages for various other applications. The implementation of this method into data collection software such as SerialEM (Mastronarde, 2005; Ni et al., 2022) and TFS/FEI Co. Tomo5 increased the ability to visualize macromolecules within cells and improved high-resolution structure determination through subtomogram averaging.

Similar to cryo-ET, microED data collection also requires a tilt-scheme with some particularities, but it can be understood as a tilt series at the reciprocal space. Upon inserting the sample into the microscope, the sample's quality is assessed, and data is acquired from the most promising diffracting crystals on the grid. Although there are recent methods for automated data collection and new user interfaces tailored for microED data collection, several scripts can be implemented into SerialEM or while using TFS/FEI Co. EPU-D.

The data collection process commences with a grid examination to identify suitable microcrystals. These crystals should exhibit distinct shapes and an appropriate spatial relationship concerning the EM grid bars and other crystals to ensure that a broad range of data can be gathered. Once a suitable crystal is identified, an initial diffraction pattern is acquired and evaluated for quality. If the diffraction pattern exhibits high quality, a continuous rotation dataset should be collected from this crystal. This involves tilting the cryo-stage to a high-tilt angle and initiating a steady, uninterrupted/continuous rotation of the stage. High-speed detectors record diffraction data while the crystal continuously tilts within the electron beam. These diffraction datasets are subsequently subjected to standard data processing workflows by indexing, merging, scaling, and phasing similar to X-ray diffraction using XDS (Kabsch, 2010), DIALS (Winter et al., 2018), iMOS-FLM (Battye et al., 2011), or HKL2000 (Otwinowski & Minor, 1997).

COLLECTING THE DATA IS JUST THE BEGINNING!

The post–data collection processing capabilities can differ significantly among facilities and users' needs. In some instances, facilities provide users with raw movie frames on a disk, placing the responsibility for all subsequent processing on the users themselves. Conversely, other facilities boast integrated processing pipelines that carry out these processes concurrently during data collection. These steps may encompass tasks such as gain normalization of images, frame motion correction, and contrast transfer function (CTF) estimation. Consequently, users receive raw movie frame data, along with aligned average frames (both weighted and unweighted), CTF estimates, and quality assessments for each image. National facilities can offer additional processing options via the web-based Appion (Lander et al., 2009) or Scipion (de la Rosa-Trevín et al., 2016), which include particle selection, stack creation, and 2D class average analysis. Ensuring that users are well informed about the processing deliverables should be an integral part of initial expectation management agreements. Furthermore, some facilities can also provide access to computational resources that provide the entire toolset for users to conduct all of their data processing by remotely accessing it, and this is what this chapter will mainly focus on (Figure 11.1j and Figure 11.2).

For SPA projects, computational resources are now almost as vital as the microscope itself. Most facilities lack the computational capacity to accommodate large-scale refinements that could potentially consume tens of thousands of hours of computer time. The computational landscape is continuously evolving, offering users a range of software packages and hardware architectures to choose from. These options span from single workstations with GPUs to institutionally hosted, multinode clusters and even cloud computing resources (Alewijnse et al., 2017; Chari & Stark, 2023; Chua et al., 2022, Cianfrocco & Leschziner, 2015; Zimanyi et al., 2022). The same rationale can be used for cryo-ET, where real-time preprocessing can be also applied. MicroED has no real-time workflow established to date, despite its similarities to diffraction acquisition in a synchrotron light source, where the procedure is well-established (Figure 11.1j).

Microscope time is often used inefficiently because the quality and value of data collection can only be assessed long after the collection process has concluded. Moreover, despite the increasingly routine nature of microscope operation and high-resolution data collection (de la Rosa-Trevín et al., 2016; Lander et al., 2009; Mastronarde, 2005; Schoehn et al., 2023; Suloway et al., 2005), these instruments remain sensitive to various factors, ranging from grid contaminants to room temperature fluctuations, and they often encounter issues during automated collection. It is not rare to have microscope downtime.

Due to these challenges, there is substantial merit in introducing automation processes similar to those already established within the X-ray community, especially in synchrotron sources. In order to mitigate financial losses and improve the effectiveness of the facility's operations, most cryo-EM facilities have been trying to implement a preprocessing workflow for on-the-fly data processing.

The data processing can be undertaken in different ways and using different software, but there is a logical sequence of procedures to obtain the reconstructed model. The details of these various workflows and strategies are not the main focus of this chapter, but they will be briefly discussed. An ideal setup to allow real-time data processing requires a dedicated fiber backbone to connect direct detectors to dedicated buffer servers for on-the-fly computation and compression. Most of the data processing–driven facilities have already planned and installed a dedicated internet backbone that boasts 10 Gbps connectivity to the server room using fiber and copper interconnects. Once data reaches the cluster or workstations, it undergoes real-time—or on-the-fly—preprocessing, which includes particle detection, frame alignment, and data-dependent compression. This workflow can be applied to SPA and STA-cryo-ET projects, but it is still in development for microED.

Initial preprocessing consists of correcting various image artifacts, such as drift and defocus, to enhance the quality of micrographs. Most high-throughput, high-end microscopes are integrated with powerful computers that can handle motion correction and CTF determination in a matter of minutes, which guides the facility staff during the data collection by showing the quality of the dataset. After that, particle picking can be done by using brute force (blob picker) or more rational strategies (template or AI-based) to identify the particles in the micrographs. Further 2D classification helps to sort and classify particles based on their 2D projections to eliminate bad particles and enhance data quality, also showing possible issues such as lack of views or preferential orientation. The core of the workflow can rely on different software packages (CryoSPARC Live, RELION-based scripts, etc) (Fernandez-Leiro & Scheres, 2017; Punjani et al., 2017; Scheres, 2012), but it also accommodates different strategies by applying particle picking within TOPAZ (Bepler et al., 2019, 2020), cryoDRGN (Zhong et al., 2021) or crYOLO (Wagner et al., 2019), for example.

After the on-the-fly analysis, the raw and preprocessed data is available for further analysis. Usually, users can have their own dedicated workstations, but if a facility wishes to create an accessible and reliable resource for its community, it should consider providing workstations with two to four GPU cards and ample local storage, capable of holding weeks' worth of data in case of system failure. Maintaining and updating the diverse array of typical processing software packages can be a complex task, requiring facilities to have appropriate support structures in place. For facilities without their own information technology staff, the SBGrid consortium offers support to the global structural biology community by deploying and maintaining software tools. The task of maintaining software can become more straightforward with the widespread adoption of containerization packages like Docker.

For the 3D reconstruction, the bias-free *ab initio* reconstruction uses the 2D class averages to generate an initial 3D model, using methods such as sinograms—or common lines—or projection matching. After the initial angular assignment, the iterative refinement improves the alignment, projection matching, and refinement using different strategies and algorithms, in order to obtain a high-resolution 3D map.

Once the map is properly obtained, it is important to start the model interpretation and validation of the solved map. New AI methods help users create a starting model or fulfill the electron density knowing the amino acid sequence (e.g., Model Angelo) (Jamali et al., 2023). In general, an atomic or near-atomic structural model is manually or automatically built on the 3D density map. The model validation indicates the accuracy and quality of the model through techniques such as Fourier shell correlation (FSC) (Afonine et al., 2018; Neumann et al., 2018; Scheres & Chen, 2012), map-to-model fitting (FSC-Q) (Ramírez-Aportela et al., 2021), and the Q-score (Pintilie et al., 2020).

After the model is determined, the 3D reconstruction and atomic model are suitable for scientific publication and analysis. The data can be shared with the scientific community through publication in scientific journals, deposition of data in public repositories such as the Protein Data Bank (or PDB, containing the atomic model), Electron Microscopy Data Bank (or EMDB, containing the reconstructed map), and the Electron Microscopy Public Image Archive (or EMPIAR, containing the raw or preprocessed data repository).

HOW TO SAVE ALL OF THIS?

Data Management Is Critical!

Since the introduction of direct detectors with high frame rates, handling the substantial volumes of data generated by high-end microscopes has become a significant challenge. These advanced facilities typically produce 1 to 5 terabytes of raw data per day per detector, which can swiftly overwhelm storage systems. A single week of data collection from a single microscope can consume up to 21 terabytes of disk space. Compounding this issue is the need to process data concurrently with data collection while simultaneously writing it to external archive drives. These multiple simultaneous operations necessitate the use of a high-performance storage system, which can be quite costly. Given the expense of high-performance storage, efficient space management is essential.

Besides raw data, preprocessing on-the-fly, and further data processing steps, all stages of the data workflow may occur concurrently with data collection. As soon as images are captured, the frame data is transferred from the camera computer to the high-performance storage. Frames are aligned at approximately the same rate they are collected, and the raw data is automatically compressed before being transferred to the storage.

Another consideration is how to transfer the data to end users. Small datasets can be sent over the internet through methods like rsync, but this approach is often susceptible to failures. Other services, such as Globus, offer data transfer with integrity and completeness checks but require an additional layer of IT implementation and licensing fees. At present, the simplest and most robust way to provide data to users is by shipping USB hard drives. Future insights into data management are also discussed under Data Is the Facility's Most Important Part!

HOW TO GET ACCESS TO ALL THESE POSSIBILITIES?

Having access to a multiuser cryo-EM facility allows users to use highly specialized and expensive equipment, including electron microscopes and sample preparation instruments. Many research institutions and individual laboratories may not have the resources to acquire and maintain these instruments. A multiuser facility allows multiple researchers and research groups to access and benefit from these expensive resources, promoting cost-sharing and efficient use of equipment. Additionally, multiuser facilities typically employ experienced staff who can provide training and support to users, helping researchers acquire the necessary skills to conduct high-quality experiments. This ensures that the technique is accessible to the broader scientific community, which brings together researchers from diverse fields, encouraging interdisciplinary collaboration and the exploration of new research avenues. This can lead to innovative discoveries and the development of novel techniques and applications. This perspective can enable researchers who might not otherwise have access to cryo-EM to conduct experiments and advance their scientific objectives. This can lead to breakthroughs in various fields and contribute to scientific progress as a whole.

Furthermore, multiuser facilities often establish and enforce quality control standards and best practices for sample preparation and data collection. This helps maintain consistency and reliability in the data generated by different users, ensuring that the results are scientifically robust and reproducible.

From a financial perspective, by sharing the cost of equipment maintenance, upgrades, and operation among multiple users or research groups, a multiuser facility can reduce the financial burden on individual researchers or institutions. This makes cryo-EM more accessible to a broader range of scientists and research projects. National facilities are freely accessible by researchers who have their projects approved, which can substantially decrease the project cost. These facilities play an important role in field development, allowing data collection using high-end instruments and facilitating access to cutting-edge microscopes and cameras.

DO YOU NEED SOME GUIDANCE?

One of the facility staff's responsibilities is to provide users with comprehensive guidance and requirements before their initial use. For instance, users should receive information about the costs and payment procedures associated with facility usage, session preparation SOPs, necessary supplies to bring, best practices for shipping samples and grids, and a preview of what to expect during and after data collection, as well as instructions for data retrieval and processing.

Maintaining an informative and regularly updated website is instrumental in delivering this information to users. Additionally, numerous facilities request or require that users and their principal investigators (PIs) sign an agreement detailing expectations, costs, and limitations before engaging in any activities. These agreements may cover aspects like authorship, confidentiality, and acknowledgment policies.

The facility staff are responsible for developing protocols and standard operating procedures (SOPs) for user access to the lab, including training and scheduling. This may include providing training programs for users on cryo-EM techniques, instrument operation, and safety procedures. Additionally, the regular maintenance of cryo-EM equipment and the establishment of service contracts with manufacturers or service providers are key to keeping the facility performing at its maximum.

The facility staff can also foster collaborations with other research groups and institutions to share resources and expertise, including outreach and educational programs to promote the use of the cryo-EM lab within the scientific community. Staff should also obtain any necessary regulatory approvals, permits, or certifications for operating a cryo-EM lab, especially if it involves the use of hazardous or BSL2 materials, and prepare reports and documentation for institutional review boards, safety committees, and funding agencies as required.

Finally, the staff is commonly involved in developing training programs and materials for new users, by providing ongoing support and training to ensure that users can operate the equipment effectively and implement quality control procedures for data collection and processing to maintain data integrity and reproducibility.

AND, HOW MUCH WOULD IT COST?

Setting up and operating a cryo-EM facility or laboratory entails considerable expenses. The initial cost of a dedicated room for any electron microscope is substantial. Generally, the cost of a room for a 300 kV cryo-EM will be considerably higher than that for a 120 kV electron microscope. As a rule of thumb, a suitable room for a 300 kV instrument may amount to approximately 20% of the microscope's purchase price. It is worth noting that refurbishing existing rooms can be as expensive as constructing a new room from the ground up. Also, it is crucial to ensure that the rooms meet the requirements for optimal instrument performance. Different microscope manufacturers may have varying room size and specification requirements, so early consultation with the EM manufacturer is essential when renovating an existing space or constructing a new one. All companies that supply high-end cryo-EM equipment offer service contracts for their instruments. While these contracts can be costly, they are essential unless the laboratory employs a qualified and experienced cryo-EM engineer. Service contracts also ensure access to replacement components, which, when purchased separately, can lead to unexpectedly high costs for spare parts and labor. A service contract minimizes downtime because it prioritizes service calls or repairs. It is strongly advisable to negotiate a service contract when acquiring a new instrument and to ensure that the contract covers associated equipment, especially water coolers, electron detectors, and energy filters whenever feasible. It is important to note that cameras are less prone to malfunctions than electron microscopes due to their sophisticated electron optics, electronics, and precision-engineered mechanics. Some labs might prefer to replace camera parts and pay for service on a case-by-case basis rather than investing in an expensive service contract. However, repairs to electron-counting cameras can be costly and time-consuming, especially when combined with an energy filter. Having a service contract ensures the smooth operation of an EM facility and the speedy restoration of a broken system to working conditions, avoiding weeks to months of downtime and user dissatisfaction.

Furthermore, each cryo-EM-related project must consider all of the consumables beyond the sample preparation (protein purification, cell culture, media, etc). As previously described, grid preparation in room-temperature or cryogenic conditions requires a series of consumables such as grids, filter paper, LN2, ethane (or other cryogenic liquids), clips, and rings. All of these have a high associated cost, not including the equipment depreciation, operational costs, and service contracts (cryo-plunger, tweezers, clipping and transfer station, nanocabs, and the microscope itself). When taking into consideration all the associated costs, cryo-EM is one of the most expensive structural methods.

Lastly, having properly trained personnel and substantial computational power is essential for image processing. Even when external users are responsible for their own data processing, the cryo-EM facility will require powerful computers for on-the-fly processing workflow. Moreover, extensive storage capacity is needed to temporarily store the large volume of data collected. The associated cost of data storage ($/TB) and data processing capabilities ($/GPU) is extremely high. GPU-based workstations and large-TB storage cost thousands of dollars and need constant supervision and maintenance. Luckily, software licenses for academia have not been an issue to date, but companies and nonacademic users must consider this as part of the total costs. Additionally, having well-trained and capable personnel helps to accelerate the data processing and to minimize loss of storage space by misprocessing the data or creating issues with unnecessary files.

Considering a small cryo-EM facility that has ~20 users, the projects can easily require 80 to 100 TB each, therefore generating 2000 TB, or 2 PB, in three to six months. Most users and facilities do not take this into consideration, mainly because of the high associated costs of the room renovation and the microscope/camera itself. However, after the facility is operational, adequate data storage is critical to keep the facility in a healthy operational status.

WHAT DO I NEED TO START A FACILITY?

Several key items need to be considered in order to build a suitable cryo-EM facility or laboratory, as detailed by previous publications (Alewijnse et al., 2017; Chua et al., 2022; Mills, 2021; Walsh et al., 2022; Zimanyi et al., 2022). Consider sustainability measures such as energy efficiency and resource management. Plan for the future growth and expansion of the cryo-EM lab as research demands evolve. Designing a cryo-EM lab is a complex and multifaceted process that requires collaboration among researchers, architects, engineers, and safety experts. It is essential to create a well-equipped, safe, and efficient environment that can support cutting-edge cryo-EM research.

The initial budget estimation for setting up the lab includes equipment costs, facility construction or renovation, staffing, and operational costs. It is imperative and also challenging to identify a variety of potential sources of funding, including grants, institutional support, and collaborations, which may be provided by the private sector or investment portfolio. In order to create an initial budget it is important to understand the full operations and needs of the future cryo-EM facility.

The initial location assessment is critical. The selection of a suitable location within your institution begins with identifying a clean, stable, and vibration-free environment. It is also important to ensure that the lab space has adequate ventilation, temperature control, and access to emergency exits. A cryo-EM facility will host expensive state-of-the-art instruments that require 24/7 power and water cooling for both data collection and safe operation/maintenance of the instruments. Several instruments within a cryo-EM facility require more than one day for on/off power switching. In addition, such power fluctuations increase the failure risk of sensitive circuit boards within the instruments. Unplanned power outages pose an additional risk of instrument malfunction, decrease the life span of certain instruments, and risk the loss of precious samples and data collection time. To mitigate these risks, backup strategies on cooling water backup systems, additional uninterrupted power supply (UPS) devices, and building emergency generators are required to protect both expensive pieces of equipment and expensive samples. The lab design must ensure that the lab has access to reliable utilities, including electrical power, water, and compressed gases. It may also consider the installation of backup generators and UPS to prevent data loss during power outages. Another important consideration is the controller room, where the water controllers, microscope chiller, and UPS should be located. This room must have sensitive temperature control (<26°C), a backup power source, and backup water cooling capacities.

The initial planning and designing of the lab layout is another critical part of the facility's design and should include the relative positioning of equipment, workstations, and storage areas. This will allow the facility staff to allocate space for sample preparation, data collection, image processing, and storage of cryo-EM grids and samples. This may also include providing fume hoods, laminar flow cabinets, and other necessary equipment for sample handling and preparation in case the project needs high biosafety levels. This helps ensure that the sample preparation area is isolated from the data collection area to prevent contamination. Moreover, it is important to ensure that the lab complies with safety regulations and standards for cryogenic materials and radiation safety. It is imperative to implement safety measures for handling cryogenic liquids and other hazardous materials. Also, provide personal protective equipment (PPE) and safety training for lab users. Furthermore, the microscope must be placed in a stable and vibration-free environment with network cabling designed to allow fast and real-time data transfer and processing.

A prospective facility manager or facility developer team must carefully select the cryogenic electron microscope that fits the institution's requirements and budget most closely. It is not always necessary to buy the most expensive and newest cryo-TEM in the market, depending on the future direction of the facility, but it is possible to make practical decisions that will ensure the best possible performance outcomes for the facility.

The decision of which cryo-EM is appropriate should be based on several factors and may depend on the setup of the entire facility. The choice of appropriate cryo-EM equipment, including electron microscopes (e.g., transmission electron microscopes or TEMs), cryo-holders, and sample preparation instruments, must consider the specifications and capabilities of the equipment, such as voltage, resolution, and automation features. From a technical perspective, the decision depends on whether the lab has the necessary infrastructure to support the equipment, including power supply, cooling systems, and data storage.

ThermoFisher Scientific (TFS) is the leading company in the market. The most popular high-end cryo-TEM in terms of global sales is the TFS Krios Series, which includes the Krios G3i and Krios G4 cryo-TEMs, known for their high stability and advanced automation. These microscopes are widely used for single-particle cryo-EM and subtomogram averaging. They offer features like autoloading and EPU (EPU, or EPU2, is software for automated data collection, previously mentioned). They are compatible with direct electron detectors, such as the Falcon 4i, for high-resolution imaging. The direct detector will be discussed below.

TFS also has a series that operates at 200 kV, suitable for high-resolution data collection and screening at small cryo-EM facilities. The Glacios (and also the newest released Glacios-2 cryo-TEM) is a 200kV cryo-TEM designed for ease of use and high-throughput screening. It is suitable for rapid sample assessment and preliminary data collection. Some of the main features include automated screening, user-friendly software, and compatibility with direct electron detectors. Analogously, the TFS Talos Arctica is another 200 kV cryo-TEM that offers versatility; it is capable of both cryo-EM and tomography experiments. It is suitable for a wide range of applications from structural biology to materials science and is compatible with both direct electron detectors and traditional CCD cameras. Additionally, the 120 kV Tecnai 12 (T12) and 200 kV F20 are versatile cryo-TEMs that are used for both life sciences and materials science applications, but these models do not have autoloader systems.

On the other hand, JEOL has been developing the new CryoARM series, including the CryoARM300 and CryoARM 300Plus, which offer automated cryo-EM solutions. These microscopes are designed for high throughput and ease of use, with features like automated data collection and grid loading. They also support various detectors and provide advanced imaging capabilities.

A few older 120 kV cryo-TEM models include the Hitachi HT7800 and HT7700 series, which are designed for a wide range of biological and materials science applications with side-entry designs. They feature high-resolution imaging capabilities and are suitable for cryo-electron tomography. These microscopes are known for their stability and ease of use. But it is important to highlight the lack of new cryogenic development in Hitachi equipments in the past decades.

Finally, TFS has developed a new low-kV cryo-TEM named Tundra, which consists of a semi-autoloader with a new direct electron detector suitable for structural biology at 100 kV. A summary of the cryo-TEMs and their potential applications are highlighted in Table 11.1.

Note that the full complement of specifications on a cryo-TEM system available to customers may change over time, particularly following the publication of the current work. Researchers and institutions should consult with microscope manufacturers and distributors for the most up-to-date information on available cryo-TEM models and their capabilities.

Because of the high price tag of any cryo-TEM, its selection and acquisition are often considered the most critical part of setting up a facility. While there may be some truth to this perception, getting a microscope is just the beginning of a long journey. This journey continues with choosing and acquiring the electron detection setup most suitable to the facility in question. Direct electron detector (DED)

TABLE 11.1
Recent and current cryo-TEMs

Microscope Manufacturer	Series Name	Model	kV_{max}	Best DED for Cryo-mode	Suitable Imaging Applications
Thermo Fischer Scientific	Krios	G3i	300	Falcon4i/Gatan K3	SPA/STA/microED
		G4	300	Falcon4i	
	Glacios	Glacios-1	200	Falcon4i	SPA/STA/microED
		Glacios-2	200	Falcon4i	
	Talos	Arctica	200	Gatan K3/K2	SPA/STA
		L120C	120	-	SPA/microED
	Tundra	Tundra	100	Falcon-C	SPA
FEI*	Tecnai	T12	200	Gatan K2	
		F20	120	Gatan K2	SPA/microED
		Spirit	120	Gatan K2	SPA/microED
		T30 Polara	300	Gatan K2	SPA/STA/microED
	Titan	Krios	300	Gatan K3/K2	SPA/STA/microED
		Halo	300	Gatan K2	SPA/STA
JEOL	JEM	2100		-	SPA
		1400+		-	SPA
		F200		Gatan K2	SPA/STA
		Z300FSC		-	SPA
		3200FS		-	SPA
		1230		-	SPA
	CryoARM	CryoARM300	300	Gatan K3	SPA/STA/microED
		CryoARM300+	300	Gatan K3	SPA/STA/microED
		CryoARM200	200	Gatan K3	SPA/STA/microED
Hitachi	HT7800		120	-	-
	HT7700		120	-	-

Note: *Formerly FEI Co., now part of Thermo Fischer Scientific.

development was one of the major developments in cryo-EM and the central pillar of the "resolution revolution." This groundbreaking development plays a pivotal role in achieving higher resolution in cryo-EM studies. DEDs enable the collection of high-quality data, resulting in atomic or near-atomic resolution structures.

The sensitivity of DEDs allows for lower electron doses during data collection, reducing radiation damage to biological specimens and preserving their native state. Direct electron detectors have a high quantum efficiency, allowing them to efficiently convert incoming electrons into measurable signals. This high sensitivity enables the detection of low-dose images, reducing radiation damage to sensitive specimens. These detectors are capable of rapidly reading out frames or movies of images, facilitating the acquisition of dynamic data and high-speed data collection for tomography and other applications. Moreover, direct electron detectors are capable of detecting single electrons, which is essential for achieving high-resolution images and accurate data collection for 3D reconstructions. The ability to capture both weak and strong signals within the same image is crucial for preserving image contrast and detail, allowing the capture of images with low noise, improving the signal-to-noise ratio in collected data, and enhancing the quality of reconstructed structures. Furthermore, many detectors can achieve high frame rates (expressed in frames or events per second), making them suitable for applications requiring rapid data acquisition. Such detectors are compatible with various electron microscopes, including transmission electron microscopes (TEMs) and scanning transmission electron microscopes (STEMs). To date, there have been several direct electron detectors. Please note that the field of electron microscopy constantly evolves; new detectors may have been developed since the time this book chapter was written.

One of the most impressive DEDs belongs to the TFS Falcon series (e.g., Falcon 2, Falcon 3, Falcon 4, and Falcon 4i). The improvements observed in Falcon 4i in comparison to Falcon 3 are amazing, as is the price. TFS also dedicated a new low-kV DED (Falcon-C) that is integrated into the new Tundra cryo-TEM. For many years, Gatan Inc. was the leading company in DEDs with the K2 series (e.g., K2 Summit, K2 Base) and K3 series. Most recently, Gatan Inc. is marketing the new Alpine DED with promising results in SPA and microED projects. Additionally, the Gatan Rio camera was used as a CCD counting model similar to the JEOL CETA camera in the previous generation of cryo-TEMs. Currently, Direct Electron is trying to deploy new DEDs in the market, such as the DE-20 and Apollo, that fit into the new TFS cryo-TEM models (Glacios-2 and Krios G4). Each of these possibilities will provide a different realm of analysis, speed, quality, and requirements. Finding the best match between the cryo-TEM, the facility needs, and the budget might be challenging, but it should be considered during the facility design process.

Finally, the design of a space for image processing and data analysis, including hosting infrastructure for high-performance computational and storage servers, is possibly the most frequently ignored aspect of a cryo-EM facility/lab design. This space must ensure secure and redundant data storage to prevent data loss and high-speed capabilities for storage and data processing.

DATA IS THE FACILITY'S MOST IMPORTANT PART!

In designing a cryo-EM facility or lab, the data storage and computational resources are often completely neglected. However, as we will describe in this section, managing data is challenging and is the longest part of a facility's service commitment to the users.

Cryo-EM data workflows have relatively high computational requirements compared to many other fields of research. NVIDIA GPUs have been found to be particularly well-suited for cryo-EM image processing workflows due to their parallel processing capabilities; tools such as CryoSPARC and RELION support the CUDA instruction set, allowing image processing steps to be completed many times faster than on a general-purpose CPU. Because the GPU is the most important computational component, the CPU and system memory requirements are typically determined by the GPU capabilities of a compute server (CryoSPARC Architecture and System Requirements, n.d.).

The amount of data generated during a facility's data collection and processing routines is gigantic. This large body of data requires significant storage resources, both for raw data and for the various stages of processed data. Raw datasets are typically a few TB (2–5 TB) in size, while processed data can be many times larger (50–100 TB). Some cameras are capable of simultaneously performing many captures in parallel and thus produce correspondingly more raw data. Both the rate at which new data is generated and the duration for which the raw and processed data must be stored should be taken into account.

To move the data around, various network topologies can be set up. These network setups are highly situational and facility-dependent. However, at the least, a high-speed network should connect the facility's microscope or camera controller (whichever writes the raw data), storage servers, and data processing nodes. Some network devices may also require access to the public internet. Datasets range from a few TB to tens of TB in size, so a 10 Gbps network is deemed a baseline networking requirement. At 1 Gbps, 20 TB of data would take between two and three days to transfer, while taking only five to six hours on a 10 Gbps connection or one to two hours at 40 Gbps. It is desirable to have an appropriately fast network so that data transfer does not create a bottleneck for the facility's operations.

While a fast internet connection is not needed for the internal operations of a facility, it may be important as a means of delivering data to external users. Without a fast internet connection, it becomes impractical for users to download data, and the only real alternative is mailing hard drives. Copying datasets to multiple hard drives and mailing them comes with a significant labor cost, which can be entirely avoided if external users are provided with a way to download their data themselves. Data-sharing platforms, such as Globus, are helpful in providing ease of access for external users.

Furthermore, a standard cryo-EM facility depends on a variety of software for data collection and processing. Linux is the platform on which most of the necessary software runs, so it is indispensable. It is necessary to have a plan for supporting both the software and the hardware of the GPU and storage servers. This plan should take into account organizational IT policies and resources as well as hardware vendors' support contract terms.

To optimize the costs and efficiency of the facility's IT needs, immediate and projected needs for computing resources must be clearly identified in order to select appropriate solutions. Even when working to minimize costs, it is important to select configurations that allow for future growth. The most expensive hardware components for a cryo-EM facility are likely to be, in descending order: networking equipment, storage servers, and compute servers.

A local 10 Gbps network requires at minimum a capable network switch and physical network links, either copper or fiber optic. However, as camera capabilities for data capture grow, so will network and storage requirements, so it is prudent to design the network for continued growth in the future. Copper cabling can provide 10 Gbps connections, but anything faster requires fiber. Therefore, it is recommended to invest in fiber optic networking infrastructure even if 10 Gbps is sufficient at the start (Figure 11.2).

The network switches should offer enough ports to connect all devices that will be part of the network. If the facility's local network must also be connected to a wider network, the bandwidth offered by the switch uplink ports should be considered. Remember that all connected devices will require appropriate network interfaces in order to connect to the network and support the highest possible connection speed.

Additionally, storage server costs are largely determined by storage capacity and hard drive prices. Solid-state storage (in SAD drives) is currently significantly more costly per TB than magnetic storage (in SAS drives). Magnetic drives are the cost-effective bulk storage option at this point. There are a number of storage technologies to choose from. Standalone ZFS servers are a popular choice due to their lower cost of entry compared to clustered file systems like Gluster or Ceph, which require multiple physical servers and the network infrastructure upon which they depend. The level of data redundancy offered by these technologies comes at the expense of capacity. It is important to consider the usable storage capacity offered by a specific configuration, not the raw capacity of the hardware. Cloud storage may be challenging to employ due to network bandwidth limitations and ingress/egress costs, making on-premises storage a more practical option.

Compute servers in cryo-EM facilities typically contain powerful GPUs, ample local disk space for data in the process, an SSD cache for software that requires one, a many-core CPU, an ample amount of system memory, and an appropriate network interface. Specific hardware requirements for each software tool can be identified through documentation, benchmarking, or user communities. A number of vendors sell turnkey cryo-EM solutions, the configurations of which are useful as a template for general requirements. Consumer-grade GPUs may be a more cost-effective solution than enterprise-grade products, but research should be performed to identify which hardware offers the optimal cost-to-performance ratio. When procuring any of the aforementioned hardware, consult with multiple hardware vendors, including some that do not specialize in turnkey cryo-EM solutions. This will aid in obtaining a competitive price, and it can be useful to compare the different configurations offered by each system builder. Sticking with standard hardware from reputable manufacturers is highly recommended, as are high-quality warranties (return merchandise authorization [RMA] or repair policies are particularly important). Hardware can and will fail, and downtime for a facility can be very costly, so it is important to plan in advance for how to fix hardware failures.

Ongoing support costs are an important part of the total cost of IT operations over time. Dedicated IT staff are typically required in order to maintain the facility's computing resources. Extensive use of standard technologies and widely supported software is encouraged in order to minimize the use of custom configurations that are often more costly to support or less frequently updated. While turnkey solutions provide a convenient toolset and system configuration, they may not come with any software support, meaning updates or additional software installations will need to be managed in-house.

REMOTE ACCESS IS KEY!

Remote work is now commonplace and can be a great productivity aid for a cryo-EM facility. What is required to enable remote access will depend on institutional resources and policies about network connectivity. It is recommended that you consult with your organization's IT resources in order to create a compliant, secure, well-supported remote access solution (Figure 11.2).

Unlike a facility's data network, the public network connection that provides remote access need not be nearly as high performing. A 1 Gbps network connection is sufficient for graphical remote sessions. Consult with your organizational IT resources about security best practices. Securely implementing remote access requires careful attention to user and access rights management. Expect remote access to require the use of a VPN client.

Cryo-EM software includes some tools that use OpenGL to render 3D visual output. This content cannot be forwarded over an X11 connection, so additional software is required in order to utilize such tools remotely. Two possible solutions to this problem are the open-source

TABLE 11.2
Relevant software

Relevant Software	Description
Linux	The operating system on which the majority of cryo-EM software runs
Windows	The operating system typically installed on microscope and camera controllers
ZFS	A file system which provides many useful features, including data redundancy across multiple disks or volumes
SMB	A network file-sharing protocol that can be used with both Windows and Linux systems
NFS	A network file-sharing protocol that is primarily used with Linux systems
Microsoft RDP	Microsoft Remote Desktop Protocol, a protocol that provides remote access to Windows systems
NoMachine	Cross-platform commercial remote desktop software that can display remotely rendered 3D graphics on Linux systems
VirtualGL	Open-source solution for displaying remotely rendered 3D graphics on Linux systems
TeamViewer	Cross-platform commercial remote desktop software that can typically function through firewalls due to how it establishes connections
CUDA	The NVIDIA computing platform that allows general-purpose processing on GPUs; is used by many cryo-EM tools
Globus	A file-sharing service that allows institutions to host their data but grants permissions through a federated authentication scheme, allowing institutions to work with each others' user credentials

VirtualGL project and the commercial NoMachine software. VirtualGL can be invoked to forward 3D graphics when running tools from the command line and is also available as part of the TurboVNC software. NoMachine is paid software that is operated similarly to other remote access software like Microsoft Remote Desktop. Additional software and descriptions are listed in Table 11.2.

Most recently, data processing cloud services have been available using cloud computing resources and services to perform data processing and analysis tasks related to cryo-EM experiments. Cloud computing provides a scalable and flexible solution for researchers and institutions to efficiently process, store, and analyze cryo-EM data without the need to invest in and maintain extensive on-premise computational infrastructure as previously described.

The workflow consists of uploading the raw images, micrographs, and metadata to cloud storage, for which providers offer scalable and secure storage solutions suitable for large datasets. Preprocessing, such as motion correction, CTF estimation, particle picking, and initial 2D classification, can be performed using cloud-based tools and software optimized for parallel processing. The cloud platforms provide access to high-performance computing (HPC) resources, including virtual machines (VMs) with powerful CPUs and GPUs. Researchers can leverage these resources for computationally intensive tasks like 3D reconstruction and refinement.

Most of the cloud-based services involve the use of specialized software packages like RELION, CryoSPARC, and others. These tools can be installed and run on cloud VMs with access to highly scalable parallel processing hardware, significantly reducing processing times for large datasets. Moreover, cloud providers offer data management and version control services, ensuring data integrity and easy access for collaborative research efforts. Several cloud providers, such as Amazon Web Services (AWS), Google Cloud Platform (GCP), and Microsoft Azure, offer specialized services and resources for scientific computing, making them popular choices for cryo-EM data processing. Research institutions and facilities can also set up private cloud environments to manage and process data efficiently (Cianfrocco et al., 2018; Li & Cianfrocco, 2022). Other cryo-EM-related companies and services, including CryoCloud (*Home | CryoCloud*, n.d.), CryoFLARE (*CryoFLARE*, n.d.), Stion (*Stion*, 2021), cryoEDU (*About – cryoEDU*, n.d.), and WEKA (*Home*, n.d.), are also often used to offload computational resources and storage.

One of the main advantages of using cloud-based services is the possibility of scaling computational resources up or down as needed, paying only for what is used. This flexibility is especially useful for handling varying workloads. After processing, results including 3D reconstructions and other analyzed data can be downloaded or transferred back to local storage for further analysis or publication.

It is important to note that the specific tools, workflows, and cloud services used for cryo-EM data processing may vary depending on the research project, the expertise of the researchers, and the available resources. Researchers and institutions should evaluate their specific needs and budget constraints when considering cryo-EM cloud data processing solutions when designing the cryo-EM facility.

GOVERNMENT-FUNDED FACILITIES ARE A GOOD RESOURCE FOR HIGH-END MICROSCOPES

In the USA, the NIH Common Fund initiated a project known as the Transformative High-Resolution Cryoelectron Microscopy Program (CryoEM) to enhance the availability of cryo-electron microscopy resources for biomedical scientists. Under this program, three National Centers for Cryoelectron Microscopy were established, and more recently, the National Network for Cryoelectron Tomography was formed, comprising four service centers. These cryo-EM and cryo-ET service centers, as listed below, have the primary goal of offering access to services related to specimen preparation and data collection, along with providing training programs. Furthermore, the CryoEM program allocated funding for Curriculum Development

Projects, which aim to create additional educational materials to assist biomedical researchers in understanding and applying cryo-EM/ET techniques, theories, and analytical methods.

This approach and financial support is also applied in Europe in facilities like Ceitec Cryo-Electron Microscopy and Tomography Core Facility (Brno, Czech Republic), Diamond The Electron Bio-Imaging Centre (eBIC, Oxfordshire, England), EMBL Cryo-Electron Microscopy Service Platform (Heidelberg, Germany), ESRF Cryo-EM Facility (Grenoble, France), NeCEN (Leiden, The Netherlands), SciLifeLab Cryo-EM National Facility (Stockholm and Umeå, Sweden), and Umeå University Umeå Core Facility for Electron Microscopy (UCEM, Umeå, Sweden). This approach is also applied in South America, in facilities such as the Brazilian Nanotechnology National Laboratory (LNNano) at the Brazilian Center for Research in Energy and Materials (CNPEM, Campinas, Brazil) and in Asia at the Chinese Academy of Science National Facility for Protein Science Shanghai Electron Microscopy facility (Shanghai, China), the Institute of Biophysics Chinese Academy of Sciences, Center for Biological Imaging (Beijing, China), the National Centre for Biological Science Cryo-EM Facility (Bangalore, India), and University of Wollongong Cryo-EM Facility (Wollongong, Australia).

Within the USA, having access to these National Labs—the National Center for CryoEM Access & Training (NCCAT), Stanford-SLAC Cryo-EM Center (S2C2), Stanford-SLAC CryoET Specimen Preparation Center (SCSC), The Pacific Northwest Center for Cryo-EM (PNCC), the Midwest Center for Cryo-Electron Tomography (MCCET), CU Boulder Center for Cryo-ET (CCET), and The National Center for In-situ Tomographic Ultramicroscopy (NCITU)—allows users with no access to a high-end cryo-TEM to collect their datasets. These resources provide access to the best equipment (microscopes and DED) suitable for the user's needs, facilitating project development. Additionally, local cryo-EM facilities can focus on sample preparation and screenings to ship the top-selected grids and conditions that will be remotely collected and processed further after the data has been transferred. This synergetic interaction is key to the future success and continuity of the development of local and national facilities and to cryo-EM field development.

HOW DO I KNOW I HAVE SUCCEEDED IN SETTING UP A CRYO-EM FACILITY?

One of the biggest challenges for a small cryo-EM facility evaluating a cryo-EM project is measuring performance. Since the equipment will be used on a daily basis to avoid unused or vacancy time, it is critical to develop some metrics that will allow the facility staff to evaluate its performance.

An increase in the number of projects and users and tracking of operation time are critical for a cryo-EM facility, especially if this is a recharge-based service facility. The ability to find clients is crucial for the financial health of the operation due to the high operational costs. This can be mitigated by minimizing vacancy time and operating as a *dynamic queue*, where the microscope and pieces of equipment for sample preparation or data analysis are constantly being used. This approach allows the facility staff to organize the operations in real time, changing samples, users, or priorities based on urgency, priorities, and needs. It is also important to reorganize microscope scheduling in case of downtime due to infrastructure or any other issues that may occur.

Another important metric is the ratio between the financial revenue and deposited structures and published manuscripts. It is obviously important for the facility to cover its operational costs, but science is taken into consideration and helps to increase the number of structures produced and papers accepted by the community.

Finally, cryo-EM facilities with an educational mission, such as those at a university, must be able to train and qualify users at various expertise levels. This pedagogical approach contributes to future generations of cryo-EM scientists and feeds the continued growth of the cryo-EM field.

From a cryo-EM user perspective, measuring success in a cryo-EM-related project involves assessing various aspects of the project's objectives, workflow, and outcomes. The primary goal of a cryo-EM project is often to generate high-quality structural data. Success can be measured by the resolution and clarity of the obtained structures. Higher resolution indicates better data quality. The continuous evaluation of the effectiveness of sample preparation methods results in well-preserved specimens that are suitable for the imaging process. Following the screening session, the sample is ready for high-resolution data collection using high-end facilities, such as national facilities. A successful project is one for which a sufficient number of images are collected without significant technical issues and the images are subsequently analyzed into a final publishable model. Consider the quality of particle picking, 2D and 3D classification, and refinement results. Lower levels of noise and better-defined structural features indicate success. If the project involves subtomogram averaging or other techniques for improving resolution, measure the degree of improvement achieved. Higher resolution or improved feature visibility is a positive outcome.

Finally, success can also be measured by the publication of research findings in reputable scientific journals and their acceptance through peer review. Published research demonstrates the project's contribution to the scientific community. Also, consider the broader impact of the project on the field of structural biology. Has it led to new insights or discoveries, advanced scientific understanding, or opened up new research directions? Moreover, the researcher should be able to answer, "Did it answer the scientific questions it set out to address?"

In summary, success in a cryo-EM project should be evaluated based on a combination of factors, including data quality, research objectives achieved, impact on the field, efficient resource utilization, and the overall contribution to scientific knowledge and advancement. Each project may have unique goals and criteria for success, so it's important to define these in advance and periodically assess progress toward those goals.

FINAL NOTES AND COMMENTS

Building a dedicated cryo-EM facility or initializing a cryo-EM project might be risky and tiring. It presents several significant challenges. First and foremost is the considerable financial investment required for acquiring and maintaining high-end cryo-EM equipment, including electron microscopes, direct detectors, and ancillary instruments. The ongoing costs for cryogens, maintenance, and technical support are substantial. Moreover, the technical complexity of cryo-EM demands skilled personnel, and recruiting and retaining experts in this specialized field can be difficult. Cryo-EM facilities must also grapple with issues related to sample preparation, as cryogenic specimens are delicate and often require precise optimization. Managing and processing the vast volumes of data generated by cryo-EM instruments pose computational and storage challenges, necessitating powerful computing infrastructure and efficient data management solutions. Furthermore, ensuring uninterrupted operation and minimizing downtime is crucial, as downtime can disrupt research schedules and impact facility sustainability. Lastly, cryo-EM facilities must stay abreast of rapidly evolving technologies and methodologies to maintain their competitiveness and relevance in the scientific community. Overcoming these challenges requires a combination of financial support, skilled personnel, robust infrastructure, and a commitment to ongoing innovation and excellence.

ACKNOWLEDGMENTS

The authors also acknowledge the Biomolecular cryoElectron Microscopy Facility at the Department of Chemistry and Biochemistry of the University of California–Santa Cruz (RRID:SCR_021755) for the scientific and technical assistance (NIH High-End Instrumentation program, S10OD02509). The authors also thank Dr. Farshad C. Azimi and Dr. Tyler D. R. Vance for their comments and suggestions on the original draft.

AUTHORS' CONTRIBUTIONS

Eric Shell - Conceptualization and writing—original draft preparation

ITS-Technology Engineering, Information Technology Services, Physical & Biological Sciences Division, University of California–Santa Cruz, Santa Cruz, CA 95064, USA

Karine Minari - Conceptualization and writing—original draft preparation

Biomolecular Cryo-Electron Microscopy Facility, University of California–Santa Cruz, Santa Cruz, CA 95064, USA

Vitor Hugo Balasco Serrão - Conceptualization and writing—original draft preparation;

Biomolecular Cryo-Electron Microscopy Facility, University of California–Santa Cruz, Santa Cruz, CA 95064, USA

Department of Chemistry and Biochemistry, University of California–Santa Cruz, Santa Cruz, CA 95064, USA

All authors equally contributed to the manuscript. All authors have read and agreed to the published version of the manuscript.

CONFLICT OF INTEREST

The authors declare no conflict of interest.

LIST OF ABBREVIATIONS

Cryo-EM Cryogenic electron microscopy
IT Information technology
SPA Single-particle analysis
STA Subtomogram averaging
Cryo-ET Cryo-electron tomography
MicroED Micro-electron diffraction
NSEM Negative-stain electron microscopy
TFS ThermoFisher Scientific
FEI Field Electron and Ion Company
Cryo-FIB Cryo-focused ion beam
Cryo-CLEM Cryo-correlative light and electron microscopy
Cryo-fLM Cryo-fluorescence light microscopy
LN2 Liquid nitrogen
Cryo-TEMs Cryo-transmission electron microscopes
TEM Transmission electron microscope
LaB_6 Lanthanum hexaboride
FEG Field emission gun
CTF Contrast transfer function
FSC Fourier shell correlation
PDB Protein Data Bank
EMDB Electron Microscopy Data Bank
EMPIAR Electron Microscopy Public Image Archive
PIs Principal investigators
SOPs Standard operating procedures
UPS Uninterrupted power supply
PPE Personal protective equipment
T12 120 kV Tecnai 12
F20 200 kV Tecnai 20
DED Direct electron detector
STEMs Scanning transmission electron microscopes
HPC High-performance computing
VMs Virtual machines
AWS Amazon Web Services
GCP Google Cloud Platform
RMA Return merchandise authorization
eBIC Electron Bio-Imaging Centre at Diamond Light Source
LNNano Nanotechnology National Laboratory
CNPEM Brazilian Center for Research in Energy and Materials

NCCAT	National Center for CryoEM Access & Training
S2C2	Stanford-SLAC Cryo-EM Center
SCSC	Stanford-SLAC CryoET Specimen Preparation Center
PNCC	Pacific Northwest Center for Cryo-EM
MCCET	Midwest Center for Cryo-Electron Tomography
CCET	CU Boulder Center for Cryo-ET
NCITU	National Center for In-situ Tomographic Ultramicroscopy

REFERENCES

About – cryoEDU. (n.d.). Retrieved October 7, 2023, from https://cryoedu.org/about/

Afonine, P. V., Klaholz, B. P., Moriarty, N. W., Poon, B. K., Sobolev, O. V., Terwilliger, T. C., Adams, P. D., & Urzhumtsev, A. (2018). New tools for the analysis and validation of cryo-EM maps and atomic models. *Acta Crystallographica Section D: Structural Biology*, *74*(9), 814–840. https://doi.org/10.1107/S2059798318009324

Alewijnse, B., Ashton, A. W., Chambers, M. G., Chen, S., Cheng, A., Ebrahim, M., Eng, E. T., Hagen, W. J. H., Koster, A. J., López, C. S., Lukoyanova, N., Ortega, J., Renault, L., Reyntjens, S., Rice, W. J., Scapin, G., Schrijver, R., Siebert, A., Stagg, S. M., … Potter, C. S. (2017). Best practices for managing large CryoEM facilities. *Journal of Structural Biology*, *199*(3), 225–236. https://doi.org/10.1016/j.jsb.2017.07.011

Battye, T. G. G., Kontogiannis, L., Johnson, O., Powell, H. R., & Leslie, A. G. W. (2011). iMOSFLM: A new graphical interface for diffraction-image processing with MOSFLM. *Acta Crystallographica Section D: Biological Crystallography*, *67*(Pt 4), 271–281. https://doi.org/10.1107/S0907444910048675

Bepler, T., Kelley, K., Noble, A. J., & Berger, B. (2020). Topaz-Denoise: General deep denoising models for cryoEM and cryoET. *Nature Communications*, *11*(1), 5208. https://doi.org/10.1038/s41467-020-18952-1

Bepler, T., Morin, A., Rapp, M., Brasch, J., Shapiro, L., Noble, A. J., & Berger, B. (2019). Positive-unlabeled convolutional neural networks for particle picking in cryo-electron micrographs. *Nature Methods*, *16*(11), 1153–1160. https://doi.org/10.1038/s41592-019-0575-8

Briegel, A., Chen, S., Koster, A. J., Plitzko, J. M., Schwartz, C. L., & Jensen, G. J. (2010). Correlated light and electron cryo-microscopy. *Methods in Enzymology*, *481*, 317–341. https://doi.org/10.1016/S0076-6879(10)81013-4

Brilot, A. F., Chen, J. Z., Cheng, A., Pan, J., Harrison, S. C., Potter, C. S., Carragher, B., Henderson, R., & Grigorieff, N. (2012). Beam-induced motion of vitrified specimen on holey carbon film. *Journal of Structural Biology*, *177*(3), 630–637. https://doi.org/10.1016/j.jsb.2012.02.003

Bu, G., & Nannenga, B. L. (2021a). MicroED sample preparation and data collection for protein crystals. *Methods in Molecular Biology (Clifton, N.J.)*, *2215*, 287–297. https://doi.org/10.1007/978-1-0716-0966-8_13

Bu, G., & Nannenga, B. L. (2021b). MicroED Sample preparation and Data Collection For Protein Crystals. In T. Gonen & B. L. Nannenga (Eds.), *CryoEM: Methods and Protocols* (pp. 287–297). Springer US. https://doi.org/10.1007/978-1-0716-0966-8_13

Chari, A., & Stark, H. (2023). Prospects and limitations of high-resolution single-particle cryo-electron microscopy. *Annual Review of Biophysics*, *52*, 391–411. https://doi.org/10.1146/annurev-biophys-111622-091300

Cheng, Y., Grigorieff, N., Penczek, P. A., & Walz, T. (2015). A primer to single-particle cryo-electron microscopy. *Cell*, *161*(3), 438–449. https://doi.org/10.1016/j.cell.2015.03.050

Chua, E. Y. D., Mendez, J. H., Rapp, M., Ilca, S. L., Tan, Y. Z., Maruthi, K., Kuang, H., Zimanyi, C. M., Cheng, A., Eng, E. T., Noble, A. J., Potter, C. S., & Carragher, B. (2022). Better, faster, cheaper: Recent advances in cryo-electron microscopy. *Annual Review of Biochemistry*, *91*, 1–32. https://doi.org/10.1146/annurev-biochem-032620-110705

Cianfrocco, M. A., Lahiri, I., DiMaio, F., & Leschziner, A. E. (2018). cryoem-cloud-tools: A software platform to deploy and manage cryo-EM jobs in the cloud. *Journal of Structural Biology*, *203*(3), 230–235. https://doi.org/10.1016/j.jsb.2018.05.014

Cianfrocco, M. A., & Leschziner, A. E. (2015). Low cost, high performance processing of single particle cryo-electron microscopy data in the cloud. *eLife*, *4*, e06664. https://doi.org/10.7554/eLife.06664

CryoFLARE—pDocumentation. (n.d.). Retrieved October 7, 2023, from https://www.cryoflare.org/documentation/index.html

CryoSPARC Architecture and System Requirements. (n.d.). Retrieved October 7, 2023, from https://guide.cryosparc.com/setup-configuration-and-management/hardware-and-system-requirements#worker-node-cluster-worker-minimum-requirements).

de la Rosa-Trevín, J. M., Quintana, A., del Cano, L., Zaldívar, A., Foche, I., Gutiérrez, J., Gómez-Blanco, J., Burguet-Castell, J., Cuenca-Alba, J., Abrishami, V., Vargas, J., Otón, J., Sharov, G., Vilas, J. L., Navas, J., Conesa, P., Kazemi, M., Marabini, R., Sorzano, C. O. S., & Carazo, J. M. (2016). Scipion: A software framework toward integration, reproducibility and validation in 3D electron microscopy. *Journal of Structural Biology*, *195*(1), 93–99. https://doi.org/10.1016/j.jsb.2016.04.010

Dobro, M. J., Melanson, L. A., Jensen, G. J., & McDowall, A. W. (2010). Plunge freezing for electron cryomicroscopy. *Methods in Enzymology*, *481*, 63–82. https://doi.org/10.1016/S0076-6879(10)81003-1

Dubochet, J., Adrian, M., Chang, J.-J., Homo, J.-C., Lepault, J., McDowall, A. W., & Schultz, P. (1988). Cryo-electron microscopy of vitrified specimens. *Quarterly Reviews of Biophysics*, *21*(2), 129–228. https://doi.org/10.1017/S0033583500004297

Dubochet, J., & McDowall, A. W. (1981). Vitrification of pure water for electron microscopy. *Journal of Microscopy*, *124*(3), 3–4. https://doi.org/10.1111/j.1365-2818.1981.tb02483.x

Faruqi, A. R., & McMullan, G. (2011). Electronic detectors for electron microscopy. *Quarterly Reviews of Biophysics*, *44*(3), 357–390. https://doi.org/10.1017/S0033583511000035

Fernandez-Leiro, R., & Scheres, S. H. W. (2017). A pipeline approach to single-particle processing in RELION. *Acta Crystallographica. Section D, Structural Biology*, *73*(Pt 6), 496–502. https://doi.org/10.1107/S2059798316019276

Glaeser, R. M., Han, B.-G., Csencsits, R., Killilea, A., Pulk, A., & Cate, J. H. D. (2016). Factors that influence the formation and stability of thin, cryo-EM specimens. *Biophysical Journal*, *110*(4), 749–755. https://doi.org/10.1016/j.bpj.2015.07.050

Grant, T., Rohou, A., & Grigorieff, N. (2018). cisTEM, user-friendly software for single-particle image processing. *eLife*, *7*, e35383. https://doi.org/10.7554/eLife.35383

Guaita, M., Watters, S. C., & Loerch, S. (2022). Recent advances and current trends in cryo-electron microscopy. *Current Opinion in Structural Biology*, *77*, 102484. https://doi.org/10.1016/j.sbi.2022.102484

Hampton, C. M., Strauss, J. D., Ke, Z., Dillard, R. S., Hammonds, J. E., Alonas, E., Desai, T. M., Marin, M., Storms, R. E., Leon, F., Melikyan, G. B., Santangelo, P. J., Spearman, P. W., & Wright, E. R. (2017). Correlated fluorescence microscopy and cryo-electron tomography of virus-infected or transfected mammalian cells. *Nature Protocols*, *12*(1), 150–167. https://doi.org/10.1038/nprot.2016.168

Han, B.-G., Armstrong, M., Fletcher, D. A., & Glaeser, R. M. (2022). Perspective: biochemical and physical constraints associated with preparing thin specimens for single-particle cryo-EM. *Frontiers in Molecular Biosciences*, *9*. https://www.frontiersin.org/articles/10.3389/fmolb.2022.864829

Home. (n.d.). WEKA. Retrieved October 7, 2023, from https://www.weka.io/

Home | CryoCloud. (n.d.). Retrieved October 7, 2023, from https://cryocloud.io//

Hylton, R. K., & Swulius, M. T. (2021). Challenges and triumphs in cryo-electron tomography. *iScience*, *24*(9), 102959. https://doi.org/10.1016/j.isci.2021.102959

Jamali, K., Käll, L., Zhang, R., Brown, A., Kimanius, D., & Scheres, S. H. W. (2023). *Automated model building and protein identification in cryo-EM maps* (p. 2023.05.16.541002). bioRxiv. https://doi.org/10.1101/2023.05.16.541002

Jun, S., Ro, H.-J., Bharda, A., Kim, S. I., Jeoung, D., & Jung, H. S. (2019). Advances in cryo-correlative light and electron microscopy: Applications for studying molecular and cellular events. *The Protein Journal*, *38*(6), 609–615. https://doi.org/10.1007/s10930-019-09856-1

Kabsch, W. (2010). XDS. *Acta Crystallographica Section D: Biological Crystallography*, *66*(2), 125–132. https://doi.org/10.1107/S0907444909047337

Kastner, B., Fischer, N., Golas, M. M., Sander, B., Dube, P., Boehringer, D., Hartmuth, K., Deckert, J., Hauer, F., Wolf, E., Uchtenhagen, H., Urlaub, H., Herzog, F., Peters, J. M., Poerschke, D., Lührmann, R., & Stark, H. (2008). GraFix: Sample preparation for single-particle electron cryomicroscopy. *Nature Methods*, *5*(1), 53–55. https://doi.org/10.1038/nmeth1139

Kelley, K., Raczkowski, A. M., Klykov, O., Jaroenlak, P., Bobe, D., Kopylov, M., Eng, E. T., Bhabha, G., Potter, C. S., Carragher, B., & Noble, A. J. (2022). Waffle method: A general and flexible approach for improving throughput in FIB-milling. *Nature Communications*, *13*(1), 1857. https://doi.org/10.1038/s41467-022-29501-3

Kühlbrandt, W. (2014). Cryo-EM enters a new era. *eLife*, *3*, e03678. https://doi.org/10.7554/eLife.03678

Lam, V., & Villa, E. (2021). Practical approaches for Cryo-FIB milling and applications for cellular cryo-electron tomography. *Methods in Molecular Biology (Clifton, N.J.)*, *2215*, 49–82. https://doi.org/10.1007/978-1-0716-0966-8_3

Lander, G. C., Stagg, S. M., Voss, N. R., Cheng, A., Fellmann, D., Pulokas, J., Yoshioka, C., Irving, C., Mulder, A., Lau, P.-W., Lyumkis, D., Potter, C. S., & Carragher, B. (2009). Appion: An integrated, database-driven pipeline to facilitate EM image processing. *Journal of Structural Biology*, *166*(1), 95–102. https://doi.org/10.1016/j.jsb.2009.01.002

Li, X., Mooney, P., Zheng, S., Booth, C. R., Braunfeld, M. B., Gubbens, S., Agard, D. A., & Cheng, Y. (2013). Electron counting and beam-induced motion correction enable near-atomic-resolution single-particle cryo-EM. *Nature Methods*, *10*(6), 584–590. https://doi.org/10.1038/nmeth.2472

Li, Y., & Cianfrocco, M. A. (2022). Cloud computing platforms to support cryo-EM structure determination. *Trends in Biochemical Sciences*, *47*(2), 103–105. https://doi.org/10.1016/j.tibs.2021.11.005

Mahamid, J., Schampers, R., Persoon, H., Hyman, A. A., Baumeister, W., & Plitzko, J. M. (2015). A focused ion beam milling and lift-out approach for site-specific preparation of frozen-hydrated lamellas from multicellular organisms. *Journal of Structural Biology*, *192*(2), 262–269. https://doi.org/10.1016/j.jsb.2015.07.012

Marko, M., Hsieh, C., Schalek, R., Frank, J., & Mannella, C. (2007). Focused-ion-beam thinning of frozen-hydrated biological specimens for cryo-electron microscopy. *Nature Methods*, *4*(3), 215–217. https://doi.org/10.1038/nmeth1014

Martynowycz, M. W., Shiriaeva, A., Clabbers, M. T. B., Nicolas, W. J., Weaver, S. J., Hattne, J., & Gonen, T. (2023). A robust approach for MicroED sample preparation of lipidic cubic phase embedded membrane protein crystals. *Nature Communications*, *14*(1), 1086. https://doi.org/10.1038/s41467-023-36733-4

Mastronarde, D. N. (2005). Automated electron microscope tomography using robust prediction of specimen movements. *Journal of Structural Biology*, *152*(1), 36–51. https://doi.org/10.1016/j.jsb.2005.07.007

Milazzo, A.-C., Cheng, A., Moeller, A., Lyumkis, D., Jacovetty, E., Polukas, J., Ellisman, M. H., Xuong, N.-H., Carragher, B., & Potter, C. S. (2011). Initial evaluation of a direct detection device detector for single particle cryo-electron microscopy. *Journal of Structural Biology*, *176*(3), 404–408. https://doi.org/10.1016/j.jsb.2011.09.002

Mills, D. J. (2021). Setting up and operating a cryo-EM laboratory. *Quarterly Reviews of Biophysics*, *54*, e2. https://doi.org/10.1017/S003358352000013X

Mu, X., Gillman, C., Nguyen, C., & Gonen, T. (2021). An overview of microcrystal electron diffraction (MicroED). *Annual Review of Biochemistry*, *90*(1), 431–450. https://doi.org/10.1146/annurev-biochem-081720-020121

Nasir, L. (2010). The checklist manifesto: How to get things right. *London Journal of Primary Care*, *3*(2), 124.

Neumann, P., Dickmanns, A., & Ficner, R. (2018). Validating resolution revolution. *Structure*, *26*(5), 785–795.e4. https://doi.org/10.1016/j.str.2018.03.004

Ni, T., Frosio, T., Mendonça, L., Sheng, Y., Clare, D., Himes, B. A., & Zhang, P. (2022). High-resolution in situ structure determination by cryo-electron tomography and subtomogram averaging using emClarity. *Nature Protocols*, *17*(2), 421–444. https://doi.org/10.1038/s41596-021-00648-5

Otwinowski, Z., & Minor, W. (1997). [20] Processing of X-ray diffraction data collected in oscillation mode. In *Methods in Enzymology* (Vol. 276, pp. 307–326). Academic Press. https://doi.org/10.1016/S0076-6879(97)76066-X

Parmenter, C. D. J., Fay, M. W., Hartfield, C., & Eltaher, H. M. (2016). Making the practically impossible "Merely difficult"—Cryogenic FIB lift-out for "Damage free" soft matter imaging. *Microscopy Research and Technique*, *79*(4), 298–303. https://doi.org/10.1002/jemt.22630

Pintilie, G., Zhang, K., Su, Z., Li, S., Schmid, M. F., & Chiu, W. (2020). Measurement of atom resolvability in cryoEM maps with Q-scores. *Nature Methods*, *17*(3), 328–334. https://doi.org/10.1038/s41592-020-0731-1

Punjani, A., Rubinstein, J. L., Fleet, D. J., & Brubaker, M. A. (2017). cryoSPARC: Algorithms for rapid unsupervised cryo-EM structure determination. *Nature Methods*, *14*(3), 290–296. https://doi.org/10.1038/nmeth.4169

Pyle, E., & Zanetti, G. (2021). Current data processing strategies for cryo-electron tomography and subtomogram averaging. *The Biochemical Journal*, *478*(10), 1827–1845. https://doi.org/10.1042/BCJ20200715

Ramírez-Aportela, E., Maluenda, D., Fonseca, Y. C., Conesa, P., Marabini, R., Heymann, J. B., Carazo, J. M., & Sorzano, C. O. S. (2021). FSC-Q: A CryoEM map-to-atomic model quality validation based on the local Fourier shell correlation. *Nature Communications*, *12*(1), 42. https://doi.org/10.1038/s41467-020-20295-w

Rice, W. J., Cheng, A., Noble, A. J., Eng, E. T., Kim, L. Y., Carragher, B., & Potter, C. S. (2018). Routine determination of ice thickness for cryo-EM grids. *Journal of Structural Biology*, *204*(1), 38–44. https://doi.org/10.1016/j.jsb.2018.06.007

Rigort, A., Bäuerlein, F. J. B., Villa, E., Eibauer, M., Laugks, T., Baumeister, W., & Plitzko, J. M. (2012). Focused ion beam micromachining of eukaryotic cells for cryoelectron tomography. Proceedings of the National Academy of Sciences of the United States of America, *109*(12), 4449–4454. https://doi.org/10.1073/pnas.1201333109

Rubino, S., Akhtar, S., Melin, P., Searle, A., Spellward, P., & Leifer, K. (2012). A site-specific focused-ion-beam lift-out method for cryo Transmission Electron Microscopy. *Journal of Structural Biology*, *180*(3), 572–576. https://doi.org/10.1016/j.jsb.2012.08.012

Sartori, A., Gatz, R., Beck, F., Rigort, A., Baumeister, W., & Plitzko, J. M. (2007). Correlative microscopy: Bridging the gap between fluorescence light microscopy and cryo-electron tomography. *Journal of Structural Biology*, *160*(2), 135–145. https://doi.org/10.1016/j.jsb.2007.07.011

Schaffer, M., Mahamid, J., Engel, B. D., Laugks, T., Baumeister, W., & Plitzko, J. M. (2017). Optimized cryo-focused ion beam sample preparation aimed at in situ structural studies of membrane proteins. *Journal of Structural Biology*, *197*(2), 73–82. https://doi.org/10.1016/j.jsb.2016.07.010

Schaffer, M., Pfeffer, S., Mahamid, J., Kleindiek, S., Laugks, T., Albert, S., Engel, B. D., Rummel, A., Smith, A. J., Baumeister, W., & Plitzko, J. M. (2019). A cryo-FIB lift-out technique enables molecular-resolution cryo-ET within native *Caenorhabditis elegans* tissue. *Nature Methods*, *16*(8), 757–762. https://doi.org/10.1038/s41592-019-0497-5

Scheres, S. H. W. (2012). RELION: Implementation of a Bayesian approach to cryo-EM structure determination. *Journal of Structural Biology*, *180*(3), 519–530. https://doi.org/10.1016/j.jsb.2012.09.006

Scheres, S. H. W., & Chen, S. (2012). Prevention of overfitting in cryo-EM structure determination. *Nature Methods*, *9*(9), 853–854. https://doi.org/10.1038/nmeth.2115

Schoehn, G., Chenavier, F., & Crépin, T. (2023). Advances in structural virology via cryo-EM in 2022. *Viruses*, *15*(6), 1315. https://doi.org/10.3390/v15061315

Schwartz, C. L., Sarbash, V. I., Ataullakhanov, F. I., McIntosh, J. R., & Nicastro, D. (2007). Cryo-fluorescence microscopy facilitates correlations between light and cryo-electron microscopy and reduces the rate of photobleaching. *Journal of Microscopy*, *227*(Pt 2), 98–109. https://doi.org/10.1111/j.1365-2818.2007.01794.x

Sexton, D. L., Burgold, S., Schertel, A., & Tocheva, E. I. (2021). Super-resolution confocal cryo-CLEM with cryo-FIB milling for in situ imaging of *Deinococcus radiodurans*. *Current Research in Structural Biology*, *4*, 1–9. https://doi.org/10.1016/j.crstbi.2021.12.001

Sgro, G. G., & Costa, T. R. D. (2018). Cryo-EM grid preparation of membrane protein samples for single particle analysis. *Frontiers in Molecular Biosciences*, *5*. https://www.frontiersin.org/articles/10.3389/fmolb.2018.00074

Stion – a Software as a Service for Cryo-EM data processing on AWS | AWS HPC Blog. (2021, August 17). https://aws.amazon.com/blogs/hpc/stion-a-saas-for-cryo-em-data-processing-on-aws/

Suloway, C., Pulokas, J., Fellmann, D., Cheng, A., Guerra, F., Quispe, J., Stagg, S., Potter, C. S., & Carragher, B. (2005). Automated molecular microscopy: The new Leginon system. *Journal of Structural Biology*, *151*(1), 41–60. https://doi.org/10.1016/j.jsb.2005.03.010

Tang, G., Peng, L., Baldwin, P. R., Mann, D. S., Jiang, W., Rees, I., & Ludtke, S. J. (2007). EMAN2: An extensible image processing suite for electron microscopy. *Journal of Structural Biology*, *157*(1), 38–46. https://doi.org/10.1016/j.jsb.2006.05.009

Tuijtel, M. W., Koster, A. J., Jakobs, S., Faas, F. G. A., & Sharp, T. H. (2019). Correlative cryo super-resolution light and electron microscopy on mammalian cells using fluorescent proteins. *Scientific Reports*, *9*(1), 1369. https://doi.org/10.1038/s41598-018-37728-8

van Driel, L. F., Valentijn, J. A., Valentijn, K. M., Koning, R. I., & Koster, A. J. (2009). Tools for correlative cryo-fluorescence microscopy and cryo-electron tomography applied to whole mitochondria in human endothelial cells. *European Journal of Cell Biology*, *88*(11), 669–684. https://doi.org/10.1016/j.ejcb.2009.07.002

Vénien-Bryan, C., & Fernandes, C. A. H. (2023). Overview of membrane protein sample preparation for single-particle cryo-electron microscopy analysis. *International Journal of Molecular Sciences*, *24*(19), 14785. https://doi.org/10.3390/ijms241914785

Wagner, F. R., Watanabe, R., Schampers, R., Singh, D., Persoon, H., Schaffer, M., Fruhstorfer, P., Plitzko, J., & Villa, E. (2020). Preparing samples from whole cells using focused-ion-beam milling for cryo-electron tomography. *Nature Protocols*, *15*(6), 2041–2070. https://doi.org/10.1038/s41596-020-0320-x

Wagner, T., Merino, F., Stabrin, M., Moriya, T., Antoni, C., Apelbaum, A., Hagel, P., Sitsel, O., Raisch, T., Prumbaum, D., Quentin, D., Roderer, D., Tacke, S., Siebolds, B., Schubert, E., Shaikh, T. R., Lill, P., Gatsogiannis, C., & Raunser, S. (2019). SPHIRE-crYOLO is a fast and accurate fully automated particle picker for cryo-EM. *Communications Biology*, *2*(1), 218. https://doi.org/10.1038/s42003-019-0437-z

Walsh, R. M., Mayer, M. L., Sun, C. H., Rawson, S., Nair, R., Sterling, S. M., & Li, Z. (2022). Practices for running a research-oriented shared cryo-EM facility. *Frontiers in Molecular Biosciences*, *9*, 960940. https://doi.org/10.3389/fmolb.2022.960940

Wan, W., & Briggs, J. a. G. (2016). Cryo-electron tomography and subtomogram averaging. *Methods in Enzymology*, *579*, 329–367. https://doi.org/10.1016/bs.mie.2016.04.014

Weissenberger, G., Henderikx, R. J. M., & Peters, P. J. (2021). Understanding the invisible hands of sample preparation for cryo-EM. *Nature Methods*, *18*(5), 463–471. https://doi.org/10.1038/s41592-021-01130-6

Winter, G., Waterman, D. G., Parkhurst, J. M., Brewster, A. S., Gildea, R. J., Gerstel, M., Fuentes-Montero, L., Vollmar, M., Michels-Clark, T., Young, I. D., Sauter, N. K., & Evans, G. (2018). DIALS: Implementation and evaluation of a new integration package. *Acta Crystallographica*

Section D: Structural Biology, *74*(2), 85–97. https://doi.org/10.1107/S2059798317017235

Xu, Y., & Dang, S. (2022). Recent technical advances in sample preparation for single-particle Cryo-EM. *Frontiers in Molecular Biosciences*, *9*. https://www.frontiersin.org/articles/10.3389/fmolb.2022.892459

Zachs, T., Schertel, A., Medeiros, J., Weiss, G. L., Hugener, J., Matos, J., & Pilhofer, M. (2020). Fully automated, sequential focused ion beam milling for cryo-electron tomography. *eLife*, *9*, e52286. https://doi.org/10.7554/eLife.52286

Zhong, E. D., Bepler, T., Berger, B., & Davis, J. H. (2021). CryoDRGN: Reconstruction of heterogeneous cryo-EM structures using neural networks. *Nature Methods*, *18*(2), 176–185. https://doi.org/10.1038/s41592-020-01049-4

Zimanyi, C. M., Kopylov, M., Potter, C. S., Carragher, B., & Eng, E. T. (2022). Broadening access to cryoEM through centralized facilities. *Trends in Biochemical Sciences*, *47*(2), 106–116. https://doi.org/10.1016/j.tibs.2021.10.007

Section III

Cryo-EM in Virology and Enzymology

High-Resolution Structures of Chaperonins and Enzymes

12 Cryo-EM Structures of Viral and Bacterial Chaperonins

Olga S. Sokolova and Tatiana B. Stanishneva-Konovalova

INTRODUCTION

The study of the mechanisms underlying the process of protein folding in the cell remains one of the major challenges in biochemistry and cell biology. Foundational discoveries made in the middle of the last century indicated that the amino acid sequence itself contains the information required for the assumption of the native structure (Anfinsen et al., 1961). Further studies also discovered the *in vivo* catalyst of protein folding, protein disulfide isomerase (Goldberger et al., 1963; Venetianer & Straub, 1963). While its catalyst activity is the ability to rearrange disulfide bonds for proteins in the endoplasmic reticulum, this raised the question of whether other cellular compartments also provide any assistance in folding. In the 1970s, such assisting protein complexes, termed heat shock proteins (hsp), were discovered in *Drosophila* and in bacterial and mammalian cells (Kelley & Schlesinger, 1978; Tissiéres et al., 1974; Yamamori et al., 1978). A first model of their action was proposed for Hsp70 (Pelham, 1986). According to this model, Hsp70 binds hydrophobic protein surfaces, which become exposed under thermal stress, and thus prevents their aggregation. This action implies Hsp70's conformational changes enabled by ATP binding and hydrolysis. Another group of assisting proteins was discovered upon investigation of T4 and λ phages, when it was found that mutations of a certain host gene restricted phage assembly (Georgopoulos et al., 1972; Takano & Kakefuda, 1972). This gene, termed *groE*, encodes a ~60 kDa GroEL protein which forms homooligomeric complexes that were first observed with negative stain EM (Hendrix, 1979; Hohn et al., 1979). Each of the complexes was composed of two heptameric rings ~13 nm in diameter with a central "hole." The second *groE* gene product, GroES, is a ~15 kDa protein, which makes donut-shaped structures ~8 nm in diameter as revealed by negative stain EM and interacts with GroEL (Chandrasekhar et al., 1986). Both GroEL and GroES were shown to be heat induced and thus could be assigned to the group of heat shock proteins (Neidhardt et al., 1981; Tilly et al., 1983). Double-ringed complexes assisting folding of other proteins were also discovered in chloroplasts (Barraclough & Ellis, 1980) and mitochondria (Cheng et al., 1989). Together with later found homologs with a similar architecture, they were allocated to the family of chaperones termed "chaperonins" (Hemmingsen et al., 1988). Chaperonins were discovered in all three kingdoms of life and were subdivided into two groups: group I unites chaperonins from bacteria, mitochondria, and chloroplasts, and group II is present in Archaea and in the eukaryotic cytosol (Bracher & Hartl, 2013; Skjærven et al., 2015). Recently several chaperonins were also found in the genomes of bacteriophages (Kurochkina et al., 2012; Marine et al., 2017; Semenyuk et al., 2016). In this chapter, we will discuss the structures of phage chaperones, most of which have been obtained using cryo-EM, as well as their similarities and differences from the chaperonins of the first and the second groups.

GroEL/GroES Complex Structure and Functional Cycle

Cryo-EM with other structural methods have provided a better understanding of the process of GroEL-assisted protein folding. To date, more than 20 structures of GroEL and its complexes obtained from cryo-EM data have been deposited into the Protein Data Bank. Among them are apo-forms and complexes with nucleotides, with GroES, and other proteins.

GroEL is a homooligomer composed of 14 subunits arranged as two stacked back-to-back rings. Each ring, a heptamer, possesses a central cavity, or the folding chamber. ATP-dependent conformational changes in GroEL subunits mediate GroES binding and closure of the folding chamber, where a misfolded protein can achieve its native conformation.

A GroEL subunit is composed of three domains: equatorial (aa 6–133 and 409–523), intermediate (aa 134–190 and 377–408), and apical (aa 191–376) (Horwich & Fenton, 2020). The equatorial domain contains an ATP-binding pocket and is responsible for most of the contacts between the adjacent subunits of the same ring and for all the contacts with subunits from the opposite ring. The intermediate domain is connected to the equatorial domain though the "lower hinge" (aa P137, G410) and to the apical domain through the "upper hinge" (G192, G375). These hinge points allow for rigid-body domain movements. The apical domain binds substrate proteins and GroES through the hydrophobic binding surface.

Stoichiometry of GroEL–GroES interaction has been debatable: either only one ring in the GroEL tetradecamer binds a GroES heptamer, resulting in the formation of an asymmetric "bullet-shaped" complex, or, alternatively, both GroEL rings can each bind a GroES heptamer, forming a "football-shaped" complex. Both types of complexes were obtained via structural methods. The X-ray method provided both "footballs" and "bullets," while cryo-EM structures were obtained only for the latter type of the complexes. Although cryo-EM yielded complexes at generally lower resolution than X-ray, it allowed researchers

DOI: 10.1201/9781003326106-15

to distinguish two different bullet-shaped complexes with ADP occupying nucleotide-binding pockets of both rings (Kudryavtseva et al., 2021). These structures were mainly different from each other in the relative orientation of their apical domains and were distinct from other X-ray and cryo-EM bullet complexes, where nucleotides were only reported in the closed ring.

A recently proposed model suggests that GroEL undergoes ring separation through the functional cycle (Yan et al., 2018). According to this model, protein folding occurs in the closed chamber of a bullet-shaped complex until seven ATP molecules are hydrolyzed. During this time another ring (which is open and nucleotide-free at this step) can bind a substrate protein. Ring separation occurs after ATP hydrolysis in the closed ring and ATP binding to the open ring. The presence of nucleotides in both rings supposedly causes steric clashes in the equatorial part of the complex, underlying negative inter-ring cooperativity. Next, ADP and GroES dissociate, the folded protein is released, and the bullet-shaped complex can reassemble with or without ring exchange between different GroEL/GroES complexes.

Despite the remaining controversy regarding the details of the functional cycle, the GroEL/GroES system is the most well-studied representative of the chaperonin family. Since recently discovered phage chaperonins are predicted to be more closely related to group I than to group II, we will compare GroEL with the known structures of chaperonins from phages EL *Pseudomonas aeruginosa*, OBP *Pseudomonas fluorescens*, and AR9 *Bacillus subtilis*.

Chaperonin from Phage EL *P. aeruginosa*

The bacterial GroEL system is sometimes used by bacteriophages to fold their capsid proteins (Georgopoulos et al., 1973). Some phages encode GroES orthologs, which can create a bigger folding cage when attached to GroEL (Ang et al., 2001; Bakkes et al., 2005; Clare et al., 2009; Hunt et al., 1997; van der Vies et al., 1994); others encode GroEL orthologs. The first phage GroEL-like protein termed gene product 146 (gp146) was found in the genome of *P. aeruginosa* bacteriophage EL (Hertveldt et al., 2005). Although it has only 25% sequence identity with GroEL, functional and physicochemical studies proved its chaperonin activity (Kurochkina et al., 2012). Unlike GroEL, however, the EL chaperonin (ϕEL) appeared to function without any co-chaperonin. Further studies demonstrated that ϕEL is unable to refold an encapsulation-dependent substrate rhodanese but prevents aggregation of denatured protein gp188 in an ATP-dependent manner (Bracher et al., 2020). The molecular weight of the complexes purified using size exclusion chromatography suggested that in the apo-state or in the presence of ADP, mostly double-ring complexes are observed, while the presence of ATP shifts the population of ϕEL complexes to the single-ring heptamers.

To investigate the switching between double-ring and single-ring conformations, structural methods were implemented. Using X-ray, two crystal forms of nucleotide-bound heptamers were obtained. The first crystal form, resolved at 4.03 Å resolution, consisted of subunits in similar conformations, termed conformation I (Bracher et al., 2020). Each subunit displayed a three-domain architecture, typical for chaperonins, and all subunits contained either ATP or ADP•BeFX in the equatorial domains. However, the apical domain made contacts with the equatorial one through its αK-αL helical hairpin, which doesn't exist in group II chaperonins and never makes such contacts in group I. The second crystal form, resolved at 3.54 Å resolution, displayed more conformational heterogeneity: Three subunits in conformation II alternated around the ring with three subunits in conformation III, with the seventh subunit being in conformation I. A pattern of alternating subunits was previously described for the apo-state of the *Pseudomonas* phage OBP chaperonin (Stanishneva-Konovalova et al., 2020). Compared to conformation I of ϕEL, the apical domain in conformation III was rotated by ~22° degrees around the point of connection to the intermediate domain. In conformation II, subunits displayed a 19°–21° outward rotation of the intermediate domains as compared to conformation I. These rotations resulted in a set of contacts between the apical domains of subunits in conformation III and intermediate domains of adjacent subunits in conformation II. Positions of intermediate domains correlated with the nucleotide density: conformation II with outward-oriented intermediate domain contained nucleotide density, corresponding to ADP, while conformation III contained density corresponding to ATP and Mg^{2+}.

Using cryo-EM, double-ring structures of apo-ϕEL as well as ϕEL in the presence of ADP or ATPγS were resolved. The tetradecameric apo-ϕEL was reconstructed at 3.45 Å resolution, allowing for the building of an atomic model and defining of intersubunit contacts (Figure 12.1).

Each ring was like the heptamer of crystal form II with alternating subunit conformations. The inter-ring contacts were similar to chaperonins of group II: Each subunit of one ring contacts only one subunit from the opposite ring through their equatorial domains. In contrast, GroEL and other group I chaperonins display staggered subunit contacts, where each subunit interacts with two subunits from the opposite ring.

ADP-bound ϕEL was also a tetradecamer but with nearly uniform conformations of subunits within each ring, thus allowing for the use of D7 symmetry. As a result, a cryo-EM density map at 5.9 Å resolution was built. All the subunits adopted conformation I as in the first crystal form described above. In the double-ring cryo-EM structure of ϕEL in the presence of ATPγS (a slowly hydrolyzing ATP analog), subunits also adopted conformation I, but the ring–ring interface was slightly changed, suggesting the breakage of several inter-ring contracts. In case of ATP binding, this effect may be more pronounced, causing the separation of the double-ring complexes into single rings (Bracher et al., 2020).

Taken together, these results allowed authors to propose a model of the ϕEL chaperonin's functional cycle. In the

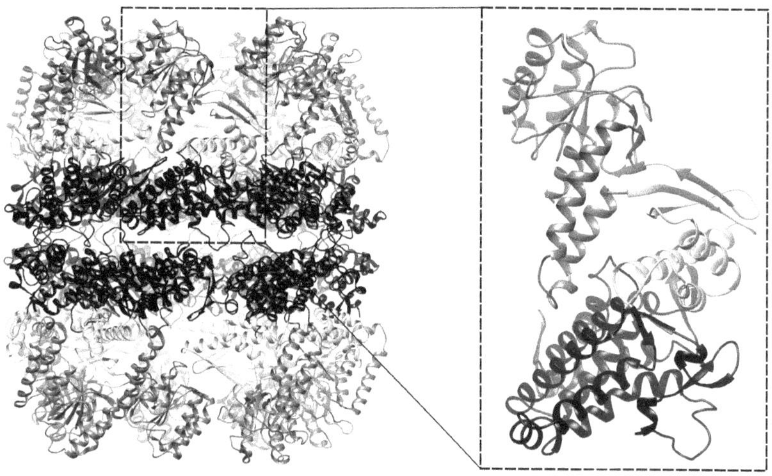

FIGURE 12.1 An atomic model of the double-ring ϕEL complex in the apo state (*left*) and of a single subunit (*right*). Equatorial, intermediate and apical domains are colored in black, light grey and dim grey, respectively.

apo-state, ϕEL is composed of two heptameric rings with two alternating subunit conformations within each ring. Upon ATP binding, all subunits acquire the same conformation where the apical domains are positioned to facilitate binding of a substrate protein. ATP binding also weakens the inter-ring interactions, causing the dissociation of the tetradecamer into two heptamers. ATP hydrolysis may happen in two steps: It is first hydrolyzed by three subunits, resulting in an alternating pattern similar to the apo-state, and then by the remaining four subunits. ADP and SP dissociation are followed by the restoration of the double-ring apo-state. This model implies there was no encapsulating state for ϕEL, which is consistent with the fact that ϕEL was unable to refold an encapsulation-dependent substrate rhodanese. This raises the possibility of ϕEL being an evolutionary precursor of chaperonins with encapsulation mechanism.

Chaperonin from Phage OBP *P. fluorescens*

Bioinformatics analysis indicated that the genome of bacteriophage OBP *P. fluorescens* also encodes a GroEL-like protein—gene product (gp) 246 (Semenyuk et al., 2016). In contrast to GroEL and ϕEL, gp246 was purified as a single heptameric ring in the apo-form. Further *in vitro* experiments proved the ability of this protein to suppress thermal protein inactivation and aggregation in the presence of ATP and without a co-chaperone. The preliminary structural studies were performed using the negatively stained samples, and they also showed the complex shaped like the single ring of GroEL. The cryo-EM method was implemented to obtain higher-resolution structures and describe the OBP chaperonin in more detail.

The structure of the nucleotide-free OBP chaperonin was resolved at 4.5 Å (deposited in EMDB under the accession number EMD-0204; atomic model deposited in PDB under the accession number 6HDD) (Stanishneva-Konovalova et al., 2020). Although such resolution does not allow for visualization of the amino acid side chains, it was enough to trace the backbone and describe the subunit conformations. The seven subunits of the ring were arranged as one unpaired subunit and three pairs, with each pair consisting of subunits in conformations termed A and B (Figure 12.2). This organization is clearly visible at the apical part of the complex resulting in the breakage of the C7 symmetry, which is present in a single ring of GroEL. In case of the OBP chaperonin, the C7 symmetry only remains at the equatorial domains. Another interesting feature of the apical part of the complex is the diminishing density of the unpaired subunit, which is caused by its movement between the A and B conformations (Figure 12.2).

Comparison of the GroEL and gp246 sequences elucidated the reason for the alternating subunit pattern. The residues G192 and G375 composing the upper hinge in GroEL are substituted to large polar residues E191 and N376 in gp246, which cannot provide efficient cis-trans switching. Thus, the apical and the intermediate domains are fixed relative to each other. The bottom hinge (residues P137, G410 in GroEL and P135, G411 in gp246) is functional and provides the existence of A and B conformations. We hypothesize that there are several salt bridges stabilizing the A–B subunit pair: E185–R227, K187–E256, and K381–E254. Although such interactions never occur throughout the functional cycle of GroEL, there are similarities for the equatorial-intermediate part of the subunits: the A subunit resembles the T state of GroEL and the B subunit resembles the Rs1 state. Importantly, while in GroEL conformational transitions are triggered by ATP binding and hydrolysis, the pattern of the OBP chaperonin described above is observed in the absence of nucleotides.

Despite similar patterns of alternating subunits of ϕEL and OBP chaperonins, they also have several differences. In the ϕEL subunit, all domains can move relative to each other, and three subunit conformations were present in the

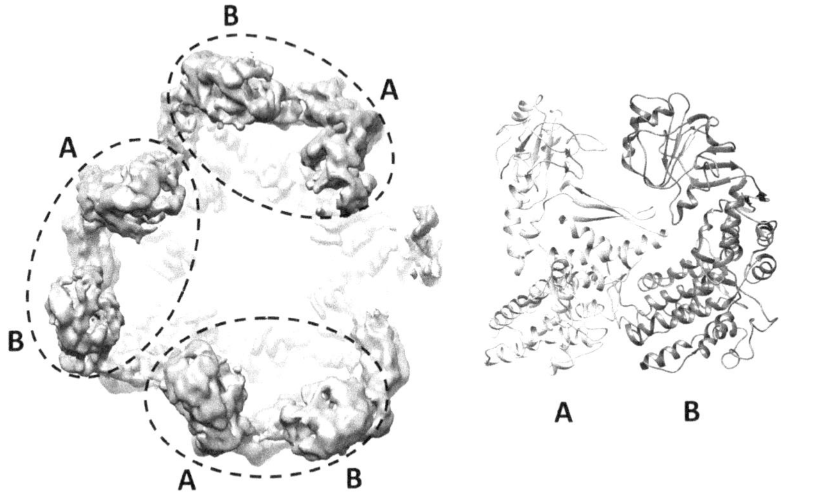

FIGURE 12.2 A cryo-EM density of the OBP chaperonin in the apo-state with A – B subunit pairs encircled (*left*) and an atomic model of A – B subunit pair (*right*).

apo-form. In OBP chaperonin, the apical-intermediate part moves as a single unit relative to the equatorial domain, providing two subunit conformations and a continuous motion between them. And more noticeably, no double-ring structures were found for OBP chaperonin, independently of the nucleotide. The addition of ADP or ATPgammaS had similar effects: Both nucleotides caused the increase in subunit mobility (presumably easing the switch between A and B states), which resulted in a cryo-EM density map with only one A–B subunit pair clearly visible.

Chaperonin from Phage AR9 *B. subtilis*

The third phage chaperonin for which the cryo-EM structure was obtained is the chaperonin from bacteriophage AR9 *B. subtilis* (Sokolova et al., 2022). Characterization by various physicochemical methods proved its chaperone activity—i.e., the ability to prevent aggregation and refold a denatured protein in an ATP-dependent manner (Semenyuk et al., 2020). Negative-stain EM showed that the apo-form is a single ring heptamer like the OBP chaperonin. Cryo-EM was implemented to study the apo-form in more detail. In contrast to the OBP chaperonin, imposing C7 symmetry helped to visualize all subunits with better resolution (4.49 Å for the symmetry-free map and 3.99 Å for the C7-symmetrical map) (Sokolova et al., 2022). This was due to the absence of an asymmetrical pattern observed in two other phage chaperonins, which makes the whole structure more like a single ring of GroEL. Nevertheless, the conformational mobility in the apo-state of AR9 chaperonin is still present, and it was elaborated using the "symmetry expansion" and three-dimensional classification in RELION (Scheres, 2016). This allowed us to describe four subunit conformations where the differences were mostly observed within the apical domains. One of them resembled the conformation of a GroEL subunit with an elevated apical domain, which occurs upon GroES binding (Sokolova et al., 2022). However, apical domain movements in GroEL are coordinated by ATP binding and hydrolysis, while the described subunit conformations of AR9 chaperonin are present in the apo-form.

DISCUSSION/CONCLUSIONS

In summary, phage chaperonins are quite different not only from bacterial and eukaryotic variants, but also from each other. According to the phylogenetic tree reconstruction, predicted phage chaperonins are closer to group I than to group II, but they cannot all be united into an individual group due to their high diversity (Semenyuk et al., 2016). However, ϕEL and OBP chaperonin can be combined into a monophyletic group. So far, structures of three phage chaperonins, from phages EL, OBP, and AR9, were obtained. Figure 12.3 illustrates their apo-states obtained by cryo-EM and compares them to GroEL.

The most obvious structural difference is that in the apo-form GroEL and ϕEL are double-ring tetradecamers, while OBP and AR9 chaperonins are single-ring heptamers. Individual subunits of each ring show a typical three-domain architecture where the equatorial domains are making most of the intra- and, for double-ring complexes, all of the inter-ring contacts. Interestingly, ϕEL and OBP chaperonin display an unusual pattern of subunit conformations within one ring. There are three pairs of subunits (two subunit conformations within a pair) and one unpaired subunit. Several contacts formed within a pair do not exist in GroEL, and they are possible in ϕEL and OBP chaperonin because of the specific orientation of the apical domains. While in GroEL the apical αK-αL helical hairpin points outward from the center of the complex, in two-phage chaperonins it is rotated downward facing the equatorial domain. In OBP chaperonin, rotation of the apical domain is disabled in contrast to ϕEL. Both these phage chaperonins appear to function without a co-chaperone like the members of the second group. However,

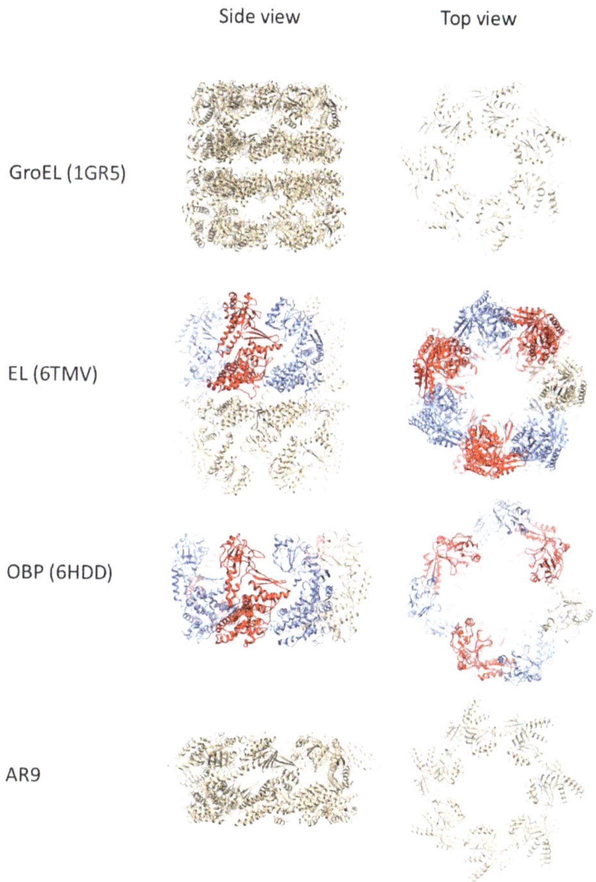

FIGURE 12.3 Atomic models of GroEL and three phage chaperonins. For EL and OBP chaperonins subunits in alternating conformations are colored red and blue.

in chaperonins of the second group, there is a helical extension in the apical domain that provides closure of the folding chamber. The described phage chaperonins are likely to perform their functions without the encapsulation of a substrate protein. According to the model of the functional cycle proposed for the ϕEL, ATP binding is necessary for the subunits to adopt the conformation with high affinity to substrate protein, which is not the case for GroEL. The question of exactly what effect the phage chaperonin has on its native substrate remains open. One possibility is that some phage proteins cannot fully rely on the host chaperonin system for their folding. Alternatively, phage chaperonins may perform an "anti-chaperonin" activity, i.e., to trap their substrate and keep it in an unfolded state until its activation is required. This might be the case for ϕEL and its putative substrate gp188, a cell wall endolysin, which should not be activated until a large number of phage particles is produced (Bracher et al., 2020).

Another interesting question is the origin of phage chaperonins. On the one hand, phage chaperonins may implement an evolutionarily more ancient folding mechanism that existed before the emergence of the encapsulation mechanism. On the other hand, they may originate from cellular chaperonins but have developed a different mechanism of function. Further structural and functional studies should answer these questions and determine the role of intrinsic chaperonins in phage reproduction.

ACKNOWLEDGMENTS

T.B.S.-K. and O.S.S. acknowledge Interdisciplinary Scientific and Educational School of Moscow University "Molecular Technologies of the Living Systems and Synthetic Biology." Authors thank RSF for support (grant 24-44-02003).

REFERENCES

Anfinsen, C. B., Haber, E., Sela, M., & White, F. H. (1961). The kinetics of formation of native ribonuclease during oxidation of the reduced polypeptide chain. *Proceedings of the National Academy of Sciences*, *47*(9), 1309–1314. https://doi.org/10.1073/pnas.47.9.1309

Ang, D., Richardson, A., Mayer, M. P., Keppel, F., Krisch, H., & Georgopoulos, C. (2001). Pseudo-T-even bacteriophage RB49 encodes CocO, a cochaperonin for GroEL, which can substitute for *Escherichia coli*'s GroES and bacteriophage T4's Gp31. *Journal of Biological Chemistry*, *276*(12), 8720–8726. https://doi.org/10.1074/jbc.M008477200

Bakkes, P. J., Faber, B. W., van Heerikhuizen, H., & van der Vies, S. M. (2005). The T4-encoded cochaperonin, gp31, has unique properties that explain its requirement for the folding of the T4 major capsid protein. *Proceedings of the National Academy of Sciences*, *102*(23), 8144–8149. https://doi.org/10.1073/pnas.0500048102

Barraclough, R., & Ellis, R. J. (1980). Protein synthesis in chloroplasts IX. Assembly of newly-synthesized large subunits into ribulose bishopshate carboxylase in isolated intact pea chloroplasts. *Biochimica et Biophysica Acta (BBA) - Nucleic Acids and Protein Synthesis*, *608*(1), 19–31. https://doi.org/10.1016/0005-2787(80)90129-X

Bracher, A., & Hartl, F. U. (2013). Chaperonins. In *Encyclopedia of Biological Chemistry* (pp. 456–460). Elsevier. https://doi.org/10.1016/B978-0-12-378630-2.00080-3

Bracher, A., Paul, S. S., Wang, H., Wischnewski, N., Hartl, F. U., & Hayer-Hartl, M. (2020). Structure and conformational cycle of a bacteriophage-encoded chaperonin. *PLoS One*, *15*, e0230090. https://doi.org/10.1101/2020.02.24.962803

Chandrasekhar, G. N., Tilly, K., Woolford, C., Hendrix, R., & Georgopoulos, C. (1986). Purification and properties of the groES morphogenetic protein of *Escherichia coli*. *The Journal of Biological Chemistry*, *261*(26), 12414–12419.

Cheng, M. Y., Hartl, F.-U., Martin, J., Pollock, R. A., Kalousek, F., Neuper, W., Hallberg, E. M., Hallberg, R. L., & Horwich, A. L. (1989). Mitochondrial heat-shock protein hsp60 is essential for assembly of proteins imported into yeast mitochondria. *Nature*, *337*(6208), 620–625. https://doi.org/10.1038/337620a0

Clare, D. K., Bakkes, P. J., van Heerikhuizen, H., van der Vies, S. M., & Saibil, H. R. (2009). Chaperonin complex with a newly folded protein encapsulated in the folding chamber. *Nature*, *457*(7225), 107–110. https://doi.org/10.1038/nature07479

Georgopoulos, C. P., Hendrix, R. W., Casjens, S. R., & Kaiser, A. D. (1973). Host participation in bacteriophage lambda head assembly. *Journal of Molecular Biology*, *76*(1), 45–60. https://doi.org/10.1016/0022-2836(73)90080-6

Georgopoulos, C. P., Hendrix, R. W., Kaiser, A. D., & Wood, W. B. (1972). Role of the host cell in bacteriophage morphogenesis: Effects of a bacterial mutation on T4 head assembly. *Nature New Biology*, *239*(89), 38–41. https://doi.org/10.1038/newbio239038a0

Goldberger, R. F., Epstein, C. J., & Anfinsen, C. B. (1963). Acceleration of reactivation of reduced bovine pancreatic ribonuclease by a microsomal system from rat liver. *The Journal of Biological Chemistry*, *238*, 628–635.

Hemmingsen, S. M., Woolford, C., van der Vies, S. M., Tilly, K., Dennis, D. T., Georgopoulos, C. P., Hendrix, R. W., & Ellis, R. J. (1988). Homologous plant and bacterial proteins chaperone oligomeric protein assembly. *Nature*, *333*(6171), 330–334. https://doi.org/10.1038/333330a0

Hendrix, R. W. (1979). Purification and properties of groE, a host protein involved in bacteriophage assembly. *Journal of Molecular Biology*, *129*(3), 375–392. https://doi.org/10.1016/0022-2836(79)90502-3

Hertveldt, K., Lavigne, R., Pleteneva, E., Sernova, N., Kurochkina, L., Korchevskii, R., Robben, J., Mesyanzhinov, V., Krylov, V. N., & Volckaert, G. (2005). Genome comparison of *Pseudomonas aeruginosa* large phages. *Journal of Molecular Biology*, *354*(3), 536–545. https://doi.org/10.1016/j.jmb.2005.08.075

Hohn, T., Hohn, B., Engel, A., Wurtz, M., & Smith, P. R. (1979). Isolation and characterization of the host protein groE involved in bacteriophage lambda assembly. *Journal of Molecular Biology*, *129*(3), 359–373. https://doi.org/10.1016/0022-2836(79)90501-1

Horwich, A. L., & Fenton, W. A. (2020). Chaperonin-assisted protein folding: A chronologue. *Quarterly Reviews of Biophysics*, *53*, e4. https://doi.org/10.1017/S0033583519000143

Hunt, J. F., van der Vies, S. M., Henry, L., & Deisenhofer, J. (1997). Structural adaptations in the specialized bacteriophage T4 co-chaperonin Gp31 expand the size of the Anfinsen cage. *Cell*, *90*(2), 361–371. https://doi.org/10.1016/S0092-8674(00)80343-8

Kelley, P. M., & Schlesinger, M. J. (1978). The effect of amino acid analogues and heat shock on gene expression in chicken embryo fibroblasts. *Cell*, *15*(4), 1277–1286. https://doi.org/10.1016/0092-8674(78)90053-3

Kudryavtseva, S. S., Pichkur, E. B., Yaroshevich, I. A., Mamchur, A. A., Panina, I. S., Moiseenko, A. V., Sokolova, O. S., Muronetz, V. I., & Stanishneva-Konovalova, T. B. (2021). Novel cryo-EM structure of an ADP-bound GroEL–GroES complex. *Scientific Reports*, *11*(1), 1–9. https://doi.org/10.1038/s41598-021-97657-x

Kurochkina, L. P., Semenyuk, P. I., Orlov, V. N., Robben, J., Sykilinda, N. N., & Mesyanzhinov, V. V. (2012). Expression and functional characterization of the first bacteriophage-encoded chaperonin. *Journal of Virology*, *86*(18), 10103–10111. https://doi.org/10.1128/JVI.00940-12

Marine, R. L., Nasko, D. J., Wray, J., Polson, S. W., & Wommack, K. E. (2017). Novel chaperonins are prevalent in the virioplankton and demonstrate links to viral biology and ecology. *The ISME Journal*, *11*(11), 2479–2491. https://doi.org/10.1038/ismej.2017.102

Neidhardt, F. C., Phillips, T. A., VanBogelen, R. A., Smith, M. W., Georgalis, Y., & Subramanian, A. R. (1981). Identity of the B56.5 protein, the A-protein, and the groE gene product of *Escherichia coli*. *Journal of Bacteriology*, *145*(1), 513–520. https://doi.org/10.1128/jb.145.1.513-520.1981

Pelham, H. R. B. (1986). Speculations on the functions of the major heat shock and glucose-regulated proteins. *Cell*, *46*(7), 959–961. https://doi.org/10.1016/0092-8674(86)90693-8

Scheres, S. H. W. (2016). Processing of structurally heterogeneous cryo-EM data in RELION. *Methods in Enzymology*, *579*, 125–157. https://doi.org/10.1016/bs.mie.2016.04.012

Semenyuk, P. I., Moiseenko, A. V., Sokolova, O. S., Muronetz, V. I., & Kurochkina, L. P. (2020). Structural and functional diversity of novel and known bacteriophage-encoded chaperonins. *International Journal of Biological Macromolecules*, *157*, 544–552. https://doi.org/10.1016/j.ijbiomac.2020.04.189

Semenyuk, P. I., Orlov, V. N., Sokolova, O. S., & Kurochkina, L. P. (2016). New GroEL-like chaperonin of bacteriophage OBP *P. fluorescens* suppresses thermal protein aggregation in an ATP-dependent manner. *Biochemical Journal*, 2383–2393. https://doi.org/10.1042/BCJ20160367

Skjærven, L., Cuellar, J., Martinez, A., & Valpuesta, J. M. (2015). Dynamics, flexibility, and allostery in molecular chaperonins. *FEBS Letters*, *589*(19), 2522–2532. https://doi.org/10.1016/j.febslet.2015.06.019

Sokolova, O. S., Pichkur, E. B., Maslova, E. S., Kurochkina, L. P., Semenyuk, P. I., Konarev, P. V., Samygina, V. R., & Stanishneva-Konovalova, T. B. (2022). Local flexibility of a new single-ring chaperonin encoded by bacteriophage AR9 *Bacillus subtilis*. *Biomedicines*, *10*(10), 2347. https://doi.org/10.3390/biomedicines10102347

Stanishneva-Konovalova, T. B., Semenyuk, P. I., Kurochkina, L. P., Pichkur, E. B., Vasilyev, A. L., Kovalchuk, M. V., Kirpichnikov, M. P., & Sokolova, O. S. (2020). Cryo-EM reveals an asymmetry in a novel single-ring viral chaperonin. *Journal of Structural Biology*, *209*(2), 107439. https://doi.org/10.1016/j.jsb.2019.107439

Takano, T., & Kakefuda, T. (1972). Involvement of a bacterial factor in morphogenesis of bacteriophage capsid. *Nature New Biology*, *239*(89), 34–37. https://doi.org/10.1038/newbio239034a0

Tilly, K., VanBogelen, R. A., Georgopoulos, C., & Neidhardt, F. C. (1983). Identification of the heat-inducible protein C15.4 as the groES gene product in *Escherichia coli*. *Journal of Bacteriology*, *154*(3), 1505–1507. https://doi.org/10.1128/jb.154.3.1505-1507.1983

Tissiéres, A., Mitchell, H. K., & Tracy, U. M. (1974). Protein synthesis in salivary glands of *Drosophila melanogaster*: Relation to chromosome puffs. *Journal of Molecular Biology*, *84*(3), 389–398. https://doi.org/10.1016/0022-2836(74)90447-1

van der Vies, S. M., Gatenby, A. A., & Georgopoulos, C. (1994). Bacteriophage T4 encodes a co-chaperonin that can substitute for *Escherichia coli* GroES in protein folding. *Nature*, *368*(6472), 654–656. https://doi.org/10.1038/368654a0

Venetianer, P., & Straub, F. B. (1963). The enzymic reactivation of reduced ribonuclease. *Biochimica et Biophysica Acta*, *67*, 166–168. https://doi.org/10.1016/0006-3002(63)91812-2

Yamamori, T., Ito, K., Nakamura, Y., & Yura, T. (1978). Transient regulation of protein synthesis in *Escherichia coli* upon shift-up of growth temperature. *Journal of Bacteriology*, *134*(3), 1133–1140. https://doi.org/10.1128/jb.134.3.1133-1140.1978

Yan, X., Shi, Q., Bracher, A., Miličić, G., Singh, A. K., Hartl, F. U., & Hayer-Hartl, M. (2018). GroEL ring separation and exchange in the chaperonin reaction. *Cell*, *172*(3), 605–617.e11. https://doi.org/10.1016/j.cell.2017.12.010

13 Cryo-EM Structure of Monkeypox Virus DNA Polymerase Holoenzyme

Yi Shi and George F Gao

INTRODUCTION

Monkeypox virus, also abbreviated as mpox virus, was firstly discovered in monkeys at a research institute in Denmark in 1959 (Magnus et al., 1959), and 11 years later, the first human mpox case was reported in a nine-month-old boy in the Democratic Republic of Congo (Ladnyj et al., 1972). Since then, the mpox disease has been mainly endemic in Central and West Africa, and in 2003, the mpox virus infection was identified in the United States for the first time outside of Africa (Hammarlund et al., 2005; Nolen et al., 2016; Pinedo et al., 2022; Rezza 2019). Prior to 2022, mpox had been considered a neglected infectious disease; only a few cases were identified outside of Africa, and most of those were associated with traveling in Africa or contact with animals there. Since May 2022, a global outbreak of mpox was identified predominantly in men who have sex with men (MSM) across the world (Ferré et al., 2022). As of October 24, 2023, a total of 90,656 human infection cases have been reported to the World Health Organization (WHO) from 115 countries/regions.

Mpox virus (MPXV) belongs to the orthopoxvirus genus of the *Poxviridae* family, which also includes vaccinia virus (VACV), cowpox virus, and variola virus (the causative virus of smallpox). It is a large, enveloped, double-stranded DNA virus that infects a wide range of host species and replicates exclusively in the cytoplasm (Falendysz et al., 2017). Many functional and molecular understandings of poxvirus are resulting from previous studies of vaccina virus, which is the prototype virus of the *Poxviridae* family. The mature poxvirus virions enter cells either via macropinocytosis and release from the endosome compartment or by direct fusion with the cell plasma membrane (Czarnecki and Traktman, 2017; Greseth and Traktman, 2022). Upon virus entry, the lateral bodies become dispersed, but the core remains intact and transports to a perinuclear position (Greseth and Traktman, 2022; Huang et al., 2022). Early transcription events will occur within the core with a proteinaceous wall that allows the entry of NTP substrates and egress of viral messenger RNAs (mRNAs), but no proteins can traffic into or out of the core. After early gene expression, the core will undergo a process of uncoating, which is complex and poorly understood. The proteinaceous wall is dissolved, and the viral DNA genome is released into the cytoplasm. Immediately, the naked parental DNA genome and the factors that are necessary to promote replication factory formation and the initiation of DNA replication will be surrounded by the endoplasmic reticulum (ER) membrane and will then form the replication factory. After robust DNA replication, as well as intermediate and late gene expression, the nascent viral genomes are encapsulated to form immature virions with nucleoid and then become mature virions through a series of maturation steps (Greseth and Traktman, 2022; Huang et al., 2022) (Figure 13.1).

The poxvirus genome is a roughly 195-kb linear duplex that is usually AT rich (~67%) and has covalently closed hairpin termini (Czarnecki and Traktman, 2017). Taking vaccinia virus as an example, the hairpin termini (104 nucleotides) contain almost exclusively A and T residues, and each hairpin terminus has 12 extrahelical residues (10 on one strand and 2 on the other) in addition to 44 fully paired bases and a four-nucleotide terminal loop (Baroudy et al., 1982; Baroudy et al., 1982; Czarnecki and Traktman, 2017). Next to the hairpin terminus is an additional 87 bp of sequence, which is found at both genomic telomeres (Baroudy et al., 1982). This 87-bp sequence contains key elements required for replication of exogenous templates and resolution for concatemeric intermediates (Merchlinsky, 1990; Du and Traktman, 1996). Subsequently after the 87-bp sequence, clusters of tandemly repeated sequences are found, and these repeats are thought to be involved in inter- and intra-genomic recombination events (Wittek and Moss, 1980) (Figure 13.1). The general scheme of the terminal sequence of the genome is conserved among poxviruses, although the precise numbers and positions of extrahelical residues and repeats could be different.

Because the poxvirus genome replication factory is physically separated from the host cell nucleus, the virus encodes its own proteins, which are required for DNA replication. Taking mpox virus as an example, a few viral proteins are known to be involved in the replication process, including the DNA polymerase holoenzyme (consisting of the catalytic polymerase F8 and the heterodimeric processivity factor A22-E4), the ATPase/primase E5, the single-stranded DNA binding protein I3, the scaffold protein H5, the DNA ligase A50, the flap endonuclease (FEN) 1-like nuclease G5, and the Holliday junction resolvase A23. Future studies are still needed to find more host and viral proteins that participate in the replication process.

Previous genetic, biochemical, and structural studies on the VACV E9-A20-D4 core replication machinery have advanced our understanding of poxvirus DNA replication. VACV E9 was recognized as a member of the B-family

DOI: 10.1201/9781003326106-16

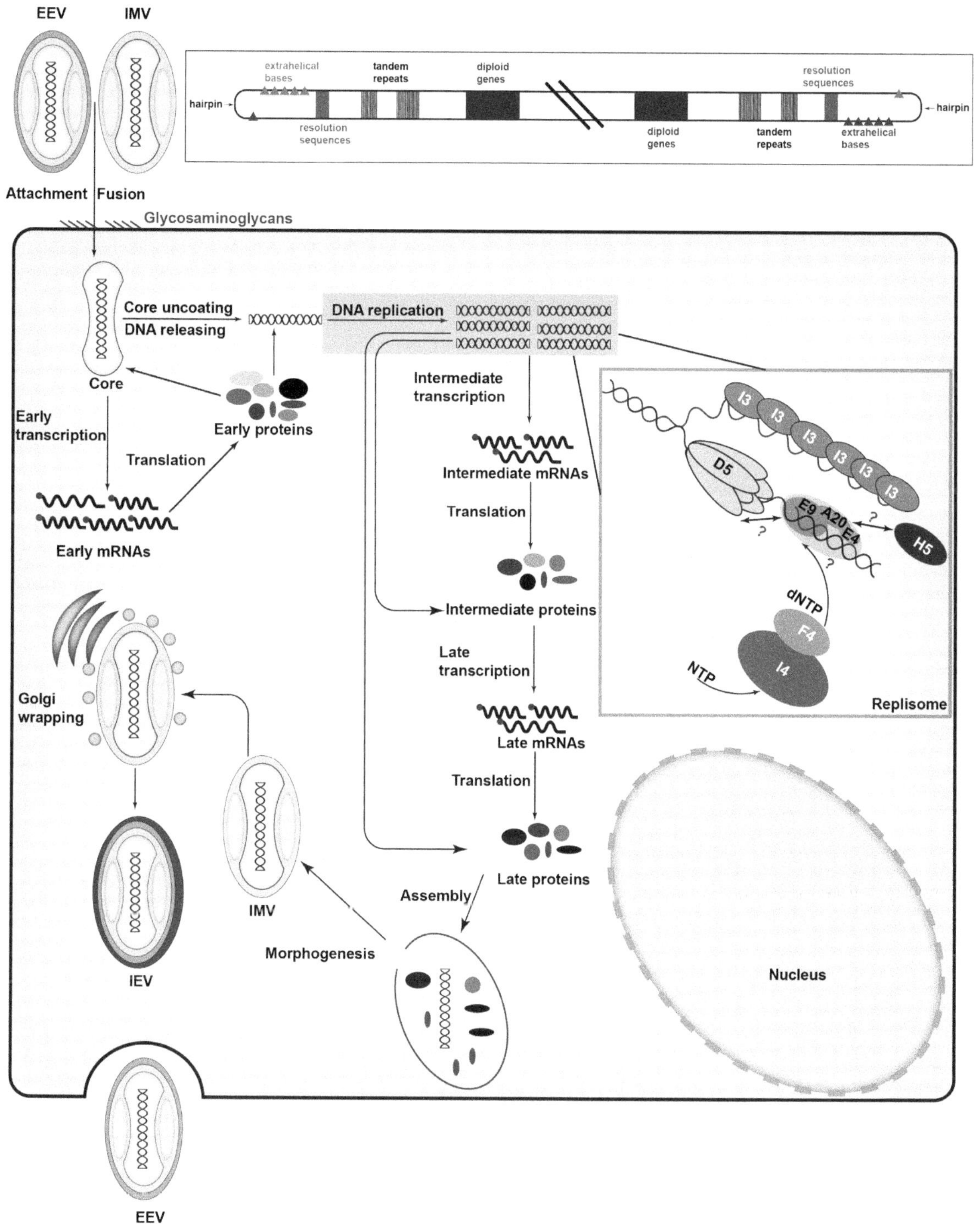

FIGURE 13.1 The life cycle of vaccinia virus. Virions could enter host cells by macropinocytosis or by direct fusion of viral membrane with the plasma membrane. Upon the virion entry, the viral core is released into cytoplasm of the host cell and undergoes morphological changes associated with activation of early transcription. The messenger RNAs (mRNAs) encoding early proteins are synthesized by DNA-dependent RNA polymerase complex carried within the virion and released from core into cytoplasm for translating. After early gene expression, the core undergoes a process of uncoating, in which the core wall dissolves, and the DNA genome is released into cytoplasm. The naked parental DNA genome is surrounded by the endoplasmic reticulum (ER) membrane and interacts with several early proteins including E9, A22, D4, H5, B1, I3 and so on to promote replication factory formation and the initiation of DNA replication. After robust DNA replication, as well as intermediate and late gene expression, the nascent viral genomes are encapsulated to form immature virions with nucleoid and then become mature virions through a series of maturation steps. EEV, extracellular enveloped virions; IMV, intracellular mature virions; IEV, intracellular enveloped virions.

DNA polymerase, possessing both 5′ to 3′ polymerizing activity and 3′ to 5′ proofreading exonuclease activity (Challberg and Englund, 1979; McDonald and Traktman, 1994 a, b)). Structural analysis has revealed the canonical features of DNA polymerases and five poxvirus-specific insertions (Czarnecki and Traktman, 2017). The E9 polymerase alone does not have processive DNA synthesis activity unless it is bound to its heterodimeric cofactor A20/D4 (Klemperer et al., 2001; Mcdonald and Traktman, 1994; McDonald et al., 1997; Stanitsa et al., 2006). Although poxvirus DNA polymerase shares many features with other B-family polymerases, the processivity factor is unique. In VACV, A20 serves as an essential bridge to link E9 and D4 together and shares no homology with viral proteins beyond poxvirus. The N-terminal domain of A20 binds to D4 (Contesto-Richefeu et al., 2016; Moss et al., 2014), and its C-terminal domain binds to one insertion in the palm domain of E9 (Bersch et al., 2021). Given that the DNA replication machinery is extremely conserved for orthopoxviruses, with a sequence identity of more than 97% between VACV and MPXV, the results obtained for VACV could also be applied to MPXV. However, we are still awaiting a reliable high-resolution structure of the replicating state of the *Orthopoxvirus* polymerase holoenzyme, and the mode of operation of the processivity factor needs to be elucidated.

BIOCHEMICAL CHARACTERIZATION OF THE PURIFIED POLYMERASE HOLOENZYME PROTEINS

Recently, the MPXV F8 polymerase and the A22-E4 heterodimer were co-expressed using the baculovirus expression system, and the homogenous F8-A22-E4 heterotrimer protein was purified for enzymatic and structural studies. Using a 38-nucleotide (nt) template DNA with a 24-nt primer DNA, the holoenzyme heterotrimer displayed weak primer extension activity in the reaction buffer with dNTP (Peng et al., 2023). Moreover, the holoenzyme possessed exonuclease activity using an enzymatic assay without dNTP substrates, and the holoenzyme could completely degrade the primer DNA in an ATP-independent manner. An exonuclease-deficient F8 mutant protein was also prepared, and the F8-mutant-A22-E4 holoenzyme did not cleave the primer-template DNA, thereby demonstrating a much stronger product band than the wild-type holoenzyme protein (Peng et al., 2023).

Another research group used the human embryonic kidney 293 freestyle (HEK293F) cells to express the MPXV DNA polymerase holoenzyme and found that the holoenzyme could exist as both dimer and monomer in its DNA-free form (Li et al., 2023). This difference might result from the lower salt concentration in the buffer. Moreover, addition of double-stranded DNA in the protein solution can shift the dimer to monomer form, exposing the DNA binding site for active function (Li et al., 2023).

OVERALL ARCHITECTURE OF F8-A22-E4 POLYMERASE HOLOENZYME

The MPXV polymerase holoenzyme consists of three protein subunits, F8, A22, and E4. The F8 structure can be traced for 1004 residues, except for the last two residues, and the classical N-terminal domain (NTD), exonuclease domain (Exo), palm domain, fingers domain, and thumb domain were observed in a closed conformation (Figure 13.2). Five "poxvirus-specific" insertion regions in MPXV F8 can also be observed, like those inserts seen in the VACV E9 structure (Czarnecki and Traktman, 2017) (Figure 13.2). The 218-residue MPXV E4 forms a compact structure with a four-stranded β-sheet surrounded by helical elements, resembling the VACV D4 structure (Burmeister et al., 2015; Contesto-Richefeu et al., 2014; Schormann et al., 2015) (Figure 13.2). The 426-residue MPXV A22 structure can be divided into three domains: the A22-NTD, middle domain (Mid), and C-terminal domain (CTD) (Figure 13.2). The Mid of A22 was aligned on the online Dali server, and it showed high structural similarities with the African swine fever virus (ASFV) DNA ligase and the bacteriophage T4 DNA ligase (Shi et al., 2018). The Mid can be further divided into two subdomains: an adenylation domain (OD) that mostly resembles the OD of ASFV DNA ligase (Chen et al., 2019), and an OB-fold domain (OB) that mostly resembles the OB of *Thermus filiformis* DNA ligase (Shi et al., 2018). Further enzymatic assay showed that the MPXV polymerase holoenzyme did not possess ordinary ligase activity similar to that of T4 ligase (Peng et al., 2023). Compared with the structures of adenylation domains from the T4 and ASFV ligases, the putative ligase active site of the A22 Mid is replaced by hydrophobic and negatively charged residues, which may prevent the binding of ATP (Peng et al., 2023). As the Mid of A22 lacks the essential DNA-binding domain, A22 may comprise a degenerative ligase domain acting simply as a flexible linker.

Upon double-stranded DNA (dsDNA) binding, the MPXV DNA polymerase holoenzyme is stabilized by pairwise interactions between the F8, A22, and E4 subunits (Li et al., 2023; Peng et al., 2023; Xu et al., 2023). A22 acts as a bridge to bind E4 and F8 via the A22-NTD and CTD, respectively (Figure 13.2). The interactions are almost identical to those of their VACV counterparts (Contesto-Richefeu et al., 2016; Moss et al., 2014; Stanitsa et al., 2006). A previous study proposed a VACV polymerase holoenzyme model with an elongated shape of the A20-D4 cofactor, leading to a ~150 Å distance between the E9 polymerase active site and the D4 DNA binding site (Sèle et al., 2013). However, in the replicating MPXV holoenzyme structure, the A22-E4 cofactor folds back, and E4 directly interacts with the Exo domain of F8 at two sites, one where Trp36 and Arg39 of E4 form hydrogen bonds and hydrophobic interactions with Phe179 and Leu278 from F8 Exo (Peng et al., 2023), and another where Asn165 of E4 forms a hydrogen bond with Asn303 from F8 Exo (Peng et al., 2023).

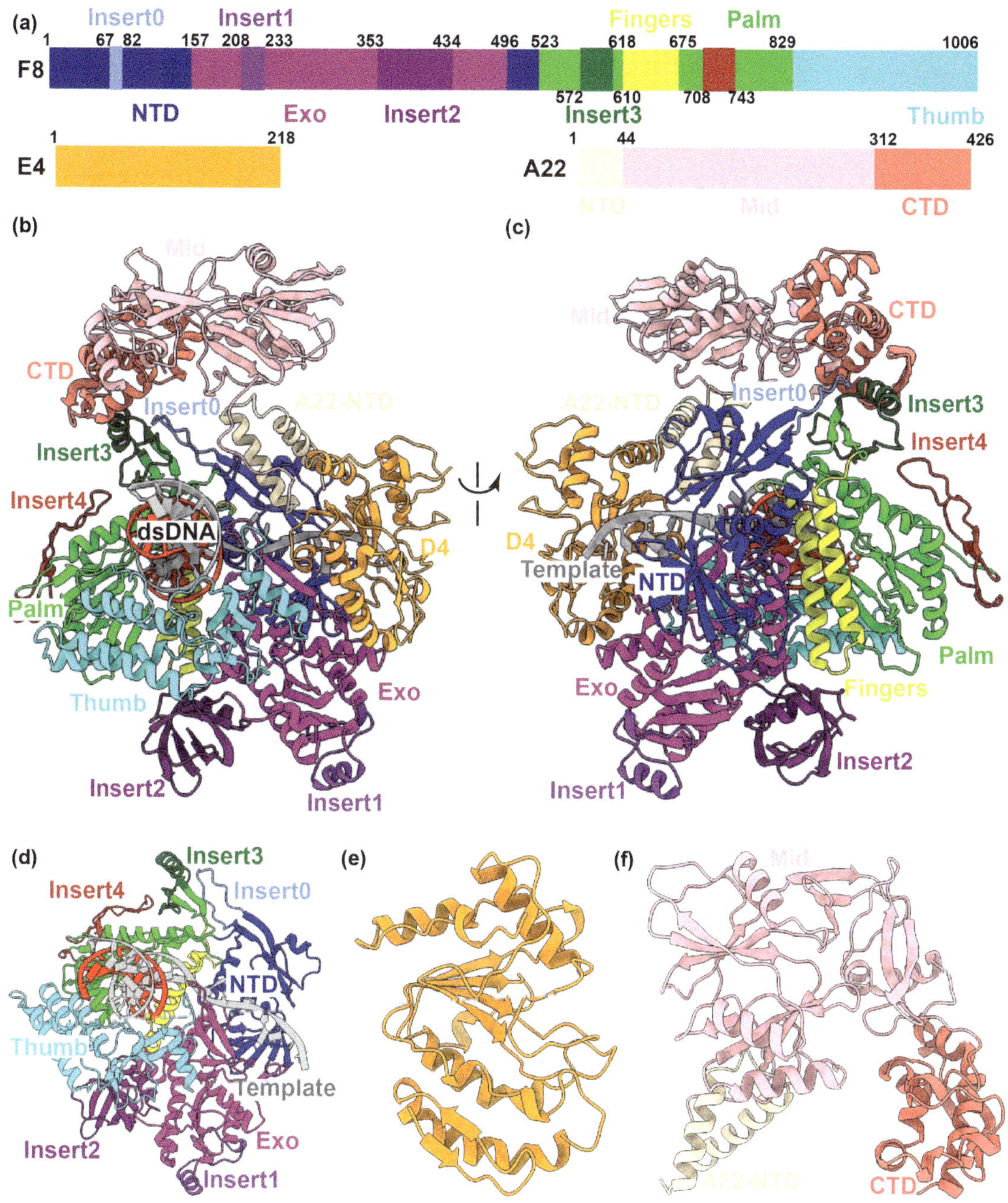

FIGURE 13.2 Overall structure of the replicating MPXV polymerase holoenzyme. (a) Schematic diagrams of the domain architecture of MPXV F8, A22, and E4. The F8 could be divided into five domains, NTD, blue; Exo, magenta; palm, green; fingers, yellow; thumb, cyan. Compared to other B family polymerases, F8 contains five inserted elements, insert 0, cornflower blue; insert 1, dark orchid; insert 2, purple; insert 3, dark green; insert 4, brown. A22 is colored by domains, NTD, wheat; Mid, pink; CTD, salmon; E4, orange; Template strand, grey; Product strand, red. (b–c) The atomic model of the replicating MPXV polymerase holoenzyme. (d–f) The structure of F8-dsDNA complex (d), E4 (e), and A22 (f).

In its DNA-free form, the holoenzyme can form a dimer structure (Li et al., 2023). For clarity, two protomers in the dimer structure are referred to as protomer A: F8/A22/E4 and protomer B: F8′/A22′/E4′ (Figure 13.3). In contrast to the DNA-binding form, the holoenzyme in the apo form undergoes dramatic conformational change in the A22/A22′ subunits. Taking protomer A as an example, the A22 protein displays an extended conformation, and the C-terminal domain of A22 interacts with the insert 3 region of the F8 subunit in the same promoter through hydrophobic interactions and hydrogen bonds (Figure 13.3). Moreover, the mid domain of A22 also contacts with the thumb domain of F8′ in the protomer B, creating an extra interface (Figure 13.3). The interface between the mid domain of A22 and

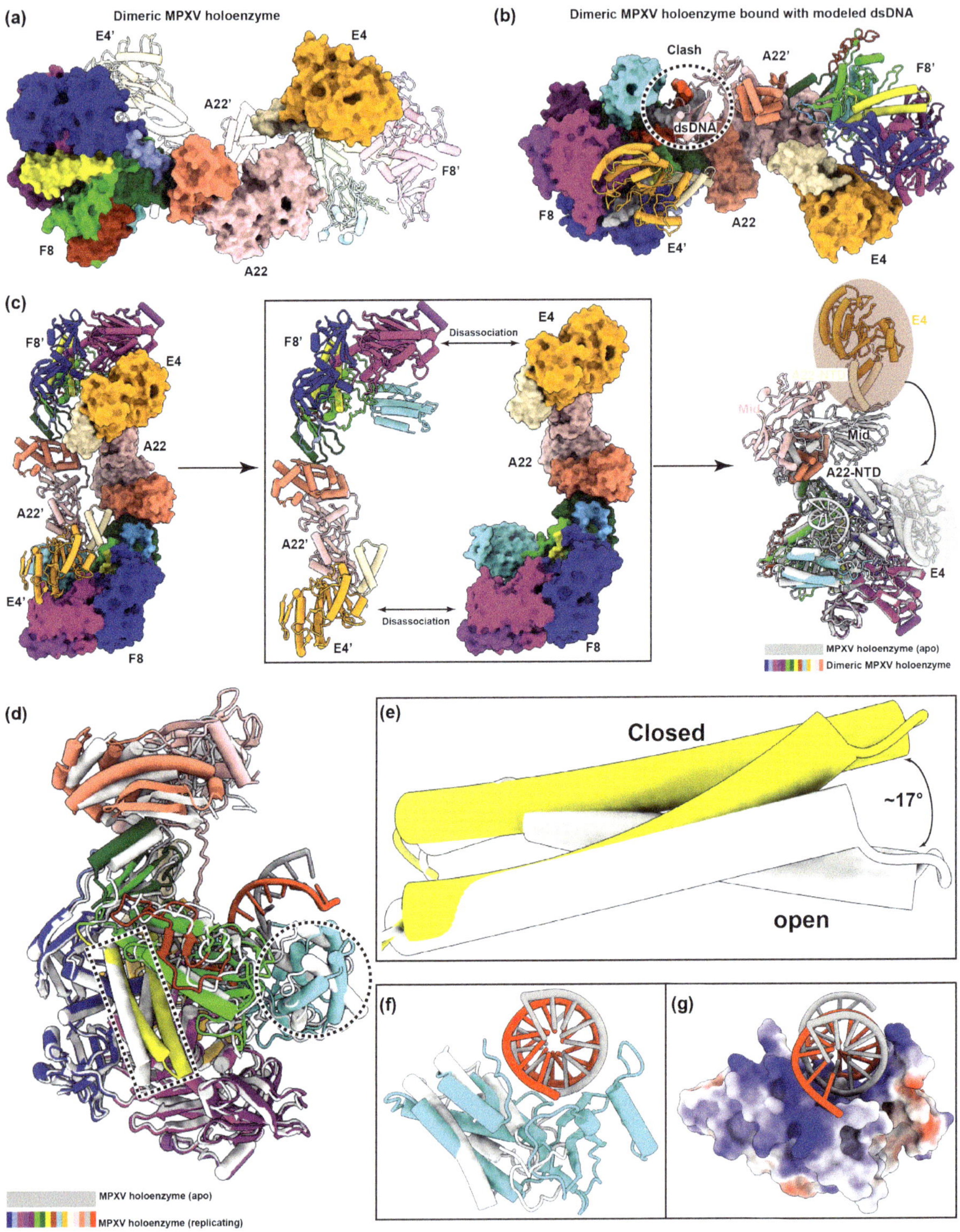

FIGURE 13.3 Conformational changes of MPXV polymerase holoenzyme. (a) The structure of dimeric MPXV holoenzyme. In the dimeric MPXV holoenzyme, the protomer adopts an extended conformation, in which the A22 interacts with the palm domain of F8′ and E4 binds with the Exo domain of F8′. (b) The modeled structure of dimeric MPXV holoenzyme in complex with dsDNA. For the MPXV holoenzyme dimer, it could not accommodate dsDNA because of the steric clash between dsDNA and the mid domain of A22. (c) The conformational changes of dimeric MPXV polymerase holoenzyme. Upon dsDNA binding, the MPXV holoenzyme dimer would disassociate into two monomers at interfaces between E4 and F8′, and E4′ and F8. Then the A22-E4 (or A22′-E4′) folds back and interacts with its own F8 (or F8′) subunit. (d) Comparison of structures of MPXV holoenzyme in apo and replicating states. (e–g) The conformation changes of the fingers domain (e) and thumb domain (f–g).

the thumb domain of F8′ possesses some flexibility, and the interaction details cannot be clearly defined.

When the dsDNA is modeled in the dimeric holoenzyme structure, an obvious steric clash can be observed between the dsDNA and the mid domain of A22 (Figure 13.3). Therefore, the dimeric holoenzyme can be recognized as an autoinhibition state, and transition from the autoinhibition state to the active state accompanies the conformational change of the A22 subunit (Figure 13.3). Previous study has proposed a putative working model that A22 undergoes a dramatic movement of the mid domain and N-terminal domain, in which the dissociation of A22′ could lead to the replacement of the interface of A22′-E4 into A22-E4, and therefore the loss of the interface between the thumb domain of F8 and A22′ will release the dsDNA binding site of F8 (Li et al., 2023). However, the E4-F8 interface has the weakest interaction force compared to the A22-E4 interface and the F8-A22 interfaces. Thus, another plausible model should be considered that the A22-E4 heterodimer can undergo dramatic conformational change and movement, and in the autoinhibited holoenzyme dimer structure, the A22-E4 from promoter A interacts with the F8′ of protomer B, while in the active holoenzyme monomer structure, the A22-E4 folds back and interacts with its own F8 subunit (Figure 13.3).

PRIMER-TEMPLATE DNA RECOGNITION BY THE POLYMERASE COMPLEX

To capture the replicating conformation of the MPXV F8-A22-E4 polymerase holoenzyme, the 3′-H modified primer/template DNA and the polymerase holoenzyme were incubated in the reaction buffer with dTTP substrate. The holoenzyme/DNA complex was resolved to ~2.8 Å (Peng et al., 2023). The EM map shows the key structural features of all proteins and DNA elements. Although the density of the 5′-end template was weak, the main chains could be traced using the unsharpened EM density map to demonstrate the template entry channel. The structure of the holoenzyme/DNA complex contains one F8, one A22, one E4, and the primer/template DNA, as well as an incoming dTTP substrate. F8, A22, and E4 form pairwise interactions with each other (Peng et al., 2023).

The structure of the MPXV polymerase holoenzyme/DNA complex contains 22-nt DNA in the template strand, 14-nt DNA in the primer strand, and the incoming dTTP, as well as a magnesium ion that may serve as a catalytic ion near the active site. The double-stranded primer-template DNA binds in a groove formed between the palm and thumb domains of F8, and the single-stranded 5′ extension of the template strand probably passes through a channel formed by the NTD and Exo of F8 and the E4 subunit in an orientation perpendicular to the DNA duplex (Peng et al., 2023). The template DNA has 12 unpaired nucleotides at the 5′ end, but only two of them are well-ordered with a defined base structure. For the remaining 10 unpaired bases, we can only trace partial phosphate–ribose backbones (C6 to T11) because of the weak EM density (Peng et al., 2023).

Upon primer-template DNA binding, F8 polymerase undergoes conformational changes, a common feature of the B family of DNA polymerases. Comparison of the MPXV F8 in this holoenzyme/DNA complex structure with the VACV apo E9 structure (Tarbouriech et al., 2017) or MPXV apo holoenzyme structure (Li et al., 2023), which has high sequence identity, shows that the finger domain rotates toward the palm domain by approximately 17° in the replicating state (Li et al., 2023; Peng et al., 2023; Tarbouriech et al., 2017) (Figure 13.3). This rotation drags the positively charged Arg634 and Lys661 of the fingers domain closer to the active site, where they can interact with the triphosphate group of incoming dNTP. The rotated fingers domain interacts with the Exo, and this interaction further stabilizes the closed conformation of the fingers domain. Moreover, the thumb domain also makes a significant rotation to wrap around the primer-template DNA duplex on its minor groove side (Peng et al., 2023) (Figure 13.3). The DNA duplex is accommodated in a positively charged groove of the thumb domain, as observed in other B-family DNA polymerases (Bergen et al., 2013; Franklin et al., 2001; Peng et al., 2023; Swan et al., 2009; Wang and Yang 2009) (Figure 13.3).

The modeled double-stranded DNA helix is formed by 14 base pairs from the primer-template DNA and maintains a B-form conformation (Figure 13.4). Extensive protein–DNA interactions are observed between the primer-template DNA and F8, with a total of 47 residues from F8 directly participating in DNA binding (within 4.0 Å distance, 29 residues to the template strand, and 18 residues to the primer strand) (Figure 13.4). Most of the key residues involved in primer-template DNA binding are highly conserved among different B-family DNA polymerases (Bergen et al., 2013; Franklin et al., 2001; Peng et al., 2023; Swan et al., 2009; Wang and Yang, 2009). Protein–DNA interactions mainly involve the phosphodiester backbone of the DNA, with many interactions directly with the phosphate groups (Figure 13.4). Interactions with the template strand phosphates are largely hydrogen bonds to the main or side chains of 13 residues from the thumb, palm, and fingers domains, Exo, and NTD of F8 (Figure 13.4), whereas the primer strand is bound by both electrostatic interactions and hydrogen bonds with the nine residues from the thumb and palm domains and Exo of F8 (Figure 13.4). There is little contact of F8 with the base pairs of primer-template DNA, except for one hydrogen bond interaction between R832 of the thumb domain and the base of T22 from the primer strand, which may be important for stabilizing the B-form conformation of the DNA duplex (Figure 13.4). This is consistent with the fact that the enzymatic activity of F8 does not rely on a specific sequence during the elongation step. In addition, residue N675 of the fingers domain forms a hydrogen bond with the base of unpaired A12 from the template strand, and this interaction may be responsible for the kinking of the single-stranded 5′ extension of the template strand near the active site.

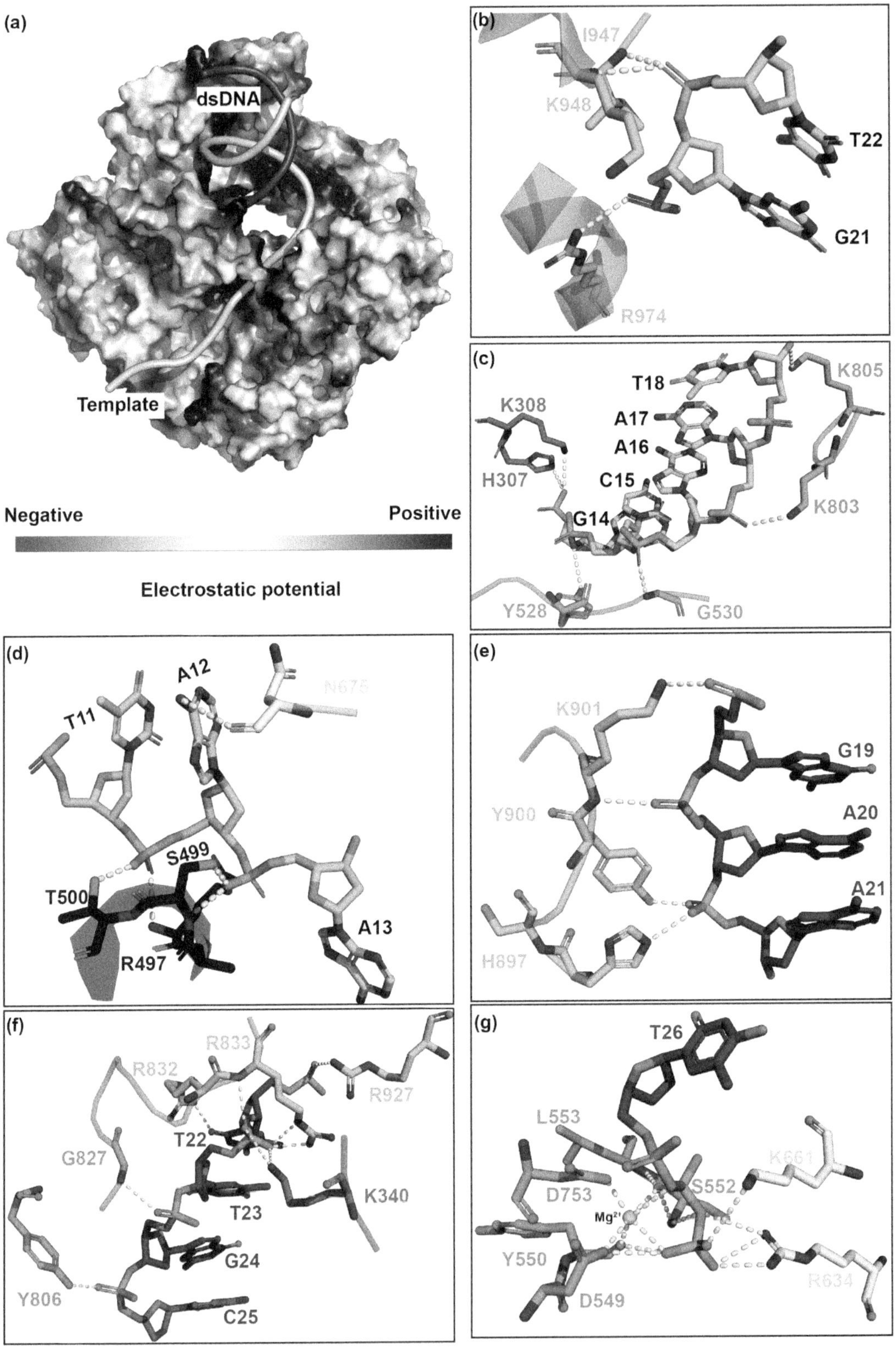

FIGURE 13.4 The interaction details between dsDNA and F8 polymerase. (a) The interface between F8 polymerase and dsDNA. The dsDNA mainly interacts with the positive channel of F8 polymerase. The F8 polymerase is shown as a surface and colored by electrostatic potential. (b–d) Major interactions between the template DNA and F8. (e–f) Major interactions between the primer DNA and F8. (g) Major interactions between the incoming dTTP and F8. The critical residues are shown as sticks, and the hydrogen bonds are indicated as dashed lines.

INTERACTIONS BETWEEN THE POLYMERASE AND THE INCOMING NUCLEOTIDE

Next to the 3′ terminus of the primer strand is the incoming dTTP, which binds to the active site of the polymerase in a manner analogous to that observed in the structures of other DNA polymerase complexes (Figure 13.4). The incoming dTTP is accommodated in a groove formed by residues from the palm and fingers domains (Figure 13.4). The two highly conserved aspartate residues, D549 (in motif A) and D753 (in motif C), together with the triphosphate tail of dTTP, coordinate one divalent metal ion (assumed to be magnesium, which has been added to the reaction buffer) (Figure 13.4). The triphosphate tail also interacts with the main chains of Y550, S552, and L553 from motif A, and the side chains of two positively charged R634 and K661 from the fingers domain (Figure 13.4).

The ribose of dTTP stacks on top of the phenyl ring of Y554 from motif A, in a manner like that previously observed with Y416 in the ternary complex structure of RB69 polymerase (Franklin et al., 2001) (Figure 13.4). There would be a steric clash between the 2′OH of ribonucleotides and Y554, hence providing a "steric gating" effect to select dNTP as the substrate. A similar steric gating effect has also been described in other DNA polymerases and was first proposed in HIV-1 RT (Gao et al., 1997; Silverman et al., 2008; Tohidi et al., 2023).

DIFFERENT OPERATION MODE OF PROCESSIVITY COFACTOR

For B-family DNA polymerases, proliferating cell nuclear antigen (PCNA) or PCNA-like proteins are required for high processivity. However, for poxviruses, including MPXV and VACV, no homologous PCNA-like proteins have been identified in the viral genome. Instead, the poxvirus-specific A22-E4 heterodimer is responsible for the high processivity of DNA replication.

A primer-extension assay using a 60-nt template DNA in the presence of heparin, which can trap the dissociated DNA polymerase from the primer-template DNA, showed that the F8 polymerase alone dissociated from the primer-template DNA after incorporating less than 14 nt, while the F8-A22-E4 holoenzyme was able to generate full-length 60-nt products with few abortive products (Peng et al., 2023). The addition of the A22-D4 heterodimer conferred processivity to F8 in a concentration-dependent manner (Peng et al., 2023). When the molar ratio of A22-D4 to F8 was 1:1, corresponding to the stoichiometry of the polymerase holoenzyme, the amount of the full-length product was almost the same as that of the product generated by the preassembled F8-A22-D4 polymerase holoenzyme (Peng et al., 2023). This indicates that the isolated A22-D4 and F8 can be efficiently assembled into functional holoenzymes to perform processive DNA synthesis. Moreover, alanine scanning of critical residues responsible for the interaction between E4 and F8 has revealed that R39A and N165A substitutions of E4 showed minor effect, and W36A reduced the synthesis of full-length products, whereas the W36A/R39A and W36A/R39A/N146A substitutions abolished the synthesis of full-length products (Peng et al., 2023). These results further confirmed the important function of the A22-E4 heterodimer in DNA replication processivity in a pure enzymatic reaction system.

As described above, the E4 cofactor interacts with the Exo of F8 polymerase, and together with the NTD of F8, they form a closed-ring channel to encircle the single-stranded template DNA (Figure 13.5). By contrast, in the yeast DNA polymerase complex, a representative of the other B-family DNA polymerases (Zheng et al., 2020), the Exo and NTD form an open semicircular channel to accommodate the single-stranded template DNA, and the trimeric PCNA ring encircles the template-product DNA duplex (Figure 13.5). This architectural difference between MPXV and yeast polymerase complexes is responsible for the different processivity mechanisms during DNA replication events. The MPXV DNA polymerase holoenzyme guarantees its high DNA replication processivity by encircling the single-stranded template DNA, which was proposed as a "forward sliding clamp" (Figure 13.5), whereas the other B-family DNA polymerase complexes possess continuous DNA replication capacity by encircling the double-stranded template-product DNA helix, which can be recognized as a "backward sliding clamp" (Figure 13.5).

DISCUSSION/CONCLUSION

The interaction between E4 and F8 could generate a ring channel that encircles the single-stranded template DNA, which is proposed to be important for high DNA replication processivity. This processivity mechanism is different from that of other B-family DNA polymerases that utilize PCNA or PCNA-like proteins to encircle the product-template DNA duplex (Lancey et al., 2020; Shamoo and Steitz, 1999; Zheng et al., 2020). The configuration of encircling the single-stranded template DNA probably allows the MPXV polymerase complex to perform continuous DNA replication by preventing template DNA disassociation from the polymerase holoenzyme. The DNA-bound and DNA-free holoenzyme structures have revealed the conformational flexibility of A22-E4 heterodimer protein. Therefore, the efficiency of polymerase holoenzyme assembly in the cellular environment requires further study, and more importantly we should investigate if the F8-A22-E4 polymerase holoenzyme has other active conformations (Shamoo and Steitz, 1999; Lancey et al., 2020).

Previous studies have implicated the unique role of the polymerase holoenzyme in DNA recombination, which involves both 3′-5′ exonuclease and DNA-joining activities (Hamilton and Evans, 2005; Hamilton et al., 2007; Gammon and Evans, 2009). Although we identified DNA ligase-like domains for the Mid of the poxvirus-specific cofactor A22, the F8-A22-E4 polymerase holoenzyme did not have canonical ligase activity in our assay. However, we cannot rule out the possibility that A22 possesses ligase activity in a different

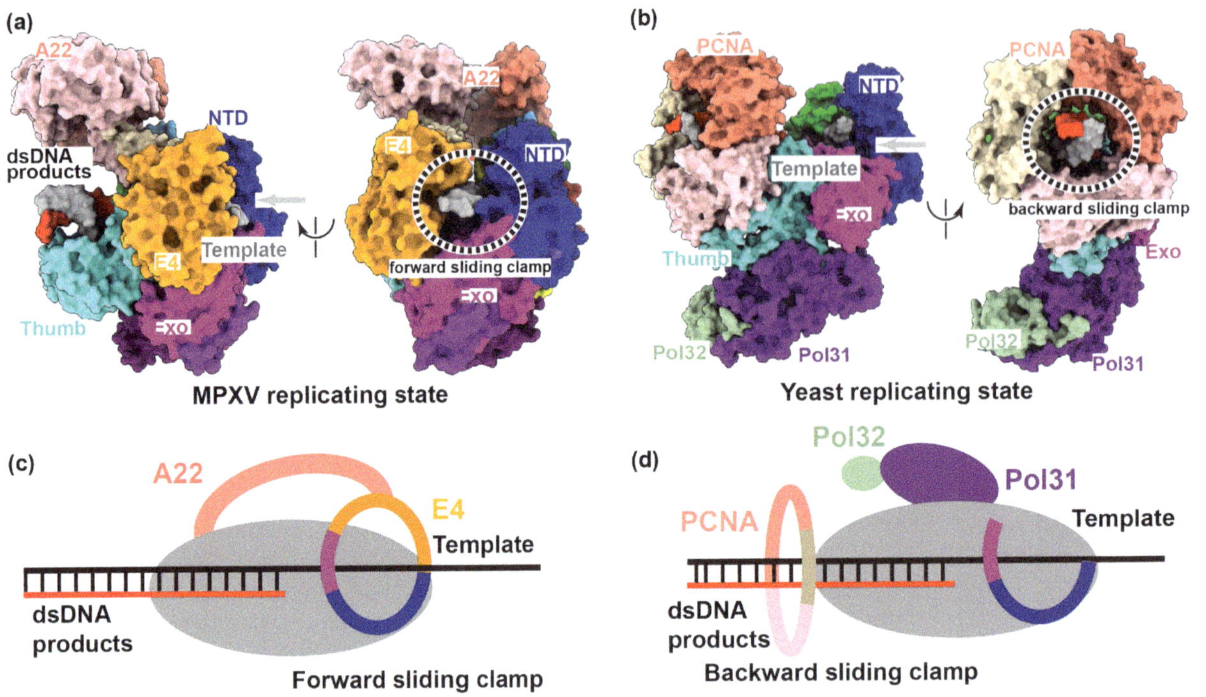

FIGURE 13.5 The unique mechanism of A22-E4 heterodimer promoting the processivity of MPXV DNA polymerase. (a–b) Comparison of the structures of the MPXV (a) and yeast (b) polymerases in complex with their own processivity factors and DNA. Polymerases and processivity factors are shown as surfaces and colored by domains, as depicted in Figure 13.2, and the primer-template DNA strands are also shown as surfaces (template DNA, grey; primer DNA, red). The processivity of yeast polymerase was strengthened by trimeric PCNA to clamp the primer-template dsDNA, recognized as a backward sliding clamp. While in the structure of MPXV polymerase holoenzyme, A22-E4 heterodimer does not interact with dsDNA. Instead, E4 located on the template DNA entry channel together with NTD and Exo domains of F8 form a forward clamp structure that would prevent the template strand from disassociating from the polymerase complex during DNA replication. (c–d) Two binding modes of processivity factors with polymerases. The processivity factors were bound with template DNA in poxvirus as "forward sliding clamp" (c) or dsDNA products in eukaryotes as "backward sliding clamp" (d).

state from the current holoenzyme structure or in an ATP-independent manner. Orthopoxviruses encode a DNA ligase that is not essential for virus replication but affects virulence and sensitivity to DNA-damaging agents (Kerr et al., 1991). Further functional dissection of the middle part of A22 and its cooperation with viral/host ligases is needed.

Poxvirus replication processes use different replication models, including self-priming, primer-dependent, and recombination models (Moss, 2013). Poxviruses, including MPXV, have linear double-stranded DNA genomes, and the termini of the two DNA strands are connected to form a continuous polynucleotide chain (Faye et al., 2018). A rolling cycle mechanism has been proposed to replicate DNA in the form of unbranched head-to-tail concatemers, which would be resolved by a Holliday junction resolvase to produce unit genomes. The proposed "forward sliding clamp" model can help in interpreting the self-priming model, as the processivity cofactor would facilitate robust continuous replication along the single-stranded template DNA unwound by the primase-helicase. Because there is also a possible presence of Okazaki fragments for *Orthopoxvirus* DNA replication (Esteban and Holowczak, 1978), other replication models should also be studied. Moreover, the working mechanism of an intact replisome, including the F8-A22-E4 holoenzyme and other replication proteins, will be a fascinating area for future studies.

ACKNOWLEDGMENTS

We thank Dr. Qi Peng for his help in preparation of the figures.

FUNDING

National Natural Science Foundation of China (NSFC) grants 81871658 (YS)

National Key Research and Development Program of China grant 2021YFC2300700 (YS)

Strategic Priority Research Program of CAS grant XDB29010000 (YS, GFG)

REFERENCES

Baroudy, B. M., Venkatesan, S. and Moss, B. (1982a). Incompletely base-paired flip-flop terminal loops link the 2 DNA strands of the vaccinia virus genome into one uninterrupted polynucleotide chain. Cell, **28**(2): 315–324.

Baroudy, B. M., Venkatesan, S. and Moss, B. (1982b). Structure and replication of vaccinia virus telomeres. Cold Spring Harbor Symposia on Quantitative Biology, **47**: 723–729.

Bergen, K., Betz, K., Welte, W., Diederichs, K. and Marx, A. (2013). Structures of KOD and 9°N DNA polymerases complexed with primer template duplex. Chembiochem, **14**(9): 1058–1062.

Bersch, B., Tarbouriech, N., Burmeister, W. P. and Iseni, F. (2021). Solution structure of the C-terminal domain of A20, the missing brick for the characterization of the interface between vaccinia virus DNA polymerase and its processivity factor. Journal of Molecular Biology, **433**(13): 167009.

Burmeister, W. P., Tarbouriech, N., Fender, P., Contesto-Richefeu, C., Peyrefitte, C. N. and Iseni, F. (2015). Crystal structure of the vaccinia virus uracil-DNA glycosylase in complex with DNA. Journal of Biological Chemistry, **290**(29): 17923–17934.

Challberg, M. D. and Englund, P. T. (1979). Purification and properties of the deoxyribonucleic-acid polymerase induced by vaccinia virus. Journal of Biological Chemistry, **254**(16): 7812–7819.

Chen, Y. Q., Liu, H. H., Yang, C., Gao, Y. Q., Yu, X., Chen, X., Cui, R. X., Zheng, L. N., Li, S. H., Li, X. H., Ma, J. B., Huang, Z., Li, J. X. and Gan, J. H. (2019). Structure of the error-prone DNA ligase of African swine fever virus identifies critical active site residues. Nature Communications, **10**: 387.

Contesto-Richefeu, C., Tarbouriech, N., Brazzolotto, X., Betzi, S., Morelli, X., Burmeister, W. P. and Iseni, F. (2014). Crystal structure of the vaccinia virus DNA polymerase holoenzyme subunit D4 in complex with the A20 N-terminal domain. Plos Pathogens, **10**(3): e1003978.

Contesto-Richefeu, C., Tarbouriech, N., Brazzolotto, X., Burmeister, W. P., Peyrefitte, C. N. and Iseni, F. (2016). Structural analysis of point mutations at the vaccinia virus A20/D4 interface. Acta Crystallographica Section F Structural Biology Communications, **72**(9): 687–691.

Czarnecki, M. W. and Traktman, P. (2017). The vaccinia virus DNA polymerase and its processivity factor. Virus Research, **234**: 193–206.

Du, S. and Traktman, P. (1996). Vaccinia virus DNA replication: Two hundred base pairs of telomeric sequence confer optimal replication efficiency on minichromosome templates. Proceedings of the National Academy of Sciences of the United States of America, **93**(18): 9693–9698.

Esteban, M. and Holowczak, J. A. (1978). Replication of vaccinia DNA in mouse L cells IV. Protein synthesis and viral DNA replication. Virology, **86**(2): 376–390.

Falendysz, E. A., Lopera, J. G., Doty, J. B., Nakazawa, Y., Crill, C., Lorenzsonn, F., Kalemba, L. S. N., Ronderos, M. D., Mejia, A., Malekani, J. M., Karem, K., Carroll, D. S., Osorio, J. E. and Rocke, T. E. (2017). Characterization of monkeypox virus infection in African rope squirrels (Funisciurus sp.). PLOS Neglected Tropical Diseases, **11**(8): e0005809.

Faye, O., Pratt, C. B. and Faye, M. (2018). Genomic characterisation of human monkeypox virus in Nigeria (vol 18, pg 246, 2018). Lancet Infectious Diseases, **18**(3): 244–244.

Ferré, V. M., Bachelard, A., Zaidi, M., Armand-Lefevre, L., Descamps, D., Charpentier, C. and Ghosn, J. (2022). Detection of monkeypox virus in anorectal swabs from asymptomatic men who have sex with men in a sexually transmitted infection screening program in Paris, France. Annals of Internal Medicine, **175**(10): 1491–1492.

Franklin, M. C., Wang, J. M. and Steitz, T. A. (2001). Structure of the replicating complex of a pol α family DNA polymerase. Cell, **105**(5): 657–667.

Gammon, D. B. and Evans, D. H. (2009). The 3′-to-5′ exonuclease activity of vaccinia virus DNA polymerase is essential and plays a role in promoting virus genetic recombination. Journal of Virology, **83**(9): 4236–4250.

Gao, G. X., Orlova, M., Georgiadis, M. M., Hendrickson, W. A. and Goff, S. P. (1997). Conferring RNA polymerase activity to a DNA polymerase: A single residue in reverse transcriptase controls substrate selection. Proceedings of the National Academy of Sciences of the United States of America, **94**(2): 407–411.

Greseth, M. D. and Traktman, P. (2022). The life cycle of the vaccinia virus genome. Annual Review of Virology, **9**: 239–259.

Hamilton, M. D. and Evans, D. H. (2005). Enzymatic processing of replication and recombination intermediates by the vaccinia virus DNA polymerase. Nucleic Acids Research, **33**(7): 2259–2268.

Hamilton, M. D., Nuara, A. A., Gammon, D. B., Buller, R. M. and Evans, D. H. (2007). Duplex strand joining reactions catalyzed by vaccinia virus DNA polymerase. Nucleic Acids Research, **35**(1): 143–151.

Hammarlund, E., Lewis, M. W., Carter, S. V., Amanna, I., Hansen, S. G., Strelow, L. I., Wong, S. W., Yoshihara, P., Hanifin, J. M. and Slifka, M. K. (2005). Multiple diagnostic techniques identify previously vaccinated individuals with protective immunity against Monkeypox. Nature Medicine, **11**(9): 1005–1011.

Huang, Y., Mu, L. and Wang, W. (2022). Monkeypox: Epidemiology, pathogenesis, treatment and prevention. Signal Transduction and Targeted Therapy, **7**(1): 373.

Kerr, S. M., Johnston, L. H., Odell, M., Duncan, S. A., Law, K. M. and Smith, G. L. (1991). Vaccinia DNA-ligase complements Saccharomyces cerevisiae Cdc9, localizes in cytoplasmic factories and affects virulence and virus sensitivity to DNA damaging agents. EMBO Journal, **10**(13): 4343–4350.

Klemperer, N., McDonald, W., Boyle, K., Unger, B. and Traktman, P. (2001). The A20R protein is a stoichiometric component of the processive form of vaccinia virus DNA polymerase. Journal of Virology, **75**(24): 12298–12307.

Ladnyj, I. D., Ziegler, P. and Kima, E. (1972). A human infection caused by monkeypox virus in Basankusu Territory, Democratic Republic of the Congo. Bulletin of the World Health Organization, **46**(5): 593–597.

Lancey, C., Tehseen, M., Raducanu, V. S., Rashid, F., Merino, N., Ragan, T. J., Savva, C. G., Zaher, M. S., Shirbini, A., Blanco, F. J., Hamdan, S. M. and De Biasio, A. (2020). Structure of the processive human Pol δ holoenzyme. Nature Communications, **11**(1): 1109.

Li, Y. N., Shen, Y. P., Hu, Z. W. and Yan, R. H. (2023). Structural basis for the assembly of the DNA polymerase holoenzyme from a monkeypox virus variant. Science Advances, **9**(16): eadg2331.

Magnus, P. V., Andersen, E. K., Petersen, K. B. and Birch-Andersen, A. (1959). A pox-like disease in cynomolgus monkeys. Acta Pathologica Microbiologica Scandinavica, **46**(2): 156–176.

McDonald, W. F., Klemperer, N. and Traktman, P. (1997). Characterization of a processive form of the vaccinia virus DNA polymerase. Virology, **234**(1): 168–175.

Mcdonald, W. F. and Traktman, P. (1994). Overexpression and purification of the vaccinia virus DNA polymerase. Protein Expression and Purification, **5**(4): 409–421.

Mcdonald, W. F. and Traktman, P. (1994). Vaccinia virus DNA polymerase. In vitro analysis of parameters affecting processivity. Journal of Biological Chemistry, **269**(49): 31190–31197.

Merchlinsky, M. (1990). Resolution of poxvirus telomeres: Processing of vaccinia virus concatemer junctions by conservative strand exchange. Journal of Virology, **64**(7): 3437–3446.

Moss, B. (2013). Poxvirus DNA replication. Cold Spring Harbor Perspectives in Biology, **5**(9): a010199.

Moss, B., Contesto-Richefeu, C., Tarbouriech, N., Brazzolotto, X., Betzi, S., Morelli, X., Burmeister, W. P. and Iseni, F. (2014). Crystal structure of the vaccinia virus DNA polymerase holoenzyme subunit D4 in complex with the A20 N-terminal domain. PLoS Pathogens, **10**(3): e1003978.

Nolen, L. D., Osadebe, L., Katomba, J., Likofata, J., Mukadi, D., Monroe, B., Doty, J., Hughes, C. M., Kabamba, J., Malekani, J., Bomponda, P. L., Lokota, J. I., Balilo, M. P., Likafi, T., Lushima, R. S., Ilunga, B. K., Nkawa, F., Pukuta, E., Karhemere, S., Tamfum, J. J. M., Nguete, B., Wemakoy, E. O., McCollum, A. M. and Reynolds, M. G. (2016). Extended human-to-human transmission during a monkeypox outbreak in the Democratic Republic of the Congo. Emerging Infectious Diseases, **22**(6): 1014–1021.

Peng, Q., Xie, Y. F., Kuai, L., Wang, H., Qi, J. X., Gao, G. F. and Shi, Y. (2023). Structure of monkeypox virus DNA polymerase holoenzyme. Science, **379**(6627): 100–105.

Pinedo, L. C., Franz, E., van den Beld, M., Van Goethem, N., Mattheus, W., Veldman, K., Bosch, T., Mughini-Gras, L. and Pijnacker, R. (2022). Changing epidemiology of *Salmonella* Enteritidis human infections in the Netherlands and Belgium, 2006 to 2019: A registry-based population study. Eurosurveillance, **27**(38): 8–18.

Rezza, G. (2019). Emergence of human monkeypox in west Africa. Lancet Infectious Diseases, **19**(8): 797–799.

Schormann, N., Banerjee, S., Ricciardi, R. and Chattopadhyay, D. (2015). Binding of undamaged double stranded DNA to vaccinia virus uracil-DNA Glycosylase. BMC Structural Biology, **15**: 10.

Sèle, C., Gabel, F., Gutsche, I., Ivanov, I., Burmeister, W. P., Iseni, F. and Tarbouriech, N. (2013). Low-resolution structure of vaccinia virus DNA replication machinery. Journal of Virology, **87**(3): 1679–1689.

Shamoo, Y. and Steitz, T. A. (1999). Building a replisome from interacting pieces: Sliding clamp complexed to a peptide from DNA polymerase and a polymerase editing complex. Cell, **99**(2): 155–166.

Shi, K., Bohl, T. E., Park, J., Zasada, A., Malik, S., Banerjee, S., Tran, V., Li, N., Yin, Z. Q., Kurniawan, F., Orellana, K. and Aihara, H. (2018). T4 DNA ligase structure reveals a prototypical ATP-dependent ligase with a unique mode of sliding clamp interaction. Nucleic Acids Research, **46**(19): 10474–10488.

Silverman, A. P., Garforth, S. J., Prasad, V. R. and Kool, E. T. (2008). Probing the active site steric flexibility of HIV-1 reverse transcriptase: Different constraints for DNA- versus RNA-templated synthesist. Biochemistry, **47**(16): 4800–4807.

Stanitsa, E. S., Arps, L. and Traktman, P. (2006). Vaccinia virus uracil DNA glycosylase interacts with the A20 protein to form a heterodimeric processivity factor for the viral DNA polymerase. Journal of Biological Chemistry, **281**(6): 3439–3451.

Swan, M. K., Johnson, R. E., Prakash, L., Prakash, S. and Aggarwal, A. K. (2009). Structural basis of high-fidelity DNA synthesis by yeast DNA polymerase δ. Nature Structural & Molecular Biology, **16**(9): 979–986.

Tarbouriech, N., Ducournau, C., Hutin, S., Mas, P. J., Man, P., Forest, E., Hart, D. J., Peyrefitte, C. N., Burmeister, W. P. and Iseni, F. (2017). The vaccinia virus DNA polymerase structure provides insights into the mode of processivity factor binding. Nature Communications, **8**: 1455.

Tohidi, N., Manshadi, S. A. D. and Hajiabdolbaghi, M. (2023). Association of polymorphism with disease progression in human immunodeficiency virus type-1 (HIV-1) infected patients. Virus Genes, **59**(6): 831–835.

Wang, F. and Yang, W. (2009). Structural insight into translesion synthesis by DNA Pol II. Cell, **139**(7): 1279–1289.

Wittek, R. and Moss, B. (1980). Tandem repeats within the inverted terminal repetition of vaccinia virus DNA. Cell, **21**(1): 277–284.

Xu, Y. X., Wu, Y. Q., Zhang, Y. Y., Fan, R. X., Yang, Y. X., Li, D. Y., Zhu, S. M., Yang, B., Zhang, Z. Y. and Dong, C. J. (2023). Cryo-EM structures of human monkeypox viral replication complexes with and without DNA duplex. Cell Research, **33**(6): 479–482.

Zheng, F. W., Georgescu, R. E., Li, H. L. and O'Donnell, M. E. (2020). Structure of eukaryotic DNA polymerase 8 bound to the PCNA clamp while encircling DNA. Proceedings of the National Academy of Sciences of the United States of America, **117**(48): 30344–30353.

14 Determining the High-Resolution Structure of the [NiFe]-hydrogenase Huc by Native Host Purification and Cryo-EM

Hari Venugopal†, Ashleigh Kropp†, Chris Greening, and Rhys Grinter

INTRODUCTION

Over the past decade, cryo-electron microscopy (cryo-EM) has emerged as a central technique for the structural analysis of macromolecules, particularly large complexes and membrane proteins (Fukuda et al., 2023; Piper et al., 2022; Yip et al., 2020). As well as revolutionizing our understanding of the structural basis for many biological processes, the use of cryo-EM has changed the types of samples prepared for structural analysis and radically decreased the quantity of sample required (Passmore and Russo, 2016; Saibil, 2022). These changes have led many researchers to isolate proteins and macromolecular complexes directly from the producing organism, which has led to stunning insights into the structure and biology of these systems (Chen et al., 2023; Gardiner et al., 2021; Johnson et al., 2021; Lauber et al., 2018; Letts et al., 2016; Qian et al., 2022; Watanabe et al., 2021).

A major focus of our research is determining the structure and function of [NiFe]-hydrogenases, a diverse superfamily of enzymes responsible for the simplest chemical reaction: the interconversion of molecular hydrogen (H_2) with two protons and two electrons ($H_2 \Leftrightarrow 2e^- + 2H^+$) (Lubitz et al., 2014). Notably, we study high-affinity [NiFe]-hydrogenases used by bacteria and archaea to oxidize the minute quantities of hydrogen present in the atmosphere as an energy source under starvation conditions (Cordero et al., 2019; Greening and Grinter, 2022; Grinter et al., 2023). The ability of microorganisms to oxidize atmospheric hydrogen is incredibly widespread, with soil microbes consuming 75% of the H_2 lost from the atmosphere and at least 22 bacterial phyla possessing the genes for atmospheric H_2 consumption (Bay et al., 2021; Conrad and Seiler, 1981; Greening and Grinter, 2022; Ortiz et al., 2021). Due to the complexity of high-affinity [NiFe]-hydrogenases, the fact that they are only expressed under starvation conditions, and their production by non-model microorganisms, until recently no enzyme capable of oxidizing the H_2 in ambient air had been isolated. All [NiFe]-hydrogenases isolated had an affinity too low to oxidize atmospheric H_2, and most were inhibited by oxygen (O_2), which is abundant in the atmosphere (Greening et al., 2014; Greening and Grinter, 2022; Shafaat et al., 2013; Wulff et al., 2014).

To isolate a high-affinity [NiFe]-hydrogenase and determine the structural basis for atmospheric H_2 oxidation, we targeted the bacterium *Mycobacterium smegmatis*. *M. smegmatis* is a soil-dwelling saprophyte that, when faced with starvation, oxidizes atmospheric H_2 via two high-affinity [NiFe]-hydrogenases, named Huc and Hhy (Cordero et al., 2019; Greening et al., 2014; Grinter et al., 2023). Inspired by the native-host purification of other macromolecular complexes, we purified chromosomally expressed Huc directly from *M. smegmatis* cells (Gardiner et al., 2021; Grinter et al., 2023; Lauber et al., 2018). Analysis of the purified enzyme demonstrated that it forms a large oligomeric complex, which is completely insensitive to inhibition by O_2 and is capable of oxidizing H_2 to well below atmospheric concentrations (Grinter et al., 2023). These properties are necessary for its function, but they are completely unprecedented for [NiFe]-hydrogenases or any other H_2-converting catalyst (Lubitz et al., 2014; Shafaat et al., 2013).

The natively purified Huc complex proved to be an excellent target for structural analysis by cryo-EM. Reconstruction of the Huc structure from cryo-EM images revealed that it forms a C4 symmetrical complex, with a flexible membrane anchoring tail. Further, focused refinement on one of the C4-symmetrical subunits, combined with extensive data processing, allowed us to determine the structure of the catalytic protomer of Huc at a very high resolution of 1.52 Å. Unlike X-ray crystallography, at this resolution, density is observed for many hydrogen atoms, which provided insights into the structure and chemistry of Huc, including the identification of a histidine with a unique D-stereoisomeric conformation (Grinter et al., 2023). These structural data, combined with in-depth biochemical and spectroscopic analysis of the Huc complex, provided us with unprecedented insight into how this enzyme can extract and oxidize H_2 from the air.

In this chapter, we present the isolation, structural determination, and analysis of the Huc structure as an example of how native host purification can be combined with structural analysis using cryo-EM to provide deep insights into the function of complex macromolecules, which would be impossible using more traditional recombinant expression and structural analysis approaches.

† These authors contributed equally to this work.

DOI: 10.1201/9781003326106-17

THE ISOLATION OF THE HUC COMPLEX USING NATIVE HOST EXPRESSION

THE CASE FOR NATIVE EXPRESSION AND PURIFICATION

Currently, most protein expression is carried out in heterologous systems, resulting in the purification of recombinant proteins (Mattanovich et al., 2012; Rosano et al., 2019; Wurm, 2004). In this process, the gene encoding the protein of interest is amplified and cloned into an extrachromosomal DNA element, where it is driven by a strong, usually chemically inducible promoter. The resulting "expression vector" is introduced into a host bacterial or eukaryotic cell line engineered for protein expression. Very often *Escherichia coli* is used as the "expression host" due to its long history as a model bacterium and the optimization of strains for this purpose (Figure 14.1a) (Rosano et al., 2019). This method often works well for proteins that do not require specialized assembly or significant posttranslational modification, leading to high protein yields (Rosano et al., 2019).

Advances in single-particle cryo-EM have enabled the structural analysis of large macromolecular complexes that cannot be produced recombinantly without significant genetic engineering and host strain optimization (Chen et al., 2023; Gardiner et al., 2021; Lauber et al., 2018; Steinhilper et al., 2022; Watanabe et al., 2021). Moreover, modern cryo-EM allows the solution of high-resolution structures using exceptionally small quantities of protein (as little as 1–10 μg), meaning that the high yields obtained from recombinant expression are no longer required for structural analysis (Bhella, 2019; Passmore and Russo, 2016). As a result, cryo-EM has enabled the structural characterization of many proteins that were inaccessible to X-ray crystallography and NMR. These developments have led to innovations in protein production for structural analysis, including a resurgence in the popularity of isolating macromolecular complexes directly from their host organism (Beckham et al., 2021; Chen et al., 2023; Lauber et al., 2018; Piper et al., 2019; Shepherd et al., 2022; Tan et al., 2022). If a macromolecular complex is abundant, it can be purified using steps that exploit its native chemistry (Figure 14.1b) (Gardiner et al., 2021; Qian et al., 2022; Tan et al., 2022; Watanabe et al., 2021). However, for low-abundance complexes, affinity tags (StrepII, Flag, Myc, Green Fluorescent Protein (GFP), His etc.) are introduced chromosomally into the gene encoding the target protein to facilitate purification (Figure 14.1c) (Beckham et al., 2021; Chen et al., 2023; Gong et al., 2018; Lauber et al., 2018; Liang et al., 2023).

A notable example of the use of this method is the isolation of the type 9 secretion system (T9SS) from *Flavobacterium johnsoniae* (Lauber et al., 2018). The T9SS is responsible for the export of large proteins (100–650 kDa) across the outer membrane of bacteria from the phylum Bacteroidota (Lasica et al., 2017; Lauber et al., 2018). It contains multiple integral membrane proteins that assemble into a large complex, making its isolation using recombinant expression challenging (Lauber et al., 2018). To isolate the T9SS complex from *F. johnsoniae,* the gene encoding SprA, the predicted channel protein for the T9SS, was chromosomally Twin-Strep tagged (Lauber et al., 2018). Purification using streptactin and size-exclusion chromatography (SEC) yielded a large complex consisting of SprA and three other proteins, two of which were known members of the T9SS and a third novel complex member (Lauber et al., 2018). Cryo-EM was used to determine the structure of the T9SS using this sample, demonstrating that SprA forms a 36-stranded transmembrane β-barrel, which acts as the export channel for the complex. The novel complex member identified during purification associates with SprA to act as a plug, preventing nonspecific import or export (Lauber et al., 2018). The cryo-EM dataset also contained the T9SS in two conformational states, providing valuable information about the mechanisms of protein export (Lauber et al., 2018).

Native expression and purification were recently employed to isolate and structurally characterize the lipid-transporting mammalian cell entry (MCE) complex from *M. smegmatis* (Chen et al., 2023). Previous attempts to characterize the MCE using recombinant expression focused on the characterization of only single proteins or fragments of the larger MCE complex, yielding limited information about its function (Asthana et al., 2021). In this recent study, MceG, one of the components of the complex, was chromosomally tagged with GFP, and the intact MCE complex was isolated using anti-GFP nanobodies (Chen et al., 2023). Mass spectrometry revealed that the MCE complex consists of nine proteins in addition to MceG (Chen et al., 2023), including LucB, a protein previously not known to form part of the complex. Cryo-EM was used to determine the structure of the MCE complex from this sample (Chen et al., 2023), revealing a structure that spans from the inner membrane to the mycobacterial outer membrane, with a length of 310 Å (Chen et al., 2023). In addition to demonstrating this remarkable complex, cryo-EM analysis revealed density corresponding to fatty acids or mycolic acids within a region of the structure that is thought to be a channel for lipids (Chen et al., 2023).

These case studies indicate how native purification can be combined with cryo-EM to determine the structures of complexes that are inaccessible to other techniques. In addition, in both cases, purification from the native host provided important new insights into the function of these complexes, including the identification of new subunits and physiologically relevant ligands (Chen et al., 2023; Lauber et al., 2018).

NATIVE EXPRESSION AND PURIFICATION OF HUC FROM *M. SMEGMATIS*

Considering the success of native expression and purification to produce complexes for structural analysis by cryo-EM, we employed this strategy to isolate the [NiFe]-hydrogenase Huc from *M. smegmatis.* Like other [NiFe]-hydrogenases, the core catalytic unit of Huc consists of large (HucL) and small (HucS) protein subunits that require

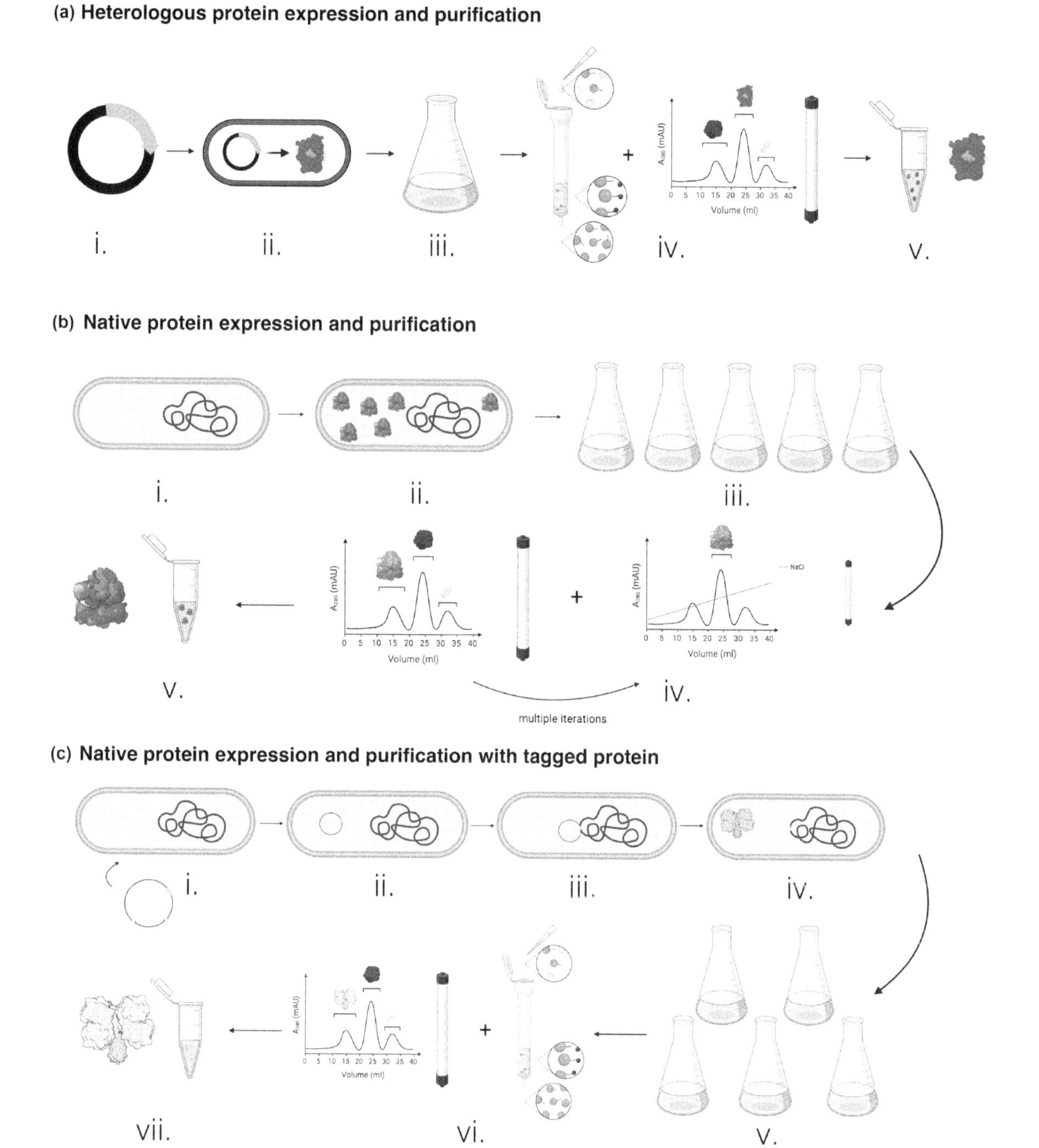

FIGURE 14.1 Heterologous and native expression strategies. (a) A scheme showing heterologous bacterial expression, (i) High-copy number expression plasmid containing the gene of interest, which is inducible with a strong promoter. (ii) *E. coli* cells are transformed with the expression vector containing the gene of interest. Cells are selected for with antibiotic resistance cassette. (iii) *E. coli* cells expressing the protein of interest are grown at large scale. (iv) Cells are harvested and lysed, and the protein is isolated through purification steps, including affinity chromatography and size exclusion chromatography (v) Pure protein, usually in high quantities (mg L^{-1}) and high purity. (b) A scheme showing native expression. (i–iii) Cells naturally expressing abundant proteins are grown in large quantities under the ideal growth conditions for the cells and expression of the protein. (iv) Cells are harvested and lysed, and protein is purified from the lysate based on the native chemistry of the molecule, such as size and charge. (v) Pure protein is isolated often with lower purity than genetically tagged proteins, but with reasonable yields. (c) A scheme showing native expression with chromosomal tagging of the gene encoding the protein of interest. (i–iv) Allelic exchange mutagenesis to genetically alter the bacterial chromosome, adding an affinity tag to the gene encoding the protein of interest. (v) The bacteria are grown in large quantities under conditions that induce production of the target protein. (vi) Cells are harvested and lysed, and the protein is isolated through purification steps, including affinity chromatography and size exclusion chromatography. (vii) Pure protein is produced of often low yield (μg/L^{-1}). (Figure created with BioRender.com.)

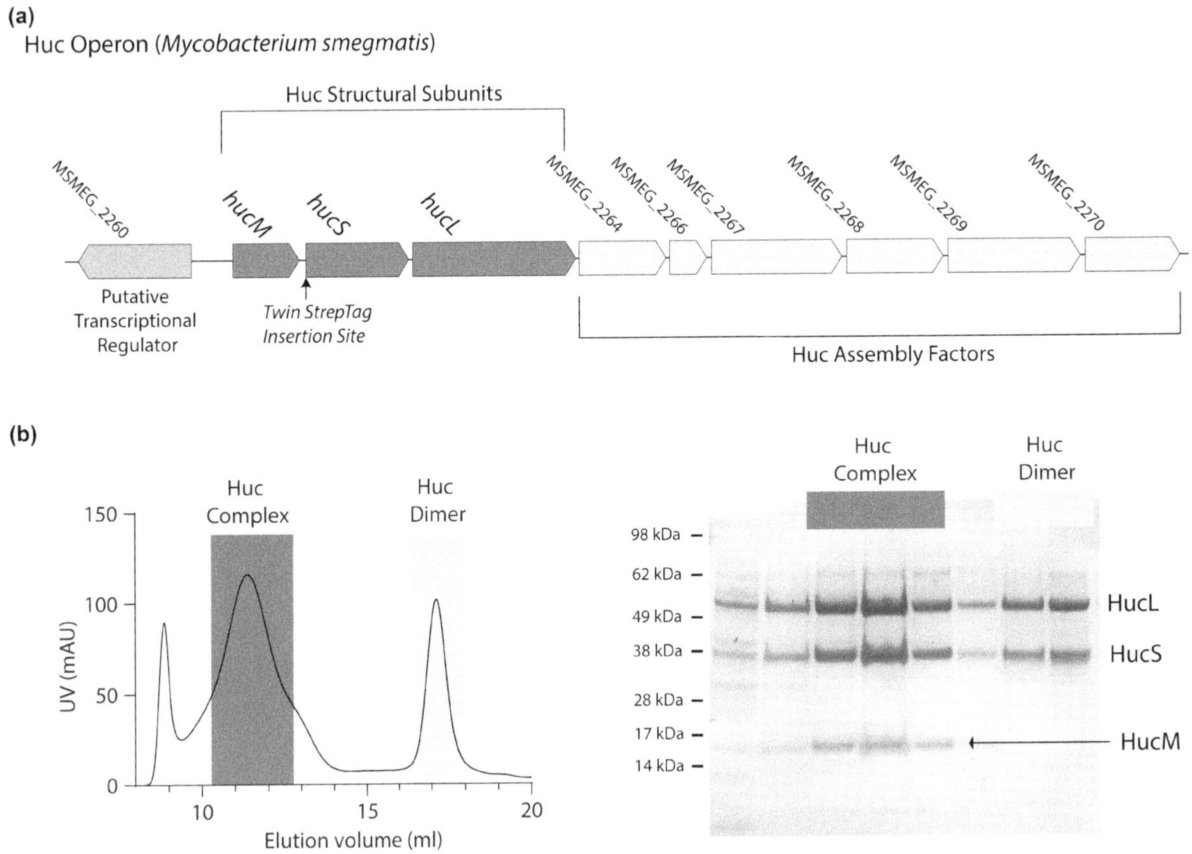

FIGURE 14.2 Isolation of the Huc complex via chromosomal tagging. (a) A schematic of the Huc operon, showing the location of Huc structural and assembly genes, and the location of the insertion of the TwinStrep affinity tag used for Huc purification. (b) An example of the SEC trace from the purification of Huc (*left*) and SDS-PAGE analysis of selected fractions (*right*), showing the tripartite nature of the Huc complex. (Adapted from Grinter et al. (2023).)

the co-expression of multiple assembly and maturation factors to form a functional enzyme (Figure 14.2a) (Berney et al., 2014; Grinter et al., 2023; Islam et al., 2019; Lubitz et al., 2014). This makes the heterologous expression of these enzymes very difficult (Fan et al., 2020). In addition, analysis by native-PAGE indicated that Huc forms a large (~900 kDa) membrane-associated complex in *M. smegmatis* cells (Cordero et al., 2019). To isolate Huc from *M. smegmatis*, we chromosomally introduced a Twin-Strep tag onto the N-terminus of the HucS subunit (Figure 14.2a). An additional barrier to Huc isolation was the low expression of the enzyme and its induction only under carbon starvation conditions (Cordero et al., 2019). However, we isolated an *M. smegmatis* strain lacking the glycerol metabolism regulator GylR, which exhibited a marked increase in Huc expression in *M. smegmatis* when grown in minimal media with glycerol as the sole carbon source (Bong et al., 2019). Even this optimized expression system yielded modest quantities of Huc with yields of 50 to 100 $\mu g/L^{-1}$ of culture media, with a culture growth time of seven days. However, it provided us with sufficient quantities of the Huc complex for biochemical and structural analysis using cryo-EM (Grinter et al., 2023).

We isolated Huc using Streptactin affinity chromatography and SEC. Although it has been shown to associate with the membrane, the enzyme was localized mostly in the soluble fraction in the Huc overexpressing strain (Cordero et al., 2019; Greening et al., 2014; Grinter et al., 2023). Huc eluted from SEC as three distinct peaks corresponding to the previously observed ~900 kDa complex, a larger peak that likely corresponds to Huc associated with membrane vesicles, and a smaller peak consistent with a dimer of the HucS and HucL subunits that form the catalytic protomer of the enzyme (Figure 14.2b). All peaks contain active hydrogenase, and we hypothesize that the smaller peak may represent the core catalytic subunit of Huc, before assembly into the Huc complex (Grinter et al., 2023). SDS-PAGE analysis of the Huc complex fraction showed that it is composed of three proteins, HucS, HucL, and a third protein that we identified by mass spectrometry as MSMEG_2261, encoded by a gene directly upstream of *hucS* in the *M. smegmatis* genome, which we designated HucM (Figure 14.2a and b). Deletion of this gene abolished the formation of the Huc complex in *M. smegmatis* cells, but not hydrogenase activity, indicating HucM is essential for oligomerization

(Grinter et al., 2023). If Huc had been produced recombinantly, it is unlikely that it would have been co-expressed with HucM, meaning this key component of its structure and the Huc complex would not have been identified. By purifying the protein from its native host, we produced the intact Huc complex, which was highly amenable to analysis by cryo-EM.

DETERMINATION OF THE HUC COMPLEX STRUCTURE BY CRYO-EM AT HIGH RESOLUTION

Initial Huc Sample Preparation and Cryo-EM Data Collection

Huc-containing cryo-EM grids were prepared according to a procedure that in our hands often works well for soluble protein in a non-detergent environment. Briefly, 300 mesh UltrAuFoil grids were used (Russo and Passmore, 2016), which were glow discharged in a PELCO glow discharger air using a current of 30 mA for 30 seconds before use. Three microliters of Huc sample was applied to the grids, and the sample was blotted and plunge frozen using a Vitrobot Mark IV, which was maintained at 100% humidity and 4°C. Grids were prepared with a Huc concentration of 0.4 and 1.0 mg/mL and imaged using a G1 Titan Krios microscope with S-FEG operated at an accelerating voltage of 300 kV and equipped with a K3 direct electron detector fitted with a Bioquantum energy filter. Both concentrations yielded high-quality ice with well-distributed particles, with the 0.4 mg/mL grids exhibiting thicker ice. We proceeded to collect datasets for structural analysis on both samples. Cryo-EM single-particle analysis (SPA) datasets show an improvement in signal-to-noise ratio when using a narrow energy slit width during energy-filtered TEM imaging (Nakane et al., 2020). This was achieved using a ThermoFischer Selectrix-X energy filter operated at a narrow-slit width of 10 eV centered on the zero-loss peak, which filtered out the plasmon peak spanning 6–20 eV in the energy loss spectrum that contributed to the noise profile of the image. The first-generation Bioquantum energy filter equipped in the G1 Titan Krios was carefully aligned to keep the zero-loss peak filter at 10 eV to achieve results similar to those shown by Nakane et al. (2020). Once stable alignments were made, periodic auto zero-loss centering (1/h) through EPU automated data collection software was sufficient to minimize and correct instabilities arising from environmental changes, thereby ensuring stable and accurate energy slit positioning during the data collection. Movies were collected at a dose rate of ~$7.5\,e^{-}\cdot pixel^{-1}\cdot s^{-1}$, with an exposure time of 6 s, yielding a total dose of $66.0\,e^{-}/Å^{-2}$ fractionated into 60 subframes. They were collected at a nominal magnification of 105,000 × in CDS mode on a K3 camera yielding a physical pixel size of 0.82 Å/pixel^{-1}. Two datasets were collected: 0.4 mg/mL composed of 2226 movies, and 1 mg/mL composed of 3113 movies (Grinter et al., 2023).

Initial Cryo-EM Image Processing and Analysis

Both the 0.4 and 1 mg/mL datasets produced high-quality images, although resolution was limited by the thicker ice of the 0.4 mg/mL dataset. Motion correction and contrast transfer function (CTF) estimation were performed using UCSF Motioncor 2 and CTFFIND 4.1.8, implemented using Relion 3.1.2 (Rohou and Grigorieff, 2015; Zheng et al., 2017; Zivanov et al., 2018). In regions of the 0.4 mg/mL grids containing thicker ice, Huc particles decorated small membrane vesicles, whereas the thinner ice of the 1.0 mg/mL grids contained largely soluble Huc particles (Figure 14.3a). In both datasets, Huc particles with a striking C4 symmetrical cloverleaf shape were observed, corresponding to a "top-down" view of the molecule, with C2 symmetrical "side views" observed as well as smaller particles corresponding to disassociated Huc complex subunits (Figure 14.3a).

The 1 mg/mL dataset was utilised for further data processing. Initially, particles corresponding to the entire Huc complex were picked with CrYOLO (Wagner et al., 2019), specifying a particle diameter of 160 Å, and using a model trained on Huc particles manually picked from a subset of 20 micrographs combined with a general model trained on low-pass filtered images. This strategy yielded 825,900 particles, which were extracted using Relion 3.1.2 and imported into CryoSparc, where reference-free 2D classification was performed, yielding high-quality class averages of the Huc complex (Figure 14.3b) (Punjani et al., 2017; Zivanov et al., 2018). *Ab initio* reconstruction was performed on particles from these classes, followed by rounds of 3D homogenous and heterogenous refinement, CTF refinement, and Bayesian polishing (in Relion 3.1.2), yielding a final set of 135,297 particles, and a final 3D reconstruction at a resolution of 2.19 Å (Figure 14.3c) (Grinter et al., 2023). Due to the flexibility of the Huc complex, the resolution of this map was better than 2 Å at the center, but worse than 4 Å at the periphery, making it difficult to discern key elements of the structure (Grinter et al., 2023). Furthermore, a flexible tail region of Huc that is observable in 2D-class averages and from the interaction with membrane vesicles in the 0.4 mg/mL dataset was not resolved (Figure 14.3a and b).

To overcome this variable map resolution, we adopted an alternative particle-picking strategy focusing on a single lobe of the cloverleaf-shaped Huc complex (Figure 14.3c), which is composed of a heterotetramer of two HucL and two HucS subunits ($HucS_2L_2$). Automatic particle picking was performed using Gautomatch 0.53 with an 80 Å particle diameter with a minimum interparticle distance set to 20 Å (Zhang et al., 2011). This strategy resulted in significant overpicking, yielding 9.3 million sets of particle coordinates, which were extracted in Relion 3.1.2 with fourfold binning to reduce dataset size and processing times (Zivanov et al., 2018). Particles were imported into CryoSparc and subjected to reference-free 2D classification. Classes containing either free or Huc complex–associated $HucS_2L_2$ particles were selected and subjected to *ab initio*

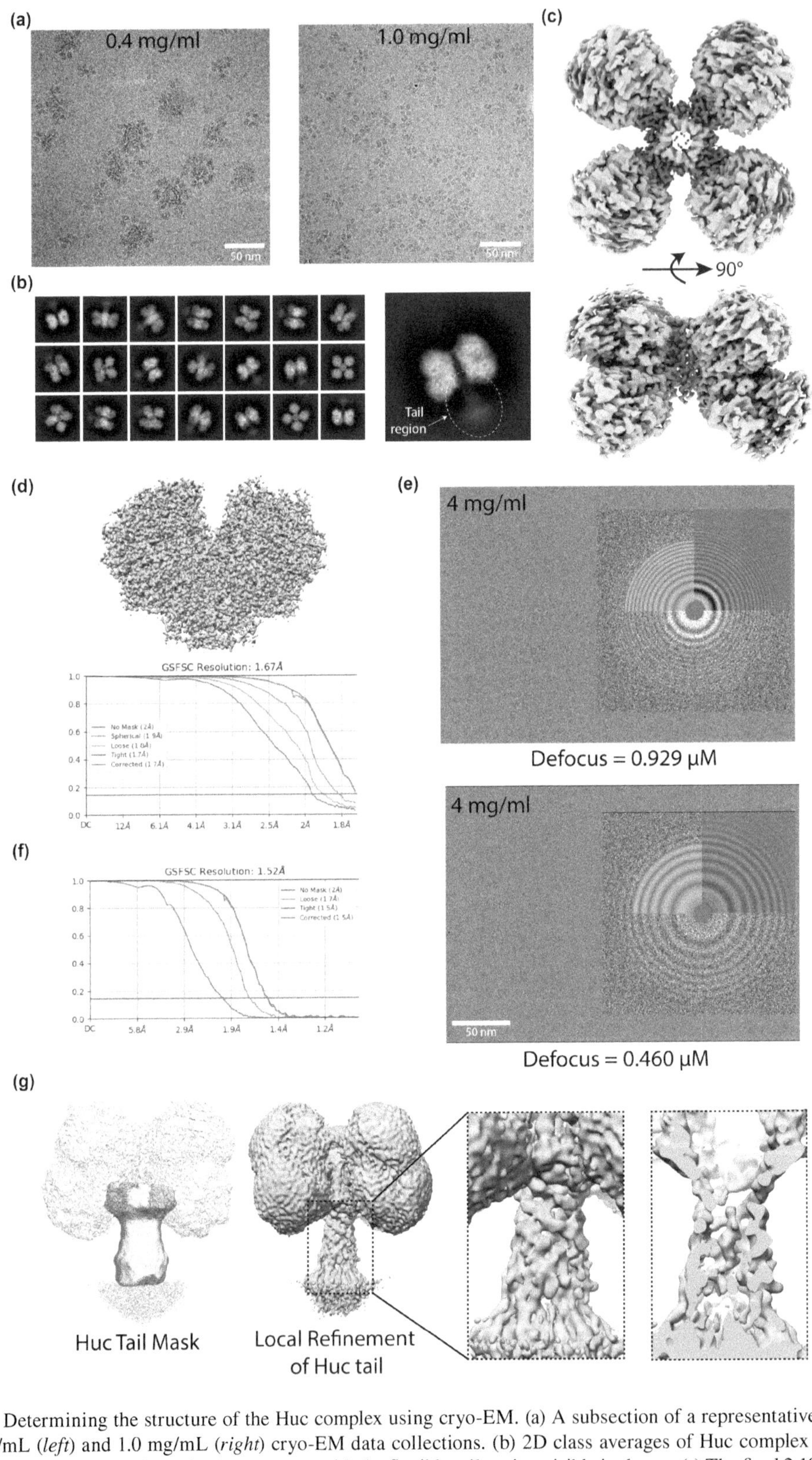

FIGURE 14.3 Determining the structure of the Huc complex using cryo-EM. (a) A subsection of a representative micrograph from the Huc 0.4 mg/mL (*left*) and 1.0 mg/mL (*right*) cryo-EM data collections. (b) 2D class averages of Huc complex particles from the 1.0 mg/mL dataset. An enlarged view of a class average with the flexible tail region visible is shown. (c) The final 2.19 Å 3D reconstruction of the main body of the Huc complex from 1.0 mg/mL dataset. (d) The final 1.67 Å reconstruction of the $HucS_2L_2$ subcomplex from 1.0 mg/mL dataset. (e) Representative low-defocus micrographs and corresponding Fourier transform showing Thon rings visible to high resolution. (g) The mask generated for local refinement of the Huc tail region (*left*), and the resulting 3D volume generated by this refinement (*right*). (Images are adapted from Grinter et al. (2023).)

reconstruction, followed by rounds of 3D heterogenous and homogenous refinement, which selected only $HucS_2L_2$ particles associated with the Huc complex (Punjani et al., 2017). These particles were reimported into Relion 3.1.2 and reextracted with no binning. Duplicated particles were removed from the dataset using a 20 Å proximity exclusion cutoff, yielding a set of 422,614 particles. CTF refinement and Bayesian polishing were performed followed by nonuniform refinement in CryoSparc, yielding a final resolution of 1.67 Å (FSC = 0.143, gold standard), essentially reaching the Nyquist limit for this dataset of 1.64 Å (Figure 14.3d) (Punjani et al., 2017).

Resolving the $HucS_2L_2$ Subcomplex at Very High Resolution

Considering that the resolution of our initial Huc reconstruction reached the Nyquist limit, we performed additional data collection to obtain a higher-resolution reconstruction. From the previous collection, we observed that a higher Huc concentration yielded thinner ice, so we prepared grids as previously, but with a Huc sample concentration of 4.0 mg/mL. Additionally, we kept data collection parameters consistent, aside from increasing the magnification to 165,000×, which resulted in a pixel size of 0.5 Å. A dataset of 9868 micrographs was collected to compensate for the smaller field of view at this magnification. The slit width was kept at 10 eV to aid signal and contrast as in the previous datasets, and a lower dose rate of $4.37\ e^{-}{\cdot}pixel{\cdot}s^{-1}$ at CDS mode was chosen. Movies were collected with an exposure time of 4 s, amounting to a total dose of $60.4\ e^{-}/Å^{-2}$ fractionated into 60 subframes. This data collection produced excellent quality micrographs, with sufficient contrast to resolve thon rings to high resolution even in low-defocus movies (Figure 14.3e). This allowed us to acquire data at a low defocus range of −1.3 and −0.3 µm, which was essential to maximize the high-resolution signal.

Initial particle picking and data processing were essentially the same as for the previous dataset, yielding 3.8 million, fourfold binned particles from initial picking and 273,568 unbinned particles after processing and reextraction. At this stage, a different processing strategy was employed. The particles were reextracted with an expanded box size of 512 pixels (256 Å) to capture the entire Huc complex and then binned fourfold. *Ab initio* reconstruction and homogenous refinement were performed in CryoSparc to adjust particle coordinates to the center of the Huc complex, before particle reextraction in Relion 3.1.2 with twofold binning and a box size of 576 pixels (Punjani et al., 2017; Zivanov et al., 2018). Duplicated particles were removed from the dataset using a 10 Å proximity exclusion cutoff, yielding a set of 179,311 particles, followed by nonuniform refinement in CryoSparc, yielding a map of the Huc complex at 2.05 Å.

To obtain the final high-resolution map of the $HucS_2L_2$ subcomplex, the 179,311 particles were reextracted unbinned with a box size of 576, subjected to Bayesian polishing, and a further round of duplicate particle removal with a 10 Å proximity exclusion cutoff, yielding a final particle set of 153,359 A. This particle set was used for homogenous refinement with C4 symmetry in CryoSparc followed by symmetry expansion and local refinement with C1 symmetry using a soft mask encompassing one lobe of the Huc map. To exploit the C2 symmetry inherent in $HucS_2L_2$, the particles and volume from the previous local refinement step were aligned to the C2 symmetry axis using the volume and particle alignment tool in CryoSparc. Per-particle and global CTF optimization and Ewald sphere correction were then performed on these particles, followed by local refinement, yielding a final resolution of 1.52 Å (Figures 14.3f and 14.4a) (Grinter et al., 2023).

Resolving the Membrane-Associated Huc Tail Region

As discussed above, 2D-class averages of the Huc complex indicated the presence of a flexible tail region extending from the main body of the Huc cloverleaf structure (Figure 14.3b). However, this region was very poorly resolved in our maps due to flexibility. To resolve this region, we performed a three-class heterologous refinement on the final high-resolution particle set, selecting particles from two reconstructions with clear density for the Huc tail region, which yielded a subset of 60,448 particles. We performed homogenous refinement, followed by CTF optimization and nonuniform refinement with C4 symmetry on this dataset, yielding a 2.05 Å resolution map. We then shifted the map fulcrum to the center of the tail region and erased the density corresponding to the body of the Huc complex and lipid molecules associated with the end of the Huc tail region. We generated a mask for the remaining tail region of this map and used it to perform masked local refinement in CryoSparc, yielding a map of the tail region of sufficient resolution to resolve individual α-helices (Figure 14.3g) (Grinter et al., 2023).

Insights into Atmospheric H_2 Oxidation from the Cryo-EM Structure of Huc

The Structure of the Huc Complex

Biochemical analysis of the purified Huc complex indicated that it is composed of three subunits, HucM, HucS, and HucL. The cryo-EM structure revealed that the Huc complex consists of 4xHucM, 8xHucS, and 8xHucL subunits, giving a total molecular mass of 819 kDa for the polypeptide components of the complex. Each of the HucS subunits contains three [3Fe4S] clusters and one bound menaquinone molecule, while the HucL subunit contains a single catalytic [NiFe]-cluster, which gives the complex a total molecular mass of 833.61 kDa (Grinter et al., 2023). The arrangement of the subunits of the complex is unique, with two HucSL units forming the heart-shaped $HucS_2L_2$ heterotetramer (Figure 14.4a). The four HucM molecules present

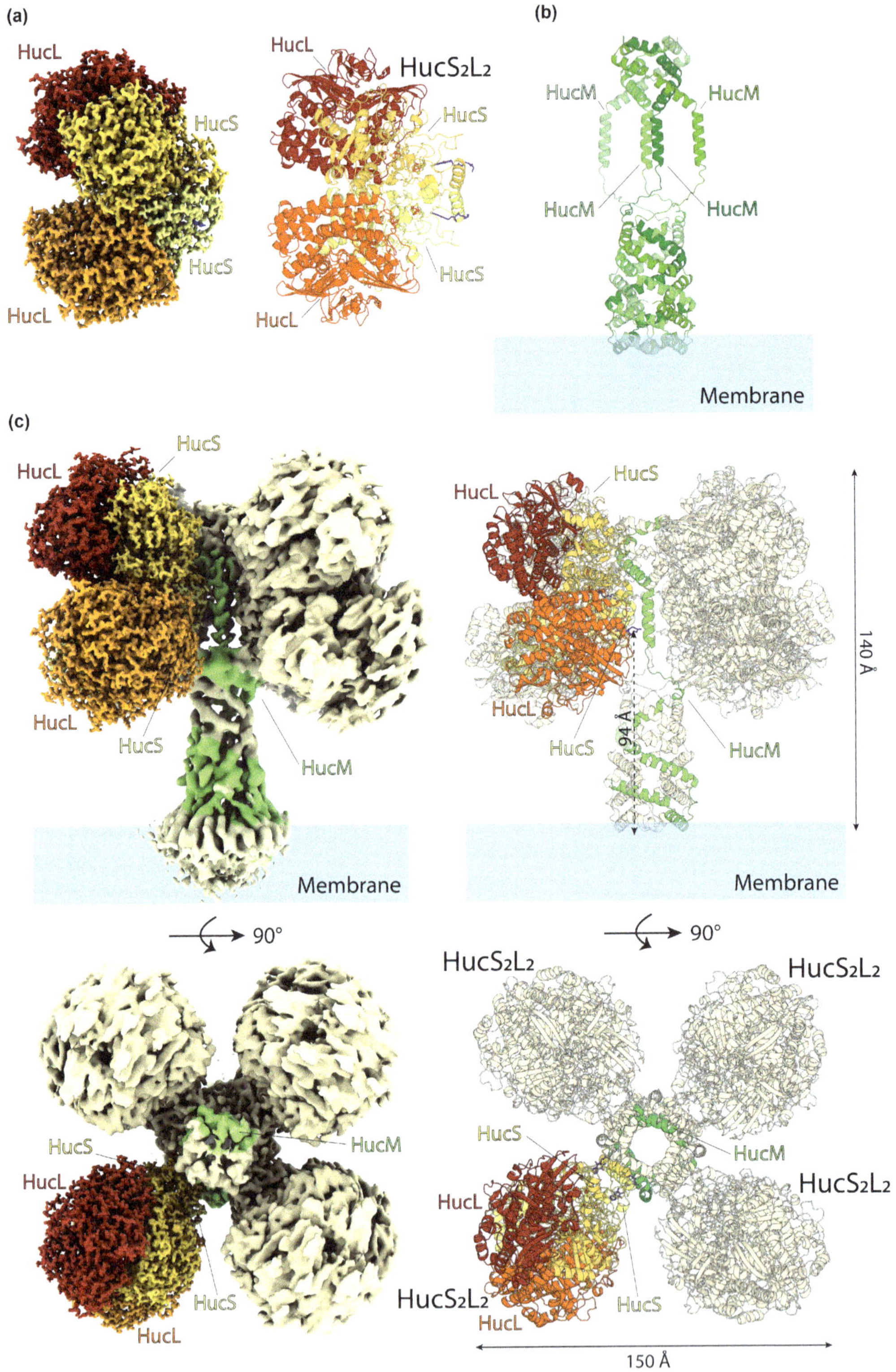

FIGURE 14.4 The cryo-EM structure of the Huc complex. (a) The 1.52 Å structure of the $HucS_2L_2$ catalytic subunit (PDB ID = 7UUR); the cryo-EM Coulomb potential map (*left*) and the refined structure (*right*) are shown. (b) The structure of the central Huc scaffold formed by a tetramer of HucM molecules (PDB ID = 7UUS). (c) The composite structure of the complete Huc complex; the cryo-EM Coulomb potential map (*left*) and the refined structure (*right*) are shown (PDB IDs = 8DVQ, 7UUR, 7UTD; EMDB IDs = EMD-27661, EMD-26802, EMD-26767). The 94 Å distance between the membrane surface and menaquinone bound to the Huc electron acceptor site is shown (*top left panel*).

in the complex have an elongated α-helical structure and intertwine to form a central scaffold for the complex with fourfold symmetry (Figure 14.4b). The four $HucS_2L_2$ subunits dock to the HucM scaffold from the Huc complex, which encloses the electron acceptor sites for the enzyme inside an internal chamber (Figure 14.4c). As is evident from the cryo-EM imaging, the complete Huc complex has a four-leafed clover shape, the $HucS_2L_2$ subunits each forming one of the lobes of the structure. An α-helical coiled-coil tail formed by the C-terminal region of HucM is connected to the cloverleaf portion of the complex via a flexible region in the structure and mediates interaction between the complex and the membrane (Figure 14.4b and c).

The Huc complex associates with the cell membrane via the HucM tail, the end of which contains a number of positively charged lysine and arginine residues that facilitate interaction with the positively charged phospholipid head groups (Grinter et al., 2023). The tetrameric coiled coil that forms this HucM tail is hollow, forming a tube that connects the membrane to the electron acceptor site contained in a cavity at the center of the complex. Intriguingly, both the interior of this tube and the cavity are lined entirely by hydrophobic residues, which are compatible with the uptake of hydrophobic molecules from the cell membrane (Figure 14.4b and c and Figure 14.5a).

Structural Insights into the O_2 Insensitivity and High Affinity of Huc

A defining feature of Huc as an [NiFe]-hydrogenase is its ability to oxidize H_2 at atmospheric concentrations (0.53 ppm), in ambient air that contains 20.95% O_2 (Grinter et al., 2023). O_2 is a potent inhibitor of most [NiFe]-hydrogenases, with trace levels of O_2 sufficient to completely inactivate some members of this family (Lu and Koo, 2019; Vincent et al., 2005). Other more O_2-tolerant [NiFe]-hydrogenases have been isolated. However, these enzymes require the presence of significant concentrations of H_2 in addition to O_2 to resist inhibition (Ludwig et al., 2009; Vincent et al., 2007). In contrast, Huc is insensitive to O_2 regardless of the concentration of H_2 present (Grinter et al., 2023). Leveraging the high-resolution cryo-EM data we obtained for Huc, we analyzed the Huc structure to gain insight into the structural basis for its remarkable O_2 insensitivity. The active site of [NiFe]-hydrogenases is buried in the heart of the enzyme, with a series of hydrophobic gas channels connecting it to the surface, which enables access to H_2 (Lubitz et al., 2014; Ogata et al., 2016). Analysis of the Huc structure revealed that these gas channels are significantly narrower than those observed in O_2-sensitive [NiFe]-hydrogenases, and in fact, the path these channels take in Huc has been rerouted compared to these hydrogenases (Figure 14.5b). Based on this analysis and on molecular dynamics simulations, we propose that the constrained Huc gas channels act as a molecular sieve, excluding O_2 while allowing H_2 to enter the enzyme's active site. These constrained gas channels may also contribute to the relatively slow catalytic rate of Huc in solution, which has an observed K_{cat} of ~7 s^{-1}, compared to other [NiFe]-hydrogenases, which can exhibit a K_{cat} of >500 s^{-1} (Grinter et al., 2023; Pershad et al., 1999; Schafer et al., 2013).

Despite extensive analysis, the structural determinants of the remarkably high affinity of Huc for H_2 are not certain. However, the high resolution of the cryo-EM structure revealed several unique features that may contribute to its ability to oxidize atmospheric H_2. First, the [NiFe]-cluster and the amino acids that surround it are conserved between Huc and other low-affinity hydrogenases, indicating that a change in the active site architecture does not dictate [NiFe]-hydrogenase affinity (Grinter et al., 2023; Ogata et al., 2015). Further, the density in the cryo-EM maps indicates that a hydroxyl (OH) molecule bridges the Ni and Fe ions in the active site (Figure 14.5c). This observation, combined with spectroscopy data, indicates that the Huc active site cluster is in a so-called Ni-B state, which is inhibitory in other [NiFe]-hydrogenases (Cracknell et al., 2009; Grinter et al., 2023; Shafaat et al., 2013). However, the Huc cryo-EM samples were prepared in ambient air, where Huc is fully active (Grinter et al., 2023). Secondly, cryo-EM density maps revealed that all the three [FeS] clusters bound by the HucS subunit have a [3Fe-4S] configuration, which has not been previously reported for an [NiFe]-hydrogenase (Grinter et al., 2023; Ogata et al., 2016). [3Fe-4S] clusters undergo their physiological redox transition at significantly higher potentials than the more common [4Fe-4S] clusters, which likely tunes Huc to oxidizing H_2 at atmospheric concentrations where the redox couple of the $H_2 \Leftrightarrow 2e^- + 2H^+$ has a significantly higher potential (Cord-Ruwisch et al., 1988; Greening and Grinter, 2022). Finally, the high resolution of the cryo-EM maps allowed us to identify that His166 of HucL is in a D-stereoisomeric configuration. The high quality of the density corresponding to the HucL amino acid backbone in this region supported the modeling of HucL His-166 as a D- rather than a typical L-isomer. Moreover, the presence of clear density for the hydrogen atom bound to the α-carbon of His-166 allowed us to unambiguously model this residue as a D-isomer (Figure 14.5d), something that would not be possible with X-ray crystallography at a resolution > ~1.0 Å (Ogata et al., 2015; Woińska et al., 2016; Yip et al., 2020). His-166 forms key interactions with two of the [3Fe-4S] clusters bound to the Huc, modifying their chemical environment and possibly tuning the affinity of the enzyme (Grinter et al., 2023).

Huc Is a Menaquinone Oxidase with a Novel Quinone Transport Mechanism

A striking discovery from the cryo-EM structure of Huc was density corresponding to menaquinone in a substrate binding site of HucS, near the electron-donating terminal [3Fe-4S] cluster (Grinter et al., 2023; Ogata et al., 2016). The high resolution of our data allowed menaquinone-9 to be unambiguously identified, and mass spectrometry confirmed it was present in the Huc complex at 96.2%

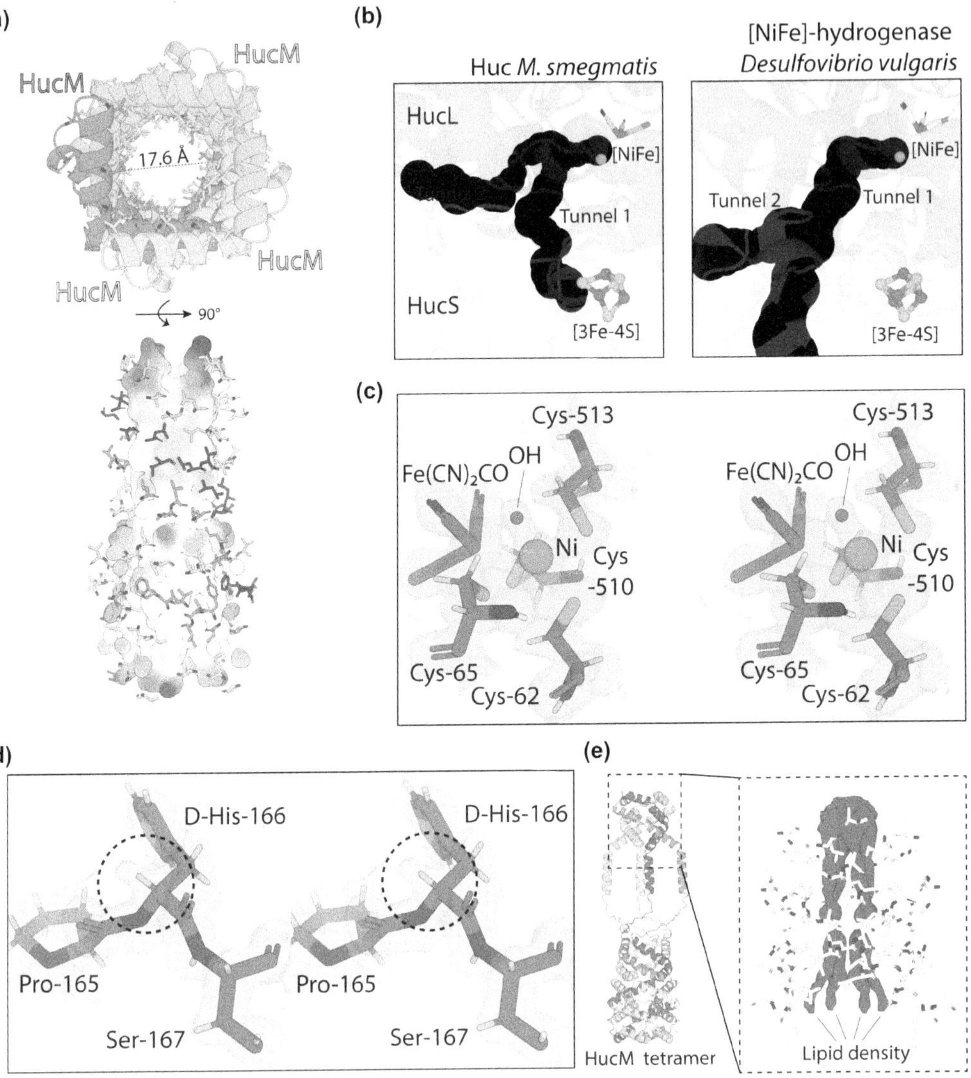

FIGURE 14.5 Key features of the Huc complex revealed by cryo-EM. (a) The C-terminal region of the HucM tetramer, showing that it forms a hollow tube capable of accommodating a menaquinone molecule (*top*), the interior of which is lined by hydrophobic amino acid side chains (*bottom*). (b) A view of the channels that provide H_2 access to the [NiFe]-hydrogenase active site for Huc (*left*) and an oxygen-sensitive [NiFe]-hydrogenase from *Desulfovibrio vulgaris* (*right*), showing that the Huc gas channels are markedly narrower. (c) A cross-eye stereo view of the Huc catalytic site, showing the [NiFe]-cluster, coordinating cysteines, and bound hydroxide as sticks/spheres with the corresponding region of the cryo-EM Coulomb potential map at 5 σ. (d) A cross-eye stereo view of D-His-166 of the HucL subunit sticks/spheres with the corresponding region of the cryo-EM Coulomb potential map at 5 σ. (e) A zoomed view of the top of the HucM tetramer showing that it is occluded by density corresponding to lipid molecules. (Images in (a) and (e) are adapted from Grinter et al. (2023).)

occupancy (Grinter et al., 2023). Menaquinone-9 is the major respiratory quinone in *M. smegmatis* and functions as an electron carrier between catabolic enzymes and the terminal respiratory complexes, generating the transmembrane proton gradient in the process (Upadhyay et al., 2015). The fact that menaquinone is bound at the Huc electron donor site indicates that Huc reduces it directly, rather than via an accessory protein, as is the case for other respiratory [NiFe]-hydrogenases studied to date (Grinter et al., 2023; Volbeda et al., 2013). This observation allowed us to determine that Huc is a hydrogen:menaquinone oxidoreductase.

While this observation confirmed that menaquinone is the electron acceptor for Huc, this raised a serious mechanistic question of how it reaches the Huc electron acceptor site. Menaquinone is a highly hydrophobic molecule that is generally confined to the cell membrane, and the electron acceptor sites of the Huc are in the interior of the complex, approximately 94 Å away from the surface of the membrane when Huc is in complex with it (Figure 14.4c). Analysis of the Huc complex structure identified features that would allow menaquinone transport from the membrane to the electron acceptor site of the enzyme. Huc forms a transient interaction with the cell membrane via a tube formed by the tetramer of HucM molecules (Figure 14.4b and c). This structure is entirely lined by hydrophobic residues and has interior dimensions capable of accommodating a menaquinone molecule (Figure 14.5a). We propose that the tube forms a hydrophobic conduit that allows menaquinone to enter the Huc complex and travel to the chamber in the center of Huc that contains the electron acceptor sites. This chamber is also lined with hydrophobic residues, and cryo-EM density indicates that at the top it is occluded by lipid molecules (Figure 14.5e). We propose that bulk solvent is excluded from the interior of the Huc complex, leading to a pseudomembrane environment that allows menaquinone to enter the enzyme for reduction to menaquinol with electrons from atmospheric H_2 oxidation. Reduced menaquinol can then return to the membrane to be oxidized by a terminal respiratory complex to generate proton motive force (Gong et al., 2018; Upadhyay et al., 2015).

CONCLUSIONS

In this work, we demonstrate the utility of affinity-tag-based native purification for isolating the Huc complex from *M. smegmatis*. This strategy was instrumental in obtaining this complex and provided key biological insights that would not have been possible using recombinant expression. The relatively modest quantities of Huc obtained using this method were no barrier to structural analysis using cryo-EM, allowing us to resolve the structure of the enzyme at record-breaking resolution (Grinter et al., 2023). This high resolution provided important insights into how Huc exhibits its extraordinary properties. We hope that this work will serve as a practical example of how this methodology can be used to determine the structure and function of other challenging macromolecular complexes.

ACKNOWLEDGEMENTS

We acknowledge the use of instruments and assistance at the Monash Ramaciotti Centre for Cryo-Electron Microscopy, a Node of Microscopy Australia. This work was supported by ARC Discovery Project Grants (DP200103074, DP230103080) (to R.G. and C.G.), a National Health and Medical Research Council Emerging Leader Grant (NHMRC) (APP1178715) (to C.G.), an NHMRC Emerging Leader Grant (APP1197376) (to R.G.), and ARC LIEF grants (LE200100045, LE120100090) for the Titan Krios Gatan K3 Camera and the Titan Krios.

REFERENCES

Asthana, P., Singh, D., Pedersen, J.S., Hynonen, M.J., Sulu, R., Murthy, A.V., Laitaoja, M., Janis, J., Riley, L.W., and Venkatesan, R. (2021). Structural insights into the substrate-binding proteins Mce1A and Mce4A from *Mycobacterium tuberculosis*. *IUCrJ*, **8**, 757–774.

Bay, S.K., Dong, X., Bradley, J.A., Leung, P.M., Grinter, R., Jirapanjawat, T., Arndt, S.K., Cook, P.L.M., LaRowe, D.E., Nauer, P.A., *et al.* (2021). Trace gas oxidizers are widespread and active members of soil microbial communities. *Nature Microbiology*, **6**, 246–256.

Beckham, K.S.H., Ritter, C., Chojnowski, G., Ziemianowicz, D.S., Mullapudi, E., Rettel, M., Savitski, M.M., Mortensen, S.A., Kosinski, J., and Wilmanns, M. (2021). Structure of the mycobacterial ESX-5 type VII secretion system pore complex. *Science Advances*, **7**, eabg9923.

Berney, M., Greening, C., Hards, K., Collins, D., and Cook, G.M. (2014). Three different [NiFe] hydrogenases confer metabolic flexibility in the obligate aerobe *Mycobacterium smegmatis*. *Environmental Microbiology*, ***16***, 318–330.

Bhella, D. (2019). Cryo-electron microscopy: An introduction to the technique, and considerations when working to establish a national facility. *Biophysical Reviews*, **11**, 515–519.

Bong, H.J., Ko, E.M., Song, S.Y., Ko, I.J., and Oh, J.I. (2019). Tripartite regulation of the glpFKD operon involved in glycerol catabolism by GylR, Crp, and SigF in *Mycobacterium smegmatis*. *Journal of Bacteriology*, **201**, e00511–00519.

Chen, J., Fruhauf, A., Fan, C., Ponce, J., Ueberheide, B., Bhabha, G., and Ekiert, D.C. (2023). Structure of an endogenous mycobacterial MCE lipid transporter. *Nature*, **620**, 445–452.

Conrad, R., and Seiler, W. (1981). Decomposition of atmospheric hydrogen by soil-microorganisms and soil enzymes. *Soil Biology & Biochemistry*, **13**, 43–49.

Cord-Ruwisch, R., Seitz, H.-J., and Conrad, R. (1988). The capacity of hydrogenotrophic anaerobic bacteria to compete for traces of hydrogen depends on the redox potential of the terminal electron acceptor. *Archives of Microbiology*, **149**, 350–357.

Cordero, P.R.F., Grinter, R., Hards, K., Cryle, M.J., Warr, C.G., Cook, G.M., and Greening, C. (2019). Two uptake hydrogenases differentially interact with the aerobic respiratory chain during mycobacterial growth and persistence. *Journal of Biological Chemistry*, **294**, 18980–18991.

Cracknell, J.A., Wait, A.F., Lenz, O., Friedrich, B., and Armstrong, F.A. (2009). A kinetic and thermodynamic understanding of O_2 tolerance in [NiFe]-hydrogenases. *Proceedings of the National Academy of Sciences, USA*, **106**, 20681–20686.

Fan, Q., Neubauer, P., Lenz, O., and Gimpel, M. (2020). Heterologous hydrogenase overproduction systems for biotechnology—An overview. *International Journal of Molecular Sciencex*, **21**, 5890.

Fukuda, Y., Stapleton, K., and Kato, T. (2023). Progress in spatial resolution of structural analysis by cryo-EM. *Microscopy*, **72**, 135–143.

Gardiner, A.T., Naydenova, K., Castro-Hartmann, P., Nguyen-Phan, T.C., Russo, C.J., Sader, K., Hunter, C.N., Cogdell, R.J., and Qian, P. (2021). The 2.4 angstrom cryo-EM structure of a heptameric light-harvesting 2 complex reveals two carotenoid energy transfer pathways. *Science Advances*, **7**, eabe4650.

Gong, H., Li, J., Xu, A., Tang, Y., Ji, W., Gao, R., Wang, S., Yu, L., Tian, C., Li, J., *et al.* (2018). An electron transfer path connects subunits of a mycobacterial respiratory supercomplex. *Science*, **362**, eaat8923.

Greening, C., Berney, M., Hards, K., Cook, G.M., and Conrad, R. (2014). A soil actinobacterium scavenges atmospheric H_2 using two membrane-associated, oxygen-dependent [NiFe] hydrogenases. *Proceedings of the National Academy of Sciences, USA*, **111**, 4257–4261.

Greening, C., and Grinter, R. (2022). Microbial oxidation of atmospheric trace gases. *Nature Reviews Microbiology*, **20**, 513–528.

Grinter, R., Kropp, A., Venugopal, H., Senger, M., Badley, J., Cabotaje, P.R., Jia, R., Duan, Z., Huang, P., Stripp, S.T., *et al.* (2023). Structural basis for bacterial energy extraction from atmospheric hydrogen. *Nature*, **615**, 541–547.

Islam, Z.F., Cordero, P.R.F., and Greening, C. (2019). Putative iron-sulfur proteins are required for hydrogen consumption and enhance survival of mycobacteria. *Frontiers in Microbiology*, **10**, 2749.

Johnson, S., Furlong, E.J., Deme, J.C., Nord, A.L., Caesar, J.J.E., Chevance, F.F.V., Berry, R.M., Hughes, K.T., and Lea, S.M. (2021). Molecular structure of the intact bacterial flagellar basal body. *Nature Microbiology*, **6**, 712–721.

Lasica, A.M., Ksiazek, M., Madej, M., and Potempa, J. (2017). The type IX secretion system (T9SS): Highlights and recent insights into its structure and function. *Frontiers in Cellular and Infection Microbiology,* **7**, 215.

Lauber, F., Deme, J.C., Lea, S.M., and Berks, B.C. (2018). Type 9 secretion system structures reveal a new protein transport mechanism. *Nature*, **564**, 77–82.

Letts, J.A., Fiedorczuk, K., and Sazanov, L.A. (2016). The architecture of respiratory supercomplexes. *Nature*, **537**, 644–648.

Liang, Y., Plourde, A., Bueler, S.A., Liu, J., Brzezinski, P., Vahidi, S., and Rubinstein, J.L. (2023). Structure of mycobacterial respiratory complex I. *Proceedings of the National Academy of Sciences, USA,* **120**, e2214949120.

Lu, Y., and Koo, J. (2019). O_2 sensitivity and H_2 production activity of hydrogenases – A review. *Biotechnology Bioengineering*, **116**, 3124–3135.

Lubitz, W., Ogata, H., Rudiger, O., and Reijerse, E. (2014). Hydrogenases. *Chemical Reviews*, **114**, 4081–4148.

Ludwig, M., Cracknell, J.A., Vincent, K.A., Armstrong, F.A., and Lenz, O. (2009). Oxygen-tolerant H_2 oxidation by membrane-bound [NiFe] hydrogenases of ralstonia species. Coping with low level H_2 in air. *Journal of Biological Chemistry*, **284**, 465–477.

Mattanovich, D., Branduardi, P., Dato, L., Gasser, B., Sauer, M., and Porro, D. (2012). Recombinant protein production in yeasts. *Methods in Molecular Biology*, **824**, 329–358.

Nakane, T., Kotecha, A., Sente, A., McMullan, G., Masiulis, S., Brown, P.M., Grigoras, I.T., Malinauskaite, L., Malinauskas, T., and Miehling, J. (2020). Single-particle cryo-EM at atomic resolution. *Nature*, **587**, 152–156.

Ogata, H., Lubitz, W., and Higuchi, Y. (2016). Structure and function of [NiFe] hydrogenases. *Journal of Biochemistry*, **160**, 251–258.

Ogata, H., Nishikawa, K., and Lubitz, W. (2015). Hydrogens detected by subatomic resolution protein crystallography in a [NiFe] hydrogenase. *Nature*, **520**, 571–574.

Ortiz, M., Leung, P.M., Shelley, G., Jirapanjawat, T., Naucr, P.A., Van Goethem, M.W., Bay, S.K., Islam, Z.F., Jordaan, K., Vikram, S., *et al.* (2021). Multiple energy sources and metabolic strategies sustain microbial diversity in Antarctic desert soils. Proceedings of the National Academy of Sciences, USA, **118**, 45.

Passmore, L.A., and Russo, C.J. (2016). Specimen preparation for high-resolution cryo-EM. *Methods in Enzymology*, **579**, 51–86.

Pershad, H.R., Duff, J.L., Heering, H.A., Duin, E.C., Albracht, S.P., and Armstrong, F.A. (1999). Catalytic electron transport in *Chromatium vinosum* [NiFe]-hydrogenase: Application of voltammetry in detecting redox-active centers and establishing that hydrogen oxidation is very fast even at potentials close to the reversible H^+/H_2 value. *Biochemistry*, **38**, 8992–8999.

Piper, S.J., Brillault, L., Rothnagel, R., Croll, T.I., Box, J.K., Chassagnon, I., Scherer, S., Goldie, K.N., Jones, S.A., Schepers, F., *et al.* (2019). Cryo-EM structures of the pore-forming a subunit from the *Yersinia entomophaga* ABC toxin. *Nature Communications*, **10**, 1952.

Piper, S.J., Johnson, R.M., Wootten, D., and Sexton, P.M. (2022). Membranes under the Magnetic Lens: A Dive into the Diverse World of Membrane Protein Structures Using Cryo-EM. *Chemical Reviews*, **122**, 13989–14017.

Punjani, A., Rubinstein, J.L., Fleet, D.J., and Brubaker, M.A. (2017). cryoSPARC: algorithms for rapid unsupervised cryo-EM structure determination. *Nature Methods*, **14**, 290–296.

Qian, P., Gardiner, A.T., Simova, I., Naydenova, K., Croll, T.I., Jackson, P.J., Nupur, Kloz, M., Cubakova, P., Kuzma, M., *et al.* (2022). 2.4 Angstrom structure of the double-ring *Gemmatimonas phototrophica* photosystem. *Science Advances*, 8, eabk3139.

Rohou, A., and Grigorieff, N. (2015). CTFFIND4: Fast and accurate defocus estimation from electron micrographs. *Journal of Structural Biology*, **192**, 216–221.

Rosano, G.L., Morales, E.S., and Ceccarelli, E.A. (2019). New tools for recombinant protein production in *Escherichia coli*: A 5-year update. *Protein Science*, **28**, 1412–1422.

Russo, C.J., and Passmore, L.A. (2016). Ultrastable gold substrates: Properties of a support for high-resolution electron cryomicroscopy of biological specimens. *Journal of Structure Biology*, **193**, 33–44.

Saibil, H.R. (2022). Cryo-EM in molecular and cellular biology. *Molecular Cell*, **82**, 274–284.

Schafer, C., Friedrich, B., and Lenz, O. (2013). Novel, oxygen-insensitive group 5 [NiFe]-hydrogenase in *Ralstonia eutropha. Applied and Environmental Microbiology*, **79**, 5137–5145.

Shafaat, H.S., Rudiger, O., Ogata, H., and Lubitz, W. (2013). [NiFe] hydrogenases: A common active site for hydrogen metabolism under diverse conditions. *Biochimica et Biophysica Acta*, **1827**, 986–1002.

Shepherd, D.C., Dalvi, S., and Ghosal, D. (2022). From cells to atoms: Cryo-EM as an essential tool to investigate pathogen biology, host–pathogen interaction, and drug discovery. *Molecular Microbiology*, **117**, 610–617.

Steinhilper, R., Höff, G., Heider, J., and Murphy, B.J. (2022). Structure of the membrane-bound formate hydrogenlyase complex from *Escherichia coli. Nature Communications*, **13**, 5395.

Tan, Y.Z., Keon, K.A., Abdelaziz, R., Imming, P., Schulze, W., Schumacher, K., and Rubinstein, J.L. (2022). Structure of V-ATPase from citrus fruit. *Structure*, **30**, 1403–1410.e4.

Upadhyay, A., Fontes, F.L., Gonzalez-Juarrero, M., McNeil, M.R., Crans, D.C., Jackson, M., and Crick, D.C. (2015). Partial saturation of menaquinone in *Mycobacterium tuberculosis*: Function and essentiality of a novel reductase, MenJ. *ACS Central Science,* **1**, 292–302.

Vincent, K.A., Parkin, A., and Armstrong, F.A. (2007). Investigating and exploiting the electrocatalytic properties of hydrogenases. *Chemical Reviews*, **107**, 4366–4413.

Vincent, K.A., Parkin, A., Lenz, O., Albracht, S.P., Fontecilla-Camps, J.C., Cammack, R., Friedrich, B., and Armstrong, F.A. (2005). Electrochemical definitions of O_2 sensitivity and oxidative inactivation in hydrogenases. *Journal of the American Chemical Society*, **127**, 18179–18189.

Volbeda, A., Darnault, C., Parkin, A., Sargent, F., Armstrong, F.A., and Fontecilla-Camps, J.C. (2013). Crystal structure of the O_2-tolerant membrane-bound hydrogenase 1 from *Escherichia coli* in complex with its cognate cytochrome b. *Structure*, **21**, 184–190.
Wagner, T., Merino, F., Stabrin, M., Moriya, T., Antoni, C., Apelbaum, A., Hagel, P., Sitsel, O., Raisch, T., Prumbaum, D., *et al.* (2019). SPHIRE-crYOLO is a fast and accurate fully automated particle picker for cryo-EM. *Communications Biology*, **2**, 1–13.
Watanabe, T., Pfeil-Gardiner, O., Kahnt, J., Koch, J., Shima, S., and Murphy, B.J. (2021). Three-megadalton complex of methanogenic electron-bifurcating and CO_2-fixing enzymes. *Science*, **373**, 1151–1155.
Woińska, M., Grabowsky, S., Dominiak, P.M., Woźniak, K., and Jayatilaka, D. (2016). Hydrogen atoms can be located accurately and precisely by x-ray crystallography. *Science Advances*, **2**, e1600192.
Wulff, P., Day, C.C., Sargent, F., and Armstrong, F.A. (2014). How oxygen reacts with oxygen-tolerant respiratory [NiFe]-hydrogenases. *Proceedings of the National Academy of Sciences, USA*, **111**, 6606–6611.
Wurm, F.M. (2004). Production of recombinant protein therapeutics in cultivated mammalian cells. *Nature Biotechnology*, **22**, 1393–1398.
Yip, K.M., Fischer, N., Paknia, E., Chari, A., and Stark, H. (2020). Atomic-resolution protein structure determination by cryo-EM. *Nature*, **587**, 157–161.
Zhang, K., Li, M., and Sun, F. (2011). Gautomatch: an efficient and convenient gpu-based automatic particle selection program. *Unpublished manuscript*.
Zheng, S.Q., Palovcak, E., Armache, J.P., Verba, K.A., Cheng, Y., and Agard, D.A. (2017). MotionCor2: Anisotropic correction of beam-induced motion for improved cryo-electron microscopy. *Nature Methods*, **14**, 331–332.
Zivanov, J., Nakane, T., Forsberg, B.O., Kimanius, D., Hagen, W.J., Lindahl, E., and Scheres, S.H. (2018). New tools for automated high-resolution cryo-EM structure determination in RELION-3. *Elife*, **7**, e42166.

15 Membrane Rupture and Lipid–Protein Interaction of γ-Hemolysin from *Staphylococcus aureus* Using Cryo-EM

Suman Mishra, Anupam Roy, Arnab Chatterjee, and Somnath Dutta

INTRODUCTION

Staphylococcus aureus is the leading cause of nosocomial and community-acquired infections, with an annual killing rate of over 100,000 people globally (Howden et al., 2023; Ikuta et al., 2022). Pore-forming toxins (PFTs) constitute the largest class of bacterial virulence factors (Los et al., 2013). These are amphipathic proteins that are secreted as water-soluble monomers, which oligomerize on host cellular membranes and subsequently puncture holes by forming a transmembrane pore complex (Iacovache et al., 2010; Peraro & van der Goot, 2016). *S. aureus* employs several such PFTs, including mono-component α-hemolysin and bicomponent PFTs such as γ-hemolysin ACB (HlgACB), Panton-Valentine leukocidin, and leukocidins AB (LukAB) and ED (LukED) (Kaneko & Kamio, 2004; Spaan et al., 2017; Tromp & van Strijp, 2020). In the case of all staphylococcal bicomponent β-PFTs, the two toxin components are classified into LukS and LukF (Alonzo & Torres, 2014). Different S components (HlgA, LukS-PV, etc.) of *S. aureus* bicomponent PFTs have 70% sequence identity, and the same also holds true for different F component toxins (HlgB, LukF-PV, etc.). Upon comparison of S components with counterpart F components, the sequence identity is ~30% (e.g., between HlgA and HlgB) (Kaneko & Kamio, 2004). Despite sharing a similar overall architecture, these bicomponent PFTs are highly specific to their respective host cellular targets (Blake et al., 2018; Lubkin et al., 2019; Spaan et al., 2017, 2014). Interestingly, HlgAB and LukED interact with Duffy antigen receptor for chemokines (DARC) on the surface of human erythrocytes (Spaan et al., 2014, 2015). The interaction of these toxin subunits with DARC is believed to trigger oligomerization and pore formation, though the mechanism is yet to be elucidated. Pore formation leads to rapid lysis of host erythrocytes. The degradation of host hemoglobin is necessary to snatch iron for *S. aureus* growth (Spaan et al., 2015). Therefore, these PFTs could also serve as an alternative target to mitigate staphylococcal pathogenesis (Peraro & van der Goot, 2016). In essence, pore formation by these PFTs can be arrested at the following checkpoints: (1) Lipid-binding resides of these toxins could be targeted to develop monoclonal antibodies or drug molecules, which disrupt lipid–toxin interaction, (2) inhibiting the interaction between toxin monomers and the respective cellular receptor, (3) targeting oligomerization after membrane binding by disrupting the molecular interactions at the interface of the two components, and (4) interfering with the post-oligomerization process of pre-stem to stem conformational transition, thus inhibiting the formation of functional transmembrane pores. HlgAB serves as an archetypical bicomponent β-PFT and could be targeted to decipher these steps of pore formation (Alessandrini et al., 2013; Ghanem et al., 2022; Olson et al., 1999; Sugawara-Tomita et al., 2002). Despite the availability of the crystal structure of octameric HlgAB (Yamashita et al., 2011), there were certain limitations since the oligomerization and pore formation was triggered by methyl pentane-diol (MPD), a common crystallizing agent (Tanaka et al., 2011). However, how this PFT interacts with lipids in its native environment remains largely unexplained. In addition, the amino latch for both toxin subunits was unstructured in the crystal structure (Yamashita et al., 2011); therefore, it was unclear whether the amino latch played any structural role during the pore formation.

In our study (Mishra et al., 2023b), we employed single-particle cryo-electron microscopy (cryo-EM) for structural characterization of HlgAB in a lipid membrane environment. We selected large unilamellar vesicles (LUVs) composed of lecithin and cholesterol as our model membrane (Alessandrini et al., 2013) for our cryo-EM-based structural studies. In this current study, we followed these major objectives: (1) to elucidate the native oligomeric state of HlgAB in liposomes, (2) to predict the residues of HlgAB engaging in binding to the membrane, (3) to elaborate the structural aspects of the amino latch in oligomerization, and (4) to capture intermediate pre-pore states.

RESULTS

Preparation of Functionally Active Toxin Oligomers in Liposomes for Cryo-EM Study

The crystal structures of γ-HL oligomer (PDBIDs: 3B07 and 4P1Y) were resolved in the absence of lipids such as phosphatidylcholine (Yamashita et al., 2011, 2014) (Figure 15.1a). This limited our current understanding of how the

DOI: 10.1201/9781003326106-18

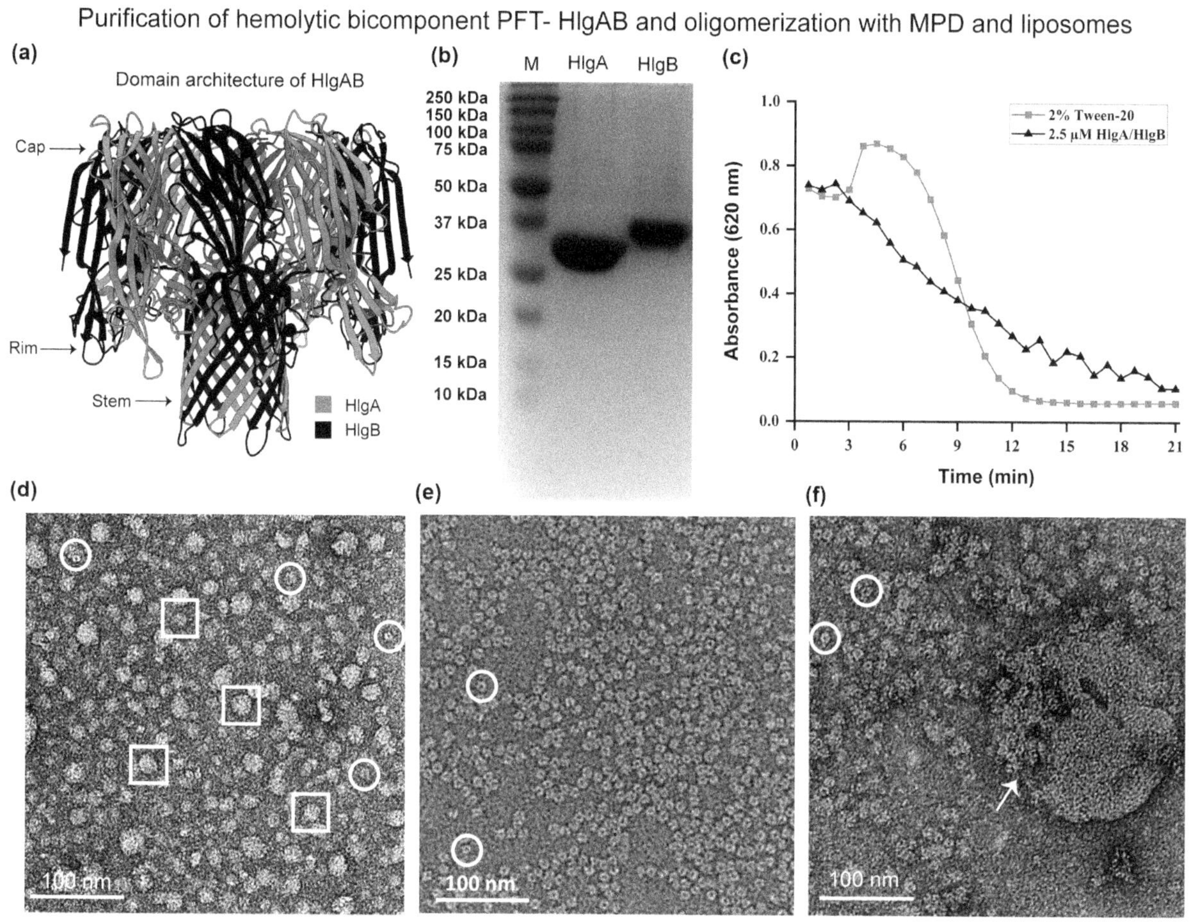

FIGURE 15.1 (a) Domain architecture of bicomponent HlgAB; cap, rim, and stem domains are highlighted with black arrows. HlgA and HlgB subunits are colored grey and black respectively (PDBID: 3B07). (b) SDS PAGE (12%) profile of SEC-purified recombinant toxins (HlgA and HlgB). Respective protein bands are shown separately with the protein molecular marker (*left*). (c) Hemolysis assay represents the lysis of rabbit erythrocytes in the presence of equimolar (2.5 μM) toxins (*solid triangle*) and with 2% detergent Tween-20 (*square box*). Adapted from Mishra et al. (2023a). (d) The transmission electron micrograph of equimolar toxins extracted from detergent (CHAPS) represents a mixture of detergent micelles (square) and a minor population of small HlgAB oligomers (encircled). (e) Negatively stained electron micrograph of homogenous oligomeric HlgAB complex in presence of 20% MPD. (f) Formation of oligomeric HlgAB complexes and lipid–protein clustering after treatment of equimolar toxins in eggPC vesicles. The scale bar is 100 nm.

toxin monomers interact with the host membrane, and factors affecting the pre-stem to stem conversion. Our structural study of HlgAB in liposomes primarily aimed to understand possible lipid-binding sites for both the toxin subunits. In addition, we also wanted to mimic the native membrane environment to investigate the interactions between the neighboring protein oligomers in the absence of detergents or model membrane systems such as lipid nanodiscs.

To achieve our targets, we employed two-step purification comprising immobilized metal affinity chromatography (IMAC) and size exclusion chromatography (SEC) to purify recombinantly expressed HlgA and HlgB subunits with a C-terminal 6xHis tag. It is essential to mention that we also purified N-terminal 6xHis tag HlgA and HlgB subunits. However, we could not identify oligomeric HlgA and HlgB in the presence of detergent or liposomes. This phenomenon suggests that N-terminal His-tag inhibits oligomerization. Thus, we strongly recommend verifying the activity of any recombinant protein sample when there is any extra tag associated with proteins, and activity may be variable based on the tag's position. However, the SDS-PAGE profile of the C-terminal 6xHis tagged two toxins illustrated purity with apparent monomeric molecular mass between 32 and 36 kDa, as expected for the monomeric subunits (Figure 15.1b), which was further verified by mass spectrometric analysis after trypsin digestion. Using an equimolar concentration of these proteins, we performed hemolysis assay with rabbit erythrocytes to ensure if the two toxin subunits were functionally active. The hemolysis assay indicated that the purified HlgA–HlgB can damage the erythrocytes which indicate the formation of transmembrane pore complexes composed of both protein subunits (Figure 15.1c). Several PFTs have been previously reported to oligomerize in the presence of solubilizing detergents (Bruggisser et al., 2022; Mueller et al., 2009), while MPD is known to trigger oligomerization for staphylococcal PFTs such as HlgAB and α-HL (Tanaka et al., 2011; Yamashita et al., 2011, 2014). We initially targeted both solubilizing detergents as well as MPD to study the formation of pore intermediates before progressing to liposomes. We screened different micellar concentrations of several solubilizing detergents such as CHAPS (3-[(3-cholamidopropyl)dimethylammonio]-1-propanesulfonate), DDM (n-dodecyl-B-D-maltoside), polysorbate 20, and octyl glucoside (OG) to see if these could trigger oligomerization. We could see small populations of oligomers, but detergent micelles were predominating, which was very difficult for cryogenic-EM-based structural studies (Figure 15.1d). Thus, we proceeded to incubate equimolar HlgA and HlgB with concentrations of MPD ranging between 20% and 30% based on previous studies (Tanaka et al., 2011; Yamashita et al., 2011). Under room temperature NS-TEM, we could observe a population of very homogenous ring-like oligomers (Figure 15.1e). This encouraged us to attempt cryo-EM study of HlgAB oligomers induced in the presence of 20% MPD. MPD is a known cryoprotectant, and its presence severely hampered optimal cryo-EM grid freezing. Removal of MPD using dialysis/buffer exchange induced severe aggregation of HlgAB oligomers. Lower concentrations of MPD, as reported previously, induced limited formation of oligomers.

However, for the study of HlgAB pore formation on model membrane, we prepared large unilamellar liposomes constituting equimolar lecithin and cholesterol. Upon incubation with both the toxin subunits, we observed drastic rupturing of the lipid vesicles, with formation of clustering of HlgAB oligomers on membrane surface (Figure 15.1f). Our studies showed that HlgAB oligomerization highly depends on protein–lipid concentration as well as incubation times of protein with lipid. We noticed that higher protein concentration and longer incubation times resulted in lipid -protein aggregations, which may be the result of membrane rupturing by pore-forming toxins (PFTs). Therefore, we recommend optimizing the lipid–protein ratio and protein incubation times for structural characterization of any types of PFT–liposome research work. However, based on our insights, optimized lipid–protein complexes were utilized for cryo-EM-based structural studies. Thus, based on our negative staining TEM and 2D class averages, we proceeded with HlgAB-treated liposomes for cryo-EM single-particle analysis.

Different Approaches to Address the Challenges in cryo-EM Study of HlgAB Proteoliposome

There are several challenges associated with the cryo-EM study of proteoliposome, including (1) optimization of vitrification conditions, since the sample constitutes both proteins and large liposomes with diameter >100 nm; (2) nonuniform distribution of liposomes in the holes due to affinity toward the surrounding carbon region; and (3) identification of appropriate liposomes with homogenously bound HlgAB oligomers at lower magnification for data collection. In addition, clustering of HlgAB oligomers also posed as a major bottleneck in our cryo-EM study. Various freezing parameters were optimized to observe the proper distribution of liposome-embedded protein complexes (Figure 15.2a and b). Simultaneously, the data acquisition pipeline was also optimized to collect a large amount of cryo-EM data of liposome-embedded HlgAB. Interestingly, we could observe a few free oligomeric pore complexes, membrane-bound oligomers, small protein-lipid clusters, an octahedral superassembly, and a square planar 2D lattice of oligomeric HlgAB (Figure 15.2c–h). Such a myriad of post-oligomerization states in the cryo-EM dataset posed an exciting challenge for cryo-EM 3D reconstruction. Due to the limited number of particles corresponding to free and membrane-bound isolated oligomeric pores, we shifted our attention to larger yet ordered clusters, such as 2D lattice and the octahedral superassembly species of oligomeric HlgAB. Though we were able to identify HlgAB's oligomeric state participating in the 2D lattice as octamers (Figure 15.3a–c), limited angular distribution affected high-resolution 3D reconstruction from 2D lattice (Figure 15.3b and c). We could observe small lipid–protein aggregation in many cryo-EM images due to longer incubations of HlgAB in liposomes (Figure 15.2d), which resulted in small lipid–protein clustering. However, the octahedral superassemblies of HlgAB oligomers were homogenous and monodispersed. Upon performing 2D classification, we could observe several distinct orientations corresponding to the superassembly state. We generated an *ab initio* 3D model without applying any symmetry, through which we were able to deduce that six octameric HlgAB complexes were associated with their rim domains and had an overall octahedral symmetry. Finally, we imposed octahedral symmetry to reconstruct HlgAB superassemblies to resolutions ranging from 3.2 to 3.5 Å (Figure 15.4a). Thus, targeting a larger assembly of ~1.8 MDa with octahedral symmetry helped us to reconstruct HlgAB to near-atomic resolution with fewer than 12,000 particles contributing to final reconstruction (Figure 15.4b). We validated our final cryo-EM map with local resolution estimation and by manual inspection of side chain fitting for both HlgA and HlgB toxin subunits. Our resultant atomic model (PDBID: 8GZ7) (Figure 15.4c) provided several novel structural insights into the phenomenon of PFT clustering, the structural role of amino latch, plausible lipid-binding residues, and the dynamic nature of the β-barrel domain.

Structural Role of the Amino Latch of HlgA in Pre-Stem to Stem Transition

The pre-stem to stem transition that drives the conversion of water-soluble monomers into a transmembrane amphiphilic state remains ambiguous. In the case of α-HL, several structural and mutagenesis studies have illustrated that the

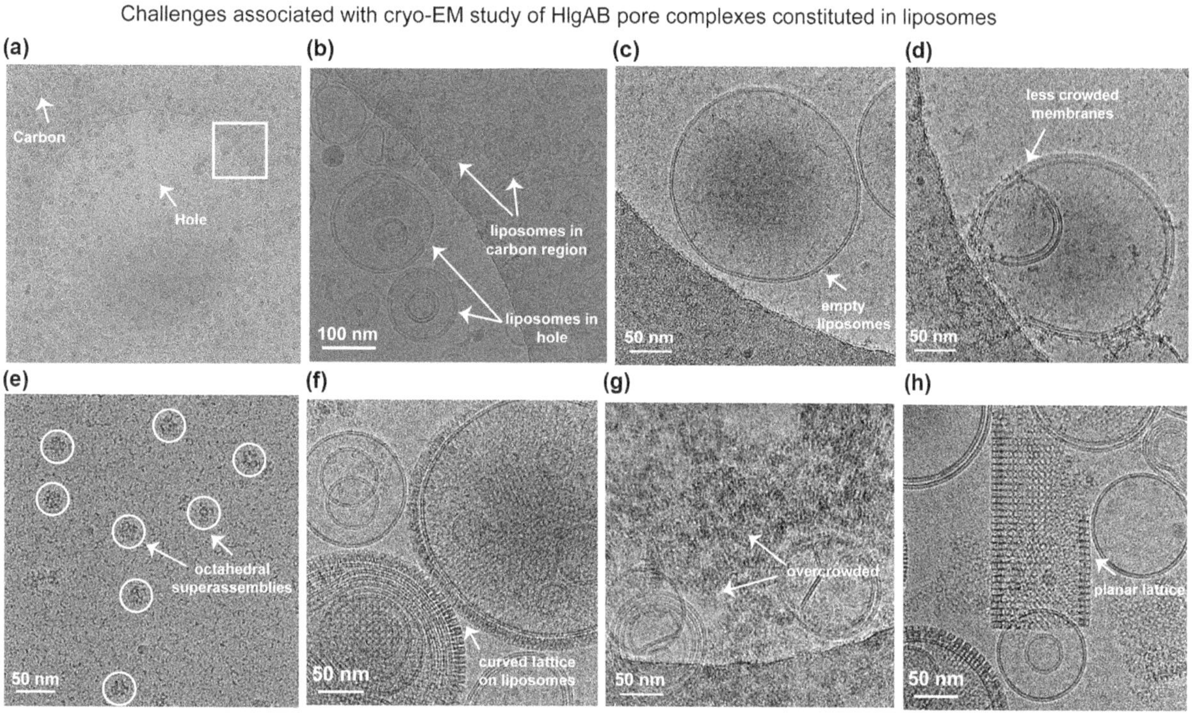

FIGURE 15.2 (a–b) Control eggPC–cholesterol large unilamellar vesicle (LUV or liposome) distribution in hole and carbon region. (c–h) Different representative cryo-EM micrographs of HlgAB octameric pore complex in presence of unilamellar vesicles as observed after the treatment of an equimolar concentration of both toxins (HlgA and HlgB) with LUVs. (c) Representative cryo-electron micrograph of an empty liposome with no embedded proteins. (d) Corresponding electron micrographs demonstrating the formation of HlgAB pore complexes on the surface of liposomal bilayer (marked with white arrows). (e) The formation of larger protein–lipid aggregates/assemblies of HlgAB with LUVs of eggPC–cholesterol (marked with white arrows). (f) Representative cryo-EM micrograph of the larger super assembled states of oligomeric HlgAB complex (encircled in white). (g) Micrograph containing 2D lattice of oligomeric pore complex on curved lipid vesicles (marked with white arrows). (h) The electron micrograph of planar 2D oligomeric lattice (indicated with white arrows). (Adapted from Mishra et al. (2023a).)

N-terminal latch plays a vital role in pore formation and its absence abrogates pore formation (Jayasinghe et al., 2006; Song et al., 1996). However, in the case of bicomponent β-PFTs, the role of the amino latch is relatively unclear (Kaneko et al., 1998; Takeda et al., 2020). Unlike α-HL, the crystal structure of γ-HL (PDBID: 3B07) had unstructured amino latches for both HlgA and HlgB subunits. Surprisingly, in our cryo-EM map, we clearly observed densities corresponding to the complete amino latch (residues K3–A11) for HlgA, while the HlgB N-terminal region (E1–K15) remained disordered. Notably, the terminal residues belonging to the amino latch of HlgA were found to be proximal to the neighboring HlgB. We aligned two protomeric subunits (HlgA and HlgB) from our cryo-EM atomic model (PDBID: 8GZ7) with the monomeric HlgA and HlgB (PDBIDs: 2QK7 and 1LKF, respectively) (Olson et al., 1999; Roblin et al., 2008) to observe the structural changes between monomeric HlgA and protomeric HlgA with respect to the amino latch. We observed that the amino latch of HlgA in our cryo-EM structure was embedded into the space between the pre-stem and β-sandwich domains for both monomeric HlgA and the nearby monomeric HlgB (Figure 15.4e). The residues K3–E5 were found closer to the nearby HlgB, while D6–A11 residues were proximal to the β-sandwich of HlgA. E5 residue of HlgA was also observed to be participating in a salt bridge interaction with N40 residue of HlgB. Thus, the N-terminal latch of HlgA could trigger pre-stem to stem transition of both HlgA and HlgB by introducing steric clashes between the pre-stem and inner leaflet of the β-sandwich in the cap domain (Figure 15.4e). The amino latch is also sequentially as well as structurally conserved across different bicomponent β-PFTs from *S. aureus*. Interestingly, it has been previously illustrated with mutagenesis studies that pore formation was not significantly hampered by an absence of the amino latch, thereby indicating that bicomponent β-PFTs might have evolved another amino latch–independent mechanism of pore formation (Jayasinghe et al., 2006; Takeda et al., 2020). Through *in silico* studies, it is now known that the monomeric toxin subunits HlgA and HlgB interact with the lipid membrane initially in a slanted fashion (Tarenzi et al., 2022). Thus, γ-HL and other bicomponent β-PFTs might have a concerted mechanism of pre-stem to stem transition that is not only driven by the steric clashes due to the amino latch but also by interaction of the pre-stem with the lipid membrane. This is supported by fuzzy lipid densities

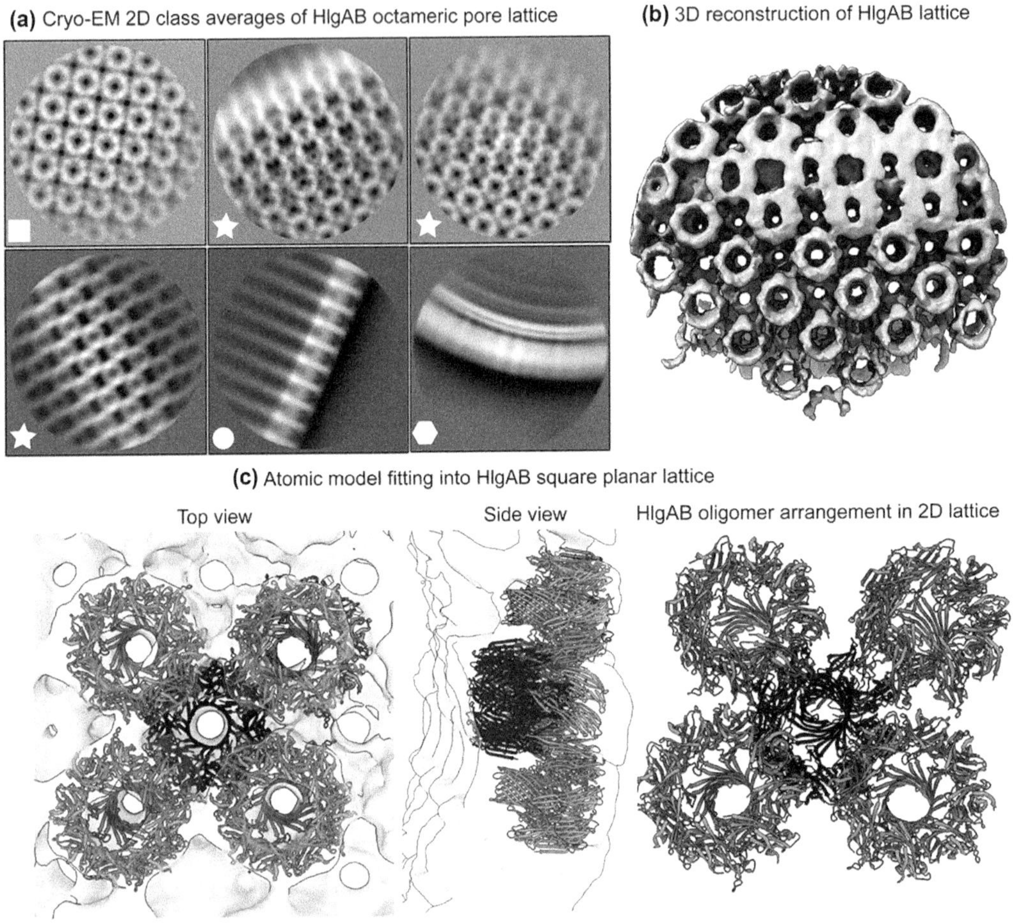

FIGURE 15.3 (a) Cryo-EM 2D class averages of square planar 2D lattice of octameric HlgAB pore complexes. Different orientations are highlighted with square (*top view*), star (*tilted views*), circle (*side view of planar 2D lattice*) and hexagon (*2D lattice over a curved liposome surface*). (b) Cryo-EM 3D reconstruction of HlgAB 2D lattice. (c) Atomic model fitted into 3D reconstruction of HlgAB 2D lattice. (Adapted from Mishra et al. (2023a).)

that we observed in the upper half of the stem domain, discussed in detail in a subsequent section.

Partial β-barrel Represents a Native pre-pore/post-pore State of Bicomponent β-PFTs

HlgAB is a typical β-PFT, i.e., the transmembrane domain that acts like a channel is constituted from a β-barrel (Peraro & van der Goot, 2016). Each toxin monomer in the octameric assembly contributes two β-strands; thus a β-barrel for octameric HlgAB comprises 16 β-strands stabilized by H-bonds and electrostatic and hydrophobic interactions. Interestingly, the upper half (closer to the rim domain) of the γ-HL β-barrel possessed many hydrophobic and aromatic residues, while the lower half (which spans the transmembrane region) was mostly enriched with Gly and Ser residues (Iacovache et al., 2010). Despite sharing a similar overall architecture with the crystal structure of HlgAB, surprisingly the β-barrel of each octameric pore complex was only partially resolved in our cryo-EM map (Figure 15.4b and c). We confirmed that the local resolution estimates were ~3.2 Å in the β-barrel domain. The unstructured regions were between the residues G114–S128 for HlgA and G120–N136 for HlgB. These residues were otherwise expected to span the lipid bilayer in a membrane-bound state. This raised an important question whether our cryo-EM structure represented a distinct state as compared to the crystal structure. Upon aligning the cryo-EM atomic structure with the wild-type crystal structure (PDBID: 3B07), we observed two important differences: (1) the β-barrel was partially resolved in our cryo-EM map, and (2) the amino latch of HlgA that was not resolved in the crystal structure was fully resolved in our cryo-EM map. These observations indicated that our cryo-EM map might correspond to a different oligomeric state as compared to the previously reported complete β-barrel pore crystal structure. Strikingly, these two unique aspects of our atomic model corroborated strongly with a previous study highlighting the mutation-stabilized pre-pore state of γ-HL (PDBID: 4P1Y) (Yamashita et al., 2014). In both these structures, the N-terminal latch of

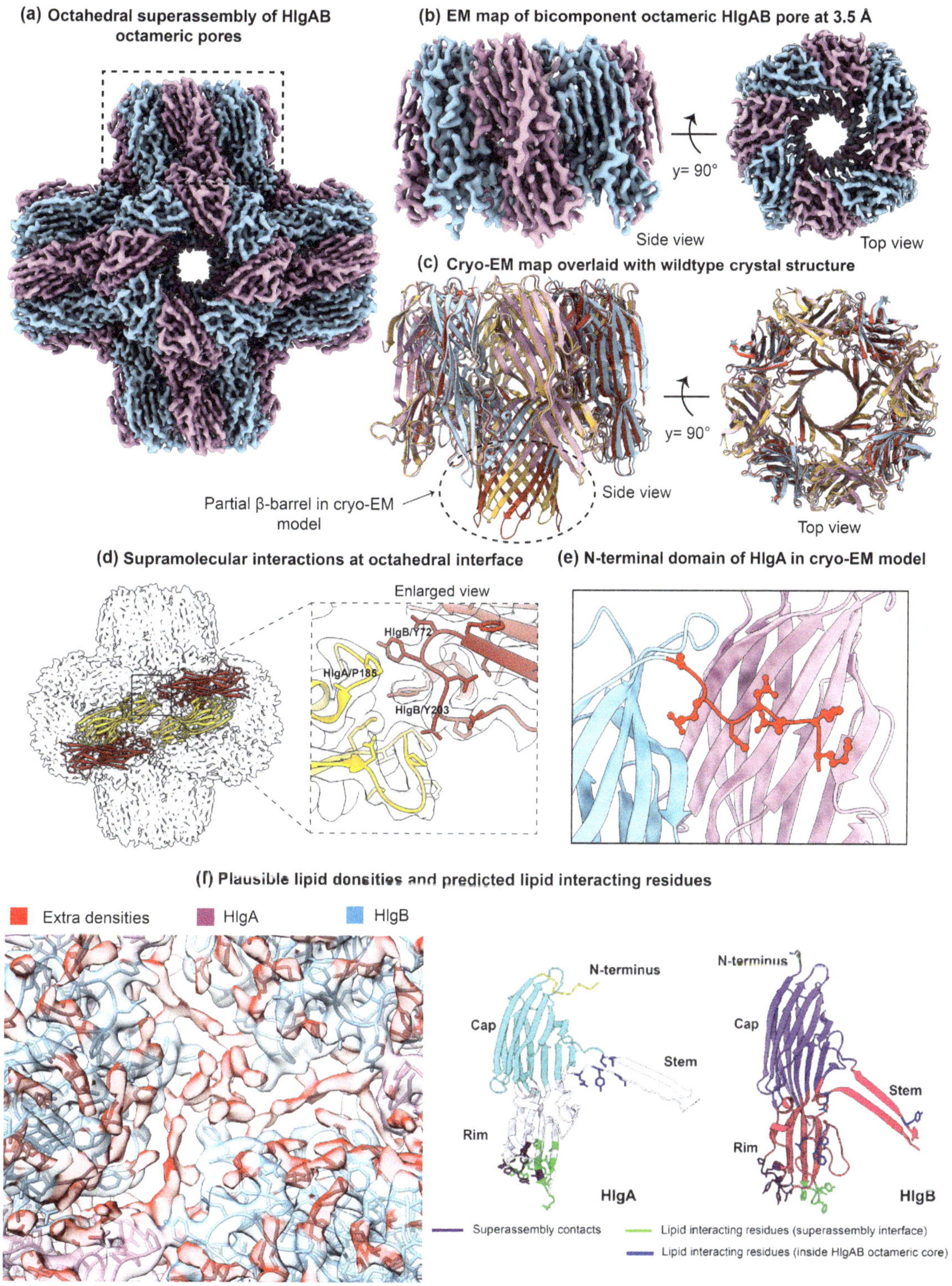

FIGURE 15.4 (a) Cryo-EM map of an octahedral assembly of six octameric HlgAB pore complexes. Alternate arrangement of the two protein components HlgA and HlgB is shown with pink and cyan. (b) Different orientations of an individual octameric HlgAB pore complex is shown in enlarged view. (c) The atomic model from this study (PDBID: 8GZ7, pink and cyan) is compared with the crystal structure with complete pore (PDBID: 3B07, goldenrod and brick red). (d) Supramolecular interactions stabilize the octahedral assembly of octameric pore complexes. Enlarged view of the rim domains of individual HlgA component (gold) and neighboring octamer's HlgB (red) is presented with a few representative residues participating in the hydrophobic interactions. (e) N-terminal latch of HlgA resolved in the cryo-EM map is highlighted in red. (f) Densities possibly corresponding to lipid/cholesterol for HlgA and HlgB. Extra densities are colored red. (Adapted from Mishra et al. (2023a).)

HlgA was resolved and found to be spanning symmetrically within the octameric pore and proximal to the neighboring HlgB subunit (Figure 15.4e). Furthermore, the residues not resolved in the β-barrel domain in our cryo-EM map were not resolved in the mutant-stabilized pre-pore state. Thus, our cryo-EM structure could represent the first wild-type pre-pore state observed for these β-PFTs without incorporation of any stabilizing mutations, thereby providing new structural evidence in favor of the existence of a pre-pore state during pore formation.

Investigating Lipid-Binding Residues for Γ-HL Toxins and Interactions Stabilizing the Octahedral Superassembly Interface

Clustering of proteins on the lipid membrane and the membrane rupturing that we reported through this study strongly agree with the previous findings of PFT aggregation in a membrane environment. The octahedral superassembly of octameric HlgAB pore complexes represents one such highly compact and stable cluster that provided us an opportunity to elaborate upon the interactions in rim domains that stabilized these larger assemblies. We observed that in our octahedral superassembly, all of the 48 toxin subunits participated in supramolecular interactions (Figure 15.4a). The rim domain of each HlgA subunit in one octameric pore complex engaged in hydrophobic interactions with the rim of the HlgB subunit belonging to another octameric pore complex inclined at 90°, and vice versa (Figure 15.4a,d). Therefore, the octahedral assembly was stabilized by 24 such HlgA–HlgB rim–rim hydrophobic interactions. The residues involved in the abovementioned hydrophobic interactions were Y68, P185, Y190, and V190 of HlgA and Y72, A202, and Y203 of HlgB (Figure 15.4d). After membrane rupturing, formation of these hydrophobic interactions in clusters of oligomers suggests that these residues might play a role in membrane interaction of the octameric pores. Indeed, the residue Y72 of HlgB from this site has previously been shown to be vital for membrane binding and its mutation abrogated hemolysis (Yokota & Kamio, 2000). Thus, the hydrophobic interactions that stabilize larger clusters of PFT oligomers were also key to experimentally pinpointing highly hydrophobic/membrane-binding regions of the toxin subunits.

Interestingly, in our cryo-EM map of octahedral super assemblies of octameric HlgAB, we could observe several fuzzy densities in the rim domains of both subunits that possibly corresponded to phosphatidylcholine/cholesterol. The resolution permitted us to identify neighboring residues to these fuzzy densities, but these densities were partial and did not allow us to completely model either phosphatidylcholine or cholesterol. We found the following residues proximal to these fuzzy densities that could play a role in binding to the membrane for both the subunits: HlgA: Y68, D101, A103, I135, Y137, Q139, D181, P185, Y237, Y239, T241, R242, R244, L245, and D248. For HlgB, the plausible lipid-interacting residues based on fuzzy extra densities were Y72, D73, Y117, Q147, N175, W177, Y180, W257, N258, F260, and Y261. Based on previous biochemical studies (Monma, 2004; Yamashita et al., 2014), HlgB's W177 and R198 (in the MPD binding site) were known to be critical for oligomerization and pore formation. In other studies (Monma, 2004; Tarenzi et al., 2022; Yokota & Kamio, 2000), the residues Y72, W257, F260, and Y261 were also found to be critical for membrane binding and lysis of erythrocytes, suggesting more lipid-binding sites in HlgB in addition to the MPD-binding site. Structural evidence for lipid binding for these other lipid-binding sites in HlgB is also provided through our study. Surprisingly, monomeric HlgA, which was previously not assumed to play any role in membrane interactions, was found to be possibly interacting with lipids in the oligomeric state. A parallel alanine mutagenesis screening study in another S component, LukS-PV, showed that a few residues were involved in membrane interaction (Laventie et al., 2014). Based on the residues we have shown proximal to the fuzzy densities, it is likely that the rim region of HlgA in HlgAB oligomer also participates in membrane binding (Figure 15.4f). Thus, studying the PFTs in the presence of a lipid environment is crucial to understanding the lipid–protein interactions of these toxins, which are so far unavailable from detergent-stabilized oligomeric pore complexes. This current study exemplifies a new method to study small pore-forming toxins in a lipid environment, and in the near future, many others PFTs will be characterized by similar approaches.

DISCUSSIONS

Bicomponent β-PFTs, though structurally similar, are functionally diverse, with unique cellular receptors for each toxin. Targeting staphylococcal virulence factors to mitigate infection would require more mechanistic insights into these PFTs. Our cryo-EM data in this study attempts to answer several pressing questions regarding γ-HL pore formation: (1) modeling of the N-terminal latch and its structural role in pore formation, (2) the structural basis of lipid binding, and (3) whether a native pre-pore state exists. We further aim to probe the mechanistic role of the amino latch and the C-terminal domains for both toxin subunits and trap other metastable pore intermediates. These attempts will further aid therapeutic intervention for *S. aureus* infections. Furthermore, this is one of the first presentations of direct evidence of structural characterization of small pore-forming toxins in a real lipid and cellular environment (RBC) using single-particle cryo-EM. Our unique approaches to characterizing the biological macromolecules in the presence of a real lipid environment will pave a new avenue to determine the structure and function of different virulence factors in near near-physiological conditions. In the near future, we anticipate a surge in structural studies of membrane-interacting proteins in near-physiological membrane environments using cryo-EM and cryo-ET. However, more optimizations are required to characterize all these virulence factors in the presence of a real cellular and intracellular environment, like in the presence of RBC, neutrophil, mitochondria, and epithelial cells. We anticipate this will be achievable in the near future due to recent advancements in cryo-EM.

METHODS

Cloning, Overexpression, and Two-Step Protein Purification of Bicomponent HlgAB

Both the HlgA and HlgB codon-optimized genes were cloned in a C-terminal His-tagged pET26b(+) vector for affinity chromatography–based purification. The recombinant genes of HlgA and HlgB were transformed into *E. coli* BL21 (DE3) strain cells (Novagen) for overexpressing both the desired proteins. The transformed cells for both genes were grown separately in Luria broth (HiMedia) to an optical density at 600 nm (OD600) of 0.6–0.8 at 37°C. Overexpression of the recombinant proteins was performed by adding 0.5 mM isopropyl-b-D-thiogalactoside (IPTG, Merck) into the cultured cells at 25°C for 16 hours. After centrifugation at 4000 rpm for 10 minutes, the pellets were then freshly dissolved in lysis buffer (buffer composition 20 mM Tris-HCl buffer pH 7.8, 150 mM NaCl) (Tris base, HiMedia). Probe sonication was performed for 30 minutes to lyse the cells and was followed by centrifugation for 40 minutes at 10,000 rpm.

The supernatant solution was incubated with pre-equilibrated Ni–nitrilotriacetic acid (NTA) agarose beads (Qiagen) to bind overnight. After passing of the flow through, the beads were washed with a wash buffer (buffer composition 20mM Tris-HCl buffer pH 7.8 and 150 mM NaCl) having a gradient imidazole (Sigma Aldrich) concentration ranging from 20 mM to 100 mM. The protein was eluted at a higher imidazole concentration of 300 mM in the wash buffer.

The eluted proteins from the Ni–NTA column were further buffer-exchanged before gel filtration chromatography and concentrated to 5 mg/mL using an Amicon® Ultra centrifugal filter with a cutoff of 10 kDa (Amicon Ultra-4 centrifugal filter unit 10K, Merck). The concentrated protein was loaded into the 10/300 GL column (Cytiva) packed with Superdex 200 beads equilibrated with buffer compositions of 20 mM Tris-HCl buffer pH 7.8, 150 mM NaCl. Peak fractions for both the components HlgA and HlgB were further loaded into 12% SDS-PAGE (Coomassie Brilliant Blue R 250, Merck) to confirm the purity. For SDS-PAGE, Precision Plus Protein™ Dual Color Standards (Bio-Rad) were used as a marker.

Rabbit Erythrocyte Lysis Assay

About 1 mL of whole blood was isolated from *Oryctolagus cuniculus* and was centrifuged several times at 3000 rpm for 5 minutes using 1X phosphate buffer saline (PBS): 137 mM sodium chloride (Qualigens), 2.7 mM potassium chloride (Sigma Aldrich), 10 mM disodium hydrogen orthophosphate (SD Fine Chemicals), and 1.8 mM potassium dihydrogen orthophosphate (Qualigens) mostly to isolate contaminant-free intact erythrocytes for hemolysis assay with bicomponent HlgAB. 1x PBS buffer solution was used to prepare 1% RBCs solution. Tween-20 detergent was used in a 2% solution condition as a positive control to calculate the changes in absorbance, and 1x PBS buffer was used as a negative control. RBCs were treated with equimolar concentrations of HlgA and HlgB (1 μM) and incubated in constant shaking conditions (160 rpm) at 37°C for a duration of 45 minutes. The optical density of the cells in different conditions was measured at 620 nm using a Varioskan Flash microplate reader (Thermo Scientific). Rabbit blood was obtained upon approval of the Institutional Animal Ethics Committee (IAEC), Indian Institute of Science, Bangalore (CAF/Ethics/881/2022).

Preparation of Large, Unilamellar Vesicles (LUVs) of eggPC and Cholesterol

Large unilamellar vesicles (LUVs) were made of eggPC and cholesterol in 1:1 molar ratio. L-α-phosphatidylcholine (egg, chicken) (Avanti Polar Lipids) lipid and cholesterol were taken for the preparation of the LUVs by the lipid extrusion method. Equimolar concentration (1:1) of 3 mg eggPC (molar mass: 770 Da) and 1.5 mg cholesterol (molar mass: 386.65 Da) were first dissolved into 1 mL chloroform and further kept in a desiccator to get a thin lipid film by evaporating the chloroform. The whitish layer of the eggPC and cholesterol mixture was further dissolved gently into 1 mL total volume of 20 mM Tris-HCl, 150 mM NaCl buffer (pH 7.8) and further incubated for 30 minutes at an above transition temperature, 45°C. The mixture lipid solution (4 mM eggPC–cholesterol) in a colloidal state was further extruded through 200 nm pore-size polycarbonate membranes (PC Membranes 0.2 μm, Avanti Polar Lipids) eight times to form liposomes. The final extruded solution containing large unilamellar vesicles (LUVs) was stored at 4°C until further use. The Avanti Polar mini-extruder apparatus contained 10 mm filter supports and syringe (Avanti Polar Lipids).

Generating Oligomeric Pore Complexes of Bicomponent HlgAB in 20% of 2-Methyl-2,4-pentanediol and EggPC-Cholesterol LUVs

Twenty percent (v/v) 2-methyl-2,4-pentanediol (MPD, Sigma Aldrich) was mixed in equimolar concentrations of HlgA and HlgB monomers (40 μM). The mixed population was incubated in a water bath at 37°C. The resultant pore complexes were loaded onto 10% SDS-PAGE without heat denaturation for the observation of SDS-stable oligomeric band and followed by observation under NS-TEM.

Furthermore, HlgA and HlgB (60 μM) were added in equimolar concentration with freshly prepared eggPC-cholesterol LUVs (4 mM) in a 1:25 protein:lipid ratio, and then incubated in water bath at 37°C. Both LUVs treated and untreated were observed under NS-TEM for the formation of oligomeric pore complexes.

Sample Preparation and Data Acquisition for Room-Temperature Negative Staining TEM (NS-TEM)

A 3 μL aqueous sample of the protein after appropriate dilution was applied to a thin film of carbon-coated copper grids (carbon film 300 mesh, copper, Electron Microscopy Sciences) which was glow discharged for 30 seconds, 20 mA before sample preparation. The sample was further stained with 2% uranyl acetate (Polysciences, Inc) for 30 seconds, followed by gentle blotting to remove excess stain from over the grid. These carbon-coated copper grids were then visualized under a room-temperature 120 kV Talos L120C transmission electron microscope equipped with bottom-mounted Ceta camera (4000 × 4000 pixels) at magnification ranging between 57,000× and 92,000× (Mishra et al., 2023a; Pramanick et al., 2021).

Image Processing for NS-TEM Micrographs

The collected micrographs were processed in EMAN 2.1. Different particle projections were manually picked using *e2projectmanager.py* followed by the extraction of particles from raw micrographs using *e2boxer.py*. Around 6000 to 7000 particles for each dataset were further classified using reference-free 2D classifications in EMAN2.1 (Tang et al., 2007), RELION2 (Scheres, 2012), and SIMPLE (Elmlund & Elmlund, 2012) programs.

Sample Preparation and Data Acquisition for Single-Particle Cryo-EM

Three microliters of LUV-treated protein complex was applied to glow-discharged Quantifoil R1.2/1.3 Cu 300 mesh holey carbon grids (Electron Microscopy Sciences) and Quantifoil R 1.2/1.3, UT, 300 mesh carbon-coated grids (Electron Microscopy Sciences). After incubation of the sample for 10 seconds on the grid at 22°C, it was blotted for 8.5 seconds, and immediately the grid was plunge-frozen into liquid ethane at approximately −180°C by a Vitrobot (FEI Mark IV) (Sengupta et al., 2021).

Around 2287 cryo-EM movies were collected using a Talos Arctica (FEI) equipped with a K2 Summit direct electron detector (counting mode) at a pixel size of 0.92 Å. The corresponding dose rate was 2 e^{-}·$Å^{2}$·s with a defocus range of −0.8 μm to −3 μm. The images were fractionated into 20 frames at 54,000× magnification.

Data Processing for Single-Particle Cryo-EM

The beam-induced motion correction was performed by MotionCor2, and the defocus parameters were estimated by CTFFIND4 (Rohou & Grigorieff, 2015). The quality of the micrographs was initially screened using cisTEM. About 1,000,000 particles were automatically picked as 2D lattice arrays using a template picker and extracted with a box size of 400 pixels (Å/pixel = 0.92). Iteratively 2D classification was performed by RELION3.1 (Scheres, 2012), obtaining 2D averages of square lattice packing in different orientations. After removing junk particles in reference-free 2D, an EM map was also generated from those classes, with aligned particles using the *relion_reconstruct* program in RELION3.1.

To reconstruct the octahedral superassembly, manual picking followed by template-based automatic picking yielded 46,595 particles from 2287 micrographs. With a box size of 400 pixels, these particles were extracted and classified rigorously to obtain the 11,634 choicest well-centered particles. A subset of these particles was used to generate an *ab initio* model without the imposition of any symmetry, which revealed an octahedral arrangement of octameric pore complexes. Octahedral symmetry was imposed to refine the cryo-EM map to a resolution of <4 Å. Another round of auto-refinement after particle polishing and CTF refinement yielded the final map at 3.5 Å resolution. Local resolution was estimated using Blocres.

Model Building and Model Validation

The atomic model of HlgAB pore complex was built using PDBID: 3B07 as the starting model in Coot. The model was further refined using PHENIX real space refinement. Model validation parameters were obtained with the comprehensive data validation and EMRinger (Barad et al., 2015) tools of PHENIX (Adams et al., 2010). Data were visualized using ChimeraX (Goddard et al., 2018) and PyMOL.

ACKNOWLEDGMENTS

We acknowledge the Department of Biotechnology, Department of Science and Technology (DST) and Science, and Ministry of Education (MoE), India, for funding and the cryo-EM facility at IISc Bangalore. We acknowledge the DBT-BUILDER Program (BT/INF/22/SP22844/2017) and DST-FIST (SR/FST/LSII-039/2015) for the national cryo-EM facility at IISc, Bangalore. We acknowledge financial support from the MoE (STARS-1/171). We also acknowledge financial support from the Science and Engineering Research Board. We acknowledge DBT-IISc partnership program phase II for the NS-TEM facility at the Biological Sciences Division, IISc, and cryo-EM data processing cluster. We acknowledge the high-performance computing cluster "Beagle" at the Biological Sciences Division, IISc. We also thank Clayton Fernando for preparing cryo-EM grids and aiding in cryo-EM data collection.

REFERENCES

Adams PD, Afonine P V, Bunkóczi G, Chen VB, Davis IW, Echols N, Headd JJ, Hung L-W, Kapral GJ, Grosse-Kunstleve RW, et al. (2010) PHENIX: a comprehensive Python-based system for macromolecular structure solution. *Acta Crystallogr D Biol Crystallogr* 66: 213–221

Alessandrini A, Viero G, Dalla Serra M, Prévost G & Facci P (2013) γ-Hemolysin oligomeric structure and effect of its formation on supported lipid bilayers: An AFM Investigation. *Biochim Biophys Acta - Biomembr* 1828: 405–411

Alonzo F & Torres VJ (2014) The bicomponent pore-forming leucocidins of *Staphylococcus aureus*. *Microbiol Mol Biol Rev* 78: 199–230

Barad BA, Echols N, Wang RY-R, Cheng Y, DiMaio F, Adams PD & Fraser JS (2015) EMRinger: Side chain-directed model and map validation for 3D cryo-electron microscopy. *Nat Methods* 12: 943–946

Blake KJ, Baral P, Voisin T, Lubkin A, Pinho-Ribeiro FA, Adams KL, Roberson DP, Ma YC, Otto M, Woolf CJ, et al. (2018) *Staphylococcus aureus* produces pain through pore-forming toxins and neuronal TRPV1 that is silenced by QX-314. *Nat Commun* 9: 37

Bruggisser J, Iacovache I, Musson SC, Degiacomi MT, Posthaus H & Zuber B (2022) Cryo-EM structure of the octameric pore of *Clostridium perfringens* β-toxin. *EMBO Rep* 23: e54856

Elmlund D & Elmlund H (2012) SIMPLE: Software for ab initio reconstruction of heterogeneous single-particles. *J Struct Biol* 180: 420–427

Ghanem N, Kanagami N, Matsui T, Takeda K, Kaneko J, Shiraishi Y, Choe CA, Uchikubo-Kamo T, Shirouzu M, Hashimoto T, et al. (2022) Chimeric mutants of staphylococcal hemolysin, which act as both one-component and two-component hemolysin, created by grafting the stem domain. *FEBS J* 289: 3505–3520

Goddard TD, Huang CC, Meng EC, Pettersen EF, Couch GS, Morris JH & Ferrin TE (2018) UCSF ChimeraX: Meeting modern challenges in visualization and analysis. *Protein Sci* 27: 14–25

Howden BP, Giulieri SG, Wong Fok Lung T, Baines SL, Sharkey LK, Lee JYH, Hachani A, Monk IR & Stinear TP (2023) *Staphylococcus aureus* host interactions and adaptation. *Nat Rev Microbiol* 21: 380–395

Iacovache I, Bischofberger M & van der Goot FG (2010) Structure and assembly of pore-forming proteins. *Curr Opin Struct Biol* 20: 241–246

Ikuta KS, Swetschinski LR, Robles Aguilar G, Sharara F, Mestrovic T, Gray AP, Davis Weaver N, Wool EE, Han C, Gershberg Hayoon A, et al. (2022) Global mortality associated with 33 bacterial pathogens in 2019: A systematic analysis for the Global Burden of Disease Study 2019. *Lancet* 400: 2221–2248

Jayasinghe L, Miles G & Bayley H (2006) Role of the amino latch of staphylococcal α-hemolysin in pore formation. *J Biol Chem* 281: 2195–2204

Kaneko J & Kamio Y (2004) Bacterial two-component and hetero-heptameric pore-forming cytolytic toxins: Structures, pore-forming mechanism, and organization of the genes. *Biosci Biotechnol Biochem* 68: 981–1003

Kaneko J, Mascarenas AL, Huda MN, Tomita T & Kamio Y (1998) An N-terminal region of LukF of staphylococcal leukocidin/γ-hemolysin crucial for the biological activity of the toxin. *Biosci Biotechnol Biochem* 62: 1465–1467

Laventie B-J, Guérin F, Mourey L, Tawk MY, Jover E, Maveyraud L & Prévost G (2014) Residues essential for panton-valentine leukocidin S component binding to its cell receptor suggest both plasticity and adaptability in its interaction surface. *PLoS One* 9: e92094

Los FCO, Randis TM, Aroian R V. & Ratner AJ (2013) Role of pore-forming toxins in bacterial infectious diseases. *Microbiol Mol Biol Rev* 77: 173–207

Lubkin A, Lee WL, Alonzo F, Wang C, Aligo J, Keller M, Girgis NM, Reyes-Robles T, Chan R, O'Malley A, et al. (2019) *Staphylococcus aureus* leukocidins target endothelial DARC to cause lethality in mice. *Cell Host Microbe* 25: 463–470.e9

Mishra S, Kundu N, Pramanick I, Kumar A, Chattopadhyay K & Dutta S (2023a) Structural insights into thermostable direct hemolysin of *Vibrio parahaemolyticus* using single-particle cryo-EM. *Proteins Struct Funct Bioinforma* 91: 137–146

Mishra S, Roy A & Dutta S (2023b) Cryo-EM-based structural insights into supramolecular assemblies of γ-hemolysin from *S. aureus* reveal the pore formation mechanism. *Structure*, *31*, 651–667.

Monma N (2004) Essential residues, W177 and R198, of LukF for phosphatidylcholine-binding and pore-formation by staphylococcal -hemolysin on human erythrocyte membranes. *J Biochem* 136: 427–431

Mueller M, Grauschopf U, Maier T, Glockshuber R & Ban N (2009) The structure of a cytolytic α-helical toxin pore reveals its assembly mechanism. *Nature* 459: 726–730

Olson R, Nariya H, Yokota K, Kamio Y & Gouaux E (1999) Crystal structure of staphylococcal LukF delineates conformational changes accompanying formation of a transmembrane channel. *Nat Struct Biol* 6: 134–140

Peraro MD & van der Goot FG (2016) Pore-forming toxins: Ancient, but never really out of fashion. *Nat Rev Microbiol* 14: 77–92

Pramanick I, Sengupta N, Mishra S, Pandey S, Girish N, Das A & Dutta S (2021) Conformational flexibility and structural variability of SARS-CoV2 S protein. *Structure* 29: 834–845.e5

Roblin P, Guillet V, Joubert O, Keller D, Erard M, Maveyraud L, Prévost G & Mourey L (2008) A covalent S-F heterodimer of leucotoxin reveals molecular plasticity of β-barrel pore-forming toxins. *Proteins Struct Funct Bioinforma* 71: 485–496

Rohou A & Grigorieff N (2015) CTFFIND4: Fast and accurate defocus estimation from electron micrographs. *J Struct Biol* 192: 216–221

Scheres SHW (2012) RELION: Implementation of a Bayesian approach to cryo-EM structure determination. *J Struct Biol* 180: 519–530

Sengupta N, Mondal AK, Mishra S, Chattopadhyay K & Dutta S (2021) Single-particle cryo-EM reveals conformational variability of the oligomeric VCC β-barrel pore in a lipid bilayer. *J Cell Biol* 220, e202102035.

Song L, Hobaugh MR, Shustak C, Cheley S, Bayley H & Gouaux JE (1996) Structure of Staphylococcal α-Hemolysin, a Heptameric Transmembrane Pore. *Science* 274: 1859–1865

Spaan AN, Reyes-Robles T, Badiou C, Cochet S, Boguslawski KM, Yoong P, Day CJ, de Haas CJC, van Kessel KPM, Vandenesch F, et al. (2015) *Staphylococcus aureus* targets the duffy antigen receptor for chemokines (DARC) to lyse erythrocytes. *Cell Host Microbe* 18: 363–370

Spaan AN, van Strijp JAG & Torres VJ (2017) Leukocidins: Staphylococcal bi-component pore-forming toxins find their receptors. *Nat Rev Microbiol* 15: 435–447

Spaan AN, Vrieling M, Wallet P, Badiou C, Reyes-Robles T, Ohneck EA, Benito Y, de Haas CJC, Day CJ, Jennings MP, et al. (2014) The staphylococcal toxins γ-haemolysin AB and CB differentially target phagocytes by employing specific chemokine receptors. *Nat Commun* 5: 5438

Sugawara-Tomita N, Tomita T & Kamio Y (2002) Stochastic assembly of two-component staphylococcal gamma-hemolysin into heteroheptameric transmembrane pores with alternate subunit arrangements in ratios of 3:4 and 4:3. *J Bacteriol* 184: 4747–4756

Takeda K, Tanaka Y & Kaneko J (2020) The N-terminal amino-latch region of Hlg2 component of staphylococcal bi-component γ-haemolysin is dispensable for prestem release to form β-barrel pores. *J Biochem* 168: 349–354

Tanaka Y, Hirano N, Kaneko J, Kamio Y, Yao M & Tanaka I (2011) 2-Methyl-2,4-pentanediol induces spontaneous assembly of staphylococcal α-hemolysin into heptameric pore structure. *Protein Sci* 20: 448–456

Tang G, Peng L, Baldwin PR, Mann DS, Jiang W, Rees I & Ludtke SJ (2007) EMAN2: An extensible image processing suite for electron microscopy. *J Struct Biol* 157: 38–46

Tarenzi T, Lattanzi G & Potestio R (2022) Membrane binding of pore-forming γ-hemolysin components studied at different lipid compositions. *Biochim Biophys Acta - Biomembranes* 1864: 183970

Tromp AT & van Strijp JAG (2020) Studying Staphylococcal leukocidins: A challenging endeavor. *Front Microbiol* 11, 611. doi: 10.3389/fmicb.2020.00611.

Yamashita D, Sugawara T, Takeshita M, Kaneko J, Kamio Y, Tanaka I, Tanaka Y & Yao M (2014) Molecular basis of transmembrane beta-barrel formation of staphylococcal pore-forming toxins. *Nat Commun* 5: 4897

Yamashita K, Kawai Y, Tanaka Y, Hirano N, Kaneko J, Tomita N, Ohta M, Kamio Y, Yao M & Tanaka I (2011) Crystal structure of the octameric pore of staphylococcal γ-hemolysin reveals the β-barrel pore formation mechanism by two components. *Proc Natl Acad Sci* 108: 17314–17319

Yokota K & Kamio Y (2000) Tyrosine72 Residue at the Bottom of Rim Domain in LukF Crucial for the Sequential Binding of the Staphylococcal γ-Hemolysin to Human Erythrocytes. *Biosci Biotechnol Biochem* 64: 2744–2747

Section IV

Cryo-EM in the Study of Membrane Biology

Probing the Structures of Membrane Proteins

16 Historical Evolution and Perspective on Cryo-EM Analysis of Membrane Proteins

Youzhong Guo

INTRODUCTION

The fundamental unit of all living things is the cell. The membrane is an essential component of a cell. The cell membrane system comparts the cytosolic environment, external environment, and intracellular organelles. The cell membrane consists primarily of lipid bilayers, membrane proteins, and other lipid-based substances. Some archaeal organisms have monolayer plasma membranes, where the isoprene chains of one phospholipid connect with the isoprene chains of a phospholipid on the opposite side of the membrane (Jain et al., 2014). Depending on how membrane proteins interact with cell membranes, they can be organized into three classes: integral membrane protein, peripheral membrane protein, and lipid-anchored membrane protein. Integral membrane proteins, such as transporters, channels, and receptors, play critical roles in facilitating the transfer of substances and biological signals across the cell membrane.

High-resolution three-dimensional atomic structural information about membrane proteins is extraordinarily beneficial in improving our understanding of their molecular mechanism of biochemical and physiological or pathological functions. Modern scientific tools, such as electron crystallography, X-ray crystallography, NMR, single-particle cryo-EM, cryo-EM tomography, neutron crystallography, and cryo-EM holography, have been used to facilitate the structural determination of macromolecules, especially proteins. Since the early 1980s, X-ray crystallography has been the predominant tool for determining membrane protein structure (Deisenhofer et al., 1984; Michel, 1982), but single-particle cryo-EM has become increasingly popular since 2013 (Liao et al., 2013). However, the determination of biological macromolecule structure did not begin with membrane proteins but with soluble proteins and DNA. The 3D structure determination of DNA double helices, myoglobin, and hemoglobin, using X-ray crystallography in the 1950s, marked the beginning of modern structural biology (Kendrew et al., 1958; Perutz, 1963; Watson & Crick, 1953).

Membrane proteins are major drug targets; more than 60% of drug targets are membrane proteins (Overington et al., 2006). However, compared with soluble proteins, there are far fewer high-quality structures available for membrane proteins because they are notoriously difficult to handle (Skrzypek et al., 2018). By their position within the lipid milieu of the cell membrane, membrane proteins are distinct from soluble proteins. Therefore, the lipid environment of membrane proteins must be considered when analyzing their structure and function. To understand why we need to consider the lipid environment, we must address the following questions (Bogdanov et al., 2008; Corin & Bowie, 2022). What roles do the lipid bilayer and its constituent elements play in membrane protein folding, assembly, and functionality? Do lipids operate as active protein folding, stability, and structure regulators, or are they just an inert solvent that helps membrane protein assembly?

There are two primary categories of transmembrane proteins: α-helical and β-barrel. Alpha-helical membrane proteins constitute the most common category of transmembrane proteins. Of all proteins in the human body, 27% are α-helical membrane proteins (Almén et al., 2009). To date, β-barrel proteins have been detected solely in the cell walls of gram-positive bacteria, chloroplast and mitochondria outer membranes, and gram-negative bacteria cell membranes; they may also be secreted as pore-forming toxins (Walther et al., 2009). It has been established that the lipid environment of α-helical membrane proteins influences their folding, stability, and function (Brady et al., 2022; Corin & Bowie, 2022; White & Wimley, 1999). Hydrophobic amino acids comprise most transmembrane (TM) portions of polypeptide chains; this adaptation allows them to partition into the nonpolar core of the lipid bilayer (White & Wimley, 1999). Although the hydrophobic thickness between phosphate groups in phospholipid bilayers of the cell membranes is maintained within 30 to 45 Å in various organisms and subcellular compartments (Mitra et al., 2004), the TM helices of the known structures vary widely in length from 14 to 36 residues (Bowie, 1997). Longer helices can be adapted by bending, kinking, or tilting their axis to the bilayer normal; shorter helices can be stabilized by tertiary contacts with nearby helices or by causing the bilayer to thin locally. The packing of the TM helices primarily forms the tertiary contacts, but folded or unstructured interhelical loops and prosthetic groups also play a role in the packing (Hong, 2015; Shelar et al., 2016). Depending on the species and subcellular location inside the cell membrane, the lipid content of cell membranes varies greatly.

Based on our understanding of the components and configurations of cell membranes and membrane proteins, several models of cell membranes have been proposed. In

DOI: 10.1201/9781003326106-20

the early 1970s, Singer and Nicolson initially proposed the "fluid mosaic" paradigm (Singer & Nicolson, 1972). Solid bodies symbolize the integral membrane proteins that diffuse at random through the fluid phase of the lipid bilayer. Integral membrane proteins are capable of forming complexes. Lipids (domains) with unique lipid compositions do not undergo phase separation, and membrane-associated structures do not exist. Following this, the modified membrane model depicts membrane domains that include an asymmetric lipid composition, glycosylated proteins and lipids, and cytoskeletons interconnected with them (Nicolson, 2013).

Specific peripheral and integral membrane proteins exhibit preferential partitioning into membrane domains or domain interfaces based on factors such as the cell tissue, developmental stage, physiological and environmental variables, disease state, and environmental context of the body (Dawidowicz, 1987; Koch et al., 2012). Additionally, even within the same subcellular compartment, the lipid content of the inner and outer leaflets of the membrane differs considerably (van Meer et al., 2008). The asymmetry of the lipid bilayer of the cell membrane results from the balanced activity of many flippases, which induce the lipid asymmetry, and scramblases, which maintain a constant lipid distribution between the two leaflets (Fadeel & Xue, 2009). Surprisingly, the calculated lateral pressure of the bilayer core is several hundred atmospheres. The lateral pressure profile is the physical origin of "curvature stress," a critical elastic property of the bilayer (Gruner, 1985). For instance, the integration of phosphatidylethanolamine (PE) into the lamellar phase of a Phosphatidylcholine (PC) bilayer diminishes the repulsion of head groups due to the small size of the PE head group and the hydrogen bonding that occurs between the ethanolamine group in PE and the phosphate group in a neighboring lipid (Gruner et al., 1988; Tate & Gruner, 1987). However, to counterbalance the reduced head group repulsion, repulsion between hydrocarbon chains will increase under the zero integrated pressure restriction without impacting the attractive line tension. A spontaneous negative curvature will occur in each monolayer leaflet due to the redistribution of the overall lateral pressure profile (Cantor, 1999).

Notably, the cell membranes, referred to as the "lamellar phase," include an abundance of non-bilayer-forming PE and Cardiolipin (CL). This suggests that membrane protein composition and functionality are significantly influenced by lateral pressure profile and curvature stress (van den Brink-van der Laan et al., 2004). Proteins encapsulated in integral membranes endure constant mechanical strain, and changes in the pressure profile can significantly impact the protein's folding, stability, and conformational equilibrium. There is a notion positing that alterations in the lateral pressure profile govern the processes underlying general anesthesia. Moreover, these alterations impact the activity of many ion and mechanosensitive channels responsible for the perception of mechanical stimuli (Milutinovic et al., 2007; Perozo et al., 2002). *Escherichia coli* LacY could potentially represent the lipid-dependent folding of membrane proteins (Guan & Kaback, 2006). Typical assembly of LacY occurs within the *E. coli* membranes containing an abundance of PE (70%, with the remainder being about 20% phosphatidylglycerol (PG) and 5% CL) (Bogdanov et al., 2008). *E. coli* LacY is organized within the inner cytoplasmic membrane as 12 transmembrane domains (TMs) connected by alternating solvent-exposed cytoplasmic and periplasmic domains with both the N- and C-termini oriented inward. Misfolding of domain P7, which is essential for the uphill energy-dependent transport of substrate, was linked to the inability of mutant cells lacking PE to perform the uphill energy-dependent transport function of LacY (Guan & Kaback, 2006). Membrane proteins are integral to the cell membrane system; their structures and functions rely on the cell membrane lipid environment.

Cryo-EM was utilized to initiate structural analysis of cell membranes in the late 1950s (Robertson, 1959; Sjostrand et al., 1958). It was J. David Robertson who identified the dark electron-dense bands as the headgroups and proteins associated with two opposite lipid monolayers (Robertson, 1959). Later, structural analysis of membrane proteins commenced. Richard Henderson and Unwin solved the first 3D structure of a membrane protein, bacterial rhodopsin, at 7 Å in the middle 1970s (Henderson & Unwin, 1975). There are several technological hurdles for high-resolution structure determination of membrane proteins. The primary obstacle is in the preparation of high-quality samples. In the case of bacterial rhodopsin, obtaining 2D crystals of superior quality is challenging. Henderson and his colleagues consistently refined the process to produce 2D crystals of rhodopsin of superior quality. By 1990, they had achieved such crystals and determined the atomic resolution of rhodopsin to be 3.5 Å. (Henderson et al., 1990). Other scientists took a different direction to solve high-resolution structures of membrane proteins. Instead of growing 2D crystals of membrane proteins in a lipid environment, they used small molecule detergents to solubilize and purify membrane proteins and stabilize them in a solution for growing 3D crystals. In this way, the first atomic-level high-resolution 3D structure of a membrane protein, the photosynthesis reaction center from the photosynthetic bacterium *Rhodopseudomonas viridis*, was solved by X-ray crystallography by Deisenhofer and colleagues in the early 1980s (Deisenhofer et al., 1985).

In this direction of 3D crystallography, many high-resolution membrane protein structures have been determined through X-ray crystallography, which is a triumph, but it is beyond the scope of this review. However, a significant hurdle is getting high-quality 3D crystals. The turning point happened in 2013, when the David Julius and Yifan Cheng groups determined the first atomic resolution single-particle cryo-EM structure of a membrane protein, the TRPV1 channel (Liao et al., 2013). This resolution revolution of single-particle cryo-EM analysis opened the door for high-resolution structure determination of membrane proteins without 2D or 3D crystals. Following the success of single-particle cryo-EM analysis of the human TRPV1 channel, many new membrane proteins resistant to crystallization have been investigated with single-particle cryo-EM analysis, according to the mpstruc database curated by the Stephen White laboratory at UC Irvine (https://blanco.biomol.uci.edu/mpstruc).

Because of the uniqueness of membrane protein associated with lipids and the importance of the protein–lipid interaction for the structure and function of membrane proteins, many efforts have been made to develop new approaches for membrane protein sample preparation. Small-molecule nonionic detergents are an approach taken from 3D membrane protein crystallography; following this direction, novel small-molecule detergents are continuously being developed, notably the successful application of LMNG (lauryl maltose neopentyl glycol) and GDN (glycoldiosgenin) (Li, 2022). DDM (n-dodecyl-β-maltoside) is the most popular and successful detergent for crystallographic and single-particle cryo-EM analysis of membrane proteins (Vénien-Bryan & Fernandes, 2023). Although small-molecule detergents are still predominantly used for membrane protein sample preparation, the limitations of detergents cannot be neglected. Small-molecule detergents often cause overdelipidation of membrane proteins during extraction from the cell membrane and following purification steps, leading to instability and even denaturing of membrane proteins (Guo, 2020; Seddon et al., 2004; Zoonens et al., 2013).

Several novel methods have been developed to overcome the limitations. (1) Amphipols. Amphipol is a type of amphiphilic polymer; after membrane protein is purified with detergents, amphibole can be used as a substitute for detergent, and usually, amphipol can better stabilize membrane proteins (Tribet et al., 1996). With this method, membrane protein can be analyzed with single-particle cryo-EM. (2) Peptidisc. Peptidisc is based on a specifically designed peptide with the amphiphilic property. Similar to amphipol, detergent in the protein samples can be removed and replaced by peptides (Carlson et al., 2018). (3) Saposin. Saposin is a small peptide that can stabilize lipid molecules after membrane protein is purified using detergent; the protein sample could be exchanged with saposin (Frauenfeld et al., 2016). (4) Proteoliposome. Proteoliposomes are often used for membrane protein functional assay. Recently, proteoliposomes have been successfully used for membrane protein structure determination. The idea was pioneered by Sigworth and colleagues and developed further by Ning Yan and colleagues (Wang & Sigworth, 2009; Yao et al., 2020). (5) Nanodisc technology. The Sligar group developed nanodisc technology. In this technology, engineered apolipoprotein and artificial lipids are used to reconstitute membrane proteins in nanodiscs (Denisov & Sligar, 2016). Although the methods mentioned above are better than detergent micelles, they are still essentially detergent-based methods and suffer from the initial interaction between membrane protein and detergents.

Several detergent-free methods have been developed to avoid detergent and keep membrane proteins associated with their native cell membrane environment. (1) SMALP. SMALP uses styrene-maleic acid copolymers to extract membrane protein and stabilize it; it has been successfully used in a few cases for membrane protein structure determination (Knowles et al., 2009; Vénien-Bryan & Fernandes, 2023). However, the incompatibility of styrene-maleic acid copolymers with lower pH conditions, divalent cations, and especially flexible transmembrane domains of membrane proteins have limited its broad application (Catalano et al., 2021; Guo, 2021). (2) Native cell membrane nanoparticle system (NCMN system). The NCMN system is composed of three sub-systems: [I] NCMN polymer library, [II] NCMN protocols for manufacturing NCMN particles of target membrane proteins, [III] NCMN analysis protocols for downstream biochemical and biophysical analysis of NCMN particles of membrane proteins (Guo, 2021). The NCMN system was set up for high-resolution structure determination and is suitable for functional assay of membrane proteins. The NCMN system has been successfully used for high-resolution structure determination of a lipid bilayer patch within a membrane protein transporter. Most recently, its power has been demonstrated by the successful determination of the high-resolution single-particle cryo-EM structure of the first full-length human integrin αIIbβ3, a challenging membrane protein drug target for anticoagulation therapy (Adair et al., 2023). (3) Cell-derived membrane vesicles. Cell-derived membrane vesicles are a relatively new strategy for membrane protein structure determination. It has been successfully used for high-resolution structure determination of the human Sol1 channel (Tao et al., 2023). (4) Cryo-focused ion beam (cryo-FIB) is a developing technology for making a thin layer of cell samples for tomography; membrane protein samples prepared in this way can be investigated using cryo-EM tomography (Berger et al., 2023).

Apart from the sample preparation using detergent or detergent-free methods, another concern is the host cells of membrane proteins. Traditionally, considering the time, cost, and convenience, we often overexpress membrane proteins heterogenously using *E. coli*, yeast, insect, Cho cells and HEK 293 cells, or even through a cell-free gene expression system (Altman et al., 1999; Yue et al., 2023; Kim et al., 2011; Nielsen, 2014; Rosano & Ceccarelli, 2014; Tan et al., 2021). With the protein–lipid interaction in mind, we need to consider expressing membrane proteins from their host cells or the original cells because the lipid environments of cells are different from different species, even different tissues within the same living organism. A single instance could illustrate the significance of the lipid environment of the host cell membrane. Human P-glycoprotein, the ATP-binding cassette transporter responsible for multidrug resistance, is located in the cytoplasmic membranes of mammals.

Significant conformational changes occur during the catalytic transport cycle of this exceptionally flexible protein or biogenesis (Moss et al., 1998; Zhang, 2001). The native host of the protein contains 12 TMs, where the cytoplasm is in contact with both the N- and C-termini. When expressed in *E. coli*, the N-terminal half of the protein adopts the architecture of the native host. However, TMs 8–12 undergo an inversion of their orientation, and TM 7 ceases to traverse the membrane. Consequently, within bacterial membranes, the nucleotide-binding domain and the complete C-terminal half are improperly oriented (Linton and Higgins, 2002). Even though topogenic signals encoded in the protein sequence of membrane proteins are the fundamental determinants of final protein organization in the membrane, the characteristics of the lipid bilayer do influence protein topology. Again, the native lipid environments must be considered when investigating the structures and function of membrane proteins.

This chapter aims to give an overall review of the evolution of cryo-EM technology and methods for membrane protein sample preparation for membrane protein structure determination in the natural cell membrane lipid environment. To achieve this goal, I selected several representative cases of application of cryo-EM technology for high-resolution membrane protein structure determination.

MODEL STUDIES OF SOME MEMBRANE PROTEINS (CASE STUDIES)

Bacteriorhodopsin

Bacteriorhodopsin, which is located on the purple cell membrane of *Halobacterium salinarum*, possesses a retinal chromophore and functions as a light-driven proton pump (Mathies et al., 1991). When light photons are received, all-*trans* retinal molecules are isomerized to 13 *cis* isomers (Braiman & Mathies, 1982). These conformational changes subsequently mediate a series of conformational changes. It is essential to get high-resolution atomic structures of bacteriorhodopsin to understand this fantastic photochemistry's active mechanism. The initial successful effort made by Richard Henderson and Nigel Unwin was to crystallize bacteriorhodopsin in 2D crystals within a lipid environment with native cell membrane purple lipids. With the 2D crystals, they used electron crystallography to determine the first-ever membrane protein structure in the world; although the resolution only reached 7 Å, all seven transmembrane helices can be solved unambiguously (Henderson & Unwin, 1975). In 1990, the improved three-dimensional map revealed the chemical details (Henderson et al., 1990). In 1996, a further improved density map allowed Henderson and colleagues to build a high-resolution structure of the bacteriorhodopsin at 3.5 Å with associated and well-ordered cell membrane lipid molecules (Grigorieff et al., 1996) (Figure 16.1a).

2D electron crystallography is a special tool for structure determination of membrane protein 2D crystals. Determining the structure of bacteriorhodopsin in a lipid environment was a tremendous success and provides an excellent example for structural biology research on membrane proteins in a near-native environment. However, successful 2D crystallization of bacteriorhodopsin was a rare case. It was challenging to get high-quality 2D crystals of membrane proteins; as history demonstrated, obtaining 3D membrane protein crystals was easier than 2D crystals. The initial 2D crystal structures revealed that the bacteriorhodopsin might be a natural trimer on the purple membrane (Henderson & Unwin, 1975). Compared to 2D crystals, 3D crystallization often requires the involvement of detergents to solubilize membrane protein from the native cell membrane environment. However, using detergents might often lead to overdelipidation of membrane protein and thus result in nonnatural protein structures. For example, bacteriorhodopsin had been crystallized as a monomer rather than a trimer (Faham & Bowie, 2002). For a review of this case and other cases, please see Guo (2021).

Mammalian Aquaporin from Eyes

Membrane holes created by aquaporin (AQP) proteins are exceptionally selective for water and other tiny neutral solutes. The most prevalent protein found in the membranes of lens fiber cells is AQP0. In addition to forming water pores, AQP0 creates the thin lens junction (Sindhu Kumari et al., 2015). Two-dimensional crystal structures of AQP0 were initially obtained from the cores of sheep lenses, diffracted to 3 Å with electron crystallography (Gonen et al., 2005), and further optimized to 1.9 Å (Gonen et al., 2005) (Figure 16.1b). The 2D crystal structure obtained using the sample from the natural sheep lenses exists as an octamer of a tetramer dimmer, indicating that it forms a junction. However, the 3D crystal of AQP0 from bovine eyes exists as a tetramer (Figure 16.1c), which differs from the 2D structures, indicating it does not form a junction (Harries et al., 2004). Although the oligomeric states are different from 2D and 3D structures, a comparison between the structures of the single subunits from these two oligomers shows that the overall structures of the individual subunit are very similar, except for the C-terminal region (Figure 16.1d). There are cleaved and uncleaved AQP0 from both the N-terminus and C-terminus; the uncleaved AQP0 forms a tetramer, and the cleaved AQP0 tends to form a junction (Engel et al., 2008). Thomas Walz and colleagues reasoned that cleavage of the C-terminus amino acids led to the formation of junctions in the eyes (Engel et al., 2008). However, position prediction using the PPM server shows that both the N-terminus and C-terminus are located in the cytosol (Figure 16.1e), and careful examination of the octamer shown in Figure 16.1b tells us that both the N-terminus and C-terminus do not directly participate in the dimerization of the tetramer. The extracellular loop region makes for direct interaction between the two tetramers within the octamer. The octamer and tetramer crystals were obtained from very different crystallization conditions. Therefore, most likely, the crystallization conditions resulted in the different oligomeric states, and the lipid molecules might be an important factor, considering the 3D crystals were obtained in a detergent micelle environment. The structural controversy suggests that the native cell membrane environment might be crucial in understanding the physiological structure and functional status.

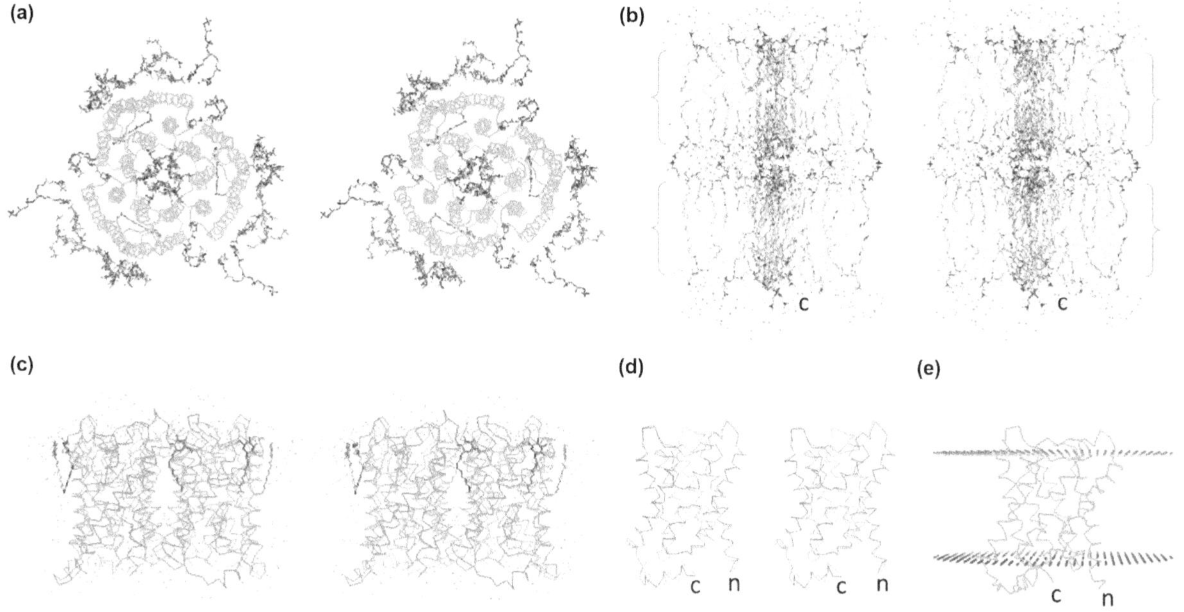

FIGURE 16.1 Membrane protein structures determined with electron crystallography. (a) 2D crystal structure of bacterial rhodopsin (PDB Code: 2BRD) determined at a resolution of 3.5 Å, in wall-eye stereo view, showing that the bacterial rhodopsin exists as a trimer with associated purple membrane lipid in 2D crystals. (b) 2D crystal structure of sheep eye aquaporin (PDB Code: 2B6O) determined at a resolution of 1.9 Å; the aquaporin exists as an octamer (dimer of tetramer from each cell). The bracketed region is the transmembrane domain. (c) 3D crystal structure of bovine eye aquaporin (PDB Code: 1YMG) determined at a resolution of 2.2 Å. (d) Comparison between 2D and 3D crystal structures shown as in panel (b) and panel (c). (e) Orientation of the aquaporin on the cell membrane predicted using PPM server (Lomize et al., 2012). The red disc represents the outer side of the plasma membrane; the blue disc represents the inner side. Note: In panels (a) and (b), the colored molecules are lipids. In panels (b), (d), and (e), n indicates the position of the N-terminus; c indicates the position of the C-terminus. In panel (b), it is obvious that the dimerization of the tetramers is mediated by the loop region of the aquaporin but not the N- or C-terminus.

Comparing the 2D crystal structure of AQP0 from sheep's eyes and the 3D crystal structures from bovines, we can see the 2D crystal structure contains many lipid molecules (Figure 16.1b); however, the 3D crystal structure only contains some detergent molecules (Figure 16.1c). In this case, unambiguously, detergents lead to over-delipidation of AQP0. It is worth pointing out that in the 2D crystals, the lipids were not native lipids from the eyes but artificial lipids reconstituted for 2D crystallization. The oligomeric and the C-terminal structural differences might result from the different molecule packing in the 2D and 3D crystals, or the different molecule packing might result from the structural differences resulting from the crystallization conditions.

Human Patched-1

During embryogenesis, the Hedgehog family of proteins transmit signals to numerous tissues and structures (Chiang et al., 1996; Dessaud et al., 2008; Ingham, 1993). The primary receptor for Hedgehog signals across the cell membrane is Human Patched-1 (Ptch1) (Ingham et al., 1991). To understand the molecular mechanism behind Hedgehog signaling, multiple groups have independently solved high-resolution structures of Ptch1 (Gong et al., 2018; Huang et al., 2022; Qi et al., 2018; Qian et al., 2019; Zhang et al., 2018). The structures determined, however, show controversies about the oligomeric states of Patched-1. Some reported that Patched-1 exists as a monomer, as shown in Figure 16.2a) (Gong et al., 2018); some reported it exists as a dimer (Huang et al., 2022), and the dimers are also different (Qi et al., 2018); some reported it exists as a tetramer as shown in (Figure 16.2b) (Qian et al., 2019). Upon careful examination of the transmembrane domain of the Patched-1 monomer and tetramer (Figure 16.2), it looks like the Patched-1 was over-delipidated during extraction and purification because there are many empty spaces among the transmembrane domains of the four protomers within the tetramer.

Lipids actively participate in the oligomerization process and the stability of the oligomers on the cell membrane. However, in the reported structures, there are no lipid molecules. Detergents can stabilize membrane proteins, but detergent molecules are different from lipids; therefore, the interaction between detergent and the transmembrane domain of the Patched-1 are also different. Gong et al. used 1% DDM to solubilize Ptch1 overexpressed in HEK293 F cells, and the exchange of the DDM to the sample produced a cryo-EM structure of a monomer (Gong et al., 2018). Huang et al. extracted *Xenopus laevis* Patch-1 using 1% DDM and reconstituted it into nanodiscs with 90% POPC (1-palmitoyl-2-oleoyl-glycero-3-phosphocholine) and 10%

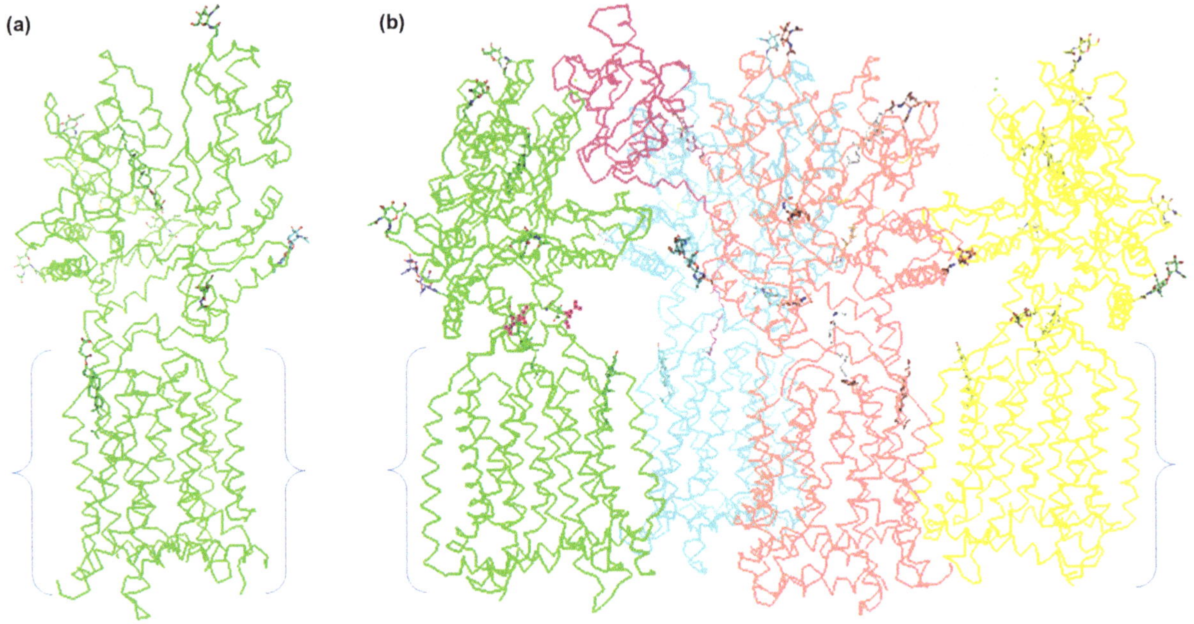

FIGURE 16.2 Controversial oligomeric states of human patched protein. (a) Human ptch1 monomer, PDB Code: 6DMB. (b) Human ptch1 tetramer, PDB Code: 6N7K.

cholesterol. Zhang et al. (2018), extracted human Patched-1 using 1% DDM from HEK293 cells, and the sample was further mixed with PMAL-C8 and removed DDM through dialysis. The sample produced a dimeric structure of human Patched-1; however, the dimer looks like it is solely mediated through the soluble domain, and the transmembrane domain of the two subunits is separated and cannot be located in a planar cell membrane bilayer, a pretty strange dimerization state (Qi et al., 2018). Qian et al. used 1% (w/v) GDN to extract and purify human Patched-1. The sample produced a tetrameric Patched-1, as shown in Figure 16.2b.

GDN is a much milder detergent than DDM; it might retain more lipids associated with the Patched-1 transmembrane domain or interact with it quite differently, producing a different oligomeric state. However, after careful examination of the tetramer, it looks like the two dimers have a stronger association with the monomer; the two dimers do not have strong interactions with each other. Therefore, the tetramer might not be in a natural oligomeric state on the native cell membrane. The diversity of the oligomeric states of the same membrane protein might reflect the perturbance by the detergent after over-dilapidation during protein extraction and purification.

E. coli Mechanosensitive Channel (MscS)

Mechanosensitive channels transform biochemical or electrical signals from mechanical forces exerted on the lipid bilayer of the cellular membrane. Among the channels, the mechanosensitive channel of small conductance (MscS) has been the subject of the most extensive research (Moller et al., 2023; Naismith and Booth, 2012). MscS exists in a minimum of two states: closed and open. MscS monitors the swift egress of solvents and solutes via a regulated gating mechanism. High-resolution MscS is necessary in order to comprehend the gating mechanism. The first crystal structure of *E.coli* MscS at 3.95 Å resolution revealed the overall shape of this protein (Bass et al., 2002) (Figure 16.3a). The structure revealed seven subunits grouped around a rotation axis of several folds parallel to the normal cell membrane. Three N-terminal transmembrane helices, designated TM1, TM2, and TM3, comprise each monomer's transmembrane domain. The crystal structure revealed that the pore's narrowest diameter is roughly 9 Å (Figure 16.3a). The inner surface of the pore was hydrophobic over its length. The original study stated that the channel was in an open state; however, it was later determined to be closed (Booth et al., 2007). The closed conformation was in line with a computational analysis that concluded the channel was dewetted, and the two rings of Leu acted as a vacuum lock (Anishkin & Sukharev, 2004). This conclusion was shared by a second independent computational analysis (Sotomayor & Schulten, 2004). However, the crystal structure does not contain any lipid molecules. Lipid–protein interactions might play a vital role in the gating of MscS. Perozo and colleagues determined a high-resolution cryo-EM structure of MscS using nanodisc technology. It identified lipid molecules associated with the transmembrane domain and revealed that pore lipids occupy the channel pore aligned with the hydrophobic inner surface (Reddy et al., 2019) (Figure 16.3b).

This is the first time we have a solid basis for understanding the role of lipids in regulating the gating of

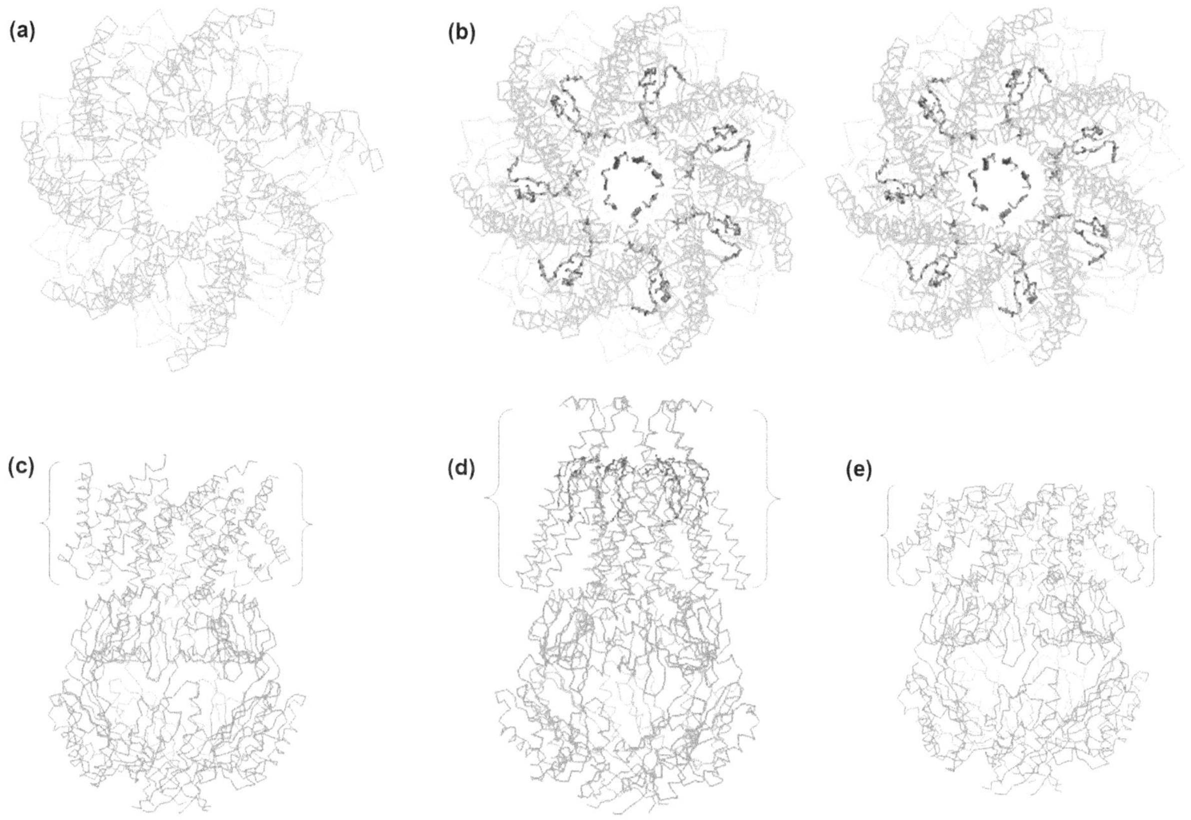

FIGURE 16.3 Structural divergence of *E.coli* mechanosensitive channel MscS. (a) Crystal structure of of MscS in detergent (PDB Code 2OAU). (b) Single-particle cryo-EM structure of MscS in nanodisc with lipid POPC:POPG (4:1) (PDB Code: 6PWN). (c) Single-particle cryo-EM structure of MscS in nanodiscs reconstituted with lipid PC-10 (PDB Code: 6VYL). (d) Single-particle cryo-EM structure of MscS in nanodiscs reconstituted with lipid PC 18 (PDB Code: 6VYK). (e) Single-particle cryo-EM structure of MscS in nanodiscs reconstituted with lipid PC-18-1 treated with β-cyclodextran (PDB Code: 6VYM).

mechanosensitive channels. However, the lipids are not directly extracted together with MscS from the *E. coli* inner membrane. Therefore, the protein–lipid interaction information from this nanodisc structure might not be adequate for understanding the working mechanism. In 2022, Thomas Walz and colleagues determined several cryo-EM structures with the nanodisc technology with different lipid components (Zhang et al., 2021) (Figure 16.3c, d, and e). With PC-10, short carbon-tailed lipid, reconstituted nanodiscs, the MscS structure shows a more helix-tilted transmembrane helix and, therefore, a compressed transmembrane domain adapted to a thin layer of lipid environment (Figure 16.3c); with PC-18, longer carbon-tailed lipid, reconstituted nanodiscs, the MscS structure shows a more or less helix-tilted transmembrane domain (Figure 16.3d); interestingly, when the nanodiscs reconstituted with PC-18-1 were treated with β-cyclodextran, the MscS structure showed a very distorted transmembrane domain (Figure 16.3e). Walz and colleagues interpreted the protein–lipid interaction to provide meaningful insight into understanding the MscS gating. However, all the lipids used for nanodisc reconstitution were artificial, different from the *E. coli* cell membrane environment. Therefore, the structural information between the interaction of protein and lipids should be more carefully interpreted.

The structural divergence suggests that the transmembrane domain is sensitive to lipid environments. Therefore, to understand the gating mechanism, it is better to get structural information on MscS with native *E. coli* cell membrane lipids. A native cell membrane nanoparticle system is a promising technology to achieve more native structures of MscS with naturally associated cell membrane lipids.

E. coli Muiti-Efflux Transporter, AcrB

In *E. coli*, AcrAB-TolC is the principal drug efflux system (Okusu et al., 1996). The homotrimeric inner membrane proton-motive force-driven H^+/drug antiporter component is called AcrB (Tam et al., 2021). During cooperative catalysis, the AcrB trimer sequentially cycles through the Loose (L), Tight (T), and Open (O) states of its three functionally interdependent protomers (Tam et al., 2021). More than 100 high-resolution structures of AcrB have been determined and deposited in the Protein Data Bank. Most structures were determined using X-ray crystallography and in a detergent micelle environment. Here, I tried to analyze three

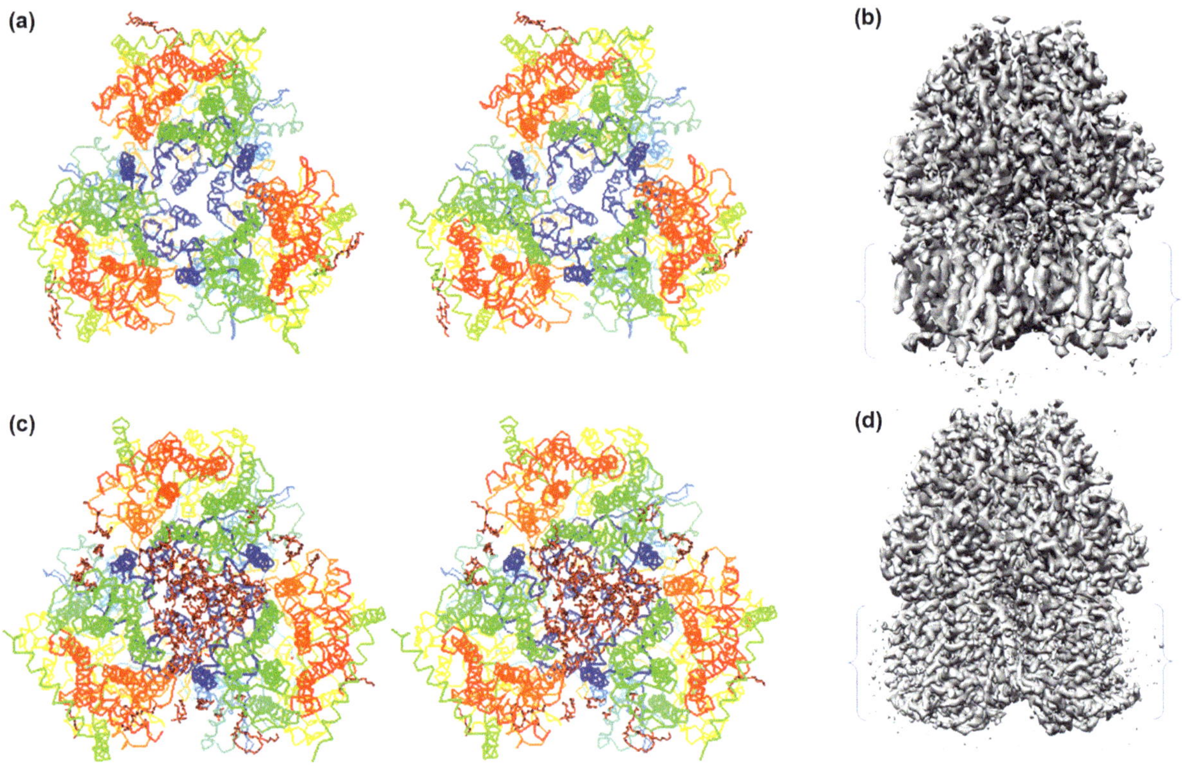

FIGURE 16.4 Divergence of structures of *E. coli* AcrB in different conditions. (a) Crystal structure of AcrB in detergent DDM (PDB Code: 1IWG) exists as a symmetric trimer at a resolution of 3.5 Å, in wall-eye stereo view. (b) Single-particle cryo-EM map of AcrB in proteoliposome exists as a symmetric trimer with a resolution of 3.9 Å (EMD-22050). The soluble domain has a higher resolution than the transmembrane domain. (c) Single-particle-cryo-EM structure of AcrB in NCMN particles exists as an asymmetric trimer with a pseudo-threefold symmetry at a resolution of 3.2 Å (PDB Code: 6BAJ, EMD-7074) in wall-eye stereo view. (d) The single-particle cryo-EM map of the structure shown in panel (c). The soluble domain and the transmembrane domain have a similar resolution.

representative structures determined in detergent, in proteoliposomes, or native cell membrane nanodiscs. Figure 16.4a shows a typical crystal structure of AcrB solved in a detergent micelle environment in a wall-eye stereo view (Murakami et al., 2002).

The three protomers exist in a pattern of a threefold symmetry. There are no solved lipid molecules in the transmembrane domain, but a detergent molecule was found associated with the transmembrane domain. However, in the transmembrane domain, the three protomers form a big cavity (Figure 16.4a). That cavity naturally is supposed to be filled with lipids; otherwise, the H^+ will leak, and the small molecules can pass through the inner membrane passively. The crystal structure was solved using an AcrB sample prepared with detergent DDM (n-dodecyl-β-D-maltopyranoside), the most widely used detergent for membrane protein structure biology. Obviously, in this sample, AcrB was over-delipidated by detergent. Figure 16.4b shows a single-particle cryo-EM structure determined using a sample of reconstituted proteoliposomes (Yao et al., 2020). In this case, AcrB was first overexpressed in *E. coli* and purified with DDM, and then the purified AcrB was reconstituted into a proteoliposome composed of POPC. This way, AcrB molecules were put into an artificial lipid environment to mimic the cell membrane lipid environment.

The proteoliposomes were imaged under cryo-EM, single-particle cryo-EM analysis was performed, and finally, a high-resolution cryo-EM map was reconstructed to a resolution of average 3.9 Å (Figure 16.4b). Symmetry analysis of the AcrB sample suggested the EM map has threefold symmetry rather than pseudo-threefold symmetry. Careful examination of the EM density tells us that the soluble domain has a much higher resolution than the transmembrane domain as bracketed (Figure 16.4b). AcrB has also been prepared as native cell membrane nanoparticles (NCMN). Single-particle cryo-EM analysis of AcrB in NCMN particles reveals a different structure (Qiu et al., 2018). The three protomers within the AcrB trimer possess a pseudo-threefold symmetry rather than a threefold symmetry. In the lipid cavity formed by the transmembrane domain, a lipid bilayer patch fully occupies the cavity and supports the transmembrane domain (Figure 16.4c). Figure 16.4d shows the EM density corresponding to the structure shown in panel 4C. Compared with the EM density map shown in Figure 16.4b, the soluble domain and transmembrane domain have very similar overall resolution of 3.2 Å.

The NCMN system is still in the early stages of development; however, it has obvious advantages. It can be conveniently used to extract membrane proteins from the native cell membrane while retaining the native associated lipid molecules with the transmembrane domain. The big disadvantage is that the NCMN particles are still a result of natural cell membrane disruption; therefore, although membrane protein could be retained with their local native lipid environment, the isolated NCMN particles are separated from the natural cell membrane system. Therefore, the environments surrounding the membrane proteins are different. This environment difference might be problematic, considering the membrane protein might interact with some cytoskeleton structures within the cell.

Human Slo1 Channel

Slo1 functions as a voltage- and Ca^{2+}-activating K^+ channel involved in hormone production, skeletal and smooth muscle contraction, audition, and neurotransmitter release (Contreras et al., 2013; Latorre et al., 2017). Slo1 functions as a tetramer of the α subunit. The structure determination of Slo1 with a detergent-solubilized protein sample yielded a fourfold symmetric tetramer (Tao & MacKinnon, 2019) (Figure 16.5a). However, single-particle structure determination of the Slo1 channel in vesicles from the total cell membrane preparation at the resolution of 3.8 Å exhibits twofold symmetry instead of fourfold symmetry (Tao et al., 2023). The symmetry differences were later found to be contributed by different Ca^{2+} concentrations but not the cell membrane environment. Slo1 channel in the plasma membrane–derived vesicles was determined to have a resolution of 2.7 Å (Figure 16.5b). The high quality of the cryo-EM map resolved the transmembrane domain at a higher resolution than previously published structures of Slo1 in detergent (Tao et al., 2023) (Figure 16.5a).

The plasma membrane environment resulted in a structure of Slo1 with a significantly greater degree of order

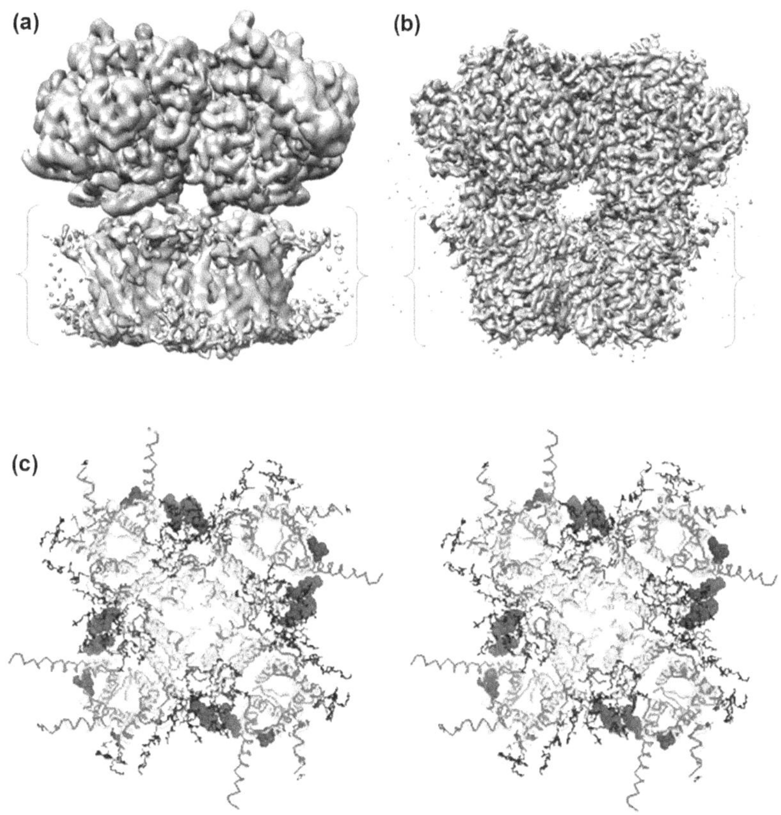

FIGURE 16.5 Diversity of the single-particle cryo-EM structures of human hsSlo1 channel. (a) Single-particle cryo-EM map of human Slo1 channel in detergent digitonin (PDB code: 6C3G and EMD-40045). The soluble domain has a higher resolution than the transmembrane domain. (b) Single-particle cryo-EM map of human Slo1 channel in native cell membrane vesicles. The soluble domain has a similar resolution to the transmembrane domain. (c) Wall-eye stereo view for comparison between the transmembrane domain of human Slo1 channel determined in detergent (PDB Code: 6V3G, at a resolution of 4.0 Å) and cell membrane lipid environment (PDB Code: 8GHF, at a resolution of 2.7 Å). In the transmembrane domain, detergent has a less resolved protein structure and much lower resolution than is determined in the cell membrane lipid environment. *Note:* The shallow grey line shows the detergent-based transmembrane domain. The dark grey line shows the cell-derived vesicle-based transmembrane domain. The black sticks show native cell membrane lipids, and the cholesterol molecules are depicted as black ball blobs.

throughout the transmembrane region, allowing for distinct characterization of previously poorly resolved areas in cryo-EM structures in detergent. It is worth noting that helices on the perimeter that have minimal interface with the rest of Slo1 and are hardly resolved in detergent structures are highly organized. The greater degree of order could be attributed to the cell plasma membrane's chemical composition and physical properties. Compared to detergent micelles, a global repositioning of transmembrane helices in the plasma membrane was discovered. The plasma membrane environment resulted in a structure of Slo1 with a significantly greater degree of order throughout the transmembrane region, allowing for distinct characterization of previously poorly resolved areas in cryo-EM structures in detergent. Figure 16.5c displays a transmembrane domain superposition: While a minor rotation is discernible, there is also an unambiguous extension of the helices outward from the channel's central axis. This excludes the selectivity filter and the pore helices, but it does entail the extension of extra helices outward from the center.

The interaction of the channel with its plasma membrane environment resulted in channel expansion and increased order. Furthermore, lipid molecules not present in the detergent structures may be detected in the transmembrane domain's surroundings and within it. This method removed membrane proteins without removing the protein from its lipid environment, resulting in a more native habitat. Cell-derived membrane vesicles have obvious advantages; they maintain membrane protein in its original lipid environment. The use of cell-derived membrane vesicles to determine structure may have limitations: (1) A tag to enhance the protein is required for native proteins with low abundance. (2) For membrane proteins deeply embedded in the cell membrane environment, fiducials such as antibodies against the target protein are required. (3) Because cryo-EM requires thin ice, lipid vesicles must be tiny enough; nevertheless, small vesicles have substantial membrane curvature, resulting in a significant bending force applied to proteins trapped in them. (4) Cell-derived membrane vesicle is not plasma membrane, but how similar they are to each other could be a limitation. However, the cell-derived vesicle method is less disruptive than detergent-based approaches for determining membrane protein structure.

CONCLUSIONS AND PERSPECTIVES

In this chapter, we analyzed representative cases of structure determination of membrane proteins, including bacteriorhodopsin, mammalian aquaporins, human Patched-1, bacterial mechanosensitive channel, MscS and multi-efflux transporter, AcrB, and human Slo1 channel. The case analysis taught us the evolution of membrane protein structural determination technology and membrane protein sample preparation. Technology development follows such an evolutionary pathway: the first is electron crystallography, the second is X-ray crystallography, and the third and current stage is single-particle cryo-EM.

The evolution of membrane protein sample preparation follows such a historical track: The first stage is to take advantage of the richness of certain naturally abundant membrane proteins, keep the membrane proteins in the native lipid environment, and produce 2D crystals. The second stage is to use detergent to purify naturally abundant membrane proteins for producing 2D and 3D crystals. The third stage is to overexpress membrane proteins and use detergent to extract the membrane proteins for making 3D crystals The fourth stage is to develop a lipid cubic phase and bicelle methods to crystallize membrane protein in the lipid environment. The fifth stage is to develop various methods to remove detergent once purified with detergent and stabilize membrane proteins for single-particle cryo-EM analysis. The sixth stage is to develop detergent-free approaches to extract membrane proteins using membrane active polymers from the cell membrane and keeping membrane proteins in the native cell membrane lipid environment. Most recently, the cell-derived vesicle method has been developed to determine membrane protein structure directly from the cell membrane vesicles. Detergent-based methods combined with single-particle cryo-EM analysis are still predominant for membrane protein structure determination; however, the limitations and disadvantages of detergent-based methods are more pronounced, especially when compared with the detergent-free approach used for the same membrane protein.

As illustrated with the case of AcrB and the human Slo1 channel, detergent-free methods, such as SMALP technology, the NCMN system, and cell-derived vesicles combined with single-particle cryo-EM analysis will be more appropriate approaches for membrane protein structure determination in the near-native cell membrane lipid environment. Hopefully, cryo-ET will be widely used for membrane protein structure determination in the natural native cell membrane system. Getting a high-resolution membrane protein structure *in situ* is still hard, especially when the protein size is small. Although we have made significant achievements, we are still far from understanding the membrane proteins' structures and functions in their native lipid environments. In other words, membrane protein structural biology is still in its infancy.

The model membrane proteins (case studies) show that protein–lipid interactions should be considered when determining membrane protein structures. Ideally, we should extract membrane proteins from their natural host cells to preserve the natural protein–lipid interactions. However, more often than not, the natural abundance of membrane proteins is inadequate for sample preparation. Therefore, we often use a heterogenous expression system to overexpress the membrane proteins using bacterial, insect, yeast, or mammalian cells. This raised another concern: If we overexpressed the target membrane protein in a heterogenous expression system, the membrane protein overexpressed might not be in their ideal cell membrane lipid environments. In this case, expressing the membrane protein in the cells closest to the original cells is preferred. For example, a human membrane protein overexpressed in HEK293 cells will be preferable to that expressed in *E. coli*, insect, or yeast cells. With current gene editing tools, such as CRISPR (Ledford, 2023), we can edit specific target membrane protein genes and add tags to them for efficient protein purification.

Cryo-EM tomography combined with cryo-FIB (focused ion beam) sample preparation is a new trend for *in situ* membrane protein structural biology (Wang et al., 2023). In practice, when the membrane protein target location has been identified by correlative microscopy through fluorescence, a thin, electron-transparent lamella can be generated using the focused ion beam by eliminating material from both above and below it. Having a thickness of 100–200 nanometers, this lamella encompasses the region of interest. Sample thinning is a critical step in the tomography workflow because the electron beam in the transmission electron microscope (TEM) can only traverse sufficiently thin materials to transmit electrons (200–300 keV). In this way, membrane proteins will be investigated in their original cell and frozen in their native states. However, this technology is still developing and is not mature enough to be used for atomic-level structure determination of membrane proteins.

Cryo-EM holography is an emerging technology for high-resolution structure determination of a protein with a single protein molecule (Longchamp et al., 2017). A low-energy electron holographic microscope (LEEH) constructed at the Max Planck Institute for Solid State Research could image a protein to a resolution of 7 Å. However, further development of LEEH and corresponding membrane protein sample preparation will be necessary to apply this technology for membrane proteins successfully.

Currently, artificial intelligence (AI) is a hot topic. AI seems to penetrate every aspect of our daily lives, including the structural biology research field. Therefore, it is necessary to evaluate the potential effect of AI in membrane protein structure prediction. The most well-known AI program is AlphaFold (Jumper et al., 2021). As mentioned, membrane proteins are distinct from soluble proteins; they need lipid environments for their natural structures and functions. AlphaFold can accurately predict the 3D structure of soluble proteins, given that the templates from the Protein Data Bank are adequate. However, we need to clarify that the principle used by artificial intelligence to predict the 3D structure of a protein is not based on the first principle laws of physics but is essentially modeling based on common structural modules. Therefore, the protein-fold problem has not been solved by AlphaFold (Moore et al., 2022). The prediction accuracy depends on the accuracy of the 3D structure template, the similarity of the common structural modules, and the sequence. AI might play an important role in predicting the 3D structure of membrane proteins; however, it cannot be successful until an adequate database of accurate experimental membrane protein structures, including protein–lipid interactions, is available.

REFERENCES

Adair, B. D., Xiong, J. P., Yeager, M., & Arnaout, M. A. (2023). Cryo-EM structures of full-length integrin αIIbβ3 in native lipids. Nature communications, 14(1), 4168. https://doi.org/10.1038/s41467-023-39763-0

Almén, M. S., Nordström, K. J., Fredriksson, R., & Schiöth, H. B. (2009). Mapping the human membrane proteome: a majority of the human membrane proteins can be classified according to function and evolutionary origin. BMC biology, 7, 50. https://doi.org/10.1186/1741-7007-7-50

Altmann, F., Staudacher, E., Wilson, I. B., & März, L. (1999). Insect cells as hosts for the expression of recombinant glycoproteins. Glycoconjugate journal, 16(2), 109–123. https://doi.org/10.1023/a:1026488408951.

Anishkin, A., & Sukharev, S. (2004). Water dynamics and dewetting transitions in the small mechanosensitive channel MscS. Biophysical journal, 86(5), 2883–2895. https://doi.org/10.1016/S0006-3495(04)74340-4

Bass, R. B., Strop, P., Barclay, M., & Rees, D. C. (2002). Crystal structure of Escherichia coli MscS, a voltage-modulated and mechanosensitive channel. Science, 298(5598), 1582–1587. https://doi.org/10.1126/science.1077945

Berger, C., Premaraj, N., Ravelli, R. B. G., Knoops, K., Lopez-Iglesias, C. and Peters, P. J. (2023). Cryo-electron tomography on focused ion beam lamellae transforms structural cell biology. Nature methods, **20**, 499–511.

Bogdanov, M., Mileykovskaya, E., & Dowhan, W. (2008). Lipids in the assembly of membrane proteins and organization of protein supercomplexes: implications for lipid-linked disorders. Sub-cellular biochemistry, 49, 197–239. https://doi.org/10.1007/978-1-4020-8831-5_8

Booth, I. R., Edwards, M. D., Black, S., Schumann, U., & Miller, S. (2007). Mechanosensitive channels in bacteria: signs of closure?. Nature reviews. Microbiology, 5(6), 431–440. https://doi.org/10.1038/nrmicro1659

Bowie J. U. (1997). Helix packing in membrane proteins. Journal of molecular biology, 272(5), 780–789. https://doi.org/10.1006/jmbi.1997.1279

Brady, R., Harris, N. J., Pellowe, G. A., Gulaidi Breen, S., & Booth, P. J. (2022). How lipids affect the energetics of co-translational alpha helical membrane protein folding. Biochemical Society transactions, 50(1), 555–567. https://doi.org/10.1042/BST20201063

Braiman, M., & Mathies, R. (1982). Resonance Raman spectra of bacteriorhodopsin's primary photoproduct: evidence for a distorted 13-cis retinal chromophore. Proceedings of the National Academy of Sciences of the United States of America, 79(2), 403–407. https://doi.org/10.1073/pnas.79.2.403

Cantor R. S. (1999). Lipid composition and the lateral pressure profile in bilayers. Biophysical journal, 76(5), 2625–2639. https://doi.org/10.1016/S0006-3495(99)77415-1

Carlson, M. L., Young, J. W., Zhao, Z., Fabre, L., Jun, D., Li, J., Li, J., Dhupar, H. S., Wason, I., Mills, A. T., Beatty, J. T., Klassen, J. S., Rouiller, I., & Duong, F. (2018). The Peptidisc, a simple method for stabilizing membrane proteins in detergent-free solution. eLife, 7, e34085. https://doi.org/10.7554/eLife.34085

Catalano, C., Ben-Hail, D., Qiu, W., Blount, P., des Georges, A., & Guo, Y. (2021). Cryo-EM Structure of Mechanosensitive Channel YnaI Using SMA2000: Challenges and Opportunities. Membranes, 11(11), 849. https://doi.org/10.3390/membranes11110849

Chiang, C., Litingtung, Y., Lee, E., Young, K. E., Corden, J. L., Westphal, H., & Beachy, P. A. (1996). Cyclopia and defective axial patterning in mice lacking Sonic hedgehog gene function. Nature, 383(6599), 407–413. https://doi.org/10.1038/383407a0

Contreras, G. F., Castillo, K., Enrique, N., Carrasquel-Ursulaez, W., Castillo, J. P., Milesi, V., Neely, A., Alvarez, O., Ferreira, G., González, C., & Latorre, R. (2013). A BK (Slo1) channel journey from molecule to physiology. Channels (Austin, Tex.), 7(6), 442–458. https://doi.org/10.4161/chan.26242

Corin, K., & Bowie, J. U. (2022). How physical forces drive the process of helical membrane protein folding. EMBO reports, 23(3), e53025. https://doi.org/10.15252embr.202153025

Dawidowicz E. A. (1987). Dynamics of membrane lipid metabolism and turnover. Annual review of biochemistry, 56, 43–61. https://doi.org/10.1146/annurev.bi.56.070187.000355

Deisenhofer, J., Epp, O., Miki, K., Huber, R., & Michel, H. (1984). X-ray structure analysis of a membrane protein complex. Electron density map at 3 A resolution and a model of the chromophores of the photosynthetic reaction center from *Rhodopseudomonas viridis*. Journal of molecular biology, 180(2), 385–398. https://doi.org/10.1016/s0022-2836(84)80011-x

Deisenhofer, J., Epp, O., Miki, K., Huber, R., & Michel, H. (1985). Structure of the protein subunits in the photosynthetic reaction centre of *Rhodopseudomonas viridis* at 3Å resolution. Nature, 318(6047), 618–624. https://doi.org/10.1038/318618a0

Denisov, I. G., & Sligar, S. G. (2016). Nanodiscs for structural and functional studies of membrane proteins. Nature structural & molecular biology, 23(6), 481–486. https://doi.org/10.1038/nsmb.3195

Dessaud, E., McMahon, A. P., & Briscoe, J. (2008). Pattern formation in the vertebrate neural tube: a sonic hedgehog morphogen-regulated transcriptional network. Development (Cambridge, England), 135(15), 2489–2503. https://doi.org/10.1242/dev.009324

Engel, A., Fujiyoshi, Y., Gonen, T., & Walz, T. (2008). Junction-forming aquaporins. Current opinion in structural biology, 18(2), 229–235. https://doi.org/10.1016/j.sbi.2007.11.003

Fadeel, B., & Xue, D. (2009). The ins and outs of phospholipid asymmetry in the plasma membrane: roles in health and disease. Critical reviews in biochemistry and molecular biology, 44(5), 264–277. https://doi.org/10.1080/10409230903193307

Faham, S., & Bowie, J. U. (2002). Bicelle crystallization: a new method for crystallizing membrane proteins yields a monomeric bacteriorhodopsin structure. Journal of molecular biology, 316(1), 1–6. https://doi.org/10.1006/jmbi.2001.5295

Frauenfeld, J., Löving, R., Armache, J. P., Sonnen, A. F., Guettou, F., Moberg, P., Zhu, L., Jegerschöld, C., Flayhan, A., Briggs, J. A., Garoff, H., Löw, C., Cheng, Y., & Nordlund, P. (2016). A saposin-lipoprotein nanoparticle system for membrane proteins. Nature methods, 13(4), 345–351. https://doi.org/10.1038/nmeth.3801

Gonen, T., Cheng, Y., Sliz, P., Hiroaki, Y., Fujiyoshi, Y., Harrison, S. C., & Walz, T. (2005). Lipid-protein interactions in double-layered two-dimensional AQP0 crystals. Nature, 438(7068), 633–638. https://doi.org/10.1038/nature04321

Gong, X., Qian, H., Cao, P., Zhao, X., Zhou, Q., Lei, J., & Yan, N. (2018). Structural basis for the recognition of Sonic Hedgehog by human Patched1. Science (New York, N.Y.), 361(6402), eaas8935. https://doi.org/10.1126/science.aas8935

Gruner S. M. (1985). Intrinsic curvature hypothesis for biomembrane lipid composition: a role for nonbilayer lipids. Proceedings of the National Academy of Sciences of the United States of America, 82(11), 3665–3669. https://doi.org/10.1073/pnas.82.11.3665

Gruner, S. M., Tate, M. W., Kirk, G. L., So, P. T., Turner, D. C., Keane, D. T., Tilcock, C. P., & Cullis, P. R. (1988). X-ray diffraction study of the polymorphic behavior of N-methylated dioleoylphosphatidylethanolamine. Biochemistry, 27(8), 2853–2866. https://doi.org/10.1021/bi00408a029

Grigorieff, N., Ceska, T. A., Downing, K. H., Baldwin, J. M., & Henderson, R. (1996). Electron-crystallographic refinement of the structure of bacteriorhodopsin. Journal of molecular biology, 259(3), 393–421. https://doi.org/10.1006/jmbi.1996.0328

Guan, L., & Kaback, H. R. (2006). Lessons from lactose permease. Annual review of biophysics and biomolecular structure, 35, 67–91. https://doi.org/10.1146/annurev.biophys.35.040405.102005

Guo Y. (2020). Be Cautious with Crystal Structures of Membrane Proteins or Complexes Prepared in Detergents. Crystals, 10(2), 86. https://doi.org/10.3390/cryst10020086

Guo Y. (2021). Detergent-free systems for structural studies of membrane proteins. Biochemical Society transactions, 49(3), 1361–1374. https://doi.org/10.1042/BST20201080

Harries, W. E., Akhavan, D., Miercke, L. J., Khademi, S., & Stroud, R. M. (2004). The channel architecture of aquaporin 0 at a 2.2-A resolution. Proceedings of the National Academy of Sciences of the United States of America, 101(39), 14045–14050. https://doi.org/10.1073/pnas.0405274101

Henderson, R., Baldwin, J. M., Ceska, T. A., Zemlin, F., Beckmann, E., & Downing, K. H. (1990). Model for the structure of bacteriorhodopsin based on high-resolution electron cryo-microscopy. Journal of molecular biology, 213(4), 899–929. https://doi.org/10.1016/S0022-2836(05)80271-2

Henderson, R., & Unwin, P. N. (1975). Three-dimensional model of purple membrane obtained by electron microscopy. Nature, 257(5521), 28–32. https://doi.org/10.1038/257028a0

Hong H. (2015). Role of lipids in folding, misfolding and function of integral membrane proteins. Advances in experimental medicine and biology, 855, 1–31. https://doi.org/10.1007/978-3-319-17344-3_1

Huang, P., Wierbowski, B. M., Lian, T., Chan, C., García-Linares, S., Jiang, J., & Salic, A. (2022). Structural basis for catalyzed assembly of the Sonic hedgehog-Patched1 signaling complex. Developmental cell, 57(5), 670–685.e8. https://doi.org/10.1016/j.devcel.2022.02.008

Ingham P. W. (1993). Localized hedgehog activity controls spatial limits of wingless transcription in the Drosophila embryo. Nature, 366(6455), 560–562. https://doi.org/10.1038/366560a0

Ingham, P. W., Taylor, A. M., & Nakano, Y. (1991). Role of the Drosophila patched gene in positional signalling. Nature, 353(6340), 184–187. https://doi.org/10.1038/353184a0

Jain, S., Caforio, A., & Driessen, A. J. (2014). Biosynthesis of archaeal membrane ether lipids. Frontiers in microbiology, 5, 641. https://doi.org/10.3389/fmicb.2014.00641

Jumper, J., Evans, R., Pritzel, A., Green, T., Figurnov, M., Ronneberger, O., Tunyasuvunakool, K., Bates, R., Žídek, A., Potapenko, A., Bridgland, A., Meyer, C., Kohl, S. A. A., Ballard, A. J., Cowie, A., Romera-Paredes, B., Nikolov, S., Jain, R., Adler, J., Back, T., … Hassabis, D. (2021). Highly accurate protein structure prediction with AlphaFold. Nature, 596(7873), 583–589. https://doi.org/10.1038/s41586-021-03819-2

Kendrew, J. C., Bodo, G., Dintzis, H. M., Parrish, R. G., Wyckoff, H., & Phillips, D. C. (1958). A three-dimensional model of the myoglobin molecule obtained by x-ray analysis. Nature, 181(4610), 662–666. https://doi.org/10.1038/181662a0

Kim, H. C., Kim, T. W., & Kim, D. M. (2011). Prolonged production of proteins in a cell-free protein synthesis system using polymeric carbohydrates as an energy source. Process Biochemistry, 46(6), 1366-1369. doi:10.1016/j.procbio.2011.03.008

Knowles, T. J., Finka, R., Smith, C., Lin, Y. P., Dafforn, T., & Overduin, M. (2009). Membrane proteins solubilized intact in lipid containing nanoparticles bounded by styrene maleic acid copolymer. Journal of the American Chemical Society, 131(22), 7484–7485. https://doi.org/10.1021/ja810046q

Koch, K., Afonin, S., Ieronimo, M., Berditsch, M., & Ulrich, A. S. (2012). Solid-state (19)F-NMR of peptides in native membranes. Topics in current chemistry, 306, 89–118. https://doi.org/10.1007/128_2011_162

Latorre, R., Castillo, K., Carrasquel-Ursulaez, W., Sepulveda, R. V., Gonzalez-Nilo, F., Gonzalez, C., & Alvarez, O. (2017). Molecular determinants of BK channel functional diversity and functioning. Physiological reviews, 97(1), 39–87. https://doi.org/10.1152/physrev.00001.2016

Ledford H. (2023). CRISPR 2.0: a new wave of gene editors heads for clinical trials. Nature, 624(7991), 234–235. https://doi.org/10.1038/d41586-023-03797-7

Liao, M., Cao, E., Julius, D., & Cheng, Y. (2013). Structure of the TRPV1 ion channel determined by electron cryo-microscopy. Nature, 504(7478), 107–112. https://doi.org/10.1038/nature12822

Linton, K. J., & Higgins, C. F. (2002). P-glycoprotein misfolds in *Escherichia coli*: evidence against alternating-topology models of the transport cycle. Molecular membrane biology, 19(1), 51–58. https://doi.org/10.1080/09687680110103622

Li S. (2022). Detergents and alternatives in cryo-EM studies of membrane proteins. Acta biochimica et biophysica Sinica, 54(8), 1049–1056. https://doi.org/10.3724/abbs.2022088

Lomize, M. A., Pogozheva, I. D., Joo, H., Mosberg, H. I., & Lomize, A. L. (2012). OPM database and PPM web server: resources for positioning of proteins in membranes. Nucleic acids research, 40(D1), D370–D376. https://doi.org/10.1093/nar/gkr703

Longchamp, J. N., Rauschenbach, S., Abb, S., Escher, C., Latychevskaia, T., Kern, K., & Fink, H. W. (2017). Imaging proteins at the single-molecule level. Proceedings of the National Academy of Sciences of the United States of America, 114(7), 1474–1479. https://doi.org/10.1073/pnas.1614519114

Mathies, R. A., Lin, S. W., Ames, J. B., & Pollard, W. T. (1991). From femtoseconds to biology: mechanism of bacteriorhodopsin's light-driven proton pump. Annual review of biophysics and biophysical chemistry, 20, 491–518. https://doi.org/10.1146/annurev.bb.20.060191.002423

Michel H. (1982). Three-dimensional crystals of a membrane protein complex. The photosynthetic reaction centre from *Rhodopseudomonas viridis*. Journal of molecular biology, 158(3), 567–572. https://doi.org/10.1016/0022-2836(82)90216-9

Milutinovic, P. S., Yang, L., Cantor, R. S., Eger, E. I., 2nd, & Sonner, J. M. (2007). Anesthetic-like modulation of a gamma-aminobutyric acid type A, strychnine-sensitive glycine, and N-methyl-d-aspartate receptors by coreleased neurotransmitters. Anesthesia and analgesia, 105(2), 386–392. https://doi.org/10.1213/01.ane.0000267258.17197.7d

Mitra, K., Ubarretxena-Belandia, I., Taguchi, T., Warren, G., & Engelman, D. M. (2004). Modulation of the bilayer thickness of exocytic pathway membranes by membrane proteins rather than cholesterol. Proceedings of the National Academy of Sciences of the United States of America, 101(12), 4083–4088. https://doi.org/10.1073/pnas.0307332101

Moller, E., Britt, M., Schams, A., Cetuk, H., Anishkin, A., & Sukharev, S. (2023). Mechanosensitive channel MscS is critical for termination of the bacterial hypoosmotic permeability response. The Journal of general physiology, 155(5), e202213168. https://doi.org/10.1085/jgp.202213168

Moore, P. B., Hendrickson, W. A., Henderson, R., & Brunger, A. T. (2022). The protein-folding problem: Not yet solved. Science (New York, N.Y.), 375(6580), 507. https://doi.org/10.1126/science.abn9422

Moss, K., Helm, A., Lu, Y., Bragin, A., & Skach, W. R. (1998). Coupled translocation events generate topological heterogeneity at the endoplasmic reticulum membrane. Molecular biology of the cell, 9(9), 2681–2697. https://doi.org/10.1091/mbc.9.9.2681

Murakami, S., Nakashima, R., Yamashita, E., & Yamaguchi, A. (2002). Crystal structure of bacterial multidrug efflux transporter AcrB. Nature, 419(6907), 587–593. https://doi.org/10.1038/nature01050

Naismith, J. H., & Booth, I. R. (2012). Bacterial mechanosensitive channels–MscS: evolution's solution to creating sensitivity in function. Annual review of biophysics, 41, 157–177. https://doi.org/10.1146/annurev-biophys-101211-113227

Nicolson G. L. (2013). Update of the 1972 Singer-Nicolson fluid-mosaic model of membrane structure. Discoveries (Craiova, Romania), 1(1), e3. https://doi.org/10.15190/d.2013.3

Nielsen K. H. (2014). Protein expression-yeast. Methods in enzymology, 536, 133–147. https://doi.org/10.1016/B978-0-12-420070-8.00012-X.

Okusu, H., Ma, D., & Nikaido, H. (1996). AcrAB efflux pump plays a major role in the antibiotic resistance phenotype of *Escherichia coli* multiple-antibiotic-resistance (Mar) mutants. Journal of bacteriology, 178(1), 306–308. https://doi.org/10.1128/jb.178.1.306-308.1996

Overington, J. P., Al-Lazikani, B., & Hopkins, A. L. (2006). How many drug targets are there?. Nature reviews. Drug discovery, 5(12), 993–996. https://doi.org/10.1038/nrd2199

Perozo, E., Cortes, D. M., Sompornpisut, P., Kloda, A., & Martinac, B. (2002). Open channel structure of MscL and the gating mechanism of mechanosensitive channels. Nature, 418(6901), 942–948. https://doi.org/10.1038/nature00992

Perutz M. F. (1963). X-ray analysis of hemoglobin. Science (New York, N.Y.), 140(3569), 863–869. https://doi.org/10.1126/science.140.3569.863

Qian, H., Cao, P., Hu, M., Gao, S., Yan, N., & Gong, X. (2019). Inhibition of tetrameric Patched1 by Sonic Hedgehog through an asymmetric paradigm. Nature communications, 10(1), 2320. https://doi.org/10.1038/s41467-019-10234-9

Qiu, W., Fu, Z., Xu, G. G., Grassucci, R. A., Zhang, Y., Frank, J., Hendrickson, W. A., & Guo, Y. (2018). Structure and activity of lipid bilayer within a membrane-protein transporter. Proceedings of the National Academy of Sciences of the United States of America, 115(51), 12985–12990. https://doi.org/10.1073/pnas.1812526115

Qi, X., Schmiege, P., Coutavas, E., Wang, J., & Li, X. (2018). Structures of human Patched and its complex with native palmitoylated sonic hedgehog. Nature, 560(7716), 128–132. https://doi.org/10.1038/s41586-018-0308-7

Reddy, B., Bavi, N., Lu, A., Park, Y., & Perozo, E. (2019). Molecular basis of force-from-lipids gating in the mechanosensitive channel MscS. eLife, 8, e50486. https://doi.org/10.7554/eLife.50486

Robertson J. D. (1959). The ultrastructure of cell membranes and their derivatives. Biochemical Society symposium, 16, 3–43.

Rosano, G. L., & Ceccarelli, E. A. (2014). Recombinant protein expression in Escherichia coli: advances and challenges. Frontiers in microbiology, 5, 172. https://doi.org/10.3389/fmicb.2014.00172.

Seddon, A. M., Curnow, P., & Booth, P. J. (2004). Membrane proteins, lipids and detergents: not just a soap opera. Biochimica et biophysica acta, 1666(1–2), 105–117. https://doi.org/10.1016/j.bbamem.2004.04.011

Shelar, A., & Bansal, M. (2016). Helix perturbations in membrane proteins assist in inter-helical interactions and optimal helix positioning in the bilayer. Biochimica et biophysica acta, 1858(11), 2804–2817. https://doi.org/10.1016/j.bbamem.2016.08.003

Sindhu Kumari, S., Gupta, N., Shiels, A., FitzGerald, P. G., Menon, A. G., Mathias, R. T., & Varadaraj, K. (2015). Role of Aquaporin 0 in lens biomechanics. Biochemical and biophysical research communications, 462(4), 339–345. https://doi.org/10.1016/j.bbrc.2015.04.138

Singer, S. J., & Nicolson, G. L. (1972). The fluid mosaic model of the structure of cell membranes. Science (New York, N.Y.), 175(4023), 720–731. https://doi.org/10.1126/science.175.4023.720

Sjostrand, F. S., Andersson-Cedergren, E., & Dewey, M. M. (1958). The ultrastructure of the intercalated discs of frog, mouse and guinea pig cardiac muscle. Journal of ultrastructure research, 1(3), 271–287. https://doi.org/10.1016/s0022-5320(58)80008-8

Skrzypek, R., Iqbal, S., & Callaghan, R. (2018). Methods of reconstitution to investigate membrane protein function. Methods (San Diego, Calif.), 147, 126–141. https://doi.org/10.1016/j.ymeth.2018.02.012

Sotomayor, M., & Schulten, K. (2004). Molecular dynamics study of gating in the mechanosensitive channel of small conductance MscS. Biophysical journal, 87(5), 3050–3065. https://doi.org/10.1529/biophysj.104.046045

Tam, H. K., Foong, W. E., Oswald, C., Herrmann, A., Zeng, H., & Pos, K. M. (2021). Allosteric drug transport mechanism of multidrug transporter AcrB. Nature communications, 12(1), 3889. https://doi.org/10.1038/s41467-021-24151-3

Tan, E., Chin, C. S. H., Lim, Z. F. S., & Ng, S. K. (2021). HEK293 Cell Line as a Platform to Produce Recombinant Proteins and Viral Vectors. Frontiers in bioengineering and biotechnology, 9, 796991. https://doi.org/10.3389/fbioe.2021.796991.

Tate, M. W., & Gruner, S. M. (1987). Lipid polymorphism of mixtures of dioleoylphosphatidylethanolamine and saturated and monounsaturated phosphatidylcholines of various chain lengths. Biochemistry, 26(1), 231–236. https://doi.org/10.1021/bi00375a031

Tao, X., & MacKinnon, R. (2019). Molecular structures of the human Slo1 K+ channel in complex with β4. eLife, 8, e51409. https://doi.org/10.7554/eLife.51409

Tao, X., Zhao, C., & MacKinnon, R. (2023). Membrane protein isolation and structure determination in cell-derived membrane vesicles. Proceedings of the National Academy of Sciences of the United States of America, 120(18), e2302325120. https://doi.org/10.1073/pnas.2302325120

Tribet, C., Audebert, R., & Popot, J. L. (1996). Amphipols: polymers that keep membrane proteins soluble in aqueous solutions. Proceedings of the National Academy of Sciences of the United States of America, 93(26), 15047–15050. https://doi.org/10.1073/pnas.93.26.15047

van den Brink-van der Laan, E., Killian, J. A., & de Kruijff, B. (2004). Nonbilayer lipids affect peripheral and integral membrane proteins via changes in the lateral pressure profile. Biochimica et biophysica acta, 1666(1–2), 275–288. https://doi.org/10.1016/j.bbamem.2004.06.010

van Meer, G., Voelker, D. R., & Feigenson, G. W. (2008). Membrane lipids: where they are and how they behave. Nature reviews. Molecular cell biology, 9(2), 112–124. https://doi.org/10.1038/nrm2330

Vénien-Bryan, C., & Fernandes, C. A. H. (2023). Overview of Membrane Protein Sample Preparation for Single-Particle Cryo-Electron Microscopy Analysis. International journal of molecular sciences, 24(19), 14785. https://doi.org/10.3390/ijms241914785

Walther, D. M., Rapaport, D., & Tommassen, J. (2009). Biogenesis of beta-barrel membrane proteins in bacteria and eukaryotes: evolutionary conservation and divergence. Cellular and molecular life sciences: CMLS, 66(17), 2789–2804. https://doi.org/10.1007/s00018-009-0029-z

Wang, L., & Sigworth, F. J. (2009). Structure of the BK potassium channel in a lipid membrane from electron cryomicroscopy. Nature, 461(7261), 292–295. https://doi.org/10.1038/nature08291

Wang, S., Zhou, H., Chen, W., Jiang, Y., Yan, X., You, H., & Li, X. (2023). CryoFIB milling large tissue samples for cryo-electron tomography. Scientific reports, 13(1), 5879. https://doi.org/10.1038/s41598-023-32716-z

Watson, J. D., & Crick, F. H. (1953). Molecular structure of nucleic acids; a structure for deoxyribose nucleic acid. Nature, 171(4356), 737–738. https://doi.org/10.1038/171737a0

White, S. H., & Wimley, W. C. (1999). Membrane protein folding and stability: physical principles. Annual review of biophysics and biomolecular structure, 28, 319–365. https://doi.org/10.1146/annurev.biophys.28.1.319

Yao, X., Fan, X., & Yan, N. (2020). Cryo-EM analysis of a membrane protein embedded in the liposome. Proceedings of the National Academy of Sciences of the United States of America, 117(31), 18497–18503. https://doi.org/10.1073/pnas.2009385117

Yue, K., Chen, J., Li, Y., & Kai, L. (2023). Advancing synthetic biology through cell-free protein synthesis. Computational and structural biotechnology journal, 21, 2899–2908. https://doi.org/10.1016/j.csbj.2023.05.003

Zhang J. T. (2001). The multi-structural feature of the multidrug resistance gene product P-glycoprotein: implications for its mechanism of action (hypothesis). Molecular membrane biology, 18(2), 145–152. https://doi.org/10.1080/09687680110048831

Zhang, Y., Bulkley, D. P., Xin, Y., Roberts, K. J., Asarnow, D. E., Sharma, A., Myers, B. R., Cho, W., Cheng, Y., & Beachy, P. A. (2018). Structural basis for cholesterol transport-like activity of the hedgehog receptor patched. Cell, 175(5), 1352–1364.e14. https://doi.org/10.1016/j.cell.2018.10.026

Zhang, Y., Daday, C., Gu, R. X., Cox, C. D., Martinac, B., de Groot, B. L., & Walz, T. (2021). Visualization of the mechanosensitive ion channel MscS under membrane tension. Nature, 590(7846), 509–514. https://doi.org/10.1038/s41586-021-03196-w

Zoonens, M., Comer, J., Masscheleyn, S., Pebay-Peyroula, E., Chipot, C., Miroux, B., & Dehez, F. (2013). Dangerous liaisons between detergents and membrane proteins. The case of mitochondrial uncoupling protein 2. Journal of the American Chemical Society, 135(40), 15174–15182. https://doi.org/10.1021/ja407424v

17 Single-Particle Cryogenic Electron Microscopy Unfolds Distinct Membrane Transporter Folds

Focus on Autoinducer-2 Exporter Family

Radhika Khera

INTRODUCTION

Membrane proteins have remained an essential and intriguing research target for decades for various scientific groups around the globe. These biomolecular machines float in the hydrophobic milieu of the fluidic biological membranes and play indispensable roles in various cellular processes across all kingdoms of life. About a quarter of the human proteome encodes for membrane proteins (Uhlén et al., 2015). Transport of molecules across the membrane, cell signaling, metabolism, and cell adhesion are some of the major examples highlighting their vital role. They are key players in fundamental biological processes like sense perception, cellular signaling, photosynthesis, and metabolism, among many others. Despite sharing a commonality in terms of their location, these proteins show high diversity in both structure and function. As these biomolecules are one of the first points of contact for any molecule traversing a biological cell or organelle, mutations or improper folding often result in fatal diseases (Marinko et al., 2019; Sanders & Myers, 2004; Sanders & Nagy, 2000; Zaucha et al., 2021).

Membrane proteins represent approximately 60% of drug targets, and a substantial drug discovery process relies on investigation of the targets' structures (Overington et al., 2006; Yin & Flynn, 2016). Thus, structural studies on these proteins are of immense interest from a therapeutic perspective. The number of accolades and scientific recognition received in membrane protein structure biology reflects the importance and difficulties in this area of research (Figure 17.1). Numerous challenges are associated with working with these insoluble proteins, and that is the primary reason for the lag in the research findings. Lower expression and lesser yields, instability of the membrane proteins, inhomogeneity of the protein preparations, and incompatibility to a particular structure biology method are some of the difficulties that impede the structural investigation of these proteins. This is reflected in the number of membrane protein structures in the Protein Data Bank (PDB), which is less than 2% of the total experimentally determined protein structures so far (Berman, 2000). Additionally, every distinct family of membrane proteins possesses its struggles and challenges, and no universal approach applies to all.

X-ray crystallography, nuclear magnetic resonance (NMR) spectroscopy, and cryo-electron microscopy (cryo-EM) are the leading protein structure biology techniques. X-ray crystallography is the oldest of them, and it requires proteins to be packed in a well-ordered crystal lattice, which is an especially tricky prerequisite for membrane proteins owing to their hydrophobic nature. This is the major reason for the delay in obtaining the first crystal structure of a membrane protein, which transpired in 1985 with the photosynthetic reaction center from *Rhodopseudomonas virdis* by Johann Deisenhofer, Robert Huber, and Hartmut Michel (Deisenhofer et al., 1985), decades after the discovery of X-ray crystallography. Despite this limitation, X-ray crystallography remains one of the principal techniques for adding atomic-resolution structures to the PDB. The major bottleneck of this method is the requirement of high amounts of the target protein and the formation of well-diffracting crystals. Previously, many membrane protein structures have been determined by using highly engineered constructs or by adding some fusion partners to stabilize the proteins and improve the crystal contacts (Ostermeier & Michel, 1997).

NMR spectroscopy is another powerful structural biology method that does not require the preparation of protein crystals for analyses, but the size limitation of <100 kDa proteins presents the primary hiccup for using this technique for several families of membrane proteins (Liang & Tamm, 2016). On the other hand, since the so-called "resolution revolution" in the last decade, cryo-EM has gone from a niche technique to become a standard structural biology method (Kühlbrandt, 2014). It is now being adopted by an increasing number of research groups. Technological developments in terms of microscope stability, the introduction of direct electron detectors (DEDs), and major improvements in image processing software have collectively led to this breakthrough in the field (Brilot et al., 2012; Li et al., 2013; McMullan et al., 2016; Scheres, 2012). Cryo-EM exists in three different flavors: single-particle cryo-EM (SPA), cryo-electron tomography (cryo-ET), and electron crystallography. Cryo-ET studies the proteins in their natural cellular environment, while SPA focuses on a near-homogenous preparation of the target protein. The aforementioned resolution revolution has greatly affected the SPA sector, and how has it

DOI: 10.1201/9781003326106-21

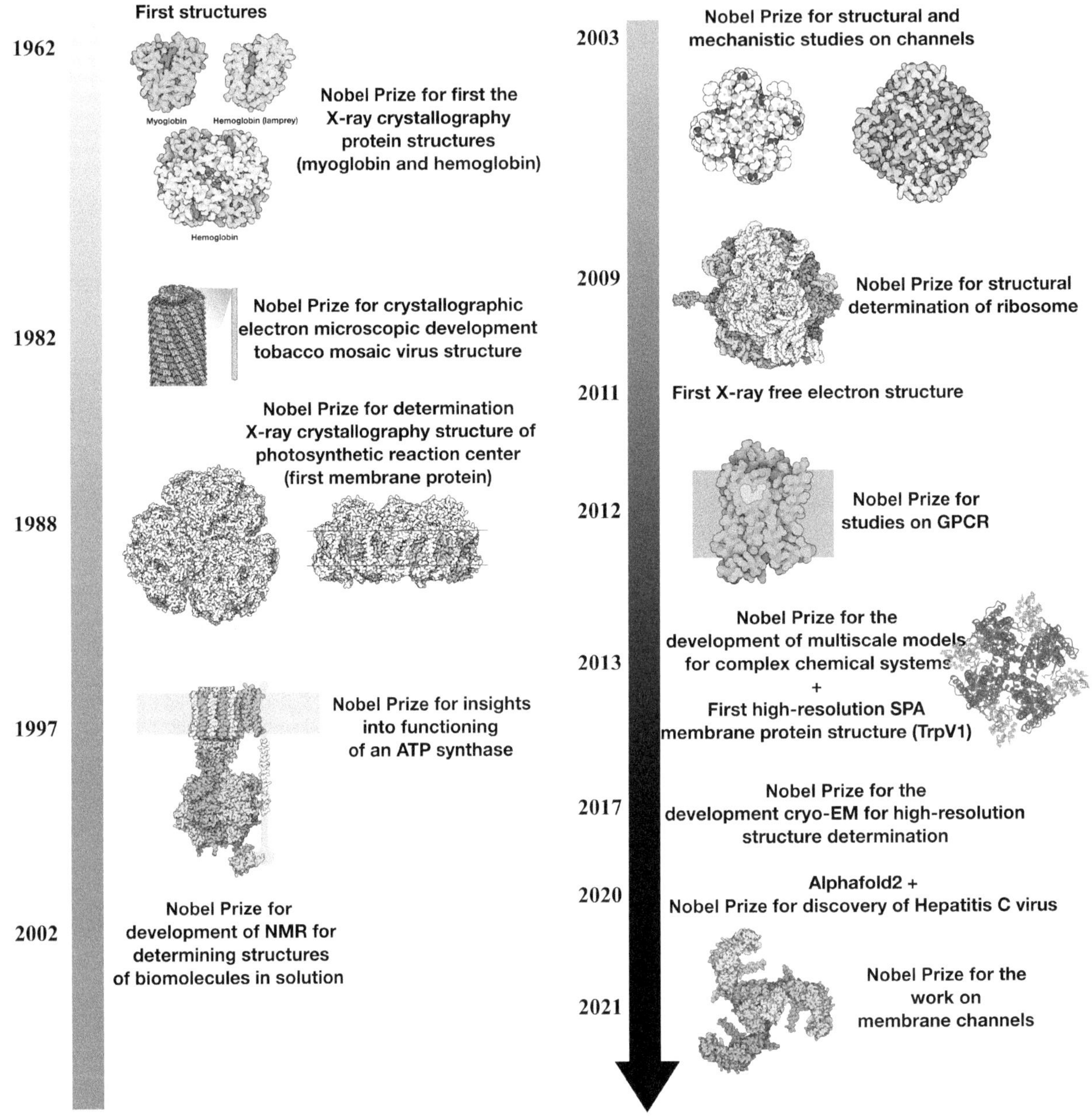

FIGURE 17.1 Major milestones in the history of structural biology. The advancements over time in the structural biology field are highlighted in the figure. The figure is prepared on similar lines as presented and celebrated by the PDB (https://www.rcsb.org/pages/about-us/history). It is also adapted and modified from a dissertation (Khera, 2023).

influenced the investigation of membrane protein structures will be the primary focus of this chapter. There is a saying in the field, "Bigger proteins (>100 kDa) are preferable for single-particle analysis," and it still has validity, but technological improvements are mitigating these barriers both in terms of investigating small membrane proteins (<100 kDa) and attaining high resolutions (around or >3Å) comparable to the X-ray structures (Dolan et al., 2022; Parker et al., 2021). The database of all the membrane protein structures is well updated and maintained by Stephen White's laboratory at UC Irvine (https://blanco.biomol.uci.edu/mpstruc/).

Although SPA has produced great progress in the membrane protein structure biology field, which has significantly reduced the timescales required to go from gene level to structure determination, certain steps in the workflow are still not completely streamlined. The optimizations are time-consuming and require back-and-forth from the lab bench to the microscope. In this chapter, we would like to go through the steps in the current single-particle workflow and discuss the obstacles one might encounter along the way in the characterization of a particular membrane protein family. This chapter also aims to highlight the developments single-particle cryo-EM has made in the field and to provide the necessary guidelines for new researchers based on previous studies. Later, we will share the case study of a novel membrane transporter family of the AI-2 exporter superfamily, focusing on its pipeline for attaining its first high-resolution structures through the single-particle approach (Khera et al., 2022). This transporter family demonstrates unique structural folds that were not observed in the previously investigated membrane transporter families. This is just one illustration among many of how cryo-EM is unraveling the membrane biology field (Baril et al., 2023; Cheng, 2018; Kühlbrandt, 2022). Finally, we conclude by highlighting the importance of complementary biophysical, genetic, and computational studies that work together with SPA to provide a complete picture of the functioning of the membrane protein family under study. We also point out that a combination of different experimental techniques (single-particle cryo-EM and *in situ* cryo-ET) and machine learning platforms such as AlphaFold2 (Jumper et al., 2021) can synergistically pave paths for answering questions that were previously considered unanswerable.

SINGLE-PARTICLE CRYO-EM WORKFLOW

Cryo-EM is no longer perceived as a complementary technique to X-ray crystallography. Instead, it is the structural biology method that can address a broader range of biological questions than the other techniques. The past decade has seen cryo-EM have a major impact on our understanding of membrane systems through both *in situ* cryo-ET and *in vitro* SPA. The initial electron microscopy work focused on negatively stained specimens and 2D electron crystallography. The high-resolution structural investigation through electron microscopy back then relied mostly on the combination of both imaging and diffraction methods (Kühlbrandt, 2013). The field has immensely progressed and gained ground since the determination of the first electron microscopy structure of the membrane protein bacteriorhodopsin at 7 Å in 1975 by Richard Henderson and Nigel Unwin and later near-atomic resolution (3.5 Å) cryo-EM structure of bacteriorhodopsin in 1990 again by Richard Henderson (Henderson et al., 1990; Henderson & Unwin, 1975). The impact of developments in the field is well reflected in the exponential growth of the number of membrane protein structures submitted to the PDB every year (Allen, 2019; Choy et al., 2021). Most of these membrane protein structures recently submitted to the PDB through cryo-EM are produced by the single-particle approach. The first near-atomic resolution structures of a membrane protein with SPA came around 2013. It was of transient receptor potential channel (TrpV1) and more structures have continued to come in ever since (Liao et al., 2013) (https://blanco.biomol.uci.edu/mpstruc/).

As mentioned earlier, in the past decade SPA has been a dominant contributor of membrane protein structures. This method needs a much smaller amount of the target protein than X-ray crystallography, and it eliminates the major bottleneck of the latter method, the requirement of well-diffracting crystals. This opened up avenues for membrane proteins that were extremely resistant to the crystallization step. SPA images the protein molecules embedded in a thin vitreous ice layer that are ideally present in different orientations; subsequently a 3D reconstruction is attained through the back projection of the averaged 2D microscopy images. SPA has reshaped the membrane structure biology workflow compared to the time when the field was exclusively dominated by X-ray crystallography. In other words, the approach toward functionally and structurally characterizing a membrane protein family is quite different now compared to the pre–resolution revolution era of cryo-EM owing to both biochemical and technological advancements. There are amazing review articles focusing on the SPA workflow for membrane proteins, but they can be a bit overwhelming, especially for newbies in the field (Piper et al., 2022; Thonghin et al., 2018). This section of the chapter aims to provide a simplified overview of the SPA workflow in structurally characterizing membrane protein families focusing on minute details in the pipeline that can sometimes be missed (Figure 17.2). The major steps in the pipeline include

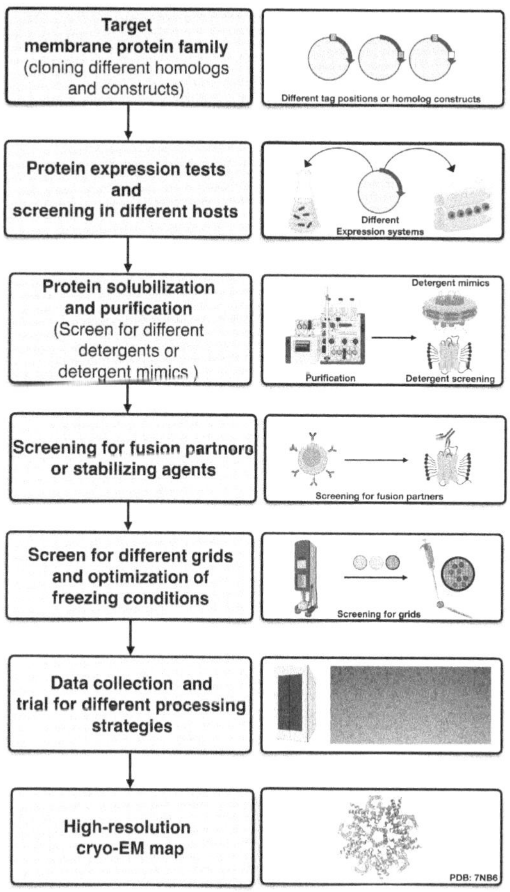

FIGURE 17.2 Single-particle cryo-EM workflow for membrane proteins. The figure describes the complete single-particle cryo-EM pipeline, beginning with cloning and preparation of different constructs and ending with the final stage of attaining a high-resolution 3D reconstruction.

protein expression and purification, screening of some stabilizing agents such as cross-linkers or fusion partners (immunoglobulins), grid freezing optimization, and lastly, single-particle data collection and processing (Figure 17.2). The points discussed in the following sections address some of the questions one might have while planning to characterize a particular membrane protein family.

Membrane Protein Expression

This step in the workflow requires a fair amount of optimization. Membrane protein expression can be achieved either through native expression or recombinant expression systems. Expression and purification from native organisms or tissues require high expression of the target protein and can be challenging. However, using native sources is the optimal choice when working with bigger membrane complexes such as those involved in respiration or photosynthesis (Graça et al., 2021; Sun et al., 2018). Since photosystems or electron transport chain complexes are composed of several subunits and encompass many cofactors and ligands, they are traditionally isolated from their native sources. Some of the protein complexes, such as bacterial secretion systems that comprise a couple of subunits, are recombinantly expressed for structural investigation like most of the other target membrane protein families (Bunduc et al., 2021; Hennell James et al., 2021).

Traditionally, in characterizing a specific membrane protein family, different homologs from several organisms would be cloned. The main idea behind this is that some homologs, especially the ones from thermophiles or archaea, would be more stable and could be relatively easier to investigate. Additionally, different constructs with a variety of affinity tags (Strep-tag II, His-tag, Flag-tag, Alpha-tag) at different locations (N- or C-terminals) are prepared to later assist in the purification of the desired protein. Chimeric constructs, where the target protein is fused with some protein partners such as EGFP (enhanced green fluorescent protein) or MBP (maltose binding protein), are also conventional in the field. These partners aid in tracking the proper folding and localization of the target protein and help in increasing the solubility in the later stage of protein extraction and purification. These fusion partners can also be used as tags for affinity purification of the target proteins. However, some tags or partners can also lead to the production of nonfunctional proteins, and one needs to be mindful of this. Beginning with several constructs in terms of homologs, different tags, tag positions, and fusion partners is a prudent approach but was often employed with X-ray crystallography as the primary structural technique. This is because only a few, or in many cases, a single one among the different constructs would reach the final stage of well-diffracting crystals. In most cases, this is not required in the SPA workflow.

The primary point to consider while choosing an expression system is the similarity with the native environment in terms of the nature of lipids and posttranslation modifications, and it should enable stable transport of the protein to the membrane. Depending upon the source organism, various expression systems (*Escherichia coli* or *Pseudomonas stutzeri*, yeast *Saccharomyces cerevisiae* or *Pichia pastoris*, insect *Spodoptera frugiperda* [Sf9], and mammalian human embryonic kidney [HEK] cells) can be picked up and investigated for the recombinant expression of the target protein (Pandey et al., 2016; Sommer et al., 2017). The cell-free expression system is also advancing considerably and can be explored. However, to the best of our knowledge, it has not yet yielded a high-resolution single-particle cryo-EM membrane protein structure (Schwarz et al., 2008). Certain parameters, such as the temperature of growth of the expressing cells, inducer concentrations, and duration of expression, can be screened and optimized, ensuring proper expression, folding, and localization of the target protein.

Protein Solubilization and Purification

After membrane protein expression, the next step is to lyse the cells and extract the target integral membrane protein from the cellular membrane environment. Traditionally, this step is carried out with the help of amphipathic detergents, which mimic the native lipid environment of the protein and attempt to bring the proteins into the solution. This is one of the critical steps in the workflow as classical detergents often lead to the denaturation and unfolding of the proteins, and it cannot be emphasized enough that the main objective of any membrane protein study is to structurally investigate a properly folded and functional protein. The detergent solubilization step, even though mandatory, can essentially strip off the native stabilizing lipids, which destabilizes the protein because lipids are vital for both the biological integrity and function of all membrane proteins.

Recognizing suitable detergent or nondetergent alternatives for protein solubilization and later structural investigation is crucial for attaining higher yields of stable proteins. However, there are no set guidelines regarding the choice of detergent. Thus, one needs to screen through various classes of detergents and nondetergent alternatives. Understanding the critical role of lipids, several nondetergent alternatives such as amphipols, cyclohexane-modified amphiphile polymer (CyclAPol), di-isobutyl maleic acid lipid particles (DIBMA), nanodiscs, native cell membrane nanoparticles (NCMN), peptidiscs, styrene-maleic acid lipid particles (SMALPs), and saposin lipoprotein nanoparticles are now available and are gaining the spotlight as they represent a detergent-free solubilization method (Angiulli et al., 2020; Guo, 2021; Knowles et al., 2009; Popot et al., 2011; Sligar & Denisov, 2021). These alternatives attempt to maintain the necessary and near-native lipid environment around the protein. They have provided several near-atomic resolution structures of different families of membrane proteins (Figure 17.3). It is essential to note that some of these alternatives, such as nanodiscs, still require a detergent solubilization step, and the detergent is exchanged at the later stage

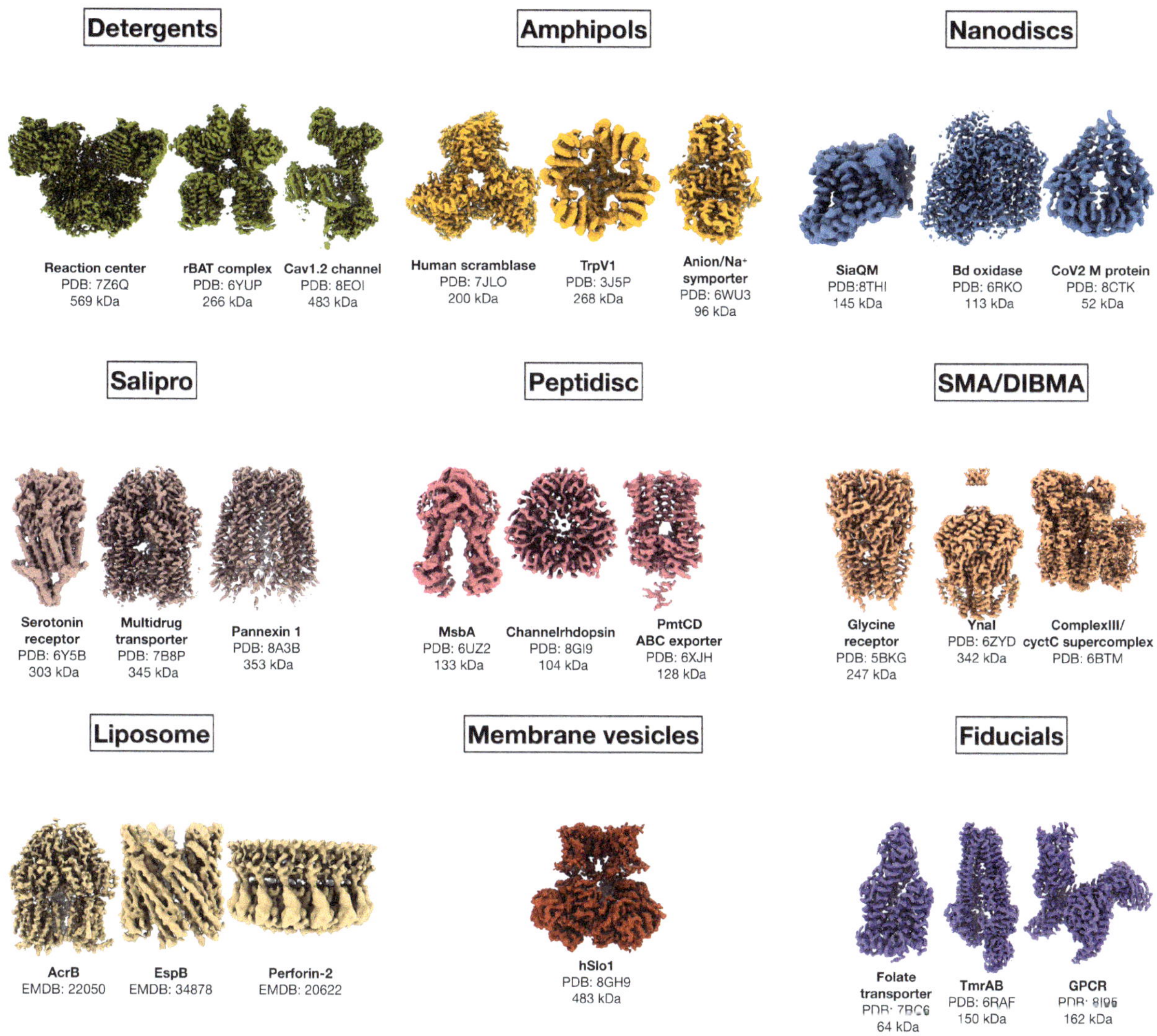

FIGURE 17.3 Single-particle cryo-EM structures determined in different detergents/detergent alternatives. This figure highlights the contributions made by single-particle cryo-EM in the field of membrane biology, providing structures of various classes of membrane proteins in different detergents or detergent alternatives. It also exhibits three illustrations (PDB: 7BC6, 6RAF, and 8I95) in purple where the structures were obtained bound to fiducials like nanobodies and scFv.

of purification. A couple of recent studies analyzed detergent usage in single-particle structures determined over a decade (2010–2019) and presented trends where certain detergents favored specific protein families. Overall, maltosides have historically been the detergent choice for membrane protein solubilization. Among the maltosides, DDM (N-dodecyl-ß-D-maltoside) is the most used detergent in protein solubilization. For eukaryotic proteins, a mix with lipidic additive cholesteryl hemisuccinate (CHS) has been observed to be beneficial (Choy et al., 2021; Le Bon et al., 2021). Glyco-diosgenin (GDN) or digitonin, neopentyl glycols with or without CHS, have also been commonly used as solubilizing detergents in several high-resolution single-particle structures.

The purification of membrane proteins is essentially like soluble proteins and requires a couple of rounds of trial and error involving a combination of different chromatography steps, and in some cases density gradients. Unlike X-ray crystallography, single-particle cryo-EM does not require an extremely homogenous sample. A high-purity protein sample is always preferable, but one can always classify out the contaminants at the data processing stage. There are some single-particle studies on crude cell extracts or everted vesicles, as illustrated in Figure 17.3 (Su et al., 2021; Tao et al., 2023). Additionally, it is crucial to perform quality control of the purified protein sample at every stage of purification. This is carried out with a variety of biochemical and functional methods like native polyacrylamide gel electrophoresis (native-PAGE), sodium dodecyl sulfate–polyacrylamide gel electrophoresis (SDS-PAGE), mass spectrometry, multiple angle light scattering coupled to size exclusion chromatography (SEC-MALS), Western blots, functional assay for the protein if available, and negative staining electron microscopy.

In most of the structural studies, the final detergent or nondetergent alternative used for single-particle sample preparation is different from the solubilization detergent. The rationales behind this are to reduce the micelle size, provide the protein with a detergent-free and near-native environment by exchanging the detergent for an alternative, and attempt to minimize the empty detergent micelle background. All the detergents have a critical micelle concentration (CMC) which typically represents the concentration above which the empty detergent micelles begin to form in the background. These empty micelles can be misinterpreted as protein molecules in the electron micrographs, especially when working with small membrane proteins. Even below nominal CMC, some detergents form micelles that generate a significant and undesirable background in the EM images (Gewering et al., 2018). Thus, screening for the optimal detergent or an alternative essentially needs to be performed at two stages, one for the solubilization and one for the final single-particle sample preparation. If the micelle background remains an issue, one could also consider exploring the GraDer technique for gently removing the excess detergent from the sample (Hauer et al., 2015). On a concluding note, the signal-to-noise ratio (SNR) in the electron micrographs is quite low. Thus, one might need to screen through a couple of buffers, salt, and detergent conditions to minimize their contribution to the background noise as much as possible. The components that severely affect the SNR of the micrographs include sucrose, glycerol, and high salt concentrations, and these should be avoided or minimized if possible.

Single-particle Sample Preparation

A typical cryo-EM grid is composed of a metal mesh (copper or gold), and it is overlaid with a thin layer of amorphous carbon or gold. The gold-based grids, even though more fragile, are known to provide better contrast images due to the low beam-induced motion of the specimens (Russo & Passmore, 2014). Normally, for cryo-EM grid preparation, 3 to 5 μL of the protein sample is applied to the grid, and later it is quickly plunge-frozen into a cryogen (liquid ethane) after blotting for vitrification (Adrian et al., 1984; Cressey & Callaway, 2017). This step can be performed manually or with a variety of commercially available devices like ThermoFischer Scientific Vitrobot, Leica GP2, Spotiton, Vitrojet, Chameleon, and Cryo-writer (Weissenberger et al., 2021). Irrespective of the device used, the primary goal is to obtain a near-ideal grid with protein molecules suspended in multiple orientations within a thin layer of vitrified ice. However, it is a challenging and time-consuming process. The major hiccups on the way include (1) denaturation at the air–water interface, (2) proteins having a strong affinity for the support film, which is a normal scenario for the membrane protein samples prepared in detergents, and (3) preferential orientation.

As mentioned before, the membrane proteins prepared in detergents have a high affinity for the support film and thus give a "halo effect," especially in thin ice (desirable) areas of the grids (Drulyte et al., 2018; Kampjut et al., 2021). The protein particles accumulate around the periphery of the hole, leaving the center empty and leading to poor distribution in the holes. This can be worked around by beginning with a concentrated protein sample such that it first saturates the film and subsequently pushes the particles into the holes. Another possible solution is to collect images in slightly thicker areas with a suitable particle distribution. Graphene oxide layered or carbon back-coated grids have also shown beautiful outcomes to circumvent this issue (Drulyte et al., 2018). Similar to the problem of poor distribution, dealing with preferential orientation issues can be quite troublesome. Screening different grids, detergents, switching to nondetergent alternatives, and the addition of fluorinated detergents like 3-([3-cholamidopropyl] dimethylammonio)-2-hydroxy-1-propanesulfonate (CHAPSO) during vitrification, and merging of a stage tilt single-particle dataset have proved promising (Chen et al., 2019; Xie et al., 2023).

In many cases, proteins get denatured at the air–water interface while freezing. This is especially observed when working with larger-membrane protein complexes. In such scenarios, a variety of things can be tried and tested. The use of stabilizing agents such as chemical cross-linkers has shown encouraging results in a couple of cases. The cross-linking step can be performed mildly through gradient fixation (GraFix) (Stark, 2010). GraFix gently cross-links the membrane complex, typically with glutaraldehyde via a density gradient. In certain cases, this method has stabilized and locked certain flexible subunits of a membrane complex that were initially not recovered in the EM density maps of the non–cross-linked preparations (Xie et al., 2023). Another way to prevent denaturation at the air–water interface is to use chemically treated grids or affinity grids (Ni-NTA, streptavidin, and antibody) for single-particle sample preparation (D'Imprima et al., 2019; Glaeser & Han, 2019; Kelly et al., 2008; Yu et al., 2016). These grids capture the protein molecules on the surface and prevent their diffusion, thereby minimizing denaturation at the air–water interface. Additionally, affinity grids concentrate a particular sample on the grids and aid in providing different angular orientations of the target protein as well. This was recently observed with the single-particle phycobilisome structure, which was investigated using streptavidin grids (Domínguez-Martín et al., 2022).

Another point worth bringing up in single-particle cryo-EM sample preparation is the use of different fusion partners such as antibodies, nanobodies, Fab or Fv fragments, ScFv fragments, megabodies, sybodies, and legobodies (Uchański et al., 2021; Wu & Rapoport, 2021; Zimmermann et al., 2020). Such immunoglobulin chaperons were also employed for X-ray crystallography studies of membrane proteins to lock the proteins in a particular conformation or to improve crystal contact. They are being used in single-particle studies for several roles: (1) to increase the size of a small membrane protein for better alignment of particles (Parker et al., 2021; Zhang et al., 2022), (2) to study the dynamics of proteins (Galazzo et al., 2022; Hofmann et al., 2019), and (3) to identify the location of a particular subunit in a bigger membrane protein complex (Brown et al., 2018). Many studies have utilized immunoglobulin libraries such as yeast surface display nanobody libraries to screen and select a suitable candidate for their target membrane protein (McMahon et al., 2018). These libraries have opened exciting avenues and eliminated the need for raising immunoglobulins in mice, rabbits, or lamas. In the best scenarios, one can obtain more than one high-affinity nanobody in the nM range. Acquiring nanobodies that bind the protein in different functional states is also feasible.

Single-particle Data Acquisition and Processing

One of the key factors that boosted the field of cryo-EM is the development of direct electron detectors that enhanced the signal-to-noise ratio in the cryo-EM images. Additionally, the images are collected as movies comprising a series of frames which are later corrected for beam-induced motion or stage drift, resulting in high-quality, unblurred images. Data collection as movies also permits discarding the later radiation-damaged frames to preserve the high-resolution information in the aligned micrographs. Once the optimal cryo-grids are obtained, the general workflow at many places is to screen the grids with a lower-end 200 keV cryo-microscope and then subsequently perform final high-resolution data collection with a 300 keV microscope. However, fast-paced technological advancements in the field have resulted in the routine use of 200 keV cryo-microscopes for the collection of high-resolution data of membrane proteins; thus, they are no longer considered just screening microscopes (Thangaratnarajah et al., 2022). The ability to use 100 to 200 keV cryo-microscopes for high-resolution data collection expands the accessibility of the technique, especially to groups that cannot afford the cost of installation and maintenance of a 300 keV microscope. The parameters to be considered for single-particle data collection include the pixel size, defocus range, total dose, and number of frames (not in electron event representation mode). Live sessions (like CryoSPARC live) help to assess the quality of the data while it is being collected and are highly recommended (Punjani et al., 2017). There are several instances where one can obtain high-resolution 3D reconstructions of the target protein from the live sessions.

There are a number of software packages that can be used for data processing, such as RELION, CryoSPARC, EMAN, and cisTEM, among many others (Grant et al., 2018; Punjani et al., 2017; Schwab et al., 2023; Tang et al., 2007). Each package has its distinct algorithm, and one should try to employ the best features of each software program in the processing workflow. Neural network–based pickers such as crYOLO and TOPAZ are beneficial in accurately picking even the low-contrast particles from the micrographs (Bepler et al., 2019; Wagner et al., 2019). The basic data processing strategy in the same order includes beam-induced motion correction, contrast transfer function (CTF) estimation, particle picking and extraction from the micrographs, 2D classification, *ab initio* model building, 3D classification, 3D refinement, local and global CTF refinements, and polishing. There are several additional steps like focused/masked classification, particle subtraction to focus on a certain subunit of the whole structure, and many more. However, they all are case-dependent and cannot be generalized. The AlphaFold2 database provides the starting models that are further built and refined in the obtained EM density maps (Jumper et al., 2021). Coot, CCP4, PHENIX, Refmac, ISOLDE, and recently developed Model Angelo help to prepare the final atomic models from the cryo-EM density maps (Afonine et al., 2018; Brown et al., 2015; Croll, 2018; Emsley et al., 2010; Jamali et al., 2023). Lastly, one should be careful while processing data to avoid the introduction of bias and overinterpretation at any step of the processing pipeline. This might produce the well-known "Einstein-from-noise" case (Henderson, 2013; Scheres & Chen, 2012).

PENTAMERIC SINGLE-PARTICLE CRYO-EM STRUCTURES OF AUTOINDUCER-2 EXPORTER (AI-2 EXPORTER) FAMILY—A CASE STUDY

So far, we have discussed how single-particle cryo-EM is boosting the field of membrane proteins and what the typical workflow looks like. Every step in taking a membrane protein from gene level to high-resolution structure determination requires excessive screening, trial and error, and optimizations. The main point is that the workflow of single-particle cryo-EM is still not as smooth as it appears to be, and it can be difficult and time-consuming. However, the ability to perform structural studies with low amounts of target protein and without the need for well-diffracting crystals is intriguing. The number of high-resolution membrane protein structures submitted every year to the PDB is escalating, and it is leading to the discovery of novel motifs and folds within the protein families. This section aims to introduce the readers to a novel membrane transporter family (Autoinducer-2 exporter family) and discusses the steps followed to attain its first high-resolution structures through the single-particle approach (Khera et al., 2022).

Quorum Sensing and autoinducer-2 Exporter Family

Membrane transporters play a vital role in both physiology and diseases. These proteins in the lipid bilayer act as gatekeepers mediating the uptake of nutrients, removal of waste products, transport of drugs, and transport of even folded proteins across the cells or organelles. Owing to their vital role, they are critical drug targets. Based on bioinformatic and phylogenetic analysis, there are a total of 15,528 transporter families or 82 superfamilies, and so far, only a small percentage of them have been structurally characterized (Saier et al., 2021).

Most of the transporter families require an energy source for their transport function (active transporters). Some of them directly rely on adenosine triphosphate (ATP) as an energy source—primary active transporters—while others depend upon the proton motive force or energy stored in the form of an electrochemical gradient across the bilayer membrane—secondary active transporters. The structural studies of the transporter families help to increase our understanding of their functions. Additionally, different conformational states of a specific transporter family protein aid in discerning the entire transport cycle of the protein. Similar structural motifs and folds are commonly observed within functionally distinct transporter families (Holm & Rosenström, 2010; Shi, 2013). Three major transporter folds are noticed among secondary active transporters: major facilitator superfamily (MFS) fold, leucine transporter (LeuT) fold, and *E. coli* Na^+/H^+ antiporter (NhaA) fold (Shi, 2013). As highlighted earlier, single-particle cryo-EM is contributing immensely to both discovering new transporter folds and investigating different dynamic conformational states.

The Autoinducer-2 exporter (AI-2 exporter) family is a large prokaryotic secondary active transporter family primarily characterized in bacteria and archaea (Rettner & Saier Jr., 2010). As the name suggests, it is an exporter of a molecule known as autoinducer-2 (AI-2). AI-2 is a universal signaling molecule of the prokaryotic signaling mechanism for quorum sensing (QS) (Bassler & Losick, 2006; Pereira et al., 2013). QS is a bacterial cell-to-cell communication mechanism that allows these primitive organisms to share information regarding their cell density and thereby adjust their gene expression. This signaling/communication is mediated by pheromones or signaling molecules known as autoinducers (self-produced) as shown in Figure 17.4a. QS is a way for these single-celled organisms to show multicellularity. It is carried out in four major steps, as demonstrated in Figure 17.4a. QS is a fascinating area, and recently it has extended from just prokaryotic communication to host–microbe communication (symbiosis) and virus communication (Ismail et al., 2016; Silpe et al., 2022). This signaling mechanism was first discovered in the findings of Tomasz (genetic competence in *Streptococcus pneumoniae*) and Nealson et al. (bioluminescence in *Vibrio fischeri*) (Nealson et al., 1970; Tomasz, 1965). Readers interested in knowing more about QS and autoinducers should see some of the several articles where this area has been well reviewed (Papenfort & Bassler, 2016; Pereira et al., 2013; Rutherford & Bassler, 2012).

As mentioned earlier, among the different discovered QS signaling molecules, AI-2 is regarded as a universal signaling molecule because it is known to mediate inter/intra-species as well as inter-kingdom signaling. AI-2 is not only recognized within the same bacterial species but also recognized by the host cells. It is known to play an important role in clinically critical processes like biofilm formation, antibiotic resistance, chemotaxis, and pathogenesis. Every QS signaling molecule like AI-2 has a QS circuit to carry out communication (Figure 17.4b and c). An exporter in the QS circuit is an essential component that exports the signaling molecules outside the cells. AI-2 exporter family proteins are the proposed exporters of AI-2. The first protein proposed to be an exporter of AI-2 was the YdgG protein from *E. coli*, which was later renamed TqsA (transporter of quorum-sensing signal AI-2) (Herzberg et al., 2006). This protein was observed to be overexpressed in *E. coli* biofilms and was later investigated with gene knockout studies. The deletion of *tqsA* leads to an increased biomass production, biofilm formation, motility, and surged intracellular concentrations of AI-2. Subsequently, the bioinformatic study of TqsA identified homologs of this membrane protein in other prokaryotes, and a new secondary active transporter family was established, the AI-2 exporter superfamily (formerly known as the PerM family) (TCDB- 2.A.86). It identified three more homologs from the model organism *E. coli*, totaling four proteins of this family in *E. coli*—TqsA, YhhT, YdiK, and PerM. At the time when the goal was set to investigate this transporter family, none of the members had been functionally or structurally characterized.

Workflow for Structural Investigation of AI-2 Exporter Family

The main goal was to investigate the AI-2 exporter family structurally and functionally. The workflow adopted for attaining single-particle high-resolution structures of AI-2 exporters is briefly discussed in this section.

Protein Expression and Purification

As highlighted before, the AI-2 exporter family is primarily a prokaryotic and archaeal family, thus *E. coli* (BL21 [DE3] or Top10 cells) was the choice for protein expression. Rooting from a crystallography background, several homologs were selected from *E. coli* (TqsA, YhhT, YdiK, PerM), *Aquifex. aeolicus* (aq_740), and some from pathogenic organisms such as *Pseudomonas aeruginosa* and *Salmonella typhimurium*. Most of the members of this family are around 38 to 40 kDa in size (Rettner & Saier Jr., 2010). Some of the homologs are bigger, and they possess some additional catalytic domains like ATP and GTP binding, purine nucleotide biosynthesis, and molecular signaling. The average size of the members of this family, ~40 kDa, was an obvious barrier to performing single-particle structural studies. Initially, both X-ray crystallography and single-particle cryo-EM were planned for the structural investigation of AI-2 exporters. All the proteins from *E. coli* and *A. aeolicus* were expressed, solubilized, and could be homogenously purified through a combination of affinity and size-exclusion chromatography. For membrane solubilization, different classes of detergents were screened, but ß-DDM worked beautifully for all the proteins (TqsA, YhhT, YdiK, PerM, and Aq_740).

Single-Particle Cryo-EM Sample Preparation

The biochemical studies (size-exclusion profiles and BN-PAGE) of the purified proteins showed some astonishing findings. The average size of the final purified proteins

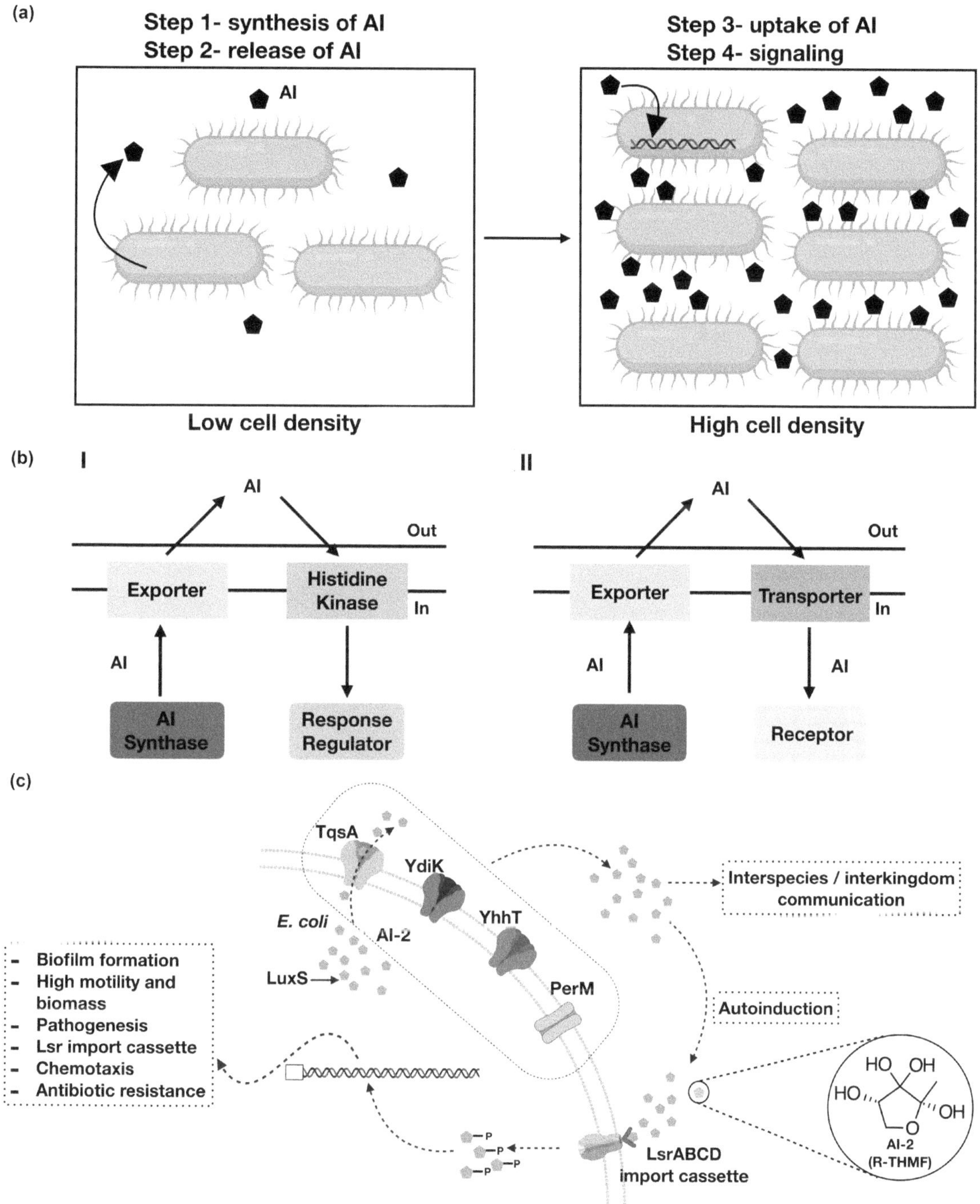

FIGURE 17.4 Quorum sensing mechanism and circuits. (a) This section of the figure describes the mechanism of QS focusing on its four major steps. QS begins with the synthesis and release of self-synthesized signaling molecules/autoinducers (AI) from the cells. Once the cell density reaches a particular threshold, the AI molecules are taken up by the cells, which signals the cells about their cell number and results in gene regulation, giving different responses such as biofilm formation. AI molecules in the figure are shown as small pentagons (Khera, 2023). (b) Two different types of QS circuits are demonstrated in this section. The QS circuit includes four major proteins that mediate QS. Type I is generally found in gram-positive bacteria, and type II is generally observed in gram-negative bacteria. Exporter is an important component of the QS circuit; it transports the AI molecules outside the cells as the second step in QS functioning (Khera, 2023). (c) This portion of the figure exhibits the QS circuit of the AI-2 (R-THMF) in *E. coli*. LuxS is the protein that synthesizes AI-2, which is then exported outside by AI-2 exporters (TqsA, YdiK, YhhT, and PerM). AI-2 is a universal signaling molecule that performs inter/intra-species as well as interkingdom signaling. Once a particular threshold is achieved, AI-2 is imported into the cells via a known ABC transporter (LsrABCD). As it is taken up by the cells, it is phosphorylated by a kinase LsrK, and subsequently it replaces the repressor LsrR to regulate gene expression (Khera, 2023; Khera et al., 2022.)

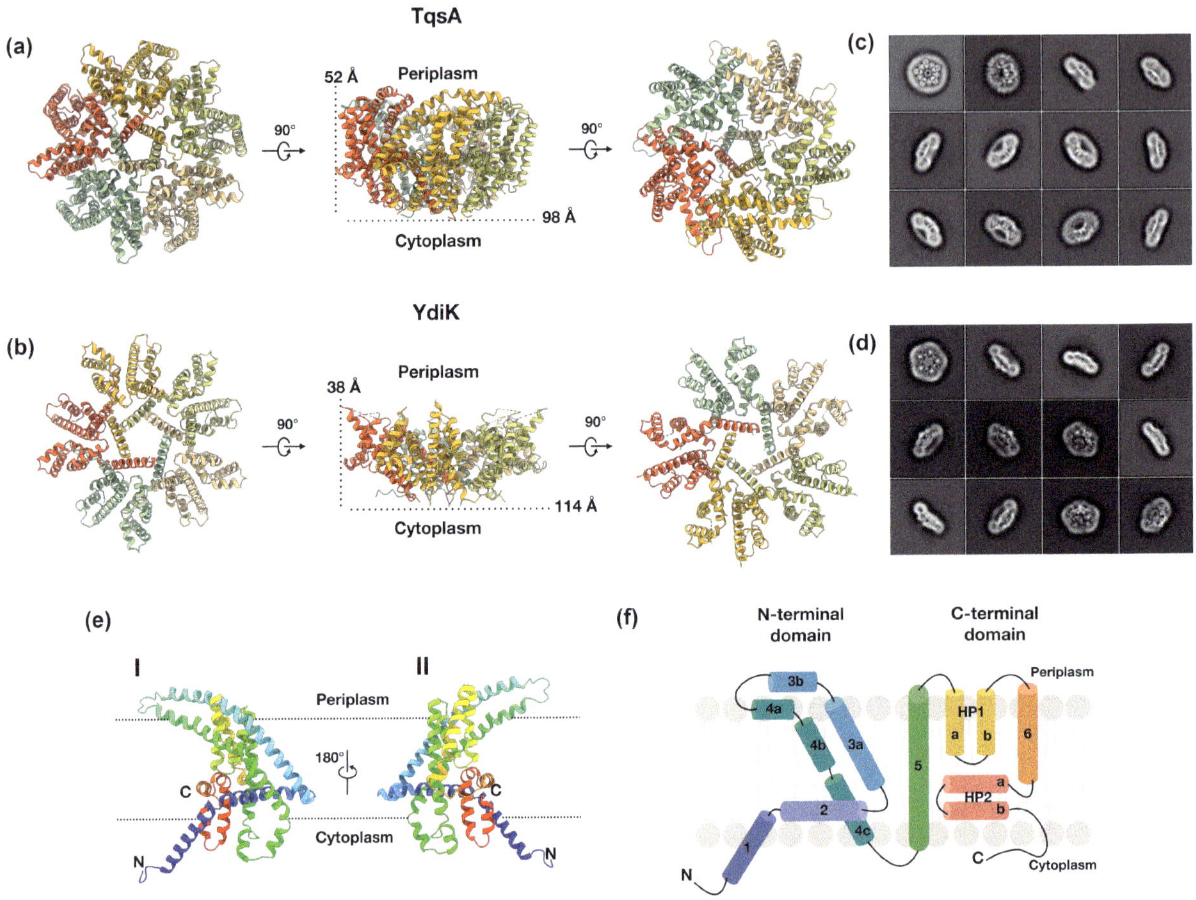

FIGURE 17.5 Structures of autoinducer-2 exporters (TqsA and YdiK) from *E. coli*. (a, b) Homo-pentameric structure of *E. coli* AI-2 exporters. The central opening that connects the cytoplasm of the cells to the periplasmic basin is visible in the structures. (c, d) 2D class average of TqsA and YdiK obtained from single-particle cryo-EM data processing. The pentameric arrangement of the two AI-2 exporters TqsA and YdiK is apparent in the classes. (e, f) Helix arrangement of TqsA protomer. Each protomer can be seen comprising two separate domains, N- and C-terminal domains. The C-terminal domain harbors two helical hairpins (HP1 and HP2), which is the suspected binding site of AI-2. (This figure is adapted and modified from [Khera et al., 2022].)

was observed to be much bigger than expected (~180 kDa). There was only one outlier, PerM, from *E. coli* (~83 kDa). There were two possibilities that accounted for a bigger size: (1) existing as higher oligomeric complexes (~4–5 oligomers), and (2) pulling down of some additional interacting protein partners. The first possibility was eventually confirmed through proteomic analysis of the purified samples.

Even after trials with different detergents and buffer conditions, these proteins didn't seem to crystallize well. As a result, the entire focus was shifted to single-particle cryo-EM. The sample quality was first confirmed with negative staining, after which different detergents and buffer conditions were screened for grid preparation. As mentioned before, ß-DDM was the membrane solubilizing detergent, but it was exchanged for other detergents (LMNG, GDN, DM) at the affinity purification step. GDN and DM-prepared samples were observed to extensively stick to the carbon, leaving the holes empty. LMNG didn't show this halo effect but formed a thicker micelle belt around the protein particles; thus, it did not result in high-resolution cryo-EM maps. After a lot of screening processes trying different grids, blotting conditions, detergents, and nanodiscs (MSP2N2), a combination of GDN-purified sample and C-flat grids finally resulted in two high-resolution cryo-EM maps of AI-2 exporters, TqsA (3.35 Å) and YdiK (2.80 Å), as shown in Figure 17.5a and b (Khera et al., 2022). GDN samples still showed a halo effect, but it was worked around by imaging at the periphery of the holes.

Data Processing and Model Building

The data-processing strategy was straightforward for the AI-2 exporters, and C5 symmetry was later used for refinements. The relaxing symmetry option can help to highlight certain differences, if any, among the protomers in such homo-oligomeric membrane proteins. Additionally, masking out the detergent belt and refining the transmembrane segments has also proven beneficial (Lee et al., 2019). The atomic models of the AI-2 exporters were *de novo* built in Coot from the initial models obtained from ArpWarp and Buccaneer from CCP4 (Chojnowski et al., 2021; Cowtan, 2006). Due to the lack of any homology or AlphaFold2 model at that time, *de novo* modeling of AI-2 exporter was time-consuming.

Homo-pentameric AI-2 Exporter Family

The pentamer oligomeric state is a rare occurrence among transporter families. It is a relatively common characteristic of membrane channels. To the best of our knowledge, the only other known pentameric transporter family is the formate/nitrite transporter (Lü et al., 2013). Each protomer of both TqsA and YdiK consists of six helical segments and two helical hairpins, and each protomer is observed to comprise two domains, N- and C-terminal domains (Figure 17.5e and f). The pentameric architecture of both structures (TqsA and YdiK) is unique, appearing as an inverted bowl facing the cytoplasm of the cell with a small cytoplasmic opening formed by helix 1 from each protomer connecting the cytoplasm to the periplasmic bowl (periplasmic basin) as shown in Figure 17.5. The structures give an illusion of a membrane channel, especially through the arrangement of helix 1 and 2 in the N-terminal domain. However, the C-terminal domain harbors a structural feature that is observed in other secondary active transporter families—helical hairpins (Figure 17.5f). Two transport mechanisms are often observed in secondary active transporters, rocking bundle and elevator-type (Garaeva & Slotboom, 2020). The transporter families with helical hairpins showed an elevator-type transport cycle, and a similar mechanism is proposed for the AI-2 exporter family as well (Garaeva & Slotboom, 2020). This proposal of the transport mechanism was further supported by a 6.4 Å map of the AI-2 exporter from *A. aeolicus* (Aq_740) (Khera et al., 2022).

Overall, this transporter family was discerned to possess many unique structural folds that were unseen before. Unfortunately, AI-2 was not recovered in the 3D reconstructions. Importantly, single-particle cryo-EM immensely contributed to its structural investigation as this family, like many other membrane proteins, was resistant to crystallization.

FURTHER STUDIES AND CONCLUSIONS

This chapter has provided an overview of the essential steps involved in determining the membrane protein structures via single-particle cryo-EM. The field is rapidly evolving, with further advancements in the algorithms of data processing software (RELION5 and 3D flex feature in CryoSPARC), development of voltage or laser phase plates for enhancing SNR in the cryo-EM images, and machine learning approaches illustrated by AlphaFold2 (Danev et al., 2014; Punjani & Fleet, 2022; Schwab et al., 2023; Schwartz et al., 2019). AlphaFold2 database has predicted the structures of all the proteins known to humankind and with its multimer platform is determining protein–protein interactions (Jumper et al., 2021). This has revolutionized the field of structural biology, particularly structure-based drug discovery.

The structures are snapshots of the proteins at certain time points while they are performing their functions within the cells. It is necessary to capture different conformation states of the same protein to get a full picture of the biological functioning of the protein. Most biological activities occur within fractions of seconds, which renders capturing other states challenging. Time-resolved studies with cryo-EM are feasible and are routinely used to study protein dynamics (Bodrug et al., 2023; Domínguez-Martín et al., 2022; Frank, 2017). This is possible with the use of different grid freezing techniques that permit the spray of substrates, inhibitors, or interacting partners right before freezing, which is performed within milliseconds (Kontziampasis et al., 2019). This is especially beneficial for studying membrane proteins with a high turnover rate. The advanced data processing algorithms further aid in classifying different states. There are several cryo-EM studies where they have obtained different snapshots of the proteins just by the addition of certain additives or illumination of the grids in the case of photosynthetic proteins (Domínguez-Martín et al., 2022; Hofmann et al., 2019; Kolbe et al., 2021; Wu et al., 2023). Time-resolved serial femtosecond crystallography is also working in the same direction to study protein dynamics (Pandey et al., 2020).

Undoubtedly, the structures of a membrane protein family provide tons of information about its functioning, but other biochemical and computational studies are vital to provide a more coherent picture. These include transport assays for membrane transporters, cell-based assays, interaction studies via surface plasmon resonance (SPR), microscale thermophoresis (MST), isothermal titration calorimetry (ITC), bio-layer interferometry (BLI), mutation or genetic studies, mass-spectrometry (MS) studies—such as cross-linking MS, hydrogen-deuterium exchange MS, lipidomic studies, molecular dynamics (MD) simulations, and many more. MD simulations beautifully complement the experimental structural and functional studies of membrane proteins in furthering our understanding of the biological function and working of the proteins in cellular environments (Domínguez-Martín et al., 2022; Hummer & Wikström, 2016; Stansfeld & Sansom, 2011; Wu et al., 2023).

This chapter celebrates the contributions made by single-particle cryo-EM in the field of membrane biology, discusses the rate-limiting steps in the workflow, and attempts to provide guidelines for overcoming the hiccups on the way. Structures of GPCRs, ion channels, and transporters are extremely beneficial for structure-based drug design platforms. The field is currently interested in and heading toward studying the membrane proteins *in situ* within their cellular environments. A combination of cryo-ET, correlative light and electron microscopy (CLEM), single-particle cryo-EM, and machine learning approaches is extremely promising. However, it has so far been restricted to large and abundant protein complexes. Its expansion to other protein families through technological improvements is highly desirable and will be a great leap forward.

ACKNOWLEDGMENTS

Most of the figures for this chapter are created with UCSF ChimeraX and bioRENDER (https://www.biorender.com/). The cryo-EM maps and coordinate files are obtained from public databases EMDB (https://www.ebi.ac.uk/emdb/) and RCSB PDB (https://www.rcsb.org/). The funding for this chapter is provided by the Howard Hughes Medical Institute, USA.

REFERENCES

Adrian, M., Dubochet, J., Lepault, J., & McDowall, A. W. (1984). Cryo-electron microscopy of viruses. *Nature*, *308*(5954), 32–36. https://doi.org/10.1038/308032a0

Afonine, P. V., Poon, B. K., Read, R. J., Sobolev, O. V., Terwilliger, T. C., Urzhumtsev, A., & Adams, P. D. (2018). Real-space refinement in PHENIX for cryo-EM and crystallography. *Acta Crystallographica. Section D, Structural Biology*, *74*(Pt 6), 531–544. https://doi.org/10.1107/S2059798318006551

Allen, J. P. (2019). Recent innovations in membrane-protein structural biology. *F1000Research*, *8*, 211. https://doi.org/10.12688/f1000research.16234.1

Angiulli, G., Dhupar, H. S., Suzuki, H., Wason, I. S., Duong Van Hoa, F., & Walz, T. (2020). New approach for membrane protein reconstitution into peptidiscs and basis for their adaptability to different proteins. *eLife*, *9*, e53530. https://doi.org/10.7554/eLife.53530

Baril, S. A., Gose, T., & Schuetz, J. D. (2023). How cryo-EM has expanded our understanding of membrane transporters. *Drug Metabolism and Disposition*, *51*(8), 904–922. https://doi.org/10.1124/dmd.122.001004

Bassler, B. L., & Losick, R. (2006). Bacterially speaking. *Cell*, *125*(2), 237–246. https://doi.org/10.1016/j.cell.2006.04.001

Bepler, T., Morin, A., Rapp, M., Brasch, J., Shapiro, L., Noble, A. J., & Berger, B. (2019). Positive-unlabeled convolutional neural networks for particle picking in cryo-electron micrographs. *Nature Methods*, *16*(11), 1153–1160. https://doi.org/10.1038/s41592-019-0575-8

Berman, H. M. (2000). The protein data bank. *Nucleic Acids Research*, *28*(1), 235–242. https://doi.org/10.1093/nar/28.1.235

Bodrug, T., Welsh, K. A., Bolhuis, D. L., Paulakonis, E., Martinez-Chacin, R. C., Liu, B., Pinkin, N., Bonacci, T., Cui, L., Xu, P., Roscow, O., Amann, S. J., Grishkovskaya, I., Emanuele, M. J., Harrison, J. S., Steimel, J. P., Hahn, K. M., Zhang, W., Zhong, E. D., … Brown, N. G. (2023). Time-resolved cryo-EM (TR-EM) analysis of substrate polyubiquitination by the RING E3 anaphase-promoting complex/cyclosome (APC/C). *Nature Structural & Molecular Biology*, *30*(11), 1663–1674. https://doi.org/10.1038/s41594-023-01105-5

Brilot, A. F., Chen, J. Z., Cheng, A., Pan, J., Harrison, S. C., Potter, C. S., Carragher, B., Henderson, R., & Grigorieff, N. (2012). Beam-induced motion of vitrified specimen on holey carbon film. *Journal of Structural Biology*, *177*(3), 630–637. https://doi.org/10.1016/j.jsb.2012.02.003

Brown, A., Long, F., Nicholls, R. A., Toots, J., Emsley, P., & Murshudov, G. (2015). Tools for macromolecular model building and refinement into electron cryo-microscopy reconstructions. *Acta Crystallographica. Section D, Biological Crystallography*, *71*(Pt 1), 136–153. https://doi.org/10.1107/S1399004714021683

Brown, Z. P., Arimori, T., Iwasaki, K., & Takagi, J. (2018). Development of a new protein labeling system to map subunits and domains of macromolecular complexes for electron microscopy. *Journal of Structural Biology*, *201*(3), 247–251. https://doi.org/10.1016/j.jsb.2017.11.006

Bunduc, C. M., Fahrenkamp, D., Wald, J., Ummels, R., Bitter, W., Houben, E. N. G., & Marlovits, T. C. (2021). Structure and dynamics of a mycobacterial type VII secretion system. *Nature*, *593*(7859), 445–448. https://doi.org/10.1038/s41586-021-03517-z

Chen, J., Noble, A. J., Kang, J. Y., & Darst, S. A. (2019). Eliminating effects of particle adsorption to the air/water interface in single-particle cryo-electron microscopy: Bacterial RNA polymerase and CHAPSO. *Journal of Structural Biology: X*, *1*, 100005. https://doi.org/10.1016/j.yjsbx.2019.100005

Cheng, Y. (2018). Single-particle cryo-EM-How did it get here and where will it go. *Science (New York, N.Y.)*, *361*(6405), 876–880. https://doi.org/10.1126/science.aat4346

Chojnowski, G., Sobolev, E., Heuser, P., & Lamzin, V. S. (2021). The accuracy of protein models automatically built into cryo-EM maps with *ARP/wARP*. *Acta Crystallographica Section D Structural Biology*, *77*(2), 142–150. https://doi.org/10.1107/S2059798320016332

Choy, B. C., Cater, R. J., Mancia, F., & Pryor, E. E. (2021). A 10-year meta-analysis of membrane protein structural biology: Detergents, membrane mimetics, and structure determination techniques. *Biochimica et Biophysica Acta (BBA) - Biomembranes*, *1863*(3), 183533. https://doi.org/10.1016/j.bbamem.2020.183533

Cowtan, K. (2006). The *Buccaneer* software for automated model building. 1. Tracing protein chains. *Acta Crystallographica Section D Biological Crystallography*, *62*(9), 1002–1011. https://doi.org/10.1107/S0907444906022116

Cressey, D., & Callaway, E. (2017). Cryo-electron microscopy wins chemistry Nobel. *Nature*, *550*(7675), 167–167. https://doi.org/10.1038/nature.2017.22738

Croll, T. I. (2018). ISOLDE: A physically realistic environment for model building into low-resolution electron-density maps. *Acta Crystallographica. Section D, Structural Biology*, *74*(Pt 6), 519–530. https://doi.org/10.1107/S2059798318002425

Danev, R., Buijsse, B., Khoshouei, M., Plitzko, J. M., & Baumeister, W. (2014). Volta potential phase plate for in-focus phase contrast transmission electron microscopy. *Proceedings of the National Academy of Sciences*, *111*(44), 15635–15640. https://doi.org/10.1073/pnas.1418377111

Deisenhofer, J., Epp, O., Miki, K., Huber, R., & Michel, H. (1985). Structure of the protein subunits in the photosynthetic reaction centre of Rhodopseudomonas viridis at 3Å resolution. *Nature*, *318*(6047), 618–624. https://doi.org/10.1038/318618a0

D'Imprima, E., Floris, D., Joppe, M., Sánchez, R., Grininger, M., & Kühlbrandt, W. (2019). Protein denaturation at the air-water interface and how to prevent it. *eLife*, *8*, e42747. https://doi.org/10.7554/eLife.42747

Dolan, K. A., Dutta, M., Kern, D. M., Kotecha, A., Voth, G. A., & Brohawn, S. G. (2022). Structure of SARS-CoV-2 M protein in lipid nanodiscs. *eLife*, *11*, e81702. https://doi.org/10.7554/eLife.81702

Domínguez-Martín, M. A., Sauer, P. V., Kirst, H., Sutter, M., Bína, D., Greber, B. J., Nogales, E., Polívka, T., & Kerfeld, C. A. (2022). Structures of a phycobilisome in light-harvesting and photoprotected states. *Nature*, *609*(7928), 835–845. https://doi.org/10.1038/s41586-022-05156-4

Drulyte, I., Johnson, R. M., Hesketh, E. L., Hurdiss, D. L., Scarff, C. A., Porav, S. A., Ranson, N. A., Muench, S. P., & Thompson, R. F. (2018). Approaches to altering particle distributions in cryo-electron microscopy sample preparation. *Acta Crystallographica. Section D, Structural Biology*, *74*(Pt 6), 560–571. https://doi.org/10.1107/S2059798318006496

Emsley, P., Lohkamp, B., Scott, W. G., & Cowtan, K. (2010). Features and development of Coot. *Acta Crystallographica. Section D, Biological Crystallography*, *66*(Pt 4), 486–501. https://doi.org/10.1107/S0907444910007493

Frank, J. (2017). Time-resolved cryo-electron microscopy: Recent progress. *Journal of Structural Biology*, *200*(3), 303–306. https://doi.org/10.1016/j.jsb.2017.06.005

Galazzo, L., Meier, G., Januliene, D., Parey, K., De Vecchis, D., Striednig, B., Hilbi, H., Schäfer, L. V., Kuprov, I., Moeller, A., Bordignon, E., & Seeger, M. A. (2022). The ABC transporter MsbA adopts the wide inward-open conformation in *E. coli* cells. *Science Advances*, *8*(41), eabn6845. https://doi.org/10.1126/sciadv.abn6845

Garaeva, A. A., & Slotboom, D. J. (2020). Elevator-type mechanisms of membrane transport. *Biochemical Society Transactions*, *48*(3), 1227–1241. https://doi.org/10.1042/BST20200290

Gewering, T., Januliene, D., Ries, A. B., & Moeller, A. (2018). Know your detergents: A case study on detergent background in negative stain electron microscopy. *Journal of Structural Biology*, *203*(3), 242–246. https://doi.org/10.1016/j.jsb.2018.05.008

Glaeser, R. M., & Han, B.-G. (2019). Streptavidin affinity grids for single-particle cryo-EM. *Microscopy and Microanalysis*, *25*(S2), 990–991. https://doi.org/10.1017/S1431927619005683

Graça, A. T., Hall, M., Persson, K., & Schröder, W. P. (2021). High-resolution model of Arabidopsis Photosystem II reveals the structural consequences of digitonin-extraction. *Scientific Reports*, *11*(1), 15534. https://doi.org/10.1038/s41598-021-94914-x

Grant, T., Rohou, A., & Grigorieff, N. (2018). cisTEM, user-friendly software for single-particle image processing. *eLife*, *7*, e35383. https://doi.org/10.7554/eLife.35383

Guo, Y. (2021). Detergent-free systems for structural studies of membrane proteins. *Biochemical Society Transactions*, *49*(3), 1361–1374. https://doi.org/10.1042/BST20201080

Hauer, F., Gerle, C., Fischer, N., Oshima, A., Shinzawa-Itoh, K., Shimada, S., Yokoyama, K., Fujiyoshi, Y., & Stark, H. (2015). GraDeR: Membrane protein complex preparation for single-particle cryo-EM. *Structure (London, England: 1993)*, *23*(9), 1769–1775. https://doi.org/10.1016/j.str.2015.06.029

Henderson, R. (2013). Avoiding the pitfalls of single particle cryo-electron microscopy: Einstein from noise. *Proceedings of the National Academy of Sciences*, *110*(45), 18037–18041. https://doi.org/10.1073/pnas.1314449110

Henderson, R., Baldwin, J. M., Ceska, T. A., Zemlin, F., Beckmann, E., & Downing, K. H. (1990). Model for the structure of bacteriorhodopsin based on high-resolution electron cryo-microscopy. *Journal of Molecular Biology*, *213*(4), 899–929. https://doi.org/10.1016/S0022-2836(05)80271-2

Henderson, R., & Unwin, P. N. T. (1975). Three-dimensional model of purple membrane obtained by electron microscopy. *Nature*, *257*(5521), 28–32. https://doi.org/10.1038/257028a0

Hennell James, R., Deme, J. C., Kjær, A., Alcock, F., Silale, A., Lauber, F., Johnson, S., Berks, B. C., & Lea, S. M. (2021). Structure and mechanism of the proton-driven motor that powers type 9 secretion and gliding motility. *Nature Microbiology*, *6*(2), 221–233. https://doi.org/10.1038/s41564-020-00823-6

Herzberg, M., Kaye, I. K., Peti, W., & Wood, T. K. (2006). YdgG (TqsA) controls biofilm formation in *Escherichia coli* K-12 through autoinducer 2 transport. *Journal of Bacteriology*, *188*(2), 587–598. https://doi.org/10.1128/JB.188.2.587-598.2006

Hofmann, S., Januliene, D., Mehdipour, A. R., Thomas, C., Stefan, E., Brüchert, S., Kuhn, B. T., Geertsma, E. R., Hummer, G., Tampé, R., & Moeller, A. (2019). Conformation space of a heterodimeric ABC exporter under turnover conditions. *Nature*, *571*(7766), 580–583. https://doi.org/10.1038/s41586-019-1391-0

Holm, L., & Rosenström, P. (2010). Dali server: Conservation mapping in 3D. *Nucleic Acids Research*, *38*(suppl_2), W545–W549. https://doi.org/10.1093/nar/gkq366

Hummer, G., & Wikström, M. (2016). Molecular simulation and modeling of complex I. *Biochimica Et Biophysica Acta*, *1857*(7), 915–921. https://doi.org/10.1016/j.bbabio.2016.01.005

Ismail, A. S., Valastyan, J. S., & Bassler, B. L. (2016). A host-produced autoinducer-2 mimic activates bacterial quorum sensing. *Cell Host & Microbe*, *19*(4), 470–480. https://doi.org/10.1016/j.chom.2016.02.020

Jamali, K., Käll, L., Zhang, R., Brown, A., Kimanius, D., & Scheres, S. H. W. (2023). *Automated model building and protein identification in cryo-EM maps* [Preprint]. Molecular Biology. https://doi.org/10.1101/2023.05.16.541002

Jumper, J., Evans, R., Pritzel, A., Green, T., Figurnov, M., Ronneberger, O., Tunyasuvunakool, K., Bates, R., Žídek, A., Potapenko, A., Bridgland, A., Meyer, C., Kohl, S. A. A., Ballard, A. J., Cowie, A., Romera-Paredes, B., Nikolov, S., Jain, R., Adler, J., … Hassabis, D. (2021). Highly accurate protein structure prediction with AlphaFold. *Nature*, *596*(7873), 583–589. https://doi.org/10.1038/s41586-021-03819-2

Kampjut, D., Steiner, J., & Sazanov, L. A. (2021). Cryo-EM grid optimization for membrane proteins. *iScience*, *24*(3), 102139. https://doi.org/10.1016/j.isci.2021.102139

Kelly, D. F., Abeyrathne, P. D., Dukovski, D., & Walz, T. (2008). The affinity grid: A pre-fabricated EM grid for monolayer purification. *Journal of Molecular Biology*, *382*(2), 423–433. https://doi.org/10.1016/j.jmb.2008.07.023

Khera, R. (2023). *Structural and functional characterization of members of the AI-2 exporter family*. Biochemie, chemie, und pharmazie der Johann Wolfgang Goethe-Universität.

Khera, R., Mehdipour, A. R., Bolla, J. R., Kahnt, J., Welsch, S., Ermler, U., Muenke, C., Robinson, C. V., Hummer, G., Xie, H., & Michel, H. (2022). Cryo-EM structures of pentameric autoinducer-2 exporter from *Escherichia coli* reveal its transport mechanism. *The EMBO Journal*, *41*(18), e109990. https://doi.org/10.15252/embj.2021109990

Knowles, T. J., Finka, R., Smith, C., Lin, Y.-P., Dafforn, T., & Overduin, M. (2009). Membrane proteins solubilized intact in lipid containing nanoparticles bounded by styrene maleic acid copolymer. *Journal of the American Chemical Society*, *131*(22), 7484–7485. https://doi.org/10.1021/ja810046q

Kolbe, F., Safarian, S., Piórek, Ż., Welsch, S., Müller, H., & Michel, H. (2021). Cryo-EM structures of intermediates suggest an alternative catalytic reaction cycle for cytochrome c oxidase. *Nature Communications*, *12*(1), 6903. https://doi.org/10.1038/s41467-021-27174-y

Kontziampasis, D., Klebl, D. P., Iadanza, M. G., Scarff, C. A., Kopf, F., Sobott, F., Monteiro, D. C. F., Trebbin, M., Muench, S. P., & White, H. D. (2019). A cryo-EM grid preparation device for time-resolved structural studies. *IUCrJ*, *6*(Pt 6), 1024–1031. https://doi.org/10.1107/S2052252519011345

Kühlbrandt, W. (2013). Introduction to electron crystallography. *Methods in Molecular Biology (Clifton, N.J.), 955*, 1–16. https://doi.org/10.1007/978-1-62703-176-9_1

Kühlbrandt, W. (2014). The resolution revolution. *Science, 343*(6178), 1443–1444. https://doi.org/10.1126/science.1251652

Kühlbrandt, W. (2022). Forty years in cryoEM of membrane proteins. *Microscopy, 71*(Supplement_1), i30–i50. https://doi.org/10.1093/jmicro/dfab041

Le Bon, C., Michon, B., Popot, J.-L., & Zoonens, M. (2021). Amphipathic environments for determining the structure of membrane proteins by single-particle electron cryo-microscopy. *Quarterly Reviews of Biophysics, 54*, e6. https://doi.org/10.1017/S0033583521000044

Lee, Y., Wiriyasermkul, P., Jin, C., Quan, L., Ohgaki, R., Okuda, S., Kusakizako, T., Nishizawa, T., Oda, K., Ishitani, R., Yokoyama, T., Nakane, T., Shirouzu, M., Endou, H., Nagamori, S., Kanai, Y., & Nureki, O. (2019). Cryo-EM structure of the human L-type amino acid transporter 1 in complex with glycoprotein CD98hc. *Nature Structural & Molecular Biology, 26*(6), 510–517. https://doi.org/10.1038/s41594-019-0237-7

Li, X., Mooney, P., Zheng, S., Booth, C. R., Braunfeld, M. B., Gubbens, S., Agard, D. A., & Cheng, Y. (2013). Electron counting and beam-induced motion correction enable near-atomic-resolution single-particle cryo-EM. *Nature Methods, 10*(6), 584–590. https://doi.org/10.1038/nmeth.2472

Liang, B., & Tamm, L. K. (2016). NMR as a tool to investigate the structure, dynamics and function of membrane proteins. *Nature Structural & Molecular Biology, 23*(6), 468–474. https://doi.org/10.1038/nsmb.3226

Liao, M., Cao, E., Julius, D., & Cheng, Y. (2013). Structure of the TRPV1 ion channel determined by electron cryo-microscopy. *Nature, 504*(7478), 107–112. https://doi.org/10.1038/nature12822

Lü, W., Du, J., Schwarzer, N. J., Wacker, T., Andrade, S. L. A., & Einsle, O. (2013). The formate/nitrite transporter family of anion channels. *Biological Chemistry, 394*(6), 715–727. https://doi.org/10.1515/hsz-2012-0339

Marinko, J. T., Huang, H., Penn, W. D., Capra, J. A., Schlebach, J. P., & Sanders, C. R. (2019). Folding and misfolding of human membrane proteins in health and disease: From single molecules to cellular proteostasis. *Chemical Reviews, 119*(9), 5537–5606. https://doi.org/10.1021/acs.chemrev.8b00532

McMahon, C., Baier, A. S., Pascolutti, R., Wegrecki, M., Zheng, S., Ong, J. X., Erlandson, S. C., Hilger, D., Rasmussen, S. G. F., Ring, A. M., Manglik, A., & Kruse, A. C. (2018). Yeast surface display platform for rapid discovery of conformationally selective nanobodies. *Nature Structural & Molecular Biology, 25*(3), 289–296. https://doi.org/10.1038/s41594-018-0028-6

McMullan, G., Faruqi, A. R., & Henderson, R. (2016). Direct electron detectors. *Methods in Enzymology, 579*, 1–17. https://doi.org/10.1016/bs.mie.2016.05.056

Nealson, K. H., Platt, T., & Hastings, J. W. (1970). Cellular control of the synthesis and activity of the bacterial luminescent system. *Journal of Bacteriology, 104*(1), 313–322. https://doi.org/10.1128/jb.104.1.313-322.1970

Ostermeier, C., & Michel, H. (1997). Crystallization of membrane proteins. *Current Opinion in Structural Biology, 7*(5), 697–701. https://doi.org/10.1016/s0959-440x(97)80080-2

Overington, J. P., Al-Lazikani, B., & Hopkins, A. L. (2006). How many drug targets are there? *Nature Reviews Drug Discovery, 5*(12), 993–996. https://doi.org/10.1038/nrd2199

Pandey, A., Shin, K., Patterson, R. E., Liu, X.-Q., & Rainey, J. K. (2016). Current strategies for protein production and purification enabling membrane protein structural biology. *Biochemistry and Cell Biology, 94*(6), 507–527. https://doi.org/10.1139/bcb-2015-0143

Pandey, S., Bean, R., Sato, T., Poudyal, I., Bielecki, J., Cruz Villarreal, J., Yefanov, O., Mariani, V., White, T. A., Kupitz, C., Hunter, M., Abdellatif, M. H., Bajt, S., Bondar, V., Echelmeier, A., Doppler, D., Emons, M., Frank, M., Fromme, R., … Schmidt, M. (2020). Time-resolved serial femtosecond crystallography at the European XFEL. *Nature Methods, 17*(1), 73–78. https://doi.org/10.1038/s41592-019-0628-z

Papenfort, K., & Bassler, B. L. (2016). Quorum sensing signal–response systems in Gram-negative bacteria. *Nature Reviews Microbiology, 14*(9), 576–588. https://doi.org/10.1038/nrmicro.2016.89

Parker, J. L., Deme, J. C., Kuteyi, G., Wu, Z., Huo, J., Goldman, I. D., Owens, R. J., Biggin, P. C., Lea, S. M., & Newstead, S. (2021). Structural basis of antifolate recognition and transport by PCFT. *Nature, 595*(7865), 130–134. https://doi.org/10.1038/s41586-021-03579-z

Pereira, C. S., Thompson, J. A., & Xavier, K. B. (2013). AI-2-mediated signalling in bacteria. *FEMS Microbiology Reviews, 37*(2), 156–181. https://doi.org/10.1111/j.1574-6976.2012.00345.x

Piper, S. J., Johnson, R. M., Wootten, D., & Sexton, P. M. (2022). Membranes under the magnetic lens: A dive into the diverse world of membrane protein structures using cryo-EM. *Chemical Reviews, 122*(17), 13989–14017. https://doi.org/10.1021/acs.chemrev.1c00837

Popot, J.-L., Althoff, T., Bagnard, D., Banères, J.-L., Bazzacco, P., Billon-Denis, E., Catoire, L. J., Champeil, P., Charvolin, D., Cocco, M. J., Crémel, G., Dahmane, T., De La Maza, L. M., Ebel, C., Gabel, F., Giusti, F., Gohon, Y., Goormaghtigh, E., Guittet, E., … Zoonens, M. (2011). Amphipols From A to Z. *Annual Review of Biophysics, 40*(1), 379–408. https://doi.org/10.1146/annurev-biophys-042910-155219

Punjani, A., & Fleet, D. (2022). 3D flexible refinement: structure and motion of flexible proteins from cryo-EM. *Microscopy and Microanalysis, 28*(S1), 1218–1218. https://doi.org/10.1017/S1431927622005074

Punjani, A., Rubinstein, J. L., Fleet, D. J., & Brubaker, M. A. (2017). cryoSPARC: Algorithms for rapid unsupervised cryo-EM structure determination. *Nature Methods, 14*(3), 290–296. https://doi.org/10.1038/nmeth.4169

Rettner, R. E., & Saier Jr., M. H. (2010). The Autoinducer-2 Exporter Superfamily. Journal of *Molecular Microbiology and Biotechnology 18*(4), 195–205. https://doi.org/10.1159/000316420

Russo, C. J., & Passmore, L. A. (2014). Ultrastable gold substrates for electron cryomicroscopy. *Science, 346*(6215), 1377–1380. https://doi.org/10.1126/science.1259530

Rutherford, S. T., & Bassler, B. L. (2012). Bacterial quorum sensing: Its role in virulence and possibilities for its control. *Cold Spring Harbor Perspectives in Medicine, 2*(11), a012427. https://doi.org/10.1101/cshperspect.a012427

Saier, M. H., Reddy, V. S., Moreno-Hagelsieb, G., Hendargo, K. J., Zhang, Y., Iddamsetty, V., Lam, K. J. K., Tian, N., Russum, S., Wang, J., & Medrano-Soto, A. (2021). The Transporter Classification Database (TCDB): 2021 update. *Nucleic Acids Research, 49*(D1), D461–D467. https://doi.org/10.1093/nar/gkaa1004

Sanders, C. R., & Myers, J. K. (2004). Disease-related misassembly of membrane proteins. *Annual Review of Biophysics and Biomolecular Structure, 33*(1), 25–51. https://doi.org/10.1146/annurev.biophys.33.110502.140348

Sanders, C. R., & Nagy, J. K. (2000). Misfolding of membrane proteins in health and disease: The lady or the tiger? *Current Opinion in Structural Biology*, *10*(4), 438–442. https://doi.org/10.1016/S0959-440X(00)00112-3

Scheres, S. H. W. (2012). RELION: Implementation of a Bayesian approach to cryo-EM structure determination. *Journal of Structural Biology*, *180*(3), 519–530. https://doi.org/10.1016/j.jsb.2012.09.006

Scheres, S. H. W., & Chen, S. (2012). Prevention of overfitting in cryo-EM structure determination. *Nature Methods*, *9*(9), 853–854. https://doi.org/10.1038/nmeth.2115

Schwab, J., Kimanius, D., Burt, A., Dendooven, T., & Scheres, S. H. W. (2023). *DynaMight: Estimating molecular motions with improved reconstruction from cryo-EM images* [Preprint]. Biophysics. https://doi.org/10.1101/2023.10.18.562877

Schwartz, O., Axelrod, J. J., Campbell, S. L., Turnbaugh, C., Glaeser, R. M., & Müller, H. (2019). Laser phase plate for transmission electron microscopy. *Nature Methods*, *16*(10), 1016–1020. https://doi.org/10.1038/s41592-019-0552-2

Schwarz, D., Dötsch, V., & Bernhard, F. (2008). Production of membrane proteins using cell-free expression systems. *PROTEOMICS*, *8*(19), 3933–3946. https://doi.org/10.1002/pmic.200800171

Shi, Y. (2013). Common folds and transport mechanisms of secondary active transporters. *Annual Review of Biophysics*, *42*, 51–72. https://doi.org/10.1146/annurev-biophys-083012-130429

Silpe, J. E., Duddy, O. P., & Bassler, B. L. (2022). Natural and synthetic inhibitors of a phage-encoded quorum-sensing receptor affect phage–host dynamics in mixed bacterial communities. *Proceedings of the National Academy of Sciences*, *119*(49), e2217813119. https://doi.org/10.1073/pnas.2217813119

Sligar, S. G., & Denisov, I. G. (2021). Nanodiscs: A toolkit for membrane protein science. *Protein Science*, *30*(2), 297–315. https://doi.org/10.1002/pro.3994

Sommer, M., Xie, H., & Michel, H. (2017). Pseudomonas stutzeri as an alternative host for membrane proteins. *Microbial Cell Factories*, *16*(1), 157. https://doi.org/10.1186/s12934-017-0771-0

Stansfeld, P. J., & Sansom, M. S. P. (2011). Molecular simulation approaches to membrane proteins. *Structure*, *19*(11), 1562–1572. https://doi.org/10.1016/j.str.2011.10.002

Stark, H. (2010). GraFix: Stabilization of fragile macromolecular complexes for single particle cryo-EM. In *Methods in Enzymology* (Vol. 481, pp. 109–126). Elsevier. https://doi.org/10.1016/S0076-6879(10)81005-5

Su, C.-C., Lyu, M., Morgan, C. E., Bolla, J. R., Robinson, C. V., & Yu, E. W. (2021). A 'Build and Retrieve' methodology to simultaneously solve cryo-EM structures of membrane proteins. *Nature Methods*, *18*(1), 69–75. https://doi.org/10.1038/s41592-020-01021-2

Sun, C., Benlekbir, S., Venkatakrishnan, P., Wang, Y., Hong, S., Hosler, J., Tajkhorshid, E., Rubinstein, J. L., & Gennis, R. B. (2018). Structure of the alternative complex III in a supercomplex with cytochrome oxidase. *Nature*, *557*(7703), 123–126. https://doi.org/10.1038/s41586-018-0061-y

Tang, G., Peng, L., Baldwin, P. R., Mann, D. S., Jiang, W., Rees, I., & Ludtke, S. J. (2007). EMAN2: An extensible image processing suite for electron microscopy. *Journal of Structural Biology*, *157*(1), 38–46. https://doi.org/10.1016/j.jsb.2006.05.009

Tao, X., Zhao, C., & MacKinnon, R. (2023). Membrane protein isolation and structure determination in cell-derived membrane vesicles. *Proceedings of the National Academy of Sciences*, *120*(18), e2302325120. https://doi.org/10.1073/pnas.2302325120

Thangaratnarajah, C., Rheinberger, J., & Paulino, C. (2022). Cryo-EM studies of membrane proteins at 200 keV. *Current Opinion in Structural Biology*, *76*, 102440. https://doi.org/10.1016/j.sbi.2022.102440

Thonghin, N., Kargas, V., Clews, J., & Ford, R. C. (2018). Cryo-electron microscopy of membrane proteins. *Methods*, *147*, 176–186. https://doi.org/10.1016/j.ymeth.2018.04.018

Tomasz, A. (1965). Control of the competent state in Pneumococcus by a hormone-like cell product: An example for a new type of regulatory mechanism in bacteria. *Nature*, *208*(5006), 155–159. https://doi.org/10.1038/208155a0

Uchański, T., Masiulis, S., Fischer, B., Kalichuk, V., López-Sánchez, U., Zarkadas, E., Weckener, M., Sente, A., Ward, P., Wohlkönig, A., Zögg, T., Remaut, H., Naismith, J. H., Nury, H., Vranken, W., Aricescu, A. R., Pardon, E., & Steyaert, J. (2021). Megabodies expand the nanobody toolkit for protein structure determination by single-particle cryo-EM. *Nature Methods*, *18*(1), 60–68. https://doi.org/10.1038/s41592-020-01001-6

Uhlén, M., Fagerberg, L., Hallström, B. M., Lindskog, C., Oksvold, P., Mardinoglu, A., Sivertsson, Å., Kampf, C., Sjöstedt, E., Asplund, A., Olsson, I., Edlund, K., Lundberg, E., Navani, S., Szigyarto, C. A.-K., Odeberg, J., Djureinovic, D., Takanen, J. O., Hober, S., … Pontén, F. (2015). Tissue-based map of the human proteome. *Science*, *347*(6220), 1260419. https://doi.org/10.1126/science.1260419

Wagner, T., Merino, F., Stabrin, M., Moriya, T., Antoni, C., Apelbaum, A., Hagel, P., Sitsel, O., Raisch, T., Prumbaum, D., Quentin, D., Roderer, D., Tacke, S., Siebolds, B., Schubert, E., Shaikh, T. R., Lill, P., Gatsogiannis, C., & Raunser, S. (2019). SPHIRE-crYOLO is a fast and accurate fully automated particle picker for cryo-EM. *Communications Biology*, *2*(1), 218. https://doi.org/10.1038/s42003-019-0437-z

Weissenberger, G., Henderikx, R. J. M., & Peters, P. J. (2021). Understanding the invisible hands of sample preparation for cryo-EM. *Nature Methods*, *18*(5), 463–471. https://doi.org/10.1038/s41592-021-01130-6

Wu, D., Mehdipour, A. R., Finke, F., Goojani, H. G., Groh, R. R., Grund, T. N., Reichhart, T. M. B., Zimmermann, R., Welsch, S., Bald, D., Shepherd, M., Hummer, G., & Safarian, S. (2023). Dissecting the conformational complexity and mechanism of a bacterial heme transporter. *Nature Chemical Biology*, *19*(8), 992–1003. https://doi.org/10.1038/s41589-023-01314-5

Wu, X., & Rapoport, T. A. (2021). Cryo-EM structure determination of small proteins by nanobody-binding scaffolds (Legobodies). *Proceedings of the National Academy of Sciences*, *118*(41), e2115001118. https://doi.org/10.1073/pnas.2115001118

Xie, H., Lyratzakis, A., Khera, R., Koutantou, M., Welsch, S., Michel, H., & Tsiotis, G. (2023). Cryo-EM structure of the whole photosynthetic reaction center apparatus from the green sulfur bacterium *Chlorobaculum tepidum*. *Proceedings of the National Academy of Sciences*, *120*(5), e2216734120. https://doi.org/10.1073/pnas.2216734120

Yin, H., & Flynn, A. D. (2016). Drugging membrane protein interactions. *Annual Review of Biomedical Engineering*, *18*(1), 51–76. https://doi.org/10.1146/annurev-bioeng-092115-025322

Yu, G., Li, K., & Jiang, W. (2016). Antibody-based affinity cryo-EM grid. *Methods*, *100*, 16–24. https://doi.org/10.1016/j.ymeth.2016.01.010

Zaucha, J., Heinzinger, M., Kulandaisamy, A., Kataka, E., Salvádor, Ó. L., Popov, P., Rost, B., Gromiha, M. M., Zhorov, B. S., & Frishman, D. (2021). Mutations in transmembrane proteins: Diseases, evolutionary insights, prediction and comparison with globular proteins. *Briefings in Bioinformatics*, *22*(3), bbaa132. https://doi.org/10.1093/bib/bbaa132

Zhang, Z., Nomura, N., Muramoto, Y., Ekimoto, T., Uemura, T., Liu, K., Yui, M., Kono, N., Aoki, J., Ikeguchi, M., Noda, T., Iwata, S., Ohto, U., & Shimizu, T. (2022). Structure of SARS-CoV-2 membrane protein essential for virus assembly. *Nature Communications*, *13*(1), 4399. https://doi.org/10.1038/s41467-022-32019-3

Zimmermann, I., Egloff, P., Hutter, C. A. J., Kuhn, B. T., Bräuer, P., Newstead, S., Dawson, R. J. P., Geertsma, E. R., & Seeger, M. A. (2020). Generation of synthetic nanobodies against delicate proteins. *Nature Protocols*, *15*(5), 1707–1741. https://doi.org/10.1038/s41596-020-0304-x

18 Rhodopsin Dimers in the Retinal Rod Outer Segment

Sahil Gulati and Krzysztof Palczewski

INTRODUCTION

G protein-coupled receptors (GPCRs) are a large family of cell membrane receptors that respond to various external signals, including light, ions, lipids, small molecules, peptides, and proteins (Mirzadegan et al., 2003; Mustafi and Palczewski, 2009). Although all GPCRs have a similar seven-transmembrane structure, maintained throughout evolution, their structural flexibility allows them to respond to a wide range of extracellular signals and affect multiple signaling pathways, depending on which G protein subtypes are bound to their respective intracellular binding sites. In addition, the discovery of noncanonical GPCR pathways and biased ligands add complexity to this scheme. This versatility has made GPCRs attractive targets for drug development and has opened possibilities for treating a wide range of diseases (Hauser et al., 2017).

GPCRs are encoded as single polypeptide chains that have an extracellular N-terminus, an intracellular C-terminus, and seven transmembrane regions (TM1–TM7) linked by three extracellular loops (ECL1-3) and three intracellular loops (ICL1-3) (Okada and Palczewski, 2001; Palczewski et al., 2000; Rasmussen et al., 2007). The rhodopsin-like (Class-A) receptor family has a structurally conserved amphipathic helical motif (helix 8) immediately after TM7 that is responsible for PDZ-domain recognition (Palczewski, 2006). The transmembrane regions arrange themselves in the plasma membrane to form a TM core that generally encompasses the ligand binding domain (Gulati et al., 2017). GPCRs share the most homology within the TM core, whereas their N-terminus and C-terminus (Kobilka, 2007), ECL2 (Peeters et al., 2011), and ICL3 (Flock et al., 2017) are the most divergent regions among different GPCR families. The intracellular region and the cytoplasmic ends of the TM core bind to downstream-signaling effector molecules such as G proteins, GPCR kinases, and arrestins Latek et al., 2012; Palczewski, 1994; Palczewski and Saari, 1997; Ridge et al., 2003). Therefore, the variability in ICLs plays a crucial role in defining effector specificity and downstream signaling events.

GPCRs can bind to various G protein subtypes, such as Gs, Gi/o, Gq, and G12/13, to set in motion specific physiological functions, including taste, vision, and synaptic transmission (Flock et al., 2017). While many studies have investigated the structure and dynamics of GPCRs as monomers, there is growing interest in the role of dimerization and oligomerization in influencing GPCR function (Liu et al., 2022; Park et al., 2004; Terrillon and Bouvier, 2004). GPCRs can form homo/heterodimers and higher-order oligomers, indicating that GPCRs may not function only as monomers (Patil et al., 2022; Yue et al., 2022; Zhao et al., 2019). This idea was first suggested by kinetic binding assays performed by Limbird et al. on frog erythrocyte membranes, which showed negative cooperativity between β-adrenergic receptor monomers (Limbird et al., 1975). Subsequent studies using various techniques such as binding assays, gel electrophoresis, immunoaffinity chromatography, chemical cross-linking, and co-immunoprecipitation have provided further evidence for this phenomenon (Ng et al., 2012). Biophysical methods such as single-molecule fluorescence, X-ray crystallography, NMR spectroscopy, and cryo-electron microscopy (cryo-EM), as well as computational methods, have provided explicit evidence for GPCR dimers and oligomers (Kasai et al., 2018; Moller et al., 2020; Patil et al., 2022; Pin and Bettler, 2016; Stenkamp, 2018; Yang et al., 2020; Yue et al., 2022; Zhao et al., 2019).

GPCR DIMERIZATION

The dimerization of GPCRs has been shown to modulate their affinity, specificity, and efficacy of ligand binding and to alter the way they activate signaling pathways (Terrillon and Bouvier, 2004). Furthermore, the conformational changes at the intradimeric GPCR interface transmit additional information to either enhance or weaken the downstream signaling activity (Dijkman et al., 2018; Herrick-Davis et al., 2005; Pellissier et al., 2011).

GPCR dimers can form through several mechanisms, including homodimerization, where two identical GPCRs associate, and heterodimerization, where two different GPCRs interact (Levoye et al., 2006). In the case of homodimerization, two protomers of the same GPCR can associate at the transmembrane (TM) domain, the intracellular domain (ICD) (Ciruela et al., 2004; Navarro et al., 2010; O'Dowd et al., 2012), or the extracellular domain (ECD) (Geng et al., 2013; Kniazeff et al., 2011; Kunishima et al., 2000; Levitz et al., 2016), with the TM domain being the best studied (Getter et al., 2019; Jastrzebska et al., 2015; Patil et al., 2022; Yue et al., 2022; Zhao et al., 2019). The TM domain–dimerization interface has been shown to involve TM1 and H8 or TM4–6 and the cytoplasmic tail of the receptor, and the interaction between the two receptor protomers has been shown to be regulated by a variety of factors, including phosphorylation and G protein interaction

DOI: 10.1201/9781003326106-22

(Berthouze et al., 2007; Guo et al., 2005, 2008; Hebert et al., 1996; Klco et al., 2003; Kota et al., 2006; Zhao et al., 2019). In the case of heterodimerization, two different GPCR monomers can be associated through a variety of mechanisms, including protein–protein interactions, hemidimerization, and domain swapping (Geng et al., 2013; Kniazeff et al., 2011; Kunishima et al., 2000; Levitz et al., 2016).

The effects of GPCR dimerization on signaling are complex and depend on the specific dimers and the signaling pathways involved. For example, the heterodimerization of the β_2-adrenergic receptor (β_2AR) with the β_1-adrenergic receptor results in an increase in cAMP signaling and an increase in the specificity of ligand binding (Zhu et al., 2005). Similarly, the heterodimerization of the dopamine D2 receptor with the dopamine D3 receptor has been shown to increase the specificity of ligand binding and to result in increased signaling through the MAPK pathway (Maggio and Millan, 2010; Scarselli et al., 2001).

The role of GPCR dimerization in disease is also becoming increasingly clear. For example, the homo- and heterodimerization of the chemokine receptors CXCR4 and CCR7 have been shown to play a role in the development of cancer and other diseases (Poltavets et al., 2021; Wang et al., 2006). Heterodimerization between prostaglandin E receptors and β_2ARs results in the uncoupling of β_2AR-mediated signaling and a diminished bronchodilator response to β_2AR agonists in airway smooth muscle cells (McGraw et al., 2006). In addition, heterodimerization of the β_2AR with mu-opioid receptors (MORs) plays a role in potentiating the aggressiveness of proliferation of human breast cancer cells by activating the glycogen synthase kinase 3 signaling (Rousseau et al., 2022). The discovery of GPCR dimerization has led to new strategies for drug discovery. For example, developing drugs that target specific GPCR dimers or oligomers rather than individual GPCRs has the potential to result in more specific and effective therapies (Akgun et al., 2013; Bhushan et al., 2004; Daniels et al., 2005; Karamitri et al., 2019; Le Naour et al., 2013; Lenard et al., 2007; Yuan et al., 2013).

Rhodopsin is the only GPCR for which the oligomeric structure has been identified using atomic force microscopy (AFM), which provided the first physiological evidence for GPCR oligomerization (Fotiadis et al., 2003; Liang et al., 2003). Similar results were obtained from native samples (Liang et al., 2003) and transmission electron microscopy, showing the rhodopsin dimer interacting with a single Gt molecule (Jastrzebska et al., 2013b). A growing body of evidence indicates that many class-A GPCRs other than rhodopsin can form concentration-dependent dimers and oligomers through transient hydrophobic interactions between the TM and extracellular domains. Studies with a variety of class-A GPCRs have documented a symmetrical intradimeric interface involving TM1 and H8 or TM4–6. Several examples of a variety of such intradimeric interfaces have been provided by X-ray crystallographic structures, including those of the neurotensin 1 receptor (Dijkman et al., 2018; Harding et al., 2009), the D_1/D_2-dopamine receptor dimer (Guo et al., 2008; Lee et al., 2004), the β_1/β_2-adrenergic receptor dimer (Dorsch et al., 2009; Hebert et al., 1996), and the C5a receptor (Klco et al., 2003). However, there are several challenges to the validity of interpreting the physiological significance of the observed dimeric interfaces, including potential crystal-packing artifacts, alterations due to protein engineering, or the use of high receptor concentrations during the crystallization process.

One distinctive feature of class-C GPCRs is that they form obligate dimers, either as homo- or heterodimers, which is essential for their function (Koehl et al., 2019a). The dimerization interfaces of class-C GPCRs are located in the TM domain as well in their large extracellular domain. Structural studies on metabotropic glutamate receptor (mGluR) (Koehl et al., 2019b; Nasrallah et al., 2021) and the γ-aminobutyric acid B receptor (GABABR) have confirmed the allosteric modulation by dimerization (Kim et al., 2020; Shen et al., 2021), further focusing attention on the relevance of dimerization in the effectuation of the physiological functions of GPCRs. Previous studies on the mGlu1 and mGlu2 receptors have led to conflicting models regarding the intersubunit-TM interfaces of their respective homodimers. While the crystal structure of mGlu1 indicates a sterol-rich TM1 interface (Wu et al., 2014), cross-linking studies of mGlu2 suggest an inactive state mediated by TM4 and TM5 that transitions to a TM6-driven interface in the presence of an agonist (Xue et al., 2015). Cryo-EM structures of the mGlu5 homodimer show that the TM domains are fully separated in the inactive state, with 21 Å between the TM5 helices of the two protomers, which are the closest transmembrane helix pair, while a TM6-driven interface is observed in the active state, with 6 Å between the TM6 domains of the two protomers (Jastrzebska et al., 2013b). Similarly, the GABABR dimeric interface is transient and changes between active (TM6) and inactive states (TM3 and TM5) (Kim et al., 2020; Koehl et al., 2019b). It remains to be confirmed whether there is a universal dimeric interface used by most GPCR dimers or if the dimerization strategies are specific to certain microenvironments and signaling pathways.

RHODOPSIN DIMERIZATION

Rhodopsin, found in the rod photoreceptor cells of the retina, is a highly light-sensitive pigment that is responsible for dim-light vision in humans (Palczewski et al., 2000). The rod cells are specialized, elongated neurons that are equipped to efficiently capture photons (Molday, 1998), and rhodopsin is located in a specialized organelle called the rod outer segment (ROS). The ROS is made up of about 1000 to 2000 stacks of bilayered disc membranes enclosed by the plasma membrane (Palczewski, 2006). Rhodopsin comprises over 90% of the total protein content of the ROS membranes (Gulati and Palczewski, 2023; Murray et al., 2009) and more than 50% of the disc total surface area (Gulati and Palczewski, 2023; Ridge and Palczewski, 2007); the remaining area is occupied by phospholipids, cholesterol, and other proteins necessary for vision (Filipek et al., 2004; Gulati and Palczewski, 2023). In the presence of light, rhodopsin's chromophore, 11-*cis*-retinal, absorbs a photon and isomerizes to all-*trans*-retinal (Gulati et al., 2017; Palczewski et al., 2000; Salom et al., 2006), causing a conformational change in the protein that initiates downstream signaling and vision (Figure 18.1). Rhodopsin is both a retinal-binding protein and a GPCR; its location in the eye makes it readily accessible to a variety of spectroscopy techniques. Moreover, it serves as a valuable prototype for the class-A GPCRs.

In recent years, much attention has been given to rhodopsin homodimerization (Gulati and Palczewski, 2023; Jastrzebska et al., 2006; Kumar et al., 2018; Milligan et al., 2019; Palczewski, 2010; Poge et al., 2021; Suda et al., 2004; Zhao et al., 2019). The precise organization of rhodopsin in the ROS is essential for proper signal transduction efficiency and function (Gulati and Palczewski, 2023; Palczewski, 2010; Poge et al., 2021). The initial structural evidence for rhodopsin dimerization came from pioneering AFM studies (Fotiadis et al., 2003, 2004), which revealed the dimeric arrangement of rhodopsin in its native environment (Figure 18.1b), making it the only GPCR whose native oligomeric state was known at the time. The AFM imaging technique demonstrated that rhodopsin assembles in rows of dimers, and this observation was confirmed by cryo-electron tomography of murine ROS in almost native conditions (Figure 18.1a) (Poge et al., 2021). The cryo-electron tomography reconstructions showed the proximity of the visual phototransduction proteins within a confined space, which is conducive to productive interactions among the key proteins involved in the propagation of the visual signal (Figure 18.1a) (Gulati and Palczewski, 2023; Poge et al., 2021). Despite a wealth of structural, biochemical, and biophysical data accumulated over the last few decades, the role of rhodopsin dimerization is still unclear, as both monomeric and dimeric structures of rhodopsin are fully capable of signal transduction. Engineered deficiency of dimerization does not lead to a complete loss of rhodopsin

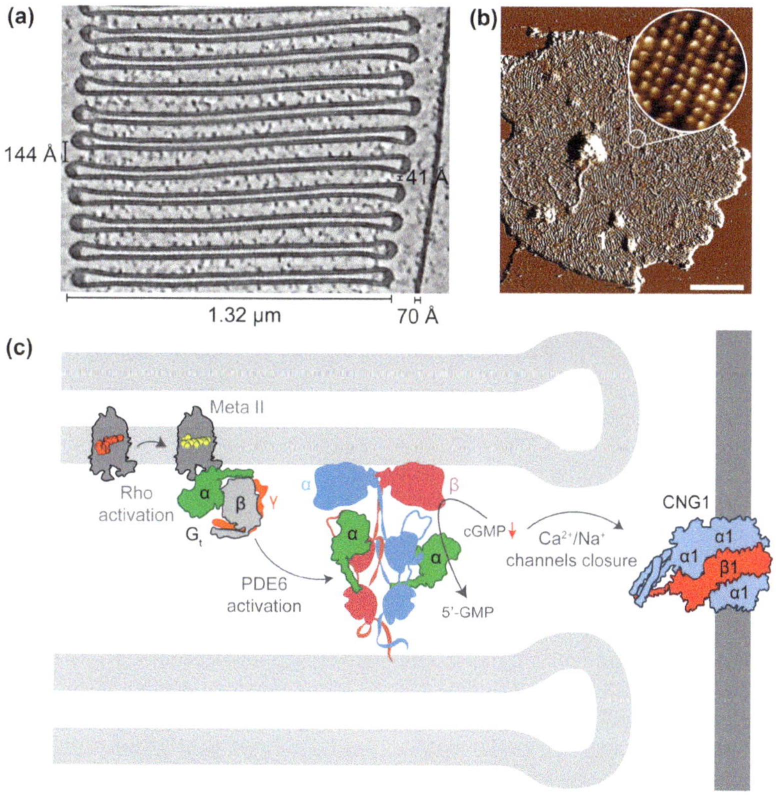

FIGURE 18.1 Organization of rhodopsin in rod outer segments (ROS) and the visual phototransduction pathway. (a) Z-slice of murine cryo-electron tomography dataset from (Poge et al., 2021), displaying distances between the ROS components. (b) Closely packed rhodopsin dimers in murine ROS, visualized with atomic force microscopy (adapted from Ref. (Fotiadis et al., 2003) and https://analyticalscience.wiley.com/do/10.1002/imaging.4681). The diameter of the round inset is 68 nm, and the scale bar is 250 nm. (c) Illustration showing steps involved in the visual phototransduction pathway. Light is detected when 11-*cis*-retinal absorbs a photon and isomerizes to all-*trans*-retinal, causing a conformational change in the protein moiety that activates transducin (Gt). Upon activation of Gt, the GTP-bound α-subunit of Gt (Gtα•GTP) dissociates from Gβγ and activates phosphodiesterase 6 (PDE6). Activated PDE6 causes the hydrolysis of cGMP to 5′-GMP, resulting in ion-channel closure, which reduces cation influx in rod cells, leading to their hyperpolarization.

functionality (Jastrzebska et al., 2015), but it could affect ROS disc formation, G protein activation, or rhodopsin deactivation *in vivo*. The inability to generate appropriate high-order rhodopsin structures in disc membranes could possibly affect the stability of ROS (Gunkel et al., 2015; Ploier et al., 2016). Various methodologies have been employed to understand the functional significance of rhodopsin dimerization, including *in silico* studies (Nemoto and Toh, 2005; Simpson et al., 2010), *in vitro* studies using synthetic peptides (Jastrzebska et al., 2015), biochemical or structural approaches (Getter et al., 2021; Jastrzebska et al., 2011, 2013a, 2013b), and *in vivo* studies (Macke et al., 1993; Ploier et al., 2016) targeting the rhodopsin dimer interface.

GPCR DIMERIZATION OBSERVED BY CRYO-EM

Cryo-EM has become a valuable tool in the study of GPCRs due to its ability to capture dynamic biological processes in near-native conditions, providing valuable insight into the structure and function of GPCRs (Garcia-Nafria and Tate, 2019; Gusach et al., 2023; Patil et al., 2022; Zhang et al., 2021). Unlike X-ray crystallography, cryo-EM does not require the formation of crystals, which can introduce artifacts such as nonphysiological conformations and packing interactions. By providing high-resolution structural information, cryo-EM has helped to elucidate the mechanisms of GPCR activation and signaling (Garcia-Nafria and Tate, 2019; Mao et al., 2020; Papasergi-Scott et al., 2023; Shen et al., 2022). Cryo-EM has enabled the visualization of GPCRs in complex with their signaling partners, such as G proteins (Gao et al., 2019; Kang et al., 2018; Tsai et al., 2019) and arrestins (Bous et al., 2022), providing a complete picture of the signaling pathways involved. Earlier crystal structures of several class-A GPCRs have shown different dimeric interfaces, such as β1AR with interface TM1–2 and H8 or with interface TM4–5 (Huang et al., 2013), MORs with interfaces at TM1–2 and H8 and TM5–TM6 (Manglik et al., 2012), and CXCR4 at TM5–6 (Wu et al., 2010). The previously discovered structure of rhodopsin dimer in nanodiscs using cryo-EM is another example of the versatility of the technique (Figure 18.2) (Zhao et al., 2019). The cryo-EM structure of the rhodopsin dimer clearly shows that the interaction between the two protomers within the dimer is facilitated by TM1 and H8 (Figure 18.2). This interface was supported in a recent crystal structure of light-activated bovine rhodopsin in lipidic cubic phase [PubMed: 36949205]. These results are also in line with previous studies, which suggested the involvement of TM1 and H8 in the dimer interface, such as mass spectral analysis of cysteine cross-linked rhodopsin in disc membranes (Knepp et al., 2012), and *in vivo* retinal peptide competition assays where peptides originated from rhodopsin domains were injected subretinally in mice to disrupt dimerization (Zhang et al., 2016). The intradimeric interface of rhodopsin encompasses a surface area of more than 300 Å^2, which is primarily composed of hydrophobic residues (Zhao et al., 2019). By comparison, the cryo-EM

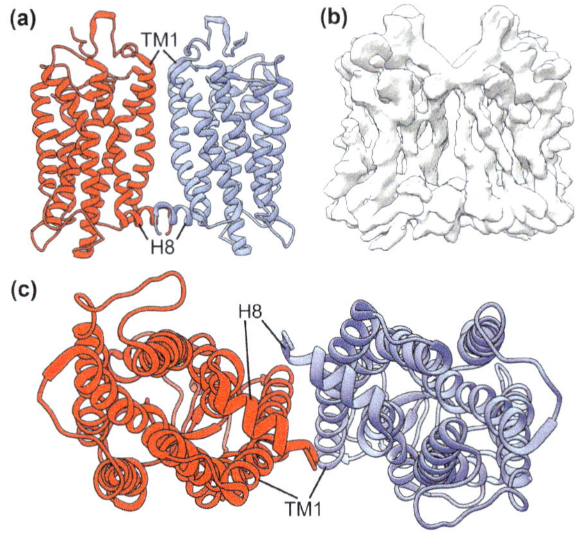

FIGURE 18.2 Rhodopsin dimer and its corresponding cryo-EM columbic map, as described in Zhao et al. (2019). (a) Rhodopsin dimer is formed by an intradimeric interface mediated by TM1 and H8 with a total interaction surface of over 300 Å^2, consisting mainly of hydrophobic residues. (b) Cryo-EM 3D reconstruction of rhodopsin dimer resolved at 4.5 Å resolution, displaying a tight packing between the two rhodopsin protomers. (c) Bottom view of rhodopsin dimer displaying an antiparallel organization between the H8 helices of the two protomers.

structure of the apelin receptor dimer in complex with Gi shows a dimeric interface consisting of TM3 and ICL1, with a buried surface area of 140 Å^2 (Yue et al., 2022). Although the apelin dimeric interface is composed of TM3 instead of TM1–2 as observed in rhodopsin, they both share a dimeric interface that is mainly governed by limited hydrophobic and π–π interactions (Yue et al., 2022; Zhao et al., 2019). This noncovalent bonding array suggests a dynamic and reversible interaction between the GPCR protomers, with a possibility of multiple interfaces of dimerization. The dimerization interface(s) could also be altered in the presence of G proteins to make room for two molecules of G protein binding with a GPCR dimer. The structure of a Ste2, a class-D GPCR, was previously resolved in complex with two molecules of G proteins using cryo-EM (Velazhahan et al., 2021). The 3D structure displayed a TM1–7 dimeric interface (Velazhahan et al., 2021). Despite the slight similarity between the dimeric interface of Ste2 and the rhodopsin dimer in terms of TM1 involvement, several interactions are mediated by TM7 of Ste2 instead of H8, as observed in the rhodopsin dimer (Velazhahan et al., 2021; Zhao et al., 2019). A cryo-EM structure of the Ste2 dimer in the absence of G protein is consistent with the expectation of changes to the dimeric interface upon G protein binding (Velazhahan et al., 2022). These changes are restricted to shifts in TM6 and TM7 only and an increase in the number of amino acid interactions between the two protomers, but the TM1–7 dimeric interface is not disturbed by G protein binding (Velazhahan et al., 2022).

Given that Ste2 needs to exist as a dimer to carry out its function (Overton and Blumer, 2000; Overton et al., 2003), it makes sense that its dimeric interface is not dynamic. On the contrary, class-A GPCRs, particularly rhodopsin, are functional in their monomeric and oligomeric states (Ernst et al., 2007; Filipek et al., 2004), suggesting the need for a dynamic interface. Further structural studies of GPCR dimers and GPCR dimer–G protein complexes are necessary to elucidate whether dimer formation (stable or transient) is necessary for the initiation of signal transduction by the various GPCRs.

Increased Gt activation efficiency of oligomeric forms of rhodopsin as compared to monomeric forms (Ernst et al., 2007; Filipek et al., 2004) suggests an alternative mode of rhodopsin–Gt coupling in which two molecules of Gt could interact simultaneously with the two rhodopsin protomers in dimeric form. This positive cooperativity has been discussed previously (McIntire, 2022) and is in line with the proposed model shown in Figure 18.3. The interactions

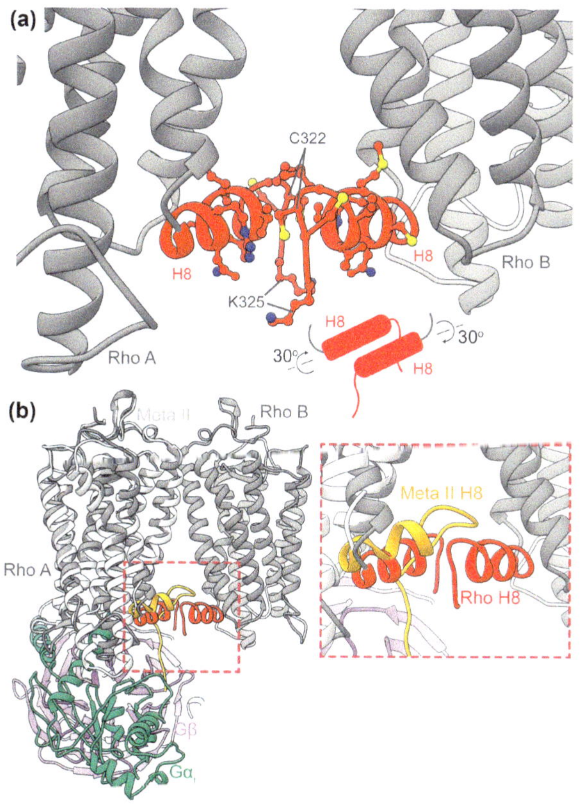

FIGURE 18.3 Interplay between H8 of the two protomers in the rhodopsin dimer. (a) As described in McIntire et al. (2022), the antiparallel organization of H8 of rhodopsin protomer A (Rho A) with H8 of protomer B (Rho B) indicates a positive cooperativity between the two protomers for G protein activation. A 30-degree rotation in H8 of Rho A induced by Gt binding results in an opposite 30-degree rotation in H8 of Rho B, priming Rho B for Gt binding. (b) Superimposition of rhodopsin with the Meta II–Gt complex, showing the conformational change in H8 of rhodopsin caused by Gt binding.

between the H8 of one protomer and the H8 of the other protomer are particularly important, as they maintain the H8 in an antiparallel orientation. A rotation in H8 of protomer A could act as a gear to produce a similar rotation in H8 of protomer B, which would likely initiate structural changes consistent with GPCR activation and accelerate the Gt activation process (Figure 18.3b). In the cryo-EM structure of the rhodopsin–Gi complex elucidated by Tsai et al. (Tsai et al., 2019), Gβ interacted with the C-tail of rhodopsin, suggesting a role for the C-tail as a signaling trigger that could interact with Gt and arrestin (Palczewski, 1994; Ridge et al., 2003). This Gβ–C-tail interaction might also suggest cross talk between rhodopsin protomers where the C-tail of rhodopsin could either mediate or inhibit the binding of the second copy of the G protein. This cross talk between the GPCR protomers could also explain the cooperative activation observed with the luteinizing hormone receptor (LHR), whereby a signaling-deficient LHR mutant was able to activate a ligand-binding-deficient LHR mutant through dimerization (Rivero-Muller et al., 2010). Allosteric communication or cross talk between two protomers of GPCRs and its role in disease progression and drug targeting has been considered previously (Maggio and Millan, 2010; Scarselli et al., 2001; Zhu et al., 2005). Although Ste2 is a class-D GPCR, the characterization of its structure is instructive. Thus, a comparison of the cryo-EM structures of the Ste2 dimer with its G protein–bound form shows the repositioning of TM6 and TM7 and displacement of one protomer by approximately 3 Å in the direction of the other protomer during the transition from the inactive state to the G protein–bound state, suggesting a cross talk between the protomers (Velazhahan et al., 2022). Similarly, the dimeric apelin receptor exhibits a decreased constitutive activity and a lower ligand efficiency in both the G protein and the β-arrestin pathway as compared to the monomeric form (Yue et al., 2022). These findings suggest a cross talk between apelin protomers for the regulation of downstream signaling.

Rhodopsin dimerization could lead to the binding of two molecules of signaling-transducer proteins to the receptor complex, resulting in the activation of distinct signaling pathways. The structural superimposition of active rhodopsin in the rhodopsin–heterotrimeric G protein complex with the rhodopsin dimer cryo-EM map shows that the interaction between one protomer of rhodopsin and Gi does not affect the intradimeric interface, suggesting that two molecules of G proteins can bind to rhodopsin dimer without obvious steric hindrance (Figure 18.4). The binding of two G protein molecules to GPCR dimers could elicit biased signaling and modulate the receptor pharmacology; this concept highlights the functional significance of understanding the dynamics of GPCR dimerization. Ste2 is the only receptor that has been structurally elucidated in its dimeric form bound to two molecules of mini-G proteins (Velazhahan et al., 2021). Although Ste2 belongs to a different class of GPCRs, its overall TM region is like class-A GPCRs except for the TM4 position and the depth and angle of α5 helix

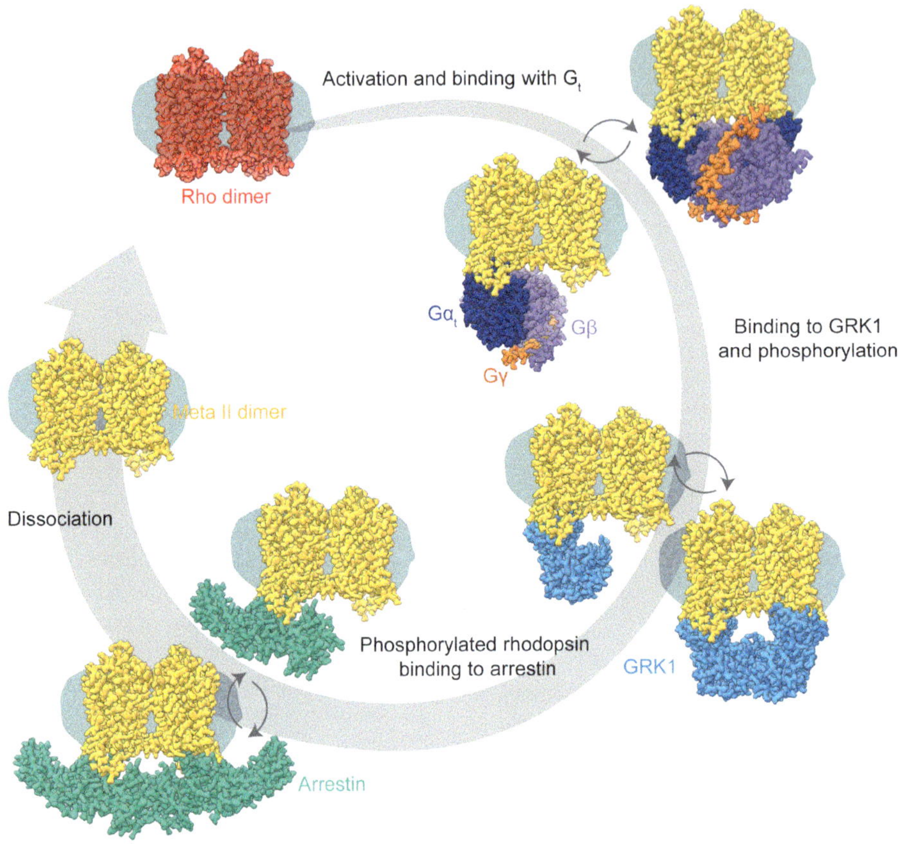

FIGURE 18.4 Illustration showing rhodopsin complexes that form during the phototransduction sequence. The initial step is photoactivation of the rhodopsin dimer to the metarhodopsin II (Meta II) dimer and its interaction with Gt (Gao et al., 2019). Due to a possible interplay between the H8 helices of the rhodopsin dimer, it is likely that there is a dynamic equilibrium with the Meta II dimer binding one or two molecules of Gt. Similarly, the complex between Meta II dimer with one or two molecules of GRK1 (Chen et al., 2021) would form a dynamic complex that produces multiple phosphorylation sites in the Meta II C-terminus. Subsequently, the Meta II dimer binds to one or two molecules of arrestin (Zhou et al., 2017), a capping protein that gets recruited upon phosphorylation of Meta II. Hydrolysis of all-*trans*-retinal causes arrestin to dissociate from the apoprotein opsin, allowing the phosphatase to access the phosphates on the C-terminus (Hsieh et al., 2022), paving the way for pigment regeneration.

insertion during GPCR–G protein complex formation (Velazhahan et al., 2021). Interestingly, the structure of the Ste2 dimer–G protein complex displayed one well-ordered mini-G protein, while the other molecule of mini-G protein was highly flexible and could not be modeled due to poor cryo-EM density. Molecular dynamics simulations revealed that the stability of the two mini-G protein molecules was mutually exclusive, i.e., one molecule was more static while the other was highly dynamic and vice versa (Velazhahan et al., 2021). This observation highlights the dynamic nature of both GPCR dimerization and G protein binding.

G protein–coupled receptor kinases (GRKs) are enzymes that phosphorylate GPCRs, leading to desensitization and internalization of the GPCRs (Gurevich et al., 2012; Maeda et al., 2003; Orban et al., 2014; Palczewski and Benovic, 1991; Palczewski and Saari, 1997; Zhao et al., 1995). Fitting two molecules of light-activated rhodopsin–GRK1 complexes (Chen et al., 2021) to the rhodopsin dimer cryo-EM map shows a spatially feasible macrocomplex (Figure 18.4). Similarly, overlaying the crystal structure of the rhodopsin–arrestin complex (Kang et al., 2015) onto the rhodopsin dimer map suggests that two molecules of arrestins could be accommodated without major steric clashes (Figure 18.4). The binding of two arrestin molecules to GPCR dimers could elicit cooperative receptor desensitization and internalization. This model will need to be tested by further structural and functional studies to determine the physiological effects of two-effector molecule binding as a result of the allosteric communication at the dimer interface.

In summary, the research on GPCR dimerization using cryo-EM has led to significant progress in understanding the molecular mechanisms of these proteins. The resolved structures of rhodopsin dimers in nanodiscs, the apelin receptor dimer, and the Ste2 dimer coupled to two mini-G proteins have provided insights into the dynamic nature of GPCR signaling and the differences imparted by the association of G proteins with the dimer. This review highlights the importance of structural studies of GPCR dimers in unraveling their functional mechanisms and providing insights for drug development. However, further research is necessary to explore the full range of dynamic behavior of GPCR dimers and to understand how this impacts their function. The study of GPCR dimers using cryo-EM is a promising area of research that has the potential to lead to new therapeutic strategies for various diseases.

ACKNOWLEDGMENTS

We thank David Salom for helpful comments on this manuscript. This research was supported, in part, by grants from the National Institutes of Health (NIH): EY030873 and EY034519. The authors acknowledge support through the Gavin Herbert Eye Institute at UCI from an unrestricted grant from Research to Prevent Blindness and from NIH core grant P30EY034070.

REFERENCES

Akgun, E., M.I. Javed, M.M. Lunzer, B.A. Smeester, A.J. Beitz, and P.S. Portoghese. 2013. Ligands that interact with putative MOR-mGluR5 heteromer in mice with inflammatory pain produce potent antinociception. *Proc Natl Acad Sci USA*. 110:11595–11599.

Berthouze, M., L. Rivail, A. Lucas, M.A. Ayoub, O. Russo, S. Sicsic, R. Fischmeister, I. Berque-Bestel, R. Jockers, and F. Lezoualc'h. 2007. Two transmembrane Cys residues are involved in 5-HT4 receptor dimerization. *Biochem Biophys Res Commun*. 356:642–647.

Bhushan, R.G., S.K. Sharma, Z. Xie, D.J. Daniels, and P.S. Portoghese. 2004. A bivalent ligand (KDN-21) reveals spinal delta and kappa opioid receptors are organized as heterodimers that give rise to delta(1) and kappa(2) phenotypes. Selective targeting of delta-kappa heterodimers. *J Med Chem*. 47:2969–2972.

Bous, J., A. Fouillen, H. Orcel, S. Trapani, X. Cong, S. Fontanel, J. Saint-Paul, J. Lai-Kee-Him, S. Urbach, N. Sibille, R. Sounier, S. Granier, B. Mouillac, and P. Bron. 2022. Structure of the vasopressin hormone-V2 receptor-beta-arrestin1 ternary complex. *Sci Adv*. 8:eabo7761.

Chen, Q., M. Plasencia, Z. Li, S. Mukherjee, D. Patra, C.L. Chen, T. Klose, X.Q. Yao, A.A. Kossiakoff, L. Chang, P.C. Andrews, and J.J.G. Tesmer. 2021. Structures of rhodopsin in complex with G-protein-coupled receptor kinase 1. *Nature*. 595:600–605.

Ciruela, F., J. Burgueno, V. Casado, M. Canals, D. Marcellino, S.R. Goldberg, M. Bader, K. Fuxe, L.F. Agnati, C. Lluis, R. Franco, S. Ferre, and A.S. Woods. 2004. Combining mass spectrometry and pull-down techniques for the study of receptor heteromerization. Direct epitope-epitope electrostatic interactions between adenosine A2A and dopamine D2 receptors. *Anal Chem*. 76:5354–5363.

Daniels, D.J., N.R. Lenard, C.L. Etienne, P.Y. Law, S.C. Roerig, and P.S. Portoghese. 2005. Opioid-induced tolerance and dependence in mice is modulated by the distance between pharmacophores in a bivalent ligand series. *Proc Natl Acad Sci USA*. 102:19208–19213.

Dijkman, P.M., O.K. Castell, A.D. Goddard, J.C. Munoz-Garcia, C. de Graaf, M.I. Wallace, and A. Watts. 2018. Dynamic tuneable G protein-coupled receptor monomer-dimer populations. *Nat Commun*. 9:1710.

Dorsch, S., K.N. Klotz, S. Engelhardt, M.J. Lohse, and M. Bunemann. 2009. Analysis of receptor oligomerization by FRAP microscopy. *Nat Methods*. 6:225–230.

Ernst, O.P., V. Gramse, M. Kolbe, K.P. Hofmann, and M. Heck. 2007. Monomeric G protein-coupled receptor rhodopsin in solution activates its G protein transducin at the diffusion limit. *Proc Natl Acad Sci USA*. 104:10859–10864.

Filipek, S., K.A. Krzysko, D. Fotiadis, Y. Liang, D.A. Saperstein, A. Engel, and K. Palczewski. 2004. A concept for G protein activation by G protein-coupled receptor dimers: the transducin/rhodopsin interface. *Photochem Photobiol Sci*. 3:628–638.

Flock, T., A.S. Hauser, N. Lund, D.E. Gloriam, S. Balaji, and M.M. Babu. 2017. Selectivity determinants of GPCR-G-protein binding. *Nature*. 545:317–322.

Fotiadis, D., Y. Liang, S. Filipek, D.A. Saperstein, A. Engel, and K. Palczewski. 2003. Atomic-force microscopy: Rhodopsin dimers in native disc membranes. *Nature*. 421:127–128.

Fotiadis, D., Y. Liang, S. Filipek, D.A. Saperstein, A. Engel, and K. Palczewski. 2004. The G protein-coupled receptor rhodopsin in the native membrane. *Febs Lett*. 564:281–288.

Gao, Y., H. Hu, S. Ramachandran, J.W. Erickson, R.A. Cerione, and G. Skiniotis. 2019. Structures of the rhodopsin-transducin complex: insights into G-protein activation. *Mol Cell*. 75:781–790 e783.

Garcia-Nafria, J., and C.G. Tate. 2019. Cryo-EM structures of GPCRs coupled to G(s), G(i) and G(o). *Mol Cell Endocrinol*. 488:1–13.

Geng, Y., M. Bush, L. Mosyak, F. Wang, and Q.R. Fan. 2013. Structural mechanism of ligand activation in human GABA(B) receptor. *Nature*. 504:254–259.

Getter, T., S. Gulati, R. Zimmerman, Y. Chen, F. Vinberg, and K. Palczewski. 2019. Stereospecific modulation of dimeric rhodopsin. *FASEB J*. 33:9526–9539.

Getter, T., A. Kemp, F. Vinberg, and K. Palczewski. 2021. Identification of small-molecule allosteric modulators that act as enhancers/disrupters of rhodopsin oligomerization. *J. Biol. Chem*. 297:101401.

Gulati, S., B. Jastrzebska, S. Banerjee, A.L. Placeres, P. Miszta, S. Gao, K. Gunderson, G.P. Tochtrop, S. Filipek, K. Katayama, P.D. Kiser, M. Mogi, P.L. Stewart, and K. Palczewski. 2017. Photocyclic behavior of rhodopsin induced by an atypical isomerization mechanism. *Proc Natl Acad Sci USA*. 114:E2608–E2615.

Gulati, S., and K. Palczewski. 2023. Structural view of G protein-coupled receptor signaling in the retinal rod outer segment. *Trends Biochem Sci*. 48:172–186.

Gunkel, M., J. Schoneberg, W. Alkhaldi, S. Irsen, F. Noe, U.B. Kaupp, and A. Al-Amoudi. 2015. Higher-order architecture of rhodopsin in intact photoreceptors and its implication for phototransduction kinetics. *Structure*. 23:628–638.

Guo, W., L. Shi, M. Filizola, H. Weinstein, and J.A. Javitch. 2005. Crosstalk in G protein-coupled receptors: changes at the transmembrane homodimer interface determine activation. *Proc Natl Acad Sci USA*. 102:17495–17500.

Guo, W., E. Urizar, M. Kralikova, J.C. Mobarec, L. Shi, M. Filizola, and J.A. Javitch. 2008. Dopamine D2 receptors form higher order oligomers at physiological expression levels. *Embo J*. 27:2293–2304.

Gurevich, E.V., J.J. Tesmer, A. Mushegian, and V.V. Gurevich. 2012. G protein-coupled receptor kinases: more than just kinases and not only for GPCRs. *Pharmacol. Ther*. 133:40–69.

Gusach, A., J. Garcia-Nafria, and C.G. Tate. 2023. New insights into GPCR coupling and dimerisation from cryo-EM structures. *Curr Opin Struct Biol*. 80:102574.

Harding, P.J., H. Attrill, J. Boehringer, S. Ross, G.H. Wadhams, E. Smith, J.P. Armitage, and A. Watts. 2009. Constitutive dimerization of the G-protein coupled receptor, neurotensin receptor 1, reconstituted into phospholipid bilayers. *Biophys J*. 96:964–973.

Hauser, A.S., M.M. Attwood, M. Rask-Andersen, H.B. Schioth, and D.E. Gloriam. 2017. Trends in GPCR drug discovery: new agents, targets and indications. *Nat Rev Drug Discov*. 16:829–842.

Hebert, T.E., S. Moffett, J.P. Morello, T.P. Loisel, D.G. Bichet, C. Barret, and M. Bouvier. 1996. A peptide derived from a beta2-adrenergic receptor transmembrane domain inhibits both receptor dimerization and activation. *J Biol Chem*. 271:16384–16392.

Herrick-Davis, K., E. Grinde, T.J. Harrigan, and J.E. Mazurkiewicz. 2005. Inhibition of serotonin 5-hydroxytryptamine2c receptor function through heterodimerization: receptor dimers bind two molecules of ligand and one G-protein. *J Biol Chem*. 280:40144–40151.

Hsieh, C.L., Y. Yao, V.V. Gurevich, and J. Chen. 2022. Arrestin facilitates rhodopsin dephosphorylation in vivo. *J Neurosci*. 42:3537–3545.

Huang, J., S. Chen, J.J. Zhang, and X.Y. Huang. 2013. Crystal structure of oligomeric beta1-adrenergic G protein-coupled receptors in ligand-free basal state. *Nat Struct Mol Biol*. 20:419–425.

Jastrzebska, B., Y. Chen, T. Orban, H. Jin, L. Hofmann, and K. Palczewski. 2015. Disruption of rhodopsin dimerization with synthetic peptides targeting an interaction interface. *J Biol Chem*. 290:25728–25744.

Jastrzebska, B., D. Fotiadis, G.F. Jang, R.E. Stenkamp, A. Engel, and K. Palczewski. 2006. Functional and structural characterization of rhodopsin oligomers. *J Biol Chem*. 281:11917–11922.

Jastrzebska, B., T. Orban, M. Golczak, A. Engel, and K. Palczewski. 2013a. Asymmetry of the rhodopsin dimer in complex with transducin. *FASEB J*. 27:1572–1584.

Jastrzebska, B., P. Ringler, D.T. Lodowski, V. Moiseenkova-Bell, M. Golczak, S.A. Muller, K. Palczewski, and A. Engel. 2011. Rhodopsin-transducin heteropentamer: three-dimensional structure and biochemical characterization. *J Struct Biol*. 176:387–394.

Jastrzebska, B., P. Ringler, K. Palczewski, and A. Engel. 2013b. The rhodopsin-transducin complex houses two distinct rhodopsin molecules. *J Struct Biol*. 182:164–172.

Kang, Y., O. Kuybeda, P.W. de Waal, S. Mukherjee, N. Van Eps, P. Dutka, X.E. Zhou, A. Bartesaghi, S. Erramilli, T. Morizumi, X. Gu, Y. Yin, P. Liu, Y. Jiang, X. Meng, G. Zhao, K. Melcher, O.P. Ernst, A.A. Kossiakoff, S. Subramaniam, and H.E. Xu. 2018. Cryo-EM structure of human rhodopsin bound to an inhibitory G protein. *Nature*. 558:553–558.

Kang, Y., X.E. Zhou, X. Gao, Y. He, W. Liu, A. Ishchenko, A. Barty, T.A. White, O. Yefanov, G.W. Han, Q. Xu, P.W. de Waal, J. Ke, M.H. Tan, C. Zhang, A. Moeller, G.M. West, B.D. Pascal, N. Van Eps, L.N. Caro, S.A. Vishnivetskiy, R.J. Lee, K.M. Suino-Powell, X. Gu, K. Pal, J. Ma, X. Zhi, S. Boutet, G.J. Williams, M. Messerschmidt, C. Gati, N.A. Zatsepin, D. Wang, D. James, S. Basu, S. Roy-Chowdhury, C.E. Conrad, J. Coe, H. Liu, S. Lisova, C. Kupitz, I. Grotjohann, R. Fromme, Y. Jiang, M. Tan, H. Yang, J. Li, M. Wang, Z. Zheng, D. Li, N. Howe, Y. Zhao, J. Standfuss, K. Diederichs, Y. Dong, C.S. Potter, B. Carragher, M. Caffrey, H. Jiang, H.N. Chapman, J.C. Spence, P. Fromme, U. Weierstall, O.P. Ernst, V. Katritch, V.V. Gurevich, P.R. Griffin, W.L. Hubbell, R.C. Stevens, V. Cherezov, K. Melcher, and H.E. Xu. 2015. Crystal structure of rhodopsin bound to arrestin by femtosecond X-ray laser. *Nature*. 523:561–567.

Karamitri, A., M.S. Sadek, A.S. Journe, F. Gbahou, R. Gerbier, M.B. Osman, S.A.M. Habib, R. Jockers, and D.P. Zlotos. 2019. O-linked melatonin dimers as bivalent ligands targeting dimeric melatonin receptors. *Bioorg Chem*. 85: 349–356.

Kasai, R.S., S.V. Ito, R.M. Awane, T.K. Fujiwara, and A. Kusumi. 2018. The class-A GPCR dopamine D2 receptor forms transient dimers stabilized by agonists: detection by single-molecule tracking. *Cell Biochem Biophys*. 76:29–37.

Kim, Y., E. Jeong, J.H. Jeong, Y. Kim, and Y. Cho. 2020. Structural basis for activation of the heterodimeric GABA(B) receptor. *J Mol Biol*. 432:5966–5984.

Klco, J.M., T.B. Lassere, and T.J. Baranski. 2003. C5a receptor oligomerization. I. Disulfide trapping reveals oligomers and potential contact surfaces in a G protein-coupled receptor. *J Biol Chem*. 278:35345–35353.

Knepp, A.M., X. Periole, S.J. Marrink, T.P. Sakmar, and T. Huber. 2012. Rhodopsin forms a dimer with cytoplasmic helix 8 contacts in native membranes. *Biochemistry*. 51: 1819–1821.

Kniazeff, J., L. Prezeau, P. Rondard, J.P. Pin, and C. Goudet. 2011. Dimers and beyond: The functional puzzles of class C GPCRs. *Pharmacol Ther*. 130:9–25.

Kobilka, B.K. 2007. G protein coupled receptor structure and activation. *Biochim Biophys Acta*. 1768:794–807.

Koehl, A., H. Hu, D. Feng, B. Sun, Y. Zhang, M.J. Robertson, M. Chu, T.S. Kobilka, T. Laeremans, J. Steyaert, J. Tarrasch, S. Dutta, R. Fonseca, W.I. Weis, J.M. Mathiesen, G. Skiniotis, and B.K. Kobilka. 2019a. Author correction: Structural insights into the activation of metabotropic glutamate receptors. *Nature*. 567:E10.

Koehl, A., H. Hu, D. Feng, B. Sun, Y. Zhang, M.J. Robertson, M. Chu, T.S. Kobilka, T. Laeremans, J. Steyaert, J. Tarrasch, S. Dutta, R. Fonseca, W.I. Weis, J.M. Mathiesen, G. Skiniotis, and B.K. Kobilka. 2019b. Structural insights into the activation of metabotropic glutamate receptors. *Nature*. 566:79–84.

Kota, P., P.J. Reeves, U.L. Rajbhandary, and H.G. Khorana. 2006. Opsin is present as dimers in COS1 cells: identification of amino acids at the dimeric interface. *Proc Natl Acad Sci USA*. 103:3054–3059.

Kumar, S., A. Lambert, J. Rainier, and Y. Fu. 2018. Disruption of rhodopsin dimerization in mouse rod photoreceptors by synthetic peptides targeting dimer interface. *Methods Mol Biol*. 1753:115–128.

Kunishima, N., Y. Shimada, Y. Tsuji, T. Sato, M. Yamamoto, T. Kumasaka, S. Nakanishi, H. Jingami, and K. Morikawa. 2000. Structural basis of glutamate recognition by a dimeric metabotropic glutamate receptor. *Nature*. 407:971–977.

Latek, D., A. Modzelewska, B. Trzaskowski, K. Palczewski, and S. Filipek. 2012. G protein-coupled receptors–recent advances. *Acta Biochim Pol*. 59:515–529.

Le Naour, M., E. Akgun, A. Yekkirala, M.M. Lunzer, M.D. Powers, A.E. Kalyuzhny, and P.S. Portoghese. 2013. Bivalent ligands that target mu opioid (MOP) and cannabinoid1 (CB1) receptors are potent analgesics devoid of tolerance. *J Med Chem*. 56:5505–5513.

Lee, S.P., C.H. So, A.J. Rashid, G. Varghese, R. Cheng, A.J. Lanca, B.F. O'Dowd, and S.R. George. 2004. Dopamine D1 and D2 receptor Co-activation generates a novel phospholipase C-mediated calcium signal. *J Biol Chem*. 279: 35671–35678.

Lenard, N.R., D.J. Daniels, P.S. Portoghese, and S.C. Roerig. 2007. Absence of conditioned place preference or reinstatement with bivalent ligands containing mu-opioid receptor agonist and delta-opioid receptor antagonist pharmacophores. *Eur J Pharmacol*. 566:75–82.

Levitz, J., C. Habrian, S. Bharill, Z. Fu, R. Vafabakhsh, and E.Y. Isacoff. 2016. Mechanism of assembly and cooperativity of homomeric and heteromeric metabotropic glutamate receptors. *Neuron*. 92:143–159.

Levoye, A., J. Dam, M.A. Ayoub, J.L. Guillaume, C. Couturier, P. Delagrange, and R. Jockers. 2006. The orphan GPR50 receptor specifically inhibits MT1 melatonin receptor function through heterodimerization. *Embo J*. 25: 3012–3023.

Liang, Y., D. Fotiadis, S. Filipek, D.A. Saperstein, K. Palczewski, and A. Engel. 2003. Organization of the G protein-coupled receptors rhodopsin and opsin in native membranes. *J Biol Chem*. 278:21655–21662.

Limbird, L.E., P.D. Meyts, and R.J. Lefkowitz. 1975. Beta-adrenergic receptors: Evidence for negative cooperativity. *Biochem Biophys Res Commun*. 64:1160–1168.

Liu, J., H. Tang, C. Xu, S. Zhou, X. Zhu, Y. Li, L. Prezeau, T. Xu, J.P. Pin, P. Rondard, W. Ji, and J. Liu. 2022. Biased signaling due to oligomerization of the G protein-coupled platelet-activating factor receptor. *Nat Commun*. 13:6365.

Macke, J.P., C.M. Davenport, S.G. Jacobson, J.C. Hennessey, F. Gonzalez-Fernandez, B.P. Conway, J. Heckenlively, R. Palmer, I.H. Maumenee, P. Sieving, et al. 1993. Identification of novel rhodopsin mutations responsible for retinitis pigmentosa: implications for the structure and function of rhodopsin. *Am J Hum Genet*. 53:80–89.

Maeda, T., Y. Imanishi, and K. Palczewski. 2003. Rhodopsin phosphorylation: 30 years later. *Prog Retin Eye Res*. 22:417–434.

Maggio, R., and M.J. Millan. 2010. Dopamine D2-D3 receptor heteromers: Pharmacological properties and therapeutic significance. *Curr Opin Pharmacol*. 10:100–107.

Manglik, A., A.C. Kruse, T.S. Kobilka, F.S. Thian, J.M. Mathiesen, R.K. Sunahara, L. Pardo, W.I. Weis, B.K. Kobilka, and S. Granier. 2012. Crystal structure of the micro-opioid receptor bound to a morphinan antagonist. *Nature*. 485:321–326.

Mao, C., C. Shen, C. Li, D.D. Shen, C. Xu, S. Zhang, R. Zhou, Q. Shen, L.N. Chen, Z. Jiang, J. Liu, and Y. Zhang. 2020. Cryo-EM structures of inactive and active GABA(B) receptor. *Cell Res*. 30:564–573.

McGraw, D.W., K.A. Mihlbachler, M.R. Schwarb, F.F. Rahman, K.M. Small, K.F. Almoosa, and S.B. Liggett. 2006. Airway smooth muscle prostaglandin-EP1 receptors directly modulate beta2-adrenergic receptors within a unique heterodimeric complex. *J Clin Invest*. 116:1400–1409.

McIntire, W.E. 2022. A model for how Gbetagamma couples Galpha to GPCR. *J Gen Physiol*. 154:e202112982.

Milligan, G., R.J. Ward, and S. Marsango. 2019. GPCR homo-oligomerization. *Curr Opin Cell Biol*. 57:40–47.

Mirzadegan, T., G. Benko, S. Filipek, and K. Palczewski. 2003. Sequence analyses of G-protein-coupled receptors: similarities to rhodopsin. *Biochemistry*. 42:2759–2767.

Molday, R.S. 1998. Photoreceptor membrane proteins, phototransduction, and retinal degenerative diseases. The Friedenwald Lecture. *Invest Ophthalmol Vis Sci*. 39:2491–2513.

Moller, J., A. Isbilir, T. Sungkaworn, B. Osberg, C. Karathanasis, V. Sunkara, E.O. Grushevskyi, A. Bock, P. Annibale, M. Heilemann, C. Schutte, and M.J. Lohse. 2020. Single-molecule analysis reveals agonist-specific dimer formation of micro-opioid receptors. *Nat Chem Biol*. 16:946–954.

Murray, A.R., S.J. Fliesler, and M.R. Al-Ubaidi. 2009. Rhodopsin: the functional significance of asn-linked glycosylation and other post-translational modifications. *Ophthalmic Genet*. 30:109–120.

Mustafi, D., and K. Palczewski. 2009. Topology of class A G protein-coupled receptors: insights gained from crystal structures of rhodopsins, adrenergic and adenosine receptors. *Mol Pharmacol*. 75:1–12.

Nasrallah, C., G. Cannone, J. Briot, K. Rottier, A.E. Berizzi, C.Y. Huang, R.B. Quast, F. Hoh, J.L. Baneres, F. Malhaire, L. Berto, A. Dumazer, J. Font-Ingles, X. Gomez-Santacana, J. Catena, J. Kniazeff, C. Goudet, A. Llebaria, J.P. Pin, K.R. Vinothkumar, and G. Lebon. 2021. Agonists and allosteric modulators promote signaling from different metabotropic glutamate receptor 5 conformations. *Cell Rep*. 36:109648.

Navarro, G., S. Ferre, A. Cordomi, E. Moreno, J. Mallol, V. Casado, A. Cortes, H. Hoffmann, J. Ortiz, E.I. Canela, C. Lluis, L. Pardo, R. Franco, and A.S. Woods. 2010. Interactions between intracellular domains as key determinants of the quaternary structure and function of receptor heteromers. *J Biol Chem*. 285:27346–27359.

Nemoto, W., and H. Toh. 2005. Prediction of interfaces for oligomerizations of G-protein coupled receptors. *Proteins*. 58:644–660.

Ng, S.Y., L.T. Lee, and B.K. Chow. 2012. Receptor oligomerization: from early evidence to current understanding in class B GPCRs. *Front Endocrinol (Lausanne)*. 3:175.

O'Dowd, B.F., X. Ji, T. Nguyen, and S.R. George. 2012. Two amino acids in each of D1 and D2 dopamine receptor cytoplasmic regions are involved in D1-D2 heteromer formation. *Biochem Biophys Res Commun*. 417:23–28.

Okada, T., and K. Palczewski. 2001. Crystal structure of rhodopsin: implications for vision and beyond. *Curr Opin Struct Biol*. 11:420–426.

Orban, T., B. Jastrzebska, and K. Palczewski. 2014. Structural approaches to understanding retinal proteins needed for vision. *Curr Opin Cell Biol*. 27:32–43.

Overton, M.C., and K.J. Blumer. 2000. G-protein-coupled receptors function as oligomers in vivo. *Curr Biol*. 10:341–344.

Overton, M.C., S.L. Chinault, and K.J. Blumer. 2003. Oligomerization, biogenesis, and signaling is promoted by a glycophorin A-like dimerization motif in transmembrane domain 1 of a yeast G protein-coupled receptor. *J Biol Chem*. 278:49369–49377.

Palczewski, K. 1994. Structure and functions of arrestins. *Protein Sci*. 3:1355–1361.

Palczewski, K. 2006. G protein-coupled receptor rhodopsin. *Annu Rev Biochem*. 75:743–767.

Palczewski, K. 2010. Oligomeric forms of G protein-coupled receptors (GPCRs). *Trends Biochem Sci*. 35:595–600.

Palczewski, K., and J.L. Benovic. 1991. G-protein-coupled receptor kinases. *Trends Biochem Sci*. 16:387–391.

Palczewski, K., T. Kumasaka, T. Hori, C.A. Behnke, H. Motoshima, B.A. Fox, I. Le Trong, D.C. Teller, T. Okada, R.E. Stenkamp, M. Yamamoto, and M. Miyano. 2000. Crystal structure of rhodopsin: A G protein-coupled receptor. *Science*. 289:739–745.

Palczewski, K., and J.C. Saari. 1997. Activation and inactivation steps in the visual transduction pathway. *Curr Opin Neurobiol*. 7:500–504.

Papasergi-Scott, M.M., G. Perez-Hernandez, H. Batebi, Y. Gao, G. Eskici, A.B. Seven, O. Panova, D. Hilger, M. Casiraghi, F. He, L. Maul, P. Gmeiner, B.K. Kobilka, P.W. Hildebrand, and G. Skiniotis. 2023. Time-resolved cryo-EM of G protein activation by a GPCR. *bioRxiv*.

Park, P.S., S. Filipek, J.W. Wells, and K. Palczewski. 2004. Oligomerization of G protein-coupled receptors: past, present, and future. *Biochemistry*. 43:15643–15656.

Patil, D.N., S. Singh, T. Laboute, T.S. Strutzenberg, X. Qiu, D. Wu, S.J. Novick, C.V. Robinson, P.R. Griffin, J.F. Hunt, T. Izard, A.K. Singh, and K.A. Martemyanov. 2022. Cryo-EM structure of human GPR158 receptor coupled to the RGS7-Gbeta5 signaling complex. *Science*. 375:86–91.

Peeters, M.C., G.J. van Westen, Q. Li, and I.J. AP. 2011. Importance of the extracellular loops in G protein-coupled receptors for ligand recognition and receptor activation. *Trends Pharmacol Sci*. 32:35–42.

Pellissier, L.P., G. Barthet, F. Gaven, E. Cassier, E. Trinquet, J.P. Pin, P. Marin, A. Dumuis, J. Bockaert, J.L. Baneres, and S. Claeysen. 2011. G protein activation by serotonin type 4 receptor dimers: evidence that turning on two protomers is more efficient. *J Biol Chem*. 286:9985–9997.

Pin, J.P., and B. Bettler. 2016. Organization and functions of mGlu and GABA(B) receptor complexes. *Nature*. 540:60–68.

Ploier, B., L.N. Caro, T. Morizumi, K. Pandey, J.N. Pearring, M.A. Goren, S.C. Finnemann, J. Graumann, V.Y. Arshavsky, J.S. Dittman, O.P. Ernst, and A.K. Menon. 2016. Dimerization deficiency of enigmatic retinitis pigmentosa-linked rhodopsin mutants. *Nat Commun*. 7: 12832.

Poge, M., J. Mahamid, S.S. Imanishi, J.M. Plitzko, K. Palczewski, and W. Baumeister. 2021. Determinants shaping the nanoscale architecture of the mouse rod outer segment. *Elife*. 10:e72817.

Poltavets, V., J.W. Faulkner, D. Dhatrak, R.J. Whitfield, S.R. McColl, and M. Kochetkova. 2021. CXCR4-CCR7 heterodimerization is a driver of breast cancer progression. *Life (Basel)*. 11:1049.

Rasmussen, S.G., H.J. Choi, D.M. Rosenbaum, T.S. Kobilka, F.S. Thian, P.C. Edwards, M. Burghammer, V.R. Ratnala, R. Sanishvili, R.F. Fischetti, G.F. Schertler, W.I. Weis, and B.K. Kobilka. 2007. Crystal structure of the human beta2 adrenergic G-protein-coupled receptor. *Nature*. 450:383–387.

Ridge, K.D., N.G. Abdulaev, M. Sousa, and K. Palczewski. 2003. Phototransduction: crystal clear. *Trends Biochem Sci*. 28:479–487.

Ridge, K.D., and K. Palczewski. 2007. Visual rhodopsin sees the light: structure and mechanism of G protein signaling. *J Biol Chem*. 282:9297–9301.

Rivero-Muller, A., Y.Y. Chou, I. Ji, S. Lajic, A.C. Hanyaloglu, K. Jonas, N. Rahman, T.H. Ji, and I. Huhtaniemi. 2010. Rescue of defective G protein-coupled receptor function in vivo by intermolecular cooperation. *Proc Natl Acad Sci USA*. 107:2319–2324.

Rousseau, B., S. Murugan, A. Palagani, and D.K. Sarkar. 2022. Beta 2 adrenergic receptor and mu opioid receptor interact to potentiate the aggressiveness of human breast cancer cell by activating the glycogen synthase kinase 3 signaling. *Breast Cancer Res*. 24:33.

Salom, D., D.T. Lodowski, R.E. Stenkamp, I. Le Trong, M. Golczak, B. Jastrzebska, T. Harris, J.A. Ballesteros, and K. Palczewski. 2006. Crystal structure of a photoactivated deprotonated intermediate of rhodopsin. *Proc Natl Acad Sci USA*. 103:16123–16128.

Scarselli, M., F. Novi, E. Schallmach, R. Lin, A. Baragli, A. Colzi, N. Griffon, G.U. Corsini, P. Sokoloff, R. Levenson, Z. Vogel, and R. Maggio. 2001. D2/D3 dopamine receptor heterodimers exhibit unique functional properties. *J Biol Chem*. 276:30308–30314.

Shen, C., C. Mao, C. Xu, N. Jin, H. Zhang, D.D. Shen, Q. Shen, X. Wang, T. Hou, Z. Chen, P. Rondard, J.P. Pin, Y. Zhang, and J. Liu. 2021. Structural basis of GABA(B) receptor-G(i) protein coupling. *Nature*. 594:594–598.

Shen, J., D. Zhang, Y. Fu, A. Chen, X. Yang, and H. Zhang. 2022. Cryo-EM structures of human bradykinin receptor-G(q) proteins complexes. *Nat Commun*. 13:714.

Simpson, L.M., B. Taddese, I.D. Wall, and C.A. Reynolds. 2010. Bioinformatics and molecular modelling approaches to GPCR oligomerization. *Curr Opin Pharmacol*. 10:30–37.

Stenkamp, R.E. 2018. Identifying G protein-coupled receptor dimers from crystal packings. *Acta Crystallogr D Struct Biol*. 74:655–670.

Suda, K., S. Filipek, K. Palczewski, A. Engel, and D. Fotiadis. 2004. The supramolecular structure of the GPCR rhodopsin in solution and native disc membranes. *Mol Membr Biol*. 21:435–446.

Terrillon, S., and M. Bouvier. 2004. Roles of G-protein-coupled receptor dimerization. *EMBO Rep*. 5:30–34.

Tsai, C.J., J. Marino, R. Adaixo, F. Pamula, J. Muehle, S. Maeda, T. Flock, N.M. Taylor, I. Mohammed, H. Matile, R.J. Dawson, X. Deupi, H. Stahlberg, and G. Schertler. 2019. Cryo-EM structure of the rhodopsin-Galphai-betagamma complex reveals binding of the rhodopsin C-terminal tail to the gbeta subunit. *Elife*. 8:e46041.

Velazhahan, V., N. Ma, G. Pandy-Szekeres, A.J. Kooistra, Y. Lee, D.E. Gloriam, N. Vaidehi, and C.G. Tate. 2021. Structure of the class D GPCR Ste2 dimer coupled to two G proteins. *Nature*. 589:148–153.

Velazhahan, V., N. Ma, N. Vaidehi, and C.G. Tate. 2022. Activation mechanism of the class D fungal GPCR dimer Ste2. *Nature*. 603:743–748.

Wang, J., L. He, C.A. Combs, G. Roderiquez, and M.A. Norcross. 2006. Dimerization of CXCR4 in living malignant cells: control of cell migration by a synthetic peptide that reduces homologous CXCR4 interactions. *Mol Cancer Ther*. 5:2474–2483.

Wu, B., E.Y. Chien, C.D. Mol, G. Fenalti, W. Liu, V. Katritch, R. Abagyan, A. Brooun, P. Wells, F.C. Bi, D.J. Hamel, P. Kuhn, T.M. Handel, V. Cherezov, and R.C. Stevens. 2010. Structures of the CXCR4 chemokine GPCR with small-molecule and cyclic peptide antagonists. *Science*. 330:1066–1071.

Wu, H., C. Wang, K.J. Gregory, G.W. Han, H.P. Cho, Y. Xia, C.M. Niswender, V. Katritch, J. Meiler, V. Cherezov, P.J. Conn, and R.C. Stevens. 2014. Structure of a class C GPCR metabotropic glutamate receptor 1 bound to an allosteric modulator. *Science*. 344:58–64.

Xue, L., X. Rovira, P. Scholler, H. Zhao, J. Liu, J.P. Pin, and P. Rondard. 2015. Major ligand-induced rearrangement of the heptahelical domain interface in a GPCR dimer. *Nat Chem Biol*. 11:134–140.

Yang, J., Z. Gong, Y.B. Lu, C.J. Xu, T.F. Wei, M.S. Yang, T.W. Zhan, Y.H. Yang, L. Lin, J. Liu, C. Tang, and W.P. Zhang. 2020. FLIM-FRET-based structural characterization of a class-A GPCR dimer in the cell membrane. *J Mol Biol*. 432:4596–4611.

Yuan, Y., C.K. Arnatt, N. El-Hage, S.M. Dever, J.C. Jacob, D.E. Selley, K.F. Hauser, and Y. Zhang. 2013. A bivalent ligand targeting the putative mu opioid receptor and chemokine receptor CCR5 heterodimers: binding affinity versus functional activities. *Medchemcomm*. 4:847–851.

Yue, Y., L. Liu, L.J. Wu, Y. Wu, L. Wang, F. Li, J. Liu, G.W. Han, B. Chen, X. Lin, R.L. Brouillette, E. Breault, J.M. Longpre, S. Shi, H. Lei, P. Sarret, R.C. Stevens, M.A. Hanson, and F. Xu. 2022. Structural insight into apelin receptor-G protein stoichiometry. *Nat Struct Mol Biol*. 29:688–697.

Zhang, T., L.H. Cao, S. Kumar, N.O. Enemchukwu, N. Zhang, A. Lambert, X. Zhao, A. Jones, S. Wang, E.M. Dennis, A. Fnu, S. Ham, J. Rainier, K.W. Yau, and Y. Fu. 2016. Dimerization of visual pigments in vivo. *Proc Natl Acad Sci USA*. 113:9093–9098.

Zhang, X., R.M. Johnson, I. Drulyte, L. Yu, A. Kotecha, R. Danev, D. Wootten, P.M. Sexton, and M.J. Belousoff. 2021. Evolving cryo-EM structural approaches for GPCR drug discovery. *Structure*. 29:963–974 e966.

Zhao, D.Y., M. Poge, T. Morizumi, S. Gulati, N. Van Eps, J. Zhang, P. Miszta, S. Filipek, J. Mahamid, J.M. Plitzko, W. Baumeister, O.P. Ernst, and K. Palczewski. 2019. Cryo-EM structure of the native rhodopsin dimer in nanodiscs. *J Biol Chem*. 294:14215–14230.

Zhao, X., K. Palczewski, and H. Ohguro. 1995. Mechanism of rhodopsin phosphorylation. *Biophys Chem*. 56:183–188.

Zhou, X.E., Y. He, P.W. de Waal, X. Gao, Y. Kang, N. Van Eps, Y. Yin, K. Pal, D. Goswami, T.A. White, A. Barty, N.R. Latorraca, H.N. Chapman, W.L. Hubbell, R.O. Dror, R.C. Stevens, V. Cherezov, V.V. Gurevich, P.R. Griffin, O.P. Ernst, K. Melcher, and H.E. Xu. 2017. Identification of phosphorylation codes for arrestin recruitment by G protein-coupled receptors. *Cell*. 170:457–469 e413.

Zhu, W.Z., K. Chakir, S. Zhang, D. Yang, C. Lavoie, M. Bouvier, T.E. Hebert, E.G. Lakatta, H. Cheng, and R.P. Xiao. 2005. Heterodimerization of beta1- and beta2-adrenergic receptor subtypes optimizes beta-adrenergic modulation of cardiac contractility. *Circ Res*. 97:244–251.

19 Cryo-EM Structures of the Fungal GPCR Dimer Ste2

Vaithish Velazhahan and Christopher G. Tate

INTRODUCTION

G protein–coupled receptors (GPCRs) are eukaryotic membrane proteins that allow cells to sense and communicate with their environment. GPCRs constitute the single largest family of membrane proteins in humans, comprising more than 800 unique receptors that mediate a broad range of physiological responses (Kobilka, 2007; Pándy-Szekeres et al., 2018). Their location on the surface of cells and their role as master switches for a variety of signal transduction cascades make GPCRs ideal drug targets, and consequently about 34% of all FDA-approved drugs mediate their actions through GPCRs (Hauser et al., 2017).

GPCRs are phylogenetically divided into six classes, denoted A to F, of which classes A, B, C, and F comprise vertebrate GPCRs while classes D and E encompass GPCRs found exclusively in fungi and amoeba, respectively (Attwood & Findlay, 1994; Kolakowski Jr, 1994). As the single largest class of cell surface receptors in fungi, GPCRs help fungi in sensing their environment by detecting a wide range of signals, such as ions, light, nutrients, proteins, and hydrophobic surfaces (Brown et al., 2018; Van Dijck et al., 2017). Such environmental sensing is critical for the regulation of metabolism, production of mycotoxins, fungal growth, sexual development, virulence, chemotropism, immune evasion, and invasiveness, allowing fungi to thrive in a wide range of natural habitats (Brown & Goldman, 2016). Fungal GPCRs are considered novel targets for drug discovery to combat the ever-growing problem of fungal pathogenesis, as deaths per annum from fungal diseases exceed those caused by tuberculosis or malaria, and only a very limited number of proteins are targeted by current antifungal drugs, none of which are GPCRs (Brown et al., 2012; Velazhahan et al., 2023; Woolhouse & Farrar, 2014).

Fungal GPCRs are classified into 10 different classes (I–X) based on homology and structural similarity (Brown et al., 2018). The wine yeast *Saccharomyces cerevisiae* has only three GPCRs, namely Ste2, Ste3, and Gpr1, which belong to classes I, II, and III, respectively. Ste2 and Ste3 are responsible for pheromone sensing and initiation of sexual reproduction in the *MAT*a and *MAT*α strains of yeast, respectively, whereas Gpr1 is involved in sensing fermentable sugars such as glucose (Xue et al., 2008).

Ste2 from *S. cerevisiae* yeast is a prototypical class I fungal GPCR. Upon stimulation with the tridecapeptide mating hormone α-factor, Ste2 catalyzes the exchange of bound GDP to GTP within Gpa1, the α-subunit of the heterotrimeric G protein. This facilitates the release of the βγ dimer (Ste4-Ste18, respectively), which then activates the p21-activated protein kinase Ste20. Subsequent activation of the Ste11-Ste7-Fus3 MAPK cascade leads to cell cycle arrest and fusion with the opposite mating type *MAT*α (Brown et al., 2018; Xue et al., 2008).

Ste2 was the first ligand-binding GPCR to be sequenced, and the genetic tractability of *S. cerevisiae* has allowed an extensive characterization of the Ste2-mediated signal transduction cascade (Burkholder & Hartwell, 1985; Naider & Becker, 2004). This has contributed enormously to our present understanding of many aspects of GPCR biology and established paradigms for understanding GPCR-mediated signaling and regulation (Alvaro & Thorner, 2016).

Over 1000 structures of vertebrate GPCRs are available (GPCRdb [Pándy-Szekeres et al., 2018]), leading to a detailed understanding of their function and activation. However, no structures of fungal GPCRs were available prior to our study, which precluded a mechanistic understanding of their function. Despite major attempts from several laboratories to crystallize Ste2 over the past two decades, no structures of Ste2 could be determined (Naider & Becker, 2020). Therefore, we developed new methodologies and strategies to determine the first high-resolution structure of Ste2. It was reasoned that electron cryomicroscopy (cryo-EM) would be the best method for the structure determination of Ste2 since cryo-EM eliminates the need for crystal formation, which typically requires small detergents and harsh conditions that may denature the protein. Cryo-EM allows the study of wild-type, unmodified GPCRs and is tolerant to receptor heterogeneity, presence of contaminants, and posttranslational modifications, such as glycosylation, which would usually interfere with the formation of diffraction-quality crystals (García-Nafría & Tate, 2021).

A major limitation of cryo-EM is the lower limit of the size of samples that could be imaged, which according to theoretical calculations is ~38 kDa (Henderson, 1995), and is roughly the size of an inactive-state monomeric GPCR (Zhang et al., 2022). While various approaches have recently been proposed to overcome this problem (Bloch et al., 2021; Robertson et al., 2022), cryo-EM is well suited for the study of active-state GPCRs coupled to transducers, such as G proteins, because the ~90 kDa heterotrimeric G protein coupled to the GPCR extends outside the detergent micelle or nanodiscs and provides a source of alignment during structure determination. The first cryo-EM structure of a GPCR–G protein complex was determined

DOI: 10.1201/9781003326106-23

in 2017, and since then cryo-EM has revolutionized GPCR structural biology, resulting in an explosion of active-state GPCR structures in complex with transducers, including all the major G protein subtypes such as $G\alpha_s$ (García-Nafría, Lee, et al., 2018), $G\alpha_i$ (Koehl et al., 2018), $G\alpha_o$ (García-Nafría, Nehmé, et al., 2018), $G\alpha_q$ (Shen et al., 2022), and $G\alpha_{13}$ (Chen et al., 2022). In addition, structures have been determined of GPCRs coupled to β-arrestins (Huang et al., 2020; Lee et al., 2020; Staus et al., 2020; Yin et al., 2019) and GPCR kinases (Chen et al., 2021).

Since active-state GPCR–G protein complexes are short-lived, different genetically engineered proteins have been developed to stabilize these complexes for structural studies. This includes nanobodies, such as Nb35 (Rasmussen et al., 2011); mini-G proteins, which contain only the GTPase domain of the Gα subunit and a number of point mutations that stabilize the nucleotide-free state of the GPCR–G protein complex (Carpenter & Tate, 2016; Nehmé et al., 2017); dominant-negative G proteins (Liang et al., 2018); and antibody fragments, such as scFv16 (Maeda et al., 2018). These engineered proteins were essential to allow the successful determination of over 569 cryo-EM structures of human GPCR–G protein complexes (as quoted by GPCRdb in June 2023 [Pándy-Szekeres et al., 2018]) in the last six years. However, these proteins were developed specifically for mammalian GPCRs, and no analogous engineered proteins were available to stabilize fungal GPCRs, such as Ste2. Therefore, using structure-based rational mutagenesis in *Escherichia coli*, we engineered mini-Gpa1, which is a miniaturized version of the fungal G protein α-subunit Gpa1 that stabilizes the active-state Ste2 complex for structural studies (Velazhahan et al., 2021). For construction of mini-Gpa1, six point mutations (G50D, E51N, A345D, E348D, M450A, V453I) were incorporated in addition to the deletion of the flexible and dynamic α-helical domain (Q66 to K299) and the N-terminal 25 amino acids (G2-V26) of Gpa1.

RESULTS

Purification and Structure Determination of the Ste2-G Protein Complex

A major challenge was obtaining sufficient quantities of purified Ste2 for structural studies. Insect cells were chosen for the expression of Ste2 since yeast cells tightly regulate the expression of their pheromone receptors, and overexpression of Ste2 in a signaling environment could lead to undesirable downstream signaling and trafficking responses (Di Segni et al., 2011; Gehret et al., 2012). Furthermore, yeast cells possess a tough cell wall and require harsh mechanical lysis methods to break it, which could denature Ste2 in the very first steps of protein isolation and increase its susceptibility to protein degradation by multiple active proteases in *S. cerevisiae*. Therefore, full-length wild-type Ste2 was overexpressed and purified from *Trichoplusia ni* (Tni5) insect cells using affinity-based and size-exclusion chromatography (Velazhahan et al., 2021). A new protocol was developed to co-express and co-purify the *S. cerevisiae* βγ subunits (Ste4-Ste18) as a complex from insect cells (Velazhahan et al., 2021). The G protein heterotrimer was assembled, and then an active-state ternary complex of Ste2 was made by mixing purified Ste2 with the G protein heterotrimer in the presence of the agonist α-factor. This active-state complex was further purified by size-exclusion chromatography, and vitrified on cryo-EM grids.

The Structure of Ste2: A Dimer Coupled to Two G Proteins

Processing of the cryo-EM images of the Ste2–G protein complex yielded a map of the entire complex at an overall resolution of 3.5 Å and another map of the Ste2 transmembrane region resolved to 3.3 Å. The cryo-EM maps featured clear density for the majority of the side chains, which facilitated *de novo* model building. The structure revealed that Ste2 forms a homodimer that couples simultaneously to two G protein heterotrimers. The density for one G-protein heterotrimer was well-resolved to about 3.5 Å, with the best density in Gpa1 that formed all contacts between the G protein and Ste2. However, the resolution of the other G protein was limited to about 8 Å, except for the C-terminal α5 helix of Gpa1 bound to the receptor. No classes consistent with a Ste2 monomer were observed during data processing despite extensive 2D and 3D classifications of the images.

Ste2 consistently behaved as a dimer in all detergents tested for its solubilization, and the simultaneous coupling of the Ste2 dimer to two G proteins was further confirmed by size-exclusion chromatography coupled to multi-angle light scattering (SEC-MALS) experiments. Our results are consistent with the biochemical observations that Ste2 exists as a dimer in yeast cells and dimerization is essential for biogenesis, signaling, and functional endocytosis (Gehret et al., 2006; Overton & Blumer, 2000; Overton et al., 2003; Yesilaltay & Jenness, 2000), with higher oligomeric states also being proposed (Stoneman et al., 2017).

Structure Determination of Ste2 in Different Conformational States

The structure of the Ste2 dimer coupled to two G proteins represents a single snapshot of Ste2 in the fully active conformation. However, GPCRs exist in a continuum of conformational states, and structural knowledge of at least some of these conformational states is necessary to understand the molecular changes that lead to activation of Ste2. Although a monomeric form of Ste2 would be too small for high-resolution studies using cryo-EM, dimeric Ste2 was calculated to have an ordered molecular weight of 67.3 kDa, which was found to be amenable for structure determination by cryo-EM.

To determine the structure of Ste2 in the agonist-bound intermediate-active state (Ste2•Ag), Ste2 was purified in the presence of the native agonist α-factor. To stabilize Ste2 in an inactive conformation (Ste2•Ant), an antagonist peptide was designed with the sequence HALQLKPGQP(Nle)Y; this is henceforth referred to as "antagonist." The antagonist peptide is derived from the agonist α-factor by deleting the first Trp and changing the third Trp to Ala, and therefore the antagonist is also known as [desTrp^{1}Ala^{3}Nle^{12}] α-factor. Conversion of Met12 to Nle (norleucine) does not affect signaling but makes the peptide more resistant to oxidation. The biochemical antagonism of this peptide and its pharmacological properties have been well-characterized previously (Blumer & Thorner, 1990; Mathew et al., 2011; Raths et al., 1988; Sommers et al., 2000).

While purification of Ste2 in the agonist-bound and antagonist-bound states were readily achieved by purifying Ste2 in the presence of the respective peptides, purification of Ste2 in the ligand-free state (Ste2•LF) was highly challenging owing to the instability of Ste2 in the absence of bound ligands. To purify Ste2 in the ligand-free state, a new methodology called PSGWAY (Pre-Stabilization of GPCRs by Weak Association) was devised that involves the addition of the purified wild-type G protein heterotrimer to stabilize ligand-free Ste2 before solubilization from cell membranes followed by dissociation of the G protein (Velazhahan et al., 2022).

Cryo-EM density maps at an overall nominal resolution of 2.7 Å and 3.1 Å were obtained for Ste2 in the antagonist-bound and ligand-free states, respectively. During processing of the cryo-EM images of Ste2 in the agonist-bound state, two distinct high-resolution classes were obtained, one resembling the antagonist-bound and ligand-free Ste2 in the inactive state (therefore called agonist-bound Ste2$^{inactive\text{-}like}$ or Ste2^{IL}•Ag) and another class resembling the G-protein heterotrimer coupled Ste2 structure in the active state (therefore called agonist-bound Ste2$^{active\text{-}like}$ or Ste2^{AL}•Ag). The other 3D classes obtained during image processing did not resolve to a high resolution due to heterogeneity; nevertheless, they were more similar to Ste2 in the active Ste2^{AL}•Ag state or G protein–coupled state (Ste2•Ag•G). Therefore, when bound to the agonist α-factor, most of the cryo-EM images of single particles (83.9%) correspond to active-like state or intermediate states resembling the active-like state, and only a minority (16.1%) of the particles correspond to a distinct inactive-like state. This is consistent with α-factor-bound Ste2 transitioning primarily into an active-like state and fits well with the paradigm that agonist binding alters the conformational equilibrium of GPCRs. Ste2 was found to exist as a dimer in all the conformational states, and no particles were observed of monomeric Ste2.

The determination of five high-resolution structures of Ste2 in different states along the entire receptor activation pathway (Figure 19.1a) with good density for the ligands (see for example Figure 19.1b) allow detailed inferences to be made about various structural aspects of Ste2 and how they change upon activation. The five conformational states of Ste2 are ligand-free (Ste2•LF), antagonist-bound (Ste2•Ant), agonist-bound inactive-like (Ste2•Ag^{IL}), agonist-bound active-like (Ste2•Ag^{AL}), and agonist-bound G protein–coupled (Ste2•Ag•G).

The Dimer Interface of Ste2

The dimer interface of Ste2 in the Ste2•Ag•G state is formed by the domain-swapped N-terminus, helices H1, H2, H7, and extracellular loop 1 (ECL1). The surface area of the dimer interface is extensive (2519 Å^2) and is over twice that of the Ste2–G protein interface (Figure 19.1c–e). The majority of contacts (56%) are mediated by the transmembrane regions H1-H2-H7, while the domain swapped N-terminus and ECL1 mediate 44% of atomic contacts. The dimer interface is stabilized by 171 van der Waals interactions and 6 hydrogen bonds, formed by 43 amino acid residues from Ste2$_A$ and 44 from Ste2$_B$ (subscripts refer to the protomers of Ste2). Substantial mutagenesis data investigating the role of 32 of the amino acid residues at the dimer interface show that 27 residues have a major effect on dimerization and/or signaling when mutated, both in the presence and absence of α-factor (Velazhahan et al., 2021).

The transmembrane α-helix H4 has previously been proposed to be involved in the formation of higher oligomeric states (Wang & Konopka, 2009). The Ste2 structures show that H4 projects from either side of the Ste2 dimer and is on the opposite side of Ste2 to the H1 dimer interface (Figure 19.1c). Therefore, it is possible that H4 could facilitate contacts with adjacent Ste2 dimers, resulting in the formation of higher-order oligomeric states of Ste2. Nevertheless, further evidence is needed to support this hypothesis.

Domain-swapping is observed in the N-terminus of Ste2. The N-terminus of Ste2$_A$ is situated on top of Ste2$_B$, where it makes contacts with ECL1$_B$. It then loops back to Ste2$_A$, allowing it to make contacts with the N-terminus of Ste2$_B$, which runs in an antiparallel direction, before it turns into H1 (Figure 19.1d). In the N-terminal region, the dimer interface is stabilized by Pro19^{D1S1x50} and Gly31^{D1S2x50}, which are the two most conserved residues in the N-terminus. Further stabilization comes from interactions made by the conserved aliphatic residues Ile/Leu/Val24^{D1S1x55} and Ile/Leu/Val36^{D1S2x55} within the antiparallel beta sheets in the N-terminus. Superscripts are the CD1 generic residue numbering system, in which the most well-conserved amino acid residue in each transmembrane α-helix is given the number Yx50, where Y is the helix number (Velazhahan et al., 2021). The dimer interface extends along the length of H1$_A$ and H1$_B$, which cross at an angle of 37°, forming a right-handed interaction. At the cytoplasmic face of Ste2, H7 extends beyond the predicted membrane surface, crossing beneath H1 to form both H7$_A$-H7$_B$ contacts as well as interactions between H7 and H1 between protomers (Figure 19.1c).

In the middle of H1, there is the sequence I^{1x43}xxGxxxGA^{1x51} (residues 53–61 of Ste2) that is highly conserved in Ste2 orthologues and contains a canonical glycophorin A dimerization

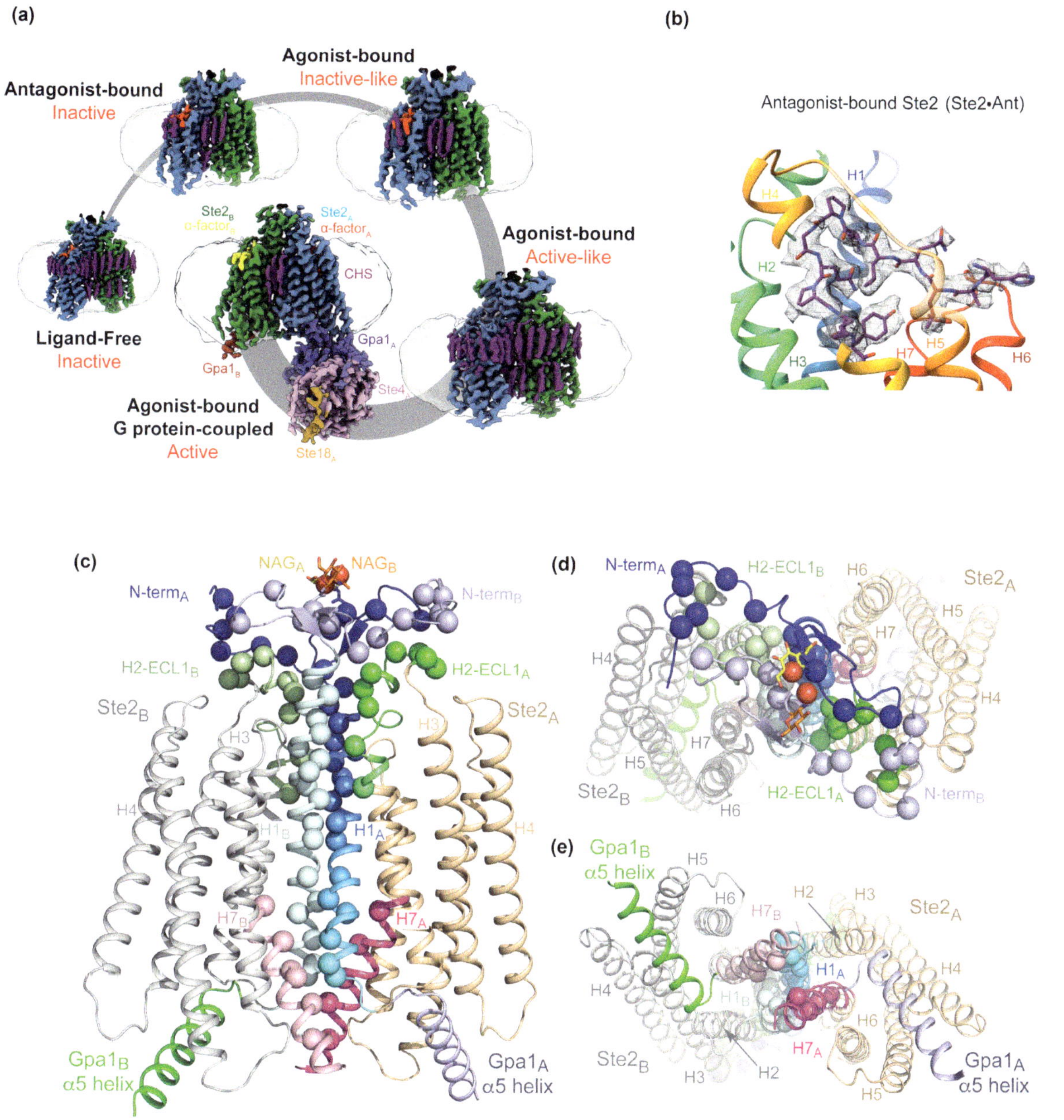

FIGURE 19.1 Cryo-EM structures of Ste2 and the dimer interface. (a) Cryo-EM density maps of Ste2 in the ligand-free state, antagonist-bound state, two agonist-bound intermediate states, and G protein–coupled state are shown. The detergent micelle is depicted in pale grey, and subunits are colored according to the key shown for the G protein–coupled state. $Ste2_A$ (blue), $Ste2_B$ (green), CHS molecules (purple), NAG molecules (black), mini-$Gpa1_A$ (purple), mini-$Gpa1_B$ (maroon), $Ste4_A$ (pink), $Ste18_A$ (orange); ligands are colored according to the protomer to which they are bound (red or yellow). Subscripts denote the protomer associated with the various subunits. Figure is adapted from (Velazhahan et al., 2023). (b) Orthosteric binding pocket of $Ste2_A$ (rainbow) depicting antagonist bound. The density encompasses a carve radius of 2 Å. Figure is taken from (Velazhahan et al., 2023). (c–e) The Ste2 dimer is shown as a cartoon (chain A, brown, chain B, grey) and the structural features of the dimer interface are labeled. The Cα atoms of residues forming contacts in the interface are depicted as spheres. (c) View parallel to the membrane; (d) extracellular surface; (e) intracellular surface. Figure is taken from (Velazhahan et al., 2022).

motif GxxxG (Overton et al., 2003). This forms a wide, shallow groove in $H1_A$ that fits into the analogous groove in $H1_B$, which is similar to that observed in other glycine zipper motif structures (Kim et al., 2005). The mutants G56A and G56A/G60A or mutants with Leu at these positions significantly reduce dimerization and greatly impair signaling responses to α-factor without affecting agonist binding (Gehret et al., 2006; Overton et al., 2003).

In all the structures, the detergent micelle surrounding Ste2 contained six ordered densities adjacent to the dimer interface (Figure 19.1a) that were assigned as the sterol cholesteryl hemisuccinate (CHS), which was present in vast molar excess during receptor purification. In addition, density corresponding to N-acetylglucosamine attached to $\text{Asn25}^{\text{D1S1x55}}$ of Ste2, a known N-glycosylation site (Mentesana & Konopka, 2001), was also found to mediate dimer contacts (Figure 19.1c).

Structural Changes Across the Dimer Interface During Receptor Activation

Ste2 was a homodimer in all the structures determined in different conformational states. As Ste2 undergoes activation, the dimer interface remains intact, but the residues making contacts as well as the surface area of the interface changes subtly. The dimer interfaces in Ste2•LF, Ste2•Ant, and Ste2^{IL}•Ag have the smallest surface areas (2166 Å^2, 2138 Å^2, and 2145 Å^2, respectively). This increases in Ste2^{AL}•Ag and Ste2•Ag•G to 2727 Å^2 and 2519 Å^2, respectively. The core of the interface dimer contacts is constituted by 21 residues in in Ste2_A in all the structures. Another seven residues mediate dimer contacts only upon activation in the Ste2^{AL}•Ag and Ste2•Ag•G states, due to the additional interactions made between the C-terminus of H7 with the cytoplasmic end of H1 in the neighboring protomer. There is also a ~3 Å relative shift of protomer B toward protomer A as Ste2 transitions from the inactive state to the active G protein–coupled state.

The Orthosteric Binding Pocket

The cryo-EM maps of the Ste2 dimer in the antagonist-bound state (Ste2•Ant) and agonist-bound state (Ste2^{IL}•Ag and Ste2^{AL}•Ag) displayed unambiguous density for two antagonist peptides and two agonist α-factor peptides, respectively, with one peptide bound to each protomer of Ste2 (Figure 19.1b). Unless noted otherwise, all discussions are on the α-factor-bound Ste2_A–G protein complex. The orthosteric binding pocket is extensive and involves residues throughout the extracellular surface of the receptor (Figure 19.2). Previous work based on extensive mutagenesis divided the α-factor into three key domains: (1) N-terminal signaling domain (Trp1, His2, Trp3, Leu4), (2) middle bend domain (Lys7, Pro8, Gly9, Gln10), and (3) C-terminal binding domain (Pro11, Met12, Tyr13) (Abel et al., 1998; Naider & Becker, 2004). The N-terminal Trp1 is situated outside the

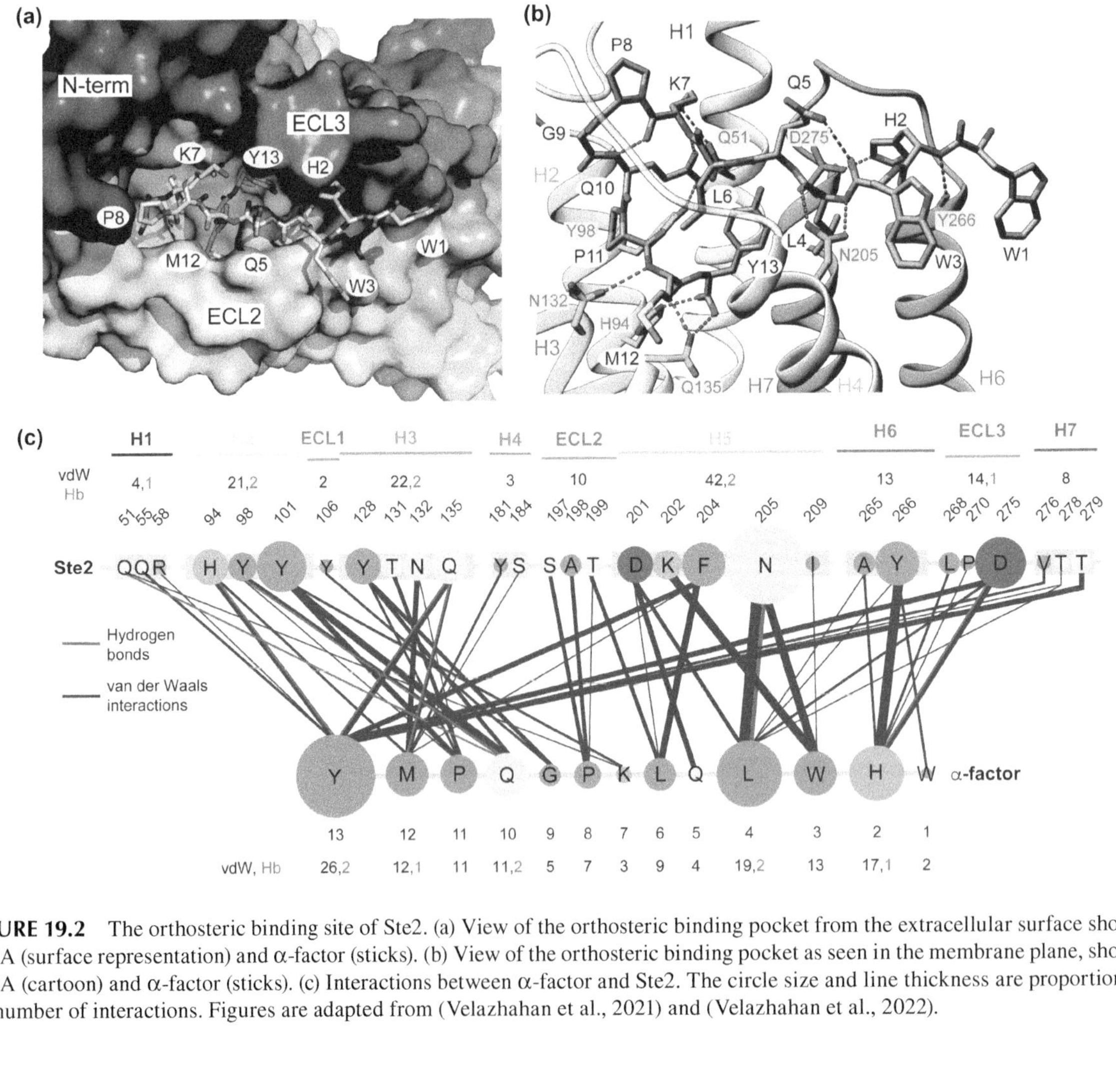

FIGURE 19.2 The orthosteric binding site of Ste2. (a) View of the orthosteric binding pocket from the extracellular surface showing Ste2A (surface representation) and α-factor (sticks). (b) View of the orthosteric binding pocket as seen in the membrane plane, showing Ste2A (cartoon) and α-factor (sticks). (c) Interactions between α-factor and Ste2. The circle size and line thickness are proportional to the number of interactions. Figures are adapted from (Velazhahan et al., 2021) and (Velazhahan et al., 2022).

orthosteric binding pocket and makes only minor contacts with the receptor (Figure 19.2b and c). In contrast, about 38% of the ligand–receptor contacts are mediated by the residues His2-Trp3-Leu4 of the α-factor that make contacts primarily to residues in H5, H6, and ECL3. The C-terminal domain Pro11-Met12-Tyr13 makes extensive contacts with residues in H1, H2, ECL1, H3 and H4, which constitute another 35% of the ligand–receptor contacts (Figure 19.2c). Twenty-three out of the 31 residues in Ste2 make contacts with α-factor, which when mutated result in major defects in ligand binding and/or signaling (Velazhahan et al., 2021).

The cryo-EM structures provide a molecular framework for understanding the key interactions between α-factor and Ste2, several of which had previously been found to be critical based on extensive mutagenesis studies. In the C-terminal binding domain of α-factor, deletion of Met12 and Tyr13 results in a peptide that does not bind (Eriotou-Bargiota et al., 1992). The negative charge of the C-terminal carboxylate of the α-factor is important as amidation of the C-terminus results in a 160-fold decrease in affinity (Zhang et al., 1997). A hydrogen bond is formed between the C-terminus of α-factor and His94^{2x56} (Figure 19.2b). A positive charge at this position is conserved in 90% of the Ste2-like fungal receptors, and it is possible that a salt bridge may form under *in vivo* acidic conditions.

The negative charge of the C-terminal carboxylate of the α-factor is also positioned optimally to interact with the conserved Arg58^{1x48}, which is adjacent to His94^{2x56}, creating an overall positively charged electrostatic field. It is notable that loss of function of Ste2 is observed when either His94^{2x56} or Arg58^{1x48} is mutated (Leavitt et al., 1999; Zuber et al., 2015). The aromaticity of Tyr13 in α-factor is also important, as analogues where Tyr is replaced with Ser or Ala do not bind to Ste2, whereas Phe13 is tolerated (Naider & Becker, 2004; Tantry et al., 2010). Tyr13 forms π-stacking interactions with Phe204^{5x28}, which when mutated (F204C and F204S) results in decreased ligand binding and signaling (Dosil et al., 1998; Lin et al., 2003). In addition, Tyr13 also engages in polar contacts with the side chain of Gln135^{3x36}, and the mutations Q135A and Q135P lead to a decrease in activity and dominant-negative phenotypes, respectively (Dosil et al., 1998; Martin et al., 2002). Met12 of α-factor makes van der Waals contacts with Asn132^{3x33}, and the mutant N132Y results in strong dominant-negative effects (Dosil et al., 1998).

Within the bend domain (Lys7-Gln10), a type II β-turn is present (Figure 19.2b), which is consistent with a biologically active conformation of α-factor based on previous NMR and biochemical studies (Gounarides et al., 1993; Shenbagamurthi et al., 1985). Substitution of Gly9 with L-Ala destabilizes the type II β-turn, and the peptide has a 350-fold decrease in affinity (Abel et al., 1998). His2-Trp3-Leu4 in the N-terminal signaling domain of the α-factor mediates most contacts (34 van der Waals, 4 hydrogen bonds) with Ste2. This region makes extensive contacts with both Asn205^{5x29} and Tyr266^{6x58}, which are essential for Ste2 signaling (Lee et al., 2006). The mutant N205A reduces signaling to less than 5% of wild-type levels (Lee et al., 2006), and the mutant Y266A exhibits a strong dominant-negative phenotype (Lee et al., 2002).

The His2 side chain of α-factor forms a hydrogen bond with Asp275^{ECL3}. This interaction could be a salt bridge under acidic conditions *in vivo* as the protonation of His2 is important for ligand recognition and binding (Blumer & Thorner, 1990; Naider & Becker, 2004). His2 also forms multiple polar contacts with the backbone carbonyl oxygen of Tyr266^{6x58} and Ala265^{6x57} and substitution of His2 with L-Ala results in a 100-fold decreased affinity (Abel et al., 1998).

The Heterotrimeric G Protein

The overall structure of the heterotrimeric G protein Gpa1-Ste4-Ste18 is similar to the structure of G_o coupled to the serotonin receptor 5-HT_{1B} (rmsd 1.7 Å over 2832 atoms) (García-Nafría, Nehmé, et al., 2018). The nucleotide binding pocket of mini-Gpa1 lacks density for GDP, indicating that GDP was not bound. Unless noted otherwise, all discussions are on the α-factor-bound $Ste2_A$–G protein complex. Only the α-subunit mini-Gpa1 of the heterotrimeric G protein makes contacts with Ste2 (Figure 19.1a), and this interface is comprised of 76 van der Waals contacts and 8 hydrogen bonds between 21 residues in mini-Gpa1 and 23 residues in Ste2 (Figure 19.3a–c). Ste2 forms interactions with only two main regions of Gpa1. This includes the β2-loop-β3, which forms 24% of all contacts, and the C-terminal α5 helix, which mediates 71% of all contacts. The loop between β2 and β3 is stacked against the second intracellular loop ICL2 of Ste2 (Figure 19.3a and b). The C-terminal α5 helix of Gpa1 packs against the cytoplasmic ends of H5, H6, and H7 and underneath H3-ICL2-H4. A slight groove exists between H3-ICL2 and H5-H6, which accommodates the side chains of the α5 helix of Gpa1 (Figure 19.3a and b). The distal end of the α5 helix does not penetrate deep into the receptor and is found juxtaposed against the intracellular end of H2. Extensive mutagenesis experiments demonstrate that 12 out of 22 residues in Ste2 making contacts with mini-Gpa1 and 11 out of 21 residues in Gpa1 forming direct contacts have implications for signaling (Velazhahan et al., 2021).

A "wavy hook" is formed by the C-terminal residues Ile469$^{H5.23}$, Gly470$^{H5.24}$, Ile471$^{H5.25}$, and Ile472$^{H5.26}$; superscripts refer to the residues position in the common Gα numbering scheme (Flock et al., 2015) similar to that observed in mammalian G proteins (García-Nafría & Tate, 2019). This region is essential for G protein coupling as mutations in Gly470-Ile472 led to strong defects in signaling and loss-of-function phenotypes (Gladue & Konopka, 2008; Kallal & Kurjan, 1997). Mutations of residues that directly interact with the wavy hook of mini-Gpa1, such as Leu247^{6x39}, Leu289^{7x49}, and Ser293^{7x53}, also impact receptor dimerization and signaling, illustrating the role of H7 in both these processes (Dube et al., 2000; Kim et al., 2009; Lin et al., 2005; Sommers et al., 2000; Umanah et al., 2011). The majority of the contacts made by the α5 helix are hydrophobic. Mutations that alter the hydrophobic regions between the receptor and the G protein led to

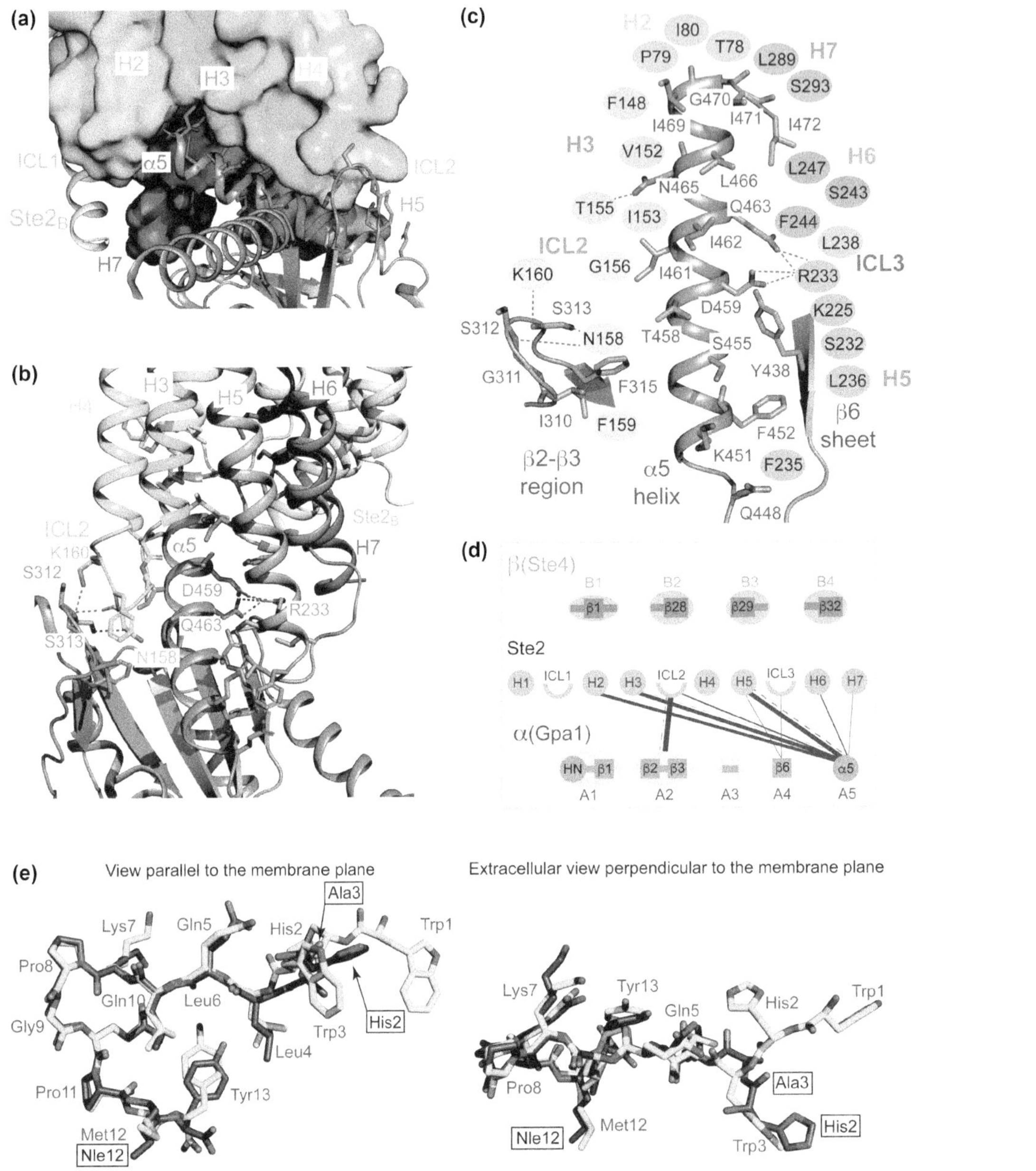

FIGURE 19.3 The G protein coupling site of Ste2. (a) Details of the interactions between the α-subunit (mini-Gpa1, cartoon) and Ste2 (surface). (b) Cartoon representation of panel (a), with side chains at the interface shown as sticks and hydrogen bonds as dashed lines. Only residues forming hydrogen bonds have been labeled. (c) Schematic of the interactions between Gpa1 (cartoon) and residues in Ste2 (ovals). Hydrogen bonds are depicted as dashed lines. (d) Interactions between structural elements in Ste2$_A$ and G protein, with the line thickness proportional to the number of contacts (solid lines, van der Waals contacts; dashed lines, hydrogen bonds). No contacts are formed between Ste2 and Ste4. (e) Superposition of ligand structures in the antagonist-bound state and G protein–coupled agonist-bound state: α-factor, pale grey; antagonist, dark grey. Figures are adapted from (Velazhahan et al., 2021).

dramatic effects. For example, the highly conserved residue Leu289^{7x49} in Ste2 forms van der Waals contacts with Ile471$^{H5.25}$ of mini-Gpa1, and the mutation L289S results in increased basal signaling, and it is an intragenic suppressor of the strong dominant negative mutant F204S. Notably, the conserved residue Leu236^{5x60} of Ste2 that forms hydrophobic contacts with Phe452$^{H5.06}$ and Tyr438$^{S6.02}$ of mini-Gpa1 is a well-characterized G protein coupling-deficient mutant when mutated to Arg or His (Schandel & Jenness, 1994; Stefan et al., 1998; Weiner et al., 1993). A salt bridge is formed between the conserved residue Arg233^{5x57} of Ste2 and Asp459$^{H5.13}$ of mini-Gpa1. Arg233^{5x57} also engages in a polar interaction with the side chain of Gln463$^{H5.17}$ of mini-Gpa1. Positive charge in this region is crucial since Ste2 with R233S/R234G mutations exhibits pheromone-resistant phenotypes (Weiner et al., 1993).

Comparison between the Ligand-Free and Antagonist-Bound States of Ste2

The structures of Ste2•LF and Ste2•Ant are virtually identical (rmsd 0.5 Å). However, a notable difference is that the extracellular end of H5 is kinked outward in Ste2•LF, while it forms a helix in Ste2•Ant. 3D variability analysis, which explores continuous conformational flexibility and motion within a 3D reconstruction (Punjani & Fleet, 2021), was performed on the particles that contributed to the Ste2•LF structure, and it was observed that the extracellular end of H5, near ECL2, was highly flexible.

A consequence of the different positions of H5 is that the side chain of Phe204^{5x28} flips inward by 180° (7 Å inward shift of the Cα atom) upon Ste2 binding the antagonist. This positions the side chain of Phe204^{5x28} to make extensive contacts to the antagonist and also to α-factor in Ste2•Ag•G (primarily to residues Leu6 and Tyr13). The importance of this conformational change is highlighted by the observation that mutation of Phe204^{5x28} and Asn205^{5x29} in this region abrogates both ligand binding and signaling (Dosil et al., 1998; Lee et al., 2006). Lys202^{5x26} flips in the opposite direction, projecting toward the detergent micelle to prevent it from clashing with the ligand. Upon ligand binding, other side chains shift to accommodate the ligand (e.g. Tyr128^{3x29}, Asn132^{5x29}, Gln135^{3x36}, Asp274^{ECL3}, Asp275^{ECL3}) while Asn205^{5x29} undergoes a slight change in its orientation to make stronger interactions with the antagonist. Mutagenesis studies indicate the essential role of all of these residues for ligand binding and/or signaling (Bajaj et al., 2007; Mathew et al., 2011; Zuber et al., 2015).

Comparison between the Antagonist-Bound and the G Protein–Coupled States

The structure of Ste2•Ant shares similarities with Ste2•Ag•G (rmsd 1.3 Å), but there are also some striking differences. The antagonist adopts a hairpin shape in the orthosteric binding pocket similar to that of the agonist α-factor; however, the antagonist interacts with fewer residues in Ste2 (20 compared to 32 in Ste2•Ag•G). Substantial differences are observed at the N-terminus of the ligands, which is congruent with the observation that the N-terminus of the ligands define their efficacy (Raths et al., 1988). The N-terminus of the antagonist adopts a linear shape with His2 forming a hydrogen bond with Asn205^{5x29} on H5 and Ala3 interacting with Thr274^{ECL3} in ECL3 (Figure 19.3e). In contrast, the N-terminus of α-factor adopts a Y-shape, which positions Trp1 and Trp3 wide apart (Figure 19.3e) with Trp3 adopting a similar position to His2 in the antagonist and resulting in interactions with H5 (Lys202^{5x26}, Asn205^{5x29}, Ile209^{5x33}), whereas Trp1 makes contacts with H6 (Ile263^{6x55}, Tyr266^{6x58}). To accommodate this conformation of the α-factor, there occurs a 5.7 Å shift of the extracellular end of H6 (measured at Cα Ser267^{6x59}), a 7.2 Å shift of the extracellular end of H7 (measured at Cα Gly273^{ECL3}), and a rearrangement of ECL3 (Figure 19.4a). This is an important feature of receptor activation, and this movement of the extracellular end of H6 is accompanied by a rotamer change of Tyr266^{6x58} that displaces a water molecule and allows it to make interactions with α-factor and Asn205^{5x29} that are not feasible in the antagonist-bound state.

The differences in the structures of the extracellular segments of Ste2•Ant and Ste2•Ag•G are accompanied by significant changes in the intracellular regions of H5, H6, and H7 (Figure 19.4). The cytoplasmic end of H6 in Ste2•Ant is pivoted away from the receptor core by 38° (12 Å shift at Cα Ser243^{6x35}) relative to its position in Ste2•Ag•G. This segment is also rotated by 170° along the helix axis. Furthermore, the intracellular end of H5 is positioned 6.6 Å away from the receptor core compared to the active state. Several notable changes occur at H7, which adopts significantly different conformations in the two different states (Figure 19.4). In the active Ste2•Ag•G structure, the intracellular end of H7 forms an α-helix and kinks outward from the receptor core by 20 Å (at Cα Asn301^{7x61}), allowing it to interact with the other Ste2 protomer within the dimer. In contrast, H7 in Ste2•Ant is rotated toward the core of the receptor by 47° compared to its position in the active state, and the residues Ala296^{7x56}-Ser303^{7x63} at the intracellular end of H7 form an unstructured coil. No density was observed for the 128 C-terminal residues in all structures, which is consistent with their prediction to be unstructured (Sonnhammer et al., 1998). The intracellular end of H7, which forms a disordered coil in Ste2•Ant, overlaps with the position of the α5 helix of the α-subunit Gpa1 in Ste2•Ag•G and therefore results in a steric block to G protein coupling. Intriguingly, two putative sterol molecules are observed at this site in H7 in the inactive states of Ste2, but this is not present in the active states; it is known that the sterol ergosterol synthesized in *S. cerevisiae* is important for Ste2 function (Morioka et al., 2013).

Agonist-Bound States of Ste2

While only a single major conformation of the Ste2 dimer was obtained from processing the cryo-EM datasets for Ste2•LF, Ste2•Ant, and Ste2•Ag•G states, two distinct conformations of Ste2 were isolated computationally from the Ste2•Ag dataset. One structure (Ste2^{IL}•Ag) was very similar to Ste2•Ant (rmsd 0.7 Å over 263 Cα atoms, protomer A aligned), and the other structure (Ste2^{AL}•Ag) was very similar to Ste2•Ag•G (rmsd 0.45 Å over 290 Cα atoms, protomer A aligned). The structures of Ste2•Ant and Ste2•Ag•G represent two extremes in the transition from an inactive state to an active state, and both Ste2^{IL}•Ag and Ste2^{AL}•Ag are consistent with being intermediate states (Figure 19.4).

The conformation of Ste2^{IL}•Ag on the extracellular surface of Ste2 is halfway between that of Ste2•Ant and Ste2•Ag•G. Binding of the α-factor causes displacement of H6 by 5.3 Å (measured at the Cα of Ser267^{6x59}); however, this is short of its relative position in Ste2•Ag•G by 0.4 Å. Agonist-induced reorientation of ECL3 is also observed. However, there is no significant displacement of the top of H7 similar to that observed in Ste2•Ag•G. On the

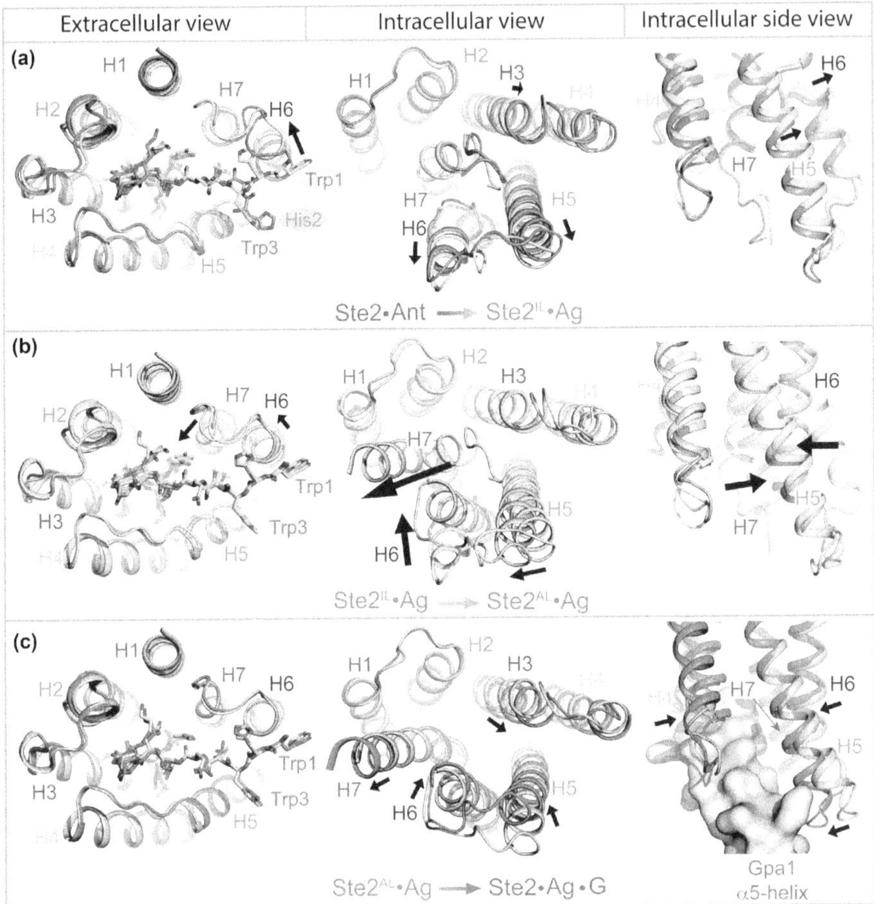

FIGURE 19.4 Structural changes upon activation of Ste2. (a–c) Superposition of Ste2•Ant, Ste2•Ag^{IL}, Ste2•Ag^{AL}, and Ste2•Ag•G with transitions between respective pairs of structures indicated by arrows. Figure is adapted from (Velazhahan et al., 2022).

cytoplasmic side of Ste2, the intracellular end of H7 forms a disordered coil similar to that observed in Ste2•Ant, and this sterically blocks G protein coupling. No major structural changes essential for G protein coupling have occurred on the intracellular surface of Ste2 (Figure 19.4a).

The major structural changes that allow G protein coupling occur during the transition from Ste2^{IL}•Ag to Ste2^{AL}•Ag (Figure 19.4b). The extracellular surface of Ste2^{AL}•Ag is virtually identical to that of Ste2•Ag•G, as the full shift of H6 by 5.7 Å and H7 by 7.2 Å (at Cα Gly273^{ECL3}) have occurred. On the intracellular side of Ste2, H7 in Ste2^{AL}•Ag kinks outward from the receptor core by 19.4 Å (at Cα Asn301^{7x61}), and H6 moves inward by 11.2 Å (at Cα Ser243^{6x35}), which are only 0.6 Å and 0.7 Å short of their respective positions in Ste2•Ag•G. There is also a 5 Å lateral movement (measured at Cα Arg234^{5x58}) of the cytoplasmic end of H5 in Ste2^{AL}•Ag.

Although the structural changes in Ste2^{AL}•Ag state unblock the G protein coupling site, additional structural changes are necessary to allow full engagement of the G protein (Figure 19.4c). While transitioning to the G protein–coupled state, Ste2^{AL}•Ag undergoes inward movements of H5 (4.7 Å at Cα Phe235^{5x59}), H6 (1.2 Å at Cα Phe244^{6x36}), and ICL3. This is accompanied by lateral movement of H3, H4, and ICL2 by 1–2 Å. This leads to the repositioning of key residues, such as Asn158^{ICL2} (1.8 Å at Cα), the side chain of Lys225^{5x49} in H5, Leu247^{6x39} in H6 (1 Å at Cα), and the side chain of Leu289^{7x49} in H7, which make major contacts with Gpa1. The essential role of these residues in signaling has previously been characterized (Velazhahan et al., 2021).

The Activation Mechanism of Ste2

Structural analysis of the five cryo-EM structures of Ste2 suggests the following activation mechanism (Figure 19.4). Binding of α-factor to the ligand-free inactive state of Ste2 (Ste2•LF) results in the reorientation of the extracellular end of H5 to allow ligand binding. There is also an outward movement of the extracellular end of H6 and a shift in the position of ECL3. This transitions the receptor from Ste2•LF to Ste2^{IL}•Ag state and is accompanied by the contraction of the orthosteric binding site by an average of 27%. The Ste2^{IL}•Ag state exists in equilibrium with the Ste2^{AL}•Ag state where the extracellular end of H7 has shifted inward and the intracellular end of H7 has moved outward from the receptor core by 20 Å, with the intracellular end of H6 moving inward by 12 Å. The structural transitions occurring on the intracellular surface of Ste2 allow the binding of the α5-helix of Gpa1, and this is accompanied by a further contraction in the ligand binding pocket with increased contacts between the agonist and H5. The cytoplasmic ends of H3, H4, and H5 move inward upon binding Gpa1, resulting in a further contraction of the G protein binding site. There is also a minor increase in the volume of the ligand binding pocket and minor changes on the extracellular face of Ste2 upon G protein coupling.

DISCUSSION

Comparison of Ste2 with Other GPCRs

There is no appreciable amino acid sequence similarity between Ste2 and mammalian GPCRs. The overall arrangement of the transmembrane helices in Ste2 was found to be similar to the active-state structures of class A, B, and F GPCRs (Figure 19.5a). However, H4 was positioned differently by over 20 Å at the intracellular side of the receptor when compared with the serotonin 5-HT_{1B}

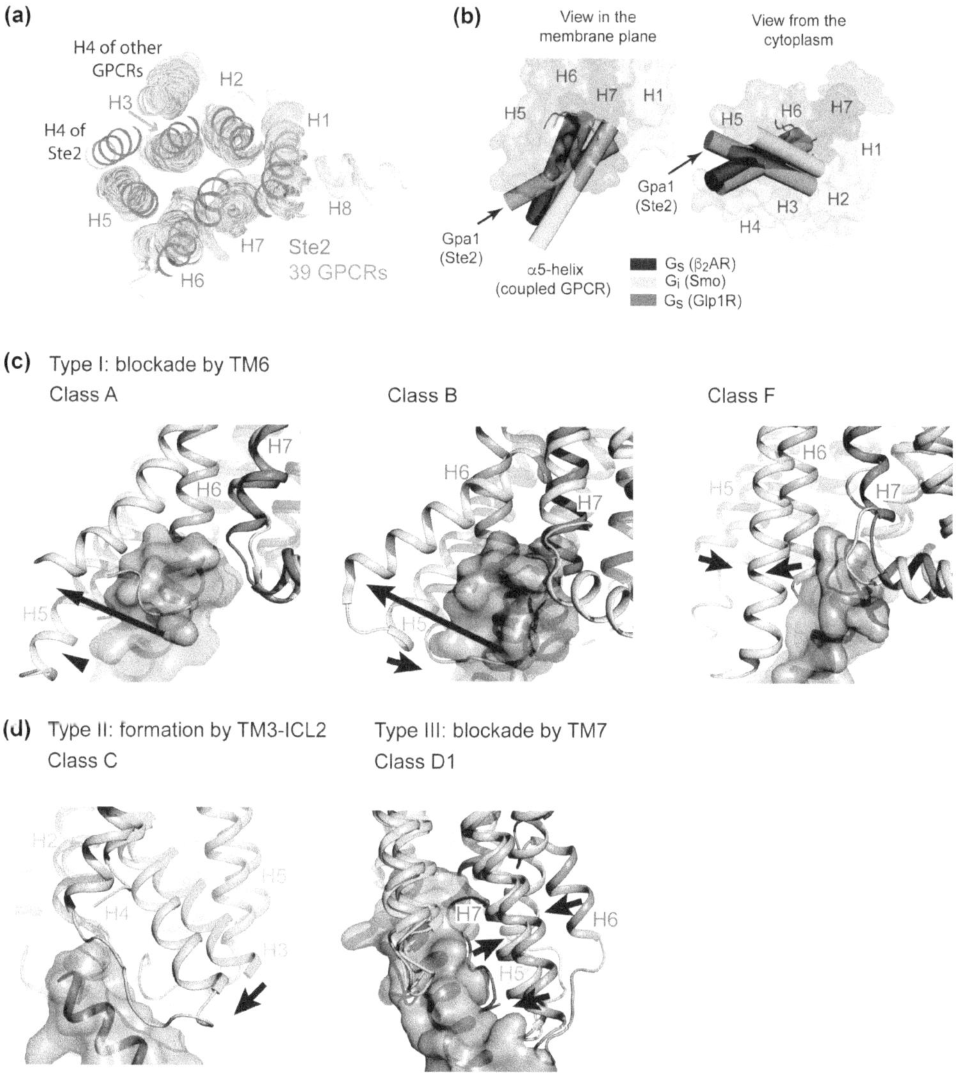

FIGURE 19.5 Comparison of Ste2 with other GPCR classes. (a) Alignment of 39 mammalian GPCRs in an active state (pale grey) with Ste2 (dark grey), viewed from the extracellular surface. (b) Positions of the α5 helices (cylinders) of coupled G proteins after global alignment of the receptors used in (a). The positions of the α5 helices are shown relative to Ste2 shown as a transparent surface. (c, d) Representative examples of GPCRs from each of the major classes are shown in the inactive state and active state coupled to a G protein (surface; only the C-terminal α5 helix is shown). Upon receptor activation, the secondary structure changes (arrows) to form the binding site for the α5 helix of the G protein α-subunit through either the outward movement of helices from the binding site and/or movement of helices inward to form the binding site. Structures shown (inactive and active states, respectively) are class A, β2-adrenoceptor (PDB IDs 2RH1 and 3SN6); class B, CRF receptor 1 (PDB IDs 4K5Y and 6P9X); class F, smoothened receptor (PDB IDs 5L7I and 6OT0); class C, GABAB receptor (PDB IDs 7EB2 and 7C7S); class D1, Ste2 receptor (PDB IDs 7QA8 and 7AD3). Figures are adapted from (Velazhahan et al., 2021) and (Velazhahan et al., 2022).

receptor–G_0 complex (García-Nafría, Nehmé, et al., 2018); PDB 6G79.

A comparison of the positions of the C-terminal α5 helix of Gpa1 coupled to Ste2 with the positions of α5 helices of mammalian G proteins coupled to class A, B, and F receptors revealed considerable differences in the angles at which they insert into the receptor when viewed both in the plane of the lipid bilayer and perpendicular to the lipid bilayer (Figure 19.5b). While the α5 helices of G proteins coupled to class A, B, and F receptors all insert to approximately the same depth (García-Nafría & Tate, 2019), the insertion of the α5 helix of Gpa1 was found to be less deep by 12 Å compared to G_0-coupled 5-HT_{1B}R (García-Nafría, Nehme, et al., 2018); PDB 6G79. This suggests a similarity between Ste2 and class C receptors where the G protein binds to a shallow cleft (Mao et al., 2020). The different insertion angle of the α5 helix of Gpa1 positions the G proteins such that the βγ (Ste4-Ste18) subunits do not interact with either Ste2 protomer.

Comparison of GPCR Activation Mechanisms across Classes

The mechanism of activation of mammalian GPCRs (classes A, B, C, and F) has been studied extensively (García-Nafría & Tate, 2019; Qi et al., 2019; Shen et al., 2021). Upon agonist binding, the GPCR undergoes conformational changes that transition it from the inactive state to an active state that allows G protein coupling (Weis & Kobilka, 2018). Based on the currently available structural knowledge of how the G protein binding site is configured to prevent coupling of the G protein in the inactive state, GPCRs could be classified into three different categories (types I–III, Figure 19.5c and d). In the case of both class A and B GPCRs (Figure 19.5c), the intracellular end of H6 forms the primary block to G protein coupling. Upon agonist binding and activation of the receptor, H6 swings away from the receptor core by 10 to 15 Å, resulting in the formation of a cleft in the cytoplasmic face of the receptor that enables binding of the α5 helix of the G protein α-subunit (García-Nafría & Tate, 2019). In class F GPCRs (Figure 19.5c), it is also the cytoplasmic end of H6 that blocks the G protein coupling site, but only a smaller movement (6 Å) of H6 is required owing to the different angle of insertion of the α5 helix of the G protein (Qi et al., 2019). As H6 presents the primary block to G protein coupling, the activation mechanism of class A, B, and F GPCRs could be grouped together under the category type I. In contrast, coupling of the G protein to class C receptors is very different (Figure 19.5d), as the binding site for the α5 helix of the G protein is formed primarily by ICL2 and a movement of H3 by 5 Å, and this allows the α5 helix to engage the periphery of the receptor (Shen et al., 2021). Thus, class C GPCRs exemplify the type II activation mechanism in GPCRs.

The structures of Ste2 determined here reveal that Ste2 has an activation mechanism distinct from any other GPCR studied thus far and is denoted type III (Figure 19.5d). In Ste2, H7 forms the primary steric block to G protein coupling. Activation of Ste2 involves an outward movement of H7, thereby removing the steric block and formation of the G protein binding site by an inward movement of H6. The inward movement of H6 by 12 Å in Ste2 upon activation is in contrast to class A GPCRs, where an outward movement of H6 to a similar extent occurs to allow G protein coupling (Figure 19.5c and d). Thus, Ste2 has evolved a fundamentally different activation mechanism to configure the movement of transmembrane helices to allow activation of the receptor.

REFERENCES

Abel, M., Zhang, Y., Lu, H. F., Naider, F., & Becker, J. (1998). Structure-function analysis of the Saccharomyces cerevisiae tridecapeptide pheromone using alanine-scanned analogs. *The Journal of peptide research*, *52*(2), 95–106.

Alvaro, C. G., & Thorner, J. (2016). Heterotrimeric G protein-coupled receptor signaling in yeast mating pheromone response. *Journal of Biological Chemistry*, *291*(15), 7788–7795.

Attwood, T., & Findlay, J. (1994). Fingerprinting G-protein-coupled receptors. *Protein Engineering, Design and Selection*, *7*(2), 195–203.

Bajaj, A., Connelly, S. M., Gehret, A. U., Naider, F., & Dumont, M. E. (2007). Role of extracellular charged amino acids in the yeast α-factor receptor. *Biochimica et Biophysica Acta (BBA)-Molecular Cell Research*, *1773*(6), 707–717.

Bloch, J. S., Mukherjee, S., Kowal, J., Filippova, E. V., Niederer, M., Pardon, E., Steyaert, J., Kossiakoff, A. A., & Locher, K. P. (2021). Development of a universal nanobody-binding Fab module for fiducial-assisted cryo-EM studies of membrane proteins. *Proceedings of the National Academy of Sciences*, *118*(47), e2115435118.

Blumer, K. J., & Thorner, J. (1990). Beta and gamma subunits of a yeast guanine nucleotide-binding protein are not essential for membrane association of the alpha subunit but are required for receptor coupling. *Proceedings of the National Academy of Sciences*, *87*(11), 4363–4367.

Brown, G. D., Denning, D. W., Gow, N. A., Levitz, S. M., Netea, M. G., & White, T. C. (2012). Hidden killers: human fungal infections. *Science Translational Medicine*, *4*(165), 165rv113.

Brown, N. A., & Goldman, G. H. (2016). The contribution of Aspergillus fumigatus stress responses to virulence and antifungal resistance. *Journal of Microbiology*, *54*, 243–253.

Brown, N. A., Schrevens, S., Van Dijck, P., & Goldman, G. H. (2018). Fungal G-protein-coupled receptors: mediators of pathogenesis and targets for disease control. *Nature Microbiology*, *3*(4), 402–414.

Burkholder, A. C., & Hartwell, L. H. (1985). The yeast α-factor receptor: structural properties deduced from the sequence of the STE2 gene. *Nucleic acids research*, *13*(23), 8463–8475.

Carpenter, B., & Tate, C. G. (2016). Engineering a minimal G protein to facilitate crystallisation of G protein-coupled receptors in their active conformation. *Protein Engineering, Design and Selection*, *29*(12), 583–594.

Chen, H., Chen, K., Huang, W., Staudt, L. M., Cyster, J. G., & Li, X. (2022). Structure of S1PR2–heterotrimeric G13 signaling complex. *Science advances*, *8*(13), eabn0067.

Chen, Q., Plasencia, M., Li, Z., Mukherjee, S., Patra, D., Chen, C.-L., Klose, T., Yao, X.-Q., Kossiakoff, A. A., & Chang, L. (2021). Structures of rhodopsin in complex with G-protein-coupled receptor kinase 1. *Nature*, *595*(7868), 600–605.

Di Segni, G., Gastaldi, S., Zamboni, M., & Tocchini-Valentini, G. P. (2011). Yeast pheromone receptor genes STE2 and STE3 are differently regulated at the transcription and poly-adenylation level. *Proceedings of the National Academy of Sciences*, *108*(41), 17082–17086.

Dosil, M., Giot, L., Davis, C., & Konopka, J. B. (1998). Dominant-negative mutations in the G-protein-coupled α-factor receptor map to the extracellular ends of the transmembrane segments. *Molecular and Cellular Biology*, *18*(10), 5981–5991.

Dube, P., DeCostanzo, A., & Konopka, J. B. (2000). Interaction between transmembrane domains five and six of the α-factor receptor. *Journal of Biological Chemistry*, *275*(34), 26492–26499.

Eriotou-Bargiota, E., Xue, C. B., Naider, F., & Becker, J. M. (1992). Antagonistic and synergistic peptide analogs of the tridecapeptide mating pheromone of Saccharomyces cerevisiae. *Biochemistry*, *31*(2), 551–557.

Flock, T., Ravarani, C. N., Sun, D., Venkatakrishnan, A. J., Kayikci, M., Tate, C. G., Veprintsev, D. B., & Babu, M. M. (2015). Universal allosteric mechanism for Gα activation by GPCRs. *Nature*, *524*(7564), 173–179.

García-Nafría, J., Lee, Y., Bai, X., Carpenter, B., & Tate, C. G. (2018). Cryo-EM structure of the adenosine A2A receptor coupled to an engineered heterotrimeric G protein. *elife*, *7*, e35946.

García-Nafría, J., Nehmé, R., Edwards, P. C., & Tate, C. G. (2018). Cryo-EM structure of the serotonin 5-HT1B receptor coupled to heterotrimeric Go. *Nature*, *558*(7711), 620–623.

García-Nafría, J., & Tate, C. G. (2019). Cryo-EM structures of GPCRs coupled to Gs, Gi and Go. *Molecular and Cellular Endocrinology*, *488*, 1–13.

García-Nafría, J., & Tate, C. G. (2021). Structure determination of GPCRs: cryo-EM compared with X-ray crystallography. *Biochemical Society Transactions*, *49*(5), 2345–2355.

Gehret, A. U., Bajaj, A., Naider, F., & Dumont, M. E. (2006). Oligomerization of the yeast α-factor receptor: implications for dominant negative effects of mutant receptors. *Journal of Biological Chemistry*, *281*(30), 20698–20714.

Gehret, A. U., Connelly, S. M., & Dumont, M. E. (2012). Functional and physical interactions among Saccharomyces cerevisiae α-factor receptors. *Eukaryotic Cell*, *11*(10), 1276–1288.

Gladue, D. P., & Konopka, J. B. (2008). Scanning mutagenesis of regions in the Gα protein Gpa1 that are predicted to interact with yeast mating pheromone receptors. *FEMS Yeast Research*, *8*(1), 71–80.

Gounarides, J., Broido, M., Becker, J., & Naider, F. (1993). Conformational analysis of [D-Ala9]. alpha.-factor and [L-Ala9]. alpha.-factor in solution and in the presence of lipid. *Biochemistry*, *32*(3), 908–917.

Hauser, A. S., Attwood, M. M., Rask-Andersen, M., Schiöth, H. B., & Gloriam, D. E. (2017). Trends in GPCR drug discovery: new agents, targets and indications. *Nature reviews Drug Discovery*, *16*(12), 829–842.

Henderson, R. (1995). The potential and limitations of neutrons, electrons and X-rays for atomic resolution microscopy of unstained biological molecules. *Quarterly Reviews of Biophysics*, *28*(2), 171–193.

Huang, W., Masureel, M., Qu, Q., Janetzko, J., Inoue, A., Kato, H. E., Robertson, M. J., Nguyen, K. C., Glenn, J. S., & Skiniotis, G. (2020). Structure of the neurotensin receptor 1 in complex with β-arrestin 1. *Nature*, *579*(7798), 303–308.

Kallal, L., & Kurjan, J. (1997). Analysis of the receptor binding domain of Gpa1p, the G (alpha) subunit involved in the yeast pheromone response pathway. *Molecular and Cellular Biology*, *17*(5), 2897–2907.

Kim, H., Lee, B.-K., Naider, F., & Becker, J. M. (2009). Identification of specific transmembrane residues and ligand-induced interface changes involved in homo-dimer formation of a yeast G protein-coupled receptor. *Biochemistry*, *48*(46), 10976–10987.

Kim, S., Jeon, T.-J., Oberai, A., Yang, D., Schmidt, J. J., & Bowie, J. U. (2005). Transmembrane glycine zippers: physiological and pathological roles in membrane proteins. *Proceedings of the National Academy of Sciences*, *102*(40), 14278–14283.

Kobilka, B. K. (2007). G protein coupled receptor structure and activation. *Biochimica et Biophysica Acta (BBA)-Biomembranes*, *1768*(4), 794–807.

Koehl, A., Hu, H., Maeda, S., Zhang, Y., Qu, Q., Paggi, J. M., Latorraca, N. R., Hilger, D., Dawson, R., & Matile, H. (2018). Structure of the μ-opioid receptor–Gi protein complex. *Nature*, *558*(7711), 547–552.

Kolakowski Jr, L. F. (1994). GCRDb: a G-protein-coupled receptor database. *Receptors & channels*, *2*(1), 1–7.

Leavitt, L., Macaluso, C., Kim, K., Martin, N., & Dumont, M. (1999). Dominant negative mutations in the α-factor receptor, a G protein-coupled receptor encoded by the STE2 gene of the yeast Saccharomyces cerevisiae. *Molecular and General Genetics MGG*, *261*, 917–932.

Lee, B.-K., Lee, Y.-H., Hauser, M., Son, C. D., Khare, S., Naider, F., & Becker, J. M. (2002). Tyr266 in the sixth transmembrane domain of the yeast α-factor receptor plays key roles in receptor activation and ligand specificity. *Biochemistry*, *41*(46), 13681–13689.

Lee, Y., Warne, T., Nehmé, R., Pandey, S., Dwivedi-Agnihotri, H., Chaturvedi, M., Edwards, P. C., García-Nafría, J., Leslie, A. G., & Shukla, A. K. (2020). Molecular basis of β-arrestin coupling to formoterol-bound β1-adrenoceptor. *Nature*, *583*(7818), 862–866.

Lee, Y.-H., Naider, F., & Becker, J. M. (2006). Interacting residues in an activated state of a G protein-coupled receptor. *Journal of Biological Chemistry*, *281*(4), 2263–2272.

Liang, Y.-L., Zhao, P., Draper-Joyce, C., Baltos, J.-A., Glukhova, A., Truong, T. T., May, L. T., Christopoulos, A., Wootten, D., & Sexton, P. M. (2018). Dominant negative G proteins enhance formation and purification of agonist-GPCR-G protein complexes for structure determination. *ACS Pharmacology & Translational Science*, *1*(1), 12–20.

Lin, J. C., Duell, K., Saracino, M., & Konopka, J. B. (2005). Identification of residues that contribute to receptor activation through the analysis of compensatory mutations in the G protein-coupled α-factor receptor. *Biochemistry*, *44*(4), 1278–1287.

Lin, J. C., Parrish, W., Eilers, M., Smith, S. O., & Konopka, J. B. (2003). Aromatic residues at the extracellular ends of transmembrane domains 5 and 6 promote ligand activation of the G protein-coupled α-factor receptor. *Biochemistry*, *42*(2), 293–301.

Maeda, S., Koehl, A., Matile, H., Hu, H., Hilger, D., Schertler, G. F., Manglik, A., Skiniotis, G., Dawson, R. J., & Kobilka, B. K. (2018). Development of an antibody fragment that stabilizes GPCR/G-protein complexes. *Nature Communications*, *9*(1), 3712.

Mao, C., Shen, C., Li, C., Shen, D.-D., Xu, C., Zhang, S., Zhou, R., Shen, Q., Chen, L.-N., & Jiang, Z. (2020). Cryo-EM structures of inactive and active GABAB receptor. *Cell Research*, *30*(7), 564–573.

Martin, N. P., Ćelić, A., & Dumont, M. E. (2002). Mutagenic mapping of helical structures in the transmembrane segments of the yeast α-factor receptor. *Journal of Molecular Biology*, *317*(5), 765–788.

Mathew, E., Bajaj, A., Connelly, S. M., Sargsyan, H., Ding, F.-X., Hajduczok, A. G., Naider, F., & Dumont, M. E. (2011). Differential interactions of fluorescent agonists and antagonists with the yeast G protein coupled receptor Ste2p. *Journal of Molecular Biology*, *409*(4), 513–528.

Mentesana, P. E., & Konopka, J. B. (2001). Mutational analysis of the role of N-glycosylation in α-factor receptor function. *Biochemistry*, *40*(32), 9685–9694.

Morioka, S., Shigemori, T., Hara, K., Morisaka, H., Kuroda, K., & Ueda, M. (2013). Effect of sterol composition on the activity of the yeast G-protein-coupled receptor Ste2. *Applied Microbiology and Biotechnology*, *97*(9), 4013–4020.

Naider, F., & Becker, J. M. (2004). The α-factor mating pheromone of Saccharomyces cerevisiae: a model for studying the interaction of peptide hormones and G protein-coupled receptors. *Peptides*, *25*(9), 1441–1463.

Naider, F., & Becker, J. M. (2020). A paradigm for peptide hormone-GPCR analyses. *Molecules*, *25*(18), 4272.

Nehme, R., Carpenter, B., Singhal, A., Strege, A., Edwards, P. C., White, C. F., Du, H., Grisshammer, R., & Tate, C. G. (2017). Mini-G proteins: Novel tools for studying GPCRs in their active conformation. *PloS one*, *12*(4), e0175642.

Overton, M. C., & Blumer, K. J. (2000). G-protein-coupled receptors function as oligomers in vivo. *Current Biology*, *10*(6), 341–344.

Overton, M. C., Chinault, S. L., & Blumer, K. J. (2003). Oligomerization, biogenesis, and signaling is promoted by a glycophorin A-like dimerization motif in transmembrane domain 1 of a yeast G protein-coupled receptor. *Journal of Biological Chemistry*, *278*(49), 49369–49377.

Pándy-Szekeres, G., Munk, C., Tsonkov, T. M., Mordalski, S., Harpsøe, K., Hauser, A. S., Bojarski, A. J., & Gloriam, D. E. (2018). GPCRdb in 2018: adding GPCR structure models and ligands. *Nucleic Acids Research*, *46*(D1), D440–D446.

Punjani, A., & Fleet, D. J. (2021). 3D variability analysis: resolving continuous flexibility and discrete heterogeneity from single particle cryo-EM. *Journal of Structural Biology*, *213*(2), 107702.

Qi, X., Liu, H., Thompson, B., McDonald, J., Zhang, C., & Li, X. (2019). Cryo-EM structure of oxysterol-bound human Smoothened coupled to a heterotrimeric Gi. *Nature*, *571*(7764), 279–283.

Rasmussen, S. G., DeVree, B. T., Zou, Y., Kruse, A. C., Chung, K. Y., Kobilka, T. S., Thian, F. S., Chae, P. S., Pardon, E., & Calinski, D. (2011). Crystal structure of the β2 adrenergic receptor–Gs protein complex. *Nature*, *477*(7366), 549–555.

Raths, S., Naider, F., & Becker, J. (1988). Peptide analogues compete with the binding of alpha-factor to its receptor in Saccharomyces cerevisiae. *Journal of Biological Chemistry*, *263*(33), 17333–17341.

Robertson, M. J., Papasergi-Scott, M. M., He, F., Seven, A. B., Meyerowitz, J. G., Panova, O., Peroto, M. C., Che, T., & Skiniotis, G. (2022). Structure determination of inactive-state GPCRs with a universal nanobody. *Nature Structural & Molecular Biology*, *29*, 1188–1195.

Schandel, K. A., & Jenness, D. D. (1994). Direct evidence for ligand-induced internalization of the yeast alpha-factor pheromone receptor. *Molecular and Cellular Biology*, *14*(11), 7245–7255.

Shen, C., Mao, C., Xu, C., Jin, N., Zhang, H., Shen, D.-D., Shen, Q., Wang, X., Hou, T., & Chen, Z. (2021). Structural basis of GABAB receptor–Gi protein coupling. *Nature*, *594*(7864), 594–598.

Shen, J., Zhang, D., Fu, Y., Chen, A., Yang, X., & Zhang, H. (2022). Cryo-EM structures of human bradykinin receptor-Gq proteins complexes. *Nature Communications*, *13*(1), 714.

Shenbagamurthi, P., Kundu, B., Raths, S., Becker, J. M., & Naider, F. (1985). Biological activity and conformational isomerism in position 9 analogs of the des-1-tryptophan, 3. beta.-cyclohexylalanine-. alpha.-factor from Saccharomyces cerevisiae. *Biochemistry*, *24*(25), 7070–7076.

Sommers, C. M., Martin, N. P., Akal-Strader, A., Becker, J. M., Naider, F., & Dumont, M. E. (2000). A limited spectrum of mutations causes constitutive activation of the yeast α-factor receptor. *Biochemistry*, *39*(23), 6898–6909.

Sonnhammer, E. L., Von Heijne, G., & Krogh, A. (1998). A hidden Markov model for predicting transmembrane helices in protein sequences. *Proceedings / ... International Conference on Intelligent Systems for Molecular Biology*, *6*, 175–182.

Staus, D. P., Hu, H., Robertson, M. J., Kleinhenz, A. L., Wingler, L. M., Capel, W. D., Latorraca, N. R., Lefkowitz, R. J., & Skiniotis, G. (2020). Structure of the M2 muscarinic receptor–β-arrestin complex in a lipid nanodisc. *Nature*, *579*(7798), 297–302.

Stefan, C. J., Overton, M. C., & Blumer, K. J. (1998). Mechanisms governing the activation and trafficking of yeast G protein-coupled receptors. *Molecular Biology of the Cell*, *9*(4), 885–899.

Stoneman, M. R., Paprocki, J. D., Biener, G., Yokoi, K., Shevade, A., Kuchin, S., & Raicu, V. (2017). Quaternary structure of the yeast pheromone receptor Ste2 in living cells. *Biochimica et Biophysica Acta (BBA)-Biomembranes*, *1859*(9), 1456–1464.

Tantry, S., Ding, F.-X., Dumont, M., Becker, J. M., & Naider, F. (2010). Binding of fluorinated phenylalanine α-factor analogues to ste2p: Evidence for a cation– π binding interaction between a peptide ligand and its cognate G protein-coupled receptor. *Biochemistry*, *49*(24), 5007–5015.

Umanah, G. K., Huang, L.-Y., Maccarone, J. M., Naider, F., & Becker, J. M. (2011). Changes in conformation at the cytoplasmic ends of the fifth and sixth transmembrane helices of a yeast G protein-coupled receptor in response to ligand binding. *Biochemistry*, *50*(32), 6841–6854.

Van Dijck, P., Brown, N. A., Goldman, G. H., Rutherford, J., Xue, C., & Van Zeebroeck, G. (2017). Nutrient sensing at the plasma membrane of fungal cells. *Microbiology Spectrum*, *5*(2). https://doi.org/10.1128/microbiolspec.funk-0031-2016

Velazhahan, V., Ma, N., Pándy-Szekeres, G., Kooistra, A. J., Lee, Y., Gloriam, D. E., Vaidchi, N., & Tate, C. G. (2021). Structure of the class D GPCR Ste2 dimer coupled to two G proteins. *Nature*, *589*(7840), 148–153.

Velazhahan, V., Ma, N., Vaidehi, N., & Tate, C. G. (2022). Activation mechanism of the class D fungal GPCR dimer Ste2. *Nature*, *603*(7902), 743–748.

Velazhahan, V., McCann, B. L., Bignell, E., & Tate, C. G. (2023). Developing novel antifungals: lessons from G protein-coupled receptors. *Trends in pharmacological sciences*, *44*(3), 162–174.

Wang, H. X., & Konopka, J. B. (2009). Identification of amino acids at two dimer interface regions of the α-factor receptor (Ste2). *Biochemistry*, *48*(30), 7132–7139.

Weiner, J., Guttierez-Steil, C., & Blumer, K. (1993). Disruption of receptor-G protein coupling in yeast promotes the function of an SST2-dependent adaptation pathway. *Journal of Biological Chemistry, 268*(11), 8070–8077.

Weis, W. I., & Kobilka, B. K. (2018). The molecular basis of G protein–coupled receptor activation. *Annual review of biochemistry, 87*, 897.

Woolhouse, M., & Farrar, J. (2014). Policy: an intergovernmental panel on antimicrobial resistance. *Nature, 509*(7502), 555–557.

Xue, C., Hsueh, Y.-P., & Heitman, J. (2008). Magnificent seven: roles of G protein-coupled receptors in extracellular sensing in fungi. *FEMS microbiology reviews, 32*(6), 1010–1032.

Yesilaltay, A., & Jenness, D. D. (2000). Homo-oligomeric complexes of the yeast α-factor pheromone receptor are functional units of endocytosis. *Molecular Biology of the Cell, 11*(9), 2873–2884.

Yin, W., Li, Z., Jin, M., Yin, Y.-L., De Waal, P. W., Pal, K., Yin, Y., Gao, X., He, Y., & Gao, J. (2019). A complex structure of arrestin-2 bound to a G protein-coupled receptor. *Cell Research, 29*(12), 971–983.

Zhang, K., Wu, H., Hoppe, N., Manglik, A., & Cheng, Y. (2022). Fusion protein strategies for cryo-EM study of G protein-coupled receptors. *Nature Communications, 13*(1), 4366.

Zhang, Y. L., LU, H. F., Becker, J. M., & Naider, F. (1997). Position one analogs of the Saccharomyces cerevisiae tridecapeptide pheromone. *The Journal of Peptide Research, 50*(5), 319–328.

Zuber, J., Danial, S. A., Connelly, S. M., Naider, F., & Dumont, M. E. (2015). Identification of destabilizing and stabilizing mutations of Ste2p, a G protein-coupled receptor in Saccharomyces cerevisiae. *Biochemistry, 54*(9), 1787–1806.

20 Human V-ATPases
Assembly, Function, and Regulation

Elizabeth Fosuah†*, Qingpeng Lin*†*, and Tian-Min Fu*

INTRODUCTION

Vacuolar-type adenosine triphosphatases (V-ATPases) are large protein complexes in many cellular organelles, including lysosomes, endosomes, and the Golgi apparatus, in eukaryotic cells.[1–4] V-ATPases play a crucial role in maintaining the acidic pH of these cellular organelles and are essential for various cellular processes such as protein degradation, membrane trafficking, endocytosis, immune response, neurotransmitter release, ion transport, and cell signaling.[5–11] Furthermore, in polarized cells like renal cells, osteoclasts, and sperm, V-ATPases are also prevalent in their cell membrane for acidifying the extracellular milieu.[12] As such, dysregulation of V-ATPases can impact cellular and organismal functions and has been associated with diseases such as osteoporosis, microbial infection, renal tubular acidosis, cancer, and neurodegeneration.[6,9]

Despite their diverse functions, V-ATPases share a common architecture comprising a cytosolic V_1 domain and a membrane-embedded V_0 domain across species. Earlier structural studies on *S. cerevisiae* (yeast) and moth V-ATPases, at up to 6 to 7 Å resolution, revealed the assembly of eukaryotic V-ATPases.[13,14] V-ATPases are organized into two main domains: the cytoplasmic V_1 complex, responsible for ATP hydrolysis, and the membrane-bound V_0 complex, responsible for proton translocation across membranes.[13] In contrast to yeast and other lower eukaryotic counterparts, mammalian V-ATPases are more complex due to the presence of multiple isoforms, resulting in an increased diversity of subunits.[15] This complexity enables the functional specialization of mammalian V-ATPases in different tissues and cellular organelles.[16,17] Recent advancements in cryo-electron microscopy (cryo-EM) have allowed the determination of the mammalian V-ATPases to atomic resolutions.[15,18,19] This chapter will delve into the assembly, function, and regulation of the human V-ATPases, shedding light on the intricate working mechanisms of mammalian V-ATPases.

STRUCTURE OF HUMAN V-ATPase

High-resolution structures of human V-ATPases reveal all the subunits in the V_1 domain and the V_0 domain[15] (Figure 20.1a). The V_1 complex is composed of subunits A, B, C, D, E, F, G, H, and the N-terminal domain of subunit a (a-NTD), while the V_0 complex is composed of subunits a, e, f (RNaseK), d, ATP6AP1, ATP6AP2, c, and c″ (Figure 20.1b). Notably, the three A and B subunits make up the AB head at the top of the V_1 complex. The three pairs of the AB complex in the V_1 complex display three different conformations, closed, semi-open, and open (Figure 20.1c).

In the V_0 complex, nine copies of subunits c (c1–c9) and one copy of subunit c″ form a cylinder-like structure called a c-ring (Figure 20.1d and e) [15]. On one side of the c-ring, three subunits, a, e, and f form a ternary complex that is tightly associated with the c-ring, constituting the functional unit for proton transfer (Figure 20.1e). Inside the c-ring, there are two accessory subunits, ATP6AP1 and ATP6AP2. On the cytosolic side of the c-ring sits subunit d, resembling a bowl to accommodate the DF stalk of the V_1 complex for torque transmission (Figure 20.1f). ATP hydrolysis by the V_1 complex is coupled with proton transfer by the V_0 complex.

ACCESSORY SUBUNITS IN THE V_O COMPLEX

The human V-ATPase contains two unique accessory subunits, ATP6AP1 and ATP6AP2, inside the c-ring within the V_0 complex (Figure 20.2a).[15] Both ATP6AP1 and ATP6AP2 contain pro-domains, which will be cleaved off by the furin protease during maturation (Figure 20.2a).[15,20,21] The mature ATP6AP2 only contains a single-pass transmembrane (TM) domain, and there is no homolog of ATP6AP2 identified in yeast (Figure 20.2a). In contrast, the mature ATP6AP1 consists of a single-pass TM domain and a luminal domain (LD), while the yeast homolog of ATP6AP1, V_0a1, only has the TM domain and lacks the LD domain (Figure 20.2a–d). The LD domain of human ATP6AP1 adopts a β-prism fold that consists of 10 β strands with β4-8 forming the flat surface and others forming the curved surface (Figure 20.2b). The TM domains of ATP6AP1 and ATP6AP2 in the c-ring interact with and stabilize subunit d, contributing to the coupling of V_1 and V_0 complexes (Figure 20.2c). Additionally, ATP6AP1 establishes extensive interactions with the c-ring, ATP6AP2, and subunit d, underscoring the functional importance of ATP6AP1 in the assembly and biogenesis of the V_0 complex (Figure 20.2c). Consistently, mutations of ATP6AP1 have been shown to be associated with various diseases, including immunodeficiency, liver

† These authors contributed equally to this work

DOI: 10.1201/9781003326106-24

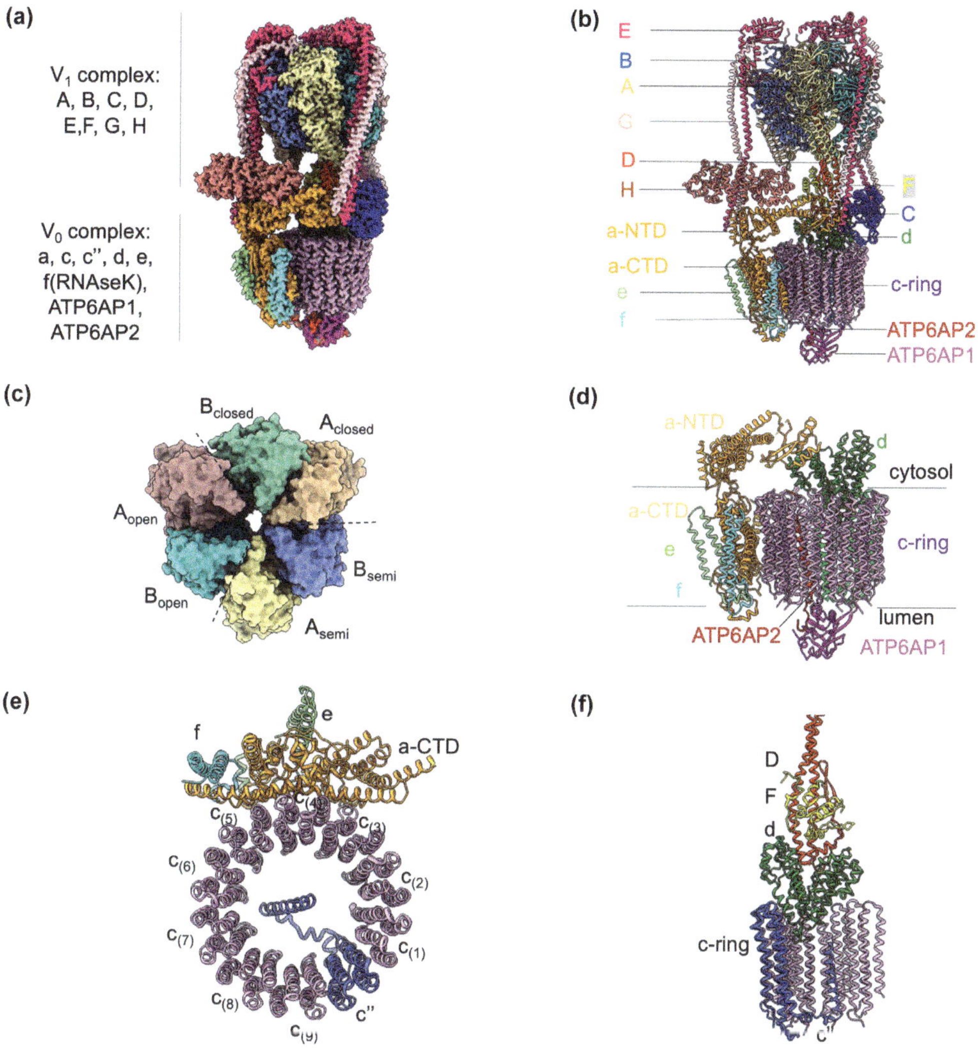

FIGURE 20.1 Structure of human V-ATPase. (a) Cryo-EM density of human V-ATPase with subunits color coded. (b) Ribbon diagram of the human V-ATPase structure with subunits color coded and labeled. (c) Surface representation of the AB heterodimers in the V_1 complex, showing the three conformations of the AB heterodimer. (d) Ribbon diagram of the V_o complex with subunits color coded. (e) Essential subunits for proton transfer. (f) Ribbon diagram to illustrate how subunits DF link with the V_o complex.

dysfunction, cancer, and neurodegeneration, as well as X-linked parkinsonism and epilepsy[21].

NONPROTEIN COMPONENTS IN V-ATPase

In addition to the protein subunits, nonprotein components, including sugars, lipids, and glycolipids, have been identified in human V-ATPases.[15] Subunits a, c, and ATP6AP1 are glycosylated in their luminal side. Specifically, six N-linked glycans were identified on the LD domain of ATP6AP1, whereas one glycosylation site each has been identified in subunits a and c[15] (Figure 20.3a). N-linked glycosylation of these subunits may be critical for the assembly, trafficking, and stabilization of human V-ATPases.[22,23] Defects in N-linked glycosylation lead to ER retention and degradation of human V-ATPases.[22] Furthermore, a molecule of dolichol pyrophosphate glycan (Dol-pp-G) has been identified to be tightly associated with the V_o complex both in human and yeast V-ATPases[15,24] (Figure 20.3b). Dol-pp-G is composed of a sugar head and an acyl tail. The hydrophilic sugar head of Dol-pp-G establishes hydrogen bonds with subunits a and e. In contrast, the hydrophobic acyl tail of Dol-pp-G inserts into a tunnel formed by the c-ring and subunit a (Figure 20.3b). As Dol-pp-G molecules are enriched in the ER, it is possible that the Dol-pp-G is incorporated into the V_o complex during V-ATPase biogenesis. In addition, many lipids, including phosphatidylcholine (PC), phosphatidylethanolamine (PE), phosphatidylserine (PS), and cholesterol (CLR), have been identified to be associated with the V_o complex[15,24] (Figure 20.3c–e). For instance, a few lipid molecules are found to bind with subunits a, e, and f (Figure 20.3c). More

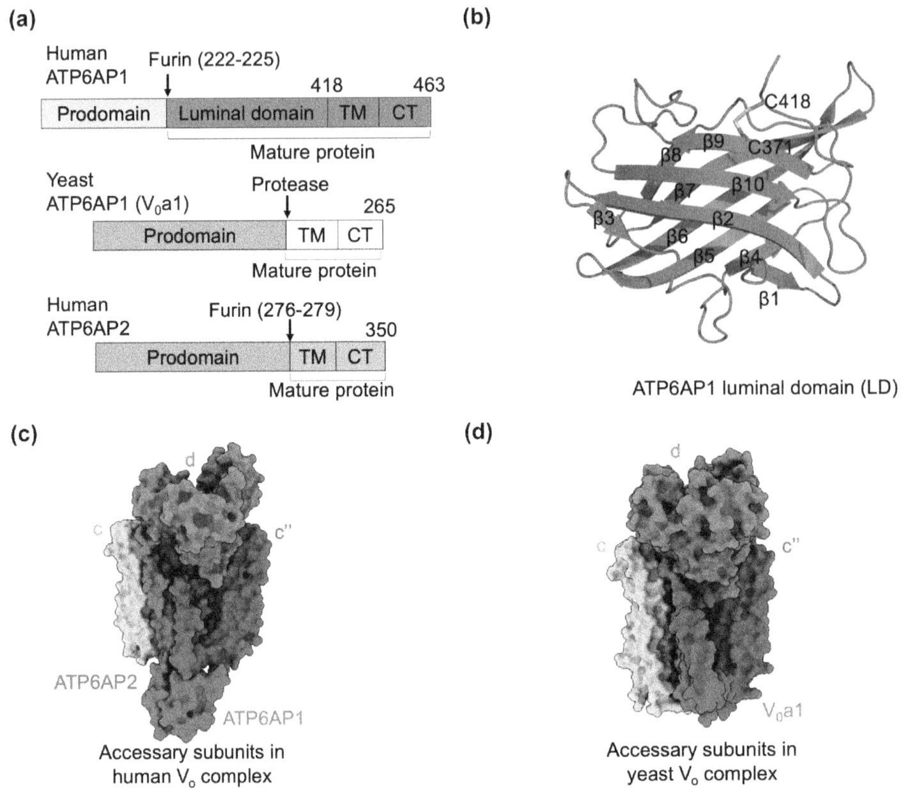

FIGURE 20.2 Accessary subunits in the V_0 complex. (a) Domain diagrams of human ATP6AP1 and ATP6AP2 and yeast ATP6AP1 (V_0a1). (b) Ribbon diagram of the luminal domain of human ATP6AP1. (c) Surface representation of ATP6AP1, ATP6AP2, c-ring, and subunit d in human V_0 complex, revealing their interactions. (d) Surface representation of V_0a1, c-ring, and subunit d in yeast V_0 complex, revealing their interactions.

strikingly, two layers of lipid molecules are located inside the c-ring and form extensive interactions with the accessory subunits (Figure 20.3d and e). These lipid molecules may play an important role in the biogenesis and function of V-ATPases. As such, the V_0 complex of the V-ATPases is a glyco-proteolipid complex (Figure 20.3f).

MECHANISM OF PROTON TRANSFER

Structures of human V-ATPases in three rotational states suggest a rotational mechanism of proton transfer.[15] As mentioned above, the three pairs of AB heterodimers are responsible for binding and hydrolyzing ATP. In each rotational state of V-ATPases, the three AB pairs in the V_1 complex display three different conformations: closed (ADP bound or ATP hydrolysis), semi-open (nucleotide-free), and open (ATP loading state) (Figure 20.4a). The conformational oscillation of AB heterodimers powered by ATP hydrolysis drives the rotation of the central DF rotor, which transmits the torque to the c-ring through connections with subunit d.

Powered by ATP hydrolysis in the V_1 complex, the c-ring rotates about 120° at the cost of one ATP molecule (Figure 20.4b). Notably, while in the F-ATP synthase each 120° rotation takes place in substeps of 80° and 40°,[25] there are no substeps observed in V-ATPases. The rotation of the c-ring is critical for proton transfer. In detail, the proton is donated to glutamate residues in the c-ring from subunit a at the cytosolic side (Figure 20.4c). After protonation, glutamate residues will rotate about 340° to arrive at the luminal side, where the proton was donated to residues in subunit a (Figure 20.4c). The continued rotation results in glutamate deprotonation and re-protonation during each revolution of the V_1-driven c-ring rotation. This results in the delivery of 10 protons at the cost of three ATP molecules.

REGULATION OF V-ATPase

The activity of V-ATPases is regulated by different factors via various mechanisms. Earlier studies have revealed that the V_1 complex and the V_0 complex can dissociate and reassemble into V-ATPases in response to nutrient availability, one important mechanism in the regulation of V-ATPases.[3,12] In addition, small molecules, including bafilomycin, concanamycin, and cadazolid, were shown to inhibit the activity of V-ATPases.[26] Moreover, recent studies have revealed a few proteins that can bind and regulate the activity of V-ATPases, including the plant toxin pean

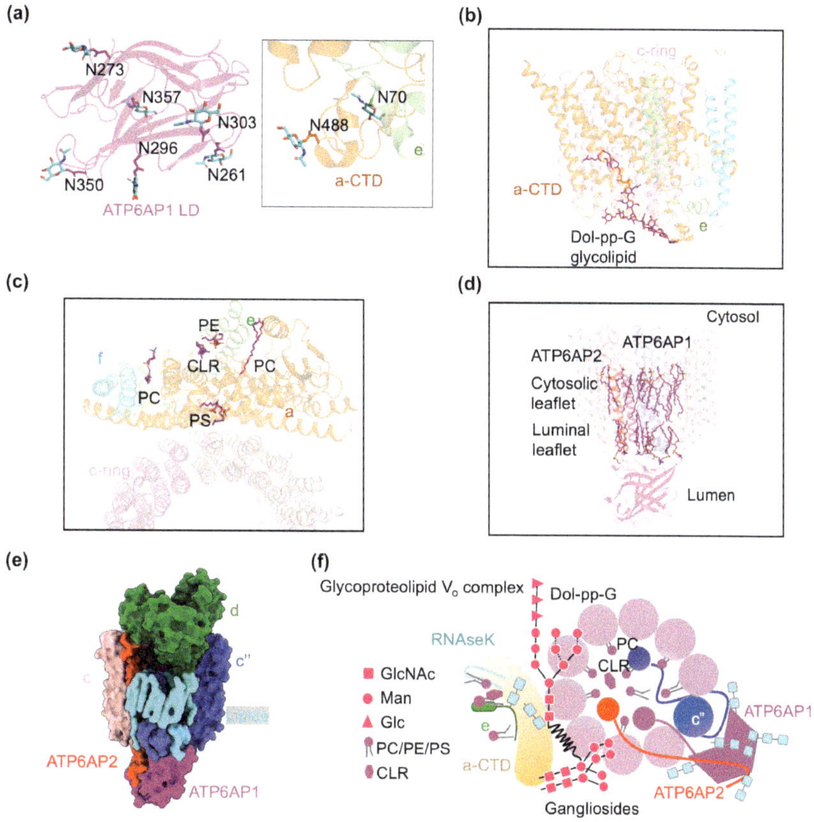

FIGURE 20.3 Nonprotein components in the V_o complex. (a) Glycosylation modifications in ATP6AP1, subunits a and e. (b) Dol-pp-Glycan in sticks associated with c-ring and subunit a. (c) Lipids tightly associated with subunits a, e, f, and c-ring. (d) Lipids within c-ring. (e) Surface representation of interactions between lipids and proteins within c-ring. (f) Diagram illustrating proteins, lipids, glycolipids, and glycans in the V_o complex.

albumin 1 subunit b, SidK from *Legionella pneumophila*, and Mammalian Enhancer-of-Akt-1-7 (mEAK-7).[27–32]

Though bafilomycin was identified as an inhibitor of V-ATPase in the 1980s, the underlying molecular mechanism was uncovered only recently through structural analysis.[33] The cryo-EM structure of V-ATPase in complex with bafilomycin showed that six copies of bafilomycin are tightly associated with the V_o complex.[33] Each bafilomycin molecule engages two neighboring subunits c in the V_o complex (Figure 20.5a). In particular, the transmembrane helix 2 of one subunit c and the transmembrane helix 4 of the neighboring subunit c form a binding site, nesting a bafilomycin molecule. Hydrophobic residues from subunits c form extensive interactions with bafilomycin (Figure 20.5b). Notably, the hydroxyl group of Y144 interacts with the 7′-hydroxyl group of bafilomycin by forming a hydrogen bond (Figure 20.5b), a binding mechanism that may be shared by other V-ATPase inhibitors like concanamycin and archazolid.[33]

Sidk, an effector protein of *Legionella pneumophila*, can bind V-ATPase to inhibit the activity of V-ATPases and thereby reduce phagosomal acidification to promote the survival of bacteria inside cells.[28] Cryo-EM structural analysis of V-ATPases in complex with Sidk has revealed that Sidk can bind to the V_1 complex of V-ATPases.[15,29] Three copies of Sidk molecules are tightly associated with subunit A in the V_1 complex (Figure 20.5c). In particular, two α-helical bundles at the N-terminus of Sidk establish extensive interactions with subunit A (Figure 20.5d). Biochemical analysis showed that Sidk inhibits the ATP hydrolysis of V-ATPases by interacting with subunit A.[29]

mEAK-7, which probably functions to regulate mTOR signaling, was identified to be a novel regulator of V-ATPases through structural analysis.[30-32,34] Cryo-EM structural analysis of V-ATPases showed that a small population of V-ATPases binds to mEAK-7[32] (Figure 20.5e). Among the three rotational states of V-ATPases, only V-ATPases in state 2 can bind to mEAK-7. More intriguingly, one copy of the mEAK-7 molecule established extensive interactions with subunits A, B, D, and E (Figure 20.5e and f). In particular, the C-terminal domain of mEAK-7 inserts into the cleft between subunit A and subunit B in semi-open conformation (Figure 20.5e). Due to the extensive interactions between mEAK-7 and V-ATPases, it was speculated that mEAK-7 may inhibit the activity of V-ATPases by blocking conformational changes of the V_1 complex. However,

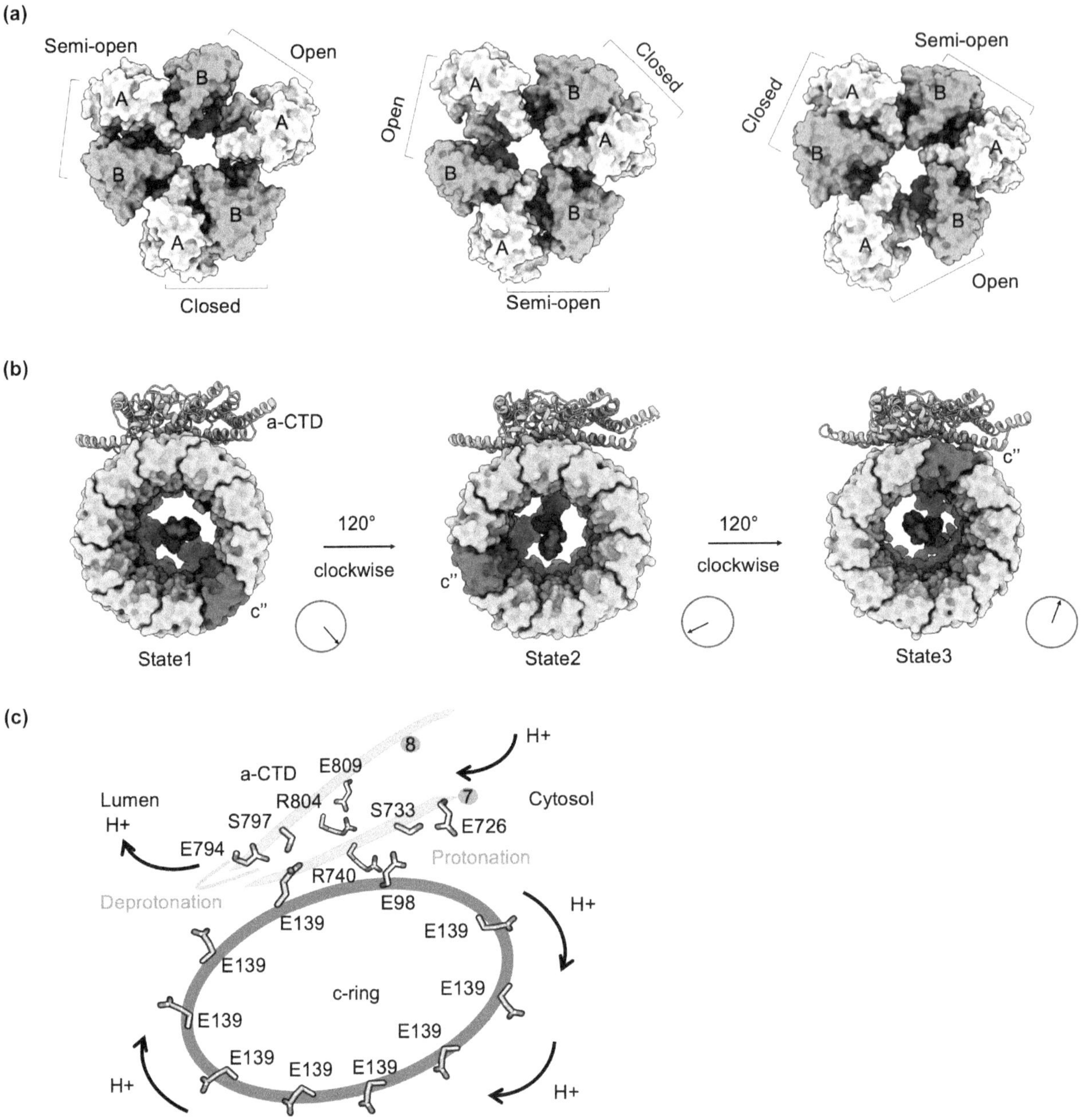

FIGURE 20.4 Mechanism of proton transfer. (a) Conformational oscillation of the V_1 complex. (b) Rotation of c-ring during proton transfer. (c) Diagram illustrating molecular details of proton transfer.

biochemical analysis revealed that mEAK-7 has limited roles in reducing ATP hydrolysis by V-ATPases.[30] The biological roles of mEAK-7 in regulating V-ATPases remain to be uncovered.

SUMMARY

With the resolution revolution of cryo-EM in the past decade, tremendous progress has been made in elucidating the assembly, function, and regulation of V-ATPases. Earlier structural studies on V-ATPases were mainly focused on yeast V-ATPases due to the mature purification protocol of yeast V-ATPases. More recently, structures of mammalian V-ATPases, either purified from mammal brains or cultured human cells, were determined to high resolution by cryo-EM.[15,18,19] In particular, the intact human V-ATPases were determined to atomic resolution, revealing that nonprotein components, including lipids, glycolipids, and glycans, are bona fide components of V-ATPases.[15] The functional roles of these nonprotein components in the assembly and function of V-ATPases are waiting to be revealed. Furthermore, we gain a comprehensive understanding of how V-ATPase

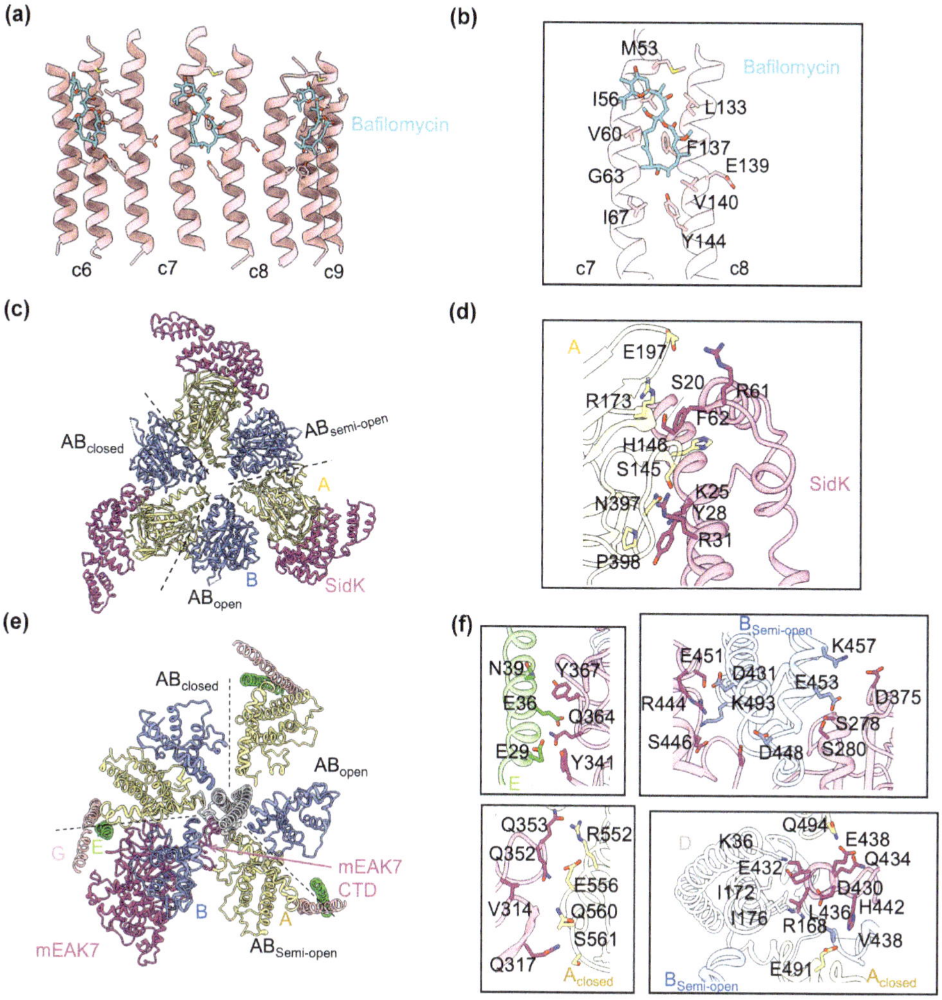

FIGURE 20.5 Regulators of V-ATPase. (a) Structure of c-ring in complex with bafilomycin (cyan). (b) Residues of c-ring coordinating bafilomycin. (c) Ribbon diagram of the V-ATPase and Sidk complex. (d) Detailed interactions between subunit A and Sidk (magenta). (e) Ribbon diagram of mEAK-7 (magenta) and V-ATPase. (f) Detailed interactions between mEAK7 (magenta) and subunits A, B, D, and E.

activity was regulated by small molecule inhibitors and binding proteins through structural analysis of V-ATPase complexes. In particular, the identification of mEAK-7 as a novel binding partner of V-ATPase represents a breakthrough discovery in the field.[32] There is a lot more work that needs to be done for us to understand the functional role of mEAK-7 in regulating V-ATPases.

REFERENCES

1. Forgac, M. Structure, function and regulation of the coated vesicle V-ATPase. *J Exp Biol* **172**, 155–169 (1992).
2. Capecci, J. & Forgac, M. The function of vacuolar ATPase (V-ATPase) a subunit isoforms in invasiveness of MCF10a and MCF10CA1a human breast cancer cells. *J Biol Chem* **288**, 32731–32741, doi:10.1074/jbc.M113.503771 (2013).
3. Vasanthakumar, T. & Rubinstein, J. L. Structure and roles of V-type ATPases. *Trends Biochem Sci* **45**, 295–307, doi:10.1016/j.tibs.2019.12.007 (2020).
4. Marshansky, V., Rubinstein, J. L. & Gruber, G. Eukaryotic V-ATPase: Novel structural findings and functional insights. *Biochim Biophys Acta* **1837**, 857–879, doi:10.1016/j.bbabio.2014.01.018 (2014).
5. Kane, P. M. The where, when, and how of organelle acidification by the yeast vacuolar H+-ATPase. *Microbiol Mol Biol Rev* **70**, 177–191, doi:10.1128/MMBR.70.1.177-191.2006 (2006).
6. Forgac, M. Vacuolar ATPases: Rotary proton pumps in physiology and pathophysiology. *Nat Rev Mol Cell Biol* **8**, 917–929, doi:10.1038/nrm2272 (2007).
7. Casey, J. R., Grinstein, S. & Orlowski, J. Sensors and regulators of intracellular pH. *Nat Rev Mol Cell Biol* **11**, 50–61, doi:10.1038/nrm2820 (2010).
8. Cruciat, C. M. *et al.* Requirement of prorenin receptor and vacuolar H+-ATPase-mediated acidification for Wnt signaling. *Science* **327**, 459–463, doi:10.1126/science.1179802 (2010).
9. Hinton, A., Bond, S. & Forgac, M. V-ATPase functions in normal and disease processes. *Pflugers Arch* **457**, 589–598, doi:10.1007/s00424-007-0382-4 (2009).

10. Pamarthy, S., Kulshrestha, A., Katara, G. K. & Beaman, K. D. The curious case of vacuolar ATPase: Regulation of signaling pathways. *Mol Cancer* **17**, 41, doi:10.1186/s12943-018-0811-3 (2018).
11. Sun-Wada, G. H. & Wada, Y. Role of vacuolar-type proton ATPase in signal transduction. *Biochim Biophys Acta* **1847**, 1166–1172, doi:10.1016/j.bbabio.2015.06.010 (2015).
12. McGuire, C., Stransky, L., Cotter, K. & Forgac, M. Regulation of V-ATPase activity. *Front Biosci (Landmark Ed)* **22**, 609–622, doi:10.2741/4506 (2017).
13. Zhao, J., Benlekbir, S. & Rubinstein, J. L. Electron cryomicroscopy observation of rotational states in a eukaryotic V-ATPase. *Nature* **521**, 241–245, doi:10.1038/nature14365 (2015).
14. Rawson, S., Harrison, M. A. & Muench, S. P. Rotating with the brakes on and other unresolved features of the vacuolar ATPase. *Biochem Soc Trans* **44**, 851–855, doi:10.1042/BST20160043 (2016).
15. Wang, L., Wu, D., Robinson, C. V., Wu, H. & Fu, T. M. Structures of a complete human V-ATPase reveal mechanisms of its assembly. *Mol Cell* **80**, 501–511, doi:10.1016/j.molcel.2020.09.029 (2020).
16. McGuire, C. M., Collins, M. P., Sun-Wada, G., Wada, Y. & Forgac, M. Isoform-specific gene disruptions reveal a role for the V-ATPase subunit a4 isoform in the invasiveness of 4T1-12B breast cancer cells. *J Biol Chem* **294**, 11248–11258, doi:10.1074/jbc.RA119.007713 (2019).
17. Hinton, A. *et al.* Function of a subunit isoforms of the V-ATPase in pH homeostasis and in vitro invasion of MDA-MB231 human breast cancer cells. *J Biol Chem* **284**, 16400–16408, doi:10.1074/jbc.M901201200 (2009).
18. Wang, R. *et al.* Cryo-EM structures of intact V-ATPase from bovine brain. *Nat Commun* **11**, 3921, doi:10.1038/s41467-020-17762-9 (2020).
19. Abbas, Y. M., Wu, D., Bueler, S. A., Robinson, C. V. & Rubinstein, J. L. Structure of V-ATPase from the mammalian brain. *Science* **367**, 1240–1246, doi:10.1126/science.aaz2924 (2020).
20. Louagie, E. *et al.* Role of furin in granular acidification in the endocrine pancreas: Identification of the V-ATPase subunit Ac45 as a candidate substrate. *Proc Natl Acad Sci USA* **105**, 12319–12324, doi:10.1073/pnas.0800340105 (2008).
21. Jansen, E. J. *et al.* ATP6AP1 deficiency causes an immunodeficiency with hepatopathy, cognitive impairment and abnormal protein glycosylation. *Nat Commun* **7**, 11600, doi:10.1038/ncomms11600 (2016).
22. Esmail, S. *et al.* N-linked glycosylation is required for vacuolar H(+)-ATPase (V-ATPase) a4 subunit stability, assembly, and cell surface expression. *J Cell Biochem* **117**, 2757–2768, doi:10.1002/jcb.25574 (2016).
23. Esmail, S. *et al.* N-linked glycosylation of a subunit isoforms is critical for vertebrate vacuolar H(+)-ATPase (V-ATPase) biosynthesis. *J Cell Biochem* **119**, 861–875, doi:10.1002/jcb.26250 (2018).
24. Roh, S. H. *et al.* Cryo-EM and MD infer water-mediated proton transport and autoinhibition mechanisms of V_o complex. *Sci Adv* **6**, eabb9605, doi:10.1126/sciadv.abb9605 (2020).
25. Kuhlbrandt, W. Structure and mechanisms of F-type ATP synthases. *Annu Rev Biochem* **88**, 515–549, doi:10.1146/annurev-biochem-013118-110903 (2019).
26. Bowman, E. J., Siebers, A. & Altendorf, K. Bafilomycins: A class of inhibitors of membrane ATPases from microorganisms, animal cells, and plant cells. *Proc Natl Acad Sci USA* **85**, 7972–7976, doi:10.1073/pnas.85.21.7972 (1988).
27. Muench, S. P. *et al.* PA1b inhibitor binding to subunits c and e of the vacuolar ATPase reveals its insecticidal mechanism. *J Biol Chem* **289**, 16399–16408, doi:10.1074/jbc.M113.541250 (2014).
28. Xu, L. *et al.* Inhibition of host vacuolar H+-ATPase activity by a legionella pneumophila effector. *PLoS Pathog* **6**, e1000822, doi:10.1371/journal.ppat.1000822 (2010).
29. Zhao, J. *et al.* Molecular basis for the binding and modulation of V-ATPase by a bacterial effector protein. *PLoS Pathog* **13**, e1006394, doi:10.1371/journal.ppat.1006394 (2017).
30. Wang, R., Qin, Y., Xie, X. S. & Li, X. Molecular basis of mEAK7-mediated human V-ATPase regulation. *Nat Commun* **13**, 3272, doi:10.1038/s41467-022-30899-z (2022).
31. Tan, Y. Z. *et al.* CryoEM of endogenous mammalian V-ATPase interacting with the TLDc protein mEAK-7. *Life Sci Alliance* **5**, doi:10.26508/lsa.202201527 (2022).
32. Wang, L., Wu, D., Robinson, C. V. & Fu, T. M. Identification of mEAK-7 as a human V-ATPase regulator via cryo-EM data mining. *Proc Natl Acad Sci USA* **119**, e2203742119, doi:10.1073/pnas.2203742119 (2022).
33. Wang, R. *et al.* Molecular basis of V-ATPase inhibition by bafilomycin A1. *Nat Commun* **12**, 1782, doi:10.1038/s41467-021-22111-5 (2021).
34. Nguyen, J. T. *et al.* Mammalian EAK-7 activates alternative mTOR signaling to regulate cell proliferation and migration. *Sci Adv* **4**, eaao5838, doi:10.1126/sciadv.aao5838 (2018).

21 Structure and Function of the Intracellular Calcium Channel Ryanodine Receptor

Marco C. Miotto

INTRODUCTION

Ryanodine receptors (RyRs) are large ~2.2 MDa ion channels located in the sarcoplasmic reticulum (SR) and endoplasmic reticulum (ER) of cells. RyRs play a crucial role in calcium (Ca^{2+}) homeostasis, releasing the Ca^{2+} stored in the SR/ER into the cytosol. This Ca^{2+} signal is vital for numerous cellular processes, including muscle contraction, synaptic transmission, proliferation, differentiation, neuronal plasticity, and other intracellular signaling pathways (Zalk, Lehnart et al. 2007). RyRs are unique because they are activated by cytosolic Ca^{2+} in a process known as Ca^{2+}-induced Ca^{2+} release (CICR). This mechanism amplifies the Ca^{2+} signal, leading to a rapid and robust increase in intracellular Ca^{2+} levels (Fill and Copello 2002). Three distinct isoforms of RyRs have been identified: RyR1, RyR2, and RyR3. RyR1 is primarily expressed in skeletal muscle and RyR2 in cardiac muscle and neurons. RyR3 is less abundant and is found in various tissues, but its exact function remains less understood (Lanner, Georgiou et al. 2010). Notably, mutations in RyRs have been associated with several muscle disorders due to dysregulated Ca^{2+} homeostasis. Given RyRs' critical roles in cellular function and disease, understanding the structure and function of RyRs is an area of active research of immense biological and clinical significance. Only after the advent of modern cryo-EM was the structure–function relationship of RyRs comprehended. Despite significant progress in the past decade (Santulli, Lewis et al. 2018, Gong, Yan et al. 2021, Hadiatullah, He et al. 2022), many aspects of RyR biology remain to be elucidated.

SECTIONS

RyRs in Muscle Cells: Excitation–Contraction Coupling

In muscle cells, RyRs are located in the terminal cisternae of the SR. RyRs are critically involved in excitation–contraction (EC) coupling, which is the physiological mechanism that translates electrical signals into mechanical contraction in muscle cells. In cardiac muscle, EC coupling is initiated by an action potential that propagates along the muscle fiber's sarcolemma and delves into the transverse tubules (T-tubules), a network of invaginations of the sarcolemma (Figure 21.1a). This electrical signal triggers the opening of voltage-gated L-type Ca^{2+} channels (Ca_V1), also known as dihydropyridine receptors (DHPR), on the T-tubule membrane. During an action potential, the opening of the $Ca_V1.2$ (the isoform expressed in cardiomyocytes) allows a small amount of extracellular Ca^{2+} to enter the cell. This Ca^{2+} influx triggers the opening of the nearby RyR2 channels, leading to the passive transport of Ca^{2+} from the SR in the process known as CICR. The resultant release of Ca^{2+} from the SR into the cytosol amplifies the initial Ca^{2+} signal, leading to an increased cytosolic Ca^{2+} concentration sufficient to initiate cardiac muscle contraction (Bers 2002). The released Ca^{2+} ions bind to troponin C on the thin filament of the muscle contractile machinery, triggering a conformational change that displaces tropomyosin from the myosin-binding sites on actin. This action exposes the sites for myosin head attachment, which initiates the power stroke of muscle contraction. Once the action potential ceases, Ca^{2+} ions are actively transported back into the SR by the action of the SR Ca^{2+}-ATPase (SERCA) and into the extracellular space by the action of Ca^{2+} transporters, allowing tropomyosin to return to its resting position and thus enabling cardiac muscle relaxation and preparing the heart for the next contraction cycle. The RyR2-mediated CICR mechanism is a crucial feature of cardiac EC coupling that ensures the heart's rhythmic and forceful contraction.

The process of EC coupling in skeletal muscle involves a similar but subtly different mechanism (Figure 21.1b). During the action potential, the activation of $Ca_V1.1$ (the isoform expressed in skeletal myocytes) opens RyR1 on the adjacent SR membrane, in a process termed "voltage-induced Ca^{2+} release." Instead of Ca^{2+} activation, a conformational change in $Ca_V1.1$ is directly transmitted to RyR1, mechanically forcing it to open and release Ca^{2+} ions that are stored in the SR into the cytosol (Rios and Pizarro 1991). The resultant release of Ca^{2+} from the SR into the cytosol amplifies the initial Ca^{2+} signal, leading to an increased cytosolic Ca^{2+} concentration sufficient to initiate muscle contraction (Bers 2002). The closing of RyR1 channels and the reuptake of Ca^{2+} into the SR by SERCA lead to a decrease in cytosolic Ca^{2+} concentration, leading to muscle relaxation (Gordon, Homsher et al. 2000). In conclusion, RyR channels serve as a central switch in EC coupling, controlling the release of Ca^{2+} from the SR and thereby the initiation of muscle contraction. Their crucial role in muscle physiology makes them a key target for studying muscle

DOI: 10.1201/9781003326106-25

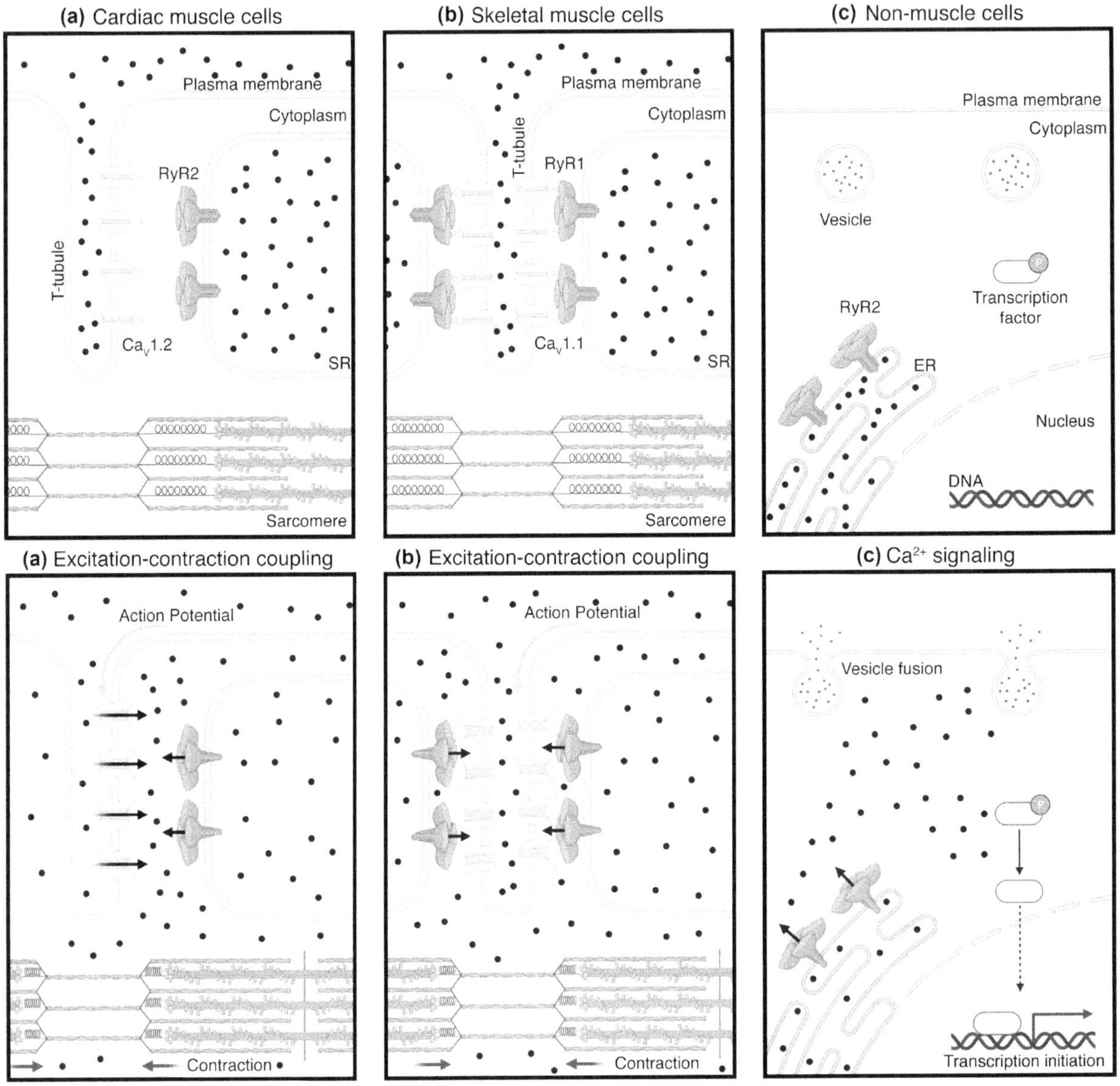

FIGURE 21.1 Function of ryanodine receptors in cells. (a–c) Representation of RyR intracellular location and biological function in Ca^{2+} homeostasis. (*Top*) RyR2 is located on the SR at the diads of cardiac muscle cells (a), on the SR at the triads of skeletal muscle cells (b), and on the ER of non-muscle cells (c). Ca^{2+} in the extracellular space and in the SR/ER lumen is depicted as black dots. (*Bottom*) In cardiac cells, the action potential of the plasma membrane activates $Ca_V1.2$, which triggers CICR by RyR2, leading to cytosolic Ca^{2+} increase and sarcomere contraction. Ca^{2+} flow is shown as black arrows. In skeletal muscle cells, the action potential of the plasma membrane activates $Ca_V1.1$, which mechanically opens RyR1, leading to cytosolic Ca^{2+} increase and sarcomere contraction. In non-muscle cells, activation of RyRs leads to cytosolic Ca^{2+} increase and activation of different pathways (vesicle fusion and content release, protein phosphorylation, and transcription activation, among others). Created with BioRender.com.

diseases associated with abnormal EC coupling. As such, gaining a deeper understanding of RyR function and regulation is critical for developing effective therapeutic strategies for these diseases.

RyRs in Non-Muscle Cells: Ca^{2+} Signaling

While RyRs are predominantly known for their role in muscle EC coupling, they also have significant functions in non-muscle cells, mediating various processes through the release of ER stored Ca^{2+} (Figure 21.1c). This aspect of RyR function is less understood but is gaining attention due to its implications in numerous biological and pathological contexts. In neurons, RyRs are implicated in synaptic plasticity, a critical mechanism for learning and memory (Chiantia, Hidisoglu et al. 2023). RyRs are present in dendritic spines of neurons, where they contribute to the Ca^{2+} dynamics necessary for long-term potentiation

(LTP), a form of synaptic plasticity. In this context, the release of Ca^{2+} from RyRs has been shown to modulate the strength of synaptic connections, which is thought to underpin memory formation (Berrout and Isokawa 2009). Moreover, RyRs are also involved in neuronal development. During the formation of the neural tube, RyR-mediated Ca^{2+} release is required for the proliferation and differentiation of neural progenitor cells. Dysregulation of this process can lead to neurodevelopmental disorders, emphasizing the importance of RyR function in the nervous system. In immune cells, RyRs play a role in regulating the immune response. For example, in T lymphocytes, RyRs are responsible for the Ca^{2+} release that is necessary for T cell activation and proliferation. In macrophages, RyRs are involved in the phagocytic process, where RyR-mediated Ca^{2+} release is necessary for the engulfment and digestion of foreign particles (Guse 2015). RyRs are also present in a variety of other non-muscle cells, including pancreatic cells, endothelial cells, and adipocytes (Olofsson, Gopel et al. 2002). In pancreatic beta cells, RyRs contribute to the regulation of insulin secretion, a process that is tightly controlled by intracellular Ca^{2+} dynamics. In endothelial cells, RyRs are involved in the regulation of vascular tone, which is crucial for blood pressure control. And in adipocytes, RyR-mediated Ca^{2+} release contributes to lipid metabolism and adipogenesis. In addition to these physiological roles, RyRs are implicated in various pathological conditions. For instance, in Alzheimer's disease, an excessive release of Ca^{2+} through neuronal RyR2 is thought to contribute to the neurodegenerative process. Similarly, in cognitive dysfunction in heart failure, hyper-adrenergic signaling leads to intracellular Ca^{2+} leak through RyR2 reducing neuron plasticity (Dridi, Liu et al. 2023). In some forms of cancer, alterations in RyRs expression or function can lead to dysregulated cell growth and proliferation. Despite the growing body of evidence on the roles of RyRs in non-muscle cells, many questions remain unanswered. For example, the mechanisms that regulate RyR activity in these cells are not fully understood. Moreover, the specific roles of the different RyR isoforms in non-muscle cells remain to be elucidated. Further research in this field will not only enhance our understanding of RyR biology but may also reveal novel therapeutic targets for a range of diseases.

RyRs and the Cryo-EM Resolution Revolution

Prior to the cryo-EM "resolution revolution," efforts to determine RyRs structure were centered on X-ray crystallography. However, due to the enormous size and dynamic nature of RyRs, they proved to be challenging subjects for crystallography. Hence, lower-resolution techniques like cryo-EM became a logical choice for studying RyR structure. The initial attempts in the 1990s provided the first low-resolution three-dimensional (3D) maps of RyRs. These studies used purified RyRs embedded in a thin layer of ice and observed under cryo-conditions using an electron microscope. Pioneering work resulted in initial 3D reconstructions of RyR1 and RyR2 at a resolution of approximately 30–40 Å (Radermacher, Rao et al. 1994, Sharma, Penczek et al. 1998). Despite the relatively low resolution, these studies were instrumental in defining the overall architecture of RyRs and their general position within the SR. These early 3D reconstructions revealed RyRs as a four-leaf clover-like structure when viewed from the cytoplasmic side, a feature that was consistent with its known tetrameric structure.

As the field progressed into the early 2000s, single-particle cryo-EM techniques began to be applied to RyRs, allowing for the imaging of the protein without the need for a carbon support layer, which led to anisotropic resolution without axis artifacts. These single-particle studies provided further insights into the overall shape of RyRs, which appeared as large mushroom-like structures with a broad "cap" region in the cytoplasm and a narrow "stalk" region forming the transmembrane pore. Advancements in cryo-EM technology, including the development of better detectors and image processing software, gradually improved the resolution of RyR images. In 2004, the structure of closed-state RyR1 was reported at a resolution of 22 Å, revealing more details about the channel's architecture being the first with isotropic resolution due to particles freely suspended in water (Samso, Wagenknecht et al. 2005). Another significant leap was made when the cryo-EM structure of RyR1 in complex with the drug ryanodine was reported at a resolution of 9.6 Å (Ludtke, Serysheva et al. 2005). In 2009 the structure of RyR1 in the open state was resolved at 10 Å (Samso, Feng et al. 2009). This considerable improvement allowed for the identification of individual protein subdomains and provided important insights into the conformational changes associated with channel gating.

These early studies also began to reveal the intricate network of interactions between RyRs and their numerous associated proteins and modulators. They showed how the binding of these modulators could induce conformational changes in RyRs, affecting their Ca^{2+} release activity. While the pre-2015 era of cryo-EM did not yield high-resolution structures of RyRs comparable to those obtained by X-ray crystallography, these studies were crucial for setting the stage for the breakthroughs that would follow. They helped define the overall architecture of RyRs, revealed their large cytoplasmic assembly, identified the approximate location of the transmembrane pore region, and provided the first insights into how RyRs might undergo conformational changes during gating. The resolution revolution in cryo-EM that began around 2015, facilitated by direct electron detector technology and advanced image processing techniques, eventually allowed for near-atomic 4–6 Å resolution structures of RyRs (Efremov, Leitner et al. 2015, Yan, Bai et al. 2015, Zalk, Clarke et al. 2015). Today, RyR structures in the 3–4 Å resolution range are routinely achieved, with records at 2.45 Å for RyR1 (Melville, Dridi et al. 2022) and 2.58 Å for RyR2 (Miotto, Weninger et al. 2022). This revolution provided unprecedented insights into the structure, function, and regulation of RyRs, enabling a deeper understanding of these critical ion channels.

RyR Cryo-EM in a Nutshell: From Purification to the Atomic Model

1. Purification of RyRs: Obtaining a pure and concentrated sample is essential for cryo-EM, making purification of RyRs a critical step. With little variation, the purification protocols can be divided into four main steps: source material harvesting, homogenization, membrane fraction separation and solubilization, and membrane protein separation and RyR purification (Figure 21.2). Currently, there are two source materials: native tissues where expression of RyRs is intrinsically high, such as skeletal muscle for RyR1 (Woll, Haji-Ghassemi et al. 2021, Melville, Kim et al. 2022) and heart for RyR2 (Gong, Chi et al. 2019); and HEK293

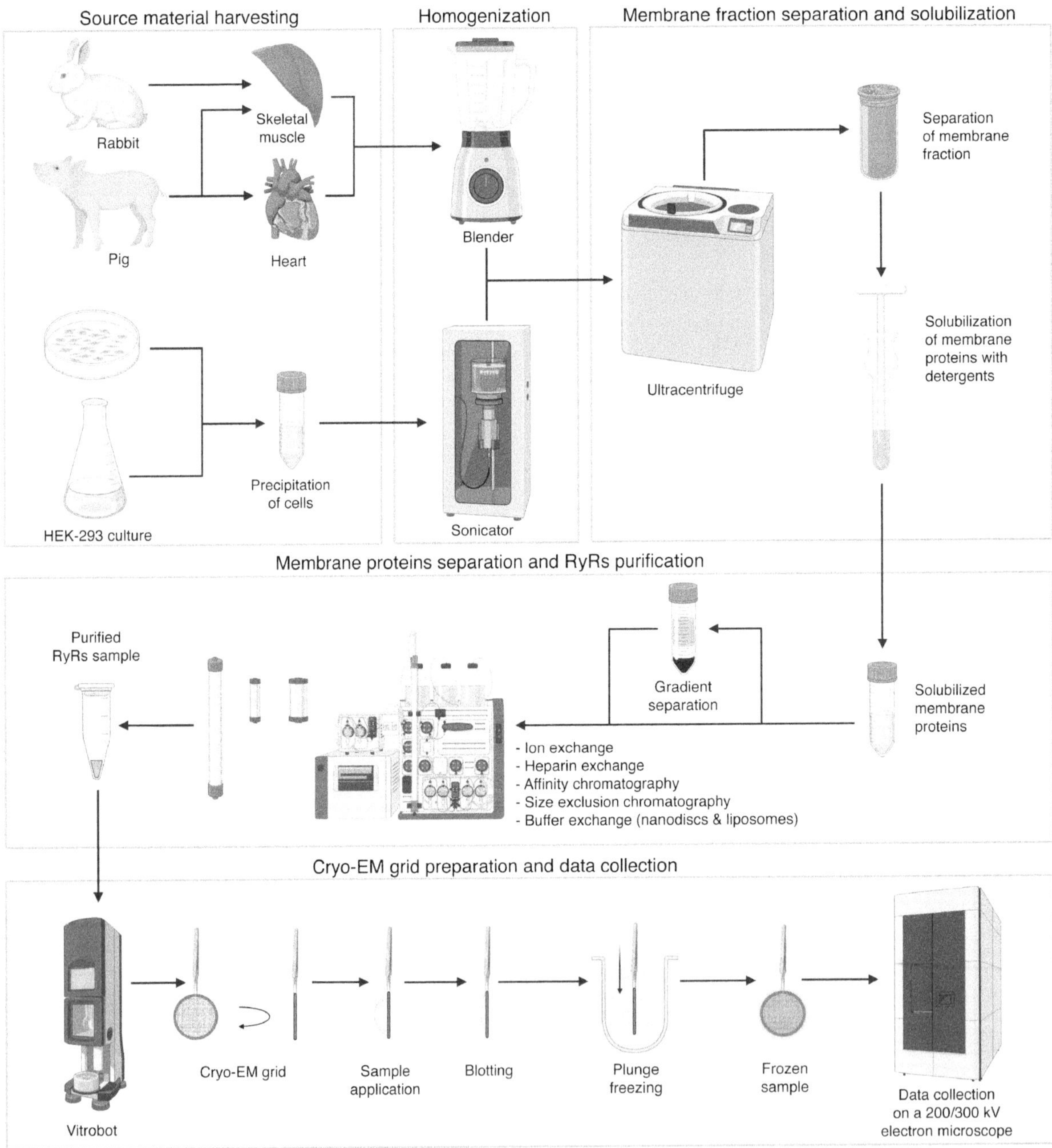

FIGURE 21.2 From purification to cryo-EM data collection. Diagram showing the sequential steps: source material harvesting, homogenization, membrane fraction separation and solubilization, membrane protein separation, RyRs purification, cryo-EM grid preparation, and cryo-EM data collection. Created with BioRender.com.

cells recombinantly expressing RyRs (Hu, Iyer et al. 2021, Kobayashi, Tsutsumi et al. 2022, Miotto, Weninger et al. 2022). The advantage of native tissues is the abundance of endogenous RyRs and the ease in obtaining them. RyR1 is usually obtained from rabbit or pig legs, where amounts as little as 20 g can lead to the purification of a good RyR1 sample for cryo-EM. RyR2 is usually obtained from a pig heart. Heart tissue is harder to work with, and purification of RyR2 is not as direct as with RyR1, leading to the need for larger amounts of source material—hence the use of a whole heart for one sample. The disadvantage of native tissues is that only the WT isoforms can be extracted. One exception was the use of a commercially available mutant pig that carries an RyR1 mutation (Woll, Haji-Ghassemi et al. 2021), but this is not usually the case as generating and maintaining mutant animals is extremely laborious and expensive. Using HEK293 cells recombinantly expressing RyRs as source material is laborious and time consuming, but it allows the expression of mutant RyRs necessary to study functionality and diseases. After homogenization of the starting material by blender or sonication, centrifugation is used to isolate the membrane fraction, where RyRs and other membrane proteins are localized, from the rest of the soluble proteins and cellular components. The membrane fraction is then incubated with detergents in order to solubilize membrane proteins, including RyRs. After this step, a series of separation steps that take advantage of the biophysical properties of RyRs are usually applied. Due to RyRs' size, sucrose gradient or other coarse centrifugal techniques are used as initial separation steps, while the more sensitive size exclusion chromatography is used as a final step. Also, due to RyRs' isolectric point of ~5.5 and negative charge at pH 7.4, anion exchange chromatography is commonly used after RyR solubilization. Finally, for complex starting materials such as heart or HEK293 cells, an extra affinity purification step is used that exploits the ability of RyRs to strongly bind to FKBPs or calstabins, efficiently separating low amounts of RyRs from complex mixtures.

2. Sample preparation: The preparation of the cryo-EM sample is a critical step in the structure determination process, which can significantly influence the resolution and quality of the final 3D reconstruction. This process begins with the application of a small sample amount, typically around 3 μL of the purified RyR sample concentrated to 1–10 g/L, onto an electron microscopy grid (Figure 21.2). The grid is a mesh-like structure, typically composed of a metal such as copper or gold, coated with a thin layer of holey supporting film, usually made of carbon or gold, which holds the sample. Before applying the sample, the grid is usually subjected to a process called "glow discharging," which involves exposing it to a low-pressure gas plasma. This process makes the grid surface hydrophilic, thus helping the sample to spread evenly across the grid surface (Erk, Delacroix et al. 1991). Care must be taken to ensure that the sample concentration is optimal—too low, and there will not be enough particles for analysis; too high, and the particles may aggregate or overlap, making it difficult to analyze. Once the sample is applied, the grid is blotted with an ashless filter paper to remove excess liquid, leaving behind a thin film of sample. Ashless filter paper is necessary to avoid Ca^{2+} impurities present in regular filter papers. Blotting parameters (blotting time, blotting force, humidity, etc.) need to be carefully optimized to achieve an ideal sample thickness (Razinkov, Dandey et al. 2016). The grid is then rapidly plunged into a cryogen, usually liquid ethane or a mixture of ethane and propane, which is kept at a temperature well below the freezing point of water. This rapid plunge freezing vitrifies the thin aqueous film, thus preserving the biological structures in their native state, free from the artifacts introduced by chemical fixation or crystal growth (Cheng, Grigorieff et al. 2015). The vitrified sample is stored in liquid nitrogen until it is loaded into the electron microscope for imaging. Throughout the process, care must be taken to avoid any warming or contamination of the sample, as these could adversely affect the quality of the images obtained.

3. Data acquisition: This stage involves the use of a cryo-electron microscope equipped with a direct electron detector to collect images. The microscope's parameters (e.g., accelerating voltage, exposure time, and dose rate) should be carefully optimized to reach a balance between obtaining high-resolution information and minimizing radiation damage to the sample (Grant and Grigorieff 2015). The focus should be set slightly underfocus to enhance phase contrast. Since RyRs are huge particles, parameters closer to defocus can be used (up to −0.5 μm). An automated data acquisition software, such as SerialEM or EPU, is typically used to acquire a large number of high-quality images, which are commonly referred to as "movies" or "micrographs" in cryo-EM parlance. This software allows for automated collection of micrographs from different regions of the grid, ensuring a comprehensive sampling of the sample. It also allows for real-time correction of drift and astigmatism, further enhancing the quality of the acquired images.

4. Image processing and refinement: The collected movies are then subjected to a series of image processing steps using specialized software packages such as RELION, cryoSPARC, or EMAN2 (Figure 21.3) (Scheres 2012, Punjani, Rubinstein et al. 2017). The initial steps include motion correction to correct for beam-induced motion and contrast transfer function (CTF) estimation to correct for phase and amplitude alterations introduced by the microscope's optics. This is followed by particle picking, either blob, template, or machine learning methods, to identify and extract individual particle images from the micrographs. These particles are then subjected to 2D classification, which sorts them into groups based on their similar views, and helps in removing bad particles (e.g., contaminants or damaged particles). From the remaining good 2D classes, initial 3D volumes can be generated. This is typically done using a method called "angular reconstitution," which determines the relative orientations of the

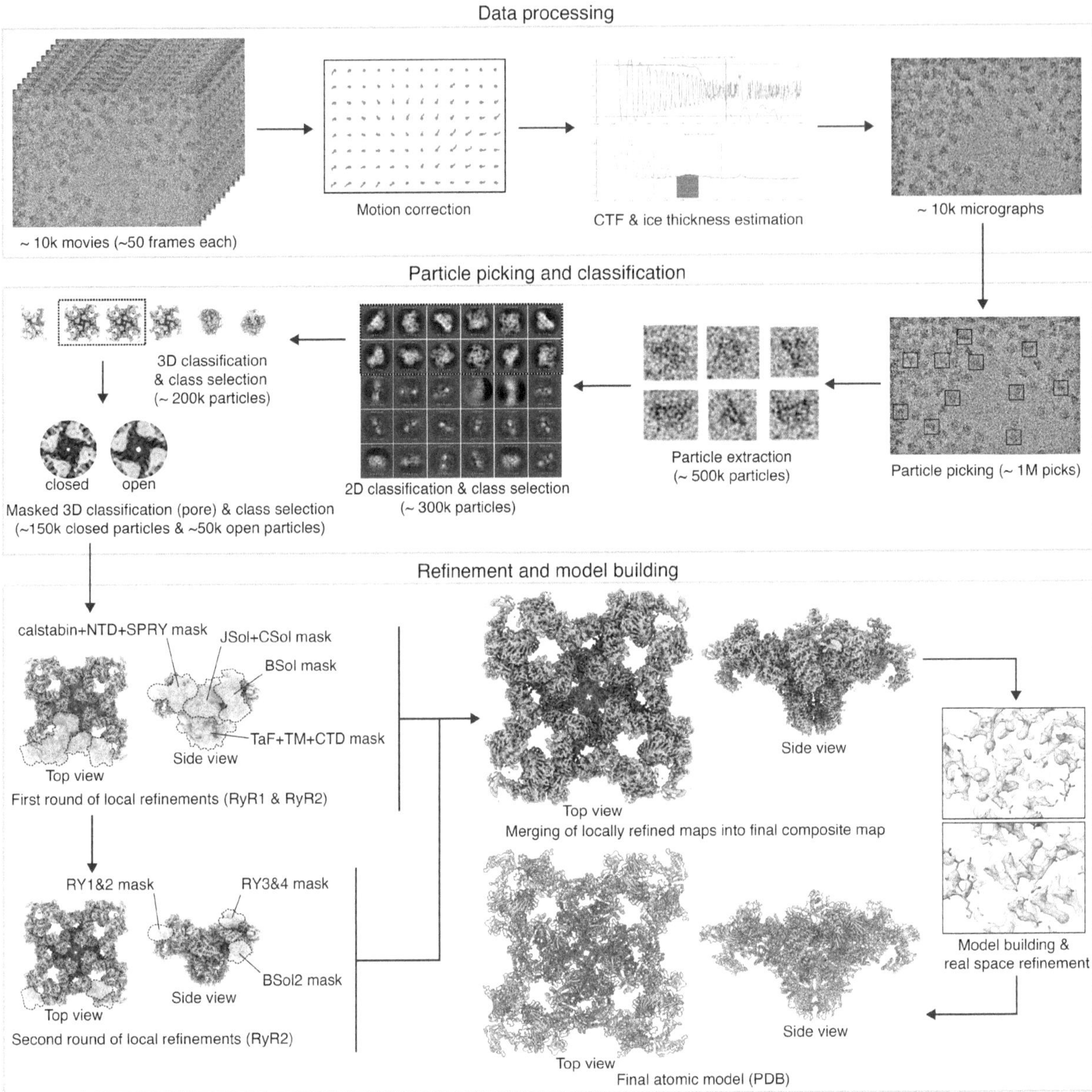

FIGURE 21.3 From cryo-EM data processing to the atomic model. Diagram showing the sequential processing steps: data processing, particle picking, particle classification, and cryo-EM map refinements with cryoSPARC, local maps merging with ChimeraX, model building with Coot, and real space refinement with PHENIX.

particles. The initial volumes are used as references for 3D classification, where particles are grouped based on their similar 3D conformations. This step is crucial for dealing with structural heterogeneity within the sample, which is common in biological samples and especially for RyRs (Scheres 2016). Then, 3D classification focused on the pore region is necessary to separate RyR particles in the closed state from those in the open state. Each 3D class is then independently refined to obtain a high-resolution 3D reconstruction. This involves iterative cycles of local 3D refinement focused on the different domains of the protein (Figure 21.3). Finally, all the local cryo-EM maps are merged, with software like ChimeraX, to get a composite cryo-EM map suitable for building the atomic model.

5. Model building and validation: Atomic models are built into the cryo-EM density maps using molecular graphics software like Coot (Figure 21.3). This process involves fitting known structures or homology models into the map, or manually building the structure based on the observed densities recently facilitated by AI software like Model-angelo. Nowadays, there are several RyR1 and

RyR2 atomic models that make model fitting a straightforward task when resolution is better than 4 Å. The atomic models are then refined against the map using real-space refinement programs like PHENIX or Rosetta, which optimize the model's fit to the density while maintaining proper stereochemistry (Adams, Afonine et al. 2010). The final models are validated using tools such as MolProbity, which checks for steric clashes and geometric outliers, and EMRinger, which assesses the fit of the model to the map and the correctness of the assigned rotameric states (Barad, Echols et al. 2015).

RyR Structure–Function Relationship

Understanding the structure–function relationship of RyRs is crucial, given the critical roles RyRs play in cellular function and the severe diseases associated with their dysfunction (Santulli, Lewis et al. 2018, Iyer, Barnakov et al. 2023). The structure and function of RyRs are influenced by a multitude of endogenous and exogenous ligands, which induce conformational changes in the receptor, modulate its activity, and regulate intracellular Ca^{2+} signaling (Gong, Yan et al. 2021, Woll and Van Petegem 2022). Structurally, all RyR isoforms are large homotetrameric proteins, with each protomer comprising around 5000 amino acids and several domains (Figure 21.4a). RyRs have a cytoplasmic region or "cytoplasmic shell" that contains the majority of the protein mass, a transmembrane region that forms the Ca^{2+} pore, and a small luminal region (Figure 21.4b). The cytoplasmic shell contains around 72% of the sequence and is mainly composed by structural solenoids, tertiary structures made of antiparallel α-helices (Figure 21.4c). This includes the N-terminal domains (NTD), the N-terminal solenoid (NSol), the SP1a/ryanodine receptor domains (SPRY), the ryanodine receptor repeats (RY repeats), the junctional solenoid (JSol), the bridging solenoid (BSol), and the shell-core linker peptide (SCLP). The remaining 28% of the sequence forms the "activation core and channel" region. This includes the core solenoid (CSol), the thumb and forefingers domain (T&F), the transmembrane domain (TM), and the C-terminal domain (CTD). Each RyR protomer has six transmembrane segments, with the channel pore being formed by the sixth segment (S6) from each of the four protomers. Although the TM is similar in structure to the voltage-gated channels, RyRs are insensitive to voltage. Gating of the channel is primarily regulated by the activation core, which senses changes in cytoplasmic Ca^{2+} levels and transmits this information to the pore region to control Ca^{2+} release.

The most critical ligand of RyRs is Ca^{2+} itself. RyRs exhibit the unique property CICR, where the binding of micromolar concentrations of Ca^{2+} triggers the opening of the channel, leading to further release of Ca^{2+} from the SR/ER. Binding of Ca^{2+} to the high-affinity site, located in the interface between the CSol and the CTD (Figure 21.4d), leads to conformational changes in the T&F domain that propagates to the TM, resulting in channel opening. There is a second low-affinity Ca^{2+} binding site that inactivates and inhibits RyRs at millimolar concentration. However, this site hasn't been structurally described, and it remains a mystery. ATP, another key modulator of RyRs, binds to a high-affinity site, located at the interface of the S6 of the pore, the T&F, and the CTD, and enhances channel activity by increasing the sensitivity of the receptor to activating Ca^{2+} (Figure 21.4d). The binding of ATP is thought to facilitate the conformational change communication between the cytoplasmic shell and the activation core with the pore. Absence of ATP usually decreases the open probability due to poor activation. Recently, a second low-affinity ATP binding site has been detected at the cleft of the RY1&2 domain (Figure 21.4e). This site was initially missed due to the lower resolution of such domain, but it has been attributed to have the ability to bind inhibitory drugs such as Rycals and Dantrolene (Melville, Dridi et al. 2022). Ligands such as xanthine, caffeine, and ryanodine also affect the structure and function of RyRs. Xanthine and caffeine bind to a low-affinity site, located in the interface between the TM and the CTD, and they increase RyR sensitivity to activating Ca^{2+}, leading to enhanced channel activity (Figure 21.4d). However, due to the low affinity and low *in vivo* concentrations of xanthine and caffeine, the physiological role remains to be addressed. Ryanodine, on the other hand, binds to a specific site in the channel pore and locks the channel in a subconductance state, effectively inhibiting channel function. Ryanodine, which is a compound produced by the South American plant *Ryania speciosa*, is the strongest known botanical insecticidal.

The cytoplasmic shell is the target of posttranslational modifications, such as phosphorylation, oxidation, and nitrosylation, and it interacts with various ligands and regulatory proteins, including calstabin or FK506-binding proteins (FKBP), calmodulin, and $Ca_V1.1$ in the case of RyR1. Other protein partners that modulate RyR function with less defined roles are STAC3, juntophillins, junctin, triadin, calsequestrin, S100A1, sorcin, Homer1, mAKAP, BIN1, calveolins, CLIC2, and presenilins, among others. For instance, calstabins or FKBPs are endogenous proteins that bind to RyRs and stabilize the closed state of the channel. The dissociation of calstabins/FKBPs from RyRs, induced by factors such as phosphorylation and other stress signals, leads to increased channel activity and has been implicated in many diseases, including heart failure (Shan, Betzenhauser et al. 2010). The calstabin binding site is a multidomain surface formed by the NSol, the SPRY1, SPRY3, and JSol domains. Differences between the closed and open states are minimal, and the structural mechanism of calstabin stabilization of RyRs remains unresolved.

Calmodulin is a Ca^{2+} binding protein that mediates the regulation of a wide range of physiological processes. In low nonactivating nanomolar Ca^{2+} concentration, apo-calmodulin activates RyR1, and in activating micromolar Ca^{2+} concentrations Ca^{2+}-bound calmodulin inhibits RyR1. On the other hand, calmodulin inhibits RyR2 in both conditions, although the affinity is higher for the Ca^{2+}-bound

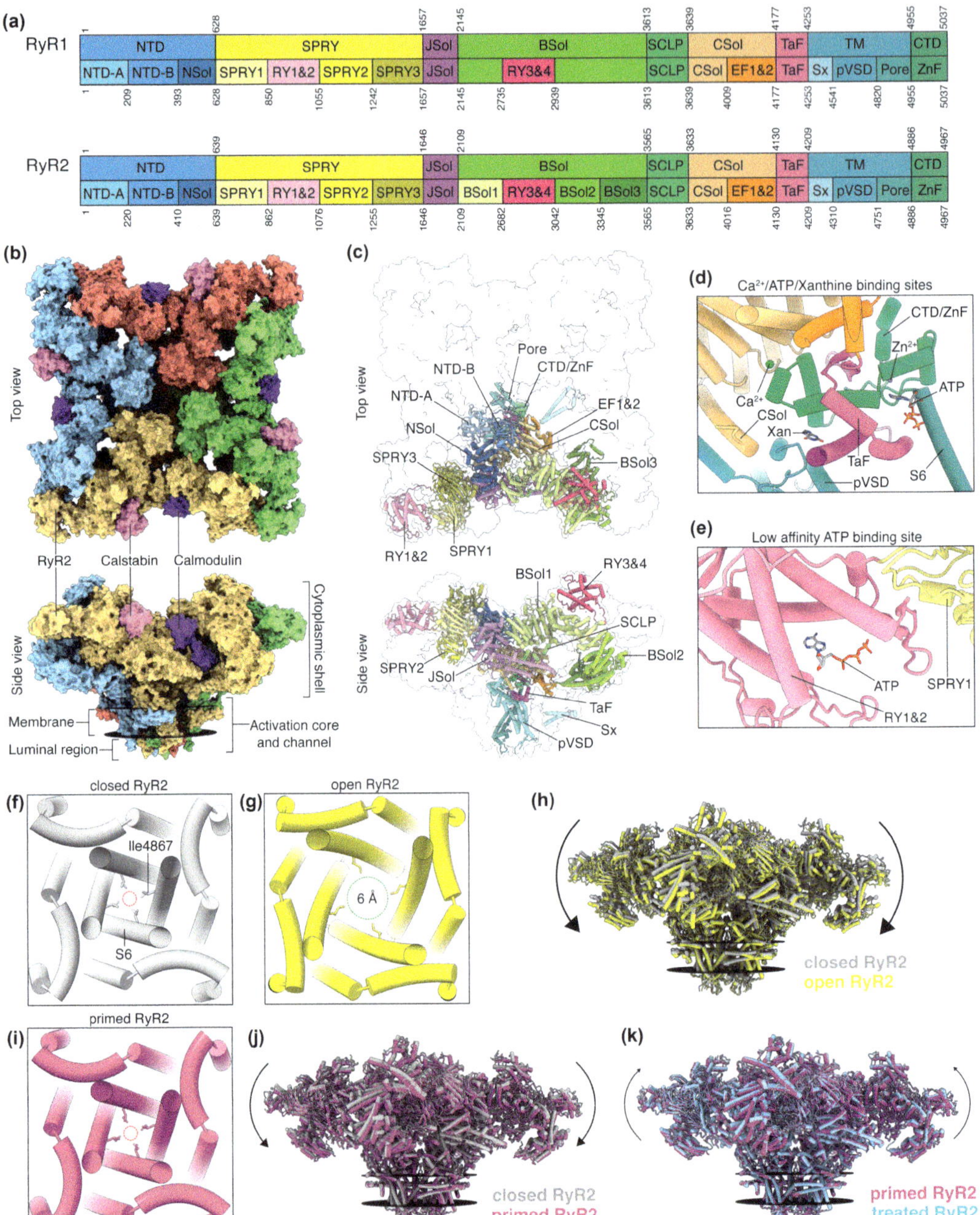

FIGURE 21.4 Structure–function relationship of ryanodine receptors. (a) Amino acid sequence of RyR1 (top) and RyR2 (bottom) divided into domains (NTD: N-terminal domain, SPRY: SP1a/ryanodine receptor domains, RY: ryanodine receptor repeats, JSol: junctional solenoid, BSol: bridging solenoid, SCLP: shell-core linker peptide, CSol: core solenoid, T&F: thumb and forefingers, TM: transmembrane domain, and CTD: C-terminal domain) and subdomains (NTD-A, NTD-B, NSol: N-terminal solenoid, SPRY1, RY1&2, SPRY2, SPRY3, RY3&4, EF1&2: EF hand, Sx: Auxiliary helix, pVSD: pseudo-voltage sensor domain, Pore, and ZnF: zinc finger motif). (b) Atomic model of RyR2 (PDB:7UA3) showing the four RyR2 protomers (yellow, green, red, and blue) and the accessory proteins calstabin (pink) and calmodulin (purple). (c) Atomic model of one protomer showing the different domains and subdomains (colors as depicted in [a]). The other protomers and accessory proteins are shown as a transparent surface. (d–e) Atomic model (PDB:7UA4) focused on the Ca^{2+}/ATP/xanthine binding sites (d) and low-affinity ATP binding site (e). (f) Atomic model focused on the pore of closed RyR2 (grey, PDB:7UA5). Residue Ile4867 in S6 is the main residue responsible for the pore constriction in the closed state (pore diameter is ~2 Å). (g) Same as (f) but of open RyR2 (yellow, PDB:7UA9). Pore diameter is ~6 Å, allowing hydrated Ca^{2+} ions to pass through. (h) Overlapped atomic models of the closed (grey, PDB:7UA5) and open (yellow, PDB:7UA9) states of RyR2 showing the cytoplasmic shell downward and outward movement. (i) Same as (f) but in the pathologic primed state (magenta, PDB:7U9X). (j–k) Overlapped atomic models of the closed (grey, PDB:7UA5), pathologic primed (magenta, PDB:7U9X), and Rycal treated (cyan, PDB:7UA1) states of RyR2 showing the cytoplasmic shell downward and outward (j), and upward and inward (k) movements, respectively. Adapted from Miotto et al. Sci. Adv. 2022. Created with ChimeraX.

form of calmodulin. Calmodulin binds to the SCLP peptide and interacts with the JSol and BSol domains, forming a complex mode of binding and regulation. The exact structural mechanism of activation and inhibition of RyRs by calmodulin is still to be determined.

As mentioned above, $Ca_V1.1$ interacts with RyR1 in the skeletal muscle cells, forming a mechanical coupling that is essential for EC coupling. This physical coupling allows for the rapid and coordinated release of Ca^{2+} from the SR following depolarization of the plasma membrane, a process essential for skeletal muscle contraction. However, the structural detail of the mechanical coupling remains a mystery.

Finally, phosphorylation and other posttranslational modifications such as nitrosylation, oxidation, and glutathionylation have been shown to activate or increase the open probability of RyRs. PKA phosphorylation of RyR2-S2808 and CaMKII phosphorylation of RyR2-S2814, both in the RY3&4 phosphorylation domain, have been shown to play crucial roles during the "flight-or-fight" response (or adrenergic response) and oxidative stress, respectively. Mice carrying the phosphomimic mutant RyR2-S2808D or the phosphorylation-resistant mutant RyR2-S2808A have shown that phosphorylation on this particular site is fundamental in the fine regulation of RyR2, and chronic phosphorylation can lead to many unwanted pathological consequences. RyRs exist in two main conformations: the closed state, where the pore is constricted and the cytoplasmic shell is in an upward and inward conformation, and the open state, where the pore is dilated and the cytoplasmic shell is in a downward and outward conformation (Figure 21.4f–h). The cryo-EM structure of the PKA phosphorylated RyR2 shows that the conformation of the cytoplasmic shell has moved from that of the closed state toward the open state, facilitating RyR2 activation by lowering the energy barrier to adopt a fully open conformation and allowing for faster heartbeat rate during the adrenergic response.

RyR Disease Variants and the Pathological Primed State

Diseases caused by mutations in the genes coding for RyRs are often referred to as "ryanodinopathies." This term is used to collectively describe a variety of neuromuscular and cardiac diseases associated with RyR dysfunction. In the context of skeletal muscle diseases, ryanodinopathies associated with mutations in the *RYR1* gene often involve malignant hyperthermia susceptibility (MHS) and a group of muscle disorders known as congenital myopathies, which include central core disease (CCD), multiminicore disease (MmD), and congenital fiber-type disproportion (CFTD), among others. MHS is a potentially lethal reaction to certain anesthetic agents characterized by a hypermetabolic state resulting in high body temperature and muscle rigidity. Mutations in the *RYR1* gene have been identified in about 50% to 70% of MHS cases. These mutations increase the sensitivity of the RyR1 channel to activation, leading to uncontrolled Ca^{2+} release (Rosenberg, Pollock et al. 2015). Dantrolene, developed in 1967, is the only approved drug to treat and prevent MHS. CCD is a congenital myopathy characterized by muscle weakness and the presence of characteristic "cores" in muscle fibers. Most CCD cases are linked to dominant mutations in the *RYR1* gene (Jungbluth 2007). MmD and CFTD are additional muscle disorders linked to *RYR1* mutations, presenting with varying degrees of muscle weakness and structural abnormalities in muscle fibers (Ferreiro, Monnier et al. 2002). In the context of cardiac diseases, ryanodinopathies associated with mutations in the *RYR2* gene typically refer to conditions like catecholaminergic polymorphic ventricular tachycardia (CPVT) and arrhythmogenic right ventricular cardiomyopathy type 2 (ARVC2). CPVT is an arrhythmogenic disorder that can lead to sudden cardiac death, especially during intense physical exercise or emotional stress. Most cases of CPVT are caused by dominant mutations in the *RYR2* gene. These mutations lead to an abnormal increase in Ca^{2+} release during sympathetic stimulation, resulting in delayed afterdepolarizations and triggered arrhythmias (Priori and Chen 2011). ARVC2 is a condition characterized by progressive replacement of the right ventricle myocardium with fibrous and fatty tissue, leading to arrhythmias and sudden cardiac death. While less common, certain mutations in the *RYR2* gene have been associated with ARVC2 (Tiso, Stephan et al. 2001). RyR1 and RyR2 dysregulation due to genetic mutations can lead to serious health conditions, though the severity and specific symptoms can vary widely among individuals with ryanodinopathies, even among those with the same mutation.

Currently, only the structures of a handful of disease variants have been resolved (Iyer, Hu et al. 2020, Woll, Haji-Ghassemi et al. 2021, Iyer, Hu et al. 2022, Miotto, Weninger et al. 2022). Interestingly, while the pore remained closed, they all showed an outward and downward movement of the cytoplasmic shell, reaching an intermediate pathological conformation known as the primed state (Figure 21.4i and j). The primed state would lower the energy barrier to reach an open state conformation, increasing Ca^{2+} sensitivity and open probability. At least for the mutants located in the cytoplasmic shell, inducing a primed state would be the common mechanism by which they promote the pathological outcomes. This would make researching drugs targeted to reverse the primed state fundamental to treat most ryanodinopathies. One drug candidate belonging to the Rycal family of drugs, called ARM210, has shown promising effects in preventing arrythmias and sudden death in CPVT models (Miotto, Weninger et al. 2022). The cryo-EM structure of the CPVT mutant RyR2-R2474S in the presence of ARM210 has been resolved, showing that binding of ARM210 reverses the pathological intermediate to a conformation closer to the closed state (Figure 21.4k). ARM210 and ATP form a complex that binds to the cleft of the RY1&2 domain. ARM210 binding induces local conformational and dynamic changes in RY1&2

that stabilize the interaction with the neighboring BSol domain. This interdomain interaction stabilizes the cytoplasmic shell, leading to a more upward and inward conformation. The RY1&2 domain and the BSol domain are in direct contact with the SPRY1 and the RY3&4 domains, which contain the calstabin binding site and the phosphorylation sites, suggesting that these factors are intrinsically connected and that regulation of RyRs is a multifaceted and complex process.

RyRs and the Membrane

RyR structures, like the majority of the membrane proteins, are resolved within detergent micelles. However, it has become a concern that detergents might affect the conformation or activity of membrane proteins. Therefore, efforts were made to resolve the structure of RyRs in lipidic environments to address this concern. Solutions of structures of RyRs reconstituted in nanodiscs (Iyer, Hu et al. 2020, Iyer, Hu et al. 2022), lipid vesicles (Melville, Kim et al. 2022), and native lipid vesicles without detergent extraction (Chen and Kudryashev 2020) have rendered structures almost identical to those obtained in detergent micelles (Figure 21.5a). This suggests that detergent does not affect RyR structure per se. However, when vesicles were present, differences in the curvature of the membrane were observed between closed and open states (Figure 21.5b) (Chen and Kudryashev 2020, Melville, Kim et al. 2022). This could be related to a physiologically relevant regulation by the membrane itself, and it cannot be disregarded that detergents might eliminate regulation of the channel by the surrounding lipid membrane. Therefore, lipidic environments are necessary to fully understand the regulation and dysfunction of RyRs.

CONCLUSIONS

In a few years' time, the cryo-EM resolution revolution opened the door to our understanding of many of the physiological and pathological mysteries around RyRs, from ligand gating to disease origins. It allowed detailed description of domains, the gating mechanism, ligand binding, and disease intermediates. However, there are still challenges to solve, including the mechanism by which $Ca_V1.1$ activates RyR1, the interaction between RyRs and the many protein partners and ligands, the activation of RyRs in nonmuscle cells, and how posttranslational modifications affect RyR structure and influence nongenetic diseases like heart failure or Alzheimer disease, among others. Cryo-EM, in conjunction with functional and physiological approaches, holds the key to answering many of these questions. Finally, cryo-ET has arisen as the technique that could revolutionize *in situ* structural biology and shed light on RyR macrostructural information such as nanoscale RyR organization, nanodomains, coupled gating, protein partner interactions in different cell types, and the membrane role in RyR structure and gating, among many others.

ACKNOWLEDGMENTS

Graphic software Biorender and ChimeraX were used for visualization. I would like to thank the lab leader Andrew R.

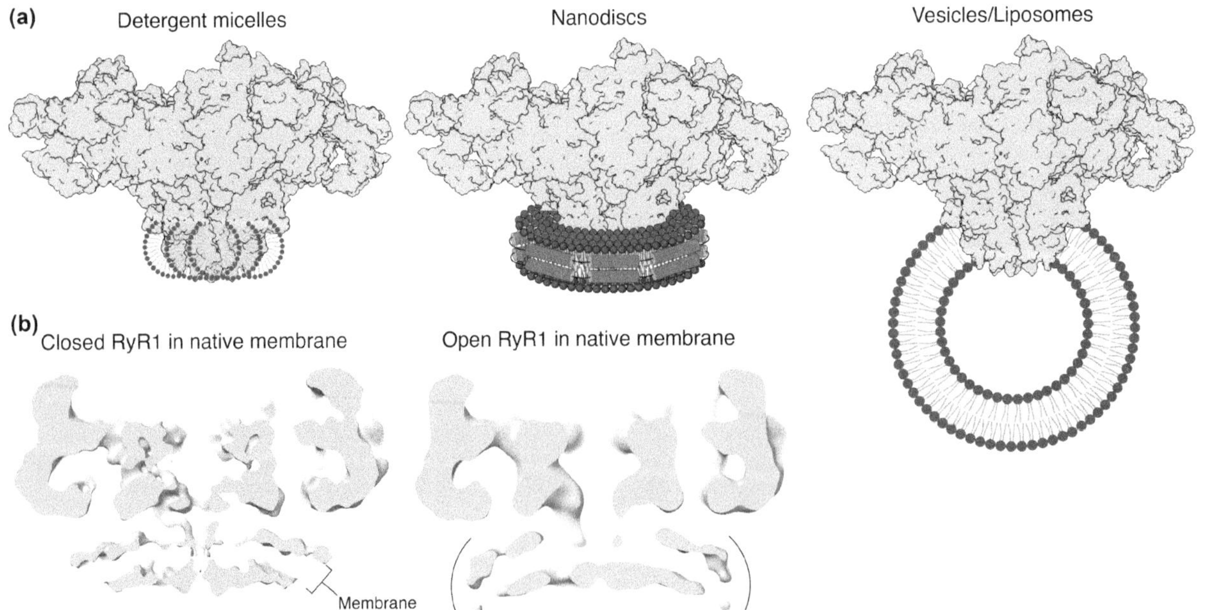

FIGURE 21.5 Detergents and lipid environments of ryanodine receptors. (a) Solubilization strategies of RyRs. Detergents and lipids are depicted as hydrophilic dark heads with hydrophobic grey tails. (b) Cryo-EM maps of RyR1 in native membrane in the closed (EMDB:EMD-10637) and open (EMDB:EMD-10638) states. Increased curvature in the open state is shown with arrows. Created with ChimeraX and BioRender.com.

Marks for leadership and funding, the current lab members Haikel Dridi, Yang Liu, Gunnar Weninger, Steve Reiken, Anetta Wronska, and Qi Yuan for discussions and collaborations, and Columbia University for research resources. Lab website: www.markslab.org

REFERENCES

Adams, P. D., P. V. Afonine, G. Bunkoczi, V. B. Chen, I. W. Davis, N. Echols, J. J. Headd, L. W. Hung, G. J. Kapral, R. W. Grosse-Kunstleve, A. J. McCoy, N. W. Moriarty, R. Oeffner, R. J. Read, D. C. Richardson, J. S. Richardson, T. C. Terwilliger and P. H. Zwart (2010). "PHENIX: a comprehensive Python-based system for macromolecular structure solution." Acta Crystallogr D Biol Crystallogr **66**(Pt 2): 213–221.

Barad, B. A., N. Echols, R. Y. Wang, Y. Cheng, F. DiMaio, P. D. Adams and J. S. Fraser (2015). "EMRinger: side chain-directed model and map validation for 3D cryo-electron microscopy." Nat Methods **12**(10): 943–946.

Berrout, J. and M. Isokawa (2009). "Homeostatic and stimulus-induced coupling of the L-type Ca2+ channel to the ryanodine receptor in the hippocampal neuron in slices." Cell Calcium **46**(1): 30–38.

Bers, D. M. (2002). "Cardiac excitation-contraction coupling." Nature **415**(6868): 198–205.

Chen, W. and M. Kudryashev (2020). "Structure of RyR1 in native membranes." EMBO Rep **21**(5): e49891.

Cheng, Y., N. Grigorieff, P. A. Penczek and T. Walz (2015). "A primer to single-particle cryo-electron microscopy." Cell **161**(3): 438–449.

Chiantia, G., E. Hidisoglu and A. Marcantoni (2023). "The role of ryanodine receptors in regulating neuronal activity and its connection to the development of Alzheimer's disease." Cells **12**(9): 1236.

Dridi, H., Y. Liu, S. Reiken, X. Liu, E. K. Argyrousi, Q. Yuan, M. C. Miotto, L. Sittenfeld, A. Meddar, R. K. Soni, O. Arancio, A. Lacampagne and A. R. Marks (2023). "Heart failure-induced cognitive dysfunction is mediated by intracellular Ca(2+) leak through ryanodine receptor type 2." Nat Neurosci **26**: 1365–1378.

Efremov, R. G., A. Leitner, R. Aebersold and S. Raunser (2015). "Architecture and conformational switch mechanism of the ryanodine receptor." Nature **517**(7532): 39–43.

Erk, I., H. Delacroix, G. Nicolas, J. L. Ranck and J. Lepault (1991). "Cryo-electron microscopy of vitrified specimens: an approach to the study of bulk specimens." J Electron Microsc Tech **18**(4): 406–410.

Ferreiro, A., N. Monnier, N. B. Romero, J. P. Leroy, C. Bonnemann, C. A. Haenggeli, V. Straub, W. D. Voss, Y. Nivoche, H. Jungbluth, A. Lemainque, T. Voit, J. Lunardi, M. Fardeau and P. Guicheney (2002). "A recessive form of central core disease, transiently presenting as multi-minicore disease, is associated with a homozygous mutation in the ryanodine receptor type 1 gene." Ann Neurol **51**(6): 750–759.

Fill, M. and J. A. Copello (2002). "Ryanodine receptor calcium release channels." Physiol Rev **82**(4): 893–922.

Gong, D., X. Chi, J. Wei, G. Zhou, G. Huang, L. Zhang, R. Wang, J. Lei, S. R. W. Chen and N. Yan (2019). "Modulation of cardiac ryanodine receptor 2 by calmodulin." Nature **572**(7769): 347–351.

Gong, D., N. Yan and H. A. Ledford (2021). "Structural basis for the modulation of ryanodine receptors." Trends Biochem Sci **46**(6): 489–501.

Gordon, A. M., E. Homsher and M. Regnier (2000). "Regulation of contraction in striated muscle." Physiol Rev **80**(2): 853–924.

Grant, T. and N. Grigorieff (2015). "Measuring the optimal exposure for single particle cryo-EM using a 2.6 A reconstruction of rotavirus VP6." Elife **4**: e06980.

Guse, A. H. (2015). "Calcium mobilizing second messengers derived from NAD." Biochim Biophys Acta **1854**(9): 1132–1137.

Hadiatullah, H., Z. He and Z. Yuchi (2022). "Structural insight into ryanodine receptor channelopathies." Front Pharmacol **13**: 897494.

Hu, Y., K. A. Iyer, A. R. Nayak, N. Kurebayashi, T. Murayama and M. Samso (2021). "Purification of recombinant wild type and mutant ryanodine receptors expressed in HEK293 cells." Bio Protoc **11**(15): e4112.

Iyer, K. A., V. Barnakov and M. Samso (2023). "Three-dimensional perspective on ryanodine receptor mutations causing skeletal and cardiac muscle-related diseases." Curr Opin Pharmacol **68**: 102327.

Iyer, K. A., Y. Hu, T. Klose, T. Murayama and M. Samso (2022). "Molecular mechanism of the severe MH/CCD mutation Y522S in skeletal ryanodine receptor (RyR1) by cryo-EM." Proc Natl Acad Sci U S A **119**(30): e2122140119.

Iyer, K. A., Y. Hu, A. R. Nayak, N. Kurebayashi, T. Murayama and M. Samso (2020). "Structural mechanism of two gain-of-function cardiac and skeletal RyR mutations at an equivalent site by cryo-EM." Sci Adv **6**(31): eabb2964.

Jungbluth, H. (2007). "Central core disease." Orphanet J Rare Dis **2**: 25.

Kobayashi, T., A. Tsutsumi, N. Kurebayashi, K. Saito, M. Kodama, T. Sakurai, M. Kikkawa, T. Murayama and H. Ogawa (2022). "Molecular basis for gating of cardiac ryanodine receptor explains the mechanisms for gain- and loss-of function mutations." Nat Commun **13**(1): 2821.

Lanner, J. T., D. K. Georgiou, A. D. Joshi and S. L. Hamilton (2010). "Ryanodine receptors: structure, expression, molecular details, and function in calcium release." Cold Spring Harb Perspect Biol **2**(11): a003996.

Ludtke, S. J., I. I. Serysheva, S. L. Hamilton and W. Chiu (2005). "The pore structure of the closed RyR1 channel." Structure **13**(8): 1203–1211.

Melville, Z., H. Dridi, Q. Yuan, S. Reiken, A. Wronska, Y. Liu, O. B. Clarke and A. R. Marks (2022). "A drug and ATP binding site in type 1 ryanodine receptor." Structure **30**(7): 1025–1034.e4.

Melville, Z., K. Kim, O. B. Clarke and A. R. Marks (2022). "High-resolution structure of the membrane-embedded skeletal muscle ryanodine receptor." Structure **30**(1): 172–180.e3.

Miotto, M. C., G. Weninger, H. Dridi, Q. Yuan, Y. Liu, A. Wronska, Z. Melville, L. Sittenfeld, S. Reiken and A. R. Marks (2022). "Structural analyses of human ryanodine receptor type 2 channels reveal the mechanisms for sudden cardiac death and treatment." Sci Adv **8**(29): eabo1272.

Olofsson, C. S., S. O. Gopel, S. Barg, J. Galvanovskis, X. Ma, A. Salehi, P. Rorsman and L. Eliasson (2002). "Fast insulin secretion reflects exocytosis of docked granules in mouse pancreatic B-cells." Pflugers Arch **444**(1–2): 43–51.

Priori, S. G. and S. R. Chen (2011). "Inherited dysfunction of sarcoplasmic reticulum Ca2+ handling and arrhythmogenesis." Circ Res **108**(7): 871–883.

Punjani, A., J. L. Rubinstein, D. J. Fleet and M. A. Brubaker (2017). "cryoSPARC: algorithms for rapid unsupervised cryo-EM structure determination." Nat Methods **14**(3): 290–296.

Radermacher, M., V. Rao, R. Grassucci, J. Frank, A. P. Timerman, S. Fleischer and T. Wagenknecht (1994). "Cryo-electron microscopy and three-dimensional reconstruction of the calcium release channel/ryanodine receptor from skeletal muscle." J Cell Biol **127**(2): 411–423.

Razinkov, I., V. Dandey, H. Wei, Z. Zhang, D. Melnekoff, W. J. Rice, C. Wigge, C. S. Potter and B. Carragher (2016). "A new method for vitrifying samples for cryoEM." J Struct Biol **195**(2): 190–198.

Rios, E. and G. Pizarro (1991). "Voltage sensor of excitation-contraction coupling in skeletal muscle." Physiol Rev **71**(3): 849–908.

Rosenberg, H., N. Pollock, A. Schiemann, T. Bulger and K. Stowell (2015). "Malignant hyperthermia: a review." Orphanet J Rare Dis **10**: 93.

Samso, M., W. Feng, I. N. Pessah and P. D. Allen (2009). "Coordinated movement of cytoplasmic and transmembrane domains of RyR1 upon gating." PLoS Biol **7**(4): e85.

Samso, M., T. Wagenknecht and P. D. Allen (2005). "Internal structure and visualization of transmembrane domains of the RyR1 calcium release channel by cryo-EM." Nat Struct Mol Biol **12**(6): 539–544.

Santulli, G., D. Lewis, A. des Georges, A. R. Marks and J. Frank (2018). "Ryanodine receptor structure and function in health and disease." Subcell Biochem **87**: 329–352.

Scheres, S. H. (2012). "RELION: implementation of a Bayesian approach to cryo-EM structure determination." J Struct Biol **180**(3): 519–530.

Scheres, S. H. (2016). "Processing of structurally heterogeneous cryo-EM data in RELION." Methods Enzymol **579**: 125–157.

Shan, J., M. J. Betzenhauser, A. Kushnir, S. Reiken, A. C. Meli, A. Wronska, M. Dura, B. X. Chen and A. R. Marks (2010). "Role of chronic ryanodine receptor phosphorylation in heart failure and beta-adrenergic receptor blockade in mice." J Clin Invest **120**(12): 4375–4387.

Sharma, M. R., P. Penczek, R. Grassucci, H. B. Xin, S. Fleischer and T. Wagenknecht (1998). "Cryoelectron microscopy and image analysis of the cardiac ryanodine receptor." J Biol Chem **273**(29): 18429–18434.

Tiso, N., D. A. Stephan, A. Nava, A. Bagattin, J. M. Devaney, F. Stanchi, G. Larderet, B. Brahmbhatt, K. Brown, B. Bauce, M. Muriago, C. Basso, G. Thiene, G. A. Danieli and A. Rampazzo (2001). "Identification of mutations in the cardiac ryanodine receptor gene in families affected with arrhythmogenic right ventricular cardiomyopathy type 2 (ARVD2)." Hum Mol Genet **10**(3): 189–194.

Woll, K. A., O. Haji-Ghassemi and F. Van Petegem (2021). "Pathological conformations of disease mutant ryanodine receptors revealed by cryo-EM." Nat Commun **12**(1): 807.

Woll, K. A. and F. Van Petegem (2022). "Calcium-release channels: structure and function of IP(3) receptors and ryanodine receptors." Physiol Rev **102**(1): 209–268.

Yan, Z., X. Bai, C. Yan, J. Wu, Z. Li, T. Xie, W. Peng, C. Yin, X. Li, S. H. W. Scheres, Y. Shi and N. Yan (2015). "Structure of the rabbit ryanodine receptor RyR1 at near-atomic resolution." Nature **517**(7532): 50–55.

Zalk, R., O. B. Clarke, A. des Georges, R. A. Grassucci, S. Reiken, F. Mancia, W. A. Hendrickson, J. Frank and A. R. Marks (2015). "Structure of a mammalian ryanodine receptor." Nature **517**(7532): 44–49.

Zalk, R., S. E. Lehnart and A. R. Marks (2007). "Modulation of the ryanodine receptor and intracellular calcium." Annu Rev Biochem **76**: 367–385.

22 Cryo-EM of Mitochondrial Membrane Proteins

Swati Yadav[†], *Shaileshanand Jha*[†], *and Kutti R. Vinothkumar*

INTRODUCTION—THE MITOCHONDRIA

In eukaryotic cells, mitochondria stand out as unique organelles due to their characteristic features and biological importance. The mitochondrion was first visualized in the 1950s using transmission electron microscopy (TEM) by the Belgian cell biologist Albert Claude and his colleagues (Claude, 1943, 1946; Ernster & Schatz, 1981). These TEM images showed mitochondria as cylindrical-shaped features measuring 1–2 µm in length and 0.1–0.5 µm in diameter. They are surrounded by a double membrane, the two parts of which are separated from each other by the intermembrane space (IMS). The inner membrane is characterized by invaginations into the mitochondrial matrix called the cristae (Ernster & Schatz, 1981). Along with the nucleus, the double membrane architecture of the mitochondria is a salient feature among the organelles in animal cells, and it plays many crucial roles, including the production of ATP. Despite mitochondria having their own genome, replication machinery, and translation apparatus, much of their proteome is encoded in the nucleus and imported into the respective locations in mitochondria. Some of the hydrophobic subunits of the electron transport chain complexes are encoded in the mitochondrial genome, and perhaps they have been retained to prevent aggregation during transport. The transport of large peptides across the mitochondrial membranes is performed by multi-subunit assemblies such as the TIM-TOM (translocase of the inner membrane–translocase of the outer membrane) complex and the SAM (signal-sequence associated motor) complex (Pfanner et al., 2019; Voos et al., 1999; Wiedemann & Pfanner, 2017). On the other hand, small molecules are transported by ion channels such as the VDAC (voltage-dependent anion channel) and porins such as mitoferrin, which is required for transporting iron across the membrane (Xian & Liou, 2021). While the inner membrane is rich in proteins involved in electron transport and energy generation, it also contains proteins that perform diverse functions such as nucleotide transport (ANT), lipid remodeling (TAFFAZIN), and cristae formation (IMMT) (Kunji et al., 2016; Mun et al., 2010; Xu et al., 2006). Thus, the mitochondrial proteins are spread across the matrix, inner membrane, IMS, and outer membrane to carry out many vital processes, including fatty acid metabolism and oxidative phosphorylation. Naturally, the mitochondria have been a source of fascination for researchers from different fields, and the dysfunction of mitochondrial proteins has been linked to several diseases, including neurodegenerative disorders, cardiovascular diseases, and cancer (Ebanks & Chakrabarti, 2022; Lee et al., 2017; Tucker & Park, 2019).

THE ELECTRON TRANSPORT CHAIN

The earliest reports of mitochondrial research can be traced back to the early 19th century, and subsequent work spanning much of the 20th century established that all living cells have a membrane potential and that the movement of ions is affected by thermodynamic and electrostatic forces. Further, the coupling of ATP synthesis to the ion gradient was postulated by Peter D. Mitchell in 1961 as the chemiosmotic hypothesis. The hypothesis states that the electrochemical gradient across the mitochondrial membrane is utilized by the ATP synthase enzyme to produce ATP (Mitchell, 1961). Over the course of the last seven decades, structural, biochemical, and biophysical studies have provided significant evidence to support the chemiosmotic hypothesis. The inner membrane of the mitochondria is densely packed, with about 46% of its area occupied by a number of proteins (Schlame, 2021), and it is the prime site of oxidative phosphorylation. The proteins involved in this process are also known as the electron transport chain complexes I, II, III, and IV (Figure 22.1). The nomenclature is based on their sequence in the series of electron transport from ubiquinone and succinate to oxygen from CI to CIV. The ion gradient generated by these complexes is utilized by ATP synthase (often called complex V) to generate ATP from ADP and P_i (Kühlbrandt, 2015).

The first step in oxidative phosphorylation is carried out by complex I (CI), also known as NADH: ubiquinone oxidoreductase. CI is an L-shaped complex with one arm embedded in the membrane and the other in the mitochondrial matrix. CI catalyzes the oxidation of NADH to NAD^+ by transferring a hydride to the flavin molecule bound close to it. The flavin molecule in turn transfers two electrons to the iron-sulfur (FeS) cluster, from where they are relayed through six FeS clusters until they reach the ubiquinone (UbQ) binding site at the intersection of the matrix and membrane arm. Here the reduction of ubiquinone to quinol takes place, and this reduction of UbQ is coupled to the transport of protons from the matrix to the IMS through multiple proton pathways in the membrane arm of CI. Mammalian CI is a large complex with a molecular weight

[†] These authors contributed equally to this work.

DOI: 10.1201/9781003326106-26

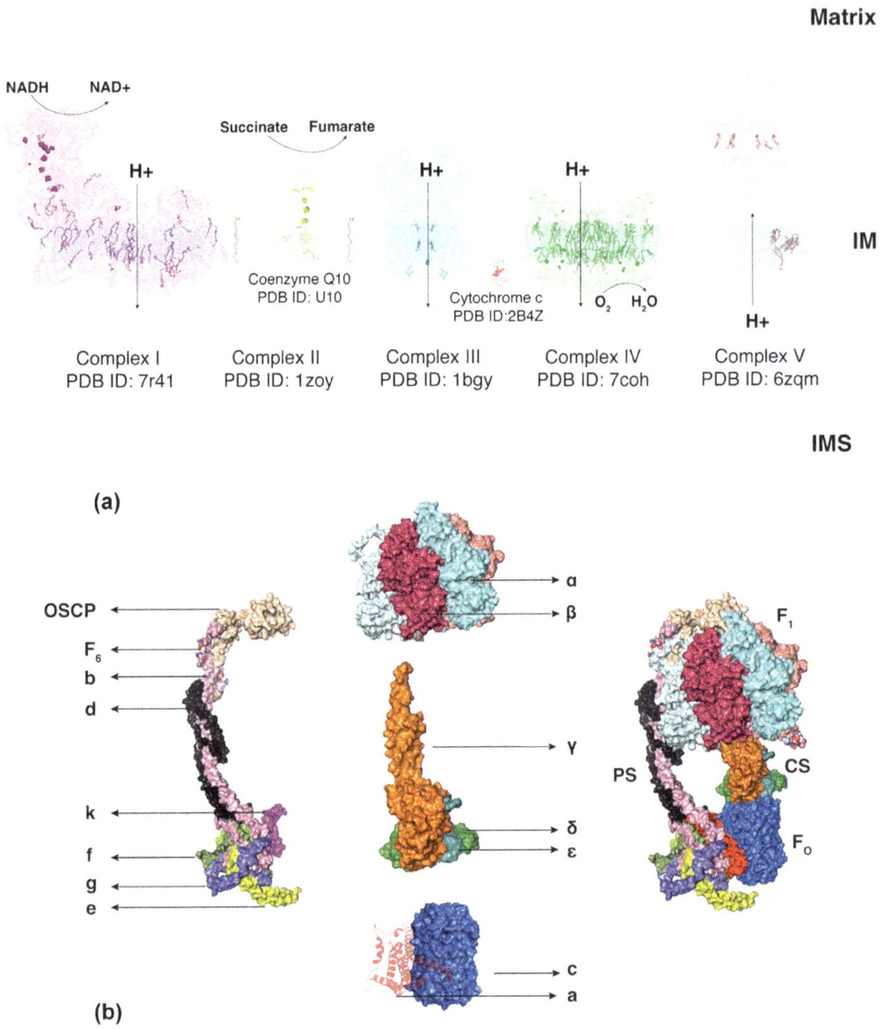

FIGURE 22.1 (a) Multi-subunit complexes in the inner mitochondrial membrane are responsible for carrying out oxidative phosphorylation. Lipids and cofactors are shown in stick and sphere representation and the protein molecule in cartoon representation. The arrows depict the direction of H^+ transport by the respective complexes. Curved arrows indicate the respective electron donors for CI and CII. (b) Salient features of the ATP synthase from *B. taurus* (Spikes et al., 2020). The F_1 and F_O domains and the central and peripheral stalk are shown, along with the fully modeled enzyme (PDB – 6zqm). The main components of the peripheral stalk include the subunits OSCP, b, F_6, and d. For completeness, here all the membrane subunits have been included in the PS (including subunit k, subunit j, and A6L, which are not visible in this view). Membrane subunits e, f, g, and A6L interact with some components of the PS. Thus, the bovine ATP synthase consists of 17 different subunits and 28 polypeptides in total (Spikes et al., 2020).

of ~1 MDa (Figure 22.1a). It comprises 14 core subunits that have all the components essential for the redox reaction and proton translocation and more than 30 supernumerary subunits (the number can vary depending on species) essential for assembly and stability (Fiedorczuk et al., 2016; Hirst, 2013; Kravchuk et al., 2022; Soufari et al., 2020; Vinothkumar et al., 2014; Zhu et al., 2016).

Complex II, also called succinate dehydrogenase, is the second enzyme in the ETC; it oxidizes succinate to fumarate, thus coupling the ETC to the Krebs cycle. The electrons from the oxidation of succinate are utilized to reduce ubiquinone. The electron transfer is linear and occurs through three FeS sites and one heme. CII is the smallest enzyme in the electron transport chain and is composed of four subunits, which form a heterodimer (Bezawork-Geleta et al., 2017; Sun et al., 2005). CII is involved in the generation of a reduced ubiquinone pool with no protons pumped.

The third enzyme in the ETC is complex III (CIII), also known as cytochrome bc_1. CIII accepts the electrons from CI and CII by catalyzing the oxidation of quinol to ubiquinone and transferring the electrons to the heme group bound to a mobile electron carrier protein, cytochrome c. CIII has two ubiquinone binding sites (Q sites), and when ubiquinol binds to the Q_o site, one electron is transferred to binuclear iron in Rieske iron-sulfur protein via heme c_1, which then reduces cytochrome c. The second electron reduces ubiquinone in the Q_i site through the heme b_L and b_H to generate a semiquinone intermediate. The binding of the second ubiquinol at the Q_o site is required for further reduction of the intermediate at the Q_i site. During this process, four protons are pumped to the IMS and two protons are also taken up from the matrix side. CIII is a homodimer composed of three core subunits per monomer containing four heme molecules, a 2Fe2S cluster, and many supernumerary subunits (Amporndanai et al., 2018; Crofts, 2021; Iwata et al., 1998; Kao & Hunte, 2022; Solmaz & Hunte, 2008).

The last step of the electron transport chain is carried out by complex IV (CIV), also known as cytochrome c oxidase (CcO). It accepts the electrons from the mobile electron carrier protein cytochrome c using its binuclear Cu_A binding site and transfers them to the heme a_3 cofactor bound in subunit I. The dioxygen species binds in between the heme a_3 and another copper site, Cu_B, where oxygen acts as the final electron acceptor and is reduced to water. The energy released in the process is utilized to pump four protons from the matrix to the intermembrane space (Rich, 2017; Tsukihara et al., 1996; Wikström et al., 2023; Wikström & Sharma, 2018; Yoshikawa et al., 1998; Zong et al., 2018).

The respiratory complexes (CI, CIII and CIV) further interact with each other and form loosely attached supercomplexes in varying stoichiometries. The presence of these supercomplexes was first observed in the Blue Native polyacrylamide gels (BN-PAGE) as they migrated at higher molecular weights corresponding to 1–2 MDa (Acín-Pérez et al., 2008). Electron cryo-microscopy (cryo-EM) has enabled the structural characterization of several of these supercomplexes in the past few years. Some of the recently observed assemblies include CIIICIV, $CIIICIV_2$, $CICIII_2CIV$, CICIIICIV, and $CICIII_2CIV_2$ (Klusch et al., 2022; Letts et al., 2016; Maldonado et al., 2021; Rathore et al., 2019; Vercellino & Sazanov, 2021, Wu et al., 2016). Several hypotheses exist to explain their presence in the inner membrane, such as facilitating substrate transfer and thereby increasing the efficiency of electron transport, reducing ROS, and regulating each other's assembly to maintain the overall stoichiometry in the inner membrane (Hirst, 2013; Klusch et al., 2022; Maldonado et al., 2021; Moe et al., 2021; Vercellino & Sazanov, 2021; Wu et al., 2016). However, a direct substrate tunnel has not yet been observed in any of these structures, but it is likely that these assemblies are an efficient way to pack a high density of these proteins in the inner membrane (Schlame, 2021).

The biochemical, biophysical, and structural studies of the ETC complexes have been pursued for many decades (summarized in Figure 22.1a. The figures were made with Chimera or ChimeraX and Pymol [DeLano, 2002; Goddard et al., 2007; Pettersen et al., 2021]), and the recent advances in cryo-EM have helped to uncover complexes from different species, the supercomplexes and their reaction processes. A complete review of the whole ETC is beyond the scope of this review as the literature on the structural details alone is vast (see reviews, Kühlbrandt, 2015; Vercellino & Sazanov, 2022). Here we focus on only one complex from the mitochondria—the ATP synthase.

STRUCTURAL FEATURES OF THE F_1F_O ATP SYNTHASES

The F_1F_O ATPases belong to the family of multi-subunit membrane protein complexes, which includes the V (vacuole) and the A (archaea) ATPases (Cross & Müller, 2004; Muench et al., 2011; Müller & Grüber, 2003; Nirody et al., 2020; Stewart et al., 2013). The F_1F_O ATP synthases comprise two major domains, the F_1 soluble domain and the F_O membrane domain. Historically, F_1 was the first intact fraction obtained and was named F_1 after "fraction 1." F_O was the domain that conferred oligomycin sensitivity to the protein and was thus named F_O where the "o" stands for oligomycin (Kagawa & Racker, 1966). The F_1 and F_O domains are connected by two structures known as the peripheral stalk (PS) and the central stalk (CS), whose names indicate their spatial locations. These four domains represent the general architecture, and the total number of subunits differs among organisms (Pedersen, 2005; Stewart et al., 2013; Walker, 2013). The binding-change mechanism to describe the mechanism of ATP synthesis by ATP synthase was proposed by Paul D. Boyer's group in the 1970s (Boyer et al., 1973; Kayalar et al., 1977). In the 1980s, John E. Walker and colleagues sequenced the genes encoding ATP synthase and identified the conserved nucleotide binding motifs (Walker A and B motifs) now commonly found in many proteins, and they eventually determined the X-ray structure of the F_1 domain (Abrahams et al., 1994; Saraste et al., 1981). For their contribution to elucidating the mechanism of ATP synthesis, they were jointly awarded the Nobel Prize for chemistry in 1997 (Boyer, 1998; Walker, 1998).

One of the well studied F_1F_O ATP synthases is from *Bos taurus*, which comprises 28 polypeptides and 17 different subunits (Spikes et al., 2020; Walker, 2013; Zhou et al., 2015). The ATP synthase was observed in the negative stain EM images of the intact mitochondria as rows of dimers as early as the 1980s (Allen et al., 1989). Structures of different domains were solved using X-ray crystallography (reviewed in [Walker, 2013; Yoshida et al., 2001]). The F_1 domain was readily isolated from the membrane fraction by chemical treatment and was found to be highly soluble and amenable to crystallization (Kamienietzky & Nelson, 1975; Tzagoloff, 1970). When the structure of the F_1 domain was solved, it was one of the largest macromolecular structures determined at that time (Abrahams et al., 1994). By the end of the 20th century, low-resolution EM maps of the full ATP synthase were obtained and the crystal structures were docked into these maps as pieces in a jigsaw puzzle (Dickson et al., 2006). Subsequently, the structures were improved to 8–15 Å resolution and the different domains could be identified distinctly. With the improvements in the detectors and data processing algorithms, now cryo-EM maps with resolutions <5 Å are obtained routinely (Courbon & Rubinstein, 2022). In this section, we describe the subunit composition and salient features of the bovine ATP synthase, which is summarized in Figure 22.1b.

The **F_1 domain** contains 3 α and 3 β subunits, which together form a hexameric structure with alternating α and β subunits (Figure 22.1b). This hexameric arrangement is also observed in the ATPases associated with diverse cellular activities such as (AAA)-ATPases, which contain a ring-shaped structure (typically hexamers) indicating a common evolutionary origin (Khan et al., 2022; Longo et al., 2020; Puchades et al., 2020). The α and β subunits are homologous, with 22% sequence identity and 37% sequence similarity in the bovine ATP synthase. They both contain the highly conserved nucleotide-binding motifs Walker A (G-X-X-X-X-G-KT) and Walker B (R/K-X-X-X-G-X-X-X-L-Q-Q-Q-D, where Q = hydrophobic amino acid) (Walker et al., 1982). However, only the β subunit is catalytically active, where ADP and P_i bind to form ATP in synthesis mode or hydrolysis of ATP to ADP and P_i in hydrolysis mode. The nucleotide-binding site on the α subunit is implicated in a regulatory role and does not perform catalysis. The ATP binding sites at the β subunit undergo large conformational change with the progress of the catalytic cycle. Three major conformational states identified for the β subunit are β_{AT}, β_{AD}, and β_E. β_{AT} corresponds to the ATP-bound tight state, β_{AD} is the ADP and P_i-bound loose state, and β_E is the empty state (Sobti et al., 2021; Stewart et al., 2014; Walker, 2013). The binding change mechanism proposed by Paul Boyer (Boyer et al., 1973) states that the three nucleotide binding sites on the β subunit bind ADP and P_i sequentially and undergo a conformational change to synthesize tightly bound ATP; this has been validated experimentally (Boyer, 1993). The enzyme can act in both synthesis and hydrolysis modes depending on factors such as proton concentration (membrane potential/hypoxia), ATP concentration (rest/stress states) and the presence of certain inhibitors (Bienert et al., 2009; Elston et al., 1998; Junesch & Gräber, 1987).

The **central stalk** couples the proton transport at the F_O domain with the catalysis at the F_1 domain. The central stalk comprises the γ subunit, which spans the entire length from the top of F_O to F_1 and is associated with δ and ε subunits at the base (Figure 22.1b). The γ subunit rotates with the rotation of the c-ring. Every 120° rotation of the γ subunit leads to the conformational change in the β subunits of the F_1. Further, substates in the 120° rotation cycle of 40° (binding dwell) and 80° (catalytic dwell) have been identified by single-molecule and cryo-EM studies (Guo & Rubinstein, 2022; Noji et al., 1997; Pinke et al., 2020). The subunits δ and ε are important in the regulation and assembly of the ATP synthase complex and are known to inhibit ATP hydrolysis (Feniouk et al., 2006; Pecina et al., 2018).

The **F_O domain** is the proton pump and is completely embedded in the inner membrane. The c subunit in the bovine enzyme is an octameric ring but varies in its oligomeric state across species (Davies & Kühlbrandt, 2018). The c-ring of the F-ATPases has two TM helices, with a conserved acidic residue in the membrane region, and is in contact with subunit a on one side and the membrane on the other (Kühlbrandt & Davies, 2016; Spikes et al., 2020). The major components of the **peripheral stalk** include the subunits OSCP (oligomycin sensitivity conferring protein), F_6, d, and b, and the membrane subunits e, f, g, and A6L can be thought of as accessories. The peripheral stalk keeps the F_1 head in place during the rotation of the c-ring and the associated central stalk. The OSCP subunit is at the top and acts as an anchor between the F_1, central, and peripheral stalk. As it is present at an accessible location in the complex, it has been implicated in the regulation of enzymatic activity by binding regulatory proteins and drugs and the formation of the much-debated permeability transition pore (Alavian et al., 2014; Giorgio et al., 2019). The subunits b, d, and F_6 form the body of the peripheral stalk and are composed of long parallel helices, which confer stiffness to the structure; toward the membrane, the C-terminus of subunit A6L adds to the peripheral stalk (Spikes et al., 2020). The hinge region of subunit b is anchored to the membrane with two helices, and they interact with membrane subunits including e, f, and g. During catalysis, the peripheral stalk is known to move laterally by 10 to 15 Å, which in turn prevents the free rotation of the catalytic domain, and this flexibility prevents enzyme wobbling during catalysis (Guo & Rubinstein, 2022; Pinke et al., 2020; Spikes et al., 2020).

As ATP synthases are important molecules in the cell, their overall architecture and mechanism is highly conserved, but the subunit composition and organization vary across species. Structures from different organisms have highlighted key differences, some of which are described in the section on dimer and higher-order assemblies and are currently an active area of research.

ELECTRON MICROSCOPY OF F_1F_O ATP SYNTHASES

Obtaining the intact structures of the full F-ATPase remained a major bottleneck for a long time. The dissociation of the F-ATPases into F_1 and F_O was extremely helpful in obtaining 3D crystals and crystal structures of the F_1 domain, F_1 domain with the central stalk, and some parts of the peripheral stalk (Stewart et al., 2014; Stock et al., 2000; Walker, 2013). It was also possible to obtain the structures of the F_1 domain with the central stalk and the c-ring but with no peripheral stalk (Stock et al., 1999; Watt et al., 2010). The intrinsic stability of the c-ring enabled the acquisition of crystal structures from different species (Kühlbrandt, 2019; Kühlbrandt & Davies, 2016), but the complete architecture of the F-ATPase remained elusive. In particular, the membrane-embedded subunit a proved to be intractable by crystallography. Thus, imaging with electrons seemed a possible way to obtain intact structures of F-ATPases. Early studies of F-ATPases were limited to negative staining and low-resolution 3D maps, which provided the overall architecture and distinct features of the enzyme (Böttcher & Gräber, 2000; Karrasch & Walker, 1999; Wilkens & Capaldi, 1998). The low-resolution cryo-EM map of bovine

ATPase was one of the major landmarks that was used to dock known crystal structures (Dickson et al., 2006; Rubinstein et al., 2003). Some of the advantages of using cryo-EM include imaging and structure determination as single molecules with minimal perturbation to complexes that might arise during crystallization required for X-ray crystallography, and the possibility of obtaining multiple structural states (Cheng, 2018; Nogales & Scheres, 2015).

3D reconstruction of macromolecules using cryo-EM relies on collecting multiple projections of the molecule of interest and determining the orientation parameters of these 2D projections to obtain a 3D map. The heart of the technique is the projection slice theorem, which states that a Fourier transform of a 2D projection represents a slice at a specific angle in the 3D Fourier transform of a given object (Bracewell, 1956). Different orientations of the macromolecule represent distinct 2D projections, which correspond to slices at distinct angles in the 3D Fourier transform. The images collected by cryo-EM are noisy and have low contrast due to many factors, including the low dose of electrons used in imaging and ice thickness. The large size of a macromolecule is advantageous because the particles have higher contrast, which makes them easy to identify. The imaging is typically performed with underfocus or defocus, which provides the contrast for visualizing macromolecules, and is corrected during image processing (Cong & Ludtke, 2010; Sigworth, 2016). The earlier detectors had a lower signal-to-noise ratio; this resulted in images that were extremely noisy, and the macromolecule of interest was often hard to identify. The components in the samples, such as free detergent micelles in membrane proteins, also contribute to the noise. Thus the pursuit of obtaining high-resolution maps for membrane proteins using cryo-EM remained difficult.

The advent of new direct electron detectors with higher signal-to noise ratios and the possibility of collecting movies rather than single images changed the way imaging was performed with an EM (Li et al., 2013; McMullan et al., 2009, 2014, 2016). These images, along with advanced algorithms, allowed to achieve higher-resolution maps and to computationally sort different populations of a given macromolecule (Grant et al., 2018; Punjani et al., 2017; Scheres, 2012). In the case of F-ATPases, the sample preparation—i.e., the biochemical preparation and the process of grid making for cryo-EM—have also played a major role in the structures that are being reported.

The inherent conformational flexibility of the F-ATPase (both in the F_1 and F_O domains) poses another challenge to achieving high resolution. To limit the heterogeneity, the use of an inhibitor protein (IF_1) that binds at the interface of α and β subunits and thus locks the enzyme in a specific conformation has been used in a few cases (Gu et al., 2019; Pinke et al., 2020; Spikes et al., 2020; Vinothkumar et al., 2016). How are the new technologies in cryo-EM helping? Perhaps the most striking are the raw images that are now obtained by cryo-EM. In Figure 22.2, the raw micrographs of the monomeric yeast and the dimeric bovine enzymes clearly reveal the particles and the distinct domains. With subsequent 2D class averages (Fig. 22.2c–e)—i.e., when similar particles are averaged to increase the signal-to-noise ratio—more features in the enzyme emerge, and often secondary structure details can be observed. For example, the central and peripheral stalk and the helices in the F_O domain are clearly seen in Figure 22.2d. We would like to emphasize that the ice thickness, particle distribution, and choice of data collection (integration vs. counting—Figure 22.2c and d) are all important factors in obtaining high resolution of the F-ATPases. Note that the 2D class averages shown here do not reflect the amount of structural heterogeneity in the sample. The need for robust 3D classification methods to separate the different states is described in the next section.

MECHANISTIC INSIGHTS OF F-ATPases FROM CRYO-EM

Several new insights have been gained from the cryo-EM structures of F-ATPases from different organisms, either as full-length monomers or as oligomers. As the F-ATPases are rotary enzymes, one of the major insights from cryo-EM structures was the possibility to resolve/classify structural states that exist in a population. In the cryo-EM image processing pipeline, the 3D classification step is used to identify and separate distinct populations existing in a given dataset that are captured when the sample is frozen rapidly in liquid ethane. In the classification step, one or more 3D references are provided, and the number of classes is defined either based on prior biochemical knowledge or by a random number provided to the software, which then performs global searches to separate conformational and compositional differences in the particles and local searches to identify minor differences in the populations (Scheres, 2016; Sigworth, 2016). In addition, the use of focused refinements and classifications, where a part of the structure is masked and the remaining part is refined, very often leads to better maps and higher resolution.

The bovine ATPase dimer is used as an example to show how 3D classification is used to separate multiple populations and reveal the heterogeneity in the enzyme present in a single sample (Figure 22.3). The preparation of the dimeric enzyme had the inhibitor protein IF_1 bound, which locked the F_1 domain in defined states, but the presence of two monomers makes the data processing more complicated as the inhibitor protein in each monomer can be bound at different interfaces, thus the need for a large dataset to separate these different populations and obtain higher-resolution maps (Spikes et al., 2020, 2021). In one approach, a mask around a monomer was used during refinement. In the second approach, both monomers were masked, extracted, and centered, and subsequent refinement of both monomers resulted in twice the number of particles. Further, focused refinement of selected regions of the enzyme allowed for the modeling of a significant portion of the enzyme that otherwise would have been difficult (Spikes et al., 2020).

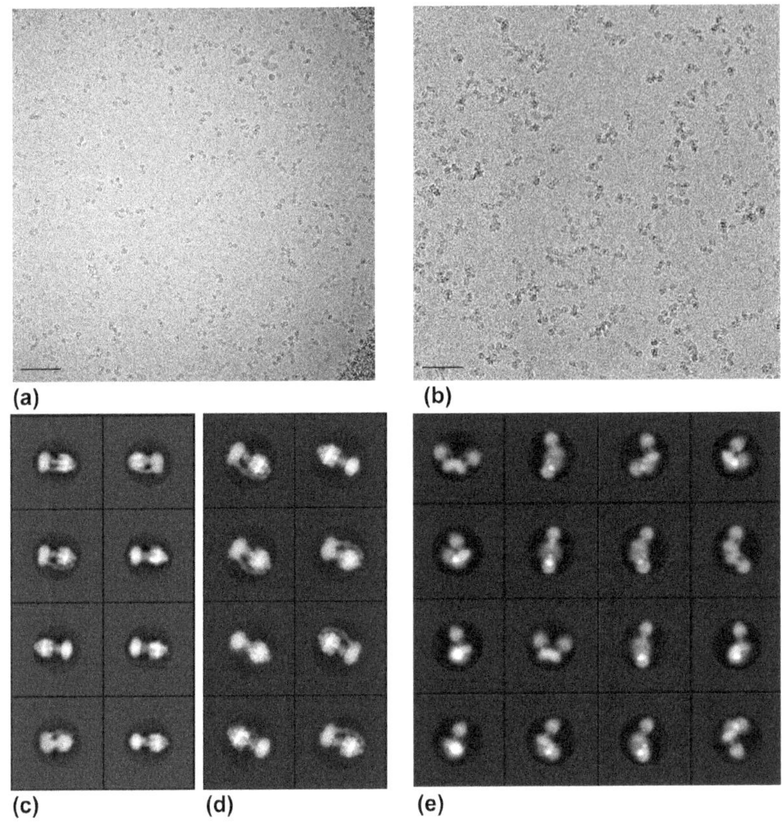

FIGURE 22.2 Cryo-EM of F-ATPases. (a–b) Micrographs of F-ATPase on ice (scale bar is 800 Å). (a) The monomer of *Pichia angusta* F-ATPase in cymal-7 (Vinothkumar et al., 2016) and (b) the bovine F-ATPase in GDN+phospholipids with Brij-35 added show the dimeric population and some higher-order oligomers (Spikes et al., 2020). (c–d) Selected reference-free 2D class averages of *P. angusta* F-ATPase are shown. (c) The classes shown were obtained with images collected in the integration mode of the Falcon 2 detector (box size is 300 pixels and sampled at 1.72 Å/pixel). (d) These classes were obtained with images collected in counting mode of the Falcon 3 detector (box size is 320 pixels and sampled at 1.38 Å/pixel). Note the higher-resolution features seen in the counting mode, in particular the TM helices in the membrane domain. (e) The 2D classes show the different views of the dimers of bovine ATPases. The box size is 384 pixels and sampled at 1.77 Å/pixel.

The lateral movement of the peripheral stalk was observed in the different populations, which represent the distinct states in the catalytic cycle. This observation was in agreement with other studies as well (Guo & Rubinstein, 2022; Pinke et al., 2020; Spikes et al., 2020).

The structure of the complete membrane domain of the F-ATPases, including the subunit a has long been sought, in particular its arrangement with respect to the c-ring. Biochemical studies have shown that some conserved residues in this subunit are important for ion transport, and a model with two half-channels for proton translocation had been proposed (Fillingame & Steed, 2014; Junge & Nelson, 2015). The earlier models of subunit a had assumed that the TM helices in this subunit follow the traditional orientation of the TM helices, i.e., perpendicular to the membrane. Thus, when cryo-EM maps of full-length F-type ATPases were obtained, it came as a surprise to learn that these long helices are highly tilted in the membrane as opposed to the conventional straight helices (Allegretti et al., 2015; Guo et al., 2017; Vinothkumar et al., 2016). Subunit a comprises six TM α-helices; four of these helices are tilted to wrap around the c-ring, and the angle of tilt differs among species (Figure 22.4). In the bovine enzyme, this angle is 30°. The horizontal helices of subunit a are characteristic features of the ATPases, and together with the c-ring they create the proton entry and exit sites at the IMS and the matrix. The ion (typically protons, sometimes sodium) translocation pathway includes the residues responsible for the proton entry (a-Glu203), exit (a-Asp149) and transport (c-Glu58) and is highly conserved across species (Kühlbrandt, 2019; Spikes et al., 2020). It is proposed that proton transfer can occur via the Grotthuss mechanism through the cluster of water molecules that form a hydrogen-bonded network in a cavity at the interface between subunit a and the c-ring, near the c-Glu58 residue (of the bovine enzyme). Additionally, several lipid molecules have been found near the proton entry site, which is believed to seal the proton pathway and prevent the proton from diffusing away (Spikes et al., 2020). Across species, subunit a has similar architecture, but the stoichiometry or subunit number in the c-ring

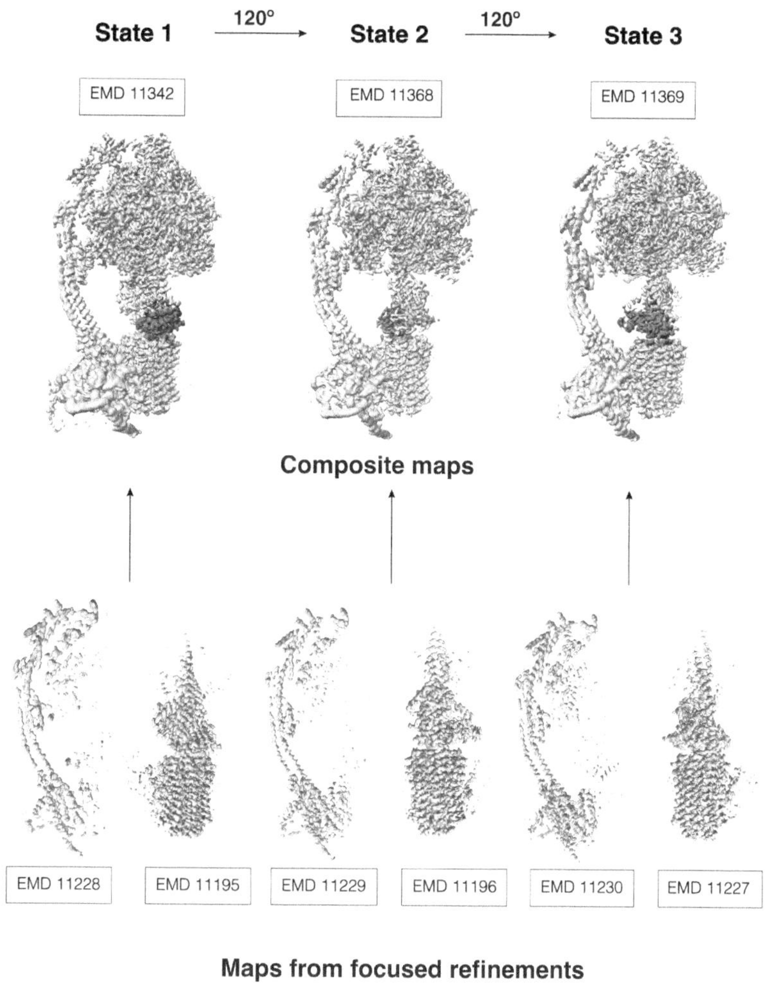

FIGURE 22.3 The different states of the bovine ATP synthase resolved as individual structures using 3D classification and focused refinement (Spikes et al., 2020, 2021). The dark grey area in the central stalk in the composite map shows the density for the subunit δ, used here to illustrate the rotation of the central stalk by ~120° along with that of the c-ring. The bottom panel shows the maps from focused refinement of the peripheral stalk and the c-ring with the central stalk.

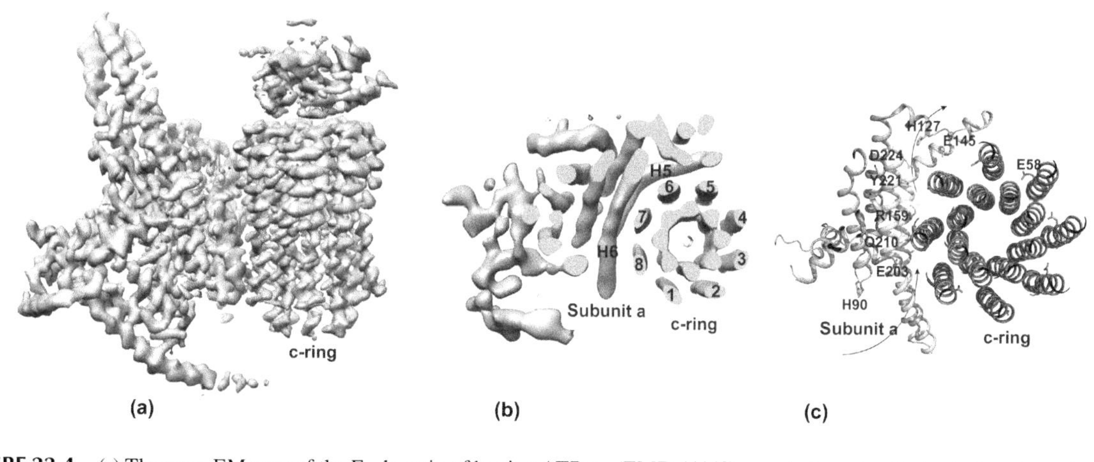

FIGURE 22.4 (a) The cryo-EM map of the F_O domain of bovine ATPase (EMD-11149) reveals the subunits, including the c-ring. (b) A section of the cryo-EM map showing the tilted helices of the subunit a and the c-ring. The helices H5 and H6 in the subunit are labeled along with the eight subunits of the c-ring. The cryo-EM map has been low-pass filtered to 6 Å for clarity. (c) The model of subunit a and c-ring with key residues labeled. The key residues of the proton entry and exit pathways (marked with arrows) are labeled (PDB–6zbb).

differs (between 8 and 17; see review by Kühlbrandt, 2019). Perhaps the arrangement of the H5/H6 helix of subunit a has evolved to follow the curvature of the c-rings and wrap around the c-rings of different dimensions, which would be difficult with conventional TM helices. Also, two features have been highlighted from the structures to be common among the ATP synthases: the distances between the proton entry and exit channels with respect to the conserved acidic residue in the c-ring, and the acidic residues in adjacent subunits of the c-ring spaced within a particular distance to be protonated/deprotonated at the same time (Kühlbrandt, 2019).

DIMER AND HIGHER-ORDER ASSEMBLIES

Following the chemiosmotic hypothesis, one of the questions that was put forward was the organization of the respiratory enzymes in the mitochondrial membrane with respect to one another (Mitchell, 1961). Two models were proposed. The first is the solid model, which assumes that these enzymes are tightly associated with each other. The other is the fluid model, where the enzymes are well separated in the membrane independent of each other (Althoff et al., 2011; Chance & Williams, 1955; Gupte et al., 1984; Hackenbrock, 1981). While the fluid model is more consistent with the chemiosmotic hypothesis, the advent of the BN-PAGE and the consequent detection of assemblies of these complexes in different detergents supported the solid model (Schägger & von Jagow, 1991; Wittig et al., 2006). The current understanding hints toward a plastic model, where a dynamic assembly or disassembly of supercomplexes occurs in response to cues such as oxygen concentration (Courbon & Rubinstein, 2022; Ramírez-Aguilar et al., 2011). Apart from the supercomplexes of the ETC, the oligomeric state of the bovine ATP synthase had been characterized early on using biochemical techniques such as BN-PAGE, density gradient centrifugation, etc., (Arnold et al., 1998). However, the recent cryo-EM advances have led to a number of novel structures, including dimers and other higher-order assemblies such as tetramer and hexamer (Gu et al., 2019; Mühleip et al., 2021). It is important to note that these oligomeric states could be artefacts of the solubilization and purification procedures, and their physiological relevance remains to be validated. While the bacterial and chloroplast ATP synthases are monomeric, the mitochondrial ATP synthases have been shown to have multiple oligomeric states that vary across species (Figure 22.5).

ATP synthase dimers have been observed in different higher organisms, and their critical role in the formation and stability of the cristae has been proposed (Davies et al., 2012). The subunits known to contribute to dimeric assembly formation and stability are, e, g, k, i/j, and c (Courbon & Rubinstein, 2022; Kühlbrandt, 2015). In the yeast ATP synthase, deleting subunits e or g leads to a complete absence of lamellar cristae, which was further correlated to the absence of dimeric ATP synthase assemblies as shown by electron cryo tomography (cryo-ET) (Davies et al., 2012). In *Podospora anserina*, a popular model system used to study aging, it has been shown that the young cultures have exclusively dimeric ATP synthase assembled in rows and forming lamellar cristae. However, as the cells grow older, the dimers start to dissociate, and the cristae structures are lost as a consequence (Daum et al., 2013). The formation of rows of ATP synthase dimers has been shown by *in vitro* reconstitution into liposomes using cryo-ET (Davies et al., 2012), and these rows of ATPase dimers have also been observed in isolated mitochondrial membranes (Davies et al., 2012) as well as in intact mitochondria (Dudkina, et al., 2010). It has been hypothesized that the assembly of the dimers in rows is an inherent feature of the protein composition and is independent of the lipid composition (Anselmi et al., 2018; Davies et al., 2012). These rows are likely to assist in the lamellar organization of the cristae and increase in the surface area available to perform oxidative phosphorylation up to 10 times (Kühlbrandt, 2019). Many of these dimeric and higher oligomeric states have been observed *in situ* with cryo-ET, and their sub-tomogram averaging has revealed the effect of oligomeric states on the cristae structure. The dimeric *Eugleozoan* enzyme induces a discoid shape in the cristae (Mühleip et al., 2019). The tetramers from *Ciliophora* form helical rows, which lead to the formation of tubular cristae (Blum et al., 2019; Mühleip et al., 2016). In the case of the *apicomplexan* hexamer, the apical bulb-shaped cristae are formed by pentagonal pyramid-shaped assemblies that are approximately 20 MDa in size (Mühleip et al., 2021).

The mitochondrial ATP synthase dimers have been classified into four types based on the angle made by the two central stalks. The type I dimers include the ones from yeast, mammals, and fungi that have a dimeric angle of 86° and assemble into long rows (Davies et al., 2011; Kühlbrandt, 2019; Strauss et al., 2008). Type II dimers have been observed in *Polytomella* sp., and their characteristic feature is the presence of ATP synthase–associated subunits that constitute the enlarged peripheral stalks. The dimeric angle in *Polytomella* sp. is 56°. The type III dimers are observed in ciliates, where the dimeric angle is close to zero, resulting in a U-shaped structure instead of the V-shape observed in mammals. Moreover, the peripheral stalks have a lateral offset, and interactions between the additional subunits at the dimeric interface in the lumen have been identified (Blum et al., 2019; Cano-Estrada et al., 2010). Cryo-ET of type III dimers from *Paramecium* shows them forming long helical rows in the membrane (Mühleip et al., 2016). Type IV dimers have been observed in *Euglena* and *Trypanosoma brucei*, a parasite that causes sleeping sickness. These dimers also have a lateral offset in their peripheral stalks and luminal subunits similar to the type III dimers. However, the dimeric angle between the monomers differs significantly and is approximately 55°. Strikingly, the *Euglean* F_1 is pyramid shaped instead of the regular globular architecture observed in all the other classes (Davies & Kühlbrandt, 2018).

The recent structures of mitochondrial ATP synthases by single-particle cryo-EM from different organisms such as apicomplexan *Toxoplasma gondii*, protozoan *Tetrahymena thermophila*, *Euglena gracilis*, and from the porcine heart have all shown different oligomeric states (Flygaard et al., 2020; Gu et al., 2019; Mühleip et al., 2019, 2021) (Figure 22.5). While some of the interacting subunits in these cases are conserved, others are unique. The structure of mammalian mitochondrial ATP synthase from the porcine heart has two V-shaped dimers having dimer–dimer interactions linked by two inhibitory factors IF_1 and stabilized by additional subunits k, g, and e that end up forming an H-shaped tetrameric assembly (Gu et al., 2019). The apicomplexan ATP synthase assembles into hexamers that are formed by three dimers and stabilized by apicomplexan-specific subunits in the luminal region. The inhibitor IF_1 is also present at the dimer interface (Mühleip et al., 2021). The type II dimer from *Polytomella* sp. is an example where the additional algae-specific subunits are incorporated into bulky and rigid peripheral stalks and act as a bridge strengthening the dimeric interface. Perhaps these additional subunits provide stability to the dimers even in relatively harsh detergents, which in other species such as yeast or mammalian ATP synthases dissociate the dimer into monomers (Cano-Estrada et al., 2010).

CONCLUSIONS AND OUTLOOK

The advances in cryo-EM have had a great impact in the study of highly dynamic and labile multi-subunit complexes, and the F_1F_O ATP synthase is a very good example (Figure 22.5). Apart from the technical advances in imaging, the advent of mild or lipid-like detergents such as lauryl maltose neopentyl glycol (LMNG) (Chae et al., 2010), digitonin, and its synthetic analogue glycol diosgenin (GDN) (Chae et al., 2012) have been important in keeping the native assembly intact, including some of the higher-order oligomers, which otherwise might dissociate due to their fragile interfaces (Kampjut et al., 2021; Li, 2022). In addition to these mild detergents, novel synthetic amphipathic polymers such as styrene maleic acid copolymer (SMA) and other derivatives are likely to play a role in the study of the ETC complexes and membrane proteins in general. These amphipathic polymers are directly added to the membranes, where they are thought to extract out portions of membrane from the bilayer, thus creating SMA lipid particles (SMALPS) in an aqueous environment and thereby completely bypassing the need for detergents (Scheidelaar et al., 2015). The alternating styrene and maleic acid groups in SMA interact with the acyl chain and head groups of the lipid molecule respectively, forming a belt of SMA around the lipids, which may also contain resident proteins. One of the major advantages that the system provides is the isolation of membrane proteins and their complexes with intact annular lipids; hence they have been called native nanodiscs (Killian, 2018; Knowles et al., 2009). However, the size of the nanodisc is constant for a particular polymer, and SMALPS are intolerant to divalent cations and sensitive to low pH (Jamshad et al., 2015; Scheidelaar et al., 2016). Thus, derivatives and variations

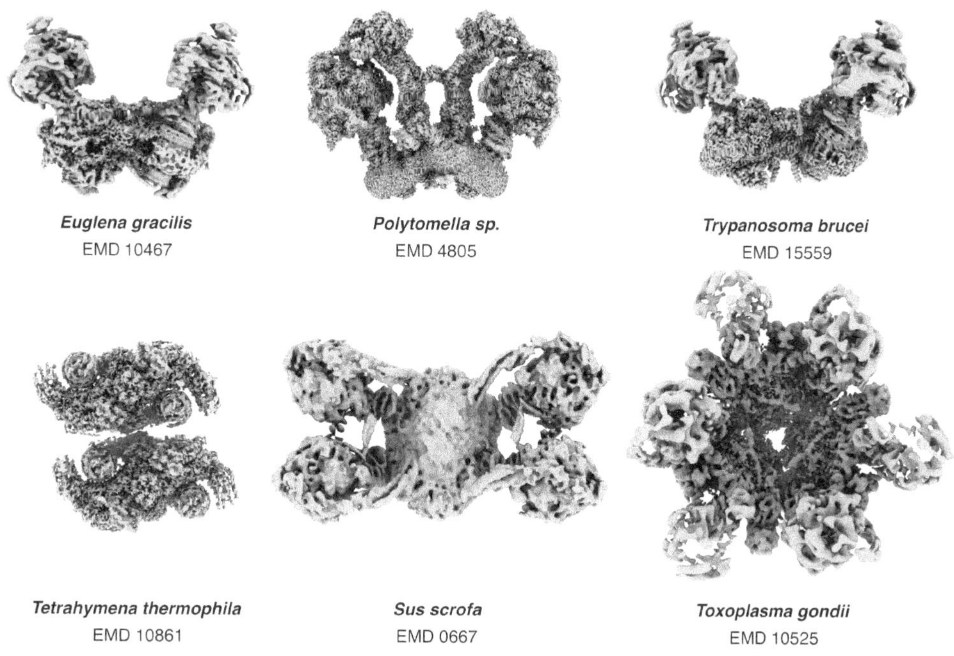

FIGURE 22.5 Higher-order structures of F_1F_O ATP synthases. The ATP synthases from different organisms have been selected here to show different oligomeric states with the dimeric form observed most commonly, and higher assemblies such as tetramer and hexamer have also been observed.

of SMA are being developed to overcome these limitations, and these include di-isobutylene maleic acid (DIBMA), which is stable at higher concentrations of divalent cations (Oluwole et al., 2017), and SMI, styrene-co-maleimide, which is stable in acidic pH as well as compatible with divalent cations (Hall et al., 2018). There is considerable promise for these kinds of nanodiscs for the structural and functional characterization of membrane proteins in the future (Killian, 2018; Krishnarjuna & Ramamoorthy, 2022; Li, 2022).

The dynamic nature of ATP synthase posed a big challenge in obtaining high-resolution structures, and with cryo-EM, it is now possible to follow its many different states as well as the diversity in the composition and architecture across different organisms (Courbon & Rubinstein, 2022). The bovine ATPase as a model system has been studied for several decades, and the recent studies on the dimeric species highlight the complexity of this enzyme, where more than 50 populations or substates have been identified (Spikes et al., 2021). These new findings keep us fascinated with how these enzymes perform their functions and assist in shaping the inner membrane of mitochondria (Mühleip et al., 2021). It is now well known that traditional 3D classification methods may not separate conformational states that are low in population or those that have very small differences. The improvements in the use of machine learning in this step have been shown to provide more discrete states, and further development of such methods and collection of larger datasets (due to detectors with faster frame rate and imaging with minimal stage movement or aberration-free image shift and fringe-free illumination) will likely provide more insights into the dynamics of these enzymes (Weis & Hagen, 2020; Zhong et al., 2021; Zivanov et al., 2020). Apart from the structural and mechanistic insights, cryo-EM is also being used in understanding the mode of small molecule binding as shown for F-ATPases from *Mycobacterium* (Guo et al., 2021; Montgomery et al., 2021; Vlasov et al., 2022). These technical developments will also aid in imaging the ATP synthases within mitochondria by cryo-ET, and perhaps in the near future, high-resolution structures of intact enzymes determined *in situ* and/or the complete annotation of the proteins in the inner membrane of the mitochondria will become realistic.

ACKNOWLEDGMENTS

SY and SJ are recipients of the PhD studentship from TIFR, and the funding is acknowledged. KRV acknowledges SERB, India, for the Ramanujan Fellowship (RJN-094/2017), DBT B-Life grant DBT/PR12422/MED/31/287/2014, and the support of the Department of Atomic Energy, Government of India, under Project Identification No. RTI4006.

DECLARATION OF INTEREST

The authors declare no competing interests.

REFERENCES

Abrahams, J. P., Leslie, A. G. W., Lutter, R., & Walker, J. E. (1994). Structure at 2.8 Â resolution of F1-ATPase from bovine heart mitochondria. *Nature, 370*(6491), 621–628. https://doi.org/10.1038/370621a0

Acín-Pérez, R., Fernández-Silva, P., Peleato, M. L., Pérez-Martos, A., & Enriquez, J. A. (2008). Respiratory active mitochondrial supercomplexes. *Molecular Cell, 32*(4), 529–539. https://doi.org/10.1016/j.molcel.2008.10.021

Alavian, K. N., Beutner, G., Lazrove, E., Sacchetti, S., Park, H. A., Licznerski, P., Li, H., Nabili, P., Hockensmith, K., Graham, M., Porter, G. A., & Jonas, E. A. (2014). An uncoupling channel within the c-subunit ring of the F1F O ATP synthase is the mitochondrial permeability transition pore. *Proceedings of the National Academy of Sciences of the United States of America, 111*(29), 10580–10585. https://doi.org/10.1073/pnas.1401591111

Allegretti, M., Davies, K. M., Klusch, N., Mills, D. J., Vonck, J., Ku, W., Bank, P. D., & Bank, P. D. (2015). Horizontal membrane-intrinsic α-helices in the stator α-subunit of an F-type ATP synthase. *Nature, 521*, 237–240. https://doi.org/10.1038/nature14185

Allen, R. D., Schroeder, C. C., & Fok, A. K. (1989). An investigation of mitochondrial inner membranes by rapid-freeze deep-etch techniques. *Journal of Cell Biology, 108*(6), 2233–2240. https://doi.org/10.1083/jcb.108.6.2233

Althoff, T., Mills, D. J., Popot, J. L., & Kühlbrandt, W. (2011). Arrangement of electron transport chain components in bovine mitochondrial supercomplex I 1 III 2 IV 1. *EMBO Journal, 30*(22). https://doi.org/10.1038/emboj.2011.324

Amporndanai, K., Johnson, R. M., O'Neill, P. M., Fishwick, C. W. G., Jamson, A. H., Rawson, S., Muench, S. P., Samar Hasnain, S., & Antonyuk, S. V. (2018). X-ray and cryo-EM structures of inhibitor-bound cytochrome bc 1 complexes for structure-based drug discovery. *IUCrJ, 5*, 200–210. https://doi.org/10.1107/S2052252518001616

Anselmi, C., Davies, K. M., & Gómez, J. D. F. (2018). Mitochondrial ATP synthase dimers spontaneously associate due to a long-range membrane-induced force. *The Journal of General Physiology, 150*(5), 763–770.

Arnold, I., Pfeiffer, K., Neupert, W., Stuart, R. A., & Schägger, H. (1998). Yeast mitochondrial F1F0-ATP synthase exists as a dimer: Identification of three dimer-specific subunits. *EMBO Journal, 17*(24), 7170–7178. https://doi.org/10.1093/emboj/17.24.7170

Bezawork-Geleta, A., Rohlena, J., Dong, L., Pacak, K., & Neuzil, J. (2017). Mitochondrial complex II: At the crossroads. *Trends in Biochemical Sciences, 42*(4), 312–325. https://doi.org/10.1016/j.tibs.2017.01.003

Bienert, R., Rombach-Riegraf, V., Diez, M., & Gräber, P. (2009). Subunit movements in single membrane-bound H+-ATP synthases from chloroplasts during ATP synthesis. *Journal of Biological Chemistry, 284*(52), 36240–36247. https://doi.org/10.1074/jbc.M109.060376

Blum, T. B., Hahn, A., Meier, T., Davies, K. M., & Kühlbrandt, W. (2019). Dimers of mitochondrial ATP synthase induce membrane curvature and self-assemble into rows. *Proceedings of the National Academy of Sciences of the United States of America, 116*(10), 4250–4255. https://doi.org/10.1073/pnas.1816556116

Böttcher, B., & Gräber, P. (2000). The structure of the H+-ATP synthase from chloroplasts and its subcomplexes as revealed by electron microscopy. *Biochimica et Biophysica Acta - Bioenergetics, 1458*(2–3), 404–416. https://doi.org/10.1016/S0005-2728(00)00090-6

Boyer, P. D. (1993). The binding change mechanism for ATP synthase - some probabilities and possibilities. *Biochimica et Biophysica Acta*, *1140*, 215–250.

Boyer, P. D. (1998). Energy, life and ATP (Nobel Lecture). *Angewandte Chemie - International Edition*, *37*(17). https://doi.org/10.1002/(SICI)1521-3773(19980918)37:17<2296::AID-ANIE2296>3.0.CO;2-W

Boyer, P. D., Cross, R. L., & Momsen, W. (1973). A new concept for energy coupling in oxidative phosphorylation based on a molecular explanation of the oxygen exchange reactions. *Proceedings of the National Academy of Sciences of the United States of America*, *70*(10), 2837–2839. https://doi.org/10.1073/pnas.70.10.2837

Bracewell, R. (1956). Strip Integration in Radio Astronomy. *Australian Journal of Physics*, *9*(2), 198. https://doi.org/10.1071/ph560198

Cano-Estrada, A., Vázquez-Acevedo, M., Villavicencio-Queijeiro, A., Figueroa-Martínez, F., Miranda-Astudillo, H., Cordeiro, Y., Mignaco, J. A., Foguel, D., Cardol, P., Lapaille, M., Remacle, C., Wilkens, S., & González-Halphen, D. (2010). Subunit-subunit interactions and overall topology of the dimeric mitochondrial ATP synthase of *Polytomella* sp. *Biochimica et Biophysica Acta - Bioenergetics*, *1797*(8), 1439–1448. https://doi.org/10.1016/j.bbabio.2010.02.024

Chae, P. S., Rasmussen, S. G. F., Rana, R. R., Gotfryd, K., Chandra, R., Goren, M. A., Kruse, A. C., Nurva, S., Loland, C. J., Pierre, Y., Drew, D., Popot, J. L., Picot, D., Fox, B. G., Guan, L., Gether, U., Byrne, B., Kobilka, B., & Gellman, S. H. (2010). Maltose-neopentyl glycol (MNG) amphiphiles for solubilization, stabilization and crystallization of membrane proteins. *Nature Methods*, *7*(12), 1003–1008. https://doi.org/10.1038/nmeth.1526

Chae, P. S., Rasmussen, S. G. F., Rana, R. R., Gotfryd, K., Kruse, A. C., Manglik, A., Cho, K. H., Nurva, S., Gether, U., Guan, L., Loland, C. J., Byrne, B., Kobilka, B. K., & Gellman, S. H. (2012). A new class of amphiphiles bearing rigid hydrophobic groups for solubilization and stabilization of membrane proteins. *Chemistry - A European Journal*, *18*(31), 9485–9490. https://doi.org/10.1002/chem.201200069

Chance, B., & Williams, G. R. (1955). Respiratory enzymes in oxidative phosphorylation. III. The steady state. *The Journal of Biological Chemistry*, *217*(1), 409–427.

Cheng, Y. (2018). Single-particle cryo-EM-How did it get here and where will it go. *Science*. https://doi.org/10.1126/science.aat4346

Claude, A. (1943). The constitution of protoplasm. *Science*, *97*(2525), 451–456. https://doi.org/10.1126/science.97.2525.451

Claude, A. (1946). Fractionation of mammalian liver cells by differential centrifugation; problems, methods, and preparation of extract. *The Journal of Experimental Medicine*, *84*, 51–59. https://doi.org/10.1084/jem.84.1.51

Cong, Y., & Ludtke, S. J. (2010). Single Particle Analysis at High Resolution. In *Methods in Enzymology Cryo-EM, Part B: 3-D Reconstruction* (1st ed., Vol. 482, pp. 211–235). Elsevier Inc. https://doi.org/10.1016/S0076-6879(10)82009-9

Courbon, G. M., & Rubinstein, J. L. (2022). CryoEM reveals the complexity and diversity of ATP synthases. *Frontiers in Microbiology*, *13*(June), 1–10. https://doi.org/10.3389/fmicb.2022.864006

Crofts, A. R. (2021). The modified Q-cycle: A look back at its development and forward to a functional model. *Biochimica et Biophysica Acta. Bioenergetics*, *1862*(8), 148417. https://doi.org/10.1016/j.bbabio.2021.148417

Cross, R. L., & Müller, V. (2004). The evolution of A-, F-, and V-type ATP synthases and ATPases: Reversals in function and changes in the H+/ATP coupling ratio. *FEBS Letters*, *576*(1–2). https://doi.org/10.1016/j.febslet.2004.08.065

Daum, B., Walter, A., Horst, A., Osiewacz, H. D., & Kühlbrandt, W. (2013). Age-dependent dissociation of ATP synthase dimers and loss of inner-membrane cristae in mitochondria. *Proceedings of the National Academy of Sciences of the United States of America*, *110*(38), 15301–15306. https://doi.org/10.1073/pnas.1305462110

Davies, K. M., Anselmi, C., Wittig, I., Faraldo-Gómez, J. D., & Kühlbrandt, W. (2012). Structure of the yeast F 1F o-ATP synthase dimer and its role in shaping the mitochondrial cristae. *Proceedings of the National Academy of Sciences of the United States of America*, *109*(34), 13602–13607. https://doi.org/10.1073/pnas.1204593109

Davies, K. M., & Kühlbrandt, W. (2018). Structure of the catalytic F1 head of the F1-Fo ATP synthase from *Trypanosoma brucei*. *Proceedings of the National Academy of Sciences of the United States of America*, *115*(13), E2906–E2907. https://doi.org/10.1073/pnas.1801103115

Davies, K. M., Strauss, M., Daum, B., Kief, J. H., Osiewacz, H. D., Rycovska, A., Zickermann, V., & Kühlbrandt, W. (2011). Macromolecular organization of ATP synthase and complex I in whole mitochondria. *Proceedings of the National Academy of Sciences of the United States of America*, *108*(34), 14121–14126. https://doi.org/10.1073/pnas.1103621108

DeLano, W. L. (2002). Pymol: An open-source molecular graphics tool. Newsletter On Protein Crystallography. https://legacy.ccp4.ac.uk/newsletters/newsletter40.pdf.

Dickson, V. K., Silvester, J. A., Fearnley, I. M., Leslie, A. G. W., & Walker, J. E. (2006). On the structure of the stator of the mitochondrial ATP synthase. *EMBO Journal*, *25*(12), 2911–2918. https://doi.org/10.1038/sj.emboj.7601177

Dudkina, N. V., Oostergetel, G. T., Lewejohann, D., Braun, H. P., & Boekema, E. J. (2010). Row-like organization of ATP synthase in intact mitochondria determined by cryo-electron tomography. *Biochimica et Biophysica Acta - Bioenergetics*, *1797*(2), 272–277. https://doi.org/10.1016/j.bbabio.2009.11.004

Ebanks, B., & Chakrabarti, L. (2022). Mitochondrial ATP synthase is a target of oxidative stress in neurodegenerative diseases. *Frontiers in Molecular Biosciences*, *9*(February), 1–7. https://doi.org/10.3389/fmolb.2022.854321

Elston, T., Wang, H., & Oster, G. (1998). Energy transduction in ATP synthase. *Nature*, *391*(January), 510–513.

Ernster, L., & Schatz, G. (1981). Mitochondria : A historical review. *The Journal of cell biology*, *91*(December), 227s–255s.

Feniouk, B. A., Suzuki, T., & Yoshida, M. (2006). The role of subunit epsilon in the catalysis and regulation of FOF1-ATP synthase. *Biochimica et Biophysica Acta - Bioenergetics*, *1757*(5–6), 326–338. https://doi.org/10.1016/j.bbabio.2006.03.022

Fiedorczuk, K., Letts, J. A., Degliesposti, G., Kaszuba, K., Skehel, M., & Sazanov, L. A. (2016). Atomic structure of the entire mammalian mitochondrial complex I. *Nature*, *538*(7625), 406–410. https://doi.org/10.1038/nature19794

Fillingame, R. H., & Steed, P. R. (2014). Half channels mediating H+ transport and the mechanism of gating in the Fo sector of Escherichia coli F1Fo ATP synthase. *Biochimica et Biophysica Acta - Bioenergetics*, *1837*(7), 1063–1068. https://doi.org/10.1016/j.bbabio.2014.03.005

Flygaard, R. K., Mühleip, A., Tobiasson, V., & Amunts, A. (2020). Type III ATP synthase is a symmetry-deviated dimer that induces membrane curvature through tetramerization. *Nature Communications*, *11*(1). https://doi.org/10.1038/s41467-020-18993-6

Giorgio, V., Fogolari, F., Lippe, G., & Bernardi, P. (2019). OSCP subunit of mitochondrial ATP synthase: Role in regulation of enzyme function and of its transition to a pore. *British Journal of Pharmacology*, *176*(22), 4247–4257. https://doi.org/10.1111/bph.14513

Goddard, T. D., Huang, C. C., & Ferrin, T. E. (2007). Visualizing density maps with UCSF Chimera. *Journal of Structural Biology*. https://doi.org/10.1016/j.jsb.2006.06.010

Grant, T., Rohou, A., & Grigorieff, N. (2018). CisTEM, user-friendly software for single-particle image processing. *ELife*. https://doi.org/10.7554/eLife.35383

Gu, J., Zhang, L., Zong, S., Guo, R., Liu, T., Yi, J., Wang, P., Zhuo, W., & Yang, M. (2019). Cryo-EM structure of the mammalian ATP synthase tetramer bound with inhibitory protein IF1. *Science*, *364*(6445). https://doi.org/10.1126/science.aaw4852

Guo, H., Bueler, S. A., & Rubinstein, J. L. (2017). Atomic model for the dimeric FO region of mitochondrial ATP synthase. *Science*, *358*(6365), 936–940. https://doi.org/10.1126/science.aao4815

Guo, H., Courbon, G. M., Bueler, S. A., Mai, J., Liu, J., & Rubinstein, J. L. (2021). Structure of mycobacterial ATP synthase bound to the tuberculosis drug bedaquiline. *Nature*, *589*(7840). https://doi.org/10.1038/s41586-020-3004-3

Guo, H., & Rubinstein, J. L. (2022). Structure of ATP synthase under strain during catalysis. *Nature Communications*, *13*(1), 2–10. https://doi.org/10.1038/s41467-022-29893-2

Gupte, S., Wu, E. S., Hoechli, L., Hoechli, M., Jacobson, K., Sowers, A. E., & Hackenbrock, C. R. (1984). Relationship between lateral diffusion, collision frequency, and electron transfer of mitochondrial inner membrane oxidation-reduction components. *Proceedings of the National Academy of Sciences of the United States of America*, *81*(9). https://doi.org/10.1073/pnas.81.9.2606

Hackenbrock, C. R. (1981). Lateral diffusion and electron transfer in the mitochondrial inner membrane. *Trends in Biochemical Sciences*, *6*(C), 151–154. https://doi.org/10.1016/0968-0004(81)90056-6

Hall, S. C. L., Tognoloni, C., Charlton, J., Bragginton, É. C., Rothnie, A. J., Sridhar, P., Wheatley, M., Knowles, T. J., Arnold, T., Edler, J., & Da, T. R. (2018). An acid-compatible co-polymer for the solubilization of membranes and proteins into lipid bilayer-containing nanoparticles. *Nanoscale*, *10*, 10609–10619. https://doi.org/10.1039/c8nr01322e

Hirst, J. (2013). Mitochondrial complex I. *Annual Review of Biochemistry*, *82*, 551–575. https://doi.org/10.1146/annurev-biochem-070511-103700

Iwata, S., Lee, J. W., Okada, K., Lee, J. K., Iwata, M., Rasmussen, B., Link, T. A., Ramaswamy, S., & Jap, B. K. (1998). Complete structure of the 11-subunit bovine mitochondrial cytochrome bc1 complex. *Science*, *281*(5373), 64–71. https://doi.org/10.1126/science.281.5373.64

Jamshad, M., Grimard, V., Idini, I., Knowles, T. J., Dowle, M. R., Schofield, N., Sridhar, P., Lin, Y., Finka, R., Wheatley, M., Thomas, O. R. T., Palmer, R. E., Overduin, M., Govaerts, C., Ruysschaert, J., Edler, K. J., & Dafforn, T. R. (2015). Structural analysis of a nanoparticle containing a lipid bilayer used for detergent-free extraction of membrane proteins. *Nano Research*, *8*(3), 774–789. https://doi.org/10.1007/s12274-014-0560-6

Junesch, U., & Gräber, P. (1987). Influence of the redox state and the activation of the chloroplast ATP synthase on proton-transport-coupled ATP synthesis/hydrolysis. *BBA - Bioenergetics*, *893*(2), 275–288. https://doi.org/10.1016/0005-2728(87)90049-1

Junge, W., & Nelson, N. (2015). ATP Synthase. *Annual Review of Biochemistry*, *84*(1), 631–657. https://doi.org/10.1146/annurev-biochem-060614-034124

Kagawa, Y & Racker, E. (1966) Partial resolution of the enzymes catalyzing oxidative phosphorylation. 8. Properties of a factor conferring oligomycin sensitivity on mitochondrial adenosine triphosphatase. Journal of Biological Chemistry, V241(10)2461-6.

Kamienietzky, A., & Nelson, N. (1975). Preparation and Properties of Chloroplasts Depleted of Chloroplast Coupling Factor 1 by Sodium Bromide Treatment. *Plant Physiology*, *55*(2). https://doi.org/10.1104/pp.55.2.282

Kampjut, D., Steiner, J., & Sazanov, L. A. (2021). Cryo-EM grid optimization for membrane proteins. *IScience*, *24*(3), 102139. https://doi.org/10.1016/j.isci.2021.102139

Kao, W.-C., & Hunte, C. (2022). Quinone binding sites of cyt bc complexes analysed by X-ray crystallography and cryogenic electron microscopy. *Biochemical Society Transactions*, *50*(2), 877–893. https://doi.org/10.1042/BST20190963

Karrasch, S., & Walker, J. E. (1999). Novel features in the structure of bovine ATP synthase. *Journal of Molecular Biology*, *290*(2), 379–384. https://doi.org/10.1006/jmbi.1999.2897

Kayalar, C., Rosing, J., & Boyer, P. D. (1977). An alternating site sequence for oxidative phosphorylation suggested by measurement of substrate binding patterns and exchange reaction inhibitions. *Journal of Biological Chemistry*, *252*(8), 2486–2491. https://doi.org/10.1016/s0021-9258(17)40484-4

Khan, Y. A., White, K. I., & Brunger, A. T. (2022). The AAA+ superfamily: A review of the structural and mechanistic principles of these molecular machines. *Critical Reviews in Biochemistry and Molecular Biology*, *57*(2), 156–187. https://doi.org/10.1080/10409238.2021.1979460

Killian, J. A. (2018). The styrene-maleic acid copolymer: A versatile tool in membrane research. *Biophysical Journal*, *114*(3). https://doi.org/10.1016/j.bpj.2017.11.2228

Klusch, N., Dreimann, M., Senkler, J., Rugen, N., Kühlbrandt, W., & Braun, H. P. (2022). Cryo-EM structure of the respiratory I + III2 supercomplex from Arabidopsis thaliana at 2 Åresolution. *Nature Plants*, *9*(1), 142–156. https://doi.org/10.1038/s41477-022-01308-6

Knowles, T. J., Finka, R., Smith, C., Lin, Y., Dafforn, T., & Overduin, M. (2009). Membrane proteins solubilized intact in lipid containing nanoparticles bounded by styrene maleic acid copolymer. *Journal of the American Chemical Society*, *131*, 7484–7485.

Kravchuk, V., Petrova, O., Kampjut, D., Wojciechowska-Bason, A., Breese, Z., & Sazanov, L. (2022). A universal coupling mechanism of respiratory complex I. *Nature*, *609*(7928), 808–814. https://doi.org/10.1038/s41586-022-05199-7

Krishnarjuna, B., & Ramamoorthy, A. (2022). Detergent-free isolation of membrane proteins and strategies to study them in a near-native membrane environment. *Biomolecules*, *12*(8). https://doi.org/10.3390/biom12081076

Kühlbrandt, W. (2015). Structure and function of mitochondrial membrane protein complexes. *BMC Biology*, *13*(1), 1–11. https://doi.org/10.1186/s12915-015-0201-x

Kühlbrandt, W. (2019). Structure and mechanisms of F-Type ATP synthases. *Annual Review of Biochemistry*, *88*. https://doi.org/10.1146/annurev-biochem-013118-110903

Kühlbrandt, W., & Davies, K. M. (2016). Rotary ATPases: A new twist to an ancient machine. *Trends in Biochemical Sciences*, *41*(1). https://doi.org/10.1016/j.tibs.2015.10.006

Kunji, E. R. S., Aleksandrova, A., King, M. S., Majd, H., Ashton, V. L., Cerson, E., Springett, R., Kibalchenko, M., Tavoulari, S., Crichton, P. G., & Ruprecht, J. J. (2016). The transport mechanism of the mitochondrial ADP/ATP carrier. *Biochimica et Biophysica Acta - Molecular Cell Research*, *1863*(10), 2379–2393. https://doi.org/10.1016/j.bbamcr.2016.03.015

Lee, S. Y., Kang, M. G., Shin, S., Kwak, C., Kwon, T., Seo, J. K., Kim, J. S., & Rhee, H. W. (2017). Architecture mapping of the inner mitochondrial membrane proteome by chemical tools in live cells. *Journal of the American Chemical Society*, *139*(10), 3651–3662. https://doi.org/10.1021/jacs.6b10418

Letts, J. A., Fiedorczuk, K., & Sazanov, L. A. (2016). The architecture of respiratory supercomplexes. *Nature*, *537*(7622), 644–648. https://doi.org/10.1038/nature19774

Li, S. (2022). Detergents and alternatives in cryo-EM studies of membrane proteins. *Acta Biochimica et Biophysica Sinica*, *54*(8), 1049–1056. https://doi.org/10.3724/abbs.2022088

Li, X., Mooney, P., Zheng, S., Booth, C. R., Braunfeld, M. B., Gubbens, S., Agard, D. A., & Cheng, Y. (2013). Electron counting and beam-induced motion correction enable near-atomic-resolution single-particle cryo-EM. *Nature Methods*, *10*(6). https://doi.org/10.1038/nmeth.2472

Longo, L. M., Jabtoiiska, J., Vyas, P., Kanade, M., Kolodny, R., Ben-Tal, N., & Tawfik, D. S. (2020). On the emergence of p-loop ntpase and rossmann enzymes from a beta-alpha-beta ancestral fragment. *ELife*, *9*. https://doi.org/10.7554/ELIFE.64415

Maldonado, M., Guo, F., Letts, J. A., & Carter, A. P. (2021). Atomic structures of respiratory complex III2, complex IV, and supercomplex III2-IV from vascular plants. *ELife*, *10*, 1–34. https://doi.org/10.7554/eLife.62047

McMullan, G., Chen, S., Henderson, R., & Faruqi, A. R. (2009). Detective quantum efficiency of electron area detectors in electron microscopy. *Ultramicroscopy*, *109*(9), 1126–1143. https://doi.org/10.1016/j.ultramic.2009.04.002

McMullan, G., Faruqi, A. R., Clare, D., & Henderson, R. (2014). Comparison of optimal performance at 300keV of three direct electron detectors for use in low dose electron microscopy. *Ultramicroscopy*, *147*, 156–163. https://doi.org/10.1016/j.ultramic.2014.08.002

McMullan, G., Faruqi, A. R., & Henderson, R. (2016). Direct Electron Detectors. In *Methods in Enzymology* (1st ed., Vol. 579, pp. 1–17). Elsevier Inc. https://doi.org/10.1016/bs.mie.2016.05.056

Mitchell P. (1961). Coupling of phosphorylation to electron and hydrogen transfer by a chemi-osmotic type of mechanism. *Nature*, *191*, 144–148.

Moe, A., Trani, J. Di, Rubinstein, J. L., & Brzezinski, P. (2021). Cryo-EM structure and kinetics reveal electron transfer by 2D diffusion of cytochrome c in the yeast III-IV respiratory supercomplex. *Proceedings of the National Academy of Sciences of the United States of America*, *118*(11), 1–9. https://doi.org/10.1073/pnas.2021157118

Montgomery, M. G., Petri, J., Spikes, T. E., & Walker, J. E. (2021). Structure of the ATP synthase from Mycobacterium smegmatis provides targets for treating tuberculosis. *Proceedings of the National Academy of Sciences of the United States of America*, *118*(47). https://doi.org/10.1073/pnas.2111899118

Muench, S. P., Trinick, J., & Harrison, M. A. (2011). Structural divergence of the rotary ATPases. *Quarterly Reviews of Biophysics*, *44*(3). https://doi.org/10.1017/S0033583510000338

Mühleip, A., Kock Flygaard, R., Ovciarikova, J., Lacombe, A., Fernandes, P., Sheiner, L., & Amunts, A. (2021). ATP synthase hexamer assemblies shape cristae of Toxoplasma mitochondria. *Nature Communications*, *12*(1). https://doi.org/10.1038/s41467-020-20381-z

Mühleip, A., McComas, S. E., & Amunts, A. (2019). Structure of a mitochondrial ATP synthase with bound native cardiolipin. *ELife*, *8*. https://doi.org/10.7554/eLife.51179

Mühleip, A. W., Joos, F., Wigge, C., Frangakis, A. S., Kühlbrandt, W., & Davies, K. M. (2016). Helical arrays of U-shaped ATP synthase dimers form tubular cristae in ciliate mitochondria. *Proceedings of the National Academy of Sciences of the United States of America*, *113*(30), 8442–8447. https://doi.org/10.1073/pnas.1525430113

Müller, V., & Grüber, G. (2003). ATP synthases: Structure, function and evolution of unique energy converters. *Cellular and Molecular Life Sciences*, *60*(3). https://doi.org/10.1007/s000180300040

Mühleip AW, Joos F, Wigge C, Frangakis AS, **Kühlbrandt W**, Davies KM (2016), Helical arrays of U-shaped ATP synthase dimers form tubular cristae in ciliate mitochondria. *Proc. Natl.* Acad Sci USA, V113, P8442-7.

Mun, J. Y., Lee, T. H., Kim, J. H., Yoo, B. H., Bahk, Y. Y., Koo, H.-S., & Han, S. S. (2010). *Caenorhabditis elegans* mitofilin homologs control the morphology of mitochondrial cristae and influence reproduction and physiology. *Journal of Cellular Physiology*, *224*(3), 748–756. https://doi.org/https://doi.org/10.1002/jcp.22177

Nirody, J. A., Budin, I., & Rangamani, P. (2020). ATP synthase: Evolution, energetics, and membrane interactions. *Journal of General Physiology*, *152*(11). https://doi.org/10.1085/JGP.201912475

Nogales, E., & Scheres, S. H. W. (2015). Cryo-EM: A Unique Tool for the Visualization of Macromolecular Complexity. *Molecular Cell*. https://doi.org/10.1016/j.molcel.2015.02.019

Noji, H., Yasuda, R., Yoshida, M., & Kinosita, K. J. (1997). Direct observation of the rotation of F1-ATPase. *Nature*, *386*(6622), 299–302. https://doi.org/10.1038/386299a0

Oluwole, A. O., Klingler, J., Babalola, J. O., Vargas, C., Pabst, G., & Keller, S. (2017). Formation of lipid-bilayer nanodiscs by diisobutylene/maleic acid (DIBMA) copolymer. *Langmuir*, *33*, 14378–14388. https://doi.org/10.1021/acs.langmuir.7b03742

Pecina, P., Nůsková, H., Karbanová, V., Kaplanová, V., Mráček, T., & Houštěk, J. (2018). Role of the mitochondrial ATP synthase central stalk subunits γ and δ in the activity and assembly of the mammalian enzyme. *Biochimica et Biophysica Acta - Bioenergetics*, *1859*(5), 374–381. https://doi.org/10.1016/j.bbabio.2018.02.007

Pedersen, P. L. (2005). Transport ATPases : Structure, motors, mechanism and medicine: A brief overview. *Journal of Bioenergetics and Biomembranes*, *37*(6). https://doi.org/10.1007/s10863-005-9470-3

Pettersen, E. F., Goddard, T. D., Huang, C. C., Meng, E. C., Couch, G. S., Croll, T. I., Morris, J. H., & Ferrin, T. E. (2021). UCSF ChimeraX: Structure visualization for researchers, educators, and developers. *Protein Science*, *30*(1). https://doi.org/10.1002/pro.3943

Pfanner, N., Warscheid, B., & Wiedemann, N. (2019). Mitochondrial proteins: From biogenesis to functional networks. *Nature Reviews Molecular Cell Biology*, *20*(5), 267–284. https://doi.org/10.1038/s41580-018-0092-0

Pinke, G., Zhou, L., & Sazanov, L. A. (2020). Cryo-EM structure of the entire mammalian F-type ATP synthase. *Nature Structural and Molecular Biology*, *27*(11), 1077–1085. https://doi.org/10.1038/s41594-020-0503-8

Puchades, C., Sandate, C. R., & Lander, G. C. (2020). The molecular principles governing the activity and functional diversity of AAA+ proteins. *Nature Reviews Molecular Cell Biology*, *21*(1), 43–58. https://doi.org/10.1038/s41580-019-0183-6

Punjani, A., Rubinstein, J. L., Fleet, D. J., & Brubaker, M. A. (2017). CryoSPARC: Algorithms for rapid unsupervised cryo-EM structure determination. *Nature Methods*, *14*(3), 290–296. https://doi.org/10.1038/nmeth.4169

Ramírez-Aguilar, S. J., Keuthe, M., Rocha, M., Fedyaev, V. V, Kramp, K., Gupta, K. J., Rasmusson, A. G., Schulze, W. X., & Van Dongen, J. T. (2011). The composition of plant mitochondrial supercomplexes changes with oxygen availability. *Journal of Biological Chemistry*, *286*(50), 43045–43053. https://doi.org/10.1074/jbc.M111.252544

Rathore, S., Berndtsson, J., Marin-Buera, L., Conrad, J., Carroni, M., Brzezinski, P., & Ott, M. (2019). Cryo-EM structure of the yeast respiratory supercomplex. *Nature Structural and Molecular Biology*, *26*(1), 50–57. https://doi.org/10.1038/s41594-018-0169-7

Rich, P. R. (2017). Mitochondrial cytochrome c oxidase: Catalysis, coupling and controversies. *Biochemical Society Transactions*, *45*(3), 813–829. https://doi.org/10.1042/BST20160139

Rubinstein, J. L., Walker, J. E., & Henderson, R. (2003). Structure of the mitochondrial ATP synthase by electron cryomicroscopy. *EMBO Journal*, *22*(23). https://doi.org/10.1093/emboj/cdg608

Saraste, M., Gay, N. J., Eberle, A., Runswick, M. J., & Walker, J. E. (1981). The atp operon: Nucleotide sequence of the genes for the γ, β, and εsubunits of Escherichia coli ATP synthase. *Nucleic Acids Research*, *9*(20), 5287–5296. https://doi.org/10.1093/nar/9.20.5287

Schägger, H., & von Jagow, G. (1991). Blue native electrophoresis for isolation of membrane protein complexes in enzymatically active form. *Analytical Biochemistry*, *199*(2), 223–231. https://doi.org/10.1016/0003-2697(91)90094-A

Scheidelaar, S., Koorengevel, M. C., Pardo, J. D., Meeldijk, J. D., Breukink, E., & Killian, J. A. (2015). Molecular Model for the solubilization of membranes into nanodisks by styrene maleic acid copolymers. *Biophysical Journal*, *108*(2), 279–290. https://doi.org/10.1016/j.bpj.2014.11.3464

Scheidelaar, S., Koorengevel, M. C., Walree, C. A. Van, Dominguez, J. J., Do, J. M., & Killian, J. A. (2016). Effect of polymer composition and pH on membrane solubilization by styrene-maleic acid copolymers. *Biophysical Journal*, *111*, 1974–1986. https://doi.org/10.1016/j.bpj.2016.09.025

Scheres, S. H. W. (2012). A bayesian view on Cryo-EM structure determination. *Journal of Molecular Biology*, *415*(2), 406–418. https://doi.org/10.1016/j.jmb.2011.11.010

Scheres, S. H. W. (2016). Processing of Structurally Heterogeneous Cryo-EM Data in RELION. In *Methods in Enzymology* (1st ed., Vol. 579, pp. 125–157). Elsevier Inc. https://doi.org/10.1016/bs.mie.2016.04.012

Schlame, M. (2021). Protein crowding in the inner mitochondrial membrane. *Biochimica et Biophysica Acta - Bioenergetics*, *1862*(1), 148305. https://doi.org/10.1016/j.bbabio.2020.148305

Sigworth, F. J. (2016). Principles of cryo-EM single-particle image processing. *Microscopy*, *65*, 57–67. https://doi.org/10.1093/jmicro/dfv370

Sobti, M., Ueno, H., Noji, H., & Stewart, A. G. (2021). The six steps of the complete F1-ATPase rotary catalytic cycle. *Nature Communications*, *12*(1). https://doi.org/10.1038/s41467-021-25029-0

Solmaz, S. R. N., & Hunte, C. (2008). Structure of complex III with bound cytochrome C in reduced state and definition of a minimal core interface for electron transfer. *Journal of Biological Chemistry*, *283*(25), 17542–17549. https://doi.org/10.1074/jbc.M710126200

Soufari, H., Parrot, C., Kuhn, L., Waltz, F., & Hashem, Y. (2020). Specific features and assembly of the plant mitochondrial complex I revealed by cryo-EM. *Nature Communications*, *11*(1). https://doi.org/10.1038/s41467-020-18814-w

Spikes, T. E., Montgomery, M. G., & Walker, J. E. (2020). Structure of the dimeric ATP synthase from bovine mitochondria. *Proceedings of the National Academy of Sciences of the United States of America*, *117*(38), 23519–23526. https://doi.org/10.1073/pnas.2013998117

Spikes, T. E., Montgomery, M. G., & Walker, J. E. (2021). Interface mobility between monomers in dimeric bovine ATP synthase participates in the ultrastructure of inner mitochondrial membranes. *Proceedings of the National Academy of Sciences of the United States of America*, *118*(8). https://doi.org/10.1073/pnas.2021012118

Stewart, A. G., Laming, E. M., Sobti, M., & Stock, D. (2014). Rotary ATPases-dynamic molecular machines. *Current Opinion in Structural Biology*, *25*. https://doi.org/10.1016/j.sbi.2013.11.013

Stewart, A. G., Sobti, M., Harvey, R. P., & Stock, D. (2013). Rotary ATPases: Models, machine elements and technical specifications. *Bioarchitecture*, *3*(1), 2–12. https://doi.org/10.4161/bioa.23301

Stock, D., Gibbons, C., Arechaga, I., Leslie, A. G. W., & Walker, J. E. (2000). The rotary mechanism of ATP synthase. *Current Opinion in Structural Biology*, *10*(6). https://doi.org/10.1016/S0959-440X(00)00147-0

Stock, D., Leslie, A. G. W., & Walker, J. E. (1999). Molecular architecture of the rotary motor in ATP synthase. *Science*, *286*(5445). https://doi.org/10.1126/science.286.5445.1700

Strauss, M., Hofhaus, G., Schröder, R. R., & Kühlbrandt, W. (2008). Dimer ribbons of ATP synthase shape the inner mitochondrial membrane. *EMBO Journal*, *27*(7), 1154–1160. https://doi.org/10.1038/emboj.2008.35

Sun, F., Huo, X., Zhai, Y., Wang, A., Xu, J., Su, D., Bartlam, M., & Rao, Z. (2005). Crystal structure of mitochondrial respiratory membrane protein Complex II. *Cell*, *121*(7), 1043–1057. https://doi.org/10.1016/j.cell.2005.05.025

Tsukihara, T., Aoyama, H., Yamashita, E., Tomizaki, T., Yamaguchi, H., Shinzawa-Itoh, K., Nakashima, R., Yaono, R., & Yoshikawa, S. (1996). The whole structure of the 13-subunit oxidized cytochrome c oxidase at 2.8 Å. *Science*, *272*(5265). https://doi.org/10.1126/science.272.5265.1136

Tucker, K., & Park, E. (2019). Cryo-EM structure of the mitochondrial protein-import channel TOM complex at near-atomic resolution. *Nature Structural and Molecular Biology*, *26*(12), 1158–1166. https://doi.org/10.1038/s41594-019-0339-2

Tzagoloff, A. (1970). Assembly of the mitochondrial membrane system. 3. Function and synthesis of the oligomycin sensitivity-conferring protein of yeast mitochondria. *Journal of Biological Chemistry*, *245*(7), 1545–1551.

Vercellino, I., & Sazanov, L. A. (2021). Structure and assembly of the mammalian mitochondrial supercomplex CIII2CIV. *Nature*, *598*(7880), 364–367. https://doi.org/10.1038/s41586-021-03927-z

Vercellino, I., & Sazanov, L. A. (2022). The assembly, regulation and function of the mitochondrial respiratory chain. *Nature Reviews Molecular Cell Biology*, *23*(2), 141–161. https://doi.org/10.1038/s41580-021-00415-0

Vinothkumar, K. R., Montgomery, M. G., Liu, S., & Walker, J. E. (2016). Structure of the mitochondrial ATP synthase from Pichia angusta determined by electron cryo-microscopy. *Proceedings of the National Academy of Sciences of the United States of America*. https://doi.org/10.1073/pnas.1615902113

Vinothkumar, K. R., Zhu, J., & Hirst, J. (2014). Architecture of mammalian respiratory complex I. *Nature*, *515*(7525), 80–84. https://doi.org/10.1038/nature13686

Vlasov, A. V, Osipov, S. D., Bondarev, N. A., Uversky, V. N., Borshchevskiy, V. I., Yanyushin, M. F., Manukhov, I. V, Rogachev, A. V, Vlasova, A. D., Ilyinsky, N. S., Kuklin, A. I., Dencher, N. A., & Gordeliy, V. I. (2022). ATP synthase FOF1 structure, function, and structure-based drug design. *Cellular and Molecular Life Sciences*, *79*(3), 1–27. https://doi.org/10.1007/s00018-022-04153-0

Voos, W., Martin, H., Krimmer, T., & Pfanner, N. (1999). Mechanisms of protein translocation into mitochondria. *Biochimica et Biophysica Acta - Reviews on Biomembranes*, *1422*(3), 235–254. https://doi.org/10.1016/S0304-4157(99)00007-6

Walker, J. E. (1998). ATP synthesis by rotary catalysis (Nobel lecture). *Angewandte Chemie - International Edition*, *37*(17). https://doi.org/10.1002/(SICI)1521-3773(19980918)37:17<2308::AID-ANIE2308>3.0.CO;2-W

Walker, J. E. (2013). The ATP synthase: The understood, the uncertain and the unknown. *Biochemical Society Transactions*, *41*(1). https://doi.org/10.1042/BST20110773

Walker, J. E., Saraste, M., Runswick, M. J., & Gay, N. J. (1982). Distantly related sequences in the alpha- and beta-subunits of ATP synthase, myosin, kinases and other ATP-requiring enzymes and a common nucleotide binding fold. *The EMBO Journal*, *1*(8), 945–951. https://doi.org/10.1002/j.1460-2075.1982.tb01276.x

Watt, I. N., Montgomery, M. G., Runswick, M. J., Leslie, A. G. W., & Walker, J. E. (2010). Bioenergetic cost of making an adenosine triphosphate molecule in animal mitochondria. *Proceedings of the National Academy of Sciences of the United States of America*, *107*(39), 16823–16827. https://doi.org/10.1073/pnas.1011099107

Weis, F., & Hagen, W. J. H. (2020). Combining high throughput and high quality for cryo-electron microscopy data collection. *Acta Crystallographica Section D: Structural Biology*, *76*, 724–728. https://doi.org/10.1107/S2059798320008347

Wiedemann, N & Pfanner, N. (2017) Mitochondrial machineries for Protein import and Assembly. *Annual Reviews of Biochemistry*, V86, 685-714.

Wikström, M., Gennis, R. B., & Rich, P. R. (2023). Structures of the intermediates in the catalytic cycle of mitochondrial cytochrome c oxidase. *Biochimica et Biophysica Acta – Bioenergetics*, *1864*(2). https://doi.org/10.1016/j.bbabio.2022.148933

Wikström, M., & Sharma, V. (2018). Proton pumping by cytochrome c oxidase – A 40 year anniversary. *Biochimica et Biophysica Acta - Bioenergetics*, *1859*(9), 692–698. https://doi.org/10.1016/j.bbabio.2018.03.009

Wilkens, S., & Capaldi, R. A. (1998). ATP synthase's second stalk comes into focus [6]. *Nature*, *393*(6680), 29. https://doi.org/10.1038/29908

Wittig, I., Braun, H. P., & Schägger, H. (2006). Blue native PAGE. *Nature Protocols*, *1*(1), 418–428. https://doi.org/10.1038/nprot.2006.62

Wu, M., Gu, J., Guo, R., Huang, Y., & Yang, M. (2016). Structure of Mammalian Respiratory Supercomplex I1III2IV1. *Cell*, *167*(6), 1598–1609.e10. https://doi.org/10.1016/j.cell.2016.11.012

Xian, H., & Liou, Y. C. (2021). Functions of outer mitochondrial membrane proteins: Mediating the crosstalk between mitochondrial dynamics and mitophagy. *Cell Death and Differentiation*, *28*(3), 827–842. https://doi.org/10.1038/s41418-020-00657-z

Xu, Y., Malhotra, A., Ren, M., & Schlame, M. (2006). The enzymatic function of tafazzin. *Journal of Biological Chemistry*, *281*(51), 39217–39224. https://doi.org/10.1074/jbc.M606100200

Yoshida, M., Muneyuki, E., & Hisabori, T. (2001). ATP synthase - A marvellous rotary engine of the cell. *Nature Reviews Molecular Cell Biology*, *2*(9), 669–677. https://doi.org/10.1038/35089509

Yoshikawa, S., Shinzawa-Itoh, K., Nakashima, R., Yaono, R., Yamashita, E., Inoue, N., Yao, M., Fei, M. J., Libeu, C. P., Mizushima, T., Yamaguchi, H., Tomizaki, T., & Tsukihara, T. (1998). Redox-coupled crystal structural changes in bovine heart cytochrome c oxidase. *Science*, *280*(5370), 1723–1729. https://doi.org/10.1126/science.280.5370.1723

Zhong, E. D., Bepler, T., Berger, B., & Davis, J. H. (2021). CryoDRGN: Reconstruction of heterogeneous cryo-EM structures using neural networks. *Nature Methods*, *18*(2), 176–185. https://doi.org/10.1038/s41592-020-01049-4

Zhou, A., Rohou, A., Schep, D. G., Bason, J. V., Montgomery, M. G., Walker, J. E., Grigorieffniko, N., & Rubinstein, J. L. (2015). Structure and conformational states of the bovine mitochondrial ATP synthase by cryo-EM. *ELife*, *4*. https://doi.org/10.7554/eLife.10180

Zhu, J., Vinothkumar, K. R., & Hirst, J. (2016). Architecture of mammalian respiratory complex I. *Nature*. https://doi.org/10.1038/nature13686

Zivanov, J., Nakane, T., & Scheres, S. H. W. (2020). Estimation of high-order aberrations and anisotropic magnification from cryo-EM data sets in RELION-3.1. *IUCrJ*, *7*. https://doi.org/10.1107/S2052252520000081

Zong, S., Wu, M., Gu, J., Liu, T., Guo, R., & Yang, M. (2018). Structure of the intact 14-subunit human cytochrome c oxidase. *Cell Research*, *28*(10), 1026–1034. https://doi.org/10.1038/s41422-018-0071-1

Section V

Cryo-EM in the Study of Molecular Biology

Structural Insights of Gene Machineries

23 Imaging Structurally Dynamic Ribosomes with Cryogenic Electron Microscopy

Samantha M. Webster, Mira B. May, Barrett M. Powell, and Joseph H. Davis

RIBOSOMES AND ELECTRON MICROSCOPY: A HISTORY INTERTWINED

The ribosome, which is responsible for protein translation in all domains of life, is a massive ribonucleoprotein complex ranging in size from ~2.5 MDa in bacteria[1] to ~4 MDa in eukaryotes[2]. This essential molecular machine is comprised of two subunits formed by interwoven RNA helices and ribosomal proteins that assemble through a dynamic, multistep biogenesis pathway[3,4]. Once assembled, the machine is itself dynamic, undergoing structural changes coupled to its function in mRNA binding and decoding, peptidyl transferase activity, and its eventual sequestration and degradation. These dynamics involve both conformational motions and regulated compositional changes, with more than 50 core proteins and hundreds of transient binders and cofactors associating with the particle during its assembly and as it functions[5].

Electron microscopy (EM), including negative stain EM, single-particle cryogenic EM, and cryogenic electron tomography (ET), have proven to be uniquely suited for studying the ribosome. Unlike X-ray crystallography, in which a requisite crystal lattice locks the ribosome in a single structural state, cryo EM observes 10^4–10^7 individual and potentially structurally heterogenous particles in a single experiment. Although such structural heterogeneity presents computational challenges for the reconstruction and refinement of cryo-EM density maps, it also provides an opportunity to see the full range of structural states populated in a sample[6]. With the advent of improved imaging technology[7,8] and 3D-reconstruction algorithms[9–11], ribosome structural dynamics are being systematically characterized in greater detail than ever before[12] (Figure 23.1).

Early studies of the ribosome are inextricably linked with the development of electron microscopy. These "small particulate component[s] of the cytoplasm" that were briefly known as "particles of Palade" were first identified in 1955 in electron micrographs of the endoplasmic reticulum of rat liver cells[13,14], though Salvador Luria had incidentally imaged ribosomes in his electron micrographs of bacteriophages years prior[15]. Early evidence for co-transcriptional ribosome assembly, which is discussed below, came from direct visualization of rRNA synthesis in rapidly growing cells using "Miller spreads"—a method that gently disperses chromatin and fixes RNA polymerases to their DNA templates and nascent transcripts[16–18]. Negative stain electron micrographs of such Miller spreads revealed dense packing of RNA polymerases along *E. coli* rRNA operons, with fibrils of increasing lengths emanating from the RNA polymerase. These fibrils were decorated with protein-dense ribosome assembly intermediates, which collectively gave rise to a "Christmas tree" morphology (Figure 23.2) that highlighted the co-transcriptional nature of ribosome assembly.

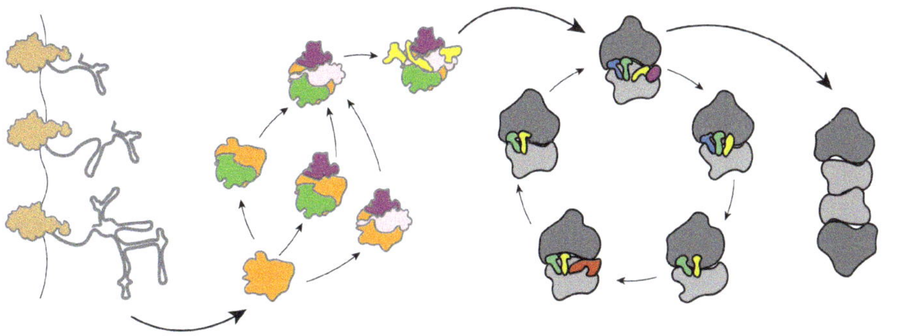

FIGURE 23.1 Selected phases from the bacterial ribosome life cycle. The ribosome life cycle involves an array of heterogenous structural states related by both parallel and sequential pathways across timescales spanning several orders of magnitude.

DOI: 10.1201/9781003326106-28

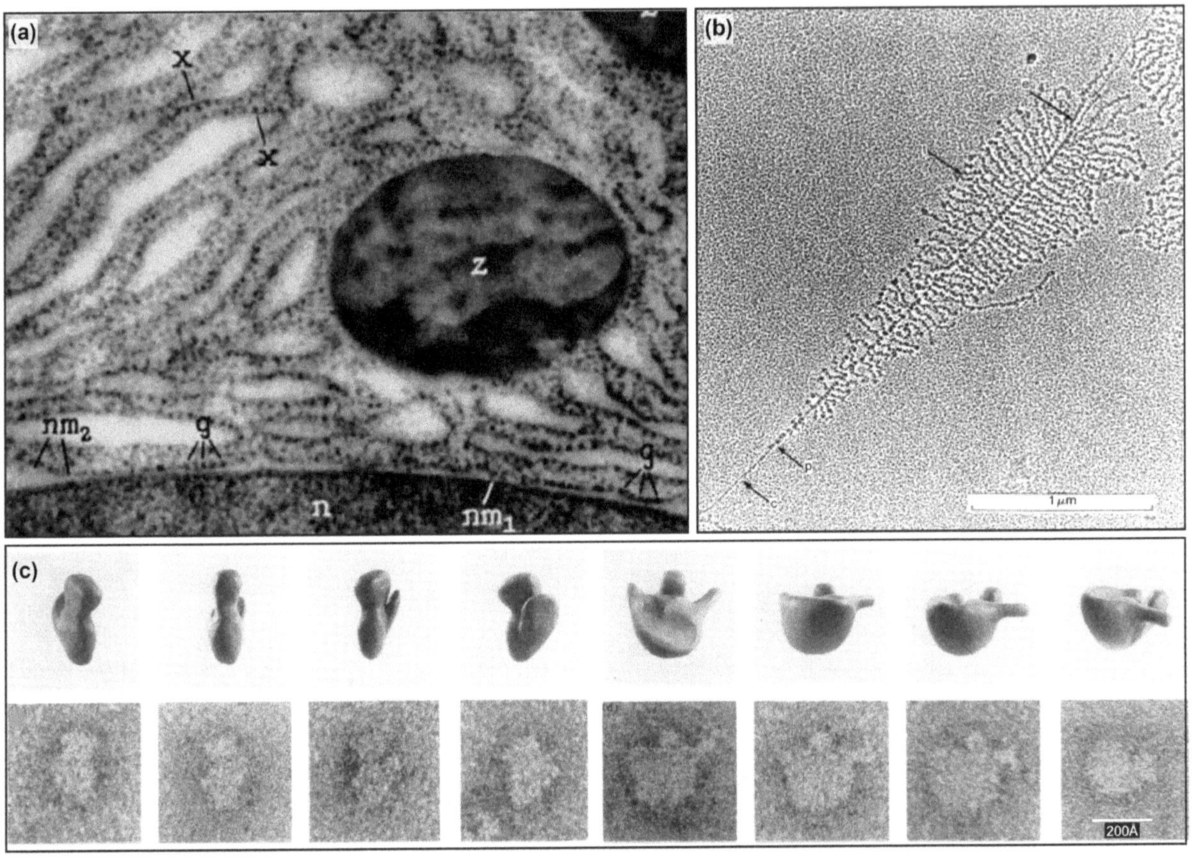

FIGURE 23.2 Early studies of ribosomes and electron microscopy. (a) Electron micrograph of endoplasmic reticulum from rat pancreatic acinar cells, first published by George Palade in 1954. *g* – Palade's "granules" *i.e.* ribosomes, *n* – nucleus, *x* – regions where granules appear to be attached to the membrane of the endoplasmic reticulum, nm_1 – nuclear membrane proper, nm_2 – endoplasmic reticulum membrane limiting the cytoplasm toward the nucleus, *z* – zymogen granule[13]. (b) Electron micrograph of a gene transcription unit prepared by the Miller spreading technique[158]. (c) The first three-dimensional models of the 30S (*left*) and 50S (*right*) subunits, with single-particle images from electron micrographs in corresponding orientations[20]. Figure panels reproduced with permission from above references.

Such applications of electron microscopy not only advanced the field's understanding of ribosomes, but many were also critical in advancing electron microscopy as a technology. Indeed, the large, dense, and abundant nature of ribosomes make them nearly ideal imaging subjects[19] and, as such, they have played key roles in the development of negative stain and cryogenic EM[20,21], image processing and 3D reconstruction methods[21–25], the development of early heterogenous classification and reconstruction algorithms[26], and the more recent development of cryo-EM and cryo-electron tomographic methods to determine structures *in situ*[27–30]. Likewise, some of the earliest examples of exhaustive structure classification methods aimed at understanding function[31,32], and the development of methods to model continuous forms of heterogeneity, were tested using datasets of ribosome assembly intermediates that exhibited vast structural heterogeneity[6,11,12,33,34].

In this chapter, we highlight how methodological advances in electron microscopy have enabled a greater understanding of the structural dynamics occurring in ribosomes as they undergo assembly and as they function, providing vignettes where electron microscopy has played a central role. Although EM has contributed to our understanding of ribosome structure–function relationships in both eukaryotes and prokaryotes, we focus on the latter, and we direct readers to excellent recent reviews of analogous studies of eukaryotes[35,36].

SNAPSHOTS OF RIBOSOME BIOGENESIS

Ribosome biogenesis is a rapid, multistep process, involving the interplay of ribosomal proteins (r-proteins) and RNA (rRNA), as well as transiently bound assembly cofactor proteins. In prokaryotic cells, much of the particle's assembly occurs co-transcriptionally, with rRNA folding coupled to r-protein binding events that help to stabilize productively folded rRNA, thereby guiding the particle's maturation[4,18,37–39]. Notably, however, pioneering work by the

Nomura and Nierhaus groups that reconstituted assembly of the 30S and 50S subunits *in vitro* showed that purified, fully transcribed rRNAs and r-proteins can self-assemble into active ribosomal subunits, which argued that co-transcriptional assembly is not strictly required for ribosome biogenesis[40–49]. Moreover, this *in vitro* approach allowed them to test the binding interdependence between r-proteins through order-of-addition experiments, and these studies provided evidence for thermodynamic coupling between many r-protein binding events. Certain proteins, termed primary binders, could bind the rRNA directly, whereas others, termed secondary and tertiary binders, exhibited stronger binding when other r-proteins were present. These experiments supported a model of hierarchical ribosome subunit assembly involving sequential binding of secondary and tertiary proteins, and parallel assembly of the primary binders.

Whereas these *in vitro* assembly reactions demonstrated the ability of subunits to assemble using only core ribosomal components, assembly is known to occur more efficiently and more rapidly in cells. Although the co-transcriptional nature of such cellular assembly is likely a dominant contributor to this improvement, a wide array of cellular assembly cofactors, including GTPases, rRNA and r-protein modification enzymes, and RNA helicases are also posited to contribute to the improved assembly efficacy observed in cells[4,38,50]. Additionally, many of these cofactors profoundly impact assembly efficacy under stress conditions, including cold-shock, suggesting roles for assembly cofactors in overcoming stress-related barriers to productive assembly[51]. Despite this broad understanding of assembly cofactor roles, there remain several areas of active inquiry, including: when in the assembly process do cofactors act; how do cofactors interact with each other; is there redundancy amongst cofactors; and how do cellular stresses impact the role of these cofactors? Moreover, basic questions about the mechanisms by which these proteins facilitate assembly remain unanswered, with postulated functions including: ensuring the correct assembly order[52–54]; expediting specific structural changes[55,56]; and proofreading specific assembly defects[12].

Answering these and related questions has proven challenging for three key reasons. First, despite its large size and complexity, the entire *E. coli* ribosome biogenesis process is completed in roughly two minutes, and ribosomal precursors account for only 2% to 5% of all ribosomes during rapid growth[57,58]. As a result, isolating and studying specific ribosomal intermediates is difficult. Second, assembly cofactors are believed to associate transiently with their ribosomal substrates, and these substrates are of low abundance in the cell, making them difficult to isolate natively. Finally, multiple parallel assembly pathways exist that produce highly heterogenous ensembles of assembly intermediates[3,31]. Approaches to isolate the relatively short-lived intermediates include: genetic systems to perturb r-protein and cofactor levels and thus stall the assembly process[31,53,59–62]; *in vitro* reconstitutions, which are inherently slow[63–65]; and biochemical approaches, such as affinity enrichment of assembly-factor bound particles[66]. Over the past 10 years, each of these strategies have been coupled to single-particle cryo-EM to characterize the structure and composition of highly heterogenous populations and, in some instances, to track dynamic structural rearrangements that occur during assembly. Here, we highlight a nonexhaustive set of such studies that leveraged single-particle cryo-EM to understand the biogenesis process.

Cryo-EM Reveals Conserved Structural Blocks in Ribosome Biogenesis

Whereas the Nierhaus experiments provided a road map to understand the consequences of withholding an r-protein from an *in vitro* large subunit assembly reaction, it was unclear how cellular assembly pathways, which are under exquisite transcriptional and translational control, and which bear an array of potentially buffering assembly cofactors, would respond to such a perturbation. To answer this question, Davis et al. depleted the early binding r-protein bL17 from cells, resulting in the accumulation of large subunit intermediates exhibiting both compositional and conformational heterogeneity. These intermediates were biochemically purified and analyzed using quantitative mass spectrometry and cryo-EM[31]—a form of *ex vivo* analysis. In one of the first applications of deep structural classification to ribosome biogenesis intermediates, this work revealed 13 distinct intermediate structures resolved at 4 to 5 Å resolution that allowed for comparison of the presence/absence of each rRNA helix and r-protein (quantified as "occupancy") across the entire range of ribosomal precursors. This quantification produced an "occupancy map" that, upon hierarchically clustering, revealed sets of similar assembly intermediate particles and sets of rRNA helices and r-proteins whose occupancy profiles were correlated across the 13 structures. Guided by these occupancy maps, the structures were then arranged into an assembly pathway that exhibited hierarchical and parallel elements, and highlighted the malleability of bacterial ribosome assembly. For example, some r-proteins that were known to bind after bL17 in wild-type cells[67] were bound to these bL17-free particles, indicating that the assembly order could be permuted when limited quantities of an early binding r-protein were available. Moreover, this approach identified five major groups of rRNA and protein "blocks" that exhibited highly correlated behavior across the accumulating intermediates and co-localized on the mature large subunit. The authors suggested that this modular block-like assembly was a result of thermodynamic cooperativity built into the assembly process, and the subsequent observation of similar assembly blocks in independently analyzed large subunit assembly intermediates is consistent with this notion[3,64]. Finally, by identifying mutually exclusive blocks among ribosome precursors, which the authors posited would not disassemble once formed, this work provided

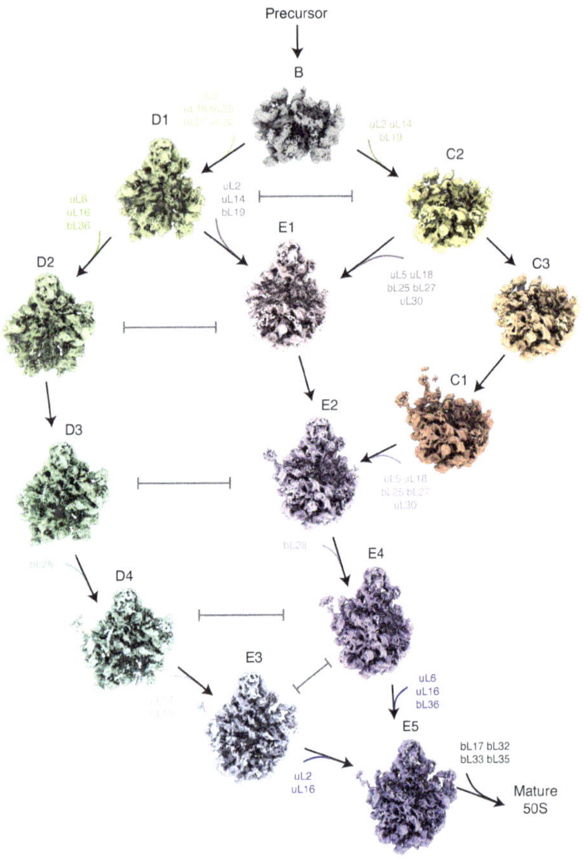

FIGURE 23.3 Parallel assembly pathways observed under bL17 depletion. The most immature observed class B is shown at the top, which is hypothesized to transition to the most mature class E5 via several branched pathways (black arrows). Incorporated r-proteins are indicated at each transition step. Adapted from Davis et. al, 2017.[31]

the first structural evidence of parallel, branched assembly pathways (Figure 23.3). Notably, more recent applications of advanced classification and reconstruction approaches, including deep hierarchical classification[32] and continuous models of structural heterogeneity, have recapitulated these original analyses[6,11,33,34].

This study additionally allowed for expansion of the classic Nierhaus assembly map under perturbed *ex vivo* conditions to include rRNA helices and r-protein binding interdependencies, and provided insights into how different assembly pathways can permute the order and coupling of these r-protein binding and rRNA docking events. Taken together, the flexibility observed in the assembly pathways in this work further highlighted the value of interrogating assembly *in vivo*. Indeed, we and others have speculated that such flexibility may be an essential and evolutionarily selected feature of bacterial ribosome assembly as it would allow assembly to proceed in the face of transient shortages in particular r-proteins or assembly cofactors, or when cells encounter environmental conditions that disfavor a particular assembly pathway[3,68,69].

More recently, the Spahn and Nikolay groups have worked to build a structural understanding of the *in vitro* reconstituted assembly process pioneered by the Nierhaus group[63,64]. Importantly, these experiments detail progressive maturation pathways based only on the inherent biochemical properties of rRNAs and r-proteins, effectively insulating the observations from the complications that could arise from interactions with assembly cofactors or regulatory cascades in cells. Interestingly, the overarching themes observed by Davis et al. *in vivo*, including multiple parallel routes of assembly, were reinforced by these *in vitro* studies. Moreover, both *in vitro* and *in vivo*, ribosomal precursors appeared to mature in a domain-wise manner, with early precursors comprised of a stable core lacking a functional peptidyl transferase center and key intersubunit bridging rRNA helices, and a significant fraction of particles lacking a well-resolved central protuberance. Of note, this reconstitution approach using purified components allowed Qin and colleagues to catalog one of the least mature assembly intermediates to date, which may represent the minimal stably folded structural domain[64]. Interestingly, similar early assembly intermediates were recently discovered both in a sample isolated from an *in vitro* co-transcriptional assembly reaction[65] and in purified from cells lacking the helicase DeaD that were grown at low temperature[70], indicating that such early assembly states can also be observed during co-transcriptional assembly *in vitro* and in particles isolated from cells.

A New Perspective on Biogenesis Cofactors

As above, cryo-EM has significantly impacted our understanding of how ribosome biogenesis cofactors affect the assembly process, with most structural studies characterizing the ribosomal particles that accumulate in the absence of a cofactor of interest, or in determining the structure of the assembly–cofactor bound complex. To date, studies have investigated the 30S cofactors RsgA/YjeQ[71–76], RbfA[72,77], RimM[59,73], RimP[55], KsgA[12,78–80], and Era[81,82], and the 50S cofactors RbgA[52,53,60,83], YphC/EngB or YsxC[61], EngA[84], RrmJ[85], ObgE[66,86], SrmB[62], and DeaD[70]. Here we highlight the biogenesis cofactors RbgA and KsgA, detailing how cryo-EM and genetic perturbations to deplete cells of these cofactors have been applied to interrogate each cofactor's mechanism of action.

Notably, given the aforementioned flexibility and inherent permutability of the *in vivo* assembly pathways, interpreting such cofactor depletion experiments is challenging[3]. Indeed, parallel assembly branches raise the possibility that the accumulating particles one observes upon cofactor depletion are not true substrates for the cofactor but rather particles that have progressed toward a thermodynamically stable state that is no longer able to interact with the factor[3,87]. It is even formally possible that such accumulated particles are not competent to mature, and are therefore off-pathway.

In analyzing the particles that accumulate in the absence of the ribosome biogenesis-associated GTPase RbgA, Jomaa et al. considered these possibilities and, by coupling pulse-labeling to quantitative mass spectrometry, they determined that the majority of the RbgA-depletion particles were in fact competent to mature to functional 70S ribosomes[53,57]. Interestingly, they further noted that the functional core (i.e., the peptidyl transferase center, PTC) in these particles was highly disordered, despite observing that the majority of the other structural elements had adopted their mature conformations. Whereas the general observation that functional domains mature late predated this work, the effect was even more pronounced in these[53,60] and related[61] particles that accumulated upon cofactor depletion. This observation has been interpreted as a quality control mechanism inherent in the assembly process that prevents premature particles from engaging in translation until the particle is completely assembled[53]—akin to issuing keys to homeowners only once the house is fully constructed, with assembly cofactors acting as building inspectors to ensure proper assembly order is maintained and particle maturation is complete before certification.

Another method for determining cofactor function is to add the protein of interest back to ribosomes purified from a cofactor-depleted strain and to then assess the changes that occur *in vitro*. Sun and colleagues employed this strategy while studying the function of KsgA[12], a methyltransferase which has long been suspected as a biogenesis cofactor involved in 30S subunit maturation[88,89]. The group characterized the ensemble of intermediates that accumulate in a cold-sensitive Δ*ksgA* strain of *E. coli* grown at low temperature, where these accumulating particles were presumed to be KsgA substrates. To assess the impact of KsgA addition to these substrates, they also analyzed structures resulting from the addition of purified KsgA to this sample. To their surprise, the KsgA-treated structures were grossly *less* mature than the particles isolated from the depletion strain, with large swaths of poorly resolved density in the head, platform, and spur domains, indicating greater conformational or compositional heterogeneity in these regions. This led to a hypothesis that KsgA may recognize a structural defect in its substrates and induce disassembly of such particles. Using cryoDRGN[6,11] to analyze particles not yet treated with KsgA, they observed that a key helix (H44) could adopt two alternative conformations—one active and the other inactive. Quantitation of the relative abundance of these conformations revealed that upon treatment, KsgA preferentially depleted the sample of H44-inactive particles. Taken together, the evidence suggested that KsgA specifically recognizes inactive 30S intermediates and, through its binding, induces large-scale structural remodeling and disassembly that may afford these inactive particles an opportunity to re-fold and progress down productive assembly (Figure 23.4). In sum, this recognition and disassembly activity was interpreted as KsgA-dependent kinetic proofreading[90] of the assembly process. In addition to illuminating a new proofreading role for KsgA, this work also highlighted the potential of next-generation cryo-EM analysis tools such as cryoDRGN for analyzing large datasets bearing extensive structural variation, and in using these data to inform our understanding of structural signaling, remodeling, and dynamics.

Despite rapid recent progress in understanding bacterial ribosome biogenesis from a structural perspective, many key questions remain. For example, what is the relative flux through different assembly pathways; how do pathways change under different environmental conditions or cofactor depletions; and do multiple cofactors work in concert on a given particle, similar to that observed in eukaryotic ribosome assembly intermediates? We expect that sophisticated biochemical and structural analysis protocols will be needed to capture these more transient stages of assembly and, as techniques improve, we anticipate drawing deeper insights into both the immature assembly intermediates and the mature ribosome as it engages in its primary function in the cell—translation.

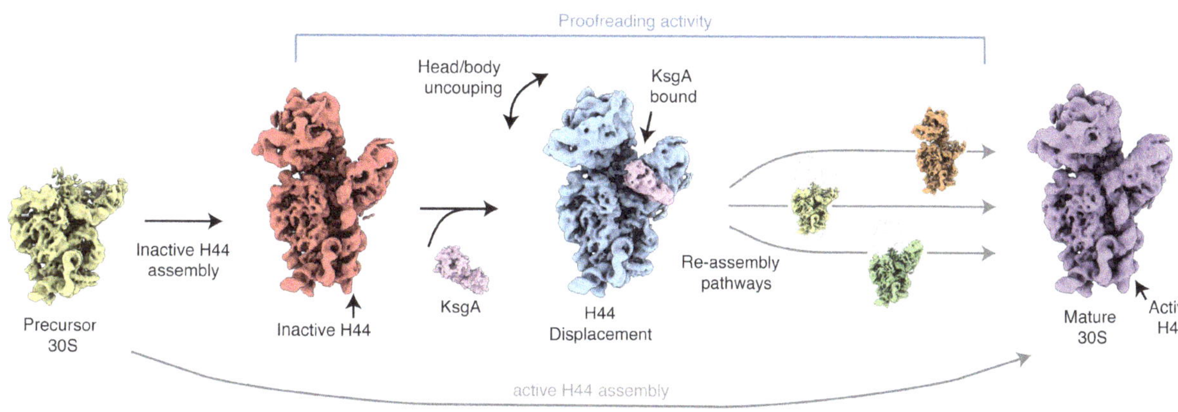

FIGURE 23.4 KsgA proofreading model. During 30S assembly, immature particles with inactive H44 are recognized by KsgA, which induces subsequent uncoupling of the head/body to provide another opportunity for subunit maturation. Adapted from Sun et al., 2022.[12]

A WINDOW INTO PROKARYOTIC TRANSLATION

The structural dynamics of translation hold two major themes: compositional dynamics, with tRNAs and GTPase initiation and elongation factors binding and dissociating from the translating ribosome; and conformational dynamics, wherein the subunits undergo movements linked to function[91,92]. For example, during translation elongation, the large (LSU) and small (SSU) subunits rock back and forth relative to each other in a distinct ratcheting motion that is critical to translocating tRNAs through the active sites[93–95], as well as a subsequent step where the SSU head rotates relative to its body in order to move along the mRNA by one codon[96].

Due to limitations in microscope and detector hardware, early cryo-EM structures of translating ribosomes were relatively low resolution[94,97–99]. Despite this, targeted studies based on previous biochemical and biophysical work[100–103] assembled desired complexes *in vitro* or stalled the translation complex pharmacologically or genetically, resulting in structures that significantly advanced our understanding of the mechanism and dynamics of translation. Such approaches provided the first visualization of ribosome-bound EF-G, a GTPase that catalyzes mRNA and tRNA translocation during polypeptide elongation[104]; facilitated discovery of the ratchet motion between the small and large subunits that occurs upon binding of EF-G[95]; and provided structures of tRNA in a hybrid state between the P- and E-sites that occurs as a prelude to tRNA translocation[105]. Early structures also complemented crystallographic approaches and led to a general understanding that the factors EF-Tu, EF-G, RF-3, and IF-2 bind in the flexible GTPase associated center (GAP) via an "induced fit" model, wherein each subsequent factor acts to stabilize the ribosome and promote productive conformational changes to advance the translation cycle[97,104,106–108].

Over the past 15 years, technical advances in cryo-EM[7,109] have produced ever-higher-resolution structures that led to new insights into ribosome structure and function. For example, the sub–3 Å resolution structure of a ribosome•EF-Tu complex bound to aminoacyl-tRNA and the antibiotic kirromycin resolved all 35 rRNA modifications[110], providing a structural framework to understand antibiotic action and resistance[110–112], and it paved the way for a recent study determining 17 high-resolution structures of antibiotic–ribosome co-complexes[113]. Further methodological improvements culminated in a 1.6 Å resolution structure of a translating ribosome that, incredibly, allowed for direct structure-guided sequencing of the rRNA[114]. These structures additionally visualized the flexible and enigmatic protein bL9 in a closed conformation and identified residues on bL9 (Glu87) and uS6 (Arg24) that stabilized this closed conformation via a salt bridge. Notably, bL9 is suspected to play a role in fidelity of translation[115], frameshifting[116,117], polysome formation[118], and quality control pathways in collided ribosomes[119,120], and the exchange between the extended conformation originally observed in crystal structures at a lattice interface[121–123] and the closed conformations observed by Fromm et al. and others may help to explain these multimodal functions.

Similarly, image analysis tools aimed at handling structural heterogeneity have begun to link mechanism to the structural ensembles within cryo-EM datasets that can be computationally resolved. Loveland et al. used such approaches to resolve six near-atomic-resolution ribosome•tRNA•EF-Tu decoding complexes, bearing cognate and near-cognate tRNA–mRNA pairs[124]. These structures established that discrimination between cognate and near-cognate tRNAs hinges on a single base on the shoulder of the 30S subunit (G530), which can assume an "ON" conformation when properly stabilized by a network of hydrogen bonds between the mRNA and tRNA bases. They further showed that movement into the "ON" conformation aids in EF-Tu activation that precipitates GTP hydrolysis and EF-Tu dissociation from the complex, which represents the final step of aminoacyl–tRNA accommodation into the LSU tRNA A-site. The ability to visualize several related complexes in such detail was critical in elucidating the mechanisms of translational fidelity and highlighted the value of accurate particle classification algorithms to not only improve resolution, but to also reveal the structural underpinnings of central biological phenomena.

Whereas the combination of biochemical data and the targeted enrichment of structural states using genetic or pharmacological approaches has allowed researchers to infer the sequence of structural progression[3], the field has long sought to directly determine time-dependent structural ensembles[125] and to visualize transitory structural states within these datasets. Pursuing these goals, Carbone et al. analyzed translational elongation as a function of time after the addition of EF-G using standard vitrification equipment[126]. Despite the relatively poor temporal resolution afforded by such plunge-freezing post-mixing, their intermediate timepoints captured transient translation states with EF-G bound, and a comparison of their structures highlighted the progression of structural states during EF-G–mediated translocation (Figure 23.5). They noted that EF-G's extended conformation both pretranslocation (structure III) and nearly post-translocation (structure IV) was consistent with a model of EF-G acting as a rigid pawl, as opposed to a flexible motor that is actively driving translocation through GTP hydrolysis. They concluded that the motion driving translocation, therefore, is inherent and spontaneous internal movement of the ribosome, including the 30S body rotation and head swivel. The study also captured a structure (structure VI) that did not have EF-G bound, but which was exclusively present in the t=24 seconds (s) dataset, indicating that it was a transient state formed directly after EF-G dissociation. The structure showed the 30S head still in the swiveled position, indicating that the reversal of the head swivel occurs after EF-G dissociation.

Building on the promise of time-resolved cryo-EM, recent innovations, including the development of fast-mixing

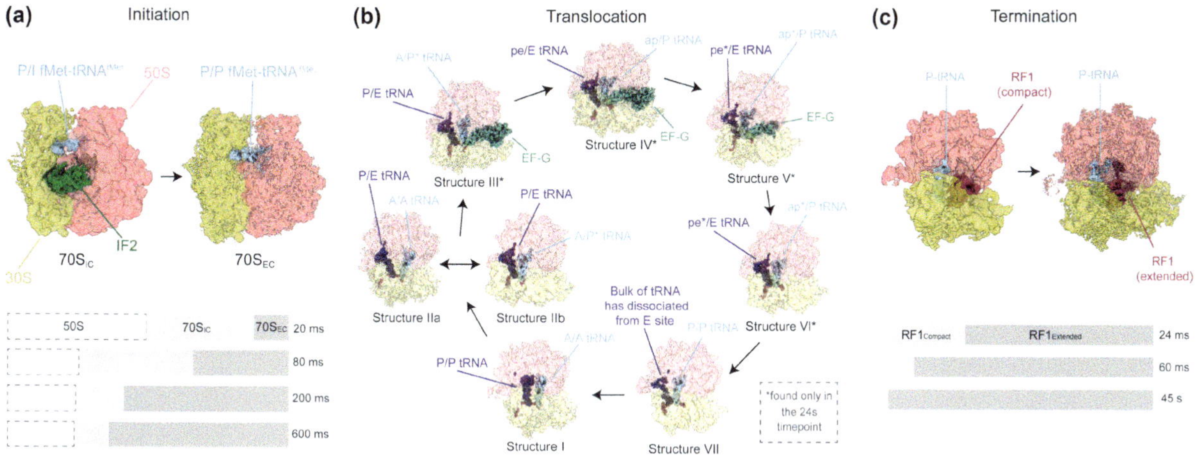

FIGURE 23.5 Time-resolved cryo-EM experiments reveal new structures at different steps of translation. (a) 30S initiation complex (IC) particles were combined with 50S particles using a mixing-spraying time resolved cryo-EM setup, resulting in four timepoints spanning 0–600 milliseconds (ms). The population of $70S_{IC}$ particles initially increased as the 50S particles bound the 30_{IC} particles, before the $70S_{IC}$ particles converted to 70S elongation complexes ($70S_{EC}$). Adapted from Kaledhonkar et al., 2019[129]. (b) EF-G was mixed with pretranslocation 70S ribosomes loaded with $tRNA^{fMet}$ in the P-site and dipeptidyl-$tRNA^{Pro}$ in the A-site. Grids were blotted at 0, 24, and 600 seconds (s). The structures show a full progression of the translocation cycle, and include many tRNA hybrid states. Structures III, IV, V, and VI were found only in the 24 s sample. Structure V shows EF-G after domains I and II have partially dissociated. Adapted from Carbone et al., 2021[126]. (c) Using a rapid-mixing cryo-EM setup, UAA-programmed release complex, (RC_0) with tripeptidyl-tRNA in the P-site was added to RF1. At 24 ms, roughly 30% of complexes were bound to RF1 in its compact form, while at 60 ms more than 90% of RF1 had converted to its extended conformation. Adapted from Fu et al., 2019[130].

microfluidic vitrification instruments,[127,128] have enabled millisecond-resolved studies of translation initiation and termination[129,130]. Kaledhonkar et al. used a rapid-mixing approach to study the later steps in bacterial translation initiation, mixing 30S initiation complexes (IC) with 50S particles to investigate mechanisms of 70S IC formation and its subsequent maturation into the 70S elongation complex (EC). Using a custom-built microfluidic device for mixing and rapid vitrification, they collected datasets spanning 20 to 600 milliseconds (ms) post-mixing. They found that the 70S IC complex abundance peaked near 80 ms, representing ~40% of the particles, and that by the end of the 600 ms more than 60% of the particles were fully converted to 70S EC complexes (Figure 23.5a). Notably, this rapid-mixing approach was vital in resolving the short-lived 70 IC complex in the absence of perturbations, and it helped to distinguish on-pathway complexes from off-pathway intermediates that could result from such perturbations. Inspection of this native 70S IC structure revealed that IF-1 had dissociated, which alleviated a steric clash with H69 of the large subunit and allowed the formation of intersubunit bridges required for translation. Moreover, in the transition from the 70S IC to the 70S EC, the authors observed dissociation of IF-2, which appeared to trigger a reverse rotation of the SSU subunit by 3 Å, thereby stabilizing the nonrotated intersubunit orientation. Finally, the combination of this rotation and the absence of IF-2 enabled the fMet-tRNA to move from the peptidyl/initiation (P/I) configuration observed in the 70S IC to the peptidyl–peptidyl (P/P) configuration observed in the 70S EC, effectively placing the fMet moiety in a peptidyl-transfer competent position within the P-site, and enabling the complex to proceed with the first round of translation elongation.

At the other end of translation, Fu et al. used a similar technique to study the structural changes in release factors (RFs) during translation termination[130]. RFs contain structural motifs involved in recognizing a stop codon, as well as a GGQ motif that is responsible for ester bond hydrolysis, which allows the nascent peptide release from the ribosome. Interestingly, crystal structures of isolated bacterial RFs are compact, with only 20 Å between the stop codon recognition and GGQ motifs, whereas ribosome-bound RFs adopt an extended conformation that spans the ~70 Å between the decoding center and the peptidyl transferase center. If the compact conformation observed crystallographically is physiologically relevant, this implies that either (1) the extended and compact conformations are in equilibrium in solution and the ribosome captures the extended form, or (2) the compact conformation binds and then extends on the ribosome. To distinguish between these possibilities, Fu et al. used a rapid-mixing technique paired with cryo-EM to study the structural progression after mixing a UAA-programmed release complex (RC_0) with RF-1. At 24 ms, 25% of ribosome-bound RF was in its compact form, in a sort of pre-accommodation state previously observed by cryo-EM in ribosomes stalled at the terminus of an mRNA lacking a stop codon[131]. This population rapidly decayed by 60 ms into a ribosome-bound extended conformation RF complex (Figure 23.5c), indicating (1) that RF-1 can bind in a compact state, and (2) that the conformation shift to the elongated state happens very rapidly upon RF binding, whereas the hydrolysis step, which acts to release the nascent polypeptide chain, occurs on a much longer timescale. The authors hypothesize that the compact form of RF may assist with rapid factor binding and more accurate stop-codon decoding.

RESPONDING TO STRESS: A GLIMPSE AT RIBOSOMAL SEQUESTRATION

Nutrient starvation and other stress can lead to translational suppression, which directs cellular resources away from the energy-intensive process of protein production and thereby promotes cell survival[132]. In *E. coli*, hibernation factors work in concert to suppress protein synthesis by binding to and blocking functional sites on the SSU, and by inducing the reversible dimerization of 70S ribosomes into a translationally inactive 100S sequestration particle[133,134]. In contrast to ribosome degradation, which has been observed in mammalian cells[135], the generation of such a particle effectively allows cells to downregulate translation during starvation, and to then rapidly reinitiate translation should nutrients become available. The first structures of this sequestration particle were generated through cryo-EM, and the visualization of these particles dimerized head-to-head through the SSU provided a structural understanding of translational repression in bacteria[136–138]. Interestingly, although hibernation factors vary between species, the process of sequestration may be conserved as a 110S sequestration particle has been identified in mammalian cells[139].

WHAT LIES AHEAD

Whereas single-particle analysis of purified ribosomes has provided key mechanistic insights into ribosome biology, there are a number of benefits to studying samples *in situ*—that is, in their native cellular environment. By avoiding cell lysis and downstream purification, one can interrogate minimally perturbed ribosomes that are likely to better retain transiently bound cofactors. Moreover, *in situ* imaging can reveal subcellular localization patterns of sub–nanometer scale features that may be associated with biological function.

Studying ribosomes *in situ* via electron microscopy is typically accomplished using cryo-electron tomography (cryo-ET), often preceded by focused ion beam (FIB) milling, which is used to thin a cellular sample to allow transmission of elastically scattered electrons. In contrast to cryo-EM, cryo-ET repeatedly images the same ~50–300 nm thick sample at different stage tilt angles, which helps to disentangle particles at different depths along any one projection axis. Such a "tilt series" can then be back-projected to reconstruct a 3D tomogram of the field of view, typically 0.05–0.5 μm^3, thereby visualizing a significant portion of small bacterial cells and a relatively small subregion of typical eukaryotic cells. From this low signal-to-noise ratio tomogram, ribosomes can be identified, extracted, aligned, and averaged to increase the signal-to-noise ratio and structural resolution in a process called subtomogram averaging (STA)[140].

Until recently, cryo-ET and STA structures rarely exceeded ~1–2 nm resolution, which limited the field's ability to glean molecular insights from the resulting structures. However, recent advancements in hardware for FIB milling and imaging[141,142], strategies for data collection[143–145], and image processing software[28,146–153] have poised structural biology for an "*in situ* resolution revolution" akin to the single-particle analysis "resolution revolution" of the early 2010s[7]. Downstream tools to analyze structural heterogeneity have also expanded, allowing identification of rare and dynamic subpopulations of ribosomes *in situ*[30,151,154–157]. Taken together, the field now enjoys faster and easier data collections, larger and higher quality datasets, more powerful data processing, and novel data analysis opportunities.

While many of these developments are recent and their synergistic effects are likely to bear fruit over the next several years, several exciting findings have already hinted at what is to come (Figure 23.6). Tegunov et al. demonstrated that STA could resolve chloramphenicol-treated *Mycoplasma pneumoniae* ribosomes at 3.5 Å resolution *in situ*[28], and Xue et al. further analyzed the same system to characterize the distribution of translational states under antibiotic and untreated conditions *in situ*, resolving 13 compositionally and conformationally distinct states that recapitulate the known translation cycle[118]. Notably, detailed analysis of native polysomes *in situ* allowed Xue et al. to directly observe protein bL9 playing a role in polysome coordination and collision avoidance. In a novel approach to *in situ* imaging and processing, Lucas et al. designed 2D template matching (2DTM), which searches individual 2D micrographs for instances of a high-resolution template[29]. In one application of 2DTM they used a *M. pneumoniae* 50S subunit structure as a "bait," identifying via local searches the variable presence of adjacent smaller 30S subunits. In principle this approach could enable quantitation of small subpopulations of ribosomes with a particular cofactor of interest bound. Finally, we have developed a deep-learning-based approach to study structural heterogeneity among subtomograms and applied it to the *M. pneumoniae* dataset first reported by Tegunov et al. In addition to recapitulating many of the aforementioned insights, this tool additionally resolved a minor population (~2%) of ribosomes associated with the membrane, thereby simultaneously visualizing a moderate-resolution structure of a 70S ribosome–SecYEG holotranslocon co-complex *in situ*[30].

With this rich toolkit to pursue structural studies *in vitro* and *in situ*, we anticipate a wave of new biologically insightful findings in the coming years. For example, one can now directly develop and test structure–function hypotheses in the native cellular environment, and subsequently elucidate the specific structural mechanism involved with ~10^{-2} second time resolution *in vitro*, all at near-atomic resolution. Such tools will be critical to address key remaining questions in the field, including: Do ribosome biogenesis, translation, and sequestration, along with their associated cofactors, exhibit specific subcellular distributions? Are such distributions perturbed in diseased states? Do the

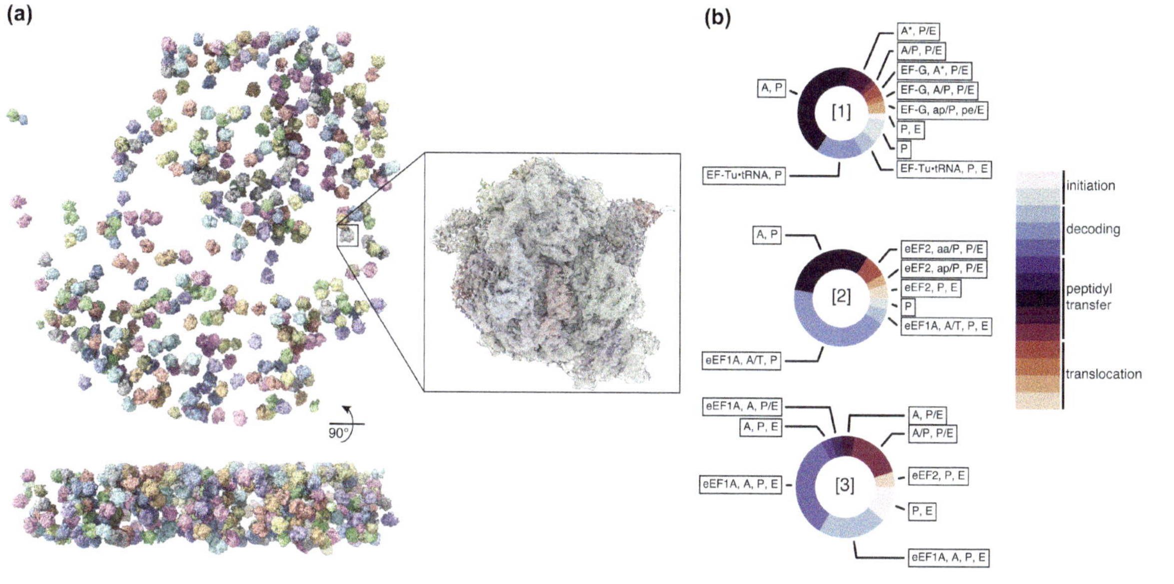

FIGURE 23.6 Cryo-ET enables new insights to ribosome translation *in situ*. (a) Illustrative example of distinct ribosome translational states identified by cryo-ET subtomogram averaging and classification, mapped back to the cell and colored by translation state. *Inset:* Example of near-atomic-resolution maps (grey, semitransparent) from cryo-ET *in situ*, with fit atomic model 7PHB (ribbon multicolor). Data were reprocessed from Xue et al., 2022.[118] (b) Distribution of translation cycle states calculated from classification results of three recent *in situ* studies, highlighting similarities in relative abundance of states associated with various stages of the translation cycle observed by Xue et al., 2022[118] [1], Hoffman et al., 2022[159] [2], and Gemmer et al., 2023[160] [3].

structural states captured *in vitro* accurately reflect states and their relative populations in the cell? We also posit that *in situ* approaches are likely to enable discovery of novel or weakly interacting cofactors not captured with traditional cryo-EM using purified samples. We eagerly await the insights that lie ahead for the study of ribosome dynamics at the intersection of structural, molecular, and cellular biology enabled by traditional cryo-EM, time-resolved cryo-EM, and cryo-ET.

ACKNOWLEDGMENTS

This work was supported by NIH grants R01-GM144542, 5T32-GM007287, and NSF-CAREER grant 2046778. Research in the Davis lab is supported by the Whitehead Family and the Sloan Foundation.

REFERENCES

1. Cate, J.H., Yusupov, M.M., Yusupova, G.Z., Earnest, T.N., and Noller, H.F. (1999). X-ray crystal structures of 70S ribosome functional complexes. Science *285*, 2095–2104.
2. Ben-Shem, A., Garreau de Loubresse, N., Melnikov, S., Jenner, L., Yusupova, G., and Yusupov, M. (2011). The structure of the eukaryotic ribosome at 3.0 A resolution. Science *334*, 1524–1529, doi: 10.1126/science.1212642.
3. Davis, J.H., and Williamson, J.R. (2017). Structure and dynamics of bacterial ribosome biogenesis. Philos Trans R Soc Lond B Biol Sci *372*, doi: 10.1098/rstb.2016.0181.
4. Shajani, Z., Sykes, M.T., and Williamson, J.R. (2011). Assembly of bacterial ribosomes. Annu Rev Biochem *80*, 501–526, doi: 10.1146/annurev-biochem-062608-160432.
5. Wilson, D.N., and Nierhaus, K.H. (2007). The weird and wonderful world of bacterial ribosome regulation. Crit Rev Biochem Mol Biol *42*, 187–219, doi: 10.1080/10409230701360843.
6. Kinman, L.F., Powell, B.M., Zhong, E.D., Berger, B., and Davis, J.H. (2023). Uncovering structural ensembles from single-particle cryo-EM data using cryoDRGN. Nat Protoc *18*, 319–339, doi: 10.1038/s41596-022-00763-x.
7. Kuhlbrandt, W. (2014). Biochemistry. The resolution revolution. Science *343*, 1443–1444, doi: 10.1126/science.1251652.
8. Faruqi, A.R., and Henderson, R. (2007). Electronic detectors for electron microscopy. Curr Opin Struct Biol *17*, 549 555, doi: 10.1016/j.sbi.2007.08.014.
9. Scheres, S.H. (2012). RELION: implementation of a Bayesian approach to cryo-EM structure determination. J Struct Biol *180*, 519–530, doi: 10.1016/j.jsb.2012.09.006.
10. Punjani, A., Rubinstein, J.L., Fleet, D.J., and Brubaker, M.A. (2017). cryoSPARC: algorithms for rapid unsupervised cryo-EM structure determination. Nat Methods *14*, 290–296, doi: 10.1038/nmeth.4169.
11. Zhong, E.D., Bepler, T., Berger, B., and Davis, J.H. (2021). CryoDRGN: reconstruction of heterogeneous cryo-EM structures using neural networks. Nat Methods *18*, 176–185, doi: 10.1038/s41592-020-01049-4.
12. Sun, J., Kinman, L.F., Jahagirdar, D., Ortega, J., Davis, J.H. (2023) KsgA facilitates ribosomal small subunit maturation by proofreading a key structural lesion. Nat Struct Mol Biol. 30, 1468-1480 doi: 10.1038/s41594-023-01078-5
13. Palade, G.E. (1955). A small particulate component of the cytoplasm. J Biophys Biochem Cytol *1*, 59–68, doi: 10.1083/jcb.1.1.59.
14. Wells, W.A. (2005). Ribosomes, or the particles of Palade. J Cell Biol, doi: 10.1083/jcb1681fta3.
15. Luria, S.E., Delbruck, M., and Anderson, T.F. (1943). Electron microscope studies of bacterial viruses. J Bacteriol *46*, 57–77, doi: 10.1128/jb.46.1.57-77.1943.

16. Miller, O.L., Jr., Hamkalo, B.A., and Thomas, C.A., Jr. (1970). Visualization of bacterial genes in action. Science *169*, 392–395, doi: 10.1126/science.169.3943.392.
17. Gotta, S.L., Miller, O.L., Jr., and French, S.L. (1991). rRNA transcription rate in Escherichia coli. J Bacteriol *173*, 6647–6649, doi: 10.1128/jb.173.20.6647-6649.1991.
18. de Narvaez, C.C., and Schaup, H.W. (1979). In vivo transcriptionally coupled assembly of Escherichia coli ribosomal subunits. J Mol Biol *134*, 1–22, doi: 10.1016/0022-2836(79)90411-x.
19. Brown, A., and Shao, S. (2018). Ribosomes and cryo-EM: a duet. Curr Opin Struct Biol *52*, 1–7, doi: 10.1016/j.sbi.2018.07.001.
20. Lake, J.A. (1976). Ribosome structure determined by electron microscopy of Escherichia coli small subunits, large subunits and monomeric ribosomes. J Mol Biol *105*, 131–139, doi: 10.1016/0022-2836(76)90200-x.
21. Frank, J., Verschoor, A., and Boublik, M. (1982). Multivariate statistical analysis of ribosome electron micrographs. L and R lateral views of the 40 S subunit from HeLa cells. J Mol Biol *161*, 107–133.
22. Verschoor, A., Frank, J., Radermacher, M., Wagenknecht, T., and Boublik, M. (1984). Three-dimensional reconstruction of the 30 S ribosomal subunit from randomly oriented particles. J Mol Biol *178*, 677–698.
23. Frank, J., Bretaudiere, J.P., Carazo, J.M., Verschoor, A., and Wagenknecht, T. (1988). Classification of images of biomolecular assemblies: a study of ribosomes and ribosomal subunits of Escherichia coli. J Microsc *150* (Pt 2), 99–115.
24. Carazo, J.M., and Frank, J. (1988). Three-dimensional matching of macromolecular structures obtained from electron microscopy: an application to the 70S and 50S E. coli ribosomal particles. Ultramicroscopy *25*, 13–22.
25. Radermacher, M., Wagenknecht, T., Verschoor, A., and Frank, J. (1987). Three-dimensional reconstruction from a single-exposure, random conical tilt series applied to the 50S ribosomal subunit of Escherichia coli. J Microsc *146* (Pt 2), 113–136.
26. Gao, H., Valle, M., Ehrenberg, M., and Frank, J. (2004). Dynamics of EF-G interaction with the ribosome explored by classification of a heterogeneous cryo-EM dataset. J Struct Biol *147*, 283–290.
27. O'Reilly, F.J., Xue, L., Graziadei, A., Sinn, L., Lenz, S., Tegunov, D., Blotz, C., Singh, N., Hagen, W.J.H., Cramer, P., et al. (2020). In-cell architecture of an actively transcribing-translating expressome. Science *369*, 554–557, doi: 10.1126/science.abb3758.
28. Tegunov, D., Xue, L., Dienemann, C., Cramer, P., and Mahamid, J. (2021). Multi-particle cryo-EM refinement with M visualizes ribosome-antibiotic complex at 3.5 A in cells. Nat Methods *18*, 186–193, doi: 10.1038/s41592-020-01054-7.
29. Lucas, B.A., Himes, B.A., Xue, L., Grant, T., Mahamid, J., and Grigorieff, N. (2021). Locating macromolecular assemblies in cells by 2D template matching with cisTEM. eLife *10*, doi: 10.7554/eLife.68946.
30. Powell, B. M., Davis, J. H. (2024) Learning structural heterogeneity from cryo-electron sub-tomograms with tomoDRGN. Nat Methods 1 -12 doi: 10.1038/s41592-024-02210-z.
31. Davis, J.H., Tan, Y.Z., Carragher, B., Potter, C.S., Lyumkis, D., and Williamson, J.R. (2016). Modular assembly of the bacterial large ribosomal subunit. Cell *167*, 1610–1622 e1615, doi: 10.1016/j.cell.2016.11.020.
32. Rabuck-Gibbons, J.N., Lyumkis, D., and Williamson, J.R. (2022). Quantitative mining of compositional heterogeneity in cryo-EM datasets of ribosome assembly intermediates. Structure *30*, 498–509 e494, doi: 10.1016/j.str.2021.12.005.
33. Chen, M., and Ludtke, S.J. (2021). Deep learning-based mixed-dimensional Gaussian mixture model for characterizing variability in cryo-EM. Nat Methods *18*, 930–936, doi: 10.1038/s41592-021-01220-5.
34. Punjani, A., and Fleet, D.J. (2021). 3D variability analysis: resolving continuous flexibility and discrete heterogeneity from single particle cryo-EM. J Struct Biol *213*, 107702, doi: 10.1016/j.jsb.2021.107702.
35. Jobe, A., Liu, Z., Gutierrez-Vargas, C., and Frank, J. (2019). New insights into ribosome structure and function. Cold Spring Harb Perspect Biol *11*, doi: 10.1101/cshperspect.a032615.
36. Klinge, S., and Woolford, J.L., Jr. (2019). Ribosome assembly coming into focus. Nat Rev Mol Cell Biol *20*, 116–131, doi: 10.1038/s41580-018-0078-y.
37. Woodson, S.A. (2011). RNA folding pathways and the self-assembly of ribosomes. Acc Chem Res *44*, 1312–1319, doi: 10.1021/ar2000474.
38. Kaczanowska, M., and Ryden-Aulin, M. (2007). Ribosome biogenesis and the translation process in *Escherichia coli*. Microbiol Mol Biol Rev *71*, 477–494, doi: 10.1128/MMBR.00013-07.
39. Duss, O., Stepanyuk, G.A., Puglisi, J.D., and Williamson, J.R. (2019). Transient protein-RNA interactions guide nascent ribosomal RNA folding. Cell *179*, 1357–1369 e1316, doi: 10.1016/j.cell.2019.10.035.
40. Traub, P., and Nomura, M. (1968). Structure and function of E. coli ribosomes. V. Reconstitution of functionally active 30S ribosomal particles from RNA and proteins. Proc Natl Acad Sci USA *59*, 777–784.
41. Traub, P., Hosokawa, K., Craven, G.R., and Nomura, M. (1967). Structure and function of E. coli ribosomes, IV. Isolation and characterization of functionally active ribosomal proteins. Proc Natl Acad Sci USA *58*, 2430–2436.
42. Mizushima, S., and Nomura, M. (1970). Assembly mapping of 30S ribosomal proteins from E. coli. Nature *226*, 1214.
43. Held, W.A., Ballou, B., Mizushima, S., and Nomura, M. (1974). Assembly mapping of 30 S ribosomal proteins from *Escherichia coli*. J Biol Chem *249*, 3103–3111.
44. Herold, M., and Nierhaus, K.H. (1987). Incorporation of six additional proteins to complete the assembly map of the 50 S subunit from Escherichia coli ribosomes. J Biol Chem *262*, 8826–8833.
45. Rohl, R., and Nierhaus, K.H. (1982). Assembly map of the large subunit (50S) of Escherichia coli ribosomes. Proc Natl Acad Sci USA *79*, 729–733.
46. Dohme, F., and Nierhaus, K.H. (1976). Total reconstitution and assembly of 50 S subunits from Escherichia coli Ribosomes in vitro. J Mol Biol *107*, 585–599.
47. Nierhaus, K.H., and Dohme, F. (1974). Total reconstitution of functionally active 50S ribosomal subunits from Escherichia coli. Proc Natl Acad Sci USA *71*, 4713–4717.
48. Nierhaus, K.H. (1991). The assembly of prokaryotic ribosomes. Biochimie *73*, 739–755, doi: 10.1016/0300-9084(91)90054-5.
49. Sieber, G., and Nierhaus, K.H. (1978). Kinetic and thermodynamic parameters of the assembly in vitro of the large subunit from Escherichia coli ribosomes. Biochemistry *17*, 3505–3511, doi: 10.1021/bi00610a013.

50. Bunner, A.E., Nord, S., Wikstrom, P.M., and Williamson, J.R. (2010). The effect of ribosome assembly cofactors on in vitro 30S subunit reconstitution. J Mol Biol *398*, 1–7, doi: 10.1016/j.jmb.2010.02.036
51. Stokes, J.M., Davis, J.H., Mangat, C.S., Williamson, J.R., and Brown, E.D. (2014). Discovery of a small molecule that inhibits bacterial ribosome biogenesis. eLife *3*, e03574, doi: 10.7554/eLife.03574.
52. Seffouh, A., Trahan, C., Wasi, T., Jain, N., Basu, K., Britton, R.A., Oeffinger, M., and Ortega, J. (2022). RbgA ensures the correct timing in the maturation of the 50S subunits functional sites. Nucleic Acids Res, doi: 10.1093/nar/gkac059.
53. Jomaa, A., Jain, N., Davis, J.H., Williamson, J.R., Britton, R.A., and Ortega, J. (2014). Functional domains of the 50S subunit mature late in the assembly process. Nucleic Acids Res *42*, 3419–3435, doi: 10.1093/nar/gkt1295.
54. Gulati, M., Jain, N., Davis, J.H., Williamson, J.R., and Britton, R.A. (2014). Functional interaction between ribosomal protein L6 and RbgA during ribosome assembly. PLoS genetics *10*, e1004694, doi: 10.1371/journal.pgen.1004694.
55. Sashital, D.G., Greeman, C.A., Lyumkis, D., Potter, C.S., Carragher, B., and Williamson, J.R. (2014). A combined quantitative mass spectrometry and electron microscopy analysis of ribosomal 30S subunit assembly in E. coli. eLife *3*, doi: 10.7554/eLife.04491.
56. Maksimova, E., Kravchenko, O., Korepanov, A., and Stolboushkina, E. (2022). Protein assistants of small ribosomal subunit biogenesis in bacteria. Microorganisms *10*, doi: 10.3390/microorganisms10040747.
57. Chen, S.S., Sperling, E., Silverman, J.M., Davis, J.H., and Williamson, J.R. (2012). Measuring the dynamics of E. coli ribosome biogenesis using pulse-labeling and quantitative mass spectrometry. Mol Biosyst *8*, 3325–3334, doi: 10.1039/c2mb25310k.
58. Lindahl, L. (1975). Intermediates and time kinetics of the in vivo assembly of Escherichia coli ribosomes. J Mol Biol *92*, 15–37. 0022-2836(75)90089-3 [pii].
59. Guo, Q., Goto, S., Chen, Y., Feng, B., Xu, Y., Muto, A., Himeno, H., Deng, H., Lei, J., and Gao, N. (2013). Dissecting the in vivo assembly of the 30S ribosomal subunit reveals the role of RimM and general features of the assembly process. Nucleic Acids Res *41*, 2609–2620, doi: 10.1093/nar/gks1256
60. Li, N., Chen, Y., Guo, Q., Zhang, Y., Yuan, Y., Ma, C., Deng, H., Lei, J., and Gao, N. (2013). Cryo-EM structures of the late-stage assembly intermediates of the bacterial 50S ribosomal subunit. Nucleic Acids Res *41*, 7073–7083, doi: 10.1093/nar/gkt423
61. Ni, X., Davis, J.H., Jain, N., Razi, A., Benlekbir, S., McArthur, A.G., Rubinstein, J.L., Britton, R.A., Williamson, J.R., and Ortega, J. (2016). YphC and YsxC GTPases assist the maturation of the central protuberance, GTPase associated region and functional core of the 50S ribosomal subunit. Nucleic Acids Res *44*, 8442–8455, doi: 10.1093/nar/gkw678.
62. Rabuck-Gibbons, J.N., Popova, A.M., Greene, E.M., Cervantes, C.F., Lyumkis, D., and Williamson, J.R. (2020). SrmB rescues trapped ribosome assembly intermediates. J Mol Biol *432*, 978–990, doi: 10.1016/j.jmb.2019.12.013.
63. Nikolay, R., Hilal, T., Qin, B., Mielke, T., Burger, J., Loerke, J., Textoris-Taube, K., Nierhaus, K.H., and Spahn, C.M.T. (2018). Structural visualization of the formation and activation of the 50S ribosomal subunit during in vitro reconstitution. Mol Cell *70*, 881–893 e883, doi: 10.1016/j.molcel.2018.05.003.
64. Qin, B., Lauer, S.M., Balke, A., Vieira-Vieira, C.H., Burger, J., Mielke, T., Selbach, M., Scheerer, P., Spahn, C.M.T., and Nikolay, R. (2023). Cryo-EM captures early ribosome assembly in action. Nat Commun *14*, 898, doi: 10.1038/s41467-023-36607-9.
65. Dong, X., Doerfel, L.K., Sheng, K., Rabuck-Gibbons, J.N., Popova, A.M., Lyumkis, D., and Williamson, J.R. (2023). Near-physiological in vitro assembly of 50S ribosomes involves parallel pathways. Nucleic Acids Res *51*, 2862–2876, doi: 10.1093/nar/gkad082.
66. Nikolay, R., Hilal, T., Schmidt, S., Qin, B., Schwefel, D., Vieira-Vieira, C.H., Mielke, T., Burger, J., Loerke, J., Amikura, K., et al. (2021). Snapshots of native pre-50S ribosomes reveal a biogenesis factor network and evolutionary specialization. Mol Cell *81*, 1200–1215 e1209, doi: 10.1016/j.molcel.2021.02.006.
67. Chen, S.S., and Williamson, J.R. (2013). Characterization of the ribosome biogenesis landscape in E. coli using quantitative mass spectrometry. J Mol Biol *425*, 767–779, doi: 10.1016/j.jmb.2012.11.040
68. Gamalinda, M., and Woolford, J.L., Jr. (2015). Paradigms of ribosome synthesis: lessons learned from ribosomal proteins. Translation (Austin) *3*, e975018, doi: 10.4161/21690731.2014.975018.
69. Gor, K., and Duss, O. (2023). Emerging quantitative biochemical, structural, and biophysical methods for studying ribosome and protein-RNA complex assembly. Biomolecules *13*, doi: 10.3390/biom13050866.
70. Sheng, K., Li, N., Rabuck-Gibbons, J.N., Dong, X., Lyumkis, D., and Williamson, J.R. (2022). Unsupervised Voxel-based Segmentation reveals a Landscape of Bacterial Ribosome Large Subunit Early Assembly. bioRxiv, doi: 10.1101/2022.11.09.515851.
71. Jomaa, A., Stewart, G., Martin-Benito, J., Zielke, R., Campbell, T.L., Maddock, J.R., Brown, E.D., and Ortega, J. (2011). Understanding ribosome assembly: the structure of in vivo assembled immature 30S subunits revealed by cryo-electron microscopy. RNA *17*, 697–709, doi: 10.1261/rna.2509811.
72. Yang, Z., Guo, Q., Goto, S., Chen, Y., Li, N., Yan, K., Zhang, Y., Muto, A., Deng, H., Himeno, H., et al. (2014). Structural insights into the assembly of the 30S ribosomal subunit in vivo: functional role of S5 and location of the 17S rRNA precursor sequence. Protein Cell *5*, 394–407, doi: 10.1007/s13238-014-0044-1.
73. Leong, V., Kent, M., Jomaa, A., and Ortega, J. (2013). Escherichia coli rimM and yjeQ null strains accumulate immature 30S subunits of similar structure and protein complement. RNA *19*, 789–802, doi: 10.1261/rna.037523.112
74. Jomaa, A., Stewart, G., Mears, J.A., Kireeva, I., Brown, E.D., and Ortega, J. (2011). Cryo-electron microscopy structure of the 30S subunit in complex with the YjeQ biogenesis factor. RNA *17*, 2026–2038, doi: 10.1261/rna.2922311.
75. Guo, Q., Yuan, Y., Xu, Y., Feng, B., Liu, L., Chen, K., Sun, M., Yang, Z., Lei, J., and Gao, N. (2011). Structural basis for the function of a small GTPase RsgA on the 30S ribosomal subunit maturation revealed by cryoelectron microscopy. Proc Natl Acad Sci USA *108*, 13100–13105, doi: 10.1073/pnas.1104645108.
76. Lopez-Alonso, J.P., Kaminishi, T., Kikuchi, T., Hirata, Y., Iturrioz, I., Dhimole, N., Schedlbauer, A., Hase, Y., Goto, S., Kurita, D., et al. (2017). RsgA couples the maturation

state of the 30S ribosomal decoding center to activation of its GTPase pocket. Nucleic Acids Res *45*, 6945–6959, doi: 10.1093/nar/gkx324.

77. Datta, P.P., Wilson, D.N., Kawazoe, M., Swami, N.K., Kaminishi, T., Sharma, M.R., Booth, T.M., Takemoto, C., Fucini, P., Yokoyama, S., and Agrawal, R.K. (2007). Structural aspects of RbfA action during small ribosomal subunit assembly. Mol Cell *28*, 434–445, doi: 10.1016/j.molcel.2007.08.026.
78. Boehringer, D., O'Farrell, H.C., Rife, J.P., and Ban, N. (2012). Structural insights into methyltransferase KsgA function in 30S ribosomal subunit biogenesis. J Biol Chem *287*, 10453–10459, doi: 10.1074/jbc.M111.318121.
79. Stephan, N.C., Ries, A.B., Boehringer, D., and Ban, N. (2021). Structural basis of successive adenosine modifications by the conserved ribosomal methyltransferase KsgA. Nucleic Acids Res *49*, 6389–6398, doi: 10.1093/nar/gkab430.
80. Schedlbauer, A., Iturrioz, I., Ochoa-Lizarralde, B., Diercks, T., Lopez-Alonso, J.P., Lavin, J.L., Kaminishi, T., Capuni, R., Dhimole, N., de Astigarraga, E., et al. (2021). A conserved rRNA switch is central to decoding site maturation on the small ribosomal subunit. Sci Adv *7*, doi: 10.1126/sciadv.abf7547.
81. Sharma, M.R., Barat, C., Wilson, D.N., Booth, T.M., Kawazoe, M., Hori-Takemoto, C., Shirouzu, M., Yokoyama, S., Fucini, P., and Agrawal, R.K. (2005). Interaction of Era with the 30S ribosomal subunit implications for 30S subunit assembly. Mol Cell *18*, 319–329, doi: 10.1016/j.molcel.2005.03.028.
82. Razi, A., Davis, J.H., Hao, Y., Jahagirdar, D., Thurlow, B., Basu, K., Jain, N., Gomez-Blanco, J., Britton, R.A., Vargas, J., et al. (2019). Role of Era in assembly and homeostasis of the ribosomal small subunit. Nucleic Acids Res *47*, 8301–8317, doi: 10.1093/nar/gkz571.
83. Seffouh, A., Jain, N., Jahagirdar, D., Basu, K., Razi, A., Ni, X., Guarne, A., Britton, R.A., and Ortega, J. (2019). Structural consequences of the interaction of RbgA with a 50S ribosomal subunit assembly intermediate. Nucleic Acids Res *47*, 10414–10425, doi: 10.1093/nar/gkz770.
84. Zhang, X., Yan, K., Zhang, Y., Li, N., Ma, C., Li, Z., Zhang, Y., Feng, B., Liu, J., Sun, Y., et al. (2014). Structural insights into the function of a unique tandem GTPase EngA in bacterial ribosome assembly. Nucleic Acids Res *42*, 13430–13439, doi: 10.1093/nar/gku1135.
85. Wang, W., Li, W., Ge, X., Yan, K., Mandava, C.S., Sanyal, S., and Gao, N. (2020). Loss of a single methylation in 23S rRNA delays 50S assembly at multiple late stages and impairs translation initiation and elongation. Proc Natl Acad Sci USA *117*, 15609–15619, doi: 10.1073/pnas.1914323117.
86. Feng, B., Mandava, C.S., Guo, Q., Wang, J., Cao, W., Li, N., Zhang, Y., Zhang, Y., Wang, Z., Wu, J., et al. (2014). Structural and functional insights into the mode of action of a universally conserved Obg GTPase. PLoS Biol *12*, e1001866, doi: 10.1371/journal.pbio.1001866.
87. Razi, A., Britton, R.A., and Ortega, J. (2017). The impact of recent improvements in cryo-electron microscopy technology on the understanding of bacterial ribosome assembly. Nucleic Acids Res *45*, 1027–1040, doi: 10.1093/nar/gkw1231.
88. Connolly, K., Rife, J.P., and Culver, G. (2008). Mechanistic insight into the ribosome biogenesis functions of the ancient protein KsgA. Mol Microbiol *70*, 1062–1075, doi: 10.1111/j.1365-2958.2008.06485.x.
89. Inoue, K., Basu, S., and Inouye, M. (2007). Dissection of 16S rRNA methyltransferase (KsgA) function in Escherichia coli. J Bacteriol *189*, 8510–8518, doi: 10.1128/JB.01259-07.
90. Hopfield, J.J. (1974). Kinetic proofreading: a new mechanism for reducing errors in biosynthetic processes requiring high specificity. Proc Natl Acad Sci USA *71*, 4135–4139, doi: 10.1073/pnas.71.10.4135.
91. Frank, J., and Gonzalez, R.L., Jr. (2010). Structure and dynamics of a processive Brownian motor: the translating ribosome. Annu Rev Biochem *79*, 381–412, doi: 10.1146/annurev-biochem-060408-173330.
92. Mitra, K., and Frank, J. (2006). Ribosome dynamics: insights from atomic structure modeling into cryo-electron microscopy maps. Annu Rev Biophys Biomol Struct *35*, 299–317.
93. Valle, M., Zavialov, A., Sengupta, J., Rawat, U., Ehrenberg, M., and Frank, J. (2003). Locking and unlocking of ribosomal motions. Cell *114*, 123–134.
94. Gao, H., Sengupta, J., Valle, M., Korostelev, A., Eswar, N., Stagg, S.M., Van Roey, P., Agrawal, R.K., Harvey, S.C., Sali, A., et al. (2003). Study of the structural dynamics of the E coli 70S ribosome using real-space refinement. Cell *113*, 789–801.
95. Frank, J., and Agrawal, R.K. (2000). A ratchet-like intersubunit reorganization of the ribosome during translocation. Nature *406*, 318–322.
96. Mohan, S., Donohue, J.P., and Noller, H.F. (2014). Molecular mechanics of 30S subunit head rotation. Proc Natl Acad Sci USA *111*, 13325–13330, doi: 10.1073/pnas.1413731111.
97. Valle, M., Zavialov, A., Li, W., Stagg, S.M., Sengupta, J., Nielsen, R.C., Nissen, P., Harvey, S.C., Ehrenberg, M., and Frank, J. (2003). Incorporation of aminoacyl-tRNA into the ribosome as seen by cryo-electron microscopy. Nat Struct Biol *10*, 899–906.
98. Seidelt, B., Innis, C.A., Wilson, D.N., Gartmann, M., Armache, J.P., Villa, E., Trabuco, L.G., Becker, T., Mielke, T., Schulten, K., et al. (2009). Structural insight into nascent polypeptide chain-mediated translational stalling. Science *326*, 1412–1415, doi: 10.1126/science.1177662.
99. Mitra, K., Schaffitzel, C., Shaikh, T., Tama, F., Jenni, S., Brooks, C.L., 3rd, Ban, N., and Frank, J. (2005). Structure of the E. coli protein-conducting channel bound to a translating ribosome. Nature *438*, 318–324, doi: 10.1038/nature04133.
100. Ermolenko, D.N., Majumdar, Z.K., Hickerson, R.P., Spiegel, P.C., Clegg, R.M., and Noller, H.F. (2007). Observation of intersubunit movement of the ribosome in solution using FRET. J Mol Biol *370*, 530–540, doi: 10.1016/j.jmb.2007.04.042.
101. Moazed, D., and Noller, H.F. (1989). Intermediate states in the movement of transfer RNA in the ribosome. Nature *342*, 142–148, doi: 10.1038/342142a0.
102. Munro, J.B., Altman, R.B., O'Connor, N., and Blanchard, S.C. (2007). Identification of two distinct hybrid state intermediates on the ribosome. Mol Cell *25*, 505–517, doi: 10.1016/j.molcel.2007.01.022.
103. Pan, D., Kirillov, S.V., and Cooperman, B.S. (2007). Kinetically competent intermediates in the translocation step of protein synthesis. Mol Cell *25*, 519–529, doi: 10.1016/j.molcel.2007.01.014.
104. Agrawal, R.K., Penczek, P., Grassucci, R.A., and Frank, J. (1998). Visualization of elongation factor G on the Escherichia coli 70S ribosome: the mechanism of translocation. Proc Natl Acad Sci USA *95*, 6134–6138.

105. Agrawal, R.K., Penczek, P., Grassucci, R.A., Burkhardt, N., Nierhaus, K.H., and Frank, J. (1999). Effect of buffer conditions on the position of tRNA on the 70 S ribosome as visualized by cryoelectron microscopy. J Biol Chem *274*, 8723–8729.
106. Allen, G.S., Zavialov, A., Gursky, R., Ehrenberg, M., and Frank, J. (2005). The cryo-EM structure of a translation initiation complex from Escherichia coli. Cell *121*, 703–712, doi: 10.1016/j.cell.2005.03.023.
107. Gao, H., Zhou, Z., Rawat, U., Huang, C., Bouakaz, L., Wang, C., Cheng, Z., Liu, Y., Zavialov, A., Gursky, R., et al. (2007). RF3 induces ribosomal conformational changes responsible for dissociation of class I release factors. Cell *129*, 929–941, doi: 10.1016/j.cell.2007.03.050.
108. Frank, J., Sengupta, J., Gao, H., Li, W., Valle, M., Zavialov, A., and Ehrenberg, M. (2005). The role of tRNA as a molecular spring in decoding, accommodation, and peptidyl transfer. FEBS Lett *579*, 959–962.
109. Bai, X.C., McMullan, G., and Scheres, S.H. (2015). How cryo-EM is revolutionizing structural biology. Trends Biochem Sci *40*, 49–57, doi: 10.1016/j.tibs.2014.10.005.
110. Fischer, N., Neumann, P., Konevega, A.L., Bock, L.V., Ficner, R., Rodnina, M.V., and Stark, H. (2015). Structure of the E. coli ribosome-EF-Tu complex at <3 A resolution by Cs-corrected cryo-EM. Nature *520*, 567–570, doi: 10.1038/nature14275.
111. Arenz, S., Nguyen, F., Beckmann, R., and Wilson, D.N. (2015). Cryo-EM structure of the tetracycline resistance protein TetM in complex with a translating ribosome at 3.9-A resolution. Proc Natl Acad Sci USA *112*, 5401–5406, doi: 10.1073/pnas.1501775112.
112. Florin, T., Maracci, C., Graf, M., Karki, P., Klepacki, D., Berninghausen, O., Beckmann, R., Vazquez-Laslop, N., Wilson, D.N., Rodnina, M.V., and Mankin, A.S. (2017). An antimicrobial peptide that inhibits translation by trapping release factors on the ribosome. Nat Struct Mol Biol *24*, 752–757, doi: 10.1038/nsmb.3439.
113. Paternoga, H., Crowe-McAuliffe, C., Bock, L.V., Koller, T.O., Morici, M., Beckert, B., Myasnikov, A.G., Grubmuller, H., Novacek, J., and Wilson, D.N. (2023). Structural conservation of antibiotic interaction with ribosomes. Nat Struct Mol Biol, doi: 10.1038/s41594-023-01047-y.
114. Fromm, S.A., O'Connor, K.M., Purdy, M., Bhatt, P.R., Loughran, G., Atkins, J.F., Jomaa, A., and Mattei, S. (2023). The translating bacterial ribosome at 1.55 A resolution generated by cryo-EM imaging services. Nat Commun *14*, 1095, doi: 10.1038/s41467-023-36742-3.
115. Herr, A.J., Nelson, C.C., Wills, N.M., Gesteland, R.F., and Atkins, J.F. (2001). Analysis of the roles of tRNA structure, ribosomal protein L9, and the bacteriophage T4 gene 60 bypassing signals during ribosome slippage on mRNA. J Mol Biol *309*, 1029–1048, doi: 10.1006/jmbi.2001.4717.
116. Smith, A.M., Costello, M.S., Kettring, A.H., Wingo, R.J., and Moore, S.D. (2019). Ribosome collisions alter frameshifting at translational reprogramming motifs in bacterial mRNAs. Proc Natl Acad Sci USA *116*, 21769–21779, doi: 10.1073/pnas.1910613116.
117. Atkins, J.F., and Bjork, G.R. (2009). A gripping tale of ribosomal frameshifting: extragenic suppressors of frameshift mutations spotlight P-site realignment. Microbiol Mol Biol Rev *73*, 178–210, doi: 10.1128/MMBR.00010-08.
118. Xue, L., Lenz, S., Zimmermann-Kogadeeva, M., Tegunov, D., Cramer, P., Bork, P., Rappsilber, J., and Mahamid, J. (2022). Visualizing translation dynamics at atomic detail inside a bacterial cell. Nature *610*, 205–211, doi: 10.1038/s41586-022-05255-2.
119. Cerullo, F., Filbeck, S., Patil, P.R., Hung, H.C., Xu, H., Vornberger, J., Hofer, F.W., Schmitt, J., Kramer, G., Bukau, B., et al. (2022). Bacterial ribosome collision sensing by a MutS DNA repair ATPase paralogue. Nature *603*, 509–514, doi: 10.1038/s41586-022-04487-6.
120. Saito, K., Kratzat, H., Campbell, A., Buschauer, R., Burroughs, A.M., Berninghausen, O., Aravind, L., Green, R., Beckmann, R., and Buskirk, A.R. (2022). Ribosome collisions induce mRNA cleavage and ribosome rescue in bacteria. Nature *603*, 503–508, doi: 10.1038/s41586-022-04416-7.
121. Noeske, J., Wasserman, M.R., Terry, D.S., Altman, R.B., Blanchard, S.C., and Cate, J.H. (2015). High-resolution structure of the Escherichia coli ribosome. Nat Struct Mol Biol *22*, 336–341, doi: 10.1038/nsmb.2994.
122. Dunkle, J.A., and Cate, J.H.D. (2011). The packing of ribosomes in crystals and polysomes. In Ribosomes: Structure, Function, and Dynamics, M.V. Rodnina, W. Wintermeyer, and R. Green, eds. (Springer Vienna), pp. 65–73, doi: 10.1007/978-3-7091-0215-2_6.
123. Yusupov, M.M., Yusupova, G.Z., Baucom, A., Lieberman, K., Earnest, T.N., Cate, J.H., and Noller, H.F. (2001). Crystal structure of the ribosome at 5.5 A resolution. Science *292*, 883–896.
124. Loveland, A.B., Demo, G., Grigorieff, N., and Korostelev, A.A. (2017). Ensemble cryo-EM elucidates the mechanism of translation fidelity. Nature *546*, 113–117, doi: 10.1038/nature22397.
125. Maeots, M.E., and Enchev, R.I. (2022). Structural dynamics: review of time-resolved cryo-EM. Acta Crystallogr D Struct Biol *78*, 927–935, doi: 10.1107/S2059798322006155.
126. Carbone, C.E., Loveland, A.B., Gamper, H.B., Jr., Hou, Y.M., Demo, G., and Korostelev, A.A. (2021). Time-resolved cryo-EM visualizes ribosomal translocation with EF-G and GTP. Nat Commun *12*, 7236, doi: 10.1038/s41467-021-27415-0.
127. Dandey, V.P., Budell, W.C., Wei, H., Bobe, D., Maruthi, K., Kopylov, M., Eng, E.T., Kahn, P.A., Hinshaw, J.E., Kundu, N., et al. (2020). Time-resolved cryo-EM using Spotiton. Nat Methods *17*, 897–900, doi: 10.1038/s41592-020-0925-6.
128. Kaledhonkar, S., Fu, Z., White, H., and Frank, J. (2018). Time-resolved cryo-electron microscopy using a microfluidic chip. Methods Mol Biol *1764*, 59–71, doi: 10.1007/978-1-4939-7759-8_4.
129. Kaledhonkar, S., Fu, Z., Caban, K., Li, W., Chen, B., Sun, M., Gonzalez, R.L., Jr., and Frank, J. (2019). Late steps in bacterial translation initiation visualized using time-resolved cryo-EM. Nature *570*, 400–404, doi: 10.1038/s41586-019-1249-5.
130. Fu, Z., Indrisiunaite, G., Kaledhonkar, S., Shah, B., Sun, M., Chen, B., Grassucci, R.A., Ehrenberg, M., and Frank, J. (2019). The structural basis for release-factor activation during translation termination revealed by time-resolved cryogenic electron microscopy. Nat Commun *10*, 2579, doi: 10.1038/s41467-019-10608-z.
131. James, N.R., Brown, A., Gordiyenko, Y., and Ramakrishnan, V. (2016). Translational termination without a stop codon. Science *354*, 1437–1440, doi: 10.1126/science.aai9127.
132. Ishihama, A. (1999). Modulation of the nucleoid, the transcription apparatus, and the translation machinery in bacteria for stationary phase survival. Genes Cells *4*, 135–143, doi: 10.1046/j.1365-2443.1999.00247.x.
133. Yamagishi, M., Matsushima, H., Wada, A., Sakagami, M., Fujita, N., and Ishihama, A. (1993). Regulation of the Escherichia coli rmf gene encoding the ribosome modulation factor: growth phase- and growth rate-dependent control. EMBO J *12*, 625–630, doi: 10.1002/j.1460-2075.1993.tb05695.x.

134. Polikanov, Y.S., Blaha, G.M., and Steitz, T.A. (2012). How hibernation factors RMF, HPF, and YfiA turn off protein synthesis. Science *336*, 915–918, doi: 10.1126/science.1218538.
135. An, H., and Harper, J.W. (2018). Systematic analysis of ribophagy in human cells reveals bystander flux during selective autophagy. Nat Cell Biol *20*, 135–143, doi: 10.1038/s41556-017-0007-x.
136. Kato, T., Yoshida, H., Miyata, T., Maki, Y., Wada, A., and Namba, K. (2010). Structure of the 100S ribosome in the hibernation stage revealed by electron cryomicroscopy. Structure *18*, 719–724, doi: 10.1016/j.str.2010.02.017.
137. Ortiz, J.O., Brandt, F., Matias, V.R., Sennels, L., Rappsilber, J., Scheres, S.H., Eibauer, M., Hartl, F.U., and Baumeister, W. (2010). Structure of hibernating ribosomes studied by cryoelectron tomography in vitro and in situ. J Cell Biol *190*, 613–621, doi: 10.1083/jcb.201005007.
138. Matzov, D., Aibara, S., Basu, A., Zimmerman, E., Bashan, A., Yap, M.F., Amunts, A., and Yonath, A.E. (2017). The cryo-EM structure of hibernating 100S ribosome dimer from pathogenic Staphylococcus aureus. Nat Commun *8*, 723, doi: 10.1038/s41467-017-00753-8.
139. Krokowski, D., Gaccioli, F., Majumder, M., Mullins, M.R., Yuan, C.L., Papadopoulou, B., Merrick, W.C., Komar, A.A., Taylor, D., and Hatzoglou, M. (2011). Characterization of hibernating ribosomes in mammalian cells. Cell Cycle *10*, 2691–2702, doi: 10.4161/cc.10.16.16844.
140. Asano, S., Engel, B.D., and Baumeister, W. (2016). In situ cryo-electron tomography: a post-reductionist approach to structural biology. J Mol Biol *428*, 332–343, doi: 10.1016/j.jmb.2015.09.030.
141. Obr, M., Hagen, W.J.H., Dick, R.A., Yu, L., Kotecha, A., and Schur, F.K.M. (2022). Exploring high-resolution cryo-ET and subtomogram averaging capabilities of contemporary DEDs. J Struct Biol *214*, 107852, doi: 10.1016/j.jsb.2022.107852.
142. Wagner, F.R., Watanabe, R., Schampers, R., Singh, D., Persoon, H., Schaffer, M., Fruhstorfer, P., Plitzko, J., and Villa, E. (2020). Preparing samples from whole cells using focused-ion-beam milling for cryo-electron tomography. Nat Protoc *15*, 2041–2070, doi: 10.1038/s41596-020-0320-x.
143. Hagen, W.J.H., Wan, W., and Briggs, J.A.G. (2017). Implementation of a cryo-electron tomography tilt-scheme optimized for high resolution subtomogram averaging. J Struct Biol *197*, 191–198, doi: 10.1016/j.jsb.2016.06.007.
144. Eisenstein, F., Yanagisawa, H., Kashihara, H., Kikkawa, M., Tsukita, S., and Danev, R. (2023). Parallel cryo electron tomography on in situ lamellae. Nat Methods *20*, 131–138, doi: 10.1038/s41592-022-01690-1.
145. Peck, A., Carter, S.D., Mai, H., Chen, S., Burt, A., and Jensen, G.J. (2022). Montage electron tomography of vitrified specimens. J Struct Biol *214*, 107860, doi: 10.1016/j.jsb.2022.107860.
146. de Teresa-Trueba, I., Goetz, S.K., Mattausch, A., Stojanovska, F., Zimmerli, C.E., Toro-Nahuelpan, M., Cheng, D.W.C., Tollervey, F., Pape, C., Beck, M., et al. (2023). Convolutional networks for supervised mining of molecular patterns within cellular context. Nat Methods *20*, 284–294, doi: 10.1038/s41592-022-01746-2.
147. Moebel, E., Martinez-Sanchez, A., Lamm, L., Righetto, R.D., Wietrzynski, W., Albert, S., Lariviere, D., Fourmentin, E., Pfeffer, S., Ortiz, J., et al. (2021). Deep learning improves macromolecule identification in 3D cellular cryo-electron tomograms. Nat Methods *18*, 1386–1394, doi: 10.1038/s41592-021-01275-4.
148. Rice, G., Wagner, T., Stabrin, M., Sitsel, O., Prumbaum, D., and Raunser, S. (2023). TomoTwin: generalized 3D localization of macromolecules in cryo-electron tomograms with structural data mining. Nat Methods *20*, 871–880, doi: 10.1038/s41592-023-01878-z.
149. Tegunov, D., and Cramer, P. (2019). Real-time cryo-electron microscopy data preprocessing with Warp. Nat Methods *16*, 1146–1152, doi: 10.1038/s41592-019-0580-y.
150. Castano-Diez, D., Kudryashev, M., Arheit, M., and Stahlberg, H. (2012). Dynamo: a flexible, user-friendly development tool for subtomogram averaging of cryo-EM data in high-performance computing environments. J Struct Biol *178*, 139–151, doi: 10.1016/j.jsb.2011.12.017.
151. Himes, B.A., and Zhang, P. (2018). emClarity: software for high-resolution cryo-electron tomography and subtomogram averaging. Nat Methods *15*, 955–961, doi: 10.1038/s41592-018-0167-z.
152. Zivanov, J., Oton, J., Ke, Z., von Kugelgen, A., Pyle, E., Qu, K., Morado, D., Castano-Diez, D., Zanetti, G., Bharat, T.A.M., et al. (2022). A Bayesian approach to single-particle electron cryo-tomography in RELION-4.0. eLife *11*, doi: 10.7554/eLife.83724.
153. Chen, M., Bell, J.M., Shi, X., Sun, S.Y., Wang, Z., and Ludtke, S.J. (2019). A complete data processing workflow for cryo-ET and subtomogram averaging. Nat Methods *16*, 1161–1168, doi: 10.1038/s41592-019-0591-8.
154. Vuillemot, R., Rouiller, I., and Jonic, S. (2023). MDTOMO method for continuous conformational variability analysis in cryo electron subtomograms based on molecular dynamics simulations. Sci Rep *13*, 10596, doi: 10.1038/s41598-023-37037-9.
155. Harastani, M., Eltsov, M., Leforestier, A., and Jonic, S. (2022). TomoFlow: analysis of continuous conformational variability of macromolecules in cryogenic subtomograms based on 3D dense optical flow. J Mol Biol *434*, 167381, doi: 10.1016/j.jmb.2021.167381.
156. Harastani, M., Eltsov, M., Leforestier, A., and Jonic, S. (2021). HEMNMA-3D: cryo electron tomography method based on normal mode analysis to study continuous conformational variability of macromolecular complexes. Front Mol Biosci *8*, 663121, doi: 10.3389/fmolb.2021.663121.
157. Frank, G.A., Bartesaghi, A., Kuybeda, O., Borgnia, M.J., White, T.A., Sapiro, G., and Subramaniam, S. (2012). Computational separation of conformational heterogeneity using cryo-electron tomography and 3D sub-volume averaging. J Struct Biol *178*, 165–176, doi: 10.1016/j.jsb.2012.01.004.
158. Macgregor, H. (2012). Oscar Miller (1925–2012)—the technological genius of chromosome science. Chromosome Res *20*, 317–318, doi: 10.1007/s10577-012-9278-z.
159. Hoffmann, P.C., Kreysing, J.P., Khusainov, I., Tuijtel, M.W., Welsch, S., and Beck, M. (2022). Structures of the eukaryotic ribosome and its translational states in situ. Nat Commun *13*, 7435, doi: 10.1038/s41467-022-34997-w.
160. Gemmer, M., Chaillet, M.L., van Loenhout, J., Cuevas Arenas, R., Vismpas, D., Grollers-Mulderij, M., Koh, F.A., Albanese, P., Scheltema, R.A., Howes, S.C., et al. (2023). Visualization of translation and protein biogenesis at the ER membrane. Nature *614*, 160–167, doi: 10.1038/s41586-022-05638-5.

24 Cryo-EM in the Study of Ribosome Biogenesis

Ankita Arora and Prem S. Kaushal

INTRODUCTION

Ribosomes are large macromolecular complexes that perform the critical cellular protein synthesis process. In the prokaryote, the ribosome is known to sediment as a 70S particle composed of two subunits: a small subunit (30S) and a large subunit (50S) (Figure 24.1). The *Escherichia coli* (*E. coli*) small subunit consists of 16S ribosomal RNA (r-RNA) 1542 nucleotides long and 21 ribosomal proteins (r-proteins).

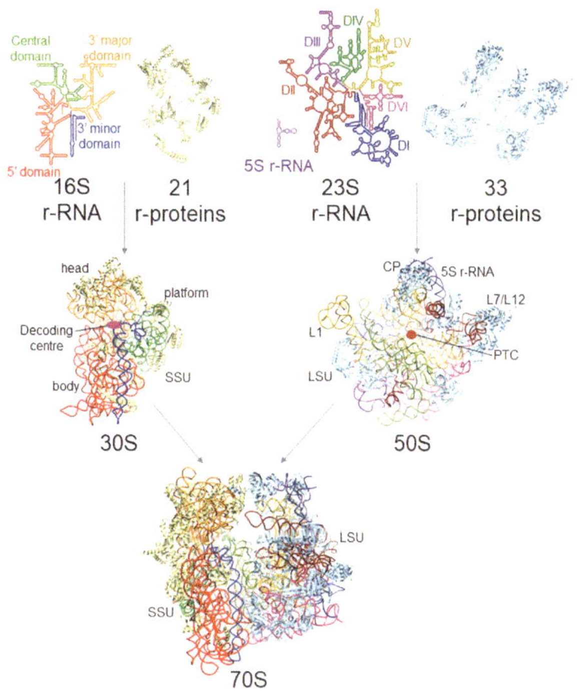

FIGURE 24.1 Flow diagram of ribosomal assembly. The SSU r-protein (khaki) and 2D map of 16S rRNA (5′ domain in red, central domain in forest green, 3′ major domain in orange, and 3′ minor domain in blue) are shown in the top left panel, and the LSU r-protein (sky blue) and 2D map of 23S (domain I in navy blue, domain II in dark red, domain III in purple, domain IV in olive drab, domain V in goldenrod, and domain VI in deep pink) and 5S rRNA (light grey) are shown in the top right panel. The assembled SSU (the 30S) is shown in the middle left panel, and the head, platform, and body of SSU are labeled. The assembled LSU (the 50S) is shown in the middle right panel, and the CP (central protuberance), L1 stalk, and L7/L12 stalk are labeled. A fully assembled 70S ribosome having both LSU and SSU is shown in the bottom panel. All figures are prepared using coordinates of PDB ID; 7N1P and program UCSF CHIMERA (https://www.cgl.ucsf.edu/chimera/).

The large subunit contains two rRNAs, 2904-nucleotide-long 23S rRNA and 120-nucleotide-long 5S rRNA, and 33 r-proteins (Figure 24.1). Each subunit harbors three tRNA binding sites, the A-site (aminoacyl), where tRNA with new incoming amino acid binds, the P-site (peptidyl), where tRNA with the growing polypeptide chain remains bound, and the E-site (exit), where deacylated tRNA remains bound to the ribosome before exiting the ribosome. The 30S ribosome subunit harbors the decoding center where mRNA decoding occurs during the protein synthesis process. The 50S ribosome subunit harbors a peptidyl transferase center, providing the site for peptide bond formation (Schmeing & Ramakrishnan, 2009) (Figure 24.1).

Assembly of both ribosomal subunits to form a functional ribosomal complex occurs by a process of ribosome biogenesis, a highly coordinated, hierarchical, and energetically expensive cellular process. Ribosome biogenesis involves a series of steps, including transcription of pre-ribosomal RNA, processing, posttranscriptional modifications, and folding of rRNA. Subsequently translation, posttranslational modifications, and folding and binding of numerous r-proteins also happens. Apart from the above steps, various assembly/biogenesis factors are also required for the maturation of both ribosomal subunits (Shajani et al., 2011). Many of these steps are coupled together and occur in parallel with rRNA transcription, which proceeds via the formation of various intermediate states having differences in RNA conformation. In bacteria, ribosome biogenesis begins with the transcription of primary rRNA transcript with the help of RNA polymerase (RNAP) (Figure 24.2). This primary transcript is cleaved by RNase III to generate three precursor rRNAs (17S, pre-23S, and 9S rRNAs) and tRNAs. Further, these precursor rRNAs are excised by various endo- and exo-nucleases to generate mature 16S, 23S, and 5S rRNAs (Figure 24.2). This mature 16S rRNA, along with small subunit r-proteins, assembles to form 30S ribosome subunits. The 23S and 5S rRNAs, along with large subunit r-proteins, assemble to form 50S ribosome subunits (Kaczanowska & Rydén-Aulin, 2007) (Figure 24.1).

In the last two decades, much progress has been made in the field to understand ribosome structure and protein synthesis. The high-resolution structures of ribosomes at different stages of protein synthesis—initiation, elongation, termination, and recycling—are now available (Rodnina, 2018; Schmeing & Ramakrishnan, 2009). However, these structures have provided little information on the process of ribosome biogenesis, due to the complex nature of this ribosome biogenesis process. Earlier studies of the last

DOI: 10.1201/9781003326106-29

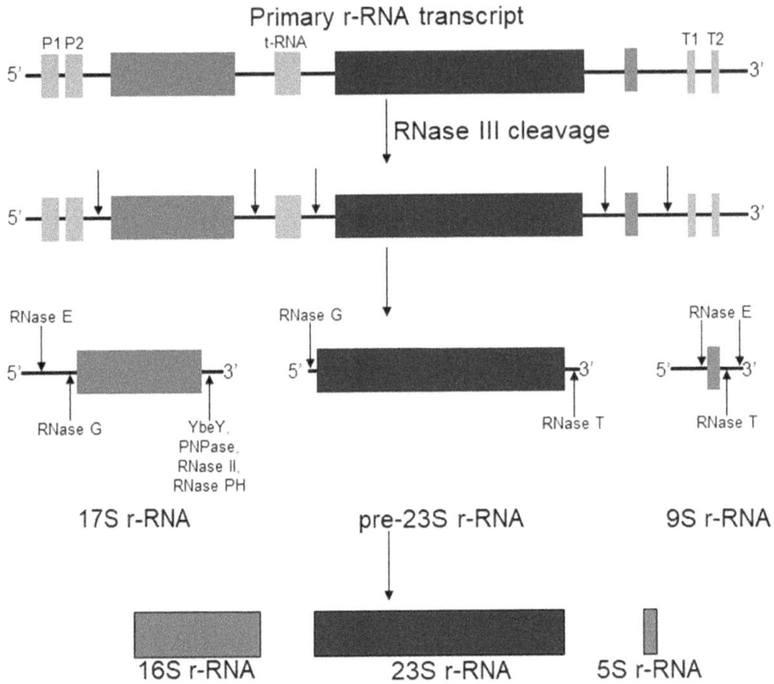

FIGURE 24.2 Schematic representation of rRNA maturation process. The process begins with the transcription of rRNA by RNA polymerase into primary transcript encoding 16S, tRNA, 23S and 5S rRNA. Further nucleolytic cleavage of the primary transcript by RNase III occurs (site of cleavage highlighted in black arrow) to produce 17S (pre-16S), pre-23S, and 9S (pre-5S) rRNA. Subsequent cleavage (highlighted in black arrow) occurs to produce mature rRNAs (16S, 23S, and 5S) by various ribonucleases (Bennison et al., 2019).

century were mainly focused on determining the various components of ribosomes. However, understanding the ribosome assembly began in the late 1960s when Nomura and Nierhaus presented the first ribosome assembly maps by reconstitution of functional ribosomes from free rRNAs and r-proteins. Later, *in vitro* reconstitution of ribosomal subunits from purified components of *E. coli* was done to classify these r-proteins into primary, secondary, and tertiary binding proteins based on their order of binding to rRNA. These reconstitution experiments were carried out in non-physiological conditions and the hierarchical binding order was called Nomura's assembly map (Nomura et al., 1969; Rohl & Nierhaus, 1982; Traub & Nomura, 1968b, 1968a, 1969a, 1969b). These studies provided a view of the process and the binding order of ribosomal proteins under *in vitro* conditions, but the structures provided by these studies were of low resolution and were often described as "blob-ology." Cryo-EM was in its infancy at this time and was not suitable for determining the structure of macromolecules at atomic resolutions (Smith & Rubinstein, 2014).

In bacteria, this process generally occurs within 2 min under rapid growth conditions and in the presence of various accessory factors, known as ribosome biogenesis/assembly factors, that assist in the process of ribosome biogenesis. In the last decade, there has been an increase in studies focusing on ribosome assembly factors and their role in ribosome biogenesis using biochemical, genetics, mass spectrometry, RNA analysis, and structural studies (Davis & Williamson, 2017). However, for most of these factors, their precise function and binding sites remained unexplored until recently.

The process of ribosome biogenesis is highly complex and dynamic, involving different conformational and compositional changes due to the incorporation of r-proteins to rRNA. These constant changes pose a major challenge for studying these intermediate states through X-ray crystallography. In recent years, with the advent of the resolution revolution in cryo-EM, studies of ribosome biogenesis have allowed visualization of this dynamic process at high resolution (Carugo & Djinović-Carugo, 2023; Kühlbrandt, 2014). Recent developments in the field of cryo-EM have helped to solve the structures at atomic resolution, thus making cryo-EM an important tool for gaining structural insights into ribosome biogenesis (Razi, Britton, et al., 2017; Smith & Rubinstein, 2014). Cryo-EM has not only abruptly increased our understanding of protein synthesis and ribosome biogenesis, but it has equally increased the understating of other cellular processes, heralding a golden era in structural biology (Carugo & Djinović-Carugo, 2023). In this chapter, we primarily focus on the role of cryo-EM in understanding the prokaryotic ribosome biogenesis process. Starting with a brief overview of the early work with the *in vitro* assembly intermediates, we follow by describing the latest advances made in understanding the assembly process. We also describe the recent cryo-EM complexes of the biogenesis factors and their role in incrementally driving the conformations into fully mature subunits.

A BRIEF ACCOUNT OF RIBOSOME ASSEMBLY

The ribosome is assembled via many coordinated steps of protein and rRNA binding, followed by conformational changes in both. In prokaryotes, ribosome assembly begins with the maturation of the 30S ribosome subunit (SSU) and the 50S ribosome subunit (LSU). These two mature subunits, the SSU and the LSU, join to form a functional 70S ribosome (Figure 24.1).

30S RIBOSOME SUBUNIT ASSEMBLY

The 16S rRNA of the 30S ribosome subunit is structurally divided into three major domains: domain I consisting of the 5′ end of rRNA, domain II formed by the central part, and domain III consisting of the 3′ end of rRNA (Noller & Woese, 1981). The 30S ribosomal subunit can be assembled under *in vitro* conditions using 16S rRNA and a mixture of 21 r-proteins (Figures 24.1 and 24.3a).

In vitro reconstitution of the 30S ribosome subunit was studied by early works of Nomura; it proceeds via a three-step reaction in the 5′–3′ direction. It begins with binding of primary (S4, S7, S8, S15, S17, S20) and secondary (S5, S6, S9, S11, S12, S13, S16, S18, and S19) r-proteins to 16S rRNA at low temperature (0°C to 15°C), resulting in sedimentation of the reconstitution intermediate (RI) particle at 21S. Changing the temperature to 42°C results in 21S intermediately undergoing conformational rearrangement to form a 26S intermediate particle (RI*). Subsequent binding of tertiary (S2, S3, S10, S14, and S21) proteins to this intermediate leads to the assembly of a fully mature 30S ribosome subunit (Held & Nomura, 1973; Mizushima & Nomura, 1970; Nomura, 1973; Traub & Nomura, 1968a, 1968b, 1969a, 1969b). The overall 30S ribosome assembly map was provided by Nomura's experiment, but there are various drawbacks with this map. First, it provided only r-protein dependency information rather than the detailed pathway. Second, it doesn't reveal the energy kinetics of protein binding. Lastly, due to a lack of structural information, the stepwise process of 16S rRNA folding and r-protein binding was also not revealed.

Further, *in vivo* studies provided an overall view of 30S ribosomal assembly by utilizing a combination of radioactive pulse labeling of ribosomes with quantitative mass spectrometry. These studies revealed the sequential order of binding of ribosomal proteins and grouped them into four categories in which the initial binders, such as S4, S6, S8, S15, S16, S17, S18, and S20 proteins, bind to the 5′ and central domains of 16S rRNA. Further, S7 and S11 proteins

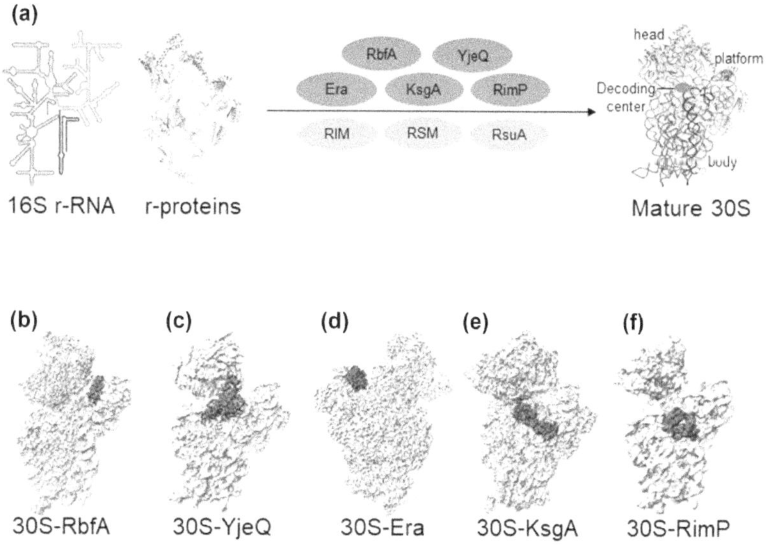

FIGURE 24.3 Assembly factors of the ribosomal small subunit. (a) The 2D diagram of partially assembled 16S rRNA along with SSU r-proteins (ribbon structure) is shown on the left, and the fully assembled 30S SSU is shown on the right. The assembly factors: RbfA, YjeQ, Era, KsgA, RimP (dark grey color circle), RIM—ribosome maturation factor (RimI, RimJ, RimM, RimN), RSM—ribosomal RNA small subunit methyltransferase (RsmB, RsmC, RsmD, RsmE, RsmF, RsmG, RsmH), RusA (light grey color circle) involved in this assembly process are listed. Light grey circle represents the factors whose structural information is not available, while the dark grey circle represents the factors whose cryo-EM structural information is available. (b) Cryo-EM structure of 30S ribosome (light grey) in complex with RbfA (dim grey) (EMDB ID; 12243). (c) Cryo-EM structure of 30S ribosome (light grey) in complex with YjeQ (dim grey) (EMDB ID; 12245). (d) Cryo-EM structure of 30S ribosome (light grey) in complex with Era (dim grey) (EMDB ID; 0484). (e) Cryo-EM structure of 30S ribosome (light grey) in complex with KsgA (dim grey) (EMDB ID; 12736). (f) Cryo-EM structure of 30S ribosome (light grey) in complex with RimP (dim grey) (EMDB ID; 12249).

attach to the central and 3′ domains, respectively, whereas S5 and S12 proteins bind to the 5′ domain. The rate of binding of S5 and S12 is slower than in other proteins, and they do not follow the 5′ to 3′ direction binding as these proteins are known to interact at the intercross of all the domains of 16S rRNA. S9, S10, S13, S14, and S19 proteins are the third set of proteins that bind to the complex. Lastly, a fourth set of binders, namely S2, S3, and S21 proteins, bind. S1, the last protein that interacts with the 16S rRNA–protein complex, completes the assembly of the 30S ribosome subunit (Chen & Williamson, 2013). Because of the lack of technology, further understanding of assembly in atomic details could not be achieved.

In vitro assembly of matured ribosomal subunits can occur from its major building blocks, such as r-proteins and rRNAs. However, it is much slower than the *in vivo* assembly process and requires non-physiological temperatures and ionic concentrations. A group of ribosome auxiliary factors/ribosome assembly factors has been identified that assist in ribosomal subunit assembly (Table 24.1) in *in vitro* conditions. These factors help in assisting ribosome assembly by preventing the kinetic traps that occur during RNA misfolding and blocking the binding of r-proteins to premature ribosomal subunits (Davis & Williamson, 2017). Therefore, these factors serve as checkpoint sensors for the ribosome assembly process (Sergeeva et al., 2014). These assembly factors belong to various types: RNA processing enzymes, protein and RNA modifying enzymes, RNA helicases, chaperones, maturation factors, and GTPases (Davis & Williamson, 2017; Kaczanowska & Rydén-Aulin, 2007; Shajani et al., 2011). A brief function of each known assembly/biogenesis factor is listed in Table 24.1.

Assembly Factors of 30S Ribosome Subunit

The cryo-EM structures of assembly factors in complex with the 30S ribosome subunit that resolved at better than 5 Å resolution are described here in detail. Other known assembly factors are listed in Table 24.1.

TABLE 24.1
Biogenesis factors involved in small ribosomal subunit maturation

	Factor	Function	EMDB	PDB	Resolution	Reference
GTPase	Era	Protects anti-SD sequences from cleavage during 3′ end processing of 16S rRNA, inhibits premature translation process	0484	6NQB	3.5 Å	(Razi et al., 2019)
	YjeQ/ RsgA	Binds to 30S ribosome subunit by interacting with h18 and h44 of decoding center	12245	7NAR	3.0 Å	(Schedlbauer et al., 2021)
r-protein modification	RimI	Acetylates the N-terminal alanine of ribosomal protein S18	-	-	-	(Connolly & Culver, 2009)
	RimJ	Acetylates the N-terminal alanine of S5 protein	-	-	-	(Shajani et al., 2011)
r-RNA modification	RsuA	Synthesis of pseudouridine at U516 of 16S rRNA	-	-	-	(Connolly & Culver, 2009)
	KsgA/ RsmA	Methylates two adjacent adenosine residues of 16S rRNA, prevents immature 30S ribosome from entering into translation cycle	12736	7O5H	3.1 Å	(Stephan et al., 2021b)
	RsmB	Methylates C967 of 16S rRNA	-	-	-	(Kaczanowska & Rydén-Aulin, 2007)
	RsmC	Methylates G1207 of 16S rRNA	-	-	-	(Kaczanowska & Rydén-Aulin, 2007)
	RsmD	Methylates G966 of 16S rRNA	-	-	-	(Kaczanowska & Rydén-Aulin, 2007)
	RsmE	Acts on fully mature 30S ribosome to methylate N3 position of U1498	-	-	-	(Connolly & Culver, 2009)
	RsmF	Methylates C1407 of 16S rRNA	-	-	-	(Kaczanowska & Rydén-Aulin, 2007)
	RsmG	Acts on fully intact 30S ribosome to methylate N7 position of G527	-	-	-	(Connolly & Culver, 2009)
	RsmH	Acts on assembled 30S ribosome to specifically methylate N4 position of C1402	-	-	-	(Kimura & Suzuki, 2009)
	RsmI	Helps in 2′-O methylation of C1402 of 16S rRNA	-	-	-	(Zhao et al., 2016)
	RsmJ	Methylates G1516 of 16S rRNA	-	-	-	(Basturea et al., 2012)
Others	RimM	Facilitates maturation of 3′ end of 16S rRNA	-	-	-	(Shajani et al., 2011)
	RimN	Promotes processing of 16S rRNA	-	-	-	(Kaczanowska & Rydén-Aulin, 2005)
	RimP	Accelerates binding of S5 and S12 protein to 16S rRNA, promotes stabilization of central pseudoknot	12249	7NAV	4.8 Å	(Schedlbauer et al., 2021)
	RbfA/ YhbB	Interacts with 3′ end of 16S rRNA and promotes maturation of h28; also prevents premature 30S ribosome from entry into translation cycle	12243	7BOH	2.8 Å	(Schedlbauer et al., 2021)

RbfA

Ribosome-binding factor A (RbfA) is a 13–15 kDa small single-domain protein, consisting of three α-helices and three β-strands with a type II K homology (KH)–domain fold topology. It is present in the majority of bacteria and archaea, as well as in eukaryotes, RbfA orthologs occur in mitochondria and chloroplasts too (Huang et al., 2003). The absence of the *rbfA* gene leads to slow growth in bacteria as well as accumulation of free ribosomal subunits and unprocessed 16S rRNA (17S rRNA). RbfA was first discovered as a multicopy suppressor of C23U mutation of 16S rRNA and is known to overcome the translational block at low temperatures, facilitating the correct folding and maturation of the 5′ end of the 16S rRNA (Datta et al., 2007).

Cryo-EM structures of the 30S ribosome subunit in complex with RbfA resolved at 2.8 Å (Figure 24.3b) reveal that RbfA's RNA-binding AxG motif interacts with the 3′ end of the 16S rRNA. Surprisingly, RbfA binding requires the 3′ end of the 16S rRNA to be in the exit channel. Further RbfA binding leads to reorganization of helix 28, which is accompanied by rearrangement of the 3′ end of the 16S rRNA from the entry to the exit of the mRNA channel. The structure showed that RbfA promotes the maturation of h28 and holds the 3′ end of the 16S rRNA near the mRNA channel's exit, in which the lower end of h28 is destabilized while the upper end is displaced. The destabilization of the lower end helps in the stabilization of h44/45 linker, and thus RbfA delays h44/45 linker folding in the fully mature 30S ribosome subunit. It also helps to prevent the premature entry of the pre-30S ribosome subunit into the translation cycle with the help of initiation factor IF3, thus acting as a gatekeeper (Schedlbauer et al., 2021; Sharma & Woodson, 2020). Also, cryo-EM analysis of pre-30S ribosome subunits accumulated in the absence of rbfA gene provided insight into the number of assembly intermediates ranging from unstructured head domain and unfolded central pseudoknot to fully resolved mature 30S subunit. The cryo-EM structures also revealed 3′ domain excessive flexibility, including in the region of h44 and h45. The study also revealed that the maturation of helix 28 is mediated by RbfA. Thus, RbfA is involved in the assembly of the 30S ribosome subunit via two distinct pathways: first in the early stage of central pseudoknot formation and later at the positioning of helix 44 in the decoding center (Maksimova et al., 2021).

YjeQ (RsgA)

Ribosomal small subunit GTPase A (RsgA) belongs to ribosome-dependent GTPase, whose activity is stimulated by the mature 30S ribosome subunit. It is a 33 kDa protein that contains three functional domains: the N-terminal OB-fold domain, the central GTPase domain having the Rossmann fold that has unique circular permutation features, and the zinc-finger-like C-terminal domain (Daigle et al., 2002).

Cryo-EM analysis of pre-30S particles from ΔyjeQ showed that ribosomal proteins such as S2 and S21 that assemble late in the process were underrepresented. Further, density corresponding to helix 44 of 16S rRNA was also distorted, along with outward displacement of the upper end. This structure provided proof that YjeQ is involved in the correct positioning of helix 44 within the decoding center at the late stage of 30S ribosome assembly (Leong et al., 2013). A recent cryo-EM structure resolved at 3 Å (Figure 24.3c) revealed the binding site of YjeQ on the mature 30S ribosome where its N-terminal interacts with decoding center helix 18 and 44, while its GTPase domain interacts with platform helix 44 and 24 and its C-terminal domain forms a bridge between platform and head (h29 and h31). YjeQ's GTPase and OB-fold restrict the mobility of helix 44 by forming a clamp at the top of it. Residue A1492 helps to monitor the geometry of codon–anticodon base pairing at the decoding center and protrudes from helix 44, forming a conformation similar to the proofreading process of codon–anticodon pairing. Further, YjeQ's OB-fold domain stabilizes A1492 position through π-stacking and hydrophobic interaction. The structure also elucidates that 70S ribosome dissociates in the presence of YjeQ as it prevents binding of premature 30S to the 50S ribosome subunit. It thus serves as a checkpoint for maturing 30S ribosome subunits for their ability to proofread (Schedlbauer et al., 2021).

Era

Era belongs to *E. coli* Ras-like protein; it is a highly conserved and essential GTPase in bacteria. It is known to play a role in numerous processes, including cell growth, cell division, and energy metabolism. It acts as a chaperone for 16S rRNA processing and 30S ribosomal subunit biogenesis. Era is a 34 kDa protein that comprises the N-terminal GTPase domain, the C-terminal KH domain, and a flexible linker 17 amino acids long. Depletion of Era leads to the accumulation of 16S precursor and increased levels of free 30S and 50S ribosomal subunits with respect to fully mature 70S ribosome subunits further inhibiting the process of translation (Tu et al., 2009). In eukaryotes, Era depletion results in arresting of the cell cycle in the G1 phase and apoptosis. Overexpression of Era is known to compensate for depletion of rbfA, suggesting that its role in the maturation of the 30S ribosomal subunit overlaps with the function of RbfA. It is also known that overexpression of KsgA suppresses the cold-sensitive mutation of Era (Inoue et al., 2003).

The 3.5 Å cryo-EM structure of Era in complex with the 30S ribosome subunit (Figure 24.3d) has revealed that the binding site of Era lies in the cleft formed between head and platform which is opposite to the face of helix 28 from RbfA. It has also revealed that Era binds with anti-Shine Dalgarno (SD) sequence at the 3′ end of 16S rRNA, thus protecting the anti-SD sequence from cleavage during the 3′ end processing of 17S rRNA. It therefore blocks the duplex formation between anti-SD of 17S rRNA and the SD sequence of mRNA, inhibiting the premature translation process. Cryo-EM studies of immature 30S ribosome accumulated during depletion of Era suggest that platform region helices 23 and 24 fail to adopt mature conformation. Also in the absence of Era, helix 44 and the decoding center are destabilized. In the presence of Era, the 70S ribosome undergoes dissociation due to destabilization of h44 and the decoding center. Era-induced destabilization of h44 further inhibits the binding of YjeQ to the 30S ribosomal subunit. The study also revealed that Era plays a crucial role in ribosome homeostasis by conversion of h44 to an immature state, including reactivation of hibernating ribosome and degradation of ribosome (Razi et al., 2019).

KsgA

KsgA is known as ribosomal RNA small subunit methyltransferase A (RsmA) that methylates two adjacent adenosine residues at positions 1518 and 1519 of 16S rRNA. KsgA is a 30 kDa protein comprising two domains, where the enzymatic N-terminal domain has a Rossmann-like fold and contains an active site for methyl transferase and a SAM binding pocket, while the C-terminal domain plays a role in rRNA recognition (O'Farrell et al., 2004). KsgA homologs are present in all three domains of life, playing a crucial role in the geometry of cells. Strains deficient in KsgA result in increased resistance to kasugamycin antibiotic, a high rate of error in the translation process, and slow growth phenotype (van Buul et al., 1984). Absence of KsgA also impairs rRNA processing during the ribosome maturation process and results in an increased amount of unprocessed 17S rRNA. Overexpression of inactive methyltransferase mutants of KsgA in ΔKsgA strains leads to accumulation of precursor 17S rRNA. The inactive KsgA associates with the pre-30S ribosome subunit and stalled 30S ribosome maturation process, preventing the premature 30S ribosome subunit from entering the translation cycle. Thus apart from the methylation process, it also serves as the crucial checkpoint during 30S ribosome biogenesis (Connolly et al., 2008).

A recent study that resolved the cryo-EM structure of KsgA bound with submethylated 30S ribosome at 3.1 Å revealed that KsgA binds to the inter-subunit interface near the decoding center and platform (Figure 24.3e). This binding site of KsgA on the 30S ribosome coincides with the initiation factor IF3 binding site and helix 69 of the large ribosomal subunit, thus ensuring that an immature 30S ribosome does not enter the translation cycle unless rRNA methylation takes place. The binding of KsgA also prevents the maturation of helix 44, which leads to a block in decoding center formation and the joining of two subunits for initiation of the translation process. The N-terminal domain of KsgA interacts with helix 45 where A1518 and A1519 residues are located, while the C-terminal binds to rRNA helices 24 and 27 near the platform region. The structure of the 30S–KsgA complex suggests that dimethylation at 16S rRNA happens one after the other without dissociation of KsgA. These methylation sites induce interaction within helices 45 and 44 in the decoding center and thus serve as another checkpoint for 30S ribosome assembly (Stephan et al., 2021a). It was also observed that overexpression of another biogenesis factor RbfA in the ΔKsgA strain leads to accumulation of 70S-like particles with impaired translational efficiency. This study revealed that the steric clash between KsgA and H69 of 23S rRNA inhibits the joining of two subunits for 70S subunit formation and also translational initiation until the methylation is finished (Connolly et al., 2008).

RimP

RimP is a 17 kDa globular protein consisting of two domains where the N-terminal domain of RimP is structurally similar to the RbfA protein while the C-terminal domain has Sm fold characteristics (Schedlbauer et al., 2020). RimP knockout is often lethal in cases of Gram-positive bacteria, while in Gram-negative bacteria it leads to slow cell growth. It is highly conserved in bacteria, but in the case of eukaryotes and archaea, RimP has not been identified yet. Absence of RimP results in a temperature-sensitive growth phenotype. An increase in the level of free 30S and 50S ribosome subunits, along with the accumulation of precursor 16S rRNA and reduced amounts of 70S ribosomes, were observed in the case of depletion of RimP at higher temperatures (Nord et al., 2009).

Pre-30S ribosome subunits accumulating in the knockout strain of RimP depict a wide range of cryo-EM 30S ribosome assembly intermediates, ranging from an early stage with immature head region to a nearly mature 30S ribosome subunit. The most characteristic feature of the depletion of RimP is the presence of an immature central pseudoknot with S5 and S12 protein depletion. Further quantitative-MS studies also suggest that RimP is known to accelerate the binding efficiency of S5 and S12 ribosomal proteins to 16S rRNA. It is also suggested that RimP plays a role in central pseudoknot stabilization at an earlier stage of 30S ribosome assembly (Sashital et al., 2014). A cryo-EM structure at 4.8 Å of RimP in complex with 30S ribosome (Figure 24.3f) reveals that RimP helps in central pseudoknot formation and interacts with helix 44 in the near vicinity of the KsgA-binding site on the 30S ribosomal subunit. RimP also prevents residues A1492 and A1493 of helix 44 from approaching S12 protein, which maintains the fidelity of the translation process. RimP along with KsgA protein delay helix 44 positioning at the 30S ribosomal subunit interface during the later stage of 30S ribosome assembly (Schedlbauer et al., 2021).

30S Ribosome Subunit Biogenesis

With the recent advancements in cryo-EM, a plethora of 30S ribosome structures bound with assembly factors such as RbfA, RimP, KsgA, and YjeQ are now available (Figure 24.3b–f). These structures have helped in visualizing the conformational landscape of the late assembly process. Prior structural analysis utilizing biochemical techniques revealed that the process of 30S ribosome assembly is robust and occurs through various parallel pathways where the body domain of the 5′ end is formed first, followed by the head and central platform domain. Finally, the minor domain of the 3′ end, along with the decoding center, is formed (Mulder et al., 2010). Earlier studies provided poor resolution of structure and were often termed "blobology" because they failed to provide the mechanistic details of the assembly landscape (Smith & Rubinstein, 2014). Advances in cryo-EM technique have helped us to gain substantial insights into the 30S ribosome assembly process, specifically the late assembly process (Mulder et al., 2010; Smith & Rubinstein, 2014). However, during the later stage of assembly, many kinetic traps, such as alternate RNA conformation, are prevalent. These traps slow down the process of assembly under physiological conditions. Thus various late assembly factors incorporate with these alternative conformations to generate a mature 30S ribosome subunit. The late assembly factors (RbfA, KsgA, YjeQ, and RimP) mainly assist in the maturation of the decoding center (Razi, Guarné, et al., 2017).

Maturation of 30S Ribosome Decoding Center

The cryo-EM structure of these late assembly factors, along with 30S ribosome, unravels the molecular mechanism of how these assembly factors help in decoding center maturation (Schedlbauer et al., 2021) (Figure 24.4). On comparing the immature 30S state with the mature state, the most prominent difference occurs in the neck region (helix 28 of 16S rRNA), where the 3′ end of 16S rRNA shifts from mRNA entry to exit channel. This swapping is often accompanied by a conformational change in helix 28 that connects the body and head regions. Immature helix 28 having nonnative secondary conformation is converted to native secondary conformation to form mature helix 28. Also, this conversion affects the conformational changes in helix 44 structure where the h28/h44 is elongated, which provides flexibility to h44 in an immature state. This h28/h44 linker is further shortened and allows h44 to adopt a more canonical position by folding the linker region to a helix-like structure and subsequently displacing it to the front of the 30S ribosome subunit. This allows the maturation of the decoding center, thus completing the maturation of the 30S ribosome subunit (Schedlbauer et al., 2021) (Figure 24.4).

These cryo-EM structures provided a model for stepwise folding of the decoding center from immature to mature 30S ribosome subunit. The model proceeds with the binding of RimP in state A with a nonnative conformation of the decoding center (Figure 24.4). In this state, the 3′ end of 16S rRNA is inside the mRNA entry channel that has immature h28, and h44 is positioned in the mRNA exit channel. This configuration allows the binding of KsgA in state B and allows methylation of A1518 and A1519 in the h45 loop (Figure 24.4). After methylation, KsgA is released in state C, which partially opens the binding site of h44 (Figure 24.4). Further, in state D, RbfA binds and swaps the position of the 3′ end from the entry to the exit channel, allowing the maturation of h28 (Figure 24.4). In state E, RimP dissociates, which allows h44 to adopt near-native conformation (Figure 24.4). Further binding of YjeQ and dissociation of RbfA takes place in state F, which allows h44 to adopt a mature conformation (Figure 24.4). Lastly, GTP hydrolysis

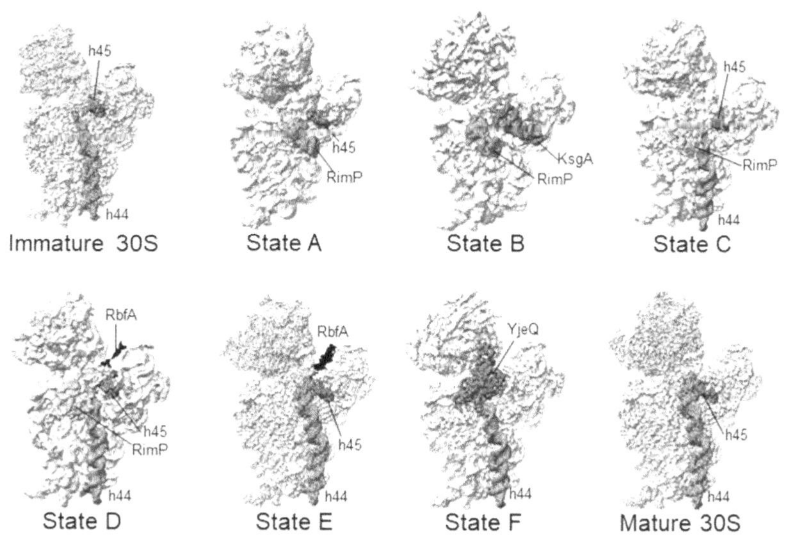

FIGURE 24.4 Cryo-EM structure of small subunit with assembly factors. The binding site of late-stage assembly factors is revealed. Immature 30S ribosome (EMDB ID; 12251) shows h28 in immature state with nonnative secondary conformation. State A (EMDB ID; 12247) has nonnative decoding center folding and shows the presence of RimP (grey) along with the disordered density of h45 (dim grey). State B (Head EMDB ID; 11771, Body EMDB ID; 11772) shows the presence of KsgA (dim grey) along with RimP. State C (EMDB ID; 12248) shows the disordered density of h45 and the nonnative conformation of h44 (grey). State D (EMDB ID; 12249) shows the binding of RbfA (black) that allows the maturation of h28. State E (EMDB ID; 12243) involves h44 to adopt near-native conformation after the dissociation of RimP. State F (EMDB ID; 12245) shows the binding of YjeQ (dim grey) that allows h44 to adopt native conformation. Mature 30S ribosome (EMDB ID; 12240) depicts the maturation of h28 along with swapping of the 3′ end of 16S rRNA from entry to exit channel.

leads to the release of YjeQ from the mature 30S ribosome conformation (Schedlbauer et al., 2021) (Figure 24.4).

50S RIBOSOME SUBUNIT ASSEMBLY

The 23S rRNA of the 50S ribosome subunit is structurally divided into three major domains, 5′ (13S), central (8S), and 3′ (12S) (Figure 24.1). The process of 50S ribosome subunit assembly is far more complex than that of the 30S ribosome subunit due to the larger size of 23S rRNA as compared to the 16S rRNA present in 30S ribosome subunit (Petrov et al., 2013). The process of 50S ribosome assembly does not occur in a manner similar to the 30S ribosome, which is assembled in the 5′ to 3′ direction. 50S ribosomal subunit assembly occurs through the incorporation of r-proteins and rRNAs (Figures 24.1 and 24.5a).

Prior studies to understand the process of 50S ribosome assembly were done in *in vitro* by reconstitution of r-RNAs and r-proteins. The *in vitro* process of 50S ribosome assembly is slow and requires non-physiological temperature and ionic concentration. Ribosome biogenesis is also studied in an *in vivo* system which is a more efficient, highly ordered, tightly regulated process and requires various biogenesis factors (Table 24.2) in contrast to *in vitro* systems. These biogenesis factors belong to a wide range of classes, such as RNA processing enzymes, protein and RNA modifying enzymes, RNA helicases, chaperones, maturation factors, and GTPases that help to accelerate the process of 50S ribosome biogenesis (Davis & Williamson, 2017; Kaczanowska & Rydén-Aulin, 2007; Shajani et al., 2011). A brief function of various biogenesis factors that assist in the assembly of 50S ribosomes are listed in Table 24.2.

Knowledge of the assembly of the large subunit was first studied by the Nierhaus group via *in vitro* reconstitution of the 50S ribosome subunit in a cell-free system that established the hierarchical order of ribosomal protein binding (Rohl & Nierhaus, 1982). *In vitro* reconstitution of functional 50S ribosome involves four steps: First, a 33S intermediate RI50(1) is produced by incubation of 23S rRNA, 5S rRNA, and 22 ribosomal proteins at 0°C in the presence of 4 mM Mg^{2+}. Subsequent incubation at 44°C leads to the production of RI*50(1), a 41S intermediate. Further addition of remaining ribosomal proteins to the 41S intermediate leads to the generation of 48S intermediate RI50(2).

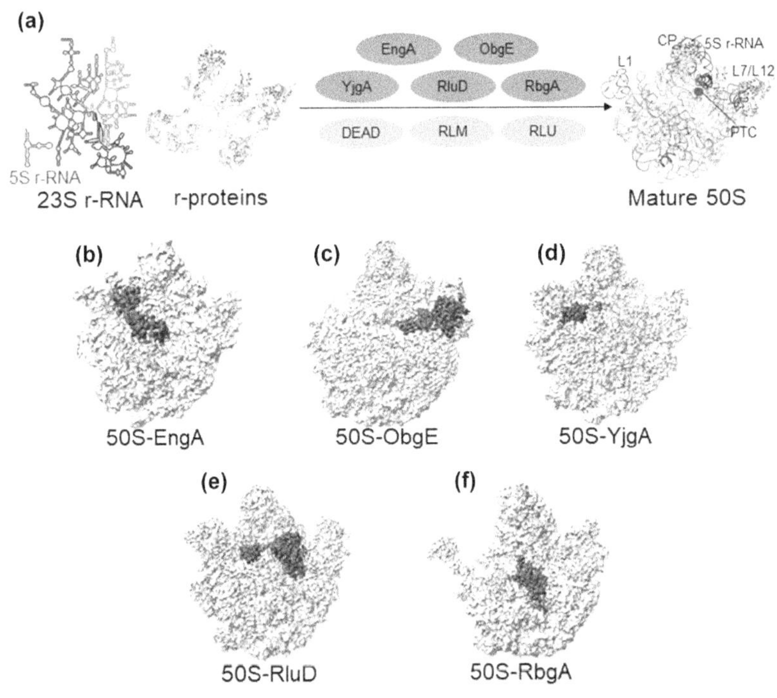

FIGURE 24.5 Assembly factors of the ribosomal large subunit. (a) The 2D diagram of the partially assembled 23S rRNA with LSU r-proteins (ribbon structure) is shown on the left, and the fully assembled 50S LSU is shown on the right. The assembly factors: EngA, ObgE, YjgA, RluD, RbgA (dark grey color circle), DEAD-box RNA helicase (SrmB, CsdA, RhlE, DbpA), RLM—23S rRNA methyltransferase (RlmA, RlmB, RlmC, RlmD, RlmE, RlmF, RlmG, RlmH, RlmI, RlmJ, RlmK/L, RlmM, RlmN), RLU—ribosomal large subunit pseudouridine synthase (RluA, RluB, RluC, RluE, RluF) (light grey color circle) involved in this assembly process are listed. Lime circle represents the factors whose cryo-EM structural information is available, while grey circle represents the factors whose structural information is not available. (b) Cryo-EM structure of 50S ribosome (light grey) in complex with EngA (dim grey) (EMDB ID; 6149). (c) Cryo-EM structure of 50S ribosome (light grey) in complex with ObgE (dim grey) (EMDB ID; 12217). (d) Cryo-EM structure of 50S ribosome (light grey) in complex with YjgA (dim grey) (EMDB ID; 12218). (e) Cryo-EM structure of 50S ribosome (light grey) in complex with RluD (dim grey) (EMDB ID; 12218). (F) Cryo-EM structure of 50S ribosome (light grey) in complex with RbgA (dim grey) (EMDB ID; 20441).

TABLE 24.2
Biogenesis factors involved in large ribosomal subunit maturation

	Factor	Function	EMDB	PDB	Resolution	Reference
GTPase	EngA	Induces conformational changes in helices 68–71 and L1 stalk, acts as chaperone to facilitate PTC assembly	6149	3J8G	5 Å	(Zhang et al., 2014)
	ObgE/ CgtA	Causes conformational changes in H89 and H38, stabilizes central protuberance and core	12217	7BL4	2.4 Å	(Nikolay et al., 2021)
	RbgA	Binds to pre-50S intermediate and induces conformational changes in H91-92 and H38, stabilizes uL6 and H80-88	20435	6PPK	3.4 Å	(Seffouh et al., 2019)
	YjgA	Mediates bending of L1 stalk	12218	7BL5	3.3 Å	(Nikolay et al., 2021)
DEAD-box RNA helicase	CsdA/ DeaD	Associates with pre-50S and mediates unwinding of 23S rRNA in ATP-dependent manner	-	-	-	(Shajani et al., 2011)
	DbpA	Binds specifically to loop of hairpin 92 of 23S rRNA and mediates its unwinding	-	-	-	(Shajani et al., 2011)
	RhlE	Mediates unwinding of 23S rRNA, plays role in interconversion of r-RNA folding intermediates that can further be processed by CsdA or SrmB	-	-	-	(Shajani et al., 2011)
	SrmB	Suppressor of temperature-sensitive mutation in ribosomal protein L24, forms complex with L4 and L24 proteins, plays role in 23S rRNA remodeling	-	-	-	(Shajani et al., 2011)
r-RNA modification	RlmA	Methylates G745 of 23S rRNA	-	-	-	(Kaczanowska & Rydén-Aulin, 2007)
	RlmB	Methylates G2251 of 23S rRNA	-	-	-	(Kaczanowska & Rydén-Aulin, 2007)
	RlmC	Methylates N5 position of U747 of 23S rRNA	-	-	-	(Kaczanowska & Rydén-Aulin, 2007)
	RlmD	Helps in formation of 5-methyl-U1939 of 23S rRNA	-	-	-	(Kaczanowska & Rydén-Aulin, 2007)
	RrmJ/ RlmE	Methylates 2′O-position of U2552 of 23S rRNA of fully assembled 50S	-	-	-	(Kaczanowska & Rydén-Aulin, 2007)
	RlmF	Methylates A1618 of 23S rRNA	-	-	-	(Sergiev et al., 2008)
	RlmG	Methylates G1835 of 23S rRNA	-	-	-	(Kaczanowska & Rydén-Aulin, 2007)
	RlmH	Methylates pseudouridine at 1915 of 23S rRNA	-	-	-	(Purta, Kaminska, et al., 2008)
	RlmI	Methylates C1962 of 23S rRNA	-	-	-	(Purta, O'Connor, et al., 2008)
	RlmJ	Methylates A2030 of 23S rRNA	-	-	-	(Golovina et al., 2012)
	RlmK/L	Methylates guanine at 2069 and 2445 of 23S rRNA	-	-	-	(Kaczanowska & Rydén-Aulin, 2007)
	RlmM	Catalyzes the 2′-O-methylation at nucleotide C2498 in 23S rRNA	-	-	-	(Purta et al., 2009)
	RlmN	Methylates A2503 of 23S rRNA	-	-	-	(Toh et al., 2008)
	RluA	Generation of pseudouridine from U746 in 23S rRNA	-	-	-	(Kaczanowska & Rydén-Aulin, 2007)
	RluB	Generation of pseudouridine from U2605 in 23S rRNA	-	-	-	(Kaczanowska & Rydén-Aulin, 2007)
	RluC	Generation of pseudouridine from U at 955, 2504, and 2580 position in 23S ribosomal RNA	-	-	-	(Kaczanowska & Rydén-Aulin, 2007)
	RluD	Generates pseudouridine at U1911, U1915, and U1917	12218	7BL5	3.3 Å	(Nikolay et al., 2021)
	RluE	Generates pseudouridine at U2457 of 23S rRNA	-	-	-	(Kaczanowska & Rydén-Aulin, 2007)
	RluF	Generates pseudouridine at U2604 of 23S rRNA	-	-	-	(Kaczanowska & Rydén-Aulin, 2007)

The functional 50S ribosome subunit is formed by heating the 48S intermediate to 50°C along with 20 mM Mg^{2+}. This study revealed the presence of three sets of ribosomal proteins based on their binding: first the primary proteins that bind to rRNA independently, then the secondary proteins and the tertiary proteins that show enhanced binding affinity in the presence of primary and secondary proteins respectively (Shajani et al., 2011). However, due to a lack of technology, structural information about assembly intermediates at atomic resolution was difficult to achieve.

Another study provided a detailed view of 50S ribosomal assembly by utilizing a combination of radioactive pulse labeling of the ribosome along with quantitative mass spectrometry. The process of 50S ribosome assembly involves the generation of six intermediates without any structural information. Ribosomal proteins L20, L21, L22, and L24 bind to the 5′ domain of 23S rRNA to produce the first intermediate. Subsequent addition of L1, L3, L4, L13, L15, L17, and L23 produce the second intermediate, which is followed by the generation of the third intermediate on the addition of L5, L18, L29, and L34. Next L14, L2, L19, and L32 proteins bind to 23S rRNA to produce the fourth intermediate, which on the subsequent addition of L6, L9, L11, L28, and L33 forms the fifth intermediate. Lastly, L30, L31, L35, and L36 ribosomal proteins bind to generate a mature 50S ribosome subunit. Assembly of 5S rRNA generally occurs at later stages as it needs prior binding of L5, L15, L18, and L25 proteins. These r-proteins are crucial as they mediate the interaction between 5S rRNA and 23S rRNA and also help in the correct formation of the central protuberance of the 50S ribosome subunit (Chen & Williamson, 2013). However, detailed information remained missing mainly because of the lack of suitable techniques such as cryo-EM.

Assembly Factors of 50S Ribosome Subunit

The cryo-EM structures of assembly factors in complex with the 50S ribosome subunit resolved at better than 5 Å resolution are described here in detail. Other known assembly factors are listed in Table 24.2.

EngA

Essential Neisseria GTP-binding protein A (EngA), also known as Der, YphC, or YfgK, 55 kDa protein, an essential GTPase in bacteria, contains two tandemly arranged GTPase domains (GD1 and GD2) that are followed by the C-terminal K-homology domain (KH domain). It was first identified in the bacteria *Neisseria gonorrhoeae* as an essential gene for its cell viability and is unique among all members of the GTPase family by its structural architecture (Hwang & Inouye, 2001). EngA is known to be a 50S ribosomal subunit assembly factor as it plays an essential role in the late steps of ribosome biogenesis. Overexpression of EngA is known to rescue aberrant ribosome profile in the Δrrmj (ribosomal RNA large subunit methyltransferase E) strain (Tan et al., 2002). Depletion of EngA leads to the accumulation of an immature large subunit (45S) that contains several missing ribosomal proteins, thus implicating its role in ribosome biogenesis.

The cryo-EM structure of 50S bound with EngA in the presence of GMP-PNP revealed that EngA binds to the inter-subunit face of the 50S ribosome subunit (Figure 24.5b), embedded deeply inside the tRNA passage, with its KH domain interacting with the peptidyl transferase center (PTC). The structure of the 50S ribosome in complex with EngA was solved at 5 Å, and it depicts that binding of EngA to the 50S ribosomal subunit leads to conformational *changes* in the 50S ribosomal subunit that result in the displacement of helices 68–71, inward movement of the L1 stalk, and loss of the L33 protein. This suggests that EngA acts as an rRNA chaperone by binding to the PTC region and facilitating its assembly (Zhang et al., 2014).

ObgE

ObgE, also known as CgtA, is a 43 kDa protein, an essential GTPase identified first in *Bacillus subtilis* as a protein encoding for sporulation. ObgE is known to be associated with various cellular activities such as cell proliferation, sporulation initiation, signal transduction, translation, antibiotic resistance, and persister formation (Verstraeten et al., 2015). ObgE depletion is known to cause defects in the growth of bacteria along with the accumulation of defective pre-50S ribosome subunits, suggesting its role in 50S ribosome biogenesis (Jiang et al., 2006). Furthermore, ObgE was found to be a phenotypic suppressor in a cold-sensitive rlmE knockout mutant, indicating its role in late 50S ribosome biogenesis downstream of the RNA methyltransferase RlmE (Tan et al., 2002). Cryo-EM study of native pre-50S ribosomes revealed a network of interacting biogenesis factors containing ObgE, YjgA, RluD, and RsfS binding at its interface. ObgE is mainly involved in the folding of H89 and further stabilizes the central protuberance and core. The cryo-EM structure of the pre-50S ribosome bound to ObgE resolved at 2.4 Å (Figure 24.5c) revealed that ObgE along with other biogenesis factor RsfS binds to the late stage of the pre-50S complex at the interface region. The binding of ObgE leads to conformational changes in H89 and H38. This provides binding sites for other biogenesis factors YjgA and RluD that lead to inward rearrangement of the L1 stalk, further promoting the binding affinity for late assembly proteins bL36 and uL16. This subsequently leads to conformational changes in H38 from a downward to an upward position in combination with conformational changes in H89 that result in hydrolysis of GTP and dissociation of ObgE along with other biogenesis factors to produce a functional 50S ribosome subunit (Nikolay et al., 2021).

YjgA

The cryo-EM structure of YjgA in complex with the 50S ribosome revealed that the side chain of the N-terminal of the protein interacts with helix 80 of the P-loop. It interacts with the lower region of the central protuberance along with the stalk base that is bent and moved away from its neutral position. This L1 stalk movement is similar when E-site tRNA interacts with it during the translation process. This bending of the L1 stalk mediated by YjgA serves as a quality control function. Cryo-EM structure with 3.3 Å (Figure 24.5d) also identifies YjgA in elongated conformation that would result in a steric clash with H68. Thus, the binding site and conformation of YjgA are subjected to various changes, depending upon the folding of H68 (Nikolay et al., 2021).

RluD

Ribosomal large subunit pseudouridine synthase D (RluD) is a 37 kDa protein involved in pseudouridine synthesis from uracil in the 1911, 1915, and 1917 positions of 23S rRNA (Wrzesinski et al., 2000). It is identified as a late biogenesis factor mainly interacting with helix 69. A cryo-EM structure of 3.3 Å (Figure 24.5e) depicting the complex of 50S ribosome bound with RluD revealed that the N-terminal of RluD interacts with YjgA protein and occupies half of H69, where the uridine residues are located, while the C-terminal region connects with a short, flexible unresolved region. The two domains are positioned in such a way that they appear to unwind H69 (Nikolay et al., 2021).

RbgA

Ribosome biogenesis GTPase A (RbgA) is a 31 kDa essential protein that plays a crucial role in 50S ribosome assembly in *Bacillus subtilis*. It is present in Gram-positive as well as some Gram-negative bacteria such as *Streptococcus pneumoniae*, *Staphylococcus aureus*, *Vibrio cholerae*, and *Neisseria meningitidis*, however, this protein is absent in *Escherichia coli*.

In *B. subtilis*, the absence of RbgA leads to the accumulation of 45S intermediate and is known to play a role in the maturation of A-, P-, and E-sites in the 50S ribosome subunit. Incorporation of RbgA to the P-site leads to maturation of H91-92 and H38 in the A-site and H93 in the P-site. It is also known to stabilize the central protuberance region. Further, two other GTPases (YphC and YsxC) with some r-proteins (L16, L27, L28, L33, L3,5 and L36) bind to stabilize H68 in the E-site, H89 in the A-site, and the folding of H69 and H71 in the P-site. Lastly, the binding of uL16 induces conformational changes that subsequently lead to the release of RbgA and thereby transformation of the intermediate to mature 50S ribosome subunit (Seffouh et al., 2019).

Cryo-EM analysis of the 45S particle accumulated in ΔRbgA has depicted that the maturation of these intermediates highly depends on the incorporation of RbgA. A cryo-EM structure resolved at 3.4 Å resolution (Figure 24.5f) provided an insight into the functional mechanics of 50S ribosome assembly, where RbgA binds to the 45S particle with the fully assembled central protuberance and unlocks the conformational acquired by H97 and H42 due to binding of uL6. Further, RbgA leads to proper folding of the functional core of the 50S ribosomal subunit. Thus, RbgA ensures that the 50S ribosome functional core is formed last, after the central protuberance has been formed (Seffouh et al., 2022).

50S Ribosome Subunit Biogenesis

A major breakthrough was made by recent work of Dong et al., 2023, where they used the iSAT (*in vitro* integrated rRNA synthesis, ribosome assembly, and translation) technique to identify early-stage intermediates of the 50S ribosome subunit assembly process in a time-resolved manner under near-physiological conditions and elucidated their intermediate structures. In this work, 13 intermediate states were identified that are formed during 50S ribosome assembly, and their structure was elucidated with the help of cryo-EM (Figure 24.6). These intermediates are arranged in chronological order, encompassing an assembly landscape that begins with the folding of 23S rRNA from domain I of the 5′ terminal to a completely mature subunit (Dong et al., 2023). The assembly landscape displays both sequential and parallel folding of domains, ensuring that the assembly proceeds efficiently.

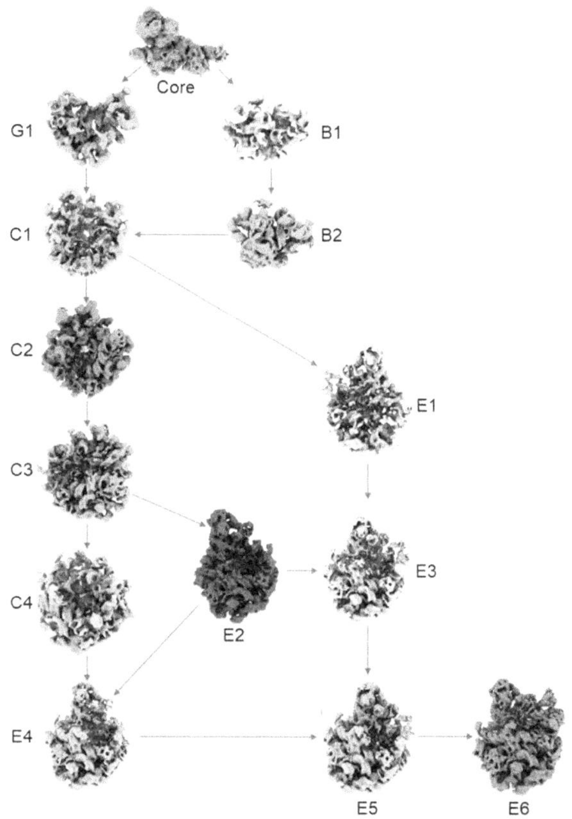

FIGURE 24.6 Cryo-EM map of *in vitro* 50S ribosome assembly intermediates from iSAT reaction. 13 assembly intermediates are arranged according to their occurrence in the assembly path. Core (EMDB ID;16509), B1 (EMDB ID; 29056), B2 (EMDB ID; 29042), G1 (EMDB ID; 29057), C1 (EMDB ID; 29058), C2 (EMDB ID; 29059), C3 (EMDB ID; 29060), C4 (EMDB ID; 29061), E1 (EMDB ID; 29062), E2 (EMDB ID; 29063), E3 (EMDB ID; 29064), E4 (EMDB ID; 29065), E5 (EMDB ID; 29066), and E6 (EMDB ID; 29067) are shown.

The first intermediate formed is Core, which is the smallest precursor in the 50S ribosome assembly. Core state mainly consists of domain I (H2-H24), some parts of domain II (H27-28) and three r-proteins (L4, L22, and L24). All of these factors are generally located at the 5′ end of 23S rRNA, supporting the fact that transcription of rRNA and assembly go hand in hand (Figure 24.6). Further, two parallel assembly pathways can occur, one through the formation of the B1 state and the other through the G1 intermediate to generate the C1 state. The assembly path through classes B1 and B2 involves sequential incorporation of the central part to domains II and III along with r-proteins corresponding to the core region (Figure 24.6). The G1 assembly pathway involves the binding of L3, L17, L19, L23, and L29 r-proteins in the base region. Once these early assembly intermediates are formed, the late assembly process starts, which is more complicated and involves various parallel pathways. Late assembly proceeds from the C1 state to the C4 state and involves maturation of H38. In the C2 state of this pathway, incorporation of L2, L14, and L32 r-proteins takes place, which subsequently involves the binding of L15 r-protein in the C3 state (Figure 24.6). From this C3 state, two parallel pathways can occur. One proceeds in the C3-C4-E4 pathway, where H68-70 assembles to give rise to the C4 state, and then assembly of the central protuberance takes place to form the E4 state. In the C3-E2-E4 pathway, an alternative conformation of these events occurs. Maturation from the E1 to the E5 pathway occurs by first assembling the central protuberance, L14, L15, and L28 r-proteins, then by incorporation of the L2 r-protein (Figure 24.6). Lastly, in the E5 state, formation of H43-44, H89-93, H71, and H80 and the incorporation of L35, L36, L16, and L6 takes place in the PTC region, which results in the generation of the E6 state, which resembles the mature 50S-like particle (Figure 24.6) (Dong et al., 2023).

50S Ribosome Early and Late Stage Assembly

The *in vitro* reconstitution of the 50S ribosome assembly intermediate has provided unprecedented structural information on the early and late stages of assembly (Nikolay et al., 2018; Qin et al., 2023). The details are described here.

Early Stages of 50S Ribosome Assembly

More recent cryo-EM study of the early assembly of the 50S ribosome has provided detailed insight into the process (Figure 24.7a) (Qin et al., 2023). The early assembly of

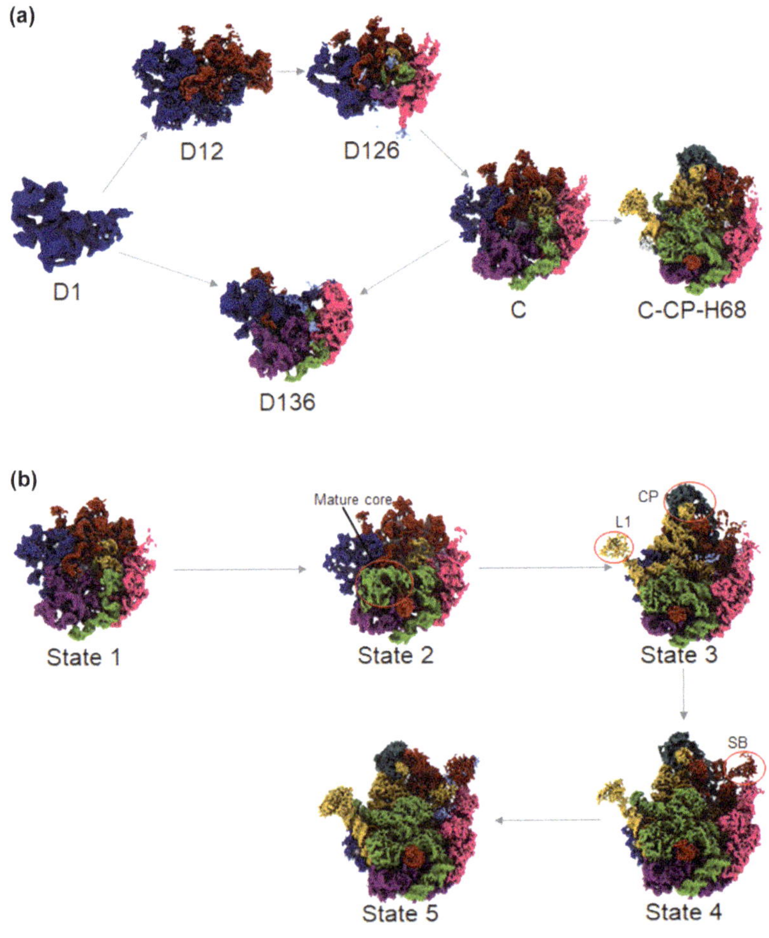

FIGURE 24.7 Large ribosomal subunit assembly intermediates. (a) Cryo-EM map of early stages of 50S ribosome assembly. Early assembly of 50S ribosome subunit occurs in a domain-wise manner. D1 (EMDB ID; 16509) represents domain I of 23S rRNA. D12 (EMDB ID; 16506) represents domain I along with domain II. D126 (EMDB ID; 16503) represents domains I, II, and VI. D136 (EMDB ID; 16504) represents domains I, III, and VI. C (EMDB ID; 16499) represents 50S ribosome core. C-CP-H68 represents 50S core, central protuberance, and helix 68 (EMDB ID; 16494). (b) Cryo-EM map of late stages of 50S ribosome assembly. Late assembly of 50S ribosome occurs in five states. State 1 (EMDB ID; 4382) shows density for core, state 2 (EMDB ID; 4381) shows a fully mature core, and state 3 (EMDB ID; 4380) shows density for central protuberance and L1 stalk. State 4 (EMDB ID; 4379) shows appearance of stalk base. State 5 (EMDB ID; 4378) resembles a fully mature 50S ribosome subunit. Here, domain I (navy blue), domain II (dark red), domain III (purple), domain IV (olive drab), domain V (goldenrod), and domain VI (deep pink) are shown.

the 50S ribosome subunit occurs in a domain-wise manner where the nucleation center begins with the interaction of uL22, uL24, and uL29 with domain I. This interaction subsequently leads to the formation of binding sites for other proteins, such as uL4 and bL23, initiating the formation of domains II and III, respectively. Once domain I is formed further, the assembly continues in a uL4-dependent manner, branching into either route of assembly and proceeding via domain I to domain II and finally to domain VI (1-2-6) or via domain I through domain III and finally to domain VI (1-3-6) (Figure 24.7a). Assembly proceeding through the 1-2-6 route involves the formation of the central protuberance (CP) due to the presence of domain II, which facilitates docking of CP-forming 5S ribonucleoprotein (RNP), leading to the formation of d12-CP or d126-CP. Alternatively, route 1-3-6 results in the formation of the core due to the presence of domain III, seeded by uL23, which allows for the incorporation of domain VI independently of domain II. Subsequently, proteins like bL17, bL32, and uL3 also contribute to the formation of domain III along with domain IV, which helps in the maturation of the 50S ribosome core. Thus, the process of early 50S ribosome assembly is characterized by the ordered arrangement of rRNA in a domain-wise fashion (Figure 24.7a) (Qin et al., 2023).

Late Stages of 50S Ribosome Assembly

Structural visualization of *in vitro* reconstitution of late 50S ribosome assembly has revealed five intermediate stages that proceed in a progressive maturation pathway (Figure 24.7b) (Nikolay et al., 2018). Two early intermediate states (state 1 and state 2) contain a well-defined core with fully folded domain I, III, and VI of 23S rRNA. However, they lack three protuberances (central protuberance, L1 protuberance, and stalk base) along with the peptidyl-transferase center (PTC). These states 1 and 2 with the stably formed core represent the 41S particle (Figure 24.7b). As the assembly proceeds further, formation of protuberances takes place from state 3 to state 5, and these states represent the 48S particle. In state 3, formation of L1 protuberance occurs with the incorporation of r-proteins (uL6, bL9, bL25, bL27, bL28) along with pre-5S complex, uL5, and uL18 (Figure 24.7b). The transition from state 3 to state 4 involves the stable incorporation of bL33 and bL35 accompanied by rotation of the central protuberance by 10° (Figure 24.7b). These structural rearrangements lead to the formation of 23S rRNA helices H43, H44, H68, and H79. Lastly, as the maturation proceeds from state 4 to state 5, the incorporation of uL16 leads to the rearrangement and formation of H89 and H38. State 5 thus represents a 48S particle that contains a full set of ribosomal proteins and a similar folding to mature 50S ribosome. However, these particles are not functionally active and require an additional step of heat activation along with a high level of Mg^{2+} concentration. Thus the final step of 50S activation is achieved by reorientation of 23S rRNA H71 that further mediates the formation of a stable H90-H93 region, therefore marking the end of the 50S ribosome assembly process (Figure 24.7b) (Nikolay et al., 2018).

SUMMARY

Ribosomes, the site of protein synthesis in the cell, are mega dalton ribonucleic protein complexes. The three-dimensional structures of the ribosomes from almost all kingdoms of life are available. The two ribosomal subunits assemble together through the process of ribosome biogenesis to form functional 70S ribosomes. The ribosome biogenesis process has been an active area of research for more than 30 years, but due to limitations of technology, earlier work on ribosome biogenesis mainly revolved around determining ribosomal components. The process of ribosome biogenesis proceeds through a series of changes in RNA conformation as well as binding of ribosomal proteins. Ribosome assembly proceeds much faster under *in vivo* conditions than *in vitro*. In the *in vitro* ribosomal assembly process, many biogenesis factors come into play to speed up the ribosome biogenesis process. Earlier studies from the assembly landscape mainly revolve around biochemical, chemical probing, and radiolabeling techniques. With recent advancements in cryo-EM, atomic-resolution structures of various ribosome assembly intermediates have emerged. As we advance into a golden era of structural biology, the high-resolution structure determination of the assembly intermediate has become straightforward. The SSU assembles with the help of various late assembly factors (RbfA, RimP, KsgA, Era, and YjeQ). The cryo-EM structures of these assembly factors bound to the ribosome have helped in understanding the maturation of the decoding center. In the case of LSU, assembly factors such as EngA, ObgE, YjgA, RluD, and RbgA have also provided mechanistic details of various intermediates of ribosome assembly. Atomic-resolution structural information from these assembly intermediates has assisted in deciphering the intricate details of the ribosome biogenesis process. Obstructing the function of an assembly factor would produce an immature ribosome that may be used as a strategy to inhibit bacterial growth. Further, the assembly intermediates may be explored as new drug targets.

REFERENCES

Basturea, G. N., Dague, D. R., Deutscher, M. P., & Rudd, K. E. (2012). YhiQ is RsmJ, the methyltransferase responsible for methylation of G1516 in 16S rRNA of *E. coli*. *Journal of Molecular Biology*, *415*(1). https://doi.org/10.1016/j.jmb.2011.10.044

Bennison, D. J., Irving, S. E., & Corrigan, R. M. (2019). The Impact of the Stringent Response on TRAFAC GTPases and Prokaryotic Ribosome Assembly. *Cells*, 8(11), 1313. https://doi.org/10.3390/cells8111313

Carugo, O., & Djinović-Carugo, K. (2023). Structural biology: A golden era. *PLoS Biology*, *21*(6). https://doi.org/10.1371/journal.pbio.3002187

Chen, S. S., & Williamson, J. R. (2013). Characterization of the ribosome biogenesis landscape in *E. Coli* using quantitative mass spectrometry. *Journal of Molecular Biology*, *425*(4). https://doi.org/10.1016/j.jmb.2012.11.040

Connolly, K., & Culver, G. (2009). Deconstructing ribosome construction. *Trends in Biochemical Sciences*, *34*(5). https://doi.org/10.1016/j.tibs.2009.01.011

Connolly, K., Rife, J. P., & Culver, G. (2008). Mechanistic insight into the ribosome biogenesis functions of the ancient protein KsgA. *Molecular Microbiology*, *70*(5). https://doi.org/10.1111/j.1365-2958.2008.06485.x

Daigle, D. M., Rossi, L., Berghuis, A. M., Aravind, L., Koonin, E. V., & Brown, E. D. (2002). YjeQ, an essential, conserved, uncharacterized protein from *Escherichia coli*, is an unusual GTPase with circularly permuted G-motifs and marked burst kinetics. *Biochemistry*, *41*(37). https://doi.org/10.1021/bi020355q

Datta, P. P., Wilson, D. N., Kawazoe, M., Swami, N. K., Kaminishi, T., Sharma, M. R., Booth, T. M., Takemoto, C., Fucini, P., Yokoyama, S., & Agrawal, R. K. (2007). Structural Aspects of RbfA Action during Small Ribosomal Subunit Assembly. *Molecular Cell*, *28*(3). https://doi.org/10.1016/j.molcel.2007.08.026

Davis, J. H., & Williamson, J. R. (2017). Structure and dynamics of bacterial ribosome biogenesis. *Philosophical Transactions of the Royal Society B: Biological Sciences*, *372*(1716). https://doi.org/10.1098/rstb.2016.0181

Dong, X., Doerfel, L. K., Sheng, K., Rabuck-Gibbons, J. N., Popova, A. M., Lyumkis, D., & Williamson, J. R. (2023). Near-physiological in vitro assembly of 50S ribosomes involves parallel pathways. *Nucleic Acids Research*, *51*(6). https://doi.org/10.1093/nar/gkad082

Golovina, A. Y., Dzama, M. M., Osterman, I. A., Sergiev, P. V., Serebryakova, M. V., Bogdanov, A. A., & Dontsova, O. A. (2012). The last rRNA methyltransferase of *E. coli* revealed: The yhiR gene encodes adenine-N6 methyltransferase specific for modification of A2030 of 23S ribosomal RNA. *RNA*, *18*(9). https://doi.org/10.1261/rna.034207.112

Held, W. A., & Nomura, M. (1973). Rate-determining step in the reconstitution of *Escherichia coli* 30S ribosomal subunits. *Biochemistry*, *12*(17). https://doi.org/10.1021/bi00741a020

Huang, Y. J., Swapna, G. V. T., Rajan, P. K., Ke, H., Xia, B., Shukla, K., Inouye, M., & Montelione, G. T. (2003). Solution NMR structure of ribosome-binding factor A (RbfA), a cold-shock adaptation protein from *Escherichia coli*. *Journal of Molecular Biology*, *327*(2). https://doi.org/10.1016/S0022-2836(03)00061-5

Hwang, J., & Inouye, M. (2001). An essential GTPase, der, containing double GTP-binding domains from *Escherichia coli* and *Thermotoga maritima*. *Journal of Biological Chemistry*, *276*(33). https://doi.org/10.1074/jbc.M104455200

Inoue, K., Alsina, J., Chen, J., & Inouye, M. (2003). Suppression of defective ribosome assembly in a *rbfA* deletion mutant by overexpression of Era, an essential GTPase in *Escherichia coli*. *Molecular Microbiology*, *48*(4). https://doi.org/10.1046/j.1365-2958.2003.03475.x

Jiang, M., Datta, K., Walker, A., Strahler, J., Bagamasbad, P., Andrews, P. C., & Maddock, J. R. (2006). The *Escherichia coli* GTPase CgtAE is involved in late steps of large ribosome assembly. *Journal of Bacteriology*, *188*(19). https://doi.org/10.1128/JB.00444-06

Kaczanowska, M., & Rydén-Aulin, M. (2005). The YrdC protein - A putative ribosome maturation factor. *Biochimica et Biophysica Acta - Gene Structure and Expression*, *1727*(2). https://doi.org/10.1016/j.bbaexp.2004.11.010

Kaczanowska, M., & Rydén-Aulin, M. (2007). Ribosome Biogenesis and the Translation Process in *Escherichia coli*. *Microbiology and Molecular Biology Reviews*, *71*(3). https://doi.org/10.1128/mmbr.00013-07

Kimura, S., & Suzuki, T. (2009). Fine-tuning of the ribosomal decoding center by conserved methyl-modifications in the *Escherichia coli* 16S rRNA. *Nucleic Acids Research*, *38*(4). https://doi.org/10.1093/nar/gkp1073

Kühlbrandt, W. (2014). Biochemistry. The resolution revolution. *Science (New York, N.Y.)*, *343*(6178). https://doi.org/10.1126/science.1251652

Leong, V., Kent, M., Jomaa, A., & Ortega, J. (2013). *Escherichia coli rimM* and *yjeQ* null strains accumulate immature 30S subunits of similar structure and protein complement. *RNA*, *19*(6). https://doi.org/10.1261/rna.037523.112

Maksimova, E. M., Korepanov, A. P., Kravchenko, O. V., Baymukhametov, T. N., Myasnikov, A. G., Vassilenko, K. S., Afonina, Z. A., & Stolboushkina, E. A. (2021). Rbfa is involved in two important stages of 30s subunit assembly: Formation of the central pseudoknot and docking of helix 44 to the decoding center. *International Journal of Molecular Sciences*, *22*(11). https://doi.org/10.3390/ijms22116140

Mizushima, S., & Nomura, M. (1970). Assembly mapping of 30S ribosomal proteins from *E. coli*. *Nature*, *226*(5252). https://doi.org/10.1038/2261214a0

Mulder, A. M., Yoshioka, C., Beck, A. H., Bunner, A. E., Milligan, R. A., Potter, C. S., Carragher, B., & Williamson, J. R. (2010). Visualizing ribosome biogenesis: Parallel assembly pathways for the 30S subunit. *Science*, *330*(6004). https://doi.org/10.1126/science.1193220

Nikolay, R., Hilal, T., Qin, B., Mielke, T., Bürger, J., Loerke, J., Textoris-Taube, K., Nierhaus, K. H., & Spahn, C. M. T. (2018). Structural visualization of the formation and activation of the 50S ribosomal subunit during in vitro reconstitution. *Molecular Cell*, *70*(5). https://doi.org/10.1016/j.molcel.2018.05.003

Nikolay, R., Hilal, T., Schmidt, S., Qin, B., Schwefel, D., Vieira-Vieira, C. H., Mielke, T., Bürger, J., Loerke, J., Amikura, K., Flügel, T., Ueda, T., Selbach, M., Deuerling, E., & Spahn, C. M. T. (2021). Snapshots of native pre-50S ribosomes reveal a biogenesis factor network and evolutionary specialization. *Molecular Cell*, *81*(6). https://doi.org/10.1016/j.molcel.2021.02.006

Noller, H. F., & Woese, C. R. (1981). Secondary Structure of 16S Ribosomal RNA. *Science*, *212*(4493), 403–411. https://doi.org/10.1126/science.6163215

Nomura, M. (1973). Assembly of bacterial ribosomes. *Science*, *179*(4076), 864–873. https://doi.org/10.1126/science.179.4076.864

Nomura, M., Mizushima, S., Ozaki, M., Traub, P., & Lowry, C. V. (1969). Structure and function of ribosomes and their molecular components. *Cold Spring Harbor Symposia on Quantitative Biology*, *34*. https://doi.org/10.1101/SQB.1969.034.01.009

Nord, S., Bylund, G. O., Lövgren, J. M., & Wikström, P. M. (2009). The RimP protein is important for maturation of the 30S ribosomal subunit. *Journal of Molecular Biology*, *386*(3). https://doi.org/10.1016/j.jmb.2008.12.076

O'Farrell, H. C., Scarsdale, J. N., & Rife, J. P. (2004). Crystal structure of KsgA, a universally conserved rRNA adenine dimethyltransferase in *Escherichia coli*. *Journal of Molecular Biology*, *339*(2). https://doi.org/10.1016/j.jmb.2004.02.068

Petrov, A. S., Bernier, C. R., Hershkovits, E., Xue, Y., Waterbury, C. C., Hsiao, C., Stepanov, V. G., Gaucher, E. A., Grover, M. A., Harvey, S. C., Hud, N. V., Wartell, R. M., Fox, G. E., & Williams, L. D. (2013). Secondary structure and domain architecture of the 23S and 5S rRNAs. *Nucleic Acids Research*, *41*(15). https://doi.org/10.1093/nar/gkt513

Purta, E., Kaminska, K. H., Kasprzak, J. M., Bujnicki, J. M., & Douthwaite, S. (2008). YbeA is the $m^3\Psi$ methyltransferase RlmH that targets nucleotide 1915 in 23S rRNA. *RNA*, *14*(10), 2234–2244. https://doi.org/10.1261/rna.1198108

Purta, E., O'Connor, M., Bujnicki, J. M., & Douthwaite, S. (2008). YccW is the m5C methyltransferase specific for 23S rRNA nucleotide 1962. *Journal of Molecular Biology*, *383*(3). https://doi.org/10.1016/j.jmb.2008.08.061

Purta, E., O'Connor, M., Bujnicki, J. M., & Douthwaite, S. (2009). YgdE is the 2′- *O*- ribose methyltransferase RlmM specific for nucleotide C2498 in bacterial 23S rRNA. *Molecular Microbiology*, *72*(5), 1147–1158. https://doi.org/10.1111/j.1365-2958.2009.06709.x

Qin, B., Lauer, S. M., Balke, A., Vieira-Vieira, C. H., Bürger, J., Mielke, T., Selbach, M., Scheerer, P., Spahn, C. M. T., & Nikolay, R. (2023). Cryo-EM captures early ribosome assembly in action. *Nature Communications*, *14*(1). https://doi.org/10.1038/s41467-023-36607-9

Razi, A., Britton, R. A., & Ortega, J. (2017). The impact of recent improvements in cryo-electron microscopy technology on the understanding of bacterial ribosome assembly. *Nucleic Acids Research*, *45*(3). https://doi.org/10.1093/nar/gkw1231

Razi, A., Davis, J. H., Hao, Y., Jahagirdar, D., Thurlow, B., Basu, K., Jain, N., Gomez-Blanco, J., Britton, R. A., Vargas, J., Guarné, A., Woodson, S. A., Williamson, J. R., & Ortega, J. (2019). Role of Era in assembly and homeostasis of the ribosomal small subunit. *Nucleic Acids Research*, *47*(15). https://doi.org/10.1093/nar/gkz571

Razi, A., Guarné, A., & Ortega, J. (2017). The cryo-EM structure of YjeQ bound to the 30S subunit suggests a fidelity checkpoint function for this protein in ribosome assembly. *Proceedings of the National Academy of Sciences of the United States of America*, *114*(17). https://doi.org/10.1073/pnas.1618016114

Rodnina, M. V. (2018). Translation in prokaryotes. *Cold Spring Harbor Perspectives in Biology*, *10*(9). https://doi.org/10.1101/cshperspect.a032664

Rohl, R., & Nierhaus, K. H. (1982). Assembly map of the large subunit (50S) of *Escherichia coli* ribosomes. *Proceedings of the National Academy of Sciences of the United States of America*, *79*(3 I). https://doi.org/10.1073/pnas.79.3.729

Sashital, D. G., Greeman, C. A., Lyumkis, D., Potter, C. S., Carragher, B., & Williamson, J. R. (2014). A combined quantitative mass spectrometry and electron microscopy analysis of ribosomal 30S subunit assembly in *E. coli*. *ELife*, *3*. https://doi.org/10.7554/eLife.04491

Schedlbauer, A., Iturrioz, I., Ochoa-Lizarralde, B., Diercks, T., Lopez-Alonso, J. P., Lavin, J. L., Kaminishi, T., Capuni, R., Dhimole, N., de Astigarraga, E., Gil-Carton, D., Fucini, P., & Connell, S. R. (2021). A conserved rRNA switch is central to decoding site maturation on the small ribosomal subunit. *Science Advances*, *7*(23). https://doi.org/10.1126/sciadv.abf7547

Schedlbauer, A., Ochoa-Lizarralde, B., Iturrioz, I., Çapuni, R., Diercks, T., de Astigarraga, E., Fucini, P., & Connell, S. R. (2020). Backbone and sidechain NMR assignments for the ribosome maturation factor RimP from *Escherichia coli*. *Biomolecular NMR Assignments*, *14*(2). https://doi.org/10.1007/s12104-020-09943-w

Schmeing, T. M., & Ramakrishnan, V. (2009). What recent ribosome structures have revealed about the mechanism of translation. *Nature*, *461*(7268). https://doi.org/10.1038/nature08403

Seffouh, A., Jain, N., Jahagirdar, D., Basu, K., Razi, A., Ni, X., Guarné, A., Britton, R. A., & Ortega, J. (2019). Structural consequences of the interaction of RbgA with a 50S ribosomal subunit assembly intermediate. *Nucleic Acids Research*, *47*(19). https://doi.org/10.1093/nar/gkz770

Seffouh, A., Trahan, C., Wasi, T., Jain, N., Basu, K., Britton, R. A., Oeffinger, M., & Ortega, J. (2022). RbgA ensures the correct timing in the maturation of the 50S subunits functional sites. *Nucleic Acids Research*, 10801–10816. https://doi.org/10.1093/nar/gkac059

Sergeeva, O. V., Sergiev, P. V., Bogdanov, A. A., & Dontsova, O. A. (2014). Ribosome: Lessons of a molecular factory construction. *Molecular Biology*, *48*(4). https://doi.org/10.1134/S0026893314040116

Sergiev, P. V., Serebryakova, M. V., Bogdanov, A. A., & Dontsova, O. A. (2008). The ybiN gene of *Escherichia coli* encodes adenine-N6 methyltransferase specific for modification of A1618 of 23 S ribosomal RNA, a methylated residue located close to the ribosomal exit tunnel. *Journal of Molecular Biology*, *375*(1). https://doi.org/10.1016/j.jmb.2007.10.051

Shajani, Z., Sykes, M. T., & Williamson, J. R. (2011). Assembly of bacterial ribosomes. *Annual Review of Biochemistry*, *80*. https://doi.org/10.1146/annurev-biochem-062608-160432

Sharma, I. M., & Woodson, S. A. (2020). RbfA and IF3 couple ribosome biogenesis and translation initiation to increase stress tolerance. *Nucleic Acids Research*, *48*(1). https://doi.org/10.1093/nar/gkz1065

Smith, M. T. J., & Rubinstein, J. L. (2014). Beyond blob-ology. *Science*, *345*(6197). https://doi.org/10.1126/science.1256358

Stephan, N. C., Ries, A. B., Boehringer, D., & Ban, N. (2021a). Structural basis of successive adenosine modifications by the conserved ribosomal methyltransferase KsgA. *Nucleic Acids Research*, *49*(11). https://doi.org/10.1093/nar/gkab430

Stephan, N. C., Ries, A. B., Boehringer, D., & Ban, N. (2021b). Structural basis of successive adenosine modifications by the conserved ribosomal methyltransferase KsgA. *Nucleic Acids Research*, *49*(11). https://doi.org/10.1093/nar/gkab430

Tan, J., Jakob, U., & Bardwell, J. C. A. (2002). Overexpression of two different GTPases rescues a null mutation in a heat-induced rRNA methyltransferase. *Journal of Bacteriology*, *184*(10). https://doi.org/10.1128/JB.184.10.2692-2698.2002

Toh, S. M., Xiong, L., Bae, T., & Mankin, A. S. (2008). The methyltransferase YfgB/RlmN is responsible for modification of adenosine 2503 in 23S rRNA. *RNA*, *14*(1). https://doi.org/10.1261/rna.814408

Traub, P., & Nomura, M. (1968a). Structure and function of *E. coli* ribosomes. V. Reconstitution of functionally active 30S ribosomal particles from RNA and proteins. *Proceedings of the National Academy of Sciences of the United States of America*, *59*(3). https://doi.org/10.1073/pnas.59.3.777

Traub, P., & Nomura, M. (1968b). Structure and functior of *Escherichia coli* ribosomes. I. Partial fractionation of the functionally active ribosomal proteins and reconstitution of artificial subribosomal particles. *Journal of Molecular Biology*, *34*(3). https://doi.org/10.1016/0022-2836(68)90182-4

Traub, P., & Nomura, M. (1969a). Structure and function of *Escherichia coli* ribosomes. VI. mechanism of assembly of 30 s ribosomes studied in vitro. *Journal of Molecular Biology*, *40*(3). https://doi.org/10.1016/0022-2836(69)90161-2

Traub, P., & Nomura, M. (1969b). Studies on the assembly of ribosomes in vitro. *Cold Spring Harbor Symposia on Quantitative Biology*, *34*. https://doi.org/10.1101/SQB.1969.034.01.010

Tu, C., Zhou, X., Tropea, J. E., Austin, B. P., Waugh, D. S., Court, D. L., & Ji, X. (2009). Structure of ERA in complex with the 3′ end of 16S rRNA: Implications for ribosome biogenesis. *Proceedings of the National Academy of Sciences of the United States of America*, *106*(35). https://doi.org/10.1073/pnas.0904032106

van Buul, C. P. J. J., Visser, W., & van Knippenberg, P. H. (1984). Increased translational fidelity caused by the antibiotic kasugamycin and ribosomal ambiguity in mutants harbouring the ksgA gene. *FEBS Letters*, *177*(1). https://doi.org/10.1016/0014-5793(84)80994-1

Verstraeten, N., Knapen, W. J., Kint, C. I., Liebens, V., Van den Bergh, B., Dewachter, L., Michiels, J. E., Fu, Q., David, C. C., Fierro, A. C., Marchal, K., Beirlant, J., Versées, W., Hofkens, J., Jansen, M., Fauvart, M., & Michiels, J. (2015). Obg and membrane depolarization are part of a microbial bet-hedging strategy that leads to antibiotic tolerance. *Molecular Cell*, *59*(1). https://doi.org/10.1016/j.molcel.2015.05.011

Wrzesinski, J., Bakin, A., Ofengand, J., & Lane, B. G. (2000). Isolation and properties of *Escherichia coli* 23S-RNA pseudouridine 1911, 1915, 1917 synthase (RluD). *IUBMB Life*, *50*(1). https://doi.org/10.1080/15216540050176566

Zhang, X., Yan, K., Zhang, Y., Li, N., Ma, C., Li, Z., Zhang, Y., Feng, B., Liu, J., Sun, Y., Xu, Y., Lei, J., & Gao, N. (2014). Structural insights into the function of a unique tandem GTPase EngA in bacterial ribosome assembly. *Nucleic Acids Research*, *42*(21). https://doi.org/10.1093/nar/gku1135

Zhao, M., Zhang, H., Liu, G., Wang, L., Wang, J., Gao, Z., Dong, Y., Zhang, L., & Gong, Y. (2016). Structural insights into the methylation of C1402 in 16S rRNA by methyltransferase RsmI. *PLoS ONE*, *11*(10). https://doi.org/10.1371/journal.pone.0163816

25 Cryo-EM Structures of Ribosomes from Human Pathogens

Implications for Drug Targeting

Ankit Dhur, Aneek Banerjee, Krishnamoorthi Srinivasan, and Jayati Sengupta

INTRODUCTION

Elucidating the three-dimensional (3D) structure of biological macromolecules is very important for understanding their key biological functions, identifying novel drug targets, and developing more specific antibiotics. Obtaining the structure of the ribosome and its subunits was a several-decades-long pursuit for several groups all over the world. The first structure of the ribosome at atomic resolution was solved in the year 2000. The structure of the ribosomal large 50S subunit from *Haloarcula marismortui* was determined using X-ray crystallography at a resolution of 2.4 Å by the group of T. A. Steitz (Ban et al., 2000). Later, in 2001, the crystal structure of the 50S subunit from *Deinococcus radiodurans* was determined at a resolution of 3.1 Å by Ada Yonath's group (Harms et al., 2001). Similarly, the structure of the ribosomal small subunit 30S was determined in the latter half of the year 2000 by two independent groups, Ada Yonath's and V. Ramakrishnan's groups at atomic resolutions of 3.3 Å and 3 Å, respectively, from *Thermus thermophilus* (Pioletti et al., 2001; Wimberly et al., 2000). The complete crystal structure of the 70S ribosome from *Thermus thermophilus* was elucidated in the year 2001 by H. F. Noller and team at a resolution of 5.5 Å (Yusupov et al., 2001). Several more crystal structures followed, and X-ray crystallography helped us understand several unknown aspects of ribosome biology, including the binding sites of several antibiotics. The first crystal structure of the eukaryotic 80S came much later in the year 2011 from *Saccharomyces cerevisiae*, which reported a resolution of 3 Å (Ben-Shem et al., 2011).

Contemporaneous with X-ray crystallography, several groups around the world were focused on the development of a new technique for structural elucidation, cryo-electron microscopy (cryo-EM) (Murata and Wolf, 2018). While X-ray crystallography had long been established as a successful technique for deciphering the structure of proteins and small molecules, it still has several limitations for deciphering the structure of large complexes (Zheng et al., 2015). Obtaining crystals of large and/or dynamic samples is a big challenge, and it may require years of optimization and standardization. Crystallization stabilizes the molecules in one or a few stable states, and hence one may miss out on the various dynamic states of the molecules, which may be crucial for understanding the physiological role of the protein/complex.

Cryo-EM is different because it enables researchers to study the structure of biomolecules in their native environment, without the need for crystallization. In this technique, the solution of proteins or other biomolecules is loaded onto grids and flash-frozen to cryogenic temperature, then bombarded with electrons to produce images of individual molecules (Callaway, 2020; Fernandez-Leiro and Scheres, 2016). The first reported cryo-EM structure of the bacterial ribosome dates back to 1991, at a resolution of ~40 Å (Frank et al., 1991). Although the resolution was a major limitation, this technique provided several important insights, especially about the ribosomes, which proved to be an excellent subject molecule for this technique (Brown and Shao, 2018). With technological advances in the field of cryo-EM, especially the development of direct detectors, cryo-EM saw a significant improvement in resolution, as cryo-EM started producing maps with resolution comparable to that of X-ray crystallography (Bai et al., 2015). This "resolution revolution" launched a golden era for cryo-EM, as the technique saw a marked increase in usage globally (Figure 25.1). Cryo-EM maps with resolution of less than 2 Å have been achieved in recent years, making it possible to analyze minute structural details of biomolecules. The number of structures solved using cryo-EM have been increasing exponentially in recent years, and scientists are preferring this technique over X-ray crystallography for determining the structure of biomolecules, especially the ribosome (Callaway, 2020).

The bacterial ribosome is a major target for most antibiotics, which act by inhibiting protein synthesis (Blanchard et al., 2010). In recent years, the rapid rise in antimicrobial resistance has posed a serious threat to the human community, and identifying novel drug targets is a pressing need. Apart from prokaryotes, various eukaryotic pathogens infect humans, causing severe, life-threatening diseases. Eukaryotic pathogens often pose a greater challenge, as the medications against them often have side effects on the human host, owing to the similarity of the molecular machinery being targeted by the drug. Understanding the detailed structure of ribosomes from different pathogenic organisms will help us to identify new targets for the development of new antimicrobials. A species-wise synopsis of the structure of the ribosome of various pathogens, and select drugs that interact with them, is given below.

DOI: 10.1201/9781003326106-30

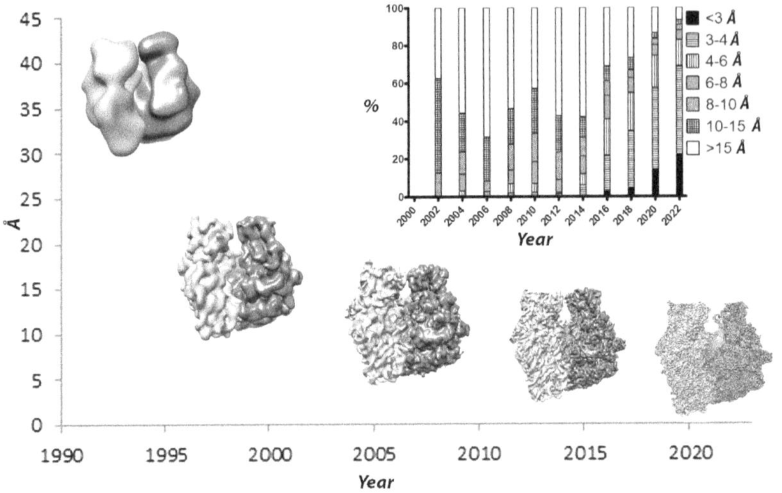

FIGURE 25.1 Advancement in achievable resolution from cryo-EM with the ribosome as a model. The journey started with a 40 Å map in 1991 followed by 15–20 Å in 1995–2000, 10–12 Å in 2000–2010, 5–8 Å in 2011–2016, and 1.55 Å in 2022. The bar graph (*inset*) represents the percentage of maps of specific resolution range deposited to the EMDB in even years since 2002, showing that 3–4 Å maps are now routinely generated.

UNDERSTANDING STRUCTURAL ASPECTS OF PATHOGENIC RIBOSOMES USING CRYO-EM

The scientific and medical community is engaged in an endless arms race against pathogenic organisms in a quest for survival. Pathogens are constantly adapting and evolving resistance against existing treatments, rendering them ineffective and leaving mankind searching for new drugs (Frieri et al., 2017). A large number of existing antibiotics and frontline drugs classically target the translation machinery (Poehlsgaard and Douthwaite, 2005). The development of structural biology tools such as cryo-EM has opened new avenues for understanding the biology of these pathogens (Herrero del Valle and Innis, 2020; Shepherd et al., 2022). Understanding the structural aspects of these molecular machines, and their interaction with various drugs, can provide us with novel insights into the working of these drugs and consequently can help us understand the evolution of resistance and pave the way toward the engineering of targeted drugs. To date, several species of pathogens belonging to multiple kingdoms have been studied in great detail using cryo-EM, some of which are summarized below.

Structural Insights into Pathogenic Bacterial Ribosomes

The bacterial ribosome is a ribonucleoprotein macromolecule with a molecular mass of approximately 2.5 MDa and a diameter of around 250 Å. It is composed of a large subunit (50S) having 5S rRNA, 23S rRNA, and approximately 31 ribosomal proteins and a small subunit 30S consisting of a 16S rRNA and approximately 21 ribosomal proteins (Kurland, 1972). The small 30S subunit is flexible and highly dynamic during the translation cycle. It functions to control the interaction of the tRNA with the mRNA and holds it in the proper reading frame. The anticodon stem loop (ASL) of the P-site tRNA directly interacts with the 30S subunit head and the platform domains (Watson et al., 2020). High-resolution cryo-EM structures of ribosomes from some of the pathogenic bacterial species are discussed below.

Escherichia

Among all the bacterial species, *Escherichia coli* (*E. coli*) is the most widely studied model organism. It colonizes in the gastrointestinal tract of humans and other mammals and is mostly nonvirulent, but in some cases it may prove to be a deadly pathogen (Clarke et al., 2003; Smith and Fratamico, 2017). Multiple structures of the *E. coli* ribosome and ribosomal subunits have been elucidated to date, including those of ribosomes trapped in various states of translation, and these have greatly improved our understanding of various aspects of protein synthesis. Cryo-EM has successfully resolved the *E. coli* ribosome at and below a resolution of 2 Å (Fromm et al., 2023; Watson et al., 2020). The structure helps us to understand the detailed positioning of the ribosomal proteins and the RNA residues along with their specific chemical interactions and modifications (Figure 25.2a).

An important ribosomal protein of the 30S subunit is the protein bS21, which has connections with the 30S subunit head domain through 13 amino acid C-terminal residues (Marzi et al., 2007; Sashital et al., 2014). The residues form an alpha helical conformation near the Shine-Dalgarno helix formed between the 3′ end of the 16S rRNA and the mRNA ribosome binding site. The Arg-Leu-Tyr (R-L-Y) residues in the C-terminal interact with the 16S rRNA helix h37 and the nucleotide A1167. Multiple sequence alignment of bS21 sequences from different bacterial phyla has shown that the C-terminal R-L-Y motif is

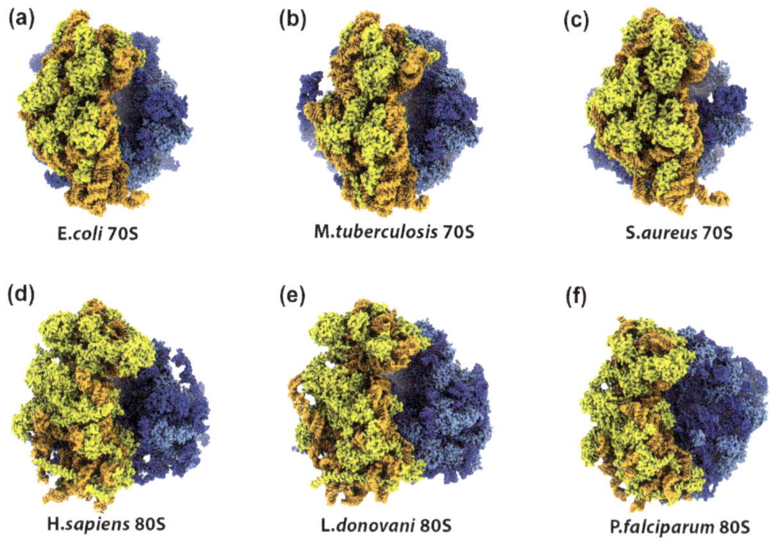

FIGURE 25.2 Visualization of the cryo-EM structures of the ribosome from various prokaryotic and eukaryotic organisms. (a) *E. coli* (3 Å); (b) *M. tuberculosis* (2.8 Å); (c) *S. aureus* (3.1 Å); (d) *H. sapiens* (3.6 Å); (e) *L. donovani* (2.9 Å); (f) *P. falciparum* (3.2 Å). Color codes are the following: LSU rRNA, dark blue; LSU protein, light blue; SSU rRNA, orange; SSU protein, yellow.

conserved in the Gammaproteobacteria phylum, while in Betaproteobacteria, bS21 possesses a Lys-Leu-Tyr (K-L-Y) motif. But no such amino acid motif is reported in other bacterial phyla.

A central component of the 30S subunit is the ribosomal protein uS11, which assembles cooperatively with the two other proteins uS6 and uS18 (Stern et al., 1988). The protein uS11 stabilizes the 16S rRNA, which forms the platform of the P-site by making direct contact with the 16S rRNA residues in a three-helix junction. The structural data have revealed the presence of a protein modification, an iso-aspartyl residue at the N119 position in the protein uS11 (David et al., 1999; Watson et al., 2020). This residue is shown to be universally conserved across bacteria, archaea, and eukaryotes and has a role in ribosome assembly and function.

The ribosomal subunit interface is an important site where large conformational changes occur during translation, which is necessary for the functioning of the ribosome. The 23S rRNA helix H69 of the 50S subunit is closely connected to the 30S subunit with the 16S rRNA helix h44 near the mRNA decoding center and the tRNA binding site. The H69 forms specific interactions with the A-site tRNA during translation, while the stem of the H69 compresses as the 30S subunit is rotated during the translational translocation event (Selmer et al., 2006). An additional bridge is formed between the 50S and the 30S subunits by 23S rRNA nucleotide G713-A718 helix H34 (Dunkle et al., 2011). The core of the 50S subunit is a lot more rigid than the 30S subunit, and hence it can be resolved at a much higher resolution. From the structural data it is evident that the protein uL16 is posttranslationally modified, with the presence of a beta-hydroxy-arginine at position 81 (Ge et al., 2012). Additionally a thio-peptide bond is present between Met82-Gly83 in the uL16, which is a posttranscriptional modification (Watson et al., 2020).

The translation machinery is largely conserved across species; hence the in-depth understanding of the *E. coli* ribosome and its complexes is fundamental to understanding global and specific properties of translation. It may also serve as a benchmark against which other bacterial ribosomes may be compared and analyzed.

Mycobacteria

Mycobacterium tuberculosis (Mtb) is the causative agent of tuberculosis (TB), which is responsible for approximately 1.5 million annual deaths worldwide and so is still a major cause of concern (Smith, 2003). Particularly with the rise of multidrug-resistant TB as well as the persistent threat of latent TB, the development of new antibiotics and the prerequisite identification of new possible targets become even more crucial.

The structure of the *Mycobacterium tuberculosis* ribosome (Mtb ribosome) is similar to that of other bacteria except it contains species-specific rRNA expansion segments (ESs) and unique ribosomal proteins (Yang et al., 2017) (Figure 25.2b). The rRNA expansion segments are found in the 5′ half of the 23S rRNA present in the 50S subunit. The H25 and H31a expansion segments are located on the solvent-exposed side of the large subunit, while the H25 leans toward the L7/L12 stalk and H31a toward the central protuberance, respectively. The expansion segments H15 and H16a are located close to the L1 stalk of the large subunit on the solvent-exposed side with their tips forming a kissing loop (Cannone et al., 2002). This expansion segment of H15 interacts and stabilizes the alpha helix of bL9. Another important expansion segment is the H54a (handle), which is 100 nt long and is present in the 23S rRNA. It is responsible for forming a new intersubunit bridge B9 by switching the interaction from rRNA helix H68-uL2 to H68-bS6 in the Mtb ribosome (Yang et al., 2017).

Mtb ribosomes are intrinsically resistant to multiple common antibiotics, including beta-lactam antibiotics such as penicillin and ampicillin. They are also capable of rapidly evolving resistance against several frontline antibiotics. Multiple structures of Mtb ribosomes bound to various mycobacterium-targeting antibiotics have been solved using cryo-EM. The structure of 70S ribosome bound to capreomycin, a drug prescribed for treating mycobacterial infections, showed that the binding position of capreomycin is near the conserved decoding center, near the helix H44 of the 16S rRNA (Zhang et al., 2022). LZD-114, a more potent inhibitor of Mtb ribosome, binds near the peptidyl transferase center in the 50S subunit by interacting with A2296, C2745, and C2848 (Yang et al., 2017).

Staphylococcus

Staphylococcus aureus is a bacterial pathogen responsible for causing acute and chronic infections, with an increasing number of strains becoming antibiotic resistant (Cheung et al., 2021). The cryo-EM structure of *S. aureus* ribosome was reported by Golubev et al. at a resolution of 3.2 Å (Golubev et al., 2020). It identifies several rRNA modifications that are unique to its ribosome (Figure 25.2c). There are altogether six modifications reported in the 23S rRNA. C1947 is positioned at the intersubunit bridge B2a, where the loop of 23S rRNA H69 contacts h44 of 16S rRNA and is modified to 2′-O-methylcytidine. The U1966 located near the A-site tRNA binding pocket is methylated; similarly, G2278 is methylated at 2′-O of ribose located near the P-site. G2472 located near the peptidyl transferase center (PTC) in the H74 of the 23S rRNA contains a methyl group at the N2 position. Another two nucleotides C2525 and A2530 located near the PTC are methylated. Similar to the 23S rRNA, the 16S rRNA also contains many modifications. The G535, located near the A-site of the 30S subunit, is methylated. This site is very close to the streptomycin binding site, and the methylation has a role in providing *S. aureus* resistance against streptomycin (Brakier-Gingras and Phoenix, 1984). C1412, A1529, and A1530 located near the P-site are dimethylated, with C1412 having a role in forming a binding pocket for P-site tRNA on the 30S subunit. These rRNA modifications at important functional sites of the ribosome can act as novel targets for the development of antibiotics and can be exploited to serve as targets against antibiotic resistance mechanisms.

Enterococcus

Enterococcus faecalis is a gram-positive bacterial pathogen showing rising cases of antibiotic resistance. E. L. Murphy et al. reported the cryo-EM structure of the 70S ribosome of *E. faecalis* at a global resolution of 2.8 Å (Murphy et al., 2020). The distribution of the rRNA structural elements and other functional elements of the ribosome is similar to the other bacterial ribosomes, whereas their conformations are different. When compared to the *E. coli* ribosome, in the 16S rRNA the major structural differences are located in the solvent-exposed side and the peripheral sections of the beak (h33a), spur (h6), and body (h7, h10, h44). Similarly, in the 23S rRNA, the structural differences can be observed for the functionally important regions of the L1 and L7/L12 stalk. Some ribosomal proteins of the *E. faecalis* ribosome differ from other bacteria, with these differences predominantly located in the peripheral regions and on the solvent-exposed side of both subunits. The C-terminal domain of UL15 is quite different, containing a number of loop regions that are shifted with respect to their locations in the *E. coli* ribosome. The structure of the bL33 protein is different from that found in *E. coli* and is quite similar to the Zn-binding paralog observed in the *T. thermophilus* ribosome.

Structural studies have shown that the antibiotic binding sites for the antibiotics commonly used to treat enterococcal infections are conserved. The binding site of linezolid is near the A-site pocket, where the base U2506 is shifted approximately 3 to 4 Å when compared to *E. coli* ribosome (Wilson, 2014). For tigecycline (Jenner et al., 2013), which inhibits the transport of aminoacylated tRNAs to the A-site by EF-Tu, the binding site is slightly altered in *E. faecalis*. The U1196 in the 23S helix 34 is replaced with an adenine, while the critical U1196, which coordinates with tigecycline, is shifted to a loop region.

Ribosome Structures from Pathogenic Eukaryotes

Targeting human pathogens may have side effects on the human host, owing to the similarity of the molecular machinery being targeted by the drug, which is highly undesirable. Hence, given the need to design antibiotics that target the pathogens specifically, we need a comprehensive idea about the structure of the human ribosome and the differences that exist between the ribosomes of humans and pathogens. High-resolution cryo-EM structures play a significant role in paving the way to understanding the minute differences.

The ribosomes of model eukaryotic organisms, such as yeasts, have been studied structurally for decades. The structure of the human ribosome has also been solved to a great resolution using cryo-EM (Figure 25.2d). The first structure of the human 80S ribosome was published by Khatter et al. at 3.6 Å. This structure highlighted the unique rRNA modifications which are absent in bacterial ribosomes (Khatter et al., 2015). These include posttranscriptional modifications like methylation on the ribose sugar and at different positions on the nitrogenous bases, as well as the presence of pseudouridine residues (Natchiar et al., 2017). A majority of these rRNA modifications are concentrated on the 18S and 28S rRNAs. The structure illustrates that there are altogether 73 sites that contain human-specific rRNA modifications predominated by 2′-O methylation and pseudouridylation. The bulge loop in the H37 near the ribosomal A-site contains two pseudouridine (ψ) residues (ψ1677 and ψ1683), and the PTC has four eukaryotic-specific ψ residues (ψ4403, 4442, 4457, and 4500). Three conserved nucleotides in eukaryotes A2363, G2364, and C2365 have 2′-OH methyl groups present near the H26-H47 junction and stabilize the H73 at the peptide exit tunnel. The ribosomal decoding center also contains specific nucleotide modifications; C1703 in the mRNA channel is methylated, and the presence of ψ1243 above the A-site and in the P-site contain a unique m1acp3Ψ1248 modification (Natchiar et al., 2017).

These studies are of supreme importance because they help us understand the translation machinery of the hosts, and they can be used in predicting whether a particular drug (against a eukaryotic pathogen) may have side effects on the host. A few persistent pathogens have been studied in this light in recent years, some of which are outlined below.

Giardia

Giardiasis, an acute diarrheal disease, is caused by the anerobic parasitic protist *Giardia lamblia*. It is mostly prevalent in developing countries (Laupland et al., 2005; Petri, 2005). The arsenal of drugs available to treat this disease often causes a wide range of side effects, which leads to patients stopping treatment and in turn the generation of antibiotic resistance (Lalle, 2010). Among the approved medicines, only paromomycin (PAR) targets the ribosome of the pathogen. However, treatment with PAR causes several side effects because of its binding to rRNA and h44 of the small subunit. These regions are highly conserved among prokaryotes, eukaryotes, and mitochondrial ribosomes (Shalev-Benami et al., 2017). A 2.75 Å cryo-EM map of *Giardia lamblia* 80S cytosolic ribosome provided very interesting observations regarding its variations. The rRNA of the *Giardia* ribosome is shorter compared to the typical prokaryotic, parasitic, and fungal rRNA (Edlind and Chakraborty, 1987). Comparing the rRNA to the human counterpart highlighted major differences (Hiregange et al., 2022). The rRNA expansion segments that are present on the head and foot regions of the small subunit are absent in *Giardia*. The large subunit also lacks any expansion segment in the solvent-exposed side. The 5.8rRNA is considerably shorter in *Giardia*, although the base pairings are conserved. The rRNA helices of the small subunit, like h6, h16, h17, h40, h41, and h44, are shorter than in its human counterpart. On the other hand, h9, h10, h21, and h39 are completely missing. In the large subunit, helices H14, H25, H30, H49, H63, H98, and H99 are shorter with significant structural differences, whereas helices H59 and H78 are absent. The regions with missing expansion segments also show variation in ribosomal proteins that stabilize those expansion segments. Two prominent ribosomal proteins, eL6 and eL28, which are present in humans are absent in *Giardia*. These proteins are responsible for stabilization of expansion segment ES7-ES15 of the human ribosome. Due to their absence in *Giardia*, the RNA–protein and protein–protein interactions of these regions vary significantly (Hiregange et al., 2022). There is also nucleotide bias in the rRNA sequences of *Giardia* with a lowering of A/U nucleotides (Eiler et al., 2020). The exact reason of such biasness is not clear yet. Another major structural difference between the *Giardia* and human ribosomes lies near the exit tunnel. The loop of the uL4 is considerably shorter in *Giardia*, which in turn results in a much wider exit tunnel as compared to its human counterpart (Majumdar et al., 2023). A thorough understanding of these significant details will pave the way for more efficient species-specific drug design.

Leishmania

Leishmania donovani is the causative agent of visceral leishmaniasis. It belongs to a distinct order of protozoan parasite called kinetoplastids, which diverged significantly from other eukaryotic species. Structural and biochemical studies of kinetoplastid cytosolic ribosomes have identified several unique features, like the unusual segmentation of 26S rRNA, a large number of sites for rRNA modification, and enlarged expansion segments (Figure 25.2e) (Eliaz et al., 2015; Hashem et al., 2013; Hernández and Cevallos, 2014). In 2016, a 2.8 Å cryo-EM map of the *Leishmania* LSU (large subunit) made it possible to visualize the unusual rRNA modifications (Shalev-Benami et al., 2016). Although *Leishmania* has a very low sequence homology with other eukaryotes, the structures of its ribosomal proteins are homologous to the already reported high-resolution structures of yeast, human, and plasmodium (Ben-Shem et al., 2011; Khatter et al., 2015; Wong et al., 2014). The two prominent differences in the leishmanial LSU are the 60 amino acid extension in the C-terminus of eL19 and the absence of eL41. These two proteins are absent in prokaryotes, and they play a significant role in the formation of bridge eB12 and eB14 in eukaryotes. The rRNA expansion segments vary considerably in the leishmanial LSU compared to other eukaryotes (Table 25.1). The variation localizes mostly in the peripheral region and most likely to interact with components that aid in translation (Shalev-Benami et al., 2016). As these regions are far away from the conserved PTC, they can be potential species-specific drug targets.

The aminoglycoside (AG) group of antibiotics is used to treat visceral leishmaniasis. Their mode of action is well studied in prokaryotes, and extrapolating that information yielded the speculation that their function was on parasitic mitochondrial ribosomes owing to the presence of adenine 1408 (Maarouf et al., 1997). Recent structural studies raised the possibility of AG binding to the cytosolic ribosome of *Leishmania*. A 2.7 Å cryo-EM map of leishmanial cytosolic ribosome bound to paromomycin made it possible to model at atomic resolution (Shalev-Benami et al., 2017). Previous studies reported the absence of eukaryotic protein eL41 in *Leishmania* and other typanosomatid ribosomes (Shalev-Benami et al., 2016). Surprisingly, the density corresponding to eL41 was observed in the PAR-binding pocket. The localization of eL41 adjacent to h44 and its interaction with several residues responsible for interaction with PAR might be of clinical significance. Visualization of the drug-binding pocket paved the way for a better understanding of its mechanism of action and laid the foundation for further research, an aspect discussed in a later part of this chapter.

TABLE 25.1
List of some of the important rRNA expansion segments of eukaryotic pathogens and their differences from human counterparts

Organism	Expansion Segment	Difference from Human
Plasmodium falciparum (Pf80S)	ES2S	Loop is shorter in Pf80S
	ES3S	Helix A is conserved but B truncated
	ES7S, ES9S, ES10S	Expanded in Pf80S
	ES12S	Helix truncated in Pf80S
	ES3L, ES4L, ES5L, SE41L	Conserved
	ES10L, ES30L	Absent in Pf80S
	ES12L	Expanded in Pf80S
	ES19L	Truncated in Pf80S
Giardia lamblia (Gl)	ES27	Highly reduced in Gl
	ES6C	Significantly reduced in Gl
	ES39L, ES7L	Absent in Gl
	ES3S, ES4S, ES5S, ES6S	Absent in Gl
Trypanosoma brucei (Tb)	ES3S, ES6S, ES7S, ES9S	Larger in Tb
	ES4L, ES7L, ES19L, ES27L, ES31L	Larger in Tb

Trypanosoma

Another kinetoplastid, *Trypanosoma brucei*, is the causative agent of African sleeping sickness. Its ribosome also possesses unique features that are absent in other eukaryotes, including humans, and can be used as organism-specific targets (Yokoyama and Suzuki, 2008). High-resolution cryo-EM maps allowed us to model the ribosome with great detail and use the information for further biological understanding. One notable feature is the presence of several expansion segments. The small ribosomal subunit (SSU) specific expansion segments like ES3S, ES6S, ES7S, and ES9S are significantly larger than in other known structures of 80S ribosomes. ES4L, ES7L, ES19L, ES27L, and ES31L belonging to the large subunit (LSU) are also considerably larger than in other known eukaryotic ribosomes (Hashem et al., 2013). An additional domain on the LSU, known as the kinetoplastid-specific domain (KSD), is situated in the vicinity of ES7L. The largest ESs, which are mostly clustered in the LSU L1 stalk and the platform region of SSU, favor the formation of several intersubunit bridges which are absent in the yeast 80S ribosome. The LSU rRNA is processed into six pieces consisting of two large ribosomal RNAs, LSU-α and LSU-β, and four short ribosomal RNAs, srRNA1 through srRNA4 (Campbell et al., 1987; Cordingley and Turner, 1980; White et al., 1986). Structural information highlights that srRNA2, srRNA3, and srRNA4 interact with the KSD to form a stable domain similar to domain VI of other eukaryotic ribosomes. The exact function of the expansion segments are not clear, but the localization of ES6S and ES7S near the mRNA exit channel makes them equipped for translation initiation regulation (Siridechadilok et al., 2005). Further research regarding ES-mediated translation regulation can open new avenues for treating kinetoplastid infections.

Plasmodium

Malaria is caused by a unicellular protozoan parasite belonging to the genus *Plasmodium*. The parasite undergoes a complex life cycle with the sexual stage in the mosquito vector and the asexual phase in the human host (Miller et al., 2002). Most of the pathological symptoms arise due to the exponential growth of *Plasmodium* within the host erythrocytes; thus it becomes crucial to understand the translational machinery. Parasitic protein synthesis occurs in three centers, the cytoplasmic ribosome, the mitochondrial ribosome, and the nonphotosynthetic plastid called the apicoplast (McFadden et al., 1996). Interestingly, *Plasmodium* species possess two distinct types of cytoplasmic ribosomes that vary in rRNA composition and are expressed in different phases of the life cycle, one specifically in the vector and another predominantly in the human host (Waters et al., 1989). Various antibiotics like macrolide azithromycin target the apicoplast ribosome and show delayed action (Dahl and Rosenthal, 2007). Treating malaria at the blood stage calls for prompt intervention, so targeting the blood-phase-specific cytoplasmic ribosome is the more logical choice. Cryo-EM plays a significant part in understanding the ribosomal differences between the host and the pathogen (Figure 25.2f). Visualizing the drug-binding sites not only allows for the generation of new drugs but also facilitates repurposing of existing antiprotozoan inhibitors (Matthews et al., 2013).

Comparing the parasitic 80S ribosome with its human counterpart highlighted the differences between them. The core structure is conserved with extensive differences in the peripheral regions arising from varied rRNA ES and protein extensions. The ESs of *P. falciparum* are generally shorter, and some (ES10L, ES20L, ES30L) are even absent in the large subunit (Wong et al., 2014). Some of the important expansion segments and their differences from human counterparts are shown in tabular form (Table 25.1). The exact functions of these segments are not clear yet. ES7E, which is highly conserved in vertebrates, facilitates the binding of SBP2 protein, which in turn recruits selenocysteine-specific tRNA and elongation factor (Kossinova et al., 2014). *Plasmodium* lacks this ES7E and any homolog of SBP2, which explains the low abundance of selenocysteine in *Plasmodium* proteins. There exist two stabilizing interactions that are absent in human ribosome: (1) The smallest ribosomal protein, eL41, bridges the two subunits, and its *Plasmodium*-specific N-terminal extension anchors the protein by accessing the 18S rRNA pocket (Ben-Shem et al., 2011). (2) The platform region of *Plasmodium* 40S forms a small bridge by interacting with regions of L1 stalk. These differences can be targeted as novel targets. A high-resolution structure of Pf80S bound to a natural antiprotozoan molecule emetine highlighted the power of cryo-EM for structure-based drug design and repurposing of existing drugs. Visualizing the binding site makes it lot easier to modify drugs for less cytotoxicity and enhanced efficiency.

STRUCTURAL CHARACTERIZATION OF ANTIBIOTIC BINDING SITES

Most naturally occurring antibiotics tend to target the bacterial ribosome, a fact which is perhaps a testament to the importance of the molecular machineries (Tenson and Mankin, 2006). However, given the universal similarity, it is important to decipher the exact binding site and mechanism of inhibition by these drugs so as to understand how these molecules specifically target and inhibit the bacteria. Broad-spectrum antibiotics that target a general range of bacteria have been in use since ages and have specific binding sites on the ribosome, targeting either the large or small ribosomal subunits (discussed earlier). Most classic broad-spectrum antibiotics function by inhibiting one of several steps of translation (Table 25.2), such as kasugamycin-mediated inhibition of translation initiation, which prevents the formation of 70S initiation complex (Schuwirth et al., 2006). Tetracycline inhibits translation elongation by binding at the base of the head of the 30S subunit and preventing the binding of aminoacyl tRNA to the A-site of the ribosome (Thaker et al., 2010). Chloramphenicol binds near the peptidyl transferase center on the large subunit, inhibiting peptide bond formation between the aminoacyl tRNA in the A-site and the peptidyl tRNA in the P-site (Svetlov et al., 2019). Erythromycin inhibits the emergence of the nascent polypeptide chain through the ribosome exit tunnel by binding near the tunnel exit on the large subunit (Weisblum, 1995).

The structures of ribosomes bound to these antibiotics have been worked on for several decades. The high-resolution structure of the ribosome from different pathogenic bacteria has helped us to accurately identify the binding sites of the different classes of antibiotics and predict their mode of action. A summary of the different groups of antibiotics, their binding sites, and available structural information is given in Table 25.2.

TABLE 25.2
Major classes of antibiotics and their mode of action

Class of Antibiotic	Name	Mode of Action	PDB ID
Oxazolidinines **PTF Inhibitor**	Linezolid Sparsomycin	Prevents tRNA accommodation	3DLL
Elfamycins **Kirromycin**	Enacyloxin Kirromycin	Prevents the release of EFTu-GDP from 70S	2BVN
EFTu Inhibitor	Pulvomycin GE2270A	Inhibits the formation of EFTu-GTP-tRNA complex	2C78 2C77
Tetracyclines	Tetracyclin Doxycycline Sarecycline	Prevents the delivery of tRNA by EFTu-GTP	4V9A 8CRX
Lincosamides **PTF Inhibitor**	Lincomycin Chloramphenicol Puromycin Sparsomycin	Prevents peptide bond formation	8A5I 6ND5
Fusidic Acid	Fusidic acid	Prevents the release of EFG-GDP	4V9L
Aminoglycosides	Paromomycin Gentamicin Streptomycin	Inhibits ribosome translocation	4V8C 4DR3
Macrolides **Ketolides**	Erythromycin Telithromycin	Plugs the peptide exit tunnel	4V7U
Thiopeptides	Thiostrepton Micrococcin	Inhibits the action of EFG-FTP	3CF5

Pathogen-specific Antibiotic-Bound Ribosome Structures

The prediction of the binding site of different antibiotics is based on the similarity of the drug to existing antibiotics with known binding sites. However, the molecular details of the binding may differ between species, which in turn causes different levels of sensitivity in different organisms. Further, these small molecules often have multiple binding sites on the ribosome, which may manifest side effects on the host. The advancement of cryo-EM has allowed for the structural elucidation of different pathogenic ribosomes, bound to their specific drugs. Atomic reconstruction of these ribosome–drug interactions can be combined with structure-based drug design to design newer and more specific antibiotics in the near future. Several such structures have been obtained in the recent past, some of which are summarized below.

The structure of the 70S ribosome of *Cutibacterium acnes* reveals the exact binding pocket of the antibiotic sarecycline (SAR) (Figure 25.3a) (Lomakin et al., 2023). SAR is a narrow-spectrum antibiotic, a third generation tetracycline class antibiotic used for the treatment of acne (Moore et al., 2019). The structure depicts that SAR has two distinct binding sites. The canonical binding site is on the 30S subunit, located in the pocket formed by helix h31 and h34 of the 16S rRNA. The second site is located on the 50S subunit within the nascent peptide exit tunnel. It interferes with the growing polypeptide chain with its ring D close to the peptidyl transferase center and the ring A extending toward the exit of the nascent polypeptide exit tunnel (Lomakin et al., 2023).

Another antibiotic, clarithromycin (CTY), a modified form of the antibiotic erythromycin, is even effective against multidrug-resistant TB. The structure of the Mtb 50S ribosome–CTY complex was solved with a global resolution of 3.3 Å (Moore et al., 2019). It describes the specific binding position of the antibiotic, at the nascent polypeptide exit tunnel (NPET) near the peptidyl transferase center (Figure 25.3b). A key residue A2062, called the gate site, is an important interacting partner of the CTY which controls protein translation (Kannan et al., 2012). It describes how alternative conformations of A2062 (*syn*- and swayed conformations) bound with water molecules plays a role in the binding affinity of the antibiotic. The interaction between CTY and the ribosome is dominated by H-bond and van der Waals (VDW) interactions. The H-bond interaction is observed between A2058 and 2′-OH of the desosamine sugar and several VDW interactions between nucleotides 2062, 2503, 2505, and 2058 around the desosamine. The A2062 in the swayed conformation forms a Hoogsteen base pair with m2A2503 and base stacking interaction with G2061. The A2062 in the *syn*-conformation forms a VDW interaction with the 6-methoxy group of CTY and is stabilized by two intramolecular C-H H-bonds with the two sugars desosamine and cladinose.

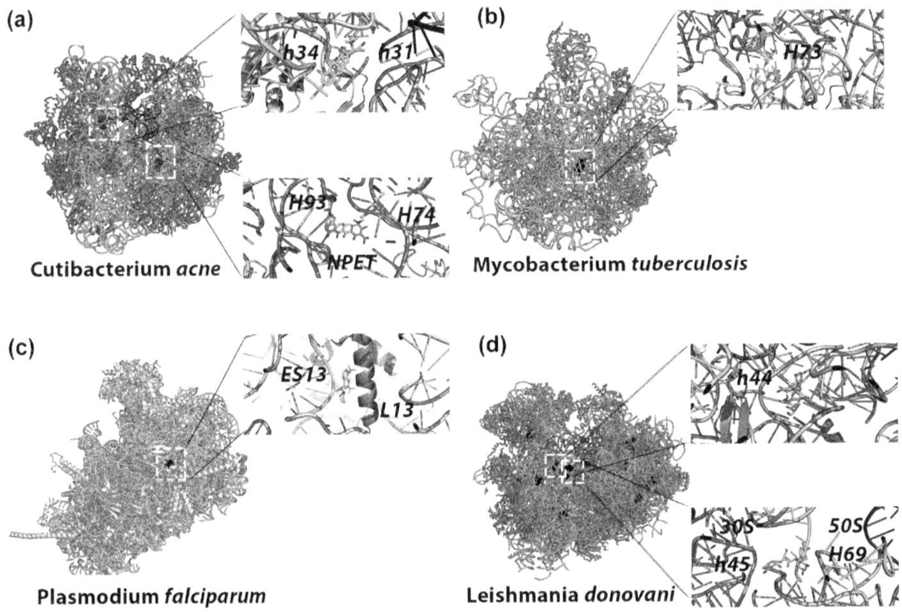

FIGURE 25.3 Identification of binding sites of pathogen-specific antibiotics on their ribosomes. (a) 70S ribosome of *Cutibacterium acnes* bound to sarecycline (PDB: 8CRX) on both 50S and 30S subunits. Inset represents the close-up view of the antibiotic binding site (30S *upper*; 50S *lower*). (b) 50S ribosomal subunit of *Mycobacterium tuberculosis* with clarithromycin (PDB: 7F0D) bound near the nascent polypeptide exit tunnel (NPET). Inset showing the antibiotic interaction region. (c) 60S ribosomal subunit of *Plasmodium falciparum* bound to mefloquine (PDB: 5UMD). The inset highlights the interaction of mefloquine with L13 and ES13 of LSU. (d) 80S ribosome of *Leishmania donovani* bound to multiple molecules of paromomycin (PDB: 6AZ3). The inset represents close-up views of the two significant interactions, one with the 40S subunit (*top*) and another with the 60S subunit (*bottom*).

The cryo-EM structure of the antimalarial Mefloquine (MFQ) bound to the 80S ribosome of the *Plasmodium falciparum* reveals its target and mode of action (Wong et al., 2017). The binding site of MFQ is located within the GTPase-associated center of the large subunit (Figure 25.3c). The binding pocket consists of the protein uL13, the sarcin-ricin loop of uL6, rRNA helix 94-5, and ES13. A second binding site was also detected which lies at the peripheral surface of the large subunit comprising the two residues Tyr290 and His294 from uL4. The high-resolution structure also serves as a platform for the development of more potent MFQ derivatives. Comparison between human cytosolic ribosome and *P. falciparum* ribosome revealed two substitutions in the drug binding pocket, (1) Glu55Ala and (2) Leu29Lys. Both of these substitutions drastically reduced the binding affinity of MFQ to the ribosome, proving these amino acid residues specific to the parasite ribosome to be critical for the stable binding of the drug. Exploiting these differences in the binding pocket, new derivatives of MFQ have been designed by introducing hydrophobic groups that can extend into the Leu59 region as well as maintain the H-bond interaction between the Glu55 and G1441 nucleotide (Wong et al., 2017). These derivatives showed increased potency compared to the parent MFQ.

The structure of the *Leishmania* ribosome in complex with the aminoglycoside paromomycin (PAR) reveals its mode of action (Shalev-Benami et al., 2017) (Figure 25.3d). PAR is a highly potent antimicrobial agent used for the treatment of visceral leishmaniasis (VL) that targets specifically the cytosolic ribosome (Wiwanitkit, 2012). The binding position of the antibiotic is on the small subunit at the tip of helix h44 at the decoding center. The binding pocket consists of two evolutionarily conserved nucleotides A2158 and A2159, evolutionarily conserved H69 of the large subunit, G626 of helix 18, and a highly conserved loop of the protein uS12. Additionally, the structure reveals the presence of two eukaryotic-specific ribosomal proteins eS30 and eL41 within the binding pocket (Shalev-Benami et al., 2017). Previous studies reported the absence of eukaryotic protein eL41 in *Leishmania* and other trypanosomatid ribosomes. Surprisingly, the density corresponding to eL41 was observed in the PAR-binding pocket. The localization of eL41 adjacent to h44 and its interaction with several residues responsible for interaction with PAR might be of clinical significance. The protein eS30 has been reported to have a role in ribosome biogenesis (Lacombe et al., 2009). The N-terminal domain of eS30 has direct contact with the PAR-binding pocket. Similarly, the structure of the *E. coli* ribosome also describes the binding position of the antibiotic paromomycin to the ribosome (Watson et al., 2020). It targets the mRNA decoding site at the A-site of the ribosome, where it establishes multiple contacts with the 16S rRNA. Ring number IV of paromomycin forms multiple contacts with the phosphate backbone of G1489 and U1490 in the 16S rRNA helix H44. High-resolution cryo-EM maps facilitate understanding of the fundamental processes and point out features that were previously not identified.

Structural Understanding of Antibiotic Resistance

The rising cases of antibiotic resistance in bacteria are becoming a global concern. There have been multiple structural studies to understand these mechanisms of resistance. Bacteria are constantly evolving several mechanisms for gaining resistance (Arenz and Wilson, 2016):

1. Target modification. Involves mutation or modifications in rRNA in the drug-binding pocket. The cryo-EM structure of the methyltransferase RmtC bound to the *E. coli* 30S subunit explains how methylation of G1405 within the ribosomal h44 in the 16S rRNA provides resistance against important aminoglycoside antibiotics (Srinivas et al., 2023). Mono- or dimethylation of A2058 carried out by Erm-methyltransferases in the 23S rRNA confers resistance against the macrolide class of antibiotics (Maravic, 2004). Methylation of A2503 by cfr methyltransferases provides resistance against streptogramin A, chloramphenicol, pleuromutilins, lincosamides, and oxazolidinones that bind at the PTC of the ribosome (Arenz and Wilson, 2016; Smith and Mankin, 2008).
2. Drug modification. The bacterium *Streptomyces hygroscopicus* is resistant to the drug HygA. The action of the enzyme phosphotransferase Hyg21, which transfers a gamma phosphate group from ATP to the ribose sugar 2′-OH of HygA, inactivates the drug (Ogawara, 2019). This phosphate group is involved in a steric clash with the nucleotide U2504 of the 23S rRNA and thus drastically reduces the drug's potency.
3. Recruitment of protection proteins. Proteins like TetO and TetM are recruited, which provides resistance to TET by competing for the binding pocket of the antibiotic on the ribosome. The cryo-EM structure of the TetM in complex with the translating ribosome reveals that the conserved Pro509 residue located in domain IV of TetM interacts with C1054 nucleotide within the 16S rRNA and hence inhibits the binding of TET on the ribosome (Arenz et al., 2015).

The 30S subunit has 11 posttranscriptionally modified nucleotides in the 16S rRNA which are instrumental in providing antibiotic resistance to bacteria. The partially methylated m7G527 located near the mRNA decoding center is responsible for providing a low level of streptomycin and neomycin resistance (Mikheil et al., 2012; Okamoto et al., 2007). The demethylation of two adenosine residues ($m_2{}^6$A1518 and $m_2{}^6$A1519) near the P-site leads to kasugamycin resistance (Ochi et al., 2009).

Exploiting this structural information will lead to the development of novel antibiotics. The recent structural understanding of the eukaryotic cytosolic and mitochondrial ribosome also has paved the way to better understand the differences with the bacterial ribosome, which will help us to develop more target-specific antibiotics with fewer side effects.

CONCLUDING REMARKS

Since the resolution revolution, cryo-EM has become a preferred structural biology tool for solving complex macromolecular structures. High-resolution maps providing atomic details help researchers understand the biological processes at a greater depth. The ribosome, the protein synthesis machine, is one of the most important cellular organelles, and targeting it with drugs is a widely used approach to treat pathogenic diseases. Visualizing the ribosomal differences between host and pathogens is crucial for designing molecules with high efficiency and minimal side effects. Cryo-EM contributes immensely in that aspect and is destined to play a pivotal role in future biomedical advances. Ribosomes of different organisms may possess characteristic structural dynamics which can be identified from high-resolution structures of multiple conformations and targeted for species-specific treatment. Many structures published in recent years have refined our understanding of the ribosome, but there is still left to uncover. A structural understanding of ribosomes from plant pathogens is also a need of the hour. Not much information is currently available on that, and we believe in coming years cryo-EM will play a crucial role in addressing that information gap.

ACKNOWLEDGMENTS

This study was primarily supported by SERB, DST (India) sponsored projects (SPF/2021/000141 and CRG/2019/001788), CSIR Niche Creating Project (NCP) MLP-139, and CSIR-Indian Institute of Chemical Biology (CSIR-IICB), Kolkata, India. AD, AB, and SK acknowledge UGC, India, DBT, India, and CSIR, India, respectively, for research fellowships.

CONFLICTS OF INTEREST

The authors declare no conflict of interest.

REFERENCES

Arenz, S., Nguyen, F., Beckmann, R., and Wilson, D.N. (2015). Cryo-EM structure of the tetracycline resistance protein TetM in complex with a translating ribosome at 3.9-Å resolution. Proceedings of the National Academy of Sciences *112*, 5401–5406.

Arenz, S., and Wilson, D.N. (2016). Blast from the past: reassessing forgotten translation inhibitors, antibiotic selectivity, and resistance mechanisms to aid drug development. Molecular Cell *61*, 3–14.

Bai, X.-C., McMullan, G., and Scheres, S.H. (2015). How cryo-EM is revolutionizing structural biology. Trends in Biochemical Sciences *40*, 49–57.

Ban, N., Nissen, P., Hansen, J., Moore, P.B., and Steitz, T.A. (2000). The complete atomic structure of the large ribosomal subunit at 2.4 A resolution. Science *289*, 905–920.

Ben-Shem, A., Garreau de Loubresse, N., Melnikov, S., Jenner, L., Yusupova, G., and Yusupov, M. (2011). The structure of the eukaryotic ribosome at 3.0 Å resolution. Science *334*, 1524–1529.

Blanchard, S.C., Cooperman, B.S., and Wilson, D.N. (2010). Probing translation with small-molecule inhibitors. Chemistry & Biology *17*, 633–645.

Brakier-Gingras, L., and Phoenix, P. (1984). The control of accuracy during protein synthesis in *Escherichia coli* and perturbations of this control by streptomycin, neomycin, or ribosomal mutations. Canadian Journal of Biochemistry and Cell Biology *62*, 231–244.

Brown, A., and Shao, S. (2018). Ribosomes and cryo-EM: a duet. Current Opinion in Structural Biology *52*, 1–7.

Callaway, E. (2020). Revolutionary cryo-EM is taking over structural biology. Nature *578*, 201–202.

Campbell, D.A., Kubo, K., Clark, C.G., and Boothroyd, J.C. (1987). Precise identification of cleavage sites involved in the unusual processing of trypanosome ribosomal RNA. Journal of Molecular Biology *196*, 113–124.

Cannone, J.J., Subramanian, S., Schnare, M.N., Collett, J.R., D'Souza, L.M., Du, Y., Feng, B., Lin, N., Madabusi, L.V., and Müller, K.M. (2002). The comparative RNA web (CRW) site: an online database of comparative sequence and structure information for ribosomal, intron, and other RNAs. BMC Bioinformatics *3*, 1–31.

Cheung, G.Y.C., Bae, J.S., and Otto, M. (2021). Pathogenicity and virulence of Staphylococcus aureus. Virulence *12*, 547–569.

Clarke, S., Haigh, R., Freestone, P., and Williams, P. (2003). Virulence of enteropathogenic *Escherichia coli*, a global pathogen. Clinical Microbiology Reviews *16*, 365–378.

Cordingley, J.S., and Turner, M.J. (1980). 6.5 S RNA; preliminary characterisation of unusual small RNAs in Trypanosoma brucei. Molecular and Biochemical Parasitology *1*, 91–96.

Dahl, E.L., and Rosenthal, P.J. (2007). Multiple antibiotics exert delayed effects against the Plasmodium falciparum apicoplast. Antimicrobial Agents and Chemotherapy *51*, 3485–3490.

David, C.L., Keener, J., and Aswad, D.W. (1999). Isoaspartate in ribosomal protein S11 of *Escherichia coli*. Journal of Bacteriology *181*, 2872–2877.

Dunkle, J.A., Wang, L., Feldman, M.B., Pulk, A., Chen, V.B., Kapral, G.J., Noeske, J., Richardson, J.S., Blanchard, S.C., and Cate, J.H. (2011). Structures of the bacterial ribosome in classical and hybrid states of tRNA binding. Science *332*, 981–984.

Edlind, T.D., and Chakraborty, P.R. (1987). Unusual ribosomal RNA of the intestinal parasite *Giardia lamblia*. Nucleic acids Research *15*, 7889–7901.

Eiler, D.R., Wimberly, B.T., Bilodeau, D.Y., Rissland, O.S., and Kieft, J.S. (2020). The *Giardia lamblia* ribosome structure reveals divergence in translation and quality control pathways. bioRxiv, 2020.2009.2030.321331.

Eliaz, D., Doniger, T., Tkacz, I.D., Biswas, V.K., Gupta, S.K., Kolev, N.G., Unger, R., Ullu, E., Tschudi, C., and Michaeli, S. (2015). Genome-wide analysis of small nucleolar RNAs of Leishmania major reveals a rich repertoire of RNAs involved in modification and processing of rRNA. RNA Biology *12*, 1222–1255.

Fernandez-Leiro, R., and Scheres, S.H. (2016). Unravelling biological macromolecules with cryo-electron microscopy. Nature *537*, 339–346.

Frank, J., Penczek, P., Grassucci, R., and Srivastava, S. (1991). Three-dimensional reconstruction of the 70S *Escherichia coli* ribosome in ice: the distribution of ribosomal RNA. The Journal of Cell Biology *115*, 597–605.

Frieri, M., Kumar, K., and Boutin, A. (2017). Antibiotic resistance. Journal of Infection and Public Health *10*, 369–378.

Fromm, S.A., O'Connor, K.M., Purdy, M., Bhatt, P.R., Loughran, G., Atkins, J.F., Jomaa, A., and Mattei, S. (2023). The translating bacterial ribosome at 1.55 A resolution generated by cryo-EM imaging services. Nature Communications *14*, 1095.

Ge, W., Wolf, A., Feng, T., Ho, C.H., Sekirnik, R., Zayer, A., Granatino, N., Cockman, M.E., Loenarz, C., Loik, N.D., *et al.* (2012). Oxygenase-catalyzed ribosome hydroxylation occurs in prokaryotes and humans. Nature Chemical Biology *8*, 960–962.

Golubev, A., Fatkhullin, B., Khusainov, I., Jenner, L., Gabdulkhakov, A., Validov, S., Yusupova, G., Yusupov, M., and Usachev, K. (2020). Cryo-EM structure of the ribosome functional complex of the human pathogen Staphylococcus aureus at 3.2 Å resolution. FEBS Letters *594*, 3551–3567.

Harms, J., Schluenzen, F., Zarivach, R., Bashan, A., Gat, S., Agmon, I., Bartels, H., Franceschi, F., and Yonath, A. (2001). High resolution structure of the large ribosomal subunit from a mesophilic eubacterium. Cell *107*, 679–688.

Hashem, Y., Des Georges, A., Fu, J., Buss, S.N., Jossinet, F., Jobe, A., Zhang, Q., Liao, H.Y., Grassucci, R.A., and Bajaj, C. (2013). High-resolution cryo-electron microscopy structure of the Trypanosoma brucei ribosome. Nature *494*, 385–389.

Hernández, R., and Cevallos, A.M. (2014). Ribosomal RNA gene transcription in trypanosomes. Parasitology Research *113*, 2415–2424.

Herrero del Valle, A., and Innis, C.A. (2020). Prospects for antimicrobial development in the cryo-EM era–a focus on the ribosome. FEMS Microbiology Reviews *44*, 793–803.

Hiregange, D.G., Rivalta, A., Bose, T., Breiner-Goldstein, E., Samiya, S., Cimicata, G., Kulakova, L., Zimmerman, E., Bashan, A., Herzberg, O., *et al.* (2022). Cryo-EM structure of the ancient eukaryotic ribosome from the human parasite *Giardia lamblia*. Nucleic Acids Research *50*, 1770–1782.

Jenner, L., Starosta, A.L., Terry, D.S., Mikolajka, A., Filonava, L., Yusupov, M., Blanchard, S.C., Wilson, D.N., and Yusupova, G. (2013). Structural basis for potent inhibitory activity of the antibiotic tigecycline during protein synthesis. Proceedings of the National Academy of Sciences of the United States of America *110*, 3812–3816.

Kannan, K., Vazquez-Laslop, N., and Mankin, A.S. (2012). Selective protein synthesis by ribosomes with a drug-obstructed exit tunnel. Cell *151*, 508–520.

Khatter, H., Myasnikov, A.G., Natchiar, S.K., and Klaholz, B.P. (2015). Structure of the human 80S ribosome. Nature *520*, 640–645.

Kossinova, O., Malygin, A., Krol, A., and Karpova, G. (2014). The SBP2 protein central to selenoprotein synthesis contacts the human ribosome at expansion segment 7L of the 28S rRNA. RNA *20*, 1046–1056.

Kurland, C. (1972). Structure and function of the bacterial ribosome. Annual Review of Biochemistry *41*, 377–408.

Lacombe, T., Garcia-Gomez, J.J., de la Cruz, J., Roser, D., Hurt, E., Linder, P., and Kressler, D. (2009). Linear ubiquitin fusion to Rps31 and its subsequent cleavage are required for the efficient production and functional integrity of 40S ribosomal subunits. Molecular Microbiology *72*, 69–84.

Lalle, M. (2010). Giardiasis in the post genomic era: treatment, drug resistance and novel therapeutic perspectives. Infectious Disorders Drug Targets *10*, 283–294.

Laupland, K.B., Bagshaw, S.M., Gregson, D.B., Kirkpatrick, A.W., Ross, T., and Church, D.L. (2005). Intensive care unit-acquired urinary tract infections in a regional critical care system. Critical Care *9*, 1–6.

Lomakin, I.B., Devarkar, S.C., Patel, S., Grada, A., and Bunick, C.G. (2023). Sarecycline inhibits protein translation in Cutibacterium acnes 70S ribosome using a two-site mechanism. Nucleic acids Research *51*, 2915–2930.

Maarouf, M., de Kouchkovsky, Y., Brown, S., Petit, P.X., and Robert-Gero, M. (1997). In vivo interference of paromomycin with mitochondrial activity of *Leishmania*. Experimental Cell Research *232*, 339–348.

Majumdar, S., Emmerich, A., Krakovka, S., Mandava, C.S., Svärd, S.G., and Sanyal, S. (2023). Insights into translocation mechanism and ribosome evolution from cryo-EM structures of translocation intermediates of *Giardia intestinalis*. Nucleic Acids Research *51*, 3436–3451.

Maravic, G. (2004). Macrolide resistance based on the Erm-mediated rRNA methylation. Current Drug Targets-Infectious Disorders *4*, 193–202.

Marzi, S., Myasnikov, A.G., Serganov, A., Ehresmann, C., Romby, P., Yusupov, M., and Klaholz, B.P. (2007). Structured mRNAs regulate translation initiation by binding to the platform of the ribosome. Cell *130*, 1019–1031.

Matthews, H., Usman-Idris, M., Khan, F., Read, M., and Nirmalan, N. (2013). Drug repositioning as a route to anti-malarial drug discovery: preliminary investigation of the in vitro anti-malarial efficacy of emetine dihydrochloride hydrate. Malaria Journal *12*, 1-11.

McFadden, G.I., Reith, M.E., Munholland, J., and Lang-Unnasch, N. (1996). Plastid in human parasites. Nature *381*, 482.

Mikheil, D.M., Shippy, D.C., Eakley, N.M., Okwumabua, O.E., and Fadl, A.A. (2012). Deletion of gene encoding methyltransferase (gidB) confers high-level antimicrobial resistance in Salmonella. The Journal of Antibiotics *65*, 185–192.

Miller, L.H., Baruch, D.I., Marsh, K., and Doumbo, O.K. (2002). The pathogenic basis of malaria. Nature *415*, 673–679.

Moore, A.Y., Charles, J.E.M., and Moore, S. (2019). Sarecycline: a narrow spectrum tetracycline for the treatment of moderate-to-severe acne vulgaris. Future Microbiology *14*, 1235–1242.

Murata, K., and Wolf, M. (2018). Cryo-electron microscopy for structural analysis of dynamic biological macromolecules. Biochimica et biophysica acta General Subjects *1862*, 324–334.

Murphy, E.L., Singh, K.V., Avila, B., Kleffmann, T., Gregory, S.T., Murray, B.E., Krause, K.L., Khayat, R., and Jogl, G. (2020). Cryo-electron microscopy structure of the 70S ribosome from *Enterococcus faecalis*. Scientific Reports *10*, 16301.

Natchiar, S.K., Myasnikov, A.G., Kratzat, H., Hazemann, I., and Klaholz, B.P. (2017). Visualization of chemical modifications in the human 80S ribosome structure. Nature *551*, 472–477.

Ochi, K., Kim, J.Y., Tanaka, Y., Wang, G., Masuda, K., Nanamiya, H., Okamoto, S., Tokuyama, S., Adachi, Y., and Kawamura, F. (2009). Inactivation of KsgA, a 16S rRNA methyltransferase, causes vigorous emergence of mutants with high-level kasugamycin resistance. Antimicrobial Agents and Chemotherapy *53*, 193–201.

Ogawara, H. (2019). Comparison of antibiotic resistance mechanisms in antibiotic-producing and pathogenic bacteria. Molecules *24*, 3430.

Okamoto, S., Tamaru, A., Nakajima, C., Nishimura, K., Tanaka, Y., Tokuyama, S., Suzuki, Y., and Ochi, K. (2007). Loss of a conserved 7-methylguanosine modification in 16S rRNA confers low-level streptomycin resistance in bacteria. Molecular Microbiology *63*, 1096–1106.

Petri, W.A. (2005). Treatment of Giardiasis. Current Treatment Options in Gastroenterology *8*, 13–17.

Pioletti, M., Schlunzen, F., Harms, J., Zarivach, R., Gluhmann, M., Avila, H., Bashan, A., Bartels, H., Auerbach, T., Jacobi, C., *et al.* (2001). Crystal structures of complexes of the small ribosomal subunit with tetracycline, edeine and IF3. The EMBO Journal *20*, 1829–1839.

Poehlsgaard, J., and Douthwaite, S. (2005). The bacterial ribosome as a target for antibiotics. Nature Reviews Microbiology *3*, 870–881.

Sashital, D.G., Greeman, C.A., Lyumkis, D., Potter, C.S., Carragher, B., and Williamson, J.R. (2014). A combined quantitative mass spectrometry and electron microscopy analysis of ribosomal 30S subunit assembly in *E. coli*. Elife *3*, e04491.

Schuwirth, B.S., Day, J.M., Hau, C.W., Janssen, G.R., Dahlberg, A.E., Cate, J.H., and Vila-Sanjurjo, A. (2006). Structural analysis of kasugamycin inhibition of translation. Nature Structural & Molecular Biology *13*, 879–886.

Selmer, M., Dunham, C.M., Murphy, F.V.t., Weixlbaumer, A., Petry, S., Kelley, A.C., Weir, J.R., and Ramakrishnan, V. (2006). Structure of the 70S ribosome complexed with mRNA and tRNA. Science *313*, 1935–1942.

Shalev-Benami, M., Zhang, Y., Matzov, D., Halfon, Y., Zackay, A., Rozenberg, H., Zimmerman, E., Bashan, A., Jaffe, C.L., Yonath, A., *et al.* (2016). 2.8-A Cryo-EM Structure of the Large Ribosomal Subunit from the Eukaryotic Parasite Leishmania. Cell Reports *16*, 288–294.

Shalev-Benami, M., Zhang, Y., Rozenberg, H., Nobe, Y., Taoka, M., Matzov, D., Zimmerman, E., Bashan, A., Isobe, T., Jaffe, C.L., *et al.* (2017). Atomic resolution snapshot of Leishmania ribosome inhibition by the aminoglycoside paromomycin. Nature Communications *8*, 1589.

Shepherd, D.C., Dalvi, S., and Ghosal, D. (2022). From cells to atoms: cryo-EM as an essential tool to investigate pathogen biology, host–pathogen interaction, and drug discovery. Molecular Microbiology *117*, 610–617.

Siridechadilok, B., Fraser, C.S., Hall, R.J., Doudna, J.A., and Nogales, E. (2005). Structural roles for human translation factor eIF3 in initiation of protein synthesis. Science *310*, 1513–1515.

Smith, I. (2003). Mycobacterium tuberculosis pathogenesis and molecular determinants of virulence. Clinical Microbiology Reviews *16*, 463–496.

Smith, J., and Fratamico, P. (2017). *Escherichia coli* as a Pathogen. In Foodborne diseases (Elsevier), pp. 189–208.

Smith, L.K., and Mankin, A.S. (2008). Transcriptional and translational control of the mlr operon, which confers resistance to seven classes of protein synthesis inhibitors. Antimicrobial Agents and Chemotherapy *52*, 1703–1712.

Srinivas, P., Nosrati, M., Zelinskaya, N., Dey, D., Comstock, L.R., Dunham, C.M., and Conn, G.L. (2023). 30S subunit recognition and G1405 modification by the aminoglycoside-resistance 16S ribosomal RNA methyltransferase RmtC. Proceedings of the National Academy of Sciences *120*, e2304128120.

Stern, S., Powers, T., Changchien, L.M., and Noller, H.F. (1988). Interaction of ribosomal proteins S5, S6, S11, S12, S18 and S21 with 16 S rRNA. Journal of Molecular Biology *201*, 683–695.

Svetlov, M.S., Plessa, E., Chen, C.-W., Bougas, A., Krokidis, M.G., Dinos, G.P., and Polikanov, Y.S. (2019). High-resolution crystal structures of ribosome-bound chloramphenicol and erythromycin provide the ultimate basis for their competition. RNA, *25*, 600–606.

Tenson, T., and Mankin, A. (2006). Antibiotics and the ribosome. Molecular Microbiology *59*, 1664–1677.

Thaker, M., Spanogiannopoulos, P., and Wright, G.D. (2010). The tetracycline resistome. Cellular and Molecular Life Sciences: CMLS *67*, 419–431.

Waters, A.P., Syin, C., and McCutchan, T.F. (1989). Developmental regulation of stage-specific ribosome populations in Plasmodium. Nature *342*, 438–440.

Watson, Z.L., Ward, F.R., Méheust, R., Ad, O., Schepartz, A., Banfield, J.F., and Cate, J.H. (2020). Structure of the bacterial ribosome at 2 Å resolution. elife *9*, e60482.

Weisblum, B. (1995). Erythromycin resistance by ribosome modification. Antimicrobial Agents and Chemotherapy *39*, 577–585.

White, T.C., Rudenko, G., and Borst, P. (1986). Three small RNAs within the 10 kb trypanosome rRNA transcription unit are analogous to domain VII of other eukaryotic 28S rRNAs. Nucleic acids Research *14*, 9471–9489.

Wilson, D.N. (2014). Ribosome-targeting antibiotics and mechanisms of bacterial resistance. Nature reviews Microbiology *12*, 35–48.

Wimberly, B.T., Brodersen, D.E., Clemons, W.M., Jr., Morgan-Warren, R.J., Carter, A.P., Vonrhein, C., Hartsch, T., and Ramakrishnan, V. (2000). Structure of the 30S ribosomal subunit. Nature *407*, 327–339.

Wiwanitkit, V. (2012). Interest in paromomycin for the treatment of visceral leishmaniasis (kala-azar). Therapeutics and Clinical Risk Management *8*, 323–328.

Wong, W., Bai, X.C., Brown, A., Fernandez, I.S., Hanssen, E., Condron, M., Tan, Y.H., Baum, J., and Scheres, S.H. (2014). Cryo-EM structure of the Plasmodium falciparum 80S ribosome bound to the anti-protozoan drug emetine. Elife *3*, e03080.

Wong, W., Bai, X.C., Sleebs, B.E., Triglia, T., Brown, A., Thompson, J.K., Jackson, K.E., Hanssen, E., Marapana, D.S., Fernandez, I.S., *et al.* (2017). Mefloquine targets the Plasmodium falciparum 80S ribosome to inhibit protein synthesis. Nature Microbiology *2*, 17031.

Yang, K., Chang, J.-Y., Cui, Z., Li, X., Meng, R., Duan, L., Thongchol, J., Jakana, J., Huwe, C.M., and Sacchettini, J.C. (2017). Structural insights into species-specific features of the ribosome from the human pathogen Mycobacterium tuberculosis. Nucleic Acids Research *45*, 10884–10894.

Yokoyama, T., and Suzuki, T. (2008). Ribosomal RNAs are tolerant toward genetic insertions: evolutionary origin of the expansion segments. Nucleic Acids Research *36*, 3539–3551.

Yusupov, M.M., Yusupova, G.Z., Baucom, A., Lieberman, K., Earnest, T.N., Cate, J.H., and Noller, H.F. (2001). Crystal structure of the ribosome at 5.5 A resolution. Science *292*, 883–896.

Zhang, W., Li, Z., Sun, Y., Cui, P., Liang, J., Xing, Q., Wu, J., Xu, Y., Zhang, W., Zhang, Y., *et al.* (2022). Cryo-EM structure of Mycobacterium tuberculosis 50S ribosomal subunit bound with clarithromycin reveals dynamic and specific interactions with macrolides. Emerging Microbes & Infections *11*, 293–305.

Zheng, H., Handing, K.B., Zimmerman, M.D., Shabalin, I.G., Almo, S.C., and Minor, W. (2015). X-ray crystallography over the past decade for novel drug discovery–where are we heading next? Expert Opinion on Drug Discovery *10*, 975–989.

Section VI

Cryo-EM in the Study of Neurobiology and Skin—Neuroreceptors to Ion Channels and Synaptic Transmission

26 Cryo-EM Structural Features of Glutamate Receptors Involved in Synaptic Neurotransmission or Brain Pathology

Nebojša Bogdanović

INTRODUCTION

Definition of Ion Channels and Their Importance in Cellular Physiology

Ion channels are integral membrane proteins that form aqueous pores that enable ions to rapidly diffuse along their electrochemical gradient. This ion flux precisely controls diverse physiological processes in electrically excitable cells. In the mammalian central nervous system, ion channels orchestrate the signaling between neurons and other cells that shapes synaptic transmission, plasticity, and information processing.

Among the most widely studied and functionally critical channels mediating brain function are glutamate receptors, which transduce the majority of excitatory neurotransmission. This chapter will elucidate the intricate architecture of glutamate receptors, uncovered through decades of research integrating genetics, biochemistry, electrophysiology, and structural biology, and how their conformational dynamics underlie receptor activation, signaling, and involvement in plasticity and disease.

Overview of Glutamate Receptors and Their Role in the Nervous System

Glutamate is the major excitatory neurotransmitter in the vertebrate brain, playing essential roles in synaptic transmission, plasticity, and development. Glutamate receptors are a family of ligand-gated ion channels that open upon binding glutamate, mediating excitatory postsynaptic currents and calcium influx that trigger downstream signaling cascades.

Milestone discoveries in the 1950s and 1960s identified glutamate as a neurotransmitter using novel tracer techniques. Early electrophysiology studies determined glutamate elicited fast excitatory postsynaptic potentials through NMDA (N-methyl-D-aspartate) and non-NMDA receptors (Young & Fagg, 1990). Starting in the late 1970s, pharmacological differentiation found the non-NMDA responses were mediated by two separable receptors, AMPA (α-amino-3-hydroxy-5-methyl-4-isoxazolepropionic acid) and kainate receptors for the selective agonists used to define them (Collingridge et al., 2009).

Molecular cloning from the late 1980s then allowed isolation of distinct receptor subunits, uncovering an unexpected diversity underlying glutamate receptor heterogeneity (Hollmann & Heinemann, 1994).

Glutamate receptors are divided into two classes: metabotropic glutamate (mGlu) receptors, which are G protein–coupled receptors modulating second messenger systems, and ionotropic glutamate receptors, which form an ion channel pore. The ionotropic glutamate receptors comprise three families named for pharmacological agonists—NMDA, AMPA, and kainate receptors (Traynelis et al., 2010).

While all mediate cation influx in response to glutamate binding, each subtype exhibits unique biophysical properties, distributions in the brain, and physiological roles. NMDA receptors require binding of the co-agonists glutamate and glycine for activation (McBain & Mayer, 1994). They exhibit high calcium permeability, voltage-dependent block by magnesium, and are critical for long-term potentiation involved in learning and memory (Paoletti & Neyton, 2007).

AMPA receptors mediate the majority of fast excitatory postsynaptic transmission in the brain. They lack voltage modulation but contain edited GluA2 subunits that render most AMPA receptors impermeable to calcium (Isaac et al., 2007). Kainate receptors display diverse pharmacological profiles and modulate network dynamics through pre- and postsynaptic positioning (Lerma, 2006).

Brief History of Research on Glutamate Receptors

In the early 1990s, molecular cloning of the first glutamate receptor subunits enabled groundbreaking structure–function analyses. Unexpected diversity emerged, as multiple genes encoding each receptor subtype were discovered along with alternatively spliced isoforms and RNA edited variants of the subunits. This multiplicity allows variable assembly of subunits into functional glutamate receptors with specialized signaling properties suited to particular neuronal circuits (Kumar & Mayer, 2013).

Recent advances in single-particle cryogenic electron microscopy (cryo-EM) have provided unprecedented atomic-resolution snapshots of glutamate receptor architecture in multiple conformational states. These breakthroughs illuminate how intricately woven modular receptor domains undergo choreographed conformational dynamics to precisely control channel gating in response to ligands

DOI: 10.1201/9781003326106-32

(Haick & Hatem, 2019). Ongoing research continues to uncover fundamental molecular mechanisms by which glutamate receptor physiology shapes excitatory synaptic transmission, plasticity, learning, memory, and brain disorders.

This chapter will first detail the modular domain structures of NMDA, AMPA, and kainate receptor subunits and highlight the diversity generated through splice variation, RNA editing, and combinatorial subunit assembly. Conformational landscapes mapped through cryo-EM studies will then be covered to understand structure–function paradigms linking receptor architecture to function.

Current knowledge on glutamate receptor activation, ion permeation, and allosteric modulation will be summarized, emphasizing the specialized roles of each subtype in synaptic signaling. Finally, involvement of dysfunctional glutamate receptor activity in plasticity disorders and neurological disease will be discussed.

This synthesis underscores decades of breakthrough discoveries on glutamate receptor structure–function relationships made possible by cross-disciplinary efforts in genetics, electrophysiology, biochemistry, and structural biology. Ongoing research promises continued insights into these essential neuronal proteins that may spur novel therapeutic approaches.

STRUCTURE OF GLUTAMATE RECEPTORS

Overview of the Different Types of Glutamate Receptors

Glutamate receptors are integral membrane proteins that function as ligand-gated ion channels, converting the chemical signal of presynaptic glutamate release into electrical and biochemical postsynaptic signaling cascades. They play essential roles mediating the excitatory synaptic transmission that underlies cognition, learning, memory, and brain development (Mayer, 2005). Three major subtypes have been extensively characterized—AMPA, NMDA, and kainate receptors—each with distinct structures, conformational dynamics, pharmacology, assembly principles, and physiological functions (Traynelis et al., 2010). Intensive research over decades combining biochemical purification, electrophysiology, molecular genetics, pharmacology, and structural biology has uncovered modular architectural elements, tremendous subunit diversity, intricately coordinated conformational mechanisms, and specialized functional attributes tailored to each receptor's synaptic roles (Sobolevsky, 2015). Recent advances in single-particle cryo-EM have enabled the determination of intact glutamate receptor structures in multiple functionally relevant conformational states at near-atomic resolution, unlocking new insights into allosteric gating mechanisms (Twomey et al., 2017). This section will outline the current understanding of glutamate receptor architecture, highlight diversity generated through splicing and subunit assembly, and compare similarities and differences in structure–function features that tailor each subtype for their roles in excitatory synaptic transmission and plasticity.

Glutamate Receptor Subtypes and Distributions

Glutamate receptors are divided into ionotropic and metabotropic classes based on structure and signaling mechanisms. Metabotropic glutamate (mGlu) receptors are G protein–coupled receptors that modulate second messenger pathways upon glutamate binding. Ionotropic glutamate receptors form an ion channel pore permeable to cations that open in response to glutamate, mediating fast excitatory postsynaptic currents. The ionotropic glutamate receptors consist of three pharmacological subtypes named according to the selective agonists historically used to differentiate their responses: AMPA, NMDA, and kainate receptors (Dingledine et al., 1999). While all three are activated by glutamate and permeable to sodium and potassium, each subtype exhibits unique signaling properties, distributions in the brain, and physiological functions.

AMPA receptors mediate the majority of fast excitatory synaptic transmission in the central nervous system. They exhibit rapid activation and deactivation kinetics in the millisecond range and lack voltage-dependent gating (Silver et al., 1996). AMPA receptors are ubiquitously expressed at high densities on excitatory neurons and are involved in synaptic strengthening underlying learning and memory (Malinow & Malenka, 2002; Shepherd & Huganir, 2007). NMDA receptors have a more specialized role in detecting coincident pre- and postsynaptic activity that can trigger various forms of synaptic plasticity through calcium influx, in addition to contributing excitation (Collingridge et al., 2013). They display slower gating kinetics, high calcium permeability, and voltage-dependent block by extracellular magnesium ions (Nowak et al., 1984). NMDA receptors are also extensively expressed but are concentrated at synaptic sites. Kainate receptors have modulatory roles shaping network dynamics through both pre- and postsynaptic positioning (Lerma, 2006). They can mediate excitatory postsynaptic currents but also modulate neurotransmitter release from presynaptic terminals (Rodríguez-Moreno & Lerma, 1998a). Kainate receptors exhibit diverse pharmacological profiles and modulate network dynamics through pre- and postsynaptic positioning (Carta et al., 2014).

STRUCTURAL FEATURES OF GLUTAMATE RECEPTORS

Modular Domain Architecture

Glutamate receptors have a conserved modular architecture consisting of discrete domains with specialized structures and functions (Kumar & Mayer, 2013). The large extracellular region is comprised of the amino-terminal domain (ATD) and the ligand-binding domain (LBD). The ATD assembles as a bilobed structure that mediates subunit dimerization (Krieger et al., 2015). The LBD adopts a similar bilobed clamshell configuration but contains a cleft between the upper (D1) and lower (D2) lobes that binds glutamate to induce conformational changes coupled to channel gating

(Furukawa et al., 2005). The transmembrane domain (TMD) is embedded within the membrane and forms the central ion channel pore, gate, and modulatory sites (Sobolevsky, 2015). The intracellular carboxyl–terminal domain (CTD) interacts with cytoplasmic scaffolding proteins and impacts membrane trafficking (Hansen et al., 2019). Linkers between the modular domains are pivotal for transducing ligand-induced cleft closure into the twisting and splaying motions that open the transmembrane ion channel (Twomey et al., 2018).

This overall architecture is conserved across ionotropic glutamate receptor subunits and subtypes. However, variations in the structural and functional properties of these modules generate differences in gating kinetics, ion permeation profiles, pharmacological modulation, and physiological roles of AMPA, NMDA, and kainate receptors tailored to their specific synaptic functions (Traynelis et al., 2010). The extracellular ATD impacts allosteric modulation and assembly. The LBD glutamate binding pockets show subtype-specific differences that determine agonist affinities and binding modes. Within the TMD, residues lining the ion channel pore and the makeup of the extracellular turret loops bestow distinct conduction properties. Alternative splicing further diversifies these elements.

Subunit Diversity and Assembly

A remarkable complexity underlying glutamate receptor structure and function arises from the diversity of genes encoding each subtype and the existence of alternatively spliced isoforms and RNA-edited variants of the subunits. This multiplicity allows variable assembly of subunits into heteromeric receptors with specialized properties adapted for particular neuronal circuits (I. H. Greger et al., 2007). Humans have four genes encoding AMPA receptor subunits designated GluA1–4 (Hollmann & Heinemann, 1994). The NMDA receptor family is encoded by three genes for GluN1 subunits, four genes for GluN2 subunits (GluN2A-D), and two genes for GluN3 subunits (GluN3A–B) (Paoletti, 2011). Kainate receptors have the greatest diversity, with five subunit genes named GluK1–5 (Jane et al., 2009). Additional splice variants and RNA editing further exponentially expand this combinatorial diversity (Plested et al., 2016).

AMPA receptors are typically comprised of heterotetramers assembled from GluA1–4 subunits in various combinations, most abundantly GluA1/GluA2 and GluA2/GluA3 (Greger et al., 2002). The edited GluA2 subunit controls key functional properties, including calcium permeability and channel conductance (Burnashev et al., 1992). NMDA receptors are obligate heterotetramers assembled from glycine-binding GluN1 subunits and glutamate-binding GluN2 subunits (Laube et al., 1998). Two GluN1 subunits form dimers with two identical or different GluN2 subunits, most commonly GluN1/GluN2A and GluN1/GluN2B (Rauner & Köhr, 2011). The GluN2 subunit isoform confers distinct pharmacological sensitivities, kinetics, and distributions (Paoletti, 2011). Kainate receptors can assemble as homomers from GluK1–3 or heteromers combining GluK1–3 with GluK4–5 (Fisher & Fisher, 2014). This subunit variability allows the construction of diverse kainate receptors. For example, GluK2/GluK5 heteromers confer distinct physiological properties (Reiner & Levitz, 2018). Overall, the modular domains of glutamate receptor subunits assemble into a multitude of functional complexes with specialized signaling properties matching their roles in synaptic transmission.

Conformational Dynamics Underlie Gating

The cryo-EM revolution in structural biology over the past decade has enabled visualization of intact glutamate receptors in multiple functional states at angstrom-level resolution, providing unprecedented molecular insights into the intricately coordinated global conformational changes that couple agonist binding to channel gating (Midgett & Madden, 2019). Agonist binding deep in the cleft between LBD bilobed layers induces cleft closure that is transduced by connecting linkers to the transmembrane helices. This causes outward pulling on the pore-lining M3 helices to splay apart the helix bundle crossing, opening the ion conduction pathway (Twomey et al., 2017). Structural rearrangements throughout the complex, including the ATD and LBD, stabilize discrete activated, deactivated, and desensitized conformations.

Recent cryo-EM structures of AMPA receptors captured open, closed, and desensitized states illuminating twist, compression, and dilation motions throughout the receptor (Twomey et al., 2018). Key movements underlying gating involve tilting, rotation, and splaying of the M3 helices and dynamic uncoupling of the LBD layers upon sustained activation (Chen & Roche, 2007). NMDA receptor structures revealed the architecture of inhibitory zinc-binding sites at ATD dimer interfaces and movements of the GluN2 LBD controlling allosteric inhibition (Karakas & Furukawa, 2014). The first intact structure of a kainate receptor, the GluK2/GluK5 heteromer, provided the framework for understanding distinctive differences in modulatory sites (Meyerson et al., 2014). This structural revolution has yielded unprecedented insight into the intricate conformational mechanisms enabling glutamate receptors to transduce ligand binding into the opening of their integral ion channel pores with exquisite speed and precision to mediate synaptic transmission.

CRYSTALLOGRAPHY AND OTHER TECHNIQUES USED TO STUDY GLUTAMATE RECEPTOR STRUCTURE

Structural Biology Techniques

X-Ray Crystallography

Early insights into glutamate receptor architecture came from X-ray crystal structures of isolated extracellular receptor domains and subunits (Armstrong & Gouaux, 2000). Crystallization of the GluA2 AMPA receptor LBD provided the first views of the bilobed clamshell configuration and identified key binding pocket residues coordinating glutamate interactions (Armstrong et al., 1998). The NMDA receptor GluN1/GluN2A LBD structure revealed the elongated heterodimer assembly and composite agonist sites (Furukawa et al., 2005). Homomeric GluK3 and heteromeric GluK2/GluK5 kainate receptor LBD structures uncovered distinctive architectures divergent from AMPA and NMDA receptor binding cores (Nanao et al., 2005). ATD constructs also yielded crystal structures delineating roles in assembly (Jin et al., 2003).

The development of stabilized receptor constructs facilitated the major achievement of elucidating intact receptor structures for the homomeric GluA2 AMPA receptor and homomeric GluK3 kainate receptors, although stabilized with antagonists in nonphysiological conformations (Sobolevsky et al., 2009). These landmark studies provided validation of the modular domain arrangement and clarified the organization of subunits surrounding the ion channel. However, crystallization constraints inherent to the technique precluded capturing conformational dynamics.

Cryogenic Electron Microscopy

The revolution in single-particle cryo-EM enabled the determination of full-length glutamate receptor structures in detergent micelles or lipid nanodiscs at near-atomic resolution, capturing multiple functionally relevant conformational states (Haick & Hatem, 2019). This transformed the understanding of glutamate receptor architecture and dynamics. Recent cryo-EM analysis of AMPA receptors in active, resting, and desensitized conformations revealed coordinated global motions throughout the modular domains that couple agonist binding to channel gating (Twomey et al., 2017). Critical movements involve vertical compression, lobe tilting, and rotations of the LBD that translate into M3 helix splaying via connecting linkers. Capturing multiple NMDA receptor states similarly illuminated mechanisms of allosteric inhibition and dynamics (Tajima et al., 2016). The GluK2/GluK5 kainate receptor structure established the framework for this understudied subtype (Meyerson et al., 2014). Ongoing advances in cryo-EM methodologies promise continued insights into the intricacies of glutamate receptor gating.

COMPARISON OF GLUTAMATE RECEPTOR STRUCTURE WITH OTHER ION CHANNELS

Similarities and Differences with Other Ion Channels

Glutamate receptors share evolutionary origins and architectural similarities with other pentameric ligand-gated ion channels, including nicotinic acetylcholine, 5-HT3 serotonin, and glycine receptors which mediate synaptic transmission in the peripheral and central nervous systems (Thompson et al., 2010). They derive from common ancestral genes and have homologous functional domains. The extracellular ATD and LBD exhibit conserved bilobed structures that facilitate assembly and ligand binding. Within the TMD, the pore-lining M2 helix in pentameric channels is analogous to the M3 helix of glutamate receptors. However, divergence has adapted each family for specialized roles.

Pentameric receptors contain five identical or homologous subunits arranged symmetrically around a central ion conduction pathway, while glutamate receptors are tetrameric complexes (Rosenmund et al., 1998). Their pores are lined exclusively by the M2 helix, whereas M3 along with the reentrant M2 loop contribute to the ion permeation pathway in glutamate receptors (Sobolevsky, 2015). Differences in the number of pore-lining elements impact conductance properties. The negatively charged pore interior attracts cations in pentameric receptors, while the neutral glutamate receptor pore allows permeation of both cations and anions. Binding sites for modulators also differ, expanding pharmacology (Armstrong & Gouaux, 2000).

In contrast to chemically gated ionotropic receptors, voltage-gated cation channels that propagate action potentials and modulate excitability lack an extracellular ligand-binding domain (Long et al., 2005). Instead they contain four homologous semi-independent voltage-sensing domains (VSD) surrounding a central pore module (Hansen et al., 2010). Conformational changes in the VSD power opening of the pore gate through electromechanical coupling (Perozo et al., 1993). The VSD contains positively charged residues on the fourth transmembrane helix, which serves a function analogous to agonist binding, but responding to changes in transmembrane voltage rather than ligands.

In summary, through divergence glutamate receptors have evolved specialized structural features that adapt them for mediating excitatory synaptic transmission, while conserving modular architecture elements shared across the ion channel superfamily. Ongoing research continues elucidating the unique aspects of glutamate receptor structure–function relationships.

Physiological Roles Tailored by Structure

Glutamate receptors, specifically AMPA, NMDA, and kainate receptors, play crucial roles in the central nervous system's excitatory neurotransmission. Their functionality is intricately tailored by their structures and interactions with various auxiliary proteins.

AMPA receptors are composed of GluA1-4 subunits and are regulated by auxiliary transmembrane proteins such as γ-TARPs, cornichons, and stargazin. These regulatory proteins associate with AMPA receptors and enhance its' conductance and modulate their response to synaptic activity, which is critical for processes like long-term potentiation, a cellular mechanism underlying learning and memory (Chen et al., 2000; Kato et al., 2010; Tomita et al., 2005). γ-TARPs regulate AMPA receptor function by enhancing conductance, facilitating surface expression, and modulating synaptic targeting and gating (Jackson et al., 2011; Menuz et al., 2008; Zheng et al., 2015). Cornichons influence AMPA receptor trafficking and gating (Schwenk et al., 2009; Shi et al., 2010), while stargazin is involved in their synaptic targeting and stabilization (Chen et al., 2000; Nicoll et al., 2006).

NMDA receptors typically consist of GluN1 and GluN2 (A-D) subunits and sometimes GluN3 (A-B) subunits. These receptors require co-agonists glutamate and glycine and exhibit a voltage-dependent magnesium block (Mayer et al., 1984; Nowak et al., 1984). The auxiliary proteins neuropilin and tolloid-like proteins 1 and 2 interact with NMDA receptors, affecting their synaptic localization and functional properties (Ng et al., 2009; Straub et al., 2011). Mutations in NMDA receptor subunits or associated proteins can lead to severe neurological disorders (Endele et al., 2010; Lesca et al., 2013). For instance, mutations in the GRIN2A gene, encoding the GluN2A subunit, are linked to epilepsy-aphasia spectrum disorders, illustrating the significant impact of structural variations on receptor function and disease (Carvill et al., 2013; Lemke et al., 2014).

Kainate receptors, composed of GluK1-5 subunits, are involved in synaptic transmission and modulation, particularly in the presynaptic regulation of neurotransmitter release (Contractor et al., 2011). Neuropilin and tolloid-like proteins 1 and 2 also interact with kainate receptors, influencing their trafficking, synaptic targeting, and functional properties (Zhang et al., 2009). Dysregulation of kainate receptors is associated with various neuropathological conditions, including epilepsy and schizophrenia (Lerma & Marques, 2013; Mulle et al., 1998). For example, mutations in the GRIK2 gene, encoding the GluK2 subunit, are implicated in intellectual disability and epilepsy, highlighting the crucial interplay between receptor structure and function (Motohashi et al., 2013).

The extensive interaction networks of glutamate receptors with auxiliary proteins like transmembrane AMPA receptor regulatory proteins, cornichons, neuropilin and tolloid-like proteins, and stargazin make it challenging to deconvolute their precise physiological functions (Jackson & Nicoll, 2011). These interactions modulate receptor trafficking, synaptic localization, gating properties, and responses to synaptic activity, contributing significantly to their physiological roles (Greger et al., 2017).

Understanding these networks is vital for elucidating the roles of glutamate receptors in synaptic transmission. This structural complexity and the intricate protein interactions of AMPA, NMDA, and kainate receptors are fundamental to their functions in the synapse, mediating excitatory synaptic transmission and playing pivotal roles in synaptic plasticity, which is essential for cognitive processes like learning and memory (Traynelis et al., 2010).

PHYSIOLOGY OF GLUTAMATE RECEPTORS

Role of Glutamate Receptors in Synaptic Transmission

Glutamate receptors play indispensable roles in mediating the majority of fast excitatory synaptic transmission in the mammalian central nervous system. Their activation underlies the signaling cascades enabling neuronal communication that underlies cognition, learning, memory, and brain development (Kerchner & Nicoll, 2008). The AMPA, NMDA, and kainate receptor subtypes each possess distinct physiological properties precisely tailored by their unique structures to fulfill specialized signaling duties across the diversity of excitatory synapses (Traynelis et al., 2010).

AMPA receptors are optimized for mediating rapid, high-fidelity excitatory transmission essential for moment-to-moment information processing in neuronal circuits (Geiger et al., 1995). Key adaptations include exceptionally fast activation and desensitization kinetics in the millisecond range conferred by their architecture to enable precise responses synchronized with presynaptic neurotransmitter release (Raman & Trussell, 1995). High single-channel conductance facilitates large, transient postsynaptic currents (Rosenmund et al., 1998). A lack of voltage-dependent gating allows unimpeded responses to repetitive activity (Silver et al., 1992). The dominance of Ca^{2+}-impermeable GluA2-containing AMPA receptors reduces the adverse effects of Ca^{2+} accumulation during repeated signaling (Sans et al., 2003).

In contrast, NMDA receptors perform unique functions in detecting coincident pre- and postsynaptic activity which can trigger various forms of longer-lasting synaptic plasticity fundamental for development, learning, and memory (Collingridge et al., 2004). Slower ion channel kinetics furnishes NMDA receptors with essential capabilities for temporal integration of synaptic inputs impinging on dendritic spines (Schiller et al., 1997). Their high calcium permeability provides a localized biochemical signal that can initiate intracellular signaling regulating synaptic strengthening and weakening (Malenka & Bear, 2004). Voltage-dependent magnesium block functionally couples NMDA receptor activation to recent depolarization (Huettner & Bean, 1988). This exquisite NMDA receptor physiology arises from distinct structural features evolved for modulatory control over plasticity induction.

Kainate receptors exhibit the greatest diversity of functional properties depending on their subunit makeup, consistent with their exceptional structural heterogeneity (Fisher & Fisher, 2014). Certain assemblies such as GluK1-3 homomers form high-conductance channels with calcium permeability, while GluK2/K5 heteromers confer low conductance and calcium impermeability (Herguedas et al., 2016). Auxiliary subunits like Neto provide further tuning of kinetics, pharmacology, and ion permeation profiles (Jackson et al., 2011). This variability suits the flexible modulatory roles of kainate receptors in fine-tuning network excitability through both pre- and postsynaptic sites (Kubo et al., 1998).

In this way, the distinct architecture of each glutamate receptor subtype encodes specialized physiological functions exquisitely tailored to enable their cooperative yet distinct signaling activities underlying the diverse dimensions of excitatory synaptic transmission and plasticity. Ongoing research continues providing insights into this structure–function relationship.

Overview of Glutamate Receptor Subtypes and Their Functions

Glutamate receptor activation by presynaptic release of glutamate triggers postsynaptic excitatory currents and second messenger signaling cascades that enable neuronal communication underlying essential CNS functions, including cognition, learning, memory, and development (Kerchner & Nicoll, 2008). The three major ionotropic glutamate receptor subtypes powering this excitatory signaling—AMPA, NMDA, and kainate receptors—each possess unique structures and biophysical properties that confer specialized physiological roles in synaptic transmission and plasticity (Traynelis et al., 2010).

AMPA receptors mediate the majority of fast excitatory synaptic transmission required for moment-to-moment electrical signaling between neurons in circuits throughout the mammalian brain (Malinow et al., 2000). Their exceptionally rapid activation and deactivation kinetics in the millisecond range enable effective transmission of temporally encoded action potential spikes (Geiger et al., 1995). High single-channel conductance allows large postsynaptic responses (Rosenmund et al., 1998). AMPA receptors lack voltage-dependent gating and exhibit only weak desensitization, enabling sustained responses to repetitive activity (Silver et al., 1992). Most contain edited GluA2 subunits that confer calcium impermeability to prevent cytotoxicity (Burnashev et al., 1992).

NMDA receptors perform more modulatory roles detecting combined pre- and postsynaptic activity, which can trigger long-term potentiation and depression, crucial mechanisms underlying learning, memory consolidation, and circuit refinement (Collingridge et al., 2004). Slower ion channel kinetics furnish NMDA receptors with capabilities for the temporal integration of successive synaptic signals impinging on dendritic spines (Schiller et al., 1997). Their high calcium permeability provides a localized biochemical signal that initiates intracellular plasticity processes (Yang et al., 2008). A voltage-dependent magnesium block functionally couples NMDA receptor opening to recent depolarization (Huettner & Bean, 1988).

Kainate receptors exhibit diverse modulatory functions shaping network dynamics through both pre- and postsynaptic locations (Lerma, 2006). At postsynapses, they can mediate excitatory transmission with distinct kinetics from AMPA receptors (Fisher & Mott, 2011). Presynaptically they regulate neurotransmitter release and inhibitory interneuron activity to tune network firing (Rodríguez-Moreno & Lerma, 1998b). Subunit combinations produce functional variability enabling specialized physiological roles (Fisher & Mott, 2011).

In summary, the unique architectures of AMPA, NMDA, and kainate receptor subtypes optimize them for cooperative yet distinct signaling functions that collectively coordinate the multifaceted dimensions of excitatory neurotransmission and plasticity. Ongoing research continues to provide insights into this structure–function relationship.

Mechanisms of Glutamate Receptor Activation

Extensive studies using biochemical, electrophysiological, pharmacological, and imaging techniques have uncovered mechanisms controlling the activation, deactivation, and desensitization kinetics of glutamate receptors that are precisely tuned by their specialized structures to match their physiological duties (Weston et al., 2006).

AMPA receptors possess exceptionally rapid activation kinetics in the microsecond range, enabling effective transmission of temporally encoded synaptic signals (Bowie et al., 1998). Structural and computational studies suggest that agonist binding induces closure of the bilobed extracellular ligand-binding domain, which pulls open the channel gate located at the bundle crossing of the pore-lining transmembrane helices, through connecting linkers like the pre-M1 and M3–S2 segments (Furukawa et al., 2005). These motions are streamlined by optimized stabilizing interactions and conformational flexibility, allowing swift induction without compromising stability (Sun et al., 2002). Alternative splicing generates variability in linker lengths and compositions, which fine-tunes activation speed (Koike et al., 1997).

In contrast, NMDA receptors exhibit slower activation kinetics in the 10 to 100 millisecond range appropriately tailored for their roles integrating synaptic inputs (Zhang et al., 2008). The weaker extent of ligand-binding domain closure and inhibitor interactions impeding cleft closure motions are key determinants imposing kinetic constraints (Gielen et al., 2008). The sophisticated conformational landscape is exquisitely tuned to furnish suitable activation rates.

AMPA and kainate receptor signaling is also shaped by rapid desensitization on a timescale of 10 to 100 milliseconds, which allows transient responses to high-frequency inputs (Andersen et al., 2006). Structural studies suggest sustained activation enables water penetration between ligand-binding domain layers, which destabilizes dimer interfaces and rearranges channel elements to close the gate and terminate ion flux (Sun et al., 2002). Alternative splicing further diversifies desensitization rates through impacts on modal transitions between open, closed, and desensitized states (Mosbacher et al., 1994).

In contrast, NMDA receptors exhibit limited desensitization essential for their functions mediating synaptic plasticity, which requires sustained responses (Zhang et al., 2008). Stabilizing interlayer and intralayer interactions in the agonist-bound activated state and high transition state barriers are structural adaptations that prevent spontaneous modal shifts to nonconducting conformations (Kussius & Popescu, 2009).

In this manner, crossover studies integrating structure–function insights have elucidated specialized mechanisms controlling glutamate receptor activation and modal gating kinetics evolved to precisely match their physiological duties.

Modulation of Glutamate Receptors by Allosteric Ligands and Posttranslational Modifications

While agonist binding to the extracellular ligand-binding domains is the primary trigger for glutamate receptor activation, their activity, kinetics, and pharmacology are powerfully influenced by binding of allosteric modulators to topographically distinct regulatory sites and by posttranslational modifications of key residues (Lee et al., 2015).

Allosteric modulators bind at interfaces between the layered domains of glutamate receptor complexes to stabilize particular conformations along the activation pathway. Benzodiazepines like GYKI52466 exert profound inhibitory effects on AMPA receptor activity by binding at a dimer interface site which enhances desensitization (Jin et al., 2003). Ifenprodil and zinc potently inhibit NMDA receptors through extracellular amino terminal domain interactions that impede agonist-induced cleft closure motions (Paoletti & Neyton, 2007). Kainate receptors have diverse allosteric sites, such as the chloride and polyamine binding pockets along the ion channel pore that provide strong electrostatic control over channel opening and closing (Braidy, 2022).

In addition to modulation by allosteric ligands, the functions of glutamate receptors are dynamically regulated by posttranslational modifications of key residues, including phosphorylation, palmitoylation, ubiquitination, and SUMOylation (Salter & Kalia, 2004). These modifications enable moment-to-moment tuning of receptor properties and signaling in response to activity. Phosphorylation of intracellular serine, threonine, and tyrosine residues by diverse kinases adjusts conductance, synaptic targeting, and membrane stabilization (Liu & Zukin, 2007). Palmitoylation of cysteine residues proximal to the ion channel pore impacts gating kinetics and pharmacology (Lan et al., 2001). Ubiquitination and SUMOylation influence receptor trafficking, stability, and proteasomal degradation (Patrick et al., 1999).

In this way, glutamate receptors possess sophisticated structural mechanisms that enable precise allosteric control over their activation states, kinetics, and signaling behaviors by both endogenous and exogenous ligands as well as posttranslational modifications of key functional elements. This affords exquisite dynamical tuning of excitatory signal transmission and plasticity. Ongoing research aimed at elucidating these intricate modulatory mechanisms promises to uncover new directions for therapeutically targeting glutamatergic signaling underlying cognition and behavior.

LIGAND BINDING AND CHANNEL GATING

Sculpting Glutamate Receptor Activation through Precisely Tailored Ligand Recognition

Glutamate receptors transduce signaling in the brain by converting the chemical signal of neurotransmitter binding into the opening of an integral ion channel pore. This activation process is exquisitely tailored for each receptor subtype through the coevolution of specialized ligand binding sites and gating mechanisms. Recent breakthroughs in structural biology have provided unprecedented glimpses into the intricate conformational landscapes enabling the selective recognition of synaptic signals.

For AMPA receptors, the endogenous neurotransmitter glutamate induces a Venus flytrap–like closure of the clamshell-shaped GluA2 ligand-binding domain (LBD), anchoring deep in the binding pocket via an intricate network of coordinating residues like arginine-485, serine-654, and threonine-655 (Armstrong et al., 1998). This energetically favorable closed-cleft conformation is stabilized like a locked trap capturing its prey. The conformational change tugs on connecting peptides, propagating structural rearrangements to open the channel gate and depolarize the cell in a beautifully choreographed motion (Sun & Gouaux, 2002).

The AMPA receptor trap can also capture synthetic agonists like the tool compound (S)-AMPA, which similarly elicits cleft closure through distinctive contacts, but shuts its door on antagonists like CNQX or GYKI52466. Structural analyses have also illuminated how AMPA receptor modulatory proteins like TARPs interact with the ligand-binding domain in a state-dependent manner to stabilize particular conformations, providing a framework for understanding agonist and antagonist selectivity (Herguedas et al., 2022). These competitors are precisely molded to occupy the binding site without inducing the downstream motions, effectively acting like a stick preventing trap closure (Jin et al., 2003).

For NMDA receptors, the Venus flytrap must bind two prey simultaneously before springing open—the endogenous neurotransmitter glutamate along with a co-agonist glycine or D-serine (Furukawa et al., 2005). Recent structural studies have provided detailed glimpses into how the coordinated binding of agonists to the bilobed NMDA receptor ligand-binding domains induces cleft closure and conformational changes coupled to ion channel opening (Zhang et al., 2021). The binding of both ligands induces closure of the clamshell-like LBDs, generating sufficient power to tug open the ion channel and excite the cell. As with AMPA receptors, this gating machinery can be jammed by antagonists like D-AP5, which occupy the glutamate site but fail to productively engage the downstream gating apparatus through steric blockade of cleft closure (Tajima et al., 2016).

Kainate receptors exhibit the greatest ingenuity in ligand recognition, containing diverse modified trap architectures tailored for specialized functions (Jane et al., 2009). Intricate disulfide linkages between cysteine residues in GluK2 and GluK5 alter the binding pocket conformation to stabilize unique inactive and active states (Lind et al., 2017). Auxiliary subunits like Neto-2 also fine-tune the interactions of the trap with its catch by interacting with the amino-terminal domain (Kato et al., 2010). Structural analyses have revealed Neto2 can bind to GluK2 receptors with variable stoichiometry, either one or two Neto2 subunits associating with tetrameric GluK2 (He et al., 2021). Furthermore, Neto2 modulation involves strategic positioning within the ion channel pore and cross-linking of ligand-binding domain lobes to stabilize particular modal conformations. This exquisite interaction network enables tailored tuning of kainate receptor signaling, sculpting a mosaic of pharmacological profiles suiting their varied modulatory roles.

In summary, atomic-level insights into the architectural ingenuity of glutamate receptor ligand-binding domains have revealed specialized structures precisely evolved to match their activation mechanisms to the demands of synaptic signaling. Further elucidation of receptor conformational landscapes promises to uncover new therapeutic strategies for manipulating these key neuronal proteins.

Orchestrating Ion Channel Opening through Conformational Choreography

The binding of agonists within the extracellular ligand-binding domains of glutamate receptors initiates a meticulously coordinated dance of protein reshaping that culminates in opening of the ion channel pore. Recent breakthroughs in structural biology have illuminated the intricate conformational choreography that enables efficient transduction of ligand-binding events into the twisting protein motions that gate the pore.

For AMPA receptors, agonist-induced closure of the ligand-binding domain clamshell exerts a tug on the adjoining linkers like the M3–D2 helix, pulling on the tips of the pore-lining helices splayed in the membrane and splaying them apart like opening a camera aperture (Sun & Gouaux, 2002). Recent insights into the unique flexibility of the GluA1 AMPA receptor subunit suggest that its conformational mobility may play a pivotal role in modulating gating kinetics and anchoring at synapses during plasticity (Zhang et al., 2023). This elegant protein motion rapidly opens the channel within microseconds of glutamate binding, enabling the flux of small signaling cations like sodium (Bowie et al., 1998). Structural studies have revealed how conformational changes initiated in the ligand-binding domain propagate through connecting linker regions to induce the twisting motions that splay open the transmembrane pore helices (Zhang et al., 2023).

However, sustained dancing can be exhausting. Prolonged AMPA receptor activation enables water penetration between the ligand-binding domains that destabilizes their closed embrace (Sun et al., 2002). This causes the intricate protein machinery to slip, with the ligand-binding and transmembrane domains uncoupling and collapsing back into inactive conformations, closing the pore. After a brief respite, the receptor can reengage its dance partners to respond to the next high-frequency synaptic signal.

For NMDA receptors, the choreography also involves coordinated motions of the ligand-binding core tugging on pore-proximal connectors like the S1–M1 linker, splaying open the channel gate (Tajima et al., 2016). However, more extensive electrostatic interactions between dancing partners in NMDA receptors provide enhanced stamina, preventing rapid exhaustion and closure. This sustains ion flux, enabling the activation of downstream signaling pathways. Recent research has uncovered how auxiliary subunits like CNIHs further slow AMPA receptor desensitization rates, likely through modulatory binding interactions, and this may provide insights into the mechanisms underlying NMDA receptors' high resistance to desensitization (Zhang et al., 2021).

Kainate receptor isoforms exhibit the greatest diversity of dance moves, with alternative splice variations introducing unique elements that influence choreography kinetics (Eibl et al., 2012). For example, the inclusion of the intracellular "flip" module speeds receptor deactivation after the removal of glutamate (Yan et al., 2005). Additionally, recent structural analyses have revealed specialized mechanisms controlling kainate receptor desensitization, involving formation of a "desensitization ring" architecture mediated by staggered helix contacts between subunits (Meyerson et al., 2016). This conformational rearrangement provides a key switch governing modal transitions between active and desensitized states. This exceptional diversity of structural elements suits the flexible modulatory roles of kainate receptors in tuning network excitation.

The integration of structural biology with biophysical studies has revealed the intricate protein conformational dynamics enabling glutamate receptors to transduce ligand binding into the gating motions that open their ion channel pores with exquisite temporal precision tailored to match their physiological functions.

The Nuances of Subunit Identities Shape Glutamate Receptor Pharmacology

The distinct structures and specialized functions of individual glutamate receptor subunits enable the assembly of heteromeric complexes with customized pharmacological profiles tuned for physiological demands. Recent high-resolution glimpses of their architectures reveal surprising intricacies underlying subtype-specific drug interactions.

For AMPA receptors, complexes containing the abundant GluA2 subunit bind the neurotransmitter glutamate within a cleft formed between two adjoining ligand-binding domain lobes, with critical arginine, serine, and threonine residues like Arginine-485 providing vital coordinating contacts (Armstrong et al., 1998). GluA2 also confers a key functional property—calcium impermeability—by neutralizing the critical Q/R/N site, essential for AMPA receptor roles mediating transient excitation (Burnashev et al., 1992). The ligand-binding core and channel gate are structurally and dynamically coupled through the M3–D2 linker, enabling efficient transduction of agonist binding into pore opening (Sobolevsky et al., 2009). Alternative splicing further diversifies signaling properties (Koike et al., 1997).

NMDA receptors require coassembly of GluN1 and GluN2 subunits into heteromeric complexes, with the GluN2 subtype providing distinctive pharmacological profiles tailored for circuit-specific roles (Paoletti, 2011). GluN1 binds the co-agonist glycine while GluN2 binds glutamate within their bilobed ligand-binding cores (Furukawa et al., 2005). For example, the GluN2B subtype confers slower channel kinetics suited for integrating synaptic inputs (Rumbaugh & Vicini, 1999). Structural elements unique to NMDA receptor subunits enable modulation by ions like zinc and magnesium to fine-tune signaling (Paoletti et al., 2013). Conformational changes elicited by agonist binding are propagated to the ion channel through specialized connector regions like the S1–M1 linker (Tajima et al., 2016).

Kainate receptors exhibit the greatest compositional variability, permitting the construction of complexes with highly diversified pharmacological profiles (Fisher & Mott, 2011). Certain combinations like GluK2/GluK5 heteromers enable modulation by auxiliary proteins that bind directly to the subunit structures, expanding functionality (Zhang et al., 2006). The Neto proteins interact with the amino-terminal domain to influence gating (Kato et al., 2010). Alternative splicing and RNA editing of the pore-lining M2 helix provide additional mechanisms for customizing ligand sensitivity (Fisher & Mott, 2011).

In this way, the intricate structures of individual glutamate receptor subunits enable combinatorial assembly of heteromeric complexes with customized pharmacological profiles. Atomic-level insights into their nuanced conformations thus promise improved therapeutic targeting.

Decades of intensive study integrating genetic, biochemical, electrophysiological, and structural techniques has uncovered the exquisite intricacy of glutamate receptor architectures, yielding specialized drug and ligand sensitivities precisely tuned to match their physiological functions. Ongoing elucidation of molecular mechanisms promises to reveal new ways of therapeutically tuning glutamate receptor signaling of underlying synaptic transmission, plasticity, learning, memory, and brain disorders.

SIGNAL TRANSDUCTION AND DOWNSTREAM EFFECTS

Overview of Glutamate Receptor–Mediated Signaling Pathways

Binding of glutamate to AMPA, NMDA, and kainate receptors triggers not only transient excitatory currents but also sophisticated signaling pathways that cooperatively modulate neuronal structure and function on multiple levels. The exquisite spatiotemporal coding enabled by different glutamate receptor subtypes activating specific downstream mechanisms orchestrates myriad facets of excitatory synaptic transmission and plasticity in the brain (Regehr, 2012).

For example, AMPA receptor activation permits the rapid influx of cations like sodium and potassium that efficiently depolarize the postsynaptic membrane, driving action potential firing to propagate signals through circuits (Choquet & Triller, 2013). This excitation encodes moment-to-moment information flow between neurons. The rapid response kinetics and calcium impermeability conferred by edited GluA2 subunits prevents adverse second messenger accumulation during repeated high-frequency signaling (Isaac et al., 2007).

In contrast, NMDA receptors can induce more sustained calcium influx when activated coincidently with postsynaptic depolarization, which then triggers localized second messenger cascades regulating synaptic strength (Greger & Esteban, 2019). Calcium binds calmodulin to activate kinases like CAMKII that phosphorylate proteins, including AMPA receptors, increasing their membrane expression during long-term potentiation (Lisman, 2017). The slower activation kinetics and voltage-dependent relief of the magnesium block makes NMDA receptors specialized for detecting concurrent pre- and postsynaptic activity patterns driving plasticity.

Kainate receptors also permit calcium influx in certain subunit configurations to modestly modulate excitability and second messenger pathways (Lerma & Marques, 2013). For example, presynaptic kainate receptors containing GluK2 enhance glutamate release probability to facilitate transmission (Rodríguez-Moreno & Lerma, 1998a). Postsynaptic GluK2-containing kainate receptors also mediate a form of metabotropic synaptic depression through calcium-dependent signaling (Melyan et al., 2002).

In this way, the signaling capabilities of ionotropic glutamate receptor subtypes are exquisitely tailored by their structures to enact downstream pathways that cooperatively shape synaptic transmission. Ongoing research continues to provide fascinating insights into this structure–function relationship.

Downstream Effects of Glutamate Receptor Activation

The remarkable signaling capabilities of AMPA, NMDA, and kainate receptor subtypes enables them to cooperatively sculpt synaptic properties essential for the hippocampal and cortical functions that underlie learning, memory formation, and higher cognition (Takeuchi et al., 2014). The diverse activity-dependent pathways coordinated by glutamate receptor signaling strengthen or weaken synapses through expression, phosphorylation, trafficking, and conductance changes that modify circuit outputs (Lisman, 2017).

For example, AMPA receptor number, subunit composition, and single-channel properties at synapses can be dynamically regulated by patterns of activity (Choquet & Triller, 2013). Incorporation of GluA1 subunits enhances calcium influx to influence dendritic spine growth and stabilization (Lu et al., 2009). Posttranslational modifications like phosphorylation also tune channel properties (Lee et al., 2003).

Sustained NMDA receptor activation provides localized calcium influx that activates numerous signaling enzymes, including CAMKII, PKC, and calcineurin (Lisman et al., 2012). These pathways induce expression, trafficking, anchoring, and phosphorylation changes to many synaptic proteins, including AMPA receptors, to enable diverse forms of bidirectional plasticity (Kessels & Malinow, 2009).

Kainate receptors shape plasticity by modulating network excitability through pre- and postsynaptic locations (Lerma & Marques, 2013). For example, presynaptic kainate receptors containing GluK2 facilitate metabotropic glutamate receptor signaling to enhance release probability during low-frequency stimulation (Rodríguez-Moreno & Lerma, 1998a).

In this way, the intricate signaling orchestrated by the specialized properties of glutamate receptor subtypes enables dynamic regulation of synapses to store information. Ongoing research elucidates how disturbances in glutamate receptor expression, composition, and signaling affects cognition in health and disease.

Role of Glutamate Receptors in Plasticity and Learning

Given the essential role of glutamate receptor–mediated signaling in enabling multiple forms of synaptic plasticity underlying learning and memory, it is perhaps unsurprising that dysfunction of AMPA, NMDA, or kainate receptor activity is implicated in the pathogenesis of diverse cognitive disorders ranging from neurodevelopmental syndromes to age-related dementia (Volk et al., 2015). Defining the specific molecular mechanisms by which altered receptor function affects synaptic homeostasis promises more targeted therapies.

For example, some evidence suggests NMDA receptor hypofunction early in development, perhaps due to altered subunit expression or scaffolding, may disrupt synaptic refinement and plasticity, contributing to autism spectrum disorder pathophysiology (Lee et al., 2015). Pharmacologically enhancing NMDA receptor signaling or dampening inhibition during critical windows may help ameliorate specific dysfunctions.

Later in life, NMDA receptor overactivation induced by beta-amyloid accumulation is implicated in the cognitive decline of Alzheimer disease. Accordingly, memantine, which selectively dampens excess NMDA activity, has shown benefits in clinical trials (Mota et al., 2014). The specific receptor subunits mediating overload could be selectively targeted.

Many types of intellectual disability arise from disrupted expression, cycling, and anchoring of glutamate receptors, which derails the signals controlling synaptic development and plasticity. Understanding these intricate molecular pathways promises pharmacological rescues by correcting composition and function. For example, modulating metabotropic glutamate receptor–mediated protein synthesis normalization shows promise in fragile X models (Suvrathan et al., 2010).

Overall, as research continues to unveil the precise balance of glutamate receptor signaling that orchestrates the synaptic connectivity changes that enable learning and cognition, new therapies can be designed to restore specific aspects of subsystem function that are disrupted in particular disorders. This promises more precise pharmacological manipulation than blanket approaches that affect all glutamate signaling equally.

DISEASES AND DISORDERS ASSOCIATED WITH GLUTAMATE RECEPTORS

Overview of Diseases and Disorders Associated with Glutamate Receptors

Given the essential roles of AMPA, NMDA, and kainate receptors in mediating excitatory synaptic transmission and plasticity, it is perhaps unsurprising that these intricate receptor systems are implicated in the pathogenesis of a broad range of neurological and psychiatric diseases when their precise signaling goes awry (Meldrum, 2000). This section will provide an overview of major disease categories linked to altered glutamate receptor activity and will discuss key mechanistic insights and therapeutic approaches.

Dysfunctional glutamate receptor signaling has been associated both with acute disorders like ischemic stroke, traumatic brain injury, and epilepsy, and with chronic neurodegenerative diseases like Alzheimer, Parkinson, and Huntington (Dong et al., 2009). Altered activity is also implicated in neurodevelopmental syndromes like autism spectrum disorder (ASD) and fragile X, major psychiatric conditions like schizophrenia, mood disorders, anxiety, and obsessive-compulsive disorders, and various forms of intellectual disability and age-related dementia (Sanacora et al., 2012).

While glutamate dysfunction is not necessarily the sole cause of these complex disorders, growing genetic, biochemical, and physiological evidence suggests that overactive or hypoactive signaling often critically contributes to the network dysfunction, impaired plasticity, excitotoxicity, and cognitive decline that is central to disease progression (Hynd et al., 2004). However, much remains unknown about the subtle molecular details that link specific alterations in glutamate receptor physiology to particular circuit vulnerabilities in distinct disease states. Ongoing research aims to map these nuanced connections to enable the development of more targeted and effective treatments.

Role of Glutamate Receptors in Neurodegenerative Diseases

Excitotoxicity from excessive glutamate receptor activation is posited to be a key driver of neuronal damage both in acute injuries like ischemic stroke, epilepsy, and traumatic brain injury, and in chronic neurodegenerative diseases like Alzheimer, Parkinson, and Huntington (Dong et al., 2009).

In ischemic stroke, for example, the dramatic release of glutamate from dying neurons, combined with the loss of energy production, leads to a toxic overload of calcium primarily through NMDA receptors, which activates injurious enzymes that degrade cell membranes and proteins while impairing mitochondrial function, triggering oxidative stress and apoptotic pathways (Lai et al., 2014). Experimental treatments that block pathological glutamate receptor overactivity show promise in reducing stroke damage—like memantine or magnesium, which can dampen NMDA signaling (Ikonomidou & Turski, 2002).

In chronic neurodegenerative conditions like Alzheimer disease, beta-amyloid–induced network dysfunction is believed to cause gradual elevation of synaptic glutamate levels, eventually overwhelming regulatory mechanisms and resulting in calcium dyshomeostasis through hyperactive NMDA and even AMPA receptor signaling (Hynd et al., 2004). This contributes to impaired plasticity, loss of dendritic spines, and destruction of vulnerable neuron populations. Accordingly, treatments that dampen glutamate-mediated excitotoxicity may ameliorate downstream pathways of damage, such as memantine, which blocks excessive NMDA activity (Parsons et al., 2007).

Overall, while triggers may differ, converging evidence points to disruptions in the glutamate receptor signaling that mediates excitotoxicity as a common driver of damage across acute CNS injuries and in age-related neurodegeneration (Dong et al., 2009). This suggests that tailored interventions that modulate receptor activity may offer neuroprotection in diverse contexts, warranting further research on the nuances.

Role of Glutamate Receptors in Psychiatric Disorders

Dysregulation of glutamate receptor signaling also contributes to psychiatric conditions like schizophrenia, mood disorders, autism spectrum disorders (ASD), and obsessive-compulsive disorders by disrupting synaptic connectivity and plasticity, especially within cortical, limbic, and striatal circuits mediating cognition, memory, and behavior (Sanacora et al., 2012).

For example, hypoactive NMDA receptor signaling caused by decreased expression or suppressed long-term potentiation is linked to schizophrenia and associated plasticity deficits that may impair cognition (Coyle, 2012). Enhanced NMDA signaling aims to ameliorate these deficiencies using agonists or positive allosteric modulators (Conn et al., 2014). On the flip side, some evidence suggests that NMDA receptor overactivity may promote compulsive behaviors by overstrengthening striatal synapses. This suggests that properly tuned activity is essential for behavioral regulation (Pittenger, 2015).

In autism spectrum disorders, abnormal composition of synaptic glutamate receptor subtypes, including excess GluN2B or GluN3A, is posited to disrupt experience-dependent synaptic refinement periods during development (Lee et al., 2015). Efforts to normalize subunit expression ratios aim to rescue appropriate strengthening and pruning for circuit maturation. Specifically, research suggests that altered subunit combinations lead to abnormal receptor kinetics that disrupt the detection of correlated pre- and postsynaptic activity patterns critical for pruning exuberant connections and strengthening productive links. For example, increased GluN2B confers slower decay kinetics that integrate inputs over broader time windows, impairing spike-timing-dependent plasticity that is important for refining circuits during critical periods (Philpot et al., 2007).

In fragile X syndrome, disrupted metabotropic glutamate receptor (mGluR) signaling due to excess protein synthesis interferes with long-term depression. This imbalance between potentiation and depression impairs synaptic pruning. Normalizing mGluR5-mediated protein synthesis shows promise in restoring this homeostatic plasticity crucial for appropriate development and learning (Suvrathan et al., 2010).

In summary, while mechanisms remain unclear, improper tuning of the precise balances and timing between excitation and inhibition required to shape circuit wiring and plasticity during critical windows appears to contribute broadly to many psychiatric conditions (Selimbeyoglu & Parvizi, 2010). Though many details remain to be elucidated, targeted modulation of associated glutamate receptor dysfunction shows promise in ameliorating symptoms of diverse disorders.

FUTURE DIRECTIONS AND EMERGING RESEARCH

Advances in Technology and Techniques for Studying Glutamate Receptors

The recent cryo-EM revolution has yielded unprecedented insights into glutamate receptor architecture, but many pressing questions remain on the intricate conformational landscapes enabling function (Lussier et al., 2015). Emerging techniques promise continued advances.

For example, phase plates enhance contrast for resolving challenging structural elements like ion channel pores in multiple states at high resolution (Lu et al., 2014). Microelectron diffraction can directly map sparsely populated conformations. Microfluidic platforms enable rapid electrophysiological analysis of disease-associated mutants identified through genome sequencing (Harada et al., 2002).

Novel super-resolution fluorescence microscopy techniques are beginning to enable direct tracking of glutamate receptors within living synapses. For instance, single-molecule tracking of genetically tagged AMPA receptors has uncovered nanodomain organization within the postsynaptic density (Noguchi et al., 2005). New genetically encoded calcium and voltage indicators can optically report glutamate receptor responses (Tønnesen et al., 2014).

Microfluidic chambers interfacing tissue sections with single cell transcriptomics and proteomics *in situ* enable the spatial mapping of glutamate receptor expression patterns throughout different brain regions and cell types (Béïque et al., 2011). Integrating such data with physiological recordings and behavioral studies will uncover new principles of receptor recruitment during learning and memory.

Overall, a new generation of structural biology, electrophysiology, optical imaging, and biochemical profiling tools promises an unprecedented multidimensional view of glutamate receptor conformational dynamics and signaling behaviors in their native synaptic environment (Makino & Malinow, 2011). Exciting times lie ahead.

Current Challenges in Understanding Glutamate Receptor Structure and Function

While recent findings have advanced our understanding substantially, many open questions and challenges remain on the intricacies of glutamate receptor physiology (Savtchouk & Volterra, 2018). How do assembly, trafficking, and anchoring proteins and posttranslational modifications construct specific combinations with particular properties suited to each synapse? What transient conformational intermediates couple agonist binding to ion channel opening? How is modal gating precisely tuned? What is the impact of neurological disease–associated mutations on structures and pathways?

Capturing high-resolution snapshots of sparsely populated transient states and mapping the full conformational landscapes that encompass transitions between multiple functional forms remains difficult but will be required to answer these questions (Sheng et al., 2012). Emerging techniques that integrate computational simulations, rapid-freezing electron microscopy, spectroscopic assays, and electrophysiology with structural methods offer promise in capturing elusive states (Lau & Zukin, 2007).

Determining the expression patterns, interactions, and signaling behaviors of specific receptor isoforms in defined neuronal and glial cell populations throughout different brain regions also remains challenging. However, emerging *in situ* structural and biochemical profiling tools promise progress by linking compositions to function with spatial precision (Herweg & Schwarz, 2018).

Overall, much remains to be learned about the intricacies of glutamate receptor physiology precisely tailored to enable specialized functional roles in particular synaptic microenvironments and neural circuits. The coming decades promise to be a golden era, unraveling these mysteries through creative new techniques.

Emerging Research on the Role of Glutamate Receptors in Health and Disease

Dysfunction of AMPA, NMDA, and kainate receptors is implicated in diverse neurological diseases, but the molecular mechanisms linking altered structure and function to pathology remain unclear (Lee et al., 1998). Do mutations trigger neurodevelopmental disorders by disrupting the receptor interactions that are essential for synaptic refinement? Does impaired control over subunit composition lead to network hyperexcitability in epilepsy? What conformational stabilization failures enable toxic ion overload in neurodegenerative diseases?

Emerging techniques promise new traction. Interrogating receptor dynamics in patient-derived neurons promises insights into disease etiologies (Sheng et al., 2012). Precise network-specific analysis of RNA and protein could identify vulnerable cell populations. Advanced longitudinal imaging of receptors *in vivo* may uncover developmental signaling timelines that go awry (Ebrahimi & Okabe, 2020).

Therapeutically, pharmacological chaperones could stabilize mutated hyperactive receptors, reducing excitation (Lau et al., 2010). Positive allosteric modulators could enhance function where subunit expression diminishes (Arendt et al., 2020). The future is bright for developing glutamate receptor drugs targeted to rescue particular structural dysfunctions that underlie specific neurological disorders.

In summary, a new generation of emerging technologies promises fresh insights into long-standing questions about glutamate receptor structure, physiology, and disease mechanisms. Creativity and cross-disciplinary collaborations will be key to overcoming current hurdles on the path to elucidating new intricacies of the glutamate receptor functions shaping the brain.

CONCLUSION

Summary of Key Points

This chapter provided an extensive overview integrating decades of research that has uncovered the elegant intricacy of glutamate receptor structure, function, and dysfunction. These essential neuronal proteins transduce presynaptic glutamate release into postsynaptic electrical and biochemical signaling cascades mediating excitatory transmission.

Their conserved modular architecture enables exquisite tailoring of physiological properties and signaling behaviors adapted for cooperative yet distinct roles across diverse excitatory synapses. Specialized ligand-binding domains recognize synaptic signals with precise pharmacology and kinetics. These are allosterically coupled through intricate conformational dynamics to ion channel pores with tailored architectures generating receptor-specific ion flux profiles.

Additional diversification arises from the assembly of homologous subunits with alternative splicing and RNA editing providing further tuning mechanisms. This heterogeneity allows nuanced control over the spatiotemporal patterning of excitation and plasticity that is essential for circuit development, cognition, and behavior. However, major gaps remain in our understanding of the precise molecular mechanisms that enable ultra-rapid activation and desensitization kinetics of glutamate receptors and how mutations trigger receptor dysfunction in neurological diseases.

When this delicate physiology goes awry, neurological and psychiatric diseases can emerge from disrupted excitation/inhibition balance, impaired plasticity, and excitotoxic damage. Defining these nuanced structure–function relationships has been made possible by ground-breaking work across disciplines.

Implications for Future Research and Therapeutics

While recent findings have provided remarkable new insights, many questions remain about receptor signaling intricacies, regulation, and involvement in synaptic computation and brain disorders. Continued cross-disciplinary research will be essential to overcome current conceptual and technical hurdles.

Emerging techniques promise unprecedented views of elusive functional transitions, integrating computational, imaging, and structural approaches. Advanced *in situ* profiling promises the ability to link receptor compositions to cell-specific physiology. Structure-guided drug design may enable the modulation of signaling with pharmacological precision to restore balances disrupted in disease. Exciting new structural biology tools like cryo-EM combined with spectroscopic and computational approaches promise insights into elusive transient conformational changes that underlie receptor physiology, while studies in patient-derived neurons may elucidate how disease-associated mutations disrupt function.

Overall, while our conception of glutamate receptor relationships has advanced remarkably, much remains unknown about how their intricacies empower cognition and go awry in brain disorders. Continued unraveling of these mysteries promises profound advances in understanding and treating neurological diseases in the exciting decades ahead.

OUTSTANDING QUESTIONS

Key Open Questions Remaining in the Field

Despite the remarkable progress in understanding glutamate receptor structure–function relationships, many fundamental questions remain about the intricacies underlying their synaptic signaling roles and involvement in neurological diseases:

- What are the conformational pathways and transient intermediates enabling ultra-rapid activation and desensitization? Cryo-EM, computational modeling, and spectroscopic tools may help capture these elusive states.
- How is receptor trafficking dynamically regulated in an activity-dependent manner? New super-resolution imaging and spatial proteomics methods promise insights.
- What are the roles of distinct receptor isoforms expressed in specific cell types? *In situ* multi-omics linking composition to physiology will illuminate differences.
- How do mutations trigger dysfunction in neuropsychiatric disorders? Studies in patient-derived neurons are beginning to unravel mechanisms.
- Can positive allosteric modulators rescue altered conformations that cause disease? Structure-based drug design offers opportunities for precision medicines.
- How can receptor combinations be tailored to restore excitation/inhibition balance? Gene therapies may enable selective editing of network elements.
- What signaling pathways shape excitatory/inhibitory synapse development? Integrative multi-omics and imaging will map out networks.
- How does disturbed glutamate signaling contribute to diverse psychiatric diseases? Circuits linking receptor dysfunction to behaviors need defining.
- Can glutamate signaling be modulated to enhance cognitive capabilities? Possibilities are emerging, but much remains unknown.

Overall, the intricacies of glutamate receptor physiology enabling synaptic computation, plasticity, learning, and memory remain only partially defined. Tackling these key open questions promises substantial advances in treating brain disorders.

REFERENCES

Andersen, T. F., Tikhonov, D., Bølcho, U., Bolshakov, K., Nelson, J., Pluteanu, F., Mellor, I., Egebjerg, J., & Strømgaard, K. (2006). Uncompetitive antagonism of AMPA receptors: Mechanistic insights from studies of polyamine toxin derivatives. *Journal of Medicinal Chemistry, 49*(18), 5414–5423. https://doi.org/10.1021/JM060606J

Arendt, K. L., Benoist, M., & Lario, A. (2020). mTORC1 inhibitor treatment of Alzheimer's disease in mice. *EMBO Molecular Medicine*, *12*(5). https://doi.org/10.15252/emmm.201911170

Armstrong, N., & Gouaux, E. (2000). Mechanisms for activation and antagonism of an AMPA-sensitive glutamate receptor: Crystal structures of the GluR2 ligand binding core. *Neuron*, *28*(1), 165–181. https://doi.org/10.1016/s0896-6273(00)00094-5

Armstrong, N., Sun, Y., Chen, G.-Q., & Gouaux, E. (1998). Structure of a glutamate-receptor ligand-binding core in complex with kainate. *Nature*, *395*(6705), 913–917. https://doi.org/10.1038/27686

Béïque, J.-C., Na, Y., Kuhl, D., Worley, P. F., & Huganir, R. L. (2011). Arc-dependent synapse-specific homeostatic plasticity. *Proceedings of the National Academy of Sciences*, *108*(2), 816–821. https://doi.org/10.1073/pnas.1017914108

Bowie, D., Lange, G. D., & Mayer, M. L. (1998). Activity-dependent modulation of glutamate receptors by polyamines. *Journal of Neuroscience*, *18*(20), 8175–8185. https://doi.org/10.1523/JNEUROSCI.18-20-08175.1998

Braidy, N., Poljak, A., Jayasena, T., Guillemin, G.J., & Sachdev, P. (2022). Ionotropic receptors in the central nervous system and neurodegenerative disease. In: Kostrzewa, R. M. (Eds.), *Handbook of Neurotoxicity*. Springer. https://doi.org/10.1007/978-3-031-15080-7_126

Burada, A. P., Vinnakota, R., & Kumar, J. (2019). Cryo-EM Structures of the Ionotropic Glutamate Receptor GluD1 Reveal a Non-Swapped Architecture. *Nature Structural & Molecular Biology*, 27, 84-91. https://doi.org/10.1038/s41594-019-0359-y

Burnashev, N., Monyer, H., Seeburg, P. H., & Sakmann, B. (1992). Divalent ion permeability of AMPA receptor channels is dominated by the edited form of a single subunit. *Neuron*, *8*(1), 189–198. https://doi.org/10.1016/0896-6273(92)90120-3

Carta, M., Fievre, S., Gorlewicz, A., & Mulle, C. (2014). Kainate receptors in the hippocampus. *European Journal of Neuroscience*, *39*(11), 1835–1844. https://doi.org/10.1111/ejn.12590

Carvill, G. L., Regan, B. M., Yendle, S. C., O'Roak, B. J., Lozovaya, N., Bruneau, N., Burnashev, N., Khan, A., Cook, J., Geraghty, E., Sadleir, L. G., Turner, S. J., Tsai, M.-H., Webster, R., Ouvrier, R., Damiano, J. A., Berkovic, S. F., Shendure, J., Hildebrand, M. S., … Mefford, H. C. (2013). GRIN2A mutations cause epilepsy-aphasia spectrum disorders. *Nature Genetics*, *45*(9), 1073–1076. https://doi.org/10.1038/ng.2727

Chen, B.-S., & Roche, K. W. (2007). Regulation of NMDA receptors by phosphorylation. *Neuropharmacology*, *53*(3), 362–368. https://doi.org/10.1016/j.neuropharm.2007.05.018

Chen, L., Chetkovich, D. M., Petralia, R. S., Sweeney, N. T., Kawasaki, Y., Wenthold, R. J., Bredt, D. S. Nicoll, R. A. (2000). Stargazin regulates synaptic targeting of AMPA receptors by two distinct mechanisms. *Nature*, *408*(6815), 936–943. https://doi.org/10.1038/35050030

Chen, S., & Gouaux, E. (2019). Structure and mechanism of AMPA receptor - auxiliary protein complexes. *Current Opinion in Structural Biology*, 54, 104-111. https://doi.org/10.1016/j.sbi.2019.01.011.

Choquet, D., & Triller, A. (2013). The dynamic synapse. *Neuron*, *80*(3), 691–703. https://doi.org/10.1016/j.neuron.2013.10.013

Collingridge, G. L., Isaac, J. T. R., & Wang, Y. (2004). Receptor trafficking and synaptic plasticity. *Nature Reviews Neuroscience*, *5*(12), 952–962. https://doi.org/10.1038/nrn1556

Collingridge, G. L., Olsen, R. W., Peters, J., & Spedding, M. (2009). A nomenclature for ligand-gated ion channels. *Neuropharmacology*, *56*(1), 2–5. https://doi.org/10.1016/j.neuropharm.2008.06.063

Collingridge, G. L., Volianskis, A., Bannister, N., France, G., Hanna, L., Mercier, M., Tidball, P., Fang, G., Irvine, M. W., Costa, B. M., Monaghan, D. T., & Jane, D. E. (2013). The NMDA receptor as a target for cognitive enhancement. *Neuropharmacology*, *64*, 13–26. https://doi.org/10.1016/j.neuropharm.2012.06.051

Conn, P. J., Lindsley, C. W., Meiler, J., Niswender, C. M., Petrovski, S., Weaver, C. D., Conn, P. J., Lindsley, C. W., Meiler, J., Niswender, C. M., Petrovski, S., & Weaver, C. D. (2014). Opportunities and challenges in the discovery of allosteric modulators of GPCRs for treating CNS disorders. *Nature Reviews Drug Discovery*, *13*(9), 692–708. https://doi.org/10.1038/nrd4399

Contractor, A., Mulle, C., & Swanson, G. T. (2011). Kainate receptors coming of age: Milestones of two decades of research. *Trends in Neurosciences*, *34*(3), 154–163. https://doi.org/10.1016/j.tins.2010.12.002

Copits, B. A., Robbins, J. S., Frausto, S., & Swanson, G. T. (2011). Synaptic targeting and functional modulation of GluK1 kainate receptors by the auxiliary neuropilin and tolloid-like (NETO) proteins. *Journal of Neuroscience*, *31*(20), 7334–7340. https://doi.org/10.1523/JNEUROSCI.0179-11.2011

Coyle, J. T. (2012). NMDA receptor and schizophrenia: a brief history. *Schizophrenia Bulletin*, *38*(5), 920–926. https://doi.org/10.1093/schbul/sbs076

Dingledine, R., Borges, K., Bowie, D., & Traynelis, S. F. (1999). The glutamate receptor ion channels. *Pharmacological Reviews*, *51*(1), 7–62.

Dong, X.-X., Wang, Y., & Qin, Z.-H. (2009). Molecular mechanisms of excitotoxicity and their relevance to pathogenesis of neurodegenerative diseases. *Acta Pharmacologica Sinica*, *30*(4), 379–387. https://doi.org/10.1038/aps.2009.24

Ebrahimi, S., & Okabe, S. (2020). Ca2+ Dynamics at the Presynaptic Active Zone. *Frontiers in Neuroscience*, *14*, 484. https://doi.org/10.3389/fnins.2020.00484

Eibl, C., Chebli, M., Kalinovich, A., Vriend, G., Rothnagel, R., Garnier, D., Wittig, I., Jahn, R., De Groot, B. L., Christine, S., Benoit, M., Sansom, M. S. P., Medina, I., Veenhoff, L. M., Hegemann, P., Orsel, B. R., Xhaard, H., Nishida, M., Greger, I. H., … Biggin, P. C. (2012). Structural and functional analysis of slack sodium-activated potassium channels. *Structure*, *20*(12), 2161–2173. https://doi.org/10.1016/j.str.2012.09.017

Endele, S., Rosenberger, G., Geider, K., Popp, B., Tamer, C., Stefanova, I., Milh, M., Kortüm, F., Fritsch, A., Pientka, F. K., Hellenbroich, Y, Kalscheuer, V. M., Kohlhase, J., Moog, U., Rappold, G., Rauch, A., Ropers, H.-H., von Spiczak, S, Tönnies, H., … Kutsche, K. (2010). Mutations in GRIN2A and GRIN2B encoding regulatory subunits of NMDA receptors cause variable neurodevelopmental phenotypes. *Nature Genetics*, *42*(11), 1021–1026. https://doi.org/10.1038/ng.677

Fisher, J. L., & Mott, D. D. (2011). Distinct functional roles of subunits within the heteromeric kainate receptor. *Journal of Neuroscience*, *31*(45), 16059–16068. https://doi.org/10.1523/JNEUROSCI.3685-11.2011

Fisher, M. T., & Fisher, J. L. (2014). Contributions of different kainate receptor subunits to the properties of recombinant homomeric and heteromeric receptors. *Neuroscience*, *278*, 70–80. https://doi.org/10.1016/j.neuroscience.2014.08.009

Furukawa, H., Singh, S. K., Mancusso, R., & Gouaux, E. (2005). Subunit arrangement and function in NMDA receptors. *Nature*, *438*(7065), 185–192. https://doi.org/10.1038/nature04089

Geiger, J. R. P., Melcher, T., Koh, D.-S., Sakmann, B., & Seeburg, P. H. (1995). Relative abundance of subunit mRNAs determines gating and Ca2+ permeability of AMPA receptors in principal neurons and interneurons in rat CNS. *Neuron*, *15*(1), 193–204. https://doi.org/10.1016/0896-6273(95)90076-4

Gielen, M., Le Goff, A., Stroebel, D., Johnson, J. W., Neyton, J., & Paoletti, P. (2008). Structural rearrangements of NR1/NR2A NMDA receptors during allosteric inhibition. *Neuron*, *57*(1), 80–93. https://doi.org/10.1016/j.neuron.2007.11.027

Greger, I. H., & Esteban, J. A. (2019). AMPA receptor biogenesis and trafficking for synapse formation and plasticity. *Frontiers in Cellular Neuroscience*, *13*, 156. https://doi.org/10.3389/fncel.2019.00156

Greger, I. H., Khatri, L., & Ziff, E. B. (2002). RNA editing at arg607 controls AMPA receptor exit from the endoplasmic reticulum. *Neuron*, *34*(5), 759–772. https://doi.org/10.1016/s0896-6273(02)00693-1

Greger, I. H., Watson, J. F., & Cull-Candy, S. G. (2017). Structural and functional architecture of AMPA-Type glutamate receptors and their auxiliary proteins. *Neuron*, *94*(4), 713–730. https://doi.org/10.1016/j.neuron.2017.04.009

Greger, I. H., Ziff, E. B., & Penn, A. C. (2007). Molecular determinants of AMPA receptor subunit assembly. *Trends in Neurosciences*, *30*(8), 407–416. https://doi.org/10.1016/j.tins.2007.06.005

Haick, A. K., & Hatem, C. L. (2019). The cryo-EM revolution in neuroscience. *F1000Research*, *8*, 564. https://doi.org/10.12688/f1000research.18963.1

Hansen, K. B., Furukawa, H., & Traynelis, S. F. (2010). Control of assembly and function of glutamate receptors by the amino-terminal domain. *Molecular Pharmacology*, *78*(4), 535–549. https://doi.org/10.1124/mol.110.067157

Hansen, K. B., Traynelis, S. F., & Ogden, K. K. (2019). Contribution of the M4 transmembrane helix to agonist efficacy in NMDA receptors. *The Journal of Physiology*, *597*(4), 885–900. https://doi.org/10.1113/JP276990

Harada, A., Teng, J., Takei, Y., Oguchi, K., & Hirokawa, N. (2002). MAP2 is required for dendrite elongation, PKA anchoring in dendrites, and proper PKA signal transduction. *Journal of Cell Biology*, *158*(3), 541–549. https://doi.org/10.1083/jcb.200110134

He, L., Sun, J., Gao, Y., Li, B., Wang, Y., Dong, Y., An, W., Li, H., Yang, B., Ge, Y., Zhang, X. C., Shi, Y. S., & Zhao, Y. (2021). Kainate receptor modulation by NETO2. *Nature*, *599*(7884), 325–329. https://doi.org/10.1038/s41586-021-03936-y

Herguedas, B., García-Nafría, J., Cais, O., & Greger, I. H. (2016). Structure and organization of heteromeric AMPA-type glutamate receptors. *Science*, *352*(6285). https://doi.org/10.1126/science.aad3873

Herguedas, B., Kohegyi, B. K., Dohrke, J. N., Watson, J. F., Zhang, D., Ho, H., Shaikh, S. A., Lape, R., Krieger, J. M., & Greger, I. H. (2022). Mechanisms underlying TARP modulation of the GluA1/2-γ8 AMPA receptor. *Nature Communications*, *13*(1). https://doi.org/10.1038/s41467-022-28404-7

Herweg, J., & Schwarz, M. K. (2018). Splice-specific Glycine Receptors: An Approach to Treat Hyperekplexia. *Frontiers in Molecular Neuroscience*, *11*, 303. https://doi.org/10.3389/fnmol.2018.00303

Hollmann, M., & Heinemann, S. (1994). Cloned glutamate receptors. *Annual Review of Neuroscience*, *17*, 31–108. https://doi.org/10.1146/annurev.ne.17.030194.000335

Huettner, J. E. (1990). Glutamate receptor channels in rat DRG neurons: Activation by kainate and quisqualate and blockade of desensitization by Con A. *Neuron*, *5*(3), 255–266. https://doi.org/10.1016/0896-6273(90)90163-A

Huettner, J. E., & Bean, B. P. (1988). Block of N-methyl-D-aspartate-activated current by the anticonvulsant MK-801: Selective binding to open channels. *Proceedings of the National Academy of Sciences*, *85*(4), 1307–1311. https://doi.org/10.1073/pnas.85.4.1307

Hynd, M. R., Scott, H. L., & Dodd, P. R. (2004). Glutamate-mediated excitotoxicity and neurodegeneration in Alzheimer's disease. *Neurochemistry International*, *45*(5), 583–595. https://doi.org/10.1016/j.neuint.2004.03.007

Ikonomidou, C., & Turski, L. (2002). Why did NMDA receptor antagonists fail clinical trials for stroke and traumatic brain injury? *The Lancet Neurology*, *1*(6), 383–386. https://doi.org/10.1016/s1474-4422(02)00164-3

Isaac, J. T. R., Ashby, M. C., & McBain, C. J. (2007). The role of the GluR2 subunit in AMPA receptor function and synaptic plasticity. *Neuron*, *54*(6), 859–871. https://doi.org/10.1016/j.neuron.2007.06.001

Jackson, A. C., Milstein, A. D., Soto, D., Farrant, M., Cull-Candy, S. G., & Nicoll, R. A. (2011). Probing TARP modulation of AMPA receptor conductance with polyamine toxins. *The Journal of Neuroscience*, *31*(20), 7511–7520. https://doi.org/10.1523/JNEUROSCI.6688-10.2011

Jackson, A. C., & Nicoll, R. A. (2011). The expanding social network of ionotropic glutamate receptors: TARPs and other transmembrane auxiliary subunits. *Neuron*, *70*(2), 178–199. https://doi.org/10.1016/j.neuron.2011.04.007

Jane, D. E., Lodge, D., & Collingridge, G. L. (2009). Kainate receptors: Pharmacology, function and therapeutic potential. *Neuropharmacology*, *56*(1), 90–113. https://doi.org/10.1016/j.neuropharm.2008.08.023

Jin, R., Banke, T. G., Mayer, M. L., Traynelis, S. F., & Gouaux, E. (2003). Structural basis for partial agonist action at ionotropic glutamate receptors. *Nature Neuroscience*, *6*(8), 803–810. https://doi.org/10.1038/nn1091

Karakas, E., & Furukawa, H. (2014). Crystal structure of a heterotetrameric NMDA receptor ion channel. *Science*, *344*(6187), 992–997. https://doi.org/10.1126/science.1251915

Kato, A. S., Gill, M. B., Ho, M. T., Yu, H., Tu, Y. H., Siuda, E. R., Wang, H.-T., Qian, Y.-W., Nisenbaum, E. S., Bredt, D. S., & Nicoll, R. A. (2010). Hippocampal AMPA receptor gating controlled by both TARP and cornichon proteins. *Neuron*, *68*(6), 1082–1096. https://doi.org/10.1016/j.neuron.2010.11.026

Kerchner, G. A., & Nicoll, R. A. (2008). Silent synapses and the emergence of a postsynaptic mechanism for LTP. *Nature Reviews Neuroscience*, *9*(11), 813–825. https://doi.org/10.1038/nrn2501

Kessels, H. W., & Malinow, R. (2009). Synaptic AMPA receptor plasticity and behavior. *Neuron*, *61*(3), 340–350. https://doi.org/10.1016/j.neuron.2009.01.015

Koike, M., Iino, M., & Ozawa, S. (1997). Blocking effect of 1-naphthyl acetyl spermine on Ca(2+)-permeable AMPA receptors in cultured rat hippocampal neurons. *Neuroscience Research*, *29*(1), 27–36. https://doi.org/10.1016/s0168-0102(97)00067-9

Krieger, J., Bahar, I., & Greger, I. H. (2015). Structure, dynamics, and allosteric potential of ionotropic glutamate receptor N-terminal domains. *Biophysical Journal*, *109*(6), 1136–1148. https://doi.org/10.1016/j.bpj.2015.07.033

Kubo, Y., Murata, Y., & Oda, S. (1998). Paradoxical enhancement of glutamate release via activation of metabotropic glutamate receptors in cerebellar Purkinje cells. *European Journal of Neuroscience*, *10*(5), 1682–1691. https://doi.org/10.1046/j.1460-9568.1998.00170.x

Kumar, J., & Mayer, M. L. (2013). Functional insights from glutamate receptor ion channel structures. *Annual Review of Physiology*, *75*, 313–337. https://doi.org/10.1146/annurev-physiol-030212-183711

Plested, A. (2016). Structural mechanisms of activation and desensitization in neurotransmitter-gated ion channels. *Nature Structural & Molecular Biology*, *23*, 494–502. https://doi.org/10.1038/nsmb.3214

Kussius, C. L., & Popescu, G. K. (2009). Kinetic basis of partial agonism at NMDA receptors. *Nature Neuroscience*, *12*(9), 1114–1120. https://doi.org/10.1038/nn.2358

Lai, T., Zhang, S., & Wang, Y. (2014). Excitotoxicity and stroke: Identifying novel targets for neuroprotection. *Progress in Neurobiology*, *115*, 157–188. https://doi.org/10.1016/j.pneurobio.2013.11.006

Lan, J.-Y., Skeberdis, V. A., Jover, T., Zheng, X., Bennett, M. V. L., & Zukin, R. S. (2001). Activation of metabotropic glutamate receptor 1 accelerates NMDA receptor trafficking. *The Journal of Neuroscience*, *21*(16), 6058–6068.

Lau, C. G., Takeuchi, K., Rodenas-Ruano, A., Takayasu, Y., Murphy, J., Bennett, M. V. L., & Zukin, R. S. (2010). Regulation of NMDA receptor trafficking by amyloid-β. *Nature Neuroscience*, *13*(8), 1051–1058. https://doi.org/10.1038/nn.2613

Lau, C. G., & Zukin, R. S. (2007). NMDA receptor trafficking in synaptic plasticity and neuropsychiatric disorders. *Nature Reviews Neuroscience*, *8*(6), 413–426. https://doi.org/10.1038/nrn2153

Laube, B., Kuhse, J., & Betz, H. (1998). Evidence for a tetrameric structure of recombinant NMDA receptors. *The Journal of Neuroscience*, *18*(8), 2954–2961. https://doi.org/10.1523/JNEUROSCI.18-08-02954.1998

Lee, E.-J., Choi, S.-Y., & Kim, E. (2015). NMDA receptor dysfunction in autism spectrum disorders. *Current Opinion in Pharmacology*, *20*, 8–13. https://doi.org/10.1016/j.coph.2014.10.007

Lee, H.-K., Kameyama, K., Huganir, R. L., & Bear, M. F. (1998). NMDA induces long-term synaptic depression and dephosphorylation of the GluR1 subunit of AMPA receptors in hippocampus. *Neuron*, *21*(5), 1151–1162. https://doi.org/10.1016/s0896-6273(00)80632-7

Lee, H.-K., Takamiya, K., Han, J.-S., Man, H.-Y., Kim, C. H., Rumbaugh, G., Yu, S., Ding, L., Petralia, R. S., Wenthold, R. J., Gallagher, M., & Huganir, R. L. (2003). Phosphorylation of the AMPA receptor GluR1 subunit is required for synaptic plasticity and retention of spatial memory. *Cell*, *112*(5), 631–643. https://doi.org/10.1016/s0092-8674(03)00122-3

Lemke, J. R., Hendrickx, R., Geider, K., Laube, B., Schwake, M., Harvey, R. J., James, V. M., Pepler, A., Steiner, I., Hörtnagel, K., Neidhardt, J., Ruf, S., Wolff, M., Bartholdi, D., Caraballo, R., Platzer, K., Suls, A., De Jonghe, P.,

Biskup, S., & Weckhuysen, S. (2014). GRIN2B mutations in West syndrome and intellectual disability with focal epilepsy. *Annals of Neurology*, *75*(1), 147–154. https://doi.org/10.1002/ana.24073

Lerma, J. (2006). Kainate receptor physiology. *Current Opinion in Pharmacology*, *6*(1), 89–97. https://doi.org/10.1016/j.coph.2005.12.002

Lerma, J., & Marques, J. M. (2013). Kainate receptors in health and disease. *Neuron*, *80*(2), 292–311. https://doi.org/10.1016/j.neuron.2013.09.045

Lesca, G., Rudolf, G., Bruneau, N., Lozovaya, N., Labalme, A., Boutry-Kryza, N., Salmi, M., Tsintsadze, T., Addis, L., Motte, J., Wright, S., Tsintsadze, V., Michel, A., Doummar, D., Lascelles, K., Strug, L., Waters, P., de Bellescize, J., Vrielynck, P., … Szepetowski, P. (2013). GRIN2A mutations in acquired epileptic aphasia and related childhood focal epilepsies and encephalopathies with speech and language dysfunction. *Nature Genetics*, *45*(9), 1061–1066. https://doi.org/10.1038/ng.2726

Lind, G. E., Mou, T. C., Tamborini, L., & Hansen, K. B. (2017). Structural basis of subunit selectivity for competitive NMDA receptor antagonists with preference for GluN2A over GluN2B subunits. *Proceedings of the National Academy of Sciences*, *114*(33), E6942–E6951. https://doi.org/10.1073/pnas.170775211

Lisman, J. (2017). Glutamatergic synapses are structurally and biochemically complex because of multiple plasticity processes: Long-term potentiation, long-term depression, short-term potentiation and scaling. *Philosophical Transactions of the Royal Society B: Biological Sciences*, *372*(1715). https://doi.org/10.1098/rstb.2016.0260

Lisman, J., Yasuda, R., & Raghavachari, S. (2012). Mechanisms of CaMKII action in long-term potentiation. *Nature Reviews Neuroscience*, *13*(3), 169–182. https://doi.org/10.1038/nrn3192

Liu, S. J., & Zukin, R. S. (2007). Ca2+-permeable AMPA receptors in synaptic plasticity and neuronal death. *Trends in Neurosciences*, *30*(3), 126–134. https://doi.org/10.1016/j.tins.2007.01.006

Long, S. B., Campbell, E. B., & Mackinnon, R. (2005). Voltage sensor of Kv1.2: Structural basis of electromechanical coupling. *Science*, *309*(5736), 903–908. https://doi.org/10.1126/science.1116270

Lu, P., Bai, X., Ma, D., Fu, Y., Luo, Y., Chanda, B., Shi, Z.-Z., Wang, Y., Yang, M., Marassi, F. M., Yao, Y., Shi, Y., Dutta, K., Li, Y., Swayze, E. E., & Lai, L. (2014). Three-dimensional structure of human γ-secretase. *Nature*, *512*(7513), 166–170. https://doi.org/10.1038/nature13567

Lu, W., Shi, Y., Jackson, A. C., Bjorgan, K., During, M. J., Sprengel, R., Seeburg, P. H., & Nicoll, R. A. (2009). Subunit arrangement and function in NMDA receptors. *Nature*, *459*(7245), 701–705. https://doi.org/10.1038/nature08016

Lussier, M. P., Sanz-Clemente, A., & Roche, K. W. (2015). Dynamic Regulation of N-Methyl-D-aspartate (NMDA) and α-Amino-3-hydroxy-5-methyl-4-isoxazolepropionic Acid (AMPA) Receptors by Posttranslational Modifications. *Journal of Biological Chemistry*, *290*(48), 28596–28603. https://doi.org/10.1074/jbc.R115.652750

Makino, H., & Malinow, R. (2011). Compartmentalized versus global synaptic plasticity on dendrites controlled by experience. *Neuron*, *72*(6), 1001–1011. https://doi.org/10.1016/j.neuron.2011.09.036

Malenka, R. C., & Bear, M. F. (2004). LTP and LTD: An embarrassment of riches. *Neuron*, *44*(1), 5–21. https://doi.org/10.1016/j.neuron.2004.09.012

Malinow, R., Mainen, Z. F., & Hayashi, Y. (2000). LTP mechanisms: From silence to four-lane traffic. *Current Opinion in Neurobiology*, *10*(3), 352–357. https://doi.org/10.1016/s0959-4388(00)00099-4

Malinow, R., & Malenka, R. C. (2002). AMPA receptor trafficking and synaptic plasticity. *Annual Review of Neuroscience*, *25*, 103–126. https://doi.org/10.1146/annurev.neuro.25.112701.142758

Mayer, M. L. (2005). Glutamate receptor ion channels. *Current Opinion in Neurobiology*, *15*(3), 282–288. https://doi.org/10.1016/j.conb.2005.05.004

Mayer, M. L., Westbrook, G. L., & Guthrie, P. B. (1984). Voltage-dependent block by Mg^{2+} of NMDA responses in spinal cord neurones. *Nature*, *309*(5965), 261–263. https://doi.org/10.1038/309261a0

McBain, C. J., & Mayer, M. L. (1994). N-methyl-D-aspartic acid receptor structure and function. *Physiological Reviews*, *74*(3), 723–760. https://doi.org/10.1152/physrev.1994.74.3.723

Meldrum, B. S. (2000). Glutamate as a neurotransmitter in the brain: Review of physiology and pathology. *The Journal of Nutrition*, *130*(4), 1007S–1015S. https://doi.org/10.1093/jn/130.4.1007S

Melyan, Z., Wheal, H. V, & Lancaster, B. (2002). Metabotropic-mediated kainate receptor regulation of IsAHP and excitability in pyramidal cells. *Neuron*, *34*(1), 107–114. https://doi.org/10.1016/s0896-6273(02)00624-4

Menuz, K., O'Brien, J. L., Karmizadegan, S., Bredt, D. S., & Nicoll, R. A. (2008). TARP redundancy is critical for maintaining AMPA receptor function. *The Journal of Neuroscience*, *28*(33), 8740–8746. https://doi.org/10.1523/JNEUROSCI.1319-08.2008

Meyerson, J. R., Kumar, J., Chittori, S., Lelyveld, V. S., Zhou, P., Gouaux, E., & Mayer, M. L. (2014). Structural mechanism of glutamate receptor activation and desensitization. *Nature*, *514*(7522), 328–334. https://doi.org/10.1038/nature13603

Meyerson, J. R., Chittori, S., Merk, A., Rao, P., Han, T. H., Serpe, M., Mayer, M. L., & Subramaniam, S. (2016). Structural basis of kainate subtype glutamate receptor desensitization. *Nature*, *537*(7621), 567–571. https://doi.org/10.1038/nature19352

Midgett, C. R., & Madden, D. R. (2019). Breaking the bottleneck: Eukaryotic membrane protein expression for high-resolution structural studies. *Journal of Biological Chemistry*, *294*(41), 15343–15363. https://doi.org/10.1016/j.jsb.2007.07.001

Mosbacher, J., Schoepfer, R., Monyer, H., Burnashev, N., Seeburg, P. H., & Ruppersberg, J. P. (1994). A molecular determinant for submillisecond desensitization in glutamate receptors. *Science*, *266*(5187), 1059–1062. https://doi.org/10.1126/science.7973663

Mota, S. I., Ferreira, I. L., & Rego, A. C. (2014). Dysfunctional synapse in Alzheimer's disease - A focus on NMDA receptors. *Neuropharmacology*, *76*, 16–26. https://doi.org/10.1016/j.neuropharm.2013.08.013

Motohashi, J., Okamoto, N., Torii, C., Kosaki, K., & Kosaki, R. (2013). Two novel mutations in GRIK2 cause developmental delay, intellectual disability with behavioral abnormalities, and epilepsy. *Congenital Anomalies*, *53*(3), 105–111. https://doi.org/10.1111/cga.12017

Mulle, C., Sailer, A., Perez-Otano, I., Dickinson-Anson, H., Castillo, P. E., Bureau, I., Maron, C., Gage, F. H., Mann, J. R., Bettler, B., & Heinemann, S. F. (1998). Altered synaptic physiology and reduced susceptibility to kainate-induced seizures in GluR6-deficient mice. *Nature*, *392*(6676), 601–605. https://doi.org/10.1038/33408

Nanao, M. H., Green, T., Stern-Bach, Y., Heinemann, S. F., & Choe, S. (2005). Structure of the kainate receptor subunit GluR6 agonist-binding domain complexed with domoic acid. *Proceedings of the National Academy of Sciences*, *102*(8), 2708–2713. https://doi.org/10.1073/pnas.0409459102

Ng, D., Pitcher, G. M., Szilard, R. K., Sertie, A., Kanisek, M., Clapcote, S. J., Lipina, T., Kalia, L. V., Joo, D., McKerlie, C., Cortez, M., Roder, J. C., Salter, M. W., & Roder, J. C. (2009). Neto1 is a novel CUB-domain NMDA receptor-interacting protein required for synaptic plasticity and learning. *PLoS Biology*, *7*(4), e1000041. https://doi.org/10.1371/journal.pbio.1000041

Noguchi, J., Matsuzaki, M., Ellis-Davies, G. C. R., & Kasai, H. (2005). Spine-neck geometry determines NMDA receptor-dependent Ca2+ signaling in dendrites. *Neuron*, *46*(4), 609–622. https://doi.org/10.1016/j.neuron.2005.03.015

Nowak, L., Bregestovski, P., Ascher, P., Herbet, A., & Prochiantz, A. (1984). Magnesium gates glutamate-activated channels in mouse central neurones. *Nature*, *307*(5950), 462–465. https://doi.org/10.1038/307462a0

Paoletti, P. (2011). Molecular basis of NMDA receptor functional diversity. *European Journal of Neuroscience*, *33*(8), 1351–1365. https://doi.org/10.1111/j.1460-9568.2011.07628.x

Paoletti, P., Bellone, C., & Zhou, Q. (2013). NMDA receptor subunit diversity: Impact on receptor properties, synaptic plasticity and disease. *Nature Reviews Neuroscience*, *14*(6), 383–400. https://doi.org/10.1038/nrn3504

Paoletti, P., & Neyton, J. (2007). NMDA receptor subunits: Function and pharmacology. *Current Opinion in Pharmacology*, *7*(1), 39–47. https://doi.org/10.1016/j.coph.2006.08.011

Parmar, M., Rawson, S., Scarff, C., Goldman, A., Dafforn, T., Muench, S., & Postis, V. L. G. (2018). Using a SMALP platform to determine a sub-nm single particle cryo-EM membrane protein structure. Biochimica et Biophysica Acta, 1860(2), 378-383. https://doi.org/10.1016/j.bbamem.2017.10.005.

Parsons, C. G., Stöffler, A., & Danysz, W. (2007). Memantine: A NMDA receptor antagonist that improves memory by restoration of homeostasis in the glutamatergic system - Too little activation is bad, too much is even worse. *Neuropharmacology*, *53*(6), 699–723. https://doi.org/10.1016/j.neuropharm.2007.07.013

Patrick, G. N., Zukerberg, L., Nikolic, M., de la Monte, S., Dikkes, P., & Tsai, L.-H. (1999). Conversion of p35 to p25 deregulates Cdk5 activity and promotes neurodegeneration. *Nature*, *402*(6762), 615–622. https://doi.org/10.1038/45159

Perozo, E., MacKinnon, R., Bezanilla, F., & Stefani, E. (1993). Gating currents from a nonconducting mutant reveal open-closed conformations in Shaker K+ channels. *Neuron*, *11*(2), 353–358. https://doi.org/10.1016/0896-6273(93)90190-3

Philpot, B. D., Cho, K. K. A., & Bear, M. F. (2007). Obligatory role of NR2A for metaplasticity in visual cortex. *Neuron*, *53*(4), 495–502. https://doi.org/10.1016/j.neuron.2007.01.027

Pittenger, C. (2015). Glutamate modulators in the treatment of obsessive-compulsive disorder. *Psychiatric Annals*, *45*(10), 506–511. https://doi.org/10.3928/00485713-20151008-06

Raman, I. M., & Trussell, L. O. (1995). The mechanism of α-amino-3-hydroxy-5-methyl-4-isoxazolepropionate receptor desensitization after removal of glutamate. *Biophysical Journal*, *68*(1), 137–146. https://doi.org/10.1016/S0006-3495(95)80168-2

Rauner, C., & Köhr, G. (2011). Triheteromeric NR1/NR2A/NR2B receptors constitute the major N-methyl-D-aspartate receptor population in adult hippocampal synapses. *Journal of Biological Chemistry*, *286*(9), 7558–7566. https://doi.org/10.1074/jbc.M110.182600

Regehr, W. G. (2012). Short-term presynaptic plasticity. *Cold Spring Harbor Perspectives in Biology*, *4*(7). https://doi.org/10.1101/cshperspect.a005702

Reiner, A., & Levitz, J. (2018). Glutamatergic signaling in the central nervous system: Ionotropic and metabotropic receptors in concert. *Neuron*, *98*(6), 1080–1098. https://doi.org/10.1016/j.neuron.2018.05.018

Rodríguez-Moreno, A., & Lerma, J. (1998a). Kainate receptor modulation of GABA release involves a metabotropic function. *Neuron*, *20*(6), 1211–1218. https://doi.org/10.1016/s0896-6273(00)80501-2

Rodríguez-Moreno, A., & Lerma, J. (1998b). Kainate receptors: Mediators of synaptic plasticity. *The Neuroscientist*, *4*(1), 15–25. https://doi.org/10.1177/107385849800400103

Rosenmund, C., Stern-Bach, Y., & Stevens, C. F. (1998). The tetrameric structure of a glutamate receptor channel. *Science*, *280*(5369), 1596–1599. https://doi.org/10.1126/science.280.5369.1596

Rumbaugh, G., & Vicini, S. (1999). Distinct synaptic and extrasynaptic NMDA receptors in developing cerebellar granule neurons. *The Journal of Neuroscience*, *19*(24), 10603–10610. https://doi.org/10.1523/JNEUROSCI.19-24-10603.1999

Salter, M. W., & Kalia, L. V. (2004). Src kinases: A hub for NMDA receptor regulation. *Nature Reviews Neuroscience*, *5*(4), 317–328. https://doi.org/10.1038/nrn1368

Sanacora, G., Treccani, G., & Popoli, M. (2012). Towards a glutamate hypothesis of depression: An emerging frontier of neuropsychopharmacology for mood disorders. *Neuropharmacology*, *62*(1), 63–77. https://doi.org/10.1016/j.neuropharm.2011.07.036

Sans, N., Vissel, B., Petralia, R. S., Wang, Y.-X., Royle, G. A., Wang, Y. T., Hardingham, G. E., Zhou, M., & Wenthold, R. J. (2003). Aberrant formation of glutamate receptor complexes in hippocampal neurons of mice lacking the GluR2 AMPA receptor subunit. *The Journal of Neuroscience*, *23*(28), 9367–9373. https://doi.org/10.1523/JNEUROSCI.23-28-09367.2003

Savtchouk, I., & Volterra, A. (2018). Gliotransmission: Beyond black-and-white. *The Journal of Neuroscience*, *38*(1), 14–25. https://doi.org/10.1523/JNEUROSCI.0017-17.2017

Schiller, J., Schiller, Y., Stuart, G., & Sakmann, B. (1997). Calcium action potentials restricted to distal apical dendrites of rat neocortical pyramidal neurons. *The Journal of Physiology*, *505*(3), 605–616. https://doi.org/10.1111/j.1469-7793.1997.605ba.x

Schwenk, J., Harmel, N., Brechet, A., Zolles, G., Berkefeld, H., Müller, C. S., Bildl, W., Baehrens, D., Hüber, B., Kulik, A., Klöcker, N., Schulte, U., & Fakler, B. (2009). High-resolution proteomics unravel architecture and molecular diversity of native AMPA receptor complexes. *Neuron*, *74*(4), 621–633. https://doi.org/10.1016/j.neuron.2012.03.034

Selimbeyoglu, A., & Parvizi, J. (2010). Electrical stimulation of the human brain: Perceptual and behavioral phenomena reported in the old and new literature. *Frontiers in Human Neuroscience*, *4*, 46. https://doi.org/10.3389/fnhum.2010.00046

Selvakumar, P., Lee, J., Khanra, N., He, C., Munguba, H., Kiese, L., Broichhagen, J., Reiner, A., Levitz, J., & Meyerson, J. R. (2021). Structural and compositional diversity in the kainate receptor family. Cell Reports, 37(4), 109891. https://doi.org/10.1016/j.celrep.2021.109891

Sheng, M., Sabatini, B. L., & Südhof, T. C. (2012). Synapses and Alzheimer's disease. *Cold Spring Harbor Perspectives in Biology*, *4*(5). https://doi.org/10.1101/cshperspect.a005777

Shepherd, J. D., & Huganir, R. L. (2007). The cell biology of synaptic plasticity: AMPA receptor trafficking. *Annual Review of Cell and Developmental Biology*, *23*, 613–643. https://doi.org/10.1146/annurev.cellbio.23.090506.123516

Shi, Y., Lu, W., Milstein, A. D., & Nicoll, R. A. (2010). The stoichiometry of AMPA receptors and TARPs varies by neuronal cell type. *Neuron*, *62*(4), 633–640. https://doi.org/10.1016/j.neuron.2010.04.008

Silver, R. A., Cull-Candy, S. G., & Takahashi, T. (1996). Non-NMDA glutamate receptor occupancy and open probability at a rat cerebellar synapse with single and multiple release sites. *The Journal of Physiology*, *494*, 231–250. https://doi.org/10.1113/jphysiol.1996.sp021560

Silver, R. A., Traynelis, S. F., & Cull-Candy, S. G. (1992). Rapid-time-course miniature and evoked excitatory currents at cerebellar synapses in situ. *Nature*, *355*(6357), 163–166. https://doi.org/10.1038/355163a0

Sobolevsky, A. I. (2015). Structure and gating of tetrameric glutamate receptors. *The Journal of Physiology*, *593*(1), 29–38. https://doi.org/10.1113/jphysiol.2013.264911

Sobolevsky, A. I., Rosconi, M. P., & Gouaux, E. (2009). X-ray structure, symmetry and mechanism of an AMPA-subtype glutamate receptor. *Nature*, *462*(7274), 745–756. https://doi.org/10.1038/nature08624

Straub, C., Zhang, W., & Howe, J. R. (2011). Neto2 modulates gating of synaptic kainate receptors and regulates synaptic transmission. *Nature Neuroscience*, *14*(7), 866–873. https://doi.org/10.1038/nn.2837

Sun, Y., Olson, R., Horning, M., Armstrong, N., Mayer, M., & Gouaux, E. (2002). Mechanism of glutamate receptor desensitization. *Nature*, *417*(6886), 245–253. https://doi.org/10.1038/417245a

Suvrathan, A., Hoeffer, C. A., Wong, H., Klann, E., & Chattarji, S. (2010). Characterization and reversal of synaptic defects in the amygdala in a mouse model of fragile X syndrome. *Proceedings of the National Academy of Sciences*, *107*(25), 11591–11596. https://doi.org/10.1073/pnas.1002262107

Tajima, N., Karakas, E., Grant, T., Simorowski, N., & Furukawa, H. (2016). Activation of NMDA receptors in rat hippocampal neurons by rapid solution exchange. *Nature*, *534*, 63–68. https://doi.org/10.1038/nature17679

Takeuchi, T., Duszkiewicz, A. J., & Morris, R. G. M. (2014). The synaptic plasticity and memory hypothesis: Encoding, storage and persistence. *Philosophical Transactions of the Royal Society B: Biological Sciences*, *369*(1633). https://doi.org/10.1098/rstb.2013.0288

Thompson, A. J., Lester, H. A., & Lummis, S. C. (2010). The structural basis of function in Cys-loop receptors. *The Quarterly Review of Biophysics*, *43*(4), 449–499. https://doi.org/10.1017/S0033583510000168

Tomita, S., Adesnik, H., Sekiguchi, M., Zhang, W., Wada, K., Howe, J. R., & Nicoll, R. A. (2005). Stargazin modulates AMPA receptor gating and trafficking by distinct domains. *Nature*, *435*(7045), 1052–1058. https://doi.org/10.1038/nature03624

Tønnesen, J., Katona, G., Rózsa, B., & Nägerl, U. V. (2014). Spine neck plasticity regulates compartmentalization of synapses. *Nature Neuroscience*, *17*(5), 678–685. https://doi.org/10.1038/nn.3682

Traynelis, S. F., Wollmuth, L. P., McBain, C. J., Menniti, F. S., Vance, K. M., Ogden, K. K., Hansen, K. B., Yuan, H., Myers, S. J., & Dingledine, R. (2010). Glutamate receptor ion channels: Structure, regulation, and function. *Pharmacological Reviews*, *62*(3), 405–496. https://doi.org/10.1124/pr.109.002451

Tsien, J. Z., Huerta, P. T., & Tonegawa, S. (1996). The essential role of hippocampal CA1 NMDA receptor-dependent synaptic plasticity in spatial memory. *Cell*, *87*(7), 1327–1338. https://doi.org/10.1016/s0092-8674(00)81827-9

Twomey, E. C., & Sobolevsky, A. I. (2017). Structural mechanisms of gating in ionotropic glutamate receptors. *Neuron*, 94(3), 494-497.e4. https://doi.org/10.1016/j.neuron.2017.04.009.

Twomey, E. C. (2018). Structural Determinants of Ionotropic Glutamate Receptor Function Revealed by Cryo-electron Microscopy. Columbia University. https://doi.org/10.7916/D8XH178F

Twomey, E. C., Vakhrushev, S. Y., Grassucci, R. A., Frank, J., & Sobolevsky, A. I. (2018). Mechanism of channel gating in the activated conformation of AMPA receptors. *Neuron*, *99*(3), 534–546.e5. https://doi.org/10.1016/j.neuron.2018.07.004

Twomey, E. C., Yelshanskaya, M. V, Grassucci, R. A., Frank, J., & Sobolevsky, A. I. (2017). Channel opening and gating mechanism in AMPA-subtype glutamate receptors. *Nature*, *549*(7670), 60–65. https://doi.org/10.1038/nature23480

Volk, L. J., Pfeiffer, B. E., Gibson, J. R., & Huber, K. M. (2015). The hippocampal glutamate neurotransmitter system in autism. *Annual Review of Pharmacology and Toxicology*, *55*, 559–573.

Vyklicky V, Korinek M, Smejkalova T, Balik A, Krausova B, Kaniakova M, Lichnerova K, Cerny J, Krusek J, Dittert I, Horak M, Vyklicky L. (2014). Structure, function, and pharmacology of NMDA receptor channels. *Physiological Research*, *64*(Suppl 1), S191–S203. https://doi.org/10.33549/physiolres.932678

Wang, J., & Furukawa, H. (2019). Dissecting diverse functions of NMDA receptors by structural biology. *Current Opinion in Structural Biology*, 54, 34-42. https://doi.org/10.1016/j.sbi.2018.12.009.

Weston, M. C., Schuck, P., Ghosal, A., Rosenmund, C., & Mayer, M. L. (2006). Conformational restriction blocks glutamate receptor desensitization. *Nature Structural & Molecular Biology*, *13*(12), 1120–1127. https://doi.org/10.1038/nsmb1173

Yan, D., Villalobos, C., & Andrade, R. (2005). TRPC1 and TRPC5 form a novel cation channel in mammalian brain. *Neuron*, *48*(5), 787–798. https://doi.org/10.1016/j.neuron.2005.10.018

Yang, Y., Wang, X., Frerking, M., & Zhou, Q. (2008). Spine expansion and stabilization associated with long-term potentiation. *The Journal of Neuroscience*, *28*(22), 5740–5751. https://doi.org/10.1523/JNEUROSCI.3998-07.2008

Young, A. B., & Fagg, G. E. (1990). Excitatory amino acid receptors in the brain: Membrane binding and receptor autoradiographic approaches. *Trends in Pharmacological Sciences*, *11*(3), 126–133. https://doi.org/10.1016/0165-6147(90)90199-I

Zhang, D., Ivica, J., Krieger, J. M., Ho, H., Yamashita, K., Stockwell, I., Baradaran, R., Cais, O., & Greger, I. H. (2023). Structural mobility tunes signalling of the GluA1 AMPA glutamate receptor. *Nature*. https://doi.org/10.1038/s41586-023-06528-0

Zhang, D., Watson, J. F., Matthews, P. M., Cais, O., & Greger, I. H. (2021). Gating and modulation of a hetero-octameric AMPA glutamate receptor. *Nature*, *594*(7863), 454–458. https://doi.org/10.1038/s41586-021-03613-0

Zhang, W., Howe, J. R., & Popescu, G. K. (2006). Single-channel properties of homomeric GluK3 kainate receptors. *Neuroscience Letters*, *402*(1–2), 92–96. https://doi.org/10.1016/j.neulet.2006.03.050

Zhang, W., Howe, J. R., & Popescu, G. K. (2008). Distinct gating modes determine the biphasic relaxation of NMDA receptor currents. *Nature Neuroscience*, *11*(12), 1373–1375. https://doi.org/10.1038/nn.2213

Zhang, W., St-Gelais, F., Grabner, C. P., Trinidad, J. C., Sumioka, A., Morimoto-Tomita, M., Kim, K. S., Straub, C., Burlingame, A. L., Howe, J. R., & Tomita, S. (2009). A transmembrane accessory subunit that modulates kainate-type glutamate receptors. *Neuron*, *61*(3), 385–396. https://doi.org/10.1016/j.neuron.2008.12.014

Zhao, Y., Chen, S., Swensen, A., Qian, W., & Gouaux, E. (2019). Architecture and subunit arrangement of native AMPA receptors elucidated by cryo-EM. Science, 364(6440), 355-362. https://doi.org/10.1126/science.aaw8250.

Zheng, W., Wen, H., Iacobucci, G., & Popescu, G. (2017). Probing the Structural Dynamics of the NMDA Receptor Activation by Coarse-Grained Modeling. *Biophysical Journal*, 112(12), 2589-2601. https://doi.org/10.1016/j.bpj.2017.04.043.

27 Cryo-EM Structures of Systemic Amyloid Fibrils

Binh An Nguyen[†], *Maria del Carmen Fernandez-Ramirez*[†], *Shumaila Afrin*[†], *and Lorena Saelices*

INTRODUCTION

Systemic amyloidosis is a group of uncommon degenerative disorders in which different proteins misfold and acquire a structure called amyloid. These amyloids, which are highly resistant to the catabolic processes of the body, deposit extracellularly in different tissues, leading to organ dysfunction and death (Gertz and Dispenzieri, 2020). Part of the scientific community has been focused on the characterization of the structural features of amyloids to better understand disease biopathology.

The term *amyloid* was firstly used in botanics, and it is derived from the Latin word *amylum* that means "starch-like." In a pathology laboratory in 1854, Rudolf Virchow used this term to refer to liver deposits that stained positively in a reaction with iodine and sulfuric acid, similar to starch (Virchow, 1854). The composition of these amyloids was discovered a few years later, when Friedrich and Kekule demonstrated their proteinaceous nature (Cohen, 1986; Friedreich and Kekulé, 1859). Although all amyloidoses are characterized by the buildup of these protein-made deposits, the precursor protein differs for each disease, and the process in which they aggregate into amyloid is also variable. Based on the current models for amyloid formation, the involved proteins need to partially or totally unfold to allow for a conformational change in a nucleation process that leads to the formation of the first amyloid assemblies (Figure 27.1). These nuclei template the incorporation of additional precursor proteins that elongate the amyloid structure into mature insoluble amyloid fibrils. These fibrils can be fragmented into smaller species or *seeds*, acting as new precursors for elongation and spreading of the amyloid deposits in a process called amyloid seeding (Figure 27.1).

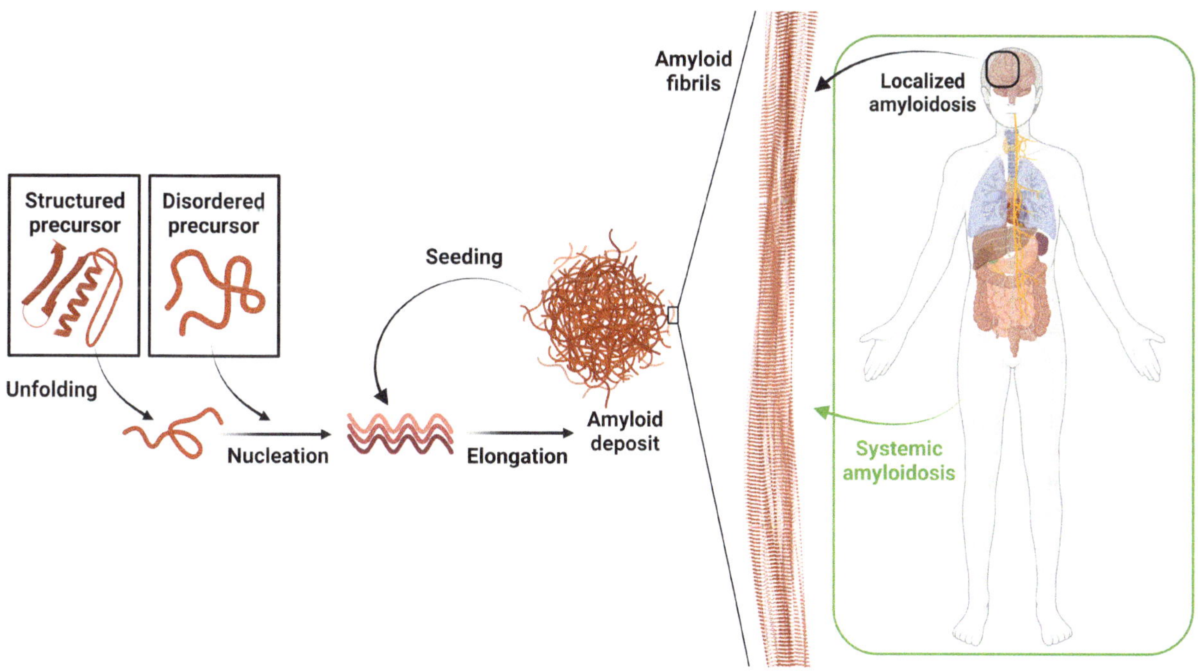

FIGURE 27.1 Schematic depiction of the mechanism of formation and deposition of systemic amyloid fibrils in the human body (created with Biorender.com). Precursor proteins may be structurally ordered or disordered and can nucleate and polymerize to form amyloid fibrils. These fibrils can deposit either locally to cause localized amyloidosis or throughout the body leading to systemic amyloidosis.

[†] These authors contributed equally on this work.

DOI: 10.1201/9781003326106-33

Disease	ATTRv amyloidosis	ATTRwt amyloidosis	AL amyloidosis	AA amyloidosis
Precursor protein	Variant transthyretin	Wild-type transthyretin	Immunoglobulin light chain	(Apo) serum amyloid A
Affected organs	Peripheral and autonomous nervous system, heart, eye, kidneys, leptomeninges	Heart mainly in males, lung, ligaments, tenosynovium	All organs, usually except central nervous system	All organs, usually except central nervous system
Systemic or localized	Systemic	Systemic	Systemic or localized	Systemic
Cryo-EM fibril structures	6sdz, 8e7d, 8e7e, 8e7h, 8e7i, 8e7j, 8pke, 8pkf, 8pkg, 8tdn, 8tdo (human hearts), 7ob4 (human vitreous humor)	7z40, 8ade, 8g9r, 8gbr, 8e7d, 8e7h (human hearts)	6ic3, 6z1o, 6z1i, 6hud (human hearts)	6mst (human kidney) 6dso (mouse spleen) 6zch (mouse spleen)
Native structure	 4tlu.pdb (V30M) (Saelices et al., 2015, JBC)	 4tlt.pdb (Saelices et al., 2015, JBC)	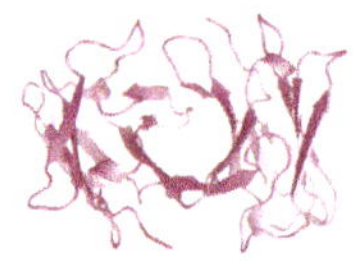 1cd0.pdb (λ6) (Pokkuluri et al., 2019, Amyloid)	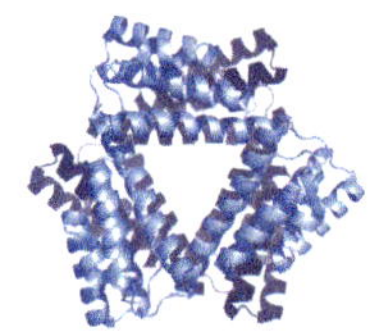 4ip9.pdb (Lu et al., 2014, PNAS)

FIGURE 27.2 A tabular presentation of ATTR, AL, and AA amyloidoses indicating their precursor proteins, primary affected organs, systemic or localized distribution, PDB IDs of their cryo-EM fibrils, and native structures. This figure also shows the native structures of amyloid precursor proteins that lead to ATTR, AL, and AA amyloidoses.

This process of protein aggregation and amyloid deposition can happen locally or systemically in localized or systemic amyloidosis, respectively (Almeida and Brito, 2020).

To date, 42 proteins have been implicated in the formation of both localized and systemic amyloidoses. Proteins causing localized amyloidosis are those that are synthesized and deposited in the same organ or tissue. These include proteins such as Aβ, tau, or α-synuclein. Proteins causing systemic amyloidosis are those that are deposited in organs or tissues that do not produce the precursor protein. These include proteins such as transthyretin, light chain immunoglobulins, or serum amyloid protein A (Buxbaum et al., 2022) (Figure 27.1). Very few of these proteins form both types (Buxbaum et al., 2022). Although caused by the deposition of different proteins, systemic amyloidoses can have overlapping clinical and pathological features. Organs commonly affected include the heart, gastrointestinal tract, kidney, liver, nerves, and eye (Muchtar et al., 2021). The present review provides a summary of the structural studies of amyloid fibrils extracted from the organs of patients suffering from systemic amyloidosis.

Advances in structural biology have been crucial in the understanding of the molecular basis of amyloidogenesis in systemic amyloidosis. In the 1900s, amyloids were observed to bind to Congo red dye, exhibiting an apple green birefringence indicative of an organized structure (Divry, 1927; Missmahl and Hartwing, 1953). In 1959, thanks to electron microscopy, it was discovered that they all share bundles of straight and rigid fibrils, whose average width ranges from 75 to 100 Å (Cohen et al., 1982; Cohen and Calkins, 1959). In 1968, X-ray fibril diffraction studies revealed ~4.7 Å and ~10 Å diffraction rings, termed a cross-β diffraction pattern, which results from the hydrogen bonding and interdigitation of β-sheets in the amyloid assembly (Fändrich, 2007; Fändrich and Schmidt, 2021; Geddes et al., 1968). Over the past several decades, X-ray crystallography and NMR spectroscopy have shed light on the inner structure of amyloid fibrils, with most success in the study of short amyloidogenic peptides (Sawaya and Eisenberg, 2013). However, the biggest revolution in the study of amyloid fibrils came with the use of cryogenic electron microscopy (cryo-EM) (Fitzpatrick et al., 2017). Cryo-EM has allowed us to visualize the atomic structure of now dozens of ex vivo amyloid fibrils (Sawaya et al., 2021), including those of systemic amyloidosis. Cryo-EM has enabled direct two-dimensional imaging and helical reconstruction to obtain the high-resolution structures of the fibrils directly extracted from systemic amyloidosis patients.

This chapter summarizes and compares the cryo-EM structures of systemic amyloid *ex vivo* fibrils modeled to date to depict their structural variabilities and commonalities. We will include fibrils from ATTR, AL, and AA amyloidoses (Figure 27.2). By presenting an overall picture of the current knowledge, we aim to offer a solid base for future advances in the field to help move us closer to understanding the molecular mechanisms behind fibril formation that lead to the different systemic amyloidoses.

CRYO-EM STRUCTURES OF AMYLOID FIBRILS FROM ATTR AMYLOIDOSIS PATIENTS

ATTR amyloidosis is a protein disorder that results in the deposition of insoluble fibrils in multiple organs throughout the body (Benson, 2012). There are two main subtypes of ATTR amyloidosis: variant (ATTRv) amyloidosis, which is caused by mutations or deletions in the transthyretin (TTR) gene, and wild-type (ATTRwt) amyloidosis, which is associated with aging (Koike et al., 2009). The currently accepted pathological pathway of ATTR amyloidosis involves the dissociation of the tetrameric transthyretin (55 kDa) and the unfolding of monomers that trigger their reassembly (Hurshman et al., 2004) into fibrils that accumulate as amyloid. This accumulation affects organs such as the heart, peripheral nerves, kidneys, gut, and eyes (Benson, 2012; Tasaki et al., 2013). The precise organs affected by amyloid deposition and clinical phenotype can vary depending on the subtype of ATTR amyloidosis and the specific mutation involved, but heart and nerves are the most common organs that are impacted by both ATTRv and ATTRwt (Benson, 2012; Wajnsztajn Yungher et al., 2020). Because previous cryo-EM studies on amyloid brain diseases suggest that a specific fibrillar folding could be related to the disease presentation (Falcon et al., 2018; Scheres et al., 2020; Schweighauser et al., 2020), structural studies of ATTR deposits could help us understand the diverse clinical phenotypes found in ATTR amyloidosis (Figure 27.3).

The first cryo-EM structure of an ex vivo ATTR fibril was published in 2019 by Schmidt et al., from a patient with the mutation ATTRv-V30M, which is the most common ATTRv variant worldwide (Schmidt et al., 2019). In 2021, Iakovleva et al. reported the structure of ATTR fibrils collected from the vitreous humor of an ATTRv-V30M patient (Iakovleva et al., 2021). In 2022, two additional cryo-EM studies were published. In the first one, Steinebrei et al. described ATTRwt fibrils from a patient with nonhereditary ATTR amyloidosis (Steinebrei et al., 2022), and in the second study, Nguyen et al. presented the structures found in three ATTRv patients carrying the mutations ATTRv-P24S and ATTRv-I84S (two patients) (Nguyen et al., 2022). In 2023, Steinebrei et al. showed three structures linked to mutations ATTRv-V20I, ATTRv-G47E, and ATTRv-V122I (Steinebrei et al., 2023). In 2024, Nguyen et al. expanded upon their previous work, providing two additional fibril structures derived from ATTRv-I84S patients (Nguyen et al., 2024).

Excluding the study from 2021, the *ex vivo* fibrils were extracted from heart, and all of them share a core with

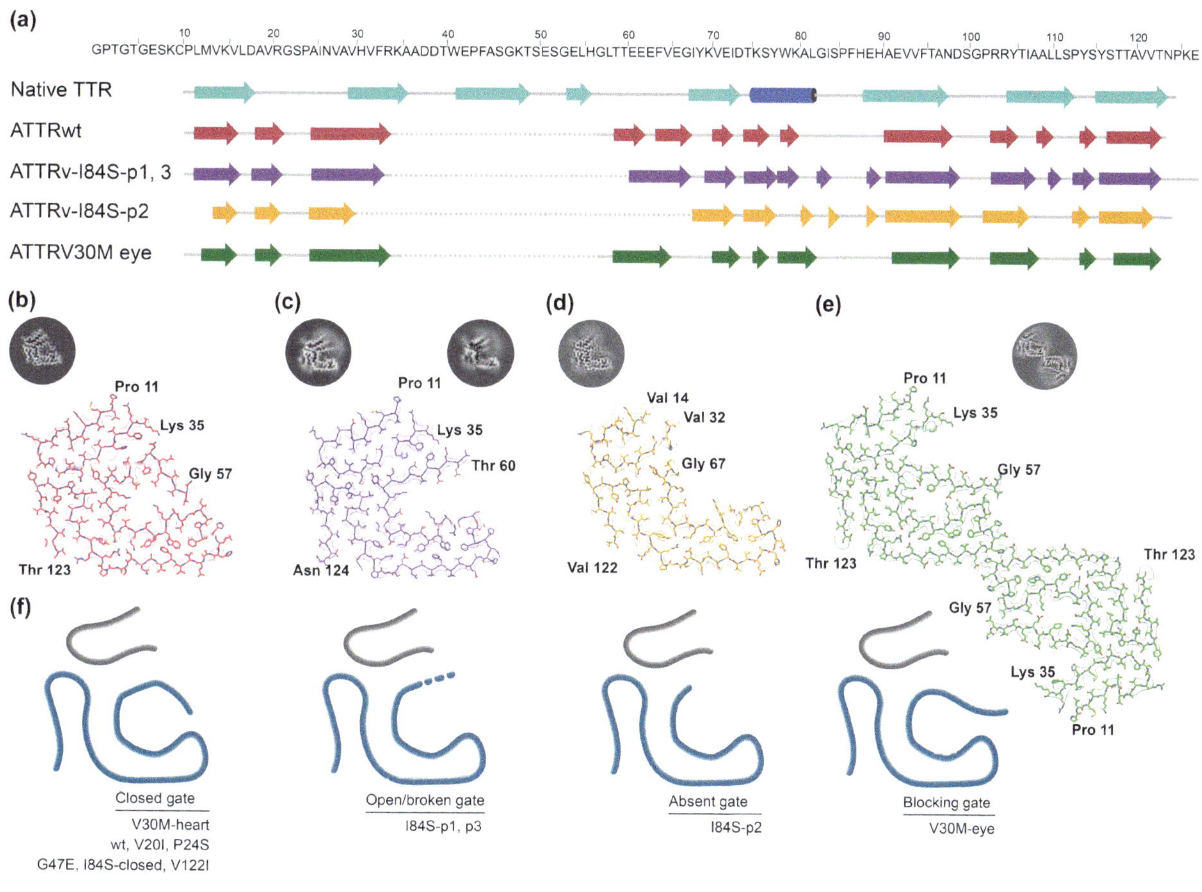

FIGURE 27.3 Structural polymorphism of ATTR fibrils. The cryo-EM studies of ATTR fibrils reveal an unprecedented polymorphism in ATTR amyloidosis patients, with the main structural differences in a polar channel ending at position glycine 67. (a) The secondary structure of fibrillary transthyretin differs from its native form. The number, length, and location of β-strands formed in ATTR fibrils is different among variants. β-strands are indicated in black arrows. (B–E) The 3D class averages and their cryo-EM density and atomic models of the structures resolved to date show alternative rearrangements in the core of the *ex vivo* ATTR fibrils from different variants. (F) Schematic illustrations of the fibrillary core structure evidence the *closed gate, open gate, broken gate, absent gate,* and *blocking gate* conformations of the polar cavity in the different variants.

similar topology composed by two transthyretin fragments: the N-terminal segment that expands from residues 14 to 32, and the C-terminal fragment that includes residues from 67 to 122. The rest of the structure is polymorphic.

The main similarities shared among all the ATTR fibrils studied to date are the absence of transthyretin fragments and the high stability of the structure acquired. Absent or weak densities are adjacent to the structured regions, indicating that residues glycine 1 to cysteine 10, alanine 36 to histidine 56, and asparagine 124 to glutamic acid 127 are disordered or absent. Although cryo-EM does not allow us to determine if every fibril layer is formed by a full-length or a fragmented protein, additional techniques such as denaturing gel electrophoresis and mass spectrometry shed light on this issue. Previous studies suggest that both full-length and fragmented transthyretin can be part of the fibrils, probably interspersed by layers (Iakovleva et al., 2021; Schmidt et al., 2019; Steinebrei et al., 2022). Cryo-EM studies also revealed that the thermodynamic stability of both ATTRwt and ATTRv fibrils is significantly higher than the ensemble of amyloid structures reported in the Amyloid Atlas from 2023 (Nguyen et al., 2024; Sawaya et al., 2021). The free energy of solvation ranges from −67.3 and −48.6 kcal/mol per chain and −0.75 and −0.60 kcal/mol per residue. The statistical comparison of these values reported the significant stability of ATTR fibrils as a result of the contribution of the three main pockets (Nguyen et al., 2024). Despite these common features, key structural differences can be found in ATTR fibrils.

There are two levels of polymorphism in ATTR fibrils. The first one is found in the polar channel, where different folds can be acquired by the fragment glycine 67 backwards, referred to as the "gate" (Figure 27.3). The closed conformation was first described in the ATTR-V30M fibrils from the heart, and it is the most common conformation, also found in ATTRwt, ATTRv-G20I, ATTRv-P24S, ATTRv-V30M, ATTRv-G47E, ATTRv-I84S (three patients), and ATTRv-V122I amyloidosis (Nguyen et al., 2022, 2024; Schmidt et al., 2019; Steinebrei et al., 2023). Notably, in each patient carrying the ATTRv-I84S mutation, two distinct morphologies were observed: one is similar to the common ATTRwt conformation, while the other differs in the gate of the polar channel, from glycine 67 backward. These polymorphic structures are illustrated in Figure 27.3C as an open gate, where a fragment from threonine 60 to glycine 67 flips by an angle of 180°; an absent gate, where a fragment prior to glycine 67 of the C-terminal is missing; and a broken gate, similar to the open gate, but the density map of the gate can only accommodate phenylalanine 64 to glycine 67 (Nguyen et al., 2024). In ATTR-V30M fibrils extracted from vitreous humor, the fragment glycine 57 to isoleucine 68 forms a long diagonal β-strand that crosses the center of the polar channel, acquiring a blocking gate conformation (Iakovleva et al., 2021). The second level of polymorphism is based on the number of protofilaments forming the fibrils. While the fibrils from the heart are composed of one single protofilament, around 80% of fibrils from the vitreous body of the eye contained multiples of them (Iakovleva et al., 2021). However, only the twisted dimers were abundant enough to enable structure determination. To summarize, all these structural variations suggest the existence of polymorphism in ATTR amyloid fibrils.

Cryo-EM studies of *ex vivo* ATTR fibrils show similarities and structural differences among ATTRv vs. ATTRwt, fibrils from eye vs. heart, and differences in fibrils from the same patient. With the limited current data, it is not possible to determine if there is an association between variant, organ, or patient and the ATTR fibril structure formed. Similarly, there is not enough information to establish an association between these diverse structures and the clinical manifestation of the patient harboring these fibrils. In order to answer these questions, more studies need to be done, increasing the sampling and including other mutations and organs.

CRYO-EM STRUCTURES OF AMYLOID FIBRILS FROM AL PATIENTS

AL amyloidosis is a rare disease characterized by the deposition and accumulation of amyloid fibrils made of immunoglobulin light chains κ or λ (Merlini, 2017). These light chains stack on top of each other to form abnormal shapes distinct from their native fold. These fibrils can deposit systemically in various tissues and organs, including the heart, kidneys, liver, peripheral nervous system, and gastrointestinal tract (Hasib Sidiqi and Gertz, 2021). Fibril deposition eventually causes damage to these organs and interferes with their normal physiological functions. The structural basis of this organ targeting is not yet clear.

The first cryo-EM structures ever described of AL amyloid fibrils were released in 2019, from subtypes λ1 and λ6 (Radamaker et al., 2019; Swuec et al., 2019). Since then, two additional cryo-EM structures from the AL fibril subtype λ3 have been reported (Radamaker et al., 2021). All these cryo-EM studies used AL fibrils extracted from cardiac tissues. Here, we review the structural information of these cryo-EM maps and their models.

AL Subtype λ1 (EMD-4452, PDB: 6IC3)

Helical reconstruction of cryo-EM data of AL subtype λ1 fibril resulted in a density map at a resolution of 3.3 Å (Radamaker et al., 2019). Details of topology parameters of this map are summarized in Figure 27.4A and B. The fibril structure is composed of 12 β-strands that extend from glycine 15 to threonine 105, featuring a disulfide bond between cysteine 22 and cysteine 89. Notably, there are three significant channels present within the structure, two of which (A and B) are hydrophilic and enclosed by residues arginine 25–tryptophan 36 and glutamine 80–glutamate 84, and by residues lysine 67–asparagine 83, respectively. The third channel (C) is hydrophobic and is enclosed by residues valine 18–cysteine 22 and cysteine 89–leucine 96. Each layer along the fibril axis is not entirely flat, and some residues form contacts with the adjacent layer; for example, strand β4 of layer i interacts with β8 of layer i and i+1. The

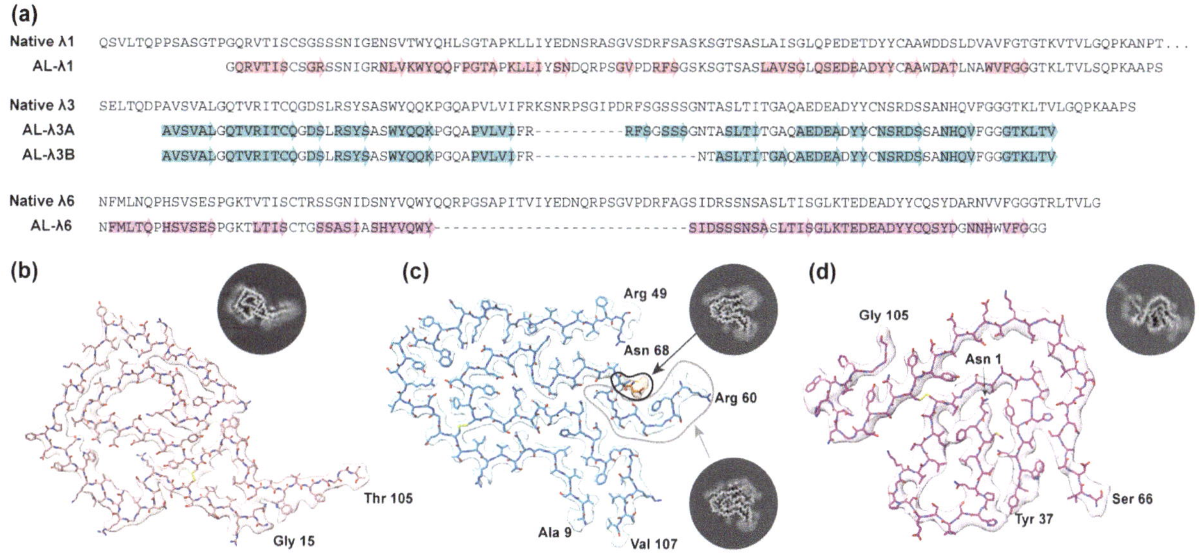

FIGURE 27.4 Schematic representation of the primary sequences, secondary sequences, and the cryo-EM density maps and models of AL fibrils. Cryo-EM studies of AL fibrils show structural polymorphism from different germlines. (A) Primary sequence alignments of AL subtypes between their native protein and the corresponding fibril protein(s). β-strands are indicated in black arrows. (B, C, and D) Subfigures of AL-λ1, AL-λ3A/B, and AL-λ6, respectively. Each subfigure contains a cross-sectional view of three-dimensional class averages of AL fibril (the dark circle), and the top view of a single fibril layer with its model. For subfigure (C), the main structural difference between the two morphologies of AL-λ3 are notated where subtype A is at the bottom and subtype B is on the top with a darker density map.

N- and C-termini of the fibril structure have fuzzy densities, indicating disordered regions. Additionally, there are extra densities near the main fibril structure, such as those near arginine 17, close to valine 100 and threonine 95, or right next to leucine 48 and tyrosine 50, which could potentially be cofactors for fibril folding or stabilization.

AL Subtype λ6 (EMD-0274, PDB: 6HUD)

Cryo-EM reconstruction of the AL subtype λ6 fibril resulted in a map with a resolution of 4.0 Å (Swuec et al., 2019). The fibril structure is composed of 9 β-strands that extend across two segments, which are connected by a disulfide bridge between cysteine 22 and cysteine 91 (Figure 27.4A and D). The N-terminal segment, spanning from aspartate 1 to tyrosine 37, contains five β-strands (β1–β5). β1 is sandwiched between β3 and β5 strands, with β2 exposed to the solution and β4 interacting with β7 of the C-terminal segment. The C-terminal segment, spanning from serine 66 to glycine 105, contains four β-strands (β6- β9) that form an arc wrapping outside the N-terminal segment. Intramolecular interactions of the N-terminal segment extend to two consecutive layers, as demonstrated by the side chain interactions between residues of β1 strand layer i, β3–4 strands of layer i+1, and β5 strand of layer i+2. In contrast, the C-terminal segment appears to be flat. Additional densities were observed after tyrosine 37 and before serine 66, which may belong to the disordered region of glutamine 38–glycine 65. A small extra density was also noted near asparagine 72.

AL Subtypes: λ3A (EMD-11031, PDB: 6Z10); λ3B (EMD-11030, PDB: 6Z1I)

Cryo-EM reconstruction of the AL subtype λ3 fibril yielded two maps, A and B, with resolutions of 3.2 and 3.4 Å, respectively (Radamaker et al., 2021). The fibril structure A comprises 12 β-strands extending across two segments featuring a disulfide bond cysteine 22–cysteine 87, with the N-terminal segment spanning from alanine 9 to arginine 49 and the C-terminal segment spanning from arginine 60 to valine 107 (Figure 27.4A and C). The N-terminal segment fold is compact without any internal channel, whereas the C-terminal segment tightly wraps around the outside of the N-terminal segment. The fibril structure B is similar to structure A, with the exception of the missing density corresponding to residues from arginine 60–glycine 67 and the extended conformation of asparagine 68–serine 71. Both structures contain a well-defined density island that was suggested to be part of an additional fibril protein fragment. Notably, both structural conformations were observed simultaneously in the same filament throughout the fibril dataset.

Folding Differences and Commonalities between Human AL Fibrils

The variability in the structure of AL amyloid fibrils (Table 27.1) may be attributed to the characteristics of immunoglobulin light chain production, including gene segment shuffling, somatic hypermutation, and posttranslational

TABLE 27.1
Structural features of fibrils in systemic amyloidoses

	Rise (Å)	Twist (°)	Pitch (nm)	Handedness
ATTRv-V30M	4.825	1.19	Not disclosed	Left
ATTRv-V30M(*)	4.72	0.55	308	Left
ATTRwt	4.82	1.23	Not disclosed	Left
ATTRv-T60A	4.79	1.26	136	Left
ATTRv-P24S	4.84	1.28	134	Left
ATTRv-V20I	4.82	−1.29	Not disclosed	Left
ATTRv-G47E	4.81	−1.24	Not disclosed	Left
ATTRv-V122I	4.81	−1.24	Not disclosed	Left
ATTRv-I84S (Closed gate)	4.81	−1.26	138	Left
ATTRv-I84S (Open gate)	4.80	−1.42	122	Left
ATTRv-I84S (Absent gate)	4.93	−1.30	137	Left
ATTRv-I84S (Broken gate)	4.96	−1.33	134	Left
SAA (human glomerular)	2.40	180.79	110	Right
SAA (human Vascular)	4.75	−1.63	110	Left
SAA feline	4.9 Å	−1.3°	135	Left
AL-λ1	4.8	−0.58	300	Left
AL-λ3A/B	4.8	−1.1	155	Left
AL-λ6	4.9	−1.6	110	Left

(*) Fibrils were extracted from vitreous humor.

modifications (Collins and Watson, 2018). These processes result in a distinct sequence of light chains for each class and a deviation from their original germline genes, while also affecting the structure, stability, and function of the light chains. Such variations can lead to polymorphism in AL amyloid structures from different patients or even within the same patient. Nevertheless, all fibrils contain a disulfide bond involving conserved cysteines with unknown implications. Generating a comprehensive structural map of AL fibrils could help us gain a better understanding of the molecular mechanisms underlying fibril formation. One particularly useful study would be to determine the structures of AL fibrils from the same organs or from different organs of the same individuals.

CRYO-EM STRUCTURE OF AMYLOID FIBRILS FROM AA AMYLOIDOSIS PATIENTS

AA amyloidosis is a form of systemic amyloidosis caused by the deposition of amyloid fibrils formed by the acute phase response protein serum amyloid A. The disease mainly affects the kidneys (Lachmann et al., 2007), spleen, and liver, leading to organ failure and death (Westermark et al., 2015). Based on the deposition site within the kidney, the disease is divided into two types, vascular, with deposits in blood vessels, and glomerular, with deposits in glomeruli. In most cases, the glomeruli are the primary site of deposition (Obici and Merlini, 2012; Röcken and Shakespeare, 2002; Westermark and Westermark, 2009). The disease occurs not only in humans but also in other species like Pekin ducks (Solomon et al., 2007), captive animals like cheetahs (Papendick et al., 1997; Zhang et al., 2008), and shelter cats (Ferri et al., 2022).

To date, we have high-resolution cryo-EM structures of *ex vivo* AA amyloid fibrils from both the glomerular and vascular variants of human kidney and from the pole region of the kidney of the domestic short-haired shelter cat (Figure 27.5) (Banerjee et al., 2022; Liberta et al., 2019; Schulte et al., 2022). The first cryo-EM ex vivo fibril structure was published in 2019 by Liberta et al. of the human glomerular variant of AA (Liberta et al., 2019), followed by the vascular variant in 2022 by Banerjee et al. (Banerjee et al., 2022). In the same year, Schulte et al. published the cryo-EM structure of feline AA fibril (Schulte et al., 2022). Here, we review the structural information from these cryo-EM structures.

Human Glomerular AA Fibril (EMD-9232, PDB-6MST)

The glomerular variant consisted of a single morphology and was resolved at the resolution of 2.7 Å (Liberta et al., 2019). Structural details are listed in Table 27.1. The fibril consisted of two identical protofilaments (referred to as equal stacks by the authors). In contrast to the native globular AA structure, which consists of only alpha helices, the fibril structure comprises seven β-strands from serine 2 to alanine 55 forming the ordered part. The fibril has a cavity lined with ionizable, charged, and polar residues, indicating that it contains water. Diffused densities are also observable outside the ordered fibril core. Some can be attributed to disordered C-terminal from residues 56–67. The identity of the remaining parts of the diffused density is unknown and

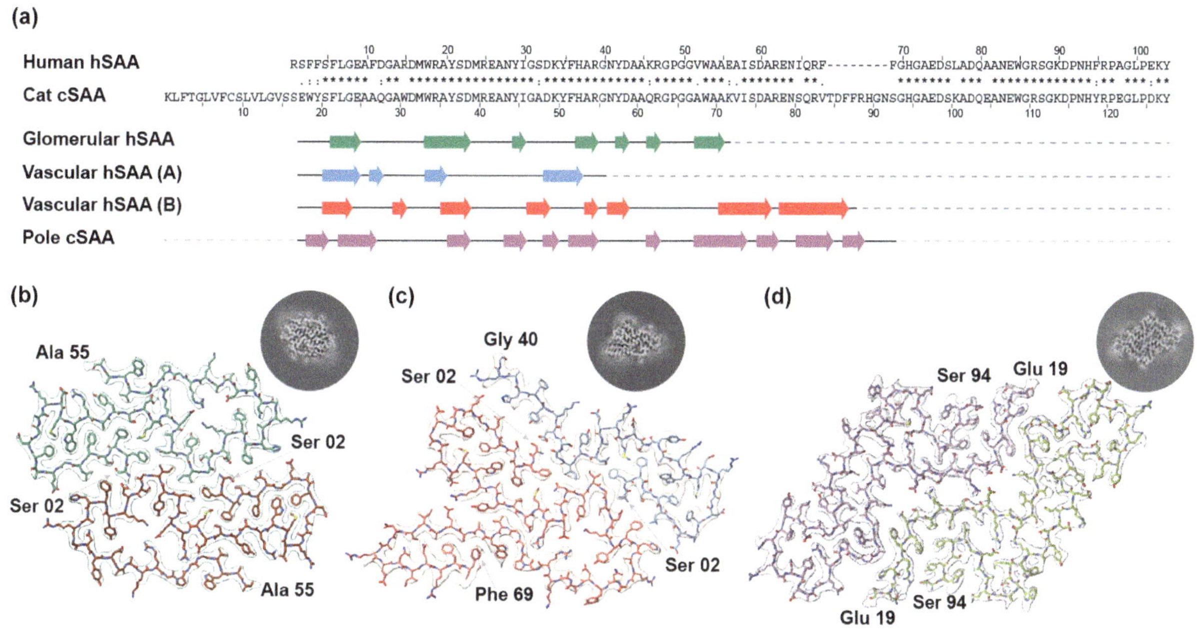

FIGURE 27.5 Schematic representation of the primary sequences, secondary sequences, and the cryo-EM density maps and models of human (vascular and glomerular) and cat AA fibrils. Cryo-EM reveals the existence of structural variations in glomerular and vascular variants of AA amyloidosis fibrils. Additionally, the structure of cat AA fibrils differs from the human AA fibrils (as expected based on sequence). (a) Primary sequence alignments of human and cat native AA proteins and the corresponding fibrils. β-strands are indicated in black arrows. (B, C and D) Subfigures of human glomerular, vascular, and cat pole region AA structures, respectively. Each subfigure contains a cross-sectional view of three-dimensional class averages of AA fibril (the dark circle), and the top view of a single fibril layer with its model. For subfigures A and B, the position of serine 2 is indicated by grey arrows.

may be attributed to several molecules that may decorate the fibril structure.

Human Vascular AA Fibrils (EMD-9232, PDB-7ZKY)

In the vascular disease variant, amyloid deposits are found in the renal medulla and the walls of the blood vessels. Two similar fibril structures from two patients of this variant were resolved at a resolution of 2.56 Å and 2.68 Å (Banerjee et al., 2022). The vascular variant has two nonidentical protofilaments or "stacks," A and B, with a C1 symmetry. Stack A comprises serine 2 to glycine 40 in the form of four cross β-sheets, while stack B contains residues serine 2 to phenylalanine 69 in the form of eight cross β-sheets. Stack A has a polar cavity that the authors speculate may be occupied by water, which is absent in stack B.

Folding Differences and Commonalities between Human AA Fibrils

Although the two variants seem to share the same fold from residue 2-53 (Figure 27.5B and C), they differ in their handedness (Banerjee et al., 2022; Liberta et al., 2019). The resolution of the structures may not justify the assignment of the handedness, but the authors performed platinum side shadowing that unambiguously defines the direction of the twist. Where the glomerular variant is right-handed, the vascular is predominantly left-handed (with a very small population of left-handed that could not be resolved). The difference in handedness may be attributed to the direction of tilt of the N-terminus β-sheets as the β-sheets are tilted upward in the glomerular variant and downward in the vascular variant relative to the central part of the fibril structure. Additionally, the hydrophobic core of the N-terminus arch from serine 2 to alanine 20 has a different packing, resulting in the flipping of the positions of phenylalanine 6 and tryptophan 18 between the two variants (Banerjee et al., 2022). In both glomerular and vascular AA cases, fibrils do not contain residue arginine 1, since the N-terminal end, which starts from residue serine 2, is buried in the densely packed core and cannot accommodate additional residues, indicating that the arginine 1 is removed before fibril formation. The two variants also differ in charge distribution and electrostatic potential (Banerjee et al., 2022). The glomerular fibril has a more uniform distribution of positive and negative charges across the fibril cross-section compared to the vascular, which results in different net charges at pH 7.0 (Banerjee et al., 2022).

Feline AA Fibrils (EMD-14726, PDB-7ZH7)

AA amyloidosis is highly prevalent in domestic short-haired cats kept in shelters, with deposits forming in the kidney, liver, spleen, and heart. The cryo-EM fibril structure of feline AA extracted from the pole region of the kidneys was resolved at a resolution of 3.3 Å (Schulte et al., 2022). The fibril structure has two identical protofilaments, with each protofilament containing residues glutamic acid 19 to serine 94 in the form of 11 β-strands, resulting in an extended hairpin structure. The fibril structure has a C2 symmetry with a left-handed twist, based on cryo-EM data processing. Although the β-arch is almost perpendicular to the fibril axis, the exposed edge and buried face-tails exhibit a slight tilt. Extra densities can be observed close to residues glycine 65 and serine 82, which the authors suggest may be attributed to ions or organic compounds.

In summary, the cryo-EM structures of AA show structural differences between human and feline AA fibrils despite a sequence similarity of around 70%. The polymorphism extends further to different deposition sites within the kidney in humans. Further studies on fibrils extracted from different organs of the same patients may shed light on the source of polymorphism, that is, whether the site drives the polymorphism or vice versa. It is also imperative to understand the clinical relevance of the polymorphism and its implications in disease prognosis and treatment, if any.

FINAL REMARKS

The growing database of cryo-EM structures of amyloid fibrils offers insights into the molecular basis of protein aggregation into amyloid fibrils and hints about nonordered regions that may play a role in the pathogenesis of the disease (Fusco et al., 2016; Hosp et al., 2017; Stewart and Radford, 2017; Zhao et al., 2019). Additionally, it demonstrates the existence of structural polymorphism in amyloid fibrils that can enable researchers to study correlations between structures and clinical presentation.

Cryo-EM structure determination of *ex* vivo amyloid fibrils has revealed relevant features that could not be described until now. We find structural polymorphism within the same genotype or even the same individual (Nguyen et al., 2024). We observe the presence of cavities lined with polar residues that could play a role in interacting with cofactors and solutes, which may stabilize or favor specific fibril structures (Gallardo et al., 2020). We also observe undefined electron densities that could be attributed to the binding of other molecules that perhaps kinetically or thermodynamically alter the formation of these fibrils, as was observed with tau fibrils from chronic traumatic encephalopathy (Falcon et al., 2019; Lövestam et al., 2022). An important limitation of the cryo-EM structural studies is the lack of information about these nonordered regions of the fibrils, since we cannot determine if the missing residues are part of the fibril or if they have been removed at some step of the self-assembly process. However, this knowledge can be achieved by exploiting a broader range of tools that include biophysical and cellular techniques (Ulamec and Radford, 2020).

Traditional and state-of-the art techniques together with the increasing advances in cryo-EM will likely provide a complete picture of the structure of amyloid fibrils and their implications in disease pathogenesis. Thanks to detailed structural data of these pathological fibrils that cryo-EM studies can provide, scientists are getting valuable information that can help them to propose mechanisms of amyloid aggregation or to rationally design structure-based diagnostic and therapeutic tools. We envision that this is the beginning of a massive effort to decipher the ins and outs of amyloid misfolding and its relationship with conformational diseases.

LIST OF ABBREVIATIONS

Aβ	Amyloid-beta
Å	Angstroms
Nm	nanometers
NMR	Nuclear magnetic resonance
Cryo-EM	Cryogenic electron microscopy
ATTRv	Variant transthyretin amyloidoses
ATTRwt	Wild-type transthyretin amyloidosis
AL	Light chain amyloidosis
AA	Serum amyloid A amyloidosis

REFERENCES

Almeida, Z.L., Brito, R.M.M., 2020. Structure and aggregation mechanisms in amyloids. Molecules. https://doi.org/10.3390/molecules25051195

Banerjee, S., Baur, J., Daniel, C., Pfeiffer, P.B., Hitzenberger, M., Kuhn, L., Wiese, S., Bijzet, J., Haupt, C., Amann, K.U., Zacharias, M., Hazenberg, B.P.C., Westermark, G.T., Schmidt, M., Fändrich, M., 2022. Amyloid fibril structure from the vascular variant of systemic AA amyloidosis. Nat. Commun. 13, 7261. https://doi.org/10.1038/s41467-022-34636-4

Benson, M.D., 2012. The Hereditary Amyloidoses. In: Picken, M.M., Herrera, G.A., Dogan, A. (Eds.), Amyloid and Related Disorders: Surgical Pathology and Clinical Correlations. Springer International Publishing, Cham, pp. 53–67. https://doi.org/10.1007/978-3-319-19294-9_5

Buxbaum, J.N., Dispenzieri, A., Eisenberg, D.S., Fändrich, M., Merlini, G., Saraiva, M.J.M., Sekijima, Y., Westermark, P., 2022. Amyloid nomenclature 2022: update, novel proteins, and recommendations by the International Society of Amyloidosis (ISA) Nomenclature Committee. Amyloid Int. J. Exp. Clin. Investig. Off. J. Int. Soc. Amyloidosis 29, 213–219. https://doi.org/10.1080/13506129.2022.2147636

Cohen, A.S., 1986. General Introduction and a Brief History of Amyloidosis. In: Marrink, J., Van Rijswijk, M.H. (Eds.), Amyloidosis. Springer, Dordrecht, pp. 3–19. https://doi.org/10.1007/978-94-009-4309-4_1

Cohen, A.S., Calkins, E., 1959. Electron microscopic observations on a fibrous component in amyloid of diverse origins. Nature 183, 1202–1203. https://doi.org/10.1038/1831202a0

Cohen, A.S., Shirahamata, T., Skinner, M., 1982. Electron microscopy of amyloid. Electron Microsc. Proteins 3, 165–205.

Collins, A.M., Watson, C.T., 2018. Immunoglobulin light chain gene rearrangements, receptor editing and the development of a self-tolerant antibody repertoire. Front. Immunol. https://doi.org/10.3389/fimmu.2018.02249

Divry, P., 1927. Sur les proprietes optiques de l'amyloide. cr soc. Bilge Biol. 97, 1808–1810.

Falcon, B., Zhang, W., Murzin, A.G., Murshudov, G., Garringer, H.J., Vidal, R., Crowther, R.A., Ghetti, B., Scheres, S.H.W., Goedert, M., 2018. Structures of filaments from Pick's disease reveal a novel tau protein fold. Nature 561, 137–140. https://doi.org/10.1038/s41586-018-0454-y

Falcon, B., Zivanov, J., Zhang, W., Murzin, A.G., Garringer, H.J., Vidal, R., Crowther, R.A., Newell, K.L., Ghetti, B., Goedert, M., Scheres, S.H.W., 2019. Novel tau filament fold in chronic traumatic encephalopathy encloses hydrophobic molecules. Nature 568, 420–423. https://doi.org/10.1038/s41586-019-1026-5

Fändrich, M., 2007. On the structural definition of amyloid fibrils and other polypeptide aggregates. Cell. Mol. Life Sci. https://doi.org/10.1007/s00018-007-7110-2

Fändrich, M., Schmidt, M., 2021. Methods to study the structure of misfolded protein states in systemic amyloidosis. Biochem. Soc. Trans. https://doi.org/10.1042/BST20201022

Ferri, F., Ferro, S., Porporato, F., Callegari, C., Guglielmetti, C., Mazza, M., Ferrero, M., Crinò, C., Gallo, E., Drigo, M., Coppola, L.M., Gerardi, G., Schulte, T.P., Ricagno, S., Vogel, M., Storni, F., Bachmann, M.F., Vogt, A.-C., Caminito, S., Mazzini, G., Lavatelli, F., Palladini, G., Merlini, G., Zini, E., 2022. AA-amyloidosis in cats (*Felis catus*) housed in shelters. bioRxiv. https://doi.org/10.1101/2022.05.04.490646

Fitzpatrick, A.W.P., Falcon, B., He, S., Murzin, A.G., Murshudov, G., Garringer, H.J., Crowther, R.A., Ghetti, B., Goedert, M., Scheres, S.H.W., 2017. Cryo-EM structures of tau filaments from Alzheimer's disease. Nature 547, 185–190. https://doi.org/10.1038/nature23002

Friedreich, N., Kekulé, A., 1859. Zur amyloidfrage. Arch. für Pathol. Anat. und Physiol. und für Klin. Med. 16. https://doi.org/10.1007/BF01945246

Fusco, G., Pape, T., Stephens, A.D., Mahou, P., Costa, A.R., Kaminski, C.F., Kaminski Schierle, G.S., Vendruscolo, M., Veglia, G., Dobson, C.M., De Simone, A., 2016. Structural basis of synaptic vesicle assembly promoted by α-synuclein. Nat. Commun. 7. https://doi.org/10.1038/ncomms12563

Gallardo, R., Ranson, N.A., Radford, S.E., 2020. Amyloid structures: much more than just a cross-β fold. Curr. Opin. Struct. Biol. https://doi.org/10.1016/j.sbi.2019.09.001

Geddes, A.J., Parker, K.D., Atkins, E.D., Beighton, E., 1968. "Cross-beta" conformation in proteins. J. Mol. Biol. 32(2), 343–358.

Gertz, M.A., Dispenzieri, A., 2020. Systemic amyloidosis recognition, prognosis, and therapy: a systematic review. J. Am. Med. Assoc. https://doi.org/10.1001/jama.2020.5493

Hasib Sidiqi, M., Gertz, M.A., 2021. Immunoglobulin light chain amyloidosis diagnosis and treatment algorithm 2021. Blood Cancer J. https://doi.org/10.1038/s41408-021-00483-7

Hosp, F., Gutiérrez-Ángel, S., Schaefer, M.H., Cox, J., Meissner, F., Hipp, M.S., Hartl, F.U., Klein, R., Dudanova, I., Mann, M., 2017. Spatiotemporal proteomic profiling of Huntington's disease inclusions reveals widespread loss of protein function. Cell Rep. 21. https://doi.org/10.1016/j.celrep.2017.10.097

Hurshman, A.R., White, J.T., Powers, E.T., Kelly, J.W., 2004. Transthyretin aggregation under partially denaturing conditions is a downhill polymerization. Biochemistry. https://doi.org/10.1021/bi0496211

Iakovleva, I., Hall, M., Oelker, M., Sandblad, L., Anan, I., Sauer-Eriksson, A.E., 2021. Structural basis for transthyretin amyloid formation in vitreous body of the eye. Nat. Commun. 12. https://doi.org/10.1038/s41467-021-27481-4

Koike, H., Ando, Y., Ueda, M., Kawagashira, Y., Iijima, M., Fujitake, J., Hayashi, M., Yamamoto, M., Mukai, E., Nakamura, T., Katsuno, M., Hattori, N., Sobue, G., 2009. Distinct characteristics of amyloid deposits in early- and late-onset transthyretin Val30Met familial amyloid polyneuropathy. J. Neurol. Sci. 287. https://doi.org/10.1016/j.jns.2009.07.028

Lachmann, H.J., Goodman, H.J.B., Gilbertson, J.A., Gallimore, J.R., Sabin, C.A., Gillmore, J.D., Hawkins, P.N., 2007. Natural History and Outcome in Systemic AA Amyloidosis. N. Engl. J. Med. 356. https://doi.org/10.1056/nejmoa070265

Liberta, F., Loerch, S., Rennegarbe, M., Schierhorn, A., Westermark, P., Westermark, G.T., Hazenberg, B.P.C., Grigorieff, N., Fändrich, M., Schmidt, M., 2019. Cryo-EM fibril structures from systemic AA amyloidosis reveal the species complementarity of pathological amyloids. Nat. Commun. 10. https://doi.org/10.1038/s41467-019-09033-z

Lövestam, S., Koh, F.A., van Knippenberg, B., Kotecha, A., Murzin, A.G., Goedert, M., Scheres, S.H., 2022. Assembly of recombinant tau into filaments identical to those of Alzheimer's disease and chronic traumatic encephalopathy. Elife 11. https://doi.org/10.7554/eLife.76494

Merlini, G., 2017. AL amyloidosis: from molecular mechanisms to targeted therapies. Hematology 2017. https://doi.org/10.1182/asheducation-2017.1.1

Missmahl, H.P., Hartwing, M., 1953. [Optical polarization studies of amyloid substance]. Virchows Arch. Pathol. Anat. Physiol. Klin. Med. 324, 489–508. https://doi.org/10.1007/BF00954791

Muchtar, E., Dispenzieri, A., Magen, H., Grogan, M., Mauermann, M., McPhail, E.D., Kurtin, P.J., Leung, N., Buadi, F.K., Dingli, D., Kumar, S.K., Gertz, M.A., 2021. Systemic amyloidosis from A (AA) to T (ATTR): a review. J. Intern. Med. https://doi.org/10.1111/joim.13169

Nguyen, B.A., Afrin, S., Singh, V., Ahmed, Y., Pedretti, R., Fernandez-Ramirez, M.D.C., Benson, M.D., Sawaya, R.M., Cao, Q., Boyer, D., Pope, A., Wydorski, P.M., Chhapra, F., Eisenberg, D.S., Saelices, L., 2022. Structural polymorphism of amyloid fibrils in cardiac ATTR amyloidosis revealed by cryo-electron microscopy. bioRxiv. https://doi.org/10.1101/2022.06.21.496949

Nguyen, B.A., Singh, V., Afrin, S., Yakubovska, A., Wang, L., Ahmed, Y., Pedretti, R., Fernandez-Ramirez, M.D.C., Singh, P., Pękała, M., Cabrera Hernandez, L.O., Kumar, S., Lemoff, A., Gonzalez-Prieto, R., Sawaya, M.R., Eisenberg, D.S., Benson, M.D., & Saelices, L., 2024. Structural polymorphism of amyloid fibrils in ATTR amyloidosis revealed by cryo-electron microscopy. Nat. Commun. 15(1), 581. https://doi.org/10.1038/s41467-024-44820-3

Obici, L., Merlini, G., 2012. AA amyloidosis: basic knowledge, unmet needs and future treatments. Swiss Med. Wkly. https://doi.org/10.4414/smw.2012.13580

Papendick, R.E., Munson, L., O'Brien, T.D., Johnson, K.H., 1997. Systemic AA amyloidosis in captive cheetahs (*Acinonyx jubatus*). Vet. Pathol. 34. https://doi.org/10.1177/030098589703400602

Radamaker, L., Baur, J., Huhn, S., Haupt, C., Hegenbart, U., Schönland, S., Bansal, A., Schmidt, M., Fändrich, M., 2021. Cryo-EM reveals structural breaks in a patient-derived amyloid fibril from systemic AL amyloidosis. Nat. Commun. 12. https://doi.org/10.1038/s41467-021-21126-2

Radamaker, L., Lin, Y.H., Annamalai, K., Huhn, S., Hegenbart, U., Schönland, S.O., Fritz, G., Schmidt, M., Fändrich, M., 2019. Cryo-EM structure of a light chain-derived amyloid fibril from a patient with systemic AL amyloidosis. Nat. Commun. 10. https://doi.org/10.1038/s41467-019-09032-0

Röcken, C., Shakespeare, A., 2002. Pathology, diagnosis and pathogenesis of AA amyloidosis. Virchows Arch. https://doi.org/10.1007/s00428-001-0582-9

Sawaya, M.R., Eisenberg, D., 2013. Amyloid Structures at the Atomic Level: Insights from Crystallography. In: Otzen, D. E. (Ed.), Amyloid Fibrils and Prefibrillar Aggregates: Molecular and Biological Properties, NJ: John Wiley & Sons, Ltd., pp. 21–38. https://doi.org/10.1002/9783527654185.ch2

Sawaya, M.R., Hughes, M.P., Rodriguez, J.A., Riek, R., Eisenberg, D.S., 2021. The expanding amyloid family: structure, stability, function, and pathogenesis. Cell. https://doi.org/10.1016/j.cell.2021.08.013

Scheres, S.H., Zhang, W., Falcon, B., Goedert, M., 2020. Cryo-EM structures of tau filaments. Curr. Opin. Struct. Biol. https://doi.org/10.1016/j.sbi.2020.05.011

Schmidt, M., Wiese, S., Adak, V., Engler, J., Agarwal, S., Fritz, G., Westermark, P., Zacharias, M., Fändrich, M., 2019. Cryo-EM structure of a transthyretin-derived amyloid fibril from a patient with hereditary ATTR amyloidosis. Nat. Commun. 10. https://doi.org/10.1038/s41467-019-13038-z

Schulte, T., Chaves-Sanjuan, A., Mazzini, G., Speranzini, V., Lavatelli, F., Ferri, F., Palizzotto, C., Mazza, M., Milani, P., Nuvolone, M., Vogt, A.-C., Vogel, M., Palladini, G., Merlini, G., Bolognesi, M., Ferro, S., Zini, E., Ricagno, S., 2022. Cryo-EM structure of ex vivo fibrils associated with extreme AA amyloidosis prevalence in a cat shelter. Nat. Commun. 13, 7041. https://doi.org/10.1038/s41467-022-34743-2

Schweighauser, M., Shi, Y., Tarutani, A., Kametani, F., Murzin, A.G., Ghetti, B., Matsubara, T., Tomita, T., Ando, T., Hasegawa, K., Murayama, S., Yoshida, M., Hasegawa, M., Scheres, S.H.W., Goedert, M., 2020. Structures of α-synuclein filaments from multiple system atrophy. Nature 585. https://doi.org/10.1038/s41586-020-2317-6

Solomon, A., Richey, T., Murphy, C.L., Weiss, D.T., Wall, J.S., Westermark, G.T., Westermark, P., 2007. Amyloidogenic potential of foie gras. Proc. Natl. Acad. Sci. U. S. A. 104. https://doi.org/10.1073/pnas.0700848104

Steinebrei, M., Baur, J., Pradhan, A., Kupfer, N., Wiese, S., Hegenbart, U., Schönland, S. O., Schmidt, M., & Fändrich, M., 2023. Common transthyretin-derived amyloid fibril structures in patients with hereditary ATTR amyloidosis. Nat. Commun. 14, 7623. https://doi.org/10.1038/s41467-023-43301-3

Steinebrei, M., Gottwald, J., Baur, J., Röcken, C., Hegenbart, U., Schönland, S., Schmidt, M., 2022. Cryo-EM structure of an ATTRwt amyloid fibril from systemic non-hereditary transthyretin amyloidosis. Nat. Commun. 13, 6398. https://doi.org/10.1038/s41467-022-33591-4

Stewart, K.L., Radford, S.E., 2017. Amyloid plaques beyond Aβ: a survey of the diverse modulators of amyloid aggregation. Biophys. Rev. https://doi.org/10.1007/s12551-017-0271-9

Swuec, P., Lavatelli, F., Tasaki, M., Paissoni, C., Rognoni, P., Maritan, M., Brambilla, F., Milani, P., Mauri, P., Camilloni, C., Palladini, G., Merlini, G., Ricagno, S., Bolognesi, M., 2019. Cryo-EM structure of cardiac amyloid fibrils from an immunoglobulin light chain AL amyloidosis patient. Nat. Commun. 10. https://doi.org/10.1038/s41467-019-09133-w

Tasaki, M., Ueda, M., Obayashi, K., Koike, H., Kitagawa, K., Ogi, Y., Jono, H., Su, Y., Suenaga, G., Oshima, T., Misumi, Y., Yoshida, M., Yamashita, T., Sobue, G., Ando, Y., 2013. Effect of age and sex differences on wild-type transthyretin amyloid formation in familial amyloidotic polyneuropathy: a proteomic approach. Int. J. Cardiol. 170. https://doi.org/10.1016/j.ijcard.2013.10.033

Ulamec, S.M., Radford, S.E., 2020. Spot the difference: function versus toxicity in amyloid fibrils. Trends Biochem. Sci. 45, 635–636. https://doi.org/10.1016/j.tibs.2020.04.007

Virchow, R., 1854. Zur cellulose -frage. Arch. für Pathol. Anat. und Physiol. und für Klin. Med. 6. https://doi.org/10.1007/BF02116546

Wajnsztajn Yungher, F., Kim, A., Boehme, A., Kleyman, I., Weimer, L.H., Maurer, M.S., Brannagan, T.H., 2020. Peripheral neuropathy symptoms in wild type transthyretin amyloidosis. J. Peripher. Nerv. Syst. 25. https://doi.org/10.1111/jns.12403

Westermark, G.T., Fändrich, M., Westermark, P., 2015. AA amyloidosis: pathogenesis and targeted therapy. Annu. Rev. Pathol. Mech. Dis. 10. https://doi.org/10.1146/annurev-pathol-020712-163913

Westermark, G.T., Westermark, P., 2009. Serum amyloid A and protein AA: molecular mechanisms of a transmissible amyloidosis. FEBS Lett. https://doi.org/10.1016/j.febslet.2009.04.026

Zhang, B., Une, Y., Fu, X., Yan, J., Ge, F., Yao, J., Sawashita, J., Mori, M., Tomozawa, H., Kametani, F., Higuchi, K., 2008. Fecal transmission of AA amyloidosis in the cheetah contributes to high incidence of disease. Proc. Natl. Acad. Sci. U. S. A. 105. https://doi.org/10.1073/pnas.0800367105

Zhao, L., Vecchi, G., Vendruscolo, M., Körner, R., Hayer-Hartl, M., Hartl, F.U., 2019. The Hsp70 chaperone system stabilizes a thermo-sensitive subproteome in *E. coli*. Cell Rep. 28. https://doi.org/10.1016/j.celrep.2019.06.081

28 Structural Insights into the 5-HT$_{3A}$ Receptor Revealed by Cryogenic Electron Microscopy (Cryo-EM)

Zhuowen Li, Nikhil Bharambe, and Sandip Basak

INTRODUCTION

DISCOVERY OF THE SEROTONIN TYPE 3 RECEPTOR (5-HT$_3$R)

Serotonin receptors are a large family of neurotransmitter receptors located in the nervous system. They respond to chemical signals such as serotonin (5-HT) and its derivatives to transfer nerve impulses across neurons. Seven classes of serotonin receptors have been identified thus far, six of which belong to GPCRs, while one class is a member of the pentameric ligand-gated ion channels (pLGICs) (Hoyer et al. 1994; Frazer and Hensler 1999). Such pLGIC-type serotonin receptors (5-HT$_3$Rs) are located on both the pre- and postsynaptic membranes of neurons, transiently conduct cations into the cytoplasm, and induce neuronal depolarization and excitation.

In the late 1950s, 5-HT$_3$R (originally named the M receptor) was initially identified as a serotonin receptor where a rapidly desensitizing aminergic impulse was detected in the submucosal neurons of the guinea pig intestine (Gaddum and Picarelli 1957). A similar effect was observed in the nicotine receptor, which was already known to be an ion channel. Subsequent electrophysiological studies using single-channel recording of excised membrane patches revealed that 5-HT$_3$ receptors are ion channels (Derkach, Surprenant, and North 1989), whereas their conductance varies in different species in a tissue-specific manner (Jackson and Yakel 1995). Initially, 5-HT$_3$R was thought to play a role in a range of functions, including emesis, cognition, and anxiety. However, the function of human 5-HT$_3$Rs in the brain remains elusive. Further investigations identified several inhibitors showing high selectivity for 5-HT$_3$R, which has been clinically proven to treat emesis (Jackson and Yakel 1995). Thus, the 5-HT$_3$ receptor has attracted significant interest in potential treatments for gastrointestinal disorders, including chemotherapy- and radiotherapy-induced nausea and vomiting.

SUBUNIT COMPOSITION OF 5-HT$_3$R AND ITS PHARMACOLOGICAL SIGNIFICANCE

mRNAs obtained from the *in vitro* transcription of genes encoding 5-HT$_3$R from mice and humans were injected into *Xenopus* oocytes, leading to the expression of functional channels and serotonin-induced currents (Maricq et al. 1991; Miyake et al. 1995). The initially identified 5-HT$_3$R subunit was termed 5-HT$_3$R subunit A, and its homopentamer was denoted as 5-HT$_{3A}$R (Maricq et al. 1991; Fletcher and Barnes 1998). A significant reduction in conductance was observed in recombinantly expressed receptors compared with native 5-HT$_3$Rs, suggesting differential subunit compositions in various tissues (Fletcher and Barnes 1998). A subsequent search for homologs of subunit A in the human genome identified the gene encoding subunit B that shared 41% protein sequence identity (Davies et al. 1999). Subunit B has also been identified in mice and rats (Hanna et al. 2000). Co-expression of subunits A and B in HEK-293 cells exhibited conductance comparable to native channels isolated directly from neurons, further supporting the presence of heteromeric compositions of physiological 5-HT$_3$R (Davies et al. 1999). Three additional subunits, C, D, and E, were identified in the human genome, but no counterparts were discovered in rodents (Niesler et al. 2003). Distinct tissue-specific subunit compositions of 5-HT$_3$Rs have been observed (Niesler et al. 2003). In humans, subunits C, D, and E are predominantly expressed in the colon and intestine, whereas subunits A and B are found in the brain, spinal cord, and intestine (Niesler et al. 2003). Subunit A can be assembled into functional homopentamers in a heterologous expression system; however, the other four subunits (B–E) require subunit A to form functional heteropentamers (Davies et al. 1999; Niesler et al. 2007). In rodents, subunits A and B are widely localized in the central nervous system (Gehlert et al. 1991; Koyama, Kondo, and Shimada 2017). Drugs targeting murine 5-HT$_3$R alleviate depression and fear (Rajkumar and Mahesh 2010; Park and Williams 2012; Bhatt et al. 2013a, 2013b). This discovery generated a considerable interest in the development of therapeutic compounds that could treat mood and anxiety disorders in humans.

FIRST STRUCTURE OF 5-HT$_{3A}$R DETERMINED BY X-RAY CRYSTALLOGRAPHY

Owing to the profound pharmacological relevance of 5-HT$_3$R, considerable functional insights have been accumulated for decades after successful 5-HT$_{3A}$R expression, mostly through the combination of mutations and

DOI: 10.1201/9781003326106-34

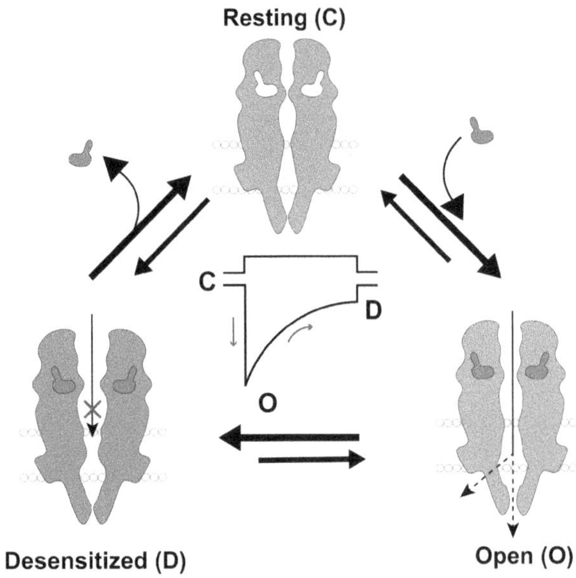

FIGURE 28.1 Schematic of the gating mechanism in pLGICs. Illustration showing the transition of three fundamental functional states of pLGICs: a resting state, a transient open state, and a desensitized state.

electrophysiological studies. Before the 5-HT$_3$R structure was revealed, the domain organization was hypothesized to resemble the other known pLGIC structures. The eukaryotic pLGICs comprise three distinct domains: the extracellular domain (ECD), houses orthosteric ligand binding sites, the transmembrane domain (TMD), which is responsible for gating and channel desensitization, and the intracellular domain (ICD), which participates in channel modulation and trafficking (Lummis 2012).

Eukaryotic pLGICs transit among three distinctive functional states: resting, open, and desensitized states (Figure 28.1). In the absence of the ligand, the receptor adopts a nonconductive conformation corresponding to the resting state. Upon agonist binding, the channel undergoes a cascade of conformational changes in ECD and TMD, leading to the transient opening of the hydrophobic gate to allow the ions to permeate. Subsequently, the channel becomes an agonist-bound nonconductive state known as the desensitization state, which exhibits the highest affinity to the agonist.

The earliest attempt to capture the structure of pLGICs was performed on the nicotinic acetylcholine receptor (nAChR), where the homogenate of the electric ray's postsynaptic membrane was crystallized into tubules with densely packed channels (Brisson and Unwin 1984). The electron crystallographic data generated a map of 6 Å resolution (Miyazawa, Fujiyoshi, and Unwin 2003), which was further improved to 4 Å after reprocessing in the following years (Unwin 2005). In the late 2000s, the structures of two prokaryotic homologs of pLGICs were reported at higher resolutions (Hilf and Dutzler 2008; Bocquet et al. 2009). Hence, the structures of eukaryotic and prokaryotic pLGICs served as a reference to solve the first X-ray structure of 5-HT$_{3A}$R (Hassaine et al. 2014). Thus far, all published structures of 5-HT$_3$R have been homopentamers composed of subunit A from *Mus musculus*. The crystal structure provides the first glimpse of the domain organization of 5-HT$_3$R (Figure 28.2A). The structure resolved the ECD at the N-terminus, which is composed of 10 β-strands (β1–β10). The TMD is composed of four α-helices (M1–M4) with the M2 helix lining the pore, and the ICD consists of a post-M3 loop, the MX helix parallel to the cytoplasmic membrane, and a membrane-associated (MA) helix preceding the M4 helix of the TMD (Figure 28.2 A). The electron density of the unstructured cytoplasmic loop in the ICD was not resolved. The ECD revealed three N-glycosylation sites in each subunit at N82, N148, and N164, which were occupied by glycans. In the ECD, the intersubunit cavity is surrounded by loops A, B, and C of the principal subunit (+) and loops D, E, and F of the complementary subunit (–) (Figure 28.2 B). Residues F199 and Y207 of loop C, W156 of loop B, Y126 of loop E, and W63 of loop D form the aromatic cage, which was predicted to stabilize serotonin with π–π stacking and cation–π interactions (Beene et al. 2002) (Figure 28.2 B). The crystal structure revealed the aromatic cage similar to the previously reported orthosteric binding site identified in homologous acetylcholine-binding protein (AChBP) (Brejc et al. 2001), suggesting that this site may be responsible for serotonin binding. Serotonin binding at the orthosteric site has been proposed to induce conformational changes that allosterically propagate to the pore-lining helices in the TMD, leading to channel opening (Grosman et al. 2000). The entire ion permeation pathway underlies all three domains. The highest constriction in the ECD was observed at residues D105 and K108. The ion permeation pathway below this constriction is lined by negatively charged residues, which facilitate the concentration of cations at the pore inlet (Hassaine et al. 2014). The aliphatic residues L260 (L9′), V264 (V13′), and I268 (I16′) at the outer half of the M2 helix and the polar residues E250 (E-1′), S253 (S-2′), and T257 (T6′) at the inner half of the M2 helix form the ion permeation pore. Among these residues, L260 (L9′) serves as the hydrophobic gate. The ICD consists of an MA helix, an MX helix, and a large, flexible loop involved in ion exit and plays a critical role in channel conductance (Kelley et al. 2003). Two possible outlets have been proposed for the ion exit mechanism, one within the ICD along the central axis and the other a lateral portal at the interface between the post-M3 loop and the MA helix (Hassaine et al. 2014) (Figure 28.1). Molecular dynamics (MD) simulations and cysteine cross-linking studies have suggested that the lateral portals are preferred over the central pore for ion exit (Yuan, Filipek, and Vogel 2016; Stuebler and Jansen 2020). In addition, three positively charged arginine residues (R416, R420, and R424) are present in the MA helix of subunit A, which substantially reduces the channel conductance in the homomeric 5-HT$_{3A}$R (Kelley et al. 2003). Mutation of these three

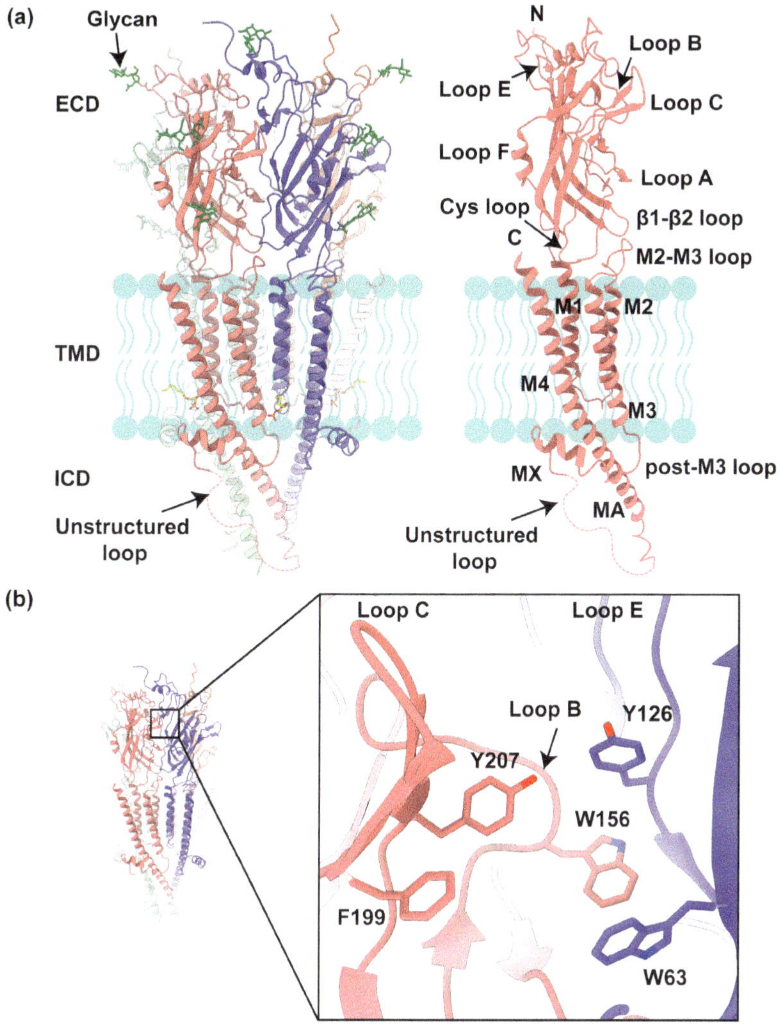

FIGURE 28.2 A representative structure of 5-HT_{3A}R and putative serotonin binding site. (A) The pentameric structure of 5-HT_{3A}R is shown in a cartoon representation (*left*). For clarity, each monomer is colored by chain. Glycans are shown as sticks (green). The important loops and helices are labeled (*right*). A schematic membrane bilayer is shown (sea green) (B) Location of the aromatic cage is highlighted in zoomed view. Residues forming aromatic cage are shown as sticks and colored by heteroatom. The principal (+) subunit is in salmon, while the complementary (–) subunit is in purple.

residues of subunit A into their counterparts in subunit B (R416Q, R420D, R424A) increased the conductivity (namely 5-HT_{3A}R-QDA mutant) to a level similar to that of the native mouse heteromeric receptor (Kelley et al. 2003).

In summary, the crystal structure of 5-HT_{3A}R revealed detailed structural insights into this channel. However, during crystallization, limited trypsin digestion was performed to remove the flexible loop of ICD and four residues of the M2–M3 loop (Hassaine et al. 2014). In addition, a nanobody (VHH15) was used during crystallization to stabilize the protein, which led to the determination of the high-resolution structure. However, two-electrode voltage-clamp recording showed that the nanobody reduced the serotonin-induced current; thus, the conformation captured in the crystal structure in complex with the nanobody might represent an inhibited nonconducting conformation and necessitates further investigation (Hassaine et al. 2014).

CRYO-EM REVEALS THE APO AND LIGAND-BOUND STRUCTURES OF 5-HT_{3A}R

The Apo and Serotonin-Bound 5-HT_{3A}R Structures Demonstrate the Conformational Transitions upon Agonist Binding

Basak et al. first reported the cryo-EM structure of apo 5-HT_{3A}R in the absence of nanobodies, which showed an overall architecture similar to that of the nanobody-bound crystal structure (Basak, Gicheru, Samanta, et al. 2018). Upon detailed analysis, an expansion is observed in all three domains of the cryo-EM structure compared with the nanobody-bound crystal structure, reflected by a radial outward twist of outer β-sheets in the ECD and an iris-like movement of the TMD and ICD (Basak, Gicheru, Samanta, et al. 2018). The ECD of apo 5-HT_{3A}R rotates clockwise compared to the crystal structure, with the highest rotation observed at loop C, loop F, and the cys loop. Several constrictions were observed at the pore-lining residues of apo 5-HT_{3A}R. The radius of the pore at the highest constriction region was narrower than that of a hydrated Na^+ ion, which confirmed that the apo structure was captured at the non-conductive conformation. Considering that the 5-HT_{3A}R structure was determined through cryo-EM without the presence of nanobodies and without undergoing enzymatic partial digestion, conformational differences compared with the nanobody-bound crystal structure suggest that the cryo-EM apo structure corresponds to a resting-like conformation.

Subsequently, two research groups independently determined the cryo-EM structures of serotonin-bound 5-HT_{3A}R (Basak, Gicheru, Rao, et al. 2018; Polovinkin et al. 2018). Two distinct serotonin-bound states, state 1 and state 2, were obtained from the same cryo-EM dataset with resolutions of 3.3 Å and 3.9 Å, respectively (Basak, Gicheru, Rao, et al. 2018). In both structures, serotonin was bound at the orthosteric binding pockets and was involved in π–π interactions with W156, Y126, and W63 (Figure 28.3 A). Compared with the apo structure, the serotonin-bound structures exhibited notable twisting of both the extracellular domain (ECD) and the transmembrane domain (TMD). This twisting involves an overall counterclockwise rotation of the ECD around the pore axis, resulting in significant repositioning of the individual interfacial loops. The conformational changes in the ECD induced by serotonin allosterically propagate to the TMD,

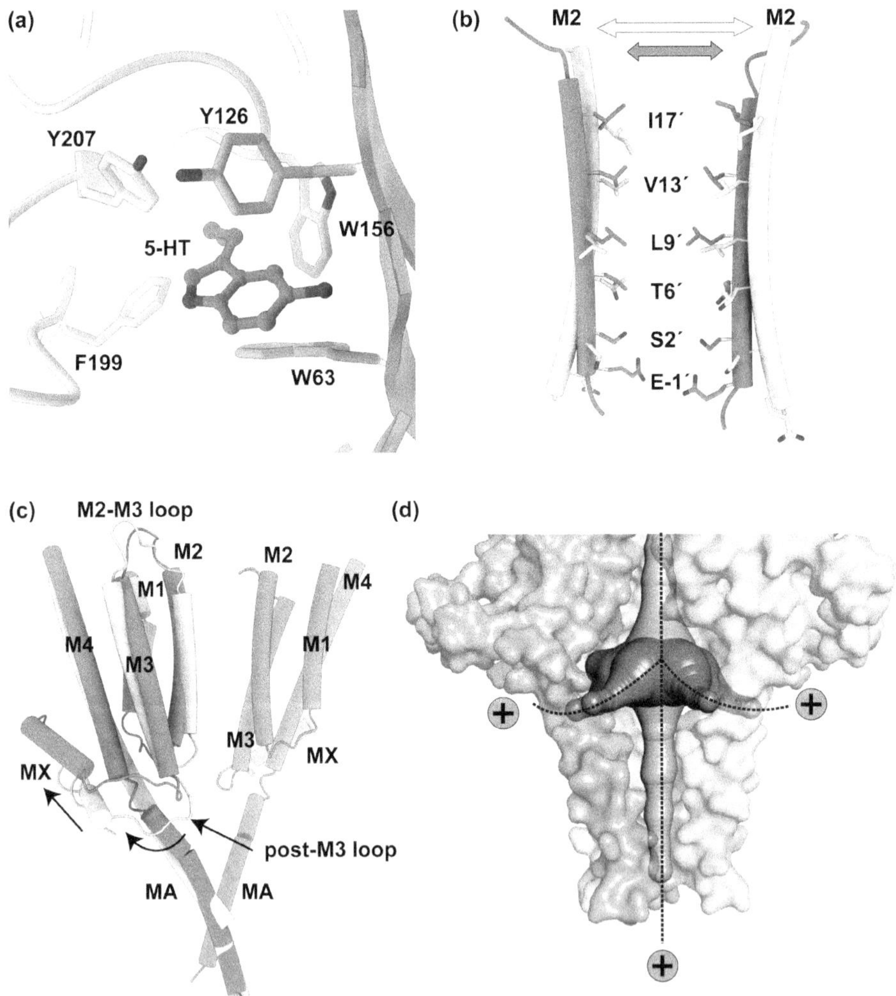

FIGURE 28.3 Serotonin binding results in channel activation. (A) Serotonin (5-HT) binds to the aromatic cage. Aromatic cage–forming residues are shown as sticks, and serotonin is shown in ball-and-stick representation. (B) The serotonin-bound structure (light grey) shows widening of the M2 helices of TMD compared to the apo structure (dark grey). Pore lining residues are shown as sticks, and widening is indicated by bidirectional arrows. (C) Superposition of the TMD and ICD of apo (light grey) and serotonin-bound structure (dark grey) is shown in cylinder representation. Arrows indicate the direction of conformational changes in the post-M3 loop and MX helix. (D) The solvent-accessible vestibule was determined using a minimum cavity radius of 2.8 Å (modified from Basak, Gicheru, Rao, et al. 2018). For clarity, one subunit is removed. Potential pathways for ion flow and exit are marked by dashed lines.

causing the transmembrane helices to expand away from the pore axis. The outward displacement of M2 coincides with a noticeable outward shift of the M2–M3 loop, which moves away from the intersubunit interface. The pore in state 1 becomes wider than the apo state due to the subtle rotation of L9′. The radius of the pore in state 1 is wider than a hydrated Na^+ ion, indicating a state initially thought to be captured in conductive conformation. However, the MD simulations of state 1 revealed an energy barrier for water molecules at L9′ and I409. Interestingly, upon the application of a transmembrane potential difference during the simulations, intermittent Na^+ ion permeation was observed (Basak, Gicheru, Rao, et al. 2018). The authors suggested that this state could be either a pre-open or a desensitized state. A major conformational rearrangement was observed in state 2 compared with state 1, particularly at the TMD and ICD, which is characterized by further tilting of M2 helices away from the pore axis, clockwise rotation of the extracellular half of the M2 helix, and drift of the peripheral MX helix. These conformational changes cause further widening of the pore (Figure 28.3B). The MD simulations revealed no energy barrier for water molecules at L9′, indicating that state 2 was captured in an open conformation. Several ion permeation events were observed throughout the MD simulations. However, an energy barrier at I409 in the ICD was still present, which suggests that ions might not be able to exit through the central pore. In state 2, the MX helix undergoes both outward and upward movement, causing the post-M3 loop to extend away from the MA–M4 helix, resulting in the formation of lateral portals (Figure 28.3 C). Interestingly, the size of these portals is wide enough to serve as pathways through which hydrated Na^+ ions could exit (Figure 28.3 D).

Concurrently, several cryo-EM structures of 5-HT$_{3A}$R were reported by Polovinkin et al., including two serotonin-bound structures (I1 and F) and one serotonin-bound structure (I2) in the presence of a positive allosteric modulator, trans-3-(4-methoxyphenyl)-N-(pentan-3-yl)acrylamide (TMPPAA). The I1 and I2 states have similar conformations with a closed pore. The conformations of both I1 and I2 resemble state 1 obtained by Basak et al. Upon extensive structural analysis along with MD simulations, the I1 and I2 states were assigned as either pre-open or desensitized state. The conformational variations in state F are much more pronounced than those in I1 and I2. The density of the ICD in state F was not observed because of flexibility. However, the entire ECD and TMD domains were modeled. The pore of state F is wider than a hydrated Na^+ ion. MD simulations assisted in assigning state F as a serotonin-bound open state with a characteristic wider pore.

Cryo-EM Reveals Multiple Antagonist-Bound 5-HT$_{3A}$R Structures

The cryo-EM structures of 5-HT$_{3A}$R in the presence of various antagonists (known as setrons) such as alosetron, ondansetron, palonosetron, tropisetron, and granisetron were obtained at high resolution (Figure 28.4) (Polovinkin et al. 2018; Basak et al. 2019; Basak et al. 2020; Zarkadas et al. 2020). The densities of setrons in each structure were clearly resolved at the orthosteric binding sites in cryo-EM maps. MD simulations further validated the stability of setron binding and its orientation (Basak et al. 2020). The overall conformation of the setron-bound structures is similar to the nanobody-bound crystal structure. Therefore, structural analysis of antagonist-bound 5-HT$_{3A}$R and

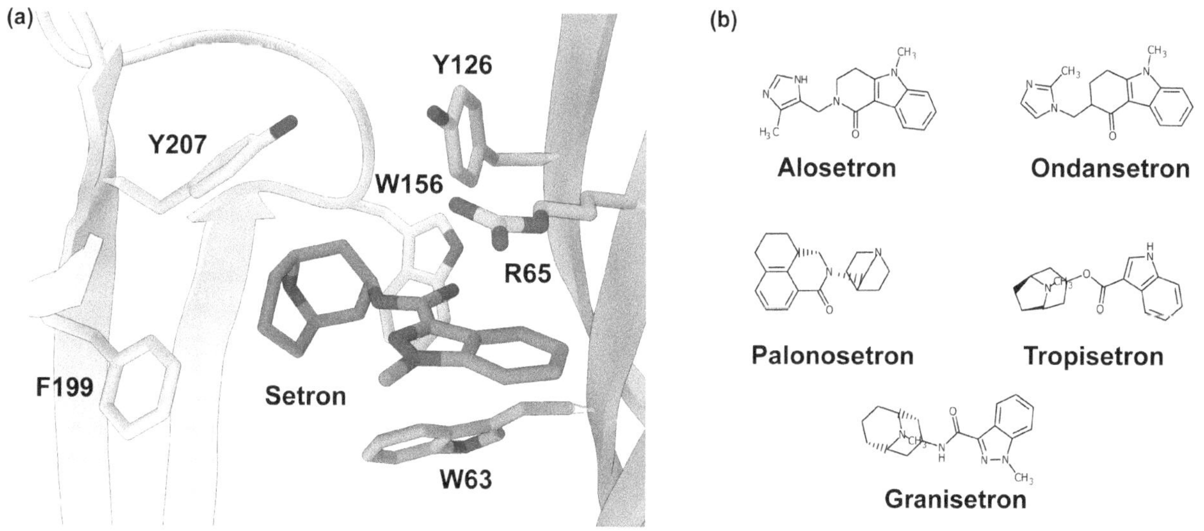

FIGURE 28.4 Antagonist binds to the orthosteric binding site of 5-HT$_{3A}$R. (A) A representative bound antagonist is shown in stick representation. Aromatic cage–forming residues and antagonist (dark grey) are shown as sticks. Only the principal (+) subunit (light grey) and the complementary (−) subunit (dark grey) are shown for clarity. (B) Chemical structures of various setrons that show anti-emetic effect are shown. These setron-bound 5-HT$_{3A}$R structures were solved using cryo-EM.

electrophysiological experiments revealed that setron stabilizes the nonconductive inhibited state.

Cryo-EM Reveals 5-HT$_{3A}$R Structures in Lipidic Environments

The first cryo-electron tomography reconstruction was achieved for 5-HT$_{3A}$R reconstituted in E. *coli* lipid vesicles at 12 Å resolution using subtomogram averaging (Kudryashev et al. 2016). Despite the limited resolution, the distribution and positioning of 5-HT$_{3A}$R on lipid vesicles were effectively illustrated. Notably, the addition of cholesterol to the sample resulted in a uniform distribution of 5-HT$_{3A}$R on vesicles, in contrast to the densely packed arrangement observed on vesicles without cholesterol. This study underscores the significance of examining the high-resolution structure of 5-HT$_{3A}$R in a lipid environment that closely resembles their native state. Recently, Zhang et al. obtained cryo-EM maps of 5-HT$_{3A}$R reconstituted in saposin-based lipid bilayer discs (5-HT$_{3A}$R-Salipro) (Frauenfeld et al. 2016; Zhang et al. 2021). A C5-symmetrical apo structure (apo-C5), an asymmetrical apo structure (apo-C1), and an asymmetrical serotonin-bound structure of 5-HT$_{3A}$R were obtained at nominal resolutions of 3.2, 3.1, and 2.8 Å, respectively (Figure 28.5 A). The conformation of the 5-HT$_{3A}$R-Salipro structures differs from that of the detergent-solubilized structures. The receptor in Salipro exhibits a significantly shorter length along the central channel axis compared with that in a detergent environment. While the pore of the apo state is closed, the serotonin-bound Salipro structure shows the pore

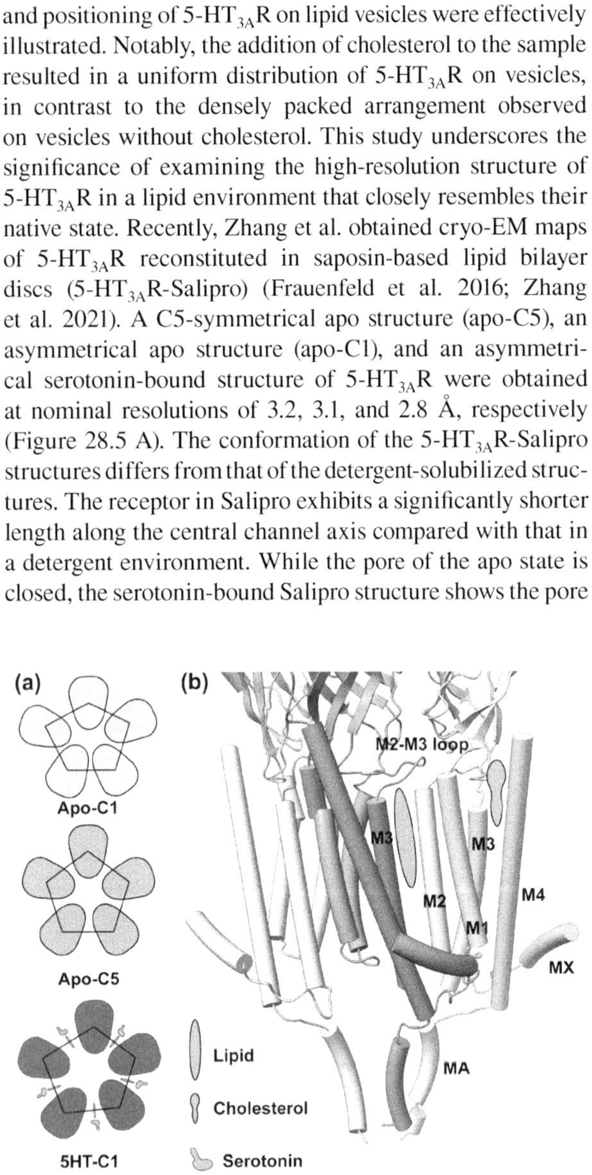

FIGURE 28.5 Asymmetric subunit arrangements of 5-HT$_{3A}$R. (A) Three structures of 5-HT$_{3A}$R reconstituted into saposin-based lipid bilayer discs: an asymmetric apo state (*top*), a symmetric apo state (*middle*), and an asymmetric agonist-bound state (*bottom*). (B) A putative location of lipid in the intersubunit interface of TMD and cholesterol near the M4 helix is shown. Two subunits were removed for clarity. TMD and ICD are shown in cylinder representation.

which is wider than the detergent solubilized open state and potentially conducts ions. MD simulations revealed ion permeation events in the serotonin-bound state. Apo-C1 was assigned an intermediate state and lies between apo-C5 and serotonin-bound models, based on the flexibility of the M2 helix during MD simulations. Several partially ordered lipid densities were observed in the Salipro 5-HT$_{3A}$R structures, which were located at the intersubunit interface (Figure 28.5 B). In addition to the lipid densities, more ordered densities were observed in the TMD near the outer M4 helix. Based on the shape and size, cholesterols were speculated to be present in those densities (Figure 28.5 B). This cholesterol was hypothesized to mediate its effect on channel function by affecting membrane packing. Although neither lipids nor cholesterol were confidently built in the models, this study provides a structural foundation for understanding how the lipid membrane can allosterically modulate the gating of 5-HT$_3$R.

CONCLUSION

Because ligand-gated ion channels are prevalent in the brain, heart, kidney, and gastrointestinal tract, 5-HT$_3$R has been a perplexing biophysical favorite and has thus received innumerable studies on its composition, functions, and pharmacological relevance. To date, extensive structural insights have been obtained on homopentameric 5-HT$_{3A}$R. The architecture and the agonist- and antagonist-triggered conformational changes have been thoroughly elucidated using X-ray crystallography and cryo-EM. The X-ray crystallography provided the first glimpse of 5-HT$_{3A}$R architecture at higher resolution and opened the door for further investigations. The cryo-EM played an important role in advancing the knowledge on ligand-induced conformational changes in 5-HT$_{3A}$R. Using cryo-EM, an apo state, several intermediate or pre-open states, two distinctively different open states, and various analogous antagonist-stabilized inhibitory states of detergent-solubilized 5-HT$_{3A}$R were assigned. MD simulations further enhance the understanding on the gating mechanism, plausible ion exit pathway, and antagonist-induced inhibition. These pioneering studies on 5-HT$_{3A}$R have provided outstanding premises for the development of newer antagonists. Occasionally, detergents strip away several annular and non-annular lipids, which play a functional role by directly interacting with the protein (Thompson and Baenziger 2020). For a comprehensive understanding of ligand-induced structural alterations in a near-native condition, cryo-EM structures of 5-HT$_{3A}$R were obtained in a lipidic environment. Notably, asymmetric conformational arrangements of subunits were observed in both apo and ligand-bound states. Such findings are consistent with earlier observations of asymmetrical channel openings in other pLGICs (Mowrey et al. 2013; Zhu et al. 2018; Ruan et al. 2018). Thus, all the insights on 5-HT$_{3A}$R gathered using structural biology methods, especially cryo-EM, lay the groundwork for exploring the physiologically significant 5-HT$_3$R heteropentamers.

REFERENCES

Basak, S., Y. Gicheru, A. Kapoor, M. L. Mayer, M. Filizola, and S. Chakrapani. 2019. 'Molecular mechanism of setron-mediated inhibition of full-length 5-HT(3A) receptor', *Nat Commun*, 10: 3225.

Basak, S., Y. Gicheru, S. Rao, M. S. P. Sansom, and S. Chakrapani. 2018. 'Cryo-EM reveals two distinct serotonin-bound conformations of full-length 5-HT(3A) receptor', *Nature*, 563: 270–74.

Basak, S., Y. Gicheru, A. Samanta, S. K. Molugu, W. Huang, M. Fuente, T. Hughes, D. J. Taylor, M. T. Nieman, V. Moiseenkova-Bell, and S. Chakrapani. 2018. 'Cryo-EM structure of 5-HT(3A) receptor in its resting conformation', *Nat Commun*, 9: 514.

Basak, S., A. Kumar, S. Ramsey, E. Gibbs, A. Kapoor, M. Filizola, and S. Chakrapani. 2020. 'High-resolution structures of multiple 5-HT(3A)R-setron complexes reveal a novel mechanism of competitive inhibition', *Elife*, 9: 24.

Beene, D. L., G. S. Brandt, W. Zhong, N. M. Zacharias, H. A. Lester, and D. A. Dougherty. 2002. 'Cation-pi interactions in ligand recognition by serotonergic (5-HT3A) and nicotinic acetylcholine receptors: the anomalous binding properties of nicotine', *Biochemistry*, 41: 10262–9.

Bhatt, S., R. Mahesh, T. Devadoss, and A. K. Jindal. 2013a. 'Antidepressant-like effect of novel 5-HT3 receptor antagonist N-n-butyl-3-ethoxyquinoxalin-2-carboxamide (6p): an approach using rodent behavioral antidepressant tests', *Indian J Pharmacol*, 45: 348–53.

Bhatt, S., R. Mahesh, T. Devadoss, and A. K. Jindal. 2013b. 'Anxiolytic-like effect of (4-benzylpiperazin-1-yl)(3-methoxyquinoxalin-2-yl)methanone (6g) in experimental mouse models of anxiety', *Indian J Pharmacol*, 45: 248–51.

Bocquet, N., H. Nury, M. Baaden, C. Le Poupon, J. P. Changeux, M. Delarue, and P. J. Corringer. 2009. 'X-ray structure of a pentameric ligand-gated ion channel in an apparently open conformation', *Nature*, 457: 111–4.

Brejc, K., W. J. van Dijk, R. V. Klaassen, M. Schuurmans, J. van Der Oost, A. B. Smit, and T. K. Sixma. 2001. 'Crystal structure of an ACh-binding protein reveals the ligand-binding domain of nicotinic receptors', *Nature*, 411: 269–76.

Brisson, A., and P. N. Unwin. 1984. 'Tubular crystals of acetyl choline receptor', *J Cell Biol*, 99: 1202–11.

Davies, P. A., M. Pistis, M. C. Hanna, J. A. Peters, J. J. Lambert, T. G. Hales, and E. F. Kirkness. 1999. 'The 5-HT3B subunit is a major determinant of serotonin-receptor function', *Nature*, 397: 359–63.

Derkach, V., A. Surprenant, and R. A. North. 1989. '5-HT3 receptors are membrane ion channels', *Nature*, 339: 706–9.

Fletcher, S., and N. M. Barnes. 1998. 'Desperately seeking subunits: are native 5-HT3 receptors really homomeric complexes?', *Trends Pharmacol Sci*, 19: 212–5.

Frauenfeld, J., R. Löving, J. P. Armache, A. F. Sonnen, F. Guettou, P. Moberg, L. Zhu, C. Jegerschöld, A. Flayhan, J. A. Briggs, H. Garoff, C. Löw, Y. Cheng, and P. Nordlund. 2016. 'A saposin-lipoprotein nanoparticle system for membrane proteins', *Nat Methods*, 13: 345–51.

Frazer, A., and J. G. Hensler. 1999. 'Chapter 13: Serotonin Receptors', in *Basic Neurochemistry: Molecular, Cellular and Medical Aspects. 6th* edition. (Lippincott-Raven: Philadelphia).

Gaddum, J. H., and Z. P. Picarelli. 1957. 'Two kinds of tryptamine receptor', *Br J Pharmacol Chemother*, 12: 323–8.

Gehlert, D. R., S. L. Gackenheimer, D. T. Wong, and D. W. Robertson. 1991. 'Localization of 5-HT3 receptors in the rat brain using [3H]LY278584', *Brain Res*, 553: 149–54.

Grosman, C., F. N. Salamone, S. M. Sine, and A. Auerbach. 2000. 'The extracellular linker of muscle acetylcholine receptor channels is a gating control element', *J Gen Physiol*, 116: 327–40.

Hanna, M. C., P. A. Davies, T. G. Hales, and E. F. Kirkness. 2000. 'Evidence for expression of heteromeric serotonin 5-HT(3) receptors in rodents', *J Neurochem*, 75: 240–7.

Hassaine, G., C. Deluz, L. Grasso, R. Wyss, M. B. Tol, R. Hovius, A. Graff, H. Stahlberg, T. Tomizaki, A. Desmyter, C. Moreau, X. D. Li, F. Poitevin, H. Vogel, and H. Nury. 2014. 'X-ray structure of the mouse serotonin 5-HT3 receptor', *Nature*, 512: 276–81.

Hilf, R. J., and R. Dutzler. 2008. 'X-ray structure of a prokaryotic pentameric ligand-gated ion channel', *Nature*, 452: 375–9.

Hoyer, D., D. E. Clarke, J. R. Fozard, P. R. Hartig, G. R. Martin, E. J. Mylecharane, P. R. Saxena, and P. P. Humphrey. 1994. 'International Union of Pharmacology classification of receptors for 5-hydroxytryptamine (Serotonin)', *Pharmacol Rev*, 46: 157–203.

Jackson, M. B., and J. L. Yakel. 1995. 'The 5-HT3 receptor channel', *Annu Rev Physiol*, 57: 447–68.

Kelley, S. P., J. I. Dunlop, E. F. Kirkness, J. J. Lambert, and J. A. Peters. 2003. 'A cytoplasmic region determines single-channel conductance in 5-HT3 receptors', *Nature*, 424: 321–4.

Koyama, Y., M. Kondo, and S. Shimada. 2017. 'Building a 5-HT3A receptor expression map in the mouse brain', *Sci Rep*, 7: 42884.

Kudryashev, M., D. Castano-Diez, C. Deluz, G. Hassaine, L. Grasso, A. Graf-Meyer, H. Vogel, and H. Stahlberg. 2016. 'The structure of the mouse serotonin 5-HT3 receptor in lipid vesicles', *Structure*, 24: 165–70.

Lummis, S. C. R. 2012. '5-HT$_3$ receptors', *Journal of Biological Chemistry*, 287: 40239–45.

Maricq, A. V., A. S. Peterson, A. J. Brake, R. M. Myers, and D. Julius. 1991. 'Primary structure and functional expression of the 5HT3 receptor, a serotonin-gated ion channel', *Science*, 254. 432–7.

Miyake, A., S. Mochizuki, Y. Takemoto, and S. Akuzawa. 1995. 'Molecular cloning of human 5-hydroxytryptamine3 receptor: heterogeneity in distribution and function among species', *Mol Pharmacol*, 48: 407–16.

Miyazawa, A., Y. Fujiyoshi, and N. Unwin. 2003. 'Structure and gating mechanism of the acetylcholine receptor pore', *Nature*, 423: 949–55.

Mowrey, D., M. H. Cheng, L. T. Liu, D. Willenbring, X. Lu, T. Wymore, Y. Xu, and P. Tang. 2013. 'Asymmetric ligand binding facilitates conformational transitions in pentameric ligand-gated ion channels', *J Am Chem Soc*, 135: 2172–80.

Niesler, B., B. Frank, J. Kapeller, and G. A. Rappold. 2003. 'Cloning, physical mapping and expression analysis of the human 5-HT3 serotonin receptor-like genes HTR3C, HTR3D and HTR3E', *Gene*, 310: 101–11.

Niesler, B., J. Walstab, S. Combrink, D. Möller, J. Kapeller, J. Rietdorf, H. Bönisch, M. Göthert, G. Rappold, and M. Brüss. 2007. 'Characterization of the novel human serotonin receptor subunits 5-HT3C,5-HT3D, and 5-HT3E', *Mol Pharmacol*, 72: 8–17.

Park, S. M., and C. L. Williams. 2012. 'Contribution of serotonin type 3 receptors in the successful extinction of cued or contextual fear conditioned responses: interactions with GABAergic signaling', *Rev Neurosci*, 23: 555–69.

Polovinkin, L., G. Hassaine, J. Perot, E. Neumann, A. A. Jensen, S. N. Lefebvre, P. J. Corringer, J. Neyton, C. Chipot, F. Dehez, G. Schoehn, and H. Nury. 2018. 'Conformational transitions of the serotonin 5-HT(3) receptor', *Nature*, 563: 275–79.

Rajkumar, R., and R. Mahesh. 2010. 'The auspicious role of the 5-HT3 receptor in depression: a probable neuronal target?', *J Psychopharmacol*, 24: 455–69.

Ruan, Y., K. Kao, S. Lefebvre, A. Marchesi, P.-J. Corringer, R. K. Hite, and S. Scheuring. 2018. 'Structural titration of receptor ion channel GLIC gating by HS-AFM', *Proc Natl Acad Sci USA*, 115: 10333–38.

Stuebler, A. G., and M. Jansen. 2020. 'Mobility of lower MA-helices for ion conduction through lateral portals in 5-HT(3A) receptors', *Biophys J*, 119: 2593–603.

Thompson, M. J., and J. E. Baenziger. 2020. 'Structural basis for the modulation of pentameric ligand-gated ion channel function by lipids', *Biochimica et Biophysica Acta (BBA) - Biomembranes*, 1862: 183304.

Unwin, N. 2005. 'Refined structure of the nicotinic acetylcholine receptor at 4A resolution', *J Mol Biol*, 346: 967–89.

Yuan, S., S. Filipek, and H. Vogel. 2016. 'A gating mechanism of the serotonin 5-HT3 receptor', *Structure*, 24: 816–25.

Zarkadas, E., H. Zhang, W. Cai, G. Effantin, J. Perot, J. Neyton, C. Chipot, G. Schoehn, F. Dehez, and H. Nury. 2020. 'The binding of palonosetron and other antiemetic drugs to the serotonin 5-HT3 receptor', *Structure*, 28: 1131–40 e4.

Zhang, Y., P. M. Dijkman, R. Zou, M. Zandl-Lang, R. M. Sanchez, L. Eckhardt-Strelau, H. Kofeler, H. Vogel, S. Yuan, and M. Kudryashev. 2021. 'Asymmetric opening of the homopentameric 5-HT(3A) serotonin receptor in lipid bilayers', *Nat Commun*, 12: 1074.

Zhu, S., C. M. Noviello, J. Teng, R. M. Walsh, J. J. Kim, and R. E. Hibbs. 2018. 'Structure of a human synaptic GABAA receptor', *Nature*, 559: 67–72.

29 Molecular Cryo-EM in the Study of Skin Tissue

Lars Norlén

INTRODUCTION

Cryogenic electron microscopy (cryo-EM) of vitreous sections (CEMOVIS) allows for the investigation of skin tissue with molecular resolution (Norlén et al., 2003; Norlén and Al-Amoudi, 2004, Al-Amoudi et al., 2005; Iwai et al., 2012; Den Hollander et al., 2016). If combined with molecular model building subject to molecular dynamics (MD) simulation and electron microscopy (EM) simulation (Lundborg et al., 2018a; Narangifard et al., 2018; Wennberg et al., 2018; Narangifard et al., 2021), CEMOVIS can be used for structure determination directly *in situ* of the skin's barrier structure (Lundborg et al., 2018a).

CRYO-ELECTRON MICROSCOPY OF VITREOUS SECTIONS (CEMOVIS) AND ITS ANALYSIS USING ATOMIC DETAIL MOLECULAR DYNAMICS MODELING AND ELECTRON MICROSCOPY SIMULATION

In cryo-EM of vitreous sections (CEMOVIS), the native tissue is preserved down to the molecular level, and the micrograph pixel intensity is directly related to the local electron density of the specimen (Dubochet et al., 1988). Biomolecules generally possess small intermolecular and intramolecular differences in electron density as they are essentially composed of atoms with similar atomic weight (carbon, nitrogen, and oxygen). However, for orderly arranged molecular assemblies, such as lipid tails and head-groups in membranes, even small differences in shape and atomic composition may be amplified because of interference effects that appear in the image phase contrast. During cryo-EM image acquisition, phase contrast is made visible using defocus.

The CEMOVIS procedure comprises the following steps: skin tissue samples, approximately 1×1 mm^2 in size and with a thickness of approximately 100–150 µm, are collected directly from living donors using a scalpel (Figure 29.1a). The samples are vitrified using a high-pressure freezer (Figure 29.1b), then cryosectioned at 140°C with a diamond blade (Figure 29.1c), transferred to a pre-cooled cryo-EM copper grid (Figure 29.1d), and mounted in a cryo-sample holder (Figure 29.1e) that is inserted into a cryo-electron microscope (Figure 29.1f). Now, a series of high-resolution cryo-electron micrographs may be collected at different microscope defocus levels (Figure 29.2).

In the analysis of cellular high-resolution cryo-EM images, atomic-detail molecular models subject to molecular dynamics (MD) simulation are employed, followed by electron microscopy (EM) simulation. The image analysis procedure comprises the following three steps: (1) construction of candidate molecular models (Figure 29.3), (2) generation of simulated electron micrographs based on these models (Figure 29.4), and finally (3) comparison of the observed cryo- electron micrographs with the simulated ones (Figure 29.5) (Iwai et al., 2012; Lundborg et al., 2018a; Narangifard et al., 2018, 2021). MD simulations may be run in GROMACS and EM simulations may be run in TEM simulator (freely available at [http://tem-simulator.sourceforge.net]).

CRYO-EM PATTERNS OF THE SKIN'S BARRIER STRUCTURE AND ITS FORMATION PROCESS

Human skin serves to uphold homeostasis by preventing water loss from the body as well as by preventing permeation of exogenous substances into the body. The major physical barrier is situated in the skin's topmost epidermal layer, the stratum corneum (Suppl. figure 1 and figure 1A in Norlén and Al-Amoudi, 2004). It consists of a lipid structure positioned intercellularly between cornified dead cells.

The formation of the skin's barrier represents the final stage of epidermal differentiation. The epidermal cells move from the basal layer with undifferentiated cells, via several layers of cells differentiating into secretory cells, to the stratum corneum with cornified dead cells. Barrier formation is initiated in the topmost secretory cell layer, the stratum granulosum (figure 1 in Al-Amoudi et al., 2005). The barrier lipids are synthesized in the endoplasmic reticulum and the Golgi apparatus and are then secreted into an extensive tubuloreticular (lamellar body) membrane system (figure 1 [white stars] in den Hollander et al., 2016) (den Hollander et al., 2016). They are finally discharged into the intercellular space separating the viable secretory cells from the dead cornified cells (figure 1 [white dotted line] in den Hollander et al., 2016). Inside the tubuloreticular (lamellar body) membrane system, the barrier lipids appear as a partly granular (green stars) and partly lamellar (blue stars) material (figure 2 in Narangifard et al., 2018). Once in the intercellular space between the cornified cells, the secreted lipids transform into laterally extended, stacked sheets (brown, pink, and yellow stars in figure 2 and Suppl. figure S1 in Narangifard et al., 2021). This process is completed at the third to fifth cornified cell layer. The stacked sheets

DOI: 10.1201/9781003326106-35

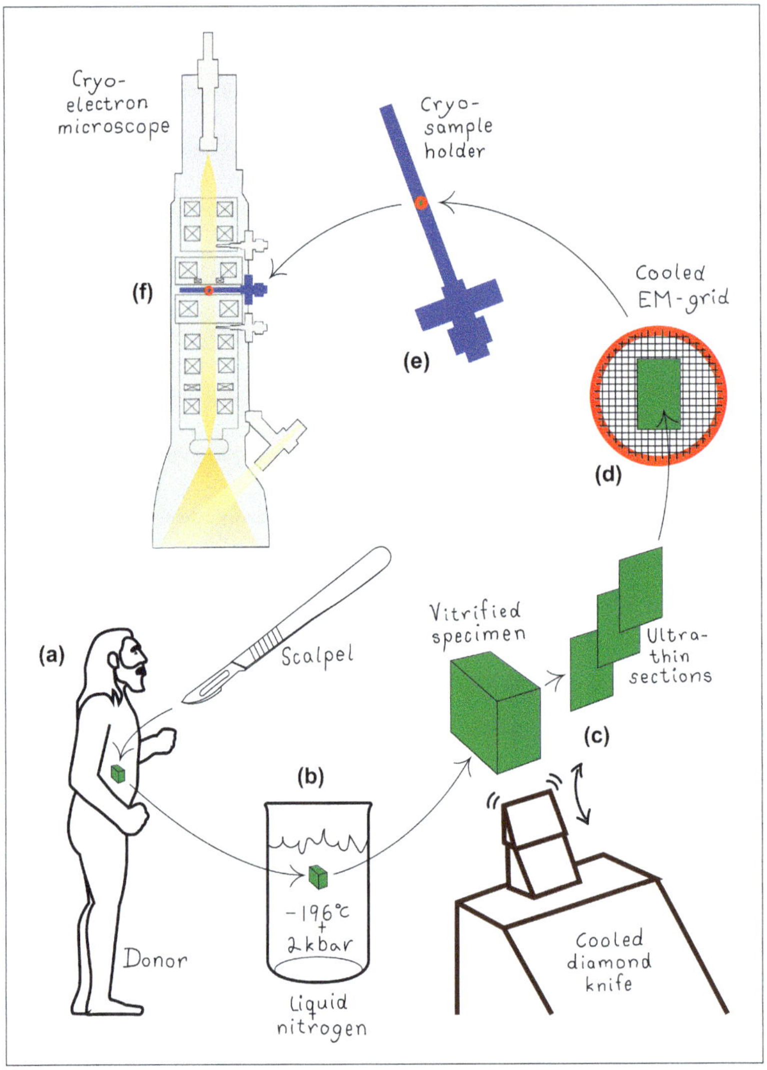

FIGURE 29.1 Specimen collection, vitrification, sectioning, and mounting for cryo-EM of vitreous sections (CEMOVIS). (a) A biological specimen is collected *in vivo* by surgical excision. (b) The specimen is vitrified by fast immersion in liquid nitrogen under pressure (2 kbar). (c) The vitrified specimen is cut by a diamond knife at cryo-conditions (–140°C) into ultrathin (30 nm) sections. (d) The cryo-sections are transferred to a cryo-EM grid, which is mounted in a cryo-EM sample holder (e) that in turn is inserted into the cryo-electron microscope (f). Adapted from Norlén (2023), with permission.

are then observed between the cornified cells throughout the stratum corneum (Suppl. figure S1 [yellow stars] in Narangifard et al., 2021).

BARRIER FORMATION

A schematic view of skin and the interface between viable and cornified epidermis, where barrier formation takes place, can be seen in figure 1 in Norlén et al., 2022.

High-Resolution Cryo-EM Patterns of the Barrier Formation Process

During epidermal differentiation, the skin's barrier structure undergoes five apparent maturation stages represented by five different cryo-EM patterns: (a) a granular pattern; (b) a lamellar pattern with 50–55 Å periodicity; (c) a lamellar pattern with 20–25 Å periodicity; (d) a lamellar pattern with 55–60 Å periodicity; and (e) a lamellar pattern with 110–120 Å periodicity) (figure 2 in Norlén et al., 2022).

The Five Stages Identified in the Barrier Formation Process

Analysis of the five cryo-EM patterns using molecular dynamics (MD) model building combined with electron microscopy (EM) simulation (see figure 3 in Norlén et al., 2022) suggests that the granular pattern represents a highly folded and highly hydrated conventional skin lipid bilayer (a) (Narangifard et al., 2018), that the lamellar

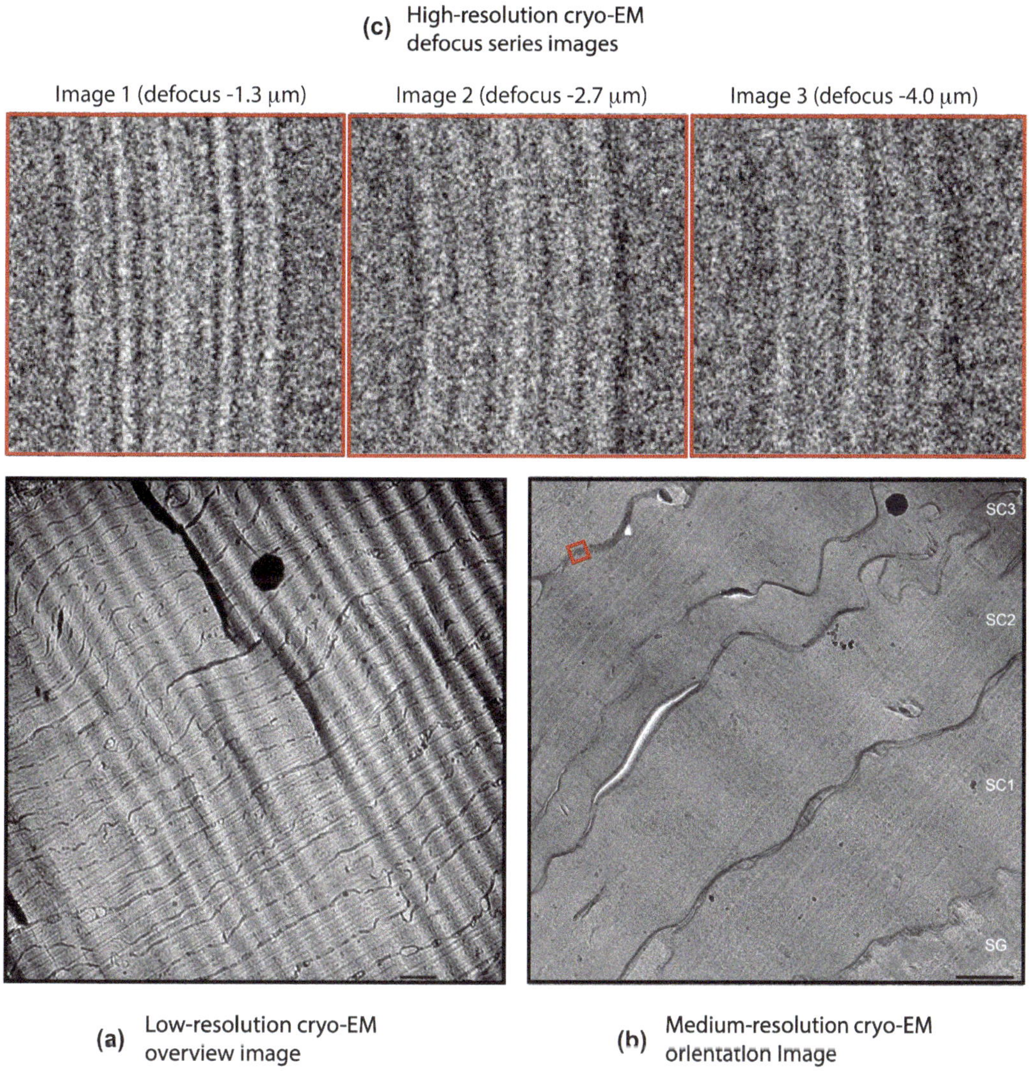

FIGURE 29.2 Collection of a CEMOVIS defocus image series. Initially, a low-resolution overview image is collected at about 1.000 to 5.000 times magnification (a), followed by a medium-resolution orientation image collected at about 5.000 to 30.000 times magnification (b), to select the specimen location (for example, red square in b) where a high-resolution defocus image series can be collected at about 80.000 to 115.000 times magnification (c). Electron-dense (black) spots in (a) and (b) correspond to surface ice contamination. Wavelike diagonal patterns in (a) and (b) are due to section compression during cutting. Scale bars in (a) 1.0 μm and in (b) 500 nm. Pixel size in (c) is 0.188 nm. SG and SC1–3 in (b): stratum granulosum and stratum corneum layer 1–3, respectively. Adapted from Norlén and Al-Amoudi 2004 (a, b) and Iwai et al., 2012 (c), with permission. Adapted from Norlén (2023), with permission.

pattern with 50–55 Å periodicity represents a stacked skin lipid monolayer with mixed hairpin and splayed ceramides (b) (Narangifard et al., 2018), that the lamellar pattern with 20–25 Å periodicity represents a stacked skin lipid monolayer with splayed ceramides and chain interdigitation (c) (Narangifard et al., 2021), that the lamellar pattern with 55–60 Å periodicity represents a stacked skin lipid bilayer with splayed ceramides and chain interdigitation (d) (Narangifard et al., 2021), and that the lamellar pattern with 110–120 Å periodicity represents a stacked skin lipid bilayer with splayed ceramides without chain interdigitation (e) (Iwai et al., 2012; Lundborg et al., 2018a).

Reorganization Steps of the Barrier Formation Process

Cryo-EM analysis of the skin's barrier formation process suggests that the formation of the skin's barrier structure starts with the secretion into the stratum corneum intercellular space of a highly folded and highly hydrated lipid bilayer (a). Cleavage of the sugar groups of glucosyl-ceramides of the lipid bilayer (a–b), followed by dehydration (b–c), provokes an unfolding, or flattening, of the folded lipid bilayer, making it collapse into stacks of tightly packed flat lipid bilayers (c) (Wennberg et al., 2018) (See figure 4 in Norlén et al., 2022).

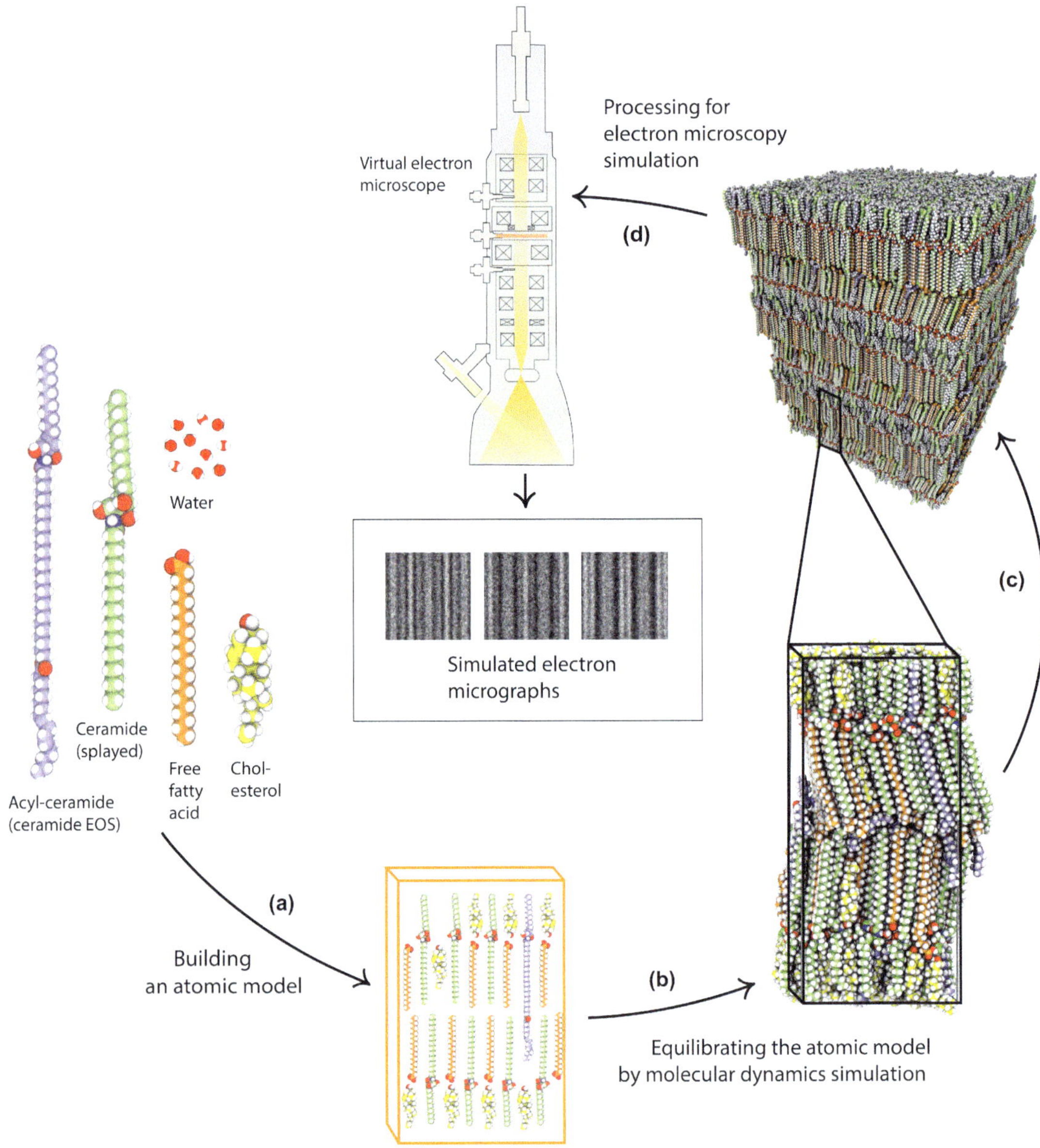

FIGURE 29.3 Molecular model building subject to molecular dynamics (MD) simulation. (a) Building an atomic model based on a known molecular composition. (b) The atomic model is equilibrated by molecular dynamics (MD) simulation, and then multiplied in three dimensions (c) before processing for electron microscopy (EM) simulation (d). Adapted from Norlén (2023), with permission.

Once shielded from water, the lipid bilayers' ceramide molecules can now start to stretch out into their preferred splayed conformation with their two hydrocarbon tails pointing in opposite directions (d). When all ceramides have become stretched out (e), the system can further relax itself and become more tightly packed by letting the ceramide molecules' short and long hydrocarbon tails, as well as the cholesterol and free fatty acid molecules, separate into different bands within the layered structure. This internal molecular rearrangement is proposed to take place via axial sliding of the ceramide molecules together with associated cholesterol and free fatty acid molecules. The axial sliding process initially causes a lipid chain interdigitation (e). Continued axial sliding results in the separation of the ceramides' short and long hydrocarbon tails, as well as of the cholesterol and free fatty acid molecules, into different bands within the layered structure (f). The axial sliding process ends with an uninterdigitation of the lipid chains (g), yielding the mature skin barrier structure (See figure 4 in Norlén et al., 2022).

BARRIER ARCHITECTURE

Cryo-EM analysis of the skin's mature barrier structure suggests that it is organized as stacked lipid bilayers of fully stretched (splayed) ceramides with cholesterol largely

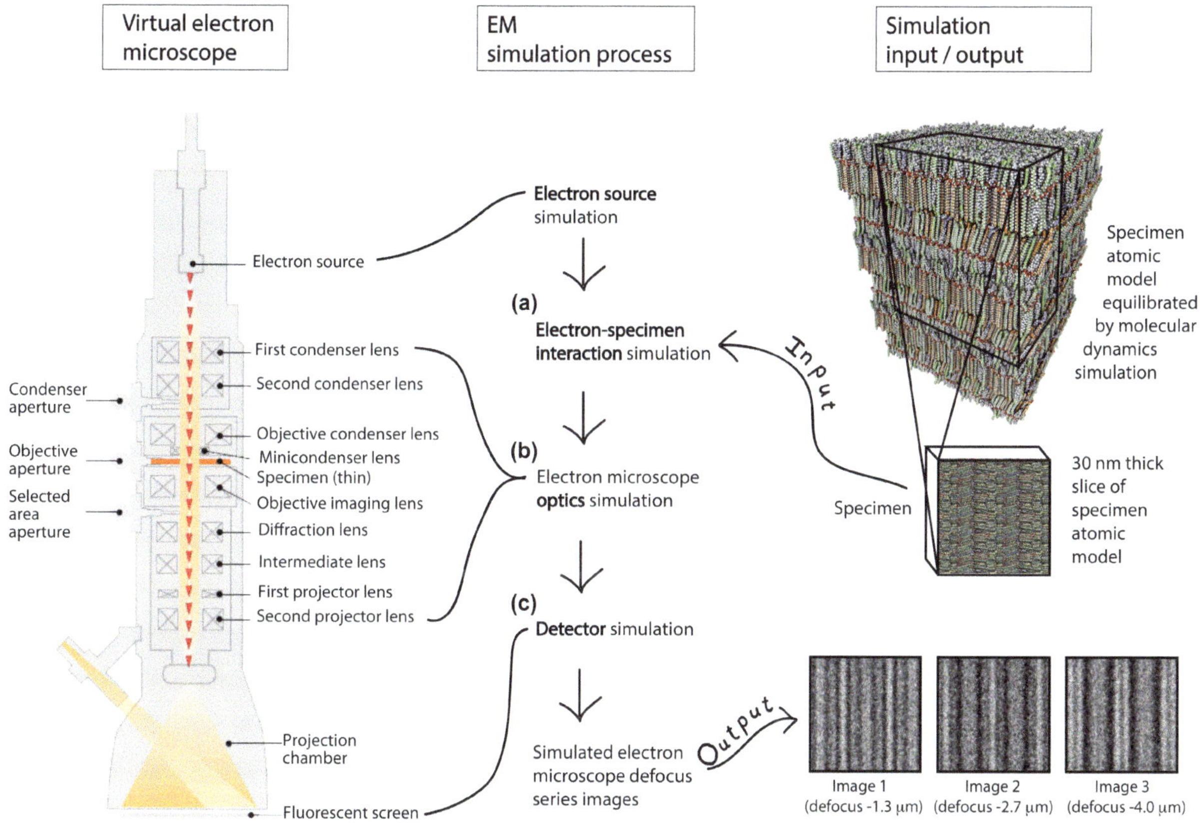

FIGURE 29.4 Electron microscopy (EM) simulation. (a) Simulated CEMOVIS defocus series images for candidate atomic models are created with a computer program simulating the interaction between the candidate atomic model's electron scattering potential map and the microscope's electron beam, (b) the optical transformation effect of the microscope's lens system, and (c) the image formation on the microscope's detector. Adapted from Norlén (2023), with permission.

associated with the ceramides' shorter sphingoid tail and with free fatty acids associated with the ceramides' longer fatty acid tail (c–d). Acyl-ceramides (ceramide EOS) (light blue) protrude their ester-bound lignoceric acid ends into the interface between opposing free fatty acid and ceramide fatty acid tail ends. The lignoceric acid ends are mobile and can extend into the opposing layer or fold back into the interface between the layers. Another feature is that the acyl-ceramides´ ester groups make the interface between the ceramides' fatty acid tail ends more polar than the interface between the ceramides' sphingoid tail ends (c–d). This arrangement offers a tight and robust barrier structure and is compatible both with cryo-EM (e–j) and with the broad:narrow:broad electron lucent ruthenium tetroxide (RuO_4) staining band pattern observed using conventional EM (See figure 5 in Norlén et al., 2022).

FUTURE PERSPECTIVE

Having obtained a near-native reference from healthy skin, it may now be possible to explore differences in lipid structure that could be at the basis of barrier malfunction in skin diseases such as atopic dermatitis, psoriasis, and the ichtyoses.

The new knowledge on the structure of the skin's permeability barrier may facilitate physics based skin permeability calculations using molecular dynamics (MD) simulation (Lundborg et al., 2018b, 2022). This may aid in predicting the properties of drugs interacting with the skin and optimizing them for topical and percutaneous delivery. In addition, it may be used for skin toxicity assessment.

The experimental approach exemplified here, which consists of obtaining *in situ* a near-native structural reference of a biomolecular complex, using cryo-EM of vitreous sections (CEMOVIS) in combination with molecular dynamics (MD) modeling and EM simulation, may be followed by more detailed investigation *in silico* and *ex situ*, using, for example, molecular dynamics simulation, X-ray crystallography, nuclear magnetic resonance spectroscopy, or single-particle cryo-EM. It may thus prove useful for exploring the molecular-level structure–function relationships of biomolecular structures inside any cell or tissue within the constraints of a near-native structural reference (Norlén et al., 2022).

CONFLICT OF INTEREST

The author states no conflict of interest.

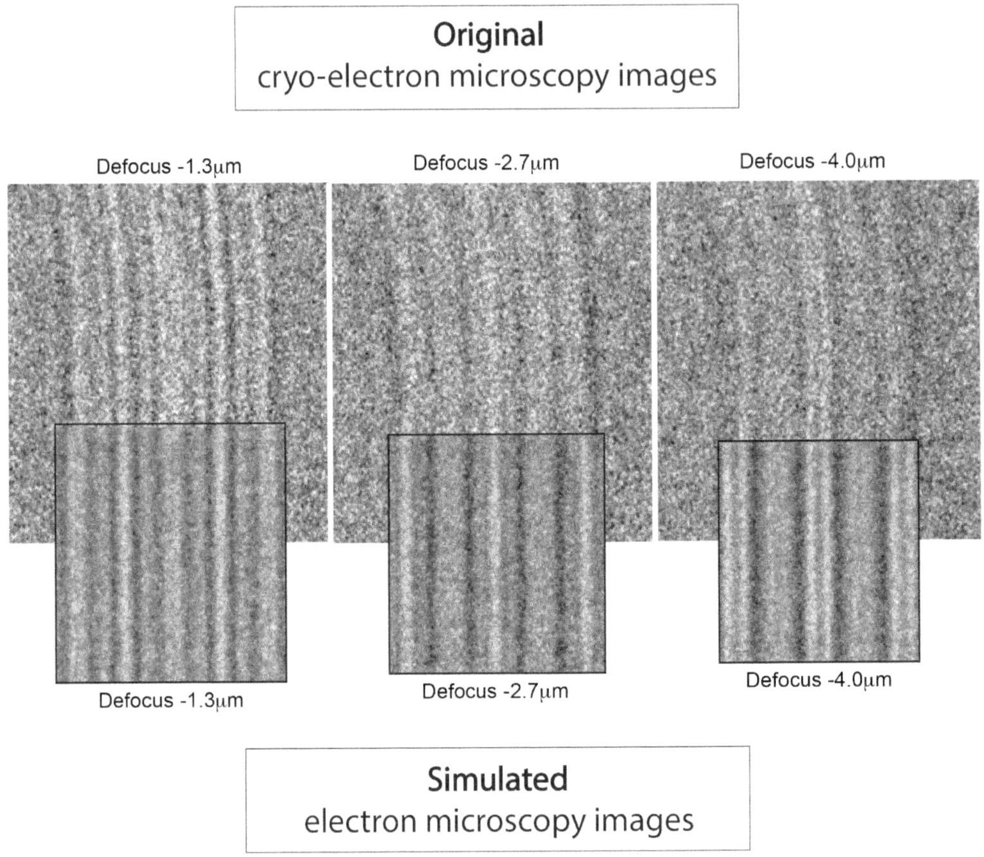

FIGURE 29.5 Comparison between simulated and original electron micrographs. The simulated electron micrographs are compared with the original cryo-electron micrographs at different microscope defocus levels. If differences are identified, the atomic model is updated and the simulation renewed. This process is repeated until optimal correspondence is attained between simulated and original electron micrographs. Adapted from Norlén (2023), with permission.

ACKNOWLEDGMENTS

The present work was made possible by the generous support from the Swedish Medical Society, Oriflame, Shiseido Japan, LEO Pharma, Wenner-Gren foundation, the Swedish eScience Research Center, the Welander foundation and ERCO Pharma. The service of the Core facility for Electron Tomography at Karolinska Institutet as well as the resources provided by the Swedish National Infrastructure for Computing (SNIC) at PDC Centre for High Performance Computing (PDCHPC), are acknowledged.

REFERENCES

Al-Amoudi, A., Chang, J.-J., Leforestier, A., McDowall, A., Michel Salamin, L., Norlén, L. et al. (2004) Cryo-electron microscopy of vitreous sections. *EMBO J.* 23(18):3583–3588.

Al-Amoudi, A., Dubochet, J., Norlén, L. (2005) Nanostructure of the epidermal extracellular space as observed by cryo-electron microscopy of vitreous sections of human skin. *J. Invest. Dermatol.* 124:764–777

den Hollander, L., Han, H.-M., de Winter, M., Svensson, L., Masich, S., Daneholt, B., Norlén, L. (2016) Skin lamellar bodies are not discrete vesicles but part of a tubuloreticular network. *ACTA Dermato-Venereol.* 96(3):303–308.

Dubochet, J., Adrian, M., Chang, J.-J., Homo, J.-C., Lepault, J., McDowall, A. W. et al. (1988) Cryo electron microscopy of vitrified specimens. *Q. Rev. Biophys.* 21:129–228.

Iwai, I., Han, H., den Hollander, L., Svensson, S., Öfverstedt, L.-G., Anwar, J. et al. (2012) The human skin barrier is organized as stacked bilayers of fully-extended ceramides with cholesterol molecules associated with the ceramide sphingoid moiety. *J. Invest. Dermatol.* 132: 2215–2225.

Lundborg, M., Narangifard, A., Wennberg, C., Lindahl, E., Norlén, L. (2018a) Human skin barrier molecular structure and function analyzed by cryo-electron microscopy and molecular dynamics simulation. *J. Struct. Biol.* 203(2):149–161.

Lundborg, M., Wennberg, C., Narangifard, A., Lindahl, E., Norlén, L. (2018b) Predicting drug permeability through skin using molecular dynamics simulation. *J. Controlled Release* 283:269–227.

Lundborg, M., Wennberg, C., Lidmar, J., Hess, B., Lindahl, E., Norlen, L. (2022) Skin permeability prediction with MD simulation sampling spatial and alchemical reaction coordinates. *Biophys. J.* 121, 3837–3849. https://doi.org/10.1016/j.bpj.2022.09.009

Narangifard, A., den Hollander, L., Iwai, I., Han, H., Wennberg, C. L., Lundborg, M. et al. (2018) Human skin barrier formation takes place via a cubic to lamellar lipid phase transition. *Exp. Cell. Res.* 366(2):139–151.

Narangifard, A., Wennberg, C. L., den Hollander, L., Iwai, I., Han, H., Lundborg, M. et al. (2021) Molecular reorganization during formation of the human skin barrier studied *in situ*. *J. Invest. Dermatol.* 141(5):1243–1253.e6.

Norlén, L. (2023), Molecular Organization of the Skin Barrier. *Acta Dermato-Venereologica, 103*, adv13356. https://doi.org/10.2340/actadv.v103.13356.

Norlén, L., Al-Amoudi, A. (2004) Stratum corneum keratin structure, function, and formation: the cubic rod-packing and membrane templating model. *J Invest Dermatol.* 123(4):715–732.

Norlén, L., Al-Amoudi, A., Dubochet, J. (2003) A cryo-transmission electron microscopy study of skin barrier formation. *J. Invest. Dermatol.* 120:555–560

Norlén, L., Lundborg, M., Wennberg, C., Narangifard, A., Daneholt, B. (2022) The Skin's Barrier: A Cryo-EM Based Overview of its Architecture and Stepwise Formation. *J. Invest. Dermatol.* 142:285–292

Wennberg, C., Narangifard, A., Lundborg, M., Lindahl, E., Norlén, L. (2018) Structural transitions in ceramide cubic phases during formation of the human skin barrier. *Biophys. J.* 114(5):1116–1127.

Section VII

Applications of Cryo-ET and Cryo-EM in the Biopharmaceutical Industry

30 Emerging Technologies in Cryo-Electron Tomography

Kanika Khanna

INTRODUCTION

> *"By the help of microscopes, there is nothing so small, as to escape our inquiry; hence there is a new visible world discovered to the understanding."*
>
> **Robert Hooke, 1665**

Embarking on a journey into the captivating realm of cell biology requires the ability to peer deep inside cells and unravel the mysteries hidden within their intricate components. Fortunately, recent light and electron microscopy developments have ignited a scientific revolution, propelling the study of cell biology to unprecedented heights.

Light microscopy allows us to observe cells in real time, uncovering their intricate morphology and dynamic interactions. Incorporating fluorophore tags into macromolecules of interest allows us to visualize and track essential cellular processes, from protein localization to organelle dynamics and intracellular signaling events (Cuny et al., 2022; Morris & Payne, 2019; Thorn, 2016). However, the resolution in light microscopy is restricted by the inherent limitations of diffraction, limiting lateral resolution to about 200 nm and axial resolution to about 500 nm (Abbe, 1873). To push the boundaries further, superresolution microscopy has emerged as a revolutionary technology capable of achieving a resolution of approximately 30–100 nm (Jacquemet et al., 2020). While this level of resolution enables the visualization of cellular structures, it falls short when it comes to delving into atomic or individual complex details.

The advent of the electron microscope in 1931, pioneered by Max Knoll and Ernst Ruska, marked a monumental breakthrough in overcoming the resolution limitations of light microscopy (Ruska, 1987). With its unprecedented resolution on the order of a few nanometers, transmission electron microscopy (TEM) has provided researchers with remarkable insights into the complex architecture and organization of cells and essentially propelled the field of cell biology in the early twentieth century.

With the advent of even newer technologies, we now have the ability to venture further, delving into the atomic scale. Single-particle cryo-electron microscopy (cryo-EM) has emerged as a routine and robust method for determining the three-dimensional structure of macromolecules, such as proteins or complexes, with near-atomic precision (Cheng, 2018; Chua et al., 2022; Guaita et al., 2022; Saibil, 2022). Technological developments that propelled single-particle cryo-EM further contributed to cryo-electron tomography (cryo-ET) development. While cryo-EM excels in determining the average structure of purified macromolecules, cryo-ET takes it further by enabling the three-dimensional reconstruction of intricate cellular components, whole cells, and even tissues to provide a more comprehensive view of complex biological structures (Hong et al., 2023; Khanna & Villa, 2022; Navarro, 2022; Turk & Baumeister, 2020; Wang et al., 2023; Young & Villa, 2023). By combining high-resolution imaging with the ability to visualize specimens in their native state, cryo-ET offers unprecedented insights into the spatial organization and macromolecular interactions within biological systems. Combining cryo-ET with subtomogram analysis and integrative modeling has unlocked new frontiers in structural biology, enabling researchers to unravel the mysteries of cellular architecture and processes with unparalleled clarity and depth (Pyle & Zanetti, 2021; Zhang, 2019).

In cryo-ET, biological samples are rapidly frozen to maintain their native context. Unlike conventional electron microscopy approaches that involve embedding samples in heavy metals, fixing or staining them, cryo-ET enables visualization of specimens in their natural, frozen state. These flash-frozen samples are then tilted inside a cryogenic TEM, allowing the capture of a series of images at different tilt angles. These 2D images are subsequently merged computationally to generate a high-resolution 3D volume called a tomogram. Cryo-ET data can provide information spanning a spectrum of resolutions. It enables the examination of ultrastructural details, capturing features measuring a few microns. It also allows investigation of the spatial distribution of macromolecules, providing insights into their arrangement at a resolution of a few nanometers. Moreover, by employing subtomogram analysis, it becomes possible to attain atomic-level details about specific molecules of interest. This versatile technique offers a comprehensive view of biological samples at various scales, enabling researchers to explore and understand their structural characteristics across different levels of detail.

This chapter will explore the latest advancements in cryo-ET, encompassing sample preparation and data acquisition. Specifically, we will discuss the application and recent developments in cryo-focused ion beam (cryo-FIB) milling to make thin cellular specimens that are suitable for high-resolution cryo-ET workflow and cryo-correlative light and electron microscopy (cryo-CLEM) to locate

DOI: 10.1201/9781003326106-37

target macromolecules for imaging. We will also discuss hardware and software developments that play an essential role in cryo-ET data acquisition. These emerging technologies significantly impact the accessibility and efficiency of the cryo-ET pipeline. By facilitating more straightforward setup and streamlined processes, they contribute to the democratization of cryo-ET, enabling a broader range of laboratories to engage with the method and address their specific biological inquiries.

SAMPLE PREPARATION: FREEZING AND PATTERNING

In cryo-ET, sample preparation is a critical step that involves depositing or growing biological samples onto electron microscopy (EM) grids. The specific approach depends on the nature of the specimen. Small molecules, viruses, bacteria, yeast, and nonadherent cells can be directly deposited onto EM grids. Adherent mammalian cells are typically cultured on gold EM grids designed to be nontoxic to cells (Lučić et al., 2008, 2013). For samples measuring less than 10 μm, vitrification can be achieved by plunge-freezing samples in a cryogen like liquid ethane or a mixture of ethane and propane. High-pressure freezing is used to vitrify thicker specimens up to 200 μm, such as larger cells, tissues, and whole organisms like *C. elegans* (McDonald & Auer, 2006). This method prevents the formation of crystalline ice and helps preserve the cellular ultrastructure.

In some instances, precise control over sample deposition on EM grids becomes crucial, particularly when seeding adherent cells. The goal is to position the feature of interest at the center of the grid square. This deliberate placement offers several advantages, such as facilitating targeted thinning through cryo-FIB milling and acquiring unhindered tilt series within the TEM. However, it is common for adherent cells to settle and spread randomly on EM grids, including metal grid bars, rendering them difficult to access for cryo-FIB milling and tomography.

Innovative strategies in micropatterning are continuously being developed to overcome such limitations and improve the control and accessibility of samples within cryo-ET workflows. Micropatterning has emerged as a valuable tool in addressing cell biological inquiries related to the cytoskeleton and intracellular organelles, primarily utilizing light microscopy techniques (D'Arcangelo & McGuigan, 2015). In recent years, this approach has been successfully integrated into cryo-ET studies for the deliberate positioning of cells on EM grids, enabling precise targeting during milling and data acquisition stages (Engel et al., 2019; Sibert et al., 2021; Toro-Nahuelpan et al., 2020).

A widely used technique for micropatterning EM grids is contactless and maskless photo-micropatterning (Toro-Nahuelpan et al., 2020). In this approach, the EM grid is initially coated with a layer of polyethylene glycol, which serves as a cell-repulsive surface. Subsequently, a UV laser precisely ablates the polyethylene glycol layer to create micropatterns on the grid. These ablated regions devoid of the cell-repulsive material can then be selectively coated with proteins that facilitate cell attachment and spreading only at these specific zones. Recently, this approach has been used to elucidate the cytoskeleton architecture of cells grown in crossbow-shaped patterns (Toro-Nahuelpan et al., 2020). In another study, researchers used photo-micropatterning to create a lattice of the extracellular matrix that facilitated the formation of cell–cell junctions at the central region of the grids for high-resolution cryo-ET imaging (Engel et al., 2021). In the future, beyond streamlining the cryo-FIB-ET workflow, photo-micropatterning holds excellent potential for investigating specific configurations of cellular architecture by manipulating cellular organelle positions (Engel et al., 2023).

SAMPLE THINNING: ADVANCES IN CRYO-FOCUSED ION BEAM MILLING WORKFLOW

One significant limitation of cryo-ET is its suitability for visualizing specimens with a maximum thickness of approximately 400–500 nm. This constraint falls short of capturing the size and complexity of many biological samples, including eukaryotic cells and even smaller bacterial cells. The underlying reason for this limitation is the mean free path of electrons in a 300 kV TEM, approximately 280 nm under cryogenic conditions (Grimm et al., 1996; Martynowycz et al., 2021). As the sample thickness exceeds this limit, multiple electron scattering events occur, leading to a decrease in the signal-to-noise ratio (SNR) in images and an increase in the susceptibility of samples to radiation damage.

High-pressure frozen samples can be thinned by a technique called cryo-electron microscopy of vitreous sections (CEMOVIS) or cryo-ultramicrotomy (Al-Amoudi et al., 2004). This method uses a diamond knife to slice thin sections from the vitrified sample, typically ranging from approximately 60 to 80 nm. These ultra-thin sections are subsequently placed on EM grids, allowing for the acquisition of tilt series consisting of multiple slices. However, cryo-ultramicrotomy has certain limitations that affect its application. Slicing biological specimens with a diamond knife can introduce compression artifacts and blade marks, which may impact the quality and interpretation of the resulting images. Additionally, cryo-ultramicrotomy is a technically challenging technique that requires specialized training and expertise, which limits its widespread applicability.

Recently, cryo-focused ion beam (cryo-FIB) milling has gained widespread adoption as a technique for thinning biological samples for cryo-ET (Figure 30.1A). Initially used in material sciences and the semiconductor industry, cryo-FIB milling is now adapted to image and thin biological specimens under cryogenic conditions (Berger, Premaraj, et al., 2023; Hutchings et al., 2023; Khanna & Villa, 2022; Lam & Villa, 2021; Marko et al., 2007; Rigort et al., 2012; Wagner et al., 2020). A cryo-FIB system comprises a dual-beam microscope that integrates an FIB and a scanning electron

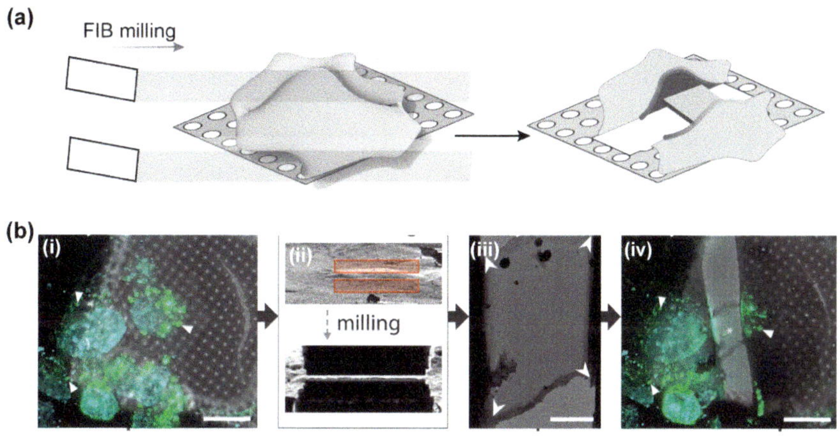

FIGURE 30.1 Cryo-FIB milling and cryo-CLEM. (A) Schematic representation of cryo-FIB milling, wherein milling patterns are used to ablate upper and lower parts of a representative mammalian cell lying on an EM grid. (B) (i) Example of a cryo-CLEM study. From left to right, nuclei and lipid droplets (cyan) were fluorescently labeled in transiently transfected mammalian cells that were transiently transfected with ABCA3-eGFP (green, marker for lamellar bodies). Scale bar: 20 µm. (ii) The cells were subsequently FIB-milled, and (iii) the lamella observed inside a TEM. Scale bar: 5 µm. (iv) Cryo-light microscopy image acquired after TEM imaging with arrows indicating the intact fluorescent signal in the cell body. Scale bar: 20 µm. (Reproduced from Klein et al., 2021 under a Creative Commons Attribution 4.0 International License.)

microscope (SEM). The FIB utilizes a focused beam of ions, typically gallium ions, which scan across the specimen and remove material through ion–sample interactions. Rectangular milling patterns are commonly employed to selectively thin the sample above and below the region of interest. This process creates a thin slice known as a lamella, typically ranging from 100 to 300 nm thick, which is suitable for subsequent steps in the cryo-ET workflow. This method has expanded the possibilities for investigating cellular structures and organelles in their native state without any fixation or perturbation, providing valuable insights into their three-dimensional organization and interactions.

Significant progress is being made in enhancing the throughput and applicability of cryo-FIB milling for a wide range of biological specimens. However, the delicate nature of thin lamellae produced through this process remains a challenge, often resulting in breakage during microscope transfers or grid handling. Particularly at cryogenic temperatures, the support film on the EM grid can crumple, exerting tension on thin lamellae and causing them to bend or fracture (Booy & Pawley, 1993). To address this issue, the implementation of "micro-expansion joints" has proven effective (Wolff et al., 2019). These joints are 1 µm wide trenches positioned approximately 3–4 µm away from the lamella on both sides. They reduce the stress on the lamella by absorbing the tension resulting from grid motion, considerably improving lamellae stability. Kelley et al. introduced another method for lamellae stress relief, resembling notches, during the milling of high-pressure frozen specimens (Kelley et al., 2022). Besides creating trenches on both sides of the lamella, this approach also carves a notch on one side, allowing the lamella to move freely and absorb any potential motion that might otherwise cause damage to the lamella during grid handling. The combined use of trenches and notches offers enhanced protection to the lamella, ensuring its structural integrity during subsequent handling and imaging processes.

The automation of the milling procedure represents another notable advancement. Cryo-FIB milling is inherently laborious and demanding, with manual milling typically producing only about four to eight lamellae on an average day. However, recent software enhancements have facilitated the automation of FIB milling, significantly expanding the instrument's usability for overnight operation (Klumpe et al., 2021; Tacke et al., 2021; Zachs et al., 2020). This automation enables the production of more than 15 to 20 lamellae without requiring user supervision, thereby considerably increasing the throughput of the procedure.

Milling Bulk Specimens

Researchers are actively investigating new approaches for performing cryo-FIB milling on thicker samples, such as tissues and whole organisms like *C. elegans*, which cannot be plunge-frozen on TEM grids. Instead, these specimens are vitrified via high-pressure freezing (HPF). One approach involves pretrimming the samples using a cryo-microtome before subsequent cryo-FIB milling of the pretrimmed sections in specialized blocks (Hsieh et al., 2014). Another technique demonstrated by Harapin et al. involves cryo-FIB milling on adult worms vitrified by HPF on TEM grids using an organic solvent like 2-methylpentane as a cryo-filler. 2-methylpentane also exposes worms in the SEM during milling, facilitating their differentiation from the background (Harapin et al., 2015). However, this method has its challenges, including long milling times, sometimes lasting a day or more, and the potential bending of lamellae in heavier samples.

An alternative approach involves a lift-out technique, where a specific region of interest (typically around 3–4 μm) is extracted from the bulk specimen through cryo-FIB milling (Klumpe et al., 2022; Mahamid et al., 2015; Schaffer et al., 2019). This extracted region is then carefully lifted out using a micromanipulator and transferred onto another EM grid within the FIB-SEM instrument. Subsequently, the lamella is further thinned to achieve a 100 to 300 nm thickness suitable for cryo-ET. Using this approach, Schaffer et al. performed structural studies on the cytosolic 80S ribosomes in *C. elegans* and could achieve a resolution of 11 to 15 Å using subtomogram analysis (Schaffer et al., 2019). While this technique shows promise for milling larger specimens like tissues and multicellular organisms, it is still in the process of being fully developed, and it currently faces some challenges. One such limitation is the relatively long milling times required, taking approximately 10 hours per lamella. It is also challenging to precisely correlate the target of interest in 3D in such large volumes using correlative electron and light microscopy methods.

An alternative method to lift-out for milling thicker specimens is the "waffle" milling technique (Kelley et al., 2022). The process involves applying the sample to the backside of the EM grid, followed by high-pressure freezing and milling grid bar lines onto the sample, which resembles a waffle when viewed in the SEM. To create lamellae in the sample, the milling is initially performed as perpendicular to the grid as possible to remove bulk material. Then, the process transitions to shallower angles and lower currents for fine milling, generating thinner lamellae. The waffle milling technique offers significant advantages compared to conventional cryo-FIB milling. One key advantage is its applicability to specimens with a wide range of thicknesses. Additionally, it effectively eliminates preferred orientation issues when freezing different types of biological samples. Kelley et al. used waffle milling to reveal the ultrastructure of the polar tube in microsporidian spores in various orientations (Kelley et al., 2022). The preferred orientation of these spores upon deposition on the EM grid had previously hindered such observations. Moreover, this technique enhances throughput, enabling the milling of stable lamellae with larger sizes.

Plasma FIB Milling

Currently, most FIB/SEM systems use gallium ions for milling samples and creating lamellae. Gallium ions possess several advantageous characteristics, including a low melting point and the ability to emit a focused beam, allowing precise control over the milling area (Forbest, 1997). However, gallium can also interact with the specimen and induce alterations in its structure. Recent studies have revealed that gallium can cause damage up to a depth of approximately 50 to 60 nm from each lamella surface (Lucas & Grigorieff, 2023; Tuijtel et al., 2023). As a result, this limitation restricts the amount of information that can be extracted from *in situ* structural biology approaches. To overcome this challenge, researchers are actively exploring alternative ion sources and techniques to minimize such adverse effects and enhance the fidelity of structural investigations.

An actively explored alternative in the field is plasma generated from gases such as oxygen, nitrogen, argon, and xenon (Berger, Dumoux, et al., 2023). When operated at higher currents, plasma beams can achieve smaller probe sizes than gallium, enabling faster milling of larger volumes. This characteristic especially proves advantageous for life-out or waffle milling when dealing with bulk specimens. As a proof of principle, Berger et al. successfully utilized argon gas to generate plasma for milling HeLa cells and achieved a remarkable 4.9 Å resolution of the 80S human ribosomes, all while fully automating the milling procedure. Furthermore, plasma FIB/SEM systems boast smaller chamber volumes, approximately one-fourth of conventional FIB/SEM systems, which allows them to achieve higher vacuums and lower redeposition rates (Berger, Dumoux, et al., 2023). Researchers are also investigating gases like xenon, which offer even faster milling capabilities. Additionally, assessing the extent of damage caused by plasma compared to gallium on samples will be essential to optimize the technique for *in situ* structural biology applications.

SAMPLE LOCALIZATION: ADVANCES IN CRYO-LIGHT MICROSCOPY

One of the most significant challenges in cryo-ET is the unambiguous identification of the molecule of interest. When observing a black-and-white tomogram, it is relatively straightforward to identify larger known macromolecular assemblies such as ribosomes, proteasomes, membranes, or structures with higher-order organizations like filaments. However, unlike in light microscopy, no equivalent reliable molecular tag akin to GFP (green fluorescent protein) allows us to precisely pinpoint smaller biological targets or those lacking recognizable assembly using *in situ* cryo-ET data. This limits the ability to study and understand these smaller, less conspicuous components at a molecular level, presenting a significant hurdle in advancing cryo-ET as a powerful technique for investigating cellular structures and processes.

In the absence of a GFP-like tag for EM, cryo-correlative light and electron microscopy (CLEM) emerges as a valuable tool to overcome this challenge (Jun et al., 2019; Zhang, 2013). Notably, when dealing with situations where only a few cells within a population express the molecule of interest, CLEM offers a useful solution. By fluorescently tagging the molecule of interest, one can identify which cells express it by imaging the plunge-frozen specimen inside a cryo-fluorescence microscope (cryo-FLM). The information about the molecule's location of interest within the cells can then guide cryo-FIB milling and cryo-ET data acquisition. FIB/SEM systems with integrated cryo-light microscopy setups are also being developed to perform localization, correlation, and milling all inside one microscope (Gorelick et al., 2019; Smeets et al., 2021; Vrbovská et al., 2023; Yang et al., 2023).

The ability to precisely correlate light microscopy features with milling in FIB/SEM is currently limited by the diffraction properties of the light microscope (Figure 30.1B). Long working distance objectives in cryo-FLM typically provide a resolution of around 300 nm in the XY direction and 500 nm in the Z direction (Dahlberg & Moerner, 2021; DeRosier, 2021). These resolution limits make it challenging to precisely align and register the two imaging modalities, leading to potential inaccuracies in locating specific biological targets for further investigation using cryo-FIB/SEM. The correlation accuracy can be improved to ~100 nm by introducing fiducial markers identifiable under both the fluorescence microscope and the SEM beam. Correlating fluorescence microscopy images with the image of a lamella under TEM remains highly challenging for locating molecules of interest for cryo-ET data acquisition unless these molecules are bound to easily identifiable features such as membranes or filaments. Nevertheless, CLEM remains a powerful approach for bridging the gap between light and electron microscopy to study biological targets within complex cellular environments. For instance, cryo-CLEM was critical to identifying LRRK2-decorated microtubules in human cells for cryo-FIB milling and subsequent structural studies (Watanabe et al., 2020).

Continuous efforts are being dedicated to refining the cryo-CLEM workflow and expanding the technique's capabilities. Notably, novel workflows are under active development, integrating cryo-superresolution microscopy with EM, such as cryo-PALM (Photoactivated Localization Microscopy), cryo-STORM (STochastic Optical Reconstruction Microscopy) (Chang et al., 2014; Dahlberg & Moerner, 2021; Li et al., 2023; Sexton et al., 2022; Wolff et al., 2019) and HOPE-SIM (Structured Illumination Microscopy) (Li et al., 2023). These advancements hold immense potential to enable correlation at a scale of a few nanometers at a single-molecule level. However, superresolution cryo-CLEM is still in the early stages of development, and various technical barriers remain to be addressed. Among these challenges, using high-intensity lasers in superresolution microscopy can lead to sample devitrification. Moreover, stable activation of fluorophores at cryogenic conditions presents a formidable obstacle. Additionally, drift in cryostage poses a significant concern, particularly in open-stage designs where the objective aperture is cooled by liquid nitrogen vapors.

Nevertheless, advancements have been encouraging, such as using specialized fluorophores like PAmKate and PA-GFP, which exhibit efficient photoactivation at room and cryogenic temperatures (Dahlberg et al., 2020). The use of single-molecule active control microscopy (SMACM) using PAmKate fusion proteins allowed Dahlberg et al. to localize polar proteins within the bacterium *Caulobacter crescentus* at the single-molecule level with an exceptional lateral precision of approximately 9 nm. As technology evolves, CLEM is expected to play an increasingly vital role in advancing our understanding of cellular structures and functions.

SAMPLE IMAGING: ADVANCES IN CRYO-ET HARDWARE AND SOFTWARE

Hardware Advances

After successfully milling lamellae containing the feature(s) of interest, the subsequent step involves acquiring tilt series within a transmission electron microscope (TEM) to generate tomograms. Apart from advancements in sample preparation instrumentation like cryo-FIB milling and cryo-CLEM, there has been rapid progress in hardware and software improvements concerning cryo-TEM data collection and image processing.

Biological samples are composed of low atomic weight materials like carbon and water and inherently possess low contrast, rendering most of the matter invisible when images are acquired in focus. To overcome this challenge, tilt series are typically obtained by defocusing the objective lens (Sigworth, 2016). While defocusing improves contrast, it comes at the expense of image resolution. The blurring effect caused by defocus reduces the ability to resolve high-resolution information in the reconstructions. Mathematically, the contrast transfer function (CTF) explains this inverse relationship between contrast and resolution in cryo-EM. Precise defocus determination allows for partial correction of CTF, contributing to enhanced resolution during data processing. Ongoing technical advancements aim to introduce phase plates that eliminate the need for defocus-based image acquisition to achieve contrast (Danev & Baumeister, 2017; Khoshouei et al., 2017; Wang & Fan, 2019). These phase plates consist of a thin carbon film with a small central aperture positioned at the back focal plane of the TEM's objective lens. By recovering low spatial frequencies, these phase plates increase contrast at low resolutions without the requirement for defocusing (Imhof et al., 2019; Sharp et al., 2016; Si et al., 2018). The development of laser phase plates is particularly promising, as they utilize a high-intensity laser beam to generate contrast without the accumulation of electrostatic charge, which is typical of carbon film–based phase plates (Schwartz et al., 2019). This innovation holds excellent potential for dose-efficient cryo-EM data collection, enabling imaging of thicker specimens without milling.

A critical technical advance that drove much of the "resolution revolution" in both single-particle cryo-EM and cryo-ET is in the field of detectors, particularly direct electron detectors (DEDs), for cryo-ET data acquisition (Kuijper et al., 2015; McMullan et al., 2016; Mendez et al., 2019). New improvements in DEDs allow for enhanced signal-to-noise ratio and higher frame rates for rapid tilt series data collection. With the current state-of-the-art cryo-EM instruments, it is possible to collect a single tilt series in less than a minute or multiple tilt series simultaneously. While such modes of data collection are not yet a standard practice owing to technical hurdles, rapid advances are being made to turn this into a reality.

Software Advances in Tilt Series Acquisition

The approach to collecting tilt series is influenced by several factors, including the location of features of interest and the sample's thickness. These factors, in turn, determine the acquisition parameters, including tilt range, tilt increment, dose, and magnification. In the traditional tilt scheme, the sample is usually tilted across a range of angles, such as −60 to +60 degrees, from the maximum tilt angle on one side to the other. However, in this scheme, a significant portion of the initial dose is used for capturing images at high tilt angles, where the sample is the thickest. As the imaging progresses to lower tilt angles, the sample has already endured considerable radiation exposure, leading to compromised resolution and a subsequent loss in image quality. Another prevalent approach to acquiring tilt series is the bidirectional scheme. In this technique, images are captured from 0 degrees, then progressing to the maximum angle on one side, for instance, +60 degrees. Subsequently, the process is repeated in the opposite direction, moving to −60 degrees. Despite the advantage of capturing initial images when the sample is thinner, this approach presents a challenge: The cumulative exposure significantly varies for the two halves of the low tilt angles. This discrepancy can result in difficulties when aligning tilt series or can even introduce artifacts in the tomographic reconstructions. Hybrid schemes often yield better results, such as acquiring images spanning from −20 to +60 degrees and then reversing the direction.

A highly effective strategy for data collection is the "dose symmetric scheme." This approach entails initiating the tilt series acquisition at 0 degrees and subsequently oscillating between positive and negative tilt angles in predefined increments (e.g., 0, 3, 6, −3, −6, and so forth) (Hagen et al., 2017). This scheme maximizes the initial dose at lower tilts, preserving the high-resolution information for subsequent subtomogram analysis (Turoňová et al., 2020).

Furthermore, apart from refining the collection strategy for individual tilt series, endeavors are underway to enhance the acquisition of multiple tilt series, thereby boosting the efficiency of data collection. An innovative method uses fast detectors in conjunction with a new single-axis tilt holder, enabling the acquisition of continuous tilt series within mere minutes. This technique, known as fast incremental single exposure (FISE), entails the camera's continuous movie acquisition while regulating sample exposure through electron beam blanking. This approach economizes time by eliminating tasks such as extensive tracking, autofocusing, and beam settling (Chreifi et al., 2019; Eisenstein et al., 2019).

Additional proposed advancements revolve around the simultaneous collection of multiple tilt series to enhance throughput. In single-particle cryo-EM, scientists opt for beam-image shift instead of stage shift to collect data, primarily due to issues associated with stage stability. This is because the stage needs time to stabilize after any movement, impeding efficiency in data acquisition. Image shifts allow the user to collect data on multiple positions within the same hole and adjacent holes simultaneously. Subsequently, modern software packages can estimate and correct the beam tilt (Cheng et al., 2018). This technique has been recently extended to cryo-ET data collection, which offers the advantages of "high-quality and high-speed collection" with the ability to achieve sub-nanometer subtomogram averages and the acquisition of each tilt series taking only a few minutes. In the BISECT (beam image-shift electron cryo-tomography) method, researchers applied image shifts to capture images of multiple areas while maintaining a specific stage position (Bouvette et al., 2021). This was achieved by incorporating relevant geometric constraints into the process and applied to an *in vitro* tomography dataset. Another approach, called "multishot tomography," allows the acquisition of multiple tilt series simultaneously along the tilt axis by applying image shifts (Khavnekar et al., 2023).

Eisenstein et al. developed another remarkable technique called parallel cryo-electron tomography or PACE-tomo, which extends the capability of acquiring tilt series in parallel from various regions on the FIB-milled lamella beyond the confines of the tilt axis (Eisenstein et al., 2023). PACE-tomo leverages beam image shifts and a sophisticated geometric model to collect tilt series concurrently. This approach offers a distinct advantage by enabling data collection from regions of the lamella that were previously inaccessible due to challenges related to focusing and tracking. Using PACE-tomo, Eisenstein et al. acquired 21 tilt series in about 1.5 hours on a single FIB-milled lamella and achieved an 8.2 Å resolution map of the eukaryotic ribosome using subtomogram analysis from ~5k particles (Eisenstein et al., 2023).

A complementary technique to PACE-tomo is montage tomography, which allows researchers to capture tilt series across almost the entirety of a lamella at high magnification without compromising resolution. So far, montage tomography has primarily been applied to resin-embedded specimens due to concerns about radiation damage in cryo-preserved samples. This is particularly relevant because the montage tomography setup necessitates repetitive exposure of different sample parts to the electron beam, which is crucial for stitching during the subsequent reconstruction process. By optimizing acquisition settings, researchers have demonstrated proof-of-concept of montage tomography to cryo-preserved specimens, including FIB-milled lamella samples (Peck et al., 2022; Yang et al., 2022). Further development of the method will allow for comprehensive imaging of more extensive fields of view. It will be particularly advantageous in investigating rare cellular events and bridging the gap between achievable resolution and the field of view captured through cryo-ET.

Examples of *In Situ* Cryo-ET Studies

Combining cryo-ET and high-resolution structures determined from purified proteins using single-particle cryo-EM is a boon for structural biology. Notably, purified protein structures obtained *in vitro* can reach exceptionally high resolutions, even down to 1.19 Å, enabling the visualization of individual amino acids and hydrogen bonds. However, these structures lack the cellular context, raising uncertainty about their conformation within a cell. On the other hand, cryo-ET-derived structures operate at lower resolutions, typically within 7 to 12 Å, limiting the visualization of individual amino acids. Nonetheless, these two techniques can synergistically provide a high-resolution perspective on protein organization within the cellular milieu when harnessed in tandem. In this context, we will delve into illustrative examples where cryo-ET has significantly enriched our understanding of intriguing biological inquiries.

Structure of Nuclear Pore Complexes

All eukaryotic cells feature a nucleus that protects the genome and is enclosed by two layers of nuclear envelope membranes that segregate chromosomes from ribosomes and metabolic enzymes in the cytoplasm. This compartmentalization necessitates the transport of mRNA produced in the nucleus to the cytoplasm for translation into protein and the import of proteins across these membranes to facilitate transcription, DNA replication, or repair. A critical element in this process is the nuclear pore complex (NPC), a large molecular weight assembly of approximately 30 nucleoporins (Nups) that serves as the gateway regulating the movement of large molecules into and out of the nucleus (Li et al., 2016). The NPC assembles into a scaffold comprising three coaxial rings encompassing a central transport channel. Understanding the NPC's structure is imperative to grasp the intricate mechanisms of nuclear import and export.

Recent observations of NPCs in their near-native state using the cryo-FIB-ET pipeline in both human and yeast cells have provided a high-resolution depiction of the complex within its native context (Akey et al., 2022; Allegretti et al., 2020; Schuller et al., 2021). A comprehensive study combining structures obtained from purified NPCs and cellular NPCs in yeast revealed that the inner ring's diameter is 58.5 nm, considerably larger than initially determined using purified proteins (~45 nm) (Akey et al., 2022, Schuller et al., 2021) (Figure 30.2A). Additionally, this investigation uncovered the existence of three diverse structural forms of NPCs within a single cell that presumably serve specialized functions in transport. This discovery underscores how a cell's environment can profoundly shape the architecture of protein complexes and the importance of visualizing cellular assemblies in their native environment in contrast to relying solely on purified proteins. In another study, Mosalaganti et al. combined AI-based structure prediction with single-particle cryo-EM maps, cryo-FIB-ET coupled with subtomogram analysis, and integrative modeling to

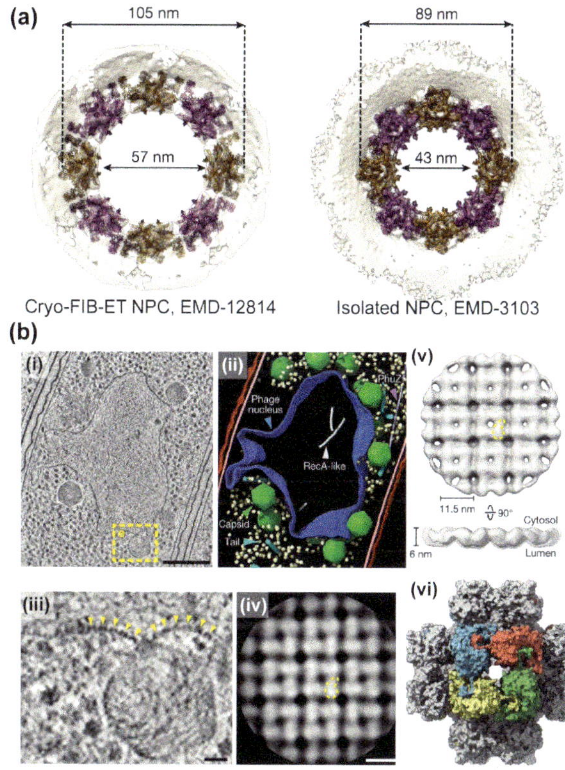

FIGURE 30.2 Examples of structures obtained using cryo-FIB-ET pipeline. (A) Comparison of the density map of the nuclear pore complex obtained from cryo-FIB-ET study in DLD-1 cells and that from single-particle analysis of isolated nuclear envelopes from HeLa cells. Diameters of the central channel and the inner ring are indicated. Reproduced from Schuller et al., 2021 under a Creative Commons Attribution 4.0 International License. (B) (i) Tomographic slice of an FIB-milled 201phi2-1 infected *Pseudomonas chlororaphis* showing the "phage nucleus." Scale bar: 250 nm. (ii) Segmentation of the tomogram in (i) with the phage nucleus (blue), capsids (green), phage tails (cyan), ribosomes (yellow), and bacterial membranes (pink, burgundy) indicated. (iii) Zoomed-in view of the inset in (i) with arrows indicating the repetitive units of the chimallin shell. Scale bar: 25 nm. (iv) Subtomogram average of the phage shell with a single comma-shaped subunit highlighted in yellow. (v) Subtomogram average of the surface of the phage nuclear shell as seen from the cytosolic and side views. (vi) *In vitro* cryo-EM reconstruction of the chimallin cubic assembly. (Reproduced from Laughlin et al., 2022 under a Creative Commons Attribution 4.0 International License.)

reveal a dynamic model of the human nuclear pore complex (Mosalaganti et al., 2022).

Structure of the Proteinaceous Shell Formed during Jumbo Phage Replication in Bacteria

Cryo-FIB milling coupled with cryo-ET was instrumental in discovering the "phage nucleus." (Chaikeeratisak et al., 2017, 2019). Certain jumbo phages (genome size > 200 kb) construct a sophisticated compartment termed the "phage nucleus" within bacteria during replication. Analogous to a eukaryotic nucleus, this specialized structure encapsulates proteins involved in phage DNA replication, repair, and transcription. Metabolic enzymes and ribosomes remain segregated in the cytoplasm. Notably, the phage nucleus serves as a safeguard, shielding phage DNA from bacterial defense mechanisms such as CRISPR/Cas9 (Mendoza et al., 2020). However, distinct from its eukaryotic counterpart, the phage nucleus is orchestrated by a protein named chimallin instead of being enveloped by a double membrane atop a protein scaffold.

Subtomogram analysis on cryo-FIB-milled tomograms of *Pseudomonas chlororaphis* cells infected by the 201phi2-1 jumbo-phage led to a 24 Å resolution reconstruction of the phage nuclear shell, unveiling its distinct square lattice structure (Laughlin et al., 2022). Concurrently, the researchers purified the chimallin protein, resulting in a 3.1 Å resolution single-particle reconstruction (Figure 30.2B). This heightened resolution facilitated the atomic modeling of the chimallin protomer, uncovering its intricate structure. Fitting the high-resolution single-particle reconstruction into the low-resolution subtomogram analysis reconstruction revealed the self-assembly mechanism of the phage shell into flexible sheets, replete with small pores measuring about 2 nm (Laughlin et al., 2022). These pores prove large enough to facilitate the passage of essential metabolites for synthesizing DNA and RNA into the nucleus.

OUTLOOK

As of August 2023, the Electron Microscopy Data Bank (EMDB) houses a collection of approximately 29,000 entries. Among these entries, around 18,000 showcase reported resolutions equal to or exceeding 5 Å. However, most of these entries stem from the resolution revolution from single-particle cryo-EM. In 2022, 85% of the maps deposited to the EMDB were generated through single-particle analysis, underscoring its dominant role in the field. In contrast, subtomogram analysis contributed to only a modest 5.7% of the entries during the same period. (These trends can be accessed online at https://www.ebi.ac.uk/emdb/statistics/emdb_entries_year.) While subtomogram analysis has yielded notably enhanced resolutions, even approaching the sub-nanometer and near-atomic scale, the cryo-ET workflow remains constrained by multiple steps and a relatively low success rate at each stage that cumulatively affect the overall throughput of the process.

Regarding sample preparation, the advancement of rapid freezing methods and the refinement of existing high-pressure freezing techniques promise to expand the scope of cryo-ET to encompass a broader spectrum of specimens while simultaneously enhancing vitrification quality. Anticipated developments, such as the availability of commercially produced micropatterned grids optimized for various cell types, offer a potential avenue for elevating throughput and seamlessly integrating with the successive stages of cryo-ET workflows.

The emergence of cutting-edge technologies, particularly in plasma FIB/SEM, presents a promising avenue to address some of these challenges. These advancements hold the potential to accelerate milling times and concurrently mitigate sample damage, thereby improving the efficiency and efficacy of the cryo-ET workflow. In this context, developing novel microscopy platforms like Arctis (Thermo Fisher Scientific) signifies a significant leap forward. These innovations offer substantial promise by integrating cryo-plasma FIB with cryo-light microscopy and incorporating a seamless autoloader mechanism for direct sample transfer between the cryo-FIB and cryo-TEM instruments.

Significant challenges persist in localizing specific macromolecules within cryo-ET data. Addressing these challenges necessitates advances in the field of cryo-CLEM, particularly through the integration of cryo-superresolution microscopy techniques. Simultaneously, endeavors are underway to devise novel molecular tags that facilitate the direct visualization of target proteins within cryo-ET datasets. One innovative approach, developed in the context of *E. coli*, involves tagging the protein of interest with ferritin (Wang et al., 2011). When cultivated with iron, iron-loaded ferritin particles become visible in the cryo-electron tomogram, enabling the precise localization of the tagged molecule. However, challenges have emerged in employing the ferritin tag. Notably, its considerable molecular weight poses the risk of protein mislocalization, while the tag's multivalent nature may lead to undesired clustering.

With the growing prospect of accumulating an increasing volume of cryo-ET data, the imperative for automation in data processing stages becomes evident. This chapter has not discussed advances in data processing and subtomogram analysis. Readers can refer to other reviews and articles on this topic (Burt et al., 2021b, 2021a; Chen et al., 2019; Hylton & Swulius, 2021; Pyle & Zanetti, 2021; Schur, 2019; Wan & Briggs, 2016; Young & Villa, 2023). Addressing the intricate challenges posed by automating tasks like tomogram reconstruction, particle picking, and subtomogram analysis is pivotal in the field. While some concepts and automation strategies have been transposed from the single-particle analysis workflow, a pressing need arises to develop novel machine-learning algorithms tailored to tasks like data segmentation, automatic reconstruction, and macromolecule localization in cryo-ET data. The rapid strides witnessed in machine learning applications, exemplified by AlphaFold2 (Jumper et al., 2021) in predicting protein structures, augur well for these advancements.

In sum, the evolving landscape of cryo-ET beckons with transformative potential as technologies, methodologies, and automation converge to usher in a new era of precision, throughput, and insight in exploring cellular biology. Addressing the current limitations of the technique holds the promise of bridging structural and cell biology and expanding the horizons of *in situ* structural biology to unlock a deeper understanding of cellular intricacies and mechanisms. Integration with other biological techniques, including genetics, biochemistry, and mass spectrometry, will further unleash the power of *in situ* cryo-ET to build comprehensive cellular landscapes at unprecedented resolution.

ACKNOWLEDGMENTS

Kanika Khanna was a Simons Foundation Awardee of the Life Sciences Research Foundation (2022-23).

REFERENCES

Abbe, E. (1873). Beiträge zur Theorie des Mikroskops und der mikroskopischen Wahrnehmung. *Archiv für Mikroskopische Anatomie*, *9*(1), 413–468. https://doi.org/10.1007/BF02956173

Akey, C. W., Singh, D., Ouch, C., Echeverria, I., Nudelman, I., Varberg, J. M., Yu, Z., Fang, F., Shi, Y., Wang, J., Salzberg, D., Song, K., Xu, C., Gumbart, J. C., Suslov, S., Unruh, J., Jaspersen, S. L., Chait, B. T., Sali, A., … Rout, M. P. (2022). Comprehensive structure and functional adaptations of the yeast nuclear pore complex. *Cell*, *185*(2), 361–378.e25. https://doi.org/10.1016/j.cell.2021.12.015

Al-Amoudi, A., Chang, J.-J., Leforestier, A., McDowall, A., Salamin, L. M., Norlén, L. P., Richter, K., Blanc, N. S., Studer, D., & Dubochet, J. (2004). Cryo-electron microscopy of vitreous sections. *The EMBO Journal*, *23*(18), 3583–3588. https://doi.org/10.1038/sj.emboj.7600366

Allegretti, M., Zimmerli, C. E., Rantos, V., Wilfling, F., Ronchi, P., Fung, H. K. H., Lee, C.-W., Hagen, W., Turoňová, B., Karius, K., Börmel, M., Zhang, X., Müller, C. W., Schwab, Y., Mahamid, J., Pfander, B., Kosinski, J., & Beck, M. (2020). In-cell architecture of the nuclear pore and snapshots of its turnover. *Nature*, *586*(7831), 796–800. https://doi.org/10.1038/s41586-020-2670-5

Berger, C., Dumoux, M., Glen, T., Yee, N. B. -y, Mitchels, J. M., Patáková, Z., Darrow, M. C., Naismith, J. H., & Grange, M. (2023). Plasma FIB milling for the determination of structures in situ. *Nature Communications*, *14*(1), 629. https://doi.org/10.1038/s41467-023-36372-9

Berger, C., Premaraj, N., Ravelli, R. B. G., Knoops, K., López-Iglesias, C., & Peters, P. J. (2023). Cryo-electron tomography on focused ion beam lamellae transforms structural cell biology. *Nature Methods*, *20*(4),499–511. https://doi.org/10.1038/s41592-023-01783-5

Booy, F. P., & Pawley, J. B. (1993). Cryo-crinkling: What happens to carbon films on copper grids at low temperature. *Ultramicroscopy*, *48*(3), 273–280. https://doi.org/10.1016/0304-3991(93)90101-3

Bouvette, J., Liu, H.-F., Du, X., Zhou, Y., Sikkema, A. P., da Fonseca Rezende e Mello, J., Klemm, B. P., Huang, R., Schaaper, R. M., Borgnia, M. J., & Bartesaghi, A. (2021). Beam image-shift accelerated data acquisition for near-atomic resolution single-particle cryo-electron tomography. *Nature Communications*, *12*(1), 1957. https://doi.org/10.1038/s41467-021-22251-8

Burt, A., Gaifas, L., Dendooven, T., & Gutsche, I. (2021a). A flexible framework for multi-particle refinement in cryo-electron tomography. *PLOS Biology*, *19*(8), e3001319. https://doi.org/10.1371/journal.pbio.3001319

Burt, A., Gaifas, L., Dendooven, T., & Gutsche, I. (2021b). Tools enabling flexible approaches to high-resolution subtomogram averaging bioRxiv. https://doi.org/10.1101/2021.01.31.428990

Chaikeeratisak, V., Khanna, K., Nguyen, K. T., Sugie, J., Egan, M. E., Erb, M. L., Vavilina, A., Nonejuie, P., Nieweglowska, E., Pogliano, K., Agard, D. A., Villa, E., & Pogliano, J. (2019). Viral capsid trafficking along treadmilling tubulin filaments in bacteria. *Cell*, *177*(7), 1771–1780.e12. https://doi.org/10.1016/j.cell.2019.05.032

Chaikeeratisak, V., Nguyen, K., Khanna, K., Brilot, A. F., Erb, M. L., Coker, J. K. C., Vavilina, A., Newton, G. L., Buschauer, R., Pogliano, K., Villa, E., Agard, D. A., & Pogliano, J. (2017). Assembly of a nucleus-like structure during viral replication in bacteria. *Science*, *355*(6321), 194–197. https://doi.org/10.1126/science.aal2130

Chang, Y.-W., Chen, S., Tocheva, E. I., Treuner-Lange, A., Löbach, S., Søgaard-Andersen, L., & Jensen, G. J. (2014). Correlated cryogenic photoactivated localization microscopy and cryo-electron tomography. *Nature Methods*, *11*(7), 737–739. https://doi.org/10.1038/nmeth.2961

Chen, M., Bell, J. M., Shi, X., Sun, S. Y., Wang, Z., & Ludtke, S. J. (2019). A complete data processing workflow for cryo-ET and subtomogram averaging. *Nature Methods*, *16*(11), 1161–1168. https://doi.org/10.1038/s41592-019-0591-8

Cheng, A., Eng, E. T., Alink, L., Rice, W. J., Jordan, K. D., Kim, L. Y., Potter, C. S., & Carragher, B. (2018). High resolution single particle cryo-electron microscopy using beam-image shift. *Journal of Structural Biology*, *204*(2), 270–275. https://doi.org/10.1016/j.jsb.2018.07.015

Cheng, Y. (2018). Single particle cryo-EM – how did it get here and where will it go. *Science (New York, N.Y.)*, *361*(6405), 876–880. https://doi.org/10.1126/science.aat4346

Chreifi, G., Chen, S., Metskas, L. A., Kaplan, M., & Jensen, G. J. (2019). Rapid tilt-series acquisition for electron cryotomography. *Journal of Structural Biology*, *205*(2), 163–169. https://doi.org/10.1016/j.jsb.2018.12.008

Chua, E. Y. D., Mendez, J. H., Rapp, M., Ilca, S. L., Tan, Y. Z., Maruthi, K., Kuang, H., Zimanyi, C. M., Cheng, A., Eng, E. T., Noble, A. J., Potter, C. S., & Carragher, B. (2022). Better, faster, cheaper: Recent advances in cryo–electron microscopy. *Annual Review of Biochemistry*, *91*(1), 1–32. https://doi.org/10.1146/annurev-biochem-032620-110705

Cuny, A. P., Schlottmann, F. P., Ewald, J. C., Pelet, S., & Schmoller, K. M. (2022). Live cell microscopy: From image to insight. *Biophysics Reviews*, *3*(2), 021302. https://doi.org/10.1063/5.0082799

Dahlberg, P. D., & Moerner, W. E. (2021). Cryogenic super-resolution fluorescence and electron microscopy correlated at the nanoscale. *Annual Review of Physical Chemistry*, *72*(1), 253–278. https://doi.org/10.1146/annurev-physchem-090319-051546

Dahlberg, P. D., Saurabh, S., Sartor, A. M., Wang, J., Mitchell, P. G., Chiu, W., Shapiro, L., & Moerner, W. E. (2020). Cryogenic single-molecule fluorescence annotations for electron tomography reveal in situ organization of key proteins in Caulobacter. *Proceedings of the National Academy of Sciences*, *117*(25), 13937–13944. https://doi.org/10.1073/pnas.2001849117

Danev, R., & Baumeister, W. (2017). Expanding the boundaries of cryo-EM with phase plates. *Current Opinion in Structural Biology*, *46*, 87–94. https://doi.org/10.1016/j.sbi.2017.06.006

D'Arcangelo, E., & McGuigan, A. P. (2015). Micropatterning strategies to engineer controlled cell and tissue architecture in vitro. *BioTechniques*, *58*(1), 13–23. https://doi.org/10.2144/000114245

DeRosier, D. J. (2021). Where in the cell is my protein? *Quarterly Reviews of Biophysics*, *54*, e9. https://doi.org/10.1017/S003358352100007X

Eisenstein, F., Danev, R., & Pilhofer, M. (2019). Improved applicability and robustness of fast cryo-electron tomography data acquisition. *Journal of Structural Biology*, *208*(2), 107–114. https://doi.org/10.1016/j.jsb.2019.08.006

Eisenstein, F., Yanagisawa, H., Kashihara, H., Kikkawa, M., Tsukita, S., & Danev, R. (2023). Parallel cryo electron tomography on in situ lamellae. *Nature Methods*, *20*(1), 131–138. https://doi.org/10.1038/s41592-022-01690-1

Engel, L., Gaietta, G., Dow, L. P., Swif, M. F., Pardon, G., Volkmann, N., Weis, W. I., Hanein, D., & Pruitt, B. L. (2019). Extracellular matrix micropatterning technology for whole cell cryogenic electron microscopy studies. *Journal of Micromechanics and Microengineering*, *29*(11), 115018. https://doi.org/10.1088/1361-6439/ab419a

Engel, L., Vasquez, C. G., Montabana, E. A., Sow, B. M., Walkiewicz, M. P., Weis, W. I., & Dunn, A. R. (2021). Lattice micropatterning for cryo-electron tomography studies of cell-cell contacts. *Journal of Structural Biology*, *213*(4), 107791. https://doi.org/10.1016/j.jsb.2021.107791

Engel, L., Zaoralova, M., Dunn, A. R., & Oliver, S. (2023). Extracellular filaments revealed by affinity capture cryo-electron tomography of lymphocytes. bioRxiv. https://doi.org/10.1101/2023.08.05.552110

Forbest, R. G. (1997). Understanding how the liquid-metal ion source works. *Vacuum*, *48*(1), 85–97. https://doi.org/10.1016/S0042-207X(96)00227-8

Gorelick, S., Buckley, G., Gervinskas, G., Johnson, T. K., Handley, A., Caggiano, M. P., Whisstock, J. C., Pocock, R., & de Marco, A. (2019). PIE-scope, integrated cryo-correlative light and FIB/SEM microscopy. *ELife*, *8*, e45919. https://doi.org/10.7554/eLife.45919

Grimm, R., Typke, D., Bärmann, M., & Baumeister, W. (1996). Determination of the inelastic mean free path in ice by examination of tilted vesicles and automated most probable loss imaging. *Ultramicroscopy*, *63*(3), 169–179. https://doi.org/10.1016/0304-3991(96)00035-6

Guaita, M., Watters, S. C., & Loerch, S. (2022). Recent advances and current trends in cryo-electron microscopy. *Current Opinion in Structural Biology*, *77*, 102484. https://doi.org/10.1016/j.sbi.2022.102484

Hagen, W. J. H., Wan, W., & Briggs, J. A. G. (2017). Implementation of a cryo-electron tomography tilt-scheme optimized for high resolution subtomogram averaging. *Journal of Structural Biology*, *197*(2), 191–198. https://doi.org/10.1016/j.jsb.2016.06.007

Harapin, J., Börmel, M., Sapra, K. T., Brunner, D., Kaech, A., & Medalia, O. (2015). Structural analysis of multicellular organisms with cryo-electron tomography. *Nature Methods*, *12*(7), 634–636. https://doi.org/10.1038/nmeth.3401

Hong, Y., Song, Y., Zhang, Z., & Li, S. (2023). Cryo-electron tomography: The resolution revolution and a surge of in situ virological discoveries. *Annual Review of Biophysics*, *52*(1), 339–360. https://doi.org/10.1146/annurev-biophys-092022-100958

Hsieh, C., Schmelzer, T., Kishchenko, G., Wagenknecht, T., & Marko, M. (2014). Practical workflow for cryo focused-ion-beam milling of tissues and cells for cryo-TEM tomography. *Journal of Structural Biology*, *185*(1), 32–41. https://doi.org/10.1016/j.jsb.2013.10.019

Hutchings, J., Basiashvili, T., & Villa, E. (2023). Cryo-FIB Milling protocol for mammalian cells. https://protocols.io/view/cryo-fib-milling-protocol-for-mammalian-cells-cxtyxnpw

Hylton, R. K., & Swulius, M. T. (2021). Challenges and triumphs in cryo-electron tomography. *IScience*, *24*(9), 102959. https://doi.org/10.1016/j.isci.2021.102959

Imhof, S., Zhang, J., Wang, H., Bui, K. H., Nguyen, H., Atanasov, I., Hui, W. H., Yang, S. K., Zhou, Z. H., & Hill, K. L. (2019). Cryo electron tomography with volta phase plate reveals novel structural foundations of the 96-nm axonemal repeat in the pathogen *Trypanosoma brucei*. *ELife*, *8*, e52058. https://doi.org/10.7554/eLife.52058

Jacquemet, G., Carisey, A. F., Hamidi, H., Henriques, R., & Leterrier, C. (2020). The cell biologist's guide to super-resolution microscopy. *Journal of Cell Science*, *133*(11), jcs240713. https://doi.org/10.1242/jcs.240713

Jumper, J., Evans, R., Pritzel, A., Green, T., Figurnov, M., Ronneberger, O., Tunyasuvunakool, K., Bates, R., Žídek, A., Potapenko, A., Bridgland, A., Meyer, C., Kohl, S. A. A., Ballard, A. J., Cowie, A., Romera-Paredes, B., Nikolov, S., Jain, R., Adler, J., … Hassabis, D. (2021). Highly accurate protein structure prediction with AlphaFold. *Nature*, *596*(7873), 583–589. https://doi.org/10.1038/s41586-021-03819-2

Jun, S., Ro, H.-J., Bharda, A., Kim, S. I., Jeoung, D., & Jung, H. S. (2019). Advances in cryo-correlative light and electron microscopy: Applications for studying molecular and cellular events. *The Protein Journal*, *38*(6), 609–615. https://doi.org/10.1007/s10930-019-09856-1

Kelley, K., Raczkowski, A. M., Klykov, O., Jaroenlak, P., Bobe, D., Kopylov, M., Eng, E. T., Bhabha, G., Potter, C. S., Carragher, B., & Noble, A. J. (2022). Waffle method: A general and flexible approach for improving throughput in FIB-milling. *Nature Communications*, *13*(1), 1857. https://doi.org/10.1038/s41467-022-29501-3

Khanna, K., & Villa, E. (2022). Revealing bacterial cell biology using cryo-electron tomography. *Current Opinion in Structural Biology*, *75*, 102419. https://doi.org/10.1016/j.sbi.2022.102419

Khavnekar, S., Wan, W., Majumder, P., Wietrzynski, W., Erdmann, P. S., & Plitzko, J. M. (2023). Multishot tomography for high-resolution in situ subtomogram averaging. *Journal of Structural Biology*, *215*(1), 107911. https://doi.org/10.1016/j.jsb.2022.107911

Khoshouei, M., Pfeffer, S., Baumeister, W., Förster, F., & Danev, R. (2017). Subtomogram analysis using the Volta phase plate. *Journal of Structural Biology*, *197*(2), 94–101. https://doi.org/10.1016/j.jsb.2016.05.009

Klein, S., Wimmer, B. H., Winter, S. L., Kolovou, A., Laketa, V., & Chlanda, P. (2021). Post-correlation on-lamella cryo-CLEM reveals the membrane architecture of lamellar bodies. *Communications Biology*, 4(1), 1–12.

Klumpe, S., Fung, H. K., Goetz, S. K., Zagoriy, I., Hampoelz, B., Zhang, X., Erdmann, P. S., Baumbach, J., Müller, C. W., Beck, M., Plitzko, J. M., & Mahamid, J. (2021). A modular platform for automated cryo-FIB workflows. *ELife*, *10*, e70506. https://doi.org/10.7554/eLife.70506

Klumpe, S., Kuba, J., Schioetz, O. H., Erdmann, P. S., Rigort, A., & Plitzko, J. M. (2022). Recent advances in gas injection system-free cryo-FIB lift-out transfer for cryo-electron tomography of multicellular organisms and tissues. *Microscopy Today*, *30*(1), 42–47. https://doi.org/10.1017/S1551929521001528

Kuijper, M., van Hoften, G., Janssen, B., Geurink, R., De Carlo, S., Vos, M., van Duinen, G., van Haeringen, B., & Storms, M. (2015). FEI's direct electron detector developments: Embarking on a revolution in cryo-TEM. *Journal of Structural Biology*, *192*(2), 179–187. https://doi.org/10.1016/j.jsb.2015.09.014

Lam, V., & Villa, E. (2021). Practical Approaches for Cryo-FIB Milling and Applications for Cellular Cryo-Electron Tomography. In T. Gonen & B. L. Nannenga (Eds.), *CryoEM: Methods and Protocols* (pp. 49–82). Springer US. https://doi.org/10.1007/978-1-0716-0966-8_3

Laughlin, T. G., Deep, A., Prichard, A. M., Seitz, C., Gu, Y., Enustun, E., Suslov, S., Khanna, K., Birkholz, E. A., Armbruster, E., McCammon, J. A., Amaro, R. E., Pogliano, J., Corbett, K. D., & Villa, E. (2022). Architecture and self-assembly of the jumbo bacteriophage nuclear shell. bioRxiv. https://doi.org/10.1101/2022.02.14.480162

Li, C., Goryaynov, A., & Yang, W. (2016). The selective permeability barrier in the nuclear pore complex. *Nucleus*, *7*(5), 430–446. https://doi.org/10.1080/19491034.2016.1238997

Li, S., Jia, X., Niu, T., Zhang, X., Qi, C., Xu, W., Deng, H., Sun, F., & Ji, G. (2023). HOPE-SIM, a cryo-structured illumination fluorescence microscopy system for accurately targeted cryo-electron tomography. *Communications Biology*, *6*(1), 474. https://doi.org/10.1038/s42003-023-04850-x

Lucas, B. A., & Grigorieff, N. (2023). Quantification of gallium cryo-FIB milling damage in biological lamellae. *Proceedings of the National Academy of Sciences*, *120*(23), e2301852120. https://doi.org/10.1073/pnas.2301852120

Lučić, V., Leis, A., & Baumeister, W. (2008). Cryo-electron tomography of cells: Connecting structure and function. *Histochemistry and Cell Biology*, *130*(2), 185–196. https://doi.org/10.1007/s00418-008-0459-y

Lučić, V., Rigort, A., & Baumeister, W. (2013). Cryo-electron tomography: The challenge of doing structural biology in situ. *Journal of Cell Biology*, *202*(3), 407–419. https://doi.org/10.1083/jcb.201304193

Mahamid, J., Schampers, R., Persoon, H., Hyman, A. A., Baumeister, W., & Plitzko, J. M. (2015). A focused ion beam milling and lift-out approach for site-specific preparation of frozen-hydrated lamellas from multicellular organisms. *Journal of Structural Biology*, *192*(2), 262–269. https://doi.org/10.1016/j.jsb.2015.07.012

Marko, M., Hsieh, C., Schalek, R., Frank, J., & Mannella, C. (2007). Focused-ion-beam thinning of frozen-hydrated biological specimens for cryo-electron microscopy. *Nature Methods*, *4*(3),215–217. https://doi.org/10.1038/nmeth1014

Martynowycz, M. W., Clabbers, M. T. B., Unge, J., Hattne, J., & Gonen, T. (2021). Benchmarking the ideal sample thickness in cryo-EM. *Proceedings of the National Academy of Sciences*, *118*(49), e2108884118. https://doi.org/10.1073/pnas.2108884118

McDonald, K. L., & Auer, M. (2006). High-pressure freezing, cellular tomography, and structural cell biology. *BioTechniques*, *41*(2), 137–143. https://doi.org/10.2144/000112226

McMullan, G., Faruqi, A. R., & Henderson, R. (2016). Chapter One—Direct Electron Detectors. In R. A. Crowther (Ed.), *Methods in Enzymology* (Vol. 579, pp. 1–17). Academic Press. https://doi.org/10.1016/bs.mie.2016.05.056

Mendez, J. H., Mehrani, A., Randolph, P., & Stagg, S. (2019). Throughput and resolution with a next-generation direct electron detector. *IUCrJ*, *6*(6), 1007–1013. https://doi.org/10.1107/S2052252519012661

Mendoza, S. D., Nieweglowska, E. S., Govindarajan, S., Leon, L. M., Berry, J. D., Tiwari, A., Chaikeeratisak, V., Pogliano, J., Agard, D. A., & Bondy-Denomy, J. (2020). A bacteriophage nucleus-like compartment shields DNA from CRISPR nucleases. *Nature*, *577*(7789), 244–248. https://doi.org/10.1038/s41586-019-1786-y

Morris, J. D., & Payne, C. K. (2019). Microscopy and Cell Biology: New Methods and New Questions. *Annual Review of Physical Chemistry*, *70*(1), 199–218. https://doi.org/10.1146/annurev-physchem-042018-052527

Mosalaganti, S., Obarska-Kosinska, A., Siggel, M., Taniguchi, R., Turoňová, B., Zimmerli, C. E., Buczak, K., Schmidt, F. H., Margiotta, E., Mackmull, M.-T., Hagen, W. J. H., Hummer, G., Kosinski, J., & Beck, M. (2022). AI-based structure prediction empowers integrative structural analysis of human nuclear pores. *Science*, *376*(6598), eabm9506. https://doi.org/10.1126/science.abm9506

Navarro, P. P. (2022). Quantitative cryo-electron tomography. *Frontiers in Molecular Biosciences*, *9*. https://www.frontiersin.org/articles/10.3389/fmolb.2022.934465

Peck, A., Carter, S. D., Mai, H., Chen, S., Burt, A., & Jensen, G. J. (2022). Montage electron tomography of vitrified specimens. *Journal of Structural Biology*, *214*(2), 107860. https://doi.org/10.1016/j.jsb.2022.107860

Pyle, E., & Zanetti, G. (2021). Current data processing strategies for cryo-electron tomography and subtomogram averaging. *Biochemical Journal*, *478*(10), 1827–1845. https://doi.org/10.1042/BCJ20200715

Rigort, A., Bäuerlein, F. J. B., Villa, E., Eibauer, M., Laugks, T., Baumeister, W., & Plitzko, J. M. (2012). Focused ion beam micromachining of eukaryotic cells for cryoelectron tomography. *Proceedings of the National Academy of Sciences of the United States of America*, *109*(12), 4449–4454. https://doi.org/10.1073/pnas.1201333109

Ruska, E. (1987). The development of the electron microscope and of electron microscopy (Nobel Lecture). *Angewandte Chemie International Edition in English*, *26*(7), 595–605. https://doi.org/10.1002/anie.198705953

Saibil, H. R. (2022). Cryo-EM in molecular and cellular biology. *Molecular Cell*, *82*(2), 274–284. https://doi.org/10.1016/j.molcel.2021.12.016

Schaffer, M., Pfeffer, S., Mahamid, J., Kleindiek, S., Laugks, T., Albert, S., Engel, B. D., Rummel, A., Smith, A. J., Baumeister, W., & Plitzko, J. M. (2019). A cryo-FIB lift-out technique enables molecular-resolution cryo-ET within native Caenorhabditis elegans tissue. *Nature Methods*, *16*(8), 757–762. https://doi.org/10.1038/s41592-019-0497-5

Schuller, A. P., Wojtynek, M., Mankus, D., Tatli, M., Kronenberg-Tenga, R., Regmi, S. G., Dip, P. V., Lytton-Jean, A. K. R., Brignole, E. J., Dasso, M., Weis, K., Medalia, O., & Schwartz, T. U. (2021). The cellular environment shapes the nuclear pore complex architecture. *Nature*, *598*(7882),667–671. https://doi.org/10.1038/s41586-021-03985-3

Schur, F. K. (2019). Toward high-resolution in situ structural biology with cryo-electron tomography and subtomogram averaging. *Current Opinion in Structural Biology*, *58*, 1–9. https://doi.org/10.1016/j.sbi.2019.03.018

Schwartz, O., Axelrod, J. J., Campbell, S. L., Turnbaugh, C., Glaeser, R. M., & Müller, H. (2019). Laser phase plate for transmission electron microscopy. *Nature Methods*, *16*(10),1016–1020. https://doi.org/10.1038/s41592-019-0552-2

Sexton, D. L., Burgold, S., Schertel, A., & Tocheva, E. I. (2022). Super-resolution confocal cryo-CLEM with cryo-FIB milling for in situ imaging of Deinococcus radiodurans. *Current Research in Structural Biology*, *4*, 1–9. https://doi.org/10.1016/j.crstbi.2021.12.001

Sharp, T. H., Koster, A. J., & Gros, P. (2016). Heterogeneous MAC initiator and pore structures in a lipid bilayer by phase-plate cryo-electron tomography. *Cell Reports*, *15*(1), 1–8. https://doi.org/10.1016/j.celrep.2016.03.002

Si, Z., Zhang, J., Shivakoti, S., Atanasov, I., Tao, C.-L., Hui, W. H., Zhou, K., Yu, X., Li, W., Luo, M., Bi, G.-Q., & Zhou, Z. H. (2018). Different functional states of fusion protein gB revealed on human cytomegalovirus by cryo electron tomography with Volta phase plate. *PLOS Pathogens*, *14*(12), e1007452. https://doi.org/10.1371/journal.ppat.1007452

Sibert, B. S., Kim, J. Y., Yang, J. E., & Wright, E. R. (2021). Micropatterning transmission electron microscopy grids to direct cell positioning within whole-cell cryo-electron tomography workflows. *JoVE (Journal of Visualized Experiments)*, *175*, e62992. https://doi.org/10.3791/62992

Sigworth, F. J. (2016). Principles of cryo-EM single-particle image processing. *Microscopy*, *65*(1), 57–67. https://doi.org/10.1093/jmicro/dfv370

Smeets, M., Bieber, A., Capitanio, C., Schioetz, O., van der Heijden, T., Effting, A., Piel, É., Lazem, B., Erdmann, P., & Plitzko, J. (2021). Integrated cryo-correlative microscopy for targeted structural investigation in situ. *Microscopy Today*, *29*(6), 20–25. https://doi.org/10.1017/S1551929521001280

Tacke, S., Erdmann, P., Wang, Z., Klumpe, S., Grange, M., Plitzko, J., & Raunser, S. (2021). A streamlined workflow for automated cryo focused ion beam milling. *Journal of Structural Biology*, *213*(3), 107743. https://doi.org/10.1016/j.jsb.2021.107743

Thorn, K. (2016). A quick guide to light microscopy in cell biology. *Molecular Biology of the Cell*, *27*(2), 219–222. https://doi.org/10.1091/mbc.e15-02-0088

Toro-Nahuelpan, M., Zagoriy, I., Senger, F., Blanchoin, L., Théry, M., & Mahamid, J. (2020). Tailoring cryo-electron microscopy grids by photo-micropatterning for in-cell structural studies. *Nature Methods*, *17*(1),50–54. https://doi.org/10.1038/s41592-019-0630-5

Tuijtel, M. W., Kreysing, J. P., Welsch, S., Hummer, G., Beck, M., & Turoňová, B. (2023). Thinner is not always better: Optimising cryo lamellae for subtomogram averaging. bioRxiv. https://doi.org/10.1101/2023.07.31.551274

Turk, M., & Baumeister, W. (2020). The promise and the challenges of cryo-electron tomography. *FEBS Letters*, *594*(20), 3243–3261. https://doi.org/10.1002/1873-3468.13948

Turoňová, B., Hagen, W. J. H., Obr, M., Mosalaganti, S., Beugelink, J. W., Zimmerli, C. E., Kräusslich, H.-G., & Beck, M. (2020). Benchmarking tomographic acquisition schemes for high-resolution structural biology. *Nature Communications*, *11*(1),876. https://doi.org/10.1038/s41467-020-14535-2

Vrbovská, V., Klumpe, S., Thompson, C., Rigort, A., Mitchels, J., Franke, T., Müllerová, M., Kasáková, A., & Hovorka, M. (2023). Fluorescence-guided cryo-lift-out using an integrated fluorescence light microscope and an optimized sample-needle attachment procedure. *Microscopy and Microanalysis*, *29*(Supplement_1), 1051–1052. https://doi.org/10.1093/micmic/ozad067.538

Wagner, F. R., Watanabe, R., Schampers, R., Singh, D., Persoon, H., Schaffer, M., Fruhstorfer, P., Plitzko, J., & Villa, E. (2020). Preparing samples from whole cells using focused-ion-beam milling for cryo-electron tomography. *Nature Protocols*, *15*(6), 2041–2070. https://doi.org/10.1038/s41596-020-0320-x

Wan, W., & Briggs, J. A. G. (2016). Chapter Thirteen—Cryo-Electron Tomography and Subtomogram Averaging. In R. A. Crowther (Ed.), *Methods in Enzymology* (Vol. 579, pp. 329–367). Academic Press. https://doi.org/10.1016/bs.mie.2016.04.014

Wang, C., Wojtynek, M., & Medalia, O. (2023). Structural investigation of eukaryotic cells: From the periphery to the interior by cryo-electron tomography. *Advances in Biological Regulation*, *87*, 100923. https://doi.org/10.1016/j.jbior.2022.100923

Wang, H.-W., & Fan, X. (2019). Challenges and opportunities in cryo-EM with phase plate. *Current Opinion in Structural Biology*, *58*, 175–182. https://doi.org/10.1016/j.sbi.2019.06.013

Wang, Q., Mercogliano, C. P., & Löwe, J. (2011). A ferritin-based label for cellular electron cryotomography. *Structure*, *19*(2), 147–154. https://doi.org/10.1016/j.str.2010.12.002

Watanabe, R., Buschauer, R., Böhning, J., Audagnotto, M., Lasker, K., Lu, T.-W., Boassa, D., Taylor, S., & Villa, E. (2020). The in situ structure of Parkinson's disease-linked LRRK2. *Cell*, *182*(6), 1508–1518.e16. https://doi.org/10.1016/j.cell.2020.08.004

Wolff, G., Limpens, R. W. A. L., Zheng, S., Snijder, E. J., Agard, D. A., Koster, A. J., & Bárcena, M. (2019). Mind the gap: Micro-expansion joints drastically decrease the bending of FIB-milled cryo-lamellae. *Journal of Structural Biology*, *208*(3), 107389. https://doi.org/10.1016/j.jsb.2019.09.006

Yang, J., Vrbovská, V., Franke, T., Sibert, B., Larson, M., Coomes, T., Rigort, A., Mitchels, J., & Wright, E. R. (2023). Precise 3D localization by integrated fluorescence microscopy (iFLM) for cryo-FIB-milling and in-situ *cryo-ET. Microscopy and Microanalysis*, *29*(Supplement_1), 1055–1057. https://doi.org/10.1093/micmic/ozad067.541

Yang, J. E., Larson, M. R., Sibert, B. S., Kim, J. Y., Parrell, D., Sanchez, J. C., Pappas, V., Kumar, A., Cai, K., Thompson, K., & Wright, E. R. (2022). Correlative cryogenic montage electron tomography for comprehensive in-situ whole-cell structural studies. bioRxiv. https://doi.org/10.1101/2021.12.31.474669

Young, L. N., & Villa, E. (2023). Bringing structure to cell biology with cryo-electron tomography. *Annual Review of Biophysics*, *52*(1), 573–595. https://doi.org/10.1146/annurev-biophys-111622-091327

Zachs, T., Schertel, A., Medeiros, J., Weiss, G. L., Hugener, J., Matos, J., & Pilhofer, M. (2020). Fully automated, sequential focused ion beam milling for cryo-electron tomography. *ELife*, *9*, e52286. https://doi.org/10.7554/eLife.52286

Zhang, P. (2013). Correlative Cryo-electron tomography and optical microscopy of cells. *Current Opinion in Structural Biology*, *23*(5),763–770. https://doi.org/10.1016/j.sbi.2013.07.017

Zhang, P. (2019). Advances in cryo-electron tomography and subtomogram averaging and classification. *Current Opinion in Structural Biology*, *58*, 249–258. https://doi.org/10.1016/j.sbi.2019.05.021

31 Cryo-Electron Tomographic Imaging of Membrane Proteins from Mitochondria and Chloroplasts

Tofayel Ahmed and Antara Saha

INTRODUCTION

Mitochondria are the energy-generating organelles in eukaryotic cells. The mitochondrial matrix is isolated from the cytoplasm by two membranes: the outer (OMM) and inner mitochondrial membranes (IMM), which differ in their protein and lipid constituents (Comte et al., 1976; Zick et al., 2009). The OMM is smooth, in contrast to the multiple invaginations that are characteristics of the IMM. The OMM is also generally considered highly permeable because it allows the passage of various ions, metabolites, nucleotides, and uncharged molecules as large as ~5 kDa (Vander Heiden et al., 2000; O'Rourke, 2000). The OMM, being in direct contact with the cytoplasmic environment, can serve as a platform for both mitochondrial protein import and the convergence of many cellular signaling events such as apoptosis (Wallace, 2012). The high dynamism of mitochondria, as evident from the continual morphological changes associated with fusion and fission, are dependent on the proteins that are harbored in the OMM (Hewitt and Whitworth, 2017). The IMM, on the other hand, forms a true barrier to the influx of materials from cytoplasm to mitochondria, and it is structurally more intriguing than the OMM, as the IMM contains many folds, called cristae (singular: crista), inside the matrix (Joubert and Puff, 2021).

On the other hand, the chloroplast is the organelle responsible for oxygenic photosynthesis in plants, algae, and cyanobacteria (Huokko et al., 2021). The chloroplast contains an extensive network of membranes known as the thylakoid that forms sheetlike compartments that in higher plants are divided into the stacked grana regions, where membranes are pressed together except the ones at grana top, bottom and margins, and the unstacked stromal thylakoid region, where all membrane surfaces are directly in contact with the fluidic stroma (Albertsson, 2001; Mechela et al., 2019; Pribil et al., 2014). Due to lateral heterogeneity in constituent membrane protein content, the granal and stromal thylakoid membranes could form specialized membrane domains that foster coordination among different energy conversion reactions (Albertsson, 2001; Andersson and Anderson, 1980). Apart from this heterogeneity, dynamism in protein constituents among the photosynthetic supercomplexes is implicated in optimal photosynthesis under limiting conditions (Li et al., 2020; Rochaix, 2014).

In the sections that follow, we summarize and discuss the topography of various membrane protein complexes in mitochondria and chloroplasts, using mainly the results obtained from conventional transmission electron microscopy (TEM) and cryo-electron tomographic (cryo-ET) studies.

EARLY TEM STUDIES ON MITOCHONDRIAL MEMBRANE

Pioneering TEM work conducted by Palade and Sjöstrand in the 1950s with plastic-embedded tissue sections provided the first glimpse of cristae structure and organization inside mitochondria (Palade, 1953; Sjöstrand, 1953; Zick et al., 2009). Palade, in his so-called baffle model, described cristae as structures that invaginate into the matrix after originating from the IMM and with broad openings at the origin (Palade, 1952, 1953) (Figure 31.1). Sjöstrand put forward a contrasting viewpoint that described cristae as barriers responsible for physically dividing the matrix into many individual compartments (Sjöstrand, 1953, 1956; Zick et al., 2009). This became known as the septa model (Figure 31.1). Between these two models, the model of Palade gained larger acceptance in the scientific community. Apart from the *in situ* TEM imaging work on mitochondria, Palade formulated a mitochondrial purification protocol by differential centrifugation and used this preparation to generate mitochondrial sections using a resin-embedding procedure that he then used for TEM imaging. Palade never published this work, but his note on the observance of membrane features associated with the OMM and IMM in this specimen was an assurance that these organelles can be studied *in vitro* (Palade, 1953). However, Palade's TEM images, low in contrast and resolution, were not suitable for examining mitochondrial membranes for constituent proteins. Soon after, studies of biochemically isolated mitochondria, as made popular by Chance and Williams (Chance and Williams, 1955), captured much attention from microscopists, and consequently TEM imaging of such mitochondrial preparations gained traction among scientists.

An extensive study that employed both isolated mitochondria and native tissue sections, was conducted by Fernandez-Moran et al. by using various organs of multiple animals (Fernandez Moran et al., 1964). The authors used a variety of negative staining procedures to rule out

DOI: 10.1201/9781003326106-38

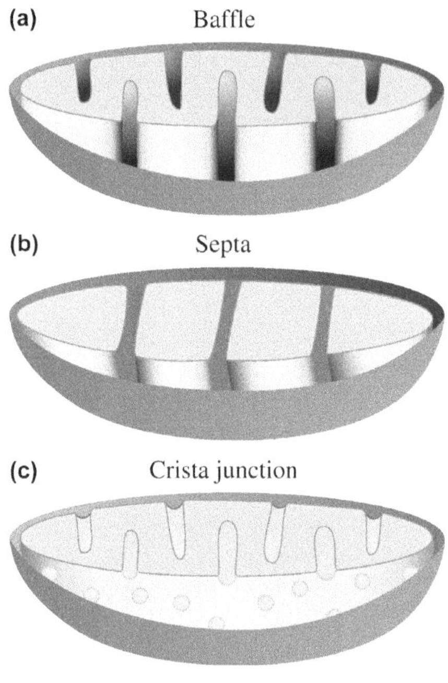

FIGURE 31.1 Models describing inner mitochondrial membrane (IMM) organization. The IMM is mainly explained by three models in the literature, namely (a) 'Baffle model', according to which cristae are formed by convoluted IMM with broad openings near their origin (Palade, 1953) (Palade, 1952), (b) 'Septa model', that describes cristae as responsible for compartmentalization of mitochondrial matrix (Sjöstrand, 1953), and (c) 'Crista junction model', according to which cristae have narrow tubular connections to mitochondrial inner boundary membrane (Daems and Wisse, 1966). Adapted from Zick et al., 2009.)

any artifactual features in the images and also to discover methods that would generate informative contrast. The authors performed imaging of mitochondrial membranes either directly from thin sections of these frozen tissues or from intact mitochondria that were isolated from them, or after membrane fractions were generated from mechanically broken mitochondria. This study was highlighted by the clear visualization and description of numerous "elementary particles" that were arranged in paired rows, as also observed and termed by Fernandez-Moran earlier, which consisted of three structural units: an 80–100 Å diameter spherical or polyhedral head, a 50 Å long and 30–40 Å wide cylindrical stalk, and a base piece of 40 × 110 Å^2 surface area (Fernandez-Moran, 1962; Fernandez Moran et al., 1964). These elementary particles were universally present in all samples studied by the authors, and despite the harsh detergent treatment of mitochondrial membranes the arrays remained intact. The studies conducted by Kagawa and Racker provided biochemical support that these particles were indeed ATP synthases (Kagawa and Racker, 1966). Thus, mitochondrial ATP synthases (mtF_1F_0), due to their conspicuous paired arrangement in cristae membranes, became the first membrane proteins to be identified unambiguously in the TEM images of mitochondria.

The initial knowledge about cristae architecture arose from TEM studies on thick tissue sections and isolated mitochondria, which provided inherently limited information because the generated images were a 2D projection of a 3D object. Another approach that was tried by scientists involved serial thin-sectioning and image reconstruction of these sections, but this technique was limited in Z-plane resolution because the sections were not thin enough to create sufficiently Z-sampled cristae images (Daems and Wisse, 1966; Winslow et al., 1991). An alternative approach involving a pair of stereo images from a thick tissue section was employed to elucidate the complexity of cristae architecture, but fine details were lost in the generated structure due to overlapping of features in stereo images (Mannella et al., 1994; Marko et al., 1988).

Much later, electron tomography studies conducted by various labs provided improved visualization of cristae 3D structure, and based on these works, the crista junction model gained traction in the community (Mannella et al., 1994, 1997; Perkins et al., 1997a, 1997b, 1998; Zick et al., 2009) (Figure 31.1). According to this model, a crista attaches to the inner boundary membrane, a stretch of IMM apposed to OMM, via a narrow pedicular structure that prevents flattening of cristae during mitochondrial swelling. This model of cristae has gained the most wide acceptance now. (Daems and Wisse, 1966; Zick et al., 2009) (Figure 31.1).

CRYO-TOMOGRAPHIC VISUALIZATION OF MEMBRANE PROTEINS IN INNER MITOCHONDRIAL MEMBRANE (IMM)

Bypassing the traditional TEM methods involving chemical fixation and resin embedding, Nicastro et al. applied cryo-ET to image frozen-hydrated mitochondria (Nicastro et al., 2000). In the volume generated for the mitochondrial matrix that is inherently noisy due to the abundance of electron-dense material, mtF_1F_0 is untraceable due to poor contrast. On the contrary, inside-out vesicles, originating from the breakage and automatic re-ligation of inner mitochondrial membrane fragments during sample preparation, had a much weaker background signal, and the dimeric rows of mtF_1F_0 could be traced on their membranes. However, detailed knowledge about the topography of mtF_1F_0 distribution in membranes, the angular relation between mtF_1F_0 monomers in a dimer, and the incurred membrane bending angle could not be elucidated due to the poor quality of the tomograms.

MITOCHONDRIAL ATP SYNTHASE (MTF_1F_0)

Moving on from a time that witnessed the wide use of traditional TEM and preliminary cryo-tomography studies to describe mitochondrial membranes, starting in 2008, a series of cryo-ET studies had attempted to dissect these membranes in molecular detail (Davies et al., 2011, 2018; Strauss et al., 2008). This happened in parallel with an improvement in electron microscopy hardware and image analysis software (Kühlbrandt, 2014; Fukuda et al., 2023). Most of these studies were conducted on either fractionated mitochondrial membranes or, whole mitochondria that were purified by differential centrifugation. In one such study, Strauss et al. generated mitochondrial membrane fragments and vesicles using rat liver and beef heart, and they performed cryo-tomographic imaging and 3D volume reconstruction (Strauss et al., 2008). Protocols for mitochondrial inner membrane vesicle generation are ubiquitous in literature; they mostly involve, in a stepwise fashion, osmotic swelling or detergent solubilization to carefully remove the outer membrane, pelleting down outer membrane-stripped mitochondria to produce mitoplasts, and applying mechanical disruption in the form of either sonication, freeze/thaw-cycles, or high pressure, to generate the vesicles (Ehmke et al., 2023). The vesicles imaged by Strauss et al. were tubular or lamellar in shape and studded by 10-nm-diameter beadlike particles on their surface near the highly curved regions (Strauss et al., 2008). These beads were readily identified as the catalytic F_1 heads of mtF_1F_0 based on their diameter and the previous descriptions of negatively stained mitochondria (Fernandez-Moran, 1962; Kagawa and Racker, 1966). The tomograms contained appropriate contrast to further identify the central stalk that is connected to the F_0 component. The F_0 component was identified from a contour map that confirmed that it was buried in the membrane. The visualization of the peripheral stalk was on hold until Davies et al. captured its density by subtomogram averaging from *P. anserina* cells (Davies et al., 2011).

Coming back to Strauss et al.'s study, it generated important knowledge about the topography of mtF_1F_0 in the IMM (Strauss et al., 2008). It revealed that the central stalks of dimeric mtF_1F_0 molecules are angled at 70 degrees in rat liver mitochondria and 55 to 95 degrees in bovine heart mitochondria, which illuminate the remarkable flexibility in dimer links which are located along the membrane curvature region (Strauss et al., 2008). In fact, based on the angular relation between the two central stalks of mtF_1F_0 dimer in different organisms, four different types of dimers have been characterized, details of which are discussed elsewhere (Kühlbrandt, 2019). Anyway, the mtF_1F_0 dimers in Strauss et al.'s study often made ~1 μ straight rows consisting of ~80 dimers with ~12 nm separation between them, and these dimers were functional. It was further shown by in-gel activity assay that not only the dimers but higher oligomers of mtF_1F_0, such as tetramers and hexamers, are capable of hydrolyzing ATP (Suhai et al., 2009). This was an indication that despite detergent solubilization that is synonymous with a harsh biochemical treatment of mitochondrial membranes, mtF_1F_0 molecules could retain their ability to hydrolyze ATP, an inherent property of such machines under *in vitro* conditions. It is interesting to speculate what functional significance might be associated with such dimeric-row arrangement, as the monomeric mtF_1F_0 is also fully capable of ATP synthesis. The answer probably lies in the advantage conferred to mitochondria to generate ATP even under limiting conditions (Strauss et al., 2008). It appears that the dimeric rows of mtF_1F_0 bend IMM to give rise to cristae structure that can act as a proton trap to locally increase proton concentration to sustain the necessary proton motive force (PMF) for ATP generation, even when membrane potential ($\Delta\psi$) remains unchanged. Strauss et al. did not observe any monomeric form of mtF_1F_0 in their tomograms. This observation contrasts with prokaryotic and chloroplastic ATP synthases (cF_1F_0) that exist as monomers. Of note, apart from the dimeric rows that appear to be ubiquitous in mitochondria, Mühleip et al. have recently captured hexameric mtF_1F_0 that are arranged in a pentagonal pyramid structure (Mühleip et al., 2021).

Respiratory Complex I

Despite clear visualization of dimeric mtF_1F_0 rows as discussed above, the possibility of artifactual dimer row generation in membrane fragments during disruptive sample preparation could not be ruled out. Davies et al. performed a subsequent cryo-ET study on intact and active mitochondria isolated from animal, plant, and fungal cells to showcase that such dimeric mtF_1F_0 rows are indeed a bona fide feature of IMM (Davies et al., 2011) (Figure 31.2). Apart from mtF_1F_0 rows, this study, probably due to an uncrowded cleaner background of smaller *P. anserina* cells, could further capture many rectangular densities that populate the flat membrane region between cristae edges. Such densities, protruding 15 nm above the membrane level, were identified as respiratory complex I or enzyme supercomplexes where respiratory complex I is a constituent member, and the confidence in such identification was derived by comparing features with proteoliposome-reconstituted complex I density and quantum dot labeling data. A few years later, better-resolved tomograms confirmed that these densities represented the matrix arm of complex I (Davies et al., 2018) (Figure 31.3). The tomograms collected by Davies et al. from 2011 further revealed that the mtF_1F_0 population in a mitochondrion is much higher than complex I, and consequently authors could obtain subtomogram averages for the former but not for the latter. A volume for the elusive peripheral stalk of mtF_1F_0 could finally be observed in such subtomogram averages (Davies et al., 2011).

Respiratory Supercomplex or Respirosomes

Supercomplex formation among mitochondrial respiratory chain components was shown earlier by blue-native polyacrylamide gel electrophoresis (BN-PAGE), negatively stained electron microscopy on gel-extracted bands, and single-particle cryo-EM on detergent-solubilized membrane (Bultema et al., 2009; Heinemeyer et al., 2007; Schägger and Pfeiffer, 2000; Seelert et al., 2006). It is now generally accepted that the individual mitochondrial electron transport chain (mETC) complexes and their supercomplexes coexist in the IMM (Acin-Perez and Enriquez, 2014; Ferna et al., 2008). Tomograms collected by Davies et al. on mitochondria

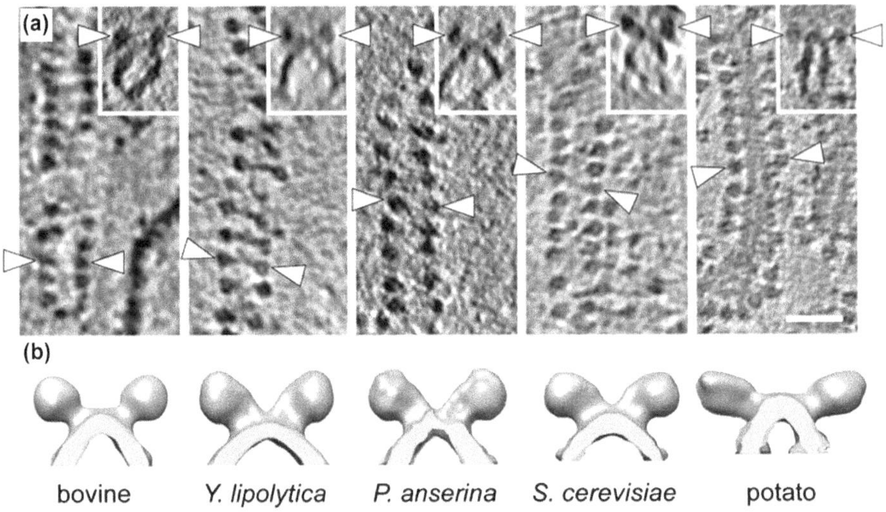

FIGURE 31.2 Dimer rows of mitochondrial ATP synthase (mtF_1F_0) from multiple organisms. (a) Left to right, tomographic slices from bovine heart, Yarrowia lipolytica, Podospora anserina, Saccharomyces cerevisiae, and potato, displaying dimer rows of mtF_1F_0 (white arrowheads) in mitochondrial inner membrane of respective organisms. Insets show dimer position with respect to membrane. (b) Subtomogram averages derived of mtF_1F_0 rows to display flexibility in angular relation between two central stalks of a mtF_1F_0 dimer. Scale bar, 50 nm. (Adapted from Davies et al., 2011)

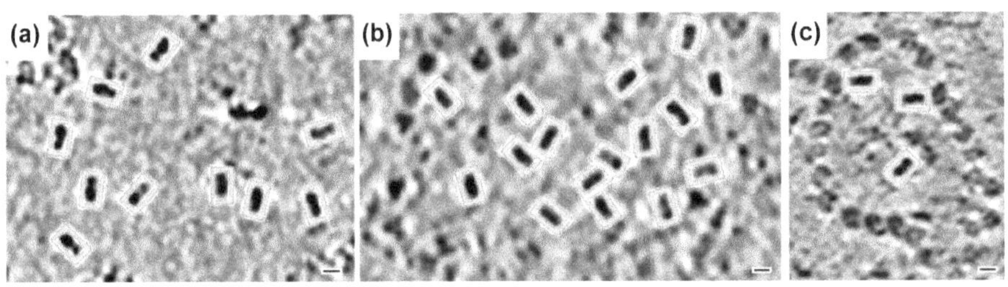

FIGURE 31.3 Complex I in mitochondrial membrane. Distribution and organization of the matrix arm of complex I are shown in gray boxes from bovine heart (a), Y. lipolytica (b), and A. officinalis (c). Respiratory strings are absent in all organisms. Scale bar, 5 nm. (Adapted from Davies et al., 2018.)

isolated from bovine, yeast, and asparagus provided direct visualization of these supercomplexes *in situ*, and they appeared to be ubiquitously present in the IMM (Davies et al., 2018). It appears that the supercomplexes can be of variable stoichiometries and compositions. For example, in the subvolumes extracted from bovine mitochondrial tomograms, supercomplex $I_1III_2IV_a$, consisting of a complex I, a dimer of complex III, and a complex IV, has the highest representation of 30% (Davies et al., 2018). This is followed by supercomplexes I_1III_2 (16%) and $I_1III_2IV_{ab}$ (10%; IV_{ab} = two separate monomers of complex IV). The yeast supercomplexes also display compositional diversity, and the respective representations are 30%, 14%, 13%, and 3% for $I_1III_2IV_{cd}$, $I_1III_2IV_c$, I_1III_2, and $I_1III_2IV_d$ supercomplexes. In marked contrast to bovine and yeast samples, asparagus IMM appears more homogenous as it displays only one type of supercomplex, a dimer of complex III (III_2), represented by 58% of subvolumes extracted from the tomograms (Davies et al., 2018). The standalone complex I densities could be respectively traced in 44%, 40%, and 42% of cases from bovine, yeast, and asparagus samples. Thus, in the above-mentioned supercomplexes from bovine and yeast, complex IV seems to attach at different interfaces such as the ones formed by the membrane arm of complex I and complex III, the peripheral arm of complex I and complex III, and the one which is adjacent to a complex III dimer.

CRYO-TOMOGRAPHIC VISUALIZATION OF MEMBRANE PROTEINS IN OUTER MITOCHONDRIAL MEMBRANE (OMM)

As portrayed in the preceding discussion, the organization and structure of macromolecular machines in the IMM are well described by mitochondrial cryo-tomograms and downstream subtomogram averages from many different species. Compared to that, the OMM is less studied.

A recent study, using biological cryo-focused ion beam (cryo-FIB) milling and cryo-ET, has provided unprecedented insight into the protein arrays adorning the OMM in the midpiece section of mammalian sperm from three different animals-pig, horse, and mouse (Ricardo Leung et al., 2021). It must be noted here that sperm cells are specialized cells that are meant to be flexible and motile. In such cells, multiple mitochondria together form a continuous sheath surrounding the central axis of the sperm midpiece, made of a microtubule-based axoneme. The OMM, therefore, must face either the plasma membrane or the axoneme. Interestingly, the tomograms revealed that the plasma membrane–facing surface is devoid of any protein array, but the axoneme-facing surface contains ordered arrays of membrane proteins in a ladderlike arrangement. While these ladders protrude from the OMM surface, they are also connected to a protein pseudo-lattice buried in the OMM. Subtomogram averaging, which could produce volumes in the 22–39 Å resolution range, revealed an S-shaped density for these ladders where a ~120 kDa homodimer of glycerol kinase (GK) could be rigid-body fitted. The average obtained for the lattice revealed a sheet of ~3–4 nm diameter pores which could be identified as VDAC molecules, an important gatekeeper of mitochondrial metabolites. Very clearly, the tomograms revealed a 2:8 ratio of GK:VDAC association where a boat of 2 GK molecules floats on top of 8 VDAC molecules. It was further revealed that the boats of GK-dimers are separated by a distance that is governed by the VDAC lattice underneath. Thus, the membrane localization and protein structures in the OMM of the midpiece section of sperm point to a conserved molecular identity in the sperm mitochondrial sheath across three mammalian species, despite varied dimensions of mitochondria and nonuniform cristae architecture when looked at closely (Ricardo Leung et al., 2021). Interestingly, the tomograms further helped in the visualization of electron-dense intramitochondrial linker complexes between adjacent mitochondria and some unknown linkers between mitochondria and outer dense filaments, but the information obtained was limited and could not be subtomogram-averaged due to either heterogeneity in the structures or sparseness of occurrence. Collectively, the information obtained from these tomograms formed a solid knowledge platform to comprehend sperm motility in molecular terms (Ricardo Leung et al., 2021).

EARLY TEM STUDIES ON THYLAKOID MEMBRANE

Conventional TEM studies provided a concrete description of thylakoid membrane organization in plant chloroplasts, revealing a bunch of stacked disc-shaped grana connected via unstacked and continuous stroma lamellae (Heslop-Harrison, 1963; Hodge et al., 1955; Sager and Palade, 1957). Concomitant with the huge diversity in thylakoid organization, different models have been proposed to describe their intricacies. The helical model describes grana as separate membrane discs that stack in a columnlike arrangement, and the discs are connected via narrow membrane protrusions of stromal thylakoid that wraps around this column in a helical fashion (Mechela et al., 2019; Mustárdy and Garab, 2003; Mustárdy et al., 2008; Paolillo, 1970). On the other hand, according to the forked membrane model, the thylakoid is a single continuous membrane that experiences bifurcation in the stroma, and the grana are where these membranes fuse. Not only that, but the granal cylinder is formed by stacking of repeating units, and each unit is rotated with respect to another around the central axis of the cylinder (Andersson and Anderson, 1980; Arvidsson and Sundby, 1999; Mechela et al., 2019). Despite multiple viewpoints, it is evident that the thylakoid architecture is governed by the proteins that adorn its membrane, and this architecture provides the necessary division of labor among the membrane system and the fluidic stroma of the chloroplast during oxygenic photosynthesis. Four large protein complexes that span such membranes are photosystems I (PSI) and II (PSII), cytochrome b_6f, and ATP synthase (Wietrzynski et al., 2020). Apart from these, light-harvesting complexes I (LHCI) and II (LHCII) are another set of entities that populate such membranes (Wietrzynski et al., 2020). Cryo-tomographic studies have provided unprecedented visualization of the membrane organization of these proteins, which has significantly improved our understanding of the photosynthetic process.

CRYO-TOMOGRAPHIC VISUALIZATION OF MEMBRANE PROTEINS IN THYLAKOID MEMBRANE

Moving on from the traditional TEM methods that involved staining, and other treatments to fix samples which significantly limit resolution, starting from 2010, unstained thylakoid materials have been studied by cryo-ET. Of late, the application of biological cryo-FIB milling along with cryo-ET has proven to be successful in mining thylakoid membranes at molecular detail.

Photosystem II (PSII) and Its Supercomplexes

Photosystem II (PSII) is responsible for splitting water to generate electrons that are later used for carbon fixation (Barber, 2012; Kato et al., 2021). Many crystallographic and single-particle cryo-EM structures of PSII have been reported from multiple sources (Athina et al., 2001; Ferreira et al., 2012; Gisriel et al., 2022; Guskov et al., 2009; Kamiya and Shen, 2003; Opatíková et al., 2023; Umena et al., 2011). A wealth of information regarding the functional mechanism of PSII that involves water channels, proton transfer, oxygen release, etc., has been revealed by these high-resolution structures. However, how PSII complexes interact with each other, and how that helps in grana stacking, was revealed by cryo-tomography studies (Daum and Kühlbrandt, 2011; Daum et al., 2010; Wietrzynski et al., 2020).

To the best of our knowledge, the first report on the visualization of native proteins in the plant thylakoid membranes by cryo-tomography was provided by Daum et al. (2010). In the tomograms captured by Daum et al., PSII was found in a dimeric form that had a distinctive rectangular shape resembling a human kidney from the direction that is perpendicular to the membrane plane. PSII densities, when viewed from the membrane plane, appeared to have protrusions that bulged into the thylakoid lumen, and these protrusions could be traced promptly at regions where two grana membranes were in close proximity across the stromal intervention. Authors indicated that the bulge in this low-resolution PSII density may harbor the oxygen-evolving complex (OEC), a cluster of manganese, calcium, and oxygen atoms forming the reaction center for water oxidation (Askerka et al., 2017; Daum et al., 2010).

The visual analysis of the tomograms revealed that PSII densities formed crystalline planar arrays of such dimers in the appressed part of thylakoid membranes and that the non-appressed thylakoid was devoid of any PSII densities (Daum et al., 2010). The PSII dimers were arranged in a head-to-tail fashion, forming a ~50-degree angle with the horizontal axis, and thus an in-plane *p*2 symmetry was formed (Figure 31.4a). Two such planar arrays from opposing membranes together formed a *p*222 symmetrical structure with a stacked appearance across stromal intervention (Figure 31.4b). Such an arrangement was prominent when authors imaged swollen spinach thylakoids generated after hypo-osmotic shock and mild detergent treatment (Daum et al., 2010).

Subtomogram averaging of the PSII and associated densities from the above-mentioned lattices produced a low-resolution map that could accommodate not only a PSII dimeric core but also two strongly associated LHCII trimers (C_2S_2 supercomplex) (Figure 31.4a). Additionally, four densities representing much smaller minor LHCs are captured inside this lattice. The subtomogram averages further confirmed the absence of any bigger supercomplex than C_2S_2 in spinach thylakoid. Each C_2S_2 density, when looked at from the direction perpendicular to the membrane plane, appeared to cast its shadow on three rows of supercomplexes in the neighboring appressed membrane (Figure 31.4c). Moreover, the space occupied by the minor LHCs in one membrane further spans two more C_2S_2 supercomplexes in the neighboring membrane, and therefore a total of five supercomplexes from one membrane are in touch with one supercomplex from the neighboring membrane (Figure 31.4c). The nature of this interaction among apposed thylakoid membranes across the stromal gap appeared to have a strong electrostatic component as the positively charged N-terminal of the LHCII trimer juxtaposed with the negatively charged patch of the opposite LHCII trimer and vice versa (Figure 31.4d and e). Other than that, interdigitation among OEC densities and charge complementarity among opposing minor LHCs (CP26 and CP29) were prominent (Daum et al., 2010). Thus, overall, these tomograms point toward a solid relation between granal stacking and electrostatic forces between the appressed membranes of grana, but the tomograms also showcase other factors that might contribute to such stacking. Such information opens new research directions that can relate granal architecture, the

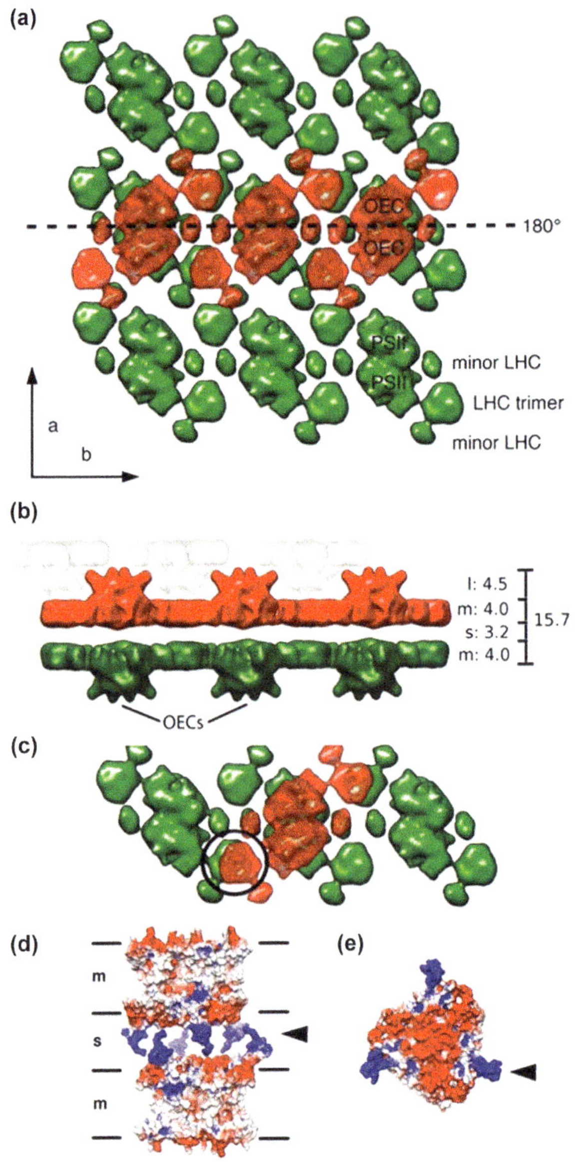

FIGURE 31.4 2D lattice of Photosystem II (PSII)-Light Harvesting Complex II (LHCII) supercomplex and their interaction in appressed grana membrane. (a) The spinach thylakoid derived 2D lattice of PSII-LHCII supercomplex along with minor LHCs are rigid-body fitted with low-pass (3-nm resolution) filtered density map of the same supercomplex earlier described by (Nield and Barber, 2006). Green and red represent densities of supercomplex lattice in apposed grana membranes separated by the stromal gap. (b) Side view of (a), PSII-LHCII supercomplexes are interdigitated in the thylakoid lumen. (c) Each PSII-LHCII supercomplex along with the associated minor LHCs interact with five rows of similar supercomplexes in the apposed thylakoid membrane across the stromal gap. (d and e) Surface charge distribution at pH 8.0 and interaction among the positive and negative patches of opposing LHCII trimers from the above-mentioned 2D lattice, across the stromal gap. (d) represents two LHCII trimers across the stromal gap, and (e) represents stroma-exposed surface of one LHCII trimer. Red and blue colors denote acidic and basic patches of proteins. (Adapted from Daum et al., 2010.)

abundance and ratios of different constituent proteins and protein complexes in granal membranes, and the local electrostatic environment (Puthiyaveetil et al., 2017; Zsiros et al., 2020).

Very recently, thylakoid membranes from *Chlamydomonas reinhardtii* were studied *in situ* by Wietrzynski et al. by generating thin lamellae using biological cryo-FIB milling (Wietrzynski et al., 2020). By virtue of using a direct electron detector camera and Volta phase plate, the authors could generate various tomograms with improved quality, enabling the identification of 75% (5094 out of 6832 densities) of the proteins embedded into many thylakoid membranes (Figure 31.5). The cryo-tomograms reveal a lack of traditional grana structure in *Chlamydomonas* thylakoid, but they contain two types of membranous regions: appressed and non-appressed (Figure 31.5). Despite lacking grana stacks, the lateral heterogeneity among these two membrane systems clearly aligns well with the protein distribution in plant thylakoid membranes described above. The stromal surface of appressed membranes does not possess any large protruding density, but as expected, density resembling PSII can be located on the luminal

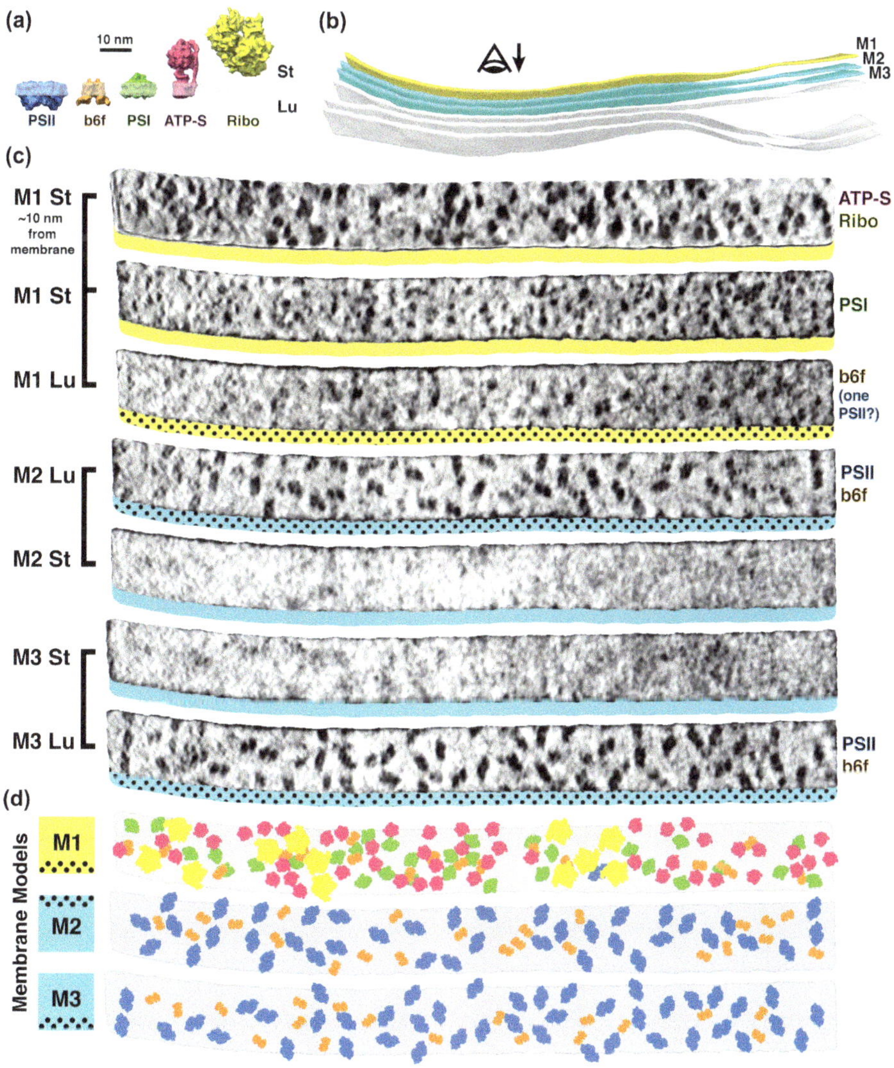

FIGURE 31.5 Distribution of photosynthetic complexes within appressed and non-appressed membranes of Chlamydomonas thylakoid. (a) Schematic representation of the four photosynthetic complexes (photosystem I and II, cytochrome b6f, and ATP synthase) along with chloroplastic ribosome to show membrane-buried region along with lumen- or stroma-exposed part, along with scale bar. (b) Three segmented thylakoid membranes are color coded to show non-appressed (yellow) and appressed (cyan) membranes. Viewing direction is denoted by an eye symbol with arrow. (c) Corresponding to the viewing direction from (b), membrane segments show the densities as they appear in field-of-view while the viewing plane is moved away starting from the stromal space toward the direction of the arrow shown in (b). Solid colors are used for representing stromal surface, and dotted color patterns for luminal surface. (d) Color coded models of macromolecules as described in (a), overlayed on their respective segmented membrane. (Adapted from Wietrzynski et al., 2020.)

surface of such membranes (Figure 31.5). The stromal thylakoid, which is inherently composed of only non-appressed membranes, completely lacks PSII densities in the lumen, but the stromal surface is populated with PSI densities (Figure 31.5). It must be borne in mind that both PSI and PSII complexes, for the most part, are buried in the membrane region, and the extensions of these complexes that are exposed on the membrane surface have similar sizes of ~3 nm and ~4 nm, respectively, but the identification of PSII is easier due to the light background in the lumen of appressed thylakoid. It was further observed that PSII did not arrange into a crystalline 2D lattice; instead it was randomly distributed in appressed thylakoid. Interestingly, such PSII complexes do not form any head-to-head contact, and they remain interdigitated across the luminal space despite having enough room in the lumen to accommodate a head-to-head arrangement.

A peculiar case of PSII organization was observed in *Phaeodactylum tricornutum*, a marine diatom, that also lacked archetypal grana architecture as opposed to plants (Levitan et al., 2019). Cryo-tomograms of such membranes revealed two distinct forms of PSII organization: noncrystalline loose cluster and 2D crystalline array (Figure 31.6). In the moderately resolved (~25 Å) subtomogram average of the former, the dimeric PSII core was adorned by additional densities in the membrane plane, which were confirmed by mass spectrometry experiments to be of fucoxanthin chlorophyll a/c (FCP) proteins implicated in light harvesting. Each asymmetric unit of the dimeric PSII core was associated with six FCP proteins. Thus, a supercomplex of 12 FCP proteins along with the PSII dimeric core was captured by Levitan et al. (Levitan et al., 2019). Authors further generated subtomogram averages of the asymmetric units of the dimeric core that spanned the above-mentioned PSII 2D crystalline array, and the resultant map resolved to a much better resolution of 10.4 Å, which could accommodate four core subunits of PSII PsbA, PsbB, PsbC, and PsbD, leaving no unaccounted density which could be assigned to any antenna protein complex. Thus, it appears that the PSII monomeric units in the crystalline and noncrystalline arrangements in *P. tricornutum* are distinct, despite having a similar domain organization (Figure 31.6). Since the densities for crystalline and noncrystalline monomeric units overlapped for most parts, it can be safely assumed that these densities represent two structural forms of the supercomplex that are in dynamic interchangeable relation. It was suggested that the crystalline array can act as a repertoire of photodamaged PSII reaction centers that oversee the replacement of photodamaged proteins with functional substitutes (Levitan et al., 2019).

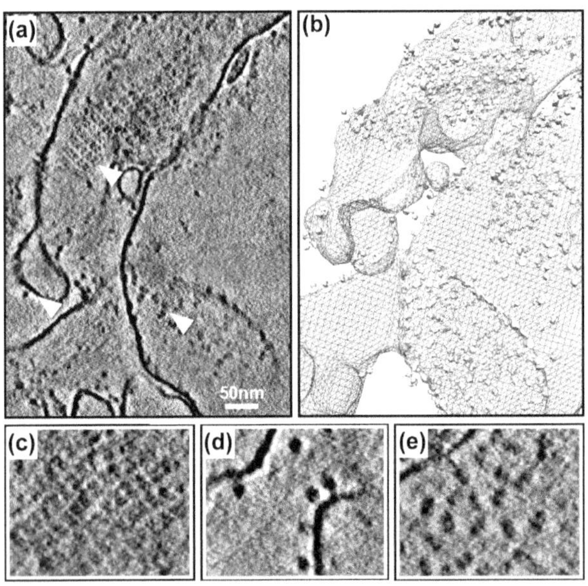

FIGURE 31.6 Organization and distribution of PSII and mtF_1F_0 in *P. tricornutum* thylakoid membrane. (a) Tomogram slice displaying organization and distribution of two types of PSII cluster: non-crystalline (bottom right) and 2D crystalline (top), and mtF_1F_0 complexes (bottom left) as indicated by white arrowheads. (b) 3D representation of the tomogram mentioned in (a), thylakoid in gray mesh, non-crystalline PSII (bottom right), 2D crystalline PSII (top), and mtF_1F_0 (bottom left) in gray surfaces. (c, d, and e) Enlarged view of the regions as pointed by the white arrowheads in (a), (c) represents 2D crystalline PSII, (d) represents mtF_1F_0 complexes, and (e) represents non-crystalline PSII. (Adapted from Levitan et al., 2019.)

Photosystem I (PSI) and Cytochrome b_6f ($Cytb_6f$)

Identification and structural analysis of chloroplastic PSI have proven to be more complicated than PSII in cryo-tomography studies. There are three possible reasons for this: The PSI core, composed of ~15 subunits, is much smaller than the PSII core which encompasses 25 to 30 subunits; eukaryotic PSI mainly exists in a monomeric form that cannot compete with PSII dimers in visibility; and the stroma-exposed PSI density is merely ~3 nm in size and does not align well for subtomogram averaging because the contrast for this region is low due to the noisy background of protein-rich stroma (Busch and Hippler, 2011; Caffarri et al., 2014). Despite such imaging challenges, Wietrzynski et al. could capture PSI densities that were exclusively populated in the non-appressed region of *Chlamydomonas* thylakoid membranes (Wietrzynski et al., 2020) (Figure 31.5). Unlike PSII, no crystalline arrangement could be traced for PSI, and PSI seemed to be rather randomly distributed all over its respective membrane plane. It seems that an absence of enough space to accommodate PSI stromal exposed density is the reason behind its absence in the appressed region of *Chlamydomonas* thylakoid (Wietrzynski et al., 2020) (Figure 31.5).

Cytochrome b_6f ($Cytb_6f$) is an integral membrane protein complex that forms a ~220 kDa functional dimer (Malone et al., 2021). The dimeric $Cytb_6f$ densities are found in both appressed and non-appressed thylakoid membranes of *Chlamydomonas*, where its luminal ~3 nm protrusion can be traced in the tomograms (Wietrzynski et al., 2020) (Figure 31.5). Wietrzynski et al. noted that Cytb$_6$f density could be unambiguously identified in the appressed membrane but not in the non-appressed region (Figure 31.5). It can be hypothesized that Cytb$_6$f is probably nontraceable in the plant thylakoid cryo-tomograms captured by Daum et al. earlier due to its smaller size in comparison to PSI, PSII, and ATP synthase (Daum and Kühlbrandt, 2011; Daum et al., 2010).

Chloroplastic ATP Synthase (cF_1F_0)

In sharp contrast to mtF_1F_0, which forms long dimer rows and are located at highly curved regions of cristae, chloroplastic ATP synthases (cF_1F_0) are monomeric, do not contribute to membrane bending in thylakoid, and are only found at the flat or slightly curved surface of non-appressed thylakoid membrane (Daum and Kühlbrandt, 2011; Daum et al., 2010; Hahn et al., 2018; Wietrzynski et al., 2020). In cryo-tomograms obtained from plant and algal thylakoids, cF_1F_0 is readily distinguished by its 10–12 nm diameter F_1 head that is connected to the membrane via a ~4 nm stalk (Daum and Kühlbrandt, 2011; Daum et al., 2010; Wietrzynski et al., 2020) (Figure 31.5). Overall, the protein complex protrudes for 15 to 16 nm above the membrane on the stromal side (Daum and Kühlbrandt, 2011; Daum et al., 2010; Wietrzynski et al., 2020) (Figure 31.5). The stator stalk of cF_1F_0 also produces a clear density that appears next to a much larger F_1 head density (Wietrzynski et al., 2020). Despite the predominant monomeric presence that accounted for 85% of cF_1F_0 densities in plant thylakoid membranes, in the remaining 15% of densities cF_1F_0 is found in small, sporadic clusters that are not only limited in size but are thought to originate from stochastic contacts among the monomeric cF_1F_0 complexes (Daum et al., 2010). In both plant and algal sources, cF_1F_0 populates only the non-appressed part of the thylakoid, and the appressed region is probably too narrow to accommodate such a big complex (Daum and Kühlbrandt, 2011; Daum et al., 2010; Wietrzynski et al., 2020) (Figure 31.5).

DISCUSSION

Mitochondria and chloroplasts are important organelles that serve essential functions inside eukaryotic cells. Both organelles contain an extensive network of membranes that are studded with a medley of protein complexes that perform important functions to sustain life. Crystallographic and high-resolution cryo-EM studies have provided detailed insights into the functional mechanism of such protein complexes. However, a complex orchestration of functions of these protein complexes is needed to generate energy in mitochondria and to produce sugar molecules in chloroplasts. Conventional TEM studies from a few decades ago and cryo-tomographic visualization in the last 15 to 20 years have incrementally improved our knowledge of this orchestration by providing contextual visualization of such complexes in native membranes. Recently developed biological cryo-FIB milling technology (Marko et al., 2007; Rigort et al., 2010) has the potential to generate thin lamellae from native whole cells, producing images in which one can peek into the topography of such membranes and the association of organellar membrane proteins under normal and stressed conditions, at molecular detail. It appears that, following in the footsteps of thylakoid exploration, cristae membranes are now about to be described with similar, if not greater, precision.

ACKNOWLEDGEMENTS

This work is supported by funding for Basic and Applied Research in Biophysics and Material Science (RSI 4002) by the Department of Atomic Energy (DAE), Govt. of India to TA. AS thanks University Grants Commission, Govt. of India for personal fellowship support.

REFERENCES

Acin-Perez, R., and Enriquez, J.A. (2014). The function of the respiratory supercomplexes: The plasticity model. BBA. *1837*, 444–450.

Albertsson, P.Å. (2001). A quantitative model of the domain structure of the photosynthetic membrane. Trends Plant Sci. *6*, 349–354.

Andersson, B., and Anderson, J.M. (1980). Lateral heterogeneity in the distribution of chlorophyll-protein complexes of the thylakoid membranes of spinach chloroplasts. BBA. *593*, 427–440.

Arvidsson, P.-O., and Sundby, C. (1999). A model for the topology of the chloroplast thylakoid membrane. Funct. Plant Biol. *26*, 687–694.

Askerka, M., Brudvig, G.W., and Batista, V.S. (2017). The O_2-evolving complex of photosystem II: Recent insights from quantum mechanics/molecular mechanics (QM/MM), extended X-ray absorption fine structure (EXAFS), and femtosecond X-ray crystallography data. Acc. Chem. Res. *50*, 41–48.

Athina, Z., Witt, H.-T., Kern, J., Fromme, P., Krauß, N., Saenger, W., and Orth, P. (2001). Crystal structure of photosystem II from synechococcus elongatus at 3.8 AÊ resolution. Nature *5648*, 739–743.

Barber, J. (2012). Photosystem II: The water-splitting enzyme of photosynthesis. Cold Spring Harb. Symp. Quant. Biol. *77*, 295–307.

Bultema, J.B., Braun, H.P., Boekema, E.J., Kouřil, R., and Kou, R. (2009). Megacomplex organization of the oxidative phosphorylation system by structural analysis of respiratory supercomplexes from potato. Biochim. Biophys. Acta. *1787*, 60–67.

Busch, A., and Hippler, M. (2011). The structure and function of eukaryotic photosystem I. Biochim. Biophys. Acta. *1807*, 864–877.

Caffarri, S., Tibiletti, T., Jennings, R.C., and Santabarbara, S. (2014). A comparison between plant photosystem I and photosystem II architecture and functioning. Curr. Protein Pept. Sci. *15*, 296–331.

Chance, B., and Williams, G.R. (1955). Respiratory enzymes in oxidative phosphorylation. II. difference spectra. J. Biol. Chem. *217*, 395–407.

Comte, J., Maïsterrena, B., and Gautheron, D.C. (1976). Lipid composition and protein profiles of outer and inner membranes from pig heart mitochondria. Comparison with microsomes. Biochim. Biophys. Acta. *419*, 271–284.

Daems, W.T., and Wisse, E. (1966). Shape and attachment of the cristae mitochondriales in mouse hepatic cell mitochondria laboratory for electron microscopy. J. Ultrastruct. Res. *16*, 123–140.

Daum, B., and Kühlbrandt, W. (2011). Electron tomography of plant thylakoid membranes. J. Exp. Bot. *62*, 2393–2402.

Daum, B., Nicastro, D., Austin, J., Richard McIntosh, J., and Kühlbrandt, W. (2010). Arrangement of photosystem II and ATP synthase in chloroplast membranes of spinach and pea. Plant Cell *22*, 1299–1312.

Davies, K.M., Strauss, M., Daum, B., Kief, J.H., Osiewacz, H.D., Rycovska, A., Zickermann, V., and Kühlbrandt, W. (2011). Macromolecular organization of ATP synthase and complex I in whole mitochondria. Proc. Natl. Acad. Sci. U. S. A. *108*, 14121–14126.

Davies, K.M., Blum, T.B., and Kühlbrandt, W. (2018). Conserved in situ arrangement of complex i and III2 in mitochondrial respiratory chain supercomplexes of mammals, yeast, and plants. Proc. Natl. Acad. Sci. U. S. A. *115*, 3024–3029.

Ehmke, L., Hause, G., and Klösgen, R.B. (2023). Preparation of physiologically active inside-out vesicles from plant inner mitochondrial membranes. Front Plant Sci.14:1216227, 1–12.

Ferna, P., Peleato, M.L., Pe, A., Acı, R., Enriquez, J.A., Bioquimica, D. De, Zaragoza, U. De, and Servet, M. (2008). Respiratory active mitochondrial supercomplexes. Mol. Cell,32, 529–539.

Fernandez-Moran, H. (1962). Cell-membrane ultrastructure: Low-temperature electron microsopy and x-ray diffraction studies of lipoprotein components in lamellar systems. Circulation *XXVI*, 1039–1065.

Fernandez Moran, H., Oda, T., Blair, P. V., and Green, D.E. (1964). A macromolecular repeating unit of mitochondrial structure and function. Correlated electron microscopic and biochemical studies of isolated mitochondria and submitochondrial particles of beef heart muscle. J. Cell Biol. *22*, 63–100.

Ferreira, K.N., Iverson, T.M., Maghlaoui, K., Barber, J., and Iwata, S. (2012). Architecture of the photosynthetic oxygen-evolving center. Biochemistry *51*, 3321–3333.

Fukuda, Y., Stapleton, K., and Kato, T. (2023). Progress in spatial resolution of structural analysis by cryo-EM. Microscopy *72*, 135–143.

Gisriel, C.J., Wang, J., Liu, J., Flesher, D.A., Reiss, K.M., Huang, H.-L., Yang, K.R., Armstrong, W.H., Gunner, M.R., Batista, V.S., et al. (2022). High-resolution cryo-electron microscopy structure of photosystem II from the mesophilic cyanobacterium, synechocystis sp. PCC 6803. Proc. Natl. Acad. Sci. *119*, e2116765118.

Guskov, A., Kern, J., Gabdulkhakov, A., Broser, M., Zouni, A., and Saenger, W. (2009). Cyanobacterial photosystem II at 2.9-Å resolution and the role of quinones, lipids, channels and chloride. Nat. Struct. Mol. Biol. *16*, 334–342.

Hahn, A., Vonck, J., Mills, D.J., Meier, T., and Kühlbrandt, W. (2018). Structure, mechanism, and regulation of the chloroplast ATP synthase. Science (80-.). *360*.

Vander Heiden, M.G., Chandel, N.S., Li, X.X., Schumacker, P.T., Colombini, M., and Thompson, C.B. (2000). Outer mitochondrial membrane permeability can regulate coupled respiration and cell survival. Proc. Natl. Acad. Sci. U. S. A. *97*, 4666–4671.

Heinemeyer, J., Braun, H., Boekema, E.J., and Kour, R. (2007). A structural model of the cytochrome c reductase/oxidase supercomplex from yeast mitochondria. J. Biol. Chem. *282*, 12240–12248.

Heslop-Harrison, J. (1963). Structure and morphogenesis of lamellar systems in grana-containing chloroplasts. Planta *60*, 243–260.

Hewitt, V.L., and Whitworth, A.J. (2017). Mitochondrial Fission and Fusion. Park. Dis. Mol. Mech. Underlying Pathol. 77–111.

Hodge, A.J., McLean, J.D., and Mercer, F. V. (1955). Ultrastructure of the lamellae and grana in the chloroplasts of ZEA mays L. J. Biophys. Biochem. Cytol. *1*, 605–613.

Huokko, T., Ni, T., Dykes, G.F., Simpson, D.M., Brownridge, P., Conradi, F.D., Beynon, R.J., Nixon, P.J., Mullineaux, C.W., Zhang, P., et al. (2021). Probing the biogenesis pathway and dynamics of thylakoid membranes. Nat. Commun. *12*, 1–14.

Joubert, F., and Puff, N. (2021). Mitochondrial cristae architecture and functions: Lessons from minimal model systems. Membranes (Basel). *11*, 1–27.

Kagawa, Y., and Racker, E. (1966). Partial Resolution of the Enzymes Catalyzing Oxidative Phosphorylation. J. Biol. Chem. *238*, 2831–2837.

Kamiya, N., and Shen, J.R. (2003). Crystal structure of oxygen-evolving photosystem II from Thermosynechococcus vulcanus at 3.7-Å resolution. Proc. Natl. Acad. Sci. U. S. A. *100*, 98–103.

Kato, K., Miyazaki, N., Hamaguchi, T., Nakajima, Y., Akita, F., Yonekura, K., and Shen, J.R. (2021). High-resolution cryo-EM structure of photosystem II reveals damage from high-dose electron beams. Commun. Biol. *4*, 382.

Kühlbrandt, W. (2014). The resolution revolution. Science (80-.). *343*, 1443–1444.

Kühlbrandt, W. (2019). Structure and mechanisms of F-type ATP synthases. Biochem. (Lond). *41*, 1–2.

Levitan, O., Chen, M., Kuang, X., Cheong, K.Y., Jiang, J., Banal, M., Nambiar, N., Gorbunov, M.Y., Ludtke, S.J., Falkowski, P.G., et al. (2019). Structural and functional analyses of photosystem II in the marine diatom Phaeodactylum tricornutum. Proc. Natl. Acad. Sci. U. S. A. *116*, 17316–17322.

Li, M., Mukhopadhyay, R., Svoboda, V., Oung, H.M.O., Mullendore, D.L., and Kirchhoff, H. (2020). Measuring the dynamic response of the thylakoid architecture in plant leaves by electron microscopy. Plant Direct *4*, 1–14.

Malone, L.A., Proctor, M.S., Hitchcock, A., Hunter, C.N., and Johnson, M.P. (2021). Cytochrome b6f – Orchestrator of photosynthetic electron transfer. Biochim. Biophys. Acta. *1862*, 148380.

Mannella, C.A., Marko, M., and Buttle, K. (1997). Reconsidering mitochondrial structure: new views of an old organelle. Trends. Biochem. Sci. *22*, 37–38.

Mannella, C.A., Marko, M., Penczek, P., Barnard, D., and Frank, J. (1994). The internal compartmentation of rat-liver mitochondria: Tomographic study using the high-voltage transmission electron microscope. Microsc. Res. Tech. *27*, 278–283.

Marko, M., Hsieh, C., Schalek, R., Frank, J., and Mannella, C. (2007). Focused-ion-beam thinning of frozen-hydrated biological specimens for cryo-electron microscopy. Nat. Methods *4*, 215–217.

Marko, M., Leith, A., and Parsons, D. (1988). Three-dimensional reconstruction of cells from serial sections and whole-cell mounts using multilevel contouring of stereo micrographs. J. Electron Microsc. Tech. *9*, 395–411.

Mechela, A., Schwenkert, S., and Soll, J. (2019). A brief history of thylakoid biogenesis. Open Biol. 9, 180237.

Mühleip, A., Kock Flygaard, R., Ovciarikova, J., Lacombe, A., Fernandes, P., Sheiner, L., and Amunts, A. (2021). ATP synthase hexamer assemblies shape cristae of Toxoplasma mitochondria. Nat. Commun. *12*, 120.

Mustárdy, L., Buttle, K., Steinbach, G., and Garab, G. (2008). The three-dimensional network of the thylakoid membranes in plants: Quasihelical model of the granum-stroma assembly. Plant Cell *20*, 2552–2557.

Mustárdy, L., and Garab, G. (2003). Granum revisited. A three-dimensional model - Where things fall into place. Trends Plant Sci. *8*, 117–122.

Nicastro, D., Frangakis, A.S., Typke, D., and Baumeister, W. (2000). Cryo-electron tomography of Neurospora mitochondria. J. Struct. Biol. *129*, 48–56.

Nield, J., and Barber, J. (2006). Refinement of the structural model for the Photosystem II supercomplex of higher plants. Biochim. Biophys. Acta - Bioenerg. 1757, 353–361.

O'Rourke, B. (2000). Pathophysiological and protective roles of mitochondrial ion channels. J. Physiol. *529*, 23–36.

Opatíková, M., Semchonok, D.A., Kopečný, D., Ilík, P., Pospíšil, P., Ilíková, I., Roudnický, P., Zeljković, S.Ć., Tarkowski, P., Kyrilis, F.L., et al. (2023). Cryo-EM structure of a plant photosystem II supercomplex with light-harvesting protein Lhcb8 and α-tocopherol. Nat. Plants *9*, 1359–1369.

Palade, G.E. (1952). The fine structure of mitochondria. Anat. Rec. *114*, 427–451.

Palade, G.E. (1953). An electron microscopic study of the mitochondrial structure. J. Histochem. Cytochem. *1*, 188–211.

Paolillo, D.J. (1970). The three-dimensional arrangement of intergranal lamellae in chloroplasts. J. Cell Sci. *6*, 243–255.

Perkins, G., Renken, C., Martone, M.E., Young, S.J., Ellisman, M., and Frey, T. (1997a). Electron tomography of neuronal mitochondria: Three-dimensional structure and organization of cristae and membrane contacts. J. Struct. Biol. *119*, 260–272.

Perkins, G.A., Renken, C.W., Song, J.Y., Frey, T.G., Young, S.J., Lamont, S., Martone, M.E., Lindsey, S., and Ellisman, M.H. (1997b). Electron tomography of large, multicomponent biological structures. J. Struct. Biol. *120*, 219–227.

Perkins, G.A., Song, J.Y., Tarsa, L., Deerinck, T.J., Ellisman, M.H., and Frey, T.G. (1998). Electron tomography of mitochondria from brown adipocytes reveals crista junctions. J. Bioenerg. Biomembr. *30*, 431–442.

Pribil, M., Labs, M., and Leister, D. (2014). Structure and dynamics of thylakoids in land plants. J. Exp. Bot. *65*, 1955–1972.

Puthiyaveetil, S., Van Oort, B., and Kirchhoff, H. (2017). Surface charge dynamics in photosynthetic membranes and the structural consequences. Nat. Plants *3*, 1–9.

Ricardo Leung, M., Chiozzi, R.Z., Roelofs, M.C., Hevler, J.F., Ravi, R.T., Maitan, P., Zhang, M., Henning, H., Bromfield, E.G., Howes, S.C., et al. (2021). In-cell structures of conserved supramolecular protein arrays at the mitochondria-cytoskeleton interface in mammalian sperm. Proc. Natl. Acad. Sci. U. S. A. *118*, 1–10.

Rigort, A., Bäuerlein, F.J.B., Leis, A., Gruska, M., Hoffmann, C., Laugks, T., Böhm, U., Eibauer, M., Gnaegi, H., Baumeister, W., et al. (2010). Micromachining tools and correlative approaches for cellular cryo-electron tomography. J. Struct. Biol. *172*, 169–179.

Rochaix, J.D. (2014). Regulation and dynamics of the light-harvesting system. Annu. Rev. Plant Biol. *65*, 287–309.

Sager, R., and Palade, G.E. (1957). Structure and development of the chloroplast in chlyamydomonous. I. the normal green cell. J. Biophys. Biochem. Cytol. *3*, 463–488.

Schägger, H., and Pfeiffer, K. (2000). Supercomplexes in the respiratory chains of yeast and mammalian mitochondria. EMBO J. *19*, 1777–1783.

Seelert, H., Reifschneider, N.H., Krause, F., Dencher, N.A., and Vonck, J. (2006). Architecture of active mammalian respiratory chain supercomplexes. J. Biol. Chem. 281, 15370–15375.

Sjöstrand, F.S. (1953). Electron microscopy of mitochondria and cytoplasmic double membranes. Nature *171*, 30–31.

Sjöstrand, F.S. (1956). The Ultrastructure of Cells as Revealed by the Electron Microscope. Int. Rev. Cytol. *5*, 455–533.

Strauss, M., Hofhaus, G., Schröder, R.R., and Kühlbrandt, W. (2008). Dimer ribbons of ATP synthase shape the inner mitochondrial membrane. EMBO J. *27*, 1154–1160.

Suhai, T., Heidrich, N.G., Dencher, N.A., and Seelert, H. (2009). Highly sensitive detection of ATPase activity in native gels. Electrophoresis *30*, 3622–3625.

Umena, Y., Kawakami, K., Shen, J.R., and Kamiya, N. (2011). Crystal structure of oxygen-evolving photosystem II at a resolution of 1.9Å. Nature *473*, 55–60.

Wallace, D.C. (2012). Mitochondria and cancer. Nat. Rev. Cancer *12*, 685–698.

Wietrzynski, W., Schaffer, M., Tegunov, D., Albert, S., Kanazawa, A., Plitzko, J.M., Baumeister, W., and Engel, B.D. (2020). Charting the native architecture of chlamydomonas thylakoid membranes with single-molecule precision. Elife *9*, 1–18.

Winslow, J.L., Hollenberg, M.J., and Lea, P.J. (1991). Resolution limit of serial sections for 3D reconstruction of tubular cristae in rat liver mitochondria. J. Electron Microsc. Tech. *18*, 241–248.

Zick, M., Rabl, R., and Reichert, A.S. (2009). Cristae formation-linking ultrastructure and function of mitochondria. Biochim. Biophys. Acta - Mol. Cell Res. *1793*, 5–19.

Zsiros, O., Ünnep, R., Nagy, G., Almásy, L., Patai, R., Székely, N.K., Kohlbrecher, J., Garab, G., Dér, A., and Kovács, L. (2020). Role of protein-water interface in the stacking interactions of granum thylakoid membranes—as revealed by the effects of Hofmeister salts. Front. Plant Sci. *11*, 1–14.

32 Functional Studies of the Actin Cytoskeleton by Cryogenic Electron Tomography

Dorit Hanein and Niels Volkmann

INTRODUCTION

The actin cytoskeleton is pivotal in orchestrating cell migration and cellular morphology across diverse eukaryotic organisms. The diverse architectural arrangements of the actin cytoskeleton contribute to various cellular functions, including protrusion, adhesion, contraction, and retraction. Branched and cross-linked networks at the leading edge form a lamellipodium that propels cell movement, while filopodia play a role in environmental sensing and in establishing initial contact between the cell and its substrate. Stress fibers and the actomyosin cortex influence cell contractility and shape changes. Focal adhesions anchor cells to their substrate, and actin-based cell-cell contacts contribute to cell adhesion within tissues.

The precise orchestration of these various actin architectures relies on a plethora of actin-associated proteins, encompassing nucleation and elongation factors, bundling proteins, and cross-linking proteins (Pollard, 2016). While high-resolution *in vitro* studies have elucidated the structures of individual actin filaments when interacting with domains or full-length proteins, comprehending the organization of physiological actin assemblies in intricate cellular environments at high resolution is crucial for deciphering the underlying mechanisms. Cryogenic electron tomography (cryo-ET) has significantly advanced our understanding of nanoscale actin organization and structure within cells.

Cryo-ET is a specialized technique in cryogenic electron transmission microscopy, involving the physical tilting of vitrified samples, typically through ±60 degrees (Baumeister, 2022). This process generates a tilt series of images that can be reconstructed into a complete three-dimensional volume (DeRosier and Klug, 1968), known as a tomogram. Notably, cryo-ET provides the unique capability to investigate biological machinery in its native cellular context within organelles or cells at high resolution (Turk and Baumeister, 2020; Bäuerlein and Baumeister, 2021; Navarro, 2022; Young and Villa, 2023). This approach avoids the disruption of interactions and context that arises when studying structures *in vitro*. To identify structures of interest, cryo-ET is often complemented by cryogenic light and electron microscopy (cryo-CLEM) (Sartori et al., 2007; Klein et al., 2021), as well as cryogenic focused ion beam (cryo-FIB) milling, which allows the creation of thin sections called lamellae from specimens that are too thick for direct vitrification or cryo-ET imaging (Rigort et al., 2012a; Mahamid et al., 2015). The analysis of data from cryo-ET tends to be challenging due to the low signal-to-noise ratio in tomograms, arising from the low scattering contrast between biomolecules and the embedding medium and the low electron dose used to image the sample without causing undue damage.

Technological developments, including direct electron detectors and Volta phase plates, coupled with advanced computational methods, have been instrumental in advancing this field (Chua et al., 2022). The current cryo-ET methodology primarily in use for investigating cytoskeletal structures relies on an exhaustive analysis of a small number of high-quality tomograms, including analysis of vectorized actin filament traces (Rigort et al., 2012b; Jasnin et al., 2013; Dimchev et al., 2021), determination of polarity (Martins et al., 2021), and subtomogram averaging (Fäßler et al., 2020; Wang et al., 2022; Tamborrini et al., 2023). These methodologies have yielded invaluable insights into the intricate organization of the actin cytoskeleton, offering high-resolution, *in situ* structures of various cytoskeletal elements. Comprehensive reviews detailing recent advances and applications in these types of studies are available for interested readers (Chakraborty et al., 2020; Schneider and Jasnin, 2022; Fäßler et al., 2023).

Despite the impressive progress attained through these approaches, their suitability for cell biology studies is often constrained by two principal factors.

1. Tomographic datasets inherently capture a relatively limited area within a single eukaryotic cell. The thorough analysis of these tomograms demands significant resources and time. Consequently, many studies rely on a small number of tomographic datasets. However, to draw robust conclusions in cell biology, it is essential to verify that observations are not isolated incidents but remain consistent across diverse cellular regions and cell types and over time.
2. The actin cytoskeleton is notably responsive to the cellular environment. Factors such as substrate stiffness and extracellular matrix composition wield a considerable influence on actin organization and cell morphology, often discernible even at the light-microscopy level. Nevertheless, due to technological limitations, a significant portion of cryo-ET studies involves cells cultured directly on electron microscopy substrates (EM grids), frequently without adequate consideration for their native environment.

DOI: 10.1201/9781003326106-39

This chapter explores alternative methodologies, emphasizing the extraction of nanoscale parameters related to actin-filament organization from comprehensive datasets comprising hundreds of tomograms. Additionally, it discusses strategies for controlling the cellular environment for cryo-ET, acknowledging the pivotal role environmental factors play in understanding the dynamic behavior of the actin cytoskeleton.

ACCELERATED FILAMENT ANALYSIS AT THE LEADING EDGE

Cell migration plays a fundamental role in many biological processes, including wound healing, immune responses, inflammation, cancer, and thrombosis. Actin cytoskeleton rearrangements that drive cell migration are primarily orchestrated by two actin nucleator classes: the Arp2/3 complex and formin family proteins. Formins assemble long, unbranched actin filaments (Pruyne et al., 2002; Zigmond et al., 2003), while the evolutionarily conserved Arp2/3 complex forms branched networks (Mullins et al., 1998; Blanchoin et al., 2000; Volkmann et al., 2001; Rouiller et al., 2008). This structural diversity enables various cellular functions, including movement, adhesion, endocytosis, and polarity establishment (Krause and Gautreau, 2014). The prevailing hypothesis proposes that Arp2/3-complex-nucleated actin networks generate protrusive forces for cell movement, supported by *in vitro* studies of *Listeria* motility (Welch et al., 1997; Loisel et al., 1999) and visualization of dendritic actin networks at the lamellipodia of moving keratocytes and fibroblasts (Svitkina and Borisy, 1999; Yang and Svitkina, 2011). However, more recent studies challenge the exclusive role of Arp2/3 complex, highlighting collaborative regulation by multiple nucleators in the intricate dynamics of actin networks during cell migration (Chesarone and Goode, 2009; Block et al., 2012; Breitsprecher et al., 2012).

Genetic knockdown or drug-induced disruption did not compromise the functionality of the Arp2/3 complex. This is likely because a background level as low as 10% of functional Arp2/3 complex appears sufficient for the maintenance of Arp2/3-dependent functions within cells (Nolen et al., 2009). Differentiated fibroblasts derived from mouse embryonic stem cells, where the *ARPC3* gene has been disrupted, lack the *ARPC3* subunit of Arp2/3 complex. Since the Arp2/3 complex requires the assembly of all seven subunits to function, the genetic elimination of *ARPC3* ensures the complete absence of a functional Arp2/3 complex in these cells, providing the opportunity to observe the consequences of complete Arp2/3 depletion (Suraneni et al., 2012).

Surprisingly, these genetically modified fibroblasts exhibit movement at rates comparable to wild-type cells, albeit with defects in persistent directional migration (Suraneni et al., 2015). Light microscopy imaging revealed that while wild-type fibroblasts have smooth lamellipodia, *ARPC3* knockout fibroblasts display large actin-based protrusions with formins mDia1 and mDia2 at their tips, enriched with the actin cross-linker fascin, similar to the thin, fingerlike filopodia of wild-type cells (Vignjevic et al., 2006; Block et al., 2008). This striking transformation and the presence of these massive filopodia-like protrusions suggest a significant molecular reorganization of actin networks in the absence of functional Arp2/3 complex.

To unravel the intricacies of these changes, an exploration into the nanoscale architecture of the corresponding actin networks becomes imperative. Lamellipodia, sheetlike membrane protrusions found at the leading edge of migrating cells, emerge as ideal subjects for high-throughput tomographic data collection, given their thin profile (<300 nm), allowing direct cryo-ET imaging without the need for labor-intensive thinning procedures such as cryo-FIB milling. The structure of Arp2/3-mediated branch junctions has been determined *in vitro* at 0.35 nm (Chou et al., 2022) and *in situ* within constitutively Rac-activated NIH-3T3 fibroblasts at 0.9 nm resolution (Fäßler et al., 2020). Previous investigations into Arp2/3-mediated actin networks in the context of actin waves in *Dictyostelium* cells (Jasnin et al., 2019) involved meticulous tracing and vectorization of actin filaments, as well as subtomogram averaging of *in situ* branch junctions, culminating in a 3.1 nm resolution reconstruction. However, this detailed characterization is notably resource-intensive and time-consuming, often limited to a few high-quality tomograms, 19 tomograms in this case (Jasnin et al., 2019).

To overcome this constraint, we implemented two innovations in the analysis of cryo-ET data of wild-type and *ARPC3* deletion mutant fibroblast cells (Anderson et al., 2016). First, we employed an efficient, accelerated automated actin tracing protocol with superior tracing performance across various tomograms and signal-to-noise ratios (Xu et al., 2015). The main advantage of this approach lies in its universality, as a single set of algorithm parameters sufficed for all collected tomograms, eliminating the need for time-consuming parameter optimization for individual tomograms. Second, we directly extracted pertinent actin network parameters from the score maps of the filament tracing algorithm, eliminating the time-consuming and error-prone step of vectorizing individual filament traces. Furthermore, we conducted the comparative analysis using a well-defined spreading assay, ensuring synchronized cell morphology at the cell edges (Suraneni et al., 2012). Correlative light microscopy, guided by fascin and actin fluorescence, facilitated the identification of regions of interest in both cell systems. This strategic approach allowed the efficient collection and analysis of 486 regions of interest from synchronized fibroblast cells, enabling a statistically meaningful quantitative comparison of wild-type and mutant cells and the detection of subtle nanoscale differences in their actin networks (Table 32.1).

In our quantitative analysis, we observed that the extensive filopodia-like protrusions (FLPs) in *ARPC3* deletion mutant fibroblasts exhibit a higher packing density compared to wild-type filopodia. Notably, our findings indicate a singular bundle type in the mutant cells, contrasting with the distinct packing densities observed in peripheral actin bundles and filopodial actin bundles in wild-type cells

TABLE 32.1
Average filament distance in regions of interest in wild-type *(WT)* and *ARPC3* deletion mutant *(ARPC$^{-/-}$)* extracted directly from the tracing algorithm score map. The volume fractions and their statistics were determined with the specified number of regions of interest (N_{ROI}) with a score cutoff calibrated to result in the known actin diameter for individual traces. The volume fraction can then be converted into an average filament distance. *FLPs* denotes filopodia-like protrusions. All differences are statistically significant with $p < 0.0001$ except the differences between *ARPC$^{-/-}$: cell periphery* and *ARPC$^{-/-}$; FLPs,* which is not statistically significant ($p = 0.0863$)

Cell/Bundle Type	Volume Fraction	Average Distance	N_{ROI}
WT; cell periphery	0.153 ± 0.049	22 nm	156
WT; filopodia	0.191 ± 0.063	20 nm	74
ARPC3$^{-/-}$; cell periphery	0.242 ± 0.077	18 nm	81
ARPC3$^{-/-}$; FLPs	0.256 ± 0.056	18 nm	175

(Anderson et al., 2016). This suggests a deficiency in location-dependent fine-tuning of actin-bundle morphology in the mutant cells. Instead, *ARPC3* knockout fibroblasts assemble massive FLPs with densely packed actin filaments, significantly larger than wild-type filopodia. Our data indicate a novel aspect of the Arp2/3 complex's role in cell motility, highlighting its involvement not only in generating dendritic actin networks and regulating filopodia but also in fine-tuning actin-bundle morphology at the cell periphery. Mathematical modeling of protrusions underscores the critical nature of this behavior in preventing the occurrence of massive, tightly packed bundles within the cell (Anderson et al., 2018b).

DYNAMIC MODULATION OF FILAMENT ORGANIZATION IN RESPONSE TO SIGNALING EVENTS

Rho GTPase proteins function as pivotal molecular switches, orchestrating critical signal transduction pathways within cells (Hall, 2012). Upon activation, these proteins engage with downstream effectors, including guanine exchange factors that catalyze GTP binding. Subsequent hydrolysis of GTP induces an inactive conformation. Since their initial characterization (Ridley and Hall, 1992; Ridley et al., 1992; Nobes and Hall, 1995), a combination of biochemical, cell biological, and genetic approaches has unveiled various effectors that mediate the impact of Rho GTPase family members on processes like actin polymerization, filament turnover, and bundling, culminating in the formation of distinct cytoskeletal actin networks. Specific Rho GTPase family members are associated with characteristic actin-based morphologies (Hall, 1998). For instance, RhoA is thought to be linked to actin stress fibers, Rac-1 to ruffles and broader protrusions, and Cdc42 to filopodia. Studies employing fluorescent biosensors to report GTPase activation states in living cells have established correlations between these states and the morphology of the actin cytoskeleton, as well as motility phenotypes, through light microscopy (Kraynov et al., 2000; Nalbant et al., 2004; Komatsu et al., 2011). These studies reveal dynamic GTPase activation patterns, particularly at the leading edge of migrating cells, where they engage in mutual regulation. Notably, Rac-1 activity fluctuates on a submicron and second scale, exerting influence over cell motility, mechanosensing, and invasion (Machacek et al., 2009).

To unravel the initial organization of actin networks prompted by Rac-1 activation at the nanometer scale, we implemented a novel correlative spatiotemporal multiscale workflow (Marston et al., 2019). This approach aimed to quantify the intricate relationship between Rac-1 activation states and the resulting functional outcomes, specifically the distinct nanoarchitecture of the cytoskeleton in regions where Rac-1 activation occurs. Unlike conventional applications of correlative light and electron microscopy that focus on integrating scales to localize entities between the two imaging modalities (Sartori et al., 2007; Schellenberger et al., 2014; Hampton et al., 2017), our methodology takes a step further. We directly connect high-resolution structural information with spatiotemporal signaling dynamics of cellular events. Our procedure involved live-cell imaging utilizing a Förster Resonance Energy Transfer (FRET)–based Rac-1 biosensor to generate spatiotemporal maps depicting Rac-1 activation dynamics at the cell periphery followed by cryo-ET to scrutinize the nanoscale structure of the same region at the time point defined by the last live-cell imaging frame.

A pivotal aspect of our analysis involved selecting regions of interest in close proximity to the cell periphery, where Rac-1 exhibits peak activity (Itoh et al., 2002; Machacek et al., 2009). This strategic choice allows for direct imaging of sufficiently thin regions, circumventing the need for cryo-FIB milling. To expedite analysis further, we implemented several computational procedures. A semiautomatic alignment procedure was introduced to accurately align the biosensor activity maps with cryo-ET data. This method relies on the visibility of features, such as the regularly spaced holes in the carbon film of the EM grids, in both imaging modalities (Anderson et al., 2018a). Achieving an accuracy between 50 and 100 nm, this alignment enables the precise identification of actin filament structures associated with different Rac-1 activity regimes. Following alignment, regions of interest can be automatically extracted using thresholding in the biosensor color maps. Subsequent filament tracing and macromolecular feature detection are then confined to these specific regions. Our accelerated feature detection approach (Xu et al., 2015) is then followed by full vectorization of individual filaments in the regions of interest, facilitating angular coherence and length analysis.

In tomograms of regions lacking high Rac-1 activity, we observed an abundance of ribosomes, diverse macromolecular machines, all three types of cytoskeletal filaments, and vesicles. In stark contrast, regions exhibiting recent high Rac-1 activity exclusively featured actin filament structures characterized by densely packed, short filaments devoid of mutual alignment or branches (Figure 32.1). These

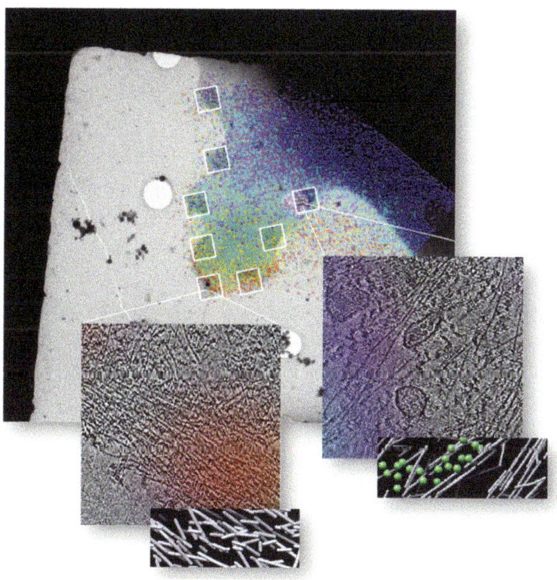

FIGURE 32.1 Correlative light and cryo-ET imaging of actin organization and Rac-1 activity. In the background, an overview cryo-EM image of a cellular region is presented with an overlaid biosensor FRET image indicating Rac-1 activity. High activity is represented in red, while low activity is in purple. The white boxes highlight areas where tomographic data was collected. In the foreground, 3 nm–thick virtual slices through the indicated tomogram regions are displayed together with the correlated biosensor signal. Protein and lipids appear dark in these slices. The insets provide a closer look, indicating traced actin filaments in grey and detected ribosomes in green. Notably, high Rac-1 activity is associated with "haystacks" of short actin filaments without ribosomes, while low Rac-1 activity is linked to long actin filaments and the presence of ribosomes. (Additional details are available in Marston et al., 2019.)

distinctive actin filament configurations showed no discernible presence of other macromolecular assemblies, such as ribosomes or other cytoskeletal filament types. Notably, 150 seconds after the cessation of high Rac-1 activity, no remnants of these actin assemblies were detectable, and the region displayed the same organizational pattern as areas with sustained low Rac-1 activity.

FILAMENT ORDER WITHIN CONTACTS BETWEEN THE CELL AND THE EXTRACELLULAR MATRIX

Cells utilize integrin cell surface receptors to establish assemblies that link the cell's cytoskeleton to the extracellular matrix (ECM). These integrin-dependent assemblies, referred to as focal adhesions (FAs), act as anchoring points where cells attach to the surrounding ECM. Cells actively exert and sense mechanical tension through FAs (Parsons et al., 2010; Eyckmans et al., 2011). Cellular tension contributes to morphogenetic movements and ECM assembly, while the mechanical characteristics of the ECM regulate cell contractility, growth, differentiation, and survival (Orr et al., 2006; Butcher et al., 2009; Costa et al., 2012; Humphrey et al., 2014). Consequently, these interconnections play pivotal roles in various biological processes, including embryonic development, wound healing, and diseases such as cancer, hypertension, and fibrosis. Talin is a central mediator of force transmission between integrins and actin filaments (Critchley, 2009). Talin's N-terminal FERM domain binds to integrin cytoplasmic domains, while its C-terminal rod domain directly links to F-actin through two actin-binding sites and indirectly through vinculin (Ziegler et al., 2008). This places talin in a crucial position for mediating force transmission between the actin cytoskeleton and the ECM, essential for cells to perceive and respond to the mechanical properties of their surroundings.

Utilizing a fluorescence tension sensor on talin in cultured cells (Kumar et al., 2016) has shown that talin is under tension in FAs. However, contrary to expectations, talin in central FAs generally experiences less tension compared to its counterparts in peripheral FAs. Tension in sizable FAs crucially relies on the actin-binding site in the central rod, while the actin-binding site at the C-terminus proves dispensable. Interestingly, talin tension diminishes in cells situated on soft substrates, aligning with the cellular stiffness sensing phenomenon where soft surfaces result in a reduction in contractility (Parsons et al., 2000). Employing talin force sensors and light microscopy, we discovered that talin tension exhibits variability not only between FAs but also locally within individual FAs (Kumar et al., 2018). To investigate these nuances at the nanometer scale, we conducted comprehensive correlative studies using both light microscopy of talin tension sensors and cellular cryo-ET on the underlying cytoskeleton, focusing on peripheral FAs in regions of the cells thin enough for direct cryo-ET imaging.

Achieving accurate alignment between tension sensor maps and cryo-ET data was realized through our semiautomated alignment procedure, leveraging the holes in the carbon film of the EM grids (Anderson et al., 2018a). Following alignment, specific tension regimes can be delineated in the tomograms by applying thresholding to the tension sensor color map. Our primary objective was to unveil the correlation between actin filament alignment and talin tension. Instead of labor-intensive processes involving filament tracing and vectorization, we opted to devise a novel method to extract filament order information directly from the tomograms. This method involved calculating two-dimensional fast Fourier transforms of overlapping boxes in the respective tomogram regions. The distance between boxes was set to a small value to emulate a sliding window. Fourier amplitudes of the contributing boxes were averaged for signal enhancement. An alignment score based on the peak shape and intensity was computed after the application of an iterative median filter to mitigate random noise effects (van der Heide et al., 2007), flattening of the background, and segmentation of the filament alignment contribution from the noise components through entropy maximization (Yen et al., 1995; Sezgin and Sankur, 2004). In regions where actin filaments exhibit precise alignment at a fixed distance and angle, distinct spots emerge in the average Fourier amplitude image derived from contributing sliding windows, resulting in high alignment scores. Any deviations from this ordered arrangement lead to peak broadening with diminished alignment score and, ultimately, complete signal loss (Figure 32.2).

Our quantitative analysis revealed a consistent pattern: In regions with elevated talin tension, actin filaments exhibited remarkable alignment, while regions with lower tension displayed significantly reduced alignment (Kumar et al., 2018). Linear regression across numerous images demonstrated a strong correlation (R = 0.70) between talin tension and local filament alignment. Even in filopodial protrusions, traditionally assumed to have well-aligned actin filaments, FAs with low talin tension exhibited comparatively lower alignment than their counterparts in FAs with high talin tension outside of filopodia.

OPTIMIZING STIFFNESS AND DIMENSIONALITY REGIMES FOR HIGH-THROUGHPUT CRYO-ET

Cells intricately sense and respond to ECM cues, a pivotal aspect in the regulation of cellular behavior. The ECM's biochemical composition, organization, and mechanical properties collectively influence cell growth, survival, migration, and differentiation, playing a crucial role in tissue development and homeostasis. Central to mechanotransduction is the force-induced restructuring of the cell's nanoscale architecture, translating into cellular signals (Katsumi et al., 2004; Schwartz, 2010; Ross et al., 2013; Humphrey et al., 2014; Ohashi et al., 2017). Experimental setups often employ gel-based ECM substrates to explore cellular responses to changes in the microenvironment. Through tunable stiffness in two- and three-dimensional models, these systems facilitate investigations into the impact of ECM composition, stiffness, and dimensionality on cell phenotype (Tse and Engler, 2010; Shinsato et al., 2020). Coupled with molecular manipulations, these systems uncovered key regulators of mechanotransduction (Pelham and Wang, 1997; Wang and Pelham, 1998; Paszek et al., 2005) and how ECM stiffness influences oncogenic transformation (Paszek et al., 2005).

While these conventional setups can capture general morphological changes observable through light microscopy, the rearrangements of cytoskeletal nanoarchitecture induced by force ultimately result from conformational

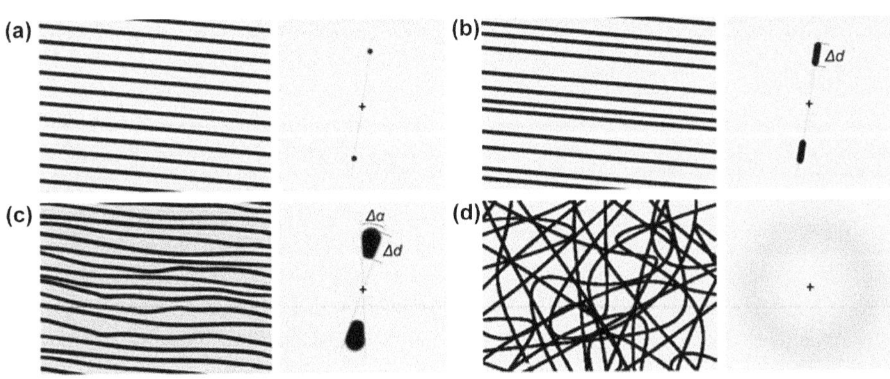

FIGURE 32.2 Schematic representation of the sliding-window Fourier amplitude averaging approach. (a) When filaments are precisely aligned at a fixed distance (left panel), distinct sharp peaks emerge in the sliding-window Fourier amplitude average image (right panel), corresponding to the specific distance between the filaments and their relative angle to the reference frame. (b) If there is a variance *(Δd)* in the distance between filaments, this variation results in broadened peaks perpendicular to the fixed angular direction, proportional to the average amount of *Δd*. (c) When the local angular relationship between filaments undergoes a variation *(Δα)*, broadening occurs around the circular pattern proportional to the average of Δα. (d) In scenarios where local relationships between filaments exhibit near-random patterns, the signal undergoes complete smearing, forming a diffuse ring. (Additional details are available in Kumar et al., 2018.)

changes in the participating molecules. To examine these processes in detail, *in situ* cryo-ET imaging is necessary. In most cryo-ET investigations with mammalian cells, the cells are cultured directly on an EM grid (Hampton et al., 2017). Occasionally, these EM grids are precoated with ECM components such as fibronectin or collagen. With prolonged cell culture studies, the cells deposit and replace the exogenous ECM proteins with their own ECM. However, under such spatial configurations, only the basal portion of the cells engages with the underlying ECM, a condition that does not accurately replicate the real-life ECM. In tissues, mammalian cells interact with three-dimensional surroundings, engaging the entire cell surface with multi-component ECMs. Furthermore, cells in the presence of gel-based ECM substrates are not directly amenable to successful vitrification (Dubochet, 2012), nor to cryo-ET imaging (Peet et al., 2019) due to the thickness of these ECM model systems, typically in the range of many micrometers (Mullen et al., 2015).

The latter obstacles can theoretically be addressed by employing high-pressure freezing and the "lift-out" method for creating electron-transparent sections (lamellae) (Mahamid et al., 2015; Schaffer et al., 2019). In this approach, a block is initially milled under cryogenic conditions from a high-pressure frozen bulk sample using cryo-FIB methodology and then extracted by cryogenically welding the milled block to a thin needle. Subsequently, the block is lifted out of the bulk sample and affixed to an EM grid to create the final lamella for cryo-ET data collection through cryo-FIB milling. However, the effective application of this technology relies heavily on the operator's skill and expertise, and even with proficient handling, success rates tend to be low, often below 30% (Schaffer et al., 2019).

We devised an alternative workflow for investigating cells within the context of gel-based ECM models that requires neither high-pressure freezing nor lift-out methodology (Gaietta et al., 2022). In fact, if the investigation is restricted to the thin cell periphery, no cryo-FIB milling is required at all. The innovation of this workflow is rooted in the use of two separate preparation modules: a hydrogel with tunable stiffness and an ECM-coated EM grid. After placing the ECM-coated EM grid onto the hydrogel, cells are plated and allowed to adhere. The basal membrane engages with the coating, mimicking common 2D cell culture practices for light microscopy studies. To emulate 3D environments, a jellifying solution of reconstituted ECM is added directly on top of the cells, enabling the apical cell membrane and its receptors to engage with the ECM. We utilized Matrigel-based reconstituted basement membrane (rBM) to replicate the ECM environment on the EM grids (Kleinman and Martin, 2005). Alternatively, other ECM model systems like tissue matrix gels or matrices extracted directly from cultured cells (Shinsato et al., 2020) can replace rBM in this workflow. Upon reaching the desired time point, the EM grids, along with the adhered cells, are gently lifted off the hydrogel. Subsequent processing follows the same procedures as cells cultured directly on EM grids, including plunge freezing and direct cryo-ET imaging of thin cell regions.

We applied this workflow to examine how ECM composition, stiffness, and dimensionality influence the nanoarchitecture of the actin cytoskeleton in MCF-10A mammary epithelial cells (Gaietta et al., 2022; Kai et al., 2022). Our focus was on the nanoarchitecture of filopodial protrusions, a structure crucial in environmental sensing. For efficient analysis, we introduced a method based on protrusion width distribution, measuring widths at defined intervals along each protrusion in the tomograms. These distributions serve as a distinctive signature of protrusion architecture (Figure 32.3a). Statistical analysis of these width distributions unveiled a significant impact of ECM parameters on protrusion nanoarchitecture (Figure 32.3b–d). Remarkably, there is no statistically significant difference between coating with rBM and the ECM molecule laminin for the protrusion nanoarchitecture. However, a statistically significant difference exists when comparing the loading of membrane compartments in cells cultured with the two coatings (Gaietta et al., 2022). Our data indicates that laminin, a prominent constituent of rBM, appears to be the primary factor driving alterations in filopodia protrusion nanoarchitecture, while other components of the rBM exert influence on other cellular pathways, such as those governing vesicle loading near the membrane and subsequent exocytosis (Kai et al., 2022). The workflow's design allowed us to gather statistical insights from nearly 200 tomograms, a feat that would require an enormous resource investment if attempted with high-pressure freezing and lift-out procedures.

CONTROLLING POSITIONING AND CELL MORPHOLOGIES IN CRYO-ET

So far, the focus of this chapter has centered on cytoskeletal features near the cell periphery, a logical strategy for exploring actin structures crucial in cell movement and adhesion. The primary advantage of examining thin cell regions lies in the ability to utilize direct cryo-ET imaging, eliminating the need for cryo-FIB milling and substantially boosting throughput by an order of magnitude or more. Nevertheless, numerous actin-based phenomena manifest in regions too thick for direct cryo-ET imaging, necessitating the use of cryo-FIB milling. Recent advancements in on-grid cryo-FIB milling have automated the process (Buckley et al., 2020; Klumpe et al., 2021; Tacke et al., 2021), resulting in significantly higher success rates and not requiring the same level of operator skill and expertise needed for the highly complex lift-out procedure mentioned earlier.

One strategy to enhance throughput in this context even further is an approach that involves utilizing micropatterned cell culture substrates to guide cell adhesion on EM grids, positioning them conveniently for cryo-FIB milling and subsequent cryo-ET data collection (Tee et al., 2015; Engel et al., 2019; Toro-Nahuelpan et al., 2020). As the size and shape of cell microenvironments play pivotal roles in regulating cell architecture, mechanics, polarity, and

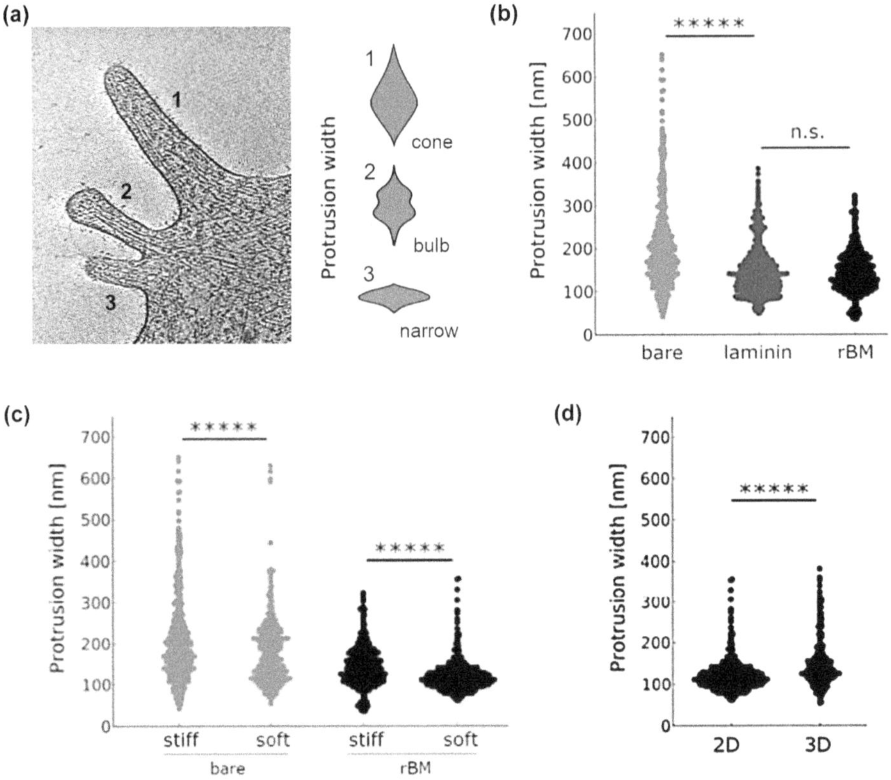

FIGURE 32.3 The impact of environmental factors on actin protrusion morphology. (a) Schematic illustrating the impact of various protrusion architectures on the distribution of protrusion width. The left panel shows a 3 nm–thick virtual slice from a cellular tomogram with three common types of protrusion geometry, including (1) a conelike shape with a wide base tapering off to the top, (2) a bulblike shape featuring a narrow stalk and a rounded bulb at the top, and (3) a protrusion with a consistently narrow width. The accompanying schematics on the right depict the width distributions as measured at defined intervals along the protrusions. The cone shape yields a broad distribution spanning various widths, the bulb shape results in a bimodal distribution corresponding to the bulb and the stalk, and the narrow shape generates a confined distribution covering only a small range of widths. Environmental factors play a significant role in influencing the protrusion width distribution. (b) Impact of extracellular matrix. *rBM* denotes reconstituted basement membrane, *bare* signifies EM grid without any coating. ***** denotes a statistically significant difference with $p < 0.00001$. (c) Impact of stiffness on the protrusion width distribution. (d) Impact of dimensionality on the protrusion width distribution. (Additional details are available in Kai et al., 2022 and in Gaietta et al., 2022.)

function (Théry, 2010; Tee et al., 2015), micropatterning of cell culture substrates is gaining prominence as a tool to alter the structure of the cell microenvironment for *in vitro* studies. As shown by light microscopy, the geometry of micropatterned substrates has proven effective in influencing the establishment of cell polarity and growth (Singhvi et al., 1994), cell viability (Singhvi et al., 1994), terminal differentiation (Watt et al., 1988), and stem cell differentiation (Lee et al., 2013).

Actin-based cell-cell contacts present a challenge for direct cryo-ET imaging due to their location in thicker cellular regions that require cryo-FIB milling. These cell-to-cell junctions play a pivotal role in both developing tissues and maintaining robust barrier function. They involve distinct epithelial assemblies between adhering cells, including tight junctions, gap junctions, and mechanical linkers like desmosomes and adherens junctions (Barone and Heisenberg, 2012). The dynamic nature of developing tissues and external loads on epithelial barriers, such as stretching in the lung or gut, subjects cell-cell contacts to forces that demand adaptive responses from these junctions. Numerous studies have highlighted how mechanical force transfer at these junctions regulates downstream epithelial function (Hartsock and Nelson, 2008). To unravel how epithelial junctions function and adapt to their environments, there is a critical need to visualize these complex structures in native cell-cell configurations at defined functional states. Achieving this task necessitates nanometer-resolution imaging of these dynamic structures.

As a first step toward this goal, we established an enhanced EM grid patterning workflow that refines and integrates previous methods to customize protein pattern sizes and seeding parameters to optimize the field of view for cryo-ET data collection and to facilitate the formation of cell doublets (Dow et al., 2022). Cell doublets represent the simplest cell-cell constructs featuring cell-cell junctions, and the precisely defined size and shape of the micropatterns enable the tuning of force-dependent functional states. We investigated the nanoarchitecture of the protein and bio-passivation layers that make up the micropatterns, employing various imaging modalities such as atomic force microscopy (AFM), cryo-ET, and scanning electron microscopy, in correlation with light microscopy (Dow et al., 2022). Our investigations revealed that the patterned ECM layer adds only a few nanometers in thickness to the EM grid's carbon film. Additionally, we observed that the edge of the pattern, appearing straight in light microscopy due to resolution limitations, exhibits significant nanometer-scale irregularities, a fact impacting physical modeling. Recognizing that substrate stiffness influences cell-cell junction activity, we also characterized the mechanical properties of the carbon films of EM grids, using AFM indentation experiments, placing them within the range of physiological epithelial tissue, such as skin or arterial walls. Within cell doublets, potential cell-to-cell contact sites for cryo-FIB milling were identified by a combination of cryogenic correlative light and electron microscopy and geometric constraints.

Our optimized workflow provides from 9 to 23 potential positions for cryo-FIB milling on each EM grid, each corresponding to a candidate cell-cell junction. Based on our current experience with the Aquilos 2 system, we can generate 15–20 lamellae from a single EM grid in a standard two-day session using semiautomated milling. This aligns well with the number of cell-cell junction candidates per grid. In approximately 90% of the successfully generated lamellae, we identified at least one cell-cell junction. Considering an estimated success rate of 60%, a single session on the Aquilos 2 system thus produces between 5 and 12 cell-cell junctions for further analysis.

CONCLUSIONS AND OUTLOOK

Cryo-ET has emerged as a crucial tool for investigating the nanoscale intricacies of actin-based cytoskeletal machinery. Coupled with advanced image analysis techniques like subtomogram averaging and vectorized actin trace analysis, it has provided crucial high-resolution insights into the *in situ* structure of the actin cytoskeleton within native environments, albeit often based on a small number of tomograms. The emergence of high-throughput techniques for analyzing actin filament structures from larger tomographic datasets is beginning to enable more robust conclusions in cell biology through the application of statistical tools ensuring that observations are not isolated incidents but rather consistent across diverse regions, cells, and time points. In conjunction with high-throughput data collection and analysis frameworks, the manipulation of environmental factors through micropatterning, control over ECM composition, and adjustment of stiffness and dimensionality are facilitating investigations into the actin cytoskeleton within more complex and biologically relevant contexts compared to conventional cell cultures on EM grids. Moreover, incorporating temporal information by analyzing data from different time points and leveraging correlative live-cell imaging adds a dynamic layer of insight. The use of functional probes, such as activation-state and force biosensors, facilitates a direct correlation between signaling events or defined forces and the nanoarchitecture of the underlying actin cytoskeleton.

While considerable progress has been made, there is still much to be accomplished to enhance throughput and fully exploit the wealth of information within cryo-ET data of actin-based processes in biologically relevant settings and specific functional states. Anticipated improvements in throughput will likely stem from advances in hardware and workflow development, particularly in the area of cryo-FIB milling and lamella production from large bulk samples. On the software front, further automation and increased efficiency in actin filament analysis are expected to facilitate a more streamlined extraction of actin network parameters and their correlation with functional data. Despite challenges related to the scarcity of training data in the cryo-ET field, machine learning methods are poised to significantly contribute to these advancements. Undoubtedly, cryo-ET will persist as a pivotal tool for investigating the structure and function of the actin cytoskeleton, emerging as the preferred method for high-resolution exploration of actin-based biological processes.

ACKNOWLEDGMENTS

DH acknowledged support from the U.S. Army Research Office, accomplished under contract W911NF-19-D-0001 for the Institute for Collaborative Biotechnologies, The Beckman Foundation Grant 2021FIB-0000000038, and The National Institutes of Health S10 Instrumentation Grant OD028727. NV acknowledges support from the Fondation pour l'audition grant FPA RD-2019-14.

REFERENCES

Anderson, K. L., Page, C., Swift, M. F., Hanein, D., and Volkmann, N. (2018a). Marker free method for accurate alignment between correlated light, cryo-light, and electron cryo-microscopy data using sample support features. J Struct Biol *201*, 46–51.

Anderson, K. L., Page, C., Swift, M. F., Suraneni, P., Janssen, M. E. W., Pollard, T. D., Li, R., Volkmann, N., and Hanein, D. (2016). Nano-scale actin-network characterization of fibroblast cells lacking functional Arp2/3 complex. J Struct Biol *197*, 312–321.

Anderson, K. L., Swift, M. F., Hanein, D., and Volkmann, N. (2018b). Does self-organized criticality drive leading edge protrusion. Biophys Rev *10*, 1571–1575.

Barone, V., and Heisenberg, C. P. (2012). Cell adhesion in embryo morphogenesis. Curr Opin Cell Biol *24*, 148–153.

Bäuerlein, F. J. B., and Baumeister, W. (2021). Towards visual proteomics at high resolution. J Mol Biol *433*, 167187.

Baumeister, W. (2022). Cryo-electron tomography: a long journey to the inner space of cells. Cell *185*, 2649–2652.

Blanchoin, L., Amann, K. J., Higgs, H. N., Marchand, J. B., Kaiser, D. A., and Pollard, T. D. (2000). Direct observation of dendritic actin filament networks nucleated by Arp2/3 complex and WASP/Scar proteins. Nature *404*, 1007–1011.

Block, J., Breitsprecher, D., Kühn, S., Winterhoff, M., Kage, F., Geffers, R., Duwe, P., Rohn, J. L., Baum, B., Brakebusch, C., Geyer, M., Stradal, T. E. B., Faix, J., and Rottner, K. (2012). FMNL2 drives actin-based protrusion and migration downstream of Cdc42. Curr Biol *22*, 1005–1012.

Block, J., Stradal, T. E. B., Hänisch, J., Geffers, R., Köstler, S. A., Urban, E., Small, J. V., Rottner, K., and Faix, J. (2008). Filopodia formation induced by active mDia2/Drf3. J Microsc *231*, 506–517.

Breitsprecher, D., Jaiswal, R., Bombardier, J. P., Gould, C. J., Gelles, J., and Goode, B. L. (2012). Rocket launcher mechanism of collaborative actin assembly defined by single-molecule imaging. Science *336*, 1164–1168.

Buckley, G., Gervinskas, G., Taveneau, C., Venugopal, H., Whisstock, J. C., and de Marco, A. (2020). Automated cryo-lamella preparation for high-throughput in-situ structural biology. J Struct Biol *210*, 107488.

Butcher, D. T., Alliston, T., and Weaver, V. M. (2009). A tense situation: forcing tumour progression. Nat Rev Cancer *9*, 108–122.

Chakraborty, S., Jasnin, M., and Baumeister, W. (2020). Three-dimensional organization of the cytoskeleton: A cryo-electron tomography perspective. Protein Sci *29*, 1302–1320.

Chesarone, M. A., and Goode, B. L. (2009). Actin nucleation and elongation factors: mechanisms and interplay. Curr Opin Cell Biol *21*, 28–37.

Chou, S. Z., Chatterjee, M., and Pollard, T. D. (2022). Mechanism of actin filament branch formation by Arp2/3 complex revealed by a high-resolution cryo-EM structureof the branch junction. Proc Natl Acad Sci USA *119*, e2206722119.

Chua, E. Y. D., Mendez, J. H., Rapp, M., Ilca, S. L., Tan, Y. Z., Maruthi, K., Kuang, H., Zimanyi, C. M., Cheng, A., Eng, E. T., Noble, A. J., Potter, C. S., and Carragher, B. (2022). Better, faster, cheaper: recent advances in cryo-electron microscopy. Annu Rev Biochem *91*, 1–32.

Costa, P., Almeida, F. V., and Connelly, J. T. (2012). Biophysical signals controlling cell fate decisions: how do stem cells really feel. Int J Biochem Cell Biol *44*, 2233–2237.

Critchley, D. R. (2009). Biochemical and structural properties of the integrin-associated cytoskeletal protein talin. Annu Rev Biophys *38*, 235–254.

DeRosier, D. J., and Klug, A. (1968). Reconstruction of three-dimensional structures from electron micrographs. Nature *217*, 130–134.

Dimchev, G., Amiri, B., Fäßler, F., Falcke, M., and Schur, F. K. (2021). Computational toolbox for ultrastructural quantitative analysis of filament networks in cryo-ET data. J Struct Biol *213*, 107808.

Dow, L. P., Gaietta, G., Kaufman, Y., Swift, M. F., Lemos, M., Lane, K., Hopcroft, M., Bezault, A., Sauvanet, C., Volkmann, N., Pruitt, B. L., and Hanein, D. (2022). Morphological control enables nanometer-scale dissection of cell-cell signaling complexes. Nat Commun *13*, 7831.

Dubochet, J. (2012). Cryo-EM–the first thirty years. J Microsc *245*, 221–224.

Engel, L., Gaietta, G., Dow, L. P., Swift, M. F., Pardon, G., Volkmann, N., Weis, W. I., Hanein, D., and Pruitt, B. L. (2019). Extracellular matrix micropatterning technology for whole cell cryogenic electron microscopy studies. J Micromech Microeng *29*, 115018.

Eyckmans, J., Boudou, T., Yu, X., and Chen, C. S. (2011). A hitchhiker's guide to mechanobiology. Dev Cell *21*, 35–47.

Fäßler, F., Dimchev, G., Hodirnau, V. V., Wan, W., and Schur, F. K. M. (2020). Cryo-electron tomography structure of Arp2/3 complex in cells reveals new insights into the branch junction. Nat Commun *11*, 6437.

Fäßler, F., Javoor, M. G., and Schur, F. K. (2023). Deciphering the molecular mechanisms of actin cytoskeleton regulation in cell migration using cryo-EM. Biochem Soc Trans *51*, 87–99.

Gaietta, G., Kai, F., Swift, M. F., Weaver, V. M., Volkmann, N., and Hanein, D. (2022). Novel cryo-tomography workflow reveals nanometer-scale responses of epithelial cells to matrix stiffness and dimensionality. Mol Biol Cell *33*, br28.

Hall, A. (1998). Rho GTPases and the actin cytoskeleton. Science *279*, 509–514.

Hall, A. (2012). Rho family GTPases. Biochem Soc Trans *40*, 1378–1382.

Hampton, C. M., Strauss, J. D., Ke, Z., Dillard, R. S., Hammonds, J. E., Alonas, E., Desai, T. M., Marin, M., Storms, R. E., Leon, F., Melikyan, G. B., Santangelo, P. J., Spearman, P. W., and Wright, E. R. (2017). Correlated fluorescence microscopy and cryo-electron tomography of virus-infected or transfected mammalian cells. Nat Protoc *12*, 150–167.

Hartsock, A., and Nelson, W. J. (2008). Adherens and tight junctions: structure, function and connections to the actin cytoskeleton. Biochim Biophys Acta *1778*, 660–669.

Humphrey, J. D., Dufresne, E. R., and Schwartz, M. A. (2014). Mechanotransduction and extracellular matrix homeostasis. Nat Rev Mol Cell Biol *15*, 802–812.

Itoh, R. E., Kurokawa, K., Ohba, Y., Yoshizaki, H., Mochizuki, N., and Matsuda, M. (2002). Activation of Rac and cdc42 video imaged by fluorescent resonance energy transfer-based single-molecule probes in the membrane of living cells. Mol Cell Biol *22*, 6582–6591.

Jasnin, M., Asano, S., Gouin, E., Hegerl, R., Plitzko, J. M., Villa, E., Cossart, P., and Baumeister, W. (2013). Three-dimensional architecture of actin filaments in Listeria monocytogenes comet tails. Proc Natl Acad Sci USA *110*, 20521–20526.

Jasnin, M., Beck, F., Ecke, M., Fukuda, Y., Martinez-Sanchez, A., Baumeister, W., and Gerisch, G. (2019). The architecture of traveling actin waves revealed by cryo-electron tomography. Structure *27*, 1211–1223.e5.

Kai, F., Ou, G., Tourdot, R. W., Stashko, C., Gaietta, G., Swift, M. F., Volkmann, N., Long, A. F., Han, Y., Huang, H. H., Northey, J. J., Leidal, A. M., Viasnoff, V., Bryant, D. M., Guo, W., Wiita, A. P., Guo, M., Dumont, S., Hanein, D., Radhakrishnan, R., and Weaver, V. M. (2022). ECM dimensionality tunes actin tension to modulate endoplasmic reticulum function and spheroid phenotypes of mammary epithelial cells. EMBO J *41*, e109205.

Katsumi, A., Orr, A. W., Tzima, E., and Schwartz, M. A. (2004). Integrins in mechanotransduction. J Biol Chem *279*, 12001–12004.

Klein, S., Wimmer, B. H., Winter, S. L., Kolovou, A., Laketa, V., and Chlanda, P. (2021). Post-correlation on-lamella cryo-CLEM reveals the membrane architecture of lamellar bodies. Commun Biol *4*, 137.

Kleinman, H. K., and Martin, G. R. (2005). Matrigel: basement membrane matrix with biological activity. Semin Cancer Biol *15*, 378–386.

Klumpe, S., Fung, H. K., Goetz, S. K., Zagoriy, I., Hampoelz, B., Zhang, X., Erdmann, P. S., Baumbach, J., Müller, C. W., Beck, M., Plitzko, J. M., and Mahamid, J. (2021). A modular platform for automated cryo-FIB workflows. Elife *10*, e70506.

Komatsu, N., Aoki, K., Yamada, M., Yukinaga, H., Fujita, Y., Kamioka, Y., and Matsuda, M. (2011). Development of an optimized backbone of FRET biosensors for kinases and GTPases. Mol Biol Cell *22*, 4647–4656.

Krause, M., and Gautreau, A. (2014). Steering cell migration: lamellipodium dynamics and the regulation of directional persistence. Nat Rev Mol Cell Biol *15*, 577–590.

Kraynov, V. S., Chamberlain, C., Bokoch, G. M., Schwartz, M. A., Slabaugh, S., and Hahn, K. M. (2000). Localized Rac activation dynamics visualized in living cells. Science *290*, 333–337.

Kumar, A., Anderson, K. L., Swift, M. F., Hanein, D., Volkmann, N., and Schwartz, M. A. (2018). Local tension on talin in focal adhesions correlates with F-actin alignment at the nanometer scale. Biophys J *115*, 1569–1579.

Kumar, A., Ouyang, M., Van den Dries, K., McGhee, E. J., Tanaka, K., Anderson, M. D., Groisman, A., Goult, B. T., Anderson, K. I., and Schwartz, M. A. (2016). Talin tension sensor reveals novel features of focal adhesion force transmission and mechanosensitivity. J Cell Biol *213*, 371–383.

Lee, J., Abdeen, A. A., Zhang, D., and Kilian, K. A. (2013). Directing stem cell fate on hydrogel substrates by controlling cell geometry, matrix mechanics and adhesion ligand composition. Biomaterials *34*, 8140–8148.

Loisel, T. P., Boujemaa, R., Pantaloni, D., and Carlier, M. F. (1999). Reconstitution of actin-based motility of Listeria and Shigella using pure proteins. Nature *401*, 613–616.

Machacek, M., Hodgson, L., Welch, C., Elliott, H., Pertz, O., Nalbant, P., Abell, A., Johnson, G. L., Hahn, K. M., and Danuser, G. (2009). Coordination of Rho GTPase activities during cell protrusion. Nature *461*, 99–103.

Mahamid, J., Schampers, R., Persoon, H., Hyman, A. A., Baumeister, W., and Plitzko, J. M. (2015). A focused ion beam milling and lift-out approach for site-specific preparation of frozen-hydrated lamellas from multicellular organisms. J Struct Biol *192*, 262–269.

Marston, D. J., Anderson, K. L., Swift, M. F., Rougie, M., Page, C., Hahn, K. M., Volkmann, N., and Hanein, D. (2019). High Rac1 activity is functionally translated into cytosolic structures with unique nanoscale cytoskeletal architecture. Proc Natl Acad Sci USA *116*, 1267–1272.

Martins, B., Sorrentino, S., Chung, W. L., Tatli, M., Medalia, O., and Eibauer, M. (2021). Unveiling the polarity of actin filaments by cryo-electron tomography. Structure 29, 488–498.e4.

Mullen, C. A., Vaughan, T. J., Billiar, K. L., and McNamara, L. M. (2015). The effect of substrate stiffness, thickness, and cross-linking density on osteogenic cell behavior. Biophys J *108*, 1604–1612.

Mullins, R. D., Heuser, J. A., and Pollard, T. D. (1998). The interaction of Arp2/3 complex with actin: nucleation, high affinity pointed end capping, and formation of branching networks of filaments. Proc Natl Acad Sci USA *95*, 6181–6186.

Nalbant, P., Hodgson, L., Kraynov, V., Toutchkine, A., and Hahn, K. M. (2004). Activation of endogenous Cdc42 visualized in living cells. Science *305*, 1615–1619.

Navarro, P. P. (2022). Quantitative cryo-electron tomography. Front Mol Biosci *9*, 934465.

Nobes, C. D., and Hall, A. (1995). Rho, rac, and cdc42 GTPases regulate the assembly of multimolecular focal complexes associated with actin stress fibers, lamellipodia, and filopodia. Cell *81*, 53–62.

Nolen, B. J., Tomasevic, N., Russell, A., Pierce, D. W., Jia, Z., McCormick, C. D., Hartman, J., Sakowicz, R., and Pollard, T. D. (2009). Characterization of two classes of small molecule inhibitors of Arp2/3 complex. Nature *460*, 1031–1034.

Ohashi, K., Fujiwara, S., and Mizuno, K. (2017). Roles of the cytoskeleton, cell adhesion and rho signalling in mechanosensing and mechanotransduction. J Biochem *161*, 245–254.

Orr, A. W., Helmke, B. P., Blackman, B. R., and Schwartz, M. A. (2006). Mechanisms of Mechanotransduction. Dev Cell *10*, 11–20.

Parsons, J. T., Horwitz, A. R., and Schwartz, M. A. (2010). Cell adhesion: integrating cytoskeletal dynamics and cellular tension. Nat Rev Mol Cell Biol *11*, 633–643.

Parsons, J. T., Martin, K. H., Slack, J. K., Taylor, J. M., and Weed, S. A. (2000). Focal adhesion kinase: a regulator of focal adhesion dynamics and cell movement. Oncogene *19*, 5606–5613.

Paszek, M. J., Zahir, N., Johnson, K. R., Lakins, J. N., Rozenberg, G. I., Gefen, A., Reinhart-King, C. A., Margulies, S. S., Dembo, M., Boettiger, D., Hammer, D. A., and Weaver, V. M. (2005). Tensional homeostasis and the malignant phenotype. Cancer Cell *8*, 241–254.

Peet, M. J., Henderson, R., and Russo, C. J. (2019). The energy dependence of contrast and damage in electron cryomicroscopy of biological molecules. Ultramicroscopy *203*, 125–131.

Pelham, R. J., and Wang, Y. (1997). Cell locomotion and focal adhesions are regulated by substrate flexibility. Proc Natl Acad Sci USA *94*, 13661–13665.

Pollard, T. D. (2016). Actin and actin-binding proteins. Cold Spring Harb Perspect Biol *8*, a018226.

Pruyne, D., Evangelista, M., Yang, C., Bi, E., Zigmond, S., Bretscher, A., and Boone, C. (2002). Role of formins in actin assembly: nucleation and barbed-end association. Science *297*, 612–615.

Ridley, A. J., and Hall, A. (1992). The small GTP-binding protein rho regulates the assembly of focal adhesions and actin stress fibers in response to growth factors. Cell *70*, 389–399.

Ridley, A. J., Paterson, H. F., Johnston, C. L., Diekmann, D., and Hall, A. (1992). The small GTP-binding protein rac regulates growth factor-induced membrane ruffling. Cell *70*, 401–410.

Rigort, A., Bäuerlein, F. J. B., Villa, E., Eibauer, M., Laugks, T., Baumeister, W., and Plitzko, J. M. (2012a). Focused ion beam micromachining of eukaryotic cells for cryoelectron tomography. Proc Natl Acad Sci USA *109*, 4449–4454.

Rigort, A., Günther, D., Hegerl, R., Baum, D., Weber, B., Prohaska, S., Medalia, O., Baumeister, W., and Hege, H.-C. (2012b). Automated segmentation of electron tomograms for a quantitative description of actin filament networks. J Struct Biol *177*, 135–144.

Ross, T. D., Coon, B. G., Yun, S., Baeyens, N., Tanaka, K., Ouyang, M., and Schwartz, M. A. (2013). Integrins in mechanotransduction. Curr Opin Cell Biol *25*, 613–618.

Rouiller, I., Xu, X. P., Amann, K. J., Egile, C., Nickell, S., Nicastro, D., Li, R., Pollard, T. D., Volkmann, N., and Hanein, D. (2008). The structural basis of actin filament branching by the Arp2/3 complex. J Cell Biol *180*, 887–895.

Sartori, A., Gatz, R., Beck, F., Rigort, A., Baumeister, W., and Plitzko, J. M. (2007). Correlative microscopy: bridging the gap between fluorescence light microscopy and cryo-electron tomography. J Struct Biol *160*, 135–145.

Schaffer, M., Pfeffer, S., Mahamid, J., Kleindiek, S., Laugks, T., Albert, S., Engel, B. D., Rummel, A., Smith, A. J., Baumeister, W., and Plitzko, J. M. (2019). A cryo-FIB lift-out technique enables molecular-resolution cryo-ET within native *Caenorhabditis elegans* tissue. Nat Methods *16*, 757–762.

Schellenberger, P., Kaufmann, R., Siebert, C. A., Hagen, C., Wodrich, H., and Grünewald, K. (2014). High-precision correlative fluorescence and electron cryo microscopy using two independent alignment markers. Ultramicroscopy *143*, 41–51.

Schneider, J., and Jasnin, M. (2022). Capturing actin assemblies in cells using in situ cryo-electron tomography. Eur J Cell Biol *101*, 151224.

Schwartz, M. A. (2010). Integrins and extracellular matrix in mechanotransduction. Cold Spring Harb Perspect Biol *2*, a005066.

Sezgin, M., and Sankur, B. (2004). Survey over image thresholding techniques and quantitative performance evaluation. J Electronic Imaging *13*, 146–166.

Shinsato, Y., Doyle, A. D., Li, W., and Yamada, K. M. (2020). Direct comparison of five different 3D extracellular matrix model systems for characterization of cancer cell migration. Cancer Rep (Hoboken) *3*, e1257.

Singhvi, R., Kumar, A., Lopez, G. P., Stephanopoulos, G. N., Wang, D. I., Whitesides, G. M., and Ingber, D. E. (1994). Engineering cell shape and function. Science *264*, 696–698.

Suraneni, P., Fogelson, B., Rubinstein, B., Noguera, P., Volkmann, N., Hanein, D., Mogilner, A., and Li, R. (2015). A mechanism of leading edge protrusion in the absence of Arp2/3 complex. Mol Biol Cell *26*, 901–912.

Suraneni, P., Rubinstein, B., Unruh, J. R., Durnin, M., Hanein, D., and Li, R. (2012). The Arp2/3 complex is required for lamellipodia extension and directional fibroblast cell migration. J Cell Biol *197*, 239–251.

Svitkina, T. M., and Borisy, G. G. (1999). Arp2/3 complex and actin depolymerizing factor/cofilin in dendritic organization and treadmilling of actin filament array in lamellipodia. J Cell Biol *145*, 1009–1026.

Tacke, S., Erdmann, P., Wang, Z., Klumpe, S., Grange, M., Plitzko, J., and Raunser, S. (2021). A streamlined workflow for automated cryo focused ion beam milling. J Struct Biol *213*, 107743.

Tamborrini, D., Wang, Z., Wagner, T., Tacke, S., Stabrin, M., Grange, M., Kho, A. L., Rees, M., Bennett, P., Gautel, M., and Raunser, S. (2023). Structure of the native myosin filament in the relaxed cardiac sarcomere. Nature *623*, 863–871.

Tee, Y. H., Shemesh, T., Thiagarajan, V., Hariadi, R. F., Anderson, K. L., Page, C., Volkmann, N., Hanein, D., Sivaramakrishnan, S., Kozlov, M. M., and Bershadsky, A. D. (2015). Cellular chirality arising from the self-organization of the actin cytoskeleton. Nat Cell Biol *17*, 445–457.

Théry, M. (2010). Micropatterning as a tool to decipher cell morphogenesis and functions. J Cell Sci *123*, 4201–4213.

Toro-Nahuelpan, M., Zagoriy, I., Senger, F., Blanchoin, L., Théry, M., and Mahamid, J. (2020). Tailoring cryo-electron microscopy grids by photo-micropatterning for in-cell structural studies. Nat Methods *17*, 50–54.

Tse, J. R., and Engler, A. J. (2010). Preparation of hydrogel substrates with tunable mechanical properties. Curr Protoc Cell Biol, *Chapter 10*, https://doi.org/10.1002/0471143030.cb1016s47.

Turk, M., and Baumeister, W. (2020). The promise and the challenges of cryo-electron tomography. FEBS Lett *594*, 3243–3261.

van der Heide, P., Xu, X. P., Marsh, B. J., Hanein, D., and Volkmann, N. (2007). Efficient automatic noise reduction of electron tomographic reconstructions based on iterative median filtering. J Struct Biol *158*, 196–204.

Vignjevic, D., Kojima, S.-I., Aratyn, Y., Danciu, O., Svitkina, T., and Borisy, G. G. (2006). Role of fascin in filopodial protrusion. J Cell Biol *174*, 863–875.

Volkmann, N., Amann, K. J., Stoilova-McPhie, S., Egile, C., Winter, D. C., Hazelwood, L., Heuser, J. E., Li, R., Pollard, T. D., and Hanein, D. (2001). Structure of Arp2/3 complex in its activated state and in actin filament branch junctions. Science *293*, 2456–2459.

Wang, Y. L., and Pelham, R. J. (1998). Preparation of a flexible, porous polyacrylamide substrate for mechanical studies of cultured cells. Methods Enzymol *298*, 489–496.

Wang, Z., Grange, M., Pospich, S., Wagner, T., Kho, A. L., Gautel, M., and Raunser, S. (2022). Structures from intact myofibrils reveal mechanism of thin filament regulation through nebulin. Science *375*, eabn1934.

Watt, F. M., Jordan, P. W., and O'Neill, C. H. (1988). Cell shape controls terminal differentiation of human epidermal keratinocytes. Proc Natl Acad Sci USA *85*, 5576–5580.

Welch, M. D., Iwamatsu, A., and Mitchison, T. J. (1997). Actin polymerization is induced by Arp2/3 protein complex at the surface of *Listeria monocytogenes*. Nature *385*, 265–269.

Xu, X. P., Page, C., and Volkmann, N. (2015). Efficient extraction of macromolecular complexes from electron tomograms based on reduced representation templates. Comput Anal Images Patterns *9256*, 423–431.

Yang, C., and Svitkina, T. (2011). Visualizing branched actin filaments in lamellipodia by electron tomography. Nat Cell Biol *13*, 1012–1013.

Yen, J. C., Chang, F. J., and Chang, S. (1995). A new criterion for automatic multilevel thresholding. IEEE Trans Image Process *4*, 370–378.

Young, L. N., and Villa, E. (2023). Bringing structure to cell biology with cryo-electron tomography. Annu Rev Biophys *52*, 573–595.

Ziegler, W. H., Gingras, A. R., Critchley, D. R., and Emsley, J. (2008). Integrin connections to the cytoskeleton through talin and vinculin. Biochem Soc Trans *36*, 235–239.

Zigmond, S. H., Evangelista, M., Boone, C., Yang, C., Dar, A. C., Sicheri, F., Forkey, J., and Pring, M. (2003). Formin leaky cap allows elongation in the presence of tight capping proteins. Curr Biol *13*, 1820–1823.

33 Cryo-EM in Drug Discovery

Youxin Kong, Maxime Killer, Madhumati Sevvana, and Chiara Rapisarda

STRUCTURAL BIOLOGY IN DRUG DISCOVERY

The development of drugs to treat human diseases usually requires the modulation of proteins or RNA by either biologics (antibodies, antibody–drug conjugates, enzymes, and pegylated proteins) or new chemical entities (small molecules, peptides, and oligonucleotides) (de la Torre et al., 2022). A drug, before reaching the clinical stage and being tested on humans, must undergo several cycles of optimization for bioavailability, affinity, potency, and safety. Before crystallography ventured out of academic labs in the 1970s (Beddell et al., 1976) and moved toward industry, the chemical discovery and optimization cycle of a drug—a time-consuming process that could take as long as 14 years (Paul et al., 2010)—was performed by blind "brute force" with high-throughput screens (Mountain, 2003). Currently, pharmaceutical companies have embraced more efficient knowledge-driven approaches by incorporating structural biology (SB) into the drug discovery pipeline right from the beginning wherever applicable.

The SB-led drug discovery process begins with the identification of a suitable target whose activity can be modulated by using a chemical or a biologic entity. This requires the availability of the three-dimensional (3D) structure of the target protein deciphered either by experimental or *in silico* approaches. The potential chemical entities are then identified by looking at naturally binding compounds, substrate or cofactor analogs, or by carrying out high-throughput or fragment-based screens (HTS) (Maveyraud et al., 2020). For therapeutic antibodies (Ab), the best binders are typically generated by inducing an immunogenic response in mice (sometimes transgenic humanized ones) to the human target protein (Green, 2014) or through the selection of synthetic human antibodies using phage libraries (Frenzel et al., 2016). The initial functional Ab binders of the target protein are biophysically characterized, and the epitopes of the shortlisted candidate Abs are structurally characterized. A virtual screen or a *de novo* design can be performed to increase the number of potential binders and enhance their druglike nature followed by an assessment of potential oral bioavailability. The ligands obtained are then optimized for absorption, distribution, metabolism, excretion, and toxicity (ADMET) and synthesized to be tested and profiled for biological activity. If all the conditions are met, one lead molecule is selected that will proceed to the clinical stage of the drug discovery pipeline (see Figure 33.1).

Crystallography is the most widely used technique for structural characterization of drug-target complexes and fragment-based drug discovery (FBDD), thanks to technological advances that have made it a high-throughput method for screening a large number of compounds simultaneously. Crystallography is, however, limited by the amount of protein required for crystallization, the requirement for the protein to form crystals, the solubility of the protein, such as membrane-embedded ones, and the impossibility of accounting for structural heterogeneity (due to posttranslational modifications) and the dynamic behaviors of proteins (Zheng et al., 2015).

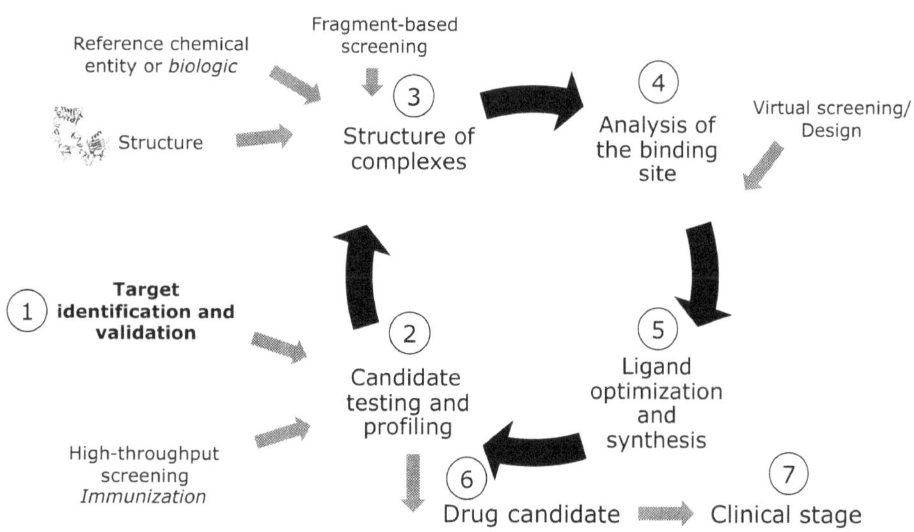

FIGURE 33.1 The stages of structure-based drug discovery.

DOI: 10.1201/9781003326106-40

To address these limitations, cryo-electron microscopy (cryo-EM) has been propelled in the pharmaceutical industry in the last eight years from a niche technique—with the pioneering infrastructure investment completed in 2017 by Genentech (U.S.) (Estevez et al., 2019) and the establishment of a consortium of five different pharma companies in Cambridge (UK) in 2016 (Sader et al., 2020)—to being a staple of structural biology–based pharma everywhere in the world (Noble, 2023). For 40 years, crystallography was leading the race in obtaining high-resolution 3D structures until recently, when cryo-EM reached the resolutions to be informational for drug design (Kühlbrandt, 2014).

Here we discuss the impact of cryo-EM on the drug discovery process, from deciphering 3D structure information of small and large molecules to specific applications on membrane proteins and vaccine design.

CRYO-EM IN THE SMALL-MOLECULE DRUG DISCOVERY PIPELINE

Small-molecule drugs are synthetic compounds or modified natural products weighing below 1000 Da. Due to their versatility for treating various diseases and accessibility for patients across economic backgrounds, small molecules dominate today's drug market by constituting as much as 90% of global sales (Makurvet, 2021). Traditionally, the binding mechanism of a small-molecule drug and its POI (protein of interest) is identified by X-ray crystallography, or in rare cases where the target is small enough (<20 kDa), by NMR. In recent years, cryo-EM has been applied to interrogate novel small-molecule-related modalities, such as protein degraders (molecular glues and PROTACs), antibody–drug conjugates (ADCs), and peptide–drug conjugates (PDCs). During a typical research stage of a drug discovery program (two to four years), cryo-EM can be adapted to provide continuous, on-demand support as it proceeds. At the exploratory stage, cryo-EM can be used to probe the target's active and allosteric sites as well as complexes with its upstream or downstream binding partners. Even low- to mid-resolution structures or 2D analyses are beneficial for accessing the target's compositional and conformational heterogeneity. To develop a first-in-class (FIC) drug, a reliable cryo-EM structure, either distinctive or complementary to previous crystal structures, can serve as a good starting point. For a fast-follow or best-in-class (BIC) drug project, a robust cryo-EM pipeline for generating compound–POI structures en masse can give the team a unique edge over competitors. At the lead identification stage, cryo-EM can generate lead–target complexes and perform fragment-based drug discovery (FBDD) as demonstrated in the case of the oncology target pyruvate kinase 2 (PKM2) (Saur et al., 2020). Even when the lead has been optimized and a drug candidate is ready to enter early development, cryo-EM can provide a retrospective structural mapping for the drug–target interactions in a timely manner.

CRYO-EM IN DRUG DISCOVERY OF ENZYMES

Due to its ability to preserve protein samples in their near-native state, cryo-EM is particularly well suited for the structure-based drug discovery (SBDD) of enzymes. One of the notable examples is the development of cytidine triphosphate synthase 1 (CTPS1)–selective inhibitors. DNA and RNA replications are dependent on a constant and tightly controlled supply of essential nucleotides such as CTP. In humans, two homologous enzymes, CTPS1 and CTPS2, catalyze the final, rate-limiting step of CTP synthesis from UTP and ammonia. Genomic studies revealed that almost all cancer cell lines rely on CTPS1 for *de novo* pyrimidine synthesis, while CTPS2 appeared to be inconsequential (https://depmap.org/portal/achilles/). In active lymphocytes, CTPS1 expression is upregulated to meet the elevated demands of nucleotides and phospholipids. In contrast, CTPS2 is used to maintain CTP levels in other tissues, serving compensatory functions outside the hematopoietic system (Martin et al., 2014). Hence, a selective inhibitor for human CTPS1 not only allows for precision targeting of tumor cells with limited systemic toxicity, but also enables specific immune suppression of lymphocyte-related disorders. Early X-ray crystallography work revealed a partial snapshot of the human CTPS1 synthetase domain (Kursula et al., 2006). In 2017, Lynch et al. discovered by cryo-EM that the two CTPS homologs achieve metabolic function by forming helical filaments of homo-tetramers (Lynch et al., 2017). Subsequent cryo-EM structures of inhibitors binding to CTPS1 and CTPS2 interrogated the differences between the two homologs across species, revealing a binding site close to the ATP hydrolysis pockets for human CTPS1 (Figure 33.2a). Despite a high similarity in polymer assembly, subtle structural changes at the monomer level, notably a one-residue substitution (I250T), holds the key to the compound's selectivity. Structure-based design of small-molecule compounds could translate into fine-tuning the two homologous enzymes' activity at a dynamic, polymer level. Further developments of these leads may faciliate a potent CTPS1-oriented drug applicable in treating cancer as well as immune disorders. Potential drug candidates targeting CTPS1 are currently in clinical trial (STP-938, Step Pharma [Asnagli et al., 2023]) and preclinical development (NTX-871, Nimbus Therapeutics, [McElwee et al., 2023]).

Another prominent example is the hunt for small-molecule drugs targeting the RAS GTPases (HRAS, KRAS or NRAS). The RAS family are the cytoplasmic initiators for the RTK (receptor tyrosine kinase)/Ras/MAPK pathway, which represents one of the best characterized signaling cascades in cell biology as well as human pathologies. Despite their importance, the RAS proteins are termed "the death star of cancer" due to their spherical shape and seemingly undruggable structure. In 2013, a breakthrough discovery identified compounds irreversibly bound to the KRAS-G12C mutant (Ostrem et al., 2013). This finding sparked a heated contest for RAS inhibitors, as SBDD of an orally viable drug to switch off a range of oncogenic

Internal

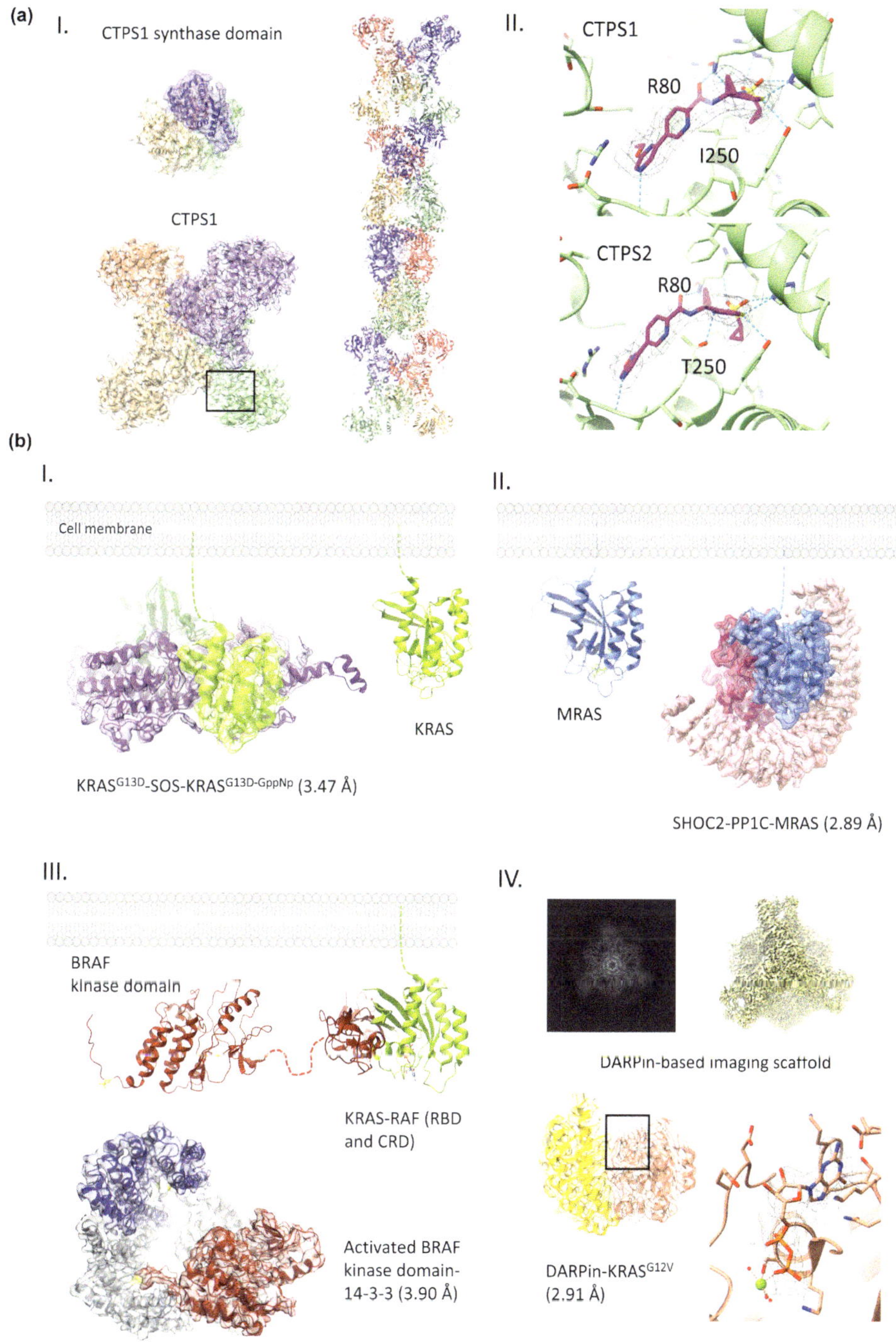

FIGURE 33.2 (a) Crystal structure of the synthase domain of hCTPS1 CTP (*top left*, PDB 2VO1), cryo-EM structure of hCTP1 tetramer bound to AMPPNP, UTP and glutaime (*bottom left*, PDB 7MGZ EMD-), and cryo-EM structure of hCTPS1 filament bound to ATP and UTP (*right*, PDB) are shown in panel I. Cryo-EM structures of hCTPS1 (*top*) and hCTP2 (*bottom*) binding to inhibior R80 (PDB 7MIF EMD-23848; PDB 7MIH EMD-23851) are shown in panel II. Hydrogen bonds are colored in cyan and the selectivity-related residue I250T is labeled in black. (b) Representative structural complexes in the RAS/RAF signaling pathway. Cryo-EM electron density maps (shown as surface) and structural models (shown as ribbon) of the KRAS-SOS complex (PDB 7KFZ EMD-22857), the SHOC2-PP1C-MRAS complex (PDB 7UPI EMD-26667), the BRAF kinase domain-14-3-3 complex (PDB 6UAN EMD-20708), and the BRAF-MEK1-14-3-3 complex (PDB 6NYB EMD-0541) are listed from panels I to III. Cryo-EM structural details of the KRASG12V-GDP (PDB 8G4F EMD-29719) binding to a DARPin-based imaging scaffold (PDB 8G3K EMD-29700) are shown in panel IV.

pathways seemed attainable (https://www.cancer.gov/tcga). However, drug resistance and other limits have begun to hamper the clinical advances, and fresh thinking is urgently needed.

Typically, RAS-mediated signaling begins with the interaction of a growth factor and its tyrosine receptor kinase on the cell surface. After being recruited to the signal transduction site, RAS proteins take on the GTP load and subsequently bind to a RAF kinase. The dephosphorylation of RAFs leads to dimerization, which is a prerequisite for downstream MAPK/ERK activation (Braicu et al., 2019). The activation of the Ras/Raf/MAPK pathway is tightly regulated by multiple checkpoints, which include trapping the RAF kinase in an inactive form before the engagement of RAS. Cryo-EM and X-ray crystallography studies in 2019 revealed the structural basis for RAF kinase (BRAF) inhibition by a 14-3-3 dimer (Kondo et al., 2019). In the same year, a cryo-EM structure of KRAS/BRAF/MEK1/14-3-3 assembly published by Pfizer pointed to a preactivation, autoinhibitory "recruitment complex" involving KRAS, suggesting a further level of regulation (Park et al., 2019). In 2022, a series of high-impact publications revealed the structural basis for SHOC2 modulation of RAS signaling, in which cryo-EM structures of the tertiary SHOC2-PP1C-MRAS complex were solved at 2.95 Å and 2.89 Å resolution (Bonsor et al., 2022; Hauseman et al., 2022; Kwon et al., 2022; Liau et al., 2022). By combining cryo-EM, X-ray crystallography, SEC-MALS-SAX, and molecular dynamics (MD) simulations, it was discovered that a criblike scaffold SHOC2 scaffold and RAS are needed to collectively drive the protein phosphatase 1 catalytic subunit (PP1C) enzyme to target RAF for dephosphorylation. These structural complexes suggest that beyond the fiercely competitive racetrack of direct RAS blockers, small-molecule drugs targeting the upstream, autoinhibitory or downstream, activation-prone complexes could offer an alternative way forward. Interestingly, cryo-EM combined with native MS revealed the existence of a tertiary complex between KRas mutants and a regulatory guanine nucleotide exchange factor Son of Sevenless (SOS), in which two KRas-G13D mutants KRasG13D and KRasG13D–GTP bound at the active and allosteric sites of SOS, respectively. This structure can provide insights into why some mutant KRas has enhanced activity, aiding the discovery of potential KRas–SOS disrupting compounds (Moghadamchargari et al., 2021). Taken together, these examples demonstrate that cryo-EM is not only instrumental in dissecting the RAS pathways but also capable of dramatically expanding the target pool in a pharmaceutical setting. Futhermore, cryo-EM has been shown to potentially perform SBDD or FBDD directly on RAS. Castells-Graells et al. used a rigidified, symmetric cage-DARPin scaffold and resolved the GDP-bound KRAS to 3 Å resolution, enabling visualization of cancer-related point mutations as well as ligands (Castells-Graells et al., 2023). As engineered scaffolds have been shown to help generate the structures for small and dynamic proteins, they can potentially be tailored to meet the drug screening demands for a wide range of disease-implicated enzymes (Martin et al., 2016; Yao et al., 2019; Yeates et al., 2020).

CRYO-EM IN DRUG DISCOVERY OF PROTEIN DEGRADERS

Protein degraders constitute an innovative class of small molecules that hijack the targeted protein degradation (TPD) pathway to selectively eliminate the proteins of interest (POI). Unlike conventional inhibitors which act on the active or allostery site, a protein degrader bridges together its target with an E3 ubiquitin ligase. After the formation of a transient tertiary complex, the target is polyubiquitinated by the E3 ligase and directed to the 26S proteasome. This proteolytic rather than inhibition-driven mechanism is a game-changer for drug discovery, as a wide range of previously considered "undruggable" proteins such as transcription factors, scaffolding proteins, and enzymes with limited structural accessibility can be targeted. Since the degrader-enabled E3 ligase recruitment is reversible and substoichiometric, protein degraders can alleviate common constraints associated with small-molecule-based therapies, including toxicity and drug resistance due to repeated, high-dosage drug administration (Chirnomas et al., 2023). Based on their modes of action, protein degraders can be broadly categorized into three modalities (Figure 33.3a). Selective hormone receptor downregulators (SERDs), such as the FDA-approved drug for HER-2 negative breast cancer elacestrant (Orserdu), bind to the estrogen receptors and induce degradation (Hernando et al., 2021). The second type of degrader, a molecular glues, binds to a cleft between the POI and/or an E3 ligase and cements the resulting tertiary complex. The third modality is termed PROTACs (proteolysis targeting chimeras), which are bifunctional, target-to-ligase tethering molecules consisting of a POI-engaging moiety (the warhead), a linker, and an E3-recruiment moiety (Lai et al., 2017).

As of 2023, three molecular glues were approved by the FDA, thalidomide (Thlomid) and its analogs pomalidomide (Pomalyst) and Ienalidomide (Revlimid). Cereblon (CRBN) appears to be the "workhorse E3 ligase" for this modality, which is utilized by not only the three drugs in the market but also most of the candidates in clinical trials. Recent studies have underscored the importance of cryo-EM in the quest for novel moleculer glue–type degraders. In 2019, Novartis revealed the structural basis of indisulam-mediated RBM39 degradation by the DCAF15 E3 ligase using X-ray crystallography (structure at 2.3 Å resolution) and cryo-EM (structure at 3.54 Å resolution) (Bussiere et al., 2020). The anticancer sulfonamide drug Indisulam slots between a structurally Complimentary pocket formed by a helix of RBM39's second RNA recognition domain and regions bolstered by both the NTD and the CTD of DCAF (Figure 33.3b). Another publication by a team in the Dana-Farber Cancer Institute solved a similar cryo-EM structure of E7820-RBM39-DCAF15 complex at 4.4 Å resolution

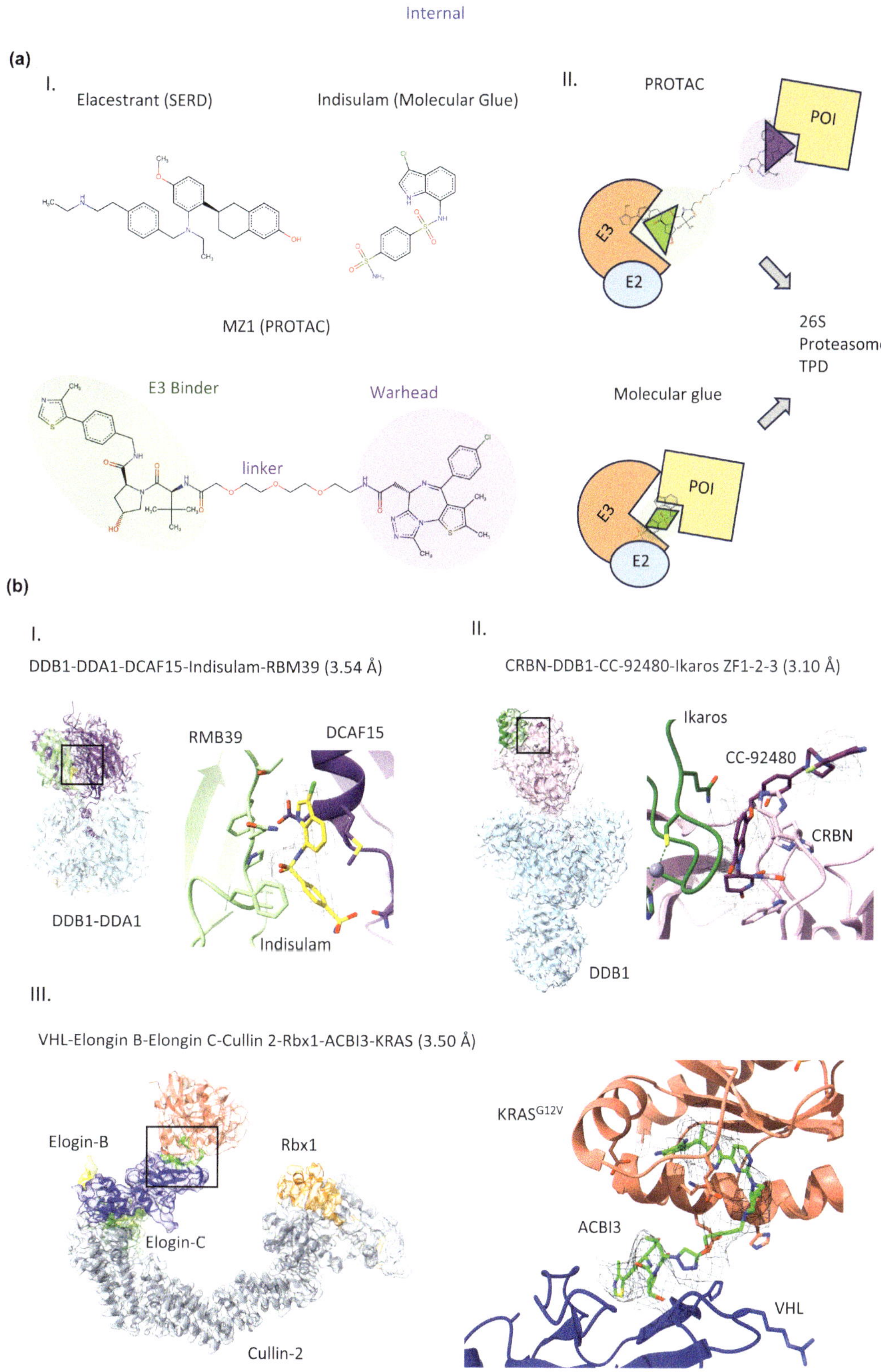

FIGURE 33.3 (a) (I) Representation of a SERD (Elacestrant/Orerdu), a molecular glue (Indisulam), and a PROTAC (MZ1) molecule. (II) Illustration for the mode of action of molecular glues and PROTACs. (b) Cryo-EM electron density maps and structural models for the RBM39- Indisulam-DDB1-DDA1-DCAF15 complex (PDB 6SJ7 EMD-10213), the CRBN-DDB1-CC-92480-Ikaros ZF1-2-3 complex (PDB 8D7Z EMD-27240), and the VHL-Elongin B-Elongin C-Cullin 2-Rbx1-ACBI3-KRAS complex (PDB 8QU8 EMD-18657) are shown in panels I to III, respectively.

(Faust et al., 2020). By using bioinformatics and MS-based proteomics, respectively, the two studies convergingly identified a new POI, RBM23, for the same CRL4/DCAF15 ubiquitylation pathway. These two publications demonstrated that structural biology can distinctively identify the degradation mechanism for a class of antitumor compounds as molecular glues, serving as the foundation for rational drug design. Another recent study elegantly demonstrated that cryo-EM can interrogate the mechanism of molecular glues and propel innovation for next-generation medicines (Watson et al., 2022). The CELMoDs (CRBN E3 ligase modulatory drugs) constitute a class of immunomodulatory imide drugs (IMiDs) used as a first-line treatment for multiple myeloma for two decades. Despite existing structural snapshots for the CRBN-DDB1-CELMoD complexes, it is crucial to understand how these drugs modulate the conformations of CRBN-DDB1 and how such modulation can be harnessed to design better compounds. In this study, the authors performed extensive rounds of iterative 3D classification and focused refinement to obtain a range of high-resolution cryo-EM structures. These structures reconstitute the mechanistic journey of CELMoDs switching on the CRBN-DDB1 complex and gearing them up for the substrate's ubiquitination. As the results drew attention to an allostery mechanism previously overlooked for molecular glues, this work highlights the unique advantage of cryo-EM in SBDD for compositionally and conformationally heterogenous degrader systems.

Since the first conception of PROTACs more than two decades ago, structural biology has played a central role in their development (Sakamoto et al., 2001). With the mode of action (MoA) for different E3 ligase–based protein degradations unveiled, pioneering biotech companies such as Arvinas, Kymera, and C4 Therapeutics were founded to bring the first PROTAC drug to the market. In 2016, two pioneering crystal structures involving the E3 ligase CRBN and its POIs were published, including the CRBN-DDB1-CC-885-GSPT1 complex at 3.6 Å resolution and the CRBN-DDB1-Lenalidomine-CK1a complex at 2.45 Å resolution (Matyskiela et al., 2016; Petzold et al., 2016). In 2017, the first structure of an engineered PROTAC, MZ1, which recruits another E3 ligase VHL (ubiquitin ligase von Hippel–Lindau) protein, was published (Gadd et al., 2017). The MZ1-Brd4^{BD2}-VHL structural complex at 2.7 Å revealed that the degrader folds in unique conformations to allow for cooperative formation of the target–E3 ligase tertiary complex. These works represent a pivotal moment for PROTAC discovery, as structure-based design for this versatile, and potentially applicable-to-all, modality is now possible. Indeed, structural and thermodynamics information obtained from the BRD-MZ1-VHL complex paved the way for the design of a trivalent PROTAC linking to two branched-out BET bromodomain to the VHL, which resulted in a more potent and long-lived degrader (Imaide et al., 2021). The field has since expanded dramatically, with pharmaceutical giants including Pfizer, AstraZeneca, Novartis, and Sanofi joining the race by partnership with biotech companies or by in-house R&D. As of 2023, more than 20 PROTAC drug candidates for cancer and autoimmune diseases are in clinical trials (Békés et al., 2022). Arvinas has recently started Phase III clinical trials for their front-runner molecule ARV-471, with results expected in the second half of 2024. Should this compound receive FDA approval, a watershed moment will arrive for the entire PROTAC R&D community.

Compared to molecular glues, cryo-EMs for PROTACs are hindered by a common obstacle: The tertiary POI-PROTAC-E3 ligase complex can be highly dynamic and hence, too heterogenous to resolve. The design of a PROTAC's linker region may result in a tremendous freedom of movement between the target and the E3 ligase. As a PROTAC acts on the complex in a tethering instead of an adhesive-like manner, subtle changes in other chemical moieties may also destabilize the whole complex. Moreover, negative or positive cooperativity exists for the PROTAC-tethered components, which may alter the equilibrium during the formation of the tertiary as well as the binary complexes (Grigglestone et al., 2021; Wurz et al., 2023). As this equilibrium affects the selectivity and degradation efficiency of the PROTAC molecule, conventional methods to stabilize multiprotein complexes for structural studies may lead to unintended results. Despite major challenges, cryo-EM of PROTAC is possible, with extensive efforts in both academia and industry moving in this direction. For instance, Arvinas has recently disclosed a cryo-EM structure of E3 ligase KLHDC2/EloB/EloC complex, which can be harnessed for the SBDD of active PROTACs. A very recent preprint used cryo-EM to discover a new modality called intramolecular bivalent glues (IBGs), which brings together the POI BRD4 and the E3 ligases DCAF11 or DCAF16 via to enhance surface complementarity and trigger ubiquitination (Hsia et al., 2023). Another preprint by Boehringer Ingelheim Pharmaceuticals and the University of Dundee revealed a 3.5 Å–resolution cryo-EM structure of KRAS in complex with the E3 ligase VHL/Elongin-B/Elongin-C/Cullin-2/Rbx1 system enabled by a PROTAC molecule ACBI3 (Popow et al., 2023) (Figure 33.3b).

Beyond identifying the binding modes of molecular glues and PROTACs, cryo-EM can be employed to meet our need for novel strategies in the quest for small-molecule-enabled TPD. It is estimated that out of more than 600 E3 ligase family members, a mere 1% have been explored (Jevtić et al., 2021). The age of data-driven drug discovery beckons our exploration into a "PROTAC-able universe" as genomic databases for all E3s in the human body are being constructed (Liu et al., 2023). Recent advancements in cryo-EM not only enabled us to delve deeper into the mechanistic details of many E3 ligase–mediated degradation machineries, but also opened up a time-resolved dimension (Bodrug et al., 2023; Pan et al., 2021; S. Zhang et al., 2022). In 2022, Genentech revealed a cryo-EM structure of a peptidic macrocyle (MC1) in complex with 26S proteasome subunit PSMD2 at 2.5 Å resolution (Bashore et al., 2023).

Chimeric compounds which conjugate the macrocycle to a BRD4 domain resulted in direct recruitment of the proteasome, demonstrating that the search for protein degraders can go beyond the canonical E3/E2 ligase landscape. Next-generation degraders such as PROTABs (proteolysis-targeting antibodies), KineTACs (cytokine receptor-targeting chimeras) and LYTACs (lysosome-targeting chimeras), are tasked with the selective degradation of cell-surface receptors (Ahn et al., 2023; Marei et al., 2022; Pance et al., 2023). With the unique advantage of cryo-EM in studying transmembrane proteins (see section below), it is reasonable to foresee its integration into the structural studies of these novel degraders. The wealth of experience accumulated in both academia and industry during the two decades of TPD discovery heralds a bright future, where more and better degraders will make it from bench to clinic.

CRYO-EM FOR LARGE MOLECULE THERAPEUTICS

Large-molecule therapeutics, also known as biologics or biopharmaceuticals, are drugs made of large (usually several thousands of daltons in size), complex molecules derived from living sources (human, animal, or microorganism). Approved biologics every year include, among others, vaccines (discussed in the section below), hormone therapies, gene therapies, antitoxins, monoclonal antibodies, and recombinant proteins. Cryo-EM has been particularly helpful in elucidating the structure and the mechanism of action of biologics because of their large size. In the case of the BIVV001, an engineered fusion protein between two Fc domains, the single chain B domain-deleted Factor VIII (FVIII) with its carrier protein, the von Willebrand factor (VWF), used currently in the clinic, cryo-EM was used to provide structural understanding of the clinical observations. The structure of the fusion complex revealed the shielding of the phospholipid binding region of FVII, thus preventing the degradation of the protein and increasing its half-life in the blood (Fuller et al., 2021).

Monoclonal antibodies (mAbs) can be used to neutralize pathogens and to agonize and antagonize certain disease-causing biochemical pathways. In contrast to small-molecule inhibitors, mAbs are highly effective therapeutic biologics because of their specific interaction properties with their respective soluble and cell-surface targets with high affinity to deliver desired functions. After approval of the first mAb in 1986 against T-cell antigen (Ag) OKT3 (now called CD3), it took more than 35 years for the FDA to reach the milestone of the 100th approved therapeutic mAb in clinic (Kaplon et al., 2023; Mullard, 2021). However, mAbs currently account for an average of more than 10 FDA approvals per year. The number of mAbs in the clinic is anticipated to further increase with the utilization of high-throughput and automation technologies in the Ab discovery and development pipelines. Among several techniques, SB is starting to have a significant impact on biologics development, from early identification of the lead molecule until its evolution to a therapeutic molecule that can be safely used in the clinic.

Antibodies are Y-shaped molecules formed by two heavy (H) and two light (L) protein chains (Harris et al., 1997; Jay et al., 2018). Each heavy and light chain has four (VH, CH1–3) and two (VL, CL) immunoglobulin domains, respectively. Each half of the Y-shaped molecule has two regions connected by a hinge: the antigen interacting Fab and the effector interacting Fc region. Fab consists of a constant (CH1, CL) and a variable region (VH, VL). The Fab VH and VL interact with epitope residues of the antigen via paratope residues formed by six CDR loops (LCDR1-3) (Figure 33.4). Antibody therapeutics come in different flavors and modalities. For example, bispecific antibodies can engage with two different antigens via each of their Fab arms, whereas nanobodies (Nb) interact with target antigens via a single VH domain (Jin et al., 2022).

3D structure information is used to decipher critical epitope–paratope information about Fab–antigen complexes and their mode of action (MoA), identify liabilities in biologics discovery pipelines, and enable rational structure-based engineering for enhanced therapeutic properties. For example, the crystal structure of dostarlimab (100th approved mAb from GSK), a therapeutic Ab against programmed cell death 1 receptor (PD-1) for the treatment of endometrial cancer, displayed the MoA to be achieved by steric occlusion of PD-L1 (PD-1 ligand) by its light chain and its high affinity induced by conformational changes within PD-1 (Park et al., 2022). As discussed earlier, though X-ray crystallography has been the method of choice to obtain high-resolution structure information about sub-100 KDa Ag–Fab complexes, the technique demands longer timelines, large amounts of reagents, and specific reagent properties like deglycosylated proteins. While high-throughput crystallization methods can be utilized for co-crystallizing protein ligand/inhibitor complexes, achieving similar throughput for Fab–antigen complexes using X-ray crystallography is often challenging because of multiple Fab binding sites and the surface properties of Fabs. Cryo-EM presents itself as a rapid and robust 3D structure determination tool by utilizing 10 times less reagent and faster data collection and processing protocols to accelerate the high-resolution structure determination of glycosylated antigen–Fab complexes. For instance, unsupervised data collection using Thermo Scientific EPU software (EPU Multigrid) has allowed for high-throughput (HT) cryo-EM–based epitope mapping of the SARS-CoV2 spike protein neutralizing Ab panel (Drulyte et al., 2022).

Small targets and their complexes (less than 50 kDa) present a challenge for cryo-EM because of low signal-to-noise ratio, lacking recognizable shape features to facilitate image alignment at low resolution. In addition to using integrated structural biology platforms for sub-100 kDa protein structure determination (Iozzo et al., 2021), methods similar to universal Nb binding Fab modules (NabFab) that rigidly bind the conserved scaffold of Nbs are routinely utilized for cryo-EM studies of smaller target Ag–Nb complexes

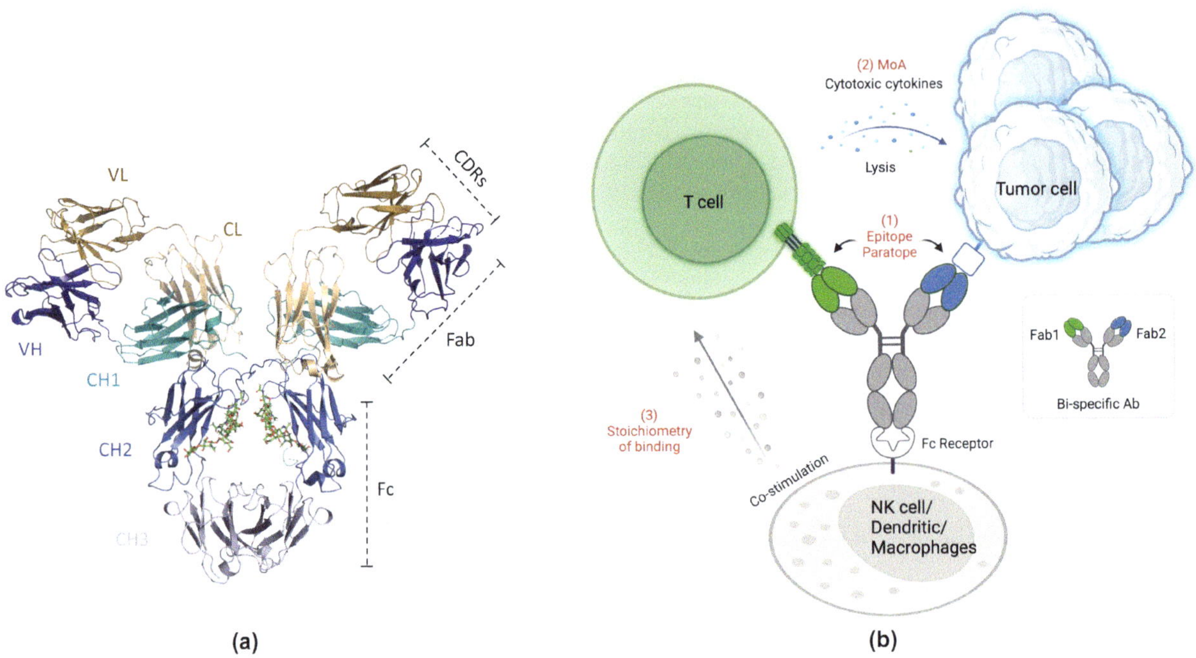

FIGURE 33.4 (a) Structure of IgG1 (PDB ID: 1IGY). (b) A bispecific Ab engaging the receptors on T cell, tumor cell, and Fc receptor on accessory cells (NK cell, dendritic or macrophages) via its Fab and Fc arms to enhance tumor cell cytotoxicity. Cryo-EM can be used for structure-based epitope/paratope mapping at atomic level (1), to decipher the stoichiometry of binding with receptors and mechanism of action at atomic level (2 and 3). Part (b) generated by BioRender.

(Bloch et al., 2021). Here, the structures of Nb-bound VcNorM and ScaDMPT, both small membrane proteins of ~50 kDa, were determined to high resolution using NabFab as a fiducial and size enhancer. Another recent study to increase the size of the Ag–Fab complex for cryo-EM studies for epitope mapping utilized bivalent engineered Fab fiducial against LAG3 (lymphocyte activation gene 3 protein). LAG3 is a major target for cancer immunotherapy, and many anti-LAG3 Abs are in clinical trials. Both LAG3 and Fab display similar Ig domain architecture, further challenging image alignment and model building at intermediate resolution. The authors therefore engineered a dimeric bivalent version of Fab favezelimab, an anti-LAG3 Fab that facilitated image processing and obtaining a high-resolution structure to 3.5 Å (Mishra et al., 2023).

In contrast to X-ray crystallography, cryo-EM can be used to extract structure information from large flexible and heterogenous complexes to decipher MoA. For example, trastuzumab and pertuzumab are mAbs that bind to distinct subdomains of human epidermal growth factor receptor 2 (HER2), which is associated with a more aggressive form of breast cancers and increased mortality rates. Although crystal structures of these mAbs in complex with HER2 provided the epitope/paratope information to shed light on how trastuzumab acts by blocking HER2 homodimerization and pertuzumab by blocking HER2 heterodimerization, the individual structures could not explain the synergistic effects of co-administering both mAbs for improved clinical outcomes. The ternary complex structure of HER2-trastuzumab-pertuzumab deciphered using cryo-EM provided key insights into the synergy in the inhibition of HER2 ligand-dependent and ligand-independent signaling pathways and opened avenues for the design of novel bispecific molecules with increased clinical efficacy (Hao et al., 2019).

High-resolution cryo-EM structure analysis in combination with other complementary techniques like HDX-MS and alanine scanning can provide mechanistic understanding of therapeutic activity. For example, a combination of these methods was utilized to understand the MoA of an investigational Ab-drug conjugate, tusamitamab, against carcinoembryonic antigen–related cell adhesion molecule 5 (CEACAM5), which is highly upregulated in certain tumor cells (Refer Anad's paper). The structure provided critical insights into molecular selectivity markers of tusamitamab against the A3-B3 domains of CEACAM5 and how it does not interact with other CEACAM homologs like CEACAM1, CEACAM6, and CEACAM8. Cryo-EM can be used to visualize multispecific antibody–antigen interactions within the context of the whole Ab. 3D reconstructions of cross-over dual variable immunoglobulin ($CODV_{Ig}$), a bispecific Ab format designed to simultaneously target interleukins IL4 and IL13 and its complexes, shed light on Ab:Ag stoichiometry, epitope–paratope interface at intermediate resolution, and the mechanism of its enhanced therapeutic potential (Fernandez-Martinez et al., 2023). Another example where high-resolution cryo-EM in combination with negative stain EM played a key role to decipher

MoA was to understand how rituximab, a therapeutic mAb in clinic for two decades, activates the complement to kill B cells. 3D structure of two rituximab Fabs in complex with its target CD20, a dimeric membrane protein on B cells, displayed that the composite epitope engagement of rituximab results in CD20 cross-linking into circular assemblies on the cell surface, allowing for complement recruitment (Rougé et al., 2020).

Over the past decade, several high-resolution 3D structures were utilized for rational structure-based engineering of therapeutic mAbs. For example, selective pH-dependent blocking of the histidine-rich interface between the extracellular domain of V-domain immunoglobulin suppressor of T-cell activation (VISTA) and its coinhibitory receptor P-selectin glycoprotein ligand-1 (PSGL-1) by mAbs were engineered to suppress VISTA-mediated immune suppression (Johnston et al., 2019). We foresee that with the availability of high-resolution cryo-EM structures of Ag–Fab/Nb complexes, the design/engineering, production, testing, and analysis of novel therapeutics in several functional assays can be increased by 100-fold in the future.

CRYO-EM FOR MEMBRANE PROTEINS

Integral membrane proteins (IMPs), such as receptors, transporters, and ion channels, are distributed throughout all organs and tissues of the human body. They localize at the surface of cells and subcellular organelles and are therefore positioned at the forefront of many biochemical processes, including cell signaling, metabolism, and sensory responses, to name a few. Defects in their regulation or activity relate to numerous diseases, so it is only logical that these proteins constitute about two-thirds of all current drug targets (Overington et al., 2006).

The most targeted class of membrane proteins are the G protein–coupled receptors (GPCRs). The GPCR superfamily, also known as seven-(pass)-transmembrane domain receptors, comprises over 800 coding genes in humans, subdivided into six classes (A to F) (Overington et al., 2006). These receptors mediate communication with the extracellular environment by sensing and translating chemical or optical stimuli into intracellular signals. Upon agonist or antagonist binding, GPCRs undergo small rearrangements allowing, or preventing, the recruitment of signal transducers such as heterotrimeric G-proteins and arrestins. Depending on the physiopathology, agonist or antagonist molecules have been developed to stabilize GPCRs in conformations which promote or inhibit the activation of signal transducers. In the past decade, structural biology has contributed to elucidate multiple mechanisms for ligand recognition in GPCRs, including orthosteric, allosteric, and bitopic receptor–ligand interactions (Kruse et al., 2013; Liu et al., 2012).

Transporters and ion channels are the two other major types of druggable membrane proteins. Ion channels trigger processes requiring electrochemical activity in our bodies. They are highly selective pores which can be regulated upon binding of a ligand (e.g., neurotransmitter, ion, nucleotide), mechanical stress (e.g., physical touch, contact), or change of membrane potential (i.e., voltage across the bilayer, electrochemical gradients). Transporters bind to small molecules and translocate them across the membrane by undergoing extremely large conformational changes (Drew et al., 2021). The energetic cost of such movements can be paid by light, ATP hydrolysis, or downhill concentration gradients. The solute carrier (SLC) superfamily forms the largest group of transporters. It includes over 400 passive and secondary active transporters classified in 66 subfamilies. SLCs mediate the disposition of nutrients and metabolites such as sugars, nucleotides, lipids, amino acids, oligopeptides, vitamins, and neurotransmitters (Pizzagalli et al., 2021).

For several decades, biochemical, biophysical, and structural studies on membrane proteins were regarded as tedious and difficult. Indeed, if over 20,000 protein structures are deposited on the PDB, only 1600 represent membrane proteins. It took 20 years to reach 100 membrane protein structures since the first high-resolution entry of the photosynthetic reaction center from *Rhodopseudomonas viridis* in 1985 (Deisenhofer et al., 1985). In 2005, an exponential growth of entries started to kick into the PDB, and since 2018, over 100 new membrane protein structures are discovered every year. We will now discuss important methodological aspects which contributed to this revolution and how cryo-EM slowly became the *go-to* structural biology applied to membrane proteins in the pharmaceutical industry, and then we will review several examples of highly relevant membrane protein cryo-EM work in the context of drug discovery.

Looking back to the early days of membrane protein structural biology, the first challenge was to obtain enough purified protein while maintaining its integrity after extraction from the host's membrane. The democratization of insect and mammalian host expression systems (Baldi et al., 2012; Vaitsopoulou et al., 2022), together with the development of various solubilization tools and screening techniques (Chatzikyriakidou et al., 2021; Denisov et al., 2016; Frauenfeld et al., 2016; Guo, 2021; Kotov et al., 2019; Privé, 2007), broadened the portfolio of targets that could be investigated. Beyond the difficulty of producing stable samples, structural biology applications are challenged by the very nature of membrane proteins.

For reproducibly crystallizable membrane protein targets, MX offers the highest potential as the crystal *physically* sorts and aligns several billions of single particles all related by symmetry operations and adopting a homogenous conformational ensemble. The three-dimensional lattice is therefore a powerful signal amplifier, and it reduces the computational cost of 3D reconstruction. The whole MX pipeline benefits from more than 60 years of hardware and software developments (Kendrew et al., 1958; Wlodawer, 2021). Therefore, routine data collection and processing can nowadays be carried out without any user intervention, and within a few minutes for each crystal. This level of automation and throughput represents a great advantage when screening a large number of small molecules into reproducibly crystallizable membrane protein targets and building pharmacophore models (Amelia et al., 2021; Congreve et al., 2014; Kotov et al., 2023).

But many IMPs are extremely challenging to crystallize (García-Nafría et al., 2021); they can be poorly stable, they can possess long and flexible loops carrying heterogenous glycosylation sites, and importantly, they often lack features outside of the detergent micelle, to promote stable crystal contacts. Given enough time and efforts, these challenges can be overcome, and there are various examples of such achievements in the literature (Billings et al., 2021; Kotov et al., 2023; Shahsavar et al., 2021). For instance, glycan heterogeneity can be dealt with by using alternative cell lines (Meuris et al., 2014) or enzymatic treatment. Poor stability—notably of GPCRs—is often increased by point mutations (Heydenreich et al., 2015). This approach has been pioneered by the Sosei Heptares group, which has built proprietary research platforms to stabilize GPCRs. These include the StaR®, CHESS, and SaBRE technologies. Conformational binders, crystallization chaperones, and chimeric constructions are also effective tools for stabilizing IMPs and promoting crystal growth (Cherezov et al., 2007; Kotov et al., 2023; Pardon et al., 2014). Finally, *in meso* crystallization methods, specially developed for membrane proteins, tend to yield type I crystal packing, which can produce higher diffraction. This method was particularly successful in elucidating the mode of action of small molecules in the GPCR field (Kenakin, 2013; Li et al., 2020). In summary, membrane protein crystallography is extremely powerful when good crystallization conditions are found, but this process can often take several months or even years of extensive troubleshooting and optimization before yielding any useful result. In the competitive industrial context, rapid and intelligent decision making is key. Cryo-EM relies on important computing resources but does not rely on a crystal. This currently places the technique as the best option to deliver impactful results in drug discovery pipelines for membrane proteins. In fact, even in the public domain, cryo-EM has now outnumbered crystallography for membrane protein annual entries in the PDB.

An iconic example that marked a pivotal shift in membrane protein structural biology toward cryo-EM is the transient receptor potential (TRP) ion channel superfamily. These channels are drug targets in numerous human diseases (Kaneko et al., 2014; Moran, 2018) and resist crystallization. Cryo-EM was game changing because it enabled researchers in less than five years to unravel atomic structures for each of the seven TRP channel subfamilies (Cheng, 2018; Diver et al., 2022).

In fact, cryo-EM structure determination of large ion channels and large ABC transporters is relatively straightforward. In contrast, inactive monomeric GPCRs and most monomeric SLCs have a molecular weight ranging between 35 kDa and 65 kDa and lack rigid extramembrane features to drive precise angular assignment. Different molecular tools (fiducials) have been used to introduce structured features and stabilize many GPCRs and SLCs.

A good example illustrating the diversity of fiducials one can use can be found in the SLC15 family, which resisted extensive crystallography trials, but for which three out of four members' high-resolution structures were recently determined by cryo-EM using various strategies (Figure 33.5). For SLC15A4, the first strategy was to use a synthetic nanobody stabilizing the transporter in its outward-facing open state and driving precise particle alignment in 2D and 3D (Custódio et al., 2023) in comparison to the naked transporter. The second strategy was to form a stable complex between SLC15A4 and the N-terminus of the signaling protein called TASL fused to EGFP. The third strategy was to insert the ALFA tag epitope (Götzke et al., 2019) between TM6 and TM7 and create a stable complex with the anti-ALFA nanobody. High-resolution structures of monomeric and dimeric SLC15A4, including an inhibitor-bound structure, could be determined using this approach (Boeszoermenyi et al., 2023; Chen et al., 2023). Taken together, these studies elucidated the dual molecular mechanism by which SLC15A4 acts both as a transporter and as a receptor (Chen et al., 2023; Custódio et al., 2023) and described the mode of action of the first anti-inflammatory small molecule targeting SLC15A4. For SLC15A1 and SLC15A2, an additional extracellular domain of 20 kDa is located between TM10 and TM11 and acts as a natural fiducial. The potential of cryo-EM dealing with large heterogenous conformational ensembles and intermediate states was also harnessed, and a detailed molecular movie of the entire transport cycle could be reconstructed, providing critical data for understanding the conformational rearrangements of the peptide and drug binding sites in various functional states (Killer et al., 2021). A high-resolution structure of SLC15A2 was also obtained in complex with an inhibitory nanobody, but this time in one unique conformation (Parker et al., 2021).

Another illustration of the possibilities available to perform cryo-EM on challenging membrane protein targets is seen in the GPCR field (Bernhard et al., 2023). While active states of GPCRs in complex with signaling proteins can easily be studied, inactive monomeric GPCRs require additional fiducials. So far, the most widely used approach is the insertion of a BRIL domain in replacement of the intracellular loop 3 (ICL3), in combination with anti-BRIL Fabs (Lees et al., 2023; Mukherjee et al., 2020; Tsutsumi et al., 2020; K. Zhang et al., 2022). A less-used approach is to replace segments of TM5 and TM6 to insert a nanobody (Nb6) binding site. With this technique, it was possible to obtain structures of neurotensin 1 receptor bound to an antagonist, of the μ-opioid receptor bound to alvimopan, and of the histamine receptor 2 bound to famotidine (Robertson et al., 2022). Finally, the recent "fuse-and-glue" hybrid approach consists in combining the BRIL insertion in ICL3 and an ALFA tag fusion to the C-terminus of the GPCR. The anti-BRIL Fab is then bound, and a molecule containing an anti-Fab nanobody and an anti-ALFA

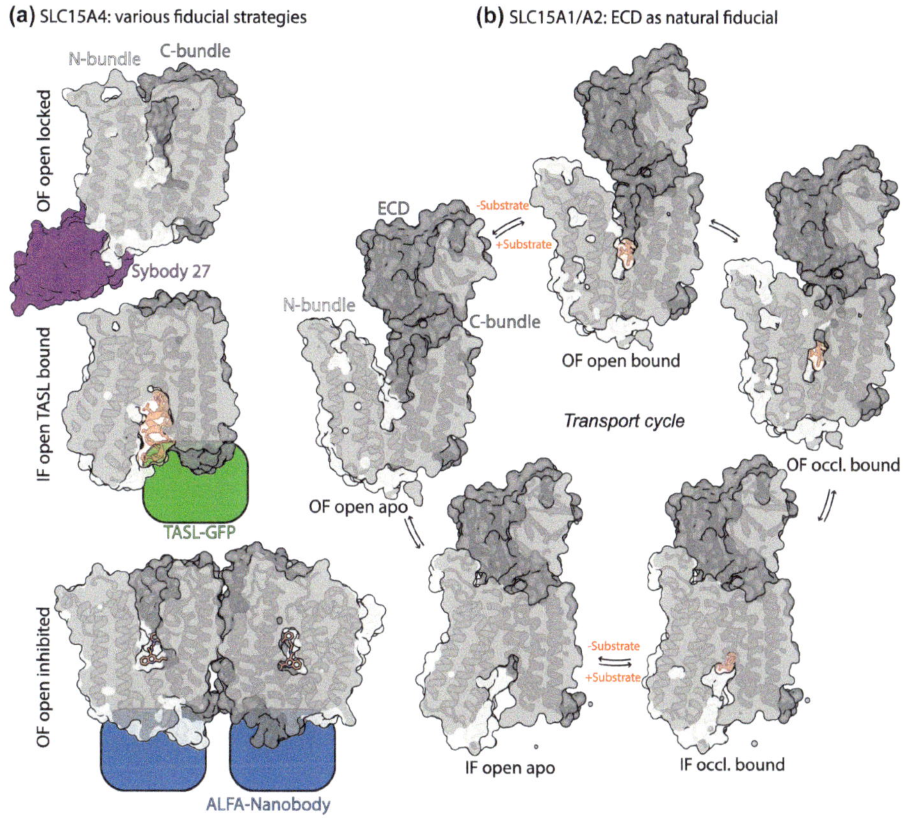

FIGURE 33.5 Strategies used in the SLC15 family. Extramembrane features allowed high-resolution cryo-EM structure determination in the SLC15 transporter family. (a) Various fiducials were used to determine the structure of SLC15A4. From top to bottom: a conformation-specific synthetic nanobody (PDB ID 8CNI), the binding domain of TASL fused to GFP (PDB ID 8JZU), and a nanobody specific to the ALFA tag introduced between the N- and C-bundles (PDB ID 8JZX). In the deposited structures, the GFP and the ALFA nanobody were not modeled in the density, although their presence was crucial to improve the precision of image alignment and therefore the final resolution of the maps. Their relative positions are indicated here by green and blue shapes, respectively. (b) SLC15A1 and SLC15A2 possess an extracellular domain (ECD) inserted between TM9 and TM10 that drives precise image alignment on its own. In this way, the transporters were captured in multiple conformational states throughout the transport cycle. (PDB IDs: OF (outward facing) open apo: 7PN1; OF open bound: 7PMX; OF occluded bound: 7PMW; IF (inward facing) occluded bound: 7PMY; IF open apo: AlphaFold PSD: P46059.

nanobody glues the anti-BRIL Fab and the ALFA tag together. This approach allowed reserchers to determine the structures of the β2-adrenergic receptor (β2AR) bound to antagonistic and agonistic ligands and the adhesion GPCR ADGRL3 in the apo state (Guo et al., 2023). For a good overview of the strategies used for GPCR structure determination by cryo-EM, we recommend the recent article from Bernhard et al. (2023).

If these tricks were proven to be successful in several cases, they often require some level of optimization of the linkers between inserted and fused domains. 3D structure prediction software can guide construct design to obtain more stable and structured objects (Guo et al., 2023; Kotov et al., 2023) and therefore help to accelerate cryo-EM structure determination.

In conclusion, it is striking to notice that every low-molecular-weight membrane protein cryo-EM structure mentioned above has benefited from the Non-Uniform Refinement (Punjani et al., 2020) or from the SIDESPLITTER (Ramlaul et al., 2020) algorithms, which have been developed to identify and downweigh high spatial frequencies in regions that are inconsistent between subsets of the data, and therefore reduce overfitting in the micelle or nanodisc region. Membrane protein structural biology has made tremendous progress in the past decade. With constant and rapid progress still very much alive in sample preparation, microscope hardware, and image analysis software, we can expect a continuous increase of cryo-EM SBDD campaigns improving drug efficacy with membrane proteins.

CRYO-EM FOR VACCINES

Vaccines are biopharmaceutical products aimed at preventing pathogens from attacking the host by eliciting a prolonged immunity against them. Traditionally, vaccines were obtained by using inactivated or attenuated viruses.

Recently, new formats have been emerging, particularly the use of recombinant proteins and nucleic acids, representing 40% of the new candidates in clinical trials in 2023 (Yue et al., 2023). So far, the most effective way to deliver nucleic acid vaccines to elicit an immune response of the host has been to use carriers, either lipid-based or made of polymeric materials, inorganic nanoparticles, or bioinspired nano-vehicles (Qin et al., 2021). In the case of lipid-based carriers, cryo-EM has been used as an imaging technique, helping researchers in academia and industry to study their biophysical properties. These properties, including lamellarity, size, shape, and polydispersity, are known to influence biodistribution, biological activity, and toxicity (Kulkarni et al., 2018). While the technique may not be as high throughput as others, such as dynamic light scattering (DLS) or size exclusion chromatography, new software is being developed to automatically detect biophysical characteristics from electron microscopy images. This software enables both quantitative and qualitative assessments of the formulation effect on these characteristics (Crawford et al., 2011).

As for canonical structure-based drug design, cryo-EM can provide information on the architecture of viral capsids (Luque et al., 2020), a snapshot of viral entry, and assembly and egress from the host cell (Klein et al., 2020; Mendonca et al., 2021). By also providing the structure of individual proteins, it can be used to create stable mutants of the antigenic protein locked into a particular conformation that promotes its neutralization. This approach proved very effective in the fight against the novel coronavirus 2019-nCoV. Based on previous structural work done on closely related betacoronaviruses (Pallesen et al., 2017), two proline mutations were quickly introduced on the fusion S protein to stabilize its prefusion conformation and obtain its structure by cryo-EM to high resolution (Wrapp et al., 2020). The stabilized prefusion state was used as the base for the development of the mRNA vaccines by Pfizer (Thorn et al., 2022), Moderna, and Curevac, as well as the adenovector by Janssen-Johnson & Johnson and the subunit vaccine by Novavax (Heinz et al., 2021). Less than a year after the discovery of the viruses, these efficacious vaccines were developed at an unprecedented speed owing to the support of the structures obtained by cryo-EM, and this created many precedents in the history of vaccinology.

Cryo-EM can be used not only to obtain the structure of individual antigens, but it is also the most efficient technique to obtain the structure of antigens interacting with protective monoclonal antibodies. After injection with an antigen, polyclonal antibodies can be isolated from either blood serum or B cells (Rush et al., 2022). Once these are purified, cryo-EM can be used to map the antibody sites, either for each antibody in isolation or more recently by mapping the binding site of all the polyclonal antibodies at the same time, to identify the most immunogenic binding sites as they evolve after repeated immunization (Bianchi et al., 2018). Once immunogenic and neutralizing epitopes have been identified, further selection is made for those that bind to specific conformations, and the antigen can be designed based on such information to prevent nonneutralizing antibody binding and promote neutralizing responses (Rush et al., 2022). Cryo-EM allows the custom design of the antigen to increase its immunogenicity using many different approaches, and this has allowed vaccinology to enter a new phase (Graham et al., 2019).

THE FUTURE OF CRYO-EM AND CONCLUSIONS

In recent years, cryo-electron microscopy has demonstrated remarkable potential in integrating into the realm of drug discovery. It is estimated that, out of a total of 757 high-end electron microscopes worldwide, at least 5% were acquired and installed by private companies or made accessible through consortiums open to the industry (Noble, 2023). Its full implementation is limited by the high entry costs and by clear bottlenecks in the sample preparation and the data processing. To address these bottlenecks, efforts are underway to develop new, more affordable microscopes (Naydenova et al., 2019). Additionally, in the field of sample preparation, several companies are striving to create more robust systems for freezing grids. These systems aim to be fully automated, to provide improved ice distribution, and to enhance overall reliability (Xu et al., 2022). Furthermore, the field of structural biology is undergoing a transformation from a three-dimensional (xyz) to a four-dimensional (xyz and time) technique. This approach enables the imaging of samples mixed together at different time points on a microsecond scale. This capability allows for the discovery of transient and metastable states that could be crucial for understanding the function of a drug (Feng et al., 2017; Kontziampasis et al., 2019; Lu et al., 2014; Maeots et al., 2020; Torino et al., 2023). New algorithms are being developed to harness machine learning for speeding up and automating data collection (Bepler et al., 2022; Kim et al., 2023; Yonekura et al., 2021) and data analysis (Chung et al., 2022). Development of high-throughput cryo-EM from the automation of sample preparation, data collection, and analysis to further software development will shorten the time from protein to structure. Computational methods for protein structure prediction such as AlphaFold or RosettaFold (Baek et al., 2021; Jumper et al., 2021) have so far not been able to replace the classical structural biology techniques, but they have instead accelerated the process of model building, becoming an invaluable tool in the hands of structural biologists.

With the increasing throughput of cryo-EM and development of AI/ML data processing methods, we can anticipate a future where cryo-EM–empowered structure-based drug discovery (SBDD) will become a routine rather than a fortuity.

REFERENCES

Ahn, G., Riley, N. M., Kamber, R. A., Wisnovsky, S., Moncayo von Hase, S., Bassik, M. C., ... Bertozzi, C. R. (2023). Elucidating the cellular determinants of targeted membrane protein degradation by lysosome-targeting chimeras. *Science*, *382*(6668), p. eadf6249. doi:10.1126/science.adf6249

Alexey Rak, Anand Kumar, Francis Duffi et al. Structural insights into epitope-paratope interactions of monoclonal antibody targeting CEACAM5-expressing tumors, 11 August 2023, PREPRINT (Version 1) available at Research Square [https://doi.org/10.21203/rs.3.rs-3235785/v1]

Amelia, T., van Veldhoven, J. P. D., Falsini, M., Liu, R., Heitman, L. H., van Westen, G. J. P., ... AP, I. J. (2021). Crystal structure and subsequent ligand design of a nonriboside partial agonist bound to the adenosine A(2A) receptor. *J Med Chem*, *64*(7), pp. 3827–3842. doi:10.1021/acs.jmedchem.0c01856

Asnagli, H., Minet, N., Pfeiffer, C., Hoeben, E., Lane, R., Laughton, D., ... Beer, P. A. (2023). CTP synthase 1 is a novel therapeutic target in lymphoma. *Hemasphere*, *7*(4), p. e864. doi:10.1097/HS9.0000000000000864

Baek, M., DiMaio, F., Anishchenko, I., Dauparas, J., Ovchinnikov, S., Lee, G. R., ... Baker, D. (2021). Accurate prediction of protein structures and interactions using a three-track neural network. *Science*, *373*(6557), pp. 871–876. doi:10.1126/science.abj8754

Baldi, L., Hacker, D. L., Meerschman, C., & Wurm, F. M. (2012). Large-scale transfection of mammalian cells. *Methods Mol Biol*, *801*, pp. 13–26. doi:10.1007/978-1-61779-352-3_2

Bashore, C., Prakash, S., Johnson, M. C., Conrad, R. J., Kekessie, I. A., Scales, S. J., ... Dueber, E. C. (2023). Targeted degradation via direct 26S proteasome recruitment. *Nat Chem Biol*, *19*(1), pp. 55–63. doi:10.1038/s41589-022-01218-w

Beddell, C. R., Goodford, P. J., Norrington, F. E., Wilkinson, S., & Wootton, R. (1976). Compounds designed to fit a site of known structure in human haemoglobin. *Br J Pharmacol*, *57*(2), pp. 201–209. doi:10.1111/j.1476-5381.1976.tb07468.x

Békés, M., Langley, D. R., & Crews, C. M. (2022). PROTAC targeted protein degraders: The past is prologue. *Nat Rev Drug Discov*, *21*(3), pp. 181–200. doi:10.1038/s41573-021-00371-6

Bepler, T., Borst, A. J., Bouvette, J., Cannone, G., Chen, S., Cheng, A., ... Potter, C. S. (2022). Smart data collection for CryoEM. *J Struct Biol*, *214*(4), p. 107913. doi:10.1016/j.jsb.2022.107913

Bernhard, S. M., & Che, T. (2023). Capturing receptor states with glue. *Nat Chem Biol*. doi:10.1038/s41589-023-01396-1

Bianchi, M., Turner, H. L., Nogal, B., Cottrell, C. A., Oyen, D., Pauthner, M., ... Hangartner, L. (2018). Electron-microscopy-based epitope mapping defines specificities of polyclonal antibodies elicited during HIV-1 BG505 envelope trimer immunization. *Immunity*, *49*(2), pp. 288–300.e8. doi:10.1016/j.immuni.2018.07.009

Billings, E., Lundquist, K., Overly, C., Srinivasan, K., & Noinaj, N. (2021). Structure determination of membrane proteins using X-ray crystallography. *Methods Mol Biol*, *2302*, pp. 101–136. doi:10.1007/978-1-0716-1394-8_7

Bloch, J. S., Mukherjee, S., Kowal, J., Filippova, E. V., Niederer, M., Pardon, E., ... Locher, K. P. (2021). Development of a universal nanobody-binding Fab module for fiducial-assisted cryo-EM studies of membrane proteins. *Proc Natl Acad Sci USA*, *118*(47). doi:10.1073/pnas.2115435118

Bodrug, T., Welsh, K. A., Bolhuis, D. L., Paulakonis, E., Martinez-Chacin, R. C., Liu, B., ... Brown, N. G. (2023). Time-resolved cryo-EM (TR-EM) analysis of substrate polyubiquitination by the RING E3 anaphase-promoting complex/cyclosome (APC/C). *Nat Struct Mol Biol*, *30*(11), pp. 1663–1674. doi:10.1038/s41594-023-01105-5

Boeszoermenyi, A., Bernaleau, L., Chen, X., Kartnig, F., Xie, M., Zhang, H., ... Superti-Furga, G. (2023). A conformation-locking inhibitor of SLC15A4 with TASL proteostatic anti-inflammatory activity. *Nat Commun*, *14*(1), p. 6626. doi:10.1038/s41467-023-42070-3

Bonsor, D. A., Alexander, P., Snead, K., Hartig, N., Drew, M., Messing, S., ... Simanshu, D. K. (2022). Structure of the SHOC2–MRAS–PP1C complex provides insights into RAF activation and Noonan syndrome. *Nat Struct Mol Biol*, *29*(10), pp. 966–977. doi:10.1038/s41594-022-00841-4

Braicu, C., Buse, M., Busuioc, C., Drula, R., Gulei, D., Raduly, L., ... Berindan-Neagoe, I. (2019). A comprehensive review on MAPK: A promising therapeutic target in cancer. *Cancers (Basel)*, *11*(10). doi:10.3390/cancers11101618

Bussiere, D. E., Xie, L., Srinivas, H., Shu, W., Burke, A., Be, C., ... Paulk, J. (2020). Structural basis of indisulam-mediated RBM39 recruitment to DCAF15 E3 ligase complex. *Nat Chem Biol*, *16*(1), pp. 15–23. doi:10.1038/s41589-019-0411-6

Castells-Graells, R., Meador, K., Arbing, M. A., Sawaya, M. R., Gee, M., Cascio, D., ... Yeates, T. O. (2023). Cryo-EM structure determination of small therapeutic protein targets at 3 Å-resolution using a rigid imaging scaffold. *Proc Natl Acad Sci USA*, *120*(37), p. e2305494120. doi:10.1073/pnas.2305494120

Chatzikyriakidou, Y., Ahn, D. H., Nji, E., & Drew, D. (2021). The GFP thermal shift assay for screening ligand and lipid interactions to solute carrier transporters. *Nat Protoc*, *16*(12), pp. 5357–5376. doi:10.1038/s41596-021-00619-w

Chen, X., Xie, M., Zhang, S., Monguió-Tortajada, M., Yin, J., Liu, C., ... Yang, M. (2023). Structural basis for recruitment of TASL by SLC15A4 in human endolysosomal TLR signaling. *Nat Commun*, *14*(1), p. 6627. doi:10.1038/s41467-023-42210-9

Cheng, Y. (2018). Single-particle cryo-EM-How did it get here and where will it go. *Science*, *361*(6405), pp. 876–880. doi:10.1126/science.aat4346

Cherezov, V., Rosenbaum, D. M., Hanson, M. A., Rasmussen, S. G., Thian, F. S., Kobilka, T. S., ... Stevens, R. C. (2007). High-resolution crystal structure of an engineered human beta2-adrenergic G protein-coupled receptor. *Science*, *318*(5854), pp. 1258–1265. doi:10.1126/science.1150577

Chirnomas, D., Hornberger, K. R., & Crews, C. M. (2023). Protein degraders enter the clinic - a new approach to cancer therapy. *Nat Rev Clin Oncol*, *20*(4), pp. 265–278. doi:10.1038/s41571-023-00736-3

Chung, J. M., Durie, C. L., & Lee, J. (2022). Artificial intelligence in cryo-electron microscopy. *Life (Basel)*, *12*(8). doi:10.3390/life12081267

Congreve, M., Dias, J. M., & Marshall, F. H. (2014). Structure-based drug design for G protein-coupled receptors. *Prog Med Chem*, *53*, pp. 1–63. doi:10.1016/b978-0-444-63380-4.00001-9

Crawford, R., Dogdas, B., Keough, E., Haas, R. M., Wepukhulu, W., Krotzer, S., ... Howell, B. J. (2011). Analysis of lipid nanoparticles by Cryo-EM for characterizing siRNA delivery vehicles. *Int J Pharm*, *403*(1-2), pp. 237–244. doi:10.1016/j.ijpharm.2010.10.025

Custódio, T. F., Killer, M., Yu, D., Puente, V., Teufel, D. P., Pautsch, A., … Löw, C. (2023). Molecular basis of TASL recruitment by the peptide/histidine transporter 1, PHT1. *Nat Commun*, *14*(1), p. 5696. doi:10.1038/s41467-023-41420-5

de la Torre, B. G., & Albericio, F. (2022). The pharmaceutical industry in 2021. An analysis of FDA drug approvals from the perspective of molecules. *Molecules*, *27*(3). doi:10.3390/molecules27031075

Deisenhofer, J., Epp, O., Miki, K., Huber, R., & Michel, H. (1985). Structure of the protein subunits in the photosynthetic reaction centre of Rhodopseudomonas viridis at 3Å resolution. *Nature*, *318*(6047), pp. 618–624. doi:10.1038/318618a0

Denisov, I. G., & Sligar, S. G. (2016). Nanodiscs for structural and functional studies of membrane proteins. *Nat Struct Mol Biol*, *23*(6), pp. 481–486. doi:10.1038/nsmb.3195

Diver, M. M., Lin King, J. V., Julius, D., & Cheng, Y. (2022). Sensory TRP channels in three dimensions. *Annu Rev Biochem*, *91*, pp. 629–649. doi:10.1146/annurev-biochem-032620-105738

Drew, D., North, R. A., Nagarathinam, K., & Tanabe, M. (2021). Structures and general transport mechanisms by the major facilitator superfamily (MFS). *Chem Rev*, *121*(9), pp. 5289–5335. doi:10.1021/acs.chemrev.0c00983

Drulyte, I., Koester, S., Lundberg, D., Eng, K., Malyutin, A., McGrath, A., … Hurdiss, D. (2022). High-throughput cryo-EM epitope mapping of SARSCoV-2 spike protein antibodies using EPU multigrid (White paper). https://assets.thermofisher.com/TFS-Assets/MSD/Reference-Materials/cryo-em-epitope-mapping-wp0031.pdf

Estevez, A., Arthur, C., Rouge, L., Kschonsak, M., Rohou, A., & Ciferri, C. (2019). Building cryo-EM at genentech to enable research and drug discovery. *Microsc Microanal*, *25*(S2), pp. 1314–1315. doi:10.1017/s143192761900730x

Faust, T. B., Yoon, H., Nowak, R. P., Donovan, K. A., Li, Z., Cai, Q., … Fischer, E. S. (2020). Structural complementarity facilitates E7820-mediated degradation of RBM39 by DCAF15. *Nat Chem Biol*, *16*(1), pp. 7–14. doi:10.1038/s41589-019-0378-3

Feng, X., Fu, Z., Kaledhonkar, S., Jia, Y., Shah, B., Jin, A., … Lin, Q. (2017). A fast and effective microfluidic spraying-plunging method for high-resolution single-particle cryo-EM. *Structure*, *25*(4), pp. 663–670 e663. doi:10.1016/j.str.2017.02.005

Fernandez-Martinez, D., Tully, M. D., Leonard, G., Mathieu, M., & Kandiah, E. (2023). Structural insights into the bi-specific cross-over dual variable antibody architecture by cryo-EM. *Sci Rep*, *13*(1), p. 8694. doi:10.1038/s41598-023-35678-4

Frauenfeld, J., Löving, R., Armache, J. P., Sonnen, A. F., Guettou, F., Moberg, P., … Nordlund, P. (2016). A saposin-lipoprotein nanoparticle system for membrane proteins. *Nat Methods*, *13*(4), pp. 345–351. doi:10.1038/nmeth.3801

Frenzel, A., Schirrmann, T., & Hust, M. (2016). Phage display-derived human antibodies in clinical development and therapy. *MAbs*, *8*(7), pp. 1177–1194. doi:10.1080/19420862.2016.1212149

Fuller, J. R., Knockenhauer, K. E., Leksa, N. C., Peters, R. T., & Batchelor, J. D. (2021). Molecular determinants of the factor VIII/von Willebrand factor complex revealed by BIVV001 cryo-electron microscopy. *Blood*, *137*(21), pp. 2970–2980. doi:10.1182/blood.2020009197

Gadd, M. S., Testa, A., Lucas, X., Chan, K. H., Chen, W., Lamont, D. J., … Ciulli, A. (2017). Structural basis of PROTAC cooperative recognition for selective protein degradation. *Nat Chem Biol*, *13*(5), pp. 514–521. doi:10.1038/nchembio.2329

García-Nafría, J., & Tate, C. G. (2021). Structure determination of GPCRs: Cryo-EM compared with X-ray crystallography. *Biochem Soc Trans*, *49*(5), pp. 2345–2355. doi:10.1042/bst20210431

Götzke, H., Kilisch, M., Martínez-Carranza, M., Sograte-Idrissi, S., Rajavel, A., Schlichthaerle, T., … Frey, S. (2019). The ALFA-tag is a highly versatile tool for nanobody-based bioscience applications. *Nat Commun*, *10*(1), p. 4403. doi:10.1038/s41467-019-12301-7

Graham, B. S., Gilman, M. S. A., & McLellan, J. S. (2019). Structure-Based Vaccine Antigen Design. *Annu Rev Med*, *70*, pp. 91–104. doi:10.1146/annurev-med-121217-094234

Green, L. L. (2014). Transgenic mouse strains as platforms for the successful discovery and development of human therapeutic monoclonal antibodies. *Curr Drug Discov Technol*, *11*(1), pp. 74–84. doi:10.2174/15701638113109990038

Grigglestone, C. E., & Yeung, K.-S. (2021). Degradation of protein kinases: Ternary complex, cooperativity, and selectivity. *ACS Med Chem Lett*, *12*(11), pp. 1629–1632. doi:10.1021/acsmedchemlett.1c00543

Guo, Q., He, B., Zhong, Y., Jiao, H., Ren, Y., Wang, Q., … Tao, Y. (2023). A method for structure determination of GPCRs in various states. *Nat Chem Biol*. doi:10.1038/s41589-023-01389-0

Guo, Y. (2021). Detergent-free systems for structural studies of membrane proteins. *Biochem Soc Trans*, *49*(3), pp. 1361–1374. doi:10.1042/bst20201080

Hao, Y., Yu, X., Bai, Y., McBride, H. J., & Huang, X. (2019). Cryo-EM structure of HER2-trastuzumab-pertuzumab complex. *PLoS One*, *14*(5), p. e0216095. doi:10.1371/journal.pone.0216095

Harris, L. J., Larson, S. B., Hasel, K. W., & McPherson, A. (1997). Refined structure of an intact IgG2a monoclonal antibody. *Biochemistry*, *36*(7), pp. 1581–1597. doi:10.1021/bi962514+

Hauseman, Z. J., Fodor, M., Dhembi, A., Viscomi, J., Egli, D., Bleu, M., … King, D. A. (2022). Structure of the MRAS-SHOC2-PP1C phosphatase complex. *Nature*, *609*(7926), pp. 416–423. doi:10.1038/s41586-022-05086-1

Heinz, F. X., & Stiasny, K. (2021). Distinguishing features of current COVID-19 vaccines: Knowns and unknowns of antigen presentation and modes of action. *NPJ Vaccines*, *6*(1), p 104. doi:10.1038/s41541-021-00369-6

Hernando, C., Ortega-Morillo, B., Tapia, M., Moragon, S., Martinez, M. T., Eroles, P., … Cejalvo, J. M. (2021). Oral selective estrogen receptor degraders (SERDs) as a novel breast cancer therapy: Present and future from a clinical perspective. *Int J Mol Sci*, *22*(15). doi:10.3390/ijms22157812

Heydenreich, F. M., Vuckovic, Z., Matkovic, M., & Veprintsev, D. B. (2015). Stabilization of G protein-coupled receptors by point mutations. *Front Pharmacol*, *6*, p. 82. doi:10.3389/fphar.2015.00082

Hsia, O., Hinterndorfer, M., Cowan, A. D., Iso, K., Ishida, T., Sundaramoorthy, R., … Ciulli, A. (2023). Targeted protein degradation via intramolecular bivalent glues. *bioRxiv*. doi:10.1101/2023.02.14.528511

Imaide, S., Riching, K. M., Makukhin, N., Vetma, V., Whitworth, C., Hughes, S. J., … Ciulli, A. (2021). Trivalent PROTACs enhance protein degradation via combined avidity and cooperativity. *Nat Chem Biol*, *17*(11), pp. 1157–1167. doi:10.1038/s41589-021-00878-4

Iozzo, Y., Qiu, Y., Xu, A., Park, A., Wendt, M., & Zhou, Y. (2021). Development of an integrated structural biology platform specialized for sub-100 kDa protein complexes to support biologics discovery and rational engineering. *Antib Ther*, *4*(4), pp. 242–251. doi:10.1093/abt/tbab025

Jay, J. W., Bray, B., Qi, Y., Igbinigie, E., Wu, H., Li, J., & Ren, G. (2018). IgG antibody 3D structures and dynamics. *Antibodies, 7*(2), p. 18. doi:10.3390/antib7020018

Jevtić, P., Haakonsen, D. L., & Rapé, M. (2021). An E3 ligase guide to the galaxy of small-molecule-induced protein degradation. *Cell Chem Biol, 28*(7), pp. 1000–1013. doi:10.1016/j.chembiol.2021.04.002

Jin, S., Sun, Y., Liang, X., Gu, X., Ning, J., Xu, Y., ... Pan, L. (2022). Emerging new therapeutic antibody derivatives for cancer treatment. *Signal Transduct Target Ther, 7*(1), p. 39. doi:10.1038/s41392-021-00868-x

Johnston, R. J., Su, L. J., Pinckney, J., Critton, D., Boyer, E., Krishnakumar, A., ... Korman, A. J. (2019). VISTA is an acidic pH-selective ligand for PSGL-1. *Nature, 574*(7779), pp. 565–570. doi:10.1038/s41586-019-1674-5

Jumper, J., Evans, R., Pritzel, A., Green, T., Figurnov, M., Ronneberger, O., ... Hassabis, D. (2021). Highly accurate protein structure prediction with AlphaFold. *Nature, 596*(7873), pp. 583–589. doi:10.1038/s41586-021-03819-2

Kaneko, Y., & Szallasi, A. (2014). Transient receptor potential (TRP) channels: A clinical perspective. *Br J Pharmacol, 171*(10), pp. 2474–2507. doi:10.1111/bph.12414

Kaplon, H., Crescioli, S., Chenoweth, A., Visweswaraiah, J., & Reichert, J. M. (2023). Antibodies to watch in 2023. *MAbs, 15*(1), p 2153410. doi:10.1080/19420862.2022.2153410

Kenakin, T. (2013). Making receptors a reality: The 2012 Nobel Prize in Chemistry. *Trends Pharmacol Sci, 34*(1), pp. 2–5. doi:10.1016/j.tips.2012.11.003

Kendrew, J. C., Bodo, G., Dintzis, H. M., Parrish, R. G., Wyckoff, H., & Phillips, D. C. (1958). A three-dimensional model of the myoglobin molecule obtained by x-ray analysis. *Nature, 181*(4610), pp. 662–666. doi:10.1038/181662a0

Killer, M., Wald, J., Pieprzyk, J., Marlovits, T. C., & Löw, C. (2021). Structural snapshots of human PepT1 and PepT2 reveal mechanistic insights into substrate and drug transport across epithelial membranes. *Sci Adv, 7*(45), p. eabk3259. doi:10.1126/sciadv.abk3259

Kim, P. T., Noble, A. J., Cheng, A., & Bepler, T. (2023). Learning to automate cryo-electron microscopy data collection with Ptolemy. *IUCrJ, 10*(Pt 1), pp. 90–102. doi:10.1107/S2052252522010612

Klein, S., Cortese, M., Winter, S. L., Wachsmuth-Melm, M., Neufeldt, C. J., Cerikan, B., ... Chlanda, P. (2020). SARS-CoV-2 structure and replication characterized by in situ cryo-electron tomography. *Nat Commun, 11*(1), p. 5885. doi:10.1038/s41467-020-19619-7

Kondo, Y., Ognjenovic, J., Banerjee, S., Karandur, D., Merk, A., Kulhanek, K., ... Kuriyan, J. (2019). Cryo-EM structure of a dimeric B-Raf:14-3-3 complex reveals asymmetry in the active sites of B-Raf kinases. *Science, 366*(6461), pp. 109–115. doi:10.1126/science.aay0543

Kontziampasis, D., Klebl, D. P., Iadanza, M. G., Scarff, C. A., Kopf, F., Sobott, F., ... White, H. D. (2019). A cryo-EM grid preparation device for time-resolved structural studies. *IUCrJ, 6*(Pt 6), pp. 1024–1031. doi:10.1107/S2052252519011345

Kotov, V., Bartels, K., Veith, K., Josts, I., Subhramanyam, U. K. T., Günther, C., ... Garcia-Alai, M. M. (2019). High-throughput stability screening for detergent-solubilized membrane proteins. *Sci Rep, 9*(1), p. 10379. doi:10.1038/s41598-019-46686-8

Kotov, V., Killer, M., Jungnickel, K. E. J., Lei, J., Finocchio, G., Steinke, J., ... Löw, C. (2023). Plasticity of the binding pocket in peptide transporters underpins promiscuous substrate recognition. *Cell Rep, 42*(8), p 112831. doi:10.1016/j.celrep.2023.112831

Kruse, A. C., Ring, A. M., Manglik, A., Hu, J., Hu, K., Eitel, K., ... Kobilka, B. K. (2013). Activation and allosteric modulation of a muscarinic acetylcholine receptor. *Nature, 504*(7478), pp. 101–106. doi:10.1038/nature12735

Kühlbrandt, W. (2014). The resolution revolution. *Science, 343*(6178), pp. 1443–1444. doi:10.1126/science.1251652

Kulkarni, J. A., Darjuan, M. M., Mercer, J. E., Chen, S., van der Meel, R., Thewalt, J. L., ... Cullis, P. R. (2018). On the formation and morphology of lipid nanoparticles containing ionizable cationic lipids and siRNA. *ACS Nano, 12*(5), pp. 4787–4795. doi:10.1021/acsnano.8b01516

Kursula, P., Flodin, S., Ehn, M., Hammarstrom, M., Schuler, H., Nordlund, P., & Stenmark, P. (2006). Structure of the synthetase domain of human CTP synthetase, a target for anticancer therapy. *Acta Crystallogr Sect F Struct Biol Cryst Commun, 62*(Pt 7), pp. 613–617. doi:10.1107/S1744309106018136

Kwon, J. J., Hajian, B., Bian, Y., Young, L. C., Amor, A. J., Fuller, J. R., ... Aguirre, A. J. (2022). Structure-function analysis of the SHOC2-MRAS-PP1C holophosphatase complex. *Nature, 609*(7926), pp. 408–415. doi:10.1038/s41586-022-04928-2

Lai, A. C., & Crews, C. M. (2017). Induced protein degradation: An emerging drug discovery paradigm. *Nat Rev Drug Discov, 16*(2), pp. 101–114. doi:10.1038/nrd.2016.211

Lees, J. A., Dias, J. M., Rajamohan, F., Fortin, J. P., O'Connor, R., Kong, J. X., ... Han, S. (2023). An inverse agonist of orphan receptor GPR61 acts by a G protein-competitive allosteric mechanism. *Nat Commun, 14*(1), p. 5938. doi:10.1038/s41467-023-41646-3

Li, D., & Caffrey, M. (2020). Structure and functional characterization of membrane integral proteins in the lipid cubic phase. *J Mol Biol, 432*(18), pp. 5104–5123. doi:10.1016/j.jmb.2020.02.024

Liau, N. P. D., Johnson, M. C., Izadi, S., Gerosa, L., Hammel, M., Bruning, J. M., ... Sudhamsu, J. (2022). Structural basis for SHOC2 modulation of RAS signalling. *Nature, 609*(7926), pp. 400–407. doi:10.1038/s41586-022-04838-3

Liu, W., Chun, E., Thompson, A. A., Chubukov, P., Xu, F., Katritch, V., ... Stevens, R. C. (2012). Structural basis for allosteric regulation of GPCRs by sodium ions. *Science, 337*(6091), pp. 232–236. doi:10.1126/science.1219218

Liu, Y., Yang, J., Wang, T., Luo, M., Chen, Y., Chen, C., ... Han, L. (2023). Expanding PROTACtable genome universe of E3 ligases. *Nat Commun, 14*(1), p. 6509. doi:10.1038/s41467-023-42233-2

Lu, Z., Barnard, D., Shaikh, T. R., Meng, X., Mannella, C. A., Yassin, A., ... Lu, T. M. (2014). Gas-assisted annular microsprayer for sample preparation for time-resolved cryo-electron microscopy. *J Micromech Microeng, 24*(11), p. 115001. doi:10.1088/0960-1317/24/11/115001

Luque, D., & Caston, J. R. (2020). Cryo-electron microscopy for the study of virus assembly. *Nat Chem Biol, 16*(3), pp. 231–239. doi:10.1038/s41589-020-0477-1

Lynch, E. M., Hicks, D. R., Shepherd, M., Endrizzi, J. A., Maker, A., Hansen, J. M., ... Kollman, J. M. (2017). Human CTP synthase filament structure reveals the active enzyme conformation. *Nat Struct Mol Biol, 24*(6), pp. 507–514. doi:10.1038/nsmb.3407

Maeots, M. E., Lee, B., Nans, A., Jeong, S. G., Esfahani, M. M. N., Ding, S., ... Enchev, R. I. (2020). Modular microfluidics enables kinetic insight from time-resolved cryo-EM. *Nat Commun, 11*(1), p. 3465. doi:10.1038/s41467-020-17230-4

Makurvet, F. D. (2021). Biologics vs. small molecules: Drug costs and patient access. *Med Drug Discov*, *9*. doi:10.1016/j.medidd.2020.100075

Marei, H., Tsai, W. K., Kee, Y. S., Ruiz, K., He, J., Cox, C., ... de Sousa, E. M. F. (2022). Antibody targeting of E3 ubiquitin ligases for receptor degradation. *Nature*, *610*(7930), pp. 182–189. doi:10.1038/s41586-022-05235-6

Martin, E., Palmic, N., Sanquer, S., Lenoir, C., Hauck, F., Mongellaz, C., ... Latour, S. (2014). CTP synthase 1 deficiency in humans reveals its central role in lymphocyte proliferation. *Nature*, *510*(7504), pp. 288–292. doi:10.1038/nature13386

Martin, T. G., Bharat, T. A., Joerger, A. C., Bai, X. C., Praetorius, F., Fersht, A. R., ... Scheres, S. H. (2016). Design of a molecular support for cryo-EM structure determination. *Proc Natl Acad Sci USA*, *113*(47), pp. E7456–E7463. doi:10.1073/pnas.1612720113

Matyskiela, M. E., Lu, G., Ito, T., Pagarigan, B., Lu, C. C., Miller, K., ... Chamberlain, P. P. (2016). A novel cereblon modulator recruits GSPT1 to the CRL4(CRBN) ubiquitin ligase. *Nature*, *535*(7611), pp. 252–257. doi:10.1038/nature18611

Maveyraud, L., & Mourey, L. (2020). Protein X-ray crystallography and drug discovery. *Molecules*, *25*(5). doi:10.3390/molecules25051030

McElwee, J. J., Kaila, N., Carreiro, S., Kumar, S., Toms, A. V., Phadke, A., ... Tummino, P. (2023). Discovery and characterization of novel inhibitors of CTP synthase 1 (CTPS1) for the treatment of autoimmune and inflammatory disease. *J Immunol*, *210*(1_Supplement), p. 165.116. doi:10.4049/jimmunol.210.Supp.165.16

Mendonca, L., Howe, A., Gilchrist, J. B., Sheng, Y., Sun, D., Knight, M. L., ... Zhang, P. (2021). Correlative multi-scale cryo-imaging unveils SARS-CoV-2 assembly and egress. *Nat Commun*, *12*(1), p. 4629. doi:10.1038/s41467-021-24887-y

Meuris, L., Santens, F., Elson, G., Festjens, N., Boone, M., Dos Santos, A., ... Callewaert, N. (2014). GlycoDelete engineering of mammalian cells simplifies N-glycosylation of recombinant proteins. *Nat Biotechnol*, *32*(5), pp. 485–489. doi:10.1038/nbt.2885

Mishra, A. K., Shahid, S., Karade, S. S., Agnihotri, P., Kolesnikov, A., Hasan, S. S., & Mariuzza, R. A. (2023). CryoEM structure of a therapeutic antibody (favezelimab) bound to human LAG3 determined using a bivalent Fab as fiducial marker. *Structure*, *31*(10), pp. 1149–1157.e1143. doi:https://doi.org/10.1016/j.str.2023.07.013

Moghadamchargari, Z., Shirzadeh, M., Liu, C., Schrecke, S., Packianathan, C., Russell, D. H., ... Laganowsky, A. (2021). Molecular assemblies of the catalytic domain of SOS with KRas and oncogenic mutants. *Proc Natl Acad Sci USA*, *118*(12). doi:10.1073/pnas.2022403118

Moran, M. M. (2018). TRP channels as potential drug targets. *Annu Rev Pharmacol Toxicol*, *58*, pp. 309–330. doi:10.1146/annurev-pharmtox-010617-052832

Mountain, V. (2003). Astex, Structural Genomix, and Syrrx. I can see clearly now: Structural biology and drug discovery. *Chem Biol*, *10*(2), pp. 95–98. doi:10.1016/s1074-5521(03)00030-9

Mukherjee, S., Erramilli, S. K., Ammirati, M., Alvarez, F. J. D., Fennell, K. F., Purdy, M. D., ... Kossiakoff, A. A. (2020). Synthetic antibodies against BRIL as universal fiducial marks for single-particle cryoEM structure determination of membrane proteins. *Nat Commun*, *11*(1), p. 1598. doi:10.1038/s41467-020-15363-0

Mullard, A. (2021). FDA approves 100th monoclonal antibody product. *Nat Rev Drug Discov*, *20*(7), pp. 491–495. doi:10.1038/d41573-021-00079-7

Naydenova, K., McMullan, G., Peet, M. J., Lee, Y., Edwards, P. C., Chen, S., ... Russo, C. J. (2019). CryoEM at 100 keV: A demonstration and prospects. *IUCrJ*, *6*(Pt 6), pp. 1086–1098. doi:10.1107/S2052252519012612

Noble, A. (2023). HighEnd CryoEM worldwide. https://www.google.com/maps/d/u/0/edit?mid=1eQ1r8BiDYfaK7D1S9EeFJEgkLggMyoaT&ll=12.86686999512855%2C0&z=2)

Ostrem, J. M., Peters, U., Sos, M. L., Wells, J. A., & Shokat, K. M. (2013). K-Ras(G12C) inhibitors allosterically control GTP affinity and effector interactions. *Nature*, *503*(7477), pp. 548–551. doi:10.1038/nature12796

Overington, J. P., Al-Lazikani, B., & Hopkins, A. L. (2006). How many drug targets are there? *Nat Rev Drug Discov*, *5*(12), pp. 993–996. doi:10.1038/nrd2199

Pallesen, J., Wang, N., Corbett, K. S., Wrapp, D., Kirchdoerfer, R. N., Turner, H. L., ... McLellan, J. S. (2017). Immunogenicity and structures of a rationally designed prefusion MERS-CoV spike antigen. *Proc Natl Acad Sci USA*, *114*(35), pp. E7348–E7357. doi:10.1073/pnas.1707304114

Pan, M., Zheng, Q., Wang, T., Liang, L., Mao, J., Zuo, C., ... Zhao, M. (2021). Structural insights into Ubr1-mediated N-degron polyubiquitination. *Nature*, *600*(7888), pp. 334–338. doi:10.1038/s41586-021-04097-8

Pance, K., Gramespacher, J. A., Byrnes, J. R., Salangsang, F., Serrano, J. C., Cotton, A. D., ... Wells, J. A. (2023). Modular cytokine receptor-targeting chimeras for targeted degradation of cell surface and extracellular proteins. *Nat Biotechnol*, *41*(2), pp. 273–281. doi:10.1038/s41587-022-01456-2

Pardon, E., Laeremans, T., Triest, S., Rasmussen, S. G., Wohlkönig, A., Ruf, A., ... Steyaert, J. (2014). A general protocol for the generation of Nanobodies for structural biology. *Nat Protoc*, *9*(3), pp. 674–693. doi:10.1038/nprot.2014.039

Park, E., Rawson, S., Li, K., Kim, B. W., Ficarro, S. B., Pino, G. G., ... Eck, M. J. (2019). Architecture of autoinhibited and active BRAF-MEK1-14-3-3 complexes. *Nature*, *575*(7783), pp. 545–550. doi:10.1038/s41586-019-1660-y

Park, U. B., Jeong, T. J., Gu, N., Lee, H. T., & Heo, Y.-S. (2022). Molecular basis of PD-1 blockade by dostarlimab, the FDA-approved antibody for cancer immunotherapy. *Biochem Biophys Res Commun*, *599*, pp. 31–37. doi:https://doi.org/10.1016/j.bbrc.2022.02.026

Parker, J. L., Deme, J. C., Wu, Z., Kuteyi, G., Huo, J., Owens, R. J., ... Newstead, S. (2021). Cryo-EM structure of PepT2 reveals structural basis for proton-coupled peptide and prodrug transport in mammals. *Sci Adv*, *7*(35). doi:10.1126/sciadv.abh3355

Paul, S. M., Mytelka, D. S., Dunwiddie, C. T., Persinger, C. C., Munos, B. H., Lindborg, S. R., & Schacht, A. L. (2010). How to improve R&D productivity: The pharmaceutical industry's grand challenge. *Nat Rev Drug Discov*, *9*(3), pp. 203–214. doi:10.1038/nrd3078

Petzold, G., Fischer, E. S., & Thomä, N. H. (2016). Structural basis of lenalidomide-induced CK1α degradation by the CRL4(CRBN) ubiquitin ligase. *Nature*, *532*(7597), pp. 127–130. doi:10.1038/nature16979

Pizzagalli, M. D., Bensimon, A., & Superti-Furga, G. (2021). A guide to plasma membrane solute carrier proteins. *FEBS J*, *288*(9), pp. 2784–2835. doi:10.1111/febs.15531

Popow, J., Farnaby, W., Gollner, A., Kofink, C., Fischer, G., Wurm, M., ... Ciulli, A. (2023). Targeting cancer with small molecule pan-KRAS degraders. *bioRxiv*. doi:10.1101/2023.10.24.563163

Privé, G. G. (2007). Detergents for the stabilization and crystallization of membrane proteins. *Methods, 41*(4), pp. 388–397. doi:10.1016/j.ymeth.2007.01.007

Punjani, A., Zhang, H., & Fleet, D. J. (2020). Non-uniform refinement: Adaptive regularization improves single-particle cryo-EM reconstruction. *Nat Methods, 17*(12), pp. 1214–1221. doi:10.1038/s41592-020-00990-8

Qin, F., Xia, F., Chen, H., Cui, B., Feng, Y., Zhang, P., ... Luo, M. (2021). A guide to nucleic acid vaccines in the prevention and treatment of infectious diseases and cancers: From basic principles to current applications. *Front Cell Dev Biol, 9*, p. 633776. doi:10.3389/fcell.2021.633776

Ramlaul, K., Palmer, C. M., Nakane, T., & Aylett, C. H. S. (2020). Mitigating local over-fitting during single particle reconstruction with SIDESPLITTER. *J Struct Biol, 211*(2), p. 107545. doi:10.1016/j.jsb.2020.107545

Robertson, M. J., Papasergi-Scott, M. M., He, F., Seven, A. B., Meyerowitz, J. G., Panova, O., ... Skiniotis, G. (2022). Structure determination of inactive-state GPCRs with a universal nanobody. *Nat Struct Mol Biol, 29*(12), pp. 1188–1195. doi:10.1038/s41594-022-00859-8

Rougé, L., Chiang, N., Steffek, M., Kugel, C., Croll, T. I., Tam, C., ... Rohou, A. (2020). Structure of CD20 in complex with the therapeutic monoclonal antibody rituximab. *Science, 367*(6483), pp. 1224–1230. doi:10.1126/science.aaz9356

Rush, S. A., Brar, G., Hsieh, C. L., Chautard, E., Rainho-Tomko, J. N., Slade, C. D., ... McLellan, J. S. (2022). Characterization of prefusion-F-specific antibodies elicited by natural infection with human metapneumovirus. *Cell Rep, 40*(12), p. 111399. doi:10.1016/j.celrep.2022.111399

Sader, K., Matadeen, R., Castro Hartmann, P., Halsan, T., & Schlichten, C. (2020). Industrial cryo-EM facility setup and management. *Acta Crystallogr D Struct Biol, 76*(Pt 4), pp. 313–325. doi:10.1107/S2059798320002223

Sakamoto, K. M., Kim, K. B., Kumagai, A., Mercurio, F., Crews, C. M., & Deshaies, R. J. (2001). Protacs: Chimeric molecules that target proteins to the Skp1-Cullin-F box complex for ubiquitination and degradation. *Proc Natl Acad Sci USA, 98*(15), pp. 8554–8559. doi:10.1073/pnas.141230798

Saur, M., Hartshorn, M. J., Dong, J., Reeks, J., Bunkoczi, G., Jhoti, H., & Williams, P. A. (2020). Fragment-based drug discovery using cryo-EM. *Drug Discov Today, 25*(3), pp. 485–490. doi:10.1016/j.drudis.2019.12.006

Shahsavar, A., Stohler, P., Bourenkov, G., Zimmermann, I., Siegrist, M., Guba, W., ... Nissen, P. (2021). Structural insights into the inhibition of glycine reuptake. *Nature, 591*(7851), pp. 677–681. doi:10.1038/s41586-021-03274-z

Thorn, C. R., Sharma, D., Combs, R., Bhujbal, S., Romine, J., Zheng, X., ... Badkar, A. (2022). The journey of a lifetime - development of Pfizer's COVID-19 vaccine. *Curr Opin Biotechnol, 78*, p. 102803. doi:10.1016/j.copbio.2022.102803

Torino, S., Dhurandhar, M., Stroobants, A., Claessens, R., & Efremov, R. G. (2023). Time-resolved cryo-EM using a combination of droplet microfluidics with on-demand jetting. *Nat Methods, 20*(9), pp. 1400–1408. doi:10.1038/s41592-023-01967-z

Tsutsumi, N., Mukherjee, S., Waghray, D., Janda, C. Y., Jude, K. M., Miao, Y., ... Garcia, K. C. (2020). Structure of human Frizzled5 by fiducial-assisted cryo-EM supports a heterodimeric mechanism of canonical Wnt signaling. *Elife, 9*. doi:10.7554/eLife.58464

Vaitsopoulou, A., Depping, P., Bill, R. M., Goddard, A. D., & Rothnie, A. J. (2022). Membrane protein production in insect cells. *Methods Mol Biol, 2507*, pp. 223–240. doi:10.1007/978-1-0716-2368-8_12

Watson, E. R., Novick, S., Matyskiela, M. E., Chamberlain, P. P., A, H. d. l. P., Zhu, J., ... Lander, G. C. (2022). Molecular glue CELMoD compounds are regulators of cereblon conformation. *Science, 378*(6619), pp. 549–553. doi:10.1126/science.add7574

Wlodawer, A. (2021). Protein crystallography: Alive and well. *FEBS J, 288*(20), pp. 5786–5787. doi:10.1111/febs.15822

Wrapp, D., Wang, N., Corbett, K. S., Goldsmith, J. A., Hsieh, C. L., Abiona, O., ... McLellan, J. S. (2020). Cryo-EM structure of the 2019-nCoV spike in the prefusion conformation. *Science, 367*(6483), pp. 1260–1263. doi:10.1126/science.abb2507

Wurz, R. P., Rui, H., Dellamaggiore, K., Ghimire-Rijal, S., Choi, K., Smither, K., ... Vaish, A. (2023). Affinity and cooperativity modulate ternary complex formation to drive targeted protein degradation. *Nature Communications, 14*(1), p. 4177. doi:10.1038/s41467-023-39904-5

Xu, Y., & Dang, S. (2022). Recent technical advances in sample preparation for single-particle cryo-EM. *Front Mol Biosci, 9*, p 892459. doi:10.3389/fmolb.2022.892459

Yao, Q., Weaver, S. J., Mock, J. Y., & Jensen, G. J. (2019). Fusion of DARPin to aldolase enables visualization of small protein by cryo-EM. *Structure, 27*(7), pp. 1148–1155 e1143. doi:10.1016/j.str.2019.04.003

Yeates, T. O., Agdanowski, M. P., & Liu, Y. (2020). Development of imaging scaffolds for cryo-electron microscopy. *Curr Opin Struct Biol, 60*, pp. 142–149. doi:https://doi.org/10.1016/j.sbi.2020.01.012

Yonekura, K., Maki-Yonekura, S., Naitow, H., Hamaguchi, T., & Takaba, K. (2021). Machine learning-based real-time object locator/evaluator for cryo-EM data collection. *Commun Biol, 4*(1), p. 1044. doi:10.1038/s42003-021-02577-1

Yue, J., Liu, Y., Zhao, M., Bi, X., Li, G., & Liang, W. (2023). The R&D landscape for infectious disease vaccines. *Nat Rev Drug Discov*. doi:10.1038/d41573-023-00119-4

Zhang, K., Wu, H., Hoppe, N., Manglik, A., & Cheng, Y. (2022). Fusion protein strategies for cryo-EM study of G protein-coupled receptors. *Nat Commun, 13*(1), p. 4366. doi:10.1038/s41467-022-32125-2

Zhang, S., Zou, S., Yin, D., Zhao, L., Finley, D., Wu, Z., & Mao, Y. (2022). USP14-regulated allostery of the human proteasome by time-resolved cryo-EM. *Nature, 605*(7910), pp. 567–574. doi:10.1038/s41586-022-04671-8

Zheng, H., Handing, K. B., Zimmerman, M. D., Shabalin, I. G., Almo, S. C., & Minor, W. (2015). X-ray crystallography over the past decade for novel drug discovery - where are we heading next? *Expert Opin Drug Discov, 10*(9), pp. 975–989. doi:10.1517/17460441.2015.1061991

34 Applications of Single-Particle Analysis and Cryo-ET in Drug Discovery

Qing Yao

INTRODUCTION

Cryo-electron microscopy (cryo-EM) has revolutionized structural biology by providing unprecedented insights into the nanoscale world of macromolecules. Within the realm of cryo-EM, there are three primary modalities that have emerged as powerful tools for investigating biological and nonbiological samples: single-particle analysis (SPA), cryo-electron tomography (cryo-ET), and microcrystal electron diffraction (microED). Each modality offers unique advantages and has found applications across a range of scientific disciplines (Figure 34.1). These cryo-EM techniques have greatly empowered scientists to explore the intricate world of macromolecules, cellular structures, and nanomaterials with unprecedented detail, fostering breakthroughs in structural biology–enabled drug design and improvement. In this chapter, we will provide a brief overview of SPA and cryo-ET and their applications in the pharmaceutical industry.

UTILIZING SINGLE-PARTICLE ANALYSIS FOR INDUSTRIAL APPLICATIONS

SPA is a transformative technique with extensive applications in biopharmaceutical research (Kuhlbrandt 2014). Its ability to determine the 3D structure of macromolecules provides crucial insights into drug discovery, vaccine development, and the study of membrane proteins (Lyumkis 2019, Bai 2021). In industrial settings, the pressure to speed

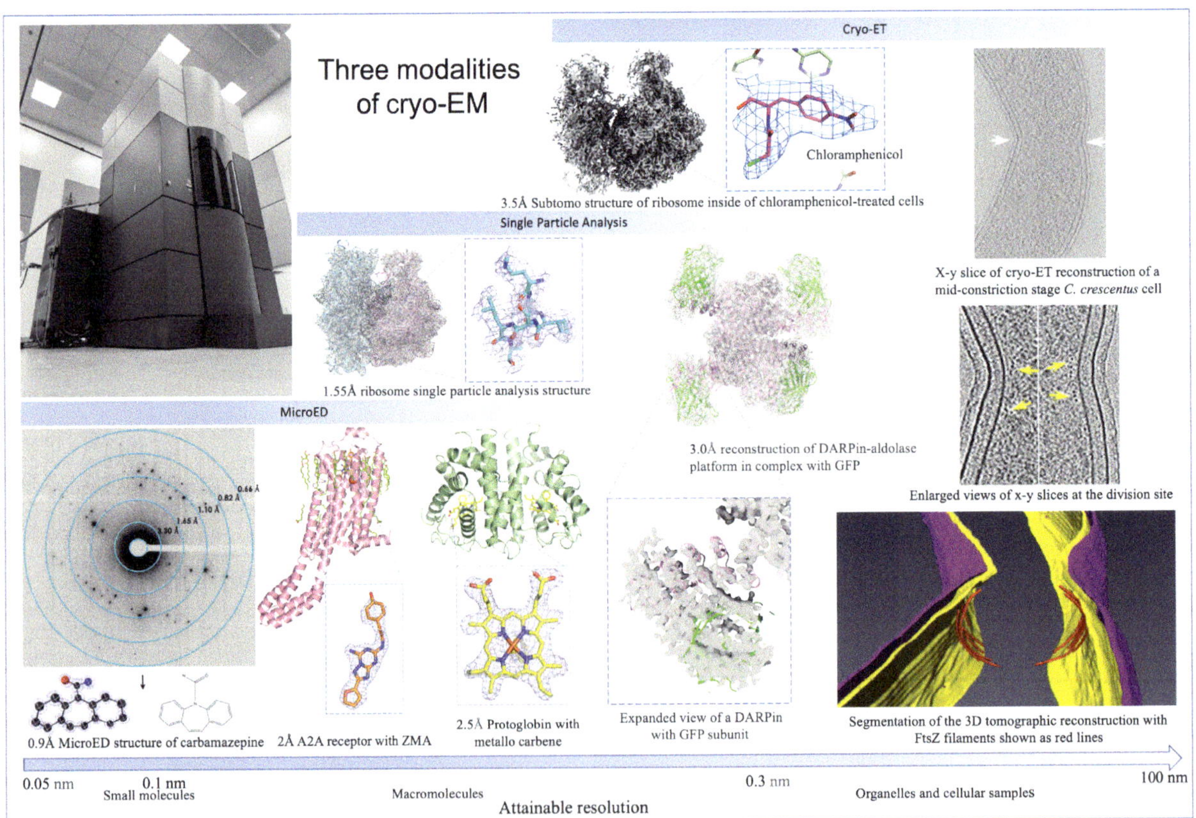

FIGURE 34.1 Three primary modalities in cryo-EM. SPA: This technique is employed to study individual particles, providing high-resolution structural information. It is routinely used in the biopharmaceutical industry. Cryo-ET: It enables three-dimensional visualization of complex structures within intact cells and tissues. MicroED: This technique is utilized for the analysis of microcrystals, offering insights into crystallography and materials science. These modalities have diverse applications across various scientific disciplines and can achieve different levels of resolution.

DOI: 10.1201/9781003326106-41

up new drug introductions to the market makes SPA a valuable tool. It maximizes the benefits of rational drug design and explores structure–activity relationships by determining the structures of large protein complexes like G protein–coupled receptors (GPCRs), membrane proteins, antibody complexes, and proteolysis targeting chimeras (PROTACs) (Paiva and Crews 2019). SPA provides crucial insights into interactions between small molecules and macromolecular assemblies, guiding compound design from hit identification to lead optimization and more. Its applications in biopharmaceutical research for drug development are widespread, including the following.

Acceleration of Structure-Based Drug Design (SBDD)

SBDD is a powerful approach in pharmaceutical research that involves designing drug molecules based on the three-dimensional structure of target proteins, leading to the development of more effective and specific therapeutic agents. X-ray crystallography has long been a key tool in SBDD (Zheng, Hou et al. 2014, van Montfort and Workman 2017). However, there are challenges in applying X-ray crystallography to every target protein. Some proteins are difficult to crystallize, and the process may introduce artifacts or alter the protein's conformation. This is where SPA has been emerging as another mainstream technique in SBDD. It has advantages in cases where proteins are challenging to crystallize or exist in multiple conformations. Moreover, SPA can provide unprecedented insights into the association and dissociation processes that underlie the dynamic nature of macromolecules (Dandey, Wei et al. 2018, Dandey, Budell et al. 2020), opening additional opportunities for drug targeting.

In industrial settings, the iterative determination of high-resolution liganded structures is crucial for successful SBDD. To fulfill this role, SPA is optimized at each step, from sample preparation to data processing, to streamline the rapid structural solutions. Despite the advancements in SPA hardware (Faruqi and Henderson 2007, Konings and Bischoff 2020, Nakane, Kotecha et al. 2020, Frechin, Holvec et al. 2023) and software (Punjani, Rubinstein et al. 2017, Grant, Rohou et al. 2018, Zivanov, Nakane et al. 2018), sample preparation continues to be a bottleneck due to various obstacles, such as sample wastage and a lack of reproducibility (Passmore and Russo 2016). Additionally, the quality of a purified specimen does not guarantee a suitable sample for data collection (Weissenberger, Henderikx et al. 2021). The current approach of traditional blotting-paper-based sample vitrification in cryo-EM setups involves a trial-and-error process influenced by various factors. These factors include the biochemical behavior of purified samples, the choice of grid material, the glow discharge technique, specific incubation conditions, and the intricacies of the blotting process, which limit the bandwidth for SPA in cryo-EM setups used in demanding industry applications.

Recent advancements in automated vitrification devices are actively addressing the challenges associated with SPA sample preparation. These devices minimize human errors and variations by automating the vitrification process, leading to more reliable and standardized results. They offer precise control over critical parameters such as temperature, humidity, and timing, ensuring optimal conditions for the vitrification of samples. For example, Prof. Bridget Carragher's laboratory (NRAMM, New York, USA) has persistently developed Spotiton (commercially known as Chameleon, TTP Labtech Ltd., Melbourn, UK), a robotic device utilizing inkjet technology to deposit picoliter droplets of samples onto self-wicking grids. The machine yielded promising results (Razinkov, Dandey et al. 2016, Dandey, Wei et al. 2018), demonstrating substantial enhancements in the efficiency and consistency of grid preparation. Moreover, Vitrojet (CryoSol Inc) is a device that utilizes inkjet printing heads with picoliter dispensing capabilities without using self-wicking grids. It is specifically designed for the purpose of spotting samples onto cryo-EM grids. By leveraging the capabilities of inkjet printing heads, Vitrojet enables researchers to dispense samples in small, controlled volumes onto the grids, ensuring reproducibility and accuracy in sample placement (Ravelli, Nijpels et al. 2020, Schotman, Weissenberger et al. 2022). These advancements in automated vitrification devices represent significant efforts to enhance the efficiency, reproducibility, and overall quality of cryo-EM sample preparation. They hold great potential for meeting the industry's need for speed in SPA.

Additionally, in recent years, SPA data collection has become standardized with automatic software such as Leginon (Suloway, Pulokas et al. 2005), SerialEM (Mastronarde 2005), and EPU (ThermoFisher Scientific). Despite significant advancements in the setup of SPA, there remains a substantial discrepancy in data collection time and efficiency compared to X-ray crystallography, falling significantly short by several orders of magnitude (Lyumkis 2019). The growing demand for automation and faster data collection is driving the focus toward the development of more intelligent and automated software solutions. These advancements in smart software aim to address the industry's needs by streamlining and optimizing the data collection process by automating various steps and incorporating advanced algorithms. For example, the EPU software Multigrid function can maximize autoloader usage by enabling users to schedule and queue multiple unattended data acquisition experiments, reducing microscope idle time and eliminating the need for manual grid switching during nights or weekends. It also facilitates faster data acquisition through the scheduling and execution of multiple single-particle analysis screening sessions. Moreover, Smart EPU Software (ThermoFisher Scientific) can dynamically adjust settings based on real-time image analysis software (EPU Quality Monitor Software or Embedded CryoSPARC Live [Punjani, Rubinstein et al. 2017]) output, optimizing focus stabilization, stage waiting time, grid-square skipping, and automated selection of high-quality holes on a grid square.

One of the paramount goals is to leverage deep learning algorithms to improve the data processing pipeline through the automation of tasks such as grid screening, ice thickness estimation, prediction of best areas for data collection, and automatic particle picking. Ultimately, concerted efforts in the coming years will establish standardized workflows that streamline the process and enable a swifter iterative approach to structurally informed drug optimization. Having discussed the emerging roles and explored the room for potential future improvement of SPA in SBDD, in the following paragraphs we will delve into a few tangible examples of how SPA can accelerate drug development.

SPA in Vaccine Development

SPA is a groundbreaking technique in vaccine development that allows scientists to shortcut the development of vaccines and monoclonal antibody therapies (Earl and Subramaniam 2016). By visualizing the three-dimensional structure of antibody/antigen complexes, SPA provides valuable insights that aid in the design of effective vaccines. The Covid-19 pandemic underscored the remarkable contribution of SPA in characterizing the ectodomain structure of coronavirus spike protein (S protein) (Henderson, Edwards et al. 2020, Walls, Park et al. 2020, Wrapp, Wang et al. 2020). The S protein is responsible for facilitating the fusion of viral and target cell membranes, allowing the virus to enter host cells (Jackson, Farzan et al. 2022). Moreover, it acts as a prominent surface antigen, prompting the production of neutralizing antibodies. Thus, S protein is a major component of the first-generation vaccines (Dai and Gao 2021). The SPA was used to solve both the prefusion and postfusion structures of the purified full-length S protein from the original Wuhan-Hu-1 strain (Cai, Zhang et al. 2020). Moreover, the structure of the S protein D614G, a mutation strain that has exhibited significantly increased infectivity and quickly became the predominant circulating strain worldwide, was also solved by SPA (Zhang, Cai et al. 2021, Zhang and Chen 2022). By comparison of the original and the mutant structures, it was found that the SARS-CoV-2 original S protein occasionally changed shape prematurely, hindering cell binding and slowing the virus. However, the D614G mutation stabilizes the S protein, increasing overall infectiousness despite weaker binding to the ACE receptor. The finding played a crucial role in developing the Covid-19 mRNA vaccine BNT162b2 (Vogel, Kanevsky et al. 2020), ensuring the conformational and antigenic integrity of its spike protein variant. This strategy was also used in several other spike-based Covid-19 vaccines, including Moderna's mRNA-1273 (Corbett, Edwards et al. 2020), both of which reported high vaccine efficacy rates, leading to global emergency use authorization. The SARS-CoV-2 case provided a vivid example of how high-resolution SPA information is critical for the rapid development of vaccines and therapeutics against emerging human pathogens.

Furthermore, in response to the prolonged Covid-19 pandemic and the emergence of vaccine-resistant variants like Omicron, scientists have sought innovative solutions to provide broader protection against the virus and potential zoonotic threats. To enhance the efficacy of the vaccine, these nanoparticles were designed to present 60 randomly arranged spike RBDs from multiple sarbecoviruses, promoting interactions with B cells through avidity effects (Cohen, van Doremalen et al. 2022). This increases the likelihood of recognizing conserved, yet typically hidden, RBD epitopes. SPA also provided detailed and high-resolution structural information about the antibody-S protein complexes. Conserved sarbecovirus epitopes were identified, and this information is crucial for understanding the interactions between the antibodies and the S protein of the virus (Fan, Cohen et al. 2022). The results are invaluable for the development of effective vaccines to combat the future spillover of animal coronaviruses and to address the current challenges posed by SARS-CoV-2 variants.

In summary, SPA has become an indispensable tool in vaccine development, significantly accelerating the process of vaccine development in several ways:

1. Structural characterization of antigens: SPA enables researchers to determine the precise atomic-level structure of viral antigens, such as the S protein structures of SARS-CoV-2. Understanding these structures is essential for designing effective vaccines, as they allow scientists to identify key regions of the virus that can elicit a strong immune response.
2. Rational vaccine design: Armed with detailed structural information from SPA, scientists can now rationally design vaccines that mimic specific antigens.
3. Vaccine candidate screening and epitope mapping: Cryo-EM helps in the early stages of vaccine development by screening various vaccine candidates and mapping the binding epitopes. This information helps gauge the effectiveness of the immune response and refine the vaccines for enhanced effectiveness.

SPA has become an indispensable tool in vaccine development, significantly accelerating the process of vaccine development. SPA will continue to be a powerful technique to fend off future public health threats.

SPA in the Studies of Membrane Proteins

SPA has significantly advanced high-resolution structural studies of membrane proteins since the first atomic structure of a membrane protein was determined using this technique in 2013 (Cao, Liao et al. 2013, Liao, Cao et al. 2013). This involves solubilizing proteins with detergents (Li 2022) or amphipols (Tribet, Audebert et al. 1996) or reconstituting them into lipid nanodiscs (Bayburt and Sligar 2010) or liposomes (Yao, Fan et al. 2020) for visualization in cryo-EM. By providing high-resolution insights into their architecture, SPA has revolutionized the structural biology of membrane proteins. In this section, we will briefly go through illustrative examples showcasing the successful application of SPA in studying membrane proteins, which are crucial players in numerous cellular processes and disease pathways.

The cryo-EM structures of the GLP-1 (glucagon-like peptide-1) receptor represent a significant breakthrough in the study of obesity and blood glucose control. The GLP-1 receptor is a type of G protein–coupled receptor (GPCR) crucially involved in regulating glucose levels and satiety (Kinzig, D'Alessio et al. 2002, Andersen, Lund et al. 2018, Wang, Wang et al. 2023). The high-resolution SPA structures have enabled researchers to visualize the GLP-1 receptor's binding pocket and overall organization. The GLP-1 receptor structure changes into an extended open shape. This change allows GLP-1 to bind more effectively and activate the receptor, leading to important biological responses (Zhang, Sun et al. 2017). This understanding has facilitated the rational design of novel GLP-1 receptor agonists with enhanced binding affinity and improved pharmacokinetic properties. These optimized agonists mimic the actions of GLP-1 for more effective and sustained weight loss and glucose control in individuals with obesity (Castro). Additionally, LY3502970 and PF-06882961, potent and selective oral small-molecule GLP-1R agonists, were developed based on the structural information. The revealed binding mechanism by SPA accounts for the compound's highly favorable pharmacological profile, highlighting its potential as a promising therapeutic agent (Kawai, Sun et al. 2020, Zhang, Belousoff et al. 2020).

SPA has also been used to determine a huge variety of membrane protein structures from many different protein classes with therapeutic interests. The other classes include ion channels, such as TRP channels (Paulsen, Armache et al. 2015, Huynh, Cohen et al. 2016, Hirschi, Herzik et al. 2017, Yin, Wu et al. 2018) and the human hERG channel (Wang and MacKinnon 2017); transporters, such as LAT1, a large amino acid transporter that serves as an anticancer therapeutic target (Yan, Zhao et al. 2019); glutamine SLC1A5 transporters, which provide novel insights into glutamate transportation across membranes (Yu, Plotnikova et al. 2019); γ-secretase, which cleaves the amyloid precursor protein and results in the aggregation of amyloid-β (Lu, Bai et al. 2014); the type VII secretion system in *Mycobacterium tuberculosis* (Bunduc, Fahrenkamp et al. 2021); and the type IVa pilus secretin required for natural competence in *Vibrio cholerae* (Weaver, Ortega et al. 2020). This pilus represents a potential druggable target and offers insight into the channel that DNA likely traverses to promote the spread of antibiotic resistance via horizontal gene transfer through natural transformation.

Collectively, the findings provide valuable insights into the mechanisms of drug action and resistance. This understanding assists in fine-tuning drug properties, including pharmacokinetics and pharmacodynamics, leading to a higher likelihood of successful drug development. Approximately 23% of the human proteome consists of membrane proteins (Almen, Nordstrom et al. 2009), and remarkably, they represent almost 60% of all FDA-approved drug targets (Yin and Flynn 2016). This prominence has led to SPA structure determination becoming a pivotal tool in drug discovery programs focused on membrane protein drugs. In this section, we provided a concise overview of recent cryo-EM highlights, demonstrating their potential applications in drug discovery and development for membrane protein targets. While numerous membrane protein cryo-EM structures have been deposited into the EMDB, listing all of them is infeasible due to their sheer number. Nonetheless, these structures have undoubtedly been becoming game-changers in membrane protein drug development research. Additionally, the rapid advancements in other cryo-EM modalities, such as cryo-ET and microED, have been becoming indispensable tools for gaining crucial structural insights into the behavior of membrane proteins within the cell or addressing the challenges posed by difficult-to-crystallize membrane proteins.

Conclusion and Perspectives

SPA has been gaining momentum as a transformative tool in drug development research by providing high-resolution structural information on biomolecules in their native state. This level of insight is critical for rational drug design, enabling the development of highly specific and effective therapeutics. Some exemplary structures solved by SPA are listed in Table 34.1. The application of SPA in drug development is constantly expanding, providing invaluable contributions to the discovery and optimization of novel drugs for various diseases. As this technology continues to advance, its impact on accelerating drug development and improving patient outcomes is expected to be profound and substantial.

CRYO-ET ADVANCEMENTS AND APPLICATIONS IN THE PHARMACEUTICAL INDUSTRY

Modern drug discovery is a highly complex and interdisciplinary endeavor that demands an in-depth understanding of collective protein dynamics and intricate cellular processes (Batool, Ahmad et al. 2019, Aplin, Milano et al. 2022). Solving protein structures offers valuable insights into the mechanisms of biological processes and plays a crucial role in the rational design of modern drug discoveries (Aplin, Milano et al. 2022). The structural biology approach has traditionally sought to gain insight in a reductionist manner by examining purified molecular components. This approach assumes that structures of isolated molecules possess enough explanatory power to comprehend the entire system (Gierasch and Gershenson 2009). This "divide and conquer" strategy has successfully explained modern biology and drug actions in terms of biophysics and biochemistry. However, the reductionist approach in structural biology has several

TABLE 34.1
Pharmacologically Significant Structures Resolved by SPA

Drug	Target	Resolution (Å)	PDB IDs and References
Emetine	Ribosome	3.2	Wong, Bai et al. 2014
Granisetron	Serotonin receptor	2.92	Basak, Gicheru et al. 2019
ZK200775 (Fanapanel)	AMPA receptor	2.97	Nakagawa 2019
Alprazolam (Xanax)	GABA receptor	3.26	Masiulis, Desai et al. 2019
Diazepam (Valium)		3.58	
Resiniferatoxin	TRP channel	2.95	Cao, Liao et al. 2013
Flecainide	Sodium channel	3.24	Jiang, Shi et al. 2020
Glibenclamide	Potassium channel	3.63	Martin, Kandasamy et al. 2017
Bay K 8644	Calcium channel	2.7	Zhao, Huang et al. 2019
Verapamil	Calcium channel	2.9	
Nifedipine	Calcium channel	2.9	
Diltiazem	Calcium channel	2.6	
Bictegravir	Intasome	2.81	Cook, Li et al. 2020, Passos, Li et al. 2020
NDI-091143	ACLY	3.67 4.38	Corbett, Edwards et al. 2020
RGT1383	GLP-1	4.2	Bai, Yu et al. 2020
Rituximab	CD20	3.3	Rouge, Chiang et al. 2020
Insulin	Insulin receptor – insulin complex	4.3	Scapin, Dandey et al. 2018
Trastuzumab and pertuzumab	HER2	4.36	Hao, Yu et al. 2019
A small molecule inhibitor	C5 component protein	3.35	Jendza, Kato et al. 2019
Fragment screening ligands	PKM2	2.6–3.2	Saur, Hartshorn et al. 2020
Virtual screening inhibitors	PRMT5	2.5–4.0	Zhou, Yadav et al. 2022
GDC-0310	NaV1.7	2.9	Kschonsak, Jao et al. 2023
AM-1473 AM-0883	TRPC6 channel	2.9–3.2	Bai, Yu et al. 2020
Ivacaftor	CFTR	3.3	Liu, Zhang et al. 2019

intrinsic limitations: it can lead to oversimplifications; it often focuses on studying individual components or processes in isolation, neglecting the complex interactions and systems-level behavior inherent in biological systems (Van Regenmortel 2004, Mazzocchi 2008).

Cryo-electron tomography (cryo-ET) is a cutting-edge imaging technique that enables scientists to visualize the intricate three-dimensional (3D) structure of biological samples with remarkable detail, approaching near-atomic resolution (Baumeister 2022, Oikonomou and Jensen 2017). It offers a powerful tool to overcome some of the limitations of reductionist approaches in biological research by holistically visualizing intact biological structures in their native state (Asano, Engel et al. 2016, Danev, Yanagisawa et al. 2019). Here, we aim to provide an overview of cryo-ET and its various applications in the pharmaceutical industry.

Introduction of Principles of Cryo-ET

Cryo-ET involves cryogenic sample preparation, transmission electron microscopy (TEM), and computational image reconstruction. The process begins with the rapid freezing of the sample, which preserves the sample's native structure by vitrifying the water molecules. The frozen sample is then imaged using TEM, where a beam of electrons is transmitted through the sample to generate images with high resolution. To obtain a three-dimensional representation, a series of images is acquired at different tilt angles, normally ranging from −60 to +60 degrees. These images, called tilt series, are then subjected to computational image reconstruction algorithms, which convert the 2D tilt series into a 3D volume (Oikonomou, Swulius et al. 2016, Oikonomou and Jensen 2017, Young and Villa 2023). The reconstructed 3D volume can be visualized and analyzed to gain insights into the structure and organization of the sample. To improve the signal-to-noise ratio and increase the clarity of the resulting 3D structure, subtomogram averaging can extract and average identical structures from multiple copies within reconstructed 3D volumes. By aligning and averaging these subvolumes, it enhances the signal-to-noise ratio, thereby revealing higher-resolution information about the individual components and their arrangement in greater detail (Wan and Briggs 2016, Leigh, Navarro et al. 2019, Scaramuzza and Castano-Diez 2021).

Cryo-ET offers a unique opportunity to study biological samples in their natural state, providing valuable information about their structure and organization. Its ability to visualize biological samples in their native state at high resolution provides unique insights into the structure and behavior of drug molecules, drug delivery systems, and biological targets. Here, we explore the impact of cryo-ET in modern pharma across different areas:

Cryo-ET in Nanoparticle and Drug Delivery Imaging and Analysis

Cryo-ET offers unique opportunities to explore the architecture and variability of nanoparticles that are challenging to investigate using other structural methods. In this overview, we focus on two exemplary studies: the cryo-ET of retroviruses and liposomal drug delivery systems like Doxil (Barenholz 2012).

Retroviruses, including the simian immunodeficiency virus (SIV) and human immunodeficiency virus (HIV), exhibit complex biological assemblies ranging from 100 nm to 200 nm in size (Fauci and Desrosiers 1997, Meissner, Talledge et al. 2022). Our understanding of their structural characteristics traditionally relied on electron microscopic studies of fixed, stained, and plastic-embedded viruses found in infected cell populations (Hockley, Wood et al. 1988, Grief, Hockley et al. 1989). Despite recent advances in SPA, the common approach used for viruses to obtain high-resolution information (Kaelber, Hryc et al. 2017) is not applicable to retroviruses. This is because retroviruses possess unique compositions and shapes and irregular features, making them unsuitable for SPA. Consequently, the analysis of retroviral ultrastructure was largely limited to qualitative descriptions based on 2D projection images, which provided only limited insights into the architecture of the viral core and the arrangement of surface envelope glycoprotein spikes. It wasn't until cryo-ET was employed to study these structures that more detailed information became available (Mak and de Marco 2018). Notably, cryo-ET applied to SIV and human HIV-infected cells led to the discovery of the viral "entry claw," a remarkable three-dimensional structure utilized by the virus to initiate cell entry (Liu, Bartesaghi et al. 2008, Mao, Wang et al. 2013, Ward and Wilson 2017). Simultaneously, the implementation of cryo-ET has started unveiling crucial details about the overall shape and distribution of envelope glycoprotein spikes within individual viruses, exemplified by SIV. These tomographic findings provide statistically relevant information regarding the variations between different viruses, and they enable the monitoring of bulk changes in parameters like spike density and distribution across diverse viral strains (Subramaniam 2006, Stewart 2017). Furthermore, an intriguing prospect lies in the possibility of averaging 3D reconstructions of individual entities, such as surface glycoproteins, to derive density maps of these protein complexes in an asymmetric context, although this approach necessitates overcoming several technical challenges for reliable and high-resolution results.

In recent years, there has been significant interest in utilizing liposomal systems for drug encapsulation. One prominent example of such a system is Doxil, the first commercially available drug in this category (Barenholz 2012). Doxil employs sterically stabilized, pegylated liposomes to encapsulate doxorubicin. These liposomes, with a nominal size of approximately 100 nm, allow the doxorubicin to selectively accumulate in tumors, as objects within this size range are believed to penetrate the tumor vasculature more effectively (Schroeder, Avnir et al. 2007). This targeted accumulation increases the local concentration of doxorubicin, enhancing its efficacy and reducing the harmful side effects of the unencapsulated form, such as tissue damage and cardiomyopathy (Chatterjee, Zhang et al. 2010).

The formation of individual Doxil liposomes involves a sequential process that includes the initial generation of the liposomal delivery system and subsequent deposition of doxorubicin within the liposomes via a pH gradient. Each of these steps can introduce variations in the chemical composition and overall size of the drug particles (Li, Hirsh et al. 1998). While the physiological effects of the drug have been extensively studied, and it has received FDA approval, cryo-ET has discovered significant variation in size and doxorubicin content among individual Doxil particles in commercially available preparations (Lengyel, Milne et al. 2008). Understanding the variability in Doxil particles is crucial for designing more efficient drug delivery systems and improving treatment efficacy. By fine-tuning the size and doxorubicin content, researchers can enhance the targeted delivery of Doxil to tumor sites, maximizing its therapeutic impact.

In conclusion, cryo-ET has revolutionized our understanding of retrovirus structures, enabling detailed insights into their unique compositions and shapes. The application of cryo-ET also holds great promise for studying envelope glycoprotein spikes and tracking variations among viral strains. Additionally, cryo-ET is critical for improving the design, efficiency, and therapeutic effects of Doxil as a liposomal drug delivery system. Taken together, cryo-ET's ability to visualize particles in their native state and provide detailed structural insights contributes to the advancement of pharmaceutical research and the development of targeted therapies.

Understanding Drug–Target Interactions Inside Cells

Cryo-ET has enabled the groundbreaking visualization of an antibiotic bound to 70S ribosomes at a remarkable resolution of 3.5 Å within cells (Tegunov, Xue et al. 2021). In the recent study, cryo-ET was utilized to analyze tilt-series data from *Mycoplasma pneumoniae* cells treated with chloramphenicol (Cm). The results obtained using cryo-ET showcased its exceptional performance in refining the 70S ribosome structure to a resolution of 3.5 Å, based on data derived from 17,890 particles obtained from 65 tomograms. The large 50S ribosomal subunit was particularly well resolved, exhibiting a higher average resolution compared to the smaller 30S subunit. The cryo-ET-generated map displayed an astonishing level of structural detail, with features such as amino acid side-chain stubs and individually resolved β-strands readily discernible (Figure 34.2). Remarkably, the presence of Cm was absent in the reconstruction obtained from untreated *M. pneumoniae* cells, emphasizing the specificity and accuracy of

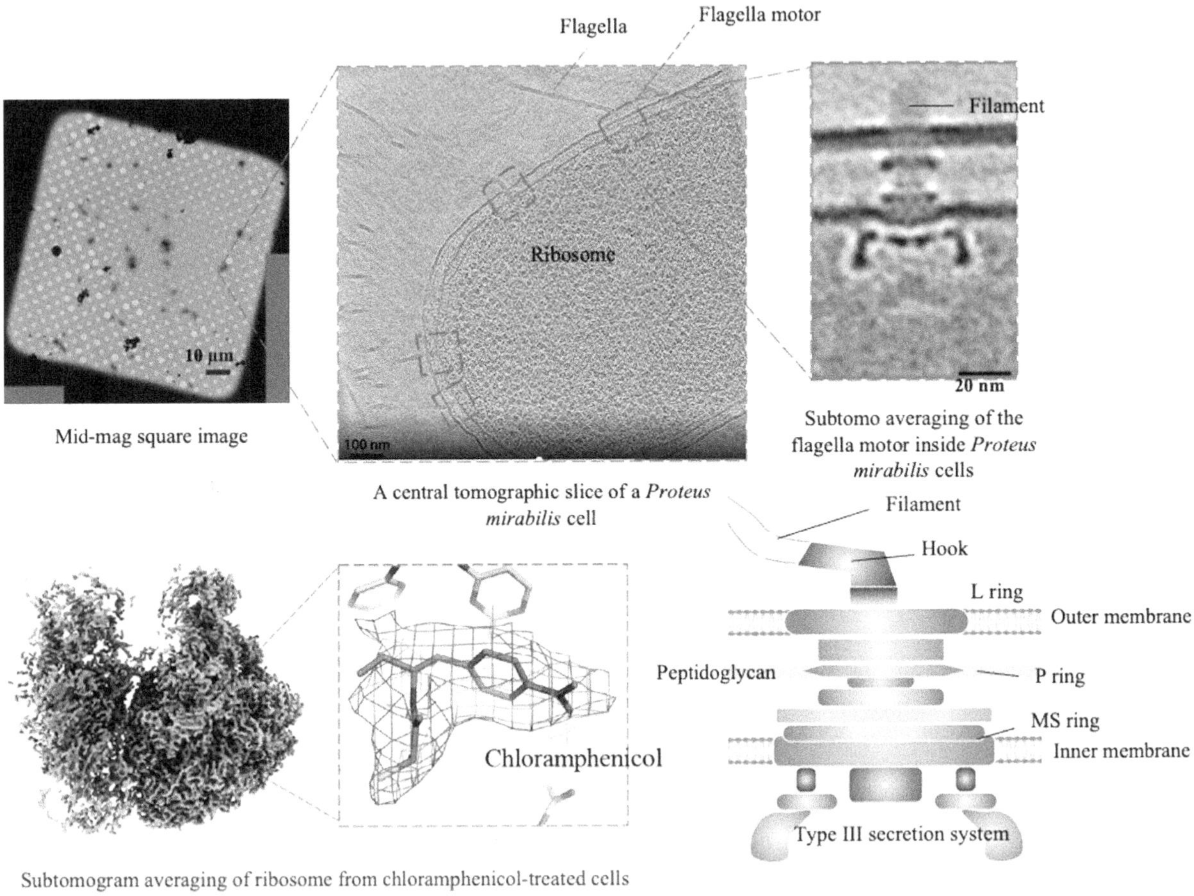

FIGURE 34.2 *In situ* structures of the bacterial flagellar motor and chloramphenicol-bound ribosome solved by cryo-ET. The upper left portion shows a mid-magnification view of bacterial cells (black dots) frozen on a square region of a cryo-EM grid. In the upper middle section, a high-magnification image of a single bacterial cell within that square is shown, with the labeling of a flagella filament and flagellar motors. The upper right shows the subtomo averaging structure of the flagellar motors. The schematic illustration of the subtomo averaged structure is shown on the lower right. The lower left portion shows the subtomo averaging structure of ribosomes obtained from chloramphenicol-treated cells with a mesh representation of the density of the chloramphenicol molecule.

cryo-ET in visualizing drug–target interactions within cellular contexts.

Cryo-ET can provide high-resolution structural information within cellular contexts, visualize drug–target interactions, and elucidate dynamic processes, holding great promise for accelerating drug discovery and development. Continued advancements in cryo-ET techniques, automation, and algorithms will undoubtedly contribute to the future of pharmaceutical applications, enabling more precise and targeted therapies for various diseases.

Target Mechanism of Action

Bacterial motility, driven by specialized structures called flagellar motors, plays a crucial role in various biological processes, including disease development and biofilm formation (Haiko and Westerlund-Wikstrom 2013, Nakamura and Minamino 2019). Understanding the structural mechanism behind bacterial motility and flagellar motor function is essential for elucidating disease pathways and designing effective strategies to combat bacterial infections (Haiko and Westerlund-Wikstrom 2013). However, the process of purification poses unique challenges and limitations. The flagellar motor is an intricate and integral part of bacterial cells embedded within the cellular envelope. Separating the flagellar motor from the rest of the cell without disrupting its structural integrity is a significant challenge (Kaplan, Tocheva et al. 2021). Given the challenges associated with purifying the flagellar motor, cryo-ET has been used to study its structure and function, allowing for the visualization of the flagellar motor in its native state within intact bacterial cells. This technique can provide valuable insights into the motor's structure, organization, and dynamics without the need for purification.

Proteus mirabilis is a gram-negative bacterium responsible for urinary tract infections in humans (Schaffer and Pearson 2015). The bacterium possesses peritrichous flagella, which contributes to its mobility and virulence. Therefore, understanding the structural mechanisms of the flagellar motor provides valuable insights for developing novel treatments that can interfere with its motility and potentially reduce its pathogenicity. A study using cryo-ET and subtomogram averaging has revealed that the flagellar motor of *P. mirabilis* shares structural similarities with the motors found in other Enterobacteriaceae, such as *Escherichia coli* and *Salmonella enterica*. However, it lacks certain periplasmic elaborations characteristic of more specialized gammaproteobacterial motors (Figure 34.2) (Kaplan, Yao et al. 2023). Additionally, the subtomogram averages did not show distinct densities corresponding to stators, indicating that the stators in the *P. mirabilis* motor are dynamic. This dynamic nature allows for the recruitment of different numbers of stators around the motor, potentially adapting to different environmental conditions and optimizing its motility. Understanding the structure and function of the flagellar motor in *P. mirabilis* is important for comprehending its pathogenicity. We can gain insights into the mechanisms underlying the movement and colonization of *P. mirabilis* in diverse environments, aiding in the development of strategies to prevent and treat infections caused by this bacterium.

Challenges and Perspective

As discussed above, cryo-ET is a powerful technique used to visualize the three-dimensional structures of biological samples at the nanometer scale. However, it comes with challenges, including the "missing wedge" problem. Due to the physical constraints of the microscope's stage tilting, it is normal to tilt the specimen from −60 to +60 degrees. As a result, there is missing information or gaps in the data collection, leading to an incomplete representation of the sample's structure. The missing information in certain orientations results in distortions, artifacts, and a loss of resolution in the final reconstructed volume (Frank 1992, Koster, Grimm et al. 1997). To overcome the missing wedge problem, various strategies have been developed. One approach involves acquiring additional tilt series with different orientations to fill in the missing information (Penczek, Marko et al. 1995, Mastronarde 1997, Lanzavecchia, Cantele et al. 2005) or using subtomogram averaging of the target with different orientations. Another approach involves employing advanced reconstruction algorithms that could partially compensate for the missing information (Paavolainen, Acar et al. 2014, Deng, Chen et al. 2016, Ding, Liu et al. 2019, Kovacs, Song et al. 2020). Despite these efforts, the missing wedge problem remains a challenge in cryo-ET. It limits the achievable resolution and introduces potential biases in the interpretation of the reconstructed density volumes.

Another significant challenge in cryo-ET is the low signal-to-noise ratio (SNR) in the recorded tilt series. Cryo-ET typically adopts a dosage range of 90–160 $e^-/Å^2$ to ensure minimal radiation damage to biological samples (Glaeser 2016), with only 1–3 $e^-/Å^2$ allocated for a single projection image in each tilt. Furthermore, for higher tilt angles, the path electrons traverse through the slab-shaped samples increases, resulting in multiple electron scattering, further reducing the SNR (Turk and Baumeister 2020). In contrast, SPA typically utilizes a dose of 40–60 $e^-/Å^2$ for a single projection image. The low SNR poses challenges for accurately visualizing and interpreting the structural details of the specimen. Additionally, sample thickness is another critical parameter in cryo-ET because it directly impacts the contrast and interpretability of the acquired image. Ideally, the sample thickness should be below 500 nm in cryo-ET (Koning, Koster et al. 2018). For thick samples, an additional step is often necessary before imaging them using TEM when dealing with thicker samples. These thick samples are typically thinned using a focused ion beam (FIB) to achieve an appropriate thickness for cryo-ET imaging, resulting in improved imaging quality and higher contrast (Berger, Premaraj et al. 2023).

Significant technical advancements and implementations have greatly enhanced cryo-ET in the past decade. The development of designated cryo-electron microscopes, FIB, phase plate technology, and direct electron detectors has resulted in substantial improvements in the speed and quality of the cryo-ET (Hong, Song et al. 2023). Despite these impressive developments, it is important to note that cryo-ET still requires considerable labor and time compared to SPA. Two key areas should be addressed to enhance the throughput and industrial applicability of cryo-ET: (1) Automation and fast tilt series acquisition. It can be achieved by implementing improvements in stage tilt behavior, adapting acquisition software, and utilizing faster cameras. For example, a pilot study showed that the speed of tilt series acquisition could be increased by approximately 10 to 50 times using the current hardware setup without sacrificing data quality (Chreifi, Chen et al. 2019). (2) Tomographic reconstruction can benefit from fully automated and integrated approaches; automatic segmentation algorithms based on neural networks can identify and segment structures of interest within tomographic volumes, streamlining the reconstruction process (Mastronarde and Held 2017, Chen, Bell et al. 2019, Zeng and Xu 2019, Sharov, Morado et al. 2021, Zheng, Wolff et al. 2022). AI-based particle-picking algorithms can recognize and extract particles from reconstructed volumes, reducing manual effort (Morado, Hu et al. 2016, Chen, Dai et al. 2017, Chung, Durie et al. 2022, Moebel, Martinez-Sanchez et al. 2022, de Teresa-Trueba, Goetz et al. 2023, Purnell, Heebner et al. 2023). These approaches involve continuous optimization of reconstruction algorithms and the improvement of AI-based particle-picking accuracy and reliability. However, there are challenges that need to be addressed, such as obtaining high-quality training datasets, fine-tuning algorithms, and dealing with imaging artifacts. AI-based cryo-ET data processing significantly increases the efficiency and accuracy of image reconstruction, particle picking, and segmentation processes. This technology can accelerate drug discovery pipelines, improve target identification, and facilitate the design of more precise and effective therapeutics.

REFERENCES

Almen, M. S., K. J. Nordstrom, R. Fredriksson and H. B. Schioth (2009). "Mapping the human membrane proteome: A majority of the human membrane proteins can be classified according to function and evolutionary origin." BMC Biol **7**: 50.

Andersen, A., A. Lund, F. K. Knop and T. Vilsboll (2018). "Glucagon-like peptide 1 in health and disease." Nat Rev Endocrinol **14**(7): 390–403.

Aplin, C., S. K. Milano, K. A. Zielinski, L. Pollack and R. A. Cerione (2022). "Evolving experimental techniques for structure-based drug design." J Phys Chem B **126**(35): 6599–6607.

Asano, S., B. D. Engel and W. Baumeister (2016). "In situ cryo-electron tomography: A post-reductionist approach to structural biology." J Mol Biol **428**(2 Pt A): 332–343.

Bai, X. C. (2021). "Seeing atoms by single-particle cryo-EM." Trends Biochem Sci **46**(4): 253–254.

Bai, Y., X. Yu, H. Chen, D. Horne, R. White, X. Wu, P. Lee, Y. Gu, S. Ghimire-Rijal, D. C. H. Lin and X. Huang (2020). "Structural basis for pharmacological modulation of the TRPC6 channel." eLife **9**: e53311.

Barenholz, Y. (2012). "Doxil(R)–the first FDA-approved nano-drug: Lessons learned." J Control Release **160**(2): 117–134.

Basak, S., Y. Gicheru, A. Kapoor, M. L. Mayer, M. Filizola and S. Chakrapani (2019). "Molecular mechanism of setron-mediated inhibition of full-length 5-HT(3A) receptor." Nat Commun **10**(1): 3225.

Batool, M., B. Ahmad and S. Choi (2019). "A structure-based drug discovery paradigm." Int J Mol Sci **20**(11): 2783.

Baumeister, W. (2022). "Cryo-electron tomography: A long journey to the inner space of cells." Cell **185**(15): 2649–2652.

Bayburt, T. H. and S. G. Sligar (2010). "Membrane protein assembly into Nanodiscs." FEBS Lett **584**(9): 1721–1727.

Berger, C., N. Premaraj, R. B. G. Ravelli, K. Knoops, C. Lopez-Iglesias and P. J. Peters (2023). "Cryo-electron tomography on focused ion beam lamellae transforms structural cell biology." Nat Methods **20**(4): 499–511.

Bunduc, C. M., D. Fahrenkamp, J. Wald, R. Ummels, W. Bitter, E. N. G. Houben and T. C. Marlovits (2021). "Structure and dynamics of a mycobacterial type VII secretion system." Nature **593**(7859): 445–448.

Cai, Y., J. Zhang, T. Xiao, H. Peng, S. M. Sterling, R. M. Walsh, Jr., S. Rawson, S. Rits-Volloch and B. Chen (2020). "Distinct conformational states of SARS-CoV-2 spike protein." Science **369**(6511): 1586–1592.

Cao, E., M. Liao, Y. Cheng and D. Julius (2013). "TRPV1 structures in distinct conformations reveal activation mechanisms." Nature **504**(7478): 113–118.

Castro, M. R. (n.d.). GLP-1 agonists: Diabetes drugs and weight loss. Mayo Clinic. https://www.mayoclinic.org/diseases-conditions/type-2-diabetes/expert-answers/byetta/faq-20057955

Chatterjee, K., J. Zhang, N. Honbo and J. S. Karliner (2010). "Doxorubicin cardiomyopathy." Cardiology **115**(2): 155–162.

Chen, M., J. M. Bell, X. Shi, S. Y. Sun, Z. Wang and S. J. Ludtke (2019). "A complete data processing workflow for cryo-ET and subtomogram averaging." Nat Methods **16**(11): 1161–1168.

Chen, M., W. Dai, S. Y. Sun, D. Jonasch, C. Y. He, M. F. Schmid, W. Chiu and S. J. Ludtke (2017). "Convolutional neural networks for automated annotation of cellular cryo-electron tomograms." Nat Methods **14**(10): 983–985.

Chreifi, G., S. Chen, L. A. Metskas, M. Kaplan and G. J. Jensen (2019). "Rapid tilt-series acquisition for electron cryotomography." J Struct Biol **205**(2): 163–169.

Chung, J. M., C. L. Durie and J. Lee (2022). "Artificial intelligence in cryo-electron microscopy." Life (Basel) **12**(8): 1267.

Cohen, A. A., N. van Doremalen, A. J. Greaney, H. Andersen, A. Sharma, T. N. Starr, J. R. Keeffe, C. Fan, J. E. Schulz, P. N. P. Gnanapragasam, L. M. Kakutani, A. P. West, Jr., G. Saturday, Y. E. Lee, H. Gao, C. A. Jette, M. G. Lewis, T. K. Tan, A. R. Townsend, J. D. Bloom, V. J. Munster and P. J. Bjorkman (2022). "Mosaic RBD nanoparticles protect against challenge by diverse sarbecoviruses in animal models." Science **377**(6606): eabq0839.

Cook, N. J., W. Li, D. Berta, M. Badaoui, A. Ballandras-Colas, A. Nans, A. Kotecha, E. Rosta, A. N. Engelman and P. Cherepanov (2020). "Structural basis of second-generation HIV integrase inhibitor action and viral resistance." Science **367**(6479): 806–810.

Corbett, K. S., D. K. Edwards, S. R. Leist, O. M. Abiona, S. Boyoglu-Barnum, R. A. Gillespie, S. Himansu, A. Schafer, C. T. Ziwawo, A. T. DiPiazza, K. H. Dinnon, S. M. Elbashir, C. A. Shaw, A. Woods, E. J. Fritch, D. R. Martinez, K. W. Bock, M. Minai, B. M. Nagata, G. B. Hutchinson, K. Wu, C. Henry, K. Bahl, D. Garcia-Dominguez, L. Ma, I. Renzi, W. P. Kong, S. D. Schmidt, L. Wang, Y. Zhang, E. Phung, L. A. Chang, R. J. Loomis, N. E. Altaras, E. Narayanan, M. Metkar, V. Presnyak, C. Liu, M. K. Louder, W. Shi, K. Leung, E. S. Yang, A. West, K. L. Gully, L. J. Stevens, N. Wang, D. Wrapp, N. A. Doria-Rose, G. Stewart-Jones, H. Bennett, G. S. Alvarado, M. C. Nason, T. J. Ruckwardt, J. S. McLellan, M. R. Denison, J. D. Chappell, I. N. Moore, K. M. Morabito, J. R. Mascola, R. S. Baric, A. Carfi and B. S. Graham (2020). "SARS-CoV-2 mRNA vaccine design enabled by prototype pathogen preparedness." Nature **586**(7830): 567–571.

Dai, L. and G. F. Gao (2021). "Viral targets for vaccines against COVID-19." Nat Rev Immunol **21**(2): 73–82.

Dandey, V. P., W. C. Budell, H. Wei, D. Bobe, K. Maruthi, M. Kopylov, E. T. Eng, P. A. Kahn, J. E. Hinshaw, N. Kundu, C. M. Nimigean, C. Fan, N. Sukomon, S. A. Darst, R. M. Saecker, J. Chen, B. Malone, C. S. Potter and B. Carragher (2020). "Time-resolved cryo-EM using Spotiton." Nat Methods **17**(9): 897–900.

Dandey, V. P., H. Wei, Z. Zhang, Y. Z. Tan, P. Acharya, E. T. Eng, W. J. Rice, P. A. Kahn, C. S. Potter and B. Carragher (2018). "Spotiton: New features and applications." J Struct Biol **202**(2): 161–169.

Danev, R., H. Yanagisawa and M. Kikkawa (2019). "Cryo-electron microscopy methodology: Current aspects and future directions." Trends Biochem Sci **44**(10): 837–848.

de Teresa-Trueba, I., S. K. Goetz, A. Mattausch, F. Stojanovska, C. E. Zimmerli, M. Toro-Nahuelpan, D. W. C. Cheng, F. Tollervey, C. Pape, M. Beck, A. Diz-Munoz, A. Kreshuk, J. Mahamid and J. B. Zaugg (2023). "Convolutional networks for supervised mining of molecular patterns within cellular context." Nat Methods **20**(2): 284–294.

Deng, Y., Y. Chen, Y. Zhang, S. Wang, F. Zhang and F. Sun (2016). "ICON: 3D reconstruction with 'missing-information' restoration in biological electron tomography." J Struct Biol **195**(1): 100–112.

Ding, G., Y. Liu, R. Zhang and H. L. Xin (2019). "A joint deep learning model to recover information and reduce artifacts in missing-wedge sinograms for electron tomography and beyond." Sci Rep **9**(1): 12803.

Earl, L. A. and S. Subramaniam (2016). "Cryo-EM of viruses and vaccine design." Proc Natl Acad Sci USA **113**(32): 8903–8905.

Fan, C., A. A. Cohen, M. Park, A. F. Hung, J. R. Keeffe, P. N. P. Gnanapragasam, Y. E. Lee, H. Gao, L. M. Kakutani, Z. Wu, H. Kleanthous, K. E. Malecek, J. C. Williams and P. J. Bjorkman (2022). "Neutralizing monoclonal antibodies elicited by mosaic RBD nanoparticles bind conserved sarbecovirus epitopes." Immunity **55**(12): 2419–2435 e2410.

Faruqi, A. R. and R. Henderson (2007). "Electronic detectors for electron microscopy." Curr Opin Struct Biol **17**(5): 549–555.

Fauci, A. S. and R. C. Desrosiers (1997). Pathogenesis of HIV and SIV. Retroviruses. J. M. Coffin, S. H. Hughes and H. E. Varmus. Cold Spring Harbor (NY).

Frank, J. (1992). Electron tomography: three-dimensional imaging with the transmission electron microscope. Plenum Press.

Frechin, L., S. Holvec, O. von Loeffelholz, I. Hazemann and B. P. Klaholz (2023). "High-resolution cryo-EM performance comparison of two latest-generation cryo electron microscopes on the human ribosome." J Struct Biol **215**(1): 107905.

Gierasch, L. M. and A. Gershenson (2009). "Post-reductionist protein science, or putting Humpty Dumpty back together again." Nat Chem Biol **5**(11): 774–777.

Glaeser, R. M. (2016). "Specimen behavior in the electron beam." Methods Enzymol **579**: 19–50.

Grant, T., A. Rohou and N. Grigorieff (2018). "cisTEM, user-friendly software for single-particle image processing." Elife **7**: e35383.

Grief, C., D. J. Hockley, C. E. Fromholc and P. A. Kitchin (1989). "The morphology of simian immunodeficiency virus as shown by negative staining electron microscopy." J Gen Virol **70**(Pt 8): 2215–2219.

Haiko, J. and B. Westerlund-Wikstrom (2013). "The role of the bacterial flagellum in adhesion and virulence." Biology (Basel) **2**(4): 1242–1267.

Hao, Y., X. Yu, Y. Bai, H. J. McBride and X. Huang (2019). "Cryo-EM structure of HER2-trastuzumab-pertuzumab complex." PLoS One **14**(5): e0216095.

Henderson, R., R. J. Edwards, K. Mansouri, K. Janowska, V. Stalls, S. M. C. Gobeil, M. Kopp, D. Li, R. Parks, A. L. Hsu, M. J. Borgnia, B. F. Haynes and P. Acharya (2020). "Controlling the SARS-CoV-2 spike glycoprotein conformation." Nat Struct Mol Biol **27**(10): 925–933.

Hirschi, M., M. A. Herzik, Jr., J. Wie, Y. Suo, W. F. Borschel, D. Ren, G. C. Lander and S. Y. Lee (2017). "Cryo-electron microscopy structure of the lysosomal calcium-permeable channel TRPML3." Nature **550**(7676): 411–414.

Hockley, D. J., R. D. Wood, J. P. Jacobs and A. J. Garrett (1988). "Electron microscopy of human immunodeficiency virus." J Gen Virol **69**(Pt 10): 2455–2469.

Hong, Y., Y. Song, Z. Zhang and S. Li (2023). "Cryo-electron tomography: The resolution revolution and a surge of in situ virological discoveries." Annu Rev Biophys **52**: 339–360.

Huynh, K. W., M. R. Cohen, J. Jiang, A. Samanta, D. T. Lodowski, Z. H. Zhou and V. Y. Moiseenkova-Bell (2016). "Structure of the full-length TRPV2 channel by cryo-EM." Nat Commun **7**: 11130.

Jackson, C. B., M. Farzan, B. Chen and H. Choe (2022). "Mechanisms of SARS-CoV-2 entry into cells." Nat Rev Mol Cell Biol **23**(1): 3–20.

Jendza, K., M. Kato, M. Salcius, H. Srinivas, A. De Erkenez, A. Nguyen, D. McLaughlin, C. Be, C. Wiesmann, J. Murphy, P. Bolduc, M. Mogi, J. Duca, A. Namil, M. Capparelli, V. Darsigny, E. Meredith, R. Tichkule, L. Ferrara, J. Heyder, F. Liu, P. A. Horton, M. J. Romanowski, M. Schirle, N. Mainolfi, K. Anderson and G. A. Michaud (2019). "A small-molecule inhibitor of C5 complement protein." Nat Chem Biol **15**(7): 666–668.

Jiang, D., H. Shi, L. Tonggu, T. M. Gamal El-Din, M. J. Lenaeus, Y. Zhao, C. Yoshioka, N. Zheng and W. A. Catterall (2020). "Structure of the cardiac sodium channel." Cell **180**(1): 122–134 e110.

Kaelber, J. T., C. F. Hryc and W. Chiu (2017). "Electron cryomicroscopy of viruses at near-atomic resolutions." Annu Rev Virol **4**(1): 287–308.

Kaplan, M., E. I. Tocheva, A. Briegel, M. J. Dobro, Y. W. Chang, P. Subramanian, A. W. McDowall, M. Beeby and G. J. Jensen (2021). "Loss of the bacterial flagellar motor switch complex upon cell lysis." mBio 12(3): e0029821.

Kaplan, M., Q. Yao and G. J. Jensen (2023). "Structure and assembly of the proteus mirabilis flagellar motor by cryo-electron tomography." Int J Mol Sci **24**(9): 8292.

Kawai, T., B. Sun, H. Yoshino, D. Feng, Y. Suzuki, M. Fukazawa, S. Nagao, D. B. Wainscott, A. D. Showalter, B. A. Droz, T. S. Kobilka, M. P. Coghlan, F. S. Willard, Y. Kawabe, B. K. Kobilka and K. W. Sloop (2020). "Structural basis for GLP-1 receptor activation by LY3502970, an orally active nonpeptide agonist." Proc Natl Acad Sci USA **117**(47): 29959–29967.

Kinzig, K. P., D. A. D'Alessio and R. J. Seeley (2002). "The diverse roles of specific GLP-1 receptors in the control of food intake and the response to visceral illness." J Neurosci **22**(23): 10470–10476.

Koning, R. I., A. J. Koster and T. H. Sharp (2018). "Advances in cryo-electron tomography for biology and medicine." Ann Anat **217**: 82–96.

Konings, S. and M. Bischoff (2020). "A Cold Field Emission Gun Optimized for Cryo-EM Applications." Microsc Microanal **26**(S2): 566–567.

Koster, A. J., R. Grimm, D. Typke, R. Hegerl, A. Stoschek, J. Walz and W. Baumeister (1997). "Perspectives of molecular and cellular electron tomography." J Struct Biol **120**(3): 276–308.

Kovacs, J., J. Song, M. Auer, J. He, W. Hunter and W. Wriggers (2020). "Correction of missing-wedge artifacts in filamentous tomograms by template-based constrained deconvolution." J Chem Inf Model **60**(5): 2626–2633.

Kschonsak, M., C. C. Jao, C. P. Arthur, A. L. Rohou, P. Bergeron, D. F. Ortwine, S. J. McKerrall, D. H. Hackos, L. Deng, J. Chen, T. Li, P. S. Dragovich, M. Volgraf, M. R. Wright, J. Payandeh, C. Ciferri and J. C. Tellis (2023). "Cryo-EM reveals an unprecedented binding site for Na(V)1.7 inhibitors enabling rational design of potent hybrid inhibitors." Elife **12**: e84151.

Kuhlbrandt, W. (2014). "Biochemistry. The resolution revolution." Science **343**(6178): 1443–1444.

Lanzavecchia, S., F. Cantele, P. L. Bellon, L. Zampighi, M. Kreman, E. Wright and G. A. Zampighi (2005). "Conical tomography of freeze-fracture replicas: A method for the study of integral membrane proteins inserted in phospholipid bilayers." J Struct Biol **149**(1): 87–98.

Leigh, K. E., P. P. Navarro, S. Scaramuzza, W. Chen, Y. Zhang, D. Castano-Diez and M. Kudryashev (2019). "Subtomogram averaging from cryo-electron tomograms." Methods Cell Biol **152**: 217–259.

Lengyel, J. S., J. L. Milne and S. Subramaniam (2008). "Electron tomography in nanoparticle imaging and analysis." Nanomedicine (Lond) **3**(1): 125–131.

Li, S. (2022). "Detergents and alternatives in cryo-EM studies of membrane proteins." Acta Biochim Biophys Sin (Shanghai) **54**(8): 1049–1056.

Li, X., D. J. Hirsh, D. Cabral-Lilly, A. Zirkel, S. M. Gruner, A. S. Janoff and W. R. Perkins (1998). "Doxorubicin physical state in solution and inside liposomes loaded via a pH gradient." Biochim Biophys Acta **1415**(1): 23–40.

Liao, M., E. Cao, D. Julius and Y. Cheng (2013). "Structure of the TRPV1 ion channel determined by electron cryo-microscopy." Nature **504**(7478): 107–112.

Liu, F., Z. Zhang, A. Levit, J. Levring, K. K. Touhara, B. K. Shoichet and J. Chen (2019). "Structural identification of a hotspot on CFTR for potentiation." Science **364**(6446): 1184–1188.

Liu, J., A. Bartesaghi, M. J. Borgnia, G. Sapiro and S. Subramaniam (2008). "Molecular architecture of native HIV-1 gp120 trimers." Nature **455**(7209): 109–113.

Lu, P., X. C. Bai, D. Ma, T. Xie, C. Yan, L. Sun, G. Yang, Y. Zhao, R. Zhou, S. H. W. Scheres and Y. Shi (2014). "Three-dimensional structure of human gamma-secretase." Nature **512**(7513): 166–170.

Lyumkis, D. (2019). "Challenges and opportunities in cryo-EM single-particle analysis." J Biol Chem **294**(13): 5181–5197.

Mak, J. and A. de Marco (2018). "Recent advances in retroviruses via cryo-electron microscopy." Retrovirology **15**(1): 23.

Mao, Y., L. Wang, C. Gu, A. Herschhorn, A. Desormeaux, A. Finzi, S. H. Xiang and J. G. Sodroski (2013). "Molecular architecture of the uncleaved HIV-1 envelope glycoprotein trimer." Proc Natl Acad Sci USA **110**(30): 12438–12443.

Martin, G. M., B. Kandasamy, F. DiMaio, C. Yoshioka and S. L. Shyng (2017). "Anti-diabetic drug binding site in a mammalian K(ATP) channel revealed by Cryo-EM." Elife **6**: e31054.

Masiulis, S., R. Desai, T. Uchanski, I. Serna Martin, D. Laverty, D. Karia, T. Malinauskas, J. Zivanov, E. Pardon, A. Kotecha, J. Steyaert, K. W. Miller and A. R. Aricescu (2019). "GABA(A) receptor signalling mechanisms revealed by structural pharmacology." Nature **565**(7740): 454–459.

Mastronarde, D. N. (1997). "Dual-axis tomography: An approach with alignment methods that preserve resolution." J Struct Biol **120**(3): 343–352.

Mastronarde, D. N. (2005). "Automated electron microscope tomography using robust prediction of specimen movements." J Struct Biol **152**(1): 36–51.

Mastronarde, D. N. and S. R. Held (2017). "Automated tilt series alignment and tomographic reconstruction in IMOD." J Struct Biol **197**(2): 102–113.

Mazzocchi, F. (2008). "Complexity in biology. Exceeding the limits of reductionism and determinism using complexity theory." EMBO Rep **9**(1): 10–14.

Meissner, M. E., N. Talledge and L. M. Mansky (2022). "Molecular biology and diversification of human retroviruses." Front Virol **2**: 872599.

Moebel, E., A. Martinez-Sanchez, L. Lamm, R. D. Righetto, W. Wietrzynski, S. Albert, D. Lariviere, E. Fourmentin, S. Pfeffer, J. Ortiz, W. Baumeister, T. Y. Peng, B. D. Engel and C. Kervrann (2022). "Deep learning improves macromolecule identification in 3D cellular cryo-electron tomograms (vol 18, pg 1386, 2021)." Nat Methods **19**(1): 129–129.

Morado, D. R., B. Hu and J. Liu (2016). "Using tomoauto: A protocol for high-throughput automated cryo-electron tomography." J Vis Exp (107): e53608.

Nakagawa, T. (2019). "Structures of the AMPA receptor in complex with its auxiliary subunit cornichon." Science **366**(6470): 1259–1263.

Nakamura, S. and T. Minamino (2019). "Flagella-driven motility of bacteria." Biomolecules **9**(7): 279.

Nakane, T., A. Kotecha, A. Sente, G. McMullan, S. Masiulis, P. M. G. E. Brown, I. T. Grigoras, L. Malinauskaite, T. Malinauskas, J. Miehling, T. Uchański, L. Yu, D. Karia, E. V. Pechnikova, E. de Jong, J. Keizer, M. Bischoff, J. McCormack, P. Tiemeijer, S. W. Hardwick, D. Y. Chirgadze, G. Murshudov, A. R. Aricescu and S. H. W. Scheres (2020). "Single-particle cryo-EM at atomic resolution." Nature **587**(7832): 152–156.

Oikonomou, C. M. and G. J. Jensen (2017). "Cellular electron cryotomography: Toward structural biology in situ." Annu Rev Biochem **86**: 873–896.

Oikonomou, C. M., M. Swulius, A. Briegel, M. Beeby, Q. Yao, Y.-W. Chang and G. Jensen (2016). Electron cryotomography. Methods in microbiology (Vol. **43**). Academic Press. 115–139.

Paavolainen, L., E. Acar, U. Tuna, S. Peltonen, T. Moriya, P. Soonsawad, V. Marjomaki, R. H. Cheng and U. Ruotsalainen (2014). "Compensation of missing wedge effects with sequential statistical reconstruction in electron tomography." PLoS One 9(10): e108978.

Paiva, S.-L. and C. M. Crews (2019). "Targeted protein degradation: Elements of PROTAC design." Curr Opin Chem Biol **50**: 111–119.

Passmore, L. A. and C. J. Russo (2016). "Specimen preparation for high-resolution cryo-EM." Methods Enzymol **579**: 51–86.

Passos, D. O., M. Li, I. K. Jozwik, X. Z. Zhao, D. Santos-Martins, R. Yang, S. J. Smith, Y. Jeon, S. Forli, S. H. Hughes, T. R. Burke, Jr., R. Craigie and D. Lyumkis (2020). "Structural basis for strand-transfer inhibitor binding to HIV intasomes." Science **367**(6479): 810–814.

Paulsen, C. E., J. P. Armache, Y. Gao, Y. Cheng and D. Julius (2015). "Structure of the TRPA1 ion channel suggests regulatory mechanisms." Nature **520**(7548): 511–517.

Penczek, P., M. Marko, K. Buttle and J. Frank (1995). "Double-tilt electron tomography." Ultramicroscopy **60**(3): 393–410.

Punjani, A., J. L. Rubinstein, D. J. Fleet and M. A. Brubaker (2017). "cryoSPARC: Algorithms for rapid unsupervised cryo-EM structure determination." Nat Methods **14**(3): 290–296.

Purnell, C., J. Heebner, M. T. Swulius, R. Hylton, S. Kabonick, M. Grillo, S. Grigoryev, F. Heberle, M. N. Waxham and M. T. Swulius (2023). "Rapid synthesis of cryo-ET data for training deep learning models." *bioRxiv*: 2023.2004.2028.538636.

Ravelli, R. B. G., F. J. T. Nijpels, R. J. M. Henderikx, G. Weissenberger, S. Thewessem, A. Gijsbers, B. Beulen, C. Lopez-Iglesias and P. J. Peters (2020). "Cryo-EM structures from sub-nl volumes using pin-printing and jet vitrification." Nat Commun **11**(1): 2563.

Razinkov, I., V. Dandey, H. Wei, Z. Zhang, D. Melnekoff, W. J. Rice, C. Wigge, C. S. Potter and B. Carragher (2016). "A new method for vitrifying samples for cryoEM." J Struct Biol **195**(2): 190–198.

Rouge, L., N. Chiang, M. Steffek, C. Kugel, T. I. Croll, C. Tam, A. Estevez, C. P. Arthur, C. M. Koth, C. Ciferri, E. Kraft, J. Payandeh, G. Nakamura, J. T. Koerber and A. Rohou (2020). "Structure of CD20 in complex with the therapeutic monoclonal antibody rituximab." Science **367**(6483): 1224–1230.

Saur, M., M. J. Hartshorn, J. Dong, J. Reeks, G. Bunkoczi, H. Jhoti and P. A. Williams (2020). "Fragment-based drug discovery using cryo-EM." Drug Discov Today **25**(3): 485–490.

Scapin, G., V. P. Dandey, Z. Zhang, W. Prosise, A. Hruza, T. Kelly, T. Mayhood, C. Strickland, C. S. Potter and B. Carragher (2018). "Structure of the insulin receptor-insulin complex by single-particle cryo-EM analysis." Nature **556**(7699): 122–125.

Scaramuzza, S. and D. Castano-Diez (2021). "Step-by-step guide to efficient subtomogram averaging of virus-like particles with Dynamo." PLoS Biol **19**(8): e3001318.

Schaffer, J. N. and M. M. Pearson (2015). "Proteus mirabilis and urinary tract infections." Microbiol Spectr **3**(5): 228–236.

Schotman, M. J. G., G. Weissenberger, R. J. M. Henderikx, F. J. T. Nijpels and B. W. A. M. M. Beulen (2022). "Controlled cryo-EM sample preparation using the VitroJet." Microsc Microanal **28**(S1): 1200–1201.

Schroeder, A., Y. Avnir, S. Weisman, Y. Najajreh, A. Gabizon, Y. Talmon, J. Kost and Y. Barenholz (2007). "Controlling liposomal drug release with low frequency ultrasound: Mechanism and feasibility." Langmuir **23**(7): 4019–4025.

Sharov, G., D. R. Morado, M. Carroni and J. M. de la Rosa-Trevin (2021). "Using RELION software within the Scipion framework." Acta Crystallogr D Struct Biol **77**(Pt 4): 403–410.

Stewart, P. L. (2017). "Cryo-electron microscopy and cryo-electron tomography of nanoparticles." Wiley Interdiscip Rev Nanomed Nanobiotechnol **9**(2).

Subramaniam, S. (2006). "The SIV surface spike imaged by electron tomography: One leg or three?" PLoS Pathog **2**(8): e91.

Suloway, C., J. Pulokas, D. Fellmann, A. Cheng, F. Guerra, J. Quispe, S. Stagg, C. S. Potter and B. Carragher (2005). "Automated molecular microscopy: The new Leginon system." J Struct Biol **151**(1): 41–60.

Tegunov, D., L. Xue, C. Dienemann, P. Cramer and J. Mahamid (2021). "Multi-particle cryo-EM refinement with M visualizes ribosome-antibiotic complex at 3.5 A in cells." Nat Methods **18**(2): 186–193.

Tribet, C., R. Audebert and J. L. Popot (1996). "Amphipols: Polymers that keep membrane proteins soluble in aqueous solutions." Proc Natl Acad Sci USA **93**(26): 15047–15050.

Turk, M. and W. Baumeister (2020). "The promise and the challenges of cryo-electron tomography." FEBS Lett **594**(20): 3243–3261.

van Montfort, R. L. M. and P. Workman (2017). "Structure-based drug design: Aiming for a perfect fit." Essays Biochem **61**(5): 431–437.

Van Regenmortel, M. H. (2004). "Reductionism and complexity in molecular biology. Scientists now have the tools to unravel biological and overcome the limitations of reductionism." EMBO Rep **5**(11): 1016–1020.

Vogel, A. B., I. Kanevsky, Y. Che, K. A. Swanson, A. Muik, M. Vormehr, L. M. Kranz, K. C. Walzer, S. Hein, A. Güler, J. Loschko, M. S. Maddur, K. Tompkins, J. Cole, B. G. Lui, T. Ziegenhals, A. Plaschke, D. Eisel, S. C. Dany, S. Fesser, S. Erbar, F. Bates, D. Schneider, B. Jesionek, B. Sänger, A.-K. Wallisch, Y. Feuchter, H. Junginger, S. A. Krumm, A. P. Heinen, P. Adams-Quack, J. Schlereth, C. Kröner, S. Hall-Ursone, K. Brasky, M. C. Griffor, S. Han, J. A. Lees, E. H. Mashalidis, P. V. Sahasrabudhe, C. Y. Tan, D. Pavliakova, G. Singh, C. Fontes-Garfias, M. Pride, I. L. Scully, T. Ciolino, J. Obregon, M. Gazi, J. Ricardo Carrion, K. J. Alfson, W. V. Kalina, D. Kaushal, P.-Y. Shi, T. Klamp, C. Rosenbaum, A. N. Kuhn, Ö. Türeci, P. R. Dormitzer, K. U. Jansen and U. Sahin (2020). "A prefusion SARS-CoV-2 spike RNA vaccine is highly immunogenic and prevents lung infection in non-human primates." *bioRxiv*: 2020.2009.2008.280818.

Walls, A. C., Y. J. Park, M. A. Tortorici, A. Wall, A. T. McGuire and D. Veesler (2020). "Structure, function, and antigenicity of the SARS-CoV-2 spike glycoprotein." Cell **181**(2): 281–292 e286.

Wan, W. and J. A. Briggs (2016). "Cryo-electron tomography and subtomogram averaging." Methods Enzymol **579**: 329–367.

Wang, J. Y., Q. W. Wang, X. Y. Yang, W. Yang, D. R. Li, J. Y. Jin, H. C. Zhang and X. F. Zhang (2023). "GLP-1 receptor agonists for the treatment of obesity: Role as a promising approach." Front Endocrinol (Lausanne) **14**: 1085799.

Wang, W. and R. MacKinnon (2017). "Cryo-EM structure of the open human ether-a-go-go-related K(+) channel hERG." Cell **169**(3): 422–430 e410.

Ward, A. B. and I. A. Wilson (2017). "The HIV-1 envelope glycoprotein structure: Nailing down a moving target." Immunol Rev **275**(1): 21–32.

Weaver, S. J., D. R. Ortega, M. H. Sazinsky, T. N. Dalia, A. B. Dalia and G. J. Jensen (2020). "CryoEM structure of the type IVa pilus secretin required for natural competence in Vibrio cholerae." Nat Commun **11**(1): 5080.

Weissenberger, G., R. J. M. Henderikx and P. J. Peters (2021). "Understanding the invisible hands of sample preparation for cryo-EM." Nat Methods **18**(5): 463–471.

Wong, W., X. C. Bai, A. Brown, I. S. Fernandez, E. Hanssen, M. Condron, Y. H. Tan, J. Baum and S. H. Scheres (2014). "Cryo-EM structure of the Plasmodium falciparum 80S ribosome bound to the anti-protozoan drug emetine." Elife **3**.

Wrapp, D., N. Wang, K. S. Corbett, J. A. Goldsmith, C. L. Hsieh, O. Abiona, B. S. Graham and J. S. McLellan (2020). "Cryo-EM Structure of the 2019-nCoV Spike in the Prefusion Conformation." *bioRxiv*.

Yan, R., X. Zhao, J. Lei and Q. Zhou (2019). "Structure of the human LAT1-4F2hc heteromeric amino acid transporter complex." Nature **568**(7750): 127–130.

Yao, X., X. Fan and N. Yan (2020). "Cryo-EM analysis of a membrane protein embedded in the liposome." Proc Natl Acad Sci USA **117**(31). 18497–18503.

Yin, H. and A. D. Flynn (2016). "Drugging membrane protein interactions." Annu Rev Biomed Eng **18**: 51–76.

Yin, Y., M. Wu, L. Zubcevic, W. F. Borschel, G. C. Lander and S. Y. Lee (2018). "Structure of the cold- and menthol-sensing ion channel TRPM8." Science **359**(6372): 237–241.

Young, L. N. and E. Villa (2023). "Bringing structure to cell biology with cryo-electron tomography." Annu Rev Biophys **52**: 573–595.

Yu, X., O. Plotnikova, P. D. Bonin, T. A. Subashi, T. J. McLellan, D. Dumlao, Y. Che, Y. Y. Dong, E. P. Carpenter, G. M. West, X. Qiu, J. S. Culp and S. Han (2019). "Cryo-EM structures of the human glutamine transporter SLC1A5 (ASCT2) in the outward-facing conformation." Elife **8**: e48120.

Zeng, X. and M. Xu (2019). "Aitom: Open-source AI platform for cryo-electron tomography data analysis." *arXiv preprint arXiv:1911.03044*.

Zhang, J., Y. Cai, T. Xiao, J. Lu, H. Peng, S. M. Sterling, R. M. Walsh, Jr., S. Rits-Volloch, H. Zhu, A. N. Woosley, W. Yang, P. Sliz and B. Chen (2021). "Structural impact on SARS-CoV-2 spike protein by D614G substitution." Science **372**(6541): 525–530.

Zhang, J. and B. Chen (2022). "Fighting SARS-CoV-2 with structural biology methods." Nat Methods **19**(4): 381–383.

Zhang, X., M. J. Belousoff, P. Zhao, A. J. Kooistra, T. T. Truong, S. Y. Ang, C. R. Underwood, T. Egebjerg, P. Senel, G. D. Stewart, Y. L. Liang, A. Glukhova, H. Venugopal, A. Christopoulos, S. G. B. Furness, L. J. Miller, S. Reedtz-Runge, C. J. Langmead, D. E. Gloriam, R. Danev, P. M. Sexton and D. Wootten (2020). "Differential GLP-1R binding and activation by peptide and non-peptide agonists." Mol Cell **80**(3): 485–500 e487.

Zhang, Y., B. Sun, D. Feng, H. Hu, M. Chu, Q. Qu, J. T. Tarrasch, S. Li, T. Sun Kobilka, B. K. Kobilka and G. Skiniotis (2017). "Cryo-EM structure of the activated GLP-1 receptor in complex with a G protein." Nature **546**(7657): 248–253.

Zhao, Y., G. Huang, J. Wu, Q. Wu, S. Gao, Z. Yan, J. Lei and N. Yan (2019). "Molecular basis for ligand modulation of a mammalian voltage-gated Ca(2+) channel." Cell **177**(6): 1495–1506 e1412.

Zheng, H., J. Hou, M. D. Zimmerman, A. Wlodawer and W. Minor (2014). "The future of crystallography in drug discovery." Expert Opin Drug Discov **9**(2): 125–137.

Zheng, S., G. Wolff, G. Greenan, Z. Chen, F. G. A. Faas, M. Barcena, A. J. Koster, Y. Cheng and D. A. Agard (2022). "AreTomo: An integrated software package for automated marker-free, motion-corrected cryo-electron tomographic alignment and reconstruction." J Struct Biol X **6**: 100068.

Zhou, W., G. P. Yadav, X. Yang, F. Qin, C. Li and Q. X. Jiang (2022). "Cryo-EM structure-based selection of computed ligand poses enables design of MTA-synergic PRMT5 inhibitors of better potency." Commun Biol **5**(1): 1054.

Zivanov, J., T. Nakane, B. O. Forsberg, D. Kimanius, W. J. Hagen, E. Lindahl and S. H. Scheres (2018). "New tools for automated high-resolution cryo-EM structure determination in RELION-3." Elife **7**: e42166.

Index

A

ABC transporters, 438
Absorption, distribution, metabolism, excretion, and toxicity (ADMET), 429
Acetylation, 76
Acquired Immune Deficiency Syndrome (AIDS), 17
Actin cytoskeleton, 17
Actin filaments, 17
Actin organization, 17
Activator kinase, 61
Adenosine triphosphate (ATP), 6, 234
Adipogenesis, 277
Aging, 14
Alignment-by-classification, 88, 89, 92, 95, 104
All-trans-retinal, 244, 248
Alosetron, 381
Alpha hemolysin, 13, 200
Alpha-amino-3-hydroxy-5-methyl-4-isoxazolepropionic acid (AMPA) receptors, 349–352, 354, 356, 359, 360
AlphaFold2, 9, 128, 131, 133, 223, 229, 237, 402, 440
Alpha-helical membrane proteins, 213
Alpha-tag, 230
Altman, Richard, 6
Alzheimer, 15, 17, 70, 277, 358
Amazon Web Services (AWS), 160
Aminoacyl-tRNA, 310, 341
Aminoglycoside, 339, 343
Amino-terminal domain, 350
Amoeba, 254
Amphibole, 214
Amphipathic phospholipid, 110
  amphipathic polymers, 295
  amphipathic proteins, 13, 200
Amphipol, 214, 230
Amyloid, 15, 367
  amyloid fibrils, 15, 367
Amyloidogenic peptides, 368
Amyloidosis, 368
Analyse factorielle des correspondances (AFC), 93
Androgen, 37
Angular reconstitution (AR), 96, 97
Anisotropic magnification, 89
Antibiotic-ribosome co-complexes, 310
Antibiotics, 15, 341
Antibody complexes, 17
  antibody evasion, 7
  antibody-drug conjugates, 430
Anti-cancer target, 11
Anti-coagulation therapy, 13, 215
Anti-Fab nanobody, 438
Antigen presentation, 11
Antimicrobials, 335
Antiprotozoan inhibitors, 340
Antiviral drugs, 7
Apoptosis, 11, 69, 76
  apoptosis execution factors, 69
Aquaporin, 13, 216, 222
Archael organisms, 213
Arrestins, 243, 247, 255, 437
Arrhythmias, 283
Arrhythmogenic right ventricular cardiomyopathy (ARVC2), 283
Artificial Intelligence (AI), 9, 10, 17, 223
  AI-based algorithms, 17
  AI-based structure prediction, 16
  AI image recognition, 144
Asymmetric viruses, 8
Atmospheric hydrogen consumption, 187
Atomic force microscopy (AM), 244, 425
Atomic model building, 14
Atopic dermatitis, 16
ATP hydrolysis, 36, 268, 270, 437
ATP synthase, 14, 17, 288, 289, 293, 295, 296
ATPases, 27, 29, 30, 31, 34, 35, 36
ATP-binding cleft, 60
  ATP-binding loop, 60
  ATP-dependent degradation, 29
ATTR fibrils, 369
Autism, 15
  Autism Spectrum Disorders (ASD), 359
Auto phosphorylation, 60, 61
Autocleavage, 48
Autoinducer-2 exporters, 13, 227, 233, 236
Auto-stigmatism, 122

B

Bacillus subtilis, 170, 172, 328, 329
Backward sliding clamp, 184
Bacterial chaperone, 12
  bacterial flagellar motor, 452
  bacterial virulence factors, 13
  bacterial viruses, 7
Bacteriophages, 305
  bacteriophage lambda, 169
  bacteriophage P22, 7
  bacteriophage T4, 7, 169
  bacteriophage T4 DNA ligase, 178
Bacteriorhodopsin, 5, 13, 73
Baker, David, 9
Band-pass filtering, 92
Baumeister, Wolfgang, 8
Beam-Image-Shift Electron Tomography (BISECT), 8
Benda, Carl, 6
Bernal, John, 3, 4
Beta-adrenergic receptor (BAR), 243, 244, 246
Beta-adrenergic receptor-mediated signalling, 244
Bioblasts, 6
Bioenergetics, 6
Biological macromolecules, 147
Biological signals, 213
Bio-molecular machines, 227
  bio-molecular structure modelling, 131
Biopharmaceutical industry, 393
Biotin-labelled DNA, 116
Biotynylation, 113
BISECT, 400
Blobology, 73, 320, 325
Blood glucose control, 17
Boyer, Paul D., 6, 289
Bragg, William, H., 3
Bragg, William, L., 3
Brain development, 350
  brain pathology, 349
Breakthrough of the year 2021, 9
Brooke, James, H., 3
Burma, Debi P., 7

C

C4 symmetrical complex, 12, 187, 191
C5 symmetry, 236
Cadmodulin, 281, 283
Calcium signalling pathways, 14
Calcium-induced calcium release (CICR), 275
Calsequestrin, 281
Calstabin, 281
Camera correlation, 88
Cancer, 14, 17, 55, 268
Capreomycin, 337
Capsaicin receptor, 6
Carbon-coated grids, 113
Cardiac muscle, 275
Cardiovascular diseases (CVD), 11, 55
Cardiomyocytes, 275
Caspar, Donald L. D., 4
Caspase-like, 35
Catecholaminergic polymorphic ventricular tachycardia (CPVT), 283
Caulbacter crescentus, 399
Ceanorabditis elegans, 396, 397, 398
Cell adhesion, 13, 227
  cell biology, 11
  cell cycle, 11
  cell cycle arrest, 69, 76
  cell cycle progression, 69
  cell signalling, 13, 227
  cell viability, 424
  cell-free gene expression system, 215
Cellular membrane biology, 5
  cellular morphology, 17
  cellular organelles, 11
  cellular physiology, 349
Central core disease (CCD), 283
Central stalk, 290
Ceramides, 389
Cereblon, 432
Chance, Britton, 6
Channel gating, 355
Channels, 13
Chaperonins, 12, 167, 169, 170, 172, 173, 232
  chaperones, 322
CHAPS, 110, 115, 201, 202, 232
Check point proteins, 69
Chemical cross-linking, 121
Chemokine receptors, 244
Chemotherapeutics, 71
Chemotropism, 254
Chimera, 88, 125, 127, 289
  chimeric constructs, 230
Chinese Hamster Ovary (CHO) cells, 13, 215
Chlamydomonas reinhardtii, 8, 413, 414
Chloramphenicol, 312, 341, 343
  chloramphenicol-bound ribosome, 452
Chloroplasts, 6, 16, 407

Cholestyryl hemi succinate (CHS), 230
Choreography, 356
Chromatin-like structures, 116
Chymotrypsin-like, 35
Ciliophora, 294
Cladinose, 342
Clarithromycin, 342
Claude, Albert, 6, 287
CMOS detector, 150
CodeOcean capsule, 133, 136
Cognition, 350
Cohn, Mildred, 6
Cold field emission (CFE), 142
Computational biology, 17
computational advancement, 11
computational methods, 131
computational tools, 12
Congenital myopathies, 283
Constraint problem, 136
Constraint programing (CP), 133
Continuous Fourier Transform (CFT), 96
Contour level, 135
Contracted tail, 7
Contractility, 421
Contrast Transfer Function correlation (CTF), 88–91, 93, 125, 126, 152, 153, 160, 193, 399
Core particle (CP), 25
Correlative light microscopy, 419
Co-transcriptional assembly, 308
Covid-19, 104
Craf kinase, 63
Crick, Francis H. C., 4
Crimean-Congo haemorrhagic fever virus, 7
CRISPR, 223
Crista, 407
Critical Micelle Concentration (CMC), 232
Crohn disease, 64
Cryo-Correlative Light and Electron Microscopy (CLEM) system, 8, 16, 17, 149, 395, 396, 398, 399, 402, 418
Cryo Arm, 122, 157, 142
Cryo-fluorescence microscopy (FLM), 162
Cryo-DRGN, 309
Cryo-EM analysis, 147, 213, 220, 214, 222, 309, 352
Cryo-EM centers, 147
Cryo-EM Coulomb potential map, 196
Cryo-EM data acquisition, 208
Cryo-EM data collection, 121, 125, 151, 191, 280
Cryo-EM data image acquisition, 12, 14
Cryo-EM data storage and processing, 12, 126, 158, 160
Cryo-EM density maps, 131, 256, 257, 269, 305
Cryo-EM facility, 12, 121, 147, 151, 156, 158, 159, 161
Cryo-EM grids, 155, 232, 396, 452
Cryo-EM holography, 8, 213, 223
Cryo-EM image processing, 14, 191
Cryo-EM images, 187, 202
Cryo-EM imaging, 195
Cryo-EM light microscopy, 398, 402
Cryo-EM projection maps, 87
Cryo-EM specimen preparation, 12, 121, 122, 147
Cryo-EM software, 159
Cryo-READ, 131, 132, 136–140
Cryo-SPARC, 49, 125, 126, 128, 153, 191, 193, 233, 237, 279, 280
Cryo-EM Maps, 7, 12, 121, 127, 128, 131, 135, 138, 139, 203–206, 220, 221, 247, 248, 292, 293, 335, 339, 340, 343
Cryo-EM of Vitreous Sections (CHEMOVIS), 385, 387, 389
Cryo-FIB milling, 13, 16, 215, 223, 312, 423, 425
Cryo-FIB-ET, 18
Cryo-fluorescence microscope (FLM), 398
Cryo-focused ion beam (Cryo-FIB), 8, 149, 150, 395–401, 410, 415
Cryogenic electron microscopy (Cryo-EM), 3, 5, 7, 9, 11, 63, 64, 65, 72, 73, 74, 76, 121, 131, 12, 147, 187, 200, 227, 237, 243, 246, 254, 255, 268, 277, 305, 308, 377, 385, 393, 395
Cryogenic electron tomography (Cryo-ET), 3, 7, 9, 10, 11, 13, 16, 63, 64, 65, 147, 206, 213, 215, 223, 227, 237, 294, 296, 305, 312, 313, 382, 393–398, 401, 402, 407, 408, 421, 418, 422, 446, 450–453
Cryo-ET data collection, 152
Cryo-tomograms, 414
Cryo-STochastic Optical Reconstruction Microscopy (STORM), 399
Cryo-transmission electron microscopy, 155, 157, 158
Cryo-ultra microtome, 150, 396
Crystal science, 3
C-terminal domains (CTDs), 45, 78, 281, 282, 351
CTF find, 90
Cutibacterium acnes, 342
Cyclic AMP signalling, 244
Cyclic GMP, 55
Cyclohexane-modified amphiphile polymer, 230
Cyclosome, 26
Cytochrome C oxidase, 289
cytochrome B6F, 414
Cytoskeleton, 214, 418
Cytosolic ribosome, 343

D

Dastinib, 61
Data acquisition, 142, 144
Data Bridge Market Research Firm, 10
de Broglie, Louis, 87
Death receptors, 69
Deep Mainmast, 12, 131–137
DeepMind, 9
Deinococcus radiodurans, 335
Deisenhofer, Johann, 6, 214
Demethylation, 343
Designer drugs, 11
Desoamine, 342
Detective quantum efficiency (DQE), 98
Detergent-free methods, 13
Deubiquitynation, 11, 29, 31, 50, 51, 52
deubiquitinases, 48
Differential phase residual (DPR), 99
Differentiation, 275
Digitonin, 231
Di-hydro-pyridine receptors (DHPR), 275
Di-isobutyl maleic acid (DIBMA), 230, 296
Dimer of dimers, 72
Dinitrophenyl (DNP), 110
Direct Electron Detectors (DEDs), 88, 157, 158, 227, 399
Disulphide bonds, 169
Double-stranded DNA, 178, 181
DNA damage, 45, 69
DNA dodecamers, 72
DNA insult, 11
DNA lesion, 113
DNA polymerase holoenzyme, 12, 176–184
DNA repair, 11, 45, 69, 72, 401, 402
DNA repair pathways, 11
DNA replication, 45, 69, 176, 177, 183, 184, 401, 422
DNA score, 133
DNA virus, 12
DNA-binding domain, 11, 70–74, 76
DNA-origami, 75
DNA-p53 complex, 75
Dopamine D2 receptor dimer, 244
Dostarlimab, 435
DRAFTER, 131
Drosophila melanogaster, 28, 33, 169
Drug discovery, 11, 17, 254, 429, 446, 449
drug candidates, 17
drug delivery imaging, 17, 451
drug targets, 213
drug-target complex structures, 17
drug-target interactions, 451
Dubochet, Jacques, 5, 73, 78
Duffy antigen receptor, 200
Dynamic Light Scattering (DLS), 440

E

E. coli mechano-sensitive Channel (MscS), 218, 219, 222
E. coli multi-eflux transporter, 219, 220, 222
E. coli Na^+/H^+ antiporter, 234
E-3 ligase, 37, 47, 69
Einstein from noise, 88, 95
Electrochemical gradient, 15
Electron crystallography, 95, 98, 213, 216, 217, 227
electron counting, 92
electron diffraction, 142
electron microscopy (EM), 3
electron microscopy data bank (EMDB), 154, 171, 402
electron microscopy micrographs, 73
electron microscopy public image archive (EMPIAR), 154
Electron transport chain (ETS), 230, 289
Electrophysiology, 15, 350
Elongation factor-G-mediated translocation, 310
elongation factor (EF)-Tu, 310
EM build, 128
EM density, 181
Endoplasmic reticulum (ER), 6, 14, 176, 177, 305
Endosomes, 14
Engineered peptide tag, 113
Engineered proteins, 255
Enhanced green fluorescent protein (EGFP), 230
Enterococcus faecalis, 338
Enzymology, 12, 167
Epidermal layer, 16
Epithelial cells, 207
Ernster, Lars, 6
Erythrocyte lysis assay, 207
Erythrocytes, 201

Escherichia coli, 7, 98, 188, 305, 312, 319, 336, 402, 453
Escherichia coli LacY, 214, 215, 219, 230
Escherichia coli large subunit (50S), 15
Escherichia coli small subunit (30S), 15
Essential Neisseria GTP-binding protein A (EngA), 328
Estrogen receptor, 37
Euglena glacilis, 294, 295
Eugleozoan, 294
Eukaryotic protein kinases (EPKs), 55, 60
Excitation Contraction (EC), 275
Excitory neurotransmission, 15
Expression vector, 188
Extra-cellular enveloped virions (EEV), 177
Extra-cellular loops (ECL), 243, 256
Extra-cellular matrix (ECM), 421, 423

F

F1F0 ATP synthase, 14, 290, 295
F8 polymerase, 181, 182, 184
FANC1-FANCD2, 45, 46, 47, 48, 50, 52
Fanconia anemia (FA), 45
FARFAR2, 131
Fast incremental single exposure (FISE), 400
Father of crystallography, 3
Father of the microscope, 4
Father of X-ray crystallography, 4
F-ATPases, 290, 292, 294
Fatty acid metabolism, 6
FFT algorithm, 92
Fibrosis, 421
Filopodia-like protrusions (FLPs), 419
Flacobacterium Johnsoniae, 188
Flagellar motors, 452
Flap endonuclease, 176
Flippases, 214
Fluid-mosaic model, 13
Forster resonance energy transfer (FRET), 420
Fourier ring correlation (FRC), 89, 99, 100, 103
Fourier ring information (FRI), 89, 92
Fourier shell correlation (FSC), 100–104, 108, 109, 154
Fourier transform, 291
Fragile-X-syndrome, 359
Fragment-based drug discovery (FBDD), 429, 430, 432
Frank, Joachim, 5, 7, 73, 78
Franklin, Rosalind, 4
Freeze-plunging specimen preparation, 87
Fucoxanthin chlorophyll proteins (FCP), 414
Fungi, 254
fungal growth, 254

G

G proteins, 259, 260
G2/M phase transition, 69
Gamma hemolysin, 13, 200
gamma proteo bacteria, 337
Gamma-amino butyric acid (GABA), 16
GABA B receptor, 244
Gene editing, 38
Gene expression, 11
Gene machineries, 14, 303
Genetic code, 15
Genotoxic events, 69
Giardia lamblica, 339
Giardiasis, 339
Git Hub, 133
Glaeser, Robert, 5
Gleevec, 61
Globus, 154
Glucagon-like peptide-1 (GLP1), 17, 449
Glutamate receptors, 15, 349, 350, 351, 359, 360
Glycine receptors, 16, 352
Glycol DiosgeniN (GDN), 214, 218, 231
Glyco-protein complex, 7
Glycosylated proteins, 214
Goldberg, Alfred L., 25
Golgi apparatus, 14, 385
Google Cloud Platform (GCP), 160
Google Colab, 133–135, 137
G-protein-coupled Receptors (GPCRs), 6, 13, 14, 17, 237, 243, 246, 254, 377, 437, 438, 447, 449
GPCR activation, 247, 264
GPCR biology, 14, 254
GPCR dimerization, 14
GPCR dimers, 254, 243, 244
GPCR hetero-trimer, 255, 256
GPCR kinases, 243, 248
GPCR monomers, 244
GPCR protomers, 246
GPCR structural biology, 14, 255
GPCR-G protein complex, 14, 248, 254, 255
G-protein coupling, 261, 262, 264
G-proteins Gi/o, 14
GRAFIX method, 121
Gram-negative bacteria, 213
Grana, 16
Granisetron, 381
Graphene oxide, 232
Graphic Processing Unit (GPU), 133
graphic processing unit cards, 153
graphic processing unit server, 151
graphic processing unit workstation, 155
graphic user interface software, 144, 145
graphic user interfaces (GUIs), 11
Green fluorescent protein (GFP), 188, 398
Green, David, 6
Grid-freezing optimization, 230
GroE, 169
GroEL, 12, 169–173
GroES, 12, 169, 170
GTPase proteins, 420
GTP hydrolysis, 310, 325
GTPases, 322, 323
GTPases domain, 328
GTPase-associated Center, 310
Guardian of the genome, 11, 69
GUOs-like Relion, 11, 125, 126, 153, 172
Gustducin, 6

H

Haemoglobin, 4
Haloarcula marismortui, 335
Halobacterium salinarium, 216
Hassabis, Demis, 9
Hauy, Abbe, 3
Heat shock protein, 70, 169
Heat-shock proteins (HSPs), 12
Helical domains, 45
Helical fibers, 4
Helix-rich C lobe, 55
Henderson, Richard, 5, 17, 73, 78, 214, 216, 229
Hepatitis C Virus (HCV), 7
Hetero-dimer, 45, 47, 181, 183, 270
Hetero-dimeric processivity factor, 176
Hetero-dimerization, 63, 243, 244
Heterogeneous protein expression, 188
Heterohexamer, 35
heteropentamers, 382
heterotrimeric G protein, 254, 259
Hierarchical ascendant classification (HAC), 94
High-speed interference contrast microscopy, 110
HlgAB pore complexes, 204, 205, 206, 208
Hodgkin, Dorothy C., 4
Holliday junction, 176, 184
Holo translocon, 312
Holoenzyme, 29
Homeostasis, 16, 25, 69, 275, 421
Homo sapiens, 33
Homo-hexamer, 34
homo-oligomerization, 61
homopentamers, 377
homotrimeric, 219
Hooke, Robert, 4, 395
HSP-90-client complex, 63
Huber, Robert, 227
Huc and Hhy, 187
Huc complex, 188–194, 197
Huc structure, 12
Huc tail region, 193
HucCS subunit, 190, 195
HucL, 190
HucM, 190, 194
HucM scaffold, 194, 196
HucSL, 193
Human embryonic kidney (HEK) 293 cells, 13, 215–217, 223, 230
HEK293F, 178
Human epidermal growth factor receptor (HER), 63, 436
Human immunodeficiency virus (HIV)-1, 8, 37
Human immunodeficiency virus (HIV)-1 RT, 183
Human integrin, 215
human patched 1 protein, 13
human platelet integrin, 13
human proteome, 227
human Slo1 channel, 13, 221
Huntington disease, 358
Hydrogenase map, 137
Hydrophobic interactions, 204, 206
Hydroxy Tryptamine (5-HT3) receptor, 16, 377, 379
Hypertension, 421

I

Ichthyoses, 16
lenalidomide, 432
Illumination microscopy, 8
Image acquisition, 122
IMAGIC system, 88, 90, 92
Immobilized metal affinity chromatography (IMAC), 76, 201
Immune disorders, 11, 55
immune evasion, 254
immunity regulation, 11
immunoglobulin, 1 (IgG1), 436
immunoglobulin G (IgG) molecule, 111
immunoprecipitation assay, 73
immunoproteasome, 25, 28

Infectious diseases, 11, 55
Information processing, 15, 349
information technology (IT) infrastructure, 12
Inhibitor protein, 291
Inner mitochondrial membrane (IMM), 407, 408, 410
In-situ Cryo-ET, 229, 423
in-situ structural biology, 284
in-situ transmission electron microscopy, 407
Integral membrane proteins, 13, 15, 213, 214
Integrative modeling, 17
Intensity transfer function, 90
Inter-membrane space (IMS), 287
Intra-cellular calcium channel, 275
intra-cellular enveloped virions (IEV), 177
intra-cellular loops (ICL), 243, 259
intra-cellular mature virions (IMV), 177
intracellular organelles, 213
Invasiveness, 254
Ion channel, 14, 15, 347, 355
ion flux, 15
ionotropic receptors, 350
Isothermal Titration Calorimetry (ITC), 237
IT infrastructure, 147

J

Julius, David, 6
Junctin, 281
Juntophillins, 281

K

Kainate receptors, 349, 352, 353, 355
Kendrew, John, 4
Kinase catalytic domain, 55
Kinase inhibitors, 61
Kinase-Signaling complexes, 11
Kinetoplastid, 340
Kinetoplastid-specific domain, 340
Knoll, Max, 4, 395
KRAS, 432
Krebs, Hans, 6, 288

L

Lake, James, 7
Lamellipodia, 419
Langmuir-Blodgett trough, 112
Lanthanum hexaboride, 162
Large Unilamellar Vesicles (LUVs), 200, 203, 207
Lauryl Maltose Neopentyl Glycol (LMNG), 214, 295
Learning and memory, 15, 350
Legionella pneumophila, 271
Legobodies, 232
Lehninger, Albert L, 6
Leishmania donovani, 339
Leishmania donovani ribosome, 343
Leucine transporter, 234
leucine-rich repeat kinase 2 (LRRK2), 64
Leukocidins (LUK), 200
Li-Fraumeni syndrome, 70
Ligand-binding domain, 355
Ligand-binding GPCR, 254
Light harvesting complexes, 411
Light microscopy, 395, 398
Lignoceric acid, 389
Lincosamides, 343
Lipid asymmetry, 214
Lipid bilayers, 13, 204, 213–215
lipid densities, 205
lipid monolayers, 111, 118, 214
lipid nanodiscs, 352, 448
lipid remodelling, 288
lipid-anchored, 13
lipidomic studies, 237
Lipid-anchored membrane protein, 213
lipid-protein aggregations, 202
lipid-protein interaction, 200
lipid-toxin interaction, 200
Liposomes, 13, 200–202, 448
Lippman, Fritz, 6
Listeria, 419
Local Dense Points (LDPs), 133
Local Information Density (LID), 103
Long-tail fibers, 7
LukF, 200
LukS, 200, 206
Luria, Salvador, 6, 305
Luteinizing Hormone Receptor (LHR), 247
Lysosomes, 14

M

Machine learning (ML), 11, 440
Macromolecular assemblies, 8, 14
Macrophages, 277
Macropinocytosis, 12, 177
Major Histocompatibility Complex (MHC) class, 1, 28
Malaria, 17
Malignant hyperthermia susceptibility (MHS), 283
Maltose binding protein (MBP), 230
Mammalian cell entry (MCE), 188
Mammalian enhancer-of-Akt-1-7 (mEAK-1), 272, 273
MAP kinases, 254
MAPK pathway, 244
Mass spectrometry, 188, 237
Matrix, 407
Maturation factors, 322
Mechano-sensitive channel protein, 13
Mefloquine, 343
Megabodies, 232
Membrane biology, 13, 229
membrane protein expression, 230
membrane protein folding, 213
membrane protein structural biology, 13
membrane proteins, 13, 17, 213–215, 222, 223, 227, 407, 448
membrane transporter folds, 227, 233
membrane vesicles, 215
membrane-anchoring tail, 12
Messenger RNA (mRNA), 15, 176, 177, 310, 325
Meta quinone oxidase, 195
Metabolic errors, 11
Metabolism, 13, 227, 254, 288
Metabotropic glutamate receptor (mGluR), 244, 349
Metalloprotease, 31
Method of the Year 2015, 5
Methyl pentane-diol (MPD), 200, 202
methyltransferases, 343
Meyerhof, Otto, 6
Michel, Hartmut, 227
Micro electron diffraction (microED), 147, 152, 153, 446
Microbial infection, 268
Microscale thermophoresis, 237
Microsporidia, 29
Microtubule-associated motor proteins (MAPs), 64
Mini-G proteins, 247
misfolding of proteins, 15
Mitchell, Peter D., 6, 288
Mitochondria, 6, 16, 287, 289, 296, 407
mitochondrial ATP synthase, 408, 409
mitochondrial biology, 6, 14
mitochondrial electron transport chain, 409
mitochondrial membrane protein complexes, 14
mitochondrial membrane proteins, 287
mitochondrial protein synthesis, 15
mitochondrial ribosome, 339
mitoribosomes, 15
Molecular biology, 14, 303
Molecular machinery, 147
molecular machines, 6, 7, 11
Molecular motors, 15
MolProbity tools, 125
Monkey PoX Virus (MPXV), 12, 176, 177
Monoclonal antibodies, 435
Moore, Peter, 7
MotionCor2, 208
Mouse Double Minute, 2(MDM2), 69, 72, 75
mRNA vaccine, 17
Multi reference alignment (MRA) procedures, 95
Multi-angle light scattering (MALS), 255
Multi-efflux transporter, 13
Multi-minicore disease (MmD), 283
Multi-omics, 15
Muscle contraction, 14, 275
Mycobacterium pneumonaie, 312
Mycobacterium smegmatis, 12, 187, 188, 190, 197
Mycobacterium tuberculosis (Mtb), 337, 449
Mycobacterium tuberculosis ribosome, 337
Mycoplasma pneumonia, 451
Myoglobin, 4
Myoviridae, 7

N

N-acetyl glucosamine, 258
Nanobodies, 16, 255
Nanodisc, 230
NASA curiosity rover, 88
Nascent polypeptide exit tunnel, 342
National Center for In-situ Tomographic Ultramicroscopy (NCITU), 161
Native cell membrane nanoparticles (NCMN) system, 13, 215, 220, 230
Native chemistry, 188
Native expression, 188
N-dodecyl-beta-maltoside (DDM), 215, 220, 231
Negative-stain electron microscopy (NSEM), 148, 150
Neisseria gonorrhoeae, 328, 329
Nerve impulses, 16
Network connectivity, 12
Neurobiology, 15, 347
neurodenerative diseases, 358
neurodegeneration, 14, 268

neurodevelopmental disorders, 360
neurological disorders, 14, 55, 70,
neuropsychiatric disorders, 15, 359–361
neuronal plasticity, 275, 277
Neuropilin, 353
Neuro-receptors, 15, 347
Neurotensin 1 receptor, 244
Neurotransmitter, 437
neurotransmitter release, 353, 354
Neutron crystallography, 213
N-glycosylation, 258, 269
Nickel-chelating lipids, 111
Nicotinic Acetyl Choline (nAch), 16, 352, 378
Nierhaus, Knud, 7, 307
Ni-Fe-hydrogenase, 187, 195, 196
Ni-nitrilotriacetic acid (NTA) agarose, 207
Nitrosylation, 281, 283
N-methyl-D-aspartate (NMDA), 349–360
Noller, Harry F., 7, 335
Nomura, Masayori, 7, 307
Northrop, John H, 4
Novel targets, 254
Novel therapeutics, 11
Novobiocine, 111
N-terminal domain (NTD), 71, 72, 78, 178, 179, 181, 184, 281, 282, 324
N-terminal solenoid, 281
Nuclear magnetic resonance (NMR), 3, 28, 72
NMR spectroscopy, 188, 213, 227, 243, 368
Nuclear pore complex (NPC), 401
Nucleoporins, 401
Nucleosome, 7
Nucleotide substitution, 29
nucleotide transport, 288
nucleotide-binding pockets, 170

O

Obesity, 14
Octahedral super-assembly, 13, 202, 205, 208
Octamer, 216
Octyl glucoside, 110, 202
Okazaki fragments, 184
Oligomeric pore complexes, 207, 187
oligomerization, 72, 200–202
Oligomycin sensitivity conferring protein (OSCP), 296
Oligopeptides, 437
Omicron variant, 7
Omicron sublineages, 7
Ondansetron, 381
Optical microscopy, 87
Ortho poxvirus, 12, 178, 184
Ortho-phenanthroline, 29
Orycolagus cuniculus, 207
Osteoporosis, 268
Outer mitochondrial membrane (OMM), 407, 410, 411
Oxazolidinones, 343
Oxidation, 281
oxidative phosphorylation, 6, 288

P

P53 structure-function studies, 11
P53-DNA interactions, 73
Palade, George, 6, 306
particles of Palade, 305
Palmitoylation, 355
palmitoyl-2-oleoyl-glycero-3-phosphocholine (POPC), 216, 220
Palonosetron, 381
Panton-Valentine leukocidin, 13
Paramecium, 294
Parkinson disease, 64, 70, 358
Paromomycin, 339, 343
Patched-1, 216, 218, 222
Pathological functions, 213
PCNA-like proteins, 183
PCR amplification, 116
PDZ domain recognition, 243
Pentameric transporter family, 237
Peptide bond formation, 341
Peptidisc, 215, 230
Peptidyl transferase, 305
peptidyl transferase center, 309, 328, 338
peptidyl site, 319
Peripheral stalk, 290
Personal protective equipment (PPE), 155
Pertuzumab, 436
Perutz, Max F., 4
P-glycoprotein, 215
Phaeodactylum tricornutum, 414
Phage nucleus, 401
phase separation, 214
Phase-contrast microscopy, 93
PHENIX, 281
Pheromone resistant phenotype, 260
Phosphatidylcholine (PC), 269
phosphatidyl cholesterol (CLR), 269
phosphatidylethanolamine (PE), 269
phosphatidylserine (PS), 269
Phosphodiesterase, 6, 245
Phosphorylation, 55, 60, 61, 76, 281, 283, 355
Photosynthesis, 13, 16, 227, 230
photosynthetic reaction center, 227
photosystems, 230
photosystem I, 411, 414
photosystem II, 411, 412, 414
Physiopathology, 437
Pichia augusta, 292
Pichia pastoris, 230
Plasma membranes, 213
Plasmodium falciparum, 339–341, 343
Plasticity, 349, 350, 354, 358, 360
Pleiotropic, 55
Pleuromutilins, 343
Podospora anserina, 294, 409
Podoviridae, 7
Polyethylene glycol, 396
Polymorphic assemblies, 9
Polysome formation, 310
Polytomella, 294, 295
Polyubiquitin, 30
Pore formation, 13
pore-forming toxins (PFTs), 13, 200–204, 206, 213
Postsynaptic Signaling, 350
Post-transcriptional modifications, 6
Post-translational modifications, 76
Power house of the cell, 6
Pox virus virions, 12
Poxviridae, 12, 176, 177
Presenelins, 281
Principal component analysis (PCA), 93, 94
Proapoptotic proteins, 69
Processome, 7
Productive catalysis, 60
Proliferation, 275
proliferating cell nuclear antigen (PCNA), 183, 184
Prostaglandin E receptors, 244
Prosthetic groups, 213
Proteasome, 11, 25–37, 398, 432, 434, 435
proteasome-interacting proteins, 11
Protein assemblies, 11
protein crystallography, 4
Protein Data Bank (PDB), 154, 171, 219, 223, 227, 229
protein disulphide isomerase, 169
Protein synthesis, 7, 14
protein folding, 12, 169
protein Kinase super family, 11
protein Kinases, 11, 55, 56
protein topology, 215
protein translational machinery, 15
protein-lipid interactions, 9, 215, 223
protein-nucleic acid (PNA) complexes, 9
Proteoliposomes, 202, 214, 220
Proteolysis, 17
proteolysis targeting chimeras (PROTACS), 432, 447
proteolysis-targeting antibodies (PROTABS), 435
Proteus mirabilis, 452, 453
Protofilaments, 64
Proton motive force, 409
Proton transfer, 272
Protonation, 259
Protrusion morphology, 424
Pseudomonas aeruginosa, 170, 234
Pseudomonas chloraphis, 401, 402
Pseudomonas fluoresces, 170, 171
Pseudomonas stutzeri, 230
Psoriasis, 16
PULCHARA, 133
Python script, 144, 145

Q

Quinone transport, 195
Quorum sensing, 13

R

Racker, Efraim, 6
Radiation exposure, 11
Ramakrishnan, Venkatraman, 7, 335
Random conical tilt (RCT), 95
Ras-like protein, 323
RCrane, 131
Receptor physiology, 15, 350, 353
receptor pharmacology, 13, 247, 356
receptor activation, 357
receptor guanylate cyclase (RGC), 55
receptor-like kinases (RLKs), 59
Recombinant expression, 187, 188
Recombination, 176, 183
Regulatory particles, 25
Release factors, 311
Relion, 191, 233, 237, 279
Renal tubular acidosis, 14, 268
Resolution Revolution, 7, 14, 277, 335
Resolvase, 176
Respiration, 6
respiratory complex I, 409
respiratory electron transport chain (ETS), 14, 17
respirosomes, 409
Retina, 14
retinal Rod Outer Segment (ROS), 243
retinal-binding protein, 14, 245
Retroviruses, 451

Rhodopeudomonas virdis, 5, 214, 227, 437
Rhodopsin, 14, 214, 244
retina outer segment membranes, 244, 245
rhodopsin chromophore, 245
rhodopsin dimerization, 244, 246–248
rhodopsin dimers, 243, 246–249
rhodopsin disc membranes, 246
rhodopsin protomer, 247
rod photoreceptor, 14
Ribo nucleotide reductase, 111
Ribo-nucleo protein (RNP) complex, 6, 305, 331, 335
Ribosomal biogenesis GTPas A (RbgA), 329
ribosomal RNA (rRNA), 15
ribosomal RNA maturation, 320
ribosomal RNA operons, 305
ribosomal RNA transcription, 319
ribosomal RNAs, 306, 307, 208, 310, 322
ribosomal sequestration, 312
ribosomal 17S RNA, 323, 324
ribosomal RNA 23 rRNA, 319, 320, 337
ribosomal RNA 16 rRNA, 15, 319, 320, 323, 326, 337
ribosomal RNA 5 rRNA, 319–321, 323, 326
Ribosomes, 5, 14, 15, 305, 310, 311, 313, 319, 336, 340, 342, 344
ribosome assembly, 306, 321, 322, 331
ribosome biogenesis, 15, 306, 307, 312, 319, 320, 324, 326–329, 343
ribosome biology, 6, 15
ribosome decoding center, 325
ribosome life cycle, 305
ribosome-binding factor A, 323
ribosomes 80S, 335, 340
Ribosomal proteins, 15, 306, 319
Giardia ribosome, 339
ribosomal large subunit pseudouridine synthase D (RluD), 328
ribosome 30S small subunit, 15, 309, 311, 319, 321
ribosome 50S large subunit, 15, 319
ribosome 70S, 319, 324, 335, 338, 342
ribosome 80S, 340, 343
ribosomes of pathogenic organisms, 15, 335
Staphylococcus aureus ribosome, 337
Thermophilus ribosome, 338
RNA structure modelling, 131, 140
RNA confirmation, 319
RNA editing, 351
RNA helices, 305
RNA mis-folding, 322
RNA Polymerase, 1 (RNP1), 113, 116, 305
RNA processing enzymes, 322
RNase III, 320
Robertson, David J., 214
Robertson, J. D., 5
Röntgen, Wilhelm, C., 3
Rosetta, 133, 135, 281
RoseTTA fold, 9, 440
Ruska, Ernst, A. F., 3, 4, 11, 395
Ruthenum tetroxide, 389
Ryanodine receptor (RyRs: RyR1, RyR2, RyR3), 275–277, 281–284
Ryanodine, 281
RyR biology, 14, 277
RyR dysfunction, 283
RyR dysregulation, 283
RyR purification, 278
ryanodinopathies, 283

S

Saccharomyces cerevisae, 7, 14, 33, 230, 254, 255, 261, 268, 335
Salmonella enterica, 7, 453
Salmonella typhimurium, 7, 234
Saposin, 214, 231
Sarcolemma, 275
Sarcoplasmic reticulum, 275
sarcosomes, 6
Sarecycline, 342
SARS-CoV-2 Spike protein, 7
Scaffolding domains, 55
Scanning electron microscope (SEM), 4
Schizophrenia, 359
Schizosaccharomyces pombe, 29, 33
Selective hormone receptor down-regulators, 432, 433
Sense perception, 13
Sequestration, 15
Serial electron microscopy, 146, 152
Serotonin receptor, 16, 352, 377, 378, 380, 381, 382
Seven helical transmembrane, 14
Sexual development, 254
Shine-Dalgarno sequence, 323
Signal transduction, 13, 245
signal peptides, 69
signaling mechanism, 13
signal-sequence-associated motor complex (SAM), 288
Signal-to-noise ratio (SNR), 98, 101, 125, 232, 291, 453
Silver-halide film, 88
Simian immunodeficiency virus (SIV), 451
Single molecule active control microscopy (SMACM), 399
Single particle analysis (SPA) Cryo-EM, 5, 6, 13, 76, 87, 88, 95, 104, 121, 123, 143, 147, 153, 191, 202, 213–215, 222, 227, 229, 234, 352, 446–450, 453
Size-exclusion chromatography (SEC), 188, 231, 255
Skin, 15, 16, 347, 385
Smith-Waterman dynamic programming, 133
Solarus plasma cleaner, 122
Solid-state storage, 159
Soluble proteins, 213
Solute carrier, 437–439
Sorcin, 281
Spatial proteomics, 15
Spermatogenesis, 28
Spermatoproteasome, 25
Spirin, Alexander, 7
Spodoptera frugiperda, 230
SpyCatcher grids, 115
Stabilization networks, 72
Standard operating procedures (SOPs), 155
Staphylococcus aureus, 13, 5, 200, 203, 206, 329
Ste2, 246, 247, 254–258, 260–263
Ste2 dimerization
Ste2-G protein complex, 255, 258
Ste2-mediated signal transduction, 254
Steitz, Thomas A, 7, 335
Stem to stem transition, 202
Stereoisomeric confirmation, 187, 188
Stratum corneum, 16
Streptavidin crystals, 111, 112
Streptavidin affinity chromatography, 190
Streptavidin affinity grids, 11, 110–117
Streptococcus pneumoniae, 234
Streptogramin A, 343
Streptomyces hygroscopicus, 343
Stroma, 16
Structural biology, 3, 4, 7, 12, 17, 73,213, 227, 228, 237, 320, 331, 350, 402, 429, 434
Structural cell biology, 8
Structural illumination microscopy (SIM), 399
Structural molecular biology, 4
Structural virology, 7
Structure-based drug design (SBDD), 15, 17, 45, 430, 432, 440, 447
Styrene maleic acid, 295
Styrene-maleic acid copolymers, 214
styrene-maleic acid lipid particles (SMALPS), 230
Sub-tomogram analysis (STA), 16, 147, 149, 398
sub-tomogram averaging, 8, 113, 294, 312
Succinate dehydrogenase, 288
Sumner, James B., 4
SUMOylation, 355
Super-resolution imaging, 15
Sybodies, 232
Synaptic transmission, 14, 15, 275, 347, 349, 353
synaptic computation, 15
Synchrotron, 10
Synuclein, 15
Systemic amyloidosis, 15

T

Targeted protein degradation, 432
Taste receptor, 6
Telomeres, 176
Tensor processing unit (TPU), 133
Tetracycline, 341
Tetrahymena thermophila, 294
Tetramer crystals, 216
Tetra-ubiquitin chain, 35
Therapeutic antibodies, 429
Thermophiles, 230
Thermus filiformis, 178
Thermus thermopilus, 7, 335
Tholidomide, 432
Thread granules, 6
Three-dimensional crystallography, 6, 214
Three-dimensional structural information, 213
three-dimensional electron diffraction (3D ED), 142, 144, 145
three-dimensional reconstruction, 87, 88, 154, 158, 160, 191, 305
three-dimensional structure, 335
Thumb and Forefingers domain, 281
Thylakoid membrane, 16, 411, 413, 415
Thymo proteasome, 25
Thyroglobulin, 113
Time-resolved serial femtosecond crystallography, 237
Tomograms, 8, 17
Tomography-guided 3D reconstruction of subcellular structures (TYGRESS), 8
TOPAZ, 153, 233

Toxin oligomers, 200
Toxin subunits, 206
Toxoplasma gondii, 294
Transducers, 255
Transfer RNA, 310
Transient receptor potential cation channel subfamily V member 1 (TRPV1), 6, 214, 229
Translation, 14, 15, 309, 310, 312, 336
  translational termination, 15
Translocase, 288
Translocation, 34
Transmembrane domain, 220, 221, 243, 268, 281, 351, 378
  transmembrane helices, 73, 213, 218, 292
  transmembrane pore complex, 13
Transmission electron microscopy (TEM), 4, 142, 145, 150, 151, 202, 223, 287, 395, 396, 399, 407, 411
Transporters, 13, 17
Transthyretin, 368
Transverse tubules, 275
Triadin, 281
Trichoplusia ni, 255
Tropisetron, 381
Troponin C, 275
Trypanosoma brucei, 28, 294, 340
Tumorigenesis, 11
  tumor suppressor protein TP53 (p53), 11, 69, 70–79
Type 9 secretion system (T9SS), 188
Tyrosine kinase-like (TKL), 55

U

Ubiquinone, 288
Ubiquitin, 11, 25–37, 45–51, 69, 6, 7, 78, 434
  ubiquitin ligase, 25, 26, 27, 434
  ubiquitin specific protease-1 (USP1), 11, 45
  ubiquitin-activating E1 enzyme, 25
  ubiquitin-conjugating E2 enzyme, 25
  ubiquitin-proteasome system, 25, 26
  ubiquitin specific protease-1 inhibitors, 11, 45
  ubiquitin specific protease-1-associated protein-1, 46, 50, 52
  ubiquitylation, 11, 25, 28, 31, 355
Unwin, Peter N. T., 5, 73, 214, 216, 229

V

Vaccine development, 17, 448
Vaccinia virus (VACV), 176, 178
Vacuolar-type adenosine triphosphatases (VATPases), 14, 268–273
VACV E9 structure, 178
  VACV polymerase, 178
van der Waals interactions, 256, 259, 260, 342
Van Heel, Marin, 7
Van Laue, M. T. F., 3
van Leeuwenhoek, Antony, 4, 11, 87
Vanilloid receptor, 6
Variant transthyretin amyloidoses (ATTRv), 374
Veesler, David, 7
VESPER, 131
Vibrio cholerae, 329, 449
  Vibrio fischeri, 234
Viral replication, 37
Virchow, Rudolf, 15
Virology, 12
Virtual Graphic Languages, 160
Virulence, 254
Visceral leishmaniasis, 343
Visual proteomics, 8
Visual transduction, 14, 245
Visualization tools, 12
Vitreous ice, 87
Vitrification, 5, 75, 150, 232, 311
Vitrobot, 122
Voltage- and calcium-activating potassium channel, 221
  voltage-dependent anion channel (VDAC), 288
  voltage-gated L-type calcium channel, 275
Von Kolliker, Rudolf, F. A., 6

W

Walker, John E., 6, 289
Ward, Andrew, 7
Water-soluble monomers, 13
Watson, James D., 4
Wilkins, Maurice H. F., 4
Wittmann, H. G., 7
World Health organization (WHO), 176

X

X-chromosome inactivation, 8
Xenopus laevis, 216
  Xenopus oocytes, 377
X-inactive specific transcript (XIST), 8
X-ray crystallography, 28, 29, 72, 98, 103, 197, 188, 213, 214, 219, 222, 227, 229, 230, 232, 234, 243, 277, 289, 291, 305, 320, 335, 357, 368, 377, 436, 447
  X-ray diffraction, 3, 4, 5, 7, 65, 142

Y

YOLO, 153
Yonath, Ada, 7, 335
YoneLocr, 142, 144
You Only Look Once (YOLO), 144

Z

Zernike phase, 90
Zero-tilt approach, 88, 95, 98
Zika, 17